A COMPENDIUM ON NONLINEAR ORDINARY DIFFERENTIAL EQUATIONS

A COMPENDIUM ON NONLINEAR ORDINARY DIFFERENTIAL EQUATIONS

P. L. SACHDEV

Department of Mathematics
Indian Institute of Science

A Wiley-Interscience Publication
JOHN WILEY & SONS, INC.
New York • Chichester • Brisbane • Toronto • Singapore • Weinheim

This text is printed on acid-free paper.

Library of Congress Cataloging in Publication Data:

Sachdev, P. L.

A compendium on nonlinear ordinary differential equations / P.L. Sachdev.

p. cm.

"A Wiley-Interscience publication."

Includes bibliographical references (p. -).

ISBN 0-471-53134-0 (cloth : acid-free paper)

1. Differential equations. I. Title.

QA372.S145 1997

515′.352—dc20 96-10204

Printed in the United States of America

10 9 8 7 6 5 4 3 2 1

TO ONE WHO INSPIRED

CONTENTS

PREFACE

I could not have accomplished the present work if I had not received help and assistance from friends, colleagues, and students, to each of whom I am deeply indebted. Dr. Neelam Gupta accompanied me in my peregrinations to many university libraries around New Jersey and New York. My friend and co-worker Dr. Varughese Philip agreed to undertake the very difficult task of checking through thousands of equations, and accomplished it with cheer and patience. I was fortunate to have Dr. Eric Lord to assist at the crucial stage of organization of the material; his rich experience in mathematical physics was of immense value in devising a sensible scheme for ordering the equations. My student Dr. B. Mayil Vaganan came to my rescue whenever I found the physical and mental effort overpowering. Mr. M. Renugopal put the entire material of the compendium in LaTeX with great care. Others who assisted me include Dr. D. Palaniappan, Dr. K. R. C. Nair, Mr. Ch. Srinivasa Rao, and Dr. R. Sarathy, and my secretary, Mrs. Sandhya. I also wish to thank Professor V. G. Tikekar for his cooperation.

My wife, Rita, patiently listened to my grumblings when I encountered hurdles of various kinds, and bore the major share of the responsibility of bringing up our two sons, Deepak and Anurag.

I would wish to record my sincere thanks to successive editors at John Wiley—Maria Taylor, Kate Roach, Elizabeth Murphy, Maria Allegra, and Jessica Downey—for their patience and cooperation during the long period the present work was in progress.

I am much obliged to the authorities of the Indian Institute of Science, in particular the Curriculum Development Cell, Ministry of Human Resources Development, Government of India, for facilities and financial assistance.

P. L. SACHDEV

Indian Institute of Science
September 1996

A COMPENDIUM ON NONLINEAR ORDINARY DIFFERENTIAL EQUATIONS

1

INTRODUCTION

I do not recollect the mystical moment when the thought to prepare this compendium captured my imagination. It was not unnatural to conceive of it after I had completed my book *Nonlinear Ordinary Differential Equations and Their Applications*, since published by Marcel Dekker (1991), but I was only vaguely aware of the task ahead, and the enormity of the effort that would be demanded of me. As I plodded on, thumbing through literally hundreds of volumes of journals, hunting out useful, interesting, known and not-so-well-known equations, I realized that the volume I had envisioned as modest in size would grow and that it could never be exhaustive. However, the search continued and the material piled up. It took a relentless effort of five years to bring this work to its present stage of completion. I ransacked mathematics sections of many libraries: the Courant Institute, NYU, Rutgers, New Jersey Institute of Technology, St. Andrews (UK), TIFR and IIT, Bombay, and the Indian Institute of Science, Bangalore. Almost all journals in applied mathematics, physics, and engineering that deal with nonlinear phenomenon were browsed through. That explains the large size of the biblography and, of course, of the compendium itself. Yet it does not seem possible to exhaust all the equations, since new ones get added to the literature almost every day. The present collection should, nevertheless, meet the needs of a large majority of scientists, engineers, and applied mathematicians.

I, like several generations of engineers, scientists, and mathematicians, had often consulted the classic collection of E. Kamke (1959) to ascertain whether a nonlinear ordinary differential equation that I encountered in my research just might be there. Despite the fact that this work is in German, it has been immensely popular outside Germany. This book, I believe, has been my principal inspiration. Although it concerns mostly linear ordinary differential equations, there are a small number (266) of nonlinear equations. The corresponding collection in English by Murphy (1960) does not go much beyond Kamke.

Since Kamke's book, research in physics, mathematics, and engineering has spawned such a large number of new and interesting equations that a compendium on the subject has long been overdue. (This gap has been widened enormously by the appearance of that ubiquitous phenomenon called chaos.) I decided that the equations should be dealt with in the following way: If an equation can be solved quickly in a closed form, the steps to arrive at the solution should be given sparsely; otherwise, a summary of the asympototics, stability, existence, or numerical results may be appended to each equation. The contributions of various authors to the same equation are generally not combined; instead, their individual results are enunciated at the same location in the compendium. Equations in an abstract setting; stiff, delay, or stochastic equations; and functional or differential-difference equations have not been included. A large majority of equations in the compendium have arisen from physical models directly or through transformations such as the similarity reduction; initial and boundary conditions have been included wherever these have been imposed. Since the compendium is pretty large as it is, there is no scope for including notes for various classes of equations, as found in Kamke's book. However, the author's book *Nonlinear Ordinary Differential Equations and Their Applications* will be found useful for elements of qualitative analysis of nonlinear ordinary differential equations.

Categorization of such a large number of equations posed some difficulty; the book of Kamke again provided useful clues. A detailed rationale for the classification of equations is given in Section 1.1. However, the basic principle adopted was to list the equations in order of increasing complexity, as they would appear to the user: One should be able to look for an equation in the same manner as one would look for a word in the dictionary. For the convenience of the reader, the equations belonging to each order—second, third, fourth, fifth, and higher—have been divided into a large number of subclasses; equations in each subclass bear a subtitle and have been ordered as explained below. The scheme given here may not be perfect but should be adequate.

There are equations, such as those of Lorenz, Van der Pol, and Painlevé, to which entire shelves of literature have been devoted. We have given only a few recent results concerning these equations. On the other hand, the present compendium includes many less known equations that the reader will find interesting, curious, or useful.

The present volume may be helpful in many ways. It should be a standard reference in the sense that Kamke's book has been for (almost precisely) five decades: An engineer or scientist can look up an equation and may find a ready-made analysis with a reference thereto or a set of similar examples of known behavior. The compendium provides a wealth of examples for teaching and exposition. Moreover, since the equations have been drawn from diverse fields, the book should cross-fertilize thinking and analysis in unrelated fields.

1.1 Instructions to the User

The equations have been ordered like the words in a dictionary; once the user has familiarized himself or herself with the system of organization that we have employed, any equation in the book can be located rapidly. The user will then find concise information about explicit solutions and the method of solution, when these can be found, and indications of the nature and behavior of solutions where explicit solutions may not be found. Also, references to the source material are given. Equations of related type will be found in close proximity.

The basis of our ordering scheme is the following hierarchy of functions of a single variable $f(x)$:

Polynomials

Rational functions (P/Q where P and Q are polynomials)

Functions involving fractional powers [e.g., $x^{3/2}$, $(1+x^2)^{1/2}$, etc.]

Functions involving unspecified powers (e.g., x^p, where p is not necessarily an integer)

Trigonometical functions

Functions involving exp

Functions involving ln

Functions involving a modulus (e.g., $|x|$, etc.)

Unspecified functions

Polynomials $a_n x^n + a_{n-1}x^{n-1} + \cdots + a_0$ are ordered according to a dictionary ordering of the "word" $a_n a_{n-1} \cdots a_0$. The "alphabet" of coefficients has the following order:

(a) Zero

(b) Positive numbers, in increasing order

(c) Negative numbers, in decreasing order

(d) Unspecified constants

Having established the ordering scheme for functions of a single variable, consider the problem of ordering functions of two variables, $f(x, y)$. We regard them as functions of y, with parameters that can be numbers or functions of x. Thus we append to the alphabet of coefficients of a polynomical in y, coefficients that are

(e) Functions of x

Similarly, functions $f(x, y, y')$ are ordered by regarding them as functions of y', with parameters that can be numbers, functions of x, or

(f) Functions of x and y

Proceeding in this way, we arrive at a principle for *ordering* a list of differential equations. To render it completely rigorous would involve more detailed considerations (e.g., does $\sin^2 x$ precede or follow $\sin x^2$?). That was not our aim. Our aim was only the pragmatic one of devising a scheme to facilitate looking up a differential equation.

For further convenience, the work has been divided into sections, each containing equations of a general type. For example:

2.1 $y'' + f(y) = 0$

2.2 $y'' + g(x)h(y) = 0$

2.3 $y'' + f(x, y) = 0$

The rule at this level of organization is that *separable functions* $g(x)h(y)$ precede inseparable functions $f(x, y)$. The scheme is less formidable than might appear from its description. Further clarification can be obtained, if necessary, simply by looking through the pages to see how it looks in practice.

Ambiguities can arise because, to a certain extent, the form of a nonlinear equation is dependent on how one chooses to write it (for example, $xy'' + 2y' + xe^y = 0$ and $y'' + 2y'/x + e^y = 0$ would occupy different positions in our scheme of ordering). In general, we have chosen to leave each equation in the form in which it appeared in the source literature. In a few instances we have deviated from this principle when it would have led to the separation of an equation from the class of equations to which it naturally belongs. The user is advised to multiply or divide by an obvious factor if an equation is not found where expected, and to try again.

No linear equations, and no first order equations, have been dealt with. The major parts of the book have been labeled 2, 3, 4, 5, 6, N to denote second order, third order, ..., sixth and higher orders, and Nth order. At the end of each of these parts are collected *systems* of simultaneous equations, of the appropriate order.

The systems of each order have been grouped according to the highest order of differentiation that occurs (the independent variable is always either x or t, with differentiation denoted by a prime or a dot, respectively). Within each group, the ordering scheme is exactly like that for single equations, except that the dependent variable and its derivatives (y, y', etc.) are now vector quantities. To facilitate the use of this scheme to look up a system of equations, the systems have been written with all terms involving *derivatives* of the dependent variables on the left-hand side and all undifferentiated terms on the right.

2

SECOND ORDER EQUATIONS

2.1 $y'' + f(y) = 0,\ f(y)$ polynomial

1. $$y'' + \mu y + y^2 - C = 0,$$

where μ and C are constants.

Put

$$\begin{aligned} y &= au(\theta) + u_1, u_1 = \left(\frac{1}{2}\right)\left\{(\mu^2 + 4C)^{1/2} - \mu\right\}, \theta = rx, \\ r &= (\mu + 2u_1)^{1/2}, a = \left(\frac{1}{2}\right) r^2 \end{aligned}$$

to obtain the canonical form of the given DE as

$$u'' + u + (1/2)u^2 = 0. \tag{1}$$

A first integral of (1) is

$$u'^2 = E - u^2 - (1/3)u^3, \tag{2}$$

where E is an integration constant related to the energy of the oscillator. The phase diagram of (2) is drawn. The closed orbit solutions, filling the region inside the loop of the separatrix, namely,

$$y_s = 3\,\mathrm{sech}^2(\theta/2) - 2, \quad E = 4/3, \tag{3}$$

are easily drawn. These closed (periodic) solutions tend asymptotically to the separatrix. Goldshtik, Hussain, and Shtern (1991)

2. $$y'' + 4y(y-1) - A = 0,$$

where A is a constant.

Multiplying the given DE by y' and integrating once, we have

$$\frac{3}{8}\left(\frac{dy}{dx}\right)^2 = -y^3 + \frac{3}{2}y^2 + 34Ay + B = (y - y_1)(y - y_2)(y_3 - y), \quad \text{say}, \tag{1}$$

where $y_1 < y_2 < y_3$. The general form of the cubic in (1) is drawn. It is evident that the only real solutions of this equation occur when $\left(\dfrac{dy}{dx}\right)^2 \geq 0$; thus the solution is either

$y = y_1$ or a nonlinear oscillation between y_2 and y_3. Two special cases of the general cubic are when $y_2 \to y_1$, giving the solitary wave, and when $y_2 \to y_3$, giving a discontinuity between y_1 and y_3. The solution can be written in terms of a Jacobian elliptic function $\mathrm{cn}(u;\nu)$ as

$$y = y_2 + (y_3 - y_2)\,\mathrm{cn}^2\left(x\left[(2/3)(y_3 - y_1)\right]^{1/2}; \nu\right),$$

where $\nu = (y_3 - y_2)/(y_3 - y_1)$. In the case $A = B = 0$, $\nu = 1$ and the solution becomes $y = (3/2)\,\mathrm{sech}^2 x$, the solitary wave. Johnson (1970)

3. $$y'' + 6y^2 + a = 0, \quad y(0) = y(\pi) = 0,$$

where $a \in R$ is a parameter.

It is shown that for all $k \in N$, there exist values $a_k < \cdots < a_1$ such that for $a < a_k$, there exist k solutions of the given Sturm-Liouville problem. Ruf and Solimini (1986)

4. $$y'' - y - (1/4)y^2 = 0, \quad -1 < x < 1, y(-1) = y(1) = 1.$$

Multiplication by y' and integration leads to

$$y'^2 = y^2 - y_0^2 + (1/6)(y^3 - y_0^3) \tag{1}$$

on using evenness of solution in x, and setting $y(0) = y_0$, as yet unknown. Separation of variables in (1) and integration yields a solution in terms of elliptic functions,

$$y(x) = y_0 - u_1\,\mathrm{sc}^2(\lambda^{-1}6^{-1/2}x; m),$$

where $u_{1,2} = -3(1 + y_0/2) \pm (1/2)[3(2 - y_0)(6 + y_0)]^{1/2}$, $\lambda^{-1} = (1/2)(-u_2)^{1/2}$, and $m = (u_1 - u_2)/(-u_2)$.

In order to satisfy the BC, we require that y_0 satisfies

$$y_0 = 1 + u_1\mathrm{sc}^2(\lambda^{-1}6^{-1/2}; m) \equiv f(y_0). \tag{2}$$

Numerical solution of (2) leads to $y_0 = 0.60850$. Hart (1980)

5. $$y'' - y - (1/4)y^2 = 0, \quad -1 < x < 1, y(-1) = 1,\ y(1) = 1.$$

(a) The given DE arises in the determination of the stationary temperature distribution in a bar whose ends $x = \pm 1$ are kept at the temperature $y = 1$, etc. It is stated from previous results that the given problem has a unique solution such that $0 \le y(x) \le 1$ and the bounds are given by

$$c_1(x) \le y((x) \le c_2(x), b_1(x) \le y(x) \le b_2(x),$$

where

$$\begin{aligned} c_1(x) &= 1 - 0.35(1 - x^2) - 0.05(1 - x^4), \\ c_2(x) &= 1 - 0.35(1 - x^2) - 0.04(1 - x^4), \end{aligned}$$

and

$$\begin{aligned} b_1(x) &= \{\cosh(3/2)^{1/2}x\}/\cosh(3/2)^{1/2}, \\ b_2(x) &= \{\cosh x\}/\{\cosh 1\}. \end{aligned}$$

These bounds are considerably improved upon, by using the maximum principle. Numerical results are also given.

(b) The given DE is autonomous. Its general solution is easily found to be

$$y(x) = (x - x_0)\ln(x - x_0) + \alpha(x - x_0),$$

where x_0 and α are arbitrary constants. Anderson and Arthurs (1982), Kruskal and Clarkson (1992)

6. $$\mathbf{y'' - y - y^2 = 0.}$$

The given DE has $(y', y) = (0, 0)$ as a saddle point, while $(y', y) = (-1, 0)$ is a center. Writing $y = -1 + \epsilon z$, we find that $z'' + z = \epsilon z^2$, $z(0) = 1$, $z'(0) = 0$. Putting $wx = \theta$, $w^{-2} = 1 - \epsilon\eta(\epsilon)$, we have

$$z'' + z = \epsilon\{\eta z + (1 - \epsilon\eta)z^2\}, \quad ' \equiv \frac{d}{d\theta}. \tag{1}$$

An equivalent integral equation form of (1) is

$$\begin{aligned} z(\theta) &= \cos\theta + \epsilon\int_0^\theta \sin(\theta - \tau)[\eta z + (1 - \epsilon\eta)z^2]d\tau, \\ z'(\theta) &= -\sin\theta + \epsilon\int_0^\theta \cos(\theta - \tau)[\eta z + (1 - \epsilon\eta)z^2]d\tau. \end{aligned} \tag{2}$$

We know that for ϵ small enough the solutions of (2) are periodic. Expanding z and η in powers of ϵ and applying the periodicity condition, we find that $\eta(\epsilon) = 0.\epsilon - (5/6)\epsilon + \epsilon^2$, with $T = 2\pi/w, T_0 = 2\pi, T_1 = 0, T_2 = 5\pi/6$. Verhulst (1990), p.143

7. $$\mathbf{y'' - (3/2)y^2 = 0, \quad y(1) = 1,\ y'(1) = 1.}$$

Writing the given DE as $y'y'' - (3/2)y^2y' = 0$, integrating twice, and using the IC, we have $y = 4(3 - x)^{-2}$. Reddick (1949), p.192

8. $$\mathbf{y'' - 2y^2 + y = 0, \quad y(0) = 1,\ y'(0) = 0.}$$

Upper and lower bounds of the solution are found. Exact solution is $y = \sec x$. Eliason (1972)

9. $$\mathbf{y'' - 6y^2 = 0.}$$

Put $p(y) = y'$, $p\dfrac{dp}{dy} = 6y^2$, $y' = \pm(4y^3 - C_1)^{1/2}$, $y = P(x + C_2)$; P is the Weierstrass P-function with invariants $g_2 = 0, g_3 = C_1$; C_1 and C_2 are arbitrary constants. Kamke (1959), p.542; Murphy (1960), p.380

10. $$\mathbf{y'' - 6y^2 + (3/2)y_0^2 = 0, \quad y(x_0) = y_0,\ y'(x_0) = 0.}$$

The solution is easily found to be

$$y = y_0 + (3/2)y_0\tan^2[(3y_0/2)^{1/2}(x - x_0)].$$

Bartashevich (1973)

11. $$\boldsymbol{y'' - 6y^2 + 4y = 0.}$$

Multiply by y' and integrate:

$$y'^2 - 4y^3 + 4y^2 + C = 0, \quad x = \int \frac{dy}{\{4y^3 - 4y^2 - C\}^{1/2}} + C_1;$$

the solution is obtained in terms of elliptic functions. C and C_1 are arbitrary constants. For $C = 0$,

$$y = \frac{1}{\sin^2(x + C_1)}.$$

Kamke (1959), p.543

12. $$\boldsymbol{y'' - k^2 y^2 = 0,}$$

where k is a constant.

Putting $y' = p$, we have $y'' = p\dfrac{dp}{dy}$; we can integrate the resulting DE twice $\left(\text{using } p = \dfrac{dy}{dx}\right)$ to obtain

$$\int \frac{dy}{\{C_1 + (2/3)k^2 y^3\}^{1/2}} + C_2 = \pm x,$$

where C_1 and C_2 are arbitrary constants. The quadrature may be evaluated explicitly. Martin and Reissner (1958), p.77

13. $$\boldsymbol{\epsilon y'' - y^2 = 0, \quad \epsilon \ll 1, \quad y(0) = \alpha, \quad y(1) = \beta.}$$

Two boundary layer solutions are constructed. They are

$$y^i = \alpha/\{1 + (\alpha/6)^{1/2}\xi\}^2 + \cdots, \xi = x\epsilon^{-1/2}$$

and

$$y^I = \beta/\{1 + (\beta/6)^{1/2}\zeta\}^2 + \cdots, \zeta = (1 - x)/\epsilon^{1/2}$$

near $x = 0$ and $x = 1$, respectively. Nayfeh (1985), p.347

14. $$\boldsymbol{y'' + k^2 y^2 = 0,}$$

where k is a constant.

Writing $y' = p, y'' = p\dfrac{dp}{dy}$, we can integrate the resulting DE twice to obtain

$$\int \frac{dy}{[C_1 - (2/3)k^2 y^3]^{1/2}} + C_2 = \pm x,$$

where C_1 and C_2 are arbitrary constants. The quadrature may be evaluated explicitly. Martin and Reissner (1958), p.77

15. $$\epsilon y'' + y^2 - 1 = 0.$$

Multiplying by y' and integrating, we get

$$(1/2)\epsilon y'^2 + (1/3)y^3 - y = E,$$

where E is a constant. The solution of this equation is $y = a\zeta(x/\delta) + b$, where $\zeta(x)$ satisfies the equation

$$\zeta'^2 = 4(1 - m^2)\zeta - 4(1 - 2m^2)\zeta^2 - 4m^2\zeta^3.$$

The coefficients a, b, and δ are related to m and ϵ according to

$$\begin{aligned} b &= (1 - 2m^2)/(m^4 - m^2 + 1)^{1/2}, \quad a = 3m^2/(m^4 - m^2 + 1)^{1/2}, \\ \delta &= [2\epsilon(m^4 - m^2 + 1)^{1/2}]^{1/2}, \quad b^3 - 3b - 3E = 0. \end{aligned}$$

More explicitly,

$$\begin{aligned} y &= [3m^2/(m^4 - m^2 + 1)^{1/2}]\,\mathrm{cn}^2[(x - x_1)/\{2\epsilon(m^4 - m^2 + 1)^{1/2}\}^{1/2}; m] \\ &\quad -(2m^2 - 1)/(m^4 - m^2 + 1)^{1/2}. \end{aligned}$$

Note that the period of $\mathrm{cn}(x)$ is $4K(m)$. The exact solution is related to asymptotic solutions, using a variational approach. Kath, Knessl, and Matkowsky (1987)

16. $$y'' - my^2 + n = 0, \quad y(0) = 0,\ y(1/2) = 1,\ y'(1/2) = 0,$$

where $m \in R$ is given. The constant n forms the eigenvalue.

The given problem occurs in viscous flow between parallel plates, and is solved in terms of elliptic functions by Tang (1967). Bespalova (1984)

17. $$y'' + y + \epsilon y^2 = 0,$$

$y(0) = A,\ y'(0) = 0$. Here ϵ is small.

A uniform perturbation solution to $O(\epsilon^3)$ is

$$\begin{aligned} y(\theta, \epsilon) &= A\cos\theta + \epsilon(A^2/6)(-3 + 2\cos\theta + \cos 2\theta) \\ &\quad +\epsilon^2(A^3/3)[-1 + (29/48)\cos\theta + (1/3)\cos 2\theta + (1/16)\cos 3\theta] + O(\epsilon^3), \end{aligned}$$

where $\theta = wx$ and $w(\epsilon) = 1 - \epsilon^2(5A^2/12) + O(\epsilon^3)$. Mickens (1981), p.39

18. $$y'' + y - \alpha y^2 = 0, \quad \alpha > 0.$$

The method of harmonic balance gives the approximate solution for small α as

$$y = c + a\cos wx,$$

where $w^2 = 1 - 2\alpha c$, $c = 1/(2\alpha) - \{[1/(2\alpha)\}(1 - 2\alpha^2 a^2)^{1/2}$, implying that the frequency-amplitude relation is

$$w = (1 - 2\alpha^2 a^2)^{1/4}, \ a < 1/(2\alpha)^{1/2}.$$

Jordan and Smith (1977), p.120

19. $$y'' - \epsilon y^2 + y - \alpha = 0,$$

where α and ϵ are positive constants.

This is the relativistic equation for the central orbit of a planet, where $y = 1/r$, and r, x are polar coordinates of the planet in the plane of its motion. The term ϵy is the Einstein correction; ϵ and α are positive constants with ϵ very small. The equilibrium point $y = \{1 + (1 - 4\epsilon\alpha)^{1/2}\}/2$ is a center according to linear approximation. Jordan and Smith (1977), p.58

20. $$y'' + w^2 y - \epsilon y^2 = 0, \quad y(0) = 0, \; y(X) = L,$$

where w, ϵ, X, and L are constants.

Introducing $\lambda = y'(0)$ and putting $\zeta = wy/L$, we get a first integral of the given system as

$$\frac{d\zeta}{dx} = \pm w \, \alpha^{1/2} P(\zeta)^{1/2}, \tag{1}$$

where $\alpha = 2\epsilon\lambda/(3w^3), P(\zeta) = \zeta^3 - (1/\alpha)\zeta^2 + 1/\alpha$; λ is yet an unknown constant. Equation (1) may be solved subject to $\zeta(0) = 0, \zeta(X) = wL/\lambda$. Assume that $\alpha < 2/3(3)^{1/2}$, so that $P(\zeta) = 0$ has three real roots, ζ_1, ζ_2, and ζ_3 ($\zeta_1 > \zeta_2 > \zeta_3$). For $\lambda > 0$, the general solution can be expressed in terms of elliptic integrals as $wx = \alpha^{-1/2}\gamma^{-1}F(\phi/\beta)$, where $F(\phi/\beta)$ is the elliptic integral, $\gamma = (1/2)(\zeta_1 - \zeta_3)^{1/2}, \sin^2\beta = (\zeta_2 - \zeta_3)/(\zeta_1 - \zeta_3)$, and $\cos^2\phi = (\zeta_1 - \zeta_2)(\zeta_1 - \zeta_3)/\{(\zeta_2 - \zeta_3)(\zeta_1 - \zeta)\}$.

Similar results may be obtained when $\lambda < 0$. Explicit approximate solution is obtained when α is small. Numerical results are depicted. Becket (1980)

21. $$y'' + y - k(1 + \epsilon y^2) = 0,$$

where $\epsilon \ll 1$ and k are parameters.

The given DE is the orbital equation of a planet about the sun. The perturbation solution with the initial conditions $y(0) = k(e + 1), y'(0) = 0$, where e is the eccentricity of the unperturbed orbit, is

$$y = k(e\cos x + 1) + \epsilon k^3\{ex\sin x + (1/2)e^2 + 1 - \{(e^2/3) + 1\}\cos x - (e^2/6)\cos 2x\} + O(\epsilon^2).$$

Jordan and Smith (1977), p.149

22. $$y'' - \epsilon y^2 + y - \mu = 0,$$

where ϵ (small) and μ are constants.

Writing $y = \mu + a\cos(\theta + \psi)$, $y' = -a\sin(\theta + \psi)$, and following the Lagrange method of averaging, we obtain, by substitution in the original DE, equations for a and ψ as functions of θ:

$$\begin{aligned} \frac{da}{d\theta} &= -\epsilon\{\mu + a\cos(\theta + \psi)\}^2 \sin(\theta + \psi), \\ \frac{d\psi}{d\theta} &= -(\epsilon/a)\{\mu + a\cos(\theta + \psi)\}^2 \cos(\theta + \psi). \end{aligned} \tag{1}$$

Averaging the RHSs of (1) with respect to θ over 0 to 2π gives

$$\frac{da}{d\theta} = 0, \quad \frac{d\psi}{d\theta} = -\epsilon\mu. \tag{2}$$

On integration of (2) and using the ICs $a(0) = a_0$, $\psi(0) = \psi_0$, we get

$$y = \mu + a_0 \cos(\theta - \epsilon\mu\theta + \psi_0) + O(\epsilon), \quad \frac{dy}{d\theta} = -a_0 \sin(\theta - \epsilon\mu\theta + \psi_0) + O(\epsilon)$$

on the time scale $O(1/\epsilon)$. Verhulst (1990), p.175

23. $\boldsymbol{y'' - (y - c)(y - 1)V - \theta y = 0, \quad |x| < L,\ y(\pm L) = 0,}$

where θ is a constant; $0 < c < 1/2$; $V > 0$ is a constant.

Stability and bifurcation of solution of the given problem are discussed. Conley and Smoller (1986)

24. $\boldsymbol{y'' + \epsilon(y - p)(y - q) + y = 0, \quad 0 < p < q,}$

where ϵ, p, and q are parameters.

Some types of relaxation oscillations are discussed. Karreman (1949)

25. $\boldsymbol{y'' - Ay^2 - By - C = 0, \quad A \neq 0,}$

$y(0) = D$, $y(X) = D$, where X, A, B, C, and D are constants.

The following result is proved regarding the given BVP: If $B^2 - 4AC \leq 0$, the BVP has either no solution, one solution, or two solutions, depending on the value of X. In case $B^2 - 4AC > 0$, let $y_- < y_+$ denote the roots of $Ay^2 + By + C = 0$. Then, if $y_- < D < y_+$, there are a finite number of solutions for each X, with this number increasing to infinity as $X \to \infty$. If either $D \leq y_-$ or $D \geq y_+$, there are either no solutions, one solution, or two solutions, depending on the values of X. Chicone (1988)

26. $\boldsymbol{y'' - Ay^2 - 2By - C = 0, \quad A > 0,}$

B and C are constants and $y(0) = \alpha$, $y'(0) = m$.

Set $w(x) = (1/6)[Ay(x) + B]$ so that the given IVP goes to

$$w'' - 6w^2 + (1/2)g_2 = 0, \tag{1}$$

$w(0) = a$, $w'(0) = b$, where $g_2 = (B^2 - AC)/3$, $a = (A\alpha + B)/6$, $b = Am/6$. Multiply (1) by w' and integrate to get

$$w'^2 = 4w^3 - g_2 w - g_3, \tag{2}$$

where $g_3 = 4a^3 - g_2 a - b^2$. Equation (2) is satisfied by the Weierstrass P-function, which has the following relevant properties. It has Laurent series representation

$$P(x) = x^{-2} + \sum_{k=2}^{\infty} C_k x^{2k-2},$$

where

$$\begin{aligned} C_2 &= g_2/20,\ C_3 = g_3/28, \\ C_k &= \frac{3}{(2k+1)(k-3)} \sum_{j=2}^{k-2} C_j C_{k-j},\ k \geq 4, \end{aligned}$$

where g_2 and g_3 are complex constants. If g_2, g_3, and x are real, $P(x)$ is also real. The function $P(x)$ is doubly periodic, and if $2w_1, 2w_3$ are the fundamental periods and $w_2 = -(w_1 + w_3)$, then $P(w_i) = e_i$ and $P'(w_i) = 0$, where e_1, e_2, e_3 are the appropriately ordered roots of the equation

$$4x^3 - g_2 x - g_3 = 0. \tag{3}$$

The function $P(x)$ satisfies the addition formula

$$P(x+w) = (1/4)[P'(x) - P'(w)]^2[P(x) - P(w)]^{-2} - [P(x) + P(w)]. \tag{4}$$

The following specific result is proved: Let e_j be the largest real root of Eqn. (3) with $e_j \leq a$ and

$$\tau = \pm \int_{e_j}^{a} (4s^3 - g_2 s - g_3)^{-1/2} ds,$$

where the sign of the integral agrees with the sign of b. Then the real solution of the given IVP is given by

$$y(x) = (1/A)[(3/2)\{P'(x+\tau)\}^2\{P(x+\tau) - e_j\}^{-2} - 6\{P(x+\tau) + e_j\} - B].$$

Heckenback and Heimes (1976)

27. $\quad y'' + (1/2)y^3 = 0.$

The given DE is a conservative Hamiltonian system, which can be integrated in terms of the Jacobi elliptic functions [see Eqn. (2.26.6)]. The motion described by it is completely regular; however, it was studied for its singularity structure as a special case of the Duffing equation. The expansion was sought in the form

$$y(x) = \sum_{j=0}^{\infty} a_j (x - x_0)^{j-1}.$$

Recursion formulas were obtained for the coefficients:

$$a_j(j+1)(j-4) = -\frac{1}{2}\sum_k \sum_\ell a_{j-k-\ell} a_k a_\ell, \quad 0 < k + \ell \leq j,\ 0 \leq k, \ell < j,$$

where $a_0 = 2i, a_1 = a_2 = 0, a_4 =$ an arbitrary constant. The arbitrary pole position x_0 and a_4 constitute two pieces of arbitrary data, consistent with a local representation of the general solution. Fournier, Levine, and Tabor (1988)

28. $\quad y'' + y' + (1/2)\, y^3 = 0$

is a constant.

The given DE is a non-Hamiltonian system with a stable spiral point at the origin to which all initial conditions tend in the limit $x \to \infty$. Although the motion in this case is regular, it is studied for its singularity structure (as a special case of the Duffing equation) in the form

$$y(x) = \sum_{j=0}^{\infty} \sum_{k=0}^{\infty} a_{jk} (x - x_0)^{j-1} [(x - x_0)^4 \ln(x - x_0)]^k.$$

For details see Eqn. (2.7.26). The given DE is not solvable in closed form. Fournier, Levine, and Tabor (1988)

29. $$\mathbf{y'' + y^3 = 0, \quad y(0) = 0.2,\ y(2) = 0.1846.}$$

The given system describes the small oscillation of a mass that is attached to two springs. The problem is solved numerically and the maximum amplitude at $x = 1$ is found to be $y(1) = 0.1960$. Kubicek and Hlavacek (1983), p.228

30. $$\mathbf{y'' + y^3 = 0.}$$

Writing $y_1 = y,\ y_2 = y'$, we have the system

$$y_1' = y_2, \quad y_2' = -y_1^3, \tag{1}$$

with the first integral $I(y_1, y_2) = (1/2)y_2^2 + (1/4)y_1^4$. Then the solution of the system (1) can be expressed in terms of Jacobi elliptic functions:

$$\begin{aligned} y_1(x) &= \frac{(y_{20}/w)\ \mathrm{sn}(wx,i) + y_{10}\ \mathrm{cn}(wx,i)\ \mathrm{dn}(wx,i)}{1 + (y_{10}/v)^2 \mathrm{sn}^2(wx,i)}, \\ y_2(x) &= \frac{y_{20}\mathrm{cn}(wx,i)\ \mathrm{dn}(wx,i)[1 - (y_{10}/v)^2\ \mathrm{sn}^2(wx,i)]}{[1 + (y_{10}/v)^2 \mathrm{sn}^2(wx,i)]^2} \\ &\quad - \frac{2y_{10}(y_{10}/v)^2\ \mathrm{sn}(wx,i)[1 + \mathrm{sn}^2(wx,i)]}{[1 + (y_{10}/v)^2\ \mathrm{sn}^2(wx,i)]^2}, \end{aligned}$$

where $w = (1/2)^{1/2}V, V = (4E)^{1/4}, E = y_{20}^2/2 + y_{10}^4/4 = I(y_1, y_2)$ with $i = (-1)^{1/2}$, and sn, cn, dn are Jacobi elliptic sine, cosine, and delta functions, respectively. The initial values are $y_{10} = y_1(x = 0)$ and $y_{20} = y_2(x = 0)$. Duarte, Euler, Moreira, and Steeb (1990)

31. $$\mathbf{y'' - y + y^3 = 0.}$$

Let $w = y'/y$; then $y = \exp[\int w dx]$ and $y(w' + w^2) - y + y^3 = 0$ or $w' + w^2 - 1 + \exp[2\int w dx] = 0$. This is a generalized Riccati equation and can be solved iteratively by choosing w in the integrand. Two methods—one using power series and continued fractions form and another employing continued fractions expansion with undetermined coefficients only—give good approximate or even exact solutions.

For example, one approximate solution is

$$y = 2^{1/2}\left[\frac{x^2}{\{14 - (133)^{1/2}\} + 1}\right]^{49/15(133)^{1/2}} \times \left[\frac{x^4}{63} + \frac{4x^2}{9} + 1\right]^{-77/60} e^{-x^2/30}.$$

This function agrees remarkably with the exact solution $2^{1/2}$ sech x. Ditto and Pickett (1988), Jordan and Smith (1977), Plaat (1971)

32. $$\mathbf{y'' + y^3 - y + \lambda = 0,}$$

where λ is a parameter.

The phase paths in the (y, y') phase plane are given by

$$(1/2)y'^2 + \lambda y + (1/4)y^4 - (1/2)y^2 = C,$$

where C is an arbitrary constant. The solution may be expressed in terms of elliptic functions. $C = 0$ describes a zero-energy surface in (y, y', λ) space. Its behavior as λ is changed may be of some interest. Jordan and Smith (1977), p.35

33. $$y'' + y^3 + \lambda y = 0, \quad 0 < x < 1, \; y(0) = y(1) = 0,$$

where λ is a real parameter.

Let y be a solution of the given problem for some $\lambda \in R$. Let $y'(0) = c > 0$. Multiplying the given DE by $2y'$ and integrating, we have

$$\frac{dy}{dx} = \left[c^2 - \lambda y^2 - \frac{1}{2}y^4\right]^{1/2} = \left[\frac{1}{2}(a^2 + y^2)(b^2 - y^2)\right]^{1/2}, \quad 0 \le x \le x_0,$$

where

$$a^2, b^2 = \pm\lambda + [\lambda^2 + 2c^2]^{1/2}, \quad a > 0, b > 0, \tag{1}$$

and where x_0 is the smallest $x \in (0,1)$ such that $y(x) = b$; such a point must exist in order that $y(1) = 0$. Thus,

$$x = \int_0^y \frac{dv}{[(1/2)(a^2 + v^2)(b^2 - v^2)]^{1/2}}, \quad 0 \le x \le x_0, \; 0 \le y(x) \le b, \tag{2}$$

and on $[0, x_0]$. This, in terms of the elliptic function sd, is equivalent to

$$y(x) = \frac{ab}{(a^2 + b^2)^{1/2}} \operatorname{sd}\left(\frac{a^2 + b^2}{2}x, \frac{b}{a^2 + b^2}\right), \tag{3}$$

which holds on $[0,1]$ because (3) satisfies the given DE and IC $y(0) = 0, y'(0) = c$ and the solution of the problem is unique.

Let $b(a^2 + b^2)^{-1/2} = \mu$; all values $\mu \in (0,1)$, and only such values are consistent with (1). Then $y(1) = 0$ if and only if

$$[a^2 + b^2]^{1/2} = 2(2)^{1/2} n K(\mu), \quad 0 < \mu < 1, \tag{4}$$

for some positive integer n. Consequently,

$$\begin{aligned} a &= 2^{3/2} n (1 - \mu^2)^{1/2} K(\mu), \; b = 2(2)^{1/2} n \mu K(\mu), \\ \lambda &= (1/2)(a^2 - b^2) = 4n^2(1 - 2\mu^2)[K(\mu)]^2. \end{aligned} \tag{5}$$

Thus, every solution with $y'(0) > 0$ has the representation (3)-(5). Fraenkel (1980)

34. $$y'' - y^3 + \lambda y = 0, \quad 0 < x < 1, \; y(0) = y(1) = 0,$$

where λ is a real parameter.

Multiplying the given DE by $2y'$ and integrating, we have

$$\frac{dy}{dx} = [c^2 - \lambda y^2 + (1/2)y^4]^{1/2} = [(1/2)(\alpha - y^2)(\beta - y^2)]^{1/2}, \quad 0 \le x \le x_0,$$

where $\alpha, \beta = \lambda \pm [\lambda^2 - 2c^2]^{1/2}$ and $y'(0) = c > 0$ and where x_0 is the smallest zero in $(0,1)$ of the quartic shown. There must be such a zero, since the condition $y(1) = 0$ requires $y'(x)$ to vanish somewhere in $(0,1)$. Now, in order that the quartic vanish at some value of y, at least one of α and β must be real and nonnegative. This implies that $\lambda \ge (2)^{1/2}c$ [since the solution corresponding to $\lambda = (2)^{1/2}c$ fails to satisfy $y(1) = 0$]. Thus we have the solution

$$x = \int_0^y \frac{dv}{[(1/2)(a^2 - v^2)(b^2 - v^2)]^{1/2}}, \quad 0 \le x \le x_0, \; 0 \le y(x) \le b, \tag{1}$$

where

$$a^2, b^2 = \lambda \pm [\lambda^2 - 2c^2]^{1/2}, a > b > 0. \tag{2}$$

The elliptic function sn is so defined that (1) is equivalent (for $0 \le x \le x_0$) to the formula

$$y(x) = b\,\mathrm{sn}(ax/(2)^{1/2}, b/a), \tag{3}$$

which holds not merely on $[0, x_0]$ but on $[0, 1]$, because this function satisfies the DE and IC $y(0) = 0, y'(0) = c$. Let $b/a = \mu$; all values $\mu \in (0, 1)$, and only such values are consistent with (2) for some λ and c. Then $y(1) = 0$ if and only if

$$a = 2(2)^{1/2} n K(\mu), \quad 0 < \mu < 1, \tag{4}$$

for some positive integer n. Consequently,

$$b = 2(2)^{1/2} n\mu K(\mu),\ \lambda = (1/2)(a^2 + b^2) = 4n^2(1 + \mu^2)[K(\mu)]^2. \tag{5}$$

Equations (3)-(5) represent the solution of the problem with $y'(0) > 0$. Fraenkel (1980)

35. $\quad$ $\mathbf{y'' - 2y - y^2 + y^3 = 0.}$

There are three equilibrium points: $y = 0, -1$, and 2. To find the solution about $y = 2$, we put $y = 2 + u$ in the given DE to get $u'' + 6u + 5u^2 + u^3 = 0$. By putting $\tau = (6)^{1/2}x$, we have

$$u'' + u + (5/6)u^2 + (1/6)u^3 = 0, \quad ' \equiv \frac{d}{d\tau}.$$

Using the method of multiple scales, etc., the solution may be found to be

$$y - 2 = u = a_0 \cos(w\tau + \beta_0) + (5/36)a_0^2[\cos(2w\tau + 2\beta_0) - 3] + \cdots,$$

where $w = 1 - (49/216)a_0^2 + \cdots$. Here a_0 and β_0 are constants. Nayfeh (1985), p.139

36. $\quad$ $\mathbf{y'' - (y - \lambda)(y^2 - \lambda) = 0,}$

where λ is a constant.

A phase plane study shows that the equilibrium points $(y = \lambda, y' = 0)$ are stable for $0 < \lambda \le 1$, unstable for $\lambda \le 0$ and $\lambda > 1$. The equilibrium points $(y = \lambda^{1/2}, y' = 0)$ are stable for $\lambda \ge 1$ and unstable for $0 \le \lambda < 1$. The points $(y = -\lambda^{1/2}, y' = 0)$ are unstable. The given DE may be solved by multiplying by y' and integrating. A second integration gives the solution in terms of elliptic functions. Jordan and Smith (1977), p.31

37. $\quad$ $\mathbf{y'' - (3/2)y^3 + (9/2)Cy^2 + 3(1 - C^2 - A_2)y + 3A_2C = 0,}$

where A_2 and C are constants.

A solution in the form is found by substitution, etc.:

$$y = \frac{A\ \mathrm{cn}^2(kx)}{1 + B\ \mathrm{cn}^2(kx)} - E.$$

Writing $y = Y + D$, D can be so chosen that $A_2 = 0$ in the given DE; now see Eqn. (2.1.38). Krishnan (1982)

38. $$y'' - (3/2)y^3 + (9/2)\lambda y^2 - 3(\lambda^2 - 1)y = 0,$$

where λ is a constant.

The given DE represents a traveling wave solution of the Boussinesq equation. The periodic solution is sought in the form

$$y(x) = \frac{AP(x)}{1 + BP(x)}, \tag{1}$$

where $P(x)$ is the Weierstrass elliptic function satisfying the DE

$$\left(\frac{dP}{dx}\right)^2 = 4P^3 - g_2 P - g_3, \tag{2}$$

with two invariants g_2 and g_3 which are assumed to be real and which satisfy $g_2^3 - 27g_3^2 > 0$; A and B are constants. Substituting (1) in the given DE and using (2), etc., we find that

$$\begin{aligned} A &= \frac{48\lambda \pm [6144\lambda^2 + 30720]^{1/2}}{12\lambda^2 + 96}, \\ B &= \frac{4 + 3\lambda A}{4(\lambda^2 - 1)}, \\ g_2 &= \frac{2(\lambda^2 - 1)}{B}, \quad g_3 = \frac{g_2}{4B}. \end{aligned} \tag{3}$$

The exact periodic solution can now be written as

$$y(x) = \frac{AP(x + \delta; g_2, g_3)}{1 + BP(x + \delta; g_2, g_3)},$$

where A, B, g_2, and g_3 are given by (3), and δ is an integration constant of (2). To obtain periodic solutions, we use

$$P(x + \delta; g_2, g_3) = e_3 + (e_2 - e_3)\ \mathrm{sn}^2\left((e_1 - e_3)^{1/2} x + \delta'\right),$$

where δ' is an arbitrary real constant and e_1, e_2, e_3 are real roots of $4y^3 - g_2 y - g_3 = 0$ with $e_1 > e_2 > e_3$. Thus, the bounded periodic solution is

$$y(x) = A\frac{e_3 + (e_2 - e_3)\ \mathrm{sn}^2\left((e_1 - e_3)^{1/2} x + \delta'\right)}{1 + B\left[e_3 + (e_2 - e_3)\ \mathrm{sn}^2\left((e_1 - e_3)^{1/2} x + \delta'\right)\right]}. \tag{4}$$

When the modulus $m = (e_2 - e_3)/(e_1 - e_3)$ of the Jacobian elliptic function is unity, i.e., $e_1 = e_2$, the period becomes infinite. Moreover, $e_1 + e_2 + e_3 = 0$ from the equation for the roots e_i. In this situation, we get instead of (4) the solitary wave solution

$$y(x) = A\frac{e_1 - (e_1 - e_3)\,\mathrm{sech}^2\{(e_1 - e_3)^{1/2} x + \delta'\}}{1 + B[e_1 - (e_1 - e_3)\,\mathrm{sech}^2\{(e_1 - e_3)^{1/2} x + \delta'\}]}.$$

Krishnan (1982a)

39. $$\mathbf{y'' - 2y^3 = 0, \quad y(1) = 1,\ y'(2) + y^2(2) = 0.}$$

The exact solution of the given BVP is $y = 1/x$. Roberts (1979), p.364

40. $$\mathbf{y'' - 2y^3 + 2y = 0, \quad y(0) = 0,\ y'(0) = 1.}$$

Putting $y' = p, y'' = (dp/dy)p$, we obtain

$$p\frac{dp}{dy} = 2y(y^2 - 1). \tag{1}$$

Integrating (1), we have $p^2/2 = y^4/2 - y^2 + c_1$. Using the IC $y(0) = 0, y'(0) = 1$, we have $c_1 = 1/2$. Therefore, $p^2 = (y^2 - 1)^2$ or

$$\frac{dy}{dx} = \pm(y^2 - 1). \tag{2}$$

Integrating (2), we obtain $\pm\tanh^{-1} y = x + c_2$. Using the IC we have $c_2 = 0$. Therefore, $y = \pm\tanh x$. Martin and Reissner (1958), p.77

41. $$\mathbf{y'' - y(2y^2 - 3\mu y + 1) = 0,}$$

where μ is a constant.

Multiplying the given DE by y', integrating, and ignoring the constant of integration, we get

$$y'^2 = y^2(1 - 2\mu y + y^2). \tag{1}$$

Writing $y = 1/\psi$ in (1), we get the linear DE

$$\psi'' - \psi + \mu = 0,$$

which is easily solved. Conte and Musette (1992)

42. $$\mathbf{y'' - ay^3 = 0,}$$

where a is a constant.

Multiply by y' and integrate:

$$y'^2 = (a/2)y^4 + C,$$

where C is an arbitrary constant. A second integration gives solutions in terms of elliptic functions. A special solution is $y = (2/a)^{1/2}[1/(x - C_1)]$, with C_1 arbitrary. Kamke (1959), p.543

43. $$\mathbf{y'' - k^2y^3 = 0.}$$

Writing $y' = p, y'' = p\dfrac{dp}{dy}$, we can integrate the resulting DE twice (using $p = dy/dx$) to obtain

$$\int \frac{dy}{[C_1 + (1/2)k^2y^4]^{1/2}} + C_2 = \pm x,$$

where C_1 and C_2 are arbitrary constants. The quadrature may be evaluated explicitly. Martin and Reissner (1958), p.77

44. $$\boldsymbol{y'' + y + \epsilon y^3 = 0, \quad y(0) = 1, \ y'(0) = 0.}$$

This anharmonic oscillator has explicit exact solution in terms of elliptic functions. Here a new perturbative method is used by writing the system as

$$y'' + y + (w^2 - 1)y^{1+2\delta} = 0, \quad y(0) = 1, \ y'(0) = 0,$$

and seeking its solution in the form

$$y^{2\delta} = 1 + \delta \ \ln(y^2) + (\delta^2/2)[\ln(y^2)]^2 + (\delta^3/6)[\ln(y^2)]^3 + \cdots .$$

Comparison is made with other methods of solutions. Bender et al. (1989)

45. $$\boldsymbol{y'' + y + \epsilon y^3 = 0, \quad y(0) = a, \ y'(0) = 0,}$$

where ϵ is not necessarily a small parameter and a is a constant.

Using an improvised Linstedt–Poincaré procedure [see Davies and James (1966)] a perturbation method gives

$$\begin{aligned} y &= a[1 - \alpha/8 - (1/64)\alpha^2 - (19/512)\alpha^3 - (13/4096)\alpha^4]\cos X \\ &\quad + a[\alpha/8 + (9/256)\alpha^3 - (7/4096)\alpha^4]\cos 3X + a[\alpha^2/64 + (19/4096)\alpha^4]\cos 5X \\ &\quad + (1/512)a\alpha^3\cos 7X + (1/4096)a\alpha^4\cos 9X + O(\alpha^5), \\ w^2 &= \frac{1}{1-3\alpha}[1 - (3/8)\alpha^2 - (51/512)\alpha^4 + O(\alpha^6)], \ \alpha = \frac{\epsilon a^2}{4 + 3\epsilon a^2}, \quad X = wx, \end{aligned}$$

where $\alpha < 1/3$ is a parameter and ϵa^2 is arbitrary. The series is shown to converge quickly, regardless of the magnitude of ϵa^2. Burton (1984)

46. $$\boldsymbol{y'' + y + \lambda y^3 = 0, \quad y(0) = A, \ y'(0) = 0,}$$

where λ is a parameter.

An exact solution of the given IVP is

$$y(x, \lambda) = \frac{2\pi A}{kK(k)} \sum_{m=1}^{\infty} a_{2m-1}\cos(2\pi(2m-1)x/P),$$

where $k = [\lambda A^2/2(1 + \lambda A^2)]^{1/2}$, $P = 4K(k)/[1 + \lambda A^2]^{1/2}$, and $K(k) = F(\pi/2; k)$ is the complete elliptic integral of the first kind. The coefficients $a_{2m-1} = q^{(2m-1)/2}/[1 + q^{2m-1}]$, where $q = \exp[-\lambda K(k')/K(k)]$, $k' = [1 - k^2]^{1/2}$. Mickens (1988)

47. $$\boldsymbol{y'' + y + \epsilon y^3 = 0,}$$

where ϵ is a small positive parameter.

Multiplying the given DE by y' and integrating, we get the first integral,

$$F(y, y') = y'^2/2 + y^2/2 + \epsilon y^4/4.$$

$F(y, y')$ = a constant gives, outside $(0,0)$, closed curves in the (y', y) plane. Introducing the transformation $y(x) = r(x)\cos(x + \psi(x)), y'(x) = -r(x)\sin(x + \psi(x))$ into the DE and using the Lagrange method, we are led to the system

$$\begin{aligned} r'(x) &= \epsilon \sin(x+\psi) r^3 \cos^3(x+\psi), \\ \psi'(x) &= \epsilon \cos(x+\psi) r^2 \cos^3(x+\psi). \end{aligned} \tag{1}$$

Averaging the RHS of (1) with respect to x over 0 to 2π, and keeping r and ψ fixed, we get

$$r_a' = 0, \quad \psi_a' = (3/8)\epsilon r_a^2, \tag{2}$$

where r_a and ψ_a denote approximate solutions. Integrating (2), we get, on substitution in expressions for $y(x)$ and $y'(x)$, the approximate solution

$$y(x) = r_0 \cos(x + (3/8)\epsilon r_0^2 x + \psi_0) + O(\epsilon),\ y'(x) = -r_0 \sin(x + (3/8)\epsilon r_0^2 x + \psi_0) + O(\epsilon)$$

on the x scale $1/\epsilon$; r_0 and ψ_0 are determined from the IC. Verhulst (1990), p.172

48. $$\boldsymbol{y'' + y + \epsilon y^3 = 0, \quad y(0) = A,\ y'(0) = 0,}$$

where $\epsilon > 0$ is small.

A naive perturbation solution is

$$y = A\cos x + \epsilon(A^3/32)[(\cos 3x - \cos x) - 12x\sin x] + O(\epsilon^2),$$

which has a secular term $(-12xA^3/32)\sin x$. However, a uniform perturbation solution is

$$\begin{aligned} y(\theta, \epsilon) &= A\cos\theta + \epsilon(A^3/32)(-\cos\theta + \cos 3\theta) \\ &\quad + \epsilon^2(A^5/1024)(23\cos\theta - 24\cos 3\theta + \cos 5\theta) + O(\epsilon^3), \end{aligned}$$

where $\theta = wx$ and $w(\epsilon) = 1 + \epsilon(3A^2/8) - \epsilon^2(57A^4/256) + O(\epsilon^3)$. Mickens (1981), p.30

49. $$\boldsymbol{y'' - y + \epsilon y^3 = 0, \quad y(0) = a,\ y'(0) = 0,}$$

where ϵ and a are constants.

Using an improvised Linstedt-Poincaré perturbation scheme, the frequency expansion is found to be

$$w^2 = [-1 + (3/4)\epsilon a^2][1 + S(\delta_i)],$$

where

$$S(\delta_i) = \alpha^2\delta_2 + \alpha^4\delta_4 + \cdots, \quad \alpha = \epsilon a^2(-4 + 3\epsilon a^2)^{-1},$$

where δ_i are chosen to annul secular terms. Here $1/3 < \alpha < 1$, corresponding to $\infty > \epsilon a^2 > 2$. The convergence of the perturbation series is slow when $\alpha \to 1$ and rapid when $\alpha \to 1/3$. Burton (1984)

50. $$\boldsymbol{y'' + (2\beta/\alpha)y^3 + 2y = 0,}$$

where α and β are constants and $\alpha\beta < 0$.

The given DE arises from the traveling wave form of the nonlinear Klein-Gordon equation $u_{tt} - u_{xx} + \alpha u + \beta u^3 = 0$. On multiplication by y' and integration, with constant of integration set equal to zero, the given DE gives

$$y'^2 + (\beta/\alpha)y^4 + 2y^2 = 0.$$

The given DE has a solitary wave solution $y = \pm[-\alpha/\beta]^{1/2}\tanh x$. Hereman, Banerjee, Korpel, and Assanto (1986)

51. $$\boldsymbol{y'' - (\beta/\alpha)y^3 - y = 0,}$$

where α and β are constants and $\alpha\beta < 0$.

The given DE arises from the traveling wave form of the nonlinear Klein-Gordon equation $u_{tt} - u_{xx} + \alpha u + \beta u^3 = 0$. If the given DE is multiplied by y' and integrated, with constant of integration set equal to zero, we have

$$y'^2 - (\beta/2\alpha)y^4 - y^2 = 0. \tag{1}$$

Equation (1) has a solitary wave solution $y = \pm[-2\alpha/\beta]^{1/2}\operatorname{sech} x$. Hereman, Banerjee, Korpel, and Assanto (1986)

52. $$\boldsymbol{y'' + w^2 y - \beta^2 y^3 = 0, \quad y(0) = A, \ y'(0) = 0,}$$

where w, β, and A are constants.

Changing the variable to $u = y + w/((3)^{1/2}\beta)$, we obtain the problem

$$u'' + (3)^{1/2}w\beta u^2 - \beta^2 u^3 = \frac{2w^3}{3(3)^{1/2}\beta}, \quad u(0) = A + \frac{w}{(3)^{1/2}\beta}, \ u'(0) = 0. \tag{1}$$

Seeking the solution of the problem in the form

$$u(x) = c_0 + c_1 \sin wx + c_2 \sin^2 wx + \cdots, \tag{2}$$

we easily check that

$$\begin{aligned} c_0 &= A + w/((3)^{1/2}\beta), \ c_1 = 0, \ 2w^2 c_2 + (3)^{1/2} w\beta c_0^2 - \beta^2 c_0^3 = 2w^3/(3(3)^{1/2}\beta), \\ c_{n+2} &= \frac{n^2}{(n+1)(n+2)}c_n - \frac{(3)^{1/2}\beta}{w(n+1)(n+2)}b_n + \frac{\beta^2}{w^2(n+1)(n+2)}\sum_{k=0}^{n} b_k c_{n-k}, \ n \geq 1, \end{aligned}$$

where $b_n = \sum_{k=0}^{n} c_k c_{n-k}$, $n \geq 0$. It is easy to check that $c_{2n+1} = 0$, $n \geq 0$, while the first few even coefficients are

$$c_0 = A + \frac{w}{(3)^{1/2}\beta}, \ c_2 = \frac{A\beta^2}{2w^2}\left(A^2 - \frac{w^2}{\beta^2}\right), \ c_4 = \frac{A\beta^4}{8w^4}\left(A^2 - \frac{w^2}{\beta^2}\right)\left(A^2 + \frac{w^2}{\beta^2}\right).$$

Hence, the solution is

$$y(x) = A + \frac{A\beta^2}{2w^2}\left(A^2 - \frac{w^2}{\beta^2}\right)\sin^2 wx + \cdots.$$

For $A = \pm w/\beta$, all even coeficients can also be shown to be zero; so $y = \pm w/\beta$ is an exact solution for this choice of A.

It is also shown that the series (2) is absolutely convergent for all x and hence (2) is solution of the given problem for all x. It is instructive to find the power series solution for (1) in the form

$$u(x) = \sum_{n=0}^{\infty} a_n x^n. \tag{3}$$

Substituting (3) in (1), etc., leads to $a_1 = a_3 = a_5 = 0, \ldots$, and

$$\begin{aligned} a_0 &= A + w/((3)^{1/2}\beta),\ 2a_2 + (3)^{1/2} w\beta a_0^2 - \beta^2 a_0^3 = 2w^3/(3(3)^{1/2}\beta), \\ a_{n+2} &= -[(3)^{1/2}\beta w/\{(n+1)(n+2)\}]d_n + [\beta^2/\{(n+1)(n+2)\}] \qquad (4) \\ &\quad (d_0 a_n + d_2 a_{n-2} + \cdots + d_n a_0),\ n = 2k, k \geq 1, \end{aligned}$$

where $d_n = a_0 a_n + a_2 a_{n-2} + \cdots + a_n a_0$. We verify from (4) that the coefficients in (4) all vanish for $A = \pm w/\beta$, again yielding the exact solution $y(x) = \pm w/\beta$. Balachandran, Thandapani, and Balasubramanian (1988)

53. $\quad y'' - (2v/9\gamma)y^3 + (v/2\gamma)y = 0,$

where γ and v are constants.

Seeking a solution in the form

$$y(x) = \sum_{n=1}^{\infty} a_n g^n, \tag{1}$$

where $g(x) = \exp[\pm i(v/2\gamma)^{1/2}x]$, we get the recurrence relation

$$\left(n^2 - 1\right) a_n + \frac{4}{9} \sum_{m=2}^{n-1} \sum_{\ell=1}^{m-1} a_{n-m} a_{m-\ell} a_\ell = 0, \quad n \geq 3,$$

leading to

$$a_{2n} = 0, \quad a_{2n+1} = \frac{(-1)^n}{2^{3n}} \left(\frac{4}{9}\right)^n a_1^{2n+1}.$$

The series (1) can be summed up to yield

$$y = \pm(3/(2)^{1/2}) \sec[(v/2\gamma)^{1/2}x + \Delta],$$

where $\Delta = (1/2)\ln(a_1/3)$. Coffey (1990)

54. $\quad y'' + w^2 y + \epsilon y^3 = 0, \quad y(0) = A,\ y'(0) = 0,$

where $\epsilon > 0$, A, and w are dimensionless parameters.

The solution of the given problem is

$$y(x) = A\,\mathrm{cn}(\lambda x, k),$$

where $\lambda = [\epsilon A^2 + w^2]^{1/2}$, $k = [\epsilon A^2/\{2(\epsilon A^2 + w^2)\}]^{1/2}$, and cn is the Jacobian elliptic function.

An approximate perturbation solution to order r is found to be

$$y = A\cos\tau + (rA/24)(\cos 3\tau - \cos\tau), \quad \tau = wx/(1-r)^{1/2},$$

where $\tau = wx/(1-r)^{1/2}, r = \lambda/(1+\lambda), \lambda = 3A^2\epsilon/(4w^2)$. These two solutions are computed and compared. Jones (1978)

55. $$y'' + \alpha y + \beta y^3 = 0,$$

where α and β are constants.

With $\dfrac{dy}{dt} = v$, we have $v\dfrac{dv}{dy} + (\alpha y + \beta y^3) = 0$. An integration yields

$$v^2 + \alpha y^2 + \beta y^4/2 = h = \text{constant}. \tag{1}$$

Near $v = 0, y = 0$, the curves (1) are all closed curves that look like ellipses. Because of symmetries in (1), the period T of motion can easily be found to be

$$T = 4\int_0^a \frac{dy}{[h - \alpha y^2 - \beta y^4/2]^{1/2}}, \tag{2}$$

where $y_{\max} = a$ is the maximum displacement where $v = 0$, so that from (1), $a^2 = [-\alpha + (\alpha^2 + 2\beta h)^{1/2}]/\beta$, where a positive sign of the radical is taken for $\beta > 0$ (hard spring) as well as for $\beta < 0$ (soft spring), since $h > 0$ and a^2 should be small and positive.

Writing $h - \alpha y^2 - \beta y^4/2 = (\beta/2)(a^2 - y^2)(b^2 + y^2)$, in which

$$(\beta/2)(-b^2 + a^2) = -\alpha, \tag{3}$$

and putting $y = a\sin\theta$, we can write (2) as

$$T = 4(2)^{1/2}\int_0^{\pi/2} \frac{d\theta}{[2\alpha + \beta a^2(1 + \sin^2\theta)]^{1/2}}, \tag{4}$$

wherein b^2 has been eliminated with the help of (3).

The solution may be drawn in the phase plane, when $\beta > 0$ or $\beta < 0$, with quite distinct portraits. Stoker (1950), p.20

56. $$y'' + (\alpha/3)y^3 - vy = 0,$$

where α and v are constants.

The given DE may be integrated directly after multiplication by y', ignoring the constant of integration and integrating once again. We illustrate a simple method which yields a solitary wave solution for many nonlinear ODEs. The linear part of this equation has two exponential solutions: $\exp[\pm Kx]$, where $K = v^{1/2}$, $v > 0$. The first term in the expansion for y may be chosen to be $g_1(x) = \exp(-Kx) = g(x)$, say. A simple normalization of the given DE is useful: Put $y = (3v/\alpha)^{1/2}\tilde{y}$; we get

$$-v\tilde{y} + v\tilde{y}^3 + \tilde{y}_{xx} = 0. \tag{1}$$

To get a particular solution (of solitary wave type), put $\tilde{y}(x) = \sum_{n=1}^{\infty} a_n g^n(x)$ into (1) and apply Cauchy's rule for the triple product appearing in it due to nonlinearity. This yields the recurrence relation

$$(n^2 - 1)a_n + \sum_{m=2}^{n-1}\sum_{\ell=1}^{m-1} a_\ell a_{m-\ell} a_{n-m} = 0, \quad n \geq 3,$$

where $a_1 > 0$ is arbitrary, $a_2 = 0, a_3 = -(a_1/2)^3, a_5 = a_1^5/2^6, a_7 = -a_1^7/2^9$. Therefore, we may write $a_{2j} = 0$, $a_{2j+1} = (-1)^j a_1^{2j+1}/2^{3j}$, $j \geq 1$. Thus, we have

$$\begin{aligned}\tilde{y} &= \sum_{n'=0}^{\infty} \frac{(-1)^{n'} a_1^{2n'+1}}{2^{3n'}} (g(x))^{2n'+1} \\ &= 2(2)^{1/2} a g(x)/(1 + a^2 g^2(x)), \quad a = \frac{a_1}{2(2)^{1/2}},\end{aligned}$$

where we have used $\sum_{n=0}^{\infty} (-1)^n x^{2n+1} = x/(1+x^2)$, $|x| < 1$. Recalling $g(x) = \exp(-kx)$, we obtain the closed-form solution $y = (6v/\alpha)^{1/2} \operatorname{sech}[v^{1/2}x]$, if $a_1 = 2^{3/2}$. Hereman, Banerjee, Korpel and Assanto (1986)

57. $$\boldsymbol{y'' + [1/(v^2 - 1)]\alpha y + [1/(v^2 - 1)]\beta y^3 = 0,}$$

where α, β, and v are constants.

The given DE describes solitary wave solutions of the nonlinear Klein-Gordon equation $y_{tt} - y_{xx} + \alpha y + \beta y^3 = 0$; it has the exact solutions (1) $\pm(-2\alpha/\beta)^{1/2} \operatorname{sech}[\{\alpha/(1-v^2)\}^{1/2} x)]$, (2) $\pm(-\alpha/\beta)^{1/2} + 2ag/(1 \mp g)$, where

$$g(x) = \exp\{[\pm 2\alpha/(-1+v^2)]^{1/2} x\}, \ \alpha > 0, v^2 > 1.$$

The solutions may be obtained directly or by series expansion in the exponential function $g(x)$. Hereman, Banerjee, Korpel, and Assanto (1986)

58. $$\boldsymbol{y'' + w^2 y + \beta y^3 = 0.}$$

With $dy/dx = v$, we have in the (y, v) plane, $dv/dy = -(w^2 y + \beta y^3/v)$, with the solution

$$v^2 + y^2(w^2 + \beta y^2/2) = C, \tag{1}$$

where C is an integration constant. For positive β, real solutions exist only for $C \geq 0$. Substituting $v = \dfrac{dy}{dx}$ in (1) and integrating, we have

$$x = \pm \int \frac{dy}{[C - y^2(w^2 + \beta y^2/2)]^{1/2}} + C_1,$$

where C_1 is another constant of integration. The solution is discussed in the phase plane. It may be found explicitly in terms of Jacobi elliptic functions. Struble (1962), p.19

59. $$\boldsymbol{y'' + w^2 y + \epsilon w^2 y^3 - f_0 = 0,}$$

where w, ϵ, f_0 are constants and $y(0) = y'(0) = 0$.

A perturbation solution is found in the form

$$\begin{aligned}y(x) &= \{f_0/w^2\}[1 - \cos w\xi] + \{\epsilon f_0^3/32w^6\}[-80 + 65\cos w\xi \\ &\quad + 16\cos 2w\xi - \cos 3w\xi] + O(\epsilon^2),\end{aligned}$$

where

$$\xi = x/[1 - (15/8)(f_0^2/w^4)\epsilon].$$

Bauer (1966)

60. $$y'' + y + \eta(a_2 y^2 + a_3 y^3) = 0, \quad \eta \ll 1,$$

where a_2 and a_3 are constants.

A perturbation solution is found, by suitably removing the secular terms, in the form

$$y(x) = y_0(x) + \eta y_1(x) + \eta^2 y_2(x) + \cdots, \quad w^2 = 1 + \eta e_1 + \eta^2 e_2 + \cdots.$$

The functions y_n and the constants e_n are found to be

$$y_0(x) = C \sin wx,$$

$$\begin{aligned} y_1(x) &= C_1 \sin(wx + r_1) - (1/2)(a_2/w^2)C^2 - (1/6)(a_2/w^2)C^2 \cos 2wx \\ &\quad - (1/32)(a_3/w^2)C^3 \sin 3wx, \\ y_2(x) &= (1/4)(a_2 a_3/w^2)C^4 + (19/96)(a_2 a_3/w^4)C^4 \cos 2wx \\ &\quad - (1/48)[a_2^2 + (3a_3 C/8)^2](C^3/w^4) \sin 3wx \\ &\quad + (1/96)[a_2 a_3/w^4]C^4 \cos 4wx + (1/32)^2[a_3^2/w^4]C^5 \sin 5wx, \\ e_1 &= (3/4)a_3 C^2, \ e_2 = [5C^2/(6w^2)][-a_2^2 + (9/320)a_3^2 C^2], \text{ etc.,} \end{aligned}$$

where

$$w^2 = 1 + \eta(3/4)a_3 C^2 + \eta^2(5/6)(C^2/w^2)(-a_2^2 + (9/320)a_3^2 C^2) + O(\eta^3).$$

Here C and r_1 are arbitrary constants. This solution is compared with an "exact" numerical solution. Hagedorn and Schäfer (1980)

61. $$y'' + w_0^2 y + by^2 + ay^3 = 0,$$

where w_0^2, b, and a are constants.

This DE may be studied in the phase plane. Putting

$$\begin{aligned} \frac{dy}{dx} &= v, \\ \frac{dv}{dx} &= -w_0^2 y - by^2 - ay^3. \end{aligned} \tag{1}$$

The singularities of the system (1) are $v = 0$,

$$y = \{-b \pm (b^2 - 4aw_0^2)^{1/2}\}/(2a).$$

Thus, there are three singularities unless $b^2 - 4aw^2 \leq 0$, in which case there is only one. For $aw_0^2 > b^2/4$, $b < 0$, a simple perturbation analysis about $(0,0)$ shows that $y = y_1, v = v_1$, where y_1 and v_1 are small, lead to

$$\frac{dv_1}{dy_1} = \frac{-w_0^2 y_1 - by_1^2 - ay_1^3}{v_1}$$

or

$$v_1^2 + w_0^2 y_1^2 + 2by_1^3/3 + ay_1^4/2 = \text{constant}.$$

Thus, (0,0) is a center. Phase portraits are drawn for the cases when (0,0) is a center, with the presence or otherwise of other singularities. Mahaffey (1976)

62. $$y'' + (\sigma_2 v/\gamma)y^3 + (v/\gamma)y^2 - (v/\gamma)y = 0,$$

where σ_2, v, and γ are constants.

The given DE arises from the traveling wave form of a combined and a modified Korteweg-deVries equation. Seeking the solution in the form

$$y(x) = \sum_{n=1}^{\infty} a_n g^n(x), \tag{1}$$

where $g(x) = e^{\pm(v/\gamma)^{1/2}x}$, we find from the given DE (after multiplying by γ) that a_1 is arbitrary. Choosing it to be positive, we have

$$a_2 = -a_1^2/3,\ a_3 = -(a_1^3/8)(\sigma_2 - 2/3),\ a_4 = (a_1^4/12)(\sigma_2 - 2/9),$$

$$a_5 = (a_1^5/64)\{\sigma_2^2 - (20/9)\sigma_2 + 20/81\},\ a_6 = -(a_1^6/64)\{\sigma_2^2 - 20/27)\sigma_2 + 4/81\},$$

and, in general,

$$(n^2 - 1)a_n + \sum_{\ell=1}^{n-1} a_{n-\ell}a_\ell + \sigma_2 \sum_{m=2}^{n-1}\sum_{\ell=1}^{m-1} a_{n-m}a_{m-\ell}a_l = 0.$$

For the special choice $\sigma_2 = -2/9$, we get

$$a_3 = a_1^3/9,\ a_4 = -a_1^4/27,\ a_5 = a_1^5/81,\ a_n = (-1)^{n-1}a_1^n/3^{n-1},\ n \geq 1.$$

Therefore, in this case,

$$\begin{aligned} y &= -3\sum_{n=1}^{\infty}(-a_1 g/3)^n = 3dg/(1 + dg), d = a_1/3 \\ &= 3de^{-Kx}/(1 + de^{-Kx}) = 3e^{-Kx/2+\Delta}/\{e^{Kx/2-\Delta} + e^{-Kx/2+\Delta}\} \\ &= (3/2)\{1 - \tanh(Kx/2 - \Delta)\}, \Delta \equiv (1/2)\ln d. \end{aligned}$$

Other solutions with a constant term in the series (1) may also be found. Coffey (1990)

63. $$y'' + Ay + 2By^3 - E = 0, \quad y(0) = 0,\ y'(0) = v_0,$$

where A, B, and E are positive constants.

Let $Y, -\overline{Y}, \overline{X}$ be the maximum, minimum, and period of the solution y, and $x_{y'}, -\tilde{x}_{y'}$ the first positive zero and the first negative zero of y'. Using elementary methods it is shown that:

(a) For fixed A, B, and E, Y and $\overline{Y}$ increase and $X, x_{y'}, \tilde{x}_{y'}$ decrease as v_0 increases for $v_0 > V_0$, a constant.

(b) For fixed v_0, A, and B, Y increases and $\overline{Y}, X, x_{y'}, \tilde{x}_{y'}$ decrease as E increases.

(c) As A or B increases (all others fixed), Y, $\overline{Y}, x_{y'}, \tilde{x}_{y'}$, and X all decrease.

Funato (1958/59)

64. $$y'' + ay^3 + by^2 + cy + d = 0,$$

where a, b, c, and d are constants.

Multiply by y' and integrate:

$$y'^2 + (1/2)ay^4 + (2/3)by^3 + 2cy^2 + 2dy + C = 0, \tag{1}$$

where C is a constant.

(a) If (1) can be written as

$$\left(\frac{dy}{dx}\right)^2 = h^2(y-\alpha)(y-\beta)(y-\gamma),$$

a cubic, then

$$y = \gamma + (\alpha - \gamma)/\operatorname{sn}^2(-hMx, k),$$

where

$$k^2 = (\beta-\gamma)/(\alpha-\gamma), M^2 = (\alpha-\gamma)/4.$$

(b) If we can write (1) as

$$\left(\frac{dy}{dx}\right)^2 = h^2(y-\alpha)(y-\beta)(y-\gamma)(y-\delta),$$

a quartic, then

$$\begin{aligned} y &= (\beta z^2 - A\alpha)/(z^2 - A), \quad z = \operatorname{sn}\{hM(x-x_0), k\}, \\ A &= (\beta-\delta)/(\alpha-\delta), \ k^2 = (\beta-\gamma)(\alpha-\delta)/\{(\alpha-\gamma)(\beta-\delta)\}, \\ M^2 &= (\beta-\delta)(\alpha-\gamma)/4. \end{aligned}$$

For the IVP, $y'' + y + \mu y^3 = 0$, $y(0) = a, y'(0) = 0$, $\mu > 0$ and a being arbitrary constants, see Kamke (1959), p.544; Murphy (1960), p.381; Ames (1968), pp.42, 55, 167; Davis (1962), p.210

65. $$\xi^2 y'' + (1/2)(y - y^3) = 0, \quad y(a) = -1, \ y(b) = 1,$$

where $0 < a < b$ and ξ is a parameter.

Multiplying the given DE by y', integrating twice, and using the BC at $x = a$, we get

$$x(y) - a = \int_{-1}^{y} \frac{\xi dy}{[c_\xi - F(Y)]^{1/2}}, \tag{1}$$

where $F(y) = (1/2)y^2 - (1/4)y^4$, and c_ξ is the constant of integration obtained from the second BC:

$$b - a = \int_{-1}^{1} \frac{\xi dy}{[c_\xi - F(Y)]^{1/2}}. \tag{2}$$

The maximum of $F(y)$ are at $y = \pm 1$ and $F(y) = F(-y), F(1) = 1/4$. Let δ be defined by $c_\xi = F(1) + \delta = 1/4 + \delta$, where $\delta > 0$ is to be determined. If we let $\psi = 1 - y$, then

$$b - a = \xi I(\delta), \tag{3}$$

where

$$I(\delta) = \int_0^2 \frac{d\psi}{[\delta + \psi^2 - \psi^3 + (1/4)\psi^4]^{1/2}}.$$

Since $I(\delta) \to \infty$ as $\delta \to 0$, $I(\delta) \to 0$ as $\delta \to \infty$, and $I(\delta)$ is a continuous function of δ, (3) can be satisfied for a unique $\delta > 0$. Hence there exists a unique solution of the given problem which is the inverse of $x(y)$, given by (1). Since c_ξ is unique, $x(y)$, and consequently $y(x)$, are unique. Caginalp and Hastings (1986)

66. $$\xi^2 y'' + (1/2)(y - y^3) + k/2 = 0, \quad y(a) = \alpha, \ y(b) = \gamma,$$

where ξ and k are parameters; a, b, α, and γ are constants.

Multiplying the given DE by y', we have

$$\xi^2 (y'^2)' + \{F(y)\}' + ky' = 0, \tag{1}$$

where $F(y) = (1/2)y^2 - (1/4)y^4$. Integrating (1), we get

$$\left(\frac{dy}{dx}\right)^2 = \xi^{-2}[c_{\xi,k} - F(y) - ky], \tag{2}$$

where $c_{\xi,k}$ is a constant.

The function $y(x)$ has an inverse, $x(y)$, provided that $y'(a) > 0$ for all $x, a \le x \le b$. Taking the positive square root of (2), we get

$$x(y) - x(\alpha) = \int_\alpha^y \frac{dy}{[c_{\xi,k} - F(y) - ky]^{1/2}}, \tag{3}$$

where $x(\alpha) = a$ by the BC. The constant $c_{\xi,k}$ is obtained from the BC $x(\gamma) = b$, i.e.,

$$b - a = \xi \int_\alpha^\gamma \frac{dy}{[c_{\xi,k} - F(y) - ky]^{1/2}}. \tag{4}$$

Equation (4) clearly defines a unique $c_{\xi,k}$ for all ξ greater than some ξ_0.

The results are summarized as follows: Let k be bounded and independent of ξ. Then the given problem has a unique, monotonic solution $y(x)$ given by the inverse of $x(y)$ in (3), in which $c_{\xi,k}$ is defined by (4). Caginalp and Hastings (1986)

67. $$y'' - y + y^4 = 0.$$

To obtain a solution about the equilibrium point $y = 1$, write $y = 1 + v$ and introduce $\tau = (3)^{1/2}x$. The new equation is $v'' + v + 2v^2 + (4/3)v^3 + \cdots = 0$ [cf. the solution of Eqn. (2.1.35)]. The solution is found to be

$$y - 1 = v = a_0 \cos(w\tau + \beta_0) + (1/3)a_0^2[\cos(2w\tau + 2\beta_0) - 3],$$

where $w = 1 - (7/6)a_0^2 + \cdots$. Here a_0 and β_0 are arbitrary constants. Nayfeh (1985), p.141

68. $$y'' + y(y-1)(y^2 + \alpha y + \beta) = 0, \alpha^2 < 4\beta,$$

where α and β are constants.

It is stated with due references that the given DE has periodic solutions. Wang (1987)

69. $$\epsilon^2 y'' + Q(y) = 0, -\infty < x < \infty,$$

where $Q(y) = y(y+1)(\beta - y)(\alpha - y)$, $0 < \beta < \alpha, y(-\infty) = 1$ and $y(\infty) = \beta$.

It is shown that the given problem has a unique solution provided that $\int_{-1}^{\beta} Q(u)du = 0$, i.e.,

$$\alpha = \beta + (3 + \beta - 2\beta^2)/5(\beta - 1) \quad \text{for} \quad 1 < \beta < 3/2.$$

This solution has the following asymptotic behavior as $|x| \to \infty$:

$$\begin{aligned} y(x) &\sim s_+ - a_+ e^{-\sigma_+ x}, \ y'(x) \sim a_+ \sigma_+ e^{-\sigma_+ x} \quad \text{as} \quad x \to \infty, \\ y(x) &\sim s_- + a_- e^{-\sigma_- x}, \ y'(x) \sim a_- \sigma_- e^{-\sigma_- x} \quad \text{as} \quad x \to -\infty, \end{aligned}$$

where

$$\sigma_- = [(1+\beta)(3\beta^2 - 2 + \beta)/(5(\beta - 1))]^{1/2}, \ \sigma_+ = (\beta + 1)[(-2\beta^2 + 3\beta)/(5(\beta - 1))]^{1/2},$$

for $1 < \beta < 3/2; s_- = -1, s_+ = \beta$; a_+ and a_- can be found in terms of β by way of a numerical integration.

Other solutions such as one with a shock are also constructed. Ward (1992)

70. $$y'' - (A_1 + A_2 y + A_3 y^2 + A_4 y^3 + A_5 y^4) = 0,$$

where A_i $(i = 1, 2, 3, 4, 5)$ are constants.

Conditions are found such that the solution of the given DE is expressible in the form

$$y(x) = \lambda P(x)/[1 + \mu P(x)],$$

where λ and μ are constants and $P(x)$ is the Weierstrass elliptic function satisfying the DE

$$p'^2 = 4p^3 - g_2 P - g_3$$

with two invariants g_2 and g_3, which are assumed to be real and satisfy $g_2^3 - 27g_3^2 > 0$. Four algebraic equations for the four unknowns λ, μ, g_2, and g_3 are obtained, and solved. Krishnan (1986)

71. $$y'' - (3/4)y^5 = 0, \quad y(0) = A^{-1/2}, \ y'(0) = (1/2)A^{-3/2}.$$

The solution of the given IVP $y(x) = (A - x)^{-1/2}$ exists only for $0 \le x < A$. Wong (1970)

72. $$y'' + y + \epsilon y^5 = 0, \quad y(0) = a, \ y'(0) = 0,$$

where ϵ and a are constants.

Using a modified Linstedt-Poincaré perturbation scheme and defining a new perturbation parameter $\alpha = \epsilon a^4/[3 + (15/8)\epsilon a^4]$, the perturbation solution is found to be

$$\begin{aligned} y(x) &= a[1 - \alpha/8 - (29/16384)\alpha^2]\cos X + a[(15/128)\alpha - (15/1024)\alpha^2]\cos 3X \\ &+ a[(1/128)\alpha + (15/1024)\alpha^2]\cos 5X + a[(55/32768)\alpha^2]\cos 7X \end{aligned}$$

$$+ \; a[(3/32768)\alpha^2]\cos 9X + O(\alpha^3),$$

$$w^2 = [1/\{1-(15/8)\alpha\}][1-(195/512)\alpha^2 + O(\alpha^3)], \; X = wx.$$

The parameter α here may at most be as large as 8/15. Burton (1984)

73. $$\boldsymbol{y'' + w_0^2 y - \epsilon y^5 = 0, \epsilon \ll 1,}$$

where w_0 is another constant.

A straightforward perturbation gives

$$\begin{aligned} y &= a\cos(w_0x+\beta) + [\epsilon a^5/(16w_0^2)][5w_0x\sin(w_0x+\beta) \\ &\quad - (5/8)\cos(3w_0x+3\beta) - (1/24)\cos(5w_0x+5\beta)] + \cdots, \end{aligned}$$

with a secular term. A first order uniform approximation by renormalization, etc., is

$$y = a\cos\left([1 - 5\epsilon a^4/(16w_0^2)]w_0x + \beta\right) + \cdots.$$

Nayfeh (1985), p.97

74. $$\boldsymbol{y'' + (1-w^2)\delta y^5 - y + y^3 = 0; \quad 0 \le w^2 < 1,}$$

where w and δ are constants.

The given DE has a solution $y = 2[1 + b(w)\cosh 2x]^{-1/2}$, where $b(w) = [1 + (16/3)\delta(1 - w^2)]^{1/2}$. Blanchard, Stubbe, and Vazquez (1988)

75. $$\boldsymbol{y'' - m^2 y + 16\pi g^2 y^3 - 48\pi^2\lambda^2 y^5 = 0,}$$

where $m, g,$ and λ are constants.

The given DE describes the classical minima in Euclidian space with renormalized parameters. It has the following five types of solutions:

(a) $y = 0$.

(b) $y = \pm[g^2/(6\pi\lambda^2) + (4g^4 - 3m^2\lambda^2)^{1/2}/(12\pi\lambda^2)]^{1/2}$.

(c) $y = \pm m e^{\pm mx}/[4\pi g^2 e^{\pm 2mx} + (2\pi m^2/k)\{k^2 + k_0^2 e^{\pm 4mx}\}]^{1/2}$, where $k = k_0 + \sqrt{k_0 - k_0^2}$, $k_0^2 = g^4/m^4 - \lambda^2/m^2 < 1$.

(d) $y = \dfrac{\pm\phi_0 \,\mathrm{sn}\left(\lambda\{\phi_1(\phi_0^2 - \phi_2)\}^{1/2}x\right)}{[4\pi(\phi_0^2\,\mathrm{sn}\,(\lambda\{\phi_1(\phi_0^2-\phi_2)\}^{1/2}x + \phi_2 - \phi_0^2))]^{1/2}}$, where sn$(x)$ is the inverse elliptic function and

$$\begin{aligned} \phi_0^2 &= [2g^2 + (4g^4 - 3m^2\lambda^2)^{1/2}]/(3\lambda^2), \\ \phi_1 &= \frac{4g^2 - (4g^4 - 3m^2\lambda^2)^{1/2} + g^2[48g^2 + 24(4g^2 - 3m^2\lambda^2)^{1/2}]^{1/2}}{6\lambda^2}, \\ \phi_2 &= \frac{4g^2 - (4g^4 - 3m^2\lambda^2)^{1/2} - g[48g^2 + 24(4g^4 - 3m^2\lambda^2)^{1/2}]^{1/2}}{6\lambda^2}. \end{aligned}$$

(e) If $g^2 = \lambda m$, we have

$$y^2(x) = \pm[(m/8\pi\lambda)\{1 \pm \tanh(mx)\}]^{1/2}.$$

Su and Chen (1987)

76. $$y'' + y + by^3 + cy^5 = 0,$$

where a, b, and $c > 0$ are constants.

With any IC, $y(0) = A, y'(0) = V_0$, the solution is periodic. A first integral with the ICs above is

$$y'^2 + y^2 + (b/2)y^4 + (c/3)y^6 = V_0^2 + A^2 + (b/2)A^4 + (c/3)A^6.$$

The solution may be expressed in terms of elliptic integrals. Mickens (1981), p.29

77. $$y'' + (v\sigma_1/\gamma)y^5 + (v/\gamma)y^3 - (v/\gamma)y = 0,$$

where σ_1, γ, and v are constants.

The given DE arises from the traveling wave form of a Korteweg-deVries equation with fifth degree nonlinearity.

A solution is sought in the form

$$y = \sum_{n=1}^{\infty} a_n g^n(x), \quad g(x) = e^{-Kx}, \ K = \sqrt{v/\gamma}.$$

It may be checked that a_1 is arbitrary. Choosing it to be positive, we find that $a_2 = 0, a_3 = -a_1^3/8, a_4 = 0,$

$$\begin{aligned} a_5 &= -(1/24)(3a_3a_1^2 + \sigma_1 a_1^5) = -(a_1^5/192)(8\sigma_1 - 3), a_6 = 0, \\ a_7 &= -(1/48)[3(a_1^2 a_5 + a_1 a_3^2) + 5\sigma_1 a_1^4 a_3] = (a_1^7/512)(8\sigma_1 - 1), \\ a_8 &= 0, a_9 = (-1/80)[3a_7a_1^2 + 6a_1a_3a_5 + \sigma_1(5a_5a_1^4 + 10a_3^2a_1^3) + a_3^3] \\ &= (a_1^9/12288)(32\sigma_1^2 - 48\sigma_1 + 3), \end{aligned}$$

$$(n^2 - 1)a_n + \sum_{m=2}^{n-1}\sum_{\ell=1}^{m-1} a_{n-m}a_{m-\ell}a_\ell + \sigma_1 \sum_{m=4}^{n-1}\sum_{\ell=3}^{m-1}\sum_{j=2}^{\ell-1}\sum_{k=1}^{j-1} a_{n-m}a_{m-\ell}a_{\ell-j}a_{j-k}a_k = 0.$$

a_{2n+1} seems to contain polynomials of degree $[n/2]$ in σ_1.

In particular, for $\sigma_1 = -3/16$, we get $a_5 = (3/2^7)a_1^5, a_7 = -15a_1^7/(3!2^9)$, and, in general, $a_{2n} = 0,\ a_{2n+1} = (-1)^n(2n-1)!!a_1^{2n+1}/(n!2^{3n})$, where $(2n-1)!! = (2n-1)(2n-3)\cdots 5.3.1.$ Therefore, in this case,

$$y = \sum_{p=1}^{\infty} a_p g^p(x) = 2\sum_{n=0}^{\infty} [(-1)^n(2n-1)!!/(n!2^n)][(a_1/2)g(x)]^{2n+1},$$

which, on using

$$1/(1+x^2)^{1/2} = \sum_{\ell=0}^{\infty} [\{(-1)^\ell(2\ell-1)!!\}/\{\ell!2^\ell\}]x^{2\ell}, \quad |x| < 1,$$

becomes

$$y = 2dg/\{1 + d^2g^2\}^{1/2}, \quad d = a_1/2, \text{ provided that } (-1)!! \equiv 1. \tag{1}$$

Equation (1) provides a closed form for y convergent for $dg < 1$ or $x > (\ln d)/K$.

Expanding y in a series in e^{Kx}, we can obtain an analogous closed form for all $-\infty < x < \infty$. Squaring (1), we have

$$y^2 = 4d^2g^2/(1+d^2g^2) = (4e^{-Kx-\Delta})/\{e^{Kx+\Delta} + e^{-Kx-\Delta}\},$$

where $\Delta = -\ln d$, so that $y^2 = 2[1 - \tanh(kx + \Delta)]$. Coffey (1990)

78. $$\boldsymbol{y'' - y + y^6 = 0.}$$

The solution about the equilibrium point $y = 1$ is found by writing $y = 1 + v$ and $\tau = (5)^{1/2}x$. We thus get $v'' + v + 3v^2 + 4v^3 + \cdots = 0$ $' \equiv d/d\tau$. See the solution of Eqn. (2.1.35). The solution is found to be

$$y - 1 = v = a_0\cos(w\tau + \beta_0) + (1/2)a_0^2[\cos(2w\tau + 2\beta_0) - 3] + \cdots,$$

where $w = 1 - (9/4)a_0^2 + \cdots$. Nayfeh (1985), p.141

79. $$\boldsymbol{y'' + y^3 + y^7 = 0.}$$

The method of time transformation is used to find an approximate periodic solution of the given DE [see Eqn. (2.2.83)]. The solution is found in the form $y = a\cos X$, where X is the transformed variable. Numerical results for various values of a are quoted and discussed with reference to other methods. Burton and Hamdan (1983)

80. $$\boldsymbol{y'' + \prod_{i=1}^{4}(y^2 - y_i^2) + y^3 - 16y = 0,}$$

where $y_i = y_{i-1} + i$, $y_0 = 2$.

The given DE is shown to have at least three isolated stable periodic solutions. Amelkin and Zhavnerchik (1988)

81. $$\boldsymbol{y'' - a^2y^n = 0, \quad y'(0) = 0, \; y(1) = 1,}$$

where a^2 is a nonzero constant and $n \geq 1$.

Writing the given BVP as

$$y'' - k^2y = a^2y^n - k^2y, k \neq 0 \text{ is a constant}, \; y'(0) = 0, \; y(1) = 1,$$

and using Green's function, the sequence

$$y^{m+1}(x) = \frac{\cosh kx}{\cosh k} + \int_0^1 g_k(x,\xi)\left[k^2y^m(\xi) - a^2[y^m(\xi)]^n\right]d\xi, \quad m = 0, 1, 2, \ldots,$$

is formed, where

$$g_k(x,\xi) = \frac{1}{k\cosh k}\begin{cases} \sinh k(1-\xi)\cosh kx, & 0 \leq x \leq \xi, \\ \cosh k\xi \sinh k(1-x), & \xi < x \leq 1. \end{cases}$$

The convergence of this sequence is proved and numerical results are provided for different values of a^2, n, and k^2. Pennline (1981)

82. $$\boldsymbol{y'' + y^{2n+1} = 0,}$$

where n is a positive integer.

Rewrite the given DE as $y' = z$, $z' = -y^{2n+1}$. This is a time-independent Hamiltonian system on

$$R^2: \; y' = \frac{\partial h(y,z)}{\partial z}, \quad z' = -\frac{\partial h(y,z)}{\partial y},$$

where $h(y,z) = \frac{1}{2}z^2 + \frac{1}{2(n+1)}y^{2n+2}$. All solutions (other than $y=0, z=0$) of this system are periodic with period tending to zero as $h = E \to \infty$. Bin (1989)

83. $$y'' - \phi^2 y^n = 0, n \geq 1, \phi^2 \neq 0, \quad y'(0) = 0, \; y(1) = 1,$$

where ϕ is a constant and n is a positive integer.

Writing the given system as

$$y'' - k^2 y = \phi^2 y^n - k^2 y, \quad y'(0) = 0, \; y(1) = 1, \tag{1}$$

and using the Green's function for the operator on the left of (1), an integral equation is formulated:

$$y(x) = \frac{\cosh kx}{\cosh k} + \int_0^1 g_k(x,\xi)\left[k^2 y(\xi) - \phi^2 [y(\xi)]^n\right] d\xi,$$

where

$$g_k(x,\xi) = \frac{1}{k\cosh k}\begin{cases} \sinh k(1-\xi)\cosh kx, & 0 \leq x \leq \xi, \\ \cosh k\xi \sinh k(1-x), & \xi < x \leq 1. \end{cases}$$

An iterative scheme

$$\begin{aligned} y^0(x) &= \frac{\cosh \phi x}{\cosh \phi} \quad \text{or} \quad \frac{1}{\cosh \phi}, \\ y^{m+1}(x) &= \frac{\cosh kx}{\cosh k} + \int_0^1 g_k(x,\xi)\left[k^2 y^m(\xi) - \phi^2 [y^m(\xi)]^n\right] d\xi, \quad m = 0,1,2,\ldots, \end{aligned}$$

is shown to converge and give a unique solution of the problem satisfying $1/\cosh\phi \leq y(x) < 1$ for all $x \in [0,1]$. Numerical results are provided for $\phi^2 = 2.25, 6.25$. Pennline (1981), De Simone and Pennline (1978)

84. $$y'' + (n+1)a^{2n} y^{2n+1} - y = 0,$$

where n is a constant.

Put $y'(x) = p(y), y'' = pp'(y)$ and integrate:

$$y' = \pm\sqrt{y^2(1 - a^{2n}y^{2n}) + C},$$

where C is an arbitrary constant. Now the variables separate. The solution is found by quadrature. Kamke (1959), p.544

85. $$y'' + w_0^2 y + \mu y^n = 0, \quad y(0) = A_0, \; y'(0) = 0,$$

where $\mu > 0$, $n \geq 3$ is an odd positive integer; $\rho \leq w_0$.

A perturbation approach with

$$y = y_0 + \mu y_1 + \mu^2 y_2 + \cdots, \quad w^2 = w_0^2 + \mu b_1 + \mu^2 b_2 + \cdots$$

is used. The zeroth order DE gives $y_0 = A_0 \cos wx$. The first order equation is

$$\begin{aligned} y_1'' + w^2 y_1 &= b_1 y_0 - y_0^n = b_1 A_0 \cos wx - A_0^n \cos^n wx \\ &= (b_1 A_0 - a_1 A_0^n) \cos wx - A_0^n \sum_{m=3, m \text{ odd}}^{n} a_m \cos wmx, \end{aligned}$$

where m is odd and $a_m = (1/\pi) \int_0^2 \cos^n wx \cos mwx d(wx)$. The usual argument to eliminate the secular term leads to $b_1 = a_1 A_0^{n-1}$.

Thus, a first order approximation to frequency is $w^2 = w_0^2 + \mu a_1 A_0^{n-1}$. The solution for y_1 is

$$y_1 = \frac{A_0^n}{w^2} \sum_{m=3, m \text{ odd}}^{n} \frac{a_m}{m^2 - 1} (\cos mwx - \cos wx).$$

The second order correction terms are given by

$$\begin{aligned} y_2'' + w^2 y_2 &= b_2 y_0 + b_1 y_1 - n y_0^{n-1} y_1 \\ &= \left[b_2 A_0 - \frac{A_0^{2n-1}}{w^2} [a_1 (1-n) + n] \sum_{m=3, m \text{ odd}}^{n} \frac{a_m}{m^2 - 1} \right] \cos wx \\ &\quad + \text{higher harmonics}. \end{aligned}$$

For the absence of secular terms,

$$b_2 = \frac{A_0^{2n-2}}{w^2} [a_1 (1-n) + n] \sum_{m=3, m \text{ odd}}^{n} \frac{a_m}{m^2 - 1}.$$

Thus, a second order approximation for frequency is

$$w^2 = w_0^2 + \mu a_1 A_0^{n-1} + \mu^2 \frac{A_0^{2n-2}}{w^2} [a_1 (1-n) + n] \sum_{m=3, m \text{ odd}}^{n} \frac{a_m}{m^2 - 1}. \tag{1}$$

Solving (1) for w^2 yields

$$w^2 = w_0^2/2 + \mu a_1 A_0^{n-1}/2 \pm \sqrt{(w_0^2 + \mu a_1 A_0^{n-1})^2/4 - \mu^2 b},$$

only a positive sign being relevant, $b = -w^2 b_2$. Ludeke and Wagner (1968)

86. $\quad \boldsymbol{y'' + \mu y - y^n = 0, \quad y(0) = y(\pi) = 0,}$

where μ is a constant.

It is proved that there is a positive solution of the given problem if and only if $\mu > 1$. Shampine (1969)

87. $\quad \boldsymbol{y'' + \beta y - y\left(a_1 y + a_3 y^3 + \cdots + a_{2N+1} y^{2N+1}\right) = 0,}$

where $a_1, \ldots, a_{2N+1}$ are nonnegative with at least one $a_i > 0$ that satisfies the "standard Neumann" situation and $\beta > 0$. Two-point boundary value problems of Dirichlet ($y(0) = 0, y(X) = 0$) or Neumann ($y'(0) = y'(X) = 0$) type are considered by geometrical methods. Chicone (1988)

2.2 $y'' + f(y) = 0$, $f(y)$ not polynomial

1. $\quad y'' + g_1/y^2 = 0, \quad g_1$ **a constant.**

The given DE describes a projectile in earth's atmosphere under the inverse square law of gravitation.

Putting $y' = p, y'' = p(dp/dy)$, we get $p(dp/dy) + g_1/y^2 = 0$. On integration we have $p^2 = C_1 + 2g_1/y$, where C_1 is an arbitrary constant. That is,

$$p = dy/dx = \pm\sqrt{C_1 + 2g_1/y}.$$

The solution is expressed in quadrature form as

$$x = \pm \int dy/\sqrt{C_1 + 2g_1/y} + C_2,$$

where C_2 is another arbitrary constant. Rabenstein (1966), p.372

2. $\quad y'' + 1/y^3 = 0.$

Multiply the given DE by $2y'$ and integate:

$$y'^2 - 1/y^2 = c_1, \quad \text{say.} \tag{1}$$

The solutions of (1) for $c_1 \neq 0$ and $c_1 = 0$, respectively, are

$$y = \pm\sqrt{c_1(x + c_2)^2 - c_1^{-1}}$$

and

$$y = \pm\sqrt{\pm 2x + c}.$$

Here c, c_1, and c_2 are arbitrary constants. Rabenstein (1972), p.47

3. $\quad y'' - 4/y^3 = 0.$

Put $y' = u$, so that $du/dx - 4/y^3 = 0$. Multiplying by $2u$, we have

$$\frac{d}{dx}u^2 = \frac{8}{y^3}\frac{dy}{dx}.$$

Integration leads to

$$u = \frac{dy}{dx} = \pm\frac{[c_1y^2 - 4]^{1/2}}{y},$$

where $y > 2/\sqrt{c_1}$ or $y < -2/\sqrt{c_1}$ and c_1 is an arbitrary constant, $c_1 > 0$. A second integration gives $c_1y^2 = (c_1x + c_1c_2)^2 + 4$, $y > 2/\sqrt{c_1}$ or $y < -2/\sqrt{c_1}$, where c_2 is another arbitrary constant. Tenebaum and Pollard (1963), p.501

4. $\quad y'' - B/y^3 + Cy = 0,$

where B and C are constants.

Multiply the given DE by y' and integrate:

$$(1/2)y'^2 + (C/2)y^2 + B/(2y^2) = D, \tag{1}$$

where D is a constant. Eliminating the term containing B from (1) and the given DE, we get

$$\frac{d^2}{dx^2}y^2 + 4Cy^2 = 4D. \tag{2}$$

Equation (2) is linear in y^2 and is easily solved. Put $4C = \nu^2$ and distinguish three cases. If we choose ICs as $y = 1, y' = 0$ at $x = 0$, then (1) gives $D = (C + B)/2 = \nu^2/8 + B/2$. Thus:

(a) $\nu^2 > 0,\quad y^2 = 1/2 + 2B/\nu^2 + [1/2 - 2(B/\nu^2)]\cos \nu x.$

(b) $\nu^2 = 0,\quad y^2 = 1 + Bx^2.$

(c) $\nu^2 = -n^2 < 0,\quad y^2 = 1/2 - 2(B/n^2) + [1/2 + 2(B/n^2)]\cosh nx.$

Ball (1962)

5. $$y'' + y - 3/\{16(1 - y)\} = 0.$$

To determine the motion about the equilibrium point $y = 1/4$, write $y = 1/4 + u(x)$ and $\tau = (2/3)^{1/2}x$. We have $u'' + u - (2/3)u^2 - (8/9)u^3 + \cdots = 0$. See the solution of Eqn. (2.1.35). The second order approximate solution is

$$y^{-1/4} = u = a_0 \cos(w\tau + \beta_0) - (a_0^2/9)[\cos(2w\tau + 2\beta_0) - 3] + \cdots,$$

where $w = 1 - (14/27)a_0^2 + \cdots$. The solution about $y = 3/4$ may be found similarly. Nayfeh (1985), p.142

6. $$y'' - M^2 y/(1 + By) = 0,\quad y(0) = 1,\ y'(1) = 0,$$

where $M = 1, B = 10; M = 4, B = 1; M = 10, B = 100$.

Numerical solution is suggested for the given problem describing diffusion in a biological flow. The problem may be solved analytically by multiplying by y' and integrating, etc. Kubicek and Hlavacek (1983), p.230

7. $$my'' + k[y - \lambda/(a - y)] = 0,$$

where m, k, λ, and a are constants.

This DE describes the attraction of current-carrying conductors. Write $y' = z$, so that $z' = (k/m)[y^2 - ay + \lambda]/(a - y)$. In the (y, z) phase plane, we have

$$\frac{dz}{dy} = \left(\frac{k}{m}\right)\frac{y^2 - ay + \lambda}{z(a - y)}. \tag{1}$$

A first integral of (1) may easily be found to be

$$(m/2)z^2 + (k/2)y^2 + k\lambda \ln(a - y) = h,$$

where h is a constant. Equation (1) has singular points $(0, a/2 - b), (0, a/2 + b)$, where $b = [a^2/4 - \lambda]^{1/2}$. If $\lambda < a^2/4$, these points are real. It is checked that the first point is stable (center) while the second is unstable (saddle point). The phase portrait is easily drawn, which may be interpreted suitably for periodic solution. Minorsky (1962), p.51

8.
$$y'' - (k/m)(y^2 - ay + \lambda)/(a - y) = 0,$$

where k, m, λ, and a are constants.

The given DE follows from the Hamiltonian

$$H(p, q) = (1/2)(p^2/m) + (1/2)kq^2 + k\lambda \ln(a - q), q < a,$$

where $p = my', q = y$. Written as the (equivalent) system

$$\dot{x}_1 = x_2, \quad \dot{x}_2 = (k/m)(x_1^2 - ax_1 + \lambda)/(a - x_1),$$

it has critical points at $x_2 = 0, x_1^2 - ax_1 + \lambda = 0$, or $x_1 = a/2 \pm (1/2)(a^2 - 4\lambda)^{1/2}$. The nature of these points is as follows:

(a) $0 < \lambda < a^2/4$. Two critical points: the minus sign gives a center while the plus sign corresponds to a saddle.

(b) $\lambda = a^2/4$. The two critical points coalesce to a degenerate critical point.

(c) $\lambda > a^2/4$. There are no critical points.

It is shown that for $0 < \lambda < a^2/4$, an infinite set of periodic solutions exists around the center point. If $\lambda < 0$, there is a critical point $x_2 = 0, x_1 = a/2 - (1/2)[a^2 - 4\lambda]^{1/2}$, which is a center; outside this critical point all motions are periodic. Verhulst (1990), p.25

9.
$$y'' + (1 - \mu)/y^2 - \mu/(1 - y)^2 = 0,$$

where μ is a small parameter.

The given DE describes the motion of a satellite. A first integral with zero energy is found to be

$$(1/2)y'^2 = (1 - \mu)/y + \mu/(1 - y).$$

A perturbation solution for small μ is

$$x = (1/3)2^{1/2}y^{3/2} + \mu 2^{1/2}[(1/3)y^{3/2} + (1/2)y^{1/2} - (1/4)\ln((1 + y^{1/2})/(1 - y^{1/2}))] + O(\mu^2),$$

except when $y \approx 1$. Jordan and Smith (1977), p.177

10.
$$y'' - 4C/(y)^{1/2} = 0,$$

where C is a constant.

Putting $y' = p$ and integrating, we get

$$p = dy/dx = [C_1 + 16Cy^{1/2}]^{1/2},$$

where C_1 is an arbitrary constant. Putting $(y)^{1/2} = z$ and $x = \sin\theta$, we get the solution in the form $h^{3/2} - 3C_1h^{1/2} = B(\sin\theta + \sin\theta_0)$, where $(h)^{1/2} = [C_1 + 16Cz]^{1/2}$, $0 < B = 3(16C)^2/4 \ll 1$, and $\sin\theta_0$ is a constant. Panayotounakos and Theocaris (1986)

11. $$\boldsymbol{y'' - 1/(1+2y)^{1/2} + 1 = 0.}$$

Put $1 + 2y = Y(x)$ so that

$$(1/2)Y^{1/2}Y'' + Y^{1/2} = 1. \tag{1}$$

Now write $Y = Z^2(x)$ so that (1) becomes

$$ZZ'' + Z'^2 = (1 - Z)/Z. \tag{2}$$

Writing $Z' = p, Z'' = p(dp/dZ)$ in (2), we get

$$Zp\frac{dp}{dZ} + p^2 = \frac{1-Z}{Z}. \tag{3}$$

Equation (3) is linear in p^2 and can therefore be integrated. The solution after that can be expressed in terms of a quadrature. Jordan and Smith (1977), p.35

12. $$\boldsymbol{y'' + w^2 y/(1+y^2)^{1/2} = 0, \quad y(0) = a,\ y'(0) = 0,}$$

where a and w are constants.

Multiplication by y' and integration lead to

$$y' = \pm\left[2w^2\left\{(1+a^2)^{1/2} - (1+y^2)\right\}^{1/2}\right]^{1/2}.$$

Thus, the trajectories in the (y, y') plane are closed ovals which are symmetric about y and y' axes.

Further, the period P of their solution is

$$P = 4\int_0^a \frac{dy}{\left[2w^2\left\{(1+a^2)^{1/2} - (1+y^2)\right\}^{1/2}\right]^{1/2}} < \infty,$$

$a < \infty$, since the only singularity in the integral at $y = a$ is integrable. Caughey (1969)

13. $$\boldsymbol{y'' + \{1 - (1+\alpha)y/(1+y^2)^{1/2}\} = 0.}$$

A first integral is

$$y' = \pm\left[2\{E - F(y)\}\right]^{1/2},$$

where $F(y) = y^2/2 - (1+\alpha)(1+y^2)^{1/2}$, and E an arbitrary constant. A phase portrait may be drawn. Hagedorn (1988), p.232

14. $$\boldsymbol{y'' + (2k/m)y - (2ka/m)y/(d^2+y^2)^{1/2} = 0.}$$

where k, m, and a are constants.

The given DE describes motion of a particle of mass m attached to the center of a stretched elastic wire. Mickens (1981), p.4

15. $$\boldsymbol{my'' + ky(y^2+\ell^2)^{-1/2}\left[(y^2+\ell^2)^{1/2} - \ell/2\right] = 0,}$$

where m, k, and ℓ are constants.

To get a first order uniform expansion for small but finite y, expand the undifferentiated term, to obtain $my'' + (1/2)ky + [k/(4\ell^2)]y^3 + \cdots = 0$ or $u'' + w_0^2 u - \alpha w_0^2 u^3 = 0$, where $w_0^2 = k/(2m), u = y/\ell, \alpha = -1/2$. Nayfeh (1985), p.106

16. $$y'' + (2\lambda/m)y[(a^2 + y^2)^{1/2} - a](a^2 + y^2)^{1/2} = 0.$$

The given DE describes motion for transverse oscillations of an elastic string of length $2a$ and stiffness λ, with a mass m attached to the midpoint; there is no gravity acting and the tension is zero in the equilibrium position. The phase portrait may be drawn for this autonomous equation. Jordan and Smith (1977), p.33

17. $$y'' - \phi^2 y^p = 0, \quad y'(0) = 0, \; y(1) = 1,$$

where ϕ is a constant; $p > 1$ is a parameter.

Multiplying the given DE by $2y'(x)$ and integrating, we have

$$y'^2 = [2\phi^2/(p+1)]\left[y^{p+1}(x) - y_0^{p+1}\right], \tag{1}$$

where $y_0 = y(0)$. An implicit solution is found by taking square root of (1) and integrating:

$$\phi x = \left[\frac{p+1}{2}\right]^{1/2} \int_{y_0}^{y} \left[\frac{1}{y^{p+1} - y_0^{p+1}}\right]^{1/2} dy. \tag{2}$$

In particular, since $y = 1$ at $x = 1$, we obtain

$$\phi = \left[\frac{p+1}{2}\right]^{1/2} \int_{y_0}^{1} \left[\frac{1}{y^{p+1} - y_0^{p+1}}\right]^{1/2} dy. \tag{3}$$

If (3) can be solved for y_0 as a function of ϕ, Eqn. (2) gives an implicit form of the solution for any ϕ. By putting $z = 1 - (y_0/y)^{p+1}$, the integral in (3) can be expressed in terms of an incomplete beta function or hypergeometric function:

$$\phi = y_0^{-(p-1)/2}[2(1 - y_0^{p+1})/(p+1)]^{1/2} F(1/2, 1/2 + q; 3/2; 1 - y_0^{p+1}), \tag{4}$$

where $q = 1/(p+1)$. Since $p > 1$, the exponent of y_0 in (4) is negative, and the hypergeometric function converges absolutely on the circle $1 - y_0^{p+1} = 1$. It is obvious that ϕ ranges from infinity to zero as y_0 ranges from zero to one. Moreover,

$$\frac{d\phi}{dy_0} = -y_0^{-(p+1)/2}[(p+1)/(2(1 - y_0^{p+1}))]^{1/2} F(-1/2, -1/2 + q, 1/2, 1 - y_0^{p+1}),$$

so that $d\phi/dy_0$ is everywhere negative and becomes infinite as y_0 approaches zero or one. Thus, (4) can be solved for y_0 as a function of ϕ for any positive ϕ. Some asymptotic results for $\phi \to 0$ or $\phi \to \infty$ are also obtained. Mehta and Aris (1971)

18. $$y'' + ky^{1+\epsilon^2} = 0,$$

where k and ϵ are constants.

A first integral is

$$y'^2 + 2ky^{2+\epsilon^2}/(2 + \epsilon^2) = c,$$

where c is a constant of integration. The solution may now be obtained in the form of a quadrature. Wilson (1971), p.33

19. $$\boldsymbol{y'' - (1 - w^2)y + y^{2p+1} = 0,}$$

where w $(0 \leq w^2 < 1)$ and p are constants.

The solution for given p is

$$y = [(p+1)(1-w^2)]^{(1/2p)} \cosh^{-1/p}[p(1-w^2)^{1/2}\, x].$$

Blanchard, Stubble, and Vazquez (1988)

20. $$\boldsymbol{y'' + y - ay^{\alpha-2} = 0,}$$

where a and α are constants.

The given DE describes motion of a particle under a central attractive force. The nontrivial equilibrium point is $y = a^{1/(3-\alpha)}, y' = 0$ for $\alpha \neq 3$; $\alpha = 3$ is the linear case. The equilibrium point is a center if $\alpha < 3$ and a saddle point if $\alpha > 3$. Jordan and Smith (1977), p.58

21. $$\boldsymbol{y'' - w^2 y - c_1 y^{1-2m} = 0,}$$

where w, c_1, and $m \neq 0, 1$ are constants.

The given DE has solutions

$$y = [a^m e^{mwx} + c_m b^m e^{-mwx}]^{1/m},$$

where a and b are arbitrary constants and $c_m = c_1/[4w^2(ab)^m(m-1)]$. Reid (1971)

22. $$\boldsymbol{y'' + \sin y = 0, \quad y(0) = 1, \; y(2\pi) = 0.}$$

This BVP is changed to the system

$$y' = z, \quad z' = -\sin y, \qquad y(0) = 1, y(2\pi) = 0,$$

which is solved numerically using an iterative approach. Vinokurov and Repnikov (1981)

23. $$\boldsymbol{y'' + \sin\, y = 0.}$$

Using the method of harmonic balance an approximate solution is $y = a\cos wx$, where a and w are related by $w^2 = 2J_1(a)/a$; here $J_1(a)$ is a Bessel function of order 1. As a result, for small a, we have $w = 1 - (1/16)a^2$. Jordan and Smith (1977), p.118

24. $$\boldsymbol{y'' + \sin\, y = 0.}$$

The DE is first written as the system

$$y_1' = y_2, \quad y_2' = -\sin y_1, \tag{1}$$

which is converted into a system with polynomial on the right-hand side by introducing $y_3 = \sin y_1$ and $y_4 = \cos y_1$:

$$y_1' = y_2, \quad y_2' = -y_3, \quad y_3' = y_2 y_4, \quad y_4' = -y_2 y_3. \tag{2}$$

For the two additional variables y_3 and y_4, IC would be related to those for y_1 and y_2. If the latter are $y_1 = y_{10}, y_2 = y_{20}$, then $y_3 = \sin y_{10}$ and $y_4 = \cos y_{10}$. A series representation for (2), namely

$$y_i = \sum_n a_{in} x^n, \tag{3}$$

leads on substitution and equating powers of x to the recursion relations

$$\begin{aligned} (n+1)a_{1,n+1} &= a_{2,n}, \ (n+1)a_{2,n+1} = -a_{3,n}, \\ (n+1)a_{3,n+1} &= \textstyle\sum_{m=0}^{n} a_{2,m}a_{4,n-m}, \ (n+1)a_{4,n+1} = -\textstyle\sum_{m=0}^{n} a_{2,m}a_{3,n-m}. \end{aligned} \tag{4}$$

The solution for the recursion is initiated with the IC, $a_{i0} = y_{i0} (i = 1, \ldots, 4)$. The relations (4) are easily programmed. Fairen, Lopez, and Conde (1988)

25. $$\boldsymbol{y'' + (mg/\ell)\sin y = 0, \quad y(a) = A, \ y(b) = B,}$$

where m, g, ℓ, a, b, A, and B are constants.

It is shown that on any interval $[a, b]$ of length less than $\pi[\ell/(mg)]^{1/2}$, the given BVP has one solution at most. Bailey, Shampine, and Waltman (1968), p.64

26. $$\boldsymbol{y'' + \mu \sin y = 0, \quad y'(0) = 0, \ y'(X) = 0,}$$

where $\mu > 0$ and X are constants.

Apart from the trivial solution, the solution considered is one for which $y(x) = 0$ exactly N times on $[0, X]$, i.e., solution with N nodes. The problem is solved geometrically. Chicone (1988)

27. $$\boldsymbol{y'' + a \sin y = 0,}$$

where a is a constant.

Multiplication by y' and integration with initial conditions $y(x_0) = \alpha, y'(x_0) = \beta$ give

$$\begin{aligned} y' &= \pm[2a\cos y + \beta^2 - 2a\cos\alpha]^{1/2}, \\ x - x_0 &= \int_\alpha^y \frac{dy}{[2a\cos y + \beta^2 - 2a\cos\alpha]^{1/2}} \end{aligned}$$

and if $\sin(y/2) = ku, k^2 = \sin^2(\alpha/2) + \beta^2/(4a)$, then

$$a^{1/2}(x - x_0) = \int_{(1/k)\sin(\alpha/2)}^{u} \frac{du}{[(1-u^2)(1-k^2u^2)]^{1/2}},$$

an elliptic integral. Kamke (1959), p.545; Murphy (1960), p.381; Ames (1968), p.52

28. $$\boldsymbol{y'' + (g/\ell) \sin y = 0,}$$

where g and ℓ are constants.

The given DE is expanded as

$$y'' + w_0^2 y - \alpha w_0^2 y^3 = 0, \quad w_0^2 = g/\ell, \ \alpha = 1/6.$$

A small nonlinearity parameter is introduced by writing $y = \epsilon^\lambda u(x)$; we have

$$u'' + w_0^2 u - \alpha\epsilon w_0^2 u^3 + \cdots = 0 \quad \text{if } \lambda = 1/2.$$

We have, with $d = 1/6, \epsilon = 1$,

$$y = a\cos\left[(1 - a^2/16)w_0 x + \beta\right] + \cdots.$$

Nayfeh (1985), p.106

29. $\mathbf{y'' + (g/a)\sin y - [Fh/(ma^2)]\sin\phi = 0,\quad F = c/(a^2 + h^2 - 2ah\cos y),}$

where h, a, ϕ, and c are constants; $h > a$.

The given DE describes the motion of a pendulum with a magnetic bob, oscillating in a vertical plane over a magnet, which repels the bob according to the inverse square law.

The phase plane study shows that $y = 0$ is a center if $c < mg(h-a)^3/h$, and a saddle point if $c > mg(h-a)^3/h$; $y = \pi$ is a center if $c > mg(h+a)^3/h$, and a saddle point if $c < mg(h+a)^3/h$; $y = \cos^{-1}[a^2 + h^2 - (ch/mg)^{2/3}/(2ah)]$ is an equilibrium if $mg(h+a)^3/h > c > mg(h-a)^3/h$; it is a center. Jordan and Smith (1977), p.33

30. $$\mathbf{y'' + \cos y = 0,\quad y(0) = y(1) = 0.}$$

It is shown by a constructive iterative method that the given problem has a unique solution. Bailey, Shampine, and Waltman (1968), p.28

31. $$\mathbf{y'' + ay - bI^2\cos y = 0,}$$

where a, b, and I are constants.

The given DE describes the angular deflection y of an electrodynamometer ammeter, where I is the steady current in the meter, and the constants a and b are positive. Expanding $\cos y$ and retaining the first few terms, the DE can be solved. Mickens (1981), p.27

32. $$\mathbf{y'' - (m^2\Omega^2 a^2/I)(\cos y - \lambda)\sin y = 0,}$$

where m, Ω, a, I, and λ are constants.

The given DE describes a rotating pendulum. Write $y' = w(x)$; then $w' = (m^2\Omega^2 a^2/I)$ $(\cos y - \lambda)\sin y$. In the (y, w) plane, we have

$$\frac{dw}{dy} = \frac{m^2\Omega^2 a^2}{Iw}(\cos y - \lambda)\sin y. \tag{1}$$

Equation (1) may be integrated. If it passes through the saddle point $y = \pm\pi, w = 0$, the integral curve is

$$w^2 = (m^2\Omega^2 a^2/I)\left[\sin^2 y + 2\lambda(\cos y + 1)\right],$$

while if it passes through $y = 0, w = 0$, it has the form

$$w^2 = (m^2\Omega^2 a^2/I)[\sin y + 2\lambda(\cos y - 1)].$$

The phase portrait of (1) is drawn. Minorsky (1962), p.48

33. $$\mathbf{y'' + \beta y - ay\tan y = 0,}$$

where β and a are positive constants.

Two-point boundary value problems with either $y(0) = y(X) = 0$ or $y'(0) = y'(X) = 0$ are considered using geometrical methods. Chicone (1988)

34. $$y'' - \sec^2 y \tan y = 0, \quad y = \pi/4,\ y' = -1 \text{ when } x = \ln 2.$$

Multiplying the given DE by y', integrating twice and using the IC, we have $x = (1/2)\ln 2 - \ln \sin y$. Reddick (1949), p.192

35. $$y'' - (2/m)[Py - k_1 \sin^{-1}(y/\ell)]/[\ell^2 - y^2]^{1/2} + \alpha y + \beta y^3 = 0,$$

where m, P, k, ℓ, α, and β are constants.

The given DE is approximated by ignoring powers of y higher than third in the expansions of the functions involved:

$$my'' + (\alpha + 2k_1/\ell^2 - 2P/\ell)y + (\beta + 4k_1/(3\ell^4) - P/\ell^3)y^3 = 0.$$

Now see Eqn. (2.1.55). A complete phase plane analysis may be carried out and different cases interpreted. Stoker (1950), p.55

36. $$y'' + e^y = 0, \quad y(0) = y(1) = 0.$$

The given BVP arises in the diffusion of heat generated by positive temperature-dependent sources. It is shown that the BVP has one and only one solution for which $|y| \leq 1$. In fact, there is one solution at most for which $|y| < \ln 8 = 2.08$. Picard's iterative method is used. Bailey, Shampine, and Waltman (1968), p.45

37. $$y'' - e^y = 0.$$

Put $y' = p, y'' = p\dfrac{dp}{dy}$, etc., and integrate twice:

$$x = \int \frac{dy}{[2e^y + C_1]^{1/2}} + C_2,$$

where C_1 and C_2 are arbitrary constants. Now put $t = [2e^y + C_1]^{1/2}$ and integrate. The BVP $y(0) = y(b) = 0$ (b a constant) for the DE above was treated by Ames (1968), p.200. For $b = 1$, an exact solution is $y(x) = -\ln 2 + 2\ln[C \sec[C(x - 1/2)/2]]$, where C is a root of $\sqrt{2} = C \sec(C/4), C \simeq 1.336056$. Kamke (1959), p.545; Ames (1968), pp.42, 200

38. $$y'' - e^y - a = 0, \quad a < 0,\ a = 0 \text{ or } a > 0.$$

A phase plane study may be carried out in the (y, y') plane. Jordan and Smith (1977), p.29

39. $$y'' - e^{-y} = 0, \quad y(0) = y'(T) = 0,$$

where $T > 0$ a constant.

Write the given DE as the system,

$$y_1' = y_2, \quad y_2' = e^{-y_1}, \quad y_1(0) = y_2(T) = 0. \tag{1}$$

A first integral of (1) is easily found to be

$$e^{-y_1} + y_2^2/2 = 1 + y_2^2(0)/2 = (1/2)a^2, \quad \text{say,} \tag{2}$$

where we use the IC. Eliminating y_1 from (2) and the second of (1), we have

$$y_2' = a^2/2 - y_2^2/2. \tag{3}$$

The solution of (3) is

$$x = \frac{1}{a}\ln\frac{[a + y_2(x)][a - y_2(0)]}{[a - y_2(x)][a + y_2(0)]}. \tag{4}$$

Putting $x = T$ in (4), we easily deduce that

$$y_2(0) = a\frac{1 - \exp(aT)}{1 + \exp(aT)}, \quad a = [2 + y_2(0)^2]^{1/2}. \tag{5}$$

Relation (5) may be analyzed graphically to find the number of solutions. There exists a value T_0 such that for $T < T_0$ there are two solutions, for $T = T_0$ one solution, and for $T > T_0$ no solutions. Daniel and Moore (1970), p.40

40. $\mathbf{2y'' - e^y = 0, \quad y(0) = 0, \; y'(0) = -1.}$

Writing the given DE as $2y'y'' = e^y y'$, integrating twice, and using the IC, we have $x = 2(e^{-y/2} - 1)$. Reddick (1949), p.192

41. $\mathbf{y'' + \delta e^y = 0, \quad y'(0) = 0, \; y(1) = 0.}$

The solution may be written as $y = 2\ln(\cosh\sigma/\cosh\sigma x)$ where $\{(1/2)\delta\}^{1/2} = \sigma\,\mathrm{sech}\,\sigma$. The given system is called the Frank–Kamenetskii problem. Adler (1991)

42. $\mathbf{y'' - \alpha e^y = 0, \quad y(0) = 0, \; y(1) = 0,}$

where α is a parameter.

The given system describes the steady-state temperature distribution in a homogeneous rod of unit thickness where the heat is generated at a rate αe^y per unit time per unit volume.

The given problem is solved numerically using a one-parameter embedding technique. The case $\alpha = 1$ is shown graphically. Kubicek and Hlavacek (1983), pp.158, 175

43. $\mathbf{y'' + 2\sinh y = 0.}$

This is a Poisson-Boltzmann equation. The given DE is replaced by an (approximate) equivalent linear equation, using harmonic balance,

$$y'' + (4I_1(a)/a)y = 0,$$

where I_1 is modified Bessel function. The approximate solution, therefore, is

$$y = a\cos wx,$$

where $w = 2[I_1(a)/a]^{1/2}$. Jordan and Smith (1977), p.115

44. $\mathbf{y'' - \sinh y = 0, \quad y(0) = 0, \; y'(0) = A.}$

A first integral of the given problem is

$$y'^2 - 4\sinh^2(y/2) = A^2. \tag{1}$$

Assuming that $y'(0) \geq 0$ ($A \geq 0$), we integrate (1) to obtain

$$y(x) = \begin{cases} 2\sinh^{-1}[\mathrm{sc}(Ax/2, \{1 - 4/A^2\}^{1/2})], & A > 2, \\ 2\sinh^{-1}(\tan x), & A = 2, \\ 2\sinh^{-1}[(A/2)\,\mathrm{sc}(x, \{1 - A^2/4\}^{1/2})], & 0 < A < 2. \end{cases}$$

Here $\mathrm{sc}(u,k) = \mathrm{sn}(u,k)/\mathrm{cn}(u,k)$, where $\mathrm{sn}(u,k)$ is the Jacobi elliptic sine function of argument u and modulus k:

$$\mathrm{sn}^{-1}(u,k) = \int_0^u \frac{dt}{[(1-t^2)(1-k^2t^2)]^{1/2}} \quad \text{and} \quad \mathrm{cn}^2(u,k) = 1 - \mathrm{sn}^2(u,k).$$

The cases $y(0) \neq 0$ and $y'(0) \leq 0$ may be treated similarly by a simple change of variable. Marshall (1979)

45. $$y'' - \lambda^2 \sinh \lambda y = 0, \quad y(0) = 0, \; y(1) = 1,$$

where λ is a constant.

The given BVP is solved numerically, using the method of successive shooting. Gaiduk (1984)

46. $$y'' - n \sinh ny = 0, \quad y(0) = 0, \; y(1) = 1,$$

where n is a parameter.

The given problem describes the confinement of a plasma column by radiation pressure. A numerical solution is obtained by using a "parameter mapping" method. Kubicek and Hlavacek (1983), pp.187, 244

47. $$y'' + e^{2y} - Ee^y - \eta = 0,$$

where E and η are constants.

Multiplying the given DE by y' and integrating, we have

$$y'^2 = -e^{2y} + 2Ee^y + 2\eta y + H, \tag{1}$$

where H is an arbitrary constant. Integration of (1) gives

$$\int \frac{d\hat{y}}{\hat{y}[-\hat{y}^2 + 2E\hat{y} + 2\eta \ln\hat{y} + H]^{1/2}} = Ax + C, \quad \hat{y} = e^y, \tag{2}$$

where C is a constant. The solution (2) is periodic, the "half-period" being given by

$$x_0/2 = \int_{y(0)}^{y_{\max}} \frac{dy}{y[-y^2 + 2Ey + 2\eta \ln y + H]^{1/2}},$$

where $y_{\max}$ and $y(0)$ are the two roots of $f(y) = -y^2 + 2Ey + 2\eta \ln y + H = 0$. Lu (1989)

48. $$y'' + 2D(1 - e^{-ry})e^{-ry} = 0,$$

where D and r are constants.

The given DE describes the Morse oscillator in modeling intramolecular atom-to-atom interaction. Putting $z = e^{-ry}$, we get

$$zz'' - z'^2 - 2Dr(1-z)z^3 = 0.$$

Now see Eqn. (2.32.9). The given DE can be changed into a system of ODE with polynomial RHS, by introducing $y_3 = \exp(-ry_1)$:

$$\frac{dy_1}{dx} = y_2, \quad \frac{dy_2}{dx} = -2Dy_3(1-y_3), \quad \frac{dy_3}{dx} = -ry_2y_3. \tag{1}$$

The first few terms of the series solutions of (1) are

$$\begin{aligned} y_1 &= y_{1,0} + y_{2,0}(x-x_0) - Dy_{3,0}(1-y_{3,0})(x-x_0)^2 \\ &\quad + (1/3)rDy_{2,0}y_{3,0}(1-2y_{3,0})(x-x_0)^3 + \cdots, \\ y_2 &= y_{2,0} - 2Dy_{3,0}(1-y_{3,0})(x-x_0) + Dy_{2,0}y_{3,0}(1-y_{3,0})(x-x_0)^2 \\ &\quad + (r/3)[6D^2y_{3,0}^3 - 2D^2y_{3,0}^2 - 4D^2y_{3,0}^4 - rDy_{2,0}^2y_{3,0} + 4rDy_{2,0}^2y_{3,0}^2](x-x_0)^3 + \cdots, \\ y_3 &= y_{3,0} - y_{3,0}y_{2,0}(x-x_0) + (1/2)r(2Dy_{3,0}^2 - 2Dy_{3,0}^3 + ry_{3,0}y_{2,0}^2)(x-x_0)^2 \\ &\quad - (1/6)r[ry_{2,0}^3y_{3,0} + 8Dy_{2,0}y_{3,0}^2 - 10Dy_{2,0}y_{3,0}^3](x-x_0)^3 + \cdots, \end{aligned}$$

with $y_{3,0} = \exp(-ry_{1,0})$. The results from the series solution and the exact numerical solution are compared and plotted. Fairen, Lopez, and Conde (1988)

49. $$y'' + \lambda e^{y/\epsilon}/(e^{y/\epsilon} + e^{1/\epsilon}) = 0, \quad 0 < x < 1,$$

where $y'(0) = 0$ and $y(1) = 0$.

The given BVP arises in a model for porous medium combustion. Using the condition $y' = 0$ at $x = 0$, we first find a first integral as

$$y'^2/2 = -\lambda\epsilon \log[1 + \exp((y-1)/\epsilon)] + \lambda\epsilon \log[1 + \exp((y_* - 1)/\epsilon)],$$

where $y_* = y(0, \epsilon)$. Another integration gives the implicit solution

$$(2\lambda\epsilon)^{1/2} x = F(y, y_*, \epsilon) = \int_y^{y_*} \frac{dy}{[\log[1 + \exp((y_* - 1)/\epsilon)] - \log\{1 + \exp((y-1)/\epsilon)\}]^{1/2}},$$

where y_* is related to ϵ, for fixed $\lambda > 0$, by $F(0, y_*, \epsilon) = (2\lambda\epsilon)^{1/2}$.

The behavior of this solution as $\epsilon \to 0$ and the qualitative features are discussed in regions $y > 1, y - 1 = O(\epsilon)$, and $0 < y < 1$, as $\epsilon \to 0$. Norbury and Stuart (1989)

50. $$y'' + ye^{-y^2} = 0.$$

For the given DE the integral condition

$$\int_0^y ye^{-y^2}\, dy \to \infty \quad \text{as} \quad y \to \pm\infty$$

is not satisfied. Therefore, the motions are not necessarily periodic. Jordan and Smith (1977), p.32

51. $$y'' + \lambda e^{-\gamma/y} = 0, \quad y'(0) = 0, \; y(1) = 0,$$

where $\lambda > 0$ and γ are constants.

Put $y = \alpha Y(x)$; α is so chosen that $Y(0) = 1$. The equation for Y is $\alpha(d^2Y/dx^2) + \lambda e^{-\gamma/(\alpha Y)} = 0$. Now setting $x = (\alpha)^{1/2}X$, $Y' = dY/dX$, we have

$$Y'' + \lambda e^{-\gamma/(\alpha Y)} = 0, \quad Y(0) = 1, \; Y'(0) = 0, \; Y[1/(\alpha)^{1/2}] = 0.$$

On integration, we have $Y'^2 = -2\lambda \int_1^Y e^{-\gamma/(\alpha t)}dt$. Another integration and the use of BC give

$$\left[\frac{1}{\alpha}\right]^{1/2} = \left[\frac{1}{2\lambda}\right]^{1/2} \int_0^1 \left[\int_Y^1 e^{-\gamma/(\alpha t)}dt\right]^{-1/2} dY \tag{1}$$

or

$$\left[\frac{2\lambda}{\gamma}\right]^{1/2} = \left[\frac{1}{\mu}\right]^{1/2} \int_0^1 \left[\int_Y^1 e^{-\mu/t}dt\right]^{-1/2} dY, \quad \mu = \gamma/\alpha. \tag{2}$$

It is now shown that given λ, γ, there exist at most two solutions α for (1) or at most two solutions μ for (2). This implies that the original BVP has two solutions for each α, where $y(0) = \alpha$. Hastings and McLeod (1985)

52. $$y'' + \lambda e^{-\gamma/y} = 0, \quad y'(0) = 0, \; y(1) = 1,$$

where $\lambda > 0$ and γ are constants.

Following the transformation in Eqn. (2.2.51) we change the BVP to

$$Y'' + e^{-\gamma/(\alpha Y)} = 0, \quad Y(0) = 1, \; Y'(0) = 0, \; Y[1/(\alpha)^{1/2}] = 1/\alpha, \; \alpha > 1,$$

which in turn leads to

$$\left[\frac{1}{\alpha}\right]^{1/2} = \left[\frac{1}{2\lambda}\right]^{1/2} \int_{1/\alpha}^1 \left[\int_Y^1 e^{-\gamma/(\alpha t)}dt\right]^{-1/2} dY. \tag{1}$$

A study of solutions of (1) shows that for the given BVP for γ sufficiently large, there exist two positive functions $K_1(\gamma), K_2(\gamma)$ such that the problem has one solution if $\lambda < K_1(\gamma)$ or $\lambda > K_2(\gamma)$, three solutions if $K_1(\gamma) < \lambda < K_2(\gamma)$, and two solutions if $\lambda = K_1(\gamma)$ or $\lambda = K_2(\gamma)$. Hastings and McLeod (1985)

53. $$y'' + \lambda(1 + \beta - y)^p e^{-\gamma/y} = 0, \quad y'(0) = 0, \; y(1) = 0,$$

where λ, β, γ are all positive and p is a nonnegative integer. Solutions are sought such that $y \leq 1 + \beta$.

Setting $y = \alpha Y, x = (\alpha)^{1/2}X$, we have for Y ($Y' = dY/dX$) the DE

$$Y'' + \lambda(1 + \beta - \alpha Y)^p e^{-\gamma/(\alpha Y)} = 0, \tag{1}$$

$$Y(0) = 1, \; Y'(0) = 0, \; Y[1/(\alpha)^{1/2}] = 0, \; \alpha = y(0) \leq 1 + \beta.$$

After two integrations of (1) and use of BC, we get

$$\left[\frac{1}{\alpha}\right]^{1/2} = \left[\frac{1}{2\lambda}\right]^{1/2} \int_0^1 \left[\int_Y^1 (1 + \beta - \alpha t)^p e^{-\gamma/(\alpha t)}dt\right]^{-1/2} dY, \tag{2}$$

$\alpha \le 1+\beta$. Considering (2), it is shown that it has, for given λ and γ, at most two solutions for α; the same is, therefore, true of the given BVP. Hastings and McLeod (1985)

54. $$y'' + \lambda(1+\beta-y)^p e^{-\gamma/y} = 0, \quad y'(0) = 0, \; y(1) = 1,$$

where λ, β, and γ are positive and p is a nonnegative integer.

Solutions (for physical reasons) are sought such that $y \le 1+\beta$. Making the substitution as in Eqn. (2.2.51), the problem is reduced to

$$Y'' + \lambda(1+\beta-\alpha Y)^p e^{-\gamma/(\alpha Y)} = 0, \tag{1}$$

$Y(0) = 1$, $Y'(0) = 0$, $Y[1/(\alpha)^{1/2}] = 1/\alpha$, $\alpha \le 1+\beta$. Since Y is clearly nonincreasing, we must have $\alpha \ge 1$. Integrating twice and using BC [see Eqn. (2.2.51)], we have

$$\left[\frac{1}{\alpha}\right]^{1/2} = \left[\frac{1}{2\lambda}\right]^{1/2} \int_{1/\alpha}^{1} \left[\int_Y^1 (1+\beta-\alpha t)^p e^{-\gamma/(\alpha t)} dt\right]^{-1/2} dY. \tag{2}$$

Considering (2), it is shown that for γ sufficiently large, β fixed, there exist two positive functions $\lambda_1(\gamma)$ and $\lambda_2(\gamma)$ such that the problem has one solution if $\lambda < \lambda_1(\gamma)$ or $\lambda > \lambda_2(\gamma)$, three solutions if $\lambda_1(\gamma) < \lambda < \lambda_2(\gamma)$, and two solutions if $\lambda = \lambda_1(\gamma)$ or $\lambda = \lambda_2(\gamma)$. Hastings and McLeod (1985)

55. $$y'' + 4\delta(1-y)\exp\{\gamma\beta y/(1+\beta y)\} = 0,$$

$y(0) = y(1) = 0$, where δ, γ, and β are constants.

The given system describes heat and mass transfer and an exothermal chemical reaction in a flat plate. The BVP is solved numerically, using a one-parameter embedding technique. Kubicek and Hlavacek (1983), p.182

56. $$y'' - \alpha y \exp[\gamma\beta(1-y)/[1+\beta/(1-y)]] = 0, \quad y'(0) = 0, \; y(1) = 1,$$

where γ, β, and α are positive constants.

The problem arises in the analysis of heat and mass transfer in a porous catalyst. A constructive iterative method using Green's function is employed to show that if $\alpha\beta \le 1$, there is a unique solution satisfying $0 \le y(x) \le 1, x \in [0,1]$. Numerical results are provided. Pennline (1984)

57. $$y'' - \phi^2 y \exp[\gamma\beta(\chi-y)/[1+\beta(\chi-y)]] = 0,$$

where $\chi = (1-S/N)y(1) + S/N$, $y(0) = 0$, $y'(1) + S(y-1) = 0$, where γ, β, ϕ, S, and N are constants.

The given problem arising in chemical engineering is solved numerically using finite difference methods. Kubicek and Hlavacek (1983), p.97

58. $$y'' - \phi^2 y \exp[\gamma\beta(1-y)/[1+\beta(1-y)]] = 0,$$

$y'(0) = 0$, $y(1) = 1$, where ϕ, γ, and β are constants.

The given BVP describes a simultaneous diffusion, heat conduction, and exothermic chemical reaction in a porous catalyst.

Changing the variable to $z = 1 - y$, we obtain the BVP with homogeneous conditions:

$$\frac{d^2z}{dx^2} = -\phi^2(1-z)\exp[(\gamma\beta z)/(1+\beta z)] \equiv R(z), \quad z'(0) = 0,\ z(1) = 0.$$

Using a Green's function approach, we write the solution as

$$z(x) = \int_0^1 G(x,t)R(z(t))dt \equiv T(z(x)),$$

where

$$G(x,t) = \begin{cases} t-1, & x \le t, \\ x-1, & x \ge t. \end{cases}$$

The iterative scheme $Z_{k+1}(x) = T(Z_k(x))$ is shown to converge. Convergence of solution is shown for $\gamma = 20$, $\beta = 0.05$, and $\phi = 0.1, 1, 2$, and 4.

Multiple solutions are found for $\gamma = 20, \beta = 0.7$, and $\phi = 0.16$. Kubicek and Hlavacek (1983), p.104

59. $$y'' - \alpha y \exp\left[\gamma\beta(1-y)/\{1+\beta(1-y)\}\right] = 0, \quad 0 \le x \le 1,$$

$y'(0) = 0, y(1) = 1$, where α, β, γ are all positive constants.

It is shown that the given problem has a solution such that $0 \le y(x) \le 1$, and that if $\gamma\beta \le 1$, then there is at most one solution $y(x)$ such that $y(x) \le 1$. Baxley (1990)

60. $$y'' + [1 + J_0(y)]y = 0,$$

where $J_0(y)$ is the Bessel function of order zero.

The given DE could describe a harmonic oscillator with position-dependent elastic constant. The DE is changed into a system with polynomial right-hand sides. Put $J_0(y) = y_3$, say, and use

$$\frac{dJ_0}{dy} = -J_1(y),\ y\frac{dJ_1(y)}{dy} = yJ_0(y) - J_1(y) \text{ and call } y_4 = J_1(y).$$

Thus, we have the equivalent system

$$\frac{dy_1}{dx} = -y_2,\ \frac{dy_2}{dx} = -(1+y_3)y_1,\ \frac{dy_3}{dx} = -y_4y_2,\ y_1\frac{dy_4}{dx} = (y_3y_1 - y_4)y_2 \tag{1}$$

with polynomial RHS. Seeking a series solution $y_i = \sum_n a_{i,n}x^n$, $i = 1, 2, 3, 4$, with suitable ICs $y_{10}, y_{20}, y_{30} = J_0(y_0), y_{40} = J_1(y_0)$, recursion relations from (1) are easily obtained and programmed. Fairen, Lopez, and Conde (1988)

61. $$y'' + y + \epsilon|y|y = 0, 0 < \epsilon << 1,$$

where ϵ is a parameter.

Using the Krylov-Bogoliubov-Mitropolsky method, the solution is found in the form

$$y = a(x,\psi)\cos\psi + \epsilon u_1(a,\psi) + O(\epsilon^2),$$

where

$$a(x,\epsilon) = A = \text{constant}, \quad \psi(x,\epsilon) = [1 + (4\epsilon/3\pi)A + (8\epsilon^2/9\pi^2)(72\Delta - 1)A^2]x + \psi_0,$$

where ψ_0 is an integration constant and $\Delta = 0.00055955$. Infinite series forms for Δ and $u_1(a, \psi)$ are given. Mickens and Ramadhani (1992)

62. $$y'' + \nu y|y|^2 - \lambda y = 0,$$

where ν and λ (real) are constants.

The given DE arises from a nonlinear Schrödinger equation. A dominant balance argument gives the local behavior near a movable singularity as $y \sim A(x - x_0)^{-1}$ as $x \to x_0$. Drazin and Johnson (1989), p.188

63. $$(1 - ic_0)y'' - (\Omega + ic_0/c_1)y + (1 + ic_0/c_1)|y^2|y = 0,$$

where c_0, Ω, and c_1 are constants, and $i = \sqrt{-1}$.

The given DE is a complex second order DE, equivalent to a fourth order real system. It is a reduced form of the Ginzburg-Landau equation. This DE is invariant under $y \to y\exp(i\theta)$ and may, therefore, be reduced to a third order equation. Exact numerical solutions are discussed; these are found to be periodic. Newton and Sirovich (1986)

64. $$y'' + \beta y - a|y|^\rho = 0, \quad \beta > 0,$$

where $a > 0, \rho \geq 2$ are constants.

Two-point boundary value problems with either $y(0) = y(X) = 0$ or $y'(0) = y'(X) = 0$ are considered, using geometrical methods. Chicone (1988)

65. $$y'' + \lambda y|y|^{p-1} = 0, \quad 0 < x < 1, y(0) = a, \; y(1) = b,$$

where a, b, and p are fixed real constants.

It is proved that if $p > 0$ and $\lambda \leq 0$, then the given problem has a unique solution. Define a solution of type n: Suppose that $\lambda > 0$, and let $\alpha = \min(|a|, |b|), \beta = \max(|a|, |b|)$; then a solution of the given problem is said to be of type n for some integer n if there exist precisely n points $x \in (0, 1)$ such that $y(x) = a$. It follows that if $a \geq 0, b \geq 0$, then any solution of the given problem is of type 0 if it is nonnegative on $(0, 1)$.

Conditions are given for $\lambda > 0$ for which the given problem has either a unique solution or more than one solution. The solutions are characterized as belonging to type 1 in each case. For example, if $p < 1$, we have the following.

(a) If $a \geq 0, b \geq 0$, and $a + b > 0$. Then the given problem has a unique nonnegative solution for all values of $\lambda, -\infty < \lambda < \infty$.

(b) If $a = b = 0$. Then the given problem has no nontrivial nonnegative solutions for all $\lambda \leq 0$, and a unique nontrivial nonnegative solution for all $\lambda > 0$.

Gilding (1987)

66. $$y'' + \lambda y|y|^{p-1} = 0, \text{ for } 0 < x < 1, y(0) = a, \; y(1) = b,$$

where a, b, and $p > 0$ are fixed real constants. The existence of the solutions of this problem is investigated for all values of $\lambda, -\infty < \lambda < \infty$. The well-known result is the following: Suppose that $p > 0$ and $\lambda \leq 0$. Then the given problem has a unique solution. Similar results are obtained for $\lambda > 0$. Gilding (1987)

67. $$y'' - y + B|y|^{2/\alpha}y = 0, \text{ for } x > 0, \quad y'(0) = 0, \; \lim_{x \to \infty} y(x) = 0,$$

where B and α are constants.

An elementary phase plane analysis shows that this BVP has a unique solution that is positive and decreasing. Stuart (1985)

68. $$y'' + (1/x^k)(|y|^{p-1}y - |y|^{q-1}y) = 0, \quad y(0) = 0,\ y'(0) > 0,$$

where $-1 \le \lim_{x\to\infty} y(x) \le 1, q > p > (N+2)(N-2), N > 2$ and $k = (2N-2)/(N-2)$.

The following results are proved:

(a) There exists a unique value $\alpha_0 > 0$ such that the solution of the given DE with $y(0) = 0, y'(0) = \alpha_0$ satisfies $y'(x) > 0$ for all $x > 0$ and $0 < \lim_{x\to\infty} y(x) \le 1$.

(b) For each integer $J \ge 1$, there exists a value $\alpha_J > 0$ such that if the solution of the given DE satisfies $(y(0), y'(0)) = (0, \alpha_J)$, then $-1 \le \lim_{x\to\infty} y(x) \le 1$ holds and the solution has exactly J isolated zeros in the interval $0 < x < \infty$.

Extensive numerical results are obtained. The method of proof consists of a topological shooting argument, suggested by numerical results. Troy (1990a)

69. $$y'' + (\beta/2)\exp(-1/|y|) = 0, \quad y(0) = y_0,\ y'(0) = 0,$$

where $\beta \ge 0$ is a constant.

It is shown that (1) the solution $y(x, y_0, \beta)$ is strictly decreasing with respect to β, and (2) the equation $y(1, y_0, \beta) - \tau = 0$ has a unique C^1 solution,

$$\beta = \beta_*(y_0, \tau),\ \tau \le y_0 < \infty.$$

Bushard (1978)

70. $$y'' = \sum_{n=1}^{\infty} c_n y^{n-1} \equiv G(y(x)).$$

If $G(w)$ is an analytic function in some neighborhood of zero and if $G'(0) = -\tau^2 < 0$, then the given DE has two distinct families of complex-valued periodic solutions of the form

$$y(x) = \sum_{n=1}^{\infty} b_n \rho^n e^{i\tau n x}, \quad b_1 = \lambda, \tag{1}$$

with period $2\pi/|\tau|$. The series (1) together with the first two derivatives converges absolutely for sufficiently small $|\lambda|$ for every positive $\rho \le 1$ and every real x.

If the function $G(w)$ satisfies the condition $G'(0) = \tau^2 > 0$, then the given DE has complex-valued solution of the form

$$y(x) = \sum_{n=1}^{\infty} b_n w^n e^{-n\tau x}, \quad b_1 = \lambda,\ \tau > 0. \tag{2}$$

The series (2) converges absolutely for sufficiently small $|\lambda|$ for every $|w| \le 1$ and every x in the interval $0 \le x < \infty$. Ifantis (1987)

71. $$y'' - f(y) + \lambda y = 0, \quad 0 < x < L, \quad y(0) = y(L) = 0,$$

where $f(0) = 0, f > 0$ on R^+ and $f(y)/y$ is decreasing.

Considering the phase plane (y, z) of the system $y' = z, z' = -\lambda y + f(y)$, the bifurcation value of λ is found such that the cases of only one solution, the origin and the two solutions, one at the origin and the other at the solution of $f(y) - \lambda y = 0$, are discussed. Liberatore and de Mottoni (1983)

72. $$y'' + yF(y^2) = 0, \quad y(0) = 0.$$

It may be shown that if $F(0) = 0$ and F is a nondecreasing function of y^2, then all solutions of the given DE are periodic and the distance between consecutive zeros decreases as $y'(0)$ increases. Nehari (1960)

73. $$y'' - \phi(y) = 0, \quad y(a) = A, \; y(b) = B,$$

where a, b, A, and B are constants.

It is proved that if $\phi(y)$ is a continuous function $\phi : R \to (-\infty, 0]$, strictly convex upward, then the given BVP has at most two solutions for any A, B and $b > a$. Stepanov (1981)

74. $$y'' + f(y) = 0, \quad y(0) = y(1) = 0,$$

where $f(y)$ is nonnegative and continuous for $0 \leq y < \infty$ and $f(0) = 0$.

Leray-Schauder degree theory is used to investigate the number of nontrivial solutions of the given two-point BVP. Guo (1984)

75. $$y'' + \lambda f(y) = 0, \quad y(0) = y(1) = 0,$$

where $\lambda > 0$ is a constant and $f \in C^2$ satisfies $f(0) < 0$.

Three distinct cases arise, and the results are summarized as follows: Let $f(0) < 0, f'(s) > 0$ for $s > 0$ and β, θ be positive real numbers that satisfy $f(\beta) = 0, F(\theta) = 0$, respectively, where $F(s) = \int_0^s f(t)dt$.

Case 1. If $f''(s) > 0$ for $s > 0$ and $\lim_{s\to\infty} f(s)/s = \infty$, then there exists $\lambda^* > 0$ such that the given BVP has a unique positive solution for $0 < \lambda \leq \lambda^*$ and has no positive solution for $\lambda > \lambda^*$. Also denoting by ρ_λ the supremum norm of the positive solution, ρ_λ increases as λ decreases, and, in particular, $\rho_{\lambda^*} = \theta, \lim_{\lambda\to 0} \rho_\lambda = \infty$.

Case 2. If $f''(s) < 0$ for $s > 0, \lim_{s\to\infty} f(s) = M$, where $0 < M \leq \infty, \lim_{s\to\infty} sf'(s) = 0$ and $f(\theta)/\theta < f'(\theta)$, then there exist λ^*, μ_1 such that $0 < \mu_1 < \lambda^*$ and the given BVP has no positive solutions for $0 < \lambda < \mu_1$ and at least one positive solution for $\lambda \geq \mu_1$. Further, the given BVP has at least two positive solutions for $\mu_1 < \lambda \leq \lambda^*$ and there exists $\mu_2 \geq \lambda^*$ such that the given BVP has a unique positive solution for $\lambda > \mu_2$. Also, $\rho_{\lambda^*} = \theta$ and $\lim_{\lambda\to\infty} \rho_\lambda = \infty$.

Case 3. If $f''(s) < 0$ for $s \in (0, s_0)$ with $s_0 > \theta, f''(s) > 0$ for $s > s_0, f(\theta)/\theta < f'(\theta)$, and if there exists $\sigma > \theta$ such that $H(\sigma) = F(\sigma) - (\sigma/2)f(\sigma) > 0, \lim_{s\to\infty} f(s)/s = \infty$ and $\lim_{s\to\infty} f(s) - \lim_{s\to\infty} sf'(s) < 0$, then there exist $\lambda^*, \lambda_1, \lambda_2$ such that $0 < \lambda_1 < \lambda^* \leq \lambda_2$ and the given BVP has a unique solution for $0 < \lambda < \lambda_1$ and no positive solution for $\lambda > \lambda_2$. Further, there exists a range for λ in (λ_1, λ^*) in which the given BVP has at least three positive solutions, and if $\lambda_2 > \lambda^*$, then the given BVP has at least two positive solutions for $\lambda \in [\lambda^*, \lambda_2)$. Also, $\rho_{\lambda^*} = \theta$ and $\lim_{\lambda\to 0} \rho_\lambda = \infty$. Bifurcation diagrams of these cases are drawn. Castro and Shivaji (1988)

76. $$y'' = f(y), \quad y(0) = y_0,\ y'(0) = y_1.$$

Special difference schemes are constructed for this autonomous system. Specific advantages of Runge–Kutta–Nystrom methods are demonstrated. Suris (1987)

77. $$y'' + f(y) = 0.$$

A thorough phase plane analysis is carried out. Lakin and Sanchez (1970), p.84

78. $$y'' + g(y) = 0.$$

It is proved that when $g(y)$ is nonlinear and every solution of the given DE oscillates with the same least period, no finite Fourier series solution exists. Obi (1980)

79. $$y'' + g(y) = 0,$$

where $g(y)$ is such that $g(0) = 0, g(y) < 0$ for $y < 0$ and $g(y) > 0$ for $y > 0$ and

$$\int_0^y g(u)du \to \infty \text{ as } y \to \pm\infty. \tag{1}$$

The motion in this case is always periodic. Counterexample: If $g(y) = y\exp(-y^2)$ so that (1) does not hold, the motions are not necessarily periodic. The given DE is autonomous. Multiplication by y' and integration leads to

$$(1/2)y'^2 - (1/2)\exp(-y^2) = C, \quad C \text{ is a constant.}$$

A second integration easily shows that the solution is not necessarily periodic. Jordan and Smith (1977), p.31

80. $$y'' - g(y) = 0.$$

The given DE is just Newton's second law of motion $my'' = g(y)$ with $m = 1$. Write this as the system

$$y' = z, \quad z' = g(y). \tag{1}$$

It is possible to obtain a complete phase description of the system (1) and hence all possible motions of the particle. The following points may be noted.

(a) The vector $V(y,z) = (z, g(y))$ has a positive y-component in the upper half-plane and a negative y-component in the lower half-plane, while on the y-axis itself the y-component vanishes. Thus, all trajectories move to the right in the upper half-plane and to the left in the lower half-plane.

(b) A trajectory that intersects the y-axis does so at right angles.

(c) The symmetry across the y-axis holds: $y = u(x), z = v(x)$ is a trajectory, thus so is $y = u(-x), z = -v(-x)$.

(d) The trajectory, therefore, describes a periodic solution, that is, it is an orbit, if and only if it intersects the y-axis in exactly two points.

(e) The given DE is exact and has the first integral

$$H(y,z) = (1/2)z^2 - \int g(y)dy = \text{kinetic energy} + \text{potential energy}.$$

This defines conservation of energy.

(f) The existence of an integral implies that the given DE has no attractor and, in particular, no asymptotically stable singular point. For the same reason all trajectories near an orbit are also orbits, so that there are no limit cycles.

The phase portrait can be easily drawn making use of the potential function $V(y) = -\int g(y)dy$. The trajectories are given by $(1/2)z^2 + V(y) = c$. If it is assumed that singular points are elementary so that $V'(y)$ and $V''(y)$ do not vanish simultaneously, the drawing of the phase portrait becomes particularly simple.

(g) The number of centers interior to an orbit of the given system exceeds by 1 the number of saddle points.

(h) The period of the solution of the given DE can be calculated without knowing the solution, but just its minimum and maximum, say α and β, respectively. Let the solution be denoted by $y = u(x), z = u'(x)$. The period p is the time (or distance) in which the representative point makes one circuit of the orbit. Therefore, $p = 2\times$ time taken to go from $(\alpha, 0)$ to $(\beta, 0)$. The upper half of the orbit satisfies

$$\frac{du}{dx} = [2\{V(\alpha) - V(u(x))\}]^{1/2}, \quad 0 \le x \le p/2, \tag{2}$$

where $u(0) = \alpha$, say. Integrating (2) from 0 to $p/2$, we get

$$\int_0^{p/2} \frac{du/dx}{[2\{V(\alpha) - V(y)\}]^{1/2}} = \frac{p}{2},$$

or

$$p = \sqrt{2}\int_\alpha^\beta \frac{dy}{[V(\alpha) - V(u(x))]^{1/2}}, \quad y = u(x).$$

Plaat (1971), pp.229-233

81. $\boldsymbol{y'' + \lambda f(y) = 0, 0 \le x \le 1, y(0) = y(1) = 0,}$

where $f(y) > 0$ for $y > 0$ and λ is a parameter.

Let x_0 be the point at which some solution $y(x)$ of the given DE assumes its maximum $||y|| = y(x_0)$. Then $y'(x) \ge 0$ on $[0, x_0]$ and $y'(x) \le 0$ on $[x_0, 1]$. Thus, the given problem may be solved as follows: Let $F(w) = \int_0^w f(v)dv$; then

$$(1/2)y'^2(x) + \lambda F(y(x)) = \lambda F(||y||),$$

$$x\sqrt{2\lambda} = \int_0^{y(x)} \frac{dw}{[F(||y||) - F(w)]^{1/2}}, \quad 0 \le x \le x_0,$$

$$(1-x)\sqrt{2\lambda} = \int_0^{y(x)} \frac{dw}{[F(||y||) - F(w)]^{1/2}}, \quad x_0 \le x \le 1. \tag{1}$$

Setting $x = x_0$ and $y(x) = ||y||$, we see that $x_0 = 1/2$ and $y(x) = y(1-x)$. That is, any solution of the given system is symmetric about $x = 1/2$. The problem is, therefore, equivalent to

$$y''(x) + \lambda f(y(x)) = 0, \quad 0 \le x \le 1/2,\ y(0) = y'(1/2) = 0. \tag{2}$$

Equation (1) may be used to construct the solution of (2). For any $\rho \in (0, r)$, define λ by

$$\sqrt{\lambda} = \sqrt{2}\int_0^{\lambda} \frac{dw}{[F(\rho) - F(w)]^{1/2}} = \sqrt{2}\rho \int_0^1 \frac{dv}{[F(\rho) - F(\rho v)]^{1/2}}.$$

Then, the equation

$$x\sqrt{2\lambda} = \int_0^y \frac{dw}{[F(\rho) - F(w)]^{1/2}}$$

defines a one-to-one relation between x and y for $0 \le x \le 1/2$ and $0 \le y \le \rho$ such that $x = 0$ iff $y = 0$ and $x = 1/2$ iff $y = \rho$. The function $y(x)$ so defined is easily seen to be twice differentiable and to satisfy (2).

A general result proved is that when $f(w)$ is a convex function of w, with $f(0) > 0$, then the given BVP has, for each $\lambda > 0$, either zero, one, or two positive solutions on the interval $(0, 1)$. Laetsch (1970)

82. $$y'' + g(y) + t = 0, \quad y(0) = y(\pi) = 0,$$

where $x \in (0, \pi), t \in R$ is a parameter and $g'(-\infty) < \infty$ and $g'(\infty) = \infty$.

It is shown that for any given $n \in N$, there exist at least n solutions of the problem if t is sufficiently negative. The proof uses variational methods. Ruf and Solimini (1986)

83. $$y'' + y + \alpha f(y) = 0,$$

where α is a parameter and $f(y)$ is a non-odd nonlinearity.

Multiply the given DE by y' and integrate from $x = 0$ to arbitrary x, with IC $y(0) = b + a, y'(0) = 0$:

$$y'^2 + y^2 - (b+a)^2 + 2\alpha \int_0^x y' f(y) dx = 0. \tag{1}$$

Transforming to the new variable $X = X(x)$ and denoting $F(x) = \dfrac{dX}{dx}$, we obtain from (1),

$$F^2 y'^2 + y^2 - (b+a)^2 + 2\alpha \int_0^X y' f(y) dX = 0, ' \equiv \frac{d}{dX}. \tag{2}$$

Now seeking the solution of (2) in the form $y(X) = b + a\cos X$, we have

$$F^2 \sin^2 X = \sin^2 X + (2b/a)(1 - \cos X) + (2\alpha/a) \int_0^X \sin X f(b + a\cos X) dX. \tag{3}$$

Equation (3) provides an algebraic relation which the differential transformation $F(X)$ must satisfy so that the response will be simply harmonic in the X domain. One next assumes that the nonlinear term $f(b + a\cos X)$ is expandable in a Fourier series

$$f(b + a\cos X) = \sum_{n=0}^{\infty} C_n \cos nX, \tag{4}$$

where the Fourier coefficients C_n are assumed to be known functions of a and b. Noting that $F(X)$ is periodic and of period 2π, one may write

$$F^2(X) = \sum_{n=0}^{\infty} G_n \cos nX, \tag{5}$$

with the coefficients G_n to be determined. Substituting (4) and (5) into (3), integrating the term containing $f(b + a\cos X)$, equating the coefficients of harmonic terms $\cos nX$ and $\sin nX$, and finally solving for the coefficients G_n as functions of Fourier coefficients C_n of $f(y)$, one arrives at the results

$$\begin{aligned} G_0 &= 1 + \alpha C_1/a,\ G_n = (4n\alpha/a)\textstyle\sum_{k=n+1,n+3,\ldots} C_k/(k^2-1), \\ b &= -\alpha\left[C_0 + \textstyle\sum_{k=2,\text{even}} C_k/(k^2-1)\right]. \end{aligned} \tag{6}$$

Note that the last of equations (6) defining the "bias" b is nonlinear since C_n are in general nonlinear functions of b. Thus, we find the X transformation

$$F^2(X) = 1 + \alpha C_1/a + \sum_{n=1}^{\infty} G_n \cos nX. \tag{7}$$

The result, that the oscillator response having a single maximum and a single minimum per cycle is such that $F(X) = dT/dx$ is always positive, is used. Thus, taking a positive root of (7), we have

$$F(X) = [1 + \alpha C_1/a]^{1/2}\left[1 + \sum_{n=1}^{\infty} H_n \cos nX\right]^{1/2}, \tag{8}$$

where $H_n \equiv G_n/(1 + \alpha C_1/a)$.

The radical in (8) will be positive for all situations resulting in periodic motion. Since $n \geq 1$, the coefficients G_n are functions of the higher harmonics $C_n, n \geq 2$, and the importance of H_n terms in $F(X)$ will depend on the magnitudes of the harmonics C_n relative to that of the fundamental harmonic C_1.

Writing $x = \int_0^X \dfrac{dX}{F(X)}$, using the form (8), and expanding the integrand in a convergent power series and integrating, we get the form

$$x(X) = \left[\frac{1}{1 + \alpha C_1/a}\right]^{1/2}\left[X\left(1 + \sum a_p A_{p0}\right) + \sum_{n=1}^{\infty} B_n \sin nX\right].$$

An expansion for the period is found. Several examples are treated. Burton and Hamdan (1983)

84. $$y'' + F'(y) + w_0^2 y = 0,$$

where $F(y) = \int_0^y f(x)dx$, $f(y)$ symmetric and negative for $|y| < \delta, \delta > 0$ and $f(y) = \alpha > 0$ otherwise. Then there exists an h such that $F(h) = F(-h) = 0$. The bounds for free oscillations of the given DE are obtained. Manaresi (1954)

85. $$y'' + w(y) + \epsilon f(y) = 0, \quad y(0) = a,\ y'(0) = 0.$$

Assuming that the reduced equation $y'' + w(y) = 0$ with the given IC has a (known) solution, the solution of the perturbed problem is found in terms of this solution. Papoulis (1958)

2.3 $y'' + g(x)h(y) = 0$

1. $$y'' - x^{-5}y^2 = 0.$$

Put $u = y/x$ and integrate twice to obtain $\psi_0 = -1/x \pm \int[1/\{(2/3)u^3 - e^{-6}\phi_0\}^{1/2}]du$, where ϕ_0 and ψ_0 are constants and the integral is expressible in terms of elliptic functions. Stephani (1989), p.72

2. $$y'' + x^{-4}y^5 = 0.$$

The given DE is related to the Emden equation in q by $y = qx$; i.e., one gets through this relation the equation $q'' + (2/x)q' + q^5 = 0$. Multiplying the given DE by x, rewriting it suitably, and integrating, we get a first integral

$$xy'^2 + y^6/(3x^3) - yy' = \text{constant}.$$

An intuitive method for such integrals is provided. Sarlet and Bahar (1980)

3. $$y'' - \beta x^{-10/3}y^{-5/3} = 0,$$

where β is a constant.

The given DE admits the first integral

$$\begin{aligned} &x^4y'^4 - 4x^3yy'^3 + (6\beta x^{2/3}y^{-2/3} + 6x^2y^2)y'^2 + (24\beta x^{-1/3}y^{1/3} - 4xy^3)y' \\ &- 3\beta x^{-4/3}y^{4/3} + 9\beta^2x^{-8/3}y^{-4/3} + y^4 = \text{constant}. \end{aligned}$$

Airault (1986)

4. $$y'' - x^ny^2 = 0.$$

Put $y = te^{-(n+2)s}$, $s = \ln x$ and the given DE reduces to $s'' + t^2s'^3 - [(n+2)(n+3)ts' - (2n+5)]s'^2 = 0$, $s' \equiv \dfrac{ds}{dt}$, which is of first order in s'. Stephani (1989), p.42.

5. $$y'' + x^{\gamma}y^{2n-1} = 0, x \geq x_0 > 0;$$

n is a positive integer and γ is a constant, $\gamma > 1 + 1/n$. It is shown that every solution of the given DE is in $L^{2n}(x_0, \infty)$. Spikes (1977)

6. $$y'' + \phi^2(x)(1-y)^{\nu} = 0, 0 < x < 1, y'(0) = 0, y(1) = 0,$$

where ν is a constant.

The existence and uniqueness of the solution of the given BVP is proved by an iterative method. Vatsya (1987)

7. $$y'' + x^4(y^5 + y) = 0, y(a) = y(b) = 0, 0 < a < b.$$

It is shown that the given BVP has only one solution. Kwong (1991)

8. $$y'' + a(x)y^{2n-1} = 0, y(0) = c, y'(0) = b,$$

where $a(x)$ is positive and nondecreasing, n is a positive integer, and c and b are constants.

It is shown that if $b = 0$, the solution of the given problem never rises above the IC; if $|c| < 1$, the sequence $\{y^{2n}(x)\}$ of iterates converges to zero for fixed x. Waltman (1963)

9. $$y'' + a(x)y^{2n+1} = 0,$$

where n is a nonnegative integer.

It is proved that if $a(x)$ is continuous and $\int^{\infty} t^{2n+1} | a(t) | dt < \infty$, then the given DE has a solution $y(x)$ with the property that $\lim_{x\to\infty} y(x)/x = a \neq 0$; the solution can then be written as $y(x) = ax + b + 0(1)$, where a and b are constants. Waltman (1964)

10. $$y'' + p(x)y^{2n+1} = 0,$$

where $p(x)$ is positive and continuous on $[0, \infty)$ and n is a positive integer.

Assuming that $p(x)$ is also locally of bounded variation on $[0, \infty)$, it is shown that for arbitrary real numbers a and b and for any $x_0 \in [0, \infty)$, the IVP $y(x_0) = a, y'(x_0) = b$ has a unique solution which exists on $[0, \infty)$. Coffman and Ullrich (1967)

11. $$y'' + f(x)y^{2n-1} = 0,$$

where n is an integer ≥ 2 and $f(x)$ is positive and continuous for $x \geq 0$.

It is proved that a necessary and sufficient condition for all solutions of the given DE to be oscillatory is that $\int_0^{\infty} xf(x)dx = \infty$. This condition is known to be necessary but not sufficient for $n = 1$. It is further shown that if $f'(x)$ is continuous and $f'(x) \leq 0$ for $x \geq 0$, a sufficient condition for nonoscillation of all solutions $\not\equiv 0$ is $\int_0^{\infty} x^{2n-1} f(x)dx < \infty$. Atkinson (1955)

12. $$y'' + f(x)y^{2n-1} = 0, f(x) > 0.$$

It is shown that the given DE has an oscillatory solution, in fact, an infinity of such solutions, if $f(x)$ satisfies $(n+1)f(x) + xf'(x) > 0$ and

$$x^{1/2} \int_x^{\infty} f(x)dx \left\{ \int_x^{\infty} du \int_u^{\infty} f(\sigma)d\sigma \right\}^{-(2n-1)/(2n-2)} < M$$

for fixed M and $x > x_0$.

According to some earlier work, the integral condition may be replaced by the condition that $x^{n+1}f(x)$ be positive and nondecreasing. Jasny (1960)

13. $$y'' - \beta xy^{-7} = 0,$$

where β is a constant.

The given DE admits the first integral $xy'^4 - yy'^3 + (2/3)\beta x^2 y^{-6} y'^2 - (1/3)\beta xy^{-5}y' - (1/12)\beta y^{-4} + (1/9)\beta^2 x^3 y^{-12} = \text{constant}$. Airault (1986)

14. $$y'' - \beta x^3 y^{-7} = 0,$$

where β is a constant.

The given DE admits a first integral $x^3 y'^4 - 3x^2 y y'^3 + [(2/3)\beta x^6 y^{-6} + 3xy^2]y'^2 - (\beta x^5 y^{-5} + y^3)y' + (1/4)\beta x^4 y^{-4} + (1/9)\beta^2 x^9 y^{-12} =$ constant. Airault (1986)

15. $$y'' - \beta x y^{-7/5} = 0,$$

where β is a constant.

The given DE admits a first integral $y'^6 + 15xy^{-2/5}y'^4 - 25\beta y^{3/5}y'^3 + 75\beta^2 x^2 y^{-4/5} y'^2 - 375\beta^2 x y^{1/5} y' + (625/2)\beta^2 y^{6/5} + 125\beta^3 x^3 y^{-6/5} =$ constant. Airault (1986)

16. $$y'' - \beta x y^{-5/3} = 0,$$

where β is a constant.

The given DE admits the first integrals $y'^4 + 6\beta x y^{-2/3} y'^2 - 18\beta y^{1/3} y' + 9\beta^2 x^2 y^{-4/3} =$ constant and xy'^4 - $yy'^3 + 6\beta x^2 y^{-2/3} y'^2 - 27\beta x y^{1/3} y' + (81/4) y^{4/3} + 9\beta^2 x^3 y^{-4/3} =$ constant. Airault (1986)

17. $$y'' - \beta x^{-7/3} y^{-5/3} = 0,$$

where β is a constant.

The given DE admits a first integral $x^3 y'^4 - 3x^2 y y'^3 + (6\beta x^{2/3} y^{-2/3} + 3xy^2) y'^2 + (15\beta x^{-1/3} y^{1/3} - y^3) y' - (3/4)\beta x^{-4/3} y^{4/3} + 9\beta^2 x^{-5/3} y^{-4/3} =$ constant and $x^4 y'^4 - 4x^3 y y'^3 + (6\beta x^{5/3} y^{-2/3} + 6x^2 y^2) y'^2 + (6\beta x^{2/3} y^{1/3} - 4xy^3) y' - 12\beta x^{-1/3} y^{4/3} + 9\beta^2 x^{-2/3} y^{-4/3} + y^4 =$ constant.

The given DE also admits a first integral of degree 6 in y': $x^6 y'^6 - 6x^5 y y'^5 + (9\beta x^{11/3} y^{-2/3} + 15x^4 y^2) y'^4 - (20x^3 y^3 + 9\beta x^{8/3} y^{1/3}) y'^3 + (15x^2 y^4 + 27\beta^2 x^{4/3} y^{-4/3} - 27\beta x^{5/3} y^{4/3}) y'^2 + (45\beta x^{2/3} y^{7/3} - 6xy^5 + 27\beta^2 x^{1/3} y^{-1/3}) y' + y^6 - 18\beta x^{-1/3} y^{10/3} + (135/2)\beta^2 x^{-2/3} y^{2/3} + 27\beta^3 x^{-1} y^{-2} =$ constant. Airault (1986)

18. $$y'' - \beta x^{-5/6} y^{-5/3} = 0,$$

where β is a constant.

The given DE admits a first integral $x^3 y'^4 - 3x^2 y y'^3 + (6\beta x^{13/6} y^{-2/3} + 3xy^2) y'^2 - (12\beta x^{7/6} y^{1/3} + y^3) y' + 6\beta x^{1/6} y^{4/3} + 9\beta^2 x^{4/3} y^{-4/3} =$ constant. Airault (1986)

19. $$y'' - \beta x^{-1/2} y^{-5/3} = 0,$$

where β is a constant.

The given DE admits the first integral $xy'^4 - yy'^3 + 6\beta x^{1/2} y^{-2/3} y'^2 + 9\beta^2 y^{-4/3} =$ constant. Airault (1986)

20. $$y'' - \beta x^2 y^{-5/3} = 0,$$

where β is a constant.

The given DE has a first integral

$$y'^4 + 6\beta x^2 y^{-2/3} y'^2 - 36\beta x y^{1/3} y' + 9\beta^2 x^4 y^{-4/3} + 27\beta y^{4/3} = \text{constant}. \tag{1}$$

If we let $\beta = -3r^4$, then any solution of the equations $y' = -3r^2xy^{-1/3} \pm 3ry^{1/3}$ or $y' = 3r^2xy^{-1/3} \pm 3r(-1)^{1/2}y^{1/3}$ is a solution of (1). Airault (1986)

21. $$y'' - \beta x^{-13/5}y^{-7/5} = 0,$$

where β is a constant.

The given DE admits a first integral $x^6y'^6 - 6x^5yy'^5 + (15\beta x^{17/5}y^{-2/5} + 15x^4y^2)y'^4 - (35\beta x^{12/5}y^{3/5} + 20x^3y^3)y'^3 + (15\beta x^{7/5}y^{8/5} + 75\beta^2x^{4/5}y^{-4/5} + 15x^2y^4)y'^2 + (15\beta x^{2/5}y^{13/5} - 6xy^5 + 225\beta^2x^{-1/5}y^{1/5})y' + y^6 - 10\beta x^{-3/5}y^{18/5} + (25/2)\beta^2x^{-6/5}y^{6/5} + 125\beta^3x^{-9/5}y^{-6/5} = \text{constant}$. Airault (1986)

22. $$y'' + ae^xy^{-1/2} = 0, \ a \text{ is constant.}$$

Write $y'' = -ae^xy^{-1/2}$ and square [see Eqn. (2.44.43)]. Kamke (1959), p.545.

23. $$y'' - x^{-1/2}/y^{3/2} = 0.$$

For the given Thomas-Fermi equation, using point transformation, first the singular solution $y = 144/x^3$ is recovered. It is suggested that using the Sommerfeld, Fermi, and regular Taylor series solutions, new solutions can be generated, employing point transformations. Meinhardt (1981)

24. $$y'' - x^{-1/2}y^{3/2} = 0, y(0) = 1, y(\infty) = 0.$$

The solution near the origin is

$$y = 1 - kx + (4/3)x^{3/2} - (2k/5)x^{5/2} + (1/3)x^3 + \cdots,$$

where k is a constant. The solution near $x = \infty$ is $y \simeq (144/x^3)(1 - A/x^s)$, $s = (73^{1/2} - 7)/2$.

It is argued that since the solution is nonanalytic at $x = \infty$, a solution by rational approximation for y is not consistent. A new function is suitably introduced and then a Padé approximation leads to an accurate solution, with an error of less than 2%. Tu (1991)

25. $$y'' - x^{-1/2}y^{3/2} = 0, y(0) = 1, y(\infty) = 0.$$

(a) The given problem is written more generally as

$$y'' - x^{-\delta}[y(x)]^{1+\delta} = 0, y(0) = 1, y(\infty) = 0.$$

The solution is sought in the form

$$y = y_0(x) + \delta y_1(x) + \delta^2 y_2(x) + \delta^3 y_3(x) + \cdots.$$

The resulting system of equations is solved iteratively. To third order the prediction for $y'(0)$, with the aid of Padé approximation, gives an error of 1.8% only. The first two terms in the perturbation expansion are

$$\begin{aligned} y_0(x) &= e^{-x}, \\ y_1(x) &= e^{-x}\int_0^x e^{2s}\{y_1'(0) + \int_0^s e^{-2t}\ln(e^{-t}/t)dt\}ds. \end{aligned}$$

The condition $y_1(\infty) = 0$ implies that

$$\begin{aligned} y_1'(0) &= -\int_0^\infty e^{-2t} \ln (e^{-t}/t) dt \\ &= 1/4 - (1/2) \ln 2 - (1/2)\gamma = -0.385181423. \end{aligned}$$

Therefore, to first order in δ, for $\delta = 1/2$, we have

$$y'(0) = -1 + (1/2)y_1'(0) = -1.192590711.$$

This value is in error by 25%. But the Padé approximation improves it after one obtains third order terms.

(b) An iterative process generated by

$$Lu_{n+1} = -x^{-1/2}(1 - x/a - u_n)^{3/2} - (3/2)x^{-1/2}u_n,$$

$Lu \equiv u'' - (3/2)x^{-1/2}u, u_{n+1}(0) = u_{n+1}(a) = 0$ is shown to have analytic forms. It converges monotonically upward if $u_0 \equiv 0$ and monotonically downward if $u_0 \equiv 1 - x/a$. The existence of a unique solution to the original problem is proved. Bender et al. (1989), Chan and Du (1986)

26. $$y'' - a^{3/2}x^{-1/2}[y]^{3/2} = 0, x \in (0,1), y(0) = 1, y(1) = 0.$$

It is proved that the solutions of the given Thomas-Fermi problem are monotonic with respect to a. Lower and upper bounds of the solution for $a = 1$ are constructed. Iterative numerical schemes for the solution are presented. Mooney (1979)

27. $$y'' - x^{-1/2}y^{3/2} = 0, x \in (0,b),$$

where $y(0) = 1, by'(b) = y(b)$.

The following results are proved:

(a) If y_1, y_2 are two distinct solutions of the given problem, then $y_1 > y_2 > 0$ or $0 < y_1 < y_2$.

(b) There exists a unique solution y of the given problem satisfying $0 \le y(x)e^{x/b}$, provided that $b \ge e^{-1/3}$. This solution is minimal in the sense that $y_1 > y$, where y_1 is any other solution of the given problem.

(c) The minimal positive solution of the given problem can be approximated by certain monotonic sequences of Picard and Newton type. The rate of convergence is discussed.

Mooney (1979)

28. $$y'' + x^{1-k}y^k = 0.$$

Using invariance properties, it may be shown that

$$\omega\{y/x^{(3-k)/(1-k)}, y'/x^{2/(1-k)}\} = C, \ k \ne 1$$

is a first integral of the given DE, for an arbitrary constant C. $\omega(\xi\eta)$ is a solution of

$$\left[\frac{2}{1-k}\eta + \xi^k\right] d\xi + \left[\left(\frac{k-3}{1-k}\right)\xi + \eta\right] d\eta = 0.$$

Jones and Ames (1967)

29.

$$\boldsymbol{y'' - x^{1-m}y^m = 0,}$$

where m is real and both x and y are positive; $m \neq -1, 1, 2, 3, \ldots$.

Assuming that $y' \neq 0$, we apply transformations $\xi = xy'/y, \eta = x^{2-m}y^m/y', t = ln \mid x \mid$ to obtain the system

$$\begin{aligned} \frac{d\xi}{dt} &= \xi[1 - \xi + \eta] = P(\xi, \eta), \\ \frac{d\eta}{dt} &= \eta[(2-m) + m\xi - \eta] = Q(\xi, \eta). \end{aligned} \tag{1}$$

A phase plane analysis of the system (1) is carried out and a detailed study of the nature of the singular points $(0,0), (1,0), (0, 2-m), \{(3-m)/(1-m), 2/(1-m)\}$ shows their diverse nature, depending on m; the results are tabulated.

The transformation $\zeta = \xi[1 - \xi + \eta]$ reduces $\dfrac{d\eta}{d\xi} = \dfrac{Q(\xi,\eta)}{P(\xi,\eta)}$ to

$$\zeta\frac{d\zeta}{d\xi} - [(4-m) + (m-3)\xi]\zeta + \xi(1-\xi)[(3-m) + (m-1)\xi] = 0, \tag{2}$$

which may easily be solved by series expansion. If $m = 5$ (and then only), Eqn. (2) admits an integrating factor of the form $[\alpha(\xi) + \beta(\xi)\zeta]^a$ with $a = 1$ or $a = -1/2$, and

$$\alpha(\xi) = (1/a)(a+1)(c)[(4-m) + (m-3)\xi]d\xi = (1/a)(a+1)c\xi(\xi - 1),$$

and $\beta(\xi) = c$, a constant of integration. Matlak (1969)

30.

$$\boldsymbol{y'' + x^\beta y^\gamma = 0, x \in (0, X),}$$

$$\boldsymbol{y(0) = y(X) = 0, y'(0) = \alpha, y(x) > 0, x \in (0, X).}$$

A necessary and sufficient condition for the existence of solution of the given Emden-Fowler equation with a free endpoint is found; a monotone iteration scheme works for the approximation of a positive solution. Iffland (1987)

31.

$$\boldsymbol{y'' \pm y^\lambda x^\sigma = 0,}$$

where λ and σ are constants.

The asymptotic solutions exist for different values of λ and σ:

1. $y'' - y^\lambda x^\sigma = 0, y(\infty) = 0, y'(\infty) = 0$. A solution exists if and only if $\sigma \geq -2$.
2. $y'' + y^\lambda x^\sigma = 0, y(0^+) = 0, y'(0^+) = \infty$. A solution exists if and only if $-2 < \sigma \leq -\lambda - 1$.
3. $y'' + y^\lambda x^\sigma = 0, y(\infty) = \infty, y'(\infty) = 0$. This problem has a solution if and only if $-2 \leq \sigma < -\lambda - 1$.

Explicit asymptotic formulas for these cases can be obtained. Taliaferro (1981)

32. $$y'' - x^\lambda y^\gamma = 0, \lambda < 0 \text{ and } \gamma \geq 1, y(a) > 0, y'(a) = 0,$$

provided that $a > 0$ when $\lambda \leq -1$, and $a \geq 0$ when $-1 < \lambda < 0$.

Upper and lower bounds of the solution are found. Eliason (1972)

33. $$y'' - x^{-\ell} y^m = 0, y(0) = 1, y(\infty) = 0,$$

where $m \geq 1, 0 < \ell < 1$.

The given DE is a special case of the more general equation (2.6.24), and hence the results for the latter apply. Kolosov and Lyubarskii (1968)

34. $$y'' - x^a y^b = 0,$$

where $a > -2, b > 0$, and $y(0) = 1$.

By a procedure similar to successive approximation, the asymptotic development of the solutions of the given DE near $x = 0$ is carried out, involving sums of terms of the type $c_{kj} x^{k+ja}$ and a remainder term of type $O(x^{n(a+2)})$, where k, j, n are integers. The coefficients $c_{kj} = c_{kj}(a, b, C)$ are given explicitly in terms of a, b, and an integration constant C. Mihailovic (1950)

35. $$y'' - x^\sigma y^n = 0,$$

where σ and n are constants.

It is easily verified that $y = cx^w$ is a solution provided that

$$w = -(\sigma + 2)/(n-1), c = [(\sigma+2)(\sigma+n+1)/(n-1)^2]^{1/(n-1)}.$$

It follows that the solution of this form exists only if $(\sigma + 2)(\sigma + n + 1) > 0$. These special solutions provide valuable clues to the structure of the set of "proper" solutions, i.e., solutions that are real and positive for some value of $t \geq t_0$. The structure of solutions will greatly be influenced by the arithmetic nature of n. Because of the presence of the term y^n, proper solutions must be positive. For certain values of $n = p/q$, say, where q is odd, y may take negative values.

Bellman proved the following results:

(a) If $\sigma + n + 1 < 0$, all positive proper solutions of the given DE have one or the other of the following asymptotic expressions:

$$y \sim \{(\sigma+2)(\sigma+n+1)/(n-1)^2\}^{1/(n-1)} x^{-\{(\sigma+2)/(n-1)\}},$$
$$y \sim a_1 x + a_2 + [1 + O(1)] a_2^n x^{\sigma+n+2}/[(\sigma+n+1)(\sigma+n+2)]$$

or

$$y \sim a_2 x + [1 + O(1)] a_2^n x^{\sigma+2}/[(\sigma+1)(\sigma+2)],$$

where a_1 and a_2 are arbitrary constants.

(b) If $\sigma + 2 < 0 < \sigma + n + 1$, every positive proper solution of the given DE has the asymptotic form $y = a + [a^n x^{\sigma+2}/\{(\sigma+1)(\sigma+2)\}](1 + O(1))$.

(c) If $\sigma + 2 < 0, \sigma + n + 1 = 0$, every positive proper solution of the given DE has the asymptotic form

$$y = a + [a^n x^{\sigma+2}/\{(\sigma+1)(\sigma+2)\}](1 + O(1)).$$

(d) If $\sigma + 2 > 0$, every positive proper solution of the given DE has the asymptotic form $y \sim cx^{-(\sigma+2)/(n-1)}$.

Bellman (1953), pp.144-151

36. $$\boldsymbol{y'' + ax^{\nu}y^n = 0,}$$

where a, ν, and n are constants.

One solution is $y = \alpha x^{\beta}$ with $\beta = (2+\nu)/(1-n)$,

$$\alpha^{n-1} = -(\nu+2)(\nu+n+1)/\{a(n-1)^2\}.$$

Writing

$$\eta(\xi) = y(x), \xi = 1/x,$$

we get

$$\xi\eta'' + 2\eta' + a\xi^{-\nu-3}\eta^n = 0.$$

Kamke (1959), p.544; Murphy (1960), p.381

37. $$\boldsymbol{y'' - \beta x^{\sigma} y^n = 0,}$$

where β, σ, and n are constants.

If $n = -2\sigma - 3$, a first integral of the given DE is $y'^2 - x^{-1}yy' - [2\beta/(n+1)]x^{-(n+3)/2}y^{n+1} = Cx^{-1}$, where C is an arbitrary constant. If $n = -\sigma - 3$, a first integral of the given DE is $y'^2 - 2x^{-1}yy' + x^{-2}y^2 - [2\beta/(n+1)]x^{-(n+3)}y^{n+1} = C_1x^{-2}$, where C_1 is an arbitrary constant.
Airault (1986)

38. $$\boldsymbol{y'' - x^{\alpha\lambda-2}y^{1+\alpha} = 0, \lambda > 0, \alpha > 0, \alpha\lambda > 1.}$$

First we note that $y = \{\lambda(\lambda+1)\}^{1/\alpha}x^{-\lambda}$ is a solution of the given DE. It is shown that the given DE with $\alpha > 0$ has a one parameter family of nontrivial bounded solutions if and only if $\beta = \alpha\lambda - 2 > -1$. It has a double series representation in the neighborhood of $x = 0$: $y(x) = \sum_{m+n\geq 0} a_{mn}x^m x^{\alpha\lambda n}, a_{00} > 0$, unless $\alpha\lambda$ is an integer.

At $x = \infty$, the solution has a representation

$$y(x) = \{\lambda(\lambda+1)\}^{1/\alpha}x^{-\lambda}(1 + Cx^{\mu/\alpha} + \cdots),$$

and therefore every nontrivial bounded solution of the given DE is asymptotic to the solution $\{\lambda(\lambda+1)\}^{1/\alpha}x^{-\lambda}$ as $x \to \infty$.

The analysis is constructive and uses Briot-Bouquet theory. Referring now to IVP $\lim_{x\to 0} y = a, \lim_{x\to 0} y' = b, 0 < a < \infty, |\, b \,| < \infty$, the solution $y = \phi(x, a, b)$ of the given DE exists and has the following properties:

(a) $\phi(x, a, b)$ admits the following double series expansion in the neighborhood of $x = 0$:

$$\phi(x, a, b) = a\left(1 + \sum_{m+n>0} \gamma_{mn}\left\{\frac{a^{\alpha}}{\lambda(\lambda+1)}x^{\alpha\lambda}\right\}^m \{(\alpha b/a)x\}^n\right).$$

(b) For each $a(0 < a < \infty)$, there exists one and only one value $\hat{b}(a)$ of b such that $y = \phi(x, a, \hat{b}(a))$ is defined and bounded for $0 < x < \infty$ together with its derivative. In the neighborhood of $x = \infty$, $\phi(x, a, \hat{b}(a))$ can be expressed in the following form: $\phi(x, a, \hat{b}(a)) = [\lambda(\lambda+1)]^{1/\alpha} x^{-\lambda} \left(1 + \sum_{n>0} c_n x^{n\mu/\alpha}\right)$, where μ is a negative eigenvalue of the matrix

$$\begin{pmatrix} 0 & \alpha \\ \lambda(\lambda+1)\alpha^2 & (2\lambda+1)\alpha \end{pmatrix}.$$

(c) If $b > \hat{b}(a)$, $\phi(x, a, b)$ has a movable singularity at $x = w(0 < w < \infty)$ and $\lim_{x\to w} \phi(x, a, b) = \infty$. In the neighborhood of $x = w$, $\phi(x, a, b)$ can be expressed as

$$\phi(x, a, b) = [2(\alpha+2)/\{\alpha^2 w^{\alpha\lambda-2}\}]^{1/\alpha}(w-x)^{-2/\alpha}$$
$$\left[1 + \sum_{m+n>0} c_{mn}(w-x)^m\{(w-x)^{2+4/\alpha}\}^n\right],$$

if $4/\alpha$ is not an integer, and as

$$\phi(x, a, b) = [2(\alpha+2)/(\alpha^2 w^{\alpha\lambda-2})]^{1/\alpha}(w-x)^{-2/\alpha}\left[1 + \sum_{m>0}(w-x)^m p_m\{\log(w-x)\}\right],$$

$p_m(\xi)$: a polynomial of ξ whose degree is at most $[m\alpha/(2\alpha+4)]$, if $4/\alpha$ is an integer. Here $w = w(a, b)$.

(d) If $b < \hat{b}(a)$, then $\lim_{x\to w} \phi(x, a, b) = 0$ for some finite positive w, and in the neighborhood of $x = w$, $\phi(x, a, b)$ is expressed as $\phi(x, a, b) = A(w-x)[1 + \sum_{m+n>0} b_{mn}(w-x)^m\ (w-x)^{\alpha n}]$. Here A and w depend on a and b, and $x = w$ is a movable branch point of the solution unless α is an integer.

Saito (1978, 1978a)

39. $$\boldsymbol{y'' + e^{\alpha\lambda x} y^{1+\alpha} = 0, -\infty < x < \infty, 0 \le y < \infty,}$$

where α and λ are positive constants and $y^{1+\alpha}$ always represents its nonnegatively valued branch.

Let $\phi(x)$ be an arbitrary solution of the given DE. Now write

$$z = -\lambda^{-2} e^{\alpha\lambda x}\phi, w = z'. \tag{1}$$

Then the given DE transforms into a first order "algebraic" DE:

$$\frac{dw}{dz} = [\alpha^2\lambda^2 z^2(z-1) + 2\alpha\lambda zw + (\alpha-1)w^2]/(\alpha zw). \tag{2}$$

Introducing a parameter s, Eqn. (2) can be written as the equivalent autonomous system

$$\frac{dz}{ds} = \alpha zw, \frac{dw}{ds} = \alpha^2\lambda^2 z^2(z-1) + 2\alpha\lambda zw + (\alpha-1)w^2. \tag{3}$$

Critical points of (3) are (0,0) and (1,0) if $\alpha \neq 1; (0, c)$, where c is an arbitrary number and (1,0) if $\alpha = 1$. Since $\phi(x) > 0$, z is always negative. So (3) is considered in the half-plane $z < 0$. With the help of (3) all solutions of (2) are found. For each $w(z)$, we determine $z(x)$ so as to satisfy $z' = w(z)$ (see the second of (1)). $\phi(x)$ can be found from the first of (1).

A complete phase plane analysis of (3) is carried out. The following results are obtained.

(a) If $\phi(x)$ denotes an arbitrary solution of the given DE, and its domain is (w', w), then w is finite and $\lim_{x\to w} \phi(x) = 0$. Moreover, if $1/\alpha$ is not integer, then $\phi(x)$ can be represented by an analytical expression of the form $\phi(x) = A(w - x)\left\{1 + \sum_{m+n>0} \phi_{mn}(w - x)^{m+n}\right\}$ in the neighborhood of $x = w$, where A and ϕ_{mn} are constants.

(b) Several asymptotic results as $x \to w'$ are also proved, with an appropriate form of asymptotic series in the neighborhood of $x = w'$. The point w' may be finite or minus infinity.

The asymptotic forms are graphically illustrated. Tsukamoto (1989)

40. $$y'' + e^{\lambda x} y^n = 0,$$

where λ and n are constants.

First define $w = -\lambda/(n-1), c = w^{2/(n-1)}$. The following cases arise.

(a) $\lambda > 0$ and $n = p/q$, where p and q are both odd; then all proper solutions are oscillatory. The proper solution is one that is real and continuous for $x \geq x_0$.

(b) $\lambda > 0$ and $n = p/q$, where p is even and q is odd; then the proper solutions form a one-dimensional manifold, all asymptotic to $-ce^{wx}$.

(c) If n is irrational or rational number with even denominator, there are no proper solutions.

(d) If $\lambda < 0$ and $n = p/q$, a rational number not of the form even/odd, the proper solutions form a two-dimensional manifold with the parameters $a_1 = \lim_{x\to\infty} y', a_0 = \lim_{x\to\infty}(y - a_1 x)$. The solution in this case can be put in the form
$$y = a_1 x + a_0 + \int_x^\infty (x - s)e^{\lambda s}(a_1 s)^n ds(1 + O(1)).$$

(e) If $\lambda < 0$ and $n = p/q$, where p is even and q is odd, then in addition to the solutions above, there is a one-dimensional manifold of proper solutions asymptotic to $-ce^{wx}$.

Bellman (1953), p.163.

41. $$y'' - e^{\alpha\lambda x} y^{1+\alpha} = 0,$$

where α and λ are positive constants.

Put $z = \psi^{-\alpha}\phi^{\alpha}, w = z'$, where $\psi(x) = \lambda^{2/\alpha}e^{-\lambda x}$ is a particular solution of the given DE and $\phi(x)$ is its general solution. Then we get

$$\frac{dw}{dz} = \{\alpha^2\lambda^2 z^2(z-1) + 2\alpha\lambda z w + (\alpha - 1)w^2\}/(\alpha w z). \tag{1}$$

Now see Eqn. (2.3.39) for an analysis of (1). Tsukamoto (1989); Tsukamoto, Mishina, and Ono (1982)

42. $$\boldsymbol{y'' + [a(x)/\{x^{2+1/\gamma}(\log x)^{\sigma}\}]y^{1+2/\gamma} = 0.}$$

Changing the variables to $s = \log x, v(s) = x^{-1/2}y(x)$, we obtain

$$v'' - (1/4)v + [a(s)/s^{\sigma}]v^{1+2/\gamma} = 0, \text{ where } a(s) \equiv a(t(s)).$$

The given DE is the same as Eqn. (2.3.51) if we write $g(x) = a(x)(\log\ x)^{-\sigma}$ therein. Kaper and Kwong (1988)

43. $$\boldsymbol{y'' + \lambda a(x)y^{\mu+1} = 0, 0 < x < 1,}$$

$\alpha y(0) - \beta y'(0) = 0, \gamma y(1) + \delta y'(1) = 0$, where $\mu > 0, a(x)$ is a positive continuous function on $(0,1)$ with $\int_0^1 a(x)dx < \infty$. The coefficients α, β, γ, and δ are assumed to satisfy $\gamma\beta + \gamma\alpha + \alpha\delta \neq 0$ along with one of the following:

(a) $\alpha \geq 0, \beta \geq 0, \gamma \geq 0, \delta \geq 0$.
(b) $\alpha > 0, \beta \geq 0, 0 < \gamma < -\alpha\delta/(\alpha + \beta), \delta < 0$.
(c) $0 < \alpha < -\gamma\beta/(\gamma + \delta), \beta < 0, \gamma > 0, \delta \geq 0$.

The following result is proved: With $y_0(x) = \mid (\alpha x + \beta) \mid$, let $\{\lambda_k, y_k\}_{k=1}^{\infty}$ be the sequence defined inductively; λ_k is the least positive eigenvalue and y_k is the corresponding positive eigenfunction of $y'' + \lambda a(x)y_{k-1}^{\mu}(x)y = 0, 0 < x < 1, \alpha y(0) - \beta y'(0) = 0, \gamma y(1) + \delta y'(1) = 0$ normalized by $\beta y_k(0) + \alpha y_k'(0) = (\text{sgn}\ \beta)(\alpha^2 + \beta^2)$. This is a positive solution (λ, y) of the given problem such that $\{\lambda_k\}$ converges to λ and $\{y_k\}$ converges to y uniformly on $[0,1]$. Moreover, the convergence is monotonic in that $0 < \lambda_1 < \lambda_2 < \cdots < \lambda$,

$$y_0(x) > y_1(x) > \cdots > y_k(x) \cdots > y(x) > 0, 0 < x < 1.$$

Luning and Perry (1974)

44. $$\boldsymbol{y'' + q(x)y^{\gamma} = 0, x \in (0, X),}$$

$$\boldsymbol{y(0) = y(X) = 0, y'(0) = \alpha, y(x) > 0, x \in (0, X),}$$

where $\gamma \geq 1, \alpha > 0$ and $q(x)$ a positive function.

A sufficient condition is given for the existence of a solution of this generalized Emden–Fowler equation, with a free endpoint. A monotone iteration scheme works for the approximation of a positive solution. Iffland (1987)

45. $$\boldsymbol{y'' - p(x)y^{1+2\epsilon} = 0, y(0) = 0, y'(0) = B, \epsilon > 0 \text{ is constant},}$$

where $p(x)$ is positive and continuous on $[0, \infty)$ and B is a positive constant.

It is proved that the given IVP has a positive, convex increasing solution $y(x)$ satisfying

$$y(x) \leq Bx\{1 - 2\epsilon B^{2\epsilon} \int_0^x t^{1+2\epsilon} p(t)dt\}^{-1/2\epsilon},$$

provided that B satisfies the inequality $2\epsilon B^{2\epsilon} \int_0^\infty x^{1+2\epsilon} p(x)dx \leq 1$. Wong (1970)

46. $\boldsymbol{y'' - g(x)y^k = 0, y(0) = A, y(b) = B,}$

where $g(x)$ is positive and continuous for $x > 0$ and is integrable over $(0, a)$ for any $a > 0$, and A, b, and B are constants.

It is shown that if $k \geq 1, b > 0$, and A and B are nonnegative, then there is exactly one solution of the BVP. The Thomas-Fermi case $k = 3/2, g(x) = x^{-1/2}$ is a special case. Bradley (1972)

47. $\boldsymbol{y'' - f(x)y^\lambda = 0, \lambda > 1,}$

where $f(x)$ is a positive continuous function.

The main result is that if

$$y(x) = O(1), f(x) \sim x^v L(x),$$

where $v > -2, L(tx)/L(x) \sim 1$ for every fixed $x > 0$ and $L(x) = O(x^\epsilon)$ for every $\epsilon > 0$, then $y(x) \sim p(x)$, where $x^{v+\lambda} L(x) p^{\lambda-1}(x) \sim (1+\lambda+v)(2+v)$. The proof is by the variation of constants method.

Other theorems furnish estimates under less restrictive assumptions of $f(x)$. Avakumovic (1947)

48. $\boldsymbol{y'' - p(x)y^{1+2\epsilon} = 0, y(0) = \alpha, y'(0) = \beta,}$

where $p(x)$ is positive and continuous on $[0, \infty)$.

The following result is proved: If there exist constants $A > 0$ and $B > 0$ such that

$$2\epsilon \int_0^\infty (A + Bx)^{1+2\epsilon} P(x)dx = B,$$

then for all α and β with $A \geq \alpha \geq 0$ and $B \geq \beta \geq 0$, the given problem has a proper continuous solution $y(x)$ that satisfies

$$y(x) \leq \ell(x) \left\{1 - 2\epsilon \int_0^x ds \ell^{-2}(s) \int_0^s \ell^{2+2\epsilon}(t)P(t)dt\right\}^{1/2\epsilon},$$

where $\ell(x) = \alpha + \beta x$. Weaker results are also proved. Wong (1970)

49. $\boldsymbol{y'' + q(x)y^\gamma = 0,}$

where $q(x) \geq 0$ and continuous on $(0, \infty)$, and $\gamma =$ quotient of odd positive integers.

Sufficient conditions for the existence of oscillatory solution in both the superlinear case $1 < \gamma$ and the sublinear case $0 < \gamma < 1$ are obtained. Heidel and Hinton (1972)

50. $$y'' - p(x)y^{\gamma} = 0, y(a) > 0, y'(a) = 0,$$

where $p(x)$ is positive and continuous on an appropriate interval and $\gamma \geq 1$. Upper and lower bounds of the solution as well as bounds for the first vertical asymptote $b > a$ of such solutions are obtained. Eliason (1972)

51. $$y'' + x^{-2-(1/\gamma)}g(x)y^{1+2/\gamma} = 0,$$

where $g(x)$ is a positive-valued function and γ is a positive constant.

It is proved that the given DE is nonoscillatory at infinity if there exists a $\sigma > 0$ such that the function $x \to g(x)(\log x)^{\sigma}$ is nonincreasing for all sufficiently large x. Kaper and Kwong (1988)

52. $$y'' - a(x)y^{\alpha} = 0,$$

where $\alpha > 0$ is a constant.

Asymptotic solutions of the given DE are found using the notions and functions of Hardy fields. Remainder terms are estimated. Klebanov (1971)

53. $$y'' - \phi(x)y^{\lambda} = 0,$$

where $\phi(x) \sim cx^{\sigma}, c > 0$.

Asymptotic solutions are found as follows.

(a) If $\lambda + \sigma < -2$ and $y(x)$ is a positive proper solution (a solution that is C^2 on $[0, \infty)$), then
   (i) either $y(x) = a + (1 + O(1))[ca^{\lambda}/\{(\sigma + 1)(\sigma + 2)\}]x^{\sigma+2}$ for some constant $a > 0$,
   or
   (ii) $y(x) = ax + b + [1 + o(1)][a^{\lambda}c/\{(\lambda + \sigma + 1)(\lambda + \sigma + 2)\}]x^{\lambda+\sigma+2}$ for some constants a and $b, a > 0$,
   or
   (iii) $y(x) \sim [(\sigma+2)(\sigma+\lambda+1)/\{c(\lambda-1)^2\}]^{1/(\lambda-1)}x^{-(\sigma+2)/(\lambda-1)}$. There exist solutions of all types.

(b) If $-2 \leq \lambda + \sigma < -1$ and $y(x)$ is a positive proper solution, then either
   (i) $y(x) = a + (1 + O(1))[ca^{\lambda}/\{(\sigma + 1)(\sigma + 2)\}]x^{\sigma+2}$ for some constants $a > 0$,
   or
   (ii) $y(x) \sim ax$ for some constant $a > 0$,
   or
   (iii) $y(x) \sim [(\sigma + 2)(\sigma + \lambda + 1)/\{c(\lambda - 1)^2\}]^{1/(\lambda-1)}x^{-(\sigma+2)/(\lambda-1)}$. All types of solutions exist.

(c) If $\lambda + \sigma \geq -1$ and $\sigma < -2$ and $y(x)$ is a positive proper solution, then
$$y(x) = a + (1 + O(1))[ca^{\lambda}/\{(\sigma + 1)(\sigma + 2)\}]x^{\sigma+2}.$$

(d) If $\sigma = -2$ and $y(x)$ is a positive proper solution, then
$$y(x) \sim [1/\{c(\lambda - 1)\log x\}]^{1/(\lambda-1)}.$$

(e) If $\sigma > -2$ and $y(x)$ is a postive proper solution, then
$$y(x) \sim [(\sigma + 2)(\sigma + \lambda + 1)/\{c(\lambda - 1)^2\}]^{1/(\lambda-1)}x^{-(\sigma+2)/(\lambda-1)}.$$

Taliaferro (1978)

54. $$y'' + x \sin y = 0, y'(0) = 0, y(0) = a, a \in R.$$

The given problem describes large deformations of a heavy cantilever by its weight. It has a unique solution. Moreover, for all $0 < a < \pi, y(x;a)$ is oscillatory over $[0,\infty)$ and $-\pi < y(x;a) < \pi$ for all $x \geq 0$. Singular behavior of the solution as $a \uparrow \pi$ is considered. The specific application, referred to above, is studied in some detail. Cheng and Hsu (1991), Hsu and Hwang (1988)

55. $$y'' + x^{-4}y^2 \cos y = 0.$$

It is shown that the given DE has solutions asymptotic to $a + bx$ as $x \to \infty$, where $b \neq 0$. Tong (1982)

56. $$y'' + (a + b \cos 2x)\sin y = 0, 0 \leq a, b \leq 8.$$

Bifurcation diagrams for two 2π-periodic and four π-periodic solutions are drawn. It is observed that for certain parameter values, the stable and unstable manifolds of a 2π-periodic solutions show transversal intersection. Batalova and Belyakova (1985)

57. $$y'' + \lambda e^{(1/2)Kx(x-1)+y} = 0, y(0) = y(1) = 0, K > 0,$$

is a parameter.

Symmetric solutions are considered. Let $v = y + (1/2)Kx(x-1) + \ln \lambda$; then v satisfies

$$v'' + e^v - K = 0, v(0) = \ln \lambda, v'(1/2) = 0, \tag{1}$$

where K is a constant.

It is now shown that the solution pair (λ, v) of (1) is a curve parameterized by $v(1/2)$. This is done by solving the IVP

$$v'' + e^v - K = 0, v(1/2) = t, v'(1/2) = 0, \tag{2}$$

which has a first integral $(1/2)v'^2 + e^v - Kv = e^t - Kt$. Since $K > 0$, it follows that v and v' are bounded; hence for every $t \in R$, (2) has a solution extending to the entire real line. Given $t \in R$, we have a solution $v(x;t)$ of (2). If we now make the identification $v(x) = v(x;t), \lambda = \exp[v(0;t)]$, we have a solution pair (λ, v) of (2) depending on the parameter t. Clèment and van Kan (1981)

58. $$y'' - (1/2)ye^{\alpha x - y} = 0, y'(\infty) = 0, y'(-\infty) = -\theta,$$

where $\theta > 0$ or the boundary conditions $y(\infty) = 0, y'(-\infty) = -\theta$. α is positive.

The given BVP arises in combustion theory. Existence and uniqueness of solution are established for all positive values of α and θ. Hastings and Poore (1983)

59. $$\boldsymbol{y'' - (1/2)ye^{\alpha x - y} = 0, y(x_0) = 1, y' \to 0 \text{ as } x \to +\infty,}$$

where x_0 is arbitrary and α is a positive constant.

It is proved that for each $\alpha > 0$ and arbitrary x_0, there exists a unique solution $y(x; x_0)$ of the given problem. Furthermore, $y \to 0$ as $x \to +\infty$. Hastings and Poore (1983)

60. $$\boldsymbol{y'' - (1/2)ye^{\alpha x - y} = 0, y(\infty) = 0, y'(-\infty) = -\theta,}$$

where α and θ are positive constants.

Let $y(x, x_0)$ denote the unique solution of the problem

$$y'' - (1/2)ye^{\alpha x - y} = 0, y(x_0) = 1,$$

$y' \to 0$ as $x \to +\infty$. Then for each $\alpha > 0$, there is an x_0 such that $\lim_{x\to\infty} y(x; x_0) = -\theta$, so that this $y(x; x_0)$ is a solution of the given problem.

It is further proved that for each $\alpha > 0$ and $\theta > 0$, the solution of the given problem is unique and $\lim_{x\to\infty} y = 0$. Hastings and Poore (1983)

61. $$\boldsymbol{y'' = k_3 e^{k_1 x^{\sigma}} e^{-k_2 y^{-\lambda}}, y(\infty) = y'(\infty) = 0,}$$

where $k_2 > 0, \lambda > 0, k_3 > 0$; σ and λ are constants.

For $\sigma > 0$ and $k_1 > 0$, it is shown that $y(x) \sim (k_2/k_1)^{1/\lambda} x^{-\sigma/\lambda}$ as $x \to \infty$. For $k_1 = 0$ or $\sigma \le 0$, it is shown that $y(x) \sim [k_2/(2\log x)]^{1/\lambda}$ as $x \to \infty$. Taliaferro (1981)

62. $$\boldsymbol{y'' - a(x)e^y = 0.}$$

If $\int_0^\infty a(x)\exp\{-cx\}dx < \infty$ for all $c > 0$, then for any $\gamma \in R$, the given DE has a unique weakly decreasing solution such that $y(0) = \gamma$. If, in addition,

$$\int_0^\infty xa(x)dx < \infty \left(\int_0^\infty xa(x)dx = \infty\right),$$

then all weakly decreasing solutions of the given DE are bounded (unbounded). If $\int_0^\infty a(x)e^{cx}dx < \infty$ for all $c > 0$ {for example, $a(x) = \exp(-kx^q)$, $k > 0, q > 1$}, then for any $\gamma \in R$, the given DE has all kinds of proper solutions satisfying $y(0) = \gamma$, as well as singular solutions.

A weakly decreasing solution is such that $\lim_{x\to\infty} y'(x) = 0$. For a proper solution, the interval of existence $[0, T_y)$ is such that $T_y = \infty$ and $\lim_{x\to\infty} y'(x)$ exists in $R \cup \{\infty\}$. Usami (1987)

63. $$\boldsymbol{y'' + e^x \sinh y = 0.}$$

The solution of the IVP $y(0) = 0, y'(0) = 1$ by Taylor series is

$$y = x - \frac{x^3}{3!} - \frac{2x^4}{4!} - \frac{3x^5}{5!} - \frac{2x^6}{6!} + \frac{17x^7}{7!} + \frac{128x^8}{8!} + \frac{549x^9}{9!} + \cdots.$$

Kamke (1959), p.545

64. $$y'' + [1/\{4x^2 \log (x+1)\}] y \log (y^2+1) = 0.$$

The given DE has a nonoscillatory exact solution $y = x^{1/2}$. Coffman and Wong (1972)

65. $$y'' + x^{-4} y^2 \text{ sgn } y = 0.$$

It is shown that the given DE has no solution that is asymptotic (as $x \to \infty$) to $a + bx$, $b \neq 0$. Chen (1987)

66. $$y'' - x^{-1/2} y^{-2} \text{ sgn } y = 0.$$

A first integral of the given DE is $xy'^2 - yy' + 2x^{1/2} \mid y \mid^{-1} = c$, where c is a constant. Trench and Bahar (1987)

67. $$y'' + x^{-\alpha} \sin x \mid y \mid^{\alpha} \text{ sgn } y = 0, 0 < \alpha < 1, x \in [x_0, \infty).$$

It is shown that the given DE is oscillatory. Onose (1983)

68. $$y'' + x^{\alpha} \sin x \mid y \mid^{1/2} (1 + \mid y \mid) \text{ sgn } y = 0.$$

It is proved that the given DE is oscillatory if $\alpha > 1/2$. Wong (1989)

69. $$y'' + [1/\{x^n (\log x)^m\}] \log(1 + \mid y \mid) = 0, x > 1.$$

It is shown that:

(a) The given DE has solutions satisfying $\lim_{x\to\infty} y'(x) = b \neq 0$ for $n > 2$, or $n = 2$ and $m > 2$.

(b) For $n = 3, m > 2$, the given DE has a solution asymptotic to $a + bx$ as $x \to \infty$ for any $a \in R$ and $b \neq 0$.

(c) For $n > 3$ or $n = 3$ and $m > 1$, and $n > 4$ or $n = 4$ and $m > 1$, the given DE has a solution satisfying $\lim_{x\to\infty} y'(x) = b \neq 0$.

Chen (1987)

70. $$y'' + f(x) \mid y \mid^{\alpha} \text{ sgn } y = 0, x > 0, 0 < \alpha < 1$$

Estimates and asymptotic formulas for solutions and their derivatives are obtained under assumptions of the type

$$\int^{\infty} x \mid f \mid dx < \infty; \int^{\infty} x^{1+\alpha} \mid f \mid dx < \infty; f \geq 0$$

and either $\int^{\infty} x f dx < \infty$ or $x^{\beta} \int^{\infty} f dx \geq$ const. > 0 for some $\beta, 0 < \beta \leq 1$. It is shown that (a) if $f(x) x^{(3+\alpha)/2}$ is nondecreasing with positive limit at $x = \infty$, then the given DE has both oscillatory and nonoscillatory solutions, but (b) if for some β on $\{0, (1-\alpha)/2\}$, $f(x) x^{(3+\alpha)/2+\beta}$ is positive, nondecreasing and bounded for $x \geq$ const. > 0, then all solutions ($\not\equiv 0$) of the given DE are nonoscillatory. A solution $y \not\equiv 0$ of the given DE is called oscillatory or nonoscillatory according as it does or does not have an unbounded set of zeros. Belhorec (1967)

71. $$y'' + a(x) \mid y \mid^{\alpha} \operatorname{sgn} y = 0, 0 < \alpha < 1,$$

where $a(x)$ may assume both positive and negative values.

It is proved that if $\lim_{x\to\infty} \sup x^{-1} \int_{x_0}^{x} \int_{x_0}^{s} \tau\alpha(\tau) d\tau ds = \infty$ for some $\beta, 0 \leq \beta \leq \alpha$, then every "proper" solution of the given DE is oscillatory. Kusa (1982)

72. $$y'' + a(x) \mid y \mid^{\gamma} \operatorname{sgn} y = 0, \gamma \neq 1, 0 < \gamma < 1,$$

where $a(x) \in C[0, \infty)$. An oscillation criterion is provided for the given equation where $a(x)$ is continuous but is not assumed to be nonnegative for all large values of x. Kwong and Wong (1983)

73. $$y'' + a(x) \mid y \mid^{n} \operatorname{sgn} y = 0,$$

$n > 1$ and $a(x) > 0$ is continuous on $0 < x < \infty$.

It is proved that there exists an oscillating solution in case $a(x)x^{(n+3)/2}$ is nondecreasing, and this result is shown to be the best possible in a certain sense. Kiguradze (1962)

74. $$y''' + p(x) \mid y \mid^{\alpha} \operatorname{sgn} y = 0, \alpha > 0, \alpha \neq 1;$$

$p : [0, +\infty) \to (0, +\infty)$ is a continuous function.

Using the oscillation-preserving transformation $t = \log x, w(t) = x^m y(x)$ (where m is an appropriate constant) together with an argument involving energy functions, a criterion is given for all nontrivial solutions of the given DE to be nonoscillatory when $0 < \alpha < 1$. Moreover, it is shown that for $\alpha > 1$, solutions of the given DE that satisfy a certain growth rate are nonoscillatory. Erbe and Rao (1987)

75. $$y'' - p(x)\{1 + q(x)\} \mid y \mid^{\sigma} \operatorname{sgn} y = 0, \sigma > 1, x \in (x_0, +\infty) \equiv \Delta,$$

where $p(x) \in C_{\Delta}^{k}(k \geq 0), p(x) \neq 0$ in $\Delta, q(x) \in C_{\Delta}, q(x) = O(1)$.

A complete asymptotic analysis of the given DE, which depends crucially on the nature of $p(x)$, is carried and six different cases distinguished. Kostin (1971)

76. $$y'' + x^{-k} f(y) = 0, x < \infty, y(x) \longrightarrow \gamma \text{ as } x \to \infty,$$

where $k > 2$ and $f(s) = s(1+ \mid s \mid^{p-1}), p = 2k - 3$.

It is stated that for $k > 2$, the given problem for every $\gamma \in R$ has a unique solution $y(x, \gamma)$. Many analytic properties of this solution are studied for different choices of γ. In particular, oscillations, extrema, and zeros are analyzed. Atkinson, Brezis, and Peletier (1990)

77. $$y'' + x^2 f(x) g(y) = 0,$$

where f has a logarithmic behavior at infinity and g is such that $0 < G'(y) \leq (1/2)yg(y)$, $y \neq 0$, where $G(y) = \int_0^y g(s)ds$. Imposing some conditions on f, it is proved that all solutions of the given DE satisfy the inequality $\int_0^\infty G\{y(t)/f(t)\}dt < +\infty$. Ben M'Barek (1988)

78. $$y'' + \lambda s(x) f(y) = 0, x \in (0,1), y(0) = y(1) = 0,$$

where (a) $s(x) : [0,1] \to R$, positive and continuous, and (b) $f(y) : R \to R$, continuously differentiable and $f(0) > 0$.

Continuation of a positive solution of the given BVP is considered. Clèment and van Kan (1981)

79. $$y'' - \phi(x) f(y) = 0, y(\infty) = y'(\infty) = 0,$$

where $\phi(x)$ and $f(y)$ are assumed to be positive and continuous.

Explicit formulas for the asymptotic behavior of solutions are obtained. Necessary conditions for the existence of the solution of the problem are given. Taliaferro (1981)

80. $$y'' - p(x) f(y) = 0, y(\alpha) = a, y'(\alpha) = b,$$

where f is continuous on $(-\infty, \infty), p(x) > 0$ for $0 \leq x < \infty$, p is continuously differentiable on $[0, \infty)$, and $p'(x)[p(x)]^{-3/2} \to 0$ as $x \to \infty$.

For fixed $a \geq 0$ and $b \geq 0$ $(a + b > 0)$, let $y(x) = y(x; \alpha)$ be the unique solution of the given DE satisfying $yf(y) > 0$ for $y \geq a$, if $a > 0$. It is then proved that if $\int_a^\infty (\int_a^x f(u)du)^{-1/2} dx < \infty$, then $y(x; \alpha)$ has a finite vertical asymptote. An asymptotic formula (valid as $\alpha \to \infty$) is given in terms of a, b, p, and f for the distance to the vertical asymptote. Implicit lower bounds for the distance to vertical asymptotes are also derived. Other choices of the parameters are also considered. Bobisud (1972)

81. $$y'' - f(x)\phi(y) = 0,$$

where f and ϕ are positive and continuous on the positive real axis.

Asymptotic formulas for $x \to \infty$ extending the results of Fowler are obtained. In fact, when Fowler's results do not apply, these theorems do. Here the functions f and ϕ are assumed to belong to an o-regularity class, possessing the form $f(x) = x^\sigma L(x)$ and $\phi(y) = y^\lambda L_1(y)$, where real numbers σ and λ are called indices of regularity and $L(x), L_1(y)$ belong to the class of slowly varying functions at infinity and at zero, respectively. More specifically, a positive continuous function L defined on (a, ∞) is said to be slowly varying at infinity if for all $t > 0, \lim_{x \to \infty}[L(tx)/L(x)] = 1$. For example, all positive functions tending to positive constants are slowly varying at infinity. Under these circumstances, precise asymptotic formulas for the behavior of the solutions are obtained. Manic and Tomic (1980)

82. $$y'' + \phi(x) f(y) = 0, y(0^+) = 0, y'(0^+) = \infty,$$

where $\phi(x)$ and $f(y)$ are assumed to be positive and continuous.

Explicit formulas for the asymptotic $(t \to 0^+)$ behavior of solutions are obtained. Existence of the solution of the given problem is also discussed. Taliaferro (1981)

83. $$y'' + \phi(x) f(y) = 0, y(\infty) = \infty, y'(\infty) = 0,$$

where $\phi(x)$ and $f(y)$ are assumed to be positive and continuous.

Explicit formulas for the behavior of the solutions as $x \to \infty$ are obtained. Existence of the solution of the given problem is also discussed. Taliaferro (1981)

84. $$y'' + a(x)f(y) = 0.$$

Let $a(x)$ be continuous, nondecreasing, and positive on $0 \le x < \infty, \lim_{x\to\infty} a(x) = \infty$, and $\log a(x)$ tends to ∞ regularly. Suppose further that $f(y)$ is a continuous, nondecreasing, odd function such that $y^{-1}f(y) = O(1)(y \to 0), y^{-1}f(y)$ is nondecreasing for $y > 0$, and $| f(y_1) - f(y_2) | \le w(| u_1 - u_2 |)$, where $w(z)$ is positive nondecreasing and $\int_0^{y_0} (w(z))^{-1} dz = \infty$. It is shown that, then, every solution of the given DE tends to 0 as $x \to \infty$. Bihari (1962)

85. $$y'' + a(x)f(y) = 0.$$

It is shown that if $yf(y) > 0, f(y')$ is integrable, and $\lim_{y\to\infty} \int_0^y f(u)du = \infty, a(x)$ is positive and nondecreasing, then all solutions of the given DE are bounded. Waltman (1963)

86. $$y'' + a(x)f(y) = 0,$$

where $a(x) \in C[0, \infty)$ and $f(y) \in C(-\infty, \infty)$ nondecreasing in y, and satisfies $yf(y) > 0$ if $y \ne 0$.

It is shown that under certain conditions, the given DE has all its continuable (solutions which exist and can be continued on some ray $[x_0, \infty)$) solutions oscillatory. Wong (1989)

87. $$y'' + \{1 + b(x)\}f(y) = 0.$$

If $b(x) \to 0$ as $x \to 0$ and $\int^\infty | b'(s) | ds < \infty$ and $f(y)$ is such that $yf(y) > 0, f(y)$ is integrable, and $\lim_{y\to\infty} \int_0^y f(u)du = \infty$, then all solutions of the given DE are bounded. Waltman (1963)

2.4 $y'' + f(x, y) = 0$, $f(x, y)$ polynomial in y

1. $$y'' + y^2 - x^2 = 0, \quad y \to \mp x \text{ as } x \to \pm\infty.$$

The following result is proved: There are two solutions $y_+(x), y_-(x)$ to the given BVP. $y_+(x) > -|x|, \forall x$, $y_-(x) < -|x|, \forall x$, and $y_+(x), y_-(x)$ satisfy the following:

(a) $y'_\pm(x) = y_\pm(-x)$ and $y'_\pm(0) = 0$ (symmetry).

(b) $\text{sign}(y'_\pm(x)) = -\text{sign}(x)$; $y_+(x)$ and $y_-(x)$ each has a single maximum at $x = 0$.

(c) $y_\pm(x) \sim \mp x + k_\pm a(|x|)$ as $x \to \pm\infty$, where $a(x) \sim (1/(2\pi^{1/2}))2^{1/12}x^{1/4} \times \exp[-(2\sqrt{2}/3)x^{3/2}]$ is the decaying solution to the Airy equation $a'' - 2xa = 0$ and $k_\pm$ are constants.

It is evident that the given DE is unchanged if $x \to -x$. Using the symmetry, one need only consider the positive half-line $x \ge 0$ since a symmetric solution of the given problem corresponds to a solution to the BVP with $y'(0) = 0, y(x) \sim -x$ as $x \to +\infty$ and a nonsymmetric solution corresponds to a pair of solutions with $y'(0) = \pm\alpha \ne 0, y(x) \sim -x$ as $x \to +\infty$. Thus, IVP

$$y'' = x^2 - y^2, \quad y(0) = \beta, \; y'(0) = \alpha,$$

α and β being constants, is considered and the results are proved with the help of several lemmas. In particular, it is shown that there is no C^1 solution to the given problem with $y(0) = 0$. Holmes (1982)

2. $$y'' - (1/2)(y^2 - x) = 0, \quad y(0) = 0, y(x) \sim \sqrt{x} \text{ as } x \to \infty.$$

The given problem is shown to have exactly two solutions such that $y(0) = 0$ and $y(x) \sim \sqrt{x}$ as $x \to \infty$. For one of these solutions, say, y_+, $y'(0) > 0$, while for the other, say, y_-, $y'(0) < 0$. A shooting argument is used. Hastings and Troy (1989)

3. $$y'' - (1/2)[y^2 - A(1 - x^2)] = 0, \quad y(-1) = y(1) = 0,$$

where $A \geq 0$ is a parameter.

The given DE appears in the problem of vertical flow of an internally heated Boussinesq fluid with viscous dissipation and pressure work.

The following result is proved: Let $N(A)$ be the number of solutions of the given IVP. Then $N(A) \to \infty$ as $A \to \infty$. A shooting argument is employed. Hastings and Troy (1989)

4. $$y'' - y^2 - x = 0.$$

Bender and Orszag (1978), p.200

5. $$y'' - y^2 + x = 0, \quad \text{either } y(0) = 0, \ y(x) \sim \sqrt{x} \text{ as } x \to +\infty \quad \text{or}$$

$$y(0) = 0, \ y(x) \sim -\sqrt{x} \text{ as } x \to \infty.$$

These problems arise in connection with studies of natural convective flows with viscous dissipation. The given DE is another version of the first Painlevé equation:

$$y'' = 6y^2 + x \quad \text{or} \quad y'' = 6y^2 - 6x.$$

It is convenient in this connection to consider the initial value problem

$$y'' = y^2 - x, \quad y(0) = y_0, \ y'(0) = y_0'.$$

Two cases arise, as was indicated by numerical integration with $y_0 = 0$ and $y_0' = \alpha$: one with $\alpha > 0$ and the other with $\alpha < 0$.

To get the asymptotic behavior near the branches $y = \pm\sqrt{x}$, we write $y = \pm\sqrt{x} + g(x)$ or $y = \pm\sqrt{x} f(x)$. Substituting in the given DE [see also Bender and Orszag (1978)], we get local linearized results as

$$\begin{aligned} y_+(x) &\sim \sqrt{x} + x^{-1/8}[c_1 \exp(\phi(x)) + c_2 \exp(-\phi(x))], & (1) \\ y_-(x) &\sim -\sqrt{x} + x^{-1/8}[c_1 \exp(\phi(x)) + c_2 \sin(\phi(x))], & (2) \end{aligned}$$

where $\phi(x) \sim (4\sqrt{2}/5)x^{5/4}$. Here, put $c_1 = 0$ in (1) to obtain the required asymptotic behavior at $x = +\infty$. Thus, we have two linearly independent solutions decaying like $x^{-1/8}$ in the neighborhood of the lower branch (2), while only one of the solutions near the upper branch is asymptotic to it as $x \to \infty$.

The continuous dependence of solutions upon initial conditions then implies that (whenever they are defined) there is a two-parameter family of solutions asymptotic to $y = -\sqrt{x}$ and a one-parameter family asymptotic to $y = \sqrt{x}$.

To get global results, Boutroux-like transformation $y = -\sqrt{x} + f/\epsilon$, $x = 1/\epsilon^2 + \sqrt{\epsilon}t$ is introduced. Letting $\mu = \epsilon^{5/2}$, we have

$$\ddot{f} + 2f - f^2 + \mu t f = O(\mu^2), \quad \cdot\cdot \equiv \frac{d^2}{dt^2}. \tag{3}$$

For $\mu \ll 1$, we have (3) as the small perturbation of the Hamiltonian system

$$\ddot{f}_0 + 2f_0 - f_0^2 = 0, \tag{4}$$

with the first integral

$$H(f_0, \dot{f}_0) = (1/2)\dot{f}_0^2 + f_0^2 - (1/3)f_0^3 = \text{constant}. \tag{5}$$

For $0 < H < 4/3$, bounded solutions of (5) are given by Jacobi elliptic functions. Perturbation solutions are obtained using properties of elliptic functions.

Shooting arguments are used to establish the uniqueness of a monotonically increasing solution and to demonstrate the existence of at least one solution with a single minimum. It is conjectured that this solution is also unique. This conjecture is supported partially by numerical arguments. Holmes and Spence (1984)

6. $$y'' - y^2 + x^2 = 0.$$

The given DE has $y = x$ and $y = -x$ as exact solutions. The approximate solutions are $y_\pm = \pm x + ka(\pm x)$, k is a (small) constant, and $a(x)$ is the bounded solution of Airy's equation $a'' - 2ax = 0$. The solution $a(x)$ has the property that

$$\begin{aligned} a(x) &\sim 2^{-11/12} x^{-1/4} \pi^{-1/2} \exp[(-2\sqrt{2}/3)x^{3/2}], \quad \text{as } x \to +\infty, \\ &\sim 2^{1/12} |x|^{-1/4} \pi^{-1/2} \cos[(2\sqrt{2}/3)|x|^{3/2} + 7\pi/4], \quad \text{as } x \to -\infty. \end{aligned}$$

From this asymptotic form it may be anticipated that in the limit $k \to 0$, the two points $(A, \phi) = (0, 3\pi/4)$ and $(A, \phi) = (0, 7\pi/4)$ correspond to solutions that are oscillatory at one end and decay exponentially to either $y = x$ or $y = -x$ at the other end. Numerical computations confirm that for very small but positive amplitudes, the range $3\pi/4 < \phi < 7\pi/4$ gives bounded solutions. A detailed study of the numerical solutions is presented. Byatt–Smith (1988)

7. $$y'' - y^2 + x^2 + c = 0,$$

where c is a constant.

The solutions of interest are those which remain finite for all finite x. The equation arises in the theory of resonant oscillations in a tank. Numerical solutions indicate that there are two bounded solutions having the property that they are even functions of x and decay exponentially to a particular solution $y_p(x, c)$ in the sense that

$$y_\pm \sim y_p(|x|, c) + k_\pm a(|x|) \quad \text{as } x \to \pm\infty, \tag{1}$$

where

$$a(x) \sim 2^{-11/12} x^{-1/4} \pi^{-1/2} \exp[(-2\sqrt{2}/3)x^{3/2}]$$

is the solution to the Airy equation $a'' - 2xa = 0$, which decays as $x \to +\infty$, and k_+ and k_- are constants which depend on c with $k_+ > 0 > k_-$, at least for $c \geq 0$. The particular solution $y_p(x,c)$ of the given DE is defined as the limit of sequences $\{y_n(x)\}$, where $y_n(x)$ is defined iteratively by $y_{n+1}(x) = [x^2 + c - y_n''(x)]^{1/2}, y_0(x) = (x^2+c)^{1/2}$. This converges for $x > x_1(c)$ and, for example, yields $y_p(x,0) \equiv x, x \geq 0$, when $c = 0$.

There are other bounded solutions of the given DE which have either the asymptotic form (1) for x tending to $+\infty$ or $-\infty$ (but not both) with a multiple k of $a(|x|)$ satisfying $k_- < k < k_+$ or oscillate about $y = -y_p$ as $x \to \pm\infty$ with asymptotic form

$$y \sim -y_p(|x|, c) + A_\pm |x|^{-1/4} \cos[(2\sqrt{2}/3)|x|^{3/2} + \phi_\pm] \quad \text{as } x \to \pm\infty,$$

where $A_+(A_-)$ and $\phi_+(\phi_-)$ are constants characterizing the behavior as $x \to +\infty$ $(-\infty)$. In this case, since the decay to y_p is also algebraic and not exponential, we can simply write

$$y \sim -|x| + A_\pm |x|^{-1/4} \cos[(2\sqrt{2}/3)|x|^{3/2} + \phi_\pm].$$

For the last class of solutions, in order to construct a solution on $-\infty < x < \infty$, we are required to solve a connection problem of finding (A_+, ϕ_+) in terms of (A_-, ϕ_-). Alternatively, we find the set of values (A_-, ϕ_-) yielding a solution of the given DE which can be connected to the asymptotic form appropriate to $x \to \infty$. Numerical and asymptotic results are provided. For the case $c = 0$, see Eqn. (2.4.6). It is further proved that if $y(x)$ is the solution of the given DE and has a maximum at $x = x_0 > 0$, then $y(x) < y(x_0)$ for all $x > x_0$. Byatt–Smith (1988)

8. $$\mathbf{y'' - y^2 - e^x = 0.}$$

(a) Changing the variables according to $y = e^{x/2}u(s), s = e^{x/4}$, one obtains the equation

$$\frac{d^2u}{ds^2} + \frac{5}{s}\frac{du}{ds} + \frac{4}{s^2}u - 16u^2 - 16 = 0,$$

whose solution for large x, i.e., s, behaves like elliptic functions in s. It may now be deduced as for Painlevé first and second transcendents that the singularities of $y(x)$ are separated by a distance proportional to $e^{-x/4}$ as $x \to \infty$. Bender and Orszag (1978), p.199

(b) The given DE is referred to by Bender and Orszag as "beyond Painlevé" since here e^x replaces x in the DE for the first Painlevé transcendent. The series in the neighborhood of a moving singularity is

$$y(t) = (1/t^2)[6 - (A/10)t^4 - (A/6)t^5 + ft^6 + (A/14)t^6 \ln t + \cdots)], \quad t = x - x_0,$$

where f and x_0 are arbitrary constants. The full asymptotic expansion is probably very complicated, involving all powers of $t^6 \ln t$.

Bender and Orszag (1978), p.164

9. $$\mathbf{y'' - (3/2)(y^2 - x) = 0.}$$

The given DE is P_I (first Painlevé equation). Writing $y(x) = \sqrt{x}u(z), z = (4/5)x^{5/4}$, we get

$$\frac{d^2u}{dz^2} = \frac{3}{2}\left(u^2 - 1\right) - \frac{1}{z}\frac{du}{dz} + \frac{4u}{25z^2}. \tag{1}$$

Equation (1) has the asymptotic form

$$u = \pm 1 - (4/75)(1/z^2) + O(1/z^4)$$

as $|z| \to \infty$ ($|x| \to \infty$). In the complex z plane, however, there are other behaviors. For example, on $\mathrm{Re}(z) = 0$,

$$u = 1 + [b_+ \exp(\sqrt{3}z) + b_- \exp(-\sqrt{3}z)]/\sqrt{z} + O(1/z^2, \exp(\pm 2\sqrt{3}z)/z),$$

and on $\mathrm{Im}(z) = 0$,

$$u = -1 + [c_+ \exp(\sqrt{3}iz) + c_- \exp(-\sqrt{3}iz)]/\sqrt{z} + O(1/z^2, \exp(\pm 2i\sqrt{3}z)/z),$$

with free parameters $b_\pm$ and $c_\pm$. The range of validity of each of these behaviors is a small strip in an angle around the associated axis. The asymptotic behaviors above are in fact degenarate forms of the elliptic function behavior valid in the quadrants bounded by each half-axis in the z-plane.

Connection problems with behaviors at $+\infty$ and $-\infty$ are studied along a large circle in the plane of the independent variable for all angles of approach to infinity. Joshi and Kruskal (1988)

10. $\quad$ $\boldsymbol{y'' - 6y^2 + x = 0.}$

This first Painlevé equation has families of solutions asymptotic to the curve

$$\sqrt{6}y = \sqrt{x}. \tag{1}$$

Let $y = y_1, y_2$ be two arbitrary regular solutions of the given DE, tending asymptotically to (1). Consider the difference function $w = y_1(x) - y_2(x)$, which is clearly a solution of the linear equation

$$w'' = 6(y_1(x) + y_2(x))w \equiv Q(x)w, \tag{2}$$

tending to zero when $x \to \infty$. It is also known that $y_i(x) = \sqrt{x/6} + O(x^{-2})$, $i = 1, 2$; hence $y_i''(x) = O(x^{-2})\,[(1/3)\sqrt{6x} + O(x^{-2})]$ and $Q''(x) = O(x^{-1/2})$. It is clear that the function $y_i'(x) > 0$ for x sufficiently large and decreases monotonically. Following the asymptotic results of Wasow (1965), it is proved that (2) has two linearly independent solutions,

$$\begin{aligned} w_1(x) &= Q^{-0.25}(x) \exp\left(-\int_a^x \sqrt{Q(x)}dx(1 + o(1))\right), \\ w_2(x) &= Q^{-0.25}(x) \exp\left(+\int_a^x \sqrt{Q(x)}dx(1 + o(1))\right). \end{aligned}$$

Therefore, $w_1(x) \to 0$ as $x \to \infty$. Hence

$$y_1(x) - y_2(x) = K_{12}x^{-1/8} \exp\left(-0.8\ (24)^{1/4}x^{1.25}\right)(1 + o(1)),$$

where K_{12} is a constant dependent on the solutions chosen. Yablonskii (1972)

11. $\quad$ $\boldsymbol{y'' - 6y^2 - x = 0.}$

This is a first Painlevé equation; closed-form solutions are not possible. See the following: Sachdev (1991), Sec. 8.5 for a general analytic treatment; Bender and Orszag (1978), p.158 for a numerical study with IC $y(0) = y'(0) = 0$ and graphical representation (DE treated here is $y'' = y^2 + x$). Kamke (1959), p.542; Murphy (1960), p.380; Davis (1962); Hille (1969); Simon (1965); Fair and Luke (1966); Kapaev (1988); Filcakova (1974); Bartashevich (1973)

12. $$\boldsymbol{y'' - 6y^2 - \lambda x = 0, \quad y(0) = 1,\ y'(0) = 0,}$$

where λ is a constant.

Numerical solution of the given IVP for $\lambda = 0, 1, 5$ are obtained. This first Painlevé equation displays movable poles. A numerical method is given which improves upon the usual method of analytic continuation. As the pole is approached, the continuous analytic continuation is halted and the solution resumed as a truncated Laurent series.

The numerical results and the error of computation are tabulated and depicted graphically. Simon (1965)

13. $$\boldsymbol{y'' - 6y^2 - a(x) = 0,}$$

where $a(x)$ is an analytic function.

Using the α-method of Painlevé, it is shown that the possible forms of the given DE, which have no critical points, may be written, by trivial changes of variables, as

$$y'' = 6y^2, \tag{1}$$

$$y'' = 6y^2 + 1/2, \tag{2}$$

$$y'' = 6y^2 + x. \tag{3}$$

Equations (1) and (2) have solutions in terms of elliptic functions as $y = \wp(x - k; 0, h)$ and $\wp(x - k; 1, h)$, where k and h are constants of integration. Equation (3) is P_I.

It is also shown that the necessary and sufficient condition for the given DE to have no movable branch points is that it have the special form $y'' = 6y^2 + a_1x + a_0$, where a_1 and a_0 are arbitrary constants. Following Ince (1956), one may show that this special form then has no movable essential singularities. Kruskal and Clarkson (1992)

14. $$\boldsymbol{y'' - 6(x + y)^2 + y = 0.}$$

Writing $y = 1/24 - \eta + w(\eta)$, $\eta = x - 1/24$, the given DE becomes

$$\frac{d^2w}{d\eta^2} = 6w^2 + \eta,$$

which is P_I. Drazin and Johnson (1989), p.188

15. $$\boldsymbol{y'' + ay^2 + (bx + c) = 0,}$$

where a, b, and c are constants.

For $b = 0$, we may integrate after multiplication by y':

$$y'^2 + (2/3)ay^3 + 2cy = C, \tag{1}$$

where C is a constant. This is integrable in terms of elliptic functions. For $b \neq 0$, put $y = \alpha\eta(\xi)$, $\xi = \beta(x + c/b)$, so that

$$\frac{d^2\eta}{d\xi^2} + (a\alpha/\beta^2)\eta^2 + \left(b/(\alpha\beta^3)\right)\xi = 0. \tag{2}$$

Choosing $a\alpha/\beta^2 = -6, b/(\alpha\beta^3) = -1$, Eqn. (2) goes to

$$\frac{d^2\eta}{d\xi^2} = 6\eta^2 + \xi,$$

the first Painlevé equation [see Eqn. (2.4.11)]. Kamke (1959), p.543; Murphy (1960), p.380

16. $\quad \epsilon y'' - y^2 + 1 + x^2 = 0, \quad y(x) = -x + O(x^{-1})$ **as** $x \to +\infty$,

where $x \in R$ and $\epsilon > 0$ is given.

The question is addressed whether the solution of the given problem is an even function of x, or equivalently, whether $y(x) = x + O(x^{-1})$ as $x \to -\infty$, or equivalently, whether $y'(0) = 0$.

The following result is proved: Let $\epsilon > 0$. There is a unique solution of the given problem on $(-\infty, \infty)$. Furthermore, (a) $y(x), y'(x), y''(x) < 0, x \geq 0$; and (b) there exists constants $B(\epsilon) \neq 0$ and $\phi(\epsilon)$ such that

$$y(x) = x + B|x|^{-1/4}\cos\left(b|x|^{3/2} + \phi\right) + O(x^{-1}) \quad \text{as } x \to -\infty,$$

where $b = b(\epsilon) = \sqrt{8/(9\epsilon)}$. Similar formulas hold for the derivatives of y.

It is shown that there exist no even solutions of the given problem for any $\epsilon > 0$. Analysis is carried partly on the real line and partly in the complex plane. Amick and Toland (1990)

17. $\quad \rho^2 y'' - y^2 + x^2 + 1 = 0, \quad -\infty < x < \infty,$

where ρ is a constant.

A connection problem is solved in the following sense. The solutions of the given DE that oscillate about $-|x|$ as $x \to \infty$ have asymptotic expansions whose leading terms are

$$y \sim -|x| + \tilde{A}_\pm |x|^{-1/4}\cos\left[2\sqrt{2}\,|x|^{3/2}/(3\rho) + \tilde{\phi}_\pm\right],$$

where $\tilde{A}_\pm$ and $\tilde{\phi}_\pm$ are constants. The connection problem is to determine the asymptotic expansion at $+\infty$ of a solution that has a given asymptotic expansion at $-\infty$. In other words, we wish to find $(\tilde{A}_+, \tilde{\phi}_+)$ as functions of $\tilde{A}_-$ and $\tilde{\phi}_-$. The given problem is related to Painlevé's first transcendent. It arises in the study of resonant oscillations of water waves.

The asymptotic behavior of the required problem is determined by solving an integral equation. Byatt-Smith (1989)

18. $\quad y'' + y + (3/K^2)\left[-\lambda y - (3/2)y^2 + (2/\pi)\cos x + (3/2)\int_\pi^\pi y^2 dx\right] = 0,$

where K is small and λ is a parameter.

The given DE represents forced water waves on shallow water near resonance. Numerical solutions are found to check the validity of some matched asymptotic solutions. Ockendon, Ockendon, and Johnson (1986)

19. $$y'' + w^2 y + \epsilon[\cos x + \cos 5x]y + \epsilon[1 + \cos 3x + \cos 7x]y^2 = 0,$$

where ϵ is a small parameter.

Lie transformation is used to find resonant frequencies of this weakly nonlinear perturbed simple harmonic oscillator. Len and Rand (1988)

20. $$y'' + a^2 y + y^2 f(x) = 0,$$

where $a \neq 0$ is a real constant.

For several types of functions $f(x)$, stability of solutions is considered. The solution is sought in the form

$$y = \epsilon[\sin(ax + \phi) + n(x, \epsilon)]$$

and it is called stable if $n(x, \epsilon)$ is bounded for all $x \geq 0$. Stability is deduced if ϵ is sufficiently small and f is constant, and if $f(x)$ is positive (negative), bounded, and increasing (decreasing) for $x \geq 0$. If $f(x)$ is monotonic and exponentially unbounded, solutions are shown to be unstable. If $f(x)$ is periodic, stability is related to a Mathieu equation. Rosenberg (1954)

21. $$y'' - p_2(x)y^2 - p_1(x)y = 0.$$

It is proved that the given DE does not contain moving critical singular points if and only if either $p_2(x)$ and $p_1(x)$ are constants or if by transformation $y = \mu(x)W + \nu(x), \tau = \phi(x)$, where

$$\begin{aligned} \mu(x) &= Cp_2^{1/5}(x),\ \phi(x) = \int_{x_0}^{x} p_2^{2/5}(x)dx, \\ \nu(x) &= \left[1/50p_2^3(x)\right]\left[6p_2^{\prime 2}(x) - 5p_2''(x)p_2(x) - 25p_1(x)p_2^2(x)\right], \end{aligned}$$

the given DE is reduced to

$$\frac{d^2W}{d\tau^2} = 6W^2 + a\tau + b,$$

where a, b, C are constants. In this case, the solution of the given DE can be expressed in terms of elementary or elliptic functions or first Painlevé transcendent. Yablonskii (1967)

22. $$y'' + (1/2)y^3 - \epsilon F(x) = 0.$$

This is a nonconservative Hamiltonian system; for sufficiently strong coupling ϵ, to an external periodic field, e.g., $F(x) = \cos \Omega x$, the solution can exhibit chaos. For singularity structure of this equation, see its more general form in Eqn. (2.7.26). Fournier, Levine, and Tabor (1988)

23. $$y'' + y^3 - \phi(x) = 0, \quad y(0) = y(1) = 0.$$

Existence of solutions of the given BVP for a certain class of $\phi(x)$ is proved. Das and Venkatesulu (1984)

24. $$y'' + y^3 - \sin x = 0.$$

A 2π-periodic solution with the condition $y(\pi/2+x) = y(\pi/2-x)$ is found by Galerkin's method. Let the nth approximation be

$$y_n(x) = b_1 \sin x + b_3 \sin(3x) + b_5 \sin(5x) + \cdots + b_{2n-1} \sin[(2n-1)x].$$

Letting $L_y = y'' + y^3 - \sin x$, we have the Galerkin equations as

$$\int_0^{2\pi} L(y_n(x)) \sin[(2k-1)x] dx = 0, \quad k = 1, 2, \ldots, n.$$

For instance, if $n = 1, y_1(x) = b_1 \sin x$ and the Galerkin's equation is

$$\begin{aligned} \int_0^{2\pi} L(y_1(x)) \sin x dx &= \int_0^{2\pi} [-b_1 \sin x + b_1^3 \sin^3 x - \sin x] \sin x dx \\ &= 2[(-\pi/2)b_1 + (3\pi/8)b_1^3 - \pi/2] = 0. \end{aligned}$$

This gives $b_1 = 1.49$, so the first approximation is $y_1(x) = (1.49...) \sin x$.

The question whether the given DE has a periodic solution, and if so, whether the Galerkin approximations are close to the periodic solutions must be investigated. Lakin and Sanchez (1970), p.145

25. $$y'' + (1 - p_1)y + y^3 - \sigma f(x) = 0,$$

where $f(x)$ is even in x and $\int_0^{2\pi} f(x) \cos x \, dx = \pi$.

It is proved that there exists a neighborhood $U \subset R^2$ of $(p_1, \sigma) = 0$ and a neighborhood of $V \subset P$ of $y = 0$ such that the only 2π-periodic solutions in V of the given DE are even functions of x if $\sigma \neq 0$. Here $P = \{h : R \to R : h \text{ is continuous and } h(x + 2\pi) = h(x)\}$. Hale and Rodrigues (1977)

26. $$y'' + (A + Bx)y + y^3 = 0,$$

where A and B are constants.

The given DE is a special case of the second Painlevé transcendent. Its solutions have no branch points, and are therefore uniform functions of x. Euler, Steeb, and Cyrus (1989)

27. $$y'' + 2y^3 - f(x) = 0.$$

Let C be the Banach space of all continuous, even, periodic functions of period 2π, with norm $\| g \| = \max_x |g(x)|$. Then the following results are proved for the given DE:

(a) If $f(x)$ is even, periodic with period 2π, twice differentiable with $\int_0^{2\pi} f(x) dx = 0$, then it is possible to find a sequence of disjoint spheres in C, in each of which the given DE has one and only one solution, even and of period 2π.

(b) Under the condition in (a), every solution of the given DE of sufficiently large energy is one of those, the existence of which is guaranteed in (a).

The methods used are functional analytic. Micheletti (1967)

28. $$y'' + 2y^3 + 2xy = 0.$$

The given DE has a solution expressible in terms of a Painlevé transcendent. It has an asymptotic form

$$y \sim \sqrt{-x} + [C_1/(-x)]^{1/4} \exp[4i(-x)^{3/2}] + [C_2/(-x)^{1/4}] \exp[-4i(-x)^{3/2}] \quad \text{as } x \to -\infty,$$

where C_2 and $C_1 = C_2^*$ are complex constants. Zakharov and Kuznetsov (1987)

29. $$y'' + 2y^3 - xy = 0.$$

Let $y(x)$ be a bounded real solution of the given DE such that as $x \to +\infty, y(x) \sim r\,\mathrm{Ai}(x)$, where r^2 is any nonnegative number. Then as $x \to -\infty$, the solution has the formal series

$$y(x) \sim d(-x)^{-1/4} \sin\theta + O(|x|^{-7/4}), \quad \theta(x) \sim (2/3)(-x)^{3/2} - (3/2)d^2 \ln(-x) + \bar{\theta} + O(|x|^{-3/2}),$$

where the constants d and $\bar{\theta}$ are determined from the nonnegative constant r by

$$d^2 = -(\sigma/\pi)\ln(1 - \sigma r^2),\ \bar{\theta} = \pi/4 - \sigma \arg[\Gamma(1 - id^2/2)] - (3/2)\sigma d^2 \ln 2.$$

The results are checked numerically. Segur and Ablowitz (1981)

30. $$y'' - (3/2)y^3 + 3(c_3/c_4 + 1/2)y^2 - (3c_2/c_4)xy$$
$$+ (k_2 - c_3/c_4 + 1)y + (3c_2/c_4)x - 3(k_1 + k_2) = 0,$$

where k_1, k_2, c_2, c_3, and c_4 are constants.

It may be checked that if $k_1 = 1 \pm c_2/(\sqrt{3}c_4), k_2 = -1$, and $c_3 = c_4$, any solution of the special Riccati equations

$$y' = \pm(\sqrt{3}/2)y^2 \mp \sqrt{3}y \pm \sqrt{3}(c_2/c_4)x \tag{1}$$

is a solution of the given DE. Equation (1) may easily be linearized. Kawamoto (1984a)

31. $$y'' - (3/2)y^3 + (9c_3/(2c_4))y^2 - (3c_2/c_4)xy$$
$$- 3(c_3^2/c_4^2 + c_5/c_4 - 1)y + (3c_2c_3/c_4^2)x - k = 0,$$

where $c_i (i = 2, \ldots, 5)$ and k are constants.

It may be checked that any solution of the special Riccati equations

$$y' \mp (\sqrt{3}/2)y^2 \pm [\sqrt{3}c_3/c_4]y \mp [\sqrt{3}c_2/c_4]x = 0 \tag{1}$$

is a solution of the given DE provided that $k = \pm\sqrt{3}c_2/c_4$ and $c_4 = c_5$. Equation (1) may easily be linearized. Kawamoto (1984a)

32. $$y'' - 2y^3 + (2 + 6/x + 6/x^2) = 0, \quad x > 1.$$

The given DE has an exact solution $y = 1 + 1/x$. Erbe, Sree Hari Rao, and Seshagiri Rao (1984)

33. $$y'' + xy - 2y^3 = 0.$$

For the given P_{II} equation, the solution is sought in the form $y(x) = dx^{-1/4}\sin[\phi_0(x)] + u(x)$, where $\phi_0(x)$ is a solution of

$$\phi_0(x) = (2/3)x^{3/2} - (3/4)d^2 \ln x + c_0 + (d^2/4)\int_{-\infty}^{x} [\cos[2\phi_0(t)]/t]dt. \tag{1}$$

With this the given DE transforms to

$$u'' + Q(x)u = g_0(x) + g_1(x)u^2 + g_2(x)u^3, \tag{2}$$

where

$$Q(x) = x - 6d^2x^{-1/2}\sin^2[\phi_0(x)],\ g_2 = 2,\ g_1 = O(x^{-1/4}),\ g_0 = O(x^{-9/4}).$$

For $y'' + Q(x)y = 0$, a fundamental system of solutions is chosen to be

$$y_{1,2} = x^{-1/4}\exp[\pm\zeta(x)][1 + O(x^{-3/2})],$$

where

$$\zeta(x) = \int_{x_0}^{x}\left[\sqrt{Q(t)} + (1/2)[(5/16)Q'^2/Q^{5/2} - (1/4)Q''/Q^{3/2}]\right]dt.$$

This leads to an asymptotic estimate for the solutions of (2) and hence the following estimates for the solutions of P_{II}: $y(x) = dx^{-1/4}\sin[\phi_0(x)] + O(x^{-7/4})$, where

$$\phi_0(x) = (2/3)x^{3/2} - (3d^2/4)\ln x + c_0 + O(x^{-3/2}),\ |d| < \sqrt{3} \quad \text{as } x \to +\infty.$$

Abdullaev (1983)

34. $$y'' - xy - 2y^3 = 0.$$

The given DE is a special case of the second Painlevé equation. It is shown that if $y(x) \sim k\,\mathrm{Ai}(x)$ as $x \to \infty$, where $-1 < k < 1$ and $\mathrm{Ai}(x)$ denotes Airy's function, then

$$y(x) \sim d|x|^{-1/4}\sin[(2/3)|x|^{3/2} - (3/4)d^2\ln|x| - c] \quad \text{as } x \to -\infty,$$

where the constants d, c depend on k. It is proved that $d^2 = -\pi^{-1}\ln(1-k^2)$. Clarkson and McLeod (1988), Lebeau and Lochak (1987), Yablonskii (1972)

35. $$y'' - 2y^3 + xy - \alpha = 0, \quad \alpha > 0 \text{ is real.}$$

The case $\alpha < 0$ may be treated by replacing $y(x)$ by $-y(x)$. This is Painlevé's second equation.

The branches $y = \phi_{1,2}(x)$ of this curve defined by the equation $2y^3 - xy + \alpha = 0$ on D_1: $(1.5)^3\sqrt{4}\alpha^2 \le x < \infty, (0.5)^3\sqrt{2}\alpha \le y < \infty$; and D_2: $0 \le x < \infty, -\infty < y \le 0$, respectively, for $\alpha \ge 0$ form asymptotes of regular solution of the given DE. Consider two pairs of regular solutions $y = y_{11}(x), y_{12}(x)$ and $y = y_{21}(x), y_{22}(x)$ of the given DE tending asymptotically to $y = \phi_1(x)$ and $y = \phi_2(x)$, respectively. We have

$$y_{1i} = \sqrt{2x}/2 - \alpha/(2x) + o(x^{-1}), \quad y_{2i} = -\sqrt{2x}/2 - \alpha/(2x) + o(x^{-1}),\ i = 1, 2.$$

Clearly,

$$(y_{k1} - y_{k2})'' = [2(y_{k1}^2 + y_{k1}y_{k2} + y_{k2}^2) - x](y_{k1} - y_{k2}) = Q_k(x)(y_{k1} - y_{k2}),$$

where $Q_k(x) = 2x + O(x^{-0.5}), k = 1,2$. It is easy to check that $Q_k'(x) = O(1), Q_k'' = O(1)$; hence using the asymptotic theorem of Wasow (1965), it is proved that

$$y_{k1}(x) - y_{k2}(x) = K_{12}^k x^{-0.25} \exp[(-2\sqrt{2}/3)x^{1.5}](1 + o(1)), \quad k = 1,2,$$

where K_{12}^k are constants depending on the solutions. Yablonskii (1972)

36. $$y'' - 2(y^3 - xy) - a_0 = 0,$$

where a_0 is a constant.

The given DE is $\mathrm{P_{II}}$. The Boutroux transformation $y(x) = \sqrt{x}\,u(z), z = (2/3)x^{3/2}$ changes the given DE to

$$u'' = 2(u^3 - u) + (a - u')/z + u/(9z^2), a = 2a_0/3. \tag{1}$$

The asymptotic behaviors for real $|z| \to \infty$ ($|x| \to \infty$) are

$$\begin{aligned} u &= \pm 1 - a/(4z) + O(1/z^2), \\ u &= a/(2z) + a(14/9 + a^2/4)/(2z^3) + O(1/z^5). \end{aligned}$$

However, when z (and x) are taken complex, we have different behaviors. For example, on $\mathrm{Im}(z) = 0$,

$$u = \pm 1 - (a/4 \pm b_+ b_-)/z + [b_+ e^{\xi} + b_- e^{-\xi}]/\sqrt{z} + O(1/z^2, e^{\pm\xi}/z),$$

and on $\mathrm{Re}(z) = 0$,

$$u = a/(2z) + 3ac_+c_-/z^2 + [c_+ e^{\psi} + c_- e^{-\psi}]/\sqrt{z} + O(1/z^3, e^{\pm\psi}/z),$$

where $\xi = 2z - [3a/4 \pm 16b_+b_-]\log z + O(1/z), \psi = \sqrt{2}iz - (3/\sqrt{2})ic_+c_- \log z + O(1/z)$, and $b_\pm, c_\pm$ are free parameters. The range of validity of each of these behaviors is a small strip in angle around the associated axis. In fact, these behaviors are degenerate forms of the general elliptic function behavior valid in the quadrants bounded by each half-axis in the z-plane.

Connection problems with behaviors at $+\infty$ and $-\infty$ are studied along a large circle in the plane of the independent variable for all angles of approach to infinity. Joshi and Kruskal (1988)

37. $$y'' - 2y^3 - xy - 1 = 0, \quad y(0) = 1, \; y'(0) = 0.$$

Writing $y = 1 + 1.5x^2u$, a rational approximation solution of the transformed system is obtained and computed. The first pole of the solution is predicted accurately. Fair and Luke (1966)

38. $$y'' - 2y^3 - xy + a = 0.$$

This is the second Painlevé equation. See the following for its analysis: Sachdev (1991), Sec. 8.6 for a general analytic treatment; Davis (1962); Rosales (1978) and Miles (1978) for the relation with the Korteweg–deVries equation, inverse scattering, and a numerical study. Kamke (1959), p.543; Murphy (1960), p.381; Fokas and Zhou (1992); Vorobev (1965); Yablonskii (1972)

39. $$y'' - 2y^3 - (Kx + H)y - K_1 = 0,$$

where K, H, and K_1 are constants.

The only movable singularities of the given DE may be shown to be poles. If $K = 0$, the given DE is solvable in terms of elliptic functions. For $K \neq 0$, a simple change of independent variable transforms it to the second Painlevé equation $y'' - 2y^3 - xy - K_2 = 0$. Bureau, 1964

40. $$y'' + ay^3 + bxy + cy + d = 0,$$

where a, b, c, and d are constants.

For $a = 0$, it is linear; for $b = 0$ it is a special case of Eqn. (10). For $a \neq 0, b \neq 0$, the given DE transforms through

$$y = \lambda\eta(\xi), \quad xi = \mu(bx + c)$$

to the second Painlevé equation,

$$\eta'' - 2\eta^3 - \xi\eta - \alpha = 0.$$

See Sachdev (1991). Kamke (1959), p.544; Murphy (1960), p.381

41. $$y'' + \Omega[(1/4)y^3 + (1 - x)y - (e^2/4)y - e] = 0, \quad y(0) = e, \ y'(0) = 0,$$

where Ω and e are constants.

The given DE arises in structural dynamics. Writing $w = 1/\Omega$, we get

$$wy'' + (1/4)y^3 + (1 - x)y - (e^2/4)y - e = 0. \tag{1}$$

A power series form of solution in w,

$$y = f_0(x) + wf_1(x) + w^2 f_2(x) + \cdots,$$

on substitution, etc., and solution of resulting algebraic relations for $f_1, f_2, \ldots$ gives

$$\begin{aligned} f_1 &= 8f_0^4[(f_0^3 - 4e)/(f_0^3 + 2e)^4], \\ f_2 &= [32f_0^7/(f_0^3 + 2e)^9](640e^3 - 2352e^2 f_0^3 + 1164ef_0^6 - 73f_0^9), \end{aligned}$$

where f_0 is a solution of the cubic equation

$$(1/4)f_0^3 + (1 - x - e^2/4)f_0 - e = 0.$$

The IC $f_0(0) = e$ is satisfied, while $y_0'(0) = 0$ is satisfied approximately for small e. Convergence of solution was obtained for $w < e^2$.

A second solution is attempted with e small:

$$y = y_1(x)e + y_2(x)e^2 + \cdots + y_n(x)e^n + \cdots.$$

The resulting DEs and IC are

$$\begin{aligned} wy_1''(x) + (1-x)y_1 &= 1, \\ wy_2''(x) + (1-x)y_2 &= 0, \\ wy_3''(x) + (1-x)y_3 &= (1/4)y_1\left(1 - y_1^2\right), \\ wy_4''(x) + (1-x)y_4 &= (1/4)y_2\left(1 - 3y_1^2\right), \end{aligned}$$

and $y_1(0) = 1, y_1'(0) = 0, y_n(0) = y_n'(0) = 0, n \geq 2$. $y_2(x)$ and $y_4(x)$ are found to be identically zero. Other linear sets of DE are solved analytically and computed. It is found that the convergence of the series is relatively good for $0 \leq x \leq 1+$ if $w > e^2$; this expansion is not good for $x > 1+$.

Another solution is sought for Eqn. (1) in the product form $y(x) = u(x)v(x)$. This is good near $x = 1$ but fails for $x > 1$. Dym and Rasmussen (1968)

42. $$\boldsymbol{y'' - 2a^2y^3 + 2abxy - b = 0,}$$

where a and b are constants. It is easily verified that every solution of Riccati equation

$$y' + ay^2 - bx = 0$$

satisfies the given DE. Substitute

$$y = (1/a)[u'(x)/u(x)]$$

to obtain

$$u''(x) - abxu = 0.$$

This is essentially Airy's equation. Kamke (1959), p.544; Murphy (1960), p.381

43. $$\boldsymbol{y'' + (\lambda/4)\left(1 - y^2\right)[y + \epsilon r(x)] = 0, \quad -1 < x < 1, \quad y'(-1) = y'(1) = 0,}$$

where the function $r(x)$ satisfies $r(-x) = -r(x)$ and $r'(x) \geq 0$ for $x \in [-1, 1]$ and $r'(0) > 0$.

Suppose that ϕ^* is a solution of the problem

$$y'' + (\lambda/4)\left(1 - y^2\right)[y + \epsilon r(x)] = 0, \quad 0 < x < 1, \quad y(0) = 0, \ y'(1) = 0. \tag{1}$$

Then, due to the symmetry property of r, the function

$$\tilde{\phi}(x) = \begin{cases} -\phi^*(-x), & -1 \leq x < 0 \\ \phi^*(x), & 0 \leq x \leq 1 \end{cases}$$

is a solution of the given BVP.

The following results are proved:

(a) For each $\lambda > 0$, the problem (1) has a unique positive solution $\phi^*(\lambda)$.
(b) $\phi^*(\lambda)$ is increasing and concave.
(c) The map $\lambda \to \phi^*(\lambda) : R^+ \to C^2[0, 1]$ is analytic.
(d) $0 \leq \lambda_1 < \lambda_2$ implies that $\phi^*(\lambda_1)(x) < \phi^*(\lambda_2)(x)$ for $x \in (0, 1]$.
(e) $\phi^*(\lambda) \to 0$ as $\lambda \to 0$ in $C^2([0, 1])$.
(f) $\lim_{\lambda \to \infty} \phi^*(\lambda)(x) = 1$ for each $x \in (0, 1]$.

Clèment and Peletier (1985)

44.
$$y'' + \left[Q_0^2 - k\delta(x)\right] y - \beta y^3 = 0,$$

where Q_0, k, and β are constants.

Minorsky's stroboscopic method is used to study the given DE, which arises in the design of the strong focusing cosmotron. Blaquiere (1956)

45.
$$y'' + w^2 y + \epsilon w^2 y^3 - f_0 e^{-\alpha x} = 0, \quad y(0) = 0, \; y'(0) = 0,$$

where ϵ, w, α, and f_0 are constants.

A perturbation solution to order $O(\epsilon^2)$ is found by suitably removing the secular term. An explicit zeroth order solution is

$$y_0(x) = \left[f_0/(\alpha^2 + w^2)\right] \left[e^{-\alpha x} + (\alpha/w)\sin wx - \cos wx\right].$$

The first order term is rather lengthy. Bauer (1966)

46.
$$y'' + w^2 y + \epsilon w^2 y^3 - f_0(1 - e^{-\alpha x}) = 0, \quad y(0) = 0, \; y'(0) = 0,$$

where w, ϵ, f_0, and α are constants.

A perturbation solution to order $O(\epsilon^2)$ is found by suitably removing the secular terms. An explicit zeroth order solution is

$$y_0(x) = \frac{f_0}{w^2}\left[1 - \frac{e^{-\alpha x}}{1 + (\alpha^2/w^2)}\right] - \frac{f_0(\alpha^2/w^2)}{w^2(1 + \alpha^2/w^2)}\cos wx - \frac{f_0(\alpha/w)}{w^2(1 + \alpha^2/w^2)}\sin wx.$$

The first order term is rather lengthy. Bauer (1966)

47.
$$y'' + w^2 y + \epsilon y^3 + \epsilon y \cos x = 0,$$

where ϵ is a small parameter.

Lie transformation is used to find resonant frequencies of this weakly nonlinear perturbed simple harmonic oscillator. Len and Rand (1988)

48.
$$y'' + y^3/x^2 - y = 0, \quad y(0) = 0, \; y(\infty) = 0.$$

A Galerkin method is used to find a solution of the given problem. Chauvette and Stenger (1975)

49.
$$y'' - y + x^{-2} y^3 = 0, \quad \lim_{x\to 0} x^{-1} y(x) = a,$$

where $a > 0$ is a constant.

It is proved that there is at most one positive value of a for which $y(x, a) > 0, 0 < x < \infty$ and $\lim_{x\to\infty} y(x, a) = 0$. Coffman (1972)

50.
$$y'' - y + y^k/x^{k-1} = 0, \quad k = 2, 3.$$

It is first easily checked that for $k = 2, y(x) \sim Ae^{-x}$ for large x and $y(x) \sim y'(0)\sinh x$

for small x. An iterative solution shows that $A = 16.0687$ and $y'(0) = 4.19169$ for a BVP with $y(0) = 0$ and $y(\infty) = 0$. For $k = 3$, corresponding values are $A = 2.71386, y'(0) = 4.33738$. Ryder (1967)

51. $$\mathbf{y'' - Ay - By^3/x^2 = 0,}$$

where A and B are constants.

An approximate solution is found as

$$y = A_m(x, k) + (k^2/4)x,$$

where A_m is the amplitude of the elliptic function with k as elliptic modulus assumed to be small. For large x, we write $y = w^{-1}$ and obtain

$$2w'^2 - ww'' - Aw^2 = Bx^{-2}, \tag{1}$$

so that as $x \to \infty$, the right-hand side of (1) is small and the solution may be approximated by

$$y \approx \lambda_1 \sin \sqrt{A}\, x + \lambda_2 \cos \sqrt{A}\, x,$$

where λ_1 and λ_2 are constants, provided that $A < 0$; for $A > 0$, corresponding hyperbolic form may be written. Dixon, Kelley, and Tuszynski (1992)

52. $$\mathbf{y'' - 2y^3/(1 + x)^6 = 0, \quad y(0) = 0, \; y'(0) = \sqrt{5}.}$$

The transformation $y(x) = (1 + x)u(t), t = (1 + x)^{-1}$ changes the problem to

$$u'' - 2u^3 = 0, \quad 0 < t < 1, \; u(1) = 0, \; u'(1) = -\sqrt{5}.$$

Integrating twice, we have $1 - x = \int_0^u (s^4 + 5)^{-1/2}\, ds = f(u)$. The solution may be expressed in terms of an elliptic function. We may obtain a bound for this solution as follows:

$$f(u) > \int_0^u \frac{ds}{s^2 + \sqrt{5}} = C \tan^{-1}(Cu),$$

where $C = 5^{-1/4}$. It follows from the montonicity of $\tan^{-1}$ that

$$Cu(x) < \tan[(1 - x)/C] \text{ or } y(x) < K(1 + x) \tan[Kx/(1 + x)],$$

where $K = 5^{1/4}$. Wong (1970)

53. $$\mathbf{y'' + Aw^2(x)y + B\lambda(x)y^3 = 0,}$$

where $w^2(x) = (1 + \Omega x)^{-\mu}, \lambda(x) = (1 + \Omega x)^{-\nu}, \mu \in R^+, \nu \in R^+$, and Ω is a parameter.

The given DE is a time-dependent anharmonic oscillator. Changing the variables according to $y = \zeta(\theta)C(x), \theta = \theta(x)$, where $d\theta/dx = 1/C^2$, we get

$$\frac{d^2\zeta}{d\theta^2} + \left[Aw^2C^4 + C^3\frac{d^2C}{dx^2}\right]\zeta + B\lambda C^6\zeta^3 = 0.$$

Now choosing $C(x) = (1 + \Omega x)^\gamma$ and different values of v, u, and γ, different asymptotic forms of the solutions to the given DE are obtained. The forms for specific choices of $C(x)$ are tabulated. Moraux, Fijalkow, and Fiex (1981)

54.
$$y'' + w^2 y + \epsilon \cos x \, y^3 = 0,$$

where w and ϵ are small parameters.

Lie transformation is used to find resonant frequencies of this weakly nonlinear perturbed simple harmonic oscillator. Len and Rand (1988)

55.
$$y'' + (1/2)\left(x^{-3/2}\sin\sqrt{x} - \cos\sqrt{x}\right) y^3 - 1/x^2 = 0.$$

Certain oscillatory properties of solutions of the given DE are discussed. Yan (1989)

56.
$$y'' + (1 + \epsilon w_1) y + \epsilon \left[s y^2 \cos x + (1-s)\cos(2x)\, y^3 \right] = 0, \quad 0 \le s \le 1.$$

When $s = 0$, the quadratic term is absent, and cubic term is resonant. When $s = 1$, the quadratic term is resonant, and the cubic term is absent. When $0 < s < 1$, both terms are resonant. The interaction of both resonances is considered. Phase plane analysis is carried out. Len and Rand (1988)

57.
$$y'' - w^2 y - \alpha^2 e^{-6wx} y^3 = 0,$$

where α and w are constants.

The given DE is shown to have a first integral

$$I = e^{2wx}(y' - wy)^2 - (1/2)\alpha^2 y^4 e^{-4wx}.$$

Ranganathan (1992)

58.
$$y'' + \alpha(x) y^3 - \beta(x) = 0, \quad y(-1) = y(1) = 0,$$

where $\alpha(x)$ and $\beta(x)$ are continuous functions on the interval $[-1, 1]$.

It is proved that if $|\alpha(x)| \le 1$ and $|\beta(x)| \le 0.6$, for $-1 \le x \le 1$, then there exists a function $y(x)$ which is twice continuously differentiable on the interval $[-1, 1]$, and which is a solution of the given BVP with $|y(x)| \le 0.7$. Locker (1970)

59.
$$y'' + y + (\alpha y^2 + \beta y^3) f(x) = 0,$$

where f is a periodic function of x, with period X, and y and $y' \in C$. f is piecewise constant. α and β are parameters. Resonance phenomena in the phase plane are studied. Bernussou, Liu, and Mira (1976)

60.
$$y'' - p_3(x) y^3 - p_2(x) y^2 - p_1(x) y = 0.$$

The given DE does not contain moving critical singular points if and only if either p_3, p_2, p_1 are constants or if by transformation $y = \mu(x)W + \nu(x), \tau = \phi(x)$, where

$$\nu(x) = -\frac{p_2(x)}{3p_3(x)}, \; \phi(x) = C\int_{x_0}^{x} p_3^{-1/3}(x)dx, \; \mu(x) = C\sqrt{p_3^{-1/3}(x)},$$

the given DE is reduced to the form

$$\frac{d^2W}{d\tau^2} - 2W^3 - (a\tau + b)W - \alpha = 0,$$

where a, b, C, and α are constants. In this case, the solution of the given DE is expressed in terms of elementary or elliptic functions or the second Painlevé transcendent. Yablonskii (1967)

61.
$$y'' + \alpha(x)y - \beta(x)y^2 + \gamma(x)y^3 = 0,$$

where α, β, and γ are measurable X-periodic functions such that if we denote by a, A, c, and C the infinimum and supremum of α and γ, respectively, and $B = ||\beta||_{L^\infty}$, then $0 < a \leq \alpha(x) \leq A < \infty, 0 < c \leq \gamma(x) \leq C < \infty$ and $B < \infty$.

It is further assumed that $m^2 < a \leq A < (m+1)^2$ for some integer $m \geq 0$ and $B^2 \leq (9/2)\delta c$, where $\delta = a - m^2$. Then the given DE has a nontrivial 2π-periodic solution. Grossinho and Sanchez (1986)

62.
$$y'' + k_1 + k_2(\cos x)y^4 = 0,$$

where $k_2 \ll 1$ and k_1 are constants.

The solutions are found when k_1 is either in the neighborhood of 0 or 1 or in the neighborhood of $1/m^2, m = 2, 4, \ldots$ or when k_1 is in the neighborhood of $1/36$ or when k_1 is in the neighborhood of $1/n^2, n = 3, 5, 7$. Hamd-Allah (1981)

63. $y'' + w^2 y + \epsilon(\cos x)\, y + \epsilon \cos(5x)\, y^2 + \epsilon \cos(12x)\, y^3 + \epsilon[1 + \cos(22x)]y^4 = 0,$

where ϵ is a small parameter.

Lie transformation is used to find resonant frequencies of this weakly nonlinear perturbed simple harmonic oscillator. Len and Rand (1988)

64.
$$y'' + (3/16)y^5 - (1/4)\epsilon x y^3 + (x^2/16 + \epsilon r/2)y + (k - r^2)y^{-3} = 0.$$

Putting $y^2 = \lambda w, x = \mu z, \lambda^4 = -1$ and $\mu = -2\epsilon\lambda$, we obtain the fourth Painlevé equation with $\epsilon^2 = 1, \alpha = 2r\epsilon\lambda^2$ and $\beta = 8(r^2 - k)$. Gromak (1987)

65.
$$y'' - 3y^5 - 2xy^3 - (x^2/4 - \nu - 1/2)y = 0, \quad y \to 0 \text{ as } x \to \infty,$$

where ν is a parameter.

If we write $y(x) = 2^{-3/4}\sqrt{w(z)}, z = x/\sqrt{2}$, the given DE goes to $\mathrm{P_{IV}}$:

$$ww'' = (1/2)w'^2 + (3/2)w^4 + 4zw^3 + 2(z^2 - \alpha)w^2 + \beta,$$

with $\alpha = 2\nu + 1$ and $\beta = 0$.

Exact bound-state solutions of the given DE with $\nu = n$, a positive integer, are obtained by using integral representation, which decay exponentially as $x \to \pm\infty$.

It is also shown that:

(a) Any solution of the given problem, for any real ν (either integer or noninteger), is asymptotic to $kD_\nu(x)$ for some k and conversely, for any k, there is a unique solution of the given DE asymptotic to $kD_\nu(x)$. If we denote this solution by $y_k(x; \nu)$, then as $x \to \infty$,

$$y_k(x; \nu) \sim kD_\nu(x) \sim kx^\nu \exp(-x^2/4).$$

Here $D_\nu(z)$ is a parabolic cylinder function, which is a solution of

$$D''_\nu(z) = (z^2/4 - \nu - 1/2)D_\nu(z),$$

satisfying $D_\nu(z) \sim z^\nu \exp(-z^2/4)$ as $x \to +\infty$ and

$$D_\nu(z) \sim [\sqrt{2\pi}/\Gamma(-\nu)]\, e^{i\pi\nu}\, z^{-\nu-1}\, e^{z^2/4} \quad \text{as } z \to -\infty,$$

provided that ν is not an integer; if $\nu = n$, a positive integer, then $D_n(z)$ is expressed in terms of Hermite functions.

(b) (a conjecture) There exists k_ν^* such that whenever $k < k_\nu^*, y_k(x;\nu)$ exists for all x and

$$(k_\nu^*)^2 = \frac{1}{2\sqrt{2\pi}\Gamma(1+\nu)}. \tag{1}$$

The result (1) for $\nu = n$, a positive integer, is proved rigorously.

Bassom et al. (1992)

66. $$y'' - 3y^5 - 2\gamma x y^3 - (x^2/4 - \nu - 1/2)y = 0,$$

where γ and ν are parameters.

It is shown that the given DE possesses a Painlevé property if and only if $\gamma = \pm 1$, since in the neighborhood of a point $x = x_0$,

$$\begin{aligned} y(x) &= (x-x_0)^{-1/2}(1/\sqrt{2} - \left(x_0/(4\sqrt{2})\right)(x-x_0) \\ &\quad + \left\{[32(\nu-\gamma) + (9\gamma^2 - 8)x_0^2 + 16]/(96\sqrt{2})\right\}(x-x_0)^2 \\ &\quad + \left\{a_3 + [(1-\gamma^2)x_0/(8\sqrt{2})]\ln(x-x_0)\right\}(x-x_0)^3 + O[(x-x_0)^4]), \end{aligned}$$

where a_3 is an arbitrary constant. The square root singularity is of no significance since it can be removed by the transformations $w(x) = y^2(x)$. At higher orders of $(x - x_0)$, higher and higher powers of $\ln(x - x_0)$ are required, so the general solution of the given DE has a movable logarithmic branch point unless $\gamma = \pm 1$ when it is equivalent to P_{IV} [see Eqn. (2.4.65)]. Bassom et al. (1992)

67. $$y'' + x^\gamma(y^5 + y^4) = 0, \quad y(a) = y(b) = 0,$$

where $-\infty < a < b < \infty$ and γ is a constant.

The given BVP has a unique solution if $-2 \le \gamma \le 2$. Kwong (1991)

68. $$y'' + k_1 y + k_2 (\cos x)\, y^5 = 0,$$

where k_1 and $k_2 \ll 1$ are constants.

Periodic solutions that are harmonic and subharmonic of even order are found. The given DE admits (a) two harmonic solutions when k_1 is in the neighborhood of zero and four when k_1 is in the neighborhood of 1; and (b) $2m$ subharmonic solutions, where m is even, when k_1 is in the neighborhood of $1/m^2$. These solutions are determined with IC $y_0 = 0.5, y'_0 = 0$ in the (k_1, k_2) plane by applying the index method. Elnaggar and Thana (1982)

69. $$y'' + x^{2/(s+1)}y - ay^{2s+1} = 0,$$

where $s = 1, 2, \ldots,$ and a is a constant.

The given DE generalizes P_{II}, which corresponds to the case $s = 1$. It is proved that if

$$|d| < \left[\frac{s+2}{|a|\sum_{j=1}^{s} j|c_j|}\right]^{1/2s},$$

then the solution of the given DE has the asymptotic form

$$y(x) = dx^{-1/(2(s+1))}\sin(\phi(x) + O(x^{-1-3/(2(s+1))})),$$

where

$$\phi(x) = \frac{1}{1/(s+1)+1}x^{1+1/(s+1)} + \frac{ad^{2s}/2(s+1)}{A_s}\ln x + c_0 + O[x^{-1-1/(s+1)}],$$

where

$$A_s = \sum_{\nu=1}^{s}(-1)^{\nu+1}2^{-2\nu}C_{s+1}^{\nu+1}C_{2\nu}^{\nu} - C_{s+1}^{1},$$
$$C_j = \sum_{\nu=j}^{s}(-1)^{\nu+1}2^{1-2\nu}C_{s+1}^{\nu+1}C_{2\nu}^{\nu-1},\ j = 1, 2, \ldots, s, \quad \text{as } x \to \infty.$$

Abdullaev (1983)

70. $y'' - y + y^n/x^{n-1} = 0, \quad n > 0,\ x \geq 0,\ y(0) = 0,\ y'(0) = \alpha < \infty,\ y(\infty) = 0,$

where α is an unknown positive parameter.

It is proved that:

(a) The given problem for $0 < n \leq 1$ has no positive solution.

(b) The given problem for $1 < n \leq 3$ has at least one positive solution.

Zhidkov and Shirikov (1964)

71. $$y'' + w^2 y + \epsilon \cos x\, y^n = 0,$$

where n even, w, and ϵ (small) are parameters.

Lie transformation is used to find resonant frequencies of this weakly nonlinear perturbed simple harmonic oscillator. Len and Rand (1988)

72. $$y'' + y^{2n+1} + \sum_{j=0}^{2n} y^j p_j(x) = 0,$$

where $n \geq 1$ and $p_j(x+1) = p_j(x), p_j \in C^\infty$.

It is proved that every solution of the given DE is bounded; that is, it exists for all $x \in R$ and $\sup_R(|y(x)| + |y'(x)|) < \infty$. Dieckerhoff and Zehnder (1987)

2.5 $y'' + f(x, y) = 0$, $f(x, y)$ not polynomial in y

1.
$$y'' + \alpha[1 + \cos 2x]y - \alpha y^{-1} = 0, \alpha > 0$$
$$y(0) = y(\pi) > 0,\ y'(0) = y'(\pi) = 0.$$

It is shown that when $0 < \alpha \leq 1$, a periodic solution exists, and for certain α with prescribed boundary conditions, there are two such solutions. Approximations to such solutions are computed. Ye and Wang (1978)

2.
$$y'' - (1/x^3)[-16/y^2 + 16/(3(8 - 1/x)^{4/3}) + 10/\{9x(8 - 1/x)^{4/3}\}] = 0,$$
$$1 < x \leq \infty,\ (2/3)y(1) - 7y'(1) = 0.$$

An exact solution is $y(x) = (8 - 1/x)^{2/3}$; it tends to 4 as $x \to \infty$. Baxley (1988)

3.
$$y'' + \{1/(4x^2)\}y - h^2/y^3 = 0,$$

where h is a constant.

An exact general solution of the given DE is

$$y = \pm\, x^{1/2}[A(\log x + B)^2 + h^2/A]^{1/2},$$

where A and B are constants. Eliezer and Gray (1976)

4.
$$y'' + w^2 x^{-4} y - K y^{-3} = 0,$$

where $K \neq 0$ and w are constants.

A first integral of the given DE is found to be

$$I = (xy' - y)^2 + w^2 y^2 x^{-2} + K y^{-2} x^2.$$

Ranganathan (1992)

5.
$$y'' + w^2 y - \{(1 + y)/(1 + y^2)\} \cos x = 0.$$

Using the Brouwer fixed-point theorem, it is shown that there exists at least one 2π-periodic solution of the given DE if $w^2 \neq 1$. Further investigation would be required to find where this periodic solution is located, whether there are other periodic solutions, and the behavior of a periodic solution if one varies w; in particular, what happens when $w = 1$, when the linear part of the DE has $\cos x$ and $\sin x$ as its solutions. Verhulst (1990), p.59

6.
$$y'' + \frac{1}{x} y^{1/3} - 2x^{-3} - x^{-4/3} = 0, \quad x > 0.$$

One exact solution of the given DE is $y = 1/x$. All solutions of the given DE are shown to satisfy the condition $\lim_{x\to\infty} \inf y(x) = 0$. Grace and Lalli (1987)

7.
$$y'' + \lambda y/(y^2 + x^2)^{1/2} = 0, \quad 0 < x < 1,\ y(0) = y'(1) = 0,$$

where λ is a constant.

Constructing a set of iterates y_k by the linear system

$$\begin{aligned} y_k'' &+ \lambda_k y_{k-1}(x)/(y_{k-1}^2(x) + x^2)^{1/2} = 0, \\ \lambda_k &= c\int_0^1 \{y_{k-1}(r)/[y_{k-1}^2(r) + r^2]^{1/2}\}dr, \qquad (1) \\ y_k(0) &= y_k'(1) = 0, y_k'(0) = c, \end{aligned}$$

it is shown that the sequence $\{y_k, \lambda_k\}$ defined by $y_0(x) = cx$, and (1), converges uniformly to the positive solution of the given problem. The convergence is monotonic:

$$c = y_0(x) > y_1(x) > \cdots > y_k(x) \cdots > y, \quad (c^2+1)^{1/2} = \lambda_1 < \lambda_2 < \cdots \lambda.$$

A similar sequence that converges in a monotonically increasing fashion to the solution is also constructed. Hence a constructive existence proof of the problem is given. Several examples are illustrated numerically. Luning and Perry (1984)

8. $$\boldsymbol{y'' - (ay^2 + bxy + cx^2 + \alpha y + \beta x + \gamma)^{-3/2} = 0,}$$

where $a \neq 0, b, c, \alpha, \beta$, and γ are constants.

Put $2au(x) = 2ay + bx + \alpha$ and find A, B, C such that

$$4aA = 4ac - b^2, \quad 2aB = 2a\beta - b\alpha, \quad 4aC = 4a\gamma - \alpha^2.$$

The DE then changes to

$$(Ax^2 + Bx + C)^{3/2}u'' = \left[au^2/(Ax^2 + Bx + C) + 1\right]^{-3/2}.$$

Write $t(x) = a(Ax^2 + Bx + C)^{-1/2}u(x)$ so that

$$(Ax^2 + Bx + C)^2 t'^2 = (B^2/4 - AC)t^2 + 2a\int (t^2/a + 1)^{-3/2}dt.$$

Now the variables separate. Kamke (1959), p.545

9. $$\boldsymbol{y'' + [(2 + \cos x + 3x \sin x)/(3x^{4/3})]\left(y + y^{1/3}\right) = 0, \quad x > 0.}$$

It is proved that the given DE is oscillatory. Blasko, Graef, Hacik, and Spikes (1990)

10. $$\boldsymbol{y'' + r^2(x)y - y^n = 0, \quad y(0) = 0 = y(\pi),}$$

where $n \geq 2$ is not necessarily an integer and where $r^2(x)$ is continuous.

The given problem arises in connection with the distribution of energy in a nuclear power reactor. It is proved that a positive solution of the given problem exists if and only if the largest eigenvalue of

$$y'' + r(x)y(x) = \lambda y(x), \quad y(0) = y(\pi) = 0,$$

is positive. Numerical procedure is also indicated. For $r^2(x) \geq \mu^2 > 1$, upper and lower bounds are also found. Shampine (1969)

11. $$\boldsymbol{y'' - x^{-2/p}[y]^{(p+2)/p} = 0,}$$

where p is real and positive is well-known Emden–Fowler equation.

Setting $y(x) = y_0(x)(1+u(x))$, where $y_0(x) = [p(p-1)]^{p/2}x^{1-p}$ is a special elementary solution of the given equation, we have

$$x^2\frac{d^2u}{dx^2} - 2(p-1)x\frac{du}{dx} - 2(p-1)u = F(u),$$

where $F(u)$ is analytic for $|u| < 1$ and $F(0) = F'(0) = 0$. Putting $x = e^{\alpha z}$, where α is an arbitrary real constant, we find the equation

$$\frac{d^2u}{dz^2} - (2p-1)\alpha\frac{du}{dz} - 2(p-1)\alpha^2 u = \alpha^2 F(u). \tag{1}$$

It is shown by applying a general theorem that Eqn. (1) has a family of solutions of the form

$$u(z) = \xi_1 + \xi_2 z + \sum_{n=2}^{\infty} b_n(\xi_1, \xi_2)z^n, \tag{2}$$

for sufficiently small $|\xi| = |\xi_1| + |\xi_2|$; this solution is written as

$$y(x) = y_0(x)\left[1 + \xi_1 + \xi_2\{(\log x)/\alpha\} + \sum_{n=2}^{\infty} b_n\{(\log x)/\alpha\}^n\right]. \tag{3}$$

The series in (3) converges absolutely for sufficiently small $|\xi|$ and every real α and x satisfying $|\log x| \leq \alpha$. By setting $\xi_1 = 0$ in (3), we obtain

$$y(x) = y_0(x)\left[1 + \xi_2(\log x)/\alpha + \sum_{n=2}^{\infty} b_n\{(\log x)/\alpha\}^n\right]. \tag{4}$$

Thus, given any real number c, we can choose α such that $\alpha c = \xi_2$ and from (4) obtain the form

$$y(x) = y_0(x)\left[1 + c\ \log x + \sum_{n=2}^{\infty} b_n c^n(\log x)^n\right]. \tag{5}$$

The series in (5) converges absolutely for every real c and sufficiently small x beyond the point 1. Ifantis (1987)

12. $$y'' + x^\gamma y^p + x^\delta y^q = 0 \quad \text{on } (a, b),$$

where $\gamma, \delta \in (-\infty, \infty)$ and $p, q \geq 1$.

A problem is posed: Classify completely the given DE according to uniqueness or nonuniqueness of BVPs with $y(a) = y(b) = 0$, or $y'(a) = y(b) = 0$, in terms of $\gamma, \delta \in (-\infty, \infty)$ and $p, q \geq 1$. Kwong (1991)

13. $$y'' + w^2 y - c_1(\sin wx \cos wx)^{m-2}y^{1-2m} = 0, \quad m \neq 0, 1,$$

where w, c_1, and m are constants.

The given DE has solutions

$$y = [a^m \cos^m wx + c_m b^m \sin^m wx]^{1/m},$$

where a and b are arbitrary constants and $c_m = c_1/[w^2(ab)^m(m-1)]$. Reid (1971)

14. $$y'' + a_0(x)y - \phi(x)y^\alpha = 0,$$

where α is a real number, $\alpha \neq 0, \alpha \neq 1$.

By introducing the transformation

$$y = v(x)z, dt = u(x)dx,$$

the given DE changes to the autonomous nonlinear form

$$\ddot{z} + k_0 z = pz^{\alpha}, \tag{1}$$

provided that

$$\phi(x)v^{\alpha-1} = pu^2, \quad vu^{1/2} = 1.$$

Here v is a solution of

$$v'' + a_0(x)v = k_0 v^{-3}. \tag{2}$$

Equation (1) may be integrated once by writing $dz/dt = q$, $d^2z/dt^2 = q$, dq/dz, etc. Therefore, the solution of the given DE may be found to be

$$\begin{aligned} y &= v(x)z(t + c_1, c_2) \\ &= v(x)z\left(\int \frac{dx}{v^2(x)} + c_1, c_2\right), \end{aligned}$$

where $z(t + c_1, c_2)$ is the general solution of (1). Sachdev (1991), p.39

15. $\boldsymbol{y'' + p(x)y - q_m(x)y^{1-2m} = 0,}$

where $m \neq 0, 1$ is real and finite.

It may be checked that

$$y = \left(u^m + c(m-1)^{-1}W^{-2}v^m\right)^{1/m}$$

is an exact solution of the given DE, where u and v are independent solutions of $y'' + p(x)y = 0$ and $q_m(x) = c(uv)^{m-2}$; W is the Wronskian: $W = W(u, v)$. Reid (1971)

16. $\boldsymbol{y'' - (1/4)y + [a(x)/x^{\sigma}]y^{1+2/\gamma} = 0,}$

where σ and γ are constants [see Eqn. (2.3.42)]. Kaper and Kwong (1988)

17. $\boldsymbol{y'' - y^n(1-y)^{\ell}/(x^k + ay^p) = 0, \quad (x \geq 0), y(0) = 1, \; y(\infty) = 0,}$

where n, ℓ, k, a, and p are constants.

For $a \geq 0, 0 < k < 2, k - \ell < 1$ and $n \geq 1$, the given problem satisfies the conditions enunciated in Eqn. (2.6.24) and so for $0 \leq y \leq 1$ and $x \geq 0$, it has at least one monotonic solution. Kolosov and Lyubarskii (1968)

18. $\boldsymbol{y'' - y - \sum_{i=1}^{m} q_i(x)y^{r_i} = 0, \quad r_i > 0,}$

where $q_i (1 \leq i \leq m)$ are continuous functions on $[0, \infty)$.

It is proved that if $r_i \leq 1$, $(1 \leq i \leq m)$, then all the solutions of the given DE satisfy the asymptotic formulas

$$\begin{aligned} y &= [\delta_1 + o(1)]e^{x} + [\delta_2 + o(1)]e^{-x}, \\ y' &= [\delta_1 + o(1)]e^{x} - [\delta_2 + o(1)]e^{-x}, \end{aligned} \tag{1}$$

where $\delta_i \, (i = 1, 2)$ are constants, as $x \to \infty$.

Even for $r_i > 1$, it can be shown that any solution of the given DE such that $|y(0)| + |y'(0)| \leq c$ is defined on $[0, \infty)$ and satisfies (1). The constant c is suitably defined. Pinto (1991)

19.
$$y'' + \sum_{k=1}^{s} b_k x^{\alpha_k}[1 + O_k(x)]y^{n_k} = 0,$$

where b_k and α_k are arbitrary real numbers; n_k $(n_1 < n_2 < \cdots < n_s)$ are nonnegative rationals with odd denominators. Such a rational is called even or odd according as its numerator is even or odd. The functions $o_k(x)$ satisfy $|o_k(x)| + |o_k'(x)| < Ax^{-c}$ for some postive A and c.

The given DE generalizes the Emden–Fowler equation. A solution is called noncontinuable if it exists at most over a finite interval. A solution is called singular in case it vanishes identically for sufficiently large x. The principal results are as follows:

(a) If n_s is odd and $b_s > 0$ or if $n_s \leq 1$, all solutions are continuable: otherwise, noncontinuable solutions exist.

(b) Every continuable solution grows no faster than a power of x except in the case $n_s = 1, b_s < 0, \alpha_s > -2$.

(c) If n_1 is odd and $b_1 > 0$, or if $n_1 \geq 1$, there are no singular solutions: otherwise, singular solutions exist.

(d) Every nonsingular solution of the given DE approaches zero no faster than some power of x except in the case $n_1 = 1, b_1 < 0, \alpha_1 > -2$.

A solution is called "ordinary" if it is continuable, nonsingular, and grows (or approaches zero) no faster than a power of x. In this case the following cases arise:

(e) When $n_s = 1, b_s < 0, \alpha_s > -2$, the given DE has a family of monotonic solutions that become infinite exponentially fast, with all other unbounded solutions being ordinary.

(f) When $n_1 = 1, b_1 < 0, \alpha_1 > -2$, the given DE has a family of monotonic solutions approaching zero, exponentially fast, with all other solutions being ordinary.

Beklemiseva (1962)

20.
$$y'' + \left[1 + x^{-1/2} \sin \beta x\right] y - \sum_{i=1}^{m} \lambda_i(x) y^{\gamma_i} = 0,$$

where $\beta \neq \pm 1, \pm 2$, $\gamma_i > 1 (1 \leq i \leq m), m \in N$, and $0 \leq \lambda_i(x)$ $(1 \leq i \leq m)$ are continuous functions over $[a, \infty), a > 0$.

It is shown that any solution $y(x, x_0, y_0)$ of the given DE such that $K|y_0| < c$ (where $K > 0$ and $c > 0$ are certain constants) is bounded for $x \to +\infty$. Medina (1991)

21.
$$y'' - \sin(2\pi y) - x + 1/5 = 0, \quad y'(0) = 0, \; y'(1) = 0.$$

It is shown that the given problem has infinitely many solutions. Saranen and Seikkala (1988)

22.
$$y'' - y - \sin y - h(x) = 0, \quad a < x < b, \; y'(a) = A, \; y'(b) = B,$$

where $h \in L_2(a, b)$ and $(b - a) < \{3(10)^{1/2}\}^{1/2}$.

It is shown that the given problem has a solution for all $A, B \in R$. The numerical solution is found for the special values $a = 0, b = 1, A = 1, B = 2$, and $h(x) = x$. Saranen and Seikkala (1988)

23.
$$\boldsymbol{y'' + \alpha^2 \sin y - \beta f(x) = 0,}$$
$$\boldsymbol{y(0) = y(2\pi),\ y'(0) = y'(2\pi).}$$

The following results are quoted: For $\alpha^2 < 1$, for every β, the BVP has just one solution. For $\alpha^2 > 1$, there can be several solutions. We obtain an approximate solution with the ansatz $y \simeq A \sin(x)$; A is determined by the equation

$$A^2 - 2\alpha^2 J_1(A) + \beta = 0$$

(J_1 is a Bessel function). The DE with the boundary condition

$$y(0) = y(\pi) = 0 \tag{1}$$

has for any α, β always at least one solution, and for $\alpha^2 < 1$, there is only one solution. The same is true for any fixed α, for all sufficiently large $|\beta|$, if in the general DE above, the function $f(x)$ is such that the solution of the boundary value problem

$$u'' = f(x), \quad u(0) = u(\pi)$$

has only a finite number of points with $u' = 0$. For a fixed β, the number of solutions of the given DE with boundary conditions (1) becomes infinite as $\alpha \to \infty$. Kamke (1959), p.546

24.
$$\boldsymbol{y'' + (mg/\ell) \sin y - g(x) = 0, \quad y(a) = A,\ y(b) = B,}$$

where m, ℓ, a, b, A, and B are constants.

It is shown that the given BVP has at least one solution if $g(x)$ is continuous and bounded. It has only one solution if the length $b - a$ of the interval $[a, b]$ is small enough, but it can have more than one solution for larger intervals. Bailey, Shampine, and Waltman (1968), p.13

25.
$$\boldsymbol{y'' - x(e^{y/x} - 1) = 0, \quad x > 0,\ y(0) = -\alpha,}$$

where $\alpha > 0$ is a given number.

The given problem describes the motion in an ionized field under the influence of the Ukawa potential.

It is proved that there exists a unique continuous bounded solution y to the given problem that is negative increasing, and concave; moreover, for some negative constant A, y is asymptotic to Ae^{-x} as $x \to \infty$. The uniqueness of the solution automatically implies uniqueness of A. It is also shown that

$$A \leq -\alpha \leq A + (A^2/4) \log 3.$$

Gross (1963)

26.
$$\boldsymbol{y'' - a(x)e^{y+\eta} - b(x)e^{-y-\eta} + f(x) = 0, \quad 0 \leq x \leq 1,\ y(0) = \mu_0,\ y(1) = \mu_1.}$$

It is proved that the given BVP has a unique solution $y = y(\eta, \mu_0, \mu_1, f, x)$ for all $\eta, \mu_0, \mu_1 \in R$, provided that $a, b, f \in C([0, 1])$ and $a, b > 0$ on $[0, 1]$. Markovich (1985)

27. $$y'' - s\left[|y|^2/(\theta^2-1) + c_1x + c_2\right]y - (\alpha - \theta^2/4)y = 0,$$

where s, θ, c_1, c_2, and α are constants.

The given DE arises from the traveling wave form of Zakharov equations. Making the change of variables

$$x = (c_1s)^{-1/3}\zeta - (1/(c_1s))(c_2s + \alpha - \theta^2/4),$$

$$y = \left[(1/2)(c_1s)^{-2/3}|s/(\theta^2-1)|\right]^{-1/2} F(\zeta),$$

we obtain

$$F'' - \zeta F - 2KF^3 = 0, ' \equiv \frac{d}{d\zeta}, \tag{1}$$

$$\text{where } K = \begin{cases} +1 \\ -1 \end{cases} \quad \text{for } s/(\theta^2-1) <> 0.$$

Equation (1) is second Painlevé if $K = 1$. Tajiri (1985)

28. $$y'' + x^\alpha|y|^\gamma \operatorname{sgn} y = x^\delta \sin x, \gamma > 0,$$

where δ is real, $x \in [0, \infty)$.

It is shown that the given DE is oscillatory for all δ, provided that $\alpha + \gamma\delta > -1$. Wong (1988)

29. $$y'' - y + y|y|^{k-1}/x^{k-1} = 0,$$

$y(0) = 0$, $\lim_{x\to\infty} y(x) = 0, y'(0)$ exists and $y(x) > 0$ on $(0, \infty)$.

This is a special case of Eqn. (2.6.16). Existence of the solution of the BVP is proven for the range $1 < k < 5$. Macki (1978)

30. $$y'' + x^\alpha(1 + \sin x)|y|^\gamma \operatorname{sgn} y - (\sin x)(1 + 1/x - 2/x^3) + (2\cos x)/x^2, 0 < \gamma < 1,$$

where α is real.

The given DE is shown to be oscillatory if $\alpha \geq -(1+\gamma)$. Wong (1988)

31. $$y'' + a(x)y + \lambda[h(x)]^{-(m+3)/2}|y|^m \operatorname{sgn} y = 0,$$

where $x \in J$, an open interval on which $h > 0$ and $h''(x)$ is continuous, and λ is an arbitrary constant; m is another constant.

It is shown that the given DE has a first integral

$$I(x, y, y') = h(x)y'^2 + [ha + (1/2)h'']y^2 + [2/(m+1)]\lambda h^{-(m+1)/2}|y|^{m+1} - h'yy', \tag{1}$$

provided that $a(x)$ is a solution of the DE

$$\left(ah^2\right)' = -hh'''/2, \tag{2}$$

or equivalently,

$$a = \alpha/h^2 - h''/(2h) + (1/4)(h'/h)^2,$$

where α is an arbitrary constant.

This result can be proved by differentiating (1) and using (2) to show that the given DE is satisfied. Trench and Bahar (1987)

32.
$$y'' + x^{\alpha} y e^{|y|^2} = e^{-x} \sin x.$$

It is shown that the given DE has oscillatory solution for $\alpha \geq -2$. This equation arises from a certain radial solution of the Klein-Gordon equation. Wong (1988)

33.
$$y'' = \begin{cases} -8 - (1 + \cos x)^3 - \cos x, & \text{if } y(x) \leq -2 \\ y^3 - (1 + \cos x)^3 - \cos x, & \text{if } |y(x)| \leq 2 \\ 8 - (1 + \cos x)^3 - \cos x, & \text{if } y(x) \geq 2, \end{cases}$$

with $y'(0) = 0, y'(1) = -\sin 1$.

The given BVP is shown to have a unique solution. Saranen and Seikkala (1988)

2.6 $y'' + f(x, y) = 0,\ f(x, y)$ general

1.
$$y'' + x^{-2} y F(x^{-1} y^2) = 0.$$

It is easy to check that the general solution of the given DE is given by $y(x) = x^{1/2} u(\log x)$, where $u(t)$ is the general solution of

$$u'' - (1/4)u + uF(u^2) = 0. \tag{1}$$

Here $F(\zeta)$ increases from 0 to ∞ as ζ varies over the same interval. Therefore, there exists a positive number c such that $F(c^2) = 1/4$. The function $u(t) \equiv c$ is evidently a solution of (1). It follows that the given DE has the nonoscillatory solution $y(x) = c\sqrt{x}$ (in fact, it has an infinity of different oscillatory solutions, as is proved by the application of a general theorem). Nehari (1960)

2.
$$y'' - x^{-3/2} f(yx^{-1/2}) = 0.$$

For $u(x) = yx^{-1/2}$,we have

$$\frac{d}{dx}(u'x)^2 = (1/2)uu' + 2f(u)u',$$

so that

$$(u'x)^2 = C_1 + (1/4)u^2 + 2\int f(u)du.$$

Separating the variables, we have

$$\int \left[C_1 + (1/4)u^2 + 2\int f(u)du\right]^{-1/2} du = C_2 + \ln|x|,$$

where C_1 and C_2 are arbitrary constants. Kamke (1959), p.547

3.
$$y'' - \left[1/\left(ax^2 + bx + c\right)^{3/2}\right] f\{y/(ax^2 + bx + c)^{1/2}\} = 0,$$

where $a, b,$ and c are constants.

With $u(x) = (ax^2 + bx + c)^{-1/2} y(x)$, the given DE goes to one with variables separable:

$$\left(ax^2 + bx + c\right)^2 u'^2 = \left[(1/4)b^2 - ac\right] u^2 + 2f(u)du + C,$$

C being an arbitrary constant. Kamke (1959), p.568; Murphy (1960), p.391

4. $$y'' + x^{-2-\nu} y F(x^{-2}y^2) = 0, \quad \nu > 0,$$

where $F(0) = 0$ and F is a nondecreasing function of its argument.

It is shown that all solutions of the given DE are nonoscillatory. An instructive case is $\nu = 2$. In this case, the general solution is of the form

$$y(x) = xu(t), \quad t = 1/a - 1/x, \tag{1}$$

where $u(t)$ is any solution of

$$\ddot{u} + uF(u^2) = 0, \tag{2}$$

for which $u(0) = 0$. It is possible to show that all solutions of (2) are periodic and that the distance between consecutive zeros decreases as $\dot{u}(0)$ increases. If a^{-1} does not coincide with a zero of $u(t)$, then (1) shows that $y \sim u(a^{-1})x$ for large x. If $u(a^{-1}) = 0$, it follows from (1) that $y(x) \to -\dot{u}(a^{-1})$ as $x \to \infty$. Thus both types of nonoscillatory solutions are permitted. Nehari (1960)

5. $$y'' + \frac{\partial}{\partial y}\left[x^{-2} y^2 g\left(x^a y^b\right)\right] = 0,$$

where a and b are real and $g \in C^1[0, \infty)$.

The general DE has several important special cases.

(a) If $a = 1, b = 0$, it is linear.

(b) If $g(u) = cu$ (c a constant), we have an Emden–Fowler equation.

(c) If $a = 0, b = 1$, and $y^2 g'(y) + 2yg(y) = f(y)$, we introduce the change of variable $t = -\log x$, and obtain a Lienard equation:

$$\frac{d^2y}{dt^2} + \frac{dy}{dt} + f(y) = 0.$$

The given DE is studied from the variational point of view. Actually, the given equation is the Euler–Lagrange equation, which determines the extremal curves for a certain variational problem, i.e., minimizing the integral

$$\int_a^b \left[\frac{1}{2}\left(\frac{dy}{dx}\right)^2 - H(x, y)\right] dx,$$

where $H(x, y) = x^{-2} y^2 g(x^a y^b)$ over all C^1 functions $y(x)$ with $y(a)$ and $y(b)$ prescribed.

For $2a + b = 0$, we get a first integral of the given DE as $yy' - xy'^2 - 2x^{-1}y^2 g(x^a y^{-2a})$ $=$ constant.

Comparison theorems are proved for the given (general) form. Benson (1974/75)

6. $$y'' + g(y) = e(x), \quad y(0) - y(2\pi) = y'(0) - y'(2\pi) = 0,$$

where $e \in C([0, 2\pi])$, $g : R \to R$ is continuous, $g(y + Y) = g(y)$ for all $y \in R$ and some $Y > 0$, and $\int_0^Y g(y)dy = 0$.

The existence of multiple solutions is proved. Mawhin and Willem (1984)

7. $$y'' + F'(y) + [w_0 + g(y)\ \sin(\Omega x + \gamma)]\, y = 0,$$

where

(a) $w_0 > 0, \Omega > 0$, and γ are constants.

(b) $F'(y)$ and $g(y)$ are continuous, and satisfy Lipschitz conditions in each finite interval.

(c) There exists a positive number Δ such that if $|y| > \Delta$, $F(y)$ has the same sign as y, and $|f(y)|$ increases monotonically without limit as $|y|$ increases.

(d) There exists a positive number N such that

$$\left(w_0^2 + |g(y)|\right)|y| < N|F(y)|, \quad |y| > \Delta.$$

It is then shown that any solution that is defined for $x = 0$ is continuable over the infinite interval $x > 0$.

Under certain other assumptions it is shown that every solution has infinitely many maxima and minima and, additionally, infinitely many zeros. Graffi (1954)

8. $$y'' + a(x)f(y) = g(x), \quad x \in [0, \infty),$$

where $a(x), g(x)$ are real-valued piecewise continuous functions on $[0, \infty)$ and $f(y)$ is a continuous nondecreasing function of $y \in (-\infty, \infty)$. Conditions on $a(x), f(y)$, and $g(x)$ are given so that all solutions are oscillatory. Wong (1988)

9. $$y'' + a(x)y + q(x)f(y) = 0.$$

It is shown that every solution of the given DE is bounded as $x \to \infty$ provided that the following conditions are met: $a(x)$ is continuous, nondecreasing, positive for $x \geq 0$, and together with the continuous function $q(x)$ satisfies $|q(x)a^{-1}(x)| < \alpha x^{-1}$ and $[q(x)a^{-1}(x)]' < \alpha x^{-2} (\alpha > 0)$ for x sufficiently large, $f(y)$ is a positive function defined for all y such that $f(u) \in \mathrm{Lip}(1)$ and $|F(u)|u^{-2}$ is bounded everywhere, where $F(u) = \int_0^u f(s)ds$. Bihari (1961)

10. $$y'' + a(x)y + q(x)f(y^2) = 0.$$

Let $a(x)$ be continuous, nondecreasing, and positive for $x \geq 0, q(x)$ continuous on $0 \leq x < \infty$ and such that

$$\int_0^\infty [a(x)]^{-1/2} x |q(x)| dx < \infty,$$

$f(y)$ continuous and nondecreasing for $y \geq 0$ and positive for $y > 0, f(0) = 0$, and

$$\int_{y_0}^\infty \left[f^2(u) + u\right]^{-1} du = \infty \quad (y_0 > 0).$$

It is shown, then, that every solution $y(x)$ of the given DE is bounded for $x \geq 0$; moreover, $A(x) = [y^2(x) + a^{-1}(x)y'^2(x)]$ tends to a finite limit as $x \to \infty$. Bihari (1961)

11. $$y'' - c^2 y^{-3} + M(x) f(y) = 0,$$

where $f(y)$ is positive for positive y, is bounded near $y = 0$, and does not approach 0 as y becomes infinite, and $M(x)$ is positive and nondecreasing for positive x.

It is shown that if M is bounded, then, as x becomes infinite, a solution $y(x)$ must either approach a positive limit y_1 or oscillate between successive maxima and minima which are themselves, respectively, decreasing. It is further shown that if M is unbounded, a solution $y(x)$ must either approach 0 as x becomes infinite or oscillate, the successive minima approaching 0. Armellini (1942)

12. $$y'' - h(\Gamma)\frac{\partial \Gamma}{\partial y} = 0,$$

where Γ is a function of x and y satisfying the PDE

$$\frac{\partial \Gamma}{\partial x} = \Gamma \frac{\partial \Gamma}{\partial y}.$$

(a) Assuming the first integral of the given DE in the form $F[y', \Gamma(x,y)] =$ constant, one checks that if $h(\Gamma) = \alpha\Gamma$, where α is a constant, then

$$\begin{aligned} \alpha &= -1/4, \ F(\hat{y}, z) = 2\hat{y}/(z + 2\hat{y}) + \log(z + 2\hat{y}) = \text{constant} \quad (1) \\ \alpha &\neq -1/4, \ F(\hat{y}, z) = (\hat{y} + r_1 z)^{2\alpha + r_2}(\hat{y} + r_2 z)^{2\alpha + r_1}, \quad (2) \end{aligned}$$

where r_1 and r_2 are two distinct roots of $r^2 - r - \alpha = 0$. In (1) and (2), $\hat{y} = y'$ and $z = \Gamma$.

(b) $h(\Gamma) = \alpha\Gamma^{-1}$, the first integral of the given DE, is

$$F(\hat{y}, z) = (1/z)e^{-\hat{y}^2/(2\alpha)} - (1/\alpha)\int^{\hat{y}} e^{u^2/(2\alpha)} du.$$

(c) $F(\hat{y}, z) = \hat{y}^4 + b(z)\hat{y}^2 + c(z)\hat{y} + d(z)$ is a first integral provided that $b = -(3/2)z^2 + Kz^{-2/3}, c = z^3 + 2Kz^{1/3}, d = -(3/16)z^4 - (5k/4)z^{4/3} + K^2/(4z^{4/3})$ and $h(z) = (3/4)z + K/(6z^{5/3})$, where K is arbitrary.

Airault (1990)

13. $$y'' + yF(y^2, x) = 0, \quad x > 0,$$

where $yF(y^2, x)$ is continuous in (y, x) for $x > 0, |y| < \infty$, and $F(t, x)$ is nonnegative.

The given DE is assumed to be either superlinear, that is, $F(t, x)$ is monotone increasing in t for every x, or it is sublinear, that is, $F(t, x)$ is monotone decreasing in t for every x. Criteria are given for equations to possess either oscillatory or nonoscillatory solutions. The methods consist primarily of comparison with linear equations and use of energy functions. Coffman and Wong (1972)

14. $$y'' + yF(y^2, x) = 0,$$

where F satisfies the following conditions:

(a) $F(t, x)$ is continuous in both t and x for $0 \le t < \infty$ and $0 < x < \infty$.

(b) $F(t, x) > 0$ for $t > 0, x > 0$.

(c) $t_2^{-\epsilon} F(t_2, x) > t_1^{-\epsilon} F(t_1, x)$ for $0 \le t_1 < t < t_2 < \infty$, fixed positive x, and some positive number ϵ.

Because of (b), $yy'' < 0$ for $y \ne 0$, i.e., all solution curves of the given DE are concave toward the horizontal axis. Accordingly, in an interval in which $y(x)$ does not change its sign, the curve $y = y(x)$ must lie between the x-axis and the tangent to the curve at any point of the interval, and it follows that a solution of the given DE for which $y(a)$ and $y'(a)$ are finite for some positive a will be continuous throughout $(0, \infty)$. Condition (c) prevents the linear equation $y'' + p(x)y = 0$ $(p(x) > 0)$ from being included in the given DE; thus only those aspects of the given DE are emphasized which are not shared by the linear DEs.

Both properly nonoscillatory solutions (which change sign at least once) and those that do not change sign in the entire interval of continuity are discussed. Several theorems are proved. We quote one: The given DE has bounded nonoscillatory solutions if and only if $\int^\infty xF(c, x)dx < \infty$ for some positive constant c. Nehari (1960)

15. $$y'' - y + yF(y^2, x) = 0,$$

where $F(\eta, x)$ satisfies the following conditions:

(a) $F(\eta, x)$ is continuous in η and x for $0 < x < \infty$ and $0 \le \eta < \infty$.

(b) $F(\eta, x) > 0$ for $\eta > 0, x > 0$.

(c) There exists a $\delta > 0$ such that for every fixed positive x and $0 \le \eta_1 < \eta_2 < \infty$, $\eta_2^{-\delta} F(\eta_2, x) > \eta_1^{-\delta} F(\eta_1, x)$.

In addition to the given condition on F, if

(d) $\lim_{x\to\infty} F(c^2, x) = 0$ for all finite c, and

(e) $\int_0^a x^{1/2-\epsilon} F(c^2 x, x)dx < \infty$ for all finite $c, 0 < a < \infty$, and for some $\epsilon \ge 0$, then the given DE has a discrete infinity of solutions $\{y_n(x)\}$, $n = 1, 2, \ldots$, whose derivatives are continuous throughout $[0, \infty)$ and are such that $y_n(x)$ has exactly $n - 1$ zeros in $(0, \infty)$. Moreover, $y_n(0) = 0, \lim x^{-1} y_n(x)$ exists as $x \to 0$ and $y_n(x) \to 0$ as $x \to \infty$ for each n.

Ryder (1967)

16. $$y'' - y + yF(y^2, x) = 0, \quad x \in (0, \infty),$$

$y(0) = 0, \lim_{x\to\infty} y(x) = 0, y'(0)$ exists, $y(x) > 0$ on $(0, \infty)$.

Existence of the solution of the singular BVP is proved under the following conditions on F :

(a) $F(\eta, x) \in C[[0, \infty) \times (0, \infty)], F(\eta, x) > 0$ for $\eta > 0$.

(b) There exists $\delta > 0$ such that for each $x > 0, \eta^{-\delta} F(\eta, x)$ is strictly increasing in η on $[0, \infty)$. In particular, $\lim_{\eta\to 0} \eta^{-\delta} F(\eta, x)$ exists for $x > 0$.

(c) $\lim_{x\to\infty} F(c^2, x) = 0$ for any constant c.

(d) For some fixed $\epsilon > 0$ (hence for all smaller values of ϵ), $\int_0^1 t^{1-\epsilon}F(c^2t,t)dt$ converges for any constant c.

Macki (1978)

17. $$y'' + y - f(y, 1/x) = 0,$$

where $f(y,u)$ is analytic near $y = u = 0$ and $f(0,u) = f(y,0) = 0$.

It is shown that the given DE possesses solutions that remain bounded as $x \to \infty$ and that every such solution $y(x)$ admits asymptotic representation of the form

$$y(x) = \gamma \sin\left[x - \{\phi(x)/(2\pi)\} \log x - \mu\right] + O(1/x),$$

where γ and μ are constants and

$$(1/2\pi)\phi(x) = 1/(2\pi\gamma) \int_0^{2\pi} f_u(\gamma \sin\sigma, 0) \sin\sigma d\sigma.$$

Ascoli (1951)

18. $$y'' - f(x,y) = 0, \quad y(0) = y(1) = 0.$$

The problem is first transformed to an integral equation

$$y(x) = \int_0^1 g_k(x,\xi)\left[k^2 y(\xi) - f(\xi, u(\xi))\right] d\xi,$$

where

$$g_k(x,\xi) = \frac{1}{k\sinh k}\begin{cases} \sinh(kx)\sinh k(1-\xi), & 0 \le x < \xi \\ \sinh k(1-x)\sinh k\xi, & \xi < x \le 1, \end{cases} \tag{1}$$

using Green's function approach for the operator $\left(\dfrac{d^2}{dx^2} - k^2\right)$. The following result by an iterative constructive method is then proved. In the given boundary value problem, let $\dfrac{\partial f}{\partial y}$ be continuous for all $x \in [0,1]$ and all u. Suppose that there exist $N > 0$ and $\delta \ge 0$ such that $0 \le \delta \le \dfrac{\partial f}{\partial y} \le N$ for all $x \in [0,1]$ and all y. Then a unique solution of the given problem exists. For $k^2 = (1/2)(\delta + N)$, it is given by the limit of the convergent sequence:

$$\begin{aligned} y^0(x) &= 0, \\ y^{m+1}(x) &= \int_0^1 g_k(x,\xi)\left[k^2 y^m(\xi) - f(\xi, y^m(x))\right] d\xi. \end{aligned}$$

Pennline (1981)

19. $$y'' - \phi(y,x) = 0, \quad x > 0, \; y(0) = -\alpha,$$

where α is a given positive number.

It is proved that if (a) ϕ is continuous in the range $y \le 0, x > 0$, (b) $\phi(y,x) \le 0$ and is nondecreasing as a function of y in the range $y \le 0, x > 0$, where, moreover, $\phi_y(y,x)$ exists and is continuous, and (c) $\phi(y,x) - y \ge 0$ and is nonincreasing as a function of y in the range $y \le 0, x > 0$, then there exists a unique, negative, bounded, continuous solution

y of the given problem. This solution is nondecreasing and concave and therefore tends to some nonpositive constant as $x \to \infty$. Gross (1963)

20. $$y'' - f(x, y) = 0,$$

where f may have singularities with respect to its first argument at $x = a$ or $x = b$ and with respect to second argument at 0.

The following results for BVPs are proved:

(a) $y(a+) = 0, y(b-) = 0, -\infty < a < b < +\infty$. If $f(x, y)$ is a nondecreasing function of y and there is $r > 0$ for which $f(x, y) \leq 0$ for $0 < y \leq r$ and

$$\text{mes}\{x \in (a, b) : \ f(x, y) < 0\} > 0, \tag{1}$$

then the condition

$$\int_a^b (x - a)(b - x)|f(x, y)|dx < +\infty \quad \text{for } 0 < y \leq r$$

is necessary and sufficient for this BVP to have a unique solution.

(b) $y(a+) = 0, y(b-) = y(x_0)$, where $-\infty < a < x_0 < b < +\infty$. If $f(x, y)$ is nondecreasing with respect to y, there is $r > 0$ such that (1) holds, and

$$\int_a^b (x - a)(b - x)|f(x, y)|dx < +\infty \quad \text{for } y > 0,$$

then this BVP has a unique solution.

(c) $y(a+) = 0, y'(b-) = 0, -\infty < a < b < +\infty$. If $f(x, y)$ is a nondecreasing function of its second argument and there is $r > 0$ such that the condition (1) are satisfied, then the condition

$$\int_a^b (x - a)|f(x, y)|dx < +\infty \quad \text{for } y > 0$$

is necessary and sufficient to ensure the solvability of this BVP.

Lomtatidze (1987)

21. $$y''(x) + F(x, y) = 0, \quad y(x) > 0, \ x \in (a, b), \ -\infty < a < b < \infty,$$

where either

$$y(a) = y(b) = 0,$$

or

$$y'(a) = y(b) = 0,$$

or

$$y(a) = y'(b) = 0.$$

Conditions are imposed on $F(x, y)$ such that the given problem has a unique solution; existence is assumed. One such condition is that $F(x, y)$ is Lipschitz continuous in y for fixed x and is superlinear, and there exists a positive concave function $\phi : (a, b) \to (0, \infty)$ such that $\phi^2 F(x, y)$ is nonincreasing in x for each fixed y. Kwong (1991)

22. $$y'' - \lambda^2 f(x, y) = 0, \quad y(0) = a, \ y(1) = b,$$

where λ^2 is sufficiently small and $a \geq b \geq 0$, and if $b = 0, a > 0$.

The existence of positive solutions of the given BVP is established under the hypothesis:

(a) f is positive and continuous on $[0,1] \times (0,\infty)$, and

$$\lim_{y \to 0+} f(x,y) = 0 \text{ for each } x \in [0,1].$$

(b) There exists a positive function g, continuous and nondecreasing on $(0,a]$, such that

(i) $f(x,y) \leq g(y)$ for $0 \leq x \leq 1,\ 0 < y \leq a$.

(ii) $\int_0^a g(u)du < \infty$.

A special case is $f(x,y) = y^\alpha$, for which $0 > \alpha > -1$. This restriction is also necessary for the case $b = 0$. Bobisud, O'Regan, and Royalty (1987)

23. $$y'' + f(x,y) = 0, \quad \alpha y(0) - \beta y'(0) = 0, \ \gamma y(1) + \delta y'(1) = 0,$$

where $f : (0,1) \times (0,\infty) \to (0,\infty)$ is continuous and decreasing in y for each fixed x and integrable on $[0,1]$ for each fixed y. A singularity condition is imposed by $\lim_{y \to 0+} f(x,y) = \infty$, uniformly on compact subsets of $(0,1)$.

An existence theorem is established. Gatica, Oliker, and Waltman (1989)

24. $$y'' - f(y,x) = 0, \quad x \geq 0, \ y(0) = 1, \ y(\infty) = 0,$$

where the following conditions on $f(y,x)$ are assumed:

(a) $f(y,x)$ is nonnegative for $0 \leq y \leq 1, x \geq 0$, continuous for $0 < y < 1$ and $x > 0$, and monotonically decreasing in x.

(b) $f(y,x) \geq y^p f_1(y,x)$, where $p \geq 0$ and $f_1(y,x)$ is nonnegative and continuous for $0 < y < 1$ and $x > 0$; $f_1(0,x) > 0$ for $x < \infty$.

(c) The function $(1/y)f(y, b(1-y))$ is locally bounded at the origin for all sufficiently small $b > 0$. The limit of $f(y, b(1-y))$ as $y \to 1$ (finite or infinite) exists.

(d) There is a function $h(y) \in L_1(0,1)$ and numbers r_1 and r_2 in $(0,2)$ such that $f(y,x) \leq h(y)[(1-y)/x]^{r_1}, 0 \leq y \leq 1$ for all sufficiently small x and

$$\lim_{x \to \infty} x^{r_2} f_1(y,x) = \infty$$

uniformly with respect to $y (0 \leq y \leq \delta)$, where δ is a positive number.

(e) $f_y(y,x)$ and $f_x(y,x)$ exist for $0 < y < 1$ and $x > 0, |f_y(y,x)|$ and $|f_x(y,x)|$ being bounded in each closed domain lying entirely within $0 < y < 1$ and $x > 0$.

Then the given singular BVP has at least one monotonic solution $y(x)$ satisfying

$$1 - x/C \leq y(x) \leq B^{1/m} x^{-1/m},$$

where m, B, C are positive constants, which are defined during the proof. The proof via integral equation formulation is iterative and uses the contraction principle. Some 0ther qualitative properties of the solution are discussed. Kolosov and Lyubarskii (1968)

25. $$y'' - f(x,y) = 0,$$

where f is a real-valued continuous function of the two variables x, y for $-\infty < x < \infty$, $a \leq y \leq b, a < b$, and f is periodic in x of period X.

Assume that the given DE satisfies the following conditions:

(a) For each pair $(A, B), a < A < b, -\infty < B < \infty$, there corresponds a unique solution $y(A, B, x)$ of the given DE that satisfies $y(A, B, 0) = A, y'(A, B, 0) = B$, and the functions $y(A, B, x), y'(A, B, x)$ are continuous functions in the triple (A, B, x) for $a < A < b, -\infty < B < \infty$ and suitable x.

(b) There are numbers $a < \alpha < \beta < b$ such that $\int_0^X f(x, y(x))dx \geq 0, \int_0^X f(x, z(x))dx \leq 0$ for all continuous functions $y(x), z(x)$ defined on $[0, X]$ that satisfy $a \leq y(x) \leq \alpha, \beta \leq z(x) \leq b$ for $0 \leq x \leq X$.

Then, if $X^2M \leq \min\{(b - \beta)/2, (\alpha - a)/2\}$, the given DE possesses at least one periodic solution of period X. Here $M = \sup\{|f(x, y)| : 0 \leq x \leq X, a \leq y \leq b\}$. Marlin and Ullrich (1968)

26. $$y'' - f(x, y) = 0.$$

It is shown that the given DE is intrinsically nonlinear; i.e., it cannot be changed by finite transformations to a linear form. Aguirre and Krause (1988)

27. $$y'' - g(x, y) = 0,$$

where $g: R \times R \to R$ is a continuous function that is X-periodic in the first variable for some $X > 0$.

Suppose that there are real numbers $M_+ \geq m_+ > m_- \geq M_-$ such that $g(x, M_\pm) \geq 0 \geq g(x, m_\pm), g(x, y) \geq 0$ if $y \leq M_-$; then it is proved that the given DE has two X-periodic solutions y_+, y_- such that $m_+ \leq y_+ \leq M_+$ and $\inf y_- \leq m_-$. In particular, $y_+ \neq y_-$. Tineo (1991)

28. $$y'' + f(x, y) = 0,$$

where $f(x, y)$ is a continuous function on $D: x \geq 0, y \in R = (-\infty, \infty)$.

Suppose that there are nonnegative continuous functions $v(x)$ for $x \geq 0$ and $g(y)$ for $y \geq 0$ such that $g(y)$ is positive and nondecreasing for $y > 0$ and $|f(x, y)| \leq v(x)g(|y|/x)$ for $x > 0$ and $y \in R$. Then it is proved that the given DE has a solution satisfying $\lim_{x\to\infty} y'(x) = b \neq 0$, or equivalently, $\lim_{x\to\infty} y(x)/x = b \neq 0$. If, in addition, $\int^\infty xv(x)dx < \infty$, then for any $a, b \in R, b \neq 0$, the given DE has a solution $y(x)$ which is asymptotic to the line $a + bx$ as $x \to \infty$ and satisfies $y' - b \in L(0, \infty)$. Further, necessary and sufficient conditions for the given DE to have a solution that is asymptotic to a line with nonzero slope are also given. Chen (1987)

29. $$y'' + f(x, y) = 0,$$

where $f(x, y)$ is continuous on $D: x \geq 0, -\infty < y < \infty$.

If there are two nonnegative continuous functions $v(x), \phi(x)$ for $x \geq 0$, and a continuous function $g(y)$ for $y \geq 0$ such that:

(a) $\int_1^\infty v(t)\phi(t)dt < \infty$,

(b) for $y > 0, g(y)$ is positive and nondecreasing,

(c) $|f(x, y)| < v(x)\phi(x)g(|y|/x)$ for $x \geq 1, -\infty < y < \infty$, then the given equation is shown to have solutions that are asymptotic to $a + bx$, where a and b are constants and $b \neq 0$. Tong (1982)

30. $$y'' + g(x, y) = 0.$$

If $g(x, y)$ is continuously differentiable for all x and y, and satisfies the following conditions, then it is proved that there exists a nontrivial 2π-periodic solution of the given DE:

(a) $g(x, y)$ is even and 2π-periodic in x, i.e., $g(x, y) = g(-x, y), g(x+2\pi, y) = g(x, y)$.

(b) $yg(x, y) > 0$ for all x and all $y \neq 0$.

(c) $g_y(x, y) > 0$ and $g_y(x, 0) > 1$ for all x, y.

(d) There exist constants β, A, B, with $\beta < 2$ such that $G(x, y) = \int_0^y g(x, s)ds \leq A|y|^\beta + B$ for all x, y.

Stevens (1980)

31. $$y'' + \phi(x, y) - f(x) = 0, \quad y(0) = y(a) = 0,$$

where $\phi(x, y)$ and $f(x)$ are continuous in the region $0 \leq x \leq a, -\infty < y < +\infty$, and $\phi(x, y)$ satisfies the additional conditions, $\phi(x, 0) = 0, |\phi(x, y_1) - \phi(x, y_2)| \leq b|y_1 - y_2|, b$ a positive constant in this region. The following result is obtained. If $a < \pi b^{-1/2}$, there exists one and only one solution of the given problem. The proof is by successive approximation, which uses a property of Green's function for the equation $y'' = 0$ over the interval $0 \leq x \leq a$. It is shown that if $f(x) = 0$, no solution not identically zero can satisfy the BC. Periodic solutions of the DE when $\phi(x, y)$ and $f(x)$ are periodic in x are also obtained, employing the general method of the Green's function. Manacorda (1946)

32. $$y'' + \lambda f(x, y) - p(x; \lambda) = 0, \quad 0 < x < \pi,$$

$y(0, \lambda) = 0 = y(\pi, \lambda)$, where λ is a positive parameter.

The function f is assumed to be smooth with the asymptotic expansion

$$f(x, y) \sim q(x)y + \beta(x)y^2 + \gamma(x)y^3 + \sum_{j=4}^{\infty} q_j(x)y^j \text{ as } y \to 0,$$

uniformly on $[0, \pi]$ with the coefficients q, β, γ, and q_j $(j \geq 4)$ being smooth functions on $[0, \pi]$.

A generalized WKB method is used to construct formal asymptotic approximations of solutions of the given nonlinear Sturm-Liouville system. Lange (1987)

33. $$y'' = f(x, y)y^{1+2\epsilon}, \epsilon > 0, \quad y(0) = \alpha, \ y'(0) = \beta, \ 0 \leq \alpha \leq A, 0 \leq \beta \leq \epsilon.$$

The existence of positive (convex) proper solutions (real valued and class C^2 over $[0, \infty)$) and their bounds are investigated. The function $f(x, y)$ is assumed to satisfy the following:

(a) $f(x, y)$ is continuous for $x \geq 0$ and $y \geq 0$,

(b) $f(x, y) > 0$ for each $x \geq 0$ and $y > 0$,

and either $f(x,y)$ is an increasing function of y for each $x \geq 0$, or $f(x,y)$ is a decreasing function of y for each $x \geq 0$. Wong (1970)

34. $$\boldsymbol{y'' - f(x,y)y^{1+2\epsilon} = 0, \quad \epsilon > 0,}$$

where f is assumed to satisfy the first two plus either the third or the fourth of the following conditions: (a) $f(x,y)$ is continuous for $x \geq 0, y \geq 0$; (b) $f(x,y) > 0$ for each $x \geq 0$ and $y > 0$; (c) $f(x,y)$ is an increasing function of y for each $x \geq 0$; and (d) $f(x,y)$ is a decreasing function of y for each $x \geq 0$.

It is proved that if $f(x,y)$ satisfies conditions (a), (b) and (c), and if δ is any positive number, then there exists a solution of the given DE that escapes to infinity as $x \to \delta^-$. Wong (1970)

35. $$\boldsymbol{\epsilon^2 y'' - x^m f(x,y) - \epsilon^2 g(x,y,\epsilon) = 0, \quad y(-1) = y(1) = 0,}$$

$0 < \epsilon \ll 1$, where f and g are suitably smooth functions, m is a positive integer.

Under certain conditions on f, an asymptotic composite solution is constructed, with boundary layer corrections at the endpoints. Kelley (1989)

36. $$\boldsymbol{y'' - [\psi(x)]^{-1}\phi(x,y) = 0, \quad 0 < x < 1,\ y(0) = a,\ y(1) = b,}$$

where $\phi : [0,1] \times R \to R$ and $\psi : [0,1] \to [0,\infty)$ are continuous with $\psi > 0$ on $(0,1]$.

For this singular BVP, existence results are proved in two parts. First, sufficient conditions on ϕ and ψ are given to imply a priori bounds on solutions of the given problem and their derivatives. Then the existence of a solution is obtained by fixed-point methods. Bobisud, O'Regan, and Royalty (1987a)

37. $$\boldsymbol{y'' - a_0 y^k - f(y,x) = 0,}$$

where a_0 is a constant, k is a natural number, and $f(y,x)$ is a polynomial in y and x, the exponent of y in the highest degree in the polynomial being $k_1 < k$. Asymptotic series are constructed, convergence conditions are found from them, and classes of equations of the given form are identified which have algebraic movable singularities.

Writing the given DE as a system,

$$\frac{dy}{dx} = z, \quad \frac{dz}{dx} = a_0 y^k + f(y,x), \tag{1}$$

asymptotic series are constructed by the method of iterations. It is then established that the solution of (1) possesses the property $y \to \infty, z \to \infty$ as $x \to x_0$ in the neighborhood of the singular point x_0, and in that neighborhood,

$$z = Ay^{(k+1)/2}\left[1 + o(y^{k_1-k})\right], \tag{2}$$

where A is determined by the exponent k and the coefficient a_0. The asymptotic series is established by successively substituting (2) in the two equations (1). The form obtained is

$$z = Ay^{(k+1)/2}\left[1 + a_1 y^{-\beta_1} + a_2 y^{-\beta_2} + \cdots + a_j y^{-\beta_j} + o(y^{-\beta})\right],$$

where a_j are uniquely determined in terms of the coefficients of the polynomial $f(y,x)$ and A, and

$$k_1 - k = -\beta_1 > -\beta_2 > -\beta_3 > \cdots > -\beta_j > -\beta.$$

Bogoslovskii (1966)

2.7 $y'' + ay' + g(x, y) = 0$

1. $$\mathbf{y'' + 0.15y' + \sin y = 0.}$$

The given DE describes a damped pendulum [see Jordan and Smith (1977), Fig. 3.26, p.97 for a phase diagram]. Jordan and Smith (1977), p.95

2. $$\mathbf{y'' + y' - y^p = 0, \quad 0 < p < 1,}$$

$$\begin{aligned} &\mathbf{y(0) = 0, y'(0) = p, \text{ and } y(x) \sim [(1-p)x]^{1/(1-p)}} \\ &\quad \mathbf{- p(1-p)^{(2p-1)/(1-p)} x^{p/(1-p)} \log x + O\left(x^{p/(1-p)}\right) \quad \text{as } x \to \infty.} \end{aligned}$$

It is shown that the given BVP has a unique solution. It is also shown that

$$\begin{aligned} \lim_{x\to 0} x^{-2/(1-p)} y(x) &= \left[\frac{1-p}{2(1+p)^{1/2}}\right]^{2/(1-p)}, \\ \lim_{x\to\infty} [(1-p)x]^{-1/(1-p)} y(x) &= 1, \end{aligned}$$

or more generally,

$$y^{1-p}(x) = (1-p)x - p\log x + C + O\left((1/x)\log x\right) \quad \text{as } x \to \infty.$$

Grundy and Peletier (1990)

3. $$\mathbf{y'' + y' - 3y^{2/3} - 6y^{1/3} = 0.}$$

The given DE for arbitrary a and $b\,(a \le b)$ has the solution of the form

$$y(x) = \begin{cases} (x-a)^3, & x \le a, \\ 0, & a \le x \le b, \\ (x-b)^3, & x \ge b. \end{cases}$$

Thus, the IVP has nonunique solutions with an IC on $y = 0$. Benson (1973)

4. $$\mathbf{y'' + y' - y^{n-1} + Ay = 0,}$$

where A is a constant.

(a) For $A = 2/9$ and $n = 4$, a kink-like solution is

$$y = \mp A^{1/2}/\{1 + \exp[\pm A^{1/2}/(x - x_0)]\}.$$

(b) For $A = 3/16$ and $n = 6$, we have the solution

$$y = -\left[\left(4/3^{1/2}\right)\{1 + \exp[\pm(1/2)(x - x_0)]\}\right]^{-1/2}.$$

Dixon, Tuszynski, and Otwinowski (1991)

5. $$y'' + 4y' + 100y + y^3 = 0, \quad y(0) = 10, \; y'(0) = 0.$$

Numerical and approximate perturbation solutions are obtained and compared graphically. The principle of harmonic balance is applied [see Eqn. (2.1.85)]. Ludeke and Wagner (1968)

6. $$y'' + 12y' + 100y + 0.05y^5 = 0, \quad y(0) = 10, \; y'(0) = 0.$$

Numerical and approximate perturbation solutions are obtained and compared graphically. The principle of harmonic balance is made use of [see Eqn. (2.1.85)]. Ludeke and Wagner (1968)

7. $$y'' + 12y' + 36y + y^5 = 0, \quad y(0) = 10, \; y'(0) = 0.$$

Numerical and approximate perturbation solutions are obtained and compared graphically. The principle of harmonic balance is made use of [see Eqn. (2.1.85)]. Ludeke and Wagner (1968)

8. $$y'' - y' - \beta y + \lambda e^y = 0, \quad 0 < x < 1, \; y(0) - y'(0) = 0, \; y'(1) = 0,$$

where β and λ are positive parameters.

Write the given problem as the system

$$y_1' = y_2, \quad y_2' = \beta y_1 + y_2 - \lambda e^{y_1}, \; y_1(0) - y_2(0) = 0, \; y_2(1) = 0. \tag{1}$$

In the phase plane (y_1, y_2), the solution of the given BVP is a trajectory of the system (1) which goes from the 45° line to the y_1-axis in unit time.

A complete phase plane analysis is carried out to show that the number of solutions of the given problem may be made arbitrarily large by appropriate choice of λ and β; the limits $\lambda \downarrow 0$ and $\beta \uparrow \infty$ are discussed in detail to illustrate this statement. Alexander (1984)

9. $$y'' - y' + \exp(-1/y) = 0, \quad y'(0) = y(0) - 1, \; y'(1) = 0.$$

The BVP describes axial mixing in tubular reaction. Writing $y - 1 = z$, we obtain a homogeneous BC:

$$z'' - z' + \exp[-1/(1+z)] = 0, \tag{1}$$
$$z'(0) = z(0), \quad z'(1) = 0. \tag{2}$$

The Green's function relevant to $z'' - z' = 0$ and (2) is

$$G(x,t) = \begin{cases} -\exp(x-t), & x \le t, \\ -1, & x \ge t. \end{cases}$$

The method of successive approximations yields

$$\begin{aligned} z_{k+1}(t) &= -\int_0^1 G(x,t)\exp[-1/\{1+z_k(t)\}]\,dt \\ &= \int_0^x \exp[-1/\{1+z_k(t)\}]\,dt + \exp(x)\int_x^1 \exp\left[-t - \frac{1}{1+z_k(t)}\right] dt. \end{aligned}$$

It is shown that approximations converge and the solution is unique in the region $z(x) > 0$. The results are shown graphically. Kubicek and Hlavacek (1983), p.107

10. $$y'' - 2y' + (1 - A^2)y - 2Fy^3 = 0,$$

where A and F are constants.

Solutions of the given DE are sought such that they also satisfy the Riccati equation,

$$y' = K_1 y^2 + K_2 y + K_3, \tag{1}$$

where $K_i\,(i = 1, 2, 3)$ are constants. Substituting (1) in the given DE and equating coefficients of y, y^2, y^3, etc., we get $K_1 = \mp\sqrt{F}, K_2 = 2/3$, and $K_3 = 0$, provided that $A^2 = 1/9$ and $F > 0$. Hence, with the help of solution of (1), the given DE has the solution

$$y = (2/3)\beta \exp\{(2/3)z\}/\{1 \pm \beta F^{1/2} \exp\{(2/3)z\},$$

where $z = \ln x$ and β is an arbitrary constant. Dixon, Kelley, and Tuszynski (1992)

11. $$y'' - 3y' - y^2 - 2y = 0.$$

Put $y'(x) = p(y)$, so that

$$p' = 3 + y(y + 2)/p.$$

The numerical solutions of this DE, for which $y \sim 2x^{-2}$ as $x \to \infty$, may be easily found. Kamke (1959), p.547

12. $$y'' - 7y' + 12y - y^{3/2} = 0.$$

The given DE is related to the Thomas–Fermi equation $\xi^{1/2}\eta'' = \eta^{3/2}$ by the transformation

$$\eta = \xi^{-3}y(x), x = \ln \xi.$$

Kamke (1959), p.547

13. $$y'' + (1 - c)y' + 2y^2 - Q = 0,$$

where c and Q are constants. The boundary conditions are $y(0) = y(2\pi/\alpha), y'(0) = y'(2\pi/\alpha), y''(0) = y''(2\pi/\alpha)$, and $\int_0^{2\pi/\alpha} y dx = 0$.

A solution of the given problem is sought in the form

$$y = \sum_{k=1}^{\infty} \left[a_k \cos(\alpha k x) + b_k \sin(\alpha k x)\right].$$

One gets $2N$ algebraic equations for $2N$ unknowns $c, a_2, \ldots, a_N, b_1, b_2, \ldots, b_N$ in a finite-dimensional case. A detailed study of this system leads to the understanding of the bifurcation phenomenon. The results are depicted on bifurcation diagrams. Attractors on a wide scale of wave numbers are studied and classified. Demekhin, Tokarev, and Shkadov (1991)

14. $$y'' - \epsilon y' + 4y(y - 1) = 0,$$

where ϵ is a parameter.

The given DE describes traveling wave solutions of the Korteweg–deVries equation, with ϵ equal to ratio of viscous to dispersive effects.

Putting $w = \dfrac{dy}{dx}$, we get

$$\frac{dw}{dy} = \{\epsilon w - 4\,y(y-1)\}/w. \tag{1}$$

Equation (1) is studied in the (w, y) phase plane. The point, $(0,0)$ is a saddle point, while $w = 0, y = 1$ is a stable node if $\epsilon \geq 4$ and a stable spiral point if $\epsilon < 4$. The solution is sketched for the case $4 > \epsilon \geq 0$ when $y = 1, w = 0$ is a spiral point; for $\epsilon = 0$, a solitary wave solution exists, and for $\epsilon \geq 4$, there is a monotonic solution. The case $\epsilon = \infty$, corresponding to a Taylor shock, is also drawn. Numerical solutions showing transitions from monotonic to oscillatory profiles for $0 \leq \epsilon \leq 4.62$ are shown graphically.

A first integral of the given DE is

$$(1/2)y'^2 = \epsilon \int_\infty^x \left(\frac{dy}{dx}\right)^2 dx - (4/3)y^3 + 2y^2, \tag{2}$$

where $y = y' = 0$ at $x = \infty$ so that the solution of (1) will be monotonic as $x \to \infty$. A straightforward expansion $y = y_0(x) + \epsilon y_1(x)$ gives

$$O(1) : \frac{1}{2}\left(\frac{dy_0}{dx}\right)^2 = 2y_0^2 - \frac{4}{3}y_0^3, \tag{3}$$

$$O(\epsilon) : \frac{dy_0}{dx}\,\frac{dy_1}{dx} = \int_\infty^x \left(\frac{dy_0}{dx}\right)^2 dx + 4y_0y_1(1-y_0). \tag{4}$$

The solution to (3) is the solitary wave

$$y_0 = (3/2)\ \mathrm{sech}^2 x, \tag{5}$$

where the peak is fixed at $x = 0$. Substituting (5) in (4) and integrating, we have

$$\begin{aligned} y_1 \;=\;& (2/5)\ \mathrm{sech}\, x\,(\tanh x - 1) + (1/20)\tanh x\,(\tanh x - 6) \\ & - (\tanh x)/\{10(1+\tanh x)\} + (3/4)x\tanh x\,\mathrm{sech}^2 x, \end{aligned} \tag{6}$$

where the condition $\left.\dfrac{dy_1}{dx}\right]_{x=0} = 0$ ensures that the peak occurs at $x = 0$ to $O(\epsilon^2)$. The expansion above becomes invalid as $\tanh x \to -1, x \to -\infty$, since (6) then approaches infinity.

The straightforward expansion above is matched to another solution expansion obtained by the method of multiple scales. A near "Taylor shock profile" is obtained by writing $x = \epsilon\tau$ so that the given DE becomes

$$\frac{1}{\epsilon^2}\frac{d^2y}{d\tau^2} + 4y(y-1) = \frac{dy}{d\tau}. \tag{7}$$

In the limit $\epsilon \to \infty$, we get the usual Taylor shock between the levels $y = 0$ and $y = 1$.

A straightforward asymptotic expansion for (7) $y = y_0(\tau) + (1/\epsilon^2)y_1(\tau) + \cdots$ gives

$$\begin{aligned} O(1) : 4y_0(y_0 - 1) &= \frac{dy_0}{d\tau}, \\ O(1/\epsilon^2) : \frac{d^2y_0}{d\tau^2} + 4y_1(2y_0 - 1) &= \frac{dy_1}{d\tau}. \end{aligned} \tag{8}$$

It is easy to solve system (8):

$$\begin{aligned} y_0 &= (1/2)[1+\tanh(-2\tau)], \\ y_1 &= \operatorname{sech}^2(-2\tau)\left[A - \ln\{\operatorname{sech}^2(-2\tau)\}\right]. \end{aligned}$$

Fixing the midpoint of the profile at $\tau = 0$ gives $A = 0$. Thus, the near-Taylor shock profile becomes

$$y = (1/2)[1+\tanh(-2\tau)] - (1/\epsilon^2)\operatorname{sech}^2(-2\tau)\ln[\operatorname{sech}^2(1-2\tau)] + \cdots.$$

Johnson (1970)

15. $$y'' + 5ay' - 6y^2 + 6a^2y = 0,$$

where a is a constant.

Put $y' = p(y), y'' = p(dp/dy)$, etc. and integrate twice:

$$y = a^2C_1^2e^{-2ax}P\,(C_1e^{-ax} + C_2, \quad 0, \quad -1)$$

in terms of the Weierstrass elliptic function. Here C_1 and C_2 are arbitrary constants. Kamke (1959), p.547

16. $$y'' + cy' - y^2 + y = 0,$$

where c is constant.

(a) The solution may be sought in the finite form

$$y(x) = A_0 + A_1p(x) + B_{-1}p'(x)/p(x), \tag{1}$$

$$p'^2(x) = b_0 + b_2p^2(x) + b_3p^3(x). \tag{2}$$

Substitution of (1)–(2) in the given DE and integration lead to the bounded solution

$$y(x) = 1/2 - (6/5)ck\tanh(kx - x_0) - (1/4)\operatorname{sech}(kx - x_0),$$

where $c^2 = 25/6, k^2 = 1/24$.

(b) A second finite (possible) form of the solution is

$$y(z) = b_3z^2p^3(z)/4, \tag{3}$$

where

$$[p'(z)]^2 = b_3p^3(z) + b_0, \tag{4}$$

where $z = e^{kx}$. Substituting (3)–(4) into the given DE, we find that if $b_3 = 4$, the solution is

$$y = e^{2kx}p(e^{kx}), \tag{5}$$

$$p''(z) = 6p^2(z), \quad k^2 = 1/6,\ c = -5k. \tag{6}$$

Equation (6) integrates to give $p(z) = P(z; 0, g_3), g_3$ being an integration constant. Here P is the Weierstrass elliptic function. Santos (1989)

17. $$\epsilon y'' - y' + y^2 - y = 0,$$

where ϵ is a parameter.

The given DE describes a tidal bore on a shallow stream. Phase plane study of the equivalent system,

$$\frac{dy}{dx} = w, \quad \epsilon\frac{dw}{dx} = y + w - y^2$$

shows that a separatrix from the saddle point at the origin reaches the other equilibrium point, $w = 0, y = 1$. Jordan and Smith (1977), p.97

18. $$y'' - \mathbf{Pe} \cdot y' - \mathbf{Da} \cdot \mathbf{Pe} \cdot y^2 = 0, \quad y(0) = 1 + (1/\mathbf{Pe})y'(0), \ y'(1) = 0,$$

where Pe and Da are constants.

The given problem arises in an isothermal tubular reactor and is solved by finite difference methods for $Pe = 5$, $Da = 1$. Kubicek and Hlavacek (1983), pp.96, 218

19. $$y'' - \epsilon(-y' + y^2) + y = 0,$$

where ϵ is a small positive parameter.

Write

$$y = r(x)\cos(x + \psi(x)), \quad y'(x) = -r(x)\sin(x + \psi(x)).$$

On substitution in the given DE and using the Lagrange method, we get

$$\begin{aligned} r'(x) &= -\epsilon r\sin^2(x + \psi) - \epsilon r^2 \sin(x + \psi) \cos^2(x + \psi), \\ \psi'(x) &= -\epsilon\cos(x + \psi)\sin(x + \psi) - \epsilon r\cos^3(x + \psi). \end{aligned} \tag{1}$$

The RHSs of (1) are 2π-periodic in x. We average x, keeping r and ψ fixed. The result is

$$r_a'(x) = -(1/2)\epsilon r_a, \quad \psi_a' = 0, \tag{2}$$

where a denotes that we have approximated r and ψ by r_a and ψ_a. The solution of (1) is $r_a(x) = r(0)e^{-\epsilon x/2}$, $\psi_a(x) = \psi(0)$. Thus, an approximate solution of $y(x)$ is $y_a(x) = r(0)e^{-\epsilon x/2}\cos(x + \psi(0))$. It is also proved that $y(x) - y_a(x) = O(\epsilon)$ on the x scale $1/\epsilon$. A phase picture of the given DE is drawn. Verhulst (1990), p.147

20. $$y'' - (\epsilon/\beta)y' - (\alpha/2\beta)y^2 + (\alpha/2\beta)y = 0,$$

where $\alpha, \beta, \epsilon > 0$ are constants.

The function $f(y) = (\alpha/2\beta)(y - y^2)$ satisfies all the requirements of the general equation (2.7.80). It has a bounded monotonic global solution. Kitada and Umehara (1988)

21. $$y'' + (b/c)y' + (a/c)y^2 + (1/c)(2a\beta - \delta)y = 0,$$

where b, c, a, β, and δ are constants.

A solution of the given DE may be found in the form $y = r[1 + \exp(\epsilon x)]^{-2}$, where $b = 5\epsilon c$ and $r = -6\epsilon^2 c/a$. Vlieg–Hultsman and Halford (1991)

22. $$y'' - (\nu/\beta)y' + (1/2\beta)y^2 - (c/\beta)y + q/\beta = 0,$$

where $\beta \neq 0, \nu, c$, and q are constants.

An exact (kink) solution of the given DE (arising from the Korteweg–deVries–Burgers equation) is

$$y = C - (12/5)\nu K \tanh(Kx + \phi_0) + 12\beta K^2 \operatorname{sech}^2(Kx + \phi_0),$$

where $K = \pm(1/10)\nu\beta$ and $c^2 = 2q + \left(\dfrac{36\nu^4}{625\beta^2}\right)$. The solution is obtained by relating the given DE to a certain Riccati equation and its solution. Kudryashov (1991)

23. $$y'' - hy' - by^2 - cy - d = 0, b \neq 0,$$

where $h, b \neq 0, c$, and d are arbitrary constants.

Put $y + (1/2)c/b = (6/b)u$; the equation for u is

$$u'' - 6u^2 - hu' - f = 0, \tag{1}$$

where $f = (1/24)(4bd - c^2)$. The solution for u with a moving pole is

$$u = 1/z^2 + (r/z) + c_0 + c_1 z + c_2 z^2 + c_3 z^3 + c_4 z^4 + \cdots,$$

where $z = x - c$ and $r = h/5$, $c_0 = -(1/12)r^2$, $c_1 = (1/12)r^3, c_2 = -(1/10)f - (7/48)r^4$, $c_3 = (11/30)fr + (79/144)r^5$, etc., and $10fr^2 + 15r^6 = 0$. Thus, either $h = 0$ in (1), leading to an equation that has a solution in terms of elliptic functions, or $f = -(3/2)r^4$, so that

$$u'' - 6u^2 - 5ru' + (3/2)r^4 = 0. \tag{2}$$

Equation (2) has solutions that are subuniform; i.e., they do not have moving algebraic singularities. Forsyth (1959), p.299

24. $$\epsilon^2 y'' + \epsilon y' + b(y + y^2) - af(x) = 0,$$

where ϵ is a small parameter, $a > 0, b > 0, f(x) = \text{sign } \sin(\pi x/w), w > 0$.

The substitution $\dfrac{dy}{dx} = z/\epsilon$ leads to the system

$$\frac{dy}{dx} = z/\epsilon, \quad \frac{dz}{dx} = (1/\epsilon)\left[-z - b(y + y^2) + af(x)\right].$$

Further introducing $\theta = x/\epsilon$ and recalling the definition of f, we arrive at the system

$$\frac{dy}{d\theta} = z, \quad \frac{dz}{d\theta} = -z - b(y + y^2) + a, \tag{1}$$

for $\theta \in [2kw/\epsilon, (2k + 1)w/\epsilon]$, where k is an integer, and at

$$\frac{dy}{d\theta} = z, \quad \frac{dz}{d\theta} = -z - b(y + y^2) - a \tag{2}$$

for $\theta \in [(2k - 1)w/\epsilon, 2kw/\epsilon]$.

The behavior of trajectories of (1)–(2) in the phase plane is studied. For small $\epsilon, 2w$-periodic solutions are found. Osipov and Pliss (1989)

25. $$y'' + \mu y' - y + y^2 - (y-1)Q \cos(wx) = 0,$$

where $\mu > 0, Q$, and w are constants.

The given DE is a dissipative Duffing equation when $Q = 0$, and has an unstable equilibrium point at $(0, 0)$ and a stable one at $(1, 0)$ in the (y, y') plane. For small $Q \neq 0$, these turn into fixed points of the Poincaré map of the system on its surface of section

$$\sum^{x_0} = \{y(x_k), y'(x_k) : \ x_k = (2\pi)/w + x_0\},$$

through which pass periodic orbits of period $2\pi/w$ in the extended three-dimensional phase space.

At both $\mu = 0$ and $Q = 0$, the hyperbolic fixed point at $(0, 0)$ has stable and unstable manifolds, W^s and W^u, respectively, which join smoothly, "embracing" the elliptic fixed point at (1, 0). For $\mu > 0$, these manifolds split, and at sufficiently high Q, they intersect each other infinitely often, thus forming a very complicated boundary for the basin of attraction around the stable fixed point on $\sum^{x_0}$. Write the given DE in the equivalent form

$$\begin{aligned} \dot{x}_1 &= x_2, \quad \dot{x}_2 = -\mu x_2 - x_1 - x_1^2 + x_1 x_4, \\ \dot{x}_3 &= w x_4, \quad \dot{x}_4 = -w x_3, \quad x_3(0) = 0, \ x_4(0) = Q. \end{aligned} \tag{1}$$

The system (1) is written as

$$\dot{x} = Ax + Mx, \tag{2}$$

where the matrix A of linear part of (2) has eigenvalues

$$\lambda_1 = -\mu/2 + (i/2)(4-\mu^2)^{1/2}, \ \lambda_2 = -\mu/2 - (i/2)(4-\mu^2)^{1/2}, \ \lambda_3 = iw, \ \lambda_4 = -iw.$$

The system (2) is transformed into a "normal" form. It is then shown exactly how the region of convergence of normal forms about a nodal fixed point is limited by the presence of singularities of solutions in the complex t-plane. This helps to obtain useful estimates of the basins of attraction of a stable fixed point of the Poincaré map, whose boundary is formed by the intersecting invariant manifolds of a second hyperbolic fixed point nearby. Bountis, Tsarouhas, and Herman (1988)

26. $$y'' + \lambda y' + (1/2)y^3 = \epsilon F(x),$$

where $F(x)$ is an analytic function.

The given system has a stable limit cycle. However, as ϵ is increased, this cycle can undergo a universal period doubling bifurcation leading to a strange attractor.

To study its singularity structure, a Laurent series of the form $y(x) = \sum_{j=0}^{\infty} a_j (x-x_0)^{j-1}$ fails because at $j = 4$, it is no longer possible to introduce an arbitrary coefficient. This problem can be handled by writing the psi series

$$y(x) = \sum_{j=0}^{\infty} \sum_{k=0}^{\infty} a_{jk}(x-x_0)^{j-1} \left((x-x_0)^4 \ln(x-x_0)\right)^k. \tag{1}$$

A tedious but straightforward computation shows that

$$\begin{aligned} &a_{jk}\left[(j-1)(j-2)+4k(2j+4k-3)\right]+a_{j-4,k+1}(k+1)(2j+8k-3) \\ &\quad + a_{j-8,k+2}(k+1)(k+2)+\lambda a_{j-1,k}(j+4k-2)+a_{j-5,k+1}(k+1) \\ &\quad = -(1/2)\sum_{p,q,r,s} a_{j-r,k-s}a_{r-p,s-q}a_{pq}+\epsilon F_{j-3}\delta_{k0}, \end{aligned}$$

where the summation is for $0 \le p \le r \le j$ and $0 \le q \le s \le k$ and where

$$F_j = \frac{1}{j!}\frac{\partial^j F}{\partial x^j}\bigg]_{x=x_0}.$$

The values of the first few coefficients are easily found to be

$$a_{00} = 2i,\ a_{10} = -i\lambda/3,\ a_{20} = -i\lambda^2/18!,\ a_{30} = -i\lambda^3/27 - \epsilon F_0/4.$$

There are "resonances" for a_{40} and a_{01}. The compatibility condition at a_{40} determines a_{01}, namely

$$a_{01} = 4i\lambda^4/135 + (\epsilon/5)\left(\lambda F_0 + F_1\right).$$

The value of the coefficient a_{01} plays a significant role in all of the subsequent analysis. x_0 and a_{40} constitute the two arbitrary data entering into the formally self-consistent expansion above. The set for which a closed set of recursion relations can be obtained are a_{0k} $(k = 0, 1, 2, \ldots)$. For these, we have

$$4k(k-1)a_{0k} + ka_{0k} + (1/2)a_{0k} = -(1/8)\sum_s\sum_q a_{0,k-s}a_{0,s-q}a_{0k},$$

where the summation is for $0 \le q \le s \le h$. Introducing the generating function

$$\theta(z) = \sum_{k=0}^{\infty} a_{0k}z^k,$$

where z is some (as yet unspecified) independent variable, the recursion relations lead to the following DE for θ:

$$16z^2\theta''(z) + 4z\theta'(z) + 2\theta(z) + (1/2)\theta^3(z) = 0, \tag{2}$$

where $' \equiv \dfrac{d}{dz}$. The DE (2) may also be obtained directly as follows. In the limit $x \to x_0$, we concentrate on the terms in psi series (1) involving powers of $(x-x_0)^4 \ln(x-x_0)$ and, therefore, make the substitution $y(x) = \dfrac{1}{x-x_0}\theta_0(z)$, where $z = (x-x_0)^4 \ln(x-x_0)$ into the DE. In the limit $x \to x_0$, it is easy to show that $\theta(z)$ again satisfies (2) provided that there is an ordering in which $|x-x_0| \ll z$.

The approach above can be thought of as a type of renormalization in that the DE (2) can be regarded as the original equation of motion rescaled in the neighborhood of a given singularity. For properties of (2), see Eqn. (2.26.6). Fournier, Levine, and Tabor (1988)

27. $$\boldsymbol{y'' + ky' + y^3 = B(x)\cos\theta(x),\ k = 0.05,}$$

where $\dfrac{d\theta}{dx} = \nu(x)$ and $B(x)$ are specified functions.

The given DE generalizes the Duffing oscillator and is treated numerically for:

(a) $\nu(x) = 1 + \alpha_\nu x, B(x) = 7.5 + \alpha_B x$.

(b) $\nu(x) = 1.05 + \gamma \sin \alpha_c x, B$ is a constant.

Different sets of constants were chosen, for which chaotic motion was observed and depicted. Moslehy and Evan-Iwanowski (1991)

28. $$\boldsymbol{y'' + \epsilon y' - y + y^3 = 0, \quad \epsilon = 0.05, 0.2.}$$

Solution trajectories are found numerically. For $\epsilon = 0.05$, the solution is shown to be very sensitive to initial conditions: the solution can go to rest either to $y = +1$ or -1 for a small difference in the initial conditions. Tongue (1986)

29. $$\boldsymbol{y'' + c_1 y' + c_2 y + y^3 = 0,}$$

where $c_1 \neq 0$ and c_2 are constants.

Put

$$y(x) = u(\xi(x))v(x) \tag{1}$$

and determine the functions ξ and v so that u satisfies

$$\frac{d^2u}{d\xi^2} + au^3 = 0, \tag{2}$$

where a is a constant. Equation (2) is integrable in terms of Jacobi elliptic functions. Substituting (1) into the given DE, we get

$$\ddot{u}\xi'^2 v + \dot{u}\left[\xi'' v + \xi'(2v' + c_1 v)\right] + u\left[v'' + c_1 v' + c_2 v\right] + u^3 v^3 = 0, \tag{3}$$

where $\dot{u} = \dfrac{du}{d\xi}$, $\xi' = \dfrac{d\xi}{dx}$, etc. Requiring that

$$v'' + c_1 v' + c_2 v = 0 \tag{4}$$

and

$$\xi'' v + \xi'(2v' + c_1 v) = 0, \tag{5}$$

we simplify (3) to

$$\ddot{u}\xi'^2 + u^3 v^2 = 0. \tag{6}$$

To obtain (2), we require further that

$$\xi'^2 = k^2 v^2, \tag{7}$$

where $k \neq 0$ is a constant, so that $k^2 = 1/a$. The solution to (4) is

$$v(x) = Ae^{r_1 x} + Be^{r_2 x}, \tag{8}$$

where A and B are constants and r_1 and r_2 are roots of $r^2 + c_1 r + c_2 = 0$. Inserting (7) into (5), we obtain for $v \neq 0$,

$$3v' + c_1 v = 0. \tag{9}$$

Inserting (8) into (9), we get

$$3r_1 + c_1 = 0, \quad 3r_2 + c_1 = 0. \tag{10}$$

Putting the values of roots r_1, r_2 into (10), we get the relation

$$2c_1^2 - 9c_2 = 0. \tag{11}$$

Under the general condition (11), we obtain the general solution of the given DE in terms of Jacobi elliptic functions.

A Painlevé analysis of the given DE also reveals that $2C_1^2 = 9C_2$ is the necessary condition for the given DE to be of Painlevé type. In fact, in this case an explicit time-dependent first integral can be found:

$$I(x, y(x), y'(x)) = \exp\{(4/3)c_1 x\}[(y' + c_1 y/3)^2 + (1/4)x^4] = \text{constant}.$$

Motion in the phase plane is studied. Euler, Steeb, and Cyrus (1989)

30. $$\boldsymbol{y'' + ky' + (a + 2q \cos 2x)y + y^3 = 0,}$$

where k, a, and q are constants.

The given DE is solved numerically. It is shown to exhibit successive subharmonic bifurcations. Transition to chaos is shown. The Poincaré section of the attractor in the chaotic region for $a = 4, q = 4.185$ is shown. Ito (1979)

31. $$\boldsymbol{y'' + \epsilon y' + y(y^2 - y - 2) = 0, \quad y(0) = 2.9,\ y'(0) = 0.1,}$$

where ϵ is a parameter.

The given DE represents oscillations in a quartic potential well in which local minima (stable equilibria) occur at $y = -1, 2$, and a saddle point occurs at $y = 0$. The initial conditions are chosen such that the initial energy is relatively large as compared with ϵ. Numerical experiments are performed with $\epsilon = 0.005, 0.001, 0.0007$ and the results compared with various asymptotic methods reported in the relevant paper. Bourland and Haberman (1990)

32. $$\boldsymbol{y'' + \gamma y' + y - y^3 = 0,}$$

where γ is a constant.

The given DE has a special kink solution in the form

$$y = \mp(1/2)\left[1 - \tanh\{(1/2^{3/2})(x - x_0)\}\right],$$

provided that $\gamma = 3/2^{1/2}$. Dixon, Tuszynski, and Otwinowski (1991)

33. $$\boldsymbol{y'' + \alpha\gamma v y' - y^3 - \lambda y^2 + y = 0, \quad \alpha\gamma = 1,}$$

where $\alpha > 0, \gamma, v$, and λ are constants.

The given DE has kink or antikink solutions

$$y = a + h/\left[1 + \exp\{f(x - x_0)\}\right] \quad \text{for } a = y_-, h = y_+ - y_- = (\lambda^2 + 4)^{1/2},$$

$$f = \pm(2 + (1/2)\lambda^2)^{1/2}, \ v = \mp\lambda/2^{1/2}.$$

It has two kink solutions or their antikinks for $a = 0, h = y_-$ or $h = y_+, 2f^2 = 1 - \lambda h > 0, \alpha\gamma v = (1 + f^2)/f$. In the above, $y_\pm = -(1/2)\lambda \pm (1 + (1/4)\lambda^2)^{1/2}$. Geicke (1988)

34. $$y'' - \theta y' - y(y-a)(y-1) - K = 0,$$

where $\theta \geq 0, a,$ and K are constants. By writing the equivalent system

$$\begin{aligned} y' &= w, \\ w' &= \theta w + (y - V)(y - W)(y - U), \quad V < W < U, \end{aligned}$$

a complete phase plane study is carried out. The singular points are $(V, 0), (W, 0), (U, 0)$. For $(U + V)/2 > W$, the following statements about the solution are proved:

(a) For $\theta = 0$, there are no solutions going from one singular point to another other than that which goes from $(V, 0)$ to $(V, 0)$.

(b) For $\theta > 0$, the only bounded solutions go from one singular point to another.

(c) For $\theta > 0$, there is a solution going from $(W, 0)$ to $(V, 0)$.

(d) For $\theta > \theta_c$, there is a solution going from $(W, 0)$ to $(V, 0)$; for $0 \leq \theta \leq \theta_c$, there is no solution going from $(W, 0)$ to $(V, 0)$. Here, $\theta_c = \lambda[U + V - 2W], \lambda = \pm 1/2^{1/2}$.

(e) For $\theta > 0$, there is no solution going from a singular point back to itself.

Casten, Cohen, and Lagerstrom (1975)

35. $$y'' + 3ay' - 2y^3 + 2a^2 y = 0,$$

where a is a constant.

Put $y' = p(y), y'' = pp'$, etc. and integrate twice:

$$y = -iaC_1 e^{-ax} \operatorname{sn}\left(k^2 = -1; C_1 e^{-ax} + C_2\right),$$

where C_1 and C_2 are arbitrary constants. Kamke (1959), p.548

36. $$y'' - 3gy' + 2g^2 y - 2y^3 = 0,$$

where g is an arbitrary constant.

Put

$$F = (y' - gy)^2 - y^4, \tag{1}$$

so that $F' = 4gF$; hence $F = e^{4gx}$. Equation (1) now is one of first order. Forsyth (1959), p.303

37. $$y'' + 2\epsilon y' + y + \alpha y^3 = 0, \quad y(0) = 1, \ y'(0) = 0,$$

where $\epsilon = 0.1, 0.2$ and $\alpha = 1, 4$, and 100.

Numerical and approximate analytic results, obtained by using certain transformations, are compared. Burton (1983)

38. $$y'' + 2\epsilon\mu y' + y + \epsilon y^3 = 0,$$

where $\epsilon \ll 1$ and μ are constants.

By using the method of multiple scales, a first order uniform solution is found to be

$$y = a_0 \exp(-\epsilon\mu x) \cos\{x - (3a_0^2/4\mu) \exp(-2\epsilon\mu x) + \beta_0\} + \cdots,$$

where a_0 and β_0 are arbitrary constants. Nayfeh (1985), pp.110-112

39. $$y'' + y - \alpha y^3 - \epsilon\delta y' - \epsilon f(\nu x) = 0,$$

where α, ν, δ are parameters, ϵ is a small parameter, and $f(\nu x)$ is a continuous periodic function of νx with period 2π. The given DE is equivalent to the system

$$y' = z, \quad z' = -y + \alpha y^3 + \epsilon[f(\nu x) + \delta z]. \tag{1}$$

A quite thorough qualitative study of the given Duffing equation is carried out both analytically and numerically.

If $\epsilon \neq 0$, the given system differs only slightly from a Hamiltonian system. In the study of this system the problems that arise relate to investigation of nonlinear resonances and the determination of stationary regions and to the establishment of the structure of the limit set in the neighborhood of the separatrix contours (the separatrix loop) of the unperturbed system ($\epsilon = 0$). Resonance energy levels are defined for the unperturbed system. The following results are proved:

(a) For arbitrary fixed $\delta \neq 0$, there is a sufficiently small constant $\epsilon_* > 0$ such that if $0 < \epsilon < \epsilon_*$, the number of δ-resonance levels of the equivalent system (1) in a region G is bounded.

(b) Corresponding to any arbitrary fixed nonzero friction coefficient δ, there is a sufficiently small positive constant $\bar{\epsilon} = \bar{\epsilon}(\delta)$ such that, for arbitrary $\epsilon \in (0, \bar{\epsilon})$, the given DE has a finite number of periodic solutions in G, and all periodic solutions except y_ϵ, which become the trivial solution for $\epsilon = 0$, are in the neighborhood of δ-resonance. If there are no δ-resonances, G contains only one periodic solution y_ϵ.

The results are supplemented and depicted through extensive numerical work. Morozov (1976)

40. $$y'' + \delta y' - \beta y + \alpha y^3 = 0,$$

where $\delta, \beta, \alpha > 0$ are constants.

The given DE is a Duffing equation with negative linear stiffness. It is changed to the system

$$\begin{aligned} \dot{x}_1 &= x_2, \\ \dot{x}_2 &= \beta x_1 - \delta x_2 - \alpha x_1^3, \end{aligned}$$

with three fixed points $(0,0)$ and $\{\pm(\beta/\alpha)^{1/2}, 0\}$, the latter two being sinks and the origin a saddle point. The system is structurally stable. For global stability, two Liapunov functions are constructed to show that all solutions starting inside the loop joining the separatrices of the saddle point approach one or the other of the sinks as $t \to +\infty$. A phase picture is drawn. Holmes (1979)

41. $$y'' + ky' + [1 + \delta \cos \Omega x)]y + \xi y^3 = 0,$$

where $k, \delta \geq 0, \Omega$, and ξ are constants.

Averaging and normal-form methods are used to obtain approximate solutions of the given DE. The results are compared with the numerical solutions. Lamarque and Stoffel (1992)

42. $$y'' + by' + y + (a - cy^2)y \cos 2x + ey^3 = 0,$$

where the parameters a, b, c, and e are assumed to be small.

For five different parametric regimes (e.g., $|a|/b > 2$ and $|c|/|e| > 3/2$), the possible number of periodic solutions, their stability, and some bounds on their amplitudes are determined. No analytic details are given. Furuya (1955)

43. $$y'' + 2\mu y' + y - \alpha y^3 + gy \cos \Omega x = 0,$$

where μ, α, g, and Ω are positive parameters.

The given DE is recast as

$$y'' + y = -\epsilon\left[2\mu y' - \alpha y^3 + gy\cos\Omega x\right].$$

Second order perturbation solutions are found and compared with digital computer solutions for $0 < \epsilon \leq 1$. Basins of attraction for different sets of parameters are found. Floquet theory is used for the prediction of some of the bifurcations. Sanchez and Nayfeh (1990)

44. $$y'' + 2ky' + \left\{(a + by^2) - 2q \cos 2x\right\} y = 0,$$

where k, a, b, and q are constants.

The given DE describes the vibration of a weighted string attached to an electromagnetically driven reed or tuning fork and subject to energy dissipation. It is studied under various simplifying assumptions; first $b = k = 0$; then $k = 0$ and b is small; finally, that k and b are small. Approximation methods are used and the results compared with experiments. McLachlan (1951)

45. $$Dy'' - y' + (y^2 - 1)(y + B) = 0, \quad y(\infty) = -1, \; y(-\infty) = 1, \; y(0) = 0,$$

where D and B are constants.

Linearizing the given DE about the values $y = -B, -1$, and 1 at the critical points and assuming exponential behavior for $x \to \pm\infty$ yield the decay rates as

$$\begin{aligned}
k_1 &= 1/(2D) \pm \left[1/(4D^2) - \{2(B+1)/D\}\right]^{1/2}, y - 1 \sim \exp(k_1 x), \\
k_{-1} &= 1/(2D) \pm \left[1/(4D^2) + \{2(B-1)/D\}\right]^{1/2}, y + 1 \sim \exp(k_{-1} x), \\
k_{-B} &= 1/(2D) \pm \left[1/(4D^2) - \{(B^2-1)/D\}\right]^{1/2}, y + B \sim \exp(k_{-B} x).
\end{aligned}$$

If $D \geq 0$, it follows that the solutions to the given problem exist for $B \geq 1$ only; then $y = -1$ is a saddle point while $y = 1$ is a saddle node if $B \leq B^*$ and an unstable spiral point if $B > B^*$, where $B^* = -1 + 1/(8D)$. Thus the solution approaches the state downstream of the "shock" monotonically if $B \leq B^*$ and in an oscillatory manner if $B > B^*$. It may also be noted that the solution for $x \to +\infty$ is algebraic rather than exponential in the limit $B = 1$:

$$y \sim -1 + 2/x - [(1+8D)/8](\ln x)/x^2 + C/x^2, \quad x \to \infty,$$

where C is an arbitrary constant. Typical results obtained by using shooting methods are depicted. Kluwick (1991)

46. $$y'' + \gamma y' - \left(Ay + By^2 + Cy^3\right) = 0,$$

where γ, A, B, and C are constants.

A particular solution of the given DE is

$$y = y_3 / \left[1 + \exp\{y_3 (C/2)^{1/2} x\}\right], \tag{1}$$

where y_1, y_2, and y_3 are the roots of the polynomial on the RHS of the given DE, namely,

$$y_1 = 0, \quad y_2, y_3 = -(1/2C)\left[B \mp (B^2 - 4AC)^{1/2}\right].$$

The solution (1) represents a kink and interpolates asymptotically between y_1 and y_3; it exists only if

$$\gamma = \frac{(C/2)y_3^2 - A}{(C/2)^{1/2} y_3}.$$

Dixon, Tuszynski, and Otwinowski (1991)

47. $$y'' + \gamma y' - a(y - y_1)(y - y_2)(y - y_3) = 0,$$

where γ, a, and $y_i\ (i = 1, 2, 3)$ are constants.

A special solution of the given DE is

$$y = y_1 + (y_3 - y_1)[1 + \exp(\mu x)]^{-1},$$

where $\mu = \pm(y_3 - y_1)(a/2)^{1/2}$; this solution exists only if

$$\gamma = \pm(a/2)^{1/2}(y_3 + y_1 - 2y_2).$$

Dixon, Tuszynski, and Otwinowski (1991)

48. $$y'' + cy' + \mu y + y^4 - y^5 = 0,$$

where c and μ are constants.

Existence of a trajectory joining $y = 0$ and $y = y_s$ such that $\mu y_s + y_s^4 - y_s^5 = 0$ is proved for some value of $c = \tilde{c}$, say. Numerical results are depicted. Powell and Tabor (1992)

49. $$y'' + 2\epsilon y' + y + \alpha y^5 = 0, \quad y(0) = 1, \ y'(0) = 0,$$

where $\epsilon = 0.2$ and $\alpha = 50$.

Numerical and approximate analytic results using the method of time transformation are compared. Burton (1983)

50. $$y'' + \gamma y' - (Ay - Cy^3 + Ey^5) = 0,$$

where γ, A, C, and E are constants.

The given DE has a special kink-like solution with the form

$$y = a/[1 + \exp(-(x/b))]^{1/2},$$

where the constants a and b may be determined by substitution, etc. Dixon, Tuszynski, and Otwinowski (1991)

51. $$\boldsymbol{y'' + vy' - P(y) = 0,}$$

where v is a constant and $P(y) = \sum_{i=0}^{5} a_{i+1} y^i$, A_{i+1} being constants.

Forms of $P(y)$ are sought for which a first integral in the form

$$y' = R(y), \quad R(y) = \sum_{k=1}^{n} a_k y^{k/2}, \ n \text{ a natural number},$$

may be found; the solution then is

$$x - x_0 = \int_{y_0}^{y} \frac{dy}{R(y)}.$$

Explicit results are obtained for the following special cases:

(a) $P(y) = \sum_{i=0}^{3} A_{i+1} y^i$.

(b) $P(y) = A_2 y + A_4 y^3 + A_6 y^5$.

(c) $P(y) = A_2 y + A_3 y^2$.

Otwinowski, Paul, and Laidlaw (1988)

52. $$\boldsymbol{y'' - PRy^n - Py' = 0, \quad n \geq 2, 0 \leq x \leq 1, \ y(0) = 1, \ y'(1) = 0,}$$

where P and R are positive constants.

The problem arises in an isothermal packed-bed chemical reactor. A constructive iterative method using a Green's function is employed to show that a unique solution exists satisfying $|y(x)| \leq e^{Px/2}, x \in [0, 1]$. Numerical results are provided. Pennline (1984)

53. $$\boldsymbol{y'' - \mathrm{Pe} \cdot y' - \mathrm{Pe} \cdot ky^n = 0,}$$

$(1/Pe)y'(0) = y(0) - 1, \ y'(1) = 0$, where Pe, n, and k are constants.

The given system describes axial mass dispersion in an isothermal tubular reactor for an nth order reaction.

Writing $y = A^{\alpha_1}\tilde{Y}, x = A^{\alpha_2}\tilde{x}$, the given system becomes

$$\tilde{Y}'' - Pe(\tilde{Y}' + \tilde{K}\tilde{y}^n) = 0, \tag{1}$$

where $\tilde{K} = kA^{\alpha_1(n-1)}, \alpha_2 = 0$ so that $x = \tilde{x}$ and $\tilde{Y}'(1) = 0$. Choosing the missing condition as $y(1) = A$, we get

$$\tilde{Y}(1) = A^{-\alpha_1} y(1) = A^{1-\alpha_1}.$$

For $\tilde{Y}(1)$ to be independent of A, we let $\alpha_1 = 1$, so that $\tilde{Y}(1) = 1$. Equation (1) is now integrated with $\tilde{Y}'(1) = 0$ and $\tilde{Y}(1) = 1$ from $x = 1$ to $x = 0$. The BC at $x = 0$ transforms to

$$\frac{A}{Pe} \frac{d\tilde{Y}}{dx}(0) - A\tilde{Y}(0) + 1 = 0,$$

giving

$$A = \left[\tilde{Y}(0) - \frac{1}{Pe} \frac{d\tilde{Y}}{dx}(0)\right]^{-1}. \tag{2}$$

Thus, the transformation is fully determined. Choosing a value of $\tilde{K}$, one integrates (1) with $\tilde{Y}(1) = 1, \tilde{Y}'(x) = 1$ from $x = 1$ to $x = 0$. The profile $y(x)$ is computed. Making use of $\tilde{Y}(0)$ and $\tilde{Y}'(0)$, the parameter A becomes known from (2). The constant k is then found from $k = \tilde{K}A^{1-n}$. The solution is now recovered from $y(x) = A\tilde{Y}(x)$.

The calculations above are performed for different values of $\tilde{K}$. The results for $n = 1/2, 2$ and $Pe = 5$ are tabulated. Kubicek and Hlavacek (1983), p.237

54. $$\mathbf{y'' + \lambda y' + \{1/(N-1)\}y^N = 0,}$$

where λ and $N > 1$ are constants.

The given DE has a leading order behavior $y(x) \sim \tau^\alpha, \alpha = 2/(1-N)$ about $\tau = x - x_0$. If one writes an expansion

$$y = \sum_{j=0}^{\infty} a_j \tau^{\alpha+j},$$

then it is easy to check that there is an incompatible resonance at $\beta = 2(1-\alpha)$ due to damping term. Making the transformation $y(x) = \lim_{\tau\to 0} \tau^\alpha \theta(z), z = \tau^\beta \ln \tau$ in the given DE, one gets the rescaled equation

$$\beta^2 z^2 \theta''(z) + \beta z \theta'(z) + \alpha(\alpha-1)\theta(z) + \{1/(N-1)\}\theta^N(z) = 0. \tag{1}$$

The transformation

$$\theta(z) = z^{-\alpha/\beta} f\left(z^{1/\beta}\right) = z^{1/(N+1)} f\left(z^{(N-1)/\{2(N+1)\}}\right)$$

yields the equation

$$f''(\xi) + \{1/(N-1)\}f^N(\xi) = 0, \quad \xi \equiv z^{1/\beta}. \tag{2}$$

Equation (2) has a first integral

$$f'^2 + \{2/(N^2-1)\}f^{N+1} = C, \tag{3}$$

which itself may be explicitly solved for certain values of N. C is an arbitrary constant. Levine and Tabor (1988)

55. $\mathbf{y'' + (k/x)y' + x^p y^n = 0, \quad y'(0) = 0, \ y(x_0) = y_0, \ y(x) \geq 0 \ \forall \, x \in [0, x_0],}$

where $k, p > -1, n > 1$, and $x_0, y_0 > 0$.

The domain of existence of the solution of the given problem is considered, with dependence on the parameters k, p, n, x_0, and y_0.

If we denote by $u(x)$ the solution of the Cauchy problem

$$y'' + (k/x)y' + x^p y^n = 0, \tag{1}$$

$y'(0) = 0$, $y(0) = 1$, we may check that

$$z(x) = \mathrm{su}\left(s^\nu x\right), \tag{2}$$

where $s \geq 0$ and $\nu = (2+p)^{-1}(n-1)$ satisfies the given DE and the conditions $z(0) = s$ and $z'(0) = 0$.

For fixed $x = x_0$, we consider $z(x_0)$ as a function of s. Let

$$y(s) = \mathrm{su}\left(s^\nu x_0\right). \tag{3}$$

Clearly, $y(0) = 0$ and $y'(0) = 1$. Let $\mu(x_0) = \max_{s\geq 0} y(s)$. Suppose that the maximum is attained for $s = s_0 < \infty$; then $s = s_0$ satisfies the equation

$$y'(s) = 0.$$

By virtue of (3) and with $\tau = s^\nu x_0$, we have

$$u(\tau) + \nu\tau u'(\tau) = 0. \tag{4}$$

Let $\tau = \tau_0$ be a solution of (4); then by using (3), we obtain

$$M(x_0) = (\tau_0/x_0)^{1/2} u(\tau_0) = Cx_0^{-(2+p)/(n-1)}, \tag{5}$$

where $C > 0$ is a constant. Hence if a solution of IVP (1) is known and a solution of (4) is unique, then (5) provides an upper bound for y_0 in the IC for which the given IVP has a solution.

Now, the following specific results are proved:

(a) If $\gamma \equiv (n-1)(k-1)/(2+p) \leq 2$, then for arbitrary $x_0 > 0$ there exists $M(x_0) > 0$ such that for $y_0 \in (0, M(x_0))$ the given IVP has two solutions, for $y_0 = M(x_0)$ it has one solution, and for $y_0 > M(x_0)$ it has no solution.

(b) If $\gamma \in \left(2, 2n + 2\{n(n-1)\}^{1/2}\right)$, then there is a sequence $M_j(x_0) > 0 (j = 1, 2, \ldots)$ such that $M_{2j-1}(x_0) > M_{2j}(x_0)$,

$$\begin{aligned} M_1(x_0) &> M_3(x_0) > \cdots > M_{2j+1}(x_0) > \cdots; \\ M_2(x_0) &< M_4(x_0) \cdots < M_{2j}(x_0) < \cdots .(M_j)_{\lim j\to\infty}(x_0) = \delta, \\ \delta &= \left[(\gamma-1)\{(2+p)/(n-1)\}^2\right]^{1/(n-1)} x_0^{-(2+p)/(n-1)}. \end{aligned}$$

For $y_0 = M_j(x_0)$ the given IVP has j solutions, for $y_0 \in [M_j(x_0), M_{j+2}(x_0)]$ it has $j+1$ solutions, and for $y_0 > M_1(x_0)$ it has no solutions.

(c) If $\gamma \geq 2n + 2[n(n-1)]^{1/2}$, then for $y_0 \in (0, \delta)$, where δ is as in (b), a solution of the given IVP is unique. For $y_0 \geq \delta$, the given IVP has no solution.

Grizan (1988)

56. $\qquad \boldsymbol{y'' + my' + f(x)y + \Phi(x)y^n - \theta(x) = 0,}$

where m and n are natural numbers.

To solve the given DE, we introduce the change of variables

$$x = \alpha(t), \quad y = \beta(t)z(t),$$

where

$$\frac{d\alpha}{dt} = \beta^2 e^{m(\alpha - t)},$$

and obtain the form

$$\ddot{z} + m\dot{z} + F(t)z + \Phi(t)z^n = \psi(t), \quad \dot{} \equiv \frac{d}{dt}, \tag{1}$$

provided that the following relations in $f(x), \Phi(x), \theta(x)$ and $F(t), \Phi(t), \psi(t)$ hold:

$$[\Phi(x)]^{-4/(n+3)}[f(x) + \{(u+2m)/(n+3)\}^2$$

$$- \{u' + m(u+2m)/(n+3)\}]e^{2mx(n-1)/(n+3)}$$

$$= [\Phi(t)]^{-4/(n+3)}[F(t) + \{(v+2m)/(n+3)\}^2$$

$$- \{(\dot{v} + m(v+2m))/(n+3)\}]e^{2mt(n-1)/(n+3)},$$

$$[\Phi(x)]^{-3/(n+3)}\theta(x)e^{2mnx/(n+3)} = [\Phi(t)]^{-3/(n+3)}\psi(t)e^{2mnt/(n+3)}, \tag{2}$$

$$u = \Phi'(x)/\Phi(x), \quad v = \dot{\Phi}(t)/\Phi(t),$$

and

$$\int \left(\Phi(x)e^{-m(n+1)x}\right)^{2/(n+3)} dx = \int \left(\Phi(t)e^{-m(n+1)t}\right)^{2/(n+3)} dt.$$

One may obtain a solution of the given DE,

$$y = [\Phi(t)/\phi(x)]^{1/(n+3)}e^{-2m(x-t)/(n+3)}z(t, C_1, C_2), \tag{3}$$

where C_1 and C_2 are arbitrary constants and z is the general integral of (1). Thus, to obtain the solution of the given form of DE, $F(t), \Phi(t)$, and $\psi(t)$ are chosen so that (1) can be solved in terms of quadratures. Then, substituting these functions into (2), we obtain two relations among the functions $f(x), \Phi(x)$, and $\theta(x)$. We may, therefore, choose one at will. It is convenient to choose $f(x)$ and then find $\Phi(x)$ and $\theta(x)$, since this choice enables us to restrict ourselves to elementary transformation. For further details, see Eqn. (2.9.36). Braude (1967)

57. $$\boldsymbol{y'' + \beta(1-\alpha)y' - \beta^2 e^{-2\beta x}\sum_{n=1}^{\infty} c_n[y]^{n-1} = 0, \quad y(\infty) = \lambda, \; y'(\infty) = 0,}$$

where $c_n =$ constant and $\beta > 0$.

The given problem has an analytic solution

$$y(x) = \lambda + \sum_{n=1}^{\infty} b_n(\lambda)e^{-\beta n x},$$

which together with its first two derivatives, converges absolutely for $0 \le x \le \infty$, provided that the problem

$$\frac{d^2y}{dz^2} + \frac{\alpha}{z}\frac{dy}{dz} = \sum_{n=1}^{\infty} c_n[y(z)]^{n-1}, c_n = \text{constant},$$

$y(0) = \lambda, \; y'(0) = 0$, has an analytic solution

$$y(z) = \lambda + \sum_{n=1}^{\infty} b_n(\lambda)z^n,$$

which together with its first two derivatives, converges absolutely for $|z| \le 1$. The result follows from the substitution $z = e^{-\beta x}$ and use of results for Eqn. (2.8.93). Ifantis (1987)

58. $$y'' + \lambda y' - (1/2)y^m + \mu y = 0,$$

where λ, μ, and m are parameters.

For $m = 2, 3$, it is easy to show that the given DE has the Painlevé property.

Using the similarity reduction $y(x) = A\{\phi_x^\beta/\phi^\beta\}f(\phi)$, $\beta = 2/(m-1)$, one finds that the function $f(\phi)/\phi^\beta \equiv F(\phi)$ satisfies the ODE

$$F'' - (1/2)A^{m-1}F^m = 0, \quad ' = \frac{d}{d\phi}, \tag{1}$$

provided that

$$-(2\beta^2 + \beta)\phi_{xx} - \lambda\beta\phi_x = 0, \tag{2}$$

$$\beta(\beta - 1)\left(\phi_{xx}/\phi_x\right)^2 + \beta\left(\phi_{xxx}/\phi_x\right) + \lambda\beta\left(\phi_{xx}/\phi_x\right) + \mu = 0. \tag{3}$$

The solution of (1) is expressible in terms of hyperelliptic integrals. Cariello and Tabor (1991)

59. $$y'' + \gamma y' - y(y^c - 1) = 0,$$

where γ and c are constants.

A kink-like solution of the given DE is

$$y = 2^{-2/c}\left[1 - \tanh[\frac{c}{2(2c+4)^{1/2}}(x - x_0)]\right]^{2/c},$$

provided that $\gamma = (c+4)/(2c+4)^{1/2}$. Dixon, Tuszynski, and Otwinowski (1991)

60. $$y'' + ay' + by^n + [(a^2 - 1)/4]y = 0,$$

where a, b, and n are constants.

With $y = \xi^\alpha\eta(\xi), \xi = e^x, \alpha = (1-a)/2$, the given DE goes to Emden's equation $\xi\eta'' + 2\eta' + b\xi^{\alpha n - \alpha - 1}\eta^n = 0$ [see Eqn. (2.8.40)]. Kamke (1959), p.548

61. $$y'' - [(3n+4)/n]y' - [2(n+1)(n+2)/n^2]y[y^{n/(n+1)} - 1] = 0.$$

The given DE is related to a generalized Thomas-Fermi DE. Put $u(y) = y'$ to obtain

$$uu' = [(3n+4)/n]u + [2(n+1)(n+2)/n^2]y[y^{n/(n+1)} - 1].$$

Let $u(y) = [(n+2)/n]y[1 - t^{n(n+1)}v(t)], y = t^{(n+1)(n+2)}$, so that

$$v'(t) = -2(n+1)^2\{t^{n(n+1)-1}v^2 - t^{n-1}\}/\{t^{n(n+1)}v - 1\}. \tag{1}$$

Kamke (1959), p.548

62. $$y'' + cy' + \mu y + p_{(n+1)/2}y^{(n+1)/2} - y^n = 0,$$

where $c, \mu, p_{(n+1)/2}$ are constants, and n is a positive integer.

A nongeneric solution of the given DE is

$$y = \left[\alpha\{\beta(\beta+1)\}^{1/2}e^{\alpha x}\right]^\beta / [1 + e^{\alpha x}]^\beta$$

$$= \quad u_s \left[e^{\alpha x}\right]^\beta / \left[1+e^{\alpha x}\right]^\beta,$$

where $\alpha\beta = \lambda$ and u_s satisfies

$$\mu u_s \quad + \quad p_{(n+1)/2} u_s^{(n+1)/2} - u_s^n = 0,$$

$$\lambda \quad = \quad \frac{1}{2}\left[\pm\{(\beta+1)^{1/2}/2\beta^{1/2}\}p_{(n+1)/2} \pm \{p_{(n+1)/2}^2(\beta+1)/(4\beta) + 2\mu/\beta\}^{1/2}\right],$$

and

$$c = \mp p_{(n+1)/2}\frac{(\beta+1)^{1/2}}{\beta}(\beta+1) \mp \frac{1}{2}\left(p_{(n+1)/2}^2 \cdot \frac{\beta+1}{4\beta} + \frac{2\mu}{\beta}\right)^{1/2}.$$

Powell and Tabor (1992)

63. $$\boldsymbol{y'' - \lambda y' - [(2\lambda - A)/4](Ay+K)\left[N^2(Ay+K)^{4(\lambda-A)/A} - 1\right] = 0,}$$

where $\lambda \neq 0, A, K,$ and N are constants.

The given DE has a first integral

$$y'^2 - (Ay+K)y' + \left[(Ay+K)^2/4\right]\left[1 - N^2(Ay+K)^{4(\lambda-A)/A}\right] = C\exp[(2\lambda - A)x],$$

where C is an arbitrary constant. Airault (1986)

64. $$\boldsymbol{y'' + 2\lambda y' + \sin y = 0,}$$

with the conditions

$$\lim_{y\to 2n\pi+} y' - \lim_{y\to 2n\pi-} y' = \alpha \ (n = 0, \pm1, \pm2, \ldots),$$

α and λ being constants.

The system describes a pendulum, capable of performing complete revolutions, which is subjected to an impulse each time it passes through its configuration of stable equilibrium.

The methods of phase plane analysis are used to obtain the following results:

(a) If $4\pi\lambda \geq \alpha + (\alpha^2 + 4\pi)^{1/2}$, there exists a unique stable oscillatory periodic motion.

(b) If $4\pi\lambda \geq \alpha \geq \pi\lambda + \pi(\lambda^2+1)^{1/2}$, there exists a periodic rotatory motion.

Aymerich (1955)

65. $$\boldsymbol{y'' + cy' + k \sin y = 0,}$$

where $c > 0$ and k are constants.

The given DE describes the motion of a pendulum with viscous damping. Putting $v = y'$, we have

$$\frac{dv}{dy} = \frac{-cv - k\sin y}{v},$$

or with $z = v/k^{1/2}, \lambda = c/k^{1/2}$, we have

$$\frac{dz}{dy} = \frac{-\lambda z - \sin y}{z}.$$

The problem posed is to determine the value of z for $y = 0$ such that y will just attain the value π. A series solution

$$z = a_1\xi + a_3\xi^3 + a_5\xi^5 + \cdots, \quad \xi = y - \pi$$

gives $a_1(a_1 + \lambda) = 1$, i.e., $a_1 = (-\lambda \pm (\lambda^2 + 4)^{1/2})/2$ and $a_3(\lambda + a_1) + 3a_3a_1 = -1/6$.

Thus, for each value of a_1, there is one value of a_3 and hence subsequent a_i. The point $p = 0, y = \pi$ is a saddle point, with two solutions entering it. The power series and numerical solution are shown to agree very well. Phase portrait is also drawn. Stoker (1950), p.62

66. $$y'' + \lambda y' + \sin y - \gamma = 0,$$

where $|\gamma| < 1$ and λ are constants.

The given DE arises in the description of pull-out torques of synchronous motors. The problem is treated in the phase plane for $\lambda = 0$; the case $\lambda \neq 0$ is handled numerically. Stoker (1950), p.72

67. $$y'' + ay' + b \sin y - c = 0,$$

where a, b, and c are constants.

The given DE occurs in connection with the synchronization of motors. Using the classical theory of Poincaré relating the behavior in the large to the behavior in the neighborhood of the singular points, asymptotic solutions are discussed. Amerio (1951)

68. $$y'' + ay' + b \sin y - p(x) = 0,$$

where a and b are constants and $p(x)$ is a continuous function on the half-line $x \geq 0$.

Conditions are found such that the solutions of the given DE satisfy either

$$\lim_{x \to \infty} y(x) = r\pi \quad \text{and} \quad \lim_{x \to \infty} y'(x) = 0$$

or

$$\lim_{x \to \infty} y(x) = (2r + 1)\pi \quad \text{and} \quad \lim_{x \to \infty} y'(x) = 0,$$

where r is a suitable integer. Andres (1992)

69. $$y'' + \delta y' + \alpha \sin y - g(ky - wx) = 0,$$

where g is a periodic function; δ, α, k, and w are constants.

The given DE describes nonlinear dynamics of a particle in a traveling electric force field. Moon (1987), p.79

70. $$y'' + ay' + (\gamma + \cos 2y) = 0,$$

where $\gamma < 1$, and a is a large constant.

Introducing $\epsilon \equiv a^{-2}, \tau = x/a, \dot{y} \equiv \dfrac{d\theta}{d\tau}$, the given DE is reduced to

$$\dot{\theta} = z, \quad \epsilon\dot{z} = -z - (\gamma + \cos 2\theta). \tag{1}$$

A multiple scales method is used to solve the system (1). Gang and Changqing (1985)

71. $$y'' + hy' + \mu(\sin x)\sin y = 0,$$

where h and μ are parameters.

This problem occurs in many physical situations, for example, synchronization of an unbalanced rotor in a vibrating machine or motion of a soliton in the field of a standing wave or the behavior of a Langmuir soliton in the background of a free ionic sound wave.

Periodic solutions such that $y(x+2\pi p) = y(x)+2\pi q$, with $p = 1, 2, \ldots; q = 0, \pm 1, \pm 2, \ldots$ or q/p a common fraction, are constructed, using Cesari's method. Writing $h = N_\ell/\mu_\ell$, where ℓ is a positive integer and N_ℓ a positive constant, the given DE becomes

$$y'' + N_\ell \mu^\ell y' + \mu \sin x \sin y = 0. \tag{1}$$

For $\mu = 0$, it has an infinite set of solutions

$$y(x) = xq/p + a, \quad |a| < 2\pi.$$

Putting $u = y - xq/p$ in (1), and applying Cesari's method, we have an auxiliary system

$$u'' = -N_\ell \mu^\ell (u' + q/p) - \mu \sin x \sin(qx/p + u) - v(a, \mu), \quad [1/(2\pi p)] \int_0^{2\pi p} u dt = a. \tag{2}$$

The system of equations (2) have solutions

$$u = u(x, a, \mu), v = v(a, \mu)$$

such that $u(x, a, 0) = a$ and $v(a, 0) = 0$, with the representations

$$\begin{aligned} u(x, a, \mu) &= a + \sum_{k=1}^{\infty} u_k(x, a)\mu^k, \\ v(a, \mu) &= \sum_{k=1}^{\infty} v_k(a)\mu^k, \end{aligned}$$

where $u_k(x, a)$ are $2\pi p$-periodic functions of x.

By a detailed analysis, domains of existence and stability of stationary and certain rotational periodic solutions of interest in application are constructed for a wide range of values of μ and h.

Stationary solutions are constructed for $0 < \mu \leq 20, 0 < h \leq 5$, while periodic solutions are treated for $0 < \mu \leq 5, 0 < h \leq 2.5$. Batalova and Bukhalova (1985)

72. $$y'' + \beta y' + (1 + A \cos \Omega x)\sin y = 0,$$

where β, A, and Ω are constants.

The given system is studied numerically. It is shown that for certain parameter intervals the pendulum described by the given DE behaves in an apparently chaotic way. This kind of motion is studied by means of the stroboscopic phase representation, the Lyapunov characteristic exponents, and the power spectrum. For fixed damping and frequency of the external modulation, the character of the motion changes as the amplitude of the driving force is varied. As an appropriate set of parameters, $\beta = 0.15$ and $\Omega = 1.56$ are chosen.

The numerical study shows that for sufficiently large values of the external perturbation, this parametrically excited mathematical pendulum with damping shows an apparently chaotic motion which has much in common with that of the Lorenz system and nonlinear oscillatory system with an additional external force. Levin and Koch (1981)

73. $$\boldsymbol{y'' + \alpha y' + \phi(x)\sin y - F(x) = 0,}$$

where $\alpha > 0$, $\phi(x) \geq a > 0$, and $F(x)$ is some smooth function.

Sufficient conditions for existence and uniqueness of a periodic solution are presented. Sun (1988)

74. $$\boldsymbol{y'' + ky' + \sin y + r\,\sin(2y) - T = 0,}$$

where k, r, and T are constants.

The given DE is related to the oscillation of the rotor of a synchronous motor with asynchronous starting.

The phase space natural to this problem is the surface of a cylinder. The existence and stability of singular points and limit cycles on the cylinder are studied. Belyustina (1955)

75. $$\boldsymbol{y'' + ay' + be^y - 2a = 0,}$$

where a and b are constants.

The given DE goes to the Emden's equation (2.8.81) through $\eta = y(x) - 2x, x = \ln\xi$, namely

$$\xi\eta'' + (a+1)\eta' + b\xi e^\eta = 0$$

[see Eqn. (2.8.81)]. Kamke (1959), p.548

76. $$\boldsymbol{y'' - \lambda y' - 2\lambda^2(N^2e^y - 1) = 0,}$$

where $\lambda \neq 0$ and N are constants.

The given DE has a first integral

$$y'^2 - 4\lambda y' + 4\lambda^2(1 - N^2e^y) = C\exp(2\lambda x),$$

where C is an arbitrary constant. Airault (1986)

77. $$\boldsymbol{y'' - Hy' - \beta Hy + BH\lambda\,\exp[y/(1+\gamma^{-1}y)] = 0,}$$

$0 < x < 1$, with $y' - Hy = 0$ at $x = 0$ and $y' = 0$ at $x = 1$, where H, β, λ, and γ are constants.

Writing $y_1 = y$ and $y_2 = y'$, we have the system

$$y_1' = y_2, \quad y_2' = H\left\{y_2 + \beta y_1 - \beta\lambda\,\exp[y_1/(1+\gamma^{-1}y_1)]\right\}, \tag{1}$$

and the boundary conditions

$$y_2(0) - Hy_1(0) = 0, \quad y_2(1) = 0.$$

It is shown that if $\beta/B\lambda > e$ and $\gamma^{-1}\beta$ is small, then the system (1) has three critical points $(y_1, y_2) = (\alpha_j, 0)$, $j = 0, 1, 2$:

$$\alpha_0 \;=\; B\lambda/\beta + O(\beta^{-2}),$$

$$\begin{aligned}\alpha_1 &= \log \beta/(B\lambda) + \log\log \beta/(B\lambda) + O(1), \\ \alpha_2 &= (B\lambda/\beta)e^{\gamma}\{1 + O(\gamma^2 e^{-\gamma})\}.\end{aligned}$$

The critical points $(\alpha_0, 0)$ and $(\alpha_2, 0)$ are always saddles, while $(\alpha_1, 0)$ is an unstable spiral point, provided that

$$\frac{1}{(1+\gamma^{-1}\alpha_1)^2} > 1 + \frac{H}{4\beta}. \tag{2}$$

The condition (2) holds when β^{-1} and γ^{-1} are small enough, for any fixed choices of H, B, and λ. Using IC, $y_1(0) = \eta, y_2(0) = H\eta, \eta$ is found such that $y_2(1;\eta) = 0$. It is shown that there are many such η. Alexander (1990)

78. $$\beta y'' - y' + pf(y) = 0,\ 0 \le x \le 1, \beta > 0,\ p > 0,\ \beta y'(0) - y(0) = 0,$$

$$y'(1) = 0 \text{ and } f(y) = (q - y)\exp[-k/(1+y)], k > 0, q > 0.$$

The given BVP arises in chemical reactor theory. The uniqueness of the solution for $k \le 4 + 4/q$ is shown for (a) sufficiently small p, (b) certain regions of p and sufficiently small β, (c) large p and sufficiently large β, and (d) fixed and sufficiently large p. Regions of points (β, p, q, k) are identified where there are at least three solutions. Numerical results are provided for $q = 1.1$ and $k = 10$. Related studies for other (k, q) values are summarized. Williams and Leggett (1982)

79. $$y'' + ay' + f(y) = 0,$$

where a is a large constant and $f(y)$ is a nonlinear function of y.

Multiple scale analysis is applied and an asymptotic solution for $a \to \infty$ is constructed; boundary and initial values are discussed. The general solution found is specified to the case $f(y) = \gamma + \cos(2y), \gamma < 1$. Gang and Changqing (1985)

80. $$y'' - 2\lambda y' + f(y) = 0, \lambda > 0, \quad -\infty < x < \infty,$$

where λ is a constant and $f(y)$ is a nonlinear function of y.

If the function f satisfies the conditions:

(a) $f(y) \in C^1([0,1])$,
(b) $f(0) = f(1) = 0$,
(c) $f'(1) < 0 < f'(0)$,
(d) $f(y) > 0, f'(0)y - f(y) \ge 0, y \in (0,1)$, and
(e) $\lambda \ge \{f'(0)\}^{1/2}$,

then there exists a global solution $y(x) \in C^3(-\infty, \infty)$ of the given DE such that $0 < y(x) < 1$ and $y'(x) > 0$ for any $x \in (-\infty, \infty)$. Kolmogoroff, Petrovskii, and Piskounoff (1937); Kitada and Umehara (1988)

81. $$y'' + (v/D)y' + hf(y) = 0, \quad 0 \le x \le b,$$

$f'(y) > 0$ on $-1 < y < y_m, f'(y) < 0$ on $y_m < y < +\infty, y' = -(v/D)y$ at $x = b$, and $y'(0) = 0$; v, D, and $h > 0$ are constants.

The given BVP arises in the dynamics of a single chemical reaction. See Eqn. (2.20.40) for a more general BVP, which contains this as a special case. Markus and Amundsun (1968)

82.
$$y'' + 2\epsilon y' + y + \alpha f(y) = 0,$$

where ϵ is small, α is unrestricted, y and $f(y)$ are $0(1)$, and $f(y)$ is an odd function of y.

Simple formulas that describe the history of the oscillation amplitude, with special attention to the influence of the nonlinearity, are derived. Burton (1983)

83.
$$y'' + ky' + g(x,y) - f(x) = 0, \quad x \in (0, 2\pi),\ y(0) = y(2\pi),\ y'(0) = y'(2\pi).$$

The growth conditions of the well-studied problem

$$y'' + g(x, u(x)) = f(x), \quad x \in (0, 2\pi),\ y(0) = y(2\pi),\ y'(0) = y'(2\pi)$$

are used to show that if the growth is restricted further depending on $K^2/4$, then the given system has at least one solution. Habets and Metzen (1989)

84.
$$\epsilon y'' + y' - g(x,y) = 0, \quad y'(0) - ay(0) = A,\ y'(1) + by(1) = B,$$

where a, b, A, and B are constants.

The existence and nonexistence of solutions of the given BVP are discussed. Iterative methods are used. Parter (1972)

85.
$$y'' + ky' + g(x,y) = 0,$$

where $g(x,y)$ and $\dfrac{\partial}{\partial y} g(x,y)$ are continuous for all $(x,y) \in R^2, g$ is periodic in x, and $k > 0$ is a constant.

The following is proved. If there exist constants a and b such that $a \le \dfrac{\partial}{\partial y} g(x,y) \le b$ holds such that there exists a closed disk B in a complex plane centered at $r = (a+b)/2$ of radius $r > (b-a)/2$ such that $\pi^2 m^2/T^2 - \pi imk/T \notin B$ holds for all $m = 0, \pm 1, \pm 2, \ldots$, then the given DE has a unique T-periodic solution which is locally, exponentially, and asymptotically stable. Lazer (1990)

2.8 $y'' + ky'/x + g(x,y) = 0$

1.
$$y'' + y'/x - \sigma\left(y - y^3/5\right) = 0,$$

where $\sigma = 1$ or $\sigma = -1$.

A solution in the phase plane is depicted using numerical integration. The solution $y = y(x)$ is also drawn. Tajiri (1983)

2.
$$y'' + y'/x - y + y^3 = 0, \quad y(0) = \alpha.$$

This nonlinear Bessel equation has the power series solution

$$y(x) = \alpha\Big[1+(1-\alpha^2)(x/2)^2+\{(1-\alpha^2)(1-3\alpha^2)/(2!)^2\}(x/2)^4 + \{(1-\alpha^2)(1-18\alpha^2+21\alpha^4)/(3!)^2\}(x/2)^6+\cdots\Big],$$

with the condition $y(0)=\alpha, \alpha$ being real. It is proved that this series converges absolutely for $|x| \le 1$ and sufficiently small $|\alpha|$. The contraction principle, however, shows that for $|\alpha| < 1/2$, the series solution is unique and satisfies the relation $|y(x)| < 1$ for $|x| \le 1$. Ifantis (1987)

3. $$y''+y'/x+[(a/2)x^2-\lambda]y+y^3=0,$$

where a and λ are constants.

It is easily seen that the asymptotic form of the solution of the given DE for large x are

$$y \simeq J_0\left\{\{a^{1/2}/[2(2^{1/2})]\}x^2\right\} \text{ for } a>0, \tag{1}$$

$$y \simeq K_0\left\{x\lambda^{1/2}\right\} \text{ for } a=0, \tag{2}$$

$$y \simeq K_0\left\{\{a^{1/2}/[2(2^{1/2})]\}x^2\right\} \text{ for } a<0. \tag{3}$$

Following the properties of Bessel functions, it is seen that the solutions corresponding to (1) are oscillatory with amplitude decaying, while those corresponding to (3) decay monotonically like $x^{-1}\exp(-ax^2/2)$. It is shown that for $a \ne 0$ the solutions corresponding to (3) are singular at $x=0$.

For $a=0$, the factor λ can be scaled out by $x \to x/\lambda^{1/2}$ and $y \to \lambda^{1/2}y$ so that the given DE becomes $y''+y'/x+y^3-y=0$. Numerical solutions of this latter equation are discussed. Rypdal, Rasmussen, and Thomsen (1985)

4. $$y''+y'/x-\nu^2y/x^2-y^3+y=0, \quad \nu>0, \ y(0)=0,$$

$$y(+\infty)=1, \ 0<y(x)<1 \text{ for } 0<x<+\infty.$$

It is proved that the given problem, called the Abrikosov problem, has at least one solution. Nagumo-Hukuhara theory for BVP for second order nonlinear ODEs is used. Iwano (1977)

5. $$y''+(1/x)y'-A^2y/x^2-2Fy^3=0,$$

where A and F are constants.

Put $y=V(\ln x)/x$ to obtain

$$V''-2V'+(1-A^2)V-2FV^3=0, \ ' \equiv \frac{d}{d(\ln x)}.$$

Now see Eqn. (2.7.10). Dixon, Kelley, and Tuszynski (1992)

6. $$y'' + y'/x + \left(1 - \Gamma^2/x^2\right) y + y - y^3 = 0; \quad y(0) = 0, \; y(\infty) = 1,$$

where Γ is a constant.

The given problem arises in the description of vortex lines in the Ginzberg–Landau model. A numerical solution was carried out for $\Gamma = 1$. As a check of the accuracy of the numerical solution, it was compared with detailed expansion of y at ∞ for $\Gamma = 1$:

$$y \sim 1 - (1/2)x^{-2} + O(x^{-4}) \quad (x \to \infty).$$

Pismen and Rubinstein (1991)

7. $$y'' + y'/x - y\left(1 + y^2\right)/x^2 = 0.$$

The given DE has an exact particular solution

$$y = 8b_0 x / \left(b_0^2 x^2 - 8\right),$$

where b_0 is an arbitrary constant.

For a general solution, one may resort to the method of continued fractions. This method gives a good approximation in a series form, which may sometimes even be summed up to yield the exact solution. Ditto and Pickett (1988)

8. $$y'' + y'/x - \alpha - \beta y^{-2} = 0, \quad y(1) = 1, \; y'(0) = 0,$$

where α and β are constants.

The given problem arises in electrohydrodynamics. First consider the case $\alpha = 0$. Putting

$$u = (x/y)y', \quad v = \left[\beta x/(2y^2)\right](y')^{-1}, \tag{1}$$

we obtain the equivalent system

$$\frac{du}{dx} = (u/x)(2v - u), \quad \frac{dv}{dx} = 2(v/x)(1 - u - v) \tag{2}$$

or in the (u, v) plane,

$$\frac{du}{dv} = \{u(2v - u)\}/\{2v(1 - u - v)\}. \tag{A}$$

The DE (A) belongs to the class studied by Jones (1953) [see Sachdev (1991)]. A local series solution of the given DE about $x = 0$ gives

$$y(x) = y_0 + \left\{\beta/(4y_0^2)\right\} x^2 - \left\{\beta^2/(32y_0^5)\right\} x^4 + \cdots, \tag{3}$$

where $y_0 = y(0)$. Using (3) in (1), we find that the BC at $x = 0$ becomes $u = 0$ when $v = 1$. At $x = 1$, the BC requires that

$$uv = (1/2)\beta. \tag{4}$$

Thus, for a given β, the integral curve $u(v)$ must terminate in the hyperbola (4). The condition $u = 0, v = 1$ leads to an indeterminate form for (A). So we put $v = 1 + v_1$ and linearize:

$$\frac{du}{dv_1} = -u/(u + v_1). \tag{5}$$

Using the IC $u = 0, v_1 = 0$, we get, on integration, $u(u/2 + v_1) = 0$. We choose

$$u = -2v_1 = 2(1 - v), \quad \text{so that } \frac{du}{dv} = -2 \text{ at } v = 1. \tag{6}$$

It is seen that the condition $u = 0, v = 1$, and (6) are independent of $y(0)$, implying that all solutions that are regular near $x = 0$ arise from a single integral curve in the (u, v) plane. By using any point (u_0, v_0) along the integral curve as the endpoint $y = 1 = x$, a solution of the given BVP corresponding to $\beta = 2u_0v_0$ is given parametrically by

$$\begin{aligned} y(v) &= \exp\left\{(1/2)\int_{v_0}^{v} \frac{u\,dv}{v(1-u-v)}\right\}, \\ x(v) &= \exp\left\{(1/2)\int_{v_0}^{v} \frac{dv}{v(1-u-v)}\right\}. \end{aligned} \tag{7}$$

The singularities in the integral in (7) are discussed.

Equation (A) is discussed in detail with respect to its singularities. Numerical results are presented. A matched asymptotic solution is also derived.

The dependence of solution on β is as follows:

(a) For $\beta < \beta_1 = 0.4152$, there is one and only one solution. This solution has a positive curvature $\dfrac{d^2y}{dx^2} \geq 0$ for $0 \leq x \leq 1$.

(b) For $\beta_1 < \beta < \beta_m = 0.7892$, there are multiple solutions for each value of β. The solutions for which $y(0) < 0.5556$ have nearly all negative curvature at $x = 1$.

(c) As $\beta \to 4/9$, the number of solutions for each β increases indefinitely. Thus, in any neighborhood of $\beta = 4/9, y(0) = 0$, there are an infinite number of solutions.

(d) If $0.7533 = \beta_2 < \beta < \beta_m$, two solutions are possible and both have positive curvature for $0 \leq x \leq 1$.

Ackerberg (1969)

9. $$\boldsymbol{y'' + y'/x + (k^2 - n^2/x^2)y - x^{-2}y^{-3} = 0,}$$

n being a nonnegative integer and $k \neq 0$.

The general solution of the given DE is

$$y^2 = aJ_n^2 + 2bJ_nY_n + cY_n^2, ac - b^2 = \beta\pi^2/4,$$

a, b, c being constants, two of them being arbitrary. $J_n(kx)$ and $Y_n(kx)$ are solutions of the Bessel equation. Ranganathan (1988)

10. $$\boldsymbol{y'' + y'/x - (K^2/x^2)(1/y^3) - 2Fy^3 = 0,}$$

where K and F are constants.

For $K = 0, y = (1/2F)^{1/2}x^{-1}$ is a solution. More general solutions may be found by writing

$$y = G(\xi)/x, \xi = \ln x,$$

so that G satisfies

$$G'' - 2G' + G - 2FG^3 = 0. \tag{1}$$

For (1), see Eqn. (2.7.46). Dixon, Kelley, and Tuszynski (1992)

11. $\boldsymbol{y'' + (1/x)y' + \lambda e^y = 0, 0 < x < 1,}$

$y'(0, \lambda) = 0, \alpha y'(1, \lambda) + y(1, \lambda) = 0$, where α and λ are parameters.

For $\alpha = 0$, the given BVP has the exact solution

$$y(x, \lambda) = \log\{(1 + \gamma)/(x^2 + \gamma)\}^2, \gamma = -1 + (4/\lambda)\{1 \pm (1 - \lambda/2)^{1/2}\}.$$

For $0 < \lambda < 2$, the given BVP has exactly two solutions:

(a) Clearly, for $\lambda = \lambda_0 = 2, y(x, 2) = \log\{2/(x^2 + 1)\}^2$, defines the limit point.

(b) For $\alpha \neq 0$, the solution of the given BVP is

$$y(x, \lambda) = \log\{(1 + \gamma)/(x^2 + \gamma)\}^2 + 4\alpha/(1 + \gamma),$$

where $\gamma = \gamma(\lambda)$ is a solution of the transcendental equation

$$\lambda = \{8\gamma/(\gamma + 1)^2\}e^{-4\alpha/(\gamma+1)}. \tag{1}$$

It is easily seen that (1) has exactly two solutions for $0 < \lambda < \lambda_0$, which coincide at the limit point $\lambda = \lambda_0$ given by (1) with

$$\gamma = 2\alpha + \{1 + 4\alpha^2\}^{1/2}.$$

Lange and Weinitschke (1991)

12. $\boldsymbol{y'' + (1/x)y' + \lambda e^y = 0, \epsilon < x < 1,}$

$y(\epsilon, \lambda; \epsilon) = \bar{y}, \alpha y'(1, \lambda; \epsilon) + y(1, \lambda; \epsilon) = 0$, where $\lambda, \bar{y}$, and ϵ are constants.

Putting $y(x) = -2 \log w(x)$, we have

$$ww'' - w'^2 + (1/x)ww' - \lambda/2 = 0. \tag{1}$$

It is easily verified that (1) has a solution of the form

$$w = Ax^a + Bx^b, \tag{2}$$

where $a = -\beta, b = 2 + \beta, \lambda = 8(1 + \beta)^2 AB$. Using the boundary conditions, we get the following equations for A and B:

$$A = -B\epsilon^{2+2\beta} + \epsilon^\beta e^{-\bar{y}/2}, \quad 2\alpha[\beta A - (2 + \beta)B] + 2(A + B)\log(A + B) = 0. \tag{3}$$

The relations (3) and $\lambda = 8(1 + \beta)^2 AB$ determine β, A, and B. For $\alpha = 0$, it is possible to find an explicit solution

$$\begin{aligned} y(x) &= -2\log\left[Ax^{-\beta} + (1 - A)x^{2+\beta}\right], \\ A &= \epsilon^\beta\left\{\left(e^{-\bar{y}/2} - \epsilon^{2+\beta}\right)/(1 - \epsilon^{2+2\beta})\right\}, \lambda = 8(1 + \beta)^2 A(1 - A), \end{aligned}$$

provided that $A + B = 1$. Lange and Weinitschke (1991)

13. $$y'' + y'/x - e^y = 0.$$

Introducing the variables $u = x^2e^y, y = xy'$, we get the equation

$$\frac{dv}{du} = \frac{1}{v+2}$$

with the integral $v^2 + 4v = 2u + C$. Choosing the constant C equal to zero for regular solution at $x = 0, y = 0$, we find that $v = (4 + 2u)^{1/2} - 2$. Here we choose a positive sign so that $v = 0$ when $u = 0$. In terms of x and y, we get

$$xy' + 2 - (4 + 2x^2e^y)^{1/2} = 0.$$

A lengthy calculation shows that the solution is

$$y = \ln\left[\{(x^2 + b^2)/(x^2 - b^2)\}^2 - 1\right] - \ln x^2 + \ln 2,$$

where b is an arbitrary constant. Dresner (1983), p.113

14. $$y'' + y'/x - e^y = 0.$$

Put $y = \psi - 2t, x = e^t$ so that the given DE changes to

$$\frac{d^2\psi}{dt^2} = \exp(\psi). \tag{1}$$

Letting $(1/2)\left(\frac{d\psi}{dt}\right)^2 = G$, we change (1) to

$$\frac{dG}{d\psi} = \exp(\psi). \tag{2}$$

Integrating (2) once, substituting for G, and integrating once again, we can obtain a general solution in terms of the original variables as

$$e^y = 2(C_1/x)^2\{\tanh^2[C_1 \ln(C_2/x)] - 1\}, \tag{3}$$

where C_1 and C_2 are arbitrary constants. Sajben (1968)

15. $$y'' + y'/x + \lambda^2(Y - y)\exp\{\gamma - \gamma/y\} = 0, 0 < x < 1,$$

$y'(0) = 0, y'(1) = L\{1 - y(1)\}, y(x) > 0$, and $Y = (1 - \sigma)y(1) + \sigma + \beta$. L, λ, σ, and β are all constants.

It is shown that the given system has no solution if $L = 0(k \neq 0)$. Perturbation solutions are obtained for the cases $\sigma = 1, |\sigma - 1| = O(1), \sigma > 1$, and $|\sigma - 1| = O(1)$. Kapila and Matkowsky (1980)

16. $$\boldsymbol{y'' + y'/x - \sinh y = 0, y'(a) = -\sigma, \sigma > 0, a > 0, y(\infty) = 0.}$$

The function $\sinh y$ satisfies the conditions of Eqn. (2.8.20). Hence, the results regarding the existence and asymptotic nature of the solution stated there hold. Krzywicki and Nadzieja (1989)

17. $$\boldsymbol{y'' + y'/x + \beta \exp(-1/|y|) = 0, \beta \geq 0,}$$

where $y'(0) = 0, y(1) = \tau; \tau$ is a constant.

The given system arises in a tubular chemical reactor. Comparison theorems regarding this problem are proved and used to find lower bounds. The comparison equation is given in Eqn. (2.2.69). This study helps reduce the set of initial values for use in the shooting method. Bushard (1978)

18. $$\boldsymbol{y'' + y'/x + \beta f(y) = 0, y'(0) = 0, y(1) = \tau.}$$

Here f is positive for y positive and increasing, $\tau \geq 0$, $\beta \geq 0$, and only nonnegative solutions are of interest. This problem arises in tubular reactor theory. It has nonunique solutions. Lower and upper bounds of solutions are found. Bushard (1978)

19. $$\boldsymbol{y'' + y'/x - f(y) = 0, y(a) = c, y'(a) = -\sigma,}$$

where $a > 0, c, \sigma > 0$ are constants. The function f is assumed to be strictly increasing, twice continuously differentiable, and vanishing at zero, $f(0) = 0$.

The solution of the IVP has the following properties:

(a) If $y > 0$, then y has at most one extremum. Integrating the DE from x_1 to x_2 gives

$$x_2 y'(x_2) - x_1 y'(x_1) = \int_{x_1}^{x_2} \mathrm{sf}(y(s))ds. \tag{1}$$

Since $f > 0$, the RHS of (1) > 0. Therefore, $y'(x_2) = y'(x_1) = 0$ would contradict.

(b) If $y > 0$ on $[a, w)$, then either $y \geq \alpha$ with some constant $\alpha > 0$ or $y(x)$ tends to zero as x goes to w and clearly in the latter case $w = \infty$.

(c) If y is negative at some $\bar{x}\epsilon[a, w)$, then y is strictly decreasing.

(d) For sufficiently large $c, y(x) > 1$ on $[a, w)$.

(e) If $y > 0$, then $y(x)$ tends to 0 or ∞ if x goes to w.

(f) If y is negative at some point, then y tends to $-\infty$ as x goes to w.

(g) If $c_1 < c_2$, then $y_{c_2} - y_{c_1} \geq c_2 - c_1$ on common intervals of existence.

Krzywicki and Nadzieja (1989)

20. $$\boldsymbol{y'' + y'/x - f(y) = 0, y'(a) = -\sigma, \sigma > 0, a > 0, y(\infty) = 0,}$$

where f is a strictly increasing, twice continuously differentiable function, vanishing at zero.

It is shown that the given BVP has exactly one solution. The following asymptotic results hold: Assuming that $y'(0) = \lambda^2, \lambda > 0$, if y is a solution of the given BVP, then

$y \sim x^{-1/2}\exp(-\lambda x)$. Moreover, if $f_1(y) < f_2(y)$ for $y \geq 0$, then $y_2(x) \leq y_1(x)$ for $x \geq a$. Krzywicki and Nadzieja (1989)

21. $$y'' + y'/x - (\nu^2/x^2)y + h(y) = 0, y(0) = 0, y(\infty) = 1,$$

$0 < y(x) < 1$ for $0 < x < \infty$ and $y(x)$ is strictly increasing for $0 < x < \infty; \nu$ is a positive constant.

Assuming that (a) $h \in C(0,1]$, (b) $h(1) = 0$, (c) $h(y) > 0$ for $0 < y < 1$, it is proved that the given DE has at least one solution $y = y(x)$ on the interval $(0, \infty)$ satisfying the given conditions. Kaminogo (1979)

22. $$y'' + (1/x)y' - (1/x^2)y - F(x, y/x) = 0, 0 < x \leq 1, y \in C^1,$$

where

$$y(0) = 0, \quad B_1Y(1) \equiv \begin{pmatrix} 1 & 0 & 0 & 0 \\ 0 & -1/3 & 0 & 1 \end{pmatrix} \begin{pmatrix} Y(1) \\ Y'(1) \end{pmatrix} = \begin{pmatrix} 1 \\ 0 \end{pmatrix},$$

where

$$F(x, y(x)/x) = \alpha \begin{pmatrix} x(\beta + [y_1(x)/x][y_2(x)/x]) \\ x[1 - \{y_1(x)/x\}^2] \end{pmatrix}$$

and α, β are parameters. Here the function $y = \begin{pmatrix} y_1 \\ y_2 \end{pmatrix}$ is a vector-valued function of dimension 2. Using the general theory for this class of problems, existence, uniqueness, and smoothness of continuous solutions are considered. Weinmüller (1984)

23. $$y'' + (2/x)y' + y^3 = 0, \int_0^{x_*} y^3x^2dx = (\pi/\sigma)^{3/2}, y'(0) = 0, y(x_*) = 0,$$

where x_* and σ are constants; x_* is the first zero of y.

Formal integration of the given DE from 0 to x_* shows that $\sigma = \pi(-x_*^2y'(x_*))^{-2/3}$. Computations show that $x_* = 1$, $y(0) \simeq 6.897, y'(1) \simeq -2.018$; thus we have $\sigma \simeq 1.967$. Ad'yutov and Zmitrenko (1991)

24. $$y'' + (2/x)y' + y^3 - y = 0, x \in (a, \infty), a \geq 0,$$

$\lim_{x\to 0} y(x, a) = a, \lim_{x\to 0} y'(x, a) = 0$.

If we put $y = x^{-1}w$, we get Eqn. (2.4.49) for w. Coffman (1972)

25. $$y'' + 2y'/x - y + y^3 = 0, y(0) = \alpha, y'(0) = 0, y(\infty) = 0, \alpha > 0.$$

It is clear that every solution of the given problem is an even function of x; therefore, writing

$$y(x) = \sum_{k=0}^{\infty} y_kx^{2k}, \tag{1}$$

and substituting in the given DE lead to the recurrence relation

$$[(2k+2)(2k+3)]y_{k+1} = y_k - \sum_{j=0}^{k} \sum_{i=0}^{j} y_iy_{j-i}y_{k-j}, k = 0, 1, 2, \ldots \tag{2}$$

We find that the assumption $y_0 = \alpha, |y_k| \leq \rho^{2k}\alpha, \rho > 0$ is valid, provided that

$$|y_{k+1}| \leq \frac{\alpha\rho^{2k}}{(2k+2)(2k+3)} + \frac{\alpha^3(k+1)(k+2)\rho^{2k}}{2(2k+2)(2k+3)} \leq \alpha(1+\alpha^2)\rho^{2k}/6 \leq \alpha\rho^{2k+2}.$$

Using the root test, we find that the series (1) converges when $|x| < 6^{1/2}/(1+\alpha^2)^{1/2}$. We may check similarly that the solution satisfying the given ICs has a Taylor series expansion at every point x_0 in the real line, which converges whenever $|x - x_0| \leq \{2/(1+\alpha^2)\}^{1/2}, y(x_0) \leq \alpha$. Therefore, we may extend this solution throughout this strip $\{x : |\text{Im } x| < [2/(1+\alpha^2)]^{1/2}\}$ by analytic continuation, provided that it remains bounded. Multiplying the given DE by y', we have $dE/dx = -2y'^2/x$, where $E = (1/2)y'^2 + V(y)$, where the Liapunov function $V(y)$ is given by $V(y) = y^4/4 - y^2/2$. It is then proved that a solution which crosses the curve $E =$ constant will remain inside this curve, and if $E < 0$, it will approach the particular point $y = +1$ or $y = -1$ which the curve $E =$ constant < 0 encloses.

To find the asymptotic, we obtain the integral equation formulation of the given DE (using variation of parameters) as

$$y(x) = Ae^{-x}/x - (1/x)\int_x^\infty t \sinh(x-t)y^3(t)dt. \tag{3}$$

For $x > 0$ sufficiently large, we may, by successive approximation, find that

$$y(x) = A(e^{-x}/x)\{1 + o(1)\}, \tag{4}$$

where $x \to \infty$.

By differentiating (3) and using arguments similar to those above, we may obtain $y'(x) = -A(e^{-x}/x)\{1 + o(1)\}$ as $x \to \infty$.

Similarly, one may find that for $y(x) \to \pm 1$ as $x \to \infty$, the asymptotic results

$$y(x) = \{\pm 1 + A\sin(2^{1/2}x)/x\}\{1 + o(1)\}$$

and

$$y'(x) = 2^{1/2}A\cos(2^{1/2}x)/x + o(1/x) \text{ as } x \to \infty.$$

It is shown that if $y(x) \not\equiv 0$ and satisfies the given DE and IC at $x = 0$ and the condition $y(x) \to 0$ as $x \to \infty$, then at each point $x_0 \geq 0$ such that $y'(x_0) = 0, y(x_0)$ satisfies $|y(x_0)| > 2^{1/2}$. Also, if $0 < \alpha \leq 2^{1/2}$, then the solution $y(x)$ with the same IC does not go to zero as $x \to \infty$.

An algorithm, using Galerkin's method, is given to solve the problem numerically. Different possibilities are treated as $x \to \infty$. Chauvette and Stenger (1975)

26. $\boldsymbol{y'' + (2/x)y' - \alpha^2 y + gy^3 = 0, y(\infty) = 0,}$

where α and g are constants.

The linear behavior shows that $y = Ag^{-1/2}e^{-\alpha x}/x$. A rough numerical integration shows that $A \sim 2.5$. The given DE may be transformed according to $t = \alpha x, \eta = g^{1/2}xy$ to the form

$$\eta''/\eta = 1 - (\eta/t)^2. \tag{1}$$

Considering $y(0) = 0$ and increasing $y'(0)$, one may conclude that there exists at least one solution with any number of nodes.

A phase plane analysis of the equation (without the $2/x$ term) and numerical results elucidate some qualitative features. Finkelstein, Lelevier, and Ruderman (1951); Rosen and Rosenstock (1952)

27. $$\boldsymbol{y'' + (2/x)y' - Ay - By^3 = 0,}$$

where A and B are constants.

Put $y = u/x$ to obtain

$$u'' - Au - (B/x^2)u^3 = 0. \tag{1}$$

Now see Eqn. (2.4.51) for (1). Dixon, Kelley, and Tuszynski (1992)

28. $$\boldsymbol{y'' + (2/x)y' + (1/x)y^3 = 0.}$$

The given DE that arises from group theoretical reduction of a nonlinear wave equation is shown to possess the Painlevé property. Duarte, Euler, Moreira, and Steeb (1990)

29. $$\boldsymbol{y'' + (2/x)y' + y^5 = 0.}$$

Multiplying the given DE by x^3, rewriting it suitably, and integrating, a first integral of the given DE is found to be

$$x^3[y'^2 + (1/3)y^6] + x^2yy' = \text{ constant.}$$

Sarlet and Bahar (1980)

30. $$\boldsymbol{y'' + (2/x)y' + y^5 = 0.}$$

A first integral of the given DE is

$$x^3y'^2 + x^2yy' + (1/3)x^3y^6 = \text{ constant.}$$

An exact solution satisfying $y(0) = \theta, y'(0) = 0$, where θ is a constant, is

$$y = \theta(1 + (\theta^4/3)x^2)^{-1/2}.$$

Crespo Da Silva (1974)

31. $$\boldsymbol{y'' + (2/x)y' + y + y^5 = 0, y'(0) = 0, y(0) = \theta,}$$

where θ is a constant.

The following results are found:

(a) In any interval $[0, x]$ in which $y(x) > 0$,

$$y(x) < (1 + \theta^4x^2/3)^{-1/2}.$$

(b) If μ_m is the mth positive zero of $y(x)$, then as $\theta \to \infty$,

$$y_x^2(\mu_{m+1}) < y_x^2(\mu_m) < \cdots < y_x^2(\mu_1) < 6(\mu_1\theta)^{-2}.$$

(c) Let y be a solution of the given problem; then, as $\theta \to \infty$,

$$y(x) = \theta(1 + \theta^4x^2/3)^{-1/2} + O\left\{\theta^{-3}[(1 + \theta^4x^2)^{1/2} - 1]\right\}.$$

Budd (1988)

32. $$y'' + 2y'/x + \alpha y^5 = 0,$$

where α is a constant.

This is a special Lane-Emden equation. Using the method of continued fractions, an intermediate integral is found to be

$$y'/y = -3^{-1}\alpha a_0^4 x/[1 + 3^{-1}\alpha(a_0^2 x)^2], \tag{1}$$

where a_0 is a constant; an integration of (1) gives the exact solution to be

$$y = a_0/[1 + 3^{-1}\alpha(a_0^2 x)^2]^{1/2},$$

where the constant of integration is put equal to zero. Ditto and Pickett (1990)

33. $$y'' + (2/x)y' - \alpha y/(y + k) = 0, \alpha, k > 0,$$

$$y'(0) = 0, y'(1) = \alpha_2[1 - y(1)] + \alpha_0/(1 + k_0) - \alpha_1 y(1)/\{y(1) + k_1\},$$

where α_i $(i = 0, 1, 2), k, k_0$ are positive constants.

The given problem models oxygen diffusion in a spherical cell with Michaelis-Menten oxygen uptake. It is proved that it has a unique nonnegative solution if

$$1 + (1/\alpha_2)\{\alpha_0/(1 + k_0) - \alpha/3 - \alpha_1\} - \alpha/3 \geq 0.$$

Garner and Shivaji (1990)

34. $$y'' + \{2/x\}y' + \{y(1 + y)\}^{3/2} = 0.$$

The given DE is the white dwarf equation, and is a special case of Eqn. (2.8.95). Since the integral $\int_0^1 \{y(1 + y)\}^{3/2} y^{-4} dy$ diverges, the solution of the given DE remains positive only for a finite value of x; thus any stellar model obtained from the white dwarf equation has a finite radius. Makino (1984)

35. $$y'' + (2/x)y' + y^n = 0,$$

where n is a parameter.

It is shown that the only group of transformations leaving the given DE invariant is $X = tx, Y = t^{-2/(n-1)}y$ $(n \neq 0, 1)$. Kranje (1951)

36. $$y'' + (2/x)y' + y^n = 0, y(0) = 1, y'(0) = 0.$$

The given DE is written as

$$y'' + (2/x)y' + [y(x)]^{1+\delta} = 0,$$

with $n = 1 + \delta$, and a solution in perturbation series is found as

$$y = y_0(x) + \delta y_1(x) + \delta^2 y_2(x) + \cdots.$$

The perturbation equations are recursively solved. The first zero of a second order perturbation solution is shown to agree reasonably well with the exact numerical value. The method requires solution of a sequence of linear equations and associated initial conditions. The first two terms of the perturbation solution are

$$\begin{aligned} y_0(x) &= (\sin x)/x, \\ y_1(x) &= [(\cos x)/(2x)] \int_0^x \ln(\sin s)ds - [(\sin x)/(2x)] \ln\{(\sin x)/x\} \\ &\quad + (3/4)\cos x + (\sin x)/(4x) - (1/2)\cos x \ln x - (\cos x/4x)\ \mathrm{Si}(2x) \\ &\quad - [(\sin x)/(4x)]\mathrm{Cin}\ 2x, \end{aligned}$$

where

$$\mathrm{Si}\ x \equiv \int_0^x [(\sin t)/t]dt, \mathrm{Cin}\ x \equiv \int_0^x [(1-\cos t)/t]dt.$$

Bender et al. (1989)

37. $$\boldsymbol{y'' + 2y'/x + y^n = 0,}$$

where n is a constant.

The given DE is invariant to the stretching group

$$y_1 = \lambda^{-2/(n-1)}y, x_1 = \lambda x.$$

Taking $u = yx^{2/(n-1)}$ and

$$v = y'x^{(n+1)/(n-1)},$$

we find that

$$\frac{dv}{du} = -\frac{(n-1)u^n + (n-3)v}{(n-1)v + 2u}. \quad (1)$$

Equation (1) is integrable for $n = 5$:

$$3uv + 3v^2 + u^6 = \text{ constant}$$

or

$$3x^2yy' + 3x^3y'^2 + x^3y^6 = \text{constant}. \quad (2)$$

If the solution sought is finite at $x = 0$ and has a zero derivative there, the constant in (2) should be zero. If we introduce $w = u^2 = xy^2$, we can separate the variables and find the solution as

$$w = 3ax/(x^2 + 3a^2),$$

or

$$y = [3a/(x^2 + 3a^2)]^{1/2},$$

where a is an arbitrary constant. For other n, the solution may be discussed in the (u, v) plane. Dresner (1983)

38. $$\boldsymbol{y'' + [2/x]y' + y^m = 0,}$$

where $0 < x < \infty, 0 < y < \infty, |y'| < \infty, m > 0$.

The given system is shown to have a solution that is positive for all $x \geq x_0$. Makino (1984)

39. $$y'' + (2/x)y' - y + y^n = 0, n > 0, x \geq 0,$$

$y(0) = y_0 < \infty, y'(0) = 0, y(\infty) = 0.$

It is proved that:

(a) The given problem for $0 < n \leq 1$ has no positive solution.

(b) The given problem has at least one positive solution for $1 < n \leq 3$. The solution for the case $n = 3/2$ is shown graphically; it is found numerically that $4.274063 < y(0) < 4.274125$.

Zhidkov and Shirikov (1964)

40. $$y'' + 2y'/x + a^\nu y^n = 0, a > 0.$$

For $n \neq 1$, the given DE goes through $y = a^{-1/(n-1)}y$ to the form

$$xy'' + 2y' + x^\nu y^n = 0, \tag{1}$$

where we have dropped the bar on y. For $\nu = 1$, (1) is Emden equation. One solution of (1) is

$$y = cx^{-\mu},$$

with $\mu = (\nu+1)/(n-1), c^{n-1} = \mu(1-\mu)$. For $y(x) = \eta(\xi)$, $\xi = 1/x$, (1) goes to

$$\eta'' + \xi^{-\nu-3}\eta^n = 0, \tag{2}$$

while, with $u(x) = xy$, it goes to

$$u'' + x^{\nu-n}u^n = 0. \tag{3}$$

In (2), if we put

$$\eta(\xi) = \xi^\mu v(t), \ t = \log \xi,$$

we get

$$v'' + (2\mu - 1)v' + \mu(\mu-1)v + v^n = 0. \tag{4}$$

Equation (4) is autonomous, so that with $v'(t) = p(v)$, we have

$$pp' + (2\mu - 1)p + \mu(\mu - 1)v + v^n = 0, \tag{5}$$

which may be solved in closed form only for some values of n.

It is shown that if $n > 0, \nu > -1$, there exists for each $c > 0$ exactly one solution for $x > 0$ for which $y(x) \to 0$ as $x \to +0$. Moreover, for $0 < x < [C^{1-n}\nu(\nu+1)]^{1/(\nu+1)}, y(x) > 0$. For such solution, $x^2y' \to 0$ as $x \to 0$. With the given IC on y and y', the solution can be obtained iteratively. For $2\nu - n + 3 > 0$, the solution is shown to have at least one positive zero. For $2\nu - n + 3 \leq 0$, the solution $y > 0$ for all $x > 0$ and $y \to 0$ as $x \to +\infty$.

For $\nu = 1$ (Emden's equation), we note the following solutions for (1):

$$n = 0 : y = C_1 + C_2/x - x^2/6,$$

$$n = 1 : y = (C_1 \sin x)/x + (C_2 \cos x)/x; C_i (i = 1, 2) \text{ are constants.}$$

Equation (5), for $\nu = 1, \mu = 2/(n-1)$, becomes

$$pp' - \{n-5)/(n-1)\}p - \{2(n-3)/(n-1)^2\}v + v^n = 0.$$

In particular,

$$\begin{aligned} pp' + p + v^3 = 0 \text{ for } n &= 3 \\ pp' = v/4 - v^5 \text{ for } n &= 5. \end{aligned} \tag{6}$$

The solution of (6) is

$$v'^2(t) = p^2 = v^2/4 - v^6/3 + C_1. \tag{7}$$

For $C_1 = 0$, (7) integrates to yield

$$y^2 = 3C/(x^2 + 3C^2),$$

where C is an arbitrary constant. It may be noted that if $y(x)$ is a solution of (1), so is $C^{2/(n-1)}y(x)$. Tables of solutions are known for $y(0) = 1, y'(0) = 0$ with $0.5 < n < 0.6$ and for $y(x_0) = 0$, $y'(x_0) = C$ in the interval $0 \leq x \leq x_0$ for the specific choices $n = 3, x_0 = 1$ and $n = 1.5, x_0 = 3.65379$. Kamke (1959), p.560

41. $$\boldsymbol{y'' + 2y'/x + a^\nu y^{2\nu+3} = 0,}$$

where a and ν are constants.

A first integral for the given DE is

$$x^3 y'^2 + x^2 yy' + \{a/(2+\nu)\}x^{\nu+2}y^{2(\nu+2)} = \text{ constant.}$$

Djukic (1973)

42. $$\boldsymbol{y'' + 2y'/x - y^n = 0.}$$

The given DE is related to the Emden DE. By eliminating the first derivative it can be transformed to $u'' = x^{1-n}u^n$. For $n = 2$, $y = 2x^{-2}$ is a solution. We may in this case transform the given DE, by writing $y(x) = x^{-2}\eta(\xi), \xi = \log x$, to $\eta'' - 3\eta' + 2\eta = \eta^2$. Now put $\eta'(\xi) = p(\eta)$ so that

$$p'(\eta) = 3 + \eta(\eta - 2)/p. \tag{1}$$

These solutions behave like $y \simeq 2x^{-2}$ as $x \to \infty$. Kamke (1959), p.559

43. $$\boldsymbol{y'' + 2y'/x - \alpha y^{2p+1} - \lambda y^{4p+1} = 0,}$$

where α, p, and λ are constants.

Using the method of continued fractions, a first integral of the given DE is found to be

$$y'(x)/y(x) = \{\alpha^2 p(2p+1)x\}[\lambda(-1+p^2) - \alpha^2 p^2 (1+2p)x^2]^{-1}. \tag{1}$$

On integration of (1) one obtains a particular integral,

$$y(x) = \left[\frac{\alpha(1+2p)(-1+p)}{\lambda(-1+p)^2 - \alpha^2 p^2 (1+2p)x^2}\right]^{1/2p},$$

with no arbitrary constants. Ditto and Pickett (1990)

44. $$y'' + 2y'/x + e^y = 0, y(0) = 0, y'(0) = 0.$$

The power series solution

$$y = -\frac{1}{6}x^2 + \frac{1}{5(4!)}x^4 - \frac{8}{21(6!)}x^6 + \frac{122}{81(8!)}x^8 - \frac{5032}{495(10!)}x^{10} + \cdots$$

is shown to be the only solution of the given problem. The domain of definition of this solution is suitably defined in a Banach space. Ifantis (1987a)

45. $$y'' + (2/x)y' - \sinh y = 0.$$

The given nonlinear spherical radial Poisson-Boltzmann equation describes the spherical distribution of the particles in a classical two-component Coulomb gas.

The following qualitative properties of solution of this DE were proved: Let x_+ denote the right boundary of the maximum interval in which a given solution y exists (x_+ may be ∞).

(a) If $y(\bar{x}) > 0$ for some $\bar{x}$, then the solution does not have a local maximum at $\bar{x}$. If $y(\bar{x}) < 0$ at the point $\bar{x}$, then $y(x)$ does not have a local minimum at $\bar{x}$.
(b) If $y(\bar{x}) > 0$ and $y'(\bar{x}) > 0$, then $y(x)$ is an increasing function in the interval $(\bar{x}, x_+)$, where x_+ denotes the right boundary of maximum interval in which a given solution exists.
(c) If for some $\bar{x}, y(\bar{x}) > 0$ and $y'(\bar{x}) > 0$, then $\lim_{x \to x_+} y(x) = \infty$.
(d) If for some $\bar{x}, y(\bar{x}) < 0$ and $y'(\bar{x}) < 0$, then $lim_{x \to x_+} y(x) = -\infty$.
(e) If y is a solution and at some $\bar{x}, y(\bar{x}) > 0$ and $y'(\bar{x}) > 0$, then x_+ is finite.
(f) $\lim_{y \to \infty} \dfrac{dx}{dy} = 0$.
(g) $\lim_{y \to \infty} \left(\dfrac{dx}{dy}\right)^3 \sinh y = 0$.

The equation was also solved numerically with $x_0 = 0.57, y_0 = 0.074$; $x_0 = 0.57, y_0 = 0.10$. Both solutions have initial slope $y'(x_0) = -0.211$. Martinov, Ouroushev, and Chelebiev (1986)

46. $$y'' + (2/x)y' - y + y|y|^{k-1} = 0,$$

$y(0) = 0, \lim_{x \to \infty} y = 0, y'(0)$ exists and $y(x) > 0$ on $(0, \infty)$.

The given DE with $k = 2$ arises in a nuclear model. Put $y = x^{-1}z(x)$ to get $z'' - z + z|z|^{k-1}/x^{k-1} = 0$. Now see Eqn. (2.5.29). Macki (1978)

47. $$y'' + (2/x)y' + f(y) = 0,$$

where $f(y)$ is positive, nondecreasing, and sufficiently regular for $y \geq 0$. Solutions are sought such that $y(x) > 0, y'(x) < 0$ in $0 < x \leq x^*$; $\lim_{x \to +0} y(x) = C < \infty$.

It is shown that $y(x)$ decreases in $0 < x < x^*$, as $f(y)$ is increased. This theorem is applied to the problem of stability of a fluid sphere under the action of its own gravity. Cimino (1956)

48. $$y'' + (2/x)y' - f(y) = 0,$$

where $f(y)$ is a strictly increasing function and $\int_{Y^*}^{\infty}(f(y))^{-1/3}dy < \infty$, where $y > y^*$, some constant.

Since these are the only conditions used in the proof of the special case $f(y) = \sinh y$, see Eqn. (2.8.45); the results quoted therein may be shown to hold for this more general equation as well. Martinov, Ouroushev, and Chelebiev (1986)

49. $$y'' + 2y'/x - y + yf(y^2) = 0.$$

Putting $y(x) = x^{-1}z(x)$, we get $z'' - z + zf(z^2/x^2) = 0$, which is a special case of Eqn. (2.6.15). Ryder (1967)

50. $$y'' + 2y'/x - 2y/x^2 - f(y) = 0,$$

where f is continuous, odd, and increasing on $(-\infty, \infty)$, with $f(+\infty) = 1, 0 < f(y) < y$ for $y > 0, f(y) = y + O(y^\alpha)$ as $y \to +0$ for some $\alpha > 1$.

The given DE is written as the system

$$y' + 2x^{-1}y = v, \quad v' = f(y). \tag{1}$$

It is shown that system (1) has a solution for $x > 0$ such that $y(x) \sim x^{-1}e^{-x}$ and $v(x) + x^{-1}e^{-x} \sim -x^{-2}e^{-x}$ as $x \to +\infty$. This is proved by using an appropriate iterative method. Prachar and Schmetterer (1956)

51. $$y'' + (2/x)y' + \sum_{m=1}^{n} a_m x^m y^{2m+5} = 0,$$

where a_m are constants and n is a positive integer.

A first integral of the given DE is

$$I = x^3y'^2 + x^2yy' + \sum_{m=1}^{n} \frac{a_m}{m+3} y^{2(m+3)} x^{m+3}.$$

Ranganathan (1992a)

52. $$y'' + (3/x)y' + (2/y^2)R^2(x, \epsilon) = 0, 0 < \epsilon < x < 1,$$

with $y(\epsilon) = 1$ or $\epsilon y'(\epsilon) + (1 - \nu)y(\epsilon) = h$ and $y(1) = S$ or $y'(1) + (1 - \nu)y(1) = H$, where ϵ, ν, S, and H are constants.

The given BVP arises in the theory of small finite deflections. An integral equation technique of solution is developed which yields the existence of positive tensile solutions for a parameter range of the boundary data. Grabmüller and Novak (1988)

53. $$y'' - y'/x + y^2 - 1/x = 0,$$

$y'(0) = 0, y''(0) > 0$.

An asymptotic series solution valid for $x \to 0+$ may easily be written out, satisfying the given IC. Bender and Orszag (1978), p.200

54. $$y'' - y'/x - B\delta x^2 e^y = 0, y'(0) = 0, y(1) = 0,$$

where δ and B are constants.

The given DE may be simplified by introducing $\xi = (Bx^2+1)^{-1}$ as the new independent variable. The solution of the BVP may be found to be $y = \ln[(8B/\delta)/(Bx^2+1)^2]$, where B is found by using the BC at $x = 1$:

$$(8B/\delta)/(B+1)^2 = 1. \tag{1}$$

It is easily checked from (1), for a given δ, that there are two distinct roots for B if $0 < \delta < 2$, leading to two solutions of the given BVP. When $\delta = 2$, there is just one root $B = 1$, yielding one solution of the BVP. For $\delta > 2$, (1) has no root and hence no solution of the BVP. Kubicek and Hlavacek (1983), p.36

55. $$\boldsymbol{\epsilon^2 y'' + \epsilon^2(N-1)y'/x + (1/2)(y-y^3) + K/2 = 0,}$$

subject to the BC $y(a) = \alpha, y(b) = \gamma, 0 < a < b < \infty$, where $f(y) \equiv y - y^3 + K$ has roots $\alpha < \beta < \gamma$. $N > 1$ is an integer, while K is a (positive) parameter.

The given DE is considered for small values of the parameter ϵ. The constant K is proportional to ϵ. This equation is a special case of phase field equations for free boundaries arising from phase transitions. It is shown that if $K = O(\epsilon)$, then any solution of the given problem must be close to a tanh function in the supremum norm, i.e.,

$$|y(X) - \tanh X/2| \le C\epsilon^{1/(1+p_0)}, X = (x - x_\epsilon)/\epsilon,$$

where $p_0 > 3^{1/2}$ and the solution y must cross the axis at

$$x_\epsilon = -[(N-1)/K](\epsilon/4)\int_{-\infty}^{\infty} \operatorname{sech}^4(t/2)dt + O(1)$$

or is within $C|\epsilon \log \epsilon|$ of one of the boundaries $x = a$ or $x = b$. Caginalp and McLeod (1986)

56. $$\boldsymbol{y'' + [(N-1)/x]y' + [1/(2\xi^2)](y-y^3) + (1/2)(k/\xi^2) = 0,}$$

$y(a) = \alpha_0, y(b) = \gamma_0$, where ξ, k, a, b, α_0, and γ_0 are constants, and $N > 1$ is an integer. Shooting arguments are advanced to show that the given problem has at least one solution for any $\xi > 0$. It is also shown that the solution must be strictly monotonic. Caginalp and Hastings (1986)

57. $$\boldsymbol{y'' + (\alpha/x)y' = -y^n, n \text{ an integer } \ge 2, y(0) = \lambda, y'(0) = 0.}$$

It is shown that for every $r > 0$ and every λ such that

$$|\lambda| \le \frac{n-1}{n}\left[\frac{2(1+\alpha)}{nr^2}\right]^{1/(n-1)},$$

the given IVP has a unique analytic solution of the form $y(x) = \lambda + \sum_{n=1}^{\infty} b_n(\lambda)x^n$, which together with its first two derivatives converges absolutely for $|x| \le r$ and satisfies the condition

$$|y(x)| \le \left[\frac{2(1+\alpha)}{nr^2}\right]^{1/(n-1)}.$$

For the special case $\alpha = 2$ it is shown that the power series solution

$$y(x) = 1 - \frac{x^2}{3!} + n\frac{x^4}{5!} + (5n - 8n^2)\frac{x^6}{(3)7!}$$

$$+ (70n - 183n^2 + 122n^3)\frac{x^8}{(9)9!}$$
$$+ (3150n - 10805n^2 + 12642n^3 - 5032n^4)\frac{x^{10}}{(45)11!} + \cdots$$

of the given equation, together with its first two derivatives, converges absolutely for $|x| \leq (2.3)^{1/2}[(n-1)/n]^n$, and satisfies the condition $y(0) = 1, y'(0) = 0$. Ifantis (1987a)

58. $$\boldsymbol{y'' + (n-1)y'/x + \lambda(1+\alpha y)^{\beta} = 0, y(1) = 0, y'(0) = 0,}$$

where n, λ, α, and β are constants, and $\alpha\beta > 0$.

Let $n - 2 > 0$ and $\lambda > 0$ be preassigned. Then positive solutions $y(x, \lambda)$ of the given system have the following properties:

(a) When $0 \leq \beta < 1$, solutions are unique.

(b) When $\beta < 0$ or $\beta > 1$, there exists a unique value $\lambda_* > 0$ such that positive solutions do not exist when $\lambda > \lambda_*$. A unique positive solution exists when $\lambda = \lambda_*$ and
$$\lambda_* > \tilde{\lambda} = \frac{\tau}{\alpha}(n - 2 - \tau),$$
where $\tau = \dfrac{2}{\beta - 1}$.

(c) When $1 < \beta \leq (2+n)/(n-2)$ and $\lambda < \lambda_*$, then there are just two solutions, one with large A $(\equiv y(0))$ and one with small A.

(d) When $\beta < 0$ or $\beta > (2+n)/(n-2), n - 2 < f(\beta), \lambda_* > \tilde{\lambda}$, and $\lambda = \tilde{\lambda}$, there are a countably infinite number of solutions. Sufficiently close to $\tilde{\lambda}$, there are a large but a finite number of positive solutions.

(e) When $\beta < 0$ or $\beta > (2+n)/(n-2), n - 2 \geq f(\beta)$ and $\lambda < \lambda_* = \tilde{\lambda}$, there is one and only one positive solution.

In the above, $\tau = \dfrac{2}{\beta - 1}$, and $f(\beta) = 4\beta/(\beta - 1) + 4[\beta/(\beta - 1)]^{1/2}$. This is a thorough study of the given problem, which uses local, global, phase plane, and asymptotic analysis, and extends in a large way the work reported in Chandrasekhar (1957). There is a limiting singular solution $y_s = (1/\alpha)[x^{-t} - 1], \lambda_s = (\tau/\alpha)(n - 2 - \tau)$, where $\tau = 2/(\beta - 1)$, satisfying $y(1) = 0$. An exact solution in the phase plane is also found for $\beta = (n+2)/(n-2)$. Joseph and Lundgren (1972)

59. $$\boldsymbol{y'' + (\alpha/x)y' - \sum_{n=1}^{\infty} \alpha_n y^{n-1} = 0,}$$

where
$$|\alpha_n| \leq \mu_1, n = 2, 3, \ldots, \alpha_1 = 0.$$

It is proved that there exists an $r > 0$ such that the given DE has a unique analytic solution which together with its first two derivatives converges absolutely for $|x| \leq r$ and satisfies the condition $y(0) = 1, y'(0) = 0$. Ifantis (1987a)

60. $$\boldsymbol{y'' + (p/x)y' + x^{q-p}y^n = 0,}$$

where $p > 1$ and $q > 0$. The qualitative properties of solutions for which $y(0) = a > 0$ are studied; in particular, the existence of zeros of such solutions is investigated. Richard (1951)

61. $$\boldsymbol{y'' + (k/x)y' + x^p y^n = 0, y'(0) = 0, y(x_0) = y_0, y(x) \geq 0, \forall x \in [0, x_0],}$$

where $k, p > -1, n > 1$, and $x_0, y_0 > 0$. Depending on

$$r \equiv (n-1)\frac{k-1}{2+p},$$

the following results are proved:

(a) If $r \leq 2$, then for an arbitrary $x_0 > 0$ there is $M(x_0) > 0$ such that for $y_0 \in (0, M(x_0))$, the given problem has two solutions, for $y_0 = M(x_0)$ it has one solution, and for $y_0 > M(x_0)$ it has no solution.

(b) If $r \in [2, 2n + 2[n(n-1)]^{1/2}]$, then there is a sequence

$$\begin{aligned} M_j(x_0) &> 0, j = 1, 2, \ldots \text{ such that } M_{2j-1}(x_0) > M_{2j}(x_0), \\ M_1(x_0) &> M_3(x_0) \cdots > M_{2j+1}(x_0) > \cdots, M_2(x_0) < M_4(x_0) < \cdots \\ &< \cdots M_{2j}(x_0) < \cdots. \text{ and } \lim_{j\to\infty} M_j(x_0) = \delta, \end{aligned}$$

where
$$\delta = [(r-1)\{(2+p)/(n-1)\}^2]^{1/(n-1)} x_0^{-(2+p)/(n-1)},$$

and for $y_0 = M_j(x_0)$ the given problem has j solutions, for $y_0 \in [M_j(x_0), M_{j+2}(x_0)]$ it has $j+1$ solutions, and for $y_0 > M_1(x_0)$ it has no solution.

(c) If $r \geq 2n + 2[n(n-1)]^{1/2}$, then for $y_0 \in (0, \delta)$, where δ is as in (b), a solution of the given problem is unique. For $y_0 \geq \delta$, the given problem has no solution.

Grizan (1988)

62. $$\boldsymbol{y'' + \rho y'/x \pm x^{\sigma-\rho}y^n = 0,}$$

where ρ, σ, and n are constants.

(a) $\rho > 1$. Set $s = (\rho-1)^{-1}x^{\rho-1}, y = (\rho-1)^{(\rho-\sigma-2)/[(\rho-1)(n-1)]}v/s$. The equation for v is

$$\frac{d^2V}{ds^2} \pm (\rho-1)^{-2}s^{\sigma_1}v^n = 0,$$

where $\sigma_1 = (\sigma+\rho)/(\rho-1) - (n+3)$. Now see Eqn. (2.3.35).

(b) $\rho < 1$. Set $s = (1-\rho)^{-1}x^{1-\rho}, y = (1-\rho)^{-(\sigma+\rho)/[(n-1)(1-\rho)]}v$. Then we have

$$\frac{d^2v}{ds^2} \pm s^{\sigma_2}v^n = 0,$$

where $\sigma_2 = (\sigma+\rho)/(1-\rho) - (n+3)$.

(c) $\rho = 1$. Set $s = \log x$. The resultant equation is

$$\frac{d^2y}{ds^2} \pm e^{(\sigma+1)s} y^n = 0.$$

Bellman (1953), p.143

63. $$y'' + (k_1/x)y' - \lambda x^{k_2} y^n = 0, n \neq 1, -3,$$

where k_1, λ, k_2, and n are constants.

A first integral of the given DE is

$$\begin{aligned} I &= T - [2\lambda/(n+1)] y^{n+1} x^{N(n+1)/2}, n \neq -1, \\ I &= T - \lambda[N \log x + 2 \log y], n = -1, \end{aligned}$$

where $\mu = [(1-n)k_1 + 2k_2 + n + 3]/[(k_1 - 1)(n+3)] = 0$ or 1,

$$N = (k_1 - 1)(\mu + 1),$$

and

$$T = x^{(\mu-1)(1-k_1)}[y'^2 x^2 + N y y' x + (1/2)\mu(\mu+1)(k_1-1)^2 y^2].$$

A solution of the given DE is

$$y^{1-n} = C x^{k_2+2}, C = \lambda(n-1)^2/[K(k_2+2)],$$

provided that $k_1 \neq 1,\ k_2 \neq -2,\ K = k_1 + k_2 + 1 + n(1-k_1) \neq 0$. Ranganathan (1992a)

64. $$y'' + (b/x)y' - c x^p y^q = 0,$$

where b, c, p, and q are constants, $0 \le b < 1, c > 0, p > -2$, and $q > 1$. Boundary conditions considered are:

(a) $y(0) = 1, B(y(a)) = 0$, where $B(y(a)) \equiv a y'(a) - y(a)$, or

(b) $y(0) = 1, \lim_{x\to\infty} y(x) = 0$, or

(c) $y(0) = 1, y(a) = 0$.

Previous work is continued: existence, uniqueness, and the dependence of the solution on the size of the interval $[0, a]$ for cases (a) and (c) are discussed. Methods are constructive; numerical examples are provided. Chan and Hon (1988)

65. $$y'' + [(2b + m + 3)/\{(m-1)x\}]y' + a x^b y^m = 0,$$

where a, b, and m are arbitrary constants, subject to $m \neq 1$ and $b \neq -2$.

A first integral here is

$$(1/2) y'^2 x^{(2b+2m+2)/(m-1)} + [(b+2)/(m-1)] y y' x^{(2b+m+3)/(m-1)}$$
$$+ [a/(m+1)] y^{m+1} x^{(b+2)(m+1)/(m-1)} = \text{constant},\ b \neq -2.$$

Crespo Da Silva (1974)

66. $$y'' + (\gamma + 1 - 2\beta)y'/x + Nx^{2\beta-1}y^{2n-1} = 0, x \geq x_0 > 0.$$

It is shown that every solution of the given DE is in $L^{2n}(x_0, \infty)$ provided that $N > 0, \gamma > 1/2n$, and $0 < \beta < (2n\gamma - 1)/(3n + 1)$. Spikes (1977)

67. $$y'' + (\nu + \alpha)y'/x + ax^{-\alpha}y^n = 0,$$

where ν, α, n, and a are constants.

A change of variables $\zeta = x^w, y = \zeta^\delta v(\zeta)$, transforms the given DE to

$$\zeta^2\ddot{v} + (2\delta + d)\zeta\dot{v} + \delta(\delta + d - 1)v + (a/w^2)\zeta^\mu v^n = 0, \tag{0}$$

$$\cdot \equiv \frac{d}{d\zeta} \text{ where } d = (w - 1 + \alpha + \nu)/w, \mu = \sigma + \delta(n - 1) + 2 - d,$$

$$\sigma = (\nu + 1 - w)/w.$$

If we choose

$$\mu = 0, 2\delta + d = 1, \tag{1}$$

we get

$$\frac{d^2v}{d\eta^2} - \delta^2 V + (a/w^2)V^n = 0, \tag{2}$$

where $n = \ln \zeta$. Conditions (1) written explicitly give

$$n = (3 - \alpha + \nu)/(\nu + \alpha - 1), w\delta = -(2 - \alpha)/(n - 1). \tag{3}$$

In view of (3) (w, δ appearing in the product form), we may choose $w \equiv 1$, so that $\delta = -(2 - \alpha)/(n - 1)$. Now (2) integrates to give

$$\ln(x/x_0) = \pm \int [V_0 + \delta^2 V^2 - [2a/(n + 1)]V^{n+1}]^{-1/2} dV, \tag{4}$$

where x_0 and V_0 are constants, and $n \neq -1$. For $n = -3, 0, 1$, (4) may be expressed in terms of elementary functions, and for $n = -7, -5, 2, 3, 5$, in terms of elliptic functions.

Choosing $V_0 = 0$ in the given DE, we may find, after some simplification, the special solution

$$y = y(0)/(1 + Ax^{2-\alpha})^{2/(n-1)},$$

$$A = a\left[\left\{\{y(0)\}^{n-1}(n - 1)^2\right\} / \{2(n + 1)(2 - \alpha)^2\}\right].$$

Another possibility of explicit solution occurs when $\delta = 1 - d$, so that equation (0) for $V = V(\zeta)$ becomes

$$\left(\zeta^{s_1}\dot{V}(\zeta)\right)^{\cdot} + (a/w^2)\zeta^{m_1}V^n(\zeta) = 0, \tag{5}$$

where

$$s_1 = 1 + (1 - s_0)/w, s_0 = \alpha + \nu, m_1 = -1 + [m_0 + 1 + (n + 1)(1 - s_0)]/w, m_0 = \nu.$$

For (5) to be immediately integrable, we require that

$$n = (3 - \nu - 2\alpha)/(\nu + \alpha - 1), \delta = -1, w = \nu + \alpha - 1,$$

and (5) becomes

$$\ddot{V} + [a/(\nu + \alpha - 1)^2]V^n(\zeta) = 0. \tag{6}$$

The solutions of (6) may be expressed in terms of beta and incomplete beta functions. Rosenau (1984)

68. $$\boldsymbol{y'' + by'/x + cx^{a-b}y^n = 0,}$$

where b, c, a, and n are constants.

Necessary and sufficient conditions for the existence of a unique solution of the given DE vanishing at a finite point and satisfying the initial condition $y(0) = c_1 > 0$ are found. Matsumoto (1944)

69. $$\boldsymbol{y'' + (b/x)y' - cx^p y^q = 0, y(0) = 1, y(a) = 0,}$$

where b, c, p, and q are constants such that $0 \le b < 1, c > 0$, $p > -2$, and $q > 1$.

Here a sequence of lower bounds as well as a sequence of upper bounds is constructed; each of these sequences is shown to converge to obtain the existence of a nonnegative solution. The monotone method and modified Bessel functions are used. Chan and Hon (1987)

70. $$\boldsymbol{y'' + [(n-1)/x]y' + C_0 x^q y^p = 0,}$$

$y(0) < \infty, y(x) > 0$ for all $x \in [0, R), y(R) = 0$, where $n > 2, C_0 > 0$, and $q > -2$ are constants.

Suppose that there exists a solution $y(x)$ of the given DE; then it is easy to check by direct substitution that $C_1 y(C_2 x)$ is also a solution of the problem if $C_1 > 0, C_2 > 0$, and $C^p = C_2^{q-2}$. In particular, if $q = [(n-2)/2][p - (n+2)/(n-2)]$, all solutions of the given problem are obtained as

$$y(x) = k[1 + 2C_0 k^{p-1}/[(p+1)(n-2)^2]x^{(n-2)(p-1)/2}]^{-2/(p-1)},$$

where $y(0) = k > 0$ is arbitrary. Yanagida (1991)

71. $$\boldsymbol{y'' + (k/x)y' + x^p y^n = 0, k, p > -1, n > 1,}$$

$$\boldsymbol{y'(0) = 0, y(0) = 1.}$$

First it is seen that if $u(x)$ is a solution of the given IVP, then

$$z(x) = su(s^\nu x),$$

where $s \ge 0, \nu = (2+p)^{-1}(n-1)$ satisfies the given DE and the conditions $z(0) = s, z'(0) = 0$.

Introducing the form $y(s) = su(s^\nu x_0)$, in the given DE we may verify that y satisfies the DE

$$y'' = (1-\gamma)y'/s + (\gamma - 1)y/s^2 - ay^n/s^2, \ ' \equiv \frac{d}{ds}, \tag{1}$$

where

$$\gamma = \frac{(n-1)(k-1)}{2+p}, \quad a = \left(\frac{n-1}{2+p}\right)^2 x_0^{2+p}.$$

Now introducing $r = \ln s$, (1) may be written as the system

$$\begin{aligned} y' &= z, \\ z' &= (2-\gamma)z + (\gamma-1)y - ay^n, \ ' \equiv \frac{d}{dr}, \end{aligned} \tag{2}$$

where $y(-\infty) = z(-\infty) = 0$.

In the phase plane, (1) becomes

$$\frac{dz}{dy} = [(2-\gamma)z + (\gamma-1)y - ay^n]/z. \tag{3}$$

For $\gamma > 1$, the origin is a saddle point for (2) and linearizing the system in the phase plane, we find that $z = y$ and $z = (1-\gamma)y$. We also have a second singular point $(\delta, 0)$, where $\delta = [a^{-1}(\gamma-1)]^{1/(n-1)}$. For $\gamma = 2$, we may solve (3) with IC $y(-\infty) = z(-\infty) = 0$ to obtain

$$y'^2(s)/2 = z^2/2 = y^2/2 - ay^{(n+1)}/(n+1). \tag{4}$$

Another integration of (4) yields

$$y(s) = s\left(1 + 2\frac{x_0^{2+p_s n-1}}{(n+1)(k-1)^2}\right)^{-2/(n-1)}.$$

Grizan (1988)

72. $$y'' + (k/x)y' + x^q y^n - f(x) = 0,$$

where $k, q > -1, n > 1, f(x) \geq 0, f \in C(R_+), R_+ = [0, +\infty)$; $y'(0) = 0, y(x_*) = 0$, and $\int_0^{x_*} x^{k+q} y^n dx = C$. Here x_* and C are constants; x_* is the first zero of the solution.

It is proved that for given k, q, n, and $f(x)$, there exists H_* such that for each positive $H = y(0) > H_*$, the given problem has a solution. Ad'yutov and Zmitrenko (1991)

73. $$y'' + [(n-1)/x]y' + \sum_{i=1}^{N} c_i x^{\gamma_i} y^{p_i} = 0, \quad y(a) = y(b) = 0,$$

$-\infty < a < b < \infty$, where $n > 2, c_i > 0$, and γ_i and p_i are given constants.

There exists a unique solution of the given BVP, if either for each i,

$$\max[1, [(n+\gamma_i)/(n-2)] - 2] \leq p_i \leq (n+\gamma_i)/(n-2),$$

or for each i,

$$\max[1, (n+\gamma_i)/(n-2)] \leq p_i \leq [(n+\gamma_i)/(n-2)] + 2.$$

Kwong (1991)

74. $$y'' + [(N-1)/x]y' + f(x)y^p = 0,$$

where $N > 2$ is an integer and p is constant.

It is proved that if $p = (N+2)/(N-2)$, then the given DE is nonoscillatory at the origin provided that there exists a $\delta > 0$ such that the function $x \to f(x)(\log(1/x))^\delta$ is nondecreasing at the origin. Kaper and Kwong (1988)

75. $$y'' + [(n-1)/x]y' + a(x)y^{-p} = 0, x \in (0,1], y'(0) = 0, y(1) = 0.$$

The following theorem is proved: Let $p \in (0,1), n \geq 2$, and a: $[0,1) \to [0,\infty)$, a continuous function such that $0 < \int_0^1 (1-x)^{-p} a(x)dx < \infty$. Then the given BVP has at least one positive solution in $C^1[0,1] \cap C^2(0,1]$. Gatica, Oliker, and Waltman (1989)

76. $$y'' + [(n-1)/x]y' + a(x)y^{-p} = 0, x \in (0,1], y'(0) = 0, y(1) = 0.$$

Let $p \in (0,1), n \geq 2$, and a:$[0,1) \to [0,\infty)$, a continuous function such that $0 < \int_0^1 (1-x)^{-p} a(x)dx < \infty$.

Then it is proved that the given boundary value problem has at least one positive solution, in $C^1[0,1] \cap C^2(0,1]$. Gatica, Oliker, and Waltman (1989)

77. $$y'' + [\gamma\sigma/x]y' + [a(x)y^{2/\gamma} - (1/4 + \alpha/x^2)]y = 0,$$

where σ, γ are constants and $\alpha = (1/2)\gamma\sigma(1-\gamma\sigma/2)$.

If we write $y(x) = x^{-(1/2)\gamma\sigma} z(x)$, we obtain Eqn. (2.5.16) for z. Kaper and Kwong (1988)

78. $$y'' + (n-1)y'/x + g(x)y + h(x)y^p = 0,$$

where $n > 2$ is not necessarily an integer, and $g(x)$ and $h(x)$ satisfy the following conditions: $g(x)$ and $h(x)$ are in $C^1(0,\infty)$, $x^{2-\sigma} g(x) \to 0$ and $a^{2-\sigma} h(x) \to 0$ as $x \to +0$ for some $\sigma > 0$.

Under the foregoing assumptions, the solution of the given DE satisfying $y(0) < \infty, y(x) > 0$ for $x \in [0,R)$ and $y(R) = 0$ is in $C([0,R)) \cap C^2((0,R))$ and satisfies $xy' \to 0$ as $x \to +0$. Here if $R \to \infty$, we also mean that $y(x) \to 0$ as $x \to +\infty$.

Under further conditions on $h(x)$ and $g(x)$ uniqueness of the solution is proved, both when R is finite and infinite. Several other qualitative properties of the solution are also discussed. Yanagida (1991)

79. $$y'' + (n-1)y'/x + \lambda e^y = 0, y(1) = y'(0) = 0,$$

where λ and n are constants.

The following uniqueness properties are proved. There exists a finite positive value λ_* depending on n such that there are:

(a) No solution when $\lambda > \lambda_* (n \geq 1)$.

(b) One solution when $\lambda = \lambda_* (n \geq 1)$.

(c) Two solutions when $0 < \lambda < \lambda_* (n = 1, 2)$.

(d) An infinite number of solutions when $\lambda = 2 (n = 3)$.

(e) A finite but large number of solutions when $|\lambda - 2| \neq 0$ is small $(n = 3)$.

(f) An infinite number of solutions when $\lambda = 2(n-2)(n < 10)$.

(g) A finite but large number of solutions when $|\lambda - 2(n-2)| \neq 0$ is small $(n < 10)$.

(h) One solution for each $\lambda < 2(n-2)(n \geq 10)$.

Local, phase plane, and global analysis are used to prove the foregoing results. A singular "similarity" solution is

$$y_s = \log(1/x^2), \lambda_s = 2(n-2).$$

Joseph and Lundgren (1972)

80. $$\boldsymbol{y'' + (\alpha/x)y' = \beta e^{\gamma y}, \gamma \neq 0, \beta \neq 0.}$$

If we put $x = rz$ and $y(rz) = f(z)$, we have

$$\frac{d^2 f}{dz^2} + \frac{\alpha}{z}\frac{df}{dz} = \beta r^2 e^{\gamma f}.$$

It is shown that the given DE has analytic solutions which together with two derivatives converges absolutely for $|x| \leq r$. For the special case $\alpha = 2, \beta = -1$, and $\gamma = 1$, the following power series solution exists:

$$y(x) = \frac{-x^2}{6} + \frac{x^4}{(5)4!} - \left[\frac{8}{(2!6!)}\right] x^6 + \left[\frac{122}{(8!8!)}\right] x^8 - \left[\frac{5032}{(495)10!}\right] x^{10} + \cdots$$

and has also been discussed by Davis (1962). Ifantis (1987a)

81. $$\boldsymbol{y'' + ay'/x + be^y = 0,}$$

where a and b are constants.

For $b(a-1) > 0$, one solution of the given DE is

$$y = \log\left[(2a-2)/(bx^2)\right]. \tag{1}$$

With $y(x) = \bar{y}(\bar{x}), x = \bar{x}|b|^{1/2}$, the given DE goes to an equation of the same form with $b = \pm 1$. Through the transformation

$$y = \eta(\xi) - 2\xi, \xi = \log x,$$

the given DE becomes autonomous:

$$\eta'' + (a-1)\eta' + be^{\eta} = 2(a-1). \tag{2}$$

Put $\eta'(\xi) = p(\eta)$ in (2) to obtain

$$pp' + (a-1)p + be^{\eta} = 2(a-1).$$

Also, if we put

$$u(t) = xy'(x), t = x^2 e^y,$$

the given DE goes to

$$t(u+2)u' + (a-1)u + bt = 0. \tag{3}$$

Through the transformation

$$y(x) = r(s) + \log[2(a-1)/(bx^2)], s = x^{1-a},$$

the given DE for $b(a-1) > 0$ goes to

$$(a-1)s^2 r'' + 2(e^r - 1) = 0. \tag{A}$$

Some special cases of the given DE may be noted:

(a) $a = 0$. A first integral of the given DE is

$$y' = \pm\beta(C_1 - e^y)^{1/2} \text{ with } \beta = (2|b|)^{1/2}.$$

Thus the cases $b > 0$ or $b < 0$ are both covered. Furthermore, one gets for $C_1 = c^2 > 0$,

$$c^2 e^{-y} = \begin{cases} \cosh^2[(1/2)c\beta(x - C_2)] & \text{for } b > 0, \\ \sinh^2[(1/2)c\beta(x - C_2)] & \text{for } b < 0. \end{cases}$$

$C_1 = -c^2 < 0$:

$$c^2 e^{-y} = \sin^2[(1/2)c\beta(x - C_2)],$$

provided that $b < 0$.

$C_1 = 0$:

$$e^{-y} = -(1/2)b(x - C_2)^2.$$

(b) $a = 1$. One gets from (3) the solution

$$u^2 + 4u + 2bt + 4C = 0, \tag{4}$$

where C is an arbitrary constant. Using the given DE along with (4), we get the Riccati equation for y':

$$2x^2 y'' = x^2 y'^2 + 2xy' + 4C,$$

which through $y' = -2v'/v$ goes to the linear Cauchy-Euler form

$$x^2 v'' - xv' + Cv = 0.$$

(c) $a = 2$. One solution in the present case is given by (1). A closed-form solution generally is not obtainable.

(d) $a \neq 0, 1$. A series solution in the neighborhood of $x = 0$ can be found.

Kamke (1959), p.562

82. $$\boldsymbol{y'' + (a/x)y' - \phi^2 y \exp[\gamma\beta(1 - y)/[1 + \beta(1 - y)]] = 0,}$$

$y'(0) = 0, y(1) = 1$, where a, ϕ, γ, and β are constants.

The numerical solution of the given BVP shows that for a given value of parameter γ and for a low value of the product $\gamma\beta$, only one solution exists. For higher values of $\gamma\beta$ in a certain range $\phi \in (\phi_1, \phi_2)$, three solutions occur. Some specific results are:

(a) $a = 0, \gamma = 20, \gamma\beta = 2, \phi = 1$; one solution and $y(0) = 0.3745$.

(b) $a = 2, \gamma = 20, \gamma\beta = 4, \phi = 2$; one solution and $y(0) = 0.0028$.

(c) $a = 0, \gamma = 20, \gamma\beta = 14, \phi = 0.16$; three solutions exist.

Kubicek and Hlavacek (1983), pp.36,92,225,242,255,261

83. $$y'' + \alpha y'/x + ax^{\beta-\alpha}e^y = 0,$$

$y'(0) = 0, y(\infty) = 0; \alpha, a,$ and β are constants.

The given BVP is converted into two IVPs using the property of invariance group. Ames and Adams (1979)

84. $$y'' + ay'/x + bx^{4-2a}e^y = 0,$$

where a and b are constants.

With $\eta(\xi) = y(x), \xi = x^{3-a}, (a \neq 3)$, the given DE goes to

$$\xi\eta'' + [2/(3-a)]\eta' + [b/(3-a)^2]\xi e^\eta = 0. \quad (1)$$

Equation (1) is a special case of Eqn. (2.8.81). Kamke (1959), p.563

85. $$y'' + (\gamma/x)y' - y + |y|^\lambda \operatorname{sgn} y = 0, y(0^+) = 0, y(+\infty) = 0,$$

where γ and λ are real and $\lambda > 0$.

A complete analysis in terms of the parameters is given. The principal results are the following:

(a) The given problem has a nontrivial solution if and only if one of the following conditions is fulfilled:

  (i) $\gamma > 1, 1 < \lambda < (\gamma + 3)/(\gamma - 1)$.
  (ii) $0 < \gamma \leq 1, \lambda > 1$.
  (iii) $\gamma = 0, \lambda > 1$.
  (iv) $\gamma < 0, \lambda > 1$.
  (v) $\gamma > 0, \lambda < 1$.

(b) All nontrivial solutions of the given problem are oscillatory when (i) or (ii) holds, have no zeros when (iii) or (iv) holds, and are oscillatory when (v) holds.

(c) In case (i) or (ii) for any nonnegative integer ℓ, there exists a solution of the given problem with exactly ℓ zeros in $(0, \infty)$; in case (iii) this problem has only two solutions, and in case (iv) or (v) for any $y_0 \in (-1, 1)$, it possesses a solution such that $y(0+) = y_0$. The proof contains several interesting oscillation and estimation arguments.

Shekhter (1986)

86. $$y'' + \gamma y'/x + y - |y|^\lambda \operatorname{sgn} y = 0, y'(0+) = 0, y(+\infty) = 0,$$

where γ and λ are real, and $\lambda > 0$.

Several qualitative properties—zeros of solutions, their monotonicity, asymptotic behavior and bounds—are considered. Singular solutions are also treated. Shekhter (1989)

87. $$y'' + \gamma y'/x + y - |y|^\lambda \operatorname{sgn} y = 0, y'(0+) = 0, y(+\infty) = 0,$$

where γ and λ are real and $\lambda > 0$.

Several qualitative properties of the solutions of the given DE, including zeros, montonicity, asymptotics, bounds, etc., are discussed. Shekhter (1989)

88. $$y'' + [(N-1)/x]y' + \lambda y + |y|^{p-1}y = 0, 0 < x < 1,$$

$y'(0) = 0, y(1) = 0$, where $N \geq 3$ is a positive integer.

Setting $\rho = \lambda^{1/2}x, v(\rho) = \lambda^{-1/(p-1)}y(x)$, we obtain

$$v'' + [(N-1)/\rho]v' + v + |v|^{p-1}v = 0, \rho > 0, v'(0) = 0 \quad (1)$$

with the boundary conditions at $x = 1$ becoming $v(R) = 0, R = \lambda^{1/2}$. Now the problem is studied by a shooting argument assuming that $v(0) = \gamma, v'(0) = 0$. For that purpose, another transformation,

$$t = [(N-2)/\rho]^{N-2}, z(t) = v(\rho),$$

changes the problem (1) to

$$z'' + t^{-k}z(1 + |z|^{p-1}) = 0, 0 < t < \infty, \ z(t) \to \gamma \text{ as } t \to \infty,$$

$$z[\{(N-2)/R\}^{N-2}] = 0, R = \lambda^{1/2},$$

where $k = 2(N-1)/(N-2), p = 2k - 3$.

It is shown that no solution of the problem exists in the neighborhood of $\lambda = 0$. Atkinson, Brezis, and Peletier (1990)

89. $$y'' + \alpha y'/x + a(\text{sgn } y)x^{\beta-\alpha}|y|^{\gamma} = 0, x \geq 0$$

(α, a, β, and γ are parameters) with BCs such as

$$y(0) = 0, \lim_{x\to\infty} y(x) = c \neq 0,$$

or

$$y(0) = 0, \lim_{x\to\infty} y'(x) = c \neq 0,$$

or

$$y(0) = c, \lim_{x\to\infty} y(x) = 0$$

are studied using simple one-parameter group properties. In all cases, boundary value problems are converted into initial value problems using the property of invariance group. An eigenvalue problem in a finite interval $y(0) = 0, y'(0) = 1, y(1) = 0$ is also studied; the eigenvalues are computed, and the results tabulated. Ames and Adams (1979)

90. $$y'' + [(N-1)/x]y' + |y|^{p-1}y - |y|^{q-1}y = 0,$$

$y(x) \to 0$ as $|x| \to \infty$, and the requirement that y remains bounded in the interval $0 \leq x < \infty$ imposes the additional constraints

$$y'(0) = 0, y = O(x^{2-N}) \text{ as } x \to \infty.$$

Here $N > 2, q > p > (N+2)(N-2)$. Putting $t = [(N-2)/x]^{N-2}$ and $u(t) = y(x)$, we get the equivalent problem

$$\begin{aligned} &u'' + (1/t^k)(|u|^{p-1}u - |u|^{q-1}u) = 0, \\ &u(0) = 0, u'(0) > 0, \text{ and } -1 \leq \lim_{t\to\infty} u(t) \leq 1, \end{aligned} \quad (1)$$

where $k = (2N-2)/(N-2)$. Now go to Eqn. (2.2.68). Troy (1990)

91. $$\boldsymbol{y'' + (n-1)y'/x + \lambda F(y) = 0, y(0) = A, y'(0) = 0, y(1) = 0,}$$

where $F(u) > 0$ when $u \geq 0$ and $F(u)$ satisfies Lipschitz condition with constant $k; n$ and λ are constants.

When n is large, the given problem may be rewritten as

$$\epsilon xy'' + y' + \mu xF(y) = 0, y(0) = A, y'(0) = 0, y(1) = 0, \tag{1}$$

where $\epsilon = 1/(n-1), \mu = \lambda/(n-1)$. Although (1) suggests that it is a singular perturbation problem, it really is not.

The following result is proved. Suppose that $y(x, A, n)$ and $\lambda(A, n)$ are positive solutions of the given problem. Let $y(x, A)$ and $\mu(A)$ solve the problem

$$\frac{dy}{dx} + \mu xF(y) = 0, y(0) = A, y(1) = 0.$$

Then, when A is fixed and n is sufficiently large,

$$|y(x, A, n) - y(x, A)|_\infty \leq k_1(A)/n$$

and

$$|\mu(A) - \lambda/n| \leq k_2(A)/n,$$

where k_1 and k_2 are independent of n. Joseph and Lundgren (1972)

92. $$\boldsymbol{y'' + (r/x)y' - y + yf(y^2) = 0, x \in (0,\infty), r > 0,}$$

$\lim_{x\to\infty} y(x) = 0$, where f satisfies certain conditions.

Provided that:

(a) $f \in C(0,\infty)$.

(b) $f(s) > 0$ if $s > 0$.

(c) $|f(s)| \leq C|s|^\sigma$, where $0 < \sigma < 2/(r-1)$ if $r > 1; \sigma > 0$ if $0 < r \leq 1$.

(d) There exists $\delta > 0$ such that $s^{-\delta}f(s)$ is strictly increasing on $(0,\infty)$, then there exists a solution of the given system for which $y \in C^2(0,\infty), y(x) > 0$ for $x \in (0,\infty)$. If $0 < r \leq 1$, or $1 < r < 5$, and $0 < \sigma \leq (5-r)/[2(r-1)]$, then also $0 < \lim_{x\to 0^+} y(x) < \infty$ and $\lim_{x\to 0^+} y'(x) = 0$.

Kurtz (1981)

93. $$\boldsymbol{y'' + (\alpha/x)y' - \beta y - G(y) = 0,}$$

where $G(y)$ is analytic in y and satisfies the conditions $G(0) = G'(0) = 0$.

From a general theorem it follows that for $\alpha \neq -n+1$, $n = 1, 2, \ldots$, the given DE has a one-parameter family of analytic solutions of the form

$$y(x) = \lambda + \sum_{n=1}^{\infty} b_n(\lambda)x^n, y(0) = \lambda, y'(0) = 0,$$

which, together with the first two derivatives, converges absolutely for $|x| \leq 1$ and sufficiently small $|\lambda|$. The dependence of the domain of analyticity on the parameter λ is studied. Ifantis (1987a)

94. $$y'' + [(n-1)/x]y' + f(y) = 0,$$

where n is a positive integer.

(a) $y'(0) = 0$; Upper and lower bounds for the solution of the given problem are found when $f(y)$ has the form $f(y) = y[1 + g(y)]$, where $g(y)$ positive, even, and monotonic increasing in y. Furthermore, asymptotic behavior of these solutions for large x and location of zeros of the solutions are also studied.

(b) $y_0 < y(0) < y_2, y'(0) = 0, y(\infty) = y_0$, where y_0, y_2, and $f(y)$ satisfy certain elaborate conditions. Using a backward shooting argument, existence of a solution of the BVP is proved; this solution is found to be "linearly unstable." The solution is also shown to be unique. Budd (1988), Klaasen and Troy (1984)

95. $$y'' + [(n-1)/x]y' + f(y) = 0, 0 < x < \infty, 0 < y < \infty, |y'| < +\infty,$$

where (a) n is a real parameter such that $n > 2$, and (b) the function $f(y)$ is defined and continuous for $y > 0$ and $f(y) > 0$ for all $y > 0$.

Suppose that the integral $\int_0^1 f(u)u^{-2(n-1)/(n-2)}du$ is convergent. Then it is shown that for any positive constant C, the given differential equation has a positive solution $y = y(x)$ which exists for sufficiently large x and satisfies $y(x) \sim Cx^{-(n-2)}$ as $x \to +\infty$. The change of variables

$$u = x^{-(n-2)}, z = -x^{n-1}\frac{dy}{dx}$$

transforms the given DE to

$$\frac{du}{dy} = (n-2)z^{-1}, \; \frac{dz}{dy} = -f(y)u^{-(n-1)/(n-2)}z^{-1}.$$

We wish to find a solution $(u(y), z(y)), 0 < y < y_0$ satisfying $u(y) \sim y/C, z(y) \to (n-2)C$ as $y \to +0$, where y_0 is a small positive number specified later. The proof is carried through an integral equation formulation and use of Schauder's fixed-point theorem. Makino (1984)

96. $$y'' - (A_1/x)y' - (A_0/x^2)y = f(x, y), 0 < x \leq 1,$$

$$B(y(0), y'(0); y(1), y'(1)) = 0,$$

where y, f are vector-valued functions of dimension n, B is a vector-valued function of dimension $m \leq n$, and A_0 and A_1 are constant $n \times n$ matrices. Existence, uniqueness, and smoothness properties of the BVP are considered. Weinmüller (1984)

97. $$y'' + \beta y'/x + f(x, y) = 0, 0 < x < 1, y'(0) = 0, y(1) = A,$$

where A is a given constant and $\beta \geq 1$.

If $\sup \dfrac{\partial f}{\partial y} < k$, where k is a first positive zero of $J_{(\beta-1)/2}(k^{1/2})$ (J_s is the Bessel function of the first kind of order s), then a unique solution is shown to exist. The linear eigenvalue problem is considered first, and then an iterative scheme is combined with maximum principle to prove the desired result. Chawla and Sivakumar (1987)

98. $$\boldsymbol{y'' + (k/x)y' + \phi(xy'/y)x^p y^n = 0,}$$

$$y'(0) = 0, y(x_0) = y_0, y(x) \geq 0, \forall x \in [0, x_0],$$

where $k, p > -1, n > 1$, and $x_0, y_0 > 0$.

The results similar to those for Eqn. (2.8.61) may be proved. Grizan (1988)

2.9 $y'' + f(x)y' + g(x, y) = 0$

1. $\boldsymbol{y'' + (x/2)y' - y/(1-p) - y^p = 0, \quad y(0) = 0, \; y'(0) = p; \; 0 < p < 1.}$

An exact solution of the given DE is

$$y = \left\{(1-p)/[2(1+p)]^{1/2}\right\}^{2/(1-p)} (-x)^{2/(1-p)}.$$

Grundy and Peletier (1990)

2. $\boldsymbol{y'' + 2xy' + y^3 - 1 = 0, y'(0) = 0, \quad y(x_*) = 0, \quad \int_0^{x_*} y^3 x^2 dx = (\pi/\sigma)^{3/2},}$

where x_* and σ are constants; x_* is the first zero of $y(x)$, which decreases monotonically from $y_0 = y(0)$ to $y(x_*) = 0$.

Conditions for the existence and uniqueness of a solution are obtained; a relation between $y(0), x_*$, and σ is obtained. See Eqn. (2.8.72) for a more general form. Ad'yutov and Zmitrenko (1991)

3. $$\boldsymbol{y'' - (x/2)y' + \lambda(e^y - 1) + (N-1)/x = 0,}$$

for $x > 0$, $k > 0$,

$$y'(0) = 0, \; y(0) = \alpha > 0, \text{ where } N \geq 2.$$

It is proved that either

$$\lim_{x \to \infty} xy' = -2\lambda$$

or

$$\lim_{x \to \infty} x^{N-1} e^{-x^2/4} \, y' = C,$$

where C is a negative constant. Budd and Qi (1989)

4. $\boldsymbol{y'' - (x/2)y' + e^y - 1 = 0, \quad y(0) = \alpha, \; y'(0) = 0, \; \alpha \in R.}$

It is shown that the given IVP has no solution which has the asymptotic property $y(x) \sim -2 \ln x + K$ as $x \to +\infty$, where K is a constant. Bebernes and Troy (1987)

5. $\boldsymbol{y'' - (x/2)y' + \lambda(e^y - 1) = 0, \quad \lambda > 0, \; y'(0) = 0, \; y(0) = \lambda > 0.}$

The following results are proved: Let $y(x)$ be a solution of the given problem, then either

(a) $\lim_{x \to \infty} xy' = -2\lambda$ called an L-solution, or

(b) $\lim_{x\to\infty} y'\, e^{x^2/4} = C < 0$, called an E-solution, where $C =$ constant < 0.

Budd and Qi (1989)

6.
$$y'' - (x/2)y' - f(y) = 0, \quad x > 0,\ y(0) = a,\ y'(0) = 0,$$

where $f(y)$ is defined for all $y \geq 0$ and satisfies the following.

There exists a $y_0 \in (0, \infty)$ such that $f(y) < 0$ if $y > y_0, f(y_0) = 0, 1 + f'(y) < 0$ and $1 + f'(y) > 0$ if $y < y_0$; and for some $\theta > 0, f(y) \leq \theta y[1 + f'(y)]$. Further, $f''(y) < 0$ if $\theta \geq 1$ and

$$(1 - \theta)[1/2 + f'(y)] > \theta y f''(y) \text{ if } \theta < 1.$$

Notice that if $a > y_0$, then $y''(0) < 0$. Under the conditions above, it is proved that if $a > y_{0'}$, then $y''(x) < 0$ as long as $y(x) > 0$. Hence it is shown that the maximal interval $\{0 < x < x_0\}$ where the positive solution of the given problem exists (when $a > y_0$) is finite. Friedman, Friedman, and McLeod (1988)

7.
$$y'' - xy' - y + y^3 = 0, \quad x \geq 0.$$

The given DE arises in nonlinear instability burst in plane parallel flow. A solution is to be such that $y(x)$ has a positive maximum at $x = 0$, and $y(x) \to 0$ monotonically as $x \to \infty$.

This means that $y'(0) = 0$ and $y'(0) < 0$ and requires that $y(0) = 1 + \alpha,\ \alpha > 0$. The following theorem is proved: Suppose that $y(x)$ is a real function of x satisfying the given DE and $y'(0) = 0,\ y(0) = 1 + \alpha,\ \alpha > 0$; then $y(x) = 0$ for some x. Hocking, Stewartson, and Stuart (1972)

8.
$$\epsilon y'' + xy' - y(1 - y) = 0, \quad -1 < x < 1,\ y(-1, \epsilon) = 0,\ y(1, \epsilon) = -1.$$

It is proved that the given BVP has a solution $y = y(x, \epsilon)$ for ϵ sufficiently small, such that $y(x, \epsilon) \to 0$ for $x \in (-1, 1) - \{0\}$ and $y(0, \epsilon) \to 3/2$ as $\epsilon \to 0$. De Santi (1987)

9.
$$\epsilon y'' + xy' + y(y^2 - 1)(2 - y) = 0, \quad -1 < x < 1,\ y(-1, \epsilon) = -1,\ y(1, \epsilon) = 1.$$

It is shown that the solution of the given BVP, $y = y(x, \epsilon)$, for ϵ sufficiently small, is such that $y(x, \epsilon) \to -1$ for x in $(-1, 0)$ and $y(x, \epsilon) \to 1$ for x in $(0, 1)$ and $y(0, \epsilon) \to s > 1$ as $\epsilon \to 0$, where s is the unique root of a certain polynomial. It may be noted that $y = -1$ and $y = 1$ are solutions of the reduced problem in respective intervals, when $\epsilon = 0$. De Santi (1987)

10.
$$\epsilon y'' + (2x + 1)y' + y^2 = 0,$$
$$y(0) = \alpha,\ y(1) = \beta,$$

where α and β are constants.

A uniformly valid first order composite solution is

$$y = \frac{\beta}{1 + (\beta/2)\ln[(2x + 1)/3]} + \left[\alpha - \frac{\beta}{1 - \left(\frac{\beta}{2}\right)\ln 3}\right]\exp(\xi) + \cdots,$$

where $\xi = x/\epsilon$. Nayfeh (1985), p.351

11.
$$\epsilon y'' - (2x+1)y' + y^2 = 0,$$
$$y(0) = \alpha, \ y(1) = \beta,$$

where α and β are constants.

A uniformly valid first order composite solution is

$$y = \frac{\alpha}{1 - (\alpha/2)\ln(2x+1)} + \left[\beta - \frac{\alpha}{1 - \left(\frac{\alpha}{2}\right)\ln 3}\right] \exp(-3\xi) + \cdots,$$

where $\xi = (1-x)/\epsilon$. Nayfeh (1985) p.355

12.
$$y'' + Rxy' - KRy^n = 0, \quad 0 \le x \le 1,$$
$$y(0) = 0, \ y(1) = 1,$$

where $n \ge 1$ is an integer; the constants K and R are positive.

The problem arises in the analysis of the stagnation point shock layer. A constructive iterative method using a Green's function is employed to show that there exists a unique solution satisfying $0 \le y(x) \le \exp[R(1-x^2)/4], x \in [0,1]$. Numerical results are provided. Pennline (1984)

13.
$$y'' + y'/x - (x/2)y' - f(y) = 0,$$

$x > 0; \ y(0) = a, \ y'(0) = 0.$

Numerical solutions show that there is no concavity of the solution of the given problem if

$$f(y) = y/(p-1) - y^p \ (1 < p < \infty) \quad \text{or} \quad f(y) = 1 - e^y.$$

There is no concavity for

$$f(y) = y^q - y/(1-q) \tag{1}$$

if $q > 0.58$; if $q < 0.58$, concavity for the solution $y(x)$ holds. Also for case (1), numerical solutions show that

$$y'' + (1/2)y' < 0.$$

Friedman, Friedman, and McLeod (1988)

14.
$$y'' + (2/x - x/2)y' + |y|^{p-1}y - y/(p-1) = 0,$$

$y(0) = \alpha \in R, \quad y'(0) = 0, \ 6 \le p \le 12.$

It is proved that there are infinitely many positive, bounded solutions of the given problem such that $\lim_{x\to\infty} y(x) = 0$. Troy (1987)

15.
$$y'' + [(n-1)/x]y' - (x/2)y' - y^q + y/(1-q) = 0,$$

$x > 0, \ n \ge 1, \ 0 < q < 1, \ y(0) = a, \ y'(0) = 0,$

$$\left\{ a > y_0, y_0 = \left[\frac{1}{1-q}\right]^{1/(1-q)} \right\}.$$

It is shown that the given problem does not have a nonnegative monotone decreasing solution in $C^2[0,\infty)$. Friedman, Friedman, and McLeod (1988)

16. $$\boldsymbol{y'' + [(N-1)/x]y' + xy'/2 - [1/(\beta-1)]y + y^\beta = 0,}$$

$$\begin{aligned} \lim_{x\to 0} y'(x) &= 0, \\ \lim_{x\to\infty} y(x) &= 0, \end{aligned}$$

$y(x) > 0$ for $x \in [0,\infty)$.

It is proved that if $n \in \{1, 2, \dots\}$ and $(\beta, N) \in S_n$, then the given BVP has at least n solutions where

$$\begin{aligned} S_0 &= \left\{(\beta, N) \in S : N \geq 10 + \frac{6}{\beta - 1}\right\}, \\ S_n &= \left\{(\beta, N) \in S \backslash \cup_{i=0}^{n-1} S_i : \beta \in \left[1, \frac{(n+1)^2}{n(n+2)}\right] \Rightarrow N \geq 2n + 8 + \frac{2}{n+1}\right. \\ &\quad \left. + \frac{4}{\beta-1} + \frac{2}{(\beta-1)(n+1)}\right\}, \ n = 1, 2, \dots, \\ S &= (1,\infty) \times [1,\infty) \backslash (S_\phi \cup S_\infty), \\ S &= \left\{(\beta, N) \in (1,\infty) \times [1,\infty) \times [1,\infty) : N \leq 2 + \frac{4}{\beta-1}\right\}, \\ S &= \left\{(\beta, N) \in (1,\infty) \times [1,\infty) \backslash S_\phi : N < 6 + \frac{4}{\beta-1} + 4\left[\frac{\beta}{\beta-1}\right]^{1/2}\right\}. \end{aligned}$$

Lepin (1990)

17. $$\boldsymbol{y'' + [(N-1)/x - x/2]y' + y/(1-p) - y^p = 0, \quad x \geq 0;}$$

$N \geq 1,\ 0 < p < 1,\ y \geq 0,\ x > 0,\ y'(0) = 0.$

It is proved that if $y(x)$ is a bounded solution of the given problem on $[0,\infty)$, it is either 0 or $k = (1-p)^{1/(1-p)}$ for all $x \geq 0$. All other bounded solutions of the problem $\to 0$ as $x \to \infty$. Peletier and Troy (1988)

18. $$\boldsymbol{y'' + [(n-1)/x]y' - (x/2)y' - y/(p-1) + y^p = 0,}$$

$x > 0,\ y(0) = a,\ y'(0) = 0$, where $[a > y_0, y_0 = \{1/(p-1)\}^{1/(p-1)}, 1 < p < \infty]$.

It is proved that if $(n-2)p \leq n$, then

$$y'' + [(p+1)/(p-1)]y'/x < 0$$

as long as $y > 0$. It is then inferred that positive global solutions of the given problem do not exist if $p(n-2) < n$; on the other hand, if $p(n-2) > n+2$, then there may exist bounded positive solutions of the given problem for $0 < x < \infty$. Friedman, Friedman, and McLeod (1988)

19. $$\boldsymbol{y'' + [(n-1)/x - x/2]y' + e^y - 1 = 0, \quad 0 < x < \infty,}$$

$y'(0) = 0,\ \lim_{x\to\infty}[1 + (1/2)xy'(x)] = 0.$

The given BVP arises in the ignition model for a high-activation-energy thermal explosion of a solid fuel in the n-dimensional unit sphere. The DE is called the Kapila–Kassoy equation. It is proved that the solution for this problem does not exist for $1 \leq n \leq 2$. For $n \in (2, 10)$, the following theorem holds: For each $n \in (2, 10)$ there is an unbounded sequence of positive numbers $\{\bar{\alpha}_m(n)\}_{m=1}^{\infty}$ such that the solutions $y(x, \bar{\alpha}_m)$ to the initial value problem with $y(0) = \alpha \in R, y'(0) = 0$, satisfy the limit condition

$$\lim_{x \to \infty} [1 + (1/2)xy'(x)] = 0.$$

Eberly and Troy (1987)

20. $$y'' + [(n-1)/x]y' - (x/2)y' + e^y - 1 = 0, \quad 0 < x < x_0,$$

$y(0) = a,\ y'(0) = 0,\ a > 0$. Set $v(t) = y(x)$, where $t = \log x$; then

$$v'' + [n - 2 - (1/2)e^{2t}]v' - e^{2t}(1 - e^v) = 0. \tag{1}$$

It is then proved from (1) that for $n \leq 2, v(t)$ is concave in $t = \log x$. It is inferred that for $n \leq 2$, there does not exist a solution for the given system for all $0 < x < \infty$ for which $y'(x)$ is uniformly bounded. Friedman, Friedman, and McLeod (1988)

21. $$y'' + [(n-1)/x]y' - (x/2)y' - f(y) = 0, \quad x > 0\ (n \geq 1),$$

$y(0) = a,\ y'(0) = 0,\ f(a) < 0$.

Concavity properties of solutions of this equation are studied. See the special cases in Eqn. (2.9.15). Friedman, Friedman, and McLeod (1988).

22. $$y'' - [1/(3-x) - 1/(3+x)]y' - y^2/(3-x) = 0,$$

$y(0) = 1,\ y'(1) = 0$.

A constructive iterative method using Green's function is employed to show that there exists a unique solution of the given problem satisfying

$$e^{v(x)}(3/4)^3 \leq y(x) \leq e^{v(x)}, x \in [0, 1],$$

where

$$e^{-v(x)} = [(3-x)/3]^{1/2}[(3+x)/3]^{1/2}.$$

Pennline (1984)

23. $$y'' + (\sin x/2x^{1/2} - 1/x)y' + xq(x)f(y) = 0,$$

where

$$q(x) = (1/2)x^{-3/2}(2 + \cos x - 2x\sin x), x \geq \pi/2$$

and f can be any of the following:

(a) $f(y) = my; y \in R$ for $m > 0$.

(b) $f(y) = ny + |y|^{\alpha} \operatorname{sgn} y, y \in R$ for $n > 0$ and $\alpha > 0$.

(c) $f(y) = y\ln^2(\mu + |y|); y \in R$ for $\mu > 1$.

(d) $f(y) = ye^{\lambda|y|}; y \in R$ for $\lambda \geq 0$.

(e) $f(y) = \sinh y; y \in R$.

Then the given DE is shown to have oscillatory solutions. Grace and Lalli (1990)

24. $$\boldsymbol{y'' + [\alpha f f'/(\beta + f^2) - f''/f']y' + [\gamma f'^2/(\beta + f^2)]y = 0,}$$

where α, β, and γ are constants and f is a function of x.

Cases are found for which the solution is obtainable in the form of quadratures. Mitrinovitch (1956)

25. $$\boldsymbol{y'' - c(x)y' - e(x)y^2 - f(x)y - g(x) = 0,}$$

where c, e, f, and g are analytic functions of x.

By a suitable change of dependent and independent variables, it becomes possible to change the given equation first to

$$y'' = 6y^2 + Kx + H,$$

where K and H are constants, and hence to one of the following forms:

(a) $y'' - 6y^2 = 0$.
(b) $y'' - 6y^2 - (1/2) = 0$.
(c) $y'' - 6y^2 - x = 0$.

Forms (a) and (b) have solutions in terms of an elliptic function, while (c) is the first Painlevé transcendent. Each of (a)–(c) has movable poles of order 2. Bureau (1964)

26. $$\boldsymbol{y'' + 3q(x)y' + [2q^2(x) + q'(x)]y - 2y^3 = 0.}$$

Changing the variables according to

$$y = u(s)V(x), \quad \ln V = -\int^x q dx, \ s = \int^x V dx,$$

we get

$$u''(s) - 2u^3 = 0, \tag{1}$$

with the first integral

$$u'^2 = u^4 + c_0, \tag{2}$$

where c_0 is integration constant. A real solution of (2) exists only if $c_0 < 0$ and is found in terms of an elliptic function:

$$u = \mp\alpha[\text{cn}\{2^{1/2}\alpha(s - s_0), (1/2)^{1/2}\}]^{-1}, \tag{3}$$

where

$$\alpha = |c_0|^{1/4} = -(1/2)^{1/2}; \tag{4}$$

the solution (3) is obviously a periodic divergent solution. Dixon, Tuszynski, and Otwinowski (1991)

27. $$\boldsymbol{y'' + f_1(x)y' + f_2(x)y + f_3(x)y^3 = 0,}$$

where $f_i\,(i = 1,\ 2,\ 3)$ are smooth functions.

A Laurent series expansion $y(x) = \sum_{j=0}^{\infty} a_j(x - x_1)^{j-n}$ shows that $n = 1, a_0^2 = -2$. Substitution into the given DE shows that at the resonance $r = 4$, a complicated condition on the derivatives of f_i and f_i' should hold; then a_4 becomes arbitrary. Special cases may be considered more easily.

(a) $f_1(x) = c_1, f_2(x) = c_2, f_3(x) = c_3, c_i$ being constants; then the resonance condition becomes

$$c_3^4\, c_1^2\, (2c_1^2 - 9c_2) = 0,$$

so that $2c_1^2 = 9c_2$ and c_3 is arbitrary.

(b) $f_1(x) = 0,\quad f_3(x) = 1,$ and $\quad f_2''(x) = 0$, i.e., $\quad f_2(x) = Ax + B$, where A and B are constants. The given DE then is

$$y'' + (A + Bx)y + y^3 = 0. \tag{1}$$

Equation (1) is a special case of the second Painlevé transcendent; the solution has no branch points.

(c) $f_2(x) = 0, f_3(x) = 1$. In this case the resonance condition becomes

$$f_1''' + 3f_1''f_1 + 2f_1'^2 + \left(\frac{10}{3}\right) f_1'f_1^2 + \frac{4}{9}f_1^4 = 0. \tag{2}$$

Equation (1) admits the particular solutions $3/x, 3/2x$. Thus, (2) admits more than one branch in the Painlevé analysis. In this case the given DE is not of Painlevé type, but weak Painlevé. It admits noninteger resonances.

(d) $f_1(x) = \dfrac{1}{4x},\quad f_2(x) = \dfrac{1}{8x^2},\quad f_3(x) = \dfrac{1}{32x^2}$. In this case the given DE is of Painlevé type and can be solved in terms of elliptic functions.

(e) $f_1(x) = \dfrac{1}{4x},\quad f_2(x) = \dfrac{1}{8x^2},\quad f_3(x) = \dfrac{-1}{8x^2}$. Writing $y(x) = x^{1/4}g(x^{1/4})$ in the given DE reduces the latter to

$$\frac{d^2g}{ds^2} = 2g^3, \quad s = x^{1/4}.$$

Hence the solution can be expressed in terms of elliptic functions.

(f) $f_1(x) = \dfrac{2}{x},\ f_2(x) = 0,\ f_3(x) = \dfrac{1}{x}$.

The given DE may be shown to be of Painlevé type. Euler, Steeb, and Cyrus (1989)

28. $$\boldsymbol{y'' + p(x)y' + q(x)y = (\lambda/y^3)\exp\left(-2\int_0^x p(x)dx\right),}$$

where λ is a parameter.

If $y_1(x)$ and $y_2(x)$ are linearly independent solutions of

$$y'' + p(x)y' + q(x)y = 0,$$

then the general solution of the given DE may be written as

$$y = [Ay_1^2 + By_2^2 + 2Cy_1y_2]^{1/2},$$

where A, B are arbitrary, and

$$C^2 - AB = \lambda\left[(y_1'y_2 - y_1y_2')^{-2}\ \exp[-2\int P(x)dx]\right].$$

Eliezer and Gray (1976)

29. $$\boldsymbol{y'' + f(x)y' - \alpha y^n \exp\left(-2\int f(x)dx\right) = 0,}$$

where α and n are constants.

The general integral of the given DE can be found to be

$$\int \left[c_1 + \frac{2\alpha}{n+1} y^{n+1}\right] dy = \theta(x, c_2),$$

where

$$\theta(x, c_2) = c_2 + \alpha^{1/2} \int \exp\left(-\int f(x)dx\right) dx.$$

Bandic (1965)

30. $$\boldsymbol{y'' + p(x)y' - Ke^{-2F} y^n = 0, \quad n \neq 1, -3,}$$

where $F = \int^x p\, dx$ and K is a constant. A first integral of the given DE is

$$\begin{aligned} I &= e^{2F} y'^2 - \frac{2K}{n+1} y^{n+1}, \quad n \neq -1 \\ &= e^{2F} y'^2 - 2K \log y, \quad n = -1. \end{aligned} \tag{1}$$

The solution for $n \neq -1$ may be found to be

$$y = [A + Bu]^{2/(1-n)}, \quad B^2 = \frac{K(n-1)^2}{2(n+1)}, \tag{2}$$

where A is an arbitrary constant and $u = \int^x e^{-F} dx$; solution (2) is obtained from (1) by putting $I = 0$. Ranganathan (1992a)

31. $$\boldsymbol{y'' + \beta(x)y' + \alpha(x)y^m = 0,}$$

where m and a are constants and

$$\frac{\alpha^{-1}(x) \exp[-2\int \beta(x)dx]}{[\int \exp\{-\int \beta(x)dx\}dx]^{(m+3)/2}} = \text{constant} = K,$$

say.

A first integral of the given DE is

$$\begin{aligned} &(K/2)y'^2 \left[\exp\left(2\int \beta(x)dx\right)\right] \left[\int \exp\left(-\int \beta(x)dx\right) dx\right] - (K/2)yy' \exp\left[\int \beta(x)dx\right] \\ &\quad + [1/(m+1)]y^{m+1} \left[\int \exp\left(-\int \beta(x)dx\right) dx\right]^{-(m+1)/2} = \text{constant}. \end{aligned}$$

Crespo Da Silva (1974)

32. $$\boldsymbol{y'' + \beta(x)y' + \alpha(x)y^m = 0, \quad m \neq -1.}$$

It is shown using an intuitive argument that if $\alpha(x)$ and $\beta(x)$ satisfy the condition

$$\left(\alpha^{-2/(m+3)}\right) \left(2\beta + \alpha^{-1}\alpha'\right) \exp\left(\frac{m-1}{m+3} \int^x \beta(x')dx'\right) = c_1,$$

where c_1 is a constant, we can find a first integral of the given DE as

$$\left(y'^2 + \frac{2\alpha}{m+1}y^{m+1}\right) \exp\left(2\int^x \beta(x')dx'\right) \left(c + c_2 \int^x \exp\left(-\int^{x'} \beta(x'')dx''\right) dx'\right)$$
$$- c_2 y y' \exp\left(\int^x \beta(x')dx'\right) = \text{constant},$$

where c and c_2 are constants. Sarlet and Bahar (1980)

33. $$\boldsymbol{y'' + [P'(x)/P(x)]y' + [Q(x)/P(x)]y^n = 0,}$$

$y(0) = a > 0$, is prescribed, n is a parameter, and $P(x)$ and $Q(x)$ are continuous functions such that $P(0) = 0$ and $P'(x) > 0, Q(x) > 0$, for $x > 0$.

Qualitative properties of solutions are studied. In particular, the existence of zeros of solutions is studied. Richard (1951)

34. $$\boldsymbol{y'' - A(x)y^n - p(x)y' = 0, \quad 0 \leq x \leq 1,}$$

where $A(x) > 0$ and $p(x)$ is continuous on $[0,1]$; the BCs are either $y(0) = 1, y(1) = 1$ or $y(0) = 1, y'(1) = 0$.

The given problem arises in stagnation point shock layer and radiation transfer for annular fins, respectively.

It is proved that each of the given problems has at least one solution and each of the solutions is nonnegative on $[0,1]$. Baxley (1990)

35. $$\boldsymbol{y'' + f(x)y' + \phi(x)y - q(x)y^n = 0.}$$

It is shown that under the transformation $x = \alpha(t), y = \beta(t)z$ the functions

$$\begin{aligned} I_1 &= f^{n+3}q^{-2}, \\ I_2 &= f^{-2}[f^{1/2}(f^{-1/2})'' - (1/2)f' + \phi] \end{aligned}$$

represent absolute invariants of the given DE. Bandic (1964)

36. $$\boldsymbol{y'' + f(x)y' + \phi(x)y + \theta(x)y^n = 0,}$$

where n is a natural number.

Introducing the new variables

$$x = \alpha(t), \quad y = \beta(t)z(t),$$

the given DE becomes

$$\ddot{z} + F(t)\dot{z} + \phi(t)z + \psi(t)z^n = 0, \quad \cdot \equiv \frac{d}{dt}, \tag{1}$$

where

$$\begin{aligned} F(t) &= \frac{\dot{\alpha}^2}{\beta}\left[\frac{2\dot{\beta}}{\dot{\alpha}^2} - \frac{\ddot{\alpha}}{\dot{\alpha}^3} + \frac{\beta}{\dot{\alpha}}f(x)\right], \\ \phi(t) &= \frac{\dot{\alpha}^2}{\beta}\left[\frac{\ddot{\beta}}{\dot{\alpha}^2} - \frac{\ddot{\alpha}\dot{\beta}}{\dot{\alpha}^3} + \frac{\dot{\beta}}{\dot{\alpha}}f(x) + \beta\phi(x)\right], \end{aligned} \tag{2}$$

where with the choice $\dot{\alpha} = \beta^2, \beta(t)$ is eliminated. The functions $\phi(t), F(t)$, and $\psi(t)$ are then related to $f(x), \phi(x)$, and $\theta(x)$ by

$$\begin{aligned} &\frac{1}{f^2(x)}\left[\phi(x) - uf(x)/2 - u'/2 + u^2/4\right] \\ &= \frac{1}{F^2(t)}\left[\phi(t) - vF(t)/2 - \dot{v}/2 + v^2/4\right]\frac{\theta(x)}{(f(x))^{(n+3)/2}} \\ &= \frac{\psi(t)}{[F(t)]^{(n+3)/2}}, \end{aligned} \tag{3}$$

where

$$u = f'(x)/f(x), \quad v = \dot{F(t)}/F(t)$$

and

$$\int f(x)dx = \int F(t)dt. \tag{4}$$

Equations (2) imply that the solution of the given DE has the form

$$y = \left[\frac{F(t)}{f(x)}\right]^{1/2} Z_1(t, C_{11}, C_{21}), \tag{5}$$

where Z_1 is the general integral of (1) and C_{11} and C_{21} are arbitrary constants.

Thus to obtain the solution of a DE of the given form, we choose $F(t), \phi(t), \psi(t)$ so that (1) can be solved by quadratures. Substituting these into (3), we obtain two relations between $f(x), \phi(x)$, and $\theta(x)$, for which (5) gives solutions for the given equation. Since (3) connect three functions $f(x), \phi(x)$, and $\theta(x)$, one may be chosen at will. It proves convenient to choose $f(x)$: $\phi(x)$ and $\theta(x)$ can then be found in an elementary way.

The transformation of the variables becomes known from (2) and (4). Braude (1967)

37. $$\boldsymbol{y'' + g(x)y' + f(y) = 0, \quad x > 0.}$$

Free boundary values for the given DE are considered. The question of the existence of a finite point p and a solution y satisfying the conditions $y'(0) = 0, y(x) > 0, 0 < x < P; y(P) = y'(P) = 0$. The existence proof is based on shooting methods. The nonnegative function g is allowed to be $O(x^{-1})$ as $x \to 0$. The classes of admissible functions f and g include functions of the type $f(y) = y^p - y^q$ and $g(x) = m/x$, where $0 \le q < p \le 1$ and $m > 0$. Kaper (1990)

38. $$\boldsymbol{y'' + f(x)y' + \mathrm{K}\left[\exp\left(-2\int f(x)dx\right)\right] h(y) = 0.}$$

The solution is given in a slightly implicit form

$$\int \left(A - 2k\int h(y)dy\right)^{-1/2} dy = \int \left(\exp\left[-\int f(x)dx\right]\right) dx + B,$$

where A and B are constants. Keckic (1975)

39. $$\boldsymbol{y'' + f(x)y' + w^2(x)y - G(x)F[k(x)y] = 0,}$$

where $f(x), w(x), G(x)$, and $F(Z)$ are arbitrary functions.

The given DE is equivalent to the pair of equations

$$\rho'' + f(x)\rho' + w^2(x)\rho = MW^2(x)/\rho^3, \tag{1}$$

$$\frac{d^2r}{d\tau^2} + Mr = F(r), \tag{2}$$

where

$$G(x) = \frac{W^2(x)}{\rho^3(x)}, \quad k(x) = \frac{1}{\rho(x)}, \quad \text{and} \quad W(x) = \exp\left(-\int^x f(\bar{x})d\bar{x}\right) \tag{3}$$

and M is an arbitrary constant.

The proof of the statement is easily carried out by using the general transformation

$$y = \rho(x)r(\tau), \quad d\tau = \mu(x)dx. \tag{4}$$

It may be noted that Eqn. (2) is autonomous and has the first integral

$$I = \frac{1}{2}\left(\frac{dr}{d\tau}\right)^2 + \frac{1}{2}Mr^2 - \int^r F(\lambda)d\lambda.$$

The transformation (4) may be regarded as a rule for finding a solution of the given DE in terms of solutions $\rho(x)$ and $r(\tau)$ of (1) and (2), respectively, under certain integrability conditions. The integrability conditions are just the relations (3) required to generate the autonomous equation (2). Thus the solution of the given DE can be written in the form

$$\begin{aligned} y &= \rho(x)r(\tau), \\ \tau &= \int_{x_0}^x \frac{W(\lambda)}{\rho^2(\lambda)}d\lambda \end{aligned}$$

provided that ρ is any particular solution of (1) and $r(\tau)$ is the general solution of (2).
Reid and Cullen (1982)

40. $$\boldsymbol{y'' + a(x)y' + b(x)y + g(x)f(y^2) - ke(x) = 0.}$$

Assume that a, b, g, and e are continuous functions of x, and k is a constant. Further assume that b is nonnegative, nondecreasing, continuously differentiable, and $b(0) \neq 0$:

(a) $\int_0^x \left[\frac{e^2(\hat{x})}{[b(\hat{x})]^{1/2}}\right] d\hat{x} \leq M \ \forall x \in [0, \infty]$.

(b) $\int_0^\infty |a(x)|dx$, $\int_0^\infty \left[\frac{|g(x)|}{[b(x)]^{1/2}}\right] dx$, and $\int_0^\infty \left[\frac{1}{[b(x)]^{1/2}}\right] dx$ are convergent.

(c) f is a continuous nondecreasing function of its argument, and $\int_0^\infty \left[\frac{1}{u^3 + f(u^2)}\right] du = \infty$. It is then shown that every solution of the given DE is bounded as $x \to \infty$.

Behzad and Mehri (1971)

41. $$\boldsymbol{y'' - p(x)y' - f(x,y) = 0,}$$

where one of the following boundary conditions is assumed:

(a) $y(0) = y_0, y(1) = y_1$.

(b) $y'(0) = 0, y(1) = y_1$.

(c) $y(0) = y_0, y'(1) = 0$.

We give some details for the first problem. Similar results are obtained for the other problems.

Let $p(x)$ be once continuously differentiable function. We first remove the gradient term from the equation by multiplying it by $\exp[-v(x)]$, where

$$v(x) = (1/2) \int_0^x p(\xi) d\xi,$$

etc. We get, in terms of u,

$$u'' = F(x, u), \tag{1}$$

where

$$F(x, u(x)) = e^{-v(x)} f(x, e^{v(x)} u(x)) + q(x) u(x),$$

$$\begin{aligned} u(x) &= e^{-v(x)} y(x), \\ q(x) &= [p(x)/2]^2 - p'(x)/2. \end{aligned}$$

The boundary conditions become $u(0) = y_0, u(1) = e^{-v(1)} y_1$. Equation (1) is replaced by the equivalent one, namely

$$u'' - k^2 u = F(x, u) - k^2 u. \tag{2}$$

Now, the BVP for (2) can be converted into an integral equation by the Green's function procedure for the operator $\dfrac{d^2}{dx^2} - k^2$:

$$u(x) = h(x) + \int_0^1 g_k(x, \xi) \left[k^2 u(\xi) - F(\xi, u(\xi))\right] d\xi, \tag{3}$$

where

$$g_k(x, \xi) = 1/(k \sinh k) \begin{cases} \sinh[k(1-\xi)] \sinh(kx), & 0 \le x < \xi, \\ \sinh[k(1-x)] \sinh(k\xi), & \xi \le x < 1, \end{cases}$$

and

$$h(x) = [y_0 \sinh[k(1-x)] + e^{-v(1)} y_1 \sinh(kx)] / \sinh k.$$

Equation (2) is looked upon as an equation of the form $u = T_k u$, where T_k is the nonlinear operator defined by the RHS of (2). For a particular choice of k, a unique fixed point of $u = T_k u$ is established in a closed subset of the space of continuous functions on $[0, 1]$ with norm $\| u \| = \max |u(x)|$, i.e., a subset of the form

$$S = \{u : m_1 \le u(x) \le m_2\}$$

for some constants m_1, m_2.

Assume that $q(x)$ occurring in the definition of (1) is nonnegative on $[0, 1]$, so that if we define

$$\delta_2 = \min_{0 \le x \le 1} q(x), \quad N_2 = \max_{0 \le x \le 1} q(x), \tag{4}$$

we shall have $\delta_2 \geq 0$ and $N_2 \geq 0$. Conditions on $q(x)$ are later weakened. The following result is then proved:

Let $\max\{|y_0|, |y_1|\} \leq M$. Suppose that there exists an $N_1 > 0$ and a $\delta_1 \geq 0$ such that $0 \leq \delta_1 \leq \dfrac{\partial f}{\partial y} \leq N_1$ for all $x \in [0, 1]$ and all y such that

$$|y(x)| \leq e^{v(x)} \max\{1, e^{-v(1)}\}M, x \in [0, 1].$$

Suppose further that

$$0 \leq f(x, y) \leq (N_1 + \delta_1)y, y \geq 0 \ \text{ and } \ (N_1 + \delta_1)y \leq f(x, y) \leq 0, y \leq 0,$$

for all $x \in [0, 1]$ and all y such that

$$|y(x)| \leq e^{v(x)} \max\{1, e^{-v(1)}\}M, x \in [0, 1].$$

Then there exists a unique solution of the given problem satisfying

$$|y(x)| \leq e^{v(x)} \max\{1, e^{-v(1)}\}M, x \in [0, 1].$$

Let $\delta = \delta_1 + \delta_2$ and $N = N_1 + N_2$, where N_2 and δ_2 are given by (4). For

$$k^2 = (\delta + N)/2,$$

the unique solution is given by

$$y(x) = e^{v(x)}u(x),$$

where $v(x)$ is given by

$$v(x) = (1/2) \int_0^x p(\xi)d\xi$$

and $u(x)$ is the limit of the convergent sequence of functions

$$\begin{aligned} u^{(0)}(x) &= h(x) = [y_0 \sinh[k(1 - x)] + e^{-v(1)}y_1 \sinh(kx)]/(\sinh k), \\ u^{(m+1)}(x) &= h(x) + \int_0^1 g_k(x, \xi)[k^2 u^m(\xi) - F(\xi, u^m(\xi))]d\xi, \quad m = 0, 1, \ldots, \end{aligned}$$

where $g_k(x, \xi)$ is defined in (3). Pennline (1984)

42. $\quad$ $\boldsymbol{y'' - g(x)y' - f(x, y) = 0.}$

The transformation

$$z = y \exp\left[-(1/2) \int_0^x g(x')dx'\right]$$

reduces the given DE to the form $z'' = h(x, z)$. Now see Eqn. (2.6.26). Aguirre and Krause (1988)

43. $\quad$ $\boldsymbol{y'' - \lambda(x)y' - F(x, y) = 0.}$

The given DE has a first integral of the form

$$y'^n + a_1(x, y)y'^{n-1} + a_2(x, y)y'^{n-2} + \cdots + a_n(x, y) = Ca_0(x, y), \tag{1}$$

where C is a constant, provided that the PDE

$$\frac{\partial A}{\partial x} + A\frac{\partial A}{\partial y} = \lambda(x)A + F(x,y) \tag{2}$$

has n solutions $A_1(x,y), A_2(x,y), \ldots, A_n(x,y)$ such that

$$\frac{\partial}{\partial y}(A_1 + \cdots + A_n) = \alpha(x),$$

where $\alpha(x)$ is a function of the variable x only. Moreover, the function a_0 in (1) is related to the function $\alpha(x)$ by the formula

$$a_0(x,y) = \exp\int [n\lambda(x) - \alpha(x)]dx. \tag{3}$$

By a certain transformation, we may choose $a_0 = 1, a_1 = 0$ in (1), and $\lambda = 0$ in the given DE. Airault (1986)

2.10 $y'' + kyy' + g(x,y) = 0$

1. $$y'' + yy' + y^3 = 0.$$

Referring to Eqn. (2.12.31), we have $f(y) = y, g(y) = y^3$ and $F(y) = y^2/2, f(y)F(y) = (1/2)y^3$, so that $g(y) > 2f(y)F(y)$ and all conditions in Eqn. (2.12.31) are satisfied with $\alpha = 2$. Hence the given DE has a center at the origin. Jordan and Smith (1977), p.333

2. $$y'' + yy' - y^3 = 0.$$

This is a special case of Eqn. (2.10.7). Kamke (1959), p.548; Murphy (1960), p.382

3. $$y'' + yy' - y^3 + ay = 0,$$

$a \neq 0$ is a constant.

This is a special case of Eqn. (2.10.7). The solution is

$$\begin{aligned} y &= (1/2)(a/3)^{1/2}\left[P'(u,12,C_1)/(P(u,12,C_1) - 1)\right], \\ u &= (x/2)(a/3)^{1/2} + C_2; \end{aligned}$$

C_1 and C_2 are arbitrary constants. P is a Weierstrass elliptic function. Kamke (1959), p.548; Murphy (1960), p.382

4. $$y'' + yy' - y^3 - f(x)y - g(x) = 0,$$

where $f(x)$ and $g(x)$ are analytic.

It may be verified that the only movable singularities of the given DE are poles provided that (a) $f'(x) + g(x) = 0$; (b) if we set $f = -12V, g = 12V'$, then the function V must satisfy

$$V'' = 6V^2 + Kx + H, \tag{1}$$

where K and H are constants. Equation (1) has solutions that are either elliptic functions or solutions of P_I. Indeed, the given equation with $f = -12V, g = 12V'$ is equivalent to the system

$$y' + y^2 = v, \tag{2}$$

$$v' = vy - 12Vy + 12V'. \tag{3}$$

On eliminating y between (2) and (3), we get

$$v'' = v^2 - 12Vv + 12V''. \tag{4}$$

The transformation $v - 6V = 6w$ takes (4) to the form

$$w'' = 6w^2 + V'' - 6V^2.$$

This DE has poles as its only movable singularities [see Eqn. (2.9.25)] only if

$$V'' = 6V^2 + Kx + H \quad \text{and} \quad w'' = 6w^2 + Kx + H. \tag{5}$$

Therefore, the general solution of

$$y'' = -y\,y' + y^3 - 12Vy + 12V',$$

where V is a particular solution of (5), is

$$y = (w' - V')/(w - V);$$

w is a solution of (5) distinct from V. Bureau (1964)

5. $$\boldsymbol{y'' + yy' - b(y' + y^2) - y^3 + (1/25)b^2 y + (1/25)b^3 = 0,}$$

where b is an arbitrary constant.

Put $y' + y^2 = 6z + \alpha$. Choose α appropriately to reduce the equation to an integrable form. See also Eqn. (2.13.9). Forsyth (1959), p.303

6. $$\boldsymbol{y'' + yy' - b(y' + y^2) - y^3 + (49/25)b^2 y + (49/25)b^3 = 0,}$$

where b is an arbitrary constant.

Put $y' + y^2 = 6z + \alpha$. Choose α appropriately to reduce it to an integrable form. See also Eqn. (2.13.9). Forsyth (1959), p.303

7. $$\boldsymbol{y'' + (y + 3a)y' - y^3 + ay^2 + 2a^2 y = 0,}$$

where a is a constant.

Put $y' = p, y'' = p\dfrac{dp}{dy}$, etc. and integrate:

$$y = C_1 e^{-ax} \frac{P'(u,0,1)}{P(u,0,1)},$$

with

$$u = \begin{cases} (C_1/a)e^{-ax} + C_2 & \text{for } a \neq 0, \\ C_1 x + C_2 & \text{for } a = 0; \end{cases}$$

C_1 and C_2 are arbitrary constants. Kamke (1959), p.548; Murphy (1960), p.382

8. $$y'' + yy' + \beta y^3 = 0.$$

For $\beta = 1/9$, the transformation $y = 3u'/u$ reduces the given DE to $u''' = 0$, yielding the solution

$$y(x) = [3(A_1 + 2A_2 x)]/[A_0 + A_1 x + A_2 x^2],$$

which involves essentially two arbitrary constants, A_1/A_0 and A_2/A_0. The analytic character (as $x \to \infty$) depends on the existence and the sign of the roots of $1 + A_1 x + A_2 x^2 = 0$ (we may put $A_0 = 1$ without loss of generality). If there is no positive root, the asymptotic behavior is $y(x) = 6/x, A_2 \neq 0$ and $y(x) = 3/x$ if $A_2 = 0$. If a positive root exists, then the solution exhibits an explosive character. For general β, we note that the given DE has elementary invariants $\xi = yx, \eta = y'x^2$. Introducing these, we obtain the system

$$\frac{d\xi}{d\theta} = w, \quad \frac{dw}{d\theta} = (3 - \xi)w + \left(\xi^2 - 2\xi - \beta\xi^3\right), \tag{1}$$

where $\theta = \log x$ and $w = \eta + \xi$. The system (1) in the (w, ξ) phase plane becomes

$$w\frac{dw}{d\xi} = (3 - \xi)w + \xi^2 - 2\xi - \beta\xi^3. \tag{2}$$

The first order DE (2) has some special exact solutions, leading to special solutions of the original equation:

(a) $w = 2\xi - (1/3)\xi^2, \beta = 1/9, \xi = 6\exp(2\theta)/[K + \exp(2\theta)], y = 6x/(K + x^2)$. Similarly,

(b) $w = -6 + 3\xi - (1/3)\xi^2, \beta = 1/9, \xi = (3K + 6\exp\theta)/(K + \exp\theta), y = 6(K/2 + x)/[-K^2/4 + (K/2 + x)^2]$.

(c) $w = \xi + a_2\xi^2, \beta = -a_2(2a_2 + 1), \xi = -\exp\theta/(K + a_2\exp\theta), y = -1/(K + a_2 x)$.

In all cases K and a_2 are arbitrary constants.

Painlevé analysis of the given DE, phase plane analysis of (2), and general qualitative behavior for general β are carried out. For example, we have the Laurent expansions

$$\begin{aligned} y(x) &= 3(x - x_0)^{-1} + a_1[1 - (a_1/3)(x - x_0) + (a_1/3)^2(x - x_0)^2 \\ &\quad - (a_1/3)^3(x - x_0)^3 + (a_1/3)^4(x - x_0)^4 - (a_1/3)^5(x - x_0)^5 + \cdots] \\ &= 3(x - x_0)^{-1} + 3a_1/[3 + a_1(x - x_0)], \end{aligned}$$

on summation, where a_1 is an arbitrary constant, and

$$\begin{aligned} y(x) &= (x - x_0)^{-1} + a_2(x - x_0)^2[1 + (a_2/21)(x - x_0)^3 \\ &\quad + [a_2^2/(3.21)]^2(x - x_0)^6 \\ &\quad + [11a_2^3/(39.21.21)](x - x_0)^9 + [4a_2^4/(39.21.21)](x - x_0)^{12} + \cdots], \end{aligned}$$

where a_2 is arbitrary. Leach, Feix, and Bouquet (1980)

9. $$y'' + 2yy' = 0, \quad y(0) = y'(0) = -1.$$

The given DE is exact. A first integral is

$$y' + y^2 = 0,$$

if the ICs are considered. Now the variables separate. The solution satisfying the IC is $y = (x-1)^{-1}$. Bender and Orszag (1978), p.34

10. $$\boldsymbol{y'' + 2yy' = 0, \quad y(0) = 1,\ y(1) = 0.5.}$$

Exact solution of the given BVP is $y = 1/(x+1)$. A numerical solution is obtained using shooting methods. The results are compared. Roberts (1979), p.356

11. $$\boldsymbol{y'' + 2yy' = 0, \quad y(0) = 1,\ y^2(1) + y'(1) = 0.}$$

Exact solution of the given BVP is $y = 1/(x+1)$. Numerical solution is obtained using shooting methods and is compared with the exact analytic solution. Roberts (1979), p.356

12. $$\boldsymbol{y'' + 2yy' = 0, \quad y'(0) + 2y(0) = 1;\ y^2(1) + y'(1) = 0.}$$

Exact solution of the given BVP is $y = 1/(x+1)$. Numerical solution is obtained using shooting methods and is compared with the exact analytic solution. Roberts (1979), p.361

13. $$\boldsymbol{y'' + 2yy' - C(x)y^2 - B(x)y - A(x) = 0,}$$

where A, B, and C are polynomials.

For the given DE it is proved that every meromorphic solution in the complex plane is either also a solution of the Riccati equation with polynomial coefficients or we have

$$B(x) = C^2(x) - C'(x)$$

and y satisfies the Riccati equation

$$y' = h(x) - C(x)y - y^2,$$

where the entire (transcendental) function $h(x)$ is defined by

$$h' = A(x) + C(x)h.$$

Steinmetz (1983)

14. $$\boldsymbol{y'' + 3yy' + y^3 = 0.}$$

The transformation $y = u'/u$ changes the given DE to $u''' = 0$ so that $u = C_0 + C_1x + C_2x^2$, where $C_i (i = 0, 1, 2)$ are arbitrary constants. The solution of the given DE is

$$y = (2x + b_1)/(x^2 + b_1x + b_0),$$

where $b_1 = C_1/C_2$ and $b_0 = C_0/C_2$. Goldstein and Braun (1973), p.101

15. $$\boldsymbol{y'' + 3yy' + y^3 + [2/\{3(3)^{1/2}\}]x^{-3} = 0.}$$

An exact solution of the given DE is

$$y = \lambda/x, \quad \lambda = 1 + (3)^{1/2}/3.$$

Erbe (1977)

16. $$\boldsymbol{y'' + 3yy' + y^3 + p(x) = 0,}$$

where $p(x) \geq 0$ and continuous on $[a, b), 0 < a < b \leq +\infty$.

This second order Riccati equation is related to a linear third order equation $u''' + p(x)u = 0$ through the transformation $y = u'/u$. Several new comparison theorems for the given Riccati equation are then obtained. Erbe (1977)

17. $$y'' + 3yy' + y^3 - y = 0.$$

Put $y = \psi'/\psi$ to obtain the linear third order DE

$$\psi''' - \psi' = 0. \tag{1}$$

Equation (1) is easily solved. Conte and Musette (1992)

18. $$y'' + 3yy' + y^3 + f(x)y - g(x) = 0.$$

The transformation $y = u'(x)/u(x)$ reduces the given DE to a linear DE of third order: $u''' + f(x)u' - g(x)u = 0$. Kamke (1959), p.550; Murphy (1960), p.383

19. $$y'' + 3yy' + y^3 - b_2(x)y - (1/2)[b_2'(x) + k] = 0,$$

where $b_2(x)$ is analytic and k is a constant.

Putting $y = w'(x)/w(x)$, we get a linear third order DE,

$$w''' - b_2(x)w' - (1/2)[b_2'(x) + k]w = 0.$$

Therefore, the given DE has poles as the only movable singularities in its solution. Bureau (1964)

20. $$y'' + 3yy' - b(y' + y^2) + y^3 - cy - f = 0,$$

where b, c, and f are arbitrary constants.

Change the variable according to $y = AY + B, x = CX$ and choose A, B, and C to relate the given DE to Eqn. (2.10.21). Forsyth (1959), p.303

21. $$y'' + [3y + f(x)]y' + y^3 + f(x)y^2 = 0.$$

The transformation $y = u'(x)/u(x)$ reduces the given DE to a linear DE of third order:

$$u''' + f(x)u'' = 0. \tag{1}$$

Equation (1) is easily integrated. Kamke (1959), p.550; Murphy (1960), p.383

22. $$y'' + (c - y)y' = 0,$$

$y \to 0$ as $x \to \infty$ and $y \to y_0(> 0)$ as $x \to -\infty$.

The given problem arises from the traveling wave form of the Burgers equation, with shock-like solution. Integrating twice and using the boundary conditions, the solution is easily found to be

$$y = (1/2)y_0[1 - \tanh(y_0x/4 + A)],$$

where A is a constant, and $y_0 = 2c$. Drazin and Johnson (1989), p.33

23. $$\mathbf{y'' - yy' + y = 0.}$$

A first integral is $y' + \ln|y' - 1| = y^2/2 + c$, where c is an arbitrary constant. Hence the solution in the phase plane may be discussed. Wilson (1971), p.40

24. $$\mathbf{y'' - yy' + y = 0.}$$

Referring to the general theorem for Eqn. (2.12.31), all conditions therein are satisfied; therefore, the origin $y = 0, y' = 0$ is a center. Jordan and Smith (1977), p.343

25. $$\mathbf{y'' - yy' - y = 0, \quad y'(-\infty) = -1, \; y(+\infty) = 0.}$$

The given DE has a first integral

$$y' = \ln(1 + y') + (1/2)y^2,$$

where we have used the condition at $x = +\infty$. The problem was then solved numerically. The numerical solution that tends to $1 - x$ as $x \to -\infty$ is depicted. Hallam and Loper (1975)

26. $$\mathbf{y'' - yy' + \sin y = 0.}$$

Referring to the general theorem for Eqn. (2.12.31), all the conditions therein are satisfied for this special case; hence the origin $y = 0, y' = 0$ is a center. Jordan and Smith (1977), p.343

27. $$\mathbf{y'' - 2yy' = 0.}$$

Putting $y' = p$, we have $p\dfrac{dp}{dy} - 2yp = 0$. Therefore, either $p = 0$, i.e., $y =$ constant or $\dfrac{dp}{dy} - 2y = 0$, which after integration twice leads respectively to $\dfrac{dy}{dx} = p = y^2 + c_1$ and

$$x = \begin{cases} \dfrac{1}{c_1^{1/2}} \tan^{-1}(y/c_1^{1/2}) + c_2 & \text{when } c_1 > 0, \\[2ex] -\dfrac{1}{y} + c_2 & \text{when } c_1 = 0, \\[2ex] \dfrac{1}{2(-c_1)^{1/2}} \ln\left[\dfrac{(y - (-c_1)^{1/2}}{y + (-c_1)^{1/2}}\right] & \text{when } c_1 < 0, \end{cases} \tag{1}$$

where c_1 and c_2 are constants. We may invert (1) to get y as a function of x. Martin and Reissner (1958), p.76

28. $$\mathbf{y'' - 2yy' - y' = 0.}$$

Putting $y' = p$ in the given autonomous DE, we have

$$p\frac{dp}{dy} - 2yp - p = 0.$$

Therefore, either $p = 0$, i.e., $y = A$, a constant, or

$$\frac{dp}{dy} - 2y - 1 = 0. \tag{1}$$

After integrating (1) once we have

$$y' = p(y) = y^2 + y + B, \tag{2}$$

where B is an arbitrary constant. The variables are now separable. The solution of (2) is

$$(2/D)[\tan^{-1}((2y+1)/D)] = x + C,$$

where $D^2 = 4B - 1$, and C is another arbitrary constant, or more explicitly,

$$y = E\tan(Ex + F) - 1/2,$$

where $E = D/2$ and $F = CE$. Zwillinger (1989), p.154

29. $$\boldsymbol{y'' - 2yy' - a\beta y' + 2\alpha y = 0,}$$

where a, α, and β are parameters.

The given DE arises from a similarity reduction of a generalized Burgers equation. The inverse transformation $y = 1/z$ leads to an Euler-Painlevé equation; see Sachdev (1991). Doyle and Englefield (1990)

30. $$\boldsymbol{y'' - 3yy' - (3ay^2 + 4a^2y + b) = 0,}$$

where a and b are constants.

With $y' = p, y'' = p\dfrac{dp}{dy}$, we have

$$pp'(y) = 3yp + b + 4a^2y + 3ay^2. \tag{1}$$

Equation (1) is an Abel DE, which may be solved explicitly in special cases. See Sachdev (1991). Kamke (1959), p.550; Murphy (1960), p.383

31. $$\boldsymbol{\epsilon y'' + 2yy' = 0, \quad y(0) = -1,\ y(1) = b.}$$

For $\epsilon = 0.05$ and for five different values of $b \simeq 1$, the solutions are found numerically and depicted. Bohe (1990)

32. $$\boldsymbol{y'' - 2(\alpha y/v - 1)y' = 0,}$$

where α and v are constants.

The given DE arises from a traveling wave form of the Burgers equation $u_t + \alpha u u_x - u_{xx} = 0$. It integrates to yield

$$y' - \alpha y^2/v + 2y = 0, \tag{1}$$

if we set constant of integration equal to zero. Equation (1) has an exact "solitary wave" solution

$$y = (v/\alpha)[1 - \tanh x]. \tag{2}$$

Hereman, Banerjee, Korpel, and Assanto (1986)

33. $\boldsymbol{y'' - 2ayy' - a = 0,}$

where a is a constant.

An integration gives the Riccati equation

$$y' - ay^2 - ax - C = 0, \tag{1}$$

where C is an arbitrary constant. The transformation $v'(x) = -ay(x)v(x)$ changes (1) to the linear equation $v'' + a^2xv + aCv = 0$. Kamke (1959), p.551; Murphy (1960), p.383

34. $\boldsymbol{y'' - \epsilon yy' + y = 0,}$

where ϵ is a small parameter.

A two-term perturbation solution with period 2π of the given DE is

$$y = a\cos x + \epsilon a^2[(1/6)\sin(2x) - (1/3)\sin x] + O(\epsilon^2).$$

The Lindstedt method is used. Jordan and Smith (1977), p.146

35. $\boldsymbol{y'' + a_0 yy' + b_0 y^3 = 0,}$

where a_0 and b_0 are constants.

As a particular case, choosing $a_0 = 3, b_0 = 1$, it is easily shown that the given DE has a unique solution of the form

$$y = (x - x_0)^{-1}\left[2 + \sum_{\sigma=1}^{\infty} \alpha_\sigma (x - x_0)^\sigma\right],$$

where α_σ are uniquely determined. Kondratenya and Yablonskii (1968)

36. $\boldsymbol{y'' + ayy' + by^3 = 0,}$

where a and b are constants.

Put $p(y) = y'$ so that $pp' + ayp + by^3 = 0$. Now, write $p(y) = y^2u(t), t = \ln y$ to obtain the DE

$$uu' + 2u^2 + au + b = 0. \tag{1}$$

Equation (1) is solved as

$$t = -\int \frac{udu}{2u^2 + au + b} + C_1, \tag{2}$$

where C_1 is an arbitrary constant. The integral (2) can be evaluated. For the original DE, one has $y'(x) = y^2u(\ln y), y' \neq 0$, so that

$$x = \int \frac{dy}{y^2u(\ln y)} + C_2,$$

where C_2 is a constant. The case $a = -4, b = 2$ can be solved explicitly. Kamke (1959), p.551

37. $\boldsymbol{y'' + (3\gamma + 1)yy' + (3\gamma - 1)y^3 + c = 0, \quad y(0) = 0,\ y'(0) = y_0,}$

where γ and c are constants.

The given system arises in the study of pellet fusion process. The transformation

$$s' = \mu y s, \quad \mu = (1/2)(\gamma+1)/(\gamma-1),$$

leads to

$$s'' = \mu s\left(y' + \mu y^2\right), \quad s''' = \mu s\left(y'' + 3\mu y\, y' + \mu^2 y^3\right).$$

The given DE for $\gamma = 5/3, \mu = 2$, linearizes to

$$s''' + 2cs = 0, \quad s(0) = 1,\ s'(0) = 0,\ s''(0) = 2y_0$$

and the solution is

$$s(x) = c_1 e^{-Fx} + e^{(F/2)x}\left[c_2\cos(3^{1/2}Fx/2) + c_3\sin(3^{1/2}Fx/2)\right],$$

where $F = (2c)^{1/3}$ and c_1, c_2, c_3 are given by

$$\begin{bmatrix} c_1 \\ c_2 \\ c_3 \end{bmatrix} = \frac{1}{3}\begin{bmatrix} 1 & 0 & 1 \\ 2 & 0 & -1 \\ 0 & 0 & 3^{1/2} \end{bmatrix}\begin{bmatrix} 1 \\ 0 \\ 2y_0F^{-2} \end{bmatrix}.$$

The solution is

$$y = \frac{s'}{2s} = \frac{-F}{2}\,\frac{c_1e^{-Fx} - c_4e^{Fx/2}\sin\{(3^{1/2}/2)Fx + \phi + \pi/3\}}{c_1e^{-Fx} + c_4e^{F/2}\sin\{(3^{1/2}/2)Fx + \phi\}},$$

where

$$c_4 = \left(c_2^2 + c_3^2\right)^{1/2}, \quad \phi = \tan^{-1}(c_2/c_3).$$

Ervin, Ames, and Adams (1984)

38. $$\boldsymbol{y'' - (ky+h)y' - ay^3 - by^2 - cy - f = 0,}$$

where a, b, c, f, k, and h are constants.

Conditions are obtained such that the integrals of this equation are subuniform, i.e., they do not have moving critical points, and that the assignment of initial arbitrary values determines the integral function, "uniform" in the vicinity of the initial point. Forsyth (1959), p.294

39. $$\boldsymbol{y'' + (1/(\beta\mu))(\alpha y - v)y' - (1/(\beta\mu^2))\left(\epsilon_0 + \epsilon_1 y + \epsilon_2 y^2 + \epsilon_3 y^3\right) = 0,}$$

where $\beta, \mu \neq 0, \alpha, v$, and $\epsilon_i (i = 0,1,2,3)$ are constants.

The given DE arises from a generalized Burgers equation with a certain type of background interaction.

An exact solution is $y = a_0 + a_1 \tanh x$, where

$$\begin{aligned} a_0 &= \frac{1}{6\epsilon_3}\left[-2\epsilon_2 \pm \left\{4\epsilon_2^2 + 12\epsilon_3\left(\alpha\mu a_1 - 2\beta\mu^2 - \epsilon_1\right)\right\}^{1/2}\right], \\ a_1 &= \left[\frac{-\alpha \pm (\alpha^2 + 8\epsilon_3\beta)^{1/2}}{2\epsilon_3}\right]\mu, \\ v &= \frac{\alpha a_0 a_1 \mu + \epsilon_2 a_1^2 + 3a_0 a_1^2 \epsilon_3}{a_1\mu}, \end{aligned}$$

and the parameter μ is determined from

$$-\epsilon_2 a_1^2 - 3a_0 a_1^2 \epsilon_3 = \epsilon_0 + \epsilon_1 a_0 + \epsilon_2 a_0^2 + \epsilon_3 a_0^3.$$

Lan and Wong (1989)

40. $$\epsilon y'' + yy' - y = 0, \quad 0 \le x \le 1, \; y(0) = \alpha, \; y(1) = \beta,$$

where $\epsilon \ll 1, \alpha$, and β are constants.

The solution of the given problem is obtained by a nonasymptotic method. It is shown that the nature of the inner solution for both left- and right-hand boundary layers depends on the roots of a transcendental equation. From sketches of this function, the location of roots can be found. For the left-hand boundary layer, depending on the relative size and signs of α and β, thirteen cases exist for the possible solutions of the transcendental equation. Of these cases, only five correspond to acceptable solutions. Similar remarks apply to the right-hand boundary layer solutions. Numerical experience with the method is also reported to confirm the theoretical analysis. Roberts (1984)

41. $$\epsilon y'' + yy' - y = 0, \quad y(0) = A(\epsilon), \; y(1) = B(\epsilon),$$

where ϵ is a parameter.

(a) Let $y(x,\epsilon)$ be a solution of the given system. Then $\min(A(\epsilon), B(\epsilon)-1) \le y(x,\epsilon) - x \le \max(A(\epsilon), B(\epsilon)-1)$. Moreover, if $A(\epsilon) \ge B(\epsilon) - 1$, then $y(x,\epsilon)$ is the unique solution of the given system.

(b) Suppose that $y(x,\epsilon)$ is a solution of the given system and $A(\epsilon) \ge 0, B(0) > 0$. If $B(0) \ge 1$, then $\lim_{\epsilon \to 0^+} y(x,\epsilon) = x + B(0) - 1$. If $0 < B(0) < 1$, then

$$\lim_{\epsilon \to 0^+} y(x,\epsilon) = \begin{cases} 0, & 0 < x \le 1 - B(0), \\ x + B(0) - 1, & 1 - B(0) \le x \le 1. \end{cases}$$

(c) Suppose that $A(0) < 0, B(\epsilon) \le 0$. If $A(0) \le -1$, then

$$\lim_{\epsilon \to 0^+} y(x,\epsilon) = A(0) + x, \quad 0 \le x < 1.$$

If $-1 < A(0) < 0$, then

$$\lim_{\epsilon \to 0^+} y(x,\epsilon) = \begin{cases} A(0) + x, & 0 \le x \le -A(0), \\ 0, & -A(0) \le x < 1. \end{cases}$$

(d) Let $y(x,\epsilon)$ be a solution of the given system. Suppose that

$$A(\epsilon) < 0 < B(\epsilon). \tag{1}$$

Then there is a unique point $C = C(\epsilon)$ such that $y(C,\epsilon) = 0$. Moreover, if

$$B(\epsilon) - A(\epsilon) < 1, \tag{2}$$

then

$$0 < y'(x,\epsilon) \le 1, \quad 0 \le x \le 1, \tag{3}$$

If $B(\epsilon) - A(\epsilon) > 1$, then $y'(x,\epsilon) \ge 1$.

(e) Suppose that there is an $\epsilon_0 > 0$ such that (1) and (2) hold for $0 < \epsilon \le \epsilon_0$. Let $y(x,\epsilon)$ be a solution of the given system. Then

$$\lim_{\epsilon\to 0^+} y(x,\epsilon) = \begin{cases} A(0)+x, & 0 \le x \le |A(0)|, \\ 0, & |A(0)| \le x \le 1-B(0), \\ B(0)-1+x, & 1-B(0) \le x \le 1. \end{cases}$$

Dorr, Parter, and Shampine (1973)

42. $$\epsilon y'' + yy' - y = 0, \quad 0 < x < 1,$$

$y(0) = A(\epsilon),\ y(1) = B(\epsilon)$.

Suppose that $y(x,\epsilon)$ is a solution of the given system and $A(\epsilon) \ge 0, B(0) > 0$. If $B(0) \ge 1$, then $y(x,\epsilon) = x + B(0) - 1, 0 < x \le 1$. If $0 < B(0) < 1,\ \epsilon \to 0^+$, then

$$\lim_{\epsilon\to 0^+} y(x,\epsilon) = \begin{cases} 0, & 0 < x \le 1-B(0), \\ x+B(0)-1, & 1-B(0) \le x \le 1. \end{cases}$$

Several other results as $\epsilon \to 0^+$ are also proved. Maximum principle methods are used. Dorr, Parter, and Shampine (1973)

43. $$\epsilon y'' + yy' - xy = 0, \quad \epsilon \ll 1,\ y(0) = \alpha,\ y(1) = \beta.$$

Inner, outer, and matched solutions for different locations of the boundary layer are discussed. Nayfeh (1985), p.340

44. $$\epsilon y'' + yy' - (2x+1)y = 0, \quad \epsilon \ll 1,\ y(0) = \alpha,\ y(1) = \beta,$$

where $y = O(1); \epsilon, \alpha$, and β are constants. Nayfeh (1985), p.359

45. $$\epsilon y'' - yy' - f(x)y = 0, \quad \epsilon \ll 1,\ y(0) = \alpha,\ y(1) = \beta,$$

where $y = O(1)$ and $f(x)$ is a known function of $x; \epsilon, \alpha$, and β are constants. Nayfeh (1985), p.359

2.11 $y'' + f(y)\,y' + g(x,y) = 0, f(y)$ polynomial

1. $$y'' + \left(3y^2 - 2\right) y' + y + y^3 = 0.$$

See Eqn. (2.12.33). All the hypotheses required of the given DE are satisfied. Hence the given DE has a stable limit cycle. Lakin and Sanchez (1970), p.93

2. $$y'' + w_0^2 y + 2\epsilon\mu y^2 y' + \epsilon\alpha y^3 = 0,$$

where $\epsilon \ll 1, w_0, \mu$, and α are constants.

By using the method of multiple scales, a first order uniform approximation is found to be

$$y = \left[a_0/[1+(1/2)\epsilon\mu a_0^2 x]^{1/2}\right] \cos\left[w_0 x + (3\alpha/(4w_0\mu)) \ln[1+(1/2)\epsilon\mu a_0^2 x] + \beta_0\right] + \cdots,$$

where a_0 and β_0 are constants. Nayfeh (1985), p.115

3. $$y'' - \epsilon\left(1-y^2\right)y' + y = 0.$$

First changing the variable to $\tau = \epsilon x$, we get the DE

$$\epsilon^2 \frac{d^2y}{d\tau^2} - \epsilon^2(1-y^2)\frac{dy}{d\tau} + y = 0.$$

New variables

$$\tau, \quad \eta = g(\tau;\epsilon) = \sum_{n=0}^{\infty} \epsilon^{n-1} g_n(\tau)$$

are introduced. Finally, the following perturbation solution is obtained:

$$\begin{aligned} y &= A_0(x)\sin(\eta+\phi_0) + \epsilon[A_1(x)\sin(\eta+\phi_1(x)) \\ &- [A_0^3(x)/32]\cos(3\eta+3\phi)] + O(\epsilon^2), \end{aligned}$$

where

$$\begin{aligned} A_0(x) &= a_0 e^{\epsilon x/2}/[1+a_0^2/4(e^{\epsilon x}-1)]^{1/2}, \\ A_1(x) &= a_1 A_0(x)\operatorname{cosec}(\bar{\phi}_1(x)-\bar{\phi}_0), \\ \phi_1(x) &= \bar{\phi}_0 + \tan^{-1}[\bar{\phi}_1 e^{\epsilon x}/A_0^2(x)], \end{aligned}$$

and

$$\eta = x + \epsilon[-(1/16)\epsilon x - (1/8)\ln[A_0(x)/A_0(0)] + (7/64)[A_0^2(x) - A_0^2(0)]] + O(\epsilon^2),$$

where $a_0, a_1, \bar{\phi}_0$, and $\bar{\phi}_1$ are constants determined by the initial conditions.

Several computer graphs of exact numerical and asymptotic (perturbation) solutions for various values of ϵ are provided. Limit cycles are obtained in the limit of large time. Davis and Alfriend (1967)

4. $$y'' + \epsilon(y^2-1)y' + y = 0,$$

where ϵ is a parameter.

Assuming a solution of the form $y = a\cos x$, the limit cycle is found to be approximately $y = 2\cos x$. This is done using an energy balance argument. The limit cycle is stable when $\epsilon > 0$ and unstable if $\epsilon < 0$.

Using an averaging method, the solution for small positive ϵ is

$$y(x) = a(x)\cos\theta(x) \equiv [2\cos(x-x_0)]/[1-(1-4/a_0^2)\exp(-\epsilon x)]^{1/2}.$$

By assuming an approximate solution $y = a\cos wx$ for ϵ small and positive and using an equivalent linear equation by harmonic balance, we have

$$\epsilon(y^2 - 1)y' \simeq \epsilon(a^2/4 - 1)y'.$$

The given DE may be approximated as

$$y'' + \epsilon[(a_0^2/4) - 1]y' + y = 0.$$

The approximate solution of the given DE, therefore, is

$$y(x) = a_0 \exp\left[-(\epsilon/2)[(a_0^2/4) - 1]x\right] \cos\{[1 - (\epsilon^2/4)(a_0^2/4 - 1)]^{1/2}x\}.$$

Here we use ICs $y(0) = a_0, y'(0) = 0$. This is a (nonperiodic) spiral solution. Jordan and Smith (1977), pp.102–104, 109, 112–113

5. $$y'' + \epsilon(y^2 - 1)y' + y = 0, \quad \epsilon > 0.$$

(The case $\epsilon < 0$ is exactly similar if we put $-x$ for x.) Here ϵ need not be small.

Referring to Eqn. (2.12.32), we have $f(y) = \epsilon(y^2-1), F(y) = \epsilon(y^3/3-y)$. The conditions of the theorem are satisfied with $a = 3^{1/2}$. The y-extremities of the limit cycle must be beyond $y = \pm 3^{1/2}$. Jordan and Smith (1977), p.338

6. $$y'' + e(y^2 - 1)y' + y = 0,$$

where e is a large parameter.

The given system is equivalent (with $\delta = 1/e$) to the equation $\delta y'' + (y^2 - 1)y' + \delta y = 0$, where δ is small. This is a singular perturbation problem. The existence of limit cycles is confirmed as in Eqn. (2.11.5). The limit cycle is analytically constructed for small δ in the phase plane (y, y'), by piecing together several approximate solutions. Jordan and Smith (1977), p.339

7. $$y'' - \alpha(1 - y^2)y' + y = 0,$$

where $\alpha > 0$ is a constant.

The periodic solutions of the given Van der Pol equation for $\alpha < 1$ are sought in the form of the Fourier series

$$y(x) = \sum_{n=-\infty}^{\infty} [A(n)\exp(\text{inwx})],$$

where the period w of the fundamental harmonic may be expressed in terms of period X by means of $w = 2\pi/X$ and $A(-n) = A^*(n)$. The asterisk denotes a complex conjugate. A successive approximation scheme is applied, and the zeroth order approximation is shown to compare well with numerical results for all values of α, even those greater than 1. Giorgini and Toebes (1971)

8. $$y'' - \lambda(1 - y^2)y' + y = 0,$$

where λ is a constant.

For this Van der Pol equation, limit cycles are generated at small amplitudes by the computer implementation of the Poincaré–Lindstedt method, and Fortran programming of the inductive algorithm yields the phase-shifting limiting cycles to graphical accuracy over the range $0 \leq \lambda \leq 1.5$. Melvin (1978)

9. $$y'' - \epsilon(1 - y^2)y' + y = 0,$$

where $\epsilon > 0$ is real.

For this Van der Pol equation, the limit cycle is found as follows. Introduce $\theta = w(\epsilon)x$, where $w(\epsilon)$ is the frequency of oscillations. The new equation is

$$w^2 \frac{d^2y}{d\theta^2} - \epsilon w(1 - y^2)\frac{dy}{d\theta} + y = 0,$$

where the normalization condition $\frac{dy}{d\theta} = 0, y > 0$ at $\theta = 0$ fixes the limit cycle in θ. The expansion for limit cycle then has the form

$$u(\theta, \epsilon) = \sum_{j=0}^{\infty} \epsilon^j u_j(\theta),$$

where the coefficient functions u_j are found to be

$$\begin{aligned} u_{2n}(\theta) &= \sum_{k=0}^{2n} y_{2n,k} \cos(2k+1)\theta \\ &= \sum_{k=0}^{2n} (-1)^{k+1} y_{2n,k} \sin[(2k+1)(\theta - \pi/2)], \end{aligned} \tag{1}$$

$$\begin{aligned} u_{2n+1}(\theta) &= \sum_{k=0}^{2n+1} y_{2n+1,k} \sin(2k+1)\theta \\ &= \sum_{k=0}^{2n+1} (-1)^k y_{2n+1,k} \cos[(2k+1)(\theta - \pi/2)]. \end{aligned}$$

The frequency w is represented by

$$w(\epsilon) = 1 + \sum_{l=1}^{\infty} w_{2l}\epsilon^{2l}. \tag{2}$$

The coefficients $y_{m,k}$ and w_{2l} in (1) and (2) are real numbers that can be determined iteratively. It was found by using Padé approximation for the series in (1) that $w(\epsilon)$ has a complex conjugate pair of branch-point singularities in the complex $-\epsilon^2$ plane at $\epsilon^2 = \tilde{R}e^{\pm i\beta}$, where $\tilde{R} = 3.420188$ and $\beta = 1.79229$.

It is shown that the moving singularities found above are caused by a family of singular complex periodic solutions. Numerical solution is used to find these singular solutions. An asymptotic description of the singular complex periodic solutions for large values of $|\epsilon|$ is given using singular perturbation theory and matched asymptotic expansions. Hunter and Tajdari (1990)

10. $$\epsilon y'' + (1 - y^2)y' + y = 0,$$

$y(\delta) = 0,\ y'(\delta) < 0$, where $\delta = \delta(\epsilon) = \epsilon^{2/3}t_0 - \epsilon|\ln\epsilon|/18$, and $t_0 = 2.3381$.

Mathematical justification for terms in the transition asymptotic expansion of relaxation oscillations of the given Van der Pol equation is presented. The construction of these terms and their formal matching is given in great detail in the book by Kevorkian and Cole (1981). MacGillivray (1990)

11. $$\epsilon y'' + (y^2 - 1)y' + y = 0, \quad 0 < \epsilon < 1.$$

This Van der Pol equation is first written as an equivalent system (with due changes in notation, $y \equiv x, x \equiv t$)

$$\epsilon \frac{dx}{dt} = -y + x - (1/3)x^3, \tag{1}$$

$$\frac{dy}{dt} = x. \tag{2}$$

The system is written in the Lienard plane as

$$(x - (1/3)x^3 - y)\frac{dy}{dx} = \epsilon x, \tag{3}$$

$$\frac{dt}{dx} = \frac{1}{x}\frac{dy}{dx}. \tag{4}$$

Near the stable branch of $y = x - x^3/3$ with $x < 0$, the solution is expanded as

$$y(x) = y_0(x) + \epsilon y_1(x) + \epsilon^2 y_2(x) + \cdots, \quad y_0(x) = x - x^3/3 \cdots.$$

Substitution in (4) yields

$$t = t_0(x) + \epsilon t_1(x) + \epsilon^2 t_2(x) + \cdots, \tag{5}$$

where

$$\begin{aligned} t_0(x) &= \ln(-x) - x^2/2, \\ t_1(x) &= \ln[1 - 1/x^2]^{1/2} + 1/(x^2 - 1) + \cdots. \end{aligned} \tag{6}$$

Rearranging (5), we obtain

$$\begin{aligned} t(x) &= \ln(-x) - x^2/2 + (\epsilon + \epsilon^2 + 2\epsilon^3 + 5\epsilon^4 + \cdots)\ln[1 - 1/x^2]^{1/2} \\ &+ (\epsilon + \epsilon^2/2 + \epsilon^3 + (5/2)\epsilon^4 + \cdots)/(x^2 - 1) + \cdots. \end{aligned}$$

The series in ϵ is found from the substitution of $y(x) = xf(\epsilon) + O(x^3)$ in (3) where $f(\epsilon)(1 - f(\epsilon)) = \epsilon$. When $x \sim -1$, we introduce $x = -1 + \epsilon^{1/3}\xi$, $y = -2/3 + \epsilon^{2/3}\eta$ in (3):

$$\epsilon \frac{d\eta}{d\xi}\left(\xi^2 - \eta - \epsilon^{1/3}\xi^3/3\right) = \epsilon(-1 + \epsilon^{1/3}\xi),$$

with an asymptotic solution

$$\eta(\xi) = \eta_0(\xi) + \epsilon^{1/3}\eta_1(\xi) + \cdots.$$

Equation (4) transforms similarly leading to the solution

$$t(\xi; \epsilon) = C_1(\epsilon) + \epsilon^{2/3}T_0(\xi) + \epsilon T_1(\xi) + \cdots.$$

After considerable detail, a matched expansion is obtained.

The solution is also found in the phase plane. Asymptotic solutions in different regions are found and suitably matched. References to numerical solution are also provided in detail. Grasman (1987), pp.65,67

12. $$y'' + \nu(y^2 - 1)y' + y - a = 0,$$

where $\nu \gg 1$ and a are constants.

The given DE is Van der Pol with a constant forcing term. This is written as a system

$$\begin{aligned} \epsilon\frac{dx}{dt} &= y - \left[\frac{1}{3}x^3 - x\right], \\ \frac{dy}{dt} &= \epsilon(-x + a), \quad 0 < \epsilon \ll 1. \end{aligned} \tag{1}$$

The main concern is with the behavior near $(x, y, a) = (1, -1, 1)$. Placing the origin at this point, we have the system

$$\begin{aligned} \epsilon\frac{dx}{dt} &= y - f(x), \quad f(x) = (1/3)x^3 + x^2, \\ \frac{dy}{dt} &= -\epsilon(x + \alpha). \end{aligned} \tag{2}$$

For $\alpha < 0$, system (2) has stable equilibrium $(\bar{x}, \bar{y}) = (-\alpha, f(-\alpha))$, which changes into an unstable point as α becomes positive. Then a periodic solution arises, being a relaxation oscillation with an orbit independent of α for $\alpha \geq \delta > 0$, with δ arbitrarily small but independent of ϵ. For $\alpha = O(1)$, transitional limit cycles are possible. They may have the shape of a duck ("canard"). A complete analysis of the given equation is available in Eckhaus (1983), who uses asymptotic analysis. Grasman (1987), p.94

13. $$y'' - \epsilon(1 - y^2)y' + y - \lambda y^2 = 0,$$

where $\lambda > 0$ and $\epsilon > 0$ is a small parameter.

For $\epsilon = 0$, we have two critical points: $(y, y') = (0, 0)$, which is a center, and $(y, y') = (1/\lambda, 0)$, which is a saddle. Moreover, the periodic solutions are located around the center and within the saddle loop. For $0 < \epsilon \ll 1$, $(0, 0)$ is an unstable saddle focus, while $(1/\lambda, 0)$ is a saddle point. Referring to a result in the form of a general expansion, it is shown that the solutions are ϵ-close to the solutions of the unperturbed problem on the time scale 1. So the periodic solutions have to be found in the region within the saddle loop (with $0(\epsilon)$ error) of the unperturbed equation. According to the Bendixson criterion, a periodic solution, if it exists, must intersect one or both of the lines $y = \pm 1$. For the saddle loop to require that solutions in the interior intersect $y = \pm 1$, we have $\lambda < 1$. Verhulst (1990), p.129

14. $$y'' + \epsilon(y^2 - 1)y' + y - \alpha y^2 = 0, \quad \epsilon \ll 1, \ \alpha \ll 1.$$

The given DE is Van der Pol with a nonlinear restoring force. By assuming an approximate form of the solution $y = c + a\cos wx$, it may be shown that the mean displacement, frequency, and amplitude are related by $c \approx 4\alpha, w \approx 1 - 4\alpha^2, a \approx 2(1 - 8\alpha^2)$. Jordan and Smith (1977), p.121

15. $$y'' + \epsilon(y^2 - 1)y' + y(1 - \beta y^2) = 0,$$

where ϵ and β are constants.

This is a combined Van der Pol-Duffing equation. Periodic solutions for a range of parameters ϵ and β are found and depicted graphically: (a) $\epsilon = 0, \beta = -8$; (b) $\epsilon = 0, \beta = 0$; (c) $\epsilon = 0, \beta = 0.2$; (d) $\epsilon = 0, \beta = -8$; (e) $\epsilon = 0, \beta = 0$; (f) $\epsilon = 0, \beta = 0.2$; (g) $\epsilon = 1, \beta = -8$; (h) $\epsilon = 1, \beta = 0$; (i) $\epsilon = 1, \beta = 0.2$; (j) $\epsilon = 10, \beta = -8$; (k) $\epsilon = 10, \beta = 0$; (l) $\epsilon = 10, \beta = 0.2$.

The results are compared with those for the Van der Pol and Duffing equations. Horvath (1975)

16. $$y'' - \epsilon(1 - y^2)y' + y^3 = 0,$$

where $0 < \epsilon \ll 1$.

A simple approach involving trigonometric functions leads to the approximate solution

$$y(x) = A(x)\cos[3^{1/2}x + \phi(x)],$$

where

$$\begin{aligned} A^2 &= 4A_0^2/[A_0^2 + (4 - A_0^2)\exp(-\epsilon x)], \\ \phi &= (1/\epsilon)(3/4)^{1/2}\log[A_0^2 + (4 - A_0^2)\exp(-\epsilon x)]. \end{aligned}$$

See Eqn. (2.22.16). Mickens and Oyedeji (1985)

17. $$y'' + e(y^2 - 1)y' + y^3 = 0.$$

Referring to the general equation (2.12.32), the given DE satisfies all the conditions required therein. Therefore, it has exactly one periodic solution. Jordan and Smith (1977), p.343

18. $$y'' + \epsilon(y^2 - 1)y' + y - \epsilon y^3 = 0,$$

where ϵ is a small parameter.

Using the energy balance method, the approximate amplitude a of the limit cycle $y = a\cos x$ is 2 and the period is $T = 2\pi[1 + (3/2)\epsilon]$. Jordan and Smith (1977), p.118

19. $$y'' - \epsilon(1 - y^2)y' + y(1 + \beta y^2) = 0,$$

where β and ϵ are parameters.

The given DE combines the Van der Pol and Duffing equations. After briefly discussing the Van der Pol and "homogeneous" Duffing equation, periodic solutions of the combined equation are discussed and shown graphically for $\epsilon = 0.1, 1, 10$ and $\beta = 0, 0.2, -8$. Other values chosen are $\epsilon = 100, \beta = -5000$ and $\epsilon = 100, \beta = -100$. Horvath (1975)

20. $$y'' + e(y^2 - 1)y' + [1 + (a + cy^2)\cos(2x)]y = 0,$$

where $e, a,$ and c are parameters.

The given DE becomes a Van der Pol equation when $a = c = 0$, and a nonlinear Mathieu equation when $e = 0$. Detailed discussion of the stable periodic solution of this DE for small parameter values is given, emphasizing in particular the relation of these solutions to those of the two simpler equations mentioned earlier. The stroboscopic method which reduces the approximate determination of the periodic solution of the given DE to that of the singular points of a certain simpler auxiliary differential system is used. Three periodic solutions are demonstrated to exist. The stability of these solutions is discussed. Minorsky (1955)

21.
$$\boldsymbol{y'' + e(y^2 - 1)y' + \tanh(ky) = 0,}$$

where $k > 0$ is a constant.

Referring to the general equation (2.12.32), the given DE satisfies all the conditions required therein for $k > 0$. Therefore, the given DE has exactly one periodic solution for $k > 0$. Jordan and Smith (1977), p.343

22.
$$\boldsymbol{\epsilon y'' - (1 - 3y^2)y' = 0, \quad y(0) = 1, \; y(1) = b,}$$

where $b \sim 1$ and $\epsilon = 0.05$.

Numerical solution shows that slight variations of $b = 1$, less than 10^{-5}, change the location of the transition layer from 0.38 to 0.73. The results are graphically shown. Bohe (1990)

23.
$$\boldsymbol{y'' + a_1 y' + a_2 y + a_3 y^2 y' = 0,}$$

where a_i $(i = 1, 2, 3)$ are constants.

Using the method of harmonic balance, an approximate solution is found as $y = A\cos wx$, where $w^2 = a_2$ and $A^2 = -4a_1/a_3$. For $a_2 > 0, a_3 > 0$, and $a_1 < 0$ we have real values of w and A. Ku and Jonnada (1971)

24.
$$\boldsymbol{y'' + \alpha y' + \gamma y^2 y' + \beta y + \delta y^3 = 0.}$$

The given free nonlinear oscillator is studied using the method of differentiable dynamics to obtain qualitative behavior. The case $\alpha, \beta < 0; \gamma, \delta > 0$ is considered in some detail; it has physical relevance as a simple model in certain flow-induced structural vibration problems in which structural nonlinearities act to maintain overall stability. The presence of local and global bifurcations is detected and their physical significance discussed. Phase plane analysis is used. Holmes and Rand (1980)

25.
$$\boldsymbol{y'' + R(A - 3By^2)y' + (A/C)y - (B/C)y^3 = 0,}$$

where R, A, B, and C are positive constants.

The given DE describes the variation in the flux y in a circuit containing an iron-core inductance coil, a resistor, and a charged condenser. Mickens (1981), p.6

26.
$$\boldsymbol{y'' - \epsilon(1 - 4y^2)y' + y(1 - \epsilon^2 y^2 + \epsilon^2 y^4) = 0, \quad \epsilon^2 < 4.}$$

A particular first integral of the given DE is

$$y' = (1 - y^2)^{1/2} + \epsilon y(1 - y^2). \tag{1}$$

Under the assumption $\epsilon^2 < 4$, we can integrate (1) as

$$y = [1 + (\epsilon/(2\mu))\tan(\mu x)]/[(\mu + \epsilon^2/(4\mu))^2 \tan^2 \mu x + [1 + (\epsilon/(2\mu))\tan(\mu x)]^2]^{1/2}, \tag{2}$$

where $\mu = (1-\epsilon^2/4)^{1/2}$, which is a periodic solution of least period $2\pi/\mu$ and with a vertex $y = 1$ at $x = 0$. To derive (2), we write (1) in the form

$$x = \int_1^y \frac{d\lambda}{(1-\lambda^2)^{1/2} + \epsilon\lambda(1-\lambda^2)}. \tag{3}$$

By putting $\lambda = \cos\theta$ and then $\mu = \tan\theta$ so that $\lambda = (1+\mu^2)^{-1/2}$, we transform (3) as

$$x = -\int_0^z \frac{d\mu}{(\mu+\epsilon/2)^2 + \mu^2}. \tag{4}$$

From the relations $\lambda = (1+\mu^2)^{-1/2}, y = (1+z^2)^{-1/2}$, we get after some elementary operations, from (4),

$$z = -\left[[\mu + (\epsilon^2/(4\mu))]\tan(\mu x)\right] / [1 + (\epsilon/(2\mu))\tan\mu x]. \tag{5}$$

Equation (5) and $y = (1+z^2)^{-1/2}$ lead at once to (2). Obi (1953)

27. $$y'' - \lambda y' + w^2 y + \beta y^3 + p y^2 y' + 2\epsilon\cos(vx)\, y = 0,$$

where λ, w, β, p, v, and ϵ (small) are constants.

When $\epsilon = 0$, the given DE exhibits a Hopf bifurcation as the bifurcation parameter λ passes through zero; it is assumed here that the natural frequency w and the external frequency v do not establish any rational relationship; i.e., the given DE is in a nonresonant region. The methods of "intrinsic" harmonic balance and multiple scales are combined to obtain a perturbation solution. Perturbed Hopf bifurcation is discussed. Huseyin and Lin (1992)

28. $$y'' + \epsilon(y-3)(y+1)y' + y = 0,$$

where $\epsilon > 0$ is small. Using the energy balance method, the approximate amplitude a of the limit cycle $y = a\cos x$ is $2(3)^{1/2}$. The limit cycle is stable. Jordan and Smith (1977), p.118

29. $$y'' + \epsilon(y-1)(y-3)y' + y = 0,$$

where ϵ is a small parameter.

Phase plane configuration and solution of an IVP with $y(0) = -4, y'(0) = 4$, and $\epsilon = 0.1$ are depicted. Karreman (1949)

30. $$y'' + [a - 2(1+a)y + 3y^2]y' + \epsilon y = 0,$$

where a and ϵ are constants.

The given DE appears in Hodgkin-Huxley theory. Putting $p = y - (1+a)/3$, we get a forced Van der Pol equation with constant forcing:

$$\frac{d^2p}{dt^2} + 3\left(p^2 - \frac{a^2-a+1}{9}\right)\frac{dp}{dt} + \epsilon p = -\epsilon\frac{1+a}{3}.$$

Casten, Cohen, and Lagerstrom (1975)

31. $$y'' - \epsilon(1 - 2\beta y - y^2)y' + y = 0,$$

where ϵ and β are constants.

The given DE is studied by the method of isoclines for the periodic solution $y = y(x)$; ϵ and β are taken to be large. It is shown that $-[\max y(x)]/[\min y(x)]$ cannot exceed 1/3. Nijenhuis (1949)

32. $$\mu y'' + (y^2 - 1 + 3\lambda^2)y' + \mu y + 2\lambda 3^{1/2} yy' = 0,$$

where μ and λ are constants.

The given DE is a model for cardiac fiber length-pressure pulsations. Writing $y = 3^{1/2}z'$, we have

$$\mu 3^{1/2} z''' - 3^{1/2}(1 - 3z'^2)z'' + \mu 3^{1/2} z' + 3\lambda^2 3^{1/2} z'' + 6(3^{1/2})\lambda z' z'' = 0. \tag{1}$$

Integrating (1) once, we get

$$\mu z'' - (1 - z'^2)z' + \mu z = c - 3\lambda^2 z' - 3\lambda z'^2. \tag{2}$$

Equation (2) can easily be transformed to the Zeeman model for cardiac pulsation if we put $c = (1 - \lambda^2)\lambda$:

$$\dot{x} = y - \lambda, \quad \mu\dot{y} = (1 - y^2)y - \mu x. \tag{3}$$

The system (3) is studied numerically in the phase plane as well as physical space. Petrov (1991).

33. $$y'' + y^3 y' + y^3 = 0.$$

Put $y' = p, y'' = p\dfrac{dp}{dy}$, etc., and integrate. A first integral is $y' - \ln|y' + 1| + y^4/4 = c$, where c is an arbitrary constant. The solution may be studied in the phase plane. Wilson (1971), p.40

34. $$y'' + y^3 y' - y^3 = 0.$$

Put $y' = p, y'' = p\dfrac{dp}{dy}$, etc., and integrate. A first integral is $y^4/4 + y' + \ln|y' - 1| = c$. Hence the solution in the phase plane may be studied. Wilson (1971), p.40

35. $$y'' + (y^3 - y)y' + y = 0.$$

It is proved that the given DE has a family of closed periodic trajectories surrounding the origin; this family, however, does not fill the entire plane. Villari and Zanolin (1988)

36. $$y'' + 4y^3 y' + y = 0.$$

Putting $z = y' + y^4$, the given DE can be changed to the system

$$\frac{dx}{dz} = -1/y, \tag{1}$$

$$\frac{dy}{dz} = (y^4 - z)/y. \tag{2}$$

By writing $Y = -2y^2$, (2) reduces to the Riccati equation

$$\frac{dY}{dz} + Y^2 = 4z. \tag{3}$$

It can be readily verified that $Y = \frac{1}{U}\frac{dU}{dz}$ is a solution of (3) whenever U satisfies the DE

$$\frac{d^2U}{dz^2} = 4zU. \tag{4}$$

A solution of (4) is

$$U = \sum_{\nu=0}^{\infty} \frac{(4/9)^{\nu} z^{3\nu+1}}{\nu!\Gamma(\nu + 4/3)}. \tag{5}$$

Indeed, if we write

$$z = -2^{-4/3}\, 3^{2/3}\, \zeta^{2/3}, \tag{6}$$

we can write (5) as

$$U = -(1/2)\, 3^{2/3}\, \zeta^{1/3}\, J_{1/3}(\zeta),$$

where $J_{1/3}$ is Bessel function of order 1/3. Therefore, a solution of (2) is

$$y = \left[-\frac{1}{2}\frac{1}{\zeta^{1/3}\, J_{1/3}(\zeta)}\frac{d}{d\zeta}\left(\zeta^{1/3}\, J_{1/3}(\zeta)\right)\frac{d\zeta}{dz}\right]^{1/2}. \tag{7}$$

Using (6) and observing that $\frac{d}{d\zeta}[\zeta^n\, J_n(\zeta)] = \zeta^n\, J_{n-1}(\zeta)$, we get

$$y = 2^{-1/3}\, 3^{1/6}\, \zeta^{1/6}\left[\frac{J_{-2/3}(\zeta)}{J_{1/3}(\zeta)}\right]^{1/2}. \tag{8}$$

From (1), (6), and (8), we get

$$\frac{dx}{d\zeta} = -\frac{1}{y}\frac{dz}{d\zeta} = (3\zeta)^{-1/2}\left[\frac{J_{1/3}(\zeta)}{J_{-2/3}(\zeta)}\right]^{1/2}. \tag{9}$$

Integrating (9), we get

$$x = 3^{-1/2}\int_{\zeta_0}^{\zeta}\left[\frac{J_{1/3}(\zeta)}{J_{-2/3}(\zeta)}\right]^{1/2}(1/\zeta^{1/2})d\zeta, \tag{10}$$

where ζ_0 is an arbitrary constant. Equations (8) and (10) give the solution.

Let C be the curve in the ζ-plane defined by

$$\zeta = t - i\cos[2(t - \pi/6 - \pi/4)], \quad 2\pi \le t < \infty.$$

It is shown that if $\zeta \to \infty$ along the curve C, then $z(\zeta) \to \infty$ and $x(\zeta)$ tends to a finite limit x^*. Smith (1953)

37. $$y'' + (a + by + cy^2 + dy^3)y' + y = 0.$$

Defining

$$F(y) = \int_0^y (a + by + cy^2 + dy^3)dy = ay + by^2/2 + cy^3/3 + dy^4/4$$

and using a general theorem, we conclude that if $b = d = 0, c > 0, a < 0$, and since the roots of coefficient polynomial are $y = \pm(a/c)^{1/2}$, there exists only one limit cycle for the given DE. Verhulst (1990), p.60

38. $$y'' + \epsilon y^4 y' + w_0^2 y = 0,$$

where $\epsilon \ll 1$ and w_0 are constants. Nayfeh (1985), p.120

39. $$y'' + \epsilon(y^4 - 1)y' + y = 0,$$

where $\epsilon > 0$ is small.

Using the energy balance method, we check that the (approximate) limit cycle solution $y = a\cos x$ is stable, with amplitude $a = 2^{3/4}$. Jordan and Smith (1977), p.117

40. $$y'' + \epsilon(y^4 - 2)y' + y = 0.$$

The amplitude $a(x)$ in the polar form of the solution $y = a\cos\theta, y' = a\sin\theta$ is given, for small ϵ, by the first order DE

$$16a'(x) = -\epsilon a(a^4 - 16). \tag{1}$$

An averaging method is used to obtain (1). Jordan and Smith (1977), p.119

41. $$y'' + (y^4 - y^2)y' + y = 0.$$

Here in the notation of Rabenstein (1966),

$$\begin{aligned} f(y) &= y^4 - y^2, \quad g(y) = y, \\ F(y) &= \int_0^y (y^4 - y^2)dy = y^5/5 - y^3/3, \\ G(y) &= \int_0^y y\,dy = y^2/2. \end{aligned}$$

So $f(y)$ and $g'(y)$ are continuous for all y and $f(y)$ is even while $g(y)$ is odd and $g(y) > 0$ for $y > 0$. $F(y) < 0$ for $0 < y < (5/3)^{1/2}$ and $F(y) > 0$ for $y > (5/3)^{1/2}$ and $F(y)$ is monotonically increasing for $y > 5/3$; Moreover, $\lim_{y\to\infty} F(y) = +\infty, \lim_{y\to\infty} G(y) = +\infty$.

Hence the given DE has a periodic solution whose closed trajectory encloses the origin of the phase plane. This periodic solution is unique in the sense that the given DE has no other closed trajectory. Moreover, every other trajectory except the point $(0,0)$ spirals toward the closed trajectory as $x \to +\infty$. Rabenstein (1966), p.405

42. $$y'' + \epsilon(y - 1)(y + 1)(y - 3)(y + 3)y' + y = 0,$$

where ϵ is a small parameter.

Using the method of Kryloff and Bogoliuboff, an approximate solution $y = a\sin(wx+\phi)$, with $a = 3.91, w = 1$, is found. It is shown to be stable. Karreman (1949)

43. $$y'' + f(y)y' + y^3 - 16y = 0,$$

where

$$f(y) = \prod_{i=0}^{4} (y^2 - y_i^2), y_i = y_{i-1} + i, y_0 = 2.$$

It is shown that the given DE has at least three isolated stable periodic solutions. Amelkin and Zhavnerchik (1989)

44. $$y'' - \epsilon(1 - y^{2n})y' + y = 0,$$

where n is a positive integer and ϵ is a parameter.

It is shown that the given DE has only one limit cycle, a circle of radius $2[(n + 1)!n!/(2n)!]^{1/2n}$. Lawden (1959), p.364

45. $$y'' - \epsilon(1 - y^{2n+2})y' + y^{2n+1} = 0,$$

where $n \geq 1$, ϵ is a parameter.

It is shown that the given DE has exactly one small "periodic oscillation" and it is $y = 0$; that is, it has no small periodic oscillation. It is also shown that the given DE has no ϵ-large periodic oscillation. So all its periodic oscillations are finite. The given DE has (save for translation in x) exactly one finite periodic oscillation and its stationary values are near $\pm(3n + 4)^{1/(2n+2)}$. Obi (1976)

46. $$y'' + [-2\alpha + (n + 1)\beta y^{n-1}]y' + w^2 y + [-\alpha + \beta y^{n-1}]^2 y = 0,$$

where α, β, n, and w are constants.

We easily find a first integral

$$y' - [\alpha + w\cot(wx + \delta)]y + \beta y^n = 0. \tag{1}$$

Equation (1) is of Bernoulli type and can be rendered linear by the transformation $Y = y^{1-n}$:

$$Y' + (n - 1)[\alpha + w\cot(wz + \delta)]Y = (n - 1)\beta. \tag{2}$$

Solving (2) and returning to the variable y, we have

$$y^{n-1}(x) = \frac{e^{\alpha(n-1)x}\sin^{n-1}(wx + \delta)}{(\sin\delta/y_0)^{n-1} + (n - 1)\beta e^{\alpha(n-1)x} I_n(x)}, \tag{3}$$

where

$$I_n(x) = \int_0^x e^{(n-1)\alpha(\tau - x)}\sin^{n-1}(w\tau + \delta)d\tau \tag{4}$$

and

$$\sin\delta/y_0 = w[(y_0' - \alpha y_0 + \beta y_0^n)^2 + w^2 y_0^2]^{-1/2}. \tag{5}$$

Several special cases of (3) may be identified.

Case 1. n is odd; $\beta > 0$ and $\alpha > 0$. In this case, $I_n(x)$ is positive definite for $x > 0$. $I_n(x)$ is given by

$$I_n(x) = F_n(x) - e^{-(n-1)\alpha x}F_n(0),$$

where

$$\begin{aligned} F_n(x) &= \frac{1}{2^{n-1}(n-1)\alpha}\begin{pmatrix} n-1 \\ \frac{n-1}{2} \end{pmatrix} + \frac{1}{2^{n-2}} \\ &\times \sum_{r=0}^{(n-1)/2-1} (-1)^{(n-1)/2+r} \begin{pmatrix} n-1 \\ r \end{pmatrix} \frac{1}{(n-1)^2\alpha^2 + (n-2r-1)^2 w^2} \\ &\times [(n-1)\alpha\cos(n-2r-1)(wx+\delta) + (n-2r-1)w\sin(n-2r-1)(wx+\delta)]. \end{aligned}$$

The initial condition comes in only through δ in Eqn. (3) as $x \to \infty$. Thus, whatever the initial conditions may be, the solutions of the given DE asymptotically approach

$$y_{1,\infty} = [(n-1)\beta F_n(x)]^{-1/(n-1)} \sin(wx+\delta).$$

The phase portrait in (y, y') plane is an isolated closed path, i.e., a limit cycle. For example, for $n = 3$, the given DE becomes

$$y'' + (-2\alpha + 4\beta y^2)y' + [w^2 + (-\alpha + \beta y^2)^2]y = 0$$

and has the limit cycle in the (y, y') plane as

$$y^2 + [1/(\alpha^2 + w^2)](y' - 2\alpha y + \beta y^3)^2 = 2\alpha/\beta.$$

Case 2. n is odd; $\beta > 0$ and $\alpha < 0$. The solution of the given DE is a modulated damped oscillation and asymptotically approaches

$$y_{2,\infty} = \left[(\sin\delta/y_0)^{n-1} - (n-1)\beta F_n(0)\right]^{-1/(n-1)} e^{\alpha x}\sin(wx+\delta).$$

Case 3. n is odd and $\beta < 0$. The denominator of the RHS of (3) becomes zero at $x = x_0$, which is given by

$$e^{\alpha(n-1)x_0} I_n(x_0) = \left[\frac{1}{(n-1)\beta}\right] (\sin\delta/y_0)^{n-1}.$$

Thus, y starting from y_0 at $x = 0$ goes to infinity at $x = x_0$.

Case 4. n even and $\alpha > 0$. In this case $e^{\alpha(n-1)x} I_n(x)$ is an oscillatory function with an increasing amplitude. Thus, $y \to \infty$ when the RHS of (3) diverges.

Case 5. n is even and $\alpha < 0$. In this case $e^{(n-1)\alpha x} I_n(x)$ is damped and oscillatory. When the RHS of (3) diverges, then $y \to \infty$ at that time. If the denominator of (3) is always positive, the solution of the given DE is a modulated damped oscillation and asymptotically approaches

$$y_{3,\infty} = e^{\alpha x}\sin(wx+\delta)[(\sin\delta/y_0)^{n-1} - (n-1)\beta G_n(0)]^{-1/(n-1)},$$

where

$$\begin{aligned} G_n(0) &= \frac{1}{2^{n-2}} \sum_{r=0}^{(n-2)/2} (-1)^{(n-2)/2+r} \begin{pmatrix} n-1 \\ r \end{pmatrix} \frac{1}{(n-1)^2\alpha^2 + (n-2r-1)^2 w^2} \\ &\times [(n-1)\alpha\sin(n-2r-1)\delta - (n-2r-1)w\cos(n-2r-1)\delta]. \end{aligned}$$

Case 6. $n \leq 0$. The solution (3) of the given DE will be formally correct, but the expression (4) for $I_n(x)$ diverges after some x. The solution may be useful before this value of x is reached. Sawada and Osawa (1978)

47. $$y'' + c(y^{2m} - k^2)y' + y^{2n-1} = 0,$$

where c and k are positive constants, and m and n are positive integers.

In the notation of Rabenstein,

$$\begin{aligned} f(y) &= c(y^{2m} - k^2), \quad g(y) = y^{2n-1}, \\ F(y) &= c\int_0^y (y^{2m} - k^2)dy = c[y^{2m+1}/(2m+1) - k^2 y], \\ G(y) &= \int_0^y y^{2n-1}dy = y^{2n}/(2n). \end{aligned}$$

Therefore, $f(y)$ and $g'(y)$ are continuous for all y and satisfy:

(a) $f(y)$ is even and $g(y)$ is odd, with $g(y) > 0$ for $y > 0$.

(b) There exists a positive constant $a = [(2m+1)k^2]^{1/(2m+1)}$ such that $F(y) < 0$ for $0 < y < a, F(y) > 0$ for $y > a$, and $F(y)$ is monotonically increasing for $y > a$.

(c) $\lim_{y\to\infty} F(y) = +\infty, \lim_{y\to\infty} G(y) = +\infty$.

Thus, the given DE has a periodic solution, whose closed trajectory encloses the origin in the phase plane. This periodic solution is unique in the sense that the given DE has no other closed trajectory. Furthermore, every other trajectory except the point $(0,0)$ spirals toward the closed trajectory. Rabenstein (1966), p.405

48. $$y'' + p(y)y' + y = 0,$$

where

$$p(y) = \sum_{i=k}^{2n} d_i y^i, 0 \le k < 2n,$$

where d_i are constants.

The given DE is written as the system

$$y' = z - P(y), \quad z' = -y,$$

where

$$P(y) = \int_0^y p(\xi)d\xi.$$

Lefschetz results for the Van der Pol equation regarding the complete (global) phase portrait for the latter are extended to the given DE. Barbalat (1955)

49. $$y'' + P(y)y' + Q(y) = 0,$$

where $P(y)$ and $Q(y)$ are polynomials of the same degree in y.

It is shown by using the small-parameter method of Painlevé that the given DE has no algebraic solutions satisfying the condition $y(x) \to \infty$ as $x \to x_0$. Logarithmic terms must appear. Kondratenya and Prolisko (1973)

50. $$y'' + (a_n y^n + a_{n-1}y^{n-1} + \cdots + a_0)y' + b_{n+1}y^{n+1} + \cdots + b_0 = 0,$$

where a_i $(i = 0, 1, \ldots, n)$ and b_j $(j = 0, 1, \ldots, n+1)$ are constants.

Using the small-parameter method of Painlevé, it is shown that it has general solutions with a nonstationary logarithmic singularity. However, the necessary and sufficient condition for the given DE to have a solution with a moving algebraic singularity satisfying $y \to \infty$ as $x \to x_0$ is that

$$\left[\frac{(n+1)^2}{n^2 a_n^2}\right] a_{n-1} b_{n+1} - \left[\frac{n+1}{n a_n}\right] b_n = 0.$$

Kondratenya and Prolisko (1973)

51. $$y'' + P(y)y' + f(x) = 0,$$

where $P(y)$ is a polynomial and $f(x)$ is holomorphic in some region D.

It is shown that the given DE has no solution with an essential singularity in D. This is shown by writing the given DE as the system

$$\frac{dy}{dx} = z - F(y), \quad \frac{dz}{dx} = -f(x), \tag{1}$$

where $F(y) = \int_0^y P(t)dt$, and integrating from the second of (1) backward, the function $z(x)$ has a finite limit when $x \to x_0 \in D$. Kondratenya (1970)

52. $$y'' + f(y)y' + g(y) - P(x) = 0,$$

where f and g are polynomials in y of degree n and m, respectively, and $P(x)$ is a regular function of the complex variable x in some domain containing the point x^*.

The types of singularities that solutions of the given DE can possess at the point x^* are discussed; the following results are proved.

(a) If $n > m$, there is an infinite family of solutions of the given DE that have an algebraic critical point at x^*. In the neighborhood of x^*, these solutions can be expressed in the form

$$y(x) = \sum_{\nu=-1}^{\infty} a_\nu (x - x^*)^{\nu/n} \text{ with } a_{-1} \neq 0.$$

(b) Suppose that $n > m$ and let γ be a contour of finite length that lies in the x-plane and has x^* as an endpoint. If $y(x)$ is a solution of the given DE that can be continued analytically along γ as far as x^* but not over it, then the singularity of $y(x)$ at x^* must be an algebraic critical point of the type discussed in (a). As a corollary, if $P(x)$ has only isolated singularities in the x-plane and $n > m$, then no solution of the given DE can have a natural boundary.

(c) Suppose that $n > m$ and let γ be any continuous curve that lies in the x-plane and has x^* as an endpoint. Let $y(x)$ be a solution of the given DE that can be continued analytically along γ as far as x^* but not over it. If the singularity of $y(x)$ at x^* is not an algebraic critical point of the type described in (a), then it must be a point of accumulation of such algebraic critical points. This means that given any circle Ω however small, with x^* as center, we can pass from a point of γ to an algebraic critical point of $y(x)$ by continuing $y(x)$ analytically along some curve that lies entirely in the circle Ω.

Smith (1953)

53. $$y'' + P(y)y' + Q(y,x) = 0,$$

where P and Q are polynomials in y of degree n and m, respectively, and the coefficients in $Q(y,x)$ are holomorphic functions of x in some region D.

The given DE may be written in the equivalent form

$$\frac{dy}{dx} = z - F(y), \quad \frac{dz}{dx} = -Q(y,x), \tag{1}$$

where $F(y) = \int_0^y P(t)dt$.

Concerning (1), the following results are proved:

(a) If $n > m$, the given DE has a one-parameter family of solutions with the property $y(x) \to \infty$ as $x \to x_0$, where x_0 is any point of D. The solutions have the form

$$y(x) = \sum_{k=-1}^{\infty} a_k(x - x_0)^{k/n}. \tag{2}$$

(b) If $x_0 \in D$ is a singularity of a solution $y(x)$ of the given DE and the function

$$z(x) = y'(x) + F(y(x))$$

does not tend to infinity for $x \to x_0$, then for $n > m$ the solution $y(x) \to \infty$ as $x \to x_0$ and is of the form (2).

Kondratenya (1970)

54. $$y'' + [P'(y) - f(x)]y' - f(x)P(y) + \phi(x) = 0,$$

where $P(y)$ is a polynomial, and $f(x)$ and $\phi(x)$ are holomorphic in some region D.

It is proved that the given DE has no solution with an essential singularity in D. Kondratenya (1970)

55. $$y'' + \epsilon(|y| - 1)y' + y = 0,$$

where $\epsilon > 0$ is small.

Using the energy balance method, the approximate amplitude a of the limit cycle $y = a\cos x$ is $3\pi/4$. The limit cycle is stable. Jordan and Smith (1977), p.118

56. $$y'' + [(n+2)by^n - 2a]y' + y[c + (by^n - a)] = 0,$$

where $a, b, c,$ and n are constants.

For any continuous function $f(y)$ we write

$$A(y) = y^{-2}\int_0^y f(\xi)d\xi. \tag{1}$$

Let $U(x)$ be any function that satisfies

$$U^{-1}U' = y^{-1}y' + A(y). \tag{2}$$

Differentiate (2) with respect to x and then add the result to the square of (2):

$$U^{-1}U'' = y^{-1}y'' + y^{-1}y'(yA' + 2A) + A^2 = y^{-1}[y'' + y'f(y) + yA^2].$$

This, together with (2), now gives

$$U^{-1}[U'' + \phi U' + \psi U] = y^{-1}[y'' + (f + \phi)y' + yA^2 + yA\phi + y\psi],$$

where $\phi(x)$ and $\psi(x)$ are any functions of x. If $U(x)$ is first chosen to satisfy the linear DE

$$U'' + \phi(x)U' + \psi(x)U = 0, \tag{3}$$

then any solution $y(x)$ of (2) satisfies

$$y'' + [f(y) + \phi(x)]y' + yA^2(y) + yA(y)\phi(x) + y\psi(x) = 0. \tag{4}$$

If, in particular, $\phi(x) = 0$ and $\psi(x) = c$, then (4) is of the form of the given DE with the last term on the left as $g(y) = y[c + A^2(y)]$. If, furthermore, $c = w^2$, a solution of (3) is $U = \cos(p + wx)$, where p is an arbitrary constant and (2) becomes

$$y' + yA(y) + yw\tan(p + wx) = 0. \tag{5}$$

When $f(y)$ is given by

$$f(y) = (n+2)by^n - 2a, g(y) = y[c + (by^n - a)^2],$$

then $A(y) = by^n - a$ by (1) and in this case (5) is Bernoulli's equation with the general solution

$$y = \cos(p + wx)[qe^{-nax} + nbe^{-nax} \int_0^x e^{na\theta} \cos^n(p + w\theta)d\theta]^{-1/n},$$

where p and q are arbitrary constants. Smith (1961)

57. $$\boldsymbol{y'' + f(y)y' + a \sin y - e(x) = 0,}$$

$y(0) - y(2\pi) = y'(0) - y'(2\pi) = 0.$

Using upper and lower solution techniques, it is shown that with 2π-periodic f and $e(x) = \bar{e} + \tilde{e}(x)$, where

$$\bar{e} = [1/(2\pi)] \int_0^{2\pi} e(t)dt, \quad \int_0^{2\pi} \tilde{e}(t)dt = 0,$$

the given system is solvable. For each $\bar{e}$, the set $R(\bar{e})$ of $\bar{e} \in R$, for which the given system is solvable, is a nonempty closed interval contained in $[-a, a]$. Mawhin and Willem (1984)

58. $$\boldsymbol{y'' + f(y)y' + g(y) - e(x) = 0,}$$

$y(0) - y(2\pi) = y'(0) - y'(2\pi) = 0$, where f and g are real continuous functions on R and $e \in L^2(0, 2\pi)$. Conditions are given to ensure at least one solution of the given problem. Mawhin and Willem (1984)

59. $$\boldsymbol{y'' + g_1'(y)y' + h(y' + g_1(y)) + g_{21}(y) - f(x) = 0,}$$

where h is a function of its argument $y' + g_1(y)$.

A graphical method for solution of the given nonautonomous class of equations is presented in which the curves represented by functions $g_1(y)$ and $h(y' + g_1(y))$ play a special role.

As examples of these equations we have the equation with nonlinear damping

$$y'' + g_3(y)y' + g_4(y) = f(x) \quad \text{with} \quad \frac{dg_1}{dy} = g_3(y)$$

and $h(y' + g_1(y)) = 0$ and the Duffing equation

$$y'' + ky' + g_2(y) = f(x) \quad \text{with} \quad g_1(y) = ky,$$

k a constant, and $h(y' + g_1(y)) = 0$. Nishikawa (1964), p.28

2.12 $y'' + f(y)y' + g(x,y) = 0, f(y)$ not polynomial

1. $$y'' - ay'/y - b/y = 0,$$

where a and b are constants.

It may be checked that the given DE may have movable singularities other than poles. Bureau (1964)

2. $$y'' + \left[(8 - 4\alpha - \alpha^2)/16 + (3/2)y/(1-y) - (5/2)y\right] y'$$
$$+ [(1-\alpha)/16]\, y - \left[\ln(1-y) + y + y^2\right]/4 = 0,$$

where α is a parameter.

The given DE is a model describing pulsating flames in the combustion of recondensed matter. Phase portrait and time evolution are found numerically for various α values ($\alpha = -5.5, -6, -7, -8$). Slow attraction to a limit cycle is observed. Frankel (1991)

3. $$y'' + \mu y'/(1-y^2) + w_0^2 y = 0,$$

where y is small; w_0 and μ are constants.

Expanding in powers of y, we have

$$y'' + w_0^2 y + \mu y'(1 + y^2 + \cdots) = 0.$$

Taking μ as a small parameter, the method of multiple scales gives a first order uniform expansion

$$y = \left[2/\{(1 + 4/a_0^2)e^{\mu x} - 1\}^{1/2}\right] \cos(w_0 x + \beta) + \cdots.$$

Nayfeh (1985), p.117

4. $$y'' - \epsilon\left[\{(1-\beta) - y^2\}/(1-y^2)\right] y' + y = 0,$$

where ϵ and β are positive constants.

The given DE generalizes the Van der Pol equation. It is shown that it has at least one stable limit cycle, and for $0 < \epsilon \ll 1$, the method of harmonic balance gives an excellent approximation to the cycle. Mickens (1987)

5. $$y'' - [6\nu/\mu_0]^{1/2} y^{-1/2} y' + 2\nu/\mu_0 = 0,$$

where ν and μ_0 are constants.

The given DE has an exact solution

$$y = (2 - 3^{1/2})(\nu/\mu_0)x^2.$$

Burde (1990)

6. $$y'' + [\beta/y^{1/2}]y' - 1/y^{1/2} + 1 = 0, \quad y(0) = y'(0) = 0,$$

where β is a parameter.

For the given autonomous equation, put $y' = p$ so that it becomes

$$\frac{dp}{dy} = [-\beta p - y^{1/2} + 1]/(y^{1/2} p). \tag{1}$$

Now let $z = y^{1/2} - 1$ in (1) to get

$$\frac{dp}{dz} = -2(z + \beta p)/p. \tag{2}$$

Equation (2) can be easily integrated by writing $p = zw$;

$$\begin{aligned} -\ln z + C &= [w/(w^2 + 2\beta w + 2)]dw \\ &= (1/2)\ln(w^2 + 2\beta w + 2) - \beta \int \left[\frac{1}{w^2 + 2\beta w + 2}\right] dw. \end{aligned} \tag{3}$$

The second integral in (3) is found to be

$$\begin{aligned} \int \left[\frac{1}{w^2 + 2\beta w + 2}\right] dw &= \left[\frac{1}{2 - \beta^2}\right]^{-1/2} \tan^{-1}\left[\frac{w + 1}{(2 - \beta^2)^{1/2}}\right], \quad \beta^2 < 2, \\ &= \frac{1}{(\beta^2 - 2)^{1/2}} \ln\left[\frac{w + \beta - (\beta^2 - 2)^{1/2}}{w + \beta + (\beta^2 - 2)^{1/2}}\right], \quad \beta^2 > 2, \\ &= -\frac{1}{(w + \beta)}, \quad \beta^2 = 2. \end{aligned}$$

The integration constant C is determined by imposing the given IC. The first integral (3) for the case $\beta^2 < 2$ may be written as

$$\begin{aligned} \frac{1}{2}\ln\left[\left(\frac{y'}{y^{1/2} - 1}\right)^2 + 2\beta\frac{y'}{y^{1/2} - 1} + 2\right] - \frac{\beta}{(2 - \beta^2)^{1/2}} \\ \times \left\{\arctan\left[\frac{y'/(y^{1/2} - 1) + \beta}{(2 - \beta^2)^{1/2}}\right] - m\pi\right\} &= -\ln|y^{1/2} - 1| + C, \end{aligned} \tag{4}$$

where

$$C = (1/2)\ln 2 - \frac{\beta}{(2 - \beta^2)^{1/2}} \tan^{-1}\left[\frac{\beta}{(2 - \beta^2)^{1/2}}\right].$$

Here, m is a positive integer, and the $m\pi$ factors arise from the branches of the inverse tangent function.

The intermediate integral (4) does not admit further solution. The Taylor series solutions for $\beta = 0$ and $\beta = 1$ about their first maxima are

$$\begin{aligned} \beta = 0: \quad y(x) &= 4 - 0.25(x - x_0)^2 + 1.30208 \times 10^{-3}(x - x_0)^4 \\ &\quad + 2.17014 \times 10^{-5}(x - x_0)^6 + 5.20736 \times 10^{-7}(x - x_0)^8 \\ &\quad + 1.50031 \times 10^{-8}(x - x_0)^{10}; \\ \beta = 1: \quad y(x) &= 1.088295 - 2.07119 \times 10^{-2}(x - x_0)^2 + 6.6179 \end{aligned}$$

$$\times\ 10^{-3}(x-x_0)^3 -\ 8.25828\ \times\ 10^{-4}(x-x_0)^4 + 3.14876$$

$$\times\ 10^{-5}(x-x_0)^5 +\ 1.37155\ \times\ 10^{-6}(x-x_0)^6 + 8.32724$$

$$\times\ 10^{-8}(x-x_0)^7 -\ 5.59307\ \times\ 10^{-9}(x-x_0)^8,$$

to a sufficient order of approximation.

The solutions are also obtained by an approximate approach called the δ-perturbation method. Abraham–Shrauner, Bender, and Zitter (1992)

7. $$\boldsymbol{y'' - 2y^{\beta}y' + (2/\beta)y = 0,}$$

where β is a parameter.

The linearized form of the given DE, namely $y_L'' + (2/\beta)y_L = 0$, has the oscillatory solution

$$y = B\cos(2/\beta)^{1/2}\, x, \tag{1}$$

where B is a constant. Oscillatory solutions having the behavior (1) at $x = \infty$ for different values of B may be constructed numerically. The solutions oscillate on the entire real line. Sachdev, Nair, and Tikekar (1988)

8. $$\boldsymbol{y'' + (ky' + 1)\sin y = 0,}$$

where $k > 0$ is a constant.

The conditions stated for the more general equation (2.12.31) are also satisfied for this special case. Therefore, for the given DE, the origin $y = 0, y' = 0$ is a center. Jordan and Smith (1977), p.343

9. $$\boldsymbol{y'' + a(1 + b\cos y)y' + \sin y = c,}$$

where a, b, and c are nonnegative constants.

It is proved that for sufficiently small a and arbitrary nonnegative b and c, the given DE has no Poincaré limit cycle which encompasses on the phase cylinder the equilibrium state, namely the trivial solution $y = \sin^{-1} c =$ constant. Pasynkova (1981)

10. $$\boldsymbol{y'' + \sin y - \epsilon y' \cos(ny) = 0, \quad n \in N,}$$

where ϵ is a parameter.

It is proved that for sufficiently small ϵ, the given DE has exactly $n-1$ coarse limit cycles in the region of oscillatory motions and no limit cycle in the region of rotary motion (i.e., saddle-type limit cycle). Chaotic behavior is also studied. Morozov (1989)

11. $$\boldsymbol{y'' + w^2 \sin y + [(4\sin^2 y)/(1 + 4(1 - \cos y))]y' = 0,}$$

where y is small but finite; w is a constant.

Expanding $\sin y$ and $\cos y$ and retaining terms to order y^3, we have

$$y'' + w^2 y - (1/6)w^2 y^3 + 4y^2 y' + \cdots = 0.$$

Putting $y = \epsilon^{1/2}u$ herein, we have

$$u'' + w^2 u + \epsilon(4u^2 u' - (1/6)w^2 u^3) + \cdots = 0. \tag{1}$$

Equation (1) is a special case of (2.11.2), with $\epsilon = 1, \mu = 2, \alpha = -(1/6)w^2$. Thus, a first order uniform approximate solution is

$$y = [a_0/(1 + a_0^2 x)^{1/2}] \cos\left[wx - (w/16)\ln(1 + a_0^2 x) + \beta_0\right] + \cdots,$$

where a_0 and β_0 are arbitrary constants. Nayfeh (1985), p.117

12. $$\boldsymbol{y'' - y'e^y = 0, \quad y(3) = 0,\ y'(3) = 1.}$$

Put $y' = u$ so that $y'' = u\dfrac{du}{dy}$, i.e., $\dfrac{du}{dy} = e^y$. Integrating twice and using the given IC, we have $y = -\log(4 - x)$. Tenebaum and Pollard (1963), p.505

13. $$\boldsymbol{y'' - (e^{\mu y} - 1)(y' + 1) = 0.}$$

The given DE describes a Lokta–Volterra system. It has the equivalent form

$$\dot{x} = 1 - e^{\mu y}, \quad \dot{y} = e^x - 1, \tag{1}$$

(where we have interchanged x and y). Equation (1) has a Hamiltonian function

$$H(x, y) = e^{\mu y}/\mu - y - 1/\mu + e^x - x - 1 \tag{2}$$

and all solutions of (1) are periodic with orbits given by $H(x, y) = h, h \geq 0$. The case $h = 0$ corresponds to the origin, which is a center. Define the amplitude a of the orbit with level $h > 0$ as the positive solution of the equation $H(a, 0) = h$.

The following result is true. The period of solutions of the given DE [and (1)] is a strictly increasing function of the amplitude with positive derivative. Furthermore, as the amplitude goes to infinity, the period goes to infinity. Hausrath and Manasevich (1991)

14. $$\boldsymbol{y'' - (\sinh y)\, y' = 0.}$$

Put $\sinh y = z$; we get a linear DE. The solution with $y'(0) = 1$ is $\sinh y = \tan x$. Reddick (1949), p.193

15. $$\boldsymbol{y'' + 2K|y|y' + ay - d = 0,}$$

where $a, K > 0, d \geq 0$ are constants.

The given DE arises in square damping problems, e.g., water flow in a hydroelectric system having a surge tank. Phase plane study is carried out; it is shown that y always tends to d/a as x tends to infinity. For small K, method of isoclines is used to obtain an approximate solution. McLachlan (1954), p.49

16. $$\boldsymbol{y'' + A|y|y' + (1/8)A^2 y^3 = 0,}$$

where A is a constant.

The given DE has the nonoscillatory solution $y = 4(Ax)^{-1}$. Benson (1973)

17. $$\boldsymbol{y'' + A|y|y' + By^3 = 0,}$$

where $A > 0, B > 0$ are constants.

All solutions of the given DE are shown to oscillate at $+\infty$ and "to the left" if $B > (1/8)A^2$. Benson (1973)

18.
$$y'' + \epsilon[|y| - 1]y' + y = 0, \quad \epsilon \ll 1.$$

Using the method of energy balance, the approximate solution of the limit cycle is $y = a\cos x$, where $a = 3\pi/4$. This limit cycle is stable. The paths close to the limit cycle are given by

$$a(\theta) = \frac{(3\pi/4)}{1 - (3\pi/4a(0))e^{\epsilon\theta/2}},$$

in the (a, θ) phase plane, where $y = a(\theta)\cos\theta, y' = a(\theta)\sin\theta$. Jordan and Smith (1977), p.118

19.
$$y'' + 2|y|y' + 2ye^{-y^2} = 0.$$

It is shown that the region of asymptotic stability for the given DE is the entire (y, y') plane. Wilson (1971), p.337

20.
$$y'' + \{(y^2 + |y| - 1)/(y^2 - |y| + 1)\}y' + y^3 = 0.$$

$$f(y) = \frac{y^2 + |y| - 1}{y^2 - |y| + 1}$$

is well shaped. $f(y) \to 1$ as $y \to \infty$.

Referring to Eqn. (2.12.32), the conditions (a)–(d) are satisfied. Therefore, a limit cycle exists. Jordan and Smith (1977), p.338

21.
$$y'' + y'|y|^p + \beta y|y|^{2p} = 0,$$

where p and β are constants.

Put $y = (1 + 2/p)^{1/p}(u'/u)^{1/p}$ to reduce the given DE to

$$u'u''' = (1 - 1/p)u''^2 + (1/p)(u'^4/u^2)[1 - \beta(p + 2)^2], \tag{1}$$

which for $\beta = (p + 2)^{-2}$ takes the readily integrable form

$$u'u''' = (1 - 1/p)u''^2. \tag{2}$$

Using (2), the solution of the given DE for this special value of β is

$$y(x) = \left[\frac{(p + 1)(p + 2)}{p}\right]^{1/p} \left[\frac{K(Kx + M)^p}{(Kx + M)^{p+1} + N}\right]^{1/p},$$

where K, M, and N are constants, of which only two are independent. Bouquet, Feix, and Leach (1991)

22.
$$y'' + f(y)y' + g(y) = 0,$$

where

$$f(y) = \begin{cases} \cos y & \text{for } y < \pi/2, \\ (y - \pi/2)[1/(y^3 + 1) + 1/y^2] & \text{for } y \geq \pi/2, \end{cases}$$

$$g(y) = \begin{cases} y & \text{for } y < 0, \\ y/(y^3+1) & \text{for } y \geq 0. \end{cases}$$

It is shown that the given system has at least one isolated stable periodic solution. Zhilevich (1987)

23. $$\epsilon y'' + [\phi(y) - 1]y' + y = 0.$$

The given DE is a modified Van der Pol equation where $\phi(y)$ is assumed to take forms other than y^2. This equation is studied as ϵ is varied from very small values (10^{-5}) to larger values (10^3), with a variety of fairly precisely defined forms for $\phi(y)$. These forms vary from the less abrupt nonlinearities $|y|^{0.8}$ and $|y|^{1.06}$ through y^2 and $|y|^{2.72}$ to the very abrupt nonlinearity $\exp(17.5(|y| - 1))$.

Matched asymptotic methods and numerical methods are used. The (numerical) experiments show that there are no dramatic changes in the solution as ϕ is varied. Thus the Van der Pol equation gives useful insight into the behavior of real oscillators. Robinson (1987)

24. $$y'' + f(y)y' + y = 0,$$

where $f(y)$ is continuous.

Sufficient conditions are given such that the given DE has a periodic solution; uniqueness of the results are also proved under further hypotheses. Sansone (1949)

25. $$y'' + [f'(y) + (1/y)f(y)]y' + w^2y + (1/y)f^2(y) = 0.$$

Consider the linear DE

$$y'' + w^2y = 0. \tag{1}$$

Introducing the new variable

$$Y = (p - iwy)(p + iwy)^{-1}, \tag{2}$$

where $p = y'$, we find that (1) is equivalent to

$$Y' = -2iwY, \tag{3}$$

which is immediately integrable. For the given nonlinear DE, let

$$p = y' + f(y). \tag{4}$$

Then it is easy to check that (3) with (2) and (4) is equivalent to the given DE. Integrating (3), we obtain

$$(y' + f(y) - iwy)(y' + f'(y) + iwy)^{-1} = Y_0e^{-2iwx}. \tag{5}$$

If $y = y_0, y' = y_0'$ for $x = 0, Y_0$ is determined as

$$Y_0 = (y_0' + f(y_0) - iwy_0)(y_0' + f(y_0) + iwy_0)^{-1} = \exp(-2i\delta), \tag{6}$$

where

$$\delta = \tan^{-1}\left[wy_0(y_0' + f(y_0))^{-1}\right].$$

The expressions (5)-(6) are equivalent to

$$y'(x) - yw\cot(wx+\delta) + f(y) = 0,$$

which is therefore the first integral of the given DE. Sawada and Osawa (1978)

26. $$\epsilon y'' + F'(y)y' + g(y) = 0,$$

where ϵ is small.

The given DE is written as the system

$$\epsilon\frac{dy}{dt} = z - F(y), \quad \frac{dz}{dt} = -g(y),$$

where $g(y)$ and $F(y)$ satisfy the following conditions:

(a) Two constants $a < 0$ and $b > 0$ exist such that $F'(a) = F'(b) = 0$ and $F'(a) > 0, F'(b) < 0$. Furthermore, the equation $F'(y) = F'(b)$ has at least one negative root. Let A be the negative number with the smallest absolute value for which $F'(A) = F'(b)$. Likewise, the equation $F'(y) = F'(a)$ has at least one positive root. Let B be the smallest positive number for which $F(B) = F(a)$.

(b) An $\epsilon > 0$ exists such that for $A - \epsilon \leq y \leq B + \epsilon$,the functions $g(y)$ and $F''(y)$ are both Lipschitz continuous and $yg(y) > 0$.

(c) For the interval $a \leq y \leq b$, the functions $g'(y)$ and $F'''(y)$ are Lipschitz continuous.

(d) Two positive constants K_1 and K_2 exist such that $F'(y) > K_1(a-y)$ for $A - \epsilon \leq y \leq a$ and $F'(y) > K_2(y-b)$ for $b < y \leq B + \epsilon$.

(e) Positive constants E_1, E_2, E_3, and E_4 can be found such that

$$E_2(a-y)^2 \leq F(a) - F(y) \leq E_1(a-y)^2,$$

$$E_4(y-b)^2 \leq F(y) - F(b) \leq E_3(b-y)^2$$

in the interval $a \leq y \leq b$.

Under the foregoing conditions, the given system has a stable periodic solution, which is approximated through the construction of a ring-shaped domain Ω with width $O(\epsilon^{2/3})$ satisfying the conditions for the Poincaré-Benedixon theorem. Also, the limit cycle within Ω is approximated asymptotically. The positive amplitude of y and the period are found to be

$$\begin{aligned}\text{Ampl}^+ &= B + \frac{\alpha}{F'(B)}\left(\frac{-2g(a)^2}{F''(a)}\right)^{1/3}\epsilon^{2/3} + \frac{1}{2F'(B)} \\ &\left[\frac{2g(B)}{F'(B)} + \frac{2}{3}g'(a)\left(\frac{-4g(a)}{F''(a)^2}\right)^{1/3} + \frac{4}{9}g(a)\frac{F'''(a)}{F''(a)^2}\right](\epsilon\ln\epsilon) + O(\epsilon),\end{aligned}$$

where $-\alpha$ is the first zero of Airy function $\alpha = 2.33811$. The negative amplitude Ampl$^-$ is obtained by replacing B by A and a by b. The time needed to go from $x = \text{Ampl}^-$ to $x = \text{Ampl}^+$ is

$$T = -\int_A^a \frac{F'(x)}{g(x)}dx + \alpha\left(\left\{\frac{2}{g(a)F''(a)}\right\}^{1/3} + \frac{1}{g(A)}\left\{\frac{-2g(b)^2}{F''(b)}\right\}^{1/3}\right)\epsilon^{2/3}$$

$$-\frac{1}{2}\left[\frac{2}{F'(b)} - \frac{2}{F'(A)} - \frac{2}{3}\frac{g'(a)}{g(a)F''(a)} + \frac{4}{9}\frac{F'''(a)}{F''(a)^2} - \frac{1}{g(A)}\left\{\frac{2}{3}g'(b)\left(\frac{-4g(b)}{F''(b)^2}\right)^{1/3} + \frac{4}{9}\frac{g(b)F'''(b)}{F''(b)^2}\right\}\right]\epsilon\ln\epsilon + O(\epsilon).$$

Ponzo and Wax (1965, 1965a)

27. $$\boldsymbol{y'' - f_1(y)y' - f_0(y) = 0,}$$

where $f_0(y)$ and $f_1(y)$ are periodic functions of period 2π.

The given DE is studied in the phase plane by introducing the variables $\frac{dy}{dx} = z, \frac{dz}{dx} = f_0(y) + f_1(y)z$, etc. The following result is proved: Suppose that $f_1(y) = \cos y$, and

$$f_0(y) = \begin{cases} \alpha_0 + \sum_{k=1}^{n/2} \alpha_{2k-1}\sin[(2k-1)y] + \alpha_{2k}\cos(2ky), & n \text{ even}, \\ \sum_{k=0}^{(n-1)/2} \alpha_{2k+1}\sin[(2k-1)y] + \alpha_{2k}\cos(2ky), & n \text{ odd}. \end{cases}$$

Then for any $\epsilon > 0$ and $\delta > 0$ there exist $\alpha_0^*, \alpha_1^*, \ldots, \alpha_n^*$ satisfying $|\alpha_i^*| < \delta$ $(i = 0, 1, \ldots, n)$, such that for $\alpha_i = \alpha_i^*$ $(i = 0, 1, \ldots, n)$ the given DE has n periodic solutions of the second kind with $y'(x) > 1/\epsilon$. Shilova (1967)

28. $$\boldsymbol{y'' + f(y)y' + g(y) = 0.}$$

It is shown that the given DE has at least one isolated periodic solution if:

(a) $f(0) < 0$.

(b) $yg(y) > 0$ for $y \neq 0$.

(c) There is a function $\phi(y) \in C^1$ such that:

  i) $\phi(0) = 0$.

  ii) $f(y) - \phi'(y) > 0\ \forall y \notin [\delta_1, \delta_2], \delta_1 < 0, \delta_2 > 0;\ f(y) - \phi'(y) = 0$ for $y = \delta_1$ and $y = \delta_2$.

  iii) The function $\Phi(y) = \phi(y) - g(y)/[f(y) - \phi'(y)]$ is either bounded above for $y < -h$, or it is bounded below for $y > h$, where $h > \max[|\delta_1|, \delta_2]$.

(d) $\lim_{y\to\infty} \phi(y) = +\infty$.

(e) $\lim_{y\to-\infty} \phi(y) = -\infty$.

Zhilevich (1987)

29. $$\boldsymbol{y'' + \phi(y)y' + X(y) = 0.}$$

The analysis generalizes that for $y'' + X(y) = 0$ in the phase plane. Writing

$$y' = f^{1/2}(y) + f(y)g(y), \tag{1}$$

and differentiating with respect to y, we have

$$\frac{dy'}{dy} = (1/2)f'f^{-1/2} + (fg)', \tag{2}$$

where the accent denotes a derivative with respect to the corresponding argument of the function. Multiplying corresponding sides of (1) and (2), we have

$$y'\frac{dy'}{dy} = (1/2)f' + f^{1/2}[(3/2)f'g + fg'] + fg(fg)'. \tag{3}$$

Eliminating $f^{1/2}$ from (3) and (1), writing y'' for $y'\dfrac{dy'}{dy}$, we get

$$y'' - [(3/2)f'g + fg']y' - (1/2)(1 - fg^2)f' = 0, \tag{4}$$

which is of the form of the given DE. These preliminaries lead to the proof of the following result:

Suppose that $f(y)$ and $g(y)$ are single valued functions that have continuous derivatives. Let a_1, a_2 with $a_2 > a_1$ be a pair of numbers such that (a) $f(a_1) = f(a_2) = 0$, (b) $f'(a_1) > 0, f'(a_2) < 0$, (c) $f(y) > 0$ for $a_1 < y < a_2$, (d) $f(y)g^2(y) \neq 1$ for $a_1 \leq y \leq a_2$ so that from (a), $f(y)g^2(y) < 1$ for $a_1 \leq y \leq a_2$.

To any such pair of numbers there corresponds a family of periodic solutions of Eqn. (4) where $f = f(y), g = g(y), f' = \dfrac{df}{dy}, g' = \dfrac{dg}{dy}$. Moreover, to this family of curves corresponds the integral curve $p^2 - 2f(y)g(y)p + f^2(y)g^2(y) - f(y) = 0, p = y'$ of the phase plane equivalent of (4); and this curve is algebraic when $f(y)$ and $g(y)$ are algebraic. Obi (1953)

30. $$\mathbf{y'' + f'(y)y' + g(y) = 0.}$$

Let $yg(y) > 0$ for all $y \neq 0, g(y)/y \to \infty$ as $y \to \infty$ and suppose that there exist constants $b, B > 0$ such that for all real $y, |f(y) - bg(y)| \leq B|y|$.

(a) If $y = y(x), y(x) \not\equiv 0$ is a solution of the given DE valid for all large x, then $y = y(x)$ is bounded and oscillatory or $y(x)$ monotonically approaches zero as $x \to \infty$.

(b) Suppose that there exist positive constants μ and λ such that

$$\lambda + 4\mu^2 \geq \mu F(y) \geq G(y) \geq \lambda > 0, \lambda > \mu^2,$$

where $F(y) = f(y)/y, G(y) = g(y)/y$. If $y = y(x), y(x) \not\equiv 0$ is any solution of the given DE valid for all large x, then $y = y(x)$ is bounded and oscillatory or $y(x)$ approaches zero monotonically as $x \to \infty$.

(c) If $yf(y) > 0$ and $yg(y) > 0$ for all $y \neq 0$ and if $g(y)/y \to \infty$ as $y \to \infty$, then any solution $y = y(x), y(x) \not\equiv 0$, of the given DE that is valid for all x is oscillatory or monotonically tends to zero as $x \to \infty$.

Utz (1957)

31. $$\mathbf{y'' + f(y)y' + g(y) = 0.}$$

It is proved that the origin $y = 0, y' = 0$ is a center for the given equation, when, in some neighborhood of the origin, f and g are continuous, and:

(a) $f(y)$ is odd, and of one sign in the half-plane $y > 0$.

(b) $g(y) > 0, y > 0$ and $g(y)$ is odd (implying that $g(0) = 0$).

(c) $g(y) > \alpha f(y)F(y)$ for $y > 0$, where $F(y) = \int_0^y f(u)du$, and $\alpha > 1$.

Jordan and Smith (1977), p.329

32. $$y'' + f(y)y' + g(y) = 0.$$

The given DE has a unique periodic solution if f and g are continuous, and:

(a) $F(y) = \int_0^y f(u)du$ is an odd function.

(b) $F(y)$ is zero only at $y = 0, y = a, y = -a$, for some $a > 0$.

(c) $F(y) \to \infty$ as $y \to \infty$ monotonically for $y > a$.

(d) $g(y)$ is an odd function, and $g(y) > 0$ for $y > 0$.

(These conditions imply that $f(y)$ is even, $f(0) < 0$ and $f(y) > 0$ for $y > a$.) Jordan and Smith (1977), p.334

33. $$y'' + f(y)y' + g(y) = 0,$$

where $f(y)$ and $g(y)$ are analytic functions, and the following hypothesis are satisfied:

(a) $g(y) = -g(-y)$.

(b) $yg(y) > 0$ for $y \neq 0$.

(c) $f(y) = f(-y)$.

(d) $f(0) < 0$.

(e) $\int_0^y f(u)du = F(y) \to \infty$ as $y \to \infty$.

(f) $F(y)$ has a unique positive root $y = a$; then the given DE has a unique stable limit cycle.

Lakin and Sanchez (1970), p.92

34. $$y'' - kf(y)y' + g(y) = 0,$$

where $k > 0$ is a parameter.

Suppose that (1) sgn $g(y)$ = sgn x, (2) $2v^2G(x) \leq [g(x)]^2$ for $-a < x < \alpha$, and (3) $f(y) \neq 0$ for $-b < x < \beta$, where $G(y) = \int_0^y g(\xi)d\xi$ and v, a, α, b, and β are positive constants; then, if $y(x)$ is a periodic solution of the given DE with least period X that has $-a \leq y(x) \leq \alpha$ for all x, then

$$X \leq v^{-1}[8 + 6\gamma^{-1/2}(J_1 + J_2) + 8\gamma^{-1/2}\max(J_1, J_2)],$$

where

$$J_1 = k\int_0^\alpha f_+(y)dy, J_2 = k\int_{-a}^0 f_+(y)dy,$$

where

$$f_+(y) = \max(0, f(y)) \quad \text{and} \quad \gamma = \min[2G(-b), 2G(\beta)].$$

The theorem is applied to the Van der Pol equation. Smith (1970)

35. $$y'' + k(y)y' + h(y) = 0,$$

where $k(y) > 0$ and $yh(y) > 0$ for nonzero y.

Putting

$$V = (1/2)y'^2 + \int_0^y h(\xi)d\xi \equiv (1/2)y'^2 + H(y),$$

we obtain the energy equation

$$\left(\frac{dV}{dy}\right)^2 - 2k^2(y)V(y) = -2k^2(y)H(y). \tag{1}$$

This formal computation is justified if $y(x)$ is a monotonic solution of the given DE. Using the first order DE (1), some comparison and oscillation theorems are proved. Conditions for the uniqueness of the IVP are derived. Benson (1973)

36. $$y'' + k(y)y' + h(y) = 0.$$

Conditions are given for oscillation and nonoscillation of solutions of the given Lienard's equation. In the oscillatory case, estimates are given for the decrement (1) of the magnitude of solution from one zero of the derivative to the next, and (2) of the magnitude of the derivative from one zero of the solution to the next. Prufer transformation is made use of. Benson (1984)

37. $$y'' + f(y)y' + g(y) = 0,$$

where f and g are continuous real functions on R, subject to conditions that ensure that the IVP for the given DE has a unique solution for every initial condition.

Sufficient conditions for the given DE are found such that it possesses no more than one nontrivial periodic solution. Albrecht and Villari (1987)

38. $$y'' + f(y)y' + g(y) = 0,$$

where $f, g : R \to R$.

The qualitative behavior from the point of view of boundedness, oscillation, and periodicity of the solution is studied. Introducing

$$F(y) = \int_0^y f(u)du, \quad G(y) = \int_0^y g(u)du,$$

the following less restrictive (compared to earlier work) conditions are imposed:

$$\lim_{y\to\infty} \sup[G(y) + F(y)] = +\infty,$$

$$\lim_{y\to-\infty} \sup[G(y) - F(y)] = +\infty.$$

$F(y)$ is assumed to be bounded below for positive y and above for y negative.

The analysis is carried out in the (y, z) plane, where

$$y' = z - F(y), z' = -g(y).$$

Villari (1987)

39. $$y'' + f(y)y' + g(y) = 0.$$

For closed trajectories surrounding $(0, 0)$, see Eqn. (2.22.19), which with some assumptions on f and g, contains the given Lienard equation as a special case. Villari and Zanolin (1988)

40. $$y'' + f(y)y' + g(y) = 0,$$

where $f(y)$ and $g(y)$ are continuous functions.

Existence of periodic solutions, based on the Poincaré-Bendixson theorem, is considered; a method using the construction of a sequence of closed curves bounding Bendixon domain is established. Amelkin and Zhavnerchik (1988)

41. $$y'' + f(y)y' + g(y) = 0,$$

where $f(y)$ and $g(y)$ are continuous for all $y, g(y)$ is monotonic and $f(y)$ is positive for $|y| > M > 0$. Also, $f(y)$ and $g(y)$ are continuous for all y and $g(0) = 0, g'(0) > 0$. It is then shown that for sufficiently large x, every solution of the DE either tends monotonically to zero as $x \to \infty$, or oscillates. If $f(0) < 2[g'(0)]^{1/2}$, all solutions oscillate; if $f(y) \geq 0$ and $f(0) > 2[g'(0)]^{1/2}$, they all show the former behavior. If $f(y) \geq 0$ and $f(0) > 0$, all solutions tend to zero (possibly oscillating). Proofs are direct and elementary. Gagliardo (1953)

42. $$y'' + \alpha(y)y' + \beta(y)y = 0,$$

where $\alpha(y) \leq 0, \ \ \beta(y) > 0$ are real functions.

It is shown that a solution $y(x)$ of the given DE which exists when $x \to \infty$ either oscillates or becomes monotone; in the latter case its limit as $x \to \infty$ is positive or negative, respectively, when y is increasing or decreasing. Utz (1956)

43. $$y'' + k(y)y' + h(y) = 0,$$

where $h(y)$ and $k(y)$ are continuous on $(-\infty, \infty)$ and $k(y)$ is positive if y is nonzero.

Put $K(y) = \int_0^y k(\xi)d\xi$ and suppose that $y(x)$ is a nonvanishing solution with nonvanishing derivative on $[a, b)$. It is clear that $\log|K(y(x))|$ is strictly monotone. Now, put $u = -\dfrac{dy}{dt}/K(y)$ and introduce a new independent variable $z = -\log|K(y(t))|$. A simple calculation shows that u satisfies the generalized Riccati equation

$$u\frac{du}{dz} = u^2 - u + F(z). \tag{1}$$

The function $F(z)$ is equal to $h_1(z)/k_1(z)K_1(z)$, where h_1, k_1, and K_1 are determined from

$$\begin{aligned}
h_1(-\log|K(y)|) &= h(y), \\
k_1(-log|K(y)|) &= k(y), \\
K_1(-log|K(y)|) &= K(y), \\
\text{i.e., } K_1(z) &= e^{-z}.
\end{aligned}$$

Making use of the "extended" Riccati equation (1), the existence, uniqueness, and continuation of solution of the given equation as well as the asymptotic behavior of the solution are studied. One such result is the following. Let $y \equiv 0$ be the global asymptotic stable solution of the given DE in $0 \le x < \infty$. Further, let $yh(y) > 0$ and $K(y) > 0$ for $y \ne 0$ and let h and k be continuous. Let

$$K(y) = \int_0^y k(\xi)d\xi,$$

and let

$$\lim_{y \to 0} \frac{h(y)}{k(y)K(y)} = C_0.$$

If $C_0 > 1/4$, then all solutions of the given DE are oscillatory. If $C_0 < 1/4$, then all solutions of the given DE are nonoscillatory. Benson (1981)

2.13 $y'' + f(x,y)y' + g(x,y) = 0$

1. $$\boldsymbol{y'' + (2/x)y' + yy' = 0.}$$

It is proved that all the solutions of the given DE satisfying IC,

$$y(\epsilon) = 0, C(1/\epsilon - C) \le y'(\epsilon) \le C(1/\epsilon + C), C \in R,$$

satisfy also the conditions $|y(x) - C(1 - \epsilon/x)| < r(x;\epsilon)$ for every $x \in (\epsilon, \infty)$, where

$$r(x;\epsilon) = \begin{cases} Ck\epsilon \ \ln(x/\epsilon) \text{ when} & x \in [\epsilon, 1/\epsilon], \\ Ck\epsilon[(1 - 2\ln\epsilon - 1/(\epsilon x)] \text{ when} & x \in [1/\epsilon, \infty), \end{cases}$$

$\epsilon < C < k \le C(1+\epsilon), k \le 1/[-\epsilon^{1/2}\ln\epsilon], 0 < \epsilon \le 1/16$. Vrdoljak (1987)

2. $$\boldsymbol{y'' + (n/x)y' + yy' = 0, 0 < \epsilon \le x < \infty,}$$

where $y(\epsilon) = 0, y \to 1$ as $x \to \infty$.

This is Lagerstrom model for incompressible viscous flow past an obstacle in $(n+1)$ dimensions; ϵ is a positive parameter, an analog of the Reynolds number. It is shown that the matched expansions for this problem have a small range of usefulness, additional terms included in the expansions making them less accurate. Far better results are obtained when a single expansion, an outer expansion, is used throughout. It is also rigorously proved that an iterative method of solution of the model equation based on outer Oseen approximation converges for all ϵ to a unique solution for all real $n > 0$. The cases $n = 1, 2$ are detailed and numerical results depicted. Hunter, Tajdari, and Boyer (1990)

3. $$\boldsymbol{y'' + (1/x)y' + \epsilon yy' = 0, \epsilon \ll 1, y(1) = 0, y(\infty) = 1.}$$

A first order composite matched asymptotic solution is

$$y = 1 + (\ln(1/\epsilon))^{-1} \int_\infty^\xi (1/\tau)e^{-\tau}d\tau + \gamma(\ln(1/\epsilon))^{-2}\ln\xi + \cdots,$$

where $\xi = \epsilon x$. Nayfeh (1985), p.332.

4. $$\boldsymbol{y'' + (1/x)y' - \epsilon yy' = 0, 1 \leq x < \infty, y(1) = 0, y(\infty) = -a,}$$

where ϵ is a small parameter and a is a positive constant.

This is a model equation for slow viscous flow past a cylinder. Inner and outer expansions are used to give a complete discussion of the given system. Hsiao (1973)

5. $$\boldsymbol{y'' + (3/x)y' + \epsilon yy' = 0, \epsilon \ll 1, y(1) = 0, y(\infty) = 1.}$$

A uniformly valid two-term perturbation solution is

$$y = 1 - 1/x^2 + (2\epsilon/3)(3/x - 4/x^2 + 1/x^3) + \cdots.$$

Nayfeh (1985), p.331

6. $$\boldsymbol{y'' + \{(2m+1)/m\}y'/x - (1/m)yy' = 0,}$$

where m is a parameter.

The given DE arises from the self-similar form of the nonlinear Landau damping equation. Several cases arise.

(a) For $m = -1/2$, the given DE reduces to $y'' + 2yy' = 0$ with the solutions

$$y_1 = a\tanh(ax - b/2), y_2 = 1/\{h_0 + x\},$$

where a, b, and h_0 are constants.

(b) $m = -1$. The given DE has a special solution

$$y = 1/x. \tag{1}$$

More generally, putting $Z = xy(x)$ in the given DE, we have

$$x^2 Z'' - xZ'(1-Z) + Z(1-Z) = 0 \tag{2}$$

and the solution (1) corresponds to $Z = 1$. Writing $u = \ln\ x$, we may change (2) to an autonomous equation and hence reduce it to a first order equation of Abel type.

Webb and Mckenzie (1991)

7. $$\boldsymbol{\epsilon[y'' + \{(n-1)/x\}y'] + yy' = 0, x > 1,}$$

with $y(1,\epsilon) = \alpha, y(\infty,\epsilon) = 1$.

Here n is a positive integer; $n = 1$ is trivially solved; hence $n > 1$. ϵ is a small positive parameter and is a fixed given constant that satisfies the condition $\alpha + 1 > 0$. The second condition can be replaced by $y(\infty,\epsilon) = \beta$, where $\beta > 0$.

Previous results for bounded interval are generalized. The multivariable method is used to provide a candidate y_0 about which the problem is perturbed. The method of successive approximation is then used to obtain existence and uniqueness in a function ball centered on y_0. It is shown that y_0 provides a uniformly valid approximation to the resulting solution in the full interval $1 \leq x \leq \infty$ for small ϵ. Smith (1975)

8. $$\boldsymbol{y'' + yy' - y^3 - P(x)(3y' + y^2) - R(x)y = S(x).}$$

This equation has been studied by Ince (1956). In the following cases, its solutions are without moving critical points (i.e., with poles as the only moving singularities).

(a) $R(x) = P'(x) - 2P^2(x), S(x) = 0.$

(b) $P(x) = q'(x)/[2q(x)], R = q''/(2q) - q'^2/q^2 - q(x), S(x) = 0.$

(c) $R(x) = P'(x) - 2P^2(x) - 12q^2(x), P(x) = q'(x)/q(x) + q(x), S(x) = -24q^3(x).$

(d) $P(x) = -2q(x)/q'(x), R(x) = -24x/q(x), S = 12/q(x), q = 4x^3 - \epsilon x^{-K}$, where $\epsilon = 0, K = 1$ or $\epsilon = 1, K$ an arbitrary constant.

(e) $P(x) = 0, R(x) = -12q(x), S = 12q'(x)$. In this case, $q(x)$ satisfies the equation $q'' = 6q^2 + x$.

Kolesnikova and Lukashevich (1972)

9.
$$\begin{gathered} \boldsymbol{y'' + yy' - y^3 - [(f'/f) + f](3y' + y^2) + \{af^2 + 3f'} \\ \boldsymbol{+ (3f'^2/f^2) - (f''/f)\}y + bf^3 = 0;} \\ \boldsymbol{f = f(x),} \end{gathered}$$

and a and b are constants.

(a) With $y(x) = \xi'\eta(\xi), \xi = \exp(\int f dx)$, we obtain $\eta'' + \eta\eta' - \eta^3 + (a-2)(\eta/\xi^2) + b/\xi^3 = 0$, which is a special case of the given DE with $f = 1/x$.

(b) For $a = 14, b = 24$, the given DE has the solution

$$y(x) = f[(\xi^3 u' + 2)/(\xi^2 u - 1)],$$

where $\xi = \exp(\int f dx)$ and $u(\xi)$ is any solution of $u'' = 6u^2$; see Eqn. (2.1.9).

Kamke (1959), p.549

10. $\boldsymbol{y'' + (y - 3f'/2f)y' - y^3 - (f'/2f)y^2 + [f + (f'^2/f^2) - f''/2f]y = 0, f = f(x).}$

(a) With $y = \xi'(x)\eta(\xi), a\xi'^2 = f(x)$, a a constant, the given DE goes to Eqn. (2.10.3) with ξ and η replacing x and y, respectively.

(b) If we write

$$u(x) = 1 + \exp\left(\int y dx\right), v(x) = 2u'' - u'f'/f - (u^2 - 1)f,$$

the given DE assumes the form

$$v'/v - (f'/f) - 2u'/(u-1) = 0,$$

so that $v = C(u-1)^2 f$, where C is a constant. Therefore, since now

$$2u'u''/f - u'^2 f'/f^2 - (u^2 - 1)u' = C(u-1)^2 u',$$

we have

$$u'^2 = f(x)[C_1(u-1)^3 + (u-1)^2 + C_2]. \tag{1}$$

C_1 and C_2 are arbitrary constants. Equation (1) can be solved in terms of elliptic functions if we write $dX = [f(x)]^{1/2} dx$.

Kamke (1959), p.549

11. $$\boldsymbol{y'' + (y + 3f)y' - y^3 + y^2 f + y(f' + 2f^2) = 0, f = f(x).}$$

(a) With $y(x) = \xi'(x)\eta(\xi)$, where $\xi(x)$ satisfies the DE $\xi'' = -f\xi'$, we get

$$\eta'' + \eta\eta' - \eta^3 = 0, \tag{1}$$

that is, the original DE with $f = 0$. Equation (1) is a special case of Eqn. (2.10.7) with $a = 0$.

(b) With $u(x) = \exp(-\int f dx), v(x) = \exp(\int y dx)$, the given DE goes to

$$\frac{d}{dx}\left[\frac{v''}{u^2v^2} - \frac{u'v'}{u^3v^2}\right] = 0.$$

It follows that

$$\left[\frac{v''}{u^2v^2}\right] - \left[\frac{u'v'}{u^3v^2}\right] = (3/2)C_1,$$

that is, $\frac{d}{dx}\frac{v'^2}{u^2} = 3C_1 v^2 v'$, so that $v'^2 = u^2(C_1 v^3 + C_2)$. The variables now separate and the solution may be found in terms of quadratures. C_1 and C_2 are arbitrary constants.

Kamke (1959), p.549; Murphy (1960), p.382

12. $\boldsymbol{y''+yy'-y^3+\left[(12x^2 - kx^{k-1})/2(4x^3 - x^k)\right](3y'+y^2)+(axy+b)/2(4x^3-x^k) = 0,}$

where $k = 0, 1, 2$, and a and b are constants.

(a) With $y(x) = \xi'(x)\eta(\xi), \xi' = (4x^3 - x^k)^{-1/2}$, we get

$$2(\eta'' + \eta\eta' - \eta^3) + [(a-24)x + k(k-1)x^{k-2}]\eta + b(4x^3 - x^k)^{1/2} = 0,$$

where x is related to ξ; the relation can be found explicitly in terms of elliptic functions for $k = 0, 1, 2$.

(b) If $k = 1, a = 48, b = -24$, then the given DE has the solution $y = (u'-1)/(u-x)$, where $u(x)$ is any solution of

$$(4x^3 - x)u'^2 = 4u^3 - u + C,$$

with C an arbitrary constant, obtainable in terms of elliptic functions.

Kamke (1959), p.566; Murphy (1960), p.391

13. $$\boldsymbol{y'' + 2yy' + f(x)y' + f'(x)y = 0.}$$

Writing $u(x) = y + (1/2)f$, we get the Riccati equation $u' + u^2 = (1/2)f' + (1/4)f^2 + C$. The usual transformation $u = Ag'/g$ changes this DE to a second order linear DE; see Sachdev (1991). Kamke (1959), p.550; Murphy (1960), p.383

14. $$\boldsymbol{y'' + 2yy' - c(x)(y' + y^2) - g(x) = 0,}$$

where $c(x)$ and $g(x)$ are analytic functions.

The given DE is equivalent to the system

$$w = y' + y^2, \tag{1}$$

$$w' = c(x)w + g(x). \tag{2}$$

Equation (2) is linear and integrable in terms of a quadrature; Eqn. (1) is Riccati and may consequently be solved by exact linearization. Bureau (1964)

15. $$y'' + 2yy' + f(x)[y' + y^2] - g(x) = 0.$$

Writing $u(x) = y' + y^2$, we obtain $u' + f(x)u = g(x)$. Thus, the original DE is reduced to a special Riccati equation and a linear DE of first order, which may be solved in the reverse order. Kamke (1959), p.550; Murphy (1960), p.383

16. $$y'' + 3yy' + y^3 - c(x)(y' + y^2) - f(x)y - g(x) = 0,$$

where $c(x), f(x)$, and $g(x)$ are analytic functions of x.

The only movable singularities of the given DE are poles. The transformation $y = v'/v$ changes it to a linear third order equation:

$$v''' - cv'' - fv' - gv = 0.$$

Bureau (1964)

17. $$y'' + 3yy' - (3/2)(q'/q)y' + y^3 - (3q'/2q)y^2 - [q''/2q - q'^2/q^2 + r]y$$
$$- (1/2)[r' + q - rq'/q] = 0,$$

where $q(x)$ and $r(x)$ are analytic functions.

On setting $y = w'/w$, we get a linear DE of third order:

$$w''' - (3/2)(q'/q)w'' - [q''/(2q) - q'^2/q^2 + r]w' - (1/2)[r' + q - q'r/q]w = 0.$$

Bureau (1964)

18. $$y'' - 2yy' + \alpha(\gamma + 1)(xy' + y) - 2\alpha\gamma\, y = 0,$$

where α and γ are constants.

The given DE arises from a similarity reduction of a generalized Burgers equation. The inverse transformation $y = 1/z$ leads to an Euler–Painlevé equation; see Sachdev (1991). Doyle and Englefield (1990)

19. $$y'' - 2yy' - \alpha(xy' + y) + 2\alpha y = 0,$$

where α is a parameter.

The given DE arises from a similarity reduction of a generalized Burgers equation. The inverse transformation $y = 1/z$ leads to an Euler–Painlevé equation; see Sachdev (1991). Doyle and Englefield (1990)

20. $$y'' - 2yy' + 2\alpha(\beta + \gamma)(xy' + y) - 4\alpha\gamma y - 2\alpha^2(1 + \beta^2)x = 0,$$

where α, β, and γ are constants.

The given DE arises from a similarity reduction of a generalized Burgers equation. The inverse transformation $y = 1/z$ leads to an Euler–Painlevé equation; see Sachdev (1991). Doyle and Englefield (1990)

21. $$\boldsymbol{y'' - 2yy' + (\lambda\gamma + \mu)(xy' + y) - 2\lambda\gamma y - 2\alpha\beta\gamma\delta x = 0,}$$

where α, β, γ, and δ are constants, and $\mu = \alpha\delta + \beta\gamma$ and $\lambda = \alpha\delta - \beta\gamma$.

The given DE arises from a similarity reduction of a generalized Burgers equation. The inverse transformation $y = 1/z$ leads to an Euler–Painlevé equation; see Sachdev (1991). Doyle and Englefield (1990)

22. $$\boldsymbol{y'' - 2yy' + 2\alpha(xy' + y) + ay' - 2\alpha y + a\alpha = 0,}$$

where a and α are parameters.

The given DE arises from a similarity reduction of a generalized Burgers equation. The inverse transformation $y = 1/z$ leads to an Euler–Painlevé equation; see Sachdev (1991). Doyle and Englefield (1990)

23. $$\boldsymbol{y'' - 2yy' + 2\alpha\delta(xy' + y) + (a\delta/\lambda)y' - 2\lambda y - 2\alpha\beta\gamma\delta x - \alpha\delta^2 a/\lambda = 0,}$$

where $\alpha, \delta, \beta, \gamma, a$, and λ are constants.

The given DE arises from a similarity reduction of a generalized Burgers equation. The inverse transformation $y = 1/z$ leads to an Euler–Painlevé equation; see Sachdev (1991). Doyle and Englefield (1990)

24. $$\boldsymbol{y'' - [3y + f(x)]y' + y^3 + f(x)y^2 = 0.}$$

This DE goes to Eqn. (2.10.21) if we first put $y(x) = -u(x)$ and then replace u and $-f$ by y and f, respectively. Kamke (1959), p.551; Murphy (1960), p.383

25. $$\boldsymbol{y'' - [(2y - 1)/(x - 1)]y' = 0.}$$

A general first integral is $(x-1)y' = y^2 + A^2$; the solution is $y = A\tan[B + A\log(x-1)]$. A first integral satisfying the IC $y = 0, y' = 0$ at $x = 0$ is $(x-1)y' = y^2$; the solution in this case is $y = 1/[A - \log(x-1)]$. Forsyth (1959), p.226

26. $$\boldsymbol{y'' - (y' - y/x)[(x - 2y)/x^2] = 0.}$$

Put $y = xv$ to get
$$x^2 v'' + xv'(2v + 1) = 0$$
after a multiplication by x. This is of Cauchy–Euler type. Putting $x = e^t$, we have
$$\ddot{v} + 2v\dot{v} = 0, \dot{} \equiv \frac{d}{dt}. \tag{1}$$
The solution of (1) is
$$\frac{dv}{dt} + v^2 = C; \tag{2}$$
C is an arbitrary constant. Equation (2) is separable. Bender and Orszag (1978), p.34

27. $$y'' - a(x)yy' - c(x)y' - e(x)y^2 - f(x)y - g(x) = 0,$$

where $a(x), c(x), \ldots, g(x)$ are analytic functions of x.

By a suitable change of variable, it can be shown that the equation of the given type with poles as their only movable singularities may be reduced to the form

$$y'' + 2yy' - c_1(x)(y' + y^2) - g_1(x) = 0,$$

provided that $f = 0$. The latter equation is equivalent to the system

$$w = y' + y^2, \tag{1}$$

$$w' = c_1(x)w + g_1(x). \tag{2}$$

Equation (2) is linear and solved in terms of a quadrature; Eqn. (1) is then a Riccati equation solvable through exact linearization. Bureau (1964)

28. $$y'' + [E_0(x) + E_1(x)y]y' + F_0(x) + F_1(x)y + F_2(x)y^2 + F_3(x)y^3 = 0,$$

where $E_0, E_1, F_0, F_1, F_2, F_3$ are given functions of the independent variable x. It is shown that this second order ODE may be "factored" into first order equations of Riccati type:

$$\begin{aligned} y' &= A_1 + B_1y + B_2z + D_1y^2, \\ z' &= A_2 + B_3y + B_4z + D_3y^2 + D_4yz, \end{aligned} \tag{A}$$

where the coefficients of the given system are related by

$$\begin{aligned} E_0 &= -B_1 - B_4 - B_2'/B_2, \\ E_1 &= -2D_1 - D_4, \\ F_0 &= A_1B_4 - A_1' + A_1B_2'/B_2 - A_2B_2, \\ F_1 &= A_1D_4 + B_1B_4 - B_1' + B_1B_2'/B_2 - B_2B_3, \\ F_2 &= B_1D_4 - B_2D_3 + B_4D_1 - D_1' + D_1B_2'/B_2, \\ F_3 &= D_1D_4, \end{aligned} \tag{1}$$

provided that $B_2(x) \neq 0$ in the range considered. Therefore, given the six coefficients in the original equation we can fit them by choosing the nine functions $\{A_1, A_2, B_1, B_2, B_3, B_4, D_1, D_3, D_4\}$ in (1). Three of these can be chosen arbitrarily. We may take them to be $A_1(x), B_1(x)$, and $B_2(x)$, with $B_2 \neq 0$; then from the system (1), we can solve for D_4, D_1:

$$\begin{aligned} D_4 &= -(1/2)[E_1 \pm (E_1^2 - 8F_3)^{1/2}], \\ D_1 &= -(1/4)[E_1 \mp (E_1^2 - 8F_3)^{1/2}]. \end{aligned} \tag{2}$$

The remaining equations of the system (1) determine B_4, A_2, B_3, and D_3. If we make the particular choice $B_1(x) \equiv b, B_2(x) \equiv 1, A_1(x) \equiv 0$, where b is independent of x, we obtain the following simple expressions for the remaining coefficients:

$$\begin{aligned} B_4 &= -E_0 - b, A_2 = -F_0, B_3 = -F_1 - b(E_0 + b), \\ D_3 &= -F_2 - D_1' + bD_4 - (b + E_0)D_1. \end{aligned}$$

These relations get simplified when $b = 0$, but it is advantageous sometimes to choose a nonzero value for b. It is important to note that even when $\{A_1, B_1, B_2\}$ have been chosen, there is still a twofold possibility in (A) corresponding to the choice of signs in (2). Chisholm and Common (1987)

29. $\boldsymbol{y'' + (E_0 + E_1 y)y' + F_0 + F_1 y + (1/3)(E_1' + E_0 E_1)y^2 + (1/9)E_1^2 y^3 = 0,}$

where E_0, E_1, F_0, F_1 are functions of x.

If the functions in the given DE can be chosen such that

$$f = (1/2)E_0, g = (1/3)E_1, k = F_0, h = F_1 - (1/2)E_0' - (1/4)E_0^2,$$

then it has a "bi-Riccati" form

$$\left(\frac{d}{dx} + f(x) + g(x)y\right)\left(\frac{d}{dx} + f(x) + g(x)y\right)y + k(x) + h(x)y = 0;$$

see Eqn. (2.13.30). Chisholm and Common (1987)

30. $\boldsymbol{\left(\frac{d}{dx} + f(x) + g(x)y\right)\left(\frac{d}{dx} + f(x) + g(x)y\right)y + k(x) + h(x)y = 0.}$

Writing

$$z = y' + fy + gy^2, \tag{1}$$

we have

$$z' + fz + gyz + k + hy = 0. \tag{2}$$

The system (1)-(2) forms a "Lie system". Chisholm and Common (1987)

31. $\boldsymbol{y'' + P(y,x)y' + Q(y,x) = 0,}$

where P and Q are polynomials of the same degree with coefficients that are holomorphic functions of x in some domain D.

It is shown using the small parameter method of Painlevé that it has no algebraic solution satisfying $y \to \infty$ as $x \to x_0$, where $x_0 \in D$. Kondratenya and Prolisko (1973)

32. $\boldsymbol{y'' + P(y,x)y' + Q(y,x) = 0,}$

where P and Q are polynomials in y of degree n and m, respectively, with coefficients that are holomorphic functions of x in a region D.

It is shown that if $y = y(x)$ is an algebraic solution of the given DE satisfying $y(x) \to \infty$ as $x \to x_0 \in D$, it has the representation

$$y(x) = \sum_{\sigma=0}^{\infty} \beta_\sigma (x - x_0)^{(-1+\sigma)/n}, \beta_0 \neq 0 \tag{1}$$

in the neighborhood of x_0 if $n \geq m - 1$; it has the representation (1) or

$$y(x) = \sum_{\sigma=0}^{\infty} \alpha_\sigma (x - x_0)^{(-1+\sigma)/(m-n-1)}$$

if $n < m - 1$ but $2n > m - 1$; finally, it has the form

$$y(x) = \sum_{\sigma=0}^{\infty} \gamma_\sigma (x - x_0)^{(-2+\sigma)/(m-1)}, \gamma_0 \neq 0$$

if $2n \leq m - 1$.

The convergence of the series is not discussed. Kondratenya and Prolisko (1973)

33. $$y'' + P(y, x)y' + Q(y, x) = 0,$$

where P and Q are polynomials in y of degree k and ϵ, respectively, and the coefficients of the polynomials are holomorphic functions of x in some domain D. Depending on whether

$$2k > \epsilon - 1 (k > 0), 2k < \epsilon - 1 (\epsilon > 1) \quad \text{or} \quad 2k = \epsilon - 1 (k > 0),$$

further conditions are obtained so that the given system has moving algebraic singularity or moving logarithmic singularity combined with an algebraic one. Forms of such solutions are written out in each case. Kondratenya and Yablonskii (1968)

34. $$y'' - Q(x, y)y' - P(x, y) = 0,$$

where P and Q are polynomials in y with rational coefficients.

It is shown that every meromorphic solution of the given DE in the complex plane is also a solution of the Riccati equation with rational coefficients or it has the form

$$y'' = p_0(x) + p_1(x)y + p_2(x)y^2 + p_3(x)y^3 + q_0(x)y' + q_1(x)yy'.$$

For $q_1(x) \not\equiv 0$, the given DE may be changed through

$$y = \mu(x)w + \nu(x),$$

where $\mu(x)$ and $\nu(x)$ are suitably chosen rational functions, to one of the forms

$$w'' + 2ww' = A(x) + B(x)w + C(x)w^2 \tag{1}$$

and

$$w'' + 3ww' = A(x) + B(x)w + C(x)w^2 + D(x)w^3. \tag{2}$$

We can change (2) with $D = -1$ to the form

$$w'' + 3ww' = A(x) + B(x)w - w^3. \tag{3}$$

Equation (3) is related to P_{IV}; see Ince (1956), p.334. For (1), see Eqn. (2.10.13). Steinmetz (1983)

35. $$y'' + (1/y^{1/2})[2xy' + 4\lambda(1 - y)] = 0, \lambda < 0,$$

$y(0) = 0, y(+\infty) = 1, 0 < y(x) < 1$ for $0 < x < +\infty$.

Assuming that the given BVP has a solution $y(x)$, it is shown that it must satisfy

$$0 < \lim_{x \to 0} y(x)/x < +\infty \text{ or } \lim_{x \to 0} y(x)/x^{4/3} = (-9\lambda)^{2/3} > 0,$$

according as $y'(0) \neq 0$ or $y'(0) = 0$. Moreover, $y'(0) \neq 0$ implies that $y'(0) > 0$;

$$+\infty \geq \lim_{x \to 0} y(x)x^{-4/3} > (-9\lambda)^{2/3}$$

implies that $y'(0) > 0$. It is also shown that the given DE has a particular local solution $y(x)$ which behaves as

$$y(x) = 1 + (1/2)C_2 x^{-2\lambda-1} \exp(-x^2)[1 + O(x^{-2})]$$

for large x. Also, we have

$$y'(x) = 2x[1 - y(x)][1 + O(x^{-2})] \text{ as } x \to +\infty,$$

where C_2 is a constant. If we set $C_2 = 0$, we have another particular solution, which behaves as

$$y(x) = 1 - C_1 x^{2\lambda}[1 + O(x^{-2})], y'(x) = -2\lambda C_1 x^{2\lambda-1}[1 + O(x^{-2})]$$

with positive C_1 for large positive x.

Existence of a solution of exponential type for the given BVP is also proved. Iwano (1977)

36. $$y'' + 2xy' - [4/(1-\alpha)]y - 4(2\delta)^{-1/2}y^{(\alpha-1)/2}y' - 4\lambda y^{\alpha} = 0,$$

where $\alpha > 1, \lambda$, and $\delta > 0$ are parameters and

$$y \sim A\exp(-x^2)H_\nu(x) \sim A\exp(-x^2)(2x)^{2\alpha_1}, x \to \infty,$$

and

$$y \to 0, \text{ as } x \to -\infty, |y| < \infty, -\infty < x < \infty, H_\nu(x)$$

is Hermite function of order $\nu = (3-\alpha)/(\alpha-1)$.

The given problem arises from the self-similar form of a (single-hump) solution of the generalized Burgers equation,

$$u_t + u^\beta u_x + \lambda u^\alpha = (\delta/2)u_{xx}.$$

It is first noted that there are special solutions of the given DE:

(a) $y = [\lambda(\alpha-1)]^{1/(1-\alpha)}$, a constant.

(b) $$y = \begin{cases} (A_+ x)^{2/(1-\alpha)}, & x > 0, \\ (-A_- x)^{2/(1-\alpha)}, & x < 0, \end{cases}$$

where

$$\begin{aligned} A_+ &= (2/\delta)^{1/2}[(\alpha-1)/(\alpha+1)][(1+\lambda\delta(1+\alpha))^{1/2}+1], \\ A_- &= (2/\delta)^{1/2}[(\alpha-1)/(\alpha+1)][(1+\lambda\delta(1+\alpha))^{1/2}-1]. \end{aligned} \quad (1)$$

The solution (1) is singular and tends, for $\alpha > 1$, to infinity as $|x| \to 0$.

Making the reciprocal transformation

$$H = \delta^{1/2} y^{(1-\alpha)/2} \tag{2}$$

(since $\alpha > 1$) in the given DE, we get

$$HH'' - 2(1+\alpha_1)H'^2 + 2xHH' - 2H^2 - 2^{3/2}H' - 2\lambda_1 = 0, \tag{3}$$

where

$$\alpha_1 = (1/2)[(3-\alpha)/(\alpha-1)], \lambda_1 = \lambda\delta(1-\alpha).$$

Equation (3) has a Taylor series solution about $x = 0$:

$$H = \sum_{n=0}^{\infty} a_n x^n, \tag{4}$$

where

$$\begin{aligned}
a_2 &= (1/a_0)[(a_0^2 + 2^{1/2}a_1 + a_1^2) + \lambda_1 + \alpha_1 a_1^2], \\
a_3 &= (1/(3a_0))[(a_0 a_1 + 2^{3/2}a_2 + 3a_1 a_2) + 4\alpha_1 a_1 a_2], \\
a_{k+2} &= 2a_k/[(k+1)(k+2)] + [2a_{k+1}/((k+2)a_0)][2^{1/2} + (\alpha_1+1)a_1] \\
&\quad + [2/((k+1)(k+2)a_0)] \sum_{i=1}^{k} [\{-(k+1-i)(k+2-i)/2\}a_i a_{k-i+2} \\
&\quad + (1+\alpha_1)(i+1)(k+1-i)a_{i+1}a_{k-i+1} + a_i a_{k-i} \\
&\quad - (k+1-i)a_{i-1}a_{k-i+1}], k = 1, 2, 3, \ldots.
\end{aligned}$$

The connection problem stated with the given DE was solved numerically. Values of the maximum amplitude A_m at infinity for a set of parameters δ, λ, and $\alpha > 1$ were found such that the solution either tends to a constant $y = y_{\max}$ as $x \to -\infty$, coinciding with the exact constant solution (a), or vanishes at $x = -\infty$ or diverges there. The solution agreed very well with the Taylor solution (3) when appropriate initial conditions $y(0)$ and $y'(0)$ were chosen and related to $H(0)$ and $H'(0)$ according to (2). Sachdev, Nair, and Tikekar (1986)

37. $$\mathbf{y'' - 2^{3/2}\delta^{-1/2}y^{\alpha}y' + 2xy' + [2(1-\alpha j)/\alpha]y = 0,}$$

where $\delta > 0, \alpha > 0$ and $j = 0, 1, 2$ are parameters.

The given DE arises from a similarity reduction of the nonplanar Burgers equation

$$u_t + u^{\alpha}u_x + ju/(2t) = (\delta/2)u_{xx}.$$

A connection problem is posed for this equation, namely, $y \sim A\exp(-x^2)H_\nu(x)$ as $x \to \infty$ and $y \to 0$ as $x \to -\infty$, and $|y| < \infty, -\infty < x < \infty$, where H_ν is a Hermite function of order $\nu = 1/\alpha - (j+1)$.

Before discussing the numerical solution of this problem, some simple analytical results may be noted.

(a) Let $\alpha = 1/(j+1)$. In this case the given DE is written as

$$y + xy' + (1/2)y'' = (2/\delta)^{1/2} y^\alpha y',$$

which on integration and use of condition $y \to 0$ as $x \to \infty$ yields

$$xy + (1/2)y' = [1/(\alpha+1)](2/\delta)^{1/2} y^{\alpha+1}. \tag{1}$$

Put $y^{-\alpha} = G$ in (1) to get

$$G' - 2\alpha x G = -[2\alpha/(\alpha+1)](2/\delta)^{1/2},$$

which on integration gives

$$G = \left[C - [2/(\alpha+1)][2\alpha/\delta]^{1/2} \int_0^{\alpha^{1/2}x} e^{-t^2} dt\right] e^{\alpha x^2},$$

where C is a constant. Thus, we have

$$y = \exp(-x^2) \left[C - \frac{2}{\alpha+1} \left[\frac{2\alpha}{\delta}\right]^{1/2} \int_0^{\alpha^{1/2}x} e^{-t^2} dt\right]^{-1/\alpha},$$

where $C = y^{-\alpha}(0)$.

(b) Also, $y = c$, a constant, is a solution if $\alpha = 1/j$.

(c) Now, we find simple results about single-hump solutions for which the solutions starting from zero at $x = \infty$ vanish either at $x = -\infty$ or at $x = x_0$, a finite number. Put $F = y^\alpha$ in the given DE to obtain

$$(1/2)FF'' - \frac{\alpha-1}{2\alpha} F'^2 + (1-\alpha j)F^2 + xFF' - \left(\frac{2}{\delta}\right)^{1/2} F^2 F' = 0. \tag{2}$$

Integrating (2) from $x = -\infty$ to $x = +\infty$ and using the conditions that y, y' and hence F, F' vanish at $x = \pm\infty$, we get

$$(2\alpha j - 1) \int_{-\infty}^{\infty} F^2 dx = \frac{1-2\alpha}{\alpha} \int_{-\infty}^{\infty} F'^2 dx$$

or

$$r \equiv \frac{\int_{-\infty}^{\infty} F^2 dx}{\int_{-\infty}^{\infty} F'^2 dx} = \frac{1-2\alpha}{\alpha(2\alpha j - 1)} > 0. \tag{3}$$

The inequality (3) implies that for $j = 0$, single-hump solutions exist for $\alpha > 1/2$; for $j = 1$ such solutions exist only for $\alpha = 1/2$. For $j = 2$, single-hump solutions exist when $1/4 < \alpha < 1/2$.

Integrating the given DE from $x = x_0$ to $x = \infty$ and assuming that $y(x_0) = 0, y'(x_0) > 0$, we get

$$\frac{\alpha[(j+1)-1]}{\alpha} \int_{x_0}^{\infty} y dx = -(1/2)y'(x_0) < 0. \tag{4}$$

Since $y > 0$ in $x_0 < x < \infty$, Eqn. (4) implies that $\alpha < 1/(j+1)$. Thus, single-hump solutions starting at $x = +\infty$ and vanishing at $x = x_0$ exist if $\alpha < 1/(j+1)$. Combining finite and (negative) infinite values of x_0, we conclude that single-hump solutions exist if $1/(j+2) < \alpha < 1/(j+1), j = 0, 1, 2$.

Numerical solution of the connection problem as well as qualitative analysis lead to certain definite conclusions. Sachdev and Nair (1987)

38. $$\boldsymbol{y'' - 2y^{\beta}y' + 2xy' = 0,}$$

where β is a parameter.

y = constant is a special solution. The linearized form of the given DE, namely, $y_L'' + 2xy_L' = 0$, has the solution $y_L = (2B/\pi^{1/2})$ erf x, where $B = y(0)$ is an amplitude parameter. Monotonical solutions joining different constants at $x = \pm\infty$ may be constructed numerically. Sachdev, Nair, and Tikekar (1988)

39. $$\boldsymbol{y'' - 2y^{\beta}y' + (1+n)xy' + [(1-n)/\beta]y = 0,}$$

where β and n are parameters.

The given DE results from the similarity reduction of the generalized Burgers equation $u_t + u^{\beta}u_x = (\delta/2)(1+t)^n u_{xx}$. The solutions of the given DE that have a single-hump form are sought. A necessary condition for that is that $n < 1$; this follows from the observation that at the maximum (if one exists), $y' = 0$ and $y'' < 0$. First we note some special cases. Choosing $\beta = (1-n)/(1+n) \equiv \beta_n$, say, we write the given DE as

$$y'' - 2y^{\beta}y' + [2/(1+\beta)](xy' + y) = 0. \tag{1}$$

Integrating (1) and assuming that $y, y' \to 0$ as $|x| \to 0$, we have

$$y' - [2/(1+\beta)]y^{1+\beta} + [2/(1+\beta)]xy = 0. \tag{2}$$

Integrating (2), we have

$$y = e^{[-1/(1+\beta)]x^2} h^{-1/\beta}(x),$$

where

$$h(x) = A - (2m)^{1/2} \text{ erf } [(m/2)^{1/2}x],$$

where

$$A = y^{-\beta}(0) \quad \text{and} \quad m = 2\beta/(1+\beta).$$

More generally, we may change the given DE to one with no fractional powers of y by writing

$$H = y^{-\beta}, \tag{3}$$

so that

$$HH'' - [(\beta+1)/\beta]H'^2 + (1+n)xHH' - (1-n)H^2 - 2H' = 0. \tag{4}$$

See Eqn. (2.32.16) for a Taylor series solution of (4).

A connection problem for the given problem is posed as follows:

$$y \sim A\phi[\alpha, 1/2; -[(1+n)/2]x^2], \text{ as } |x| \to \infty,$$
$$|y| < \infty, -\infty < x < \infty; \alpha = [1/(2\beta)][(1-n)/(1+n)],$$

where the function ϕ is confluent hypergeometric function with the given arguments. This function vanishes for $|x| \to \infty$ only if $|n| < 1$. A numerical solution of this problem is attempted. The following conclusions are drawn.

(a) $0 < n < 1, \beta \geq 1$. The problem has solutions that either tend to zero as $|x| \to \infty$, or tend to zero at $x = \infty$ but tend to a nonzero value as $x \to -\infty$. These kinds of solutions are also found for $-1 < n < 0, \beta \geq \beta_n$.

(b) $-1 < n < 0, \beta < \beta_n$. In this case the limiting (linear) solution as $x \to \infty$ is oscillatory. The nonlinear solutions starting from these linear solutions for large x are such that either they are nonoscillatory at other end, namely $x \to -\infty$, or they are oscillatory there also. The variations of the solutions with changes in the parameters are shown in figures. The cases $n = 1$ and $n = -1$ are treated separately; see Eqns. (2.13.38) and (2.12.7).

Sachdev, Nair, and Tikekar (1988)

40. $$y'' + [x/(2m)]y^{1/m-1}y' = 0,$$

$y(x_0) = 0, y'(x_0) = (\alpha/2)x_0, y(\infty) = \beta^m$, where m, x_0, α, and β are constants.

The solution of the given problem is sought such that it belongs to $C^1[x_0, \infty) \cap C^2(x_0, \infty)$ and is positive for $x > x_0$. The problem is first solved for any fixed $x_0 < 0$ and then x_0 is chosen such that the solution satisfies $v(\infty) = \beta^m$ as well. The proof of existence of the solution is by integral equation formulation. A unique solution is shown to exist for $x_0 < 0$. Wu (1985)

41. $$y'' + \sin y - \epsilon(\cos ny + a)(1 + c \sin \nu x)y' = 0,$$

where ϵ, a, c, and ν are parameters.

For $c = 0$, it is shown that the given DE has exactly n limit cycles at $a = (-1)^n/(4n^2 - 1)$. Of these, $n-1$ lie in the region of oscillatory motions and one on the boundary between the oscillatory and rotary regions (i.e., saddle-type limit cycle). If n is odd, the saddle-type limit cycle is stable if $\epsilon > 0$. Using this fact, the chaotic dynamics of the full equation for $c \neq 0$ is studied. The role of limit cycles in the formation of nontrivial attracting sets is demonstrated. It is shown that for the given equation, attracting sets may possibly contain stable points with a large period—these are called quasi-attractors or chaos. Numerical results are depicted. Morozov (1989)

42. $$y'' + (x/2)B'(y)y' = 0, x > x_0, y(x_0) = 0,$$

$y'(x_0) = -(\alpha/2)x_0, \lim_{x\to\infty} y(x) = \beta$; α and β are constants.

It is proved that the given free boundary value problem has a unique solution $\{y(x), x_0\}$ provided that $B \in C[0, +\infty) \cap C'(0, +\infty)$ and $B(y)$ is strictly increasing and concave. Yan (1988)

43. $$y'' + 2xF(y)y' = 0, y(0) = 1, y(\infty) = 0,$$

where $F(y)$ is positive and C^1.

It is shown that the given BVP has at least one solution: For $0 < x < \infty$ the solution is monotonic decreasing and tends asymptotically to 0 as $x \to \infty$, and its derivative is negative, continuously differentiable, and monotonic increasing to 0 as $x \to +\infty$. Lee (1971/72)

44. $$y'' + F(x,y)y' = 0, 0 \leq x \leq a, y(0) = 0, y(a) = \lambda,$$

where a and λ are constants.

For the given BVP, the following result is proved. Let $F(x,y)$ satisfy the following conditions:

(a) It is continuous in the rectangle $Q: 0 \leq x \leq a, 0 \leq y \leq b$.

(b) It has a continuous derivative $F_y(x, y)$ in this rectangle.

(c) It satisfies the conditions $|F(x, y)| \leq Mx^{\ell}, |F_y(x, y)| \leq M'x^{\ell'}$ in Q. Then if

$$q = \lambda \frac{2M'}{(\ell' + 1)u^{l+1}} \frac{E(\ell, \ell', a)}{G(\ell, a)} < 1,$$

where

$$E(\ell, \ell', a) = \int_0^{\mu a} t^{\ell'+1} \exp(t^{\ell+1}) dt, G(\ell, a) = \int_0^{\mu a} \exp(-t^{\ell+1}) dt,$$

$\mu = [M/(\ell+1)]^{1/(\ell+1)}$, the given BVP has a unique solution $y_*(x)$. This solution can be found as the limit of the sequence $u_0(x) = 0, u_{k+1}(k) = A(u_k|x)(k = 0, 1, 2, \ldots.)$, where $A(u|x)$ is the operator in Eqn. (N.1.14), $u(x) = A(u|x)$,

$$A(u|x) = \lambda \frac{\int_0^x \exp[-B(u|t)]dt}{\int_0^a \exp[-B(u|t)]dt}, \quad B(u|x) = \int_0^x F(t, u(t)) dt,$$

and the error of the kth approximation is described by the inequality

$$|y_*(x) - u_k(x)| \leq \lambda[q^k/(1 - q)]a, k = 0, 1, \ldots.$$

If $\ell = \ell' = 0$, a condition for the convergence of the sequence of approximations to the solution of the given BVP is

$$\lambda M'a < f(Ma), f(\xi) = (1/2)[\xi(1 - \exp(-\xi))]/[1 - (1 - \xi)\exp(\xi)];$$

if $\ell = 0$ and $\ell' = 1$, the condition is

$$\lambda M'a < f[(M/2)^{1/2}a], f(\xi) = [\pi^{1/2}/2][\exp(\xi)][\exp(\xi^2) - 1].$$

The results improve if $F(x, y)$ is positive in Q.

Pykhteev and Myachina (1972)

45. $$y'' + f(x, y)y' + g(x, y) = 0.$$

If $g_y - f_x = Xf - X^2 - X'$, where $X = X(x)$, calculate $F(x) = \exp(\int X(x)dx), G_x = gF, G_y = (f - X)F$. Then solve the first order DE $F(x)y' + G(x, y) = 0$. Kamke (1959), p.551; Murphy (1960), p.384

46. $$\epsilon y'' + \alpha(x, y, \epsilon)y' - \gamma(x, y, \epsilon) = 0, \; a < x < b, \; y(a) = A(\epsilon), y'(b) = B(\epsilon),$$

where the functions $\alpha(x, y, \epsilon), \gamma(x, y, \epsilon)$ are continuously differentiable w.r.t. (x, y) in any region of the form

$$R(k) \equiv \{(x, y, \epsilon)|a \leq x \leq b, |y| \leq k, 0 \leq \epsilon \leq 1\}$$

and the function $A(\epsilon), B(\epsilon)$ are continuous for $0 \leq \epsilon \leq 1$. The following result holds with reference to the "reduced" problem: Suppose that there are three positive constants ϵ_0, M, and α_0 such that:

(a) For $0 < \epsilon \leq \epsilon_0$, there is a solution of the given system.

(b) $|\gamma[x, y(x, \epsilon), \epsilon]| \leq M$.

(c) $0 < \alpha_0 \leq \alpha[x, y(x, \epsilon), \epsilon]$.

Then there is a function $u(x)$ that satisfies the reduced equation

$$\alpha[x, u(x), 0]u'(x) = \gamma[x, u(x), 0], u(b) = B(0). \tag{1}$$

Moreover, $u(x)$ is the only solution of (1) that also satisfies

$$|u(x)| \leq k_0 \equiv \max(\bar{A}, \bar{B}) + (M/\alpha_0)[(b - a)(b - a + 1) + 1].$$

Evidently, for any $\delta, 0 < \delta < b - a$,

$$\lim_{\epsilon \to 0+} \max_{a+\delta \leq x \leq b} [|y(x, \epsilon) - u(x)| + |y'(x, \epsilon) - u'(x)|] = 0.$$

Maximum principle methods are applied. Dorr, Parter, and Shampine (1973)

47. $$\boldsymbol{y'' + y' \int_{-\infty}^{x} y^2 dx + y^3 + [1/(2\nu)](xy' + y) = 0, y(-\infty) = 0,}$$

where ν is a parameter.

The given DE has a first integral

$$xy + 2\nu \left[y' + y \int_{-\infty}^{x} y^2 dx\right] = 0 \tag{1}$$

satisfying $y(-\infty) = 0$. Dividing (1) by y and differentiating, we have

$$1 + 2\nu[(y'/y)' + y^2] = 0. \tag{2}$$

Multiplying (2) by y'/y and integrating, we obtain

$$(y'/y)^2 = -(y^2 + \nu^{-1}\ln y) + C, \tag{3}$$

where C is an arbitrary constant; y is assumed to be positive; since $y(-x) = y(x)$ in (3), it suffices to consider $x \geq 0$. Now specifying IC $y(0) = 1, y'(0) = 0$, we solve (3) as a quadrature:

$$\int_y^1 \left[\frac{1}{y(1 - y^2 - \nu^{-1}\ln y)^{1/2}}\right] dy = x. \tag{4}$$

To obtain (approximate) explicit form for (4), we consider two cases:

(a) $\nu \gg 1$. Then (4) is readily integrated to give

$$y = \operatorname{sech} x. \tag{5}$$

(b) $\nu \ll 1$. The integrand in (4) is approximated as

$$\int_y^1 \left[\frac{1}{y(-\nu^{-1}\ln y)^{1/2}}\right] dy = x,$$

with explicit form

$$y = e^{-x^2/4\nu}. \tag{6}$$

It is easy to check from (3)-(6) that $y(x) < \operatorname{sech} x$ and

$$e^{-(x^2/4\nu + |x|)} < y(x) < e^{-x^2/4\nu}.$$

The approximate solutions are compared with exact numerical solutions and depicted. The problem arises in Alfven wave propagation. Matsuno (1991)

2.14 $y'' + ay'^2 + g(x,y)y' + h(x,y) = 0$

1. $$\boldsymbol{y'' + y'^2 + 4 = 0.}$$

Put $y' = p(x)$, so that

$$\frac{dp}{dx} + p^2 + 4 = 0. \tag{1}$$

The change $p = w'(x)/w(x)$ makes (1) linear: $w'' + 4w = 0$. Hence the solution may be written as

$$y = \ln \cos(2x + C_1) + C_2;$$

C_1 and C_2 are arbitrary constants. Reddick (1949), p.191.

2. $$\boldsymbol{y'' + y'^2 - 1 = 0.}$$

Put $y' = v, \dfrac{dv}{dx} = 1 - v^2$. $v = \pm 1$ are special solutions. Therefore, $y' = \pm 1$. On integration, we get $y = \pm x + C_1$, where C_1 is an arbitrary constant. The other possibility, $dv/dx = 1 - v^2$, on integrating, and using $v = \dfrac{dy}{dx}$ and integrating again, gives

$$y = \ln |e^x + C_2 e^{-x}| + C_3,$$

where C_2 and C_3 are arbitrary constants. Rabenstein (1966), p.373

3. $$\boldsymbol{y'' + y'^2 - y' = 0.}$$

Put $y' = p, y'' = p\dfrac{dp}{dy}$; we have $p\dfrac{dp}{dy} + p^2 - p = 0$. Therefore, either $p = 0$, i.e., $y =$ constant, or

$$\frac{dp}{p-1} + dy = 0,$$

$$\text{i.e., } p - 1 = \exp(C_1 - y), \tag{1}$$

$$\frac{dy}{dx} = 1 + \exp(C_1 - y). \tag{2}$$

Hence,

$$x = \int \frac{1}{1 + \exp(C_1 - y)} dy + C_2,$$

where C_1 and C_2 are arbitrary constants. Rabenstein (1966), p.373

4. $$\boldsymbol{y'' + y'^2 - yy' = 0.}$$

Putting

$$\frac{dy}{dx} = p, \quad \frac{d^2y}{dx^2} = p\frac{dp}{dy},$$

we have

$$p\frac{dp}{dy} + p^2 - py = 0.$$

Therefore either $p = 0$, i.e., $y =$ constant or $\dfrac{dp}{dy} = -p + y$. This linear DE is solved:

$$e^y p = \int y e^y dy + c_1 = e^y(y-1) + c_1. \tag{1}$$

Integrating (1), with $p = \dfrac{dy}{dx}$, we have

$$\int \frac{dy}{y - 1 + c_1 e^{-y}} + c_2 = x,$$

where c_1 and c_2 are arbitrary constants. Martin and Reissner (1958), p.77

5. $$\boldsymbol{y'' + y'^2 - yy'/(y+1)^2 = 0, y(1/2) = 0, y'(1/2) = 1.}$$

The solution is found to be $(y+1)^2 = 2x$. To find it, put $(y+1)^2 = z$, etc. Reddick (1949), p.192

6. $$\boldsymbol{y'' \pm y'^2/2 + y = 0.}$$

Put $y' = p, y'' = p\dfrac{dp}{dy}$, etc., to obtain

$$\frac{d}{dy}(p^2) \pm p^2 = -2y. \tag{1}$$

Integrating (1) w.r.t. y we have

$$p^2/2 = Ae^{\mp y} \mp y + 1, \tag{2}$$

where A is a constant. Another integration of (2) gives

$$x = \frac{1}{2^{1/2}} \int \left[Ae^{\mp y} \mp y + 1\right]^{-1/2} dy + B,$$

where B is a constant. Lawden (1959), p.317

7. $$\boldsymbol{y'' - (1/2)y'^2 - (3/2)y^3 - 4xy^2 - 2(x^2 + \alpha)y - \beta/y = 0,}$$

where α and β are constants.

A method to linearize IVP for P_{IV} is given. The procedure involves formulating a Riemann–Hilbert boundary value problem on intersecting lines for the inverse monodromy problem. This boundary value problem is reduced to a sequence of standard problems on single lines in a certain range of parameter space. Schlesinger transformations allow one to cover the parametric space completely. Special solutions of P_{IV} are constructed from special cases of the Riemann problem as well. Fokas, Mugan, and Ablowitz (1988)

8. $$\boldsymbol{y'' - (1/x)y' - (1/2)y'^2 = 0.}$$

An exact solution is

$$y = -2\ln[1 + x^2/(4R^2)],$$

where R is a constant. McVittie (1933)

9. $$y'' - y'^2 + w^2 = 0,$$

where w is a real constant.

Putting

$$Y = [1/(2w)]e^{2wx}, X = -e^{-y+wx},$$

we obtain the reduced equation $\dfrac{d^2Y}{dX^2} = 0$, with the solution $Y = C_1X + C_2$ or

$$\overline{C}_1 e^{-y-wx} + \overline{C}_2 e^{-2wx} = 1,$$

where C_i and $\overline{C}_i$ $(i = 1, 2)$ are arbitrary constants. Sarlet, Mahomed, and Leach (1987)

10. $$y'' - y'^2 + y = 0.$$

The given DE has polynomial solutions

$$y = x^2/4 + kx + k^2 - 1/2$$

for any constant k.

It is shown that if

$$k_1 = e^{-2y_0}(y_0'^2 - y_0 - 1/2),$$

a solution of the given DE with IC $y = y_0, y' = y_0'$ and valid for large x is periodic if $-1/2 < k_1 < 0$ and nonoscillatory if $k_1 \geq 0$. The polynomial solutions occur for $k_1 = 0$. Utz (1969)

11. $$y'' - y'^2 - [2y^2 + a(x)]y' - y^4 - 2y^3 - b_1(x)y^2 - b_2(x)y - b_3(x) = 0,$$

where $a(x)$ and $b_j(x)$ are holomorphic in a region D.

This becomes a special case of Eqn. (2.14.12) if suitable transformations are made. A series solution of the form

$$y = \sum_{n=0}^{\infty} y_n(x - x_0)^{(n-\alpha)/\beta},$$

where $\alpha, \beta > 0$ are integers, is constructed. The convergence of the series is established by transition via appropriate transformations to a Briot-Bouquet system.

In particular, if $b_1(x) = 2 + a(x)$ and $b_2(x) = a(x) + 1$, then by the transformation

$$\beta' = \beta^2 + [a(x) + 1]\beta + b_3(x), \tag{1}$$

the given DE reduces to the Riccati equation

$$y' = -y^2 - y + \beta. \tag{2}$$

The Riccati system (1)–(2) would imply the existence of single-valued movable singularities of the given DE for the present special case. However, the system (1)–(2) does not have poles as the only movable singularities. This is quickly seen by using the Painlevé method of small parameters. Write the given system with the present special choice of b_1 and b_2 as

$$\begin{aligned} \frac{dy}{dx} &= z, \\ \frac{dz}{dx} &= z^2 + z[2y^2 + a(x)] + y^4 + 2y^3 + [2 + a(x)]y^2 + [1 + a(x)]y + b_3(x). \end{aligned} \tag{3}$$

Introducing the parameter λ through

$$y = Y, z = \lambda^{-1}Z, x = x_0 + \lambda X$$

in (3) and then putting $\lambda = 0$, we have the simplified system

$$\frac{dY}{dX} = Z, \quad \frac{dZ}{dX} = Z^2,$$

with the solution

$$Z = -1/(X+C), Y = -\ln(X+C),$$

where C is an arbitrary constant. Hence the existence of multivalued singularities is demonstrated. Bogoslovskii (1972)

12. $$\boldsymbol{y'' - y'^2 - y' \sum_{j=0}^{k} a_j(x) y^{k-j} - \sum_{j=0}^{s} b_j(x) y^{s-j} = 0.}$$

The structure of the solution of the given DE is found at the point $x_0 \in D = G \backslash w_1 \cup w_2 \cup w_3 \cup w_4$, where G is the extended complex plane, w_i is the set of singular points of coefficients $a_0(x)$ and $b_0(x)$, and $x = \infty$. The solution of the given DE is constructed with the property that $z \equiv y' \to \infty$ as $x \to x_0 \in D$.

We replace the given DE by the equivalent system

$$dy = \frac{dz}{z[1 + \phi(z^{-1}, y, x)]}, \tag{1}$$

$$dx = \frac{dz}{z^2[1 + \phi(z^{-1}, y, x)]}, \tag{2}$$

where

$$\phi = z^{-1} \sum_{j=0}^{k} \sum_{\nu=0}^{\infty} a_{j\nu}(x - x_0)^{\nu} y^{k-j} + z^{-2} \sum_{j=0}^{2k} \sum_{\nu=0}^{\infty} b_{j\nu}(x - x_0)^{\nu} y^{2k-j}. \tag{3}$$

Here we have set $s = 2k$, without loss of generality.

The solution to (1)–(3) will be sought as $z \to \infty$, starting from the approximated system

$$\frac{dy}{dz} = \frac{1}{z}, \quad \frac{dx}{dz} = \frac{1}{z^2}, \tag{4}$$

namely

$$y = \ln z + C, \quad x = x_0 - z^{-1}, \tag{5}$$

where C is an arbitrary constant, and $x_0 \in D$. As $z \to \infty$, in view of (5),

$$1 + \phi(z^{-1}, y, x) \sim 1 + O(z^{-1} \ln^k z);$$

therefore,

$$\begin{aligned} dy &= z^{-1}[1 + O(z^{-1} \ln^k z)]dz, \\ dx &= z^{-2}[1 + O(z^{-1} \ln^k z)]dz. \end{aligned} \tag{6}$$

Integrating (6) along some path $L(\infty, 0)$ and using the estimate

$$\int_{\infty}^{z} z^{-m} O[z^{-\ell} \ln^s z] dz \sim O[z^{-m-\ell+1} \ln^s z]$$

(m, ℓ, and s are positive integers), we obtain a first approximation to the solution of the system (1)–(2),

$$\begin{aligned} y &= \ln z + C + O(z^{-1} \ln^k z), \\ x &= x_0 - z^{-1} + O(z^{-2} \ln^k z), \end{aligned} \tag{7}$$

where C is the same constant as in (5). The series (7) is generalized as

$$\begin{aligned} y &= \ln z + C + \textstyle\sum_{n=1}^{\infty} \left[\sum_{m=kn}^{0} A_{nm} \ln^m z\right] z^{-n}, \\ x &= x_0 - z^{-1} \left[1 + \textstyle\sum_{n=1}^{\infty} \left(\sum_{m=kn}^{0} B_{nm} \ln^m z\right) z^{-n}\right]. \end{aligned} \tag{8}$$

The recurrence relation for A_{nm} and B_{nm} are obtained and the convergence of series in (8) proved. This is achieved by introducing the variables

$$y = \ln z + C + u, x = x_0 - z^{-1}(1 + v),$$

$$\xi = \ln^{-1} z, \quad \eta = z^{-1} \ln^k z,$$

where $\xi \to 0$, $\eta \to 0$ as $z \to \infty$, forming a PDE for $u(\xi, \eta)$ and $v(\xi, \eta)$, finding their solution about $\xi = 0, \eta = 0$ and proving the convergence of the relevant series. The solution in the form

$$y = \sum_{n=0}^{\infty} y_n (x - x_0)^{(n-\alpha)/\beta},$$

where $\alpha, \beta > 0$ are integers, is found in the following cases:

(a) $\alpha = 1, \beta = p - 1, x_0^{2p} = -\dfrac{1}{(p-1)^2 b_{00}}, s > 2k, s = 2p, p \neq 1.$

(b) $\alpha = 2, \beta = 2p - 1, x_0^{2p-1} = -\dfrac{4}{(2p-1)^2 b_{00}}, s > 2k, s = 2p + 1.$

(c) $\alpha = 1, \beta = k - 1, x_0^{k-1} = \dfrac{1}{(k-1) a_{00}}, s < 2k, k \neq 1.$

(d) $\alpha = 1, \beta = s - k - 1, x_0^{s-k-1} = \dfrac{a_{00}}{(s-k-1) b_{00}}, k + 1 < s < 2k, k \neq 1.$

(e) $\alpha = 1, \beta = k-1, (k-1)^2 b_{00} x_0^{2k-2} - (k-1) a_{00} x_0^{k-1} + 1 = 0$ if $a_{00}^2 \neq 4 b_{00}; s = 2k, k \neq 1$

The other coefficients are uniquely found for each case by recursion formulas. The convergence of the series in appropriate domains is proved by constructing a relevant system of Briot and Bouquet equations. Bogoslovskii (1972)

13. $$\boldsymbol{y'' - y'^2 - \sum_{j=0}^{p} a_j(x) y^{p-j} y' - \sum_{j=0}^{k} b_j(x) y^{k-j} = 0,}$$

where $a_j(x)$ and $b_j(x)$ are holomorphic in some domain D.

It was shown in an earlier paper by the authors that the given DE has a one-parameter family of solutions which have the property that as $x \to x_0, y \to \ln z + C$, when $z = y' \to \infty$ along paths bounded with respect to the argument. Algebraic solutions were also shown to exist. In the paper referred to here, all subset of equations of the given one are singled out which at a point $x_0 \in D$ have a one-parameter family of algebraic solutions having the property that $y \to \infty,\ z = y' \to \infty$ as $x \to x_0$. Bogoslovskii and Ostroumov (1981)

14.
$$y'' + \epsilon y'^2 + y = 0,$$

$y(0) = A, y'(0) = 0$. Here ϵ is small.

A uniform perturbation solution to $O(\epsilon^3)$ is

$$\begin{aligned} y(\theta,\epsilon) &= A\cos\theta + \epsilon(A^2/6)(-3 + 4\cos\theta - \cos 2\theta) + \epsilon^2(A^3/3) \\ &\quad \times [-2 + (61/24)\cos\theta - (2/3)\cos 2\theta + (1/8)\cos 3\theta] + O(\epsilon^3), \end{aligned}$$

where $\theta = wx$ and

$$w(\epsilon) = 1 - \epsilon^2(A^2/6) + O(\epsilon^3).$$

Mickens (1981), p.42

15.
$$y'' + ay'^2 + by = 0,$$

where a and b are constants.

Put $v(y) = y'^2$ so that we get a linear DE $v' + 2av + 2by = 0$ with the solution

$$v(y) = y'^2 = Ce^{-2ay} + (b/2a^2)(1 - 2ay) \equiv Y, x = C_1 + \int Y^{-1/2}dy,$$

where C and C_1 are arbitrary constants. Kamke (1959), p.551; Murphy (1960), p.384

16.
$$y'' + \epsilon y'^2 + \epsilon y^2 + y = 0, \epsilon \ll 1.$$

The method of multiple scales gives the solution as

$$y = a\cos(wx + \beta) - \epsilon a^2 + \cdots,$$

where $w = 1 - \epsilon^2 a^2 + \cdots$, and a and β are constants. Nayfeh (1985), p.145

17.
$$y'' + \epsilon\alpha_4 y'^2 + w_0^2 y + \epsilon\alpha_2 y^2 = 0, \epsilon \ll 1,$$

where w_0, α_2, and α_4 are constants.

The solution to second order approximation, by the method of multiple scales, is

$$y = a\cos(wx + \beta_0) + \epsilon a^2 \left[\frac{\alpha_2 - w_0^2\alpha_4}{6w_0^2}\cos(2wx + 2\beta_0) - \frac{\alpha_2 + w_0^2\alpha_4}{2w_0^2}\right] + \cdots,$$

where

$$w = w_0 - \left[\epsilon^2 \frac{5\alpha_2^2 + 5w_0^2\alpha_2\alpha_4 + 2w_0^4\alpha_4^2}{12w_0^3}\right] a^2 + \cdots,$$

a and β being arbitrary constants. Nayfeh (1985), p.147

18. $$y'' + \epsilon y'^2 + \epsilon^2 y^3 + y = 0, \epsilon \ll 1.$$

Using the method of multiple scales, the solution may be found to be

$$y = a\cos(wx + \beta_0) - (1/6)\epsilon a^2[\cos(2wx + 2\beta_0) + 3] + \cdots,$$

where $w = 1 + (5/24)\epsilon^2 a^2 + \cdots$. Nayfeh (1985), p.144

19. $$y'' + ay'^2 + b\sin y = 0,$$

where a and b are constants.

This is the pendulum DE with quadratic resistance. Writing $v(y) = y'^2$, we get

$$v' + 2av + 2b\sin y = 0, \tag{1}$$

with the integral

$$v(y) = [y'(x)]^2 = Ce^{-2ay} + [2b/(4a^2 + 1)](\cos y - 2a\sin y);$$

we have a case of separable variables. C is an arbitrary constant. Kamke (1959), p.553; Murphy (1960), p.384

20. $$\epsilon y'' + y'^2 - 1 = 0, \quad y(0) = A, y(1) = B,$$

with $0 < B - A < 1$.

The limit solution ($\epsilon \to 0$) is the angular solution

$$y(x)_{\text{lim}} = \begin{cases} A - x, & 0 \le x \le [1 - (B - A)]/2, \\ B + (x - 1), & [1 - (B - A)]/2 \le x \le 1. \end{cases}$$

If $1 \le B - A$, this limit solution is

$$y_{\text{lim}} = B + (t - 1).$$

Dorr, Parter, and Shampine (1973)

21. $$\epsilon y'' + y'^2 - y = 0, 0 < x < 1, \epsilon \ll 1,$$

$-y'(0, \epsilon) = A, y(1, \epsilon) = B > 0$; A and B are constants.

A detailed singular perturbation analysis that depends on the solution of the reduced problem $y = y'^2$, subject to either right or left boundary condition, is carried out. Several possibilities arise which are summarized in the (A, B) plane. Howes (1977)

22. $$\epsilon y'' + y'^2 - y^3 = 0, 0 < x < 1, y(0, \epsilon) = A, y(1, \epsilon) = B.$$

The asymptotic forms of the solution for small ϵ are discussed, as an application of the general theory. See Eqn. (2.16.68). Howes (1978)

23. $$\epsilon y'' + y'^2 + yy' = 0,$$

$y(0) = A, y(1) = B, A < B$.

Explicit solution of the reduced equation when $\epsilon = 0$ are obtained:

(a) $B \geq 0, y(x) = B, 0 < x \leq 1$.

(b) $B < 0$ and $Be < A$; then

$$y(x) = \begin{cases} A, & 0 \leq x \leq \ln(Be/A), \\ Be^{1-x}, & \ln(Be/A) \leq x \leq 1. \end{cases}$$

(c) $B < 0$ and $Be \geq A, y(x) = Be^{1-x}$. These solutions are then related to those of the full equation. Maximum principle methods are used.

Dorr, Parter, and Shampine (1973)

24. $$\epsilon y'' - y'^2 - y^2 = 0, 0 < x < 1, y(0,\epsilon) = A, y(1,\epsilon) = B.$$

This BVP can be solved exactly. See Eqn. (2.16.68). The asymptotic forms of the solution for small ϵ are discussed as an application of the general theory presented. Howes (1978)

25. $$\epsilon y'' + \mu y'^2 - y = 0, 0 < x < 1, \epsilon \ll 1,$$

$-y'(0,\epsilon,\mu) = A, y(1,\epsilon,\mu) = B > 0$, where the parameter μ is assumed to be small and positive, and both ϵ and μ tend to zero in an interrelated fashion.

A singular perturbation analysis is carried out; the nature of solution depends crucially on the relation between ϵ and μ and the corresponding reduced problems when $\epsilon \to 0^+$. Howes (1977)

26. $$y'' + \epsilon(y'^2/3 - y') + y = 0,$$

where ϵ is a small parameter. For the given Rayleigh equation approximate solution is found to be $y = 2\cos x$, while the frequency is $1 + O(\epsilon^2)$. Averaging methods are used. Jordan and Smith (1977), p.107

27. $$y'' - ay'^2 - by' - c = 0.$$

We can make a preliminary rescaling of y to make $a = 1$. Having done so, a transformation of the form $y = Y + (1/2)bx$ will eliminate y' term. So we may solve the equation

$$y'' = y'^2 + c.$$

Now see Eqn. (2.14.9). Conditions and symmetries are found such that the given DE may be linearized. Sarlet, Mahomed, and Leach (1987)

28. $$y'' + ay'^2 + by' + cy = 0;$$

a, b, and c are constants.

Put $v(y) = y'(x)$ to obtain an Abel DE

$$vv' + av^2 + bv + cy = 0;$$

see Sachdev (1991), p.30. Kamke (1959), p.553; Murphy (1960), p.384

29. $$\epsilon y'' - y'^2 + 2xy' - y = 0, -1 < x < 1,$$

$y(-1, \epsilon) = A, y(1, \epsilon) = B.$

The asymptotic forms of the solution for small ϵ are discussed, as an application of the general theory presented. Howes (1978)

30. $$y'' + (a/x)y' + byy' + cy'^2 = 0, y(\epsilon) = 0, y(\infty) = C,$$

where a, b, c, ϵ, and C are constants; $a, b, c \in R$ on the interval $I(u, v), u \geq -\infty, v < 0$, or $u > 0, v \leq +\infty$.

Rather complicated conditions for the existence of solutions of BVP are given. Vrdoljak (1987)

2.15 $y'' + ky'^2/y + g(x, y)y' + h(x, y) = 0$

1. $$y'' + y'^2/y + (1/x)y' - y/2x^2 = 0.$$

Use of the substitution

$$t = xy'/y$$

reduces the given DE to first order; hence the solution is

$$y^2 = (C/x)(x^2 + D),$$

where C and D are arbitrary constants. Murphy (1992)

2. $$y'' - (1/2)y'^2/y - 4y^2 - 2xy = 0.$$

Put $y = z^2$ to obtain P_{II}:

$$z'' - 2z^3 - xz = 0.$$

Bureau (1964)

3. $$y'' - (1/2)y'^2/y - 4y^2 - (Kx + H)y = 0,$$

where K and H are constants.

The following cases arise:

(a) $K = H = 0$. The equation reduces to

$$y'' - (1/2)y'^2/y - 4y^2 = 0.$$

It has the first integral

$$y'^2 = y(4y^2 + K_1), \qquad (1)$$

where K_1 is an arbitrary constant. Equation (1) has solutions expressible in terms of elliptic functions. Alternatively, one may put $y = z^2$ and determine z from $z'' = 2z^3$.

(b) $K = 0, H \neq 0$. A linear transformation $X = ax + b$ and $Y = cy$ can be used to change the given DE to the form

$$y'' - (1/2)y'^2/y - 4y^2 - 2y = 0. \quad (2)$$

Equation (2) has the first integral

$$y'^2 = 4y(y^2 + y + K_1), \quad (3)$$

where K_1 is an arbitrary constant. Equation (3) may be integrated in terms of elliptic functions.

(c) $KH \neq 0$. A simple change of variables, as in (b), transforms the given DE to

$$y'' - (1/2)y'^2/y - 4y^2 - 2xy = 0,$$

which on setting $y = z^2$ becomes

$$z'' - 2z^3 - xz = 0. \quad (4)$$

(4) is a special case of P_{II}.

Bureau (1964)

4. $$\mathbf{y'' - (1/2)y'^2/y + 1/(2y) = 0.}$$

The given DE has a first integral

$$y'^2 = 1 + 2Ky, \quad (1)$$

where K is an arbitrary constant. On differentiating (1), one gets

$$y'' = K. \quad (2)$$

Now integrating (2), we have

$$y = (K/2)x^2 + K_1x + K_2, \quad (3)$$

where K_1 and K_2 are arbitrary constants. Differentiating (3) and substituting in (1), we get the relation between K_1 and K_2:

$$K_1^2 = 1 + 2KK_2.$$

Bureau (1964)

5. $$\mathbf{y'' - (1/2)y'^2/y + 1/(2y) - 4y^2 - 2Hy = 0,}$$

where $H \neq 0$ is a constant.

The given DE has a first integral

$$y'^2 = 4y^3 + 4Hy^2 + eH_1y + 1,$$

where H_1 is a constant; hence the solution can be found in terms of elliptic functions. Bureau (1964)

6. $$\boldsymbol{y'' - (1/2)y'^2/y - 4y^2 - (Kx + H)y + 1/(2y) = 0,}$$

where K and H are constants.

$K = 0$. We get

$$y'' = (1/2)y'^2/y - 1/(2y) + 4y^2 + Hy. \tag{1}$$

Equation (1) may be shown to have a first integral

$$y'^2 = 4y^3 + 2Hy^2 + 2H_1y + 1. \tag{2}$$

Equation (2) is integrable in terms of elliptic functions. Bureau (1964)

7. $$\boldsymbol{y'' - y'^2/(2y) + 1/(2y) - 4Hy^2 + xy = 0,}$$

where $H \neq 0$ is a constant.

The given DE is equivalent to the system

$$\begin{aligned} y' &= 2vy - 1, \\ 4Hy &= 2v' + 2v^2 + x. \end{aligned} \tag{1}$$

On eliminating y from (1), we have

$$v'' = 2v^3 + xv - 2H - 1/2,$$

so that v is a solution of the second Painlevé equation. Hence y can be found from the second equation of (1). Bureau (1964)

8. $$\boldsymbol{y'' - y'^2/(2y) - (3/2)y^3 - 4K_1y^2 - 2K_2y + K/(2y) = 0,}$$

where K_1, K_2, and K are constants.

Multiplying by $2y'/y$ and noting that

$$2y'y''/y - y'^3/y^2 = \frac{d}{dx}(y'^2/y),$$

one obtains the first integral

$$y'^2 = y^4 + 4K_1y^3 + 4K_2y^2 + K + K_3y, \tag{1}$$

where K_3 is an arbitrary constant. Equation (1) is integrable in terms of elliptic functions. Bureau (1964)

9. $$\boldsymbol{y'' - y'^2/(2y) - (3/2)y^3 - 4xy^2 - 2(x^2 - \alpha)y - \beta/y = 0.}$$

The given DE is P_{IV}. The following results are proved:

(a) If $y = y(x)$ is a solution of the given DE for fixed α and β, then the mapping $(T, \wedge)$, where

$$\begin{array}{rcl} T: & & y \to \bar{y}, \wedge : (\alpha, \beta) \to (\bar{\alpha}, \bar{\beta}), \bar{y} = R(y)/(2\epsilon y), \\ R(y) & = & y' - q - 2\epsilon xy - \epsilon y^2, \epsilon^2 = 1, q^2 = -2\beta, \end{array}$$

yields a solution $y(x)$ of the given DE with parameter values $\bar{\alpha}$ and $\bar{\beta}$. The transformation can be used successively.

(b) The given DE has a one-parameter family of solutions expressible in terms of Weber–Hermite functions if either

$$(1) \quad \beta = -2(\alpha\epsilon + 2n - 1)^2$$

or

$$(2) \quad \beta = -2n^2, \text{ where } n \in N.$$

(c) The given DE has rational solution if and only if either

$$\alpha = n_1, \beta = -2(1 + 2n_3 - n_1)^2, n_1, n_3 \in Z,$$

or

$$\alpha = n_1, \beta = -2(6n_3 - 3n_1 + \epsilon)^2/9, n_1, n_3 \in Z.$$

For fixed α and β in these expressions, there is only one solution.

Gromak (1987)

10. $$\boldsymbol{y'' - (1/2y)y'^2 - (3/2)y^3 - 4xy^2 - 2(x^2 - \alpha)y - \beta/y = 0.}$$

The given DE is Painlevé's fourth transcendent. It has some special elementary solutions. It is easy to verify that all solutions of the Riccati equation

$$y' = y^2 + 2xy - 2(1 + \alpha) \tag{1}$$

are also solutions of the given DE if $\beta + 2(1+\alpha)^2 = 0$. Also, all solutions of the Riccati equation

$$y' = -y^2 - 2xy + 2(\alpha - 1) \tag{2}$$

are solutions of the given DE if $\beta + 2(\alpha - 1)^2 = 0$. Setting $y = -2(\alpha + 1)/v$ in (1) we get

$$\frac{dv}{dx} = -v^2 - 2xv + 2(1 + \alpha). \tag{3}$$

Comparing (2) and (3), we see that the solution $y = y_\alpha$ of (1) with any α will lead to a solution $y_{\alpha_1} = -2(\alpha + 1)/y_\alpha$ of (2) for the value $\alpha_1 = \alpha + 2$, and conversely. Therefore, it is sufficient to consider (1). Letting $y = -u'/u$, (1) becomes

$$u'' - 2xu' - 2(1 + \alpha)u = 0. \tag{4}$$

The general solution of (4) for all nonintegral values of α is expressible in terms of Weber–Hermite functions. When α is an integer, (4) admits a particular solution in the form of a Hermite polynomial $u = H_\alpha(x)$. It follows that for any integral value of the parameter α such that either

$$\beta + 2(\alpha+1)^2 = 0 \quad \text{or} \quad \beta + 2(1-\alpha)^2 = 0,$$

the given DE has a solution that is rational in x. All poles of this solution may be verified to be simple and lie on the real or imaginary axis if x is considered to be complex, depending on the sign of the parameter α. These poles are symmetric about the origin.

The given DE also has other special elementary solutions. For example, for $\alpha = 1$, $\beta = -8/9$, the solution is

$$y = -(2/3)x + 1/x;$$

for $\alpha = -1, \beta = -8/9$, there is a solution $y = -(2/3)x - 1/x$. Lukashevich (1965)

11. $$\boldsymbol{y'' - [1/(2y)]y'^2 - (3/2)y^3 - 4xy^2 - 2(x^2 - \alpha)y - \beta/y = 0,}$$

where α and β are constants.

The given P_{IV} equation is studied by solving the Riemann–Hilbert problem. It is shown that the Cauchy problems for this equation admit, in general, global meromorphic (in x) solutions. Furthermore, for special relations among the monodromy data and for x on Stokes lines, these solutions are bounded for finite x. Fokas and Zhou (1992)

12. $$\boldsymbol{y'' + (2/x)y' - (y^{-1}/2)y'^2 - [2L(L+1)/x^2]y - (4m/h^2)[V(x) - E]y = 0,}$$

where L, m, h, and E are constants; $V(x)$ is a given function of x.

The given DE has a solution $y = (au_1 + bu_2)^2$, where u_1 and u_2 are linearly independent solutions of the linear DE

$$u'' + (2/x)u' - \frac{L(L+1)u}{x^2} - [2m/h^2](V(x) - E)u = 0$$

and a and b are arbitrary constants. The statement may be verified by direct substitution.

If we multiply the given DE by $y(x)$ and differentiate, we get

$$\begin{aligned} y''' &+ (2/x)y'' - (2/x^2)y' + [2/(xy)]y'^2 \\ &- [4L(L+1)/x^2](y' - y/x) - (4m/h^2)[2(V(x) - E)y' + V'(x)y] = 0. \quad (1) \end{aligned}$$

Eliminate $2y'^2/(xy)$ from (1) and the given DE to get a linear DE of third order for y:

$$\begin{aligned} y''' &+ (6/x)y'' + (6/x^2)y' - [4L(L+1)/x^2](y' - y/x) \\ &- (4m/h^2)[2[V(x) - E]y' + 4(V(x) - E)y/x + V'(x)y] = 0. \end{aligned}$$

Burt and Reid (1973), Kostin (1971a)

13. $$\boldsymbol{y'' - (1/2)y'^2/y - b_1(x)y^2 - a_1(x)y' - b_2(x)y = 0,}$$

where a_1, b_1, and b_2 are analytic functions of x.

Using the transformation $y = \lambda(x)u, t = \phi(x)$, it is possible to reduce this DE to a form such that $a_1 = 0$ and $b_1 = 4$. Thus, starting from

$$y'' - (1/2)y'^2/y - 4y^2 - b_2(x)y = 0 \quad (1)$$

and setting $y = z^{-2}$, $\dfrac{dz}{dx} = 1 + uz$ in (1), we get

$$zu' = 3u - (b_2/2)z + zu^2.$$

To obtain the condition for poles to be the only movable singularities, set $u = \beta z + \gamma z^2 + z^3 v$ and determine β, γ by

$$4\beta = b_2, \frac{d\beta}{dx} = \gamma, \quad \frac{d\gamma}{dx} = 0;$$

therefore, γ is a constant and $b_2 = Kx + H; K$ and H are arbitrary constants. Thus, the given DE reduces to the form

$$y'' - (1/2)y'^2/y = 4y^2 + (Kx + H)y.$$

Now, go to Eqn. (2.15.3). Bureau (1964)

14. $$\boldsymbol{y'' - (1/2)y'^2/y + yy' - q(x)y' - (1/2)y^3 + 2q(x)y^2 - 3[q'(x) + q^2(x)/2]y + [72Hr^2(x)]/y = 0,}$$

where H is constant and $r'(x)/r = q$.

Setting

$$12w = y' + y^2 + 3qy, \tag{1}$$

we obtain

$$y = \frac{6(w^2 - Hr^2)}{w' - qw}. \tag{2}$$

Eliminating y from (1) and (2), we obtain

$$w'' = 6w^2 + (q' + q^2)w - 6Hr^2.$$

Because $q' + q^2 = r''/r$, we have

$$w'' = 6w^2 + (r''/r)w - 6Hr^2.$$

Now put $12g = r''/r, w = v - g$ to obtain the canonical form

$$v'' = 6v^2 + g''(x) - 6g^2 - 6Hr^2.$$

Bureau (1964)

15. $$\boldsymbol{y'' - (2/3)y'^2/y + (2/3)yy' - (2/3)q(x)y' - (2/3)y^3 + (10/3)q(x)y^2 - [4q'(x) + (8/3)q^2(x)]y = 0,}$$

where $q(x)$ is a solution of

$$q' + q^2 = Kx + H,$$

where K and H are arbitrary constants. The general solution of the given DE is

$$y = (w' - q' + w^2 - q^2)/(w - q),$$

where w is a solution of

$$w'' = 2w^3 - 2(Kx + H)w + K. \tag{1}$$

Equation (1) is easily related to the second Painlevé equation. Bureau (1964)

16. $\boldsymbol{y'' - (2/3)y'^2/y + (2/3)yy' - (2/3)q(x)y' - ry'/y - (2/3)y^3 + (10/3)q(x)y^2}$

$$\boldsymbol{- [4q'(x) + r + (8/3)q^2]y - [2q(x)r(x) - 3r'(x)] + 3r^2(x)/y = 0,}$$

where $q(x)$ is given by

$$q'' = 2q^3 - (Kx + H)q + K_1, \tag{1}$$

K, H, K_1 being constants, and $yr(x)$ by

$$3r = -(Kx + H) + 2q'(x) + 2q^2(x).$$

The only movable singularities of the given DE can be shown to be poles; its general solution is given by

$$y = (w' - q' + w^2 - q^2)/(w - q),$$

where q and w ($\neq q$) are, respectively, a particular solution and the general solution of (1). Bureau (1964)

17. $$\boldsymbol{y'' - (1/2)y'^2/y + 2yy' + y^3/2 + 1/(2y) - f(x)y = 0,}$$

where f is an arbitrary function of x.

Setting

$$y' = -y^2 + 1 + 2vy, \tag{1}$$

we get the equation

$$v' + v^2 = (1/2)(f - 1). \tag{2}$$

Thus, $y(x)$ is determined by the two Riccati equations (1)-(2). Now, set $y = w'/w$ in (1) so that $2vw' = w'' - w$; Eqn. (2) becomes

$$2w'w''' = w''^2 - w^2 + 2fw'^2. \tag{3}$$

Upon differentiating (3), we get the linear DE

$$w^{iv} - 2fw'' - f'w' + w = 0. \tag{4}$$

Hence the given DE has poles as the only movable singularities in its solution. Bureau (1964)

18. $$\boldsymbol{y'' - (3/4)y'^2/y - 3y^2 - b_2(x)y - b_3(x) = 0,}$$

where $b_2(x)$ and $b_3(x)$ are analytic functions.

This is a special case of Eqn. (2.15.20). Bureau (1964)

19. $$\boldsymbol{y'' - (3/4)y'^2/y - a_1(x)y' - b_2(x)y - b_3(x) = 0,}$$

where $a_1(x), b_2(x)$, and $b_3(x)$ are analytic functions.

By the transformation $y = \lambda(x)u, t = \phi(x)$, it is possible to reduce the given DE to a form such that $a_1 = b_2 = 0$. Further for the given DE to have poles as its movable singularities, $b_3 = K$, a constant. Thus, we assume the form

$$y'' - (3/4)y'^2/y - K = 0,$$

where K is an arbitrary constant. The case $K = 0$ is a special case discussed in Eqn. (2.15.46). For $K \neq 0$, set $y = -Ku$ and find the canonical form

$$u'' = (3/4)u'^2/u - 1. \tag{1}$$

Setting $u = z^2$ in (1), we find that

$$2zz'' = z'^2 - 1. \tag{2}$$

Upon differentiating (2), we have $z''' = 0$; therefore, z is a quadratic in x and hence y may be found. Bureau (1964)

20. $$\boldsymbol{y'' - (3/4)y'^2/y - b_1(x)y^2 - b_3(x) - a_1(x)y' - b_2(x)y = 0,}$$

where a_1, b_1, b_2, and b_3 are analytic functions of x.

By a transformation $y = \lambda(x)u, t = \phi(x)$, it is possible to change the DE to a form such that $a_1 = 0, b_1 = 3$; therefore, we may start with

$$y'' - (3/4)y'^2/y - 3y^2 - b_3 - b_2y = 0.$$

Now, set $y = z^{-2}, z' = 1 + uz$ and determine u from

$$zu' = 2u - (1/2)(b_2z + b_3z^3) + (z/2)u^2. \tag{1}$$

To get the condition that poles are the only movable singularities in the solutions of (1), we set $u = \beta z + z^2v$, and analyze; we obtain $2\beta = b_2, \beta' = 0$; therefore, $b_2 = 2k$. Thus, we write the given DE in the form

$$y'' - (3/4)y'^2/y - 3y^2 - 2Ky - H = 0, \tag{2}$$

where K and H are arbitrary constants. Equation (2) has the first integral

$$y'^2 = 4[y^3 + 2Ky^2 - Hy + K_1y^{3/2}],$$

where K_1 is an arbitrary constant. Setting $y = z^2$, we obtain

$$z'^2 = z^4 + 2Kz^2 - H + K_1z. \tag{3}$$

Equation (3) is solvable in terms of elliptic functions. Bureau (1964)

21. $$\boldsymbol{y'' - (3/4)y'^2/y + (3/2)yy' + y^{3/4} - b_2(x)y - k = 0,}$$

where $b_2(x)$ is an arbitrary analytic function of x and k is a constant.

The given DE is equivalent to the system

$$y' + y^2 = 2vy, y = k/[2v' + v^2 - b_2(x)]. \tag{1}$$

On eliminating y from the system (1), we get

$$v'' + 3vv' + v^3 - b_2 v - (1/2)[b_2'(x) + k] = 0.$$

On setting $v = w'(x)/w$, we get a linear third order equation

$$w''' - b_2(x)w'(x) - (1/2)[b_2'(x) + k]w = 0.$$

Therefore, the given DE has poles as the only movable singularities in its solution. Bureau (1964)

22. $\mathbf{y'' - (3/4)y'^2/y + (3/2)yy' + y^{3/4} - [q'(x)/(2q(x))](y' + y^2) - r(x)y - q(x) = 0,}$

where $q(x)$ and $r(x)$ are arbitrary analytic functions of x.

The given DE is equivalent to the system

$$y' + y^2 = 2vy, y[2v' + v^2 - (q'/q)v - r] = q. \tag{1}$$

On eliminating y from the system (1), we get

$$\begin{aligned} v'' + 3vv' - (3/2)(q'/q)v' + v^3 - (3q'/(2q))v^2 - [q''/(2q) \\ - q'^2/q^2 + r]v - (1/2)(r' + q - q'r/q) = 0. \end{aligned} \tag{2}$$

On setting $v = w'/w$, Eqn. (2) becomes linear of third order:

$$w''' - (3/2)(q'/q)w'' - [q''/(2q) - q'^2/q^2 + r]w' - (1/2)[r' + q - q'r/q]w = 0.$$

Therefore, the given DE has poles as the only movable singularities in its solution. Bureau (1964)

23. $$\mathbf{y'' - (3/4)y'^2/y - 6q'(x)y'/y - 3y^2 - 12q(x)y + 12q''(x) + 36q'(x)^2/y = 0,}$$

where q is a solution of the equation

$$q'' = 6q^2 + Kx + H,$$

where K and H are constants.

It may be shown that the given DE has poles as its only movable singularities. The given DE is equivalent to the system

$$\begin{aligned} y' &+ 12q' = -2vy, \\ 3y &= -2v' + v^2 - 12q, \end{aligned} \tag{1}$$

which shows that v is stable if y is stable, and conversely. Eliminating y from (1), we get an equation for v:

$$v'' = -vv' + v^3 - 12qv + 12q'$$

[see Eqn. (2.10.4)]. Bureau (1964)

24.
$$y'' - y'^2/y - H_1 e^{Kx} y^2 = H_1 e^{Kx},$$

where H_1, K_1, and K_2 are constants.

It may be checked that the given DE has poles as its only movable singularity, and is, after suitable transformations, a special case of P_{III} [see Eqn. (2.15.28)]. Bureau (1964)

25.
$$y'' - y'^2/y - H_1 e^{Kx} y^2 - H_3 e^{Kx} - (H_4/y) e^{2Kx} = 0,$$

where H_1, H_3, H_4, K_1, and K_2 are constants.

This DE may be reduced to P_{III} by a suitable change of variables. Bureau (1964)

26.
$$y'' - y'^2/y + 4y^3 - 2Ce^{-2x} + 2\delta y^2 = 2,$$

where c and δ are constants.

Put $Y = e^x y, X = e^{-x}$; the given DE changes to P_{III}. Bountis, Ramani, Grammaticos, and Dorizzi (1984)

27.
$$y'' - y'^2/y - H_1 y^3 - H_4/y - H_2 y^2 - H_3 = 0,$$

where $H_1, \ldots, H_4$ are constants.

The given DE can be solved in terms of elliptic functions [see Eqn. (2.15.28)]. Bureau (1964)

28.
$$y'' - y'^2/y - H_1 e^{2Kx} y^3 - H_2 e^{Kx} y^2 - H_3 e^{Kx} - (H_4/y) e^{2Kx} = 0,$$

where $H_1, \ldots, H_4, K_1$, and K_2 are constants.

The transformation $y \to \lambda(x) y$, with λ defined by

$$\lambda^2 e^{(K_1 - K_2)x} = 1 \quad \text{and} \quad m = (K_1 + K_2)/2,$$

takes the given DE to

$$y'' - y'^2/y - e^{2mx}(H_1 y^3 + H_4/y) - e^{mx}(H_2 y^2 + H_3) = 0 \cdot y. \tag{1}$$

Two cases arise.

(a) $m = 0$. The substitution $u = y'(x)/y(x)$ changes (1) to

$$y\,\frac{du}{dx} = H_1 y^3 + H_2 y^2 + H_3 + H_4/y$$

or

$$u\,\frac{du}{dy} = H_1 y + H_2 + H_3/y^2 + H_4/y^3. \tag{2}$$

Therefore, by integration of (2), we have

$$u^2 = H_1 y^2 + 2H_2 y - 2H_3/y - H_4/y^2 + H_5$$

or

$$y'^2(x) = H_1 y^4 + H_2 y^3 + H_5 y^2 - 2H_3 y - H_4. \tag{3}$$

Equation (3) can be integrated in terms of elliptic functions.

(b) $m \neq 0$. The linear transformation $t = e^{mx}/m$ takes the DE (1) to the canonical form

$$y'' - y'^2/y + y'/x - H_1 y^3 - H_4/y - [1/(mx)](H_2 y^2 + H_3) = 0, H_1 H_4 \neq 0. \quad (4)$$

Equation (4) is of the form P_{III}.

Bureau (1964)

29. $$\boldsymbol{y'' - (y'^2/y) - K y^{1/3}(y-1)e^x = 0,}$$

where K is a constant.

Putting

$$y = v^3, v = e^{-x/4} w(\tau), \tau = e^{3x/4},$$

we get

$$w'' = (w'^2/w) - w'/\tau + K_1 w^2/\tau - K_1/w, \quad (1)$$

where K_1 is another constant. Equation (1) is a special case of P_{III}. Leonovich (1984)

30. $$\boldsymbol{y'' - (1/y)y'^2 + (1/x)y' - (1/x)(\alpha y^2 + \beta) - \gamma y^3 - \delta/y = 0,}$$

where α, β, γ, and δ are constants.

The given DE is the third Painlevé transcendent. Its special explicit solutions may be found as follows. Putting $x = e^t, y = v e^{kt}, k$ a constant, we get

$$v'' + (2k-1)v' + k(k-1)v = (1/v)(v' + kv)^2 - (v' + kv)$$
$$+ \alpha v^2 e^{(k+1)t} + \beta e^{(1-k)t} + \gamma v^3 e^{(2k+2)t} + (\delta/v) e^{(2-2k)t}. \quad (1)$$

Choosing $k = 1, \alpha = \gamma = 0$, we get from (1)

$$v v'' = v'^2 + \beta v + \delta. \quad (2)$$

Put $v' = u$ so that Eqn. (2) becomes

$$v u \frac{du}{dv} = u^2 + \beta v + \delta, \quad (3)$$

the solution of (3) may be written as

$$u = \pm(C_1 v^2 - 2\beta v - \delta)^{1/2}$$

and hence

$$\int \frac{dv}{(C_1 v^2 - 2\beta v - \delta)^{1/2}} = \pm t + C_2,$$

where C_1 and C_2 are arbitrary constants. With $y = ve^t, x = e^t$, we obtain the solution in quadrature form. Similarly, choosing $k = -1, \beta = \delta = 0$ in (1), we get

$$v v'' = v'^2 + \alpha v^3 + \gamma v^4. \quad (4)$$

Now, put $v' = u$ in (4) so that

$$v u \frac{du}{dv} = u^2 + \alpha v^3 + \gamma v^4. \quad (5)$$

Equation (5) can be integrated to yield

$$u = \pm v[\gamma v^2 + 2\alpha v + C_1']^{1/2},$$

leading to

$$\int \frac{dv}{v[\gamma v^2 + 2\alpha v + C_1]^{1/2}} = \pm t + C_2, \tag{6}$$

where C_1 and C_2 are arbitrary constants. With $y = ve^{-t}, x = e^t$, we have the special solution (6).

Another special circumstance is when $\beta + [(\alpha - 2a)/a]b = 0$, $\gamma = a^2 \neq 0, \delta + b^2 = 0$. Then it is easy to verify that all solutions of the Riccati equation

$$\frac{dy}{dx} = ay^2 + \left[\frac{\alpha - a}{ax}\right] y + b \tag{7}$$

are also solutions of the given DE. Setting $y = -u'/(au)$, the Riccati equation (7) becomes

$$u'' + \left[\frac{a - \alpha}{ax}\right] u' + abu = 0. \tag{8}$$

Equation (8) can be solved in terms of Bessel functions. In particular, if $\alpha = (2n+1)a$, where n is an arbitrary integer, its solution is

$$u = \left(\frac{2}{\pi}\right)^{1/2} x^{n+1/2} \left[(-1)^n C_1'' \frac{d^n}{(\tau d\tau)^n} \left(\frac{\sin\tau}{\tau}\right) C_2'' \frac{d^n}{(\tau d\tau)^n} \left(\frac{\cos\tau}{\tau}\right)\right],$$

where $\tau = x(ab)^{1/2}$; C_1'' and C_2'' are arbitrary constants.

The given DE may also have other solutions. For example, if $\beta = \gamma = 0$, and $\alpha\delta \neq 0$, then $y = hx^{1/3}$, where h is a root of $\alpha h^3 + \delta = 0$. Similarly, if $\alpha = \delta = 0, \beta\gamma \neq 0$, then the functions $y = hx^{-1/3}$, where $rh^3 + \beta = 0$, are also solutions. Moreover, if we let $y = 1/y_1$, the given DE becomes

$$y_1'' - (1/y_1)y_1'^2 + (1/x)y_1' + (1/x)(\beta y_1^2 + \alpha) + \delta y_1^3 + \gamma y_1 = 0. \tag{9}$$

Equation (9) coincides with the given DE if we set

$$\beta = -\alpha_1, \alpha = -\beta_1, \delta = -\gamma_1, \gamma = -\delta_1.$$

Thus if $y(\alpha, \beta, \gamma, \delta)$ is a solution, then $y = 1/y_1(\alpha_1, \beta_1, \gamma_1, \delta_1)$ is also a solution, where $\alpha_1, \beta_1, \gamma_1$, and δ_1 are known in terms of α, β, γ, and δ. Lukashevich (1965)

31. $\boldsymbol{y'' + p(x)y' + q(x)y \ln y - y'^2/y = 0.}$

If we put $y = e^{U(x)}$, then U satisfies the linear DE

$$U''(x) + p(x)U'(x) + q(x)U(x) = 0. \tag{1}$$

The solution of (1), therefore, leads to the solution of the given DE. Finch (1989)

32. $$y'' + r(x)y' + q(x)y \log y - y^{-1}y'^2 - g(x)y = 0.$$

The substitution $u = \log y$ changes the given DE to the linear inhomogeneous DE

$$u'' + r(x)u' + q(x)u = g(x).$$

Klamkin and Reid (1976)

33. $$y'' + r(x)y' + q(x)y \log y - y^{-1}y'^2 - g(x)y(\log y)^a = 0,$$

where a is a constant.

Put $Y = \log y$; we obtain a simpler nonlinear equation

$$Y'' + r(x)Y' + q(x)Y = g(x)Y^a. \tag{1}$$

Equation (1) can be solved explicitly in some cases. For example, if $a = -3$ and $g(x) = c$, it becomes a Pinney-type equation. Klamkin and Reid (1976)

34. $$y'' + a_1(x)y' + [1/(\beta+1)]a_0(x)y + \beta y'^2/y$$
$$+ vu^2F(y^{\beta+1}/v)y^{-\beta} - [1/(\beta+1)]y^{-\beta}f(x) = 0,$$

where β is a constant, and the linear part

$$L(y) = y'' + a_1(x)y' + [a_0(x)/(\beta+1)]y$$

is such that it can be changed by

$$y^{\beta+1} = v(x)z, dt = u(x)dx \tag{1}$$

to one with constant coefficients

$$z''(t) + b_1z'(t) + b_0(z),$$

and the function F is arbitrary.

Under these circumstances, the given DE goes to

$$z''(t) + b_1z'(t) + b_0z + (\beta+1)F(z) = f(x(t))u^{-2}v^{-1}. \tag{2}$$

If $f(x) \equiv 0$, then the given DE has the solution $y = (\rho v)^{1/(\beta+1)}$, where v is a solution of $L(v) - b_0u^2v = 0$ and ρ satisfies the equation $b_0\rho + (\beta+1)F(\rho) = 0$. The proof is facilitated by writing $y = w^{1/(\beta+1)}$. Berkovich (1971)

35. $$y'' + a_1(x)y' + a_0(x)y \ln y - y'^2/y + vu^2yF[(\ln y)/v] - yf(x) = 0,$$

where F is an arbitrary function and the nonlinear part of the equation

$$L(y) = y'' + a_1(x)y' + a_0(x)y \ln y$$

can be reduced to

$$z''(t) + b_1z'(t) + b_0(z),$$

with constant coefficients, by

$$\ln y = v(x)z, dt = u(x)dx. \tag{1}$$

It may be checked by direct substitution that the given DE changes by (1) to

$$z''(t) + b_1 z'(t) + b_0 z + F(z) = f(x(t))u^{-2}v^{-1}. \tag{2}$$

If $f(x) \equiv 0$, then the given DE has the solution $y = \exp(\rho v)$, where v is a solution of

$$v'' + a_1(x)v' + a_0(x)v - b_0 u^2 v = 0;$$

the function $u(x)$ is known, and ρ is a constant that satisfies the equation $b_0\rho + F(\rho) = 0$. The proof is made easier if we make the change of variable $y = \exp(w)$ in the given equation. Berkovich (1971)

36. $$\boldsymbol{y'' - y'^2/y + q'(x)yy' + 1/y + q''(x)y^2 - q(x) = 0,}$$

where $q(x)$ is analytic.

We easily find that the given DE is equivalent to the system

$$y' + q'(x)y^2 - 1 + q(x)y = vy, \tag{1}$$

$$yv'(x) = v. \tag{2}$$

On eliminating y between (1) and (2), we have

$$v'' + vv' - q'(x)v - q(x)v' = 0,$$

which integrates to yield the Riccati equation

$$v' + (1/2)v^2 - qv = H,$$

where H is an arbitrary constant. Setting $v = 2w'(x)/w(x)$, we get the linear DE

$$w'' - q(x)w' - (H/2)w = 0.$$

Thus, the original solution $y(x)$ is determined by

$$y^{-1} = \frac{v'(x)}{v} = \frac{w''(x)}{w'(x)} - \frac{w'(x)}{w(x)}$$

and has poles as its only movable singularities. Bureau (1964)

37. $$\boldsymbol{y'' - y'^2/y - r(x)y'/y - y^2 + r'(x) = 0.}$$

The given DE is reducible to a Riccati equation and has poles as its only movable singularities [see Eqn. (2.15.38)]. Bureau (1964)

38. $$\boldsymbol{y'' - y'^2/y - q(x)y'/y - He^{Kx}y^2 + q'(x) = 0,}$$

where H and K are constants; $q(x)$ is an analytic function.

Put $y = \lambda(x)u$, where $\lambda(x)He^{Kx} = 1$ and set $r = q(x)/\lambda(x)$, then the given DE becomes

$$u'' - u'^2/u - ru'/u - u^2 + r'(x) = 0. \tag{1}$$

Now, with $r = R'(x)$ and $v = (u' + R')/u$, Eqn. (1) is equivalent to the system

$$\frac{u'}{u} + \frac{R'}{u} = v, v'(x) = u. \tag{2}$$

Eliminating u between the relations (2), we get

$$v'' + R' = vv',$$

so that on integration, we have the Riccati equation

$$v' + R = (1/2)v^2 + H_1.$$

Thus, u and hence y has poles as its only movable singularities. Bureau (1964)

39. $$\boldsymbol{y'' - y'^2/y - a(x)yy' - a_2(x)y'/y - a'(x)y^2 + a_2'(x) = 0.}$$

It is easy to check that the given DE is equivalent to the system

$$y' - a(x)y^2 = v, \tag{1}$$

$$v + a_2(x) = Ky, \tag{2}$$

where K is a constant. Thus, y is determined by the Riccati equation

$$y' = a(x)y^2 + Ky - a_2(x), \tag{3}$$

and so has poles as its only movable singularities. Bureau (1964)

40. $$\boldsymbol{y'' - (5/4)y^{-1}y'^2 + (1/4)wy^2 = 0,}$$

where w is a constant.

The general solution of the given DE is

$$y = (ax^2 + 2bx + c)^{-2}, ac - b^2 = w/16,$$

two of the constants a, b, and c being arbitrary. Ranganathan (1988)

41. $$\boldsymbol{y'' - 3y'^2/(2y) + \epsilon y'/y - \epsilon y' - 2y = 0,}$$

where ϵ is a parameter.

The given DE arises from the Van der Pol equation by the transformation $y = z^{-2}$, where $z'' - \epsilon(1 - z^2)z' + z = 0$.

Looking for the dominant behavior in the given DE and writing $y \sim A(x - x_0)^m$, two cases arise:

(a) $y \sim A(x - x_0)^{-2}$, where A is arbitrary and leading terms arise from the first two terms of the given DE.

(b) $y \sim (2/3)\epsilon(x - x_0)$, where the dominant terms arise from the second and third terms in the equation.

For case (a), it is easy to check that there is a Laurent series

$$y(x) = A(x - x_0)^{-2} + a_{-1}(x - x_0)^{-1} + a_0 + a_1(x - x_0) + \cdots,$$

where

$$a_{-1} = -\epsilon A, a_0 = (4 + 5\epsilon^2)A/12, a_1 = \epsilon[2 - (\epsilon^2 + 4)A]/12, \ldots,$$

where x_0 and A are arbitrary.

For case (b), if we write

$$y = (2/3)\epsilon(x - x_0) + \cdots + a_r(x - x_0)^{1+r} + \cdots \tag{1}$$

and substitute in the given DE, we find that $(r+1)(r - 3/2) = f(a_j)$, where $f(a_j)$ is a function of coefficients of lower orders, i.e., $0 \le j \le r$. There are two resonances: $r = -1$ corresponds to x_0 while $r = 3/2$ gives rise to the second arbitrary constant. The first few terms in the series (1) are found as follows:

$$y = (2/3)\epsilon\tau - (2/3)\epsilon^2\tau^2 + B\tau^{5/2} + (4/9)(\epsilon^3 + 2\epsilon)\tau^3 + \cdots, \tag{2}$$

where $\tau \equiv x - x_0$ and B is an arbitrary constant. The series (2) shows that the solution has an algebraic branch point. Thus, the given DE and hence the Van der Pol equation do not enjoy the Painlevé property. Bountis (1983)

42. $$\boldsymbol{y'' + [(\alpha^2 - 1)/[4(n-1)]](y/x^2) - ny^{-1}y'^2 - \beta x^{-(\alpha+1)/2}y^n = 0,}$$

where $\beta \neq 0, n \neq 1, \alpha \neq 0, 1$.

The given DE has a solution

$$y = [(a + Bx + 2cx^\alpha)x^{(1-\alpha)/2}]^{1/(1-n)},$$

where $B = \beta(1-n)/(1-\alpha)$, and a and c are arbitrary constants. Ranganathan (1988)

43. $$\boldsymbol{y'' + q(x)y - ny'^2y^{-1} - f(x)y^n = 0, n \neq 1,}$$

where

$$\begin{aligned} q(x) &= 3D(T_1T_2)^{-1}\frac{2a_0^2x^2 + 4a_0b_0x + a_0c_0 + b_0^2}{4(n-1)}, \\ D &= a_0c_0 - b_0^2 \neq 0, \\ T_1 &= a_0x + b_0, \\ T_2 &= a_0x^2 + 2b_0x + c_0, \\ \text{and } f(x) &= \beta T_1^{-3/2}T_2^{-3/4}, \beta \neq 0. \end{aligned}$$

The general solution of the given DE is

$$y = \left[(a + Bx + 2cT_2^{1/2})u\right]^{1/(1-n)}, u = T_1^{-1/2}T_2^{1/4},$$

where $B = (n-1)\beta/D$; a and c are arbitrary constants. Ranganathan (1988)

44. $\boldsymbol{y'' + ny^{-1}y'^2 + (N-1)y'/x + [1/[2(n+1)]]xy^{-n}y' + [1/(n+1)]y^{1-n} = 0,}$

where N is a positive integer, and $n \neq -1$ is a parameter.

The given DE arises from a similarity reduction of a nonlinear diffusion equation. It has a first integral

$$xy^n y' + \frac{(N-2)y^{n+1}}{n+1} + \left[\frac{1}{2(n+1)}\right] x^2 y + \alpha = 0, \tag{1}$$

where α is an arbitrary constant. If $\alpha = 0$ and $n \neq 0$, (1) is a Bernoulli equation and may be solved in a closed form:

$$\begin{aligned} y &= x^{(2-N)/(n+1)} \left[\frac{n}{2(nN+2)} \left[\beta - x^{(nN+2)/(n+1)}\right]\right]^{1/n}, n \neq 0, n \neq -2/N; \\ y &= \beta x^{2-N} e^{-x^2/4}, n = 0; \\ y &= \left[\frac{1}{N-2}\, x^2 \ln(x/\beta)\right]^{-N/2}, n = -2/N. \end{aligned}$$

Other cases for $\alpha \neq 0$ are as follows.

(a) $n = -1/2$. Equation (1), on putting $y = g^2$, becomes

$$2x\,\frac{dg}{dx} + 2(N-2)g + x^2 g^2 = -\alpha. \tag{2}$$

Writing $g = [2/(xq)]\,\dfrac{dq}{dx}$ in (2), we have

$$x\,\frac{d^2q}{dx^2} + (N-3)\,\frac{dq}{dx} + \frac{\alpha}{4} xq = 0,$$

so that

$$q = x^{(4-N)/2}\left[AJ_\nu(\alpha^{1/2}x/2) + BY_\nu(\alpha^{1/2}x/2)\right],$$

where A and B are arbitrary constants and $\nu = 2 - N/2$. Hence,

$$\begin{aligned} g &= (\alpha^{1/2}/x)\left[\beta J_{\nu-1}(\alpha^{1/2}x/2) + (1-\beta)Y_{\nu-1}(\alpha^{1/2}x/2)\right] \\ &\quad \times \left[\beta J_\nu(\alpha^{1/2}x/2) + (1-\beta)Y_\nu(\alpha^{1/2}x/2)\right]^{-1}, \end{aligned} \tag{3}$$

where β is an arbitrary constant. For $N = 3, 1$, (3) can be written in a much simpler form. For

$$N = 3, g = -(\alpha^{1/2}/x)\tan[(\alpha^{1/2}/2)(x - x_0)]$$

or

$$g = (\gamma^{1/2}/x)\tanh[(\gamma^{1/2}/2)(x - x_0)],$$

where $\gamma = -\alpha$, and x_0 is an arbitrary constant. For $N = 1$, (3) can be shown to reduce to

$$g = \frac{\alpha}{2 - \alpha^{1/2}x\cot[(1/2)\alpha^{1/2}(x - x_0)]} \tag{4}$$

or

$$g = \frac{-\gamma}{2 - \gamma^{1/2}x\coth[(1/2)\gamma^{1/2}(x - x_0)]}, \gamma = -\alpha. \tag{5}$$

For $\alpha \to 0$, with $x_0 = -\dfrac{a^3\alpha}{12}$, (4) becomes $g = \dfrac{6x}{x^3 + a^3}$, while if $x_0 \to -\infty$ with $\gamma > 0$, (5) becomes $g = \gamma/(\gamma^{1/2}x - 2)$.

(b) $n = (1/2)N - 2$. If we write $y = x^{(2-N)/(n+1)}g$, (1) becomes

$$-\left[\frac{1}{2(n+1)}\right] x^{(2n+4-N)/(n+1)} g = x^{3-N} g^n \frac{dg}{dx} + \alpha. \tag{6}$$

When $n = N/2 - 2$, (6) is separable and has the solution

$$\int_{g_0}^{g} \frac{g^{(N/2-2)}}{g + (n-2)\alpha}\, dg = -[1/(N-2)^2]x^{N-2},$$

where g_0 is an arbitrary constant.

King (1990)

45. $\boldsymbol{y'' + p(x)y' + q(x)y - \mu y'^2 y^{-1} = 0, \mu \neq 1.}$

The general solution of the given DE is

$$y = [Au + Bv]^{1/(1-\mu)},$$

where u and v are two linearly independent solutions of

$$y'' + p(x)y' + (1-\mu)q(x)y = 0,$$

and A and B are arbitrary constants. Ranganathan (1988)

46. $\boldsymbol{y'' - (1 - 1/n)y'^2/y - a_1(x)y' - b_2(x)y = 0,}$

where n is a positive or negative integer and a_1 and b_2 are analytic functions.

Using a transformation $y = \lambda(x)u, t = \phi(x)$ it is possible to reduce the given DE to a form such that $a_1 = b_2 = 0$, so that we may consider

$$y'' - (1 - 1/n)y'^2/y = 0. \tag{1}$$

Equation (1) has the general solution $y = (Kx + H)^n$, with poles as the only movable singularities; K and H are arbitrary constants. Bureau (1964)

47. $\boldsymbol{y'' + p(x)y' + kr(x)y - (1-\ell)y'^2 y^{-1} - f(x)y^{1-2m\ell} = 0,}$

where k, ℓ, and m are nonzero constants such that $k\ell = 1$.

It is shown that with $\phi(u,v)$ a homogeneous function of u and v,

$$y = [\phi(u,v)]^{k/m}, v = uw, \phi(u,v) = u^m\psi(w), \psi(w) = \phi(l,w) \tag{1}$$

is a solution provided that

$$f(x) = km^{-2}Q_0, Q_0 = 2mu^{2m-1}\psi\psi'u'w' + u^{2m}w'^2[m\psi\psi'' + (1-m)\psi'^2] + mu^{2m}\psi\psi'[w'' + p(x)w'],$$

where $\psi' = \dfrac{d}{dw}(\psi)$. In the above u is a solution of $y'' + p(x)y' + r(x)y = 0$ and $w(x)$ is an arbitrary function. The accent denotes a derivative with respect to the respective argument. Ranganathan (1988)

48. $$\boldsymbol{y'' + p(x)y' - \mu y'^2 y^{-1} - f(x)y^n = 0, \mu \neq 1, n \neq 1.}$$

The given DE has a solution

$$y = (bF_1^2 + AF_1 + B)^{1/(1-\mu)}, F_1(x) = \int^x e^{-\int^x p(x)dx} dx$$

provided that

$$f(x) = \beta(bF_1^2 + AF_1 + B)^{(n-\mu)/(\mu-1)} e^{-2F}; \beta \neq 0, F = \int^x p(x)dx,$$

where A and B are constants, and $(1-\mu)\beta = 2b$. Ranganathan (1988)

49. $$\boldsymbol{y'' + p(x)y' - ny'^2 y^{-1} - \beta e^{-2F(x)} y^n = 0, \beta \neq 0, n \neq 1,}$$

where $F(x) = \int^x p(x)dx$.

The given DE has a solution

$$y = \left[bF_1^2 + AF_1 + B\right]^{1/(1-n)},$$

where

$$F_1 = \int^x e^{-F(x)} dx, F(x) = \int^x p(x)dx, 2b = \beta(1-n),$$

and A and B are arbitrary constants. Ranganathan (1988)

50. $$\boldsymbol{y'' + p(x)y' + q(x)y - (1/4)(n+3)y'^2 y^{-1} - \beta \exp(-2F) y^n = 0,}$$

where $\beta \neq 0, n \neq 1$, are constants, and $F(x) = \int^x p(x)dx$.

The following result is quoted: The general solution of the given equation is

$$y = (au^2 + 2buv + cv^2)^{2/(1-n)}, ac - b^2 = (1-n)\beta/(4C_w^2),$$

where a, b, and c are constants, two of which are arbitrary and u and v are two linearly independent solutions of

$$y'' + p(x)y' + [(1-n)/4]q(x)y = 0.$$

The constant C_w appears in the Abel formula:

$$W(u,v)\exp(F(x)) = C_w,$$

where $W(u,v) = vu' - uv'$ is the Wronskian. Ranganathan (1987)

51. $$\boldsymbol{y'' + p(x)y' + q(x)y - \mu y'^2 y^{-1} - f(x)y^n = 0,}$$

where μ and n are constants not equal to 1.

$$y = [(bF_1 + a)u]^{1/(1-\mu)}, F_1 = \int^x e^{-\int^x p(x)dx} dx$$

is a solution of the given DE provided that

$$f(x) = \beta[(bF_1 + a)u]^{(n-\mu)/(\mu-1)} e^{-F} u', \beta \neq 0,$$
$$F(x) = \int^x p(x)dx, (1-\mu)\beta = 2b,$$

where a and b are constants, and u is a solution of

$$y'' + p(x)y' + (1-\mu)q(x)y = 0.$$

Ranganathan (1988)

52. $$\boldsymbol{y'' - (1-\ell)y^{-1}y'^2 + r(x)y' + kq(x)y - [\beta q(x) + kg(x)]y^{1-\ell} = 0.}$$

The transformation $u = y^\ell - \ell\beta$ changes the given DE to linear form: $u'' + r(x)u' + q(x)u = g(x)$, provided that $k\ell = 1$. Klamkin and Reid (1976)

53. $$\boldsymbol{y'' + r(x)y' + kq(x)y - (1-\ell)y^{-1}y'^2 - kg(x)y^{1-\ell+a\ell} = 0,}$$

where a and ℓ are constants.

The substitution $Y = y^\ell$, changes the given DE to the simpler form

$$Y'' + r(x)Y' + q(x)Y = g(x)Y^a,$$

provided that $k\ell = 1$. Now see Eqn. (2.15.33). Klamkin and Reid (1976)

54. $$\boldsymbol{y'' - (1-\ell)y^{-1}y'^2 + r(x)y' - [kq(x)(1-y^\ell) + g(x)]y^{\ell-1} = 0.}$$

The substitution $u = (y^\ell - 1)/\ell$ transforms the given DE to the linear form $u'' + r(x)u' + q(x)u = g(x)$, provided that $k\ell = 1$. Klamkin and Reid (1976)

55. $$\boldsymbol{y'' + p(x)y' + q(x)y - ny'^2y^{-1} - f(x)y^n = 0, n \neq 1.}$$

The given DE has a solution

$$y = [(bF_1 + a)u]^{1/(1-n)},$$

where u is a solution of

$$y'' + p(x)y' + (1-\mu)q(x)y = 0,$$

with $\mu = n$ and

$$f(x) = \beta e^{-F}u', \beta \neq 0, F(x) = \int^x p(x)dx, F_1 = \int^x e^{-F(x)}dx, 2b = \beta(1-n),$$

and a is an arbitrary constant. Ranganathan (1988)

56. $$\boldsymbol{y'' - (1-1/n)y'^2/y - (1+2/n)yy' + (1/n)y^3 - Cy = 0,}$$

where n is an integer and C is a constant.

If we put $y = -u'/u$ and $u' = v^n$, we may verify that

$$v = \sin(C_1x + C_2),$$

where $C_1 = (-C/n)^{1/2}$ and C_2 is an arbitrary constant. For $n \neq 0, -2, -1, \ell = 0, 1, \ldots, u$ is meromorphic, so y has the Painlevé property. Martynov (1985)

57. $$\boldsymbol{y'' - (1 - 1/n)y'^2/y + [(n+2)/n]yy' + [(n-2)/n]y'/y}$$
$$\boldsymbol{+ y^3/n + 1/(ny) - b(x)(y' + y^2 - 1) - f(x)y = 0,}$$

where $f(x)$ and $b(x)$ are arbitrary analytic functions, and n is an integer.

Set $y' = -y^2 + 1 + nvy$ so that

$$v' + v^2 = bv + (1/n)(f - 2/n).$$

Now, putting $y = u'/u, v = w'/w$, we obtain

$$w'' - bw' - (1/n)(f - 2/n)w = 0, \tag{1}$$

$$u'' - n(w'/w)u' - u = 0. \tag{2}$$

Since (1) is linear, its singular points are fixed. Let $x = x_0$ be a simple movable zero of $w(x)$; then $x = x_0$ is a regular singular point of (2). The roots of the corresponding indicial equation are 0 and $(n+1)$; therefore, $u(x)$ may have logarithmic term except if a certain condition is satisfied. In that exceptional case, (2) has two regular solutions

$$\begin{aligned} u_1(x) &= (x - x_0)^{n+1}[1 + \cdots], \\ u_2(x) &= 1 + (x - x_0)^2/[2(1-n)] + \cdots, \end{aligned}$$

to which correspond, respectively,

$$\begin{aligned} y_1(x) &= (n+1)/(x - x_0) + \cdots \\ y_2(x) &= (x - x_0)/(1-n) + \cdots. \end{aligned}$$

Bureau (1964)

58. $$\boldsymbol{y'' - (1 - 1/n)y'^2/y - a(x)yy' + [na^2(x)/(n+2)^2]y^3}$$
$$\boldsymbol{- a_1(x)y' - b_1(x)y^2 - b_2(x)y = 0,}$$

where a, a_1, b_1, and b_2 are analytic functions of x and n is an integer. By the transformation $y = \lambda(x)u, t = \phi(x)$, it is possible to change the given DE to a form in which $a_1 = b_2 = 0$; therefore, we may assume that the given DE has the form

$$y'' - (1 - 1/n)y'^2/y - ayy' + [na^2/(n+2)^2]y^3 - b_1 y^2 = 0. \tag{1}$$

It is possible to check that the condition for (1) to have poles as the only movable singularities is that $b_1 = [n/(n+2)]a'$, where a as in (1) is an arbitrary analytic function of x. Then (1) becomes

$$y'' - (1 - 1/n)y'^2/y - ayy' + [na^2/(n+2)^2]y^3 - [n/(n+2)]a'(x)y^2 = 0. \tag{2}$$

To solve (2), put

$$y'/y - nay/(n+2) = nv(x);$$

then

$$v'(x) + v^2 = 0,$$

so that

$$v = 1/(x + H).$$

The solution y is determined by the Bernoulli equation

$$y' = [n/(x + H)]y + nay^2/(n + 2).$$

Setting $y = 1/w$, we get the linear DE

$$w' + nw/(x + H) + na/(n + 2) = 0,$$

which on setting $(x + H)^n w = W$ becomes

$$W' + [na/(n + 2)](x + H)^n = 0. \tag{3}$$

Equation (3) is easily integrated. The original DE, under assumptions made here, therefore, has poles as its only movable singularity in its solutions. Bureau (1964)

59. $$\boldsymbol{y'' - (1 - 1/n)y'^2/y - a(x)yy' - a_1(x)y' + [na^2(x)/(n + 2)^2]y^3 - [n/(n + 2)][a'(x) - a_1(x)a(x)]y^2 - b_2(x)y = 0,}$$

where a, a_1, and b_1 are analytic functions of x.

Making the transformations $y'/y - nay/(n + 2) = nv$ and $v = w'(x)/w$, we get the linear DE for w:

$$w'' - a_1 w' - (b_2/n)w = 0;$$

therefore, y is solution of Bernoulli's equation

$$y' = (nw'/w)y + [na/(n + 2)]y^2$$

or

$$u' + [na/(n + 2)]w^n = 0 \text{ if } y = w^n/u.$$

Therefore, the given DE has poles as the only movable singularities in its solutions. Bureau (1964)

60. $$\boldsymbol{y'' - (1 - 1/n)y'^2/y + [(n - 2)/n]y'/y + 1/(ny) - b_2 y = 0,}$$

where $b_2 = b_2(x)$ is an arbitrary function and n is an integer.

Writing

$$y' = 1 + nvy,$$

we check that v satisfies the Riccati equation

$$v' + v^2 = (1/n)b_2,$$

which is linearized by writing $v = w'(x)/w(x)$; we have the linearized equation

$$w'' - (b_2/n)w = 0.$$

Thus, y is given by

$$y = w^n \left[K + \int \frac{dx}{w^n} \right].$$

The given DE has poles as its only movable singularities in its solutions provided that b_2 is a constant. Bureau (1964)

61. $$y'' - (1 - 1/n)y'^2/y - ayy' - a_2 y'/y - by^3 - b_4/y - F(x,y) = 0,$$

where $F(x,y) = a_1 y' + b_1 y^2 + b_2 y + b_3$, n is a positive or negative y integer; $a, a_2, b, b_4, a_1, b_1, b_2$, and b_3 are analytic functions.

Conditions on a's and b's are found such that the given DE has poles as its only movable singularities. This leads to 20 subcases, which are individually treated. We also list them individually. Bureau (1964)

62. $$n\alpha^2 y'' - (\alpha/x)(B - n\alpha)y' + \alpha^2 y'^2/y + (A/x^2)y + (1/xy)y' - 1/x^2 = 0,$$

where n, α, B, and A are constants; $y(0) = 0,\ y'(0) = -c$. For convenience, c is chosen as 1.

The given DE is nonautonomous, but is equidimensional. If we make the change of variables $x = -e^{-z}$ and write the resulting equation as a system, we obtain

$$\begin{aligned} \psi' &= \theta, \\ \psi\theta' &= (1/n\alpha^2)(\theta + \psi - \alpha^2\theta^2 - A\psi^2 - \alpha B\theta\psi), \end{aligned} \tag{1}$$

where $' \equiv \dfrac{d}{dz}$ and $\psi = y(x(z))$. The initial conditions now become

$$\psi(z) \sim e^{-z}, \theta(z) \sim -e^{-z} \text{ as } z \to \infty. \tag{2}$$

The system (1), being singular at $\psi = 0$, is changed by the transformation $\dfrac{d\tau}{dz} = 1/\rho(z)$, so that the system in the new variables $\rho(\tau) \equiv \psi(z(\tau))$ and $\sigma(\tau) \equiv \theta(z(\tau))$ becomes

$$\begin{aligned} \dot\rho &= \rho\sigma, \\ \dot\sigma &= (1/n\alpha^2)(\sigma + \rho - \alpha^2\sigma^2 - A\rho^2 - \alpha B\sigma\rho), \end{aligned} \tag{3}$$

where $\cdot = \dfrac{d}{d\tau}$. Since $\tau(z) \sim e^z$ as $z \to \infty$, the conditions (2) become $\rho(\tau) \sim 1/\tau, \sigma(\tau) \sim -1/\tau$ as $\tau \to \infty$. The system (3), with another minor change of variables, is studied in the phase plane to solve a focusing problem for the porous medium equation. Aronson and Graveleau (1993)

63. $$y'' - (1 - 1/n)y'^2/y - ay'/y^k - b/y^{2k-1} = 0,$$

where a and b are constants.

The necessary conditions for the solution of the given DE about any movable singularity to be single-valued are $k = 1$ and $n \neq 1$. It may be verified that for $k > 1$, there is a solution of the given DE

$$y^k = (k/h)(x + c), \tag{1}$$

where c is an arbitrary constant, and h is a solution of $P(h) = 1/n - k - ah - bh^2 = 0$. The solution of (1) is in general not single-valued. Bureau (1964)

64. $$y'' + py^{-1}y'^2 - [a_0/[(p+1)|a|^2]]y^{-p}y' = 0,$$

where p, a_0, and a are constants.

The given DE arises from the similarity reduction of a nonlinear diffusion equation in n-spatial variables. Branson and Steeb (1983)

65. $$y'' + py^{-1}y'^2 + [n/(2x)]y' + [1/[4(p+1)]]y^{-p}y' = 0,$$

where n, p are arbitrary constants.

The given DE results from the similarity reduction of a nonlinear diffusion equation in n-space variables. Branson and Steeb (1983)

2.16 $y'' + f(y)y'^2 + g(x,y)y' + h(x,y) = 0$

1. $$y'' + [y^2 + y'^2]y = 0.$$

A first integral is easily found to be

$$[y'^2 + y^2 - 1]\exp(y^2) = \text{ constant.}$$

Assuming the approximate solution as $y = a\cos wx$, the method of harmonic balance gives

$$w^2 = 3a^2/(4 - a^2) \text{ for } a < 2.$$

An exact solution is $y = \cos x$. Jordan and Smith (1977), p.120

2. $$y'' + yy'^2 + q^2y^3 = 0,$$

where q is a constant.

The given DE has exact solutions $y = \pm\sin qx$ and $y = \pm\cos qx$. In fact, it may be shown that all solutions of the given DE have similar oscillatory properties. See Eqn. (2.16.5). Utz (1967)

3. $$y'' + yy'^2 - q^2y^3 = 0,$$

where q is a constant.

Two exact solutions of the given DE are $y = \pm\cosh qx$. Utz (1967)

4. $$y'' + yy'^2 - q^2y^{2n-1} = 0,$$

where q is a constant and n is a positive integer.

It is shown that no solution of the given DE is oscillatory. Utz (1969)

5. $$y'' + yy'^2 + yf(y) = 0,$$

where $f(y)$ is an even integrable function and $f(y) > 0$ for $y = 0$.

It is shown that if $f(y)$ also satisfies the condition that $\int_0^y uf(u)du \to \infty$ as $y \to \infty$ and if the solutions of the given DE are unique, and valid for all x, then they are oscillatory (if $y \not\equiv 0$) and periodic. Utz (1967)

6.
$$\boldsymbol{y'' + [y^3 + 13y^2 + 2y - 1/4]y' + (y - 1/4)y'^2 + y = 0.}$$

It is shown that the given DE admits at least one nonperiodic solution. Guidorizzi (1993)

7.
$$\boldsymbol{y'' - yy'^2 = 0.}$$

Put
$$y' = p, y'' = p\frac{dp}{dy};$$
we have
$$p\frac{dp}{dy} - yp^2 = 0.$$
Therefore, either $p = 0$, i.e., $y =$ constant or
$$\frac{dp}{dy} - yp = 0. \tag{1}$$

Integrating (1) twice $\left(\text{using } p = \frac{dp}{dy}\right)$, we get, respectively,
$$\frac{dp}{dy} = p = \frac{1}{C_1}e^{y^2/2}, x = c_1 \int e^{-y^2/2}dy + c_2,$$
where c_1 and c_2 are arbitrary constants. Martin and Reissner (1958), p.77

8.
$$\boldsymbol{y'' - yy'^2 + q^2y^3 = 0,}$$
where q is a constant.

The given DE has exact solutions $y = \pm \sinh qx$. Utz (1967)

9.
$$\boldsymbol{y'' - yy'^2 + q^2y^{2n-1} = 0,}$$
where q is a constant.

Putting $y' = v, y'' = v\frac{dv}{dy}$ and then $z = v^2$, one easily finds that
$$v\frac{dv}{dy} - yv^2 + q^2y^{2n-1} = 0$$
or
$$v^2(y) = z(y) = -2e^{y^2}\int_{y_0}^{y} q^2u^{2n-1}e^{-u^2}du + e^{-y_0^2}z(y_0)e^{y^2}.$$
By introducing
$$\int_{u_0}^{u}\left[-2q^2u^{2n-1}e^{-u^2}\right]du = F(u) - F(u_0), \tag{1}$$

where

$$F(u) = q^2 e^{-u^2} \left[u^{2n-2} + (n-1)u^{2n-4} + (n-1)(n-2)u^{2n-6} \cdots \right].$$

Thus, using (1), we have

$$v^2(y) = e^{y^2} F(y) + \left[v^2(y_0) e^{-y_0^2} - F(y_0) \right] e^{y^2}. \tag{2}$$

Equation (2) may be examined in the (y, v) plane, to arrive at the result that if

$$k = y_0' \, e^{-y_0^2} - F(y_0),$$

a solution of the given DE with IC $y = y_0, y' = y_0'$ is periodic if $-q^2 n! < k < 0$, and nonoscillatory for $k \leq 0$. Utz (1969)

10. $$\boldsymbol{y'' + ayy'^2 + by = 0,}$$

where a and b are constants.

The given DE is autonomous. With $p(y) = y'(x)$, we obtain the Bernoulli equation

$$pp' + ayp^2 + by = 0.$$

Put $p^2 = P$, solve the linear DE in P, and hence solve the first order DE $dy/dx = p = P^{1/2}$. Kamke (1959), p.553

11. $$\boldsymbol{y'' - \epsilon yy'^2 + w_0^2 y = 0, \epsilon \ll 1.}$$

A straightforward perturbation gives

$$y = a\cos(w_0 x + \beta) + (1/32)\epsilon a^3 \left[4w_0 x \sin(w_0 x + \beta) + \cos(3w_0 x + 3\beta) \right] + \cdots$$

with a secular term. Renormalization leads to a uniform first order solution as

$$y = a\cos\left[w_0(1 - \epsilon a^2/8)x + \beta \right] + \cdots.$$

Here β is an arbitrary constant. Nayfeh (1985), p.85

12. $$\boldsymbol{y'' + w_0^2 y + \epsilon\alpha_4 y'^2 + \epsilon^2 \alpha_5 yy'^2 = 0, \epsilon \ll 1.}$$

The solution to a second order approximation on using the method of multiple scales is

$$y = a\cos(wx + \beta_0) - (\epsilon\alpha_4 a^2/6)\left[\cos(2wx + 2\beta_0) + 3\right] + \cdots,$$

where

$$w = w_0 + \frac{\epsilon^2(3\alpha_5 - 4\alpha_4^2)w_0 a^2}{24} + \cdots.$$

Here a and β_0 are arbitrary constants. Nayfeh (1985), p.147

13. $$\boldsymbol{y'' + \beta yy'^2 + y(1 + \alpha y^2) = 0,}$$

where α and β are constants.

Put

$$y' = p, y'' = p\frac{dp}{dy};$$

we get

$$p\frac{dp}{dy} + \beta y p^2 + y(1 + \alpha y^2) = 0.$$

Now we put $p^2 = z$ to obtain

$$\frac{dz}{dy} + 2\beta y z = -2y(1 + \alpha y^2). \tag{1}$$

The general solution of (1) for $\beta \neq 0$ is found to be

$$z = Ce^{-\beta y^2} - \frac{\alpha}{\beta}y^2 - \frac{\alpha}{\beta^2} - \frac{1}{\beta},$$

where C is a constant of integration. Therefore,

$$\frac{dy}{dx} = p = \pm\left[Ce^{-\beta y^2} - \frac{\alpha}{\beta}y^2 + \frac{\alpha}{\beta^2} - \frac{1}{\beta}\right]^{1/2}.$$

The solution is now obtained by quadrature. Kane (1966)

14. $$\boldsymbol{y'' + w_0^2 y + \epsilon\alpha_2 y^2 + \epsilon^2\alpha_3 y^3 + \epsilon\alpha_4 y'^2 + \epsilon^2\alpha_5 y y'^2 = 0, \epsilon \ll 1.}$$

The solution, to second order approximation, by the method of multiple scales is

$$y = a\cos(wx + \beta_0) + \epsilon a^2\left[\frac{\alpha_2 - w_0^2\alpha_4}{6w_0^2}\cos(2wx + 2\beta_0) - \frac{\alpha_2 + w_0^2\alpha_4}{2w_0^2}\right] + \cdots,$$

where

$$w = w_0 + \left[\frac{\epsilon^2(9w_0^2\alpha_3 + 3w_0^4\alpha_5 - 10\alpha_2^2 - 10w_0^2\alpha_2\alpha_4 - 4w_0^4\alpha_4^2)}{24w_0^3}\right]a^2\ldots,$$

and a and β_0 are arbitrary constants. Nayfeh (1985), p.147

15. $$\boldsymbol{y'' + (\delta y^2 + \beta y)y' + \alpha y y'^2 + y = 0,}$$

where $\delta \geq 0, \alpha > 0$, and β are given.

The following results are proved:

(a) If $\delta = 0$ and $\beta^2 - 4\alpha = 0$, then every nontrivial solution is periodic; if $\delta = 0$ and $\beta^2 - 4\alpha \geq 0$, then every nontrivial solution passing by $(x_0, y_0) \in \Omega_{\alpha,\beta}$ is periodic.

(b) If $\delta > 0$ and $\beta^2 - 4\alpha > 0$, then every nontrivial solution is oscillating and $(y(x), y'(x))$ approaches the origin when $x \to \infty$. If $\delta > 0$ and $\beta^2 - 4 \geq 0$, then every nontrivial solution $y(x)$ passing by $(x_0, y_0) \in \Omega_{\alpha,\beta}$ is oscillating and $(y(x), y'(x))$ approaches the origin when $x \to +\infty$.

The open set $\Omega_{\alpha,\beta}$ is defined as follows. Let $\alpha, \beta \in R$ with $\alpha > 0$; then

$$\begin{aligned}\Omega_{\alpha,\beta} &= R^2 \text{ if } \beta^2 - 4\alpha < 0,\\ \Omega_{\alpha,\beta} &= \{(x,y) \in R^2 | y > \rho_1\} \text{ if } \beta > 0 \text{ and } \beta^2 - 4\alpha \geq 0\end{aligned}$$

and

$$= \{(x,y) \in R^2 | y < \rho_1\} \text{ if } \beta < 0 \text{ and } \alpha^2 - 4\alpha \geq 0,$$

where

$$\rho_1 = [-\beta + (\beta^2 - 4\alpha)^{1/2}]/2\alpha \quad \text{and} \quad \rho_2 = [-\beta - (\beta^2 - 4\alpha^2)^{1/2}]/2\alpha.$$

Guidorizzi (1993)

16. $\mathbf{y'' + \alpha y y'^2 + 2\alpha y^2 y'/x + y'/x - y/x^2 + (1 + \alpha/x^2)y^3 - y = 0, x \geq 0,}$

where α is a parameter.

The given DE describes TM waves in a nonlinear medium. A numerical investigation is carried out for bounded solutions of the given DE, which decrease at infinity. It is shown that there are no such solutions for $\alpha > 0.145$.

By way of analysis, the given DE is shown to have two types of expansions of $y(x)$ as $x \to 0$:

$$y(x) = R_1 x + \sum_{n=1}^{\infty} R_{2n+1} x^{2n+1}, R_{2n+1} = R_{2n+1}(R_1, \alpha) \tag{1}$$

and

$$y(x) = R_0 + \sum_{n=1}^{\infty} R_{2n} x^{2n}, R_{2n} = R_{2n}(R_0), \alpha R_0^2 = 1. \tag{2}$$

The eigenvalues R_1 that depend on α and R_0 are determined from the condition that $y(x)$ tend to modified Bessel function $K_1(x)$ as $x \to \infty$, the solution of the linearized equation. Although the physical requirement is that $\alpha > 0$, numerical integration shows that solutions corresponding to (1) exist for $\alpha < 0$, but as α is reduced, the solutions spread rapidly so that prescribed asymptotic behavior is achieved at increasingly larger values of x_0. For $\alpha > 0$, there exist boundary values α_N in whose vicinity $R_1 \to \infty$: $\alpha_0 \sim 0.145, \alpha_1 =\sim 0.077, \alpha_2 = 0.050$. Some sketchy results are shown in figures. Gisin (1990)

17. $\mathbf{y'' + (y^6 - 1)y' + (y^3 + y - 1)y'^2 + y^5 = 0.}$

It is shown that the given DE admits at least one nontrivial periodic solution. Guidorizzi (1993)

18. $\mathbf{y'' + y'^2/(1 - y) - \phi^2 y(1 - y) = 0, y'(0) = 0,}$

$y(1) = 1$, where ϕ is a constant.

The given problem arises in chemical engineering and is solved by finite difference methods. Kubicek and Hlavacek (1983), p.96

19. $\mathbf{y'' - [a/(ay + b)]y'^2 = 0,}$

where a and b are constants.

The solutions of the given DE about any movable singularity are single valued. It has a first integral

$$y' = K(ay + b), \tag{1}$$

where K is a constant. Equation (1) is linear and is easily solved. Bureau (1964)

20. $$y'' - [1/(2y) + 1/(y-1)]y'^2 = 0.$$

The solution of the given DE is

$$y = \tanh^2(Hx + H_1),$$

where H and H_1 are constants. Bureau (1964)

21. $$y'' - [1/(2y) + 1/(y-1)]y'^2 - He^{kx}y = 0,$$

where H and K are constants.

On writing $t = e^{Kx}$, etc., the given DE can be reduced to a special case of Painlevé's fifth equation. Bureau (1964)

22. $$y'' - (1/2)[(1/y) + 1/(y-1)]y'^2 - Ky(y-1)e^x = 0,$$

where K is a constant.

The transformation $y = 1/(1-v), \tau = e^x$, takes the given DE to the following special case of $\mathrm{P_V}$:

$$v'' = [1/(2v) + 1/(v-1)]v'^2 - v'/\tau + Kv/\tau, \quad ' \equiv \frac{d}{d\tau}.$$

Leonovich (1984)

23. $$y'' - [1/(2y) + 1/(y-1)]y'^2 - y[k/(y-1) + \ell] = 0,$$

where k and ℓ are constants. Bureau (1964)

24. $$y'' - [1/(2y) + 1/(y-1)]y'^2 - y(y-1)[e(y-1) + \ell/(y-1)] = 0,$$

where e, k, and ℓ are constants.

The given DE is a special case of Eqn. (2.16.25). Bureau (1964)

25. $$y'' - [1/(2y) + 1/(y-1)]y'^2 - y(y-1)[e(y-1) + (g/y^2)(1-y) + \ell/(y-1)] = 0,$$

where e and g are constants, and $\ell = \ell(x)$ is an analytic function of x.

If $\ell(x) = He^{Kx}$, where H and K are constants, the given DE has solutions, in which the only movable singularities are poles. Bureau (1964)

26. $$y'' - [1/(2y) + 1/(y-1)]y'^2 - y(y-1)[H(y-1) + H_1(y-1)/y^2 + H_2/(y-1) + 2H_3/(y-1)^2] = 0,$$

where H, H_1, H_2, and H_3 are constants.

The given DE has poles as its only movable singularities. It has a first integral

$$y'^2 - 2y(y-1)^2[Hy - H_1/y - H_2/(y-1) - H_3/(y-1)^2 + H_4] = 0,$$

where H_4 is an arbitrary constant. The solution now is obtainable through a quadrature. Bureau (1964)

27.
$$\boldsymbol{y'' - [1/(2y) + 1/(y-1)]y'^2 - (y-1)^2[H_2 y + H_3/y] - H_1 x y - H x^2 y(y+1)/(y-1) = 0,}$$

where H_1, H_2, H_3, and H are constants.

A change of variable $t = \phi(x)$ can be found such that the given DE becomes Painlevé's fifth equation:

$$\begin{aligned} y'' &= [1/(2y) + 1/(y-1)]y'^2 - y'/x + [(y-1)^2/x^2][H_2 y + H_3/y] \\ &\quad + H_1 y/x + Hy(y+1)/(y-1). \end{aligned}$$

Bureau (1964)

28.
$$\boldsymbol{y'' - [(1/y) + 1/\{2(y-1)\}]y'^2 - K y^{1/2}(y-1)e^x = 0,}$$

where K is a constant.

The transformation $y = \left(\dfrac{1+v}{1-v}\right)^2$, $\tau = e^x$, reduces the given DE to the following special case of a fifth Painlevé equation:

$$v'' = [1/(2v) + 1/(v+1)]v'^2 - v'/\tau + Kv/\tau, \quad ' \equiv \frac{d}{d\tau}.$$

Leonovich (1984)

29.
$$\boldsymbol{y'' - [1/2y + 1/(y-1)]y'^2 + (1/x)y' - [(y-1)^2/x^2](\alpha y + \beta/y) - \gamma y/x - \delta y(y+1)/(y-1) = 0,}$$

where α, β, γ, and δ are constants.

The given DE is P_V. It has special explicit solutions for the following choices:

(a) $\alpha = \beta = 0, \gamma^2 + 2\delta = 0$, then $y = C\exp[\pm(-2\delta)^{1/2}x]$, where C is an arbitrary constant.

(b) $\delta = \gamma = 0; \alpha, \beta$ are arbitrary.

Thus, the solution may be written as

$$\int \frac{dy}{(y-1)(\alpha y^2 + C_1 y - \gamma)^{1/2}} = \pm 2^{1/2}(C_2 + \ln x),$$

where C_1 and C_2 are arbitrary constants. Lukashevich (1965)

30.
$$\boldsymbol{y'' - [1/(2y) + 1/(y-1)]y'^2 + y'/x - (y-1)^2(\alpha y + \beta/y)/x^2 - \gamma y/x - \delta y(y+1)/(y-1) = 0,}$$

where α, β, γ, and δ are constants.

The given DE is P_V. Writing $y = \tanh^2(u/2)$, it is changed to

$$x(xu')' = \left(\tanh\frac{u}{2} + \beta\tanh^{-3}\frac{u}{2}\right)\cosh^{-2}\frac{u}{2} + \frac{\gamma}{2}x\sinh u - \frac{\delta}{4}x^2\sinh 2u. \tag{1}$$

Series expansions of solutions about fixed regular singular points are found. Some asymptotic results regarding the number of poles are also derived. Shimomura (1987)

31.
$$y'' - [1/(2y) + 1/(y-1)]y'^2 + (1/x)y' - [(y-1)^2/x^2](\alpha y + \beta/y) - \gamma y/x - \delta y(y+1)/(y-1) = 0,$$

where α, β, γ, and δ are constants.

A method to linearize IVP for P_V is given. The procedure involves formulating a Riemann–Hilbert boundary value problem on intersecting lines for the inverse monodromy problem. This boundary value problem is reduced to a sequence of standard problems on single lines in a certain range of parameter space. Schlesinger transformations allow one to cover the parametric space completely. Special solutions of P_V are constructed from special cases of the Riemann problem as well. Fokas, Mugan, and Ablowitz (1988)

32.
$$y'' - [1/(2y) + 1/(y-1)]y'^2 + (1/x)y' - [(y-1)^2/x^2](\alpha y + \beta/y) - (\gamma/x)y - \delta y(y+1)/(y-1) = 0,$$

where α, β, γ, and δ are constants.

For the given DE that is P_V, an initial value problem with $y(x_0) = y_0, y'(x_0) = y_1$, and x_0 near 0 is solved. It is also indicated that

$$[\{y(x) - 1\}x^{-w}]^{\pm 1}, 0 \le w < 1$$

and

$$[(\tilde{y}(x) - 1)^{-1}x^{w}]^{\pm 1}, 0 < w \le 1$$

are bounded on suitably chosen curves tending to $x = 0$. Convergent series solutions are constructed in the neighborhood of $x = 0$. Shimomura (1982,1982a)

33.
$$y'' - [1/(2y) + 1/(y-1)]y'^2 + y'/x - (\alpha/x^2)(y-1)^2 y - \gamma y/x + \delta y(y+1)/(y-1) = 0,$$

where α, γ, and δ are real constants.

The given DE is a special case of P_V. Let $y = Y_0(x)$ be a solution of the given DE subject to the ICs $Y_0(x_0) = y_0, Y_0'(x_0) = y_1$, where x_0, y_0, and y_1 are real constants satisfying $x_0 > 0, y_0 \ne 0, 1$. Several cases of $Y_0(x)$ are treated for which it is oscillatory; an asymptotic representation of $Y_0(x)$ is given as $x \to +\infty$. For example, when $\delta < 0, 0 < y_0 < 1$, it is shown that

$$\begin{aligned} Y_0(x) &= r_1[1 + O(x^{-1})]x^{-1}\cos^2[(-\delta/2)^{1/2}x - \left[(\gamma/4)\,(2/(-\delta))^{1/2}\right. \\ &\quad \left. - \; (-\delta/2)^{1/2}r_1]\log x + \theta_1 + O(x^{-1})\right] \quad \text{as } x \to +\infty, \end{aligned}$$

where r_1 and θ_1 are real constants depending on (x_0, y_0, y_1) and satisfying $r_1 > 0, 0 \le \theta_1 < 2\pi$. Other asymptotic representations for other sets of parameters are given. Shimomura (1987)

34.
$$y'' - [1/(2y) + 1/(y-1)]y'^2 + (1/x)y' - (1/x^2)(y-1)^2(\alpha y + \beta/y) - (\gamma/x)y - \delta y(y+1)/(y-1) = 0,$$

where α, β, γ, and δ are constants.

For the given DE, which is P_V, a two-parameter family of solutions near the point at infinity is constructed; this point is a (fixed) irregular singular point. The general solution behaves as a damped oscillator on a half-line. The domain of convergence of the series solution is clearly defined. Takuno (1983)

35. $$\mathbf{y'' - [1/(2y) + 1/(y-1)]y'^2 - 2[(cy+d)/(y-1)]y' - [(y-1)^2/2](q^2 y - r^2/y) - 2[c^2 - d^2 - c' - d']y = 0,}$$

where $c = c(x)$, $d = d(x)$, $q(x)$, and $r(x)$ are analytic functions such that $q' = 2cq$, $r' + 2dr = 0$. With

$$-2w = \frac{y'}{y-1} + 2\left[\frac{c+d}{y-1}\right] y + qy, \tag{1}$$

the given DE yields

$$\frac{y-1}{4y} = -\frac{w' + 2wd}{4w^2 - r^2}. \tag{2}$$

The system (1) and (2) is equivalent to the original DE. Eliminating y, we obtain an equation for w. For convenience, write $A = w' + 2dw, B = 4w^2 - r^2$; then after some simplification we get

$$B' + 4A' + 4Aq + 2(d-c)(B+4A) = 0 \tag{3}$$

or

$$w'' = -2ww' + a_1 w' + a_2 w^2 + a_3 w + a_4$$

with suitable coefficients $a_1, \ldots, a_4$. To integrate (3) we observe that since $q' = 2cq$, one also has

$$Aq = qw' + 2dqw = \frac{d}{dx}(qw) + 2(d-c)qw;$$

hence on setting

$$4Z = B + 4A + 4qw = 4[w' + w^2 + w(q+2d) - r^2/4], \tag{4}$$

(3) becomes $Z' = 2(c-d)Z$. Therefore, Z is determined by $Z = Kqr$ (see the definition of c and d) and hence w by the Riccati equation (4). Bureau (1964)

36. $$\mathbf{y'' - [1/(2y) + 1/(y-1)]y'^2 - [(a(x)y - b(x))/(y-1)]y' - \ell(x)y = 0,}$$

where $a(x), b(x)$, and $\ell(x)$ are analytic. The given DE is a special case of Eqn. (2.16.35). Bureau (1964)

37. $$\mathbf{y'' - (2/3)[1/y + 1/(y-1)]y'^2 = 0.}$$

The given DE, on setting, $2y = 1 + u$, etc., admits a first integral

$$u'^3 = K(u^2 - 1)^2, \tag{1}$$

where K is an arbitrary constant. Equation (1) can be solved in terms of elliptic functions. Bureau (1964)

38. $$\mathbf{y'' - (3/4)[1/y + 1/(y-1)]y'^2 - ay' = 0,}$$

where a is a constant.

Making a change of variable $x = \phi(t)$, etc., it is possible to set $a = 0$. A first integral of the resulting equation is

$$y'^4 = Ky^3(y-1)^3, \tag{1}$$

where K is an arbitrary constant. The solution of (1) is expressible in terms of elliptic functions. Bureau (1964)

39. $$\mathbf{y'' - (3/4)[1/y + 1/(y-1)]y'^2 - ay'}$$
$$\mathbf{- y(y-1)[4d^2(2y-1) + (k/y) + \ell/(y-1)] = 0,}$$

where d, k, and ℓ are constants.

By a change of variable $x = \phi(t)$, it is possible to make the coefficient a of y' zero. We may, therefore, assume that to be the case; d, k, and ℓ are constants. Now to solve the given DE with $a = 0$, we put $y = Aw^{-1}$, where A is a constant, and follow the approach for Eqn. (2.16.40). Bureau (1964)

40. $\mathbf{y'' - [1/(2y) + 3/\{4(y-1)\}]y'^2 - (y-1)[4d^2(2/y-1) + ky - \ell y/(y-1)] = 0.}$

To solve the given DE, we first note that the DE

$$w'' = \left[\frac{1}{2w} + \frac{3}{4(w-1)}\right] w'^2$$

has an integral

$$w'^2 = Kw(w-1)^{3/2}, \tag{1}$$

where K is an arbitrary constant. Now making the constant K depend on w, we get an equation for $K(w)$ by substitution in the given DE (where we put $y = w$):

$$\frac{dK}{dw} = \left[\frac{2}{(w-1)^{1/2}}\right] \{4d^2(2-w)/w^2 + K - l/(w-1)\}.$$

Set

$$w = 1 + u^2, K(u) = K(1+u^2); \tag{2}$$

then

$$\frac{dK}{du} = 4\left[\frac{4d^2(1-u^2)}{(1+u^2)^2} + K - l/u^2\right]$$

and

$$K(u) = 4\left[ku + \frac{\ell}{u} + \frac{4d^2u}{1+u^2} + H\right],$$

where H is an arbitrary constant. Thus, u is an elliptic function [see (1) and (2)]. The function y satisfies the first order DE

$$y'^2 = K(y)y(y-1)^{3/2}.$$

Bureau (1964)

41. $$\mathbf{y'' - [2/(3y) + 1/\{2(y-1)\}]y'^2 - d(x)y' = 0,}$$

where $d(x)$ is an analytic function.

Making a change of variable $x \to \phi(x)$, it is possible to remove the term $d(x)y'$ so that the given DE has the form

$$y'' - \left[\frac{2}{(3y)} + \frac{1}{2(y-1)}\right] y'^2 = 0,$$

with the first integral

$$y'^6 = ky^4(y-1)^3, \tag{1}$$

where k is a constant. Equation (1) is solvable in terms of elliptic functions. Bureau (1964)

42. $$\mathbf{y'' - [2/(3y) + 1/\{2(y-1)\}]y'^2 - 3\ell y/(y-1) - \ell y = 0,}$$

where ℓ is a constant.

Put $y = w^{-1}$ to obtain

$$w'' = \left[\frac{4}{(3w)} + \frac{1}{2w(w-1)}\right] w'^2 + 3\ell w^2/(w-1) - \ell w.$$

On setting $w = z^3$, we have

$$z'' = \left[\frac{2}{z} + \frac{3}{2z(z^3-1)}\right] z'^2 + \ell z^4/(z^3-1) - (l/3)z. \tag{1}$$

The integral of (1) may be found by the method of variation of parameters. To this end, we put

$$z'^2 = z(z^3-1)K(z), \tag{2}$$

so that by substituting in (1), we get

$$\frac{dK}{dz} = -(2/3)\ell\frac{d}{dz}[z/(z^3-1)].$$

Solving for $K(z)$ and substituting in (2), we have

$$z'^2 = K_1 z(z^3-1) - (2/3)\ell z^2,$$

where K_1 is an arbitrary constant. z is, therefore, an elliptic function. Hence w and therefore y can be obtained. Bureau (1964)

43. $$\mathbf{y'' - [a/(ay+b) + c/(cy+d)]y'^2 = 0,}$$

where $a, b, c,$ and d are constants.

The solutions of the given DE about any movable singularity are single valued. This DE has the first integral

$$y' = K(ay+b)(cy+d), \tag{1}$$

where K is a constant. Equation (1) is of separable form and, therefore, easily solved. Bureau (1964)

44. $$\mathbf{y'' - [(1+1/n)a/(ay+b) + (1-1/n)c/(cy+d)]y'^2 = 0,}$$

where $a, b, c,$ and d are constants and $n > 1$ is an integer.

The solutions of the given DE about any movable singularity are single valued. It has the first integral

$$y'^n = K(ay+b)^{n+1}(cy+d)^{n-1}, \tag{1}$$

where K is a constant. The variables in (1) are separable; the solution for some n may be expressed in terms of elliptic integrals. Bureau (1964)

45. $$\mathbf{y'' + 2y'/x + [y/(a^2 - y^2)][y'^2 + (1 - y^2/a^2)^2] = (2y/x^2)(1 - y^2/a^2)}$$
$$\mathbf{+ (3y/a^2)[(1 - R^2/a^2)(1 - y^2/a^2)]^{1/2}, x \in [0, R], y(0) = 0, y(R) = 2,}$$

where $R^3 = 2a^2, R \geq 2$

Under the assumption $R > 2^{4/3}$, the uniqueness of the solution of the given problem is proved in the class of functions

$$y \in C^0[0, R] \cap C^2(0, R), 0 \leq y(x) \leq 2, x \in [0, R].$$

Moiseev and Sadovnichii (1989)

46. $$\mathbf{y'' - (y'^2/2)[1/y + 1/(y-1) + 1/(y-H)]}$$
$$\mathbf{- y(y-1)(y-H)[e/y^2 - f/(y-1)^2 + g/(y-H)^2 + h] = 0,}$$

where H, e, f, g, and h are constants.

To integrate the given DE, we use the method of variation of parameters. A first integral is found to be

$$y'^2 = 2y(y-1)(y-H)[K_1 + hy - e/y + f/(y-1) - g/(y-H)],$$

where K_1 is an arbitrary constant. The solution y is therefore expressible in terms of elliptic functions. The result also holds if one or more of the constants e, f, g, h is zero. Bureau (1964)

47. $$\mathbf{y'' - [(1/2)\{a/(ay+b) + c/(cy+d)\} + e/(ey+f)]y'^2 = 0,}$$

where a, b, c, d, e, and f are constants.

The solution of the given DE about any movable singularity may be shown to be single valued. It has the first integral

$$y'^2 = K(ay+b)(cy+d)(ey+f)^2,$$

where K is an arbitrary constant. The variables here also are separable; the solution may be expressed in terms of elliptic integrals or some elementary functions. Bureau (1964)

48. $$\mathbf{y'' - [(1/2)a/(ay+b) + (3/4)\{c/(cy+d) + e/(ey+f)\}]y'^2 = 0,}$$

where a, b, c, d, e, and f are constants.

The solution of the given DE about any movable singularity may be shown to be single valued. It has the first integral

$$y'^4 = K(ay+b)^2(cy+d)^3(ey+f)^3,$$

where K is a constant. The variables separate, and the solution may be expressed in the form of a quadrature. Bureau (1964)

49. $$\mathbf{y'' - [(2/3)\{a/(ay+b) + c/(cy+d) + e/(ey+f)\}]y'^2 = 0,}$$

where a, b, c, d, e, and f are constants.

The solution of the given DE about any movable singularity may be shown to be single valued. It has the first integral

$$y'^3 = K(ay+b)^2(cy+d)^2(ey+f)^2,$$

where K is an arbitrary constant. The solution by separating the variables may be expressed in terms of elliptic or elementary functions. Bureau (1964)

50. $$\mathbf{y'' - [(5/6)a/(ay+b) + (2/3)c/(cy+d) + (1/2)e/(ey+f)]y'^2 = 0,}$$

where a, b, c, d, e, and f are constants.

The solution of the given DE about any movable singularity may be shown to be single valued. It has the first integral

$$y'^6 = K(ay+b)^5(cy+d)^4(ey+f)^3,$$

where K is an arbitrary constant. The variables separate and the solution may be written in the form of a quadrature. Bureau (1964)

51. $$\mathbf{y'' - (1/2)\{a/(ay+b) + c/(cy+d) + e/(ey+f) + g/(gy+h)\}y'^2 = 0,}$$

where a, b, c, d, e, f, g, and h are constants.

The solution of the given DE about any movable singularity may be shown to be single valued. It has the first integral

$$y'^2 = K(ay+b)(cy+d)(ey+f)(gy+h), \tag{1}$$

where K is an arbitrary constant. Equation (1) is separable; the solution may be expressed in terms of elliptic functions. Bureau (1964)

52. $$\mathbf{y'' - \sum_{k=1}^{p}[A_k/(y-a_k)]y'^2 = 0,}$$

where $a_k, k = 1, \ldots, p$ are constants, and $A_k = 1 - 1/n_k$, where $n_k \neq 0$ is a positive or negative integer. It may then be shown that the conditions enumerated above are necessary so that the solution about any movable singularity is single valued. Thus, one must find the set of integers n, n_k $(k = 1, 2, \ldots, p)$ different from zero, satisfying $\sum_{k=1}^{p}(1-1/n_k) = 1-1/n$. Bureau (1964)

53. $$\mathbf{y'' - [\lambda y/(1+\lambda y^2)]y'^2 + \alpha y/(1+\lambda y^2) = 0,}$$

where λ and α are constants.

The given DE has an exact solution $y = A\sin(wx+\theta)$, provided that $(1+\lambda A^2)w^2 - \alpha = 0$; w and θ are arbitrary constants. Mickens (1981), p.29

54. $\boldsymbol{y'' - [y/(y^2+2\delta x^2)]y'^2 + (1/x)[(y^2-2\delta x^2)/(y^2+2\delta x^2)]y' - (1/x)\theta_1(y,x) = 0,}$

where

$$\begin{aligned}\theta_1(y,x) &= 2\delta x - \gamma - (a/x)(2\delta x^2+y^2) - \frac{y}{x(2\delta x^2+y^2)}[\gamma x - \delta x^2 - (a+c)y - y^2/2]^2 \\ &+ \left[\frac{4\delta x}{2\delta x^2+y^2} - (1/x)(a+c+y)\right][\gamma x - (a+c)y - \delta x^2 - y^2/2],\end{aligned}$$

where a, c, γ, and δ ($\neq 0$) are constants.

Using the transformation $w = (y+kx)/(y-kx)$, where $k^2 + 2\delta = 0$, we obtain

$$\begin{aligned}w'' &- \frac{3w-1}{2w(w-1)}w'^2 + (1/x)w' - (\alpha^*/x^2)w(w-1)^2 \\ &- (\beta^*/x^2)(w-1)^2/w - (\gamma^*/x)w - \frac{\delta^* w(w+1)}{w-1} = 0,\end{aligned} \tag{1}$$

where

$$\begin{aligned}\alpha^* &= -\frac{1}{16\delta}[r + k(1-a-c)]^2, \\ \beta^* &= \frac{1}{16\delta}[r - k(1-a-c)]^2, \\ \gamma^* &= k(a-c), \\ \delta^* &= \delta.\end{aligned}$$

The form (1) is the same as Painlevé's fifth equation with specific values of the parameters. Gromak (1975)

55. $$\boldsymbol{y'' + \frac{(b-1)y'-1}{(2y+1)\{(b+1)y+b\}}y' - \sigma(y-y_0)\frac{(b+1)y+b}{b(2y+1)} = 0;}$$

$y'(0) = -1/2, y'(x_1) = 1/(2b)$, where b, y_0, σ are constants.

The given BVP arises in the investigation of the minority carrier distribution in the base of a semiconductor structure. A numerical (iterative) algorithm is given to solve this problem. This algorithm uses quasilinearization, pivotal condensation, and continuation of the solution with respect to a parameter. Grigorenko, Smirnova, and Tai (1975)

56. $$\boldsymbol{\begin{aligned}y'' + [k/y^n - n/y]y'^2 &+ [a(x) + b(x)y^{1-n}]y' \\ &+ [h(x)y + d(x)y^n + (1/4)b^2y^{2-n}/k] = 0,\end{aligned}}$$

where a, b, d are all arbitrary functions of x, and n, k are arbitrary constants. The function $h(x)$ is given by

$$h(x) = (1/2)[b'(x) + ab + (1/2)(n-1)b^2/k]/k.$$

Using general theory given by the author, a first integral of the given DE is found to be

$$y'(x) = -(1/2)by/k + \left[c/(1 + c\int u\,dx) - A/u\right]y^n/k, \tag{1}$$

where $u = 1/r^2$ with $r(x)$ satisfying

$$r''(x) - (1/2)r[a' + (1/2)(n-1)b'/k + (1/2)\{a + (1/2)(n-1)b/k\}^2 - 2kd] = 0,$$

where c is a constant. Equation (1), being of Bernoulli type, can be integrated:

$$y^{1-n} = C + \left[(1-n)\int \left\{(c/\left(k\left(1+c\int u dx\right)\right) - A/(ku)\right\} B dx\right]/B, \tag{2}$$

where $B(x) = \exp[\{(1-n)/2k\}\int b dx]$. Equation (2) constitutes the general integral of the given DE involving constants C and c. Finch (1989)

57. $$\mathbf{y'' + 2y'^2 \tan y = 0.}$$

A first integral of the given DE is

$$y' = c_1 \cos^2 y. \tag{1}$$

On integration of (1), we get $\tan y = c_1 x + c_2$. c_1 and c_2 are arbitrary constants. Rabenstein (1972), p.47

58. $$\mathbf{y'' + \cot y\, y'^2 = 0, y(0) = \pi/3, y'(0) = -2/3^{1/2}.}$$

The solution is $y = \cos^{-1}(x + 1/2)$, obtained by putting

$$y' = p, y'' = p\frac{dp}{dy}$$

and integrating twice, etc. Wilson (1971), p.40

59. $$\mathbf{y'' - 2y'^2 \cot y - \sin y \cos y = 0,}$$

which is the equation of geodesics on the sphere.

Using Lie point symmetries, two first integrals are found to be

$$\begin{aligned} &- y'(\sin x)/(\sin^2 y) - \cos x \cot y = \phi_0, \\ &- y'(\cos x)/(\sin^2 y) + \sin x \cot y = \psi_0, \end{aligned} \tag{1}$$

where ϕ_0 and ψ_0 are constants. Eliminating y' from (1), we get the solution as $\cot y = \psi_0 \sin x - \phi_0 \cos x$. Stephani (1989), p.78.

60. $$\mathbf{y'' + (e^y - 2)y' + (y^3 - y^2 - 1)y'^2 + y^3 = 0.}$$

It is shown that the given DE admits at least one nontrivial periodic solution. Guidorizzi (1993)

61. $$\mathbf{y'' - e^{-y} y'^2 = 0.}$$

Put

$$y' = p, y'' = p\frac{dp}{dy}$$

to get

$$p\frac{dp}{dy} - e^{-y}p^2 = 0.$$

A first integral is

$$\ln p = -e^{-y} + C \text{ or } p = \exp\{C - e^{-y}\},$$

where C is an arbitrary constant. Hence integrate by separating the variables. Bender and Orszag (1978), p.34.

62. $$y'' - y'^2 \tanh y = 0.$$

A first integral of the given DE is

$$\ln y' - \ln \cosh y = \ln c_1$$

or

$$y' = c_1 \cosh y. \tag{1}$$

On integration of (1), we have $\arctan(\sinh y) = c_1 x + c_2$. c_1 and c_2 are arbitrary constants. Rabenstein (1972), p.47.

63. $$y'' - aZ(y) - bT(y) - A(y)y'^2 = 0,$$

where a and b are constants; Z, T, and A are functions of y.

The given DE has the general solution $y = F(u)$, where u satisfies

$$u'' = au + bu^3 \tag{1}$$

if and only if Z, T, and A satisfy

$$\dot{Z} - AZ = 1, Z\dot{T} - (3 + AZ)T = 0, \quad \cdot \equiv \frac{d}{dy}.$$

The function F is found from the relationship $u^2 = T(F)/Z(F)$.

The more general nonlinear DE of the given type can be solved in terms of solutions of (1). de Spautz and Lerman (1967)

64. $$y'' + g(y)y'^2 + f(y) = 0,$$

where $g(y)$ and $f(y)$ are defined and continuous for all y.

Assume that $g(y)$ and $f(y)$ are odd functions and $yf(y) > 0$ for $y \neq 0$. Define $G(y) = e^{2\int_0^y g(u)du}$ and assume that $\int_0^\infty f(u)G(u)du$ is finite and hence positive. Then depending on the particular conditions $y(x_0) = y_0, y'(x_0) = y_0'$, the solutions of the given DE for all large x may be oscillatory or nonoscillatory. Utz (1969)

65. $$y'' + g(y)y'^2 + f(x)y' = 0.$$

The given DE is called Liouville's equation. It can be made exact by multiplying by $1/y'$:

$$y''/y' + g(y)y' + f(x) = 0. \tag{1}$$

Integrating (1), we have

$$\ln y' + Y + X = C_0, \tag{2}$$

where $Y = \int g(y)dy$ and $X = \int f(x)dx$, or

$$y'e^Y = C_1 e^{-X}. \tag{3}$$

Integrating (3), we obtain the general solution of the given DE as

$$\int e^{Y(y)}dy = \int C_1 e^{-X(x)}dx + C_2.$$

Goldstein and Braun (1973), p.98

66. $$\boldsymbol{y'' + r(x)y' + q(x)Z(y) - A(y)y'^2 - g(x)z(y) = 0,}$$

where $(dZ/dy) - A(y)Z = 1$.

Putting $Z(y) = z(y)u(y)$, where

$$z(y) = \exp\left(\int^y A(u)du\right) \quad \text{and} \quad u(y) = \beta + \int^y \exp\left(-\int^u A(t)dt\right)du, \tag{1}$$

we can change the given equation to the linear form

$$\frac{d^2u}{dx^2} + r(x)\frac{du}{dx} + q(x)u = g(x). \tag{2}$$

Thus, solving for u from (2), we may solve the implicit equation (1) for $y(x)$. Klamkin and Reid (1976)

67. $$\boldsymbol{y'' + r(x)y' + q(x)Z(y) - A(y)y'^2 - g(x)z(y)[u(y)]^a = 0,}$$

where Z is found from

$$\frac{dZ}{dy} - A(y)Z = 1,$$

where $a \neq 0$ is a constant.

The given DE can be reduced to a simpler form

$$\frac{d^2u}{dx^2} + r(x)\frac{du}{dx} + q(x)u = g(x)u^a, a \neq 0,$$

through the transformation $Z(y) = z(y)u(y)$, where

$$z(y) = \exp\left(\int^y A(u)du\right) \quad \text{and} \quad u(y) = \beta + \int^y \exp\left(-\int^u A(t)dt\right)du.$$

Klamkin and Reid (1976)

68. $$\boldsymbol{y'' + f(y)y'^2 + g(y)y' + h(y) = 0.}$$

The given DE is autonomous. Put $p(y) = y'(x)$ so that

$$pp' + f(y)p^2 + g(y)p + h(y) = 0. \tag{1}$$

If $g = 0$, Eqn. (1) becomes Bernoulli's equation. If $h = 0$, it is linear. In general, it is an Abel DE. Kamke (1959), p.554

69. $$y'' + f_1(y)y' + f_2(y)y'^2 + g(y) = 0,$$

where:

(a) f_1, f_2, and g are of class C^1.

(b) $yg(y) > 0$ for $y \neq 0$.

(c) $\int_0^\infty g(y)dy = +\infty = \int_0^{-\infty} g(y)dy$.

Varieties of conditions are given on f_1, f_2, and g so that the given DE has either oscillatory solutions approach $(y, y') = (0, 0)$ or has periodic solutions. Guidorizzi (1993)

70. $$y'' + \phi(y)y'^2 + f(x)y' + g(x)\psi(y) = 0.$$

If $\phi(y) = [1 - \psi'(y)]/\psi(y), F(x) = \int f(x)dx$ and $g(x) = e^{-2F(x)}[\pm\exp(2\int e^{-F(x)}dx) - \nu^2]$, then by writing $\eta(\xi) = \exp(\int [1/\psi(y)]dy), \xi = \exp(\int e^{-F(x)}dx)$, we get the Bessel equation $\xi^2\eta'' + \xi\eta' + (\pm\xi^2 - r\nu^2)\eta = 0$, which is exactly solvable. Kamke (1959), p.554

71. $$y'' - [f'(y)/f(y)]y'^2 + g(x)y' + h(x)f(y) = 0.$$

With $y(x) = \eta(\xi)$, where $\xi = \xi(x)$ is a solution of the linear DE $\xi'' + g(x)\xi' + h(x) = 0$, we get

$$\xi'^2\eta'' - \{f'(\eta)/f(\eta)\}\xi'^2\eta'^2 + h(x)\{f(\eta) - \eta'] = 0. \tag{1}$$

Equation (1) is, for example, satisfied by the solution of $\eta'(\xi) = f(\eta)$. The original DE may be solved by solving the system

$$y'(x) = uf(y), u'(x) = -g(x)u - h(x). \tag{2}$$

The second equation of the system (2) may be solved first; it is then substituted into the first to obtain $y'(x)$ and hence $y(x)$. Kamke (1959), p.554.

2.17 $y'' + f(x, y)y'^2 + g(x, y)y' + h(x, y) = 0$

1. $$y'' + 2xy'^2 = 0.$$

Put $y' = v$; then

$$\frac{dv}{dx} + 2xv^2 = 0. \tag{1}$$

Equation (1) has $v = 0$ and

$$y' = v = \frac{1}{x^2 - C_1} \tag{2}$$

as solutions. $v = 0$ gives $y =$ constant. If the arbitrary constant $C_1 = -k^2 < 0$, then integrating (2), we have

$$y = \int vdx = (1/k)\tan^{-1}(x/k) + C_2,$$

where C_2 is an arbitrary constant. If $C_1 = k^2 > 0$, we have the solution

$$y = \frac{1}{2k} \ln |(x-k)/(x+k)| + C_2.$$

When $C_1 = 0$ in (2), we integrate to get $y = -1/x + C_3$. C_2 and C_3 are arbitrary constants. Rabenstein (1966), p.371

2. $$\boldsymbol{y'' - xy'^2 = 0.}$$

Put $\frac{dy}{dx} = u$, $\frac{du}{dx} = xu^2$, so that if $u \neq 0$, we have the integral

$$u = -2/(x^2 + c). \tag{1}$$

Two cases arise : $c = c_1^2$ or $c = -c_1^2$. Integrating (1) with $u = \frac{dy}{dx}$, we have either

$$y = -2[(1/c_1)\tan^{-1}(x/c_1) + c_2] \text{ if } c > 0$$

or

$$y = -\frac{1}{c_1} \log \frac{x - c_1}{x + c_1} + c_2 \text{ if } c < 0.$$

Here c_2 is another constant. For the special case $c = 0$, the solution is $y = 2/x + C, x \neq 0$, where C is another constant. Tenebaum and Pollard (1963), p.502

3. $$\boldsymbol{\epsilon y'' + xy'^2 - y = 0, -1 < x < 1,}$$

$y(-1,\epsilon) = -1, y(1,\epsilon) = B \leq 0.$

It is shown that the given problem has a unique solution $y = y(x,\epsilon)$ for which

$$\min(x,0) + B\exp\left[-\epsilon^{-1/2}(1-x)\right] - \epsilon^{1/2}\gamma \leq y(x,\epsilon) \leq \min(x,0), -1 \leq x \leq 1,$$

where γ is a positive constant independent of ϵ. Howes (1978a)

4. $$\boldsymbol{\epsilon y'' + xy'^2 - y = 0, -1 < x < 1,}$$

$y(-1,\epsilon) = A, y(1,\epsilon) = B, A < -1, B \leq 0.$

It is shown that the given problem has a unique solution $y(x,\epsilon)$ satisfying

$$\begin{aligned} y(x,\epsilon) &= u_L(x) + O\left(\exp\left[-\epsilon^{-1}\int_x^0 h_1(s)ds\right]\right) + O(\epsilon^{1/2}), -1 \leq x \leq 0, \\ y(x,\epsilon) &= O\left(\exp[-\epsilon^{-1/2}x]\right) + O\left(|B|\exp[-\epsilon^{-1/2}(1-x)]\right) + O(\epsilon^{1/2}), 0 \leq x \leq 1, \end{aligned}$$

where

$$u_L(x) = -\left[(-A)^{1/2} - 1 + (-x)^{1/2}\right]^2, -1 \leq x \leq 0$$

and

$$h_1(x) = 2\left[(-A^{1/2} - 1)(-x)^{1/2}\right], -1 \leq x \leq 0.$$

Howes (1978a)

5. $$\epsilon y'' + x^2 y'^2 - y^2 = 0, -1 < x < 1,$$

$y(-1, \epsilon) = A, y(1, \epsilon) = B, -B < A < 0.$

It is shown that the given problem has a solution $y(x, \epsilon)$ satisfying

$$u(x) - \epsilon^{1/2}\gamma \leq y(x, \epsilon) \leq u(x), -1 \leq x \leq 1,$$

where

$$u(x) = \begin{cases} -Ax^{-1}, & -1 \leq x \leq x_0, \\ Bx, & x_0 \leq x \leq 1, \end{cases}$$

and γ is a positive constant independent of ϵ. Howes (1978a)

6. $$\epsilon y'' + xy'^2 - xy' - y = 0, -1 < x < 1,$$

$y(-1, \epsilon) = A > 0,\ y(1, \epsilon) = B < 0.$

It is shown that the given problem has a unique solution $y = y(x, \epsilon)$ satisfying

$$B \exp\left[-(1-\delta)\epsilon^{-1}(1-x)\right] - \epsilon\gamma \leq y(x, \epsilon) \leq A \exp\left[-(1-\delta)\epsilon^{-1}(1+x)\right] + \epsilon\gamma, -1 \leq x \leq 1,$$

where $0 < \delta < 1$ and γ is a positive constant independent of ϵ. Howes (1978a)

7. $$\epsilon y'' - xy'^2 + xy' = 0, -1 < x < 1,$$

$y(-1, \epsilon) = A,\ y(1, \epsilon) = B, A > B.$

It is shown that for each $\epsilon > 0$, a solution $y = y(t, \epsilon)$ exists and satisfies

$$A - (A - B)\exp[-(2\epsilon)^{-1}x^2] \leq y(x, \epsilon) \leq A, \qquad -1 \leq x \leq 0,$$

$$B \leq y(x, \epsilon) \leq B + (A - B)\exp[-(2\epsilon)^{-1}x^2], \qquad 0 \leq x \leq 1.$$

Howes (1978a)

8. $$\epsilon y'' - xy'^2 - x^2 y' - y = 0, -1 < x < 1,$$

$y(-1, \epsilon) = 0,\ y(1, \epsilon) = B \geq 0.$

It is shown that the given problem has a unique solution $y = y(x, \epsilon)$ for which

$$0 \leq y(x, \epsilon) \leq B \exp[-k\epsilon^{-1}(1 - x)] + \epsilon\gamma, -1 \leq x \leq 1,$$

with $0 < k < 1$ and γ is a positive constant independent of ϵ. Howes (1978a)

9. $$\epsilon y'' - xy'^2 + yy' = 0, -1 < x < 1,$$

$y(-1, \epsilon) = A,\ y(1, \epsilon) = B, A > 0, B < 0, A \neq -B.$

It is shown that the given problem has a solution $y(x, \epsilon)$ for small enough $\epsilon > 0$, satisfying

$$u(x) \leq y(x, \epsilon) \leq u(x) + \epsilon^{1/2}\gamma, \qquad -1 \leq x \leq 1, \text{if} - A < B,$$

$$u(x) - \epsilon^{1/2}\gamma \leq y(x, \epsilon) \leq u(x), \qquad -1 \leq x \leq 1, \text{if} - A > B,$$

where

$$u(x) = \begin{cases} -Ax, & -1 \le x \le 0, \\ Bx, & 0 \le x \le 1, \end{cases}$$

and γ is a positive constant independent of ϵ. Howes (1978a)

10.

$$\boldsymbol{\epsilon y'' - xy'^2 - yy' = 0, -1 < x < 1,}$$

$y(-1,\epsilon) = A,\ y(1,\epsilon) = B, A = -B > 0.$

It is shown that the given problem has a solution satisfying $y(x,\epsilon) \to A, -1 \le x < 0, y(x,\epsilon) \to B, 0 < x \le 1$, as $\epsilon \to 0^+$. Howes (1978a)

11.

$$\boldsymbol{\epsilon y'' + x^3 y'^2 - y^3 = 0, -1 < x < 1,}$$

$y(-1,\epsilon) = A < 0,\ y(1,\epsilon) = B \le 0.$

It is shown that the given problem has a solution $y = y(x,\epsilon)$ for which

$$u(x) + B[1 + (2\epsilon)^{-1/2}(1-x)]^{-1} - \epsilon^{1/6}\gamma \le y(x,\epsilon) \le u(x) + \epsilon^{1/6}\gamma, -1 \le x \le 1,$$

where

$$u(x) = \begin{cases} x\left[(-x)^{1/2} - ((-A)^{1/2} - 1) + 1\right]^{-2}, & -1 \le x \le 0, \\ 0, & 0 \le x \le 1, \end{cases}$$

where γ is a positive constant independent of ϵ. Howes (1978a)

12.

$$\boldsymbol{y'' - (1+x)^{-2} y'^2 = 0, y(0) = y'(0) = 1.}$$

Put $y' = p(x)$, to get

$$p'(x) = (1+x)^{-2}p^2. \tag{1}$$

On integration, we get from (1)

$$\frac{dy}{dx} = p = \frac{1+x}{1 - C(1+x)}; \tag{2}$$

C is an arbitrary constant.

An integration of (2) is easily performed, and initial conditions satisfied. Bender and Orszag (1978), p.34

13.

$$\boldsymbol{y'' + 4y + 4\epsilon[(1/16)y'^2 \sin 2x - \sin 2x] = 0,}$$

where ϵ is a small positive parameter.

Assuming that

$$\begin{aligned} y(x) &= u(x)\cos 2x + (1/2)v(x)\sin 2x, \\ y'(x) &= -2u(x)\sin 2x + v(x)\cos 2x, \end{aligned}$$

substituting into the given DE, and using the Lagrange method of averaging, we get

$$\begin{aligned} u' &= 2\epsilon \sin 2x \left[(1/16)(-2u\sin 2x + v\cos 2x)^2 \sin 2x - \sin 2x\right], \\ v' &= -4\epsilon \cos 2x \left[(1/16)(-2u\sin 2x + v\cos 2x)^2 \sin 2x - \sin 2x\right]. \end{aligned} \tag{1}$$

Averaging the RHSs of system (1) over $0 - 2\pi$ yields the approximate system

$$u_a' = \epsilon\left[(3/16)u_a^2 + v_a^2/64 - 1\right], v_a' = (1/8)\epsilon u_a v_a. \tag{2}$$

The system (2) shows that the periodic solutions of the original DE correspond with the critical points $(u_a, v_a) = (0, \pm 8)$, (unstable), $((4/3)3^{1/2}, 0)$ (unstable), and $(-(4/3)3^{1/2}, 0)$ (stable). Verhulst (1990), p.175

14. $$\boldsymbol{y'' - e^{-x}y'^2 = 0.}$$

Putting $e^x = t, y' = \dfrac{dy}{dt}\,t, y'' = \left(t\,\dfrac{d^2y}{dt^2} + \dfrac{dy}{dt}\right)t$. Therefore,

$$t\,\frac{d^2y}{dt^2} - \left(\frac{dy}{dt}\right)^2 + \frac{dy}{dt} = 0. \tag{1}$$

Put $\dfrac{dy}{dt} = v, t\,\dfrac{dv}{dt} = v^2 - v$; $v = 1$ is a solution, i.e., $\dfrac{dy}{dt} = 1$; therefore, $y = t + c_1 = e^x + c_1$. The other case, $\dfrac{dt}{t} = \dfrac{dv}{v(v-1)}$, on integration and using $e^x = t$ gives

$$y = C_2 \ln|e^x + C_2| + C_3.$$

C_1, C_2, and C_3 are arbitrary constants. Rabenstein (1966), p.373

15. $$\boldsymbol{\epsilon y'' - |x|y'^2 + 2xy' - y = 0, -1 < x < 1,}$$

$y(-1, \epsilon) = A,\ y(b, \epsilon) = B, A, B \geq 1.$

It is shown that the given problem has a unique solution $y = y(x, \epsilon)$ for which

$$\begin{aligned}|x| \leq y(x,\epsilon) \ \leq\ & |x| + (A-1)\exp[-\epsilon^{-1/2}(1+x)] \\ & + (B-1)\exp[-\epsilon^{-1/2}(1-x)] + \epsilon^{-1/2}\gamma, -1 \leq x \leq 1,\end{aligned}$$

where γ is a positive constant independent of ϵ. Howes (1978)

16. $$\boldsymbol{y'' - (x - y)y'^2 = 0.}$$

Put $T = \ln(x - y), S(x, y) = y$, from which one obtains

$$y' = \frac{S'}{e^T + S'},\quad S' = \frac{y'(x-y)}{1 - y'},$$

leading to

$$S''(T) = S'(T) + S'^3(T) + e^T S'^2(T). \tag{1}$$

A first integral of the given DE is

$$S'^2 = \frac{e^{2(T-\phi_0)}}{1 - e^{2(T-\phi_0)}},$$

which on integration yields

$$S + \psi_0 = \pm \arcsin\,(e^{T-\phi_0}),$$

where ϕ_0 and ψ_0 are constants. Stephani (1989), p.72

17. $$\boldsymbol{y'' - y'^2/y + y'/x - (1/x)[\alpha y^2 + \beta] - \gamma y^3 - \delta/y = 0.}$$

The given DE is P_V. Putting $y = x^{-1}e^{-u}, x^2 = t$, it is transformed into

$$t(tu')' = -(1/4)\left[\alpha e^{-u} + \beta t e^{u} + \gamma e^{-2u} + \delta t^2 e^{2u}\right], \quad ' \equiv \frac{d}{dt}. \tag{1}$$

Series solution of (1) about fixed regular points is discussed. For $\alpha = \beta = 0$, some asymptotic results regarding the zeros and poles are given. Shimomura (1987)

18. $$\boldsymbol{y'' - [1/(y-2)]a(x)y'^2 = 0, 0 \le x \le 1, y(0) = A, y(1) = B,}$$

where $a(x)$ is a continuous positive function.

It is proved that the given problem has at least one solution $\phi(x)$ satisfying $B \le \phi(x) \le A$ on $[0, 1]$. Baxley (1990)

19. $$\boldsymbol{y'' - (1/2)[1/y + 1/(y-1) + 1/(y-x)]y'^2 + [1/x + 1/(x-1) + 1/(y-x)]y' - [y(y-1)(y-x)/(x^2(x-1)^2)] \times \left[\alpha + \beta x/y^2 + \gamma(x-1)/(y-1)^2 + \delta x(x-1)/(y-x)^2\right] = 0,}$$

where α, β, γ, and δ are constants.

The given DE is a sixth Painlevé equation. It has fixed singularities at $x = 0, 1, \infty$. It is known that the given DE is invariant under the change of variables

$$\xi = 1 - x, \quad \eta = 1 - y, \quad \beta = -\gamma', \quad \gamma = -\beta',$$

or

$$\xi = 1/x, \quad \eta = y/x, \quad \gamma = 1/2 - \delta', \quad \delta = 1/2 - \gamma'.$$

Therefore, it suffices to consider solutions about the fixed singular point $x = 0$.

Families of solutions of the given DE such that $y(x_0) = y_0, y'(x_0) = y_1$ are constructed following the method of Boutroux. It is also shown that either $[y(x)]^{-1}x^w$ is bounded on a curve C extending from $x = x_0$ and tending to $x = 0$ or $y(x)x^{-w'}$ is bounded on a suitably chosen curve C' tending to $x = 0$. Here $0 \le w < 1, 0 < w' \le 1$.

Convergent series solutions are constructed in the neighborhood of $x = 0$. Shimomura (1982,1982a)

20. $$\boldsymbol{y'' - (1/2)\left[1/y + 1/(y-1) + 1/(y-x)\right]y'^2 + \left[1/x + 1/(x-1) + 1/(y-x)\right]y' + \left[y(y-1)(y-x)/x^2(x-1)^2\right]\left[\alpha + \beta x/y^2 + \gamma(x-1)/(y-1)^2 + \delta x(x-1)/(y-x)^2\right] = 0,}$$

where α, β, γ, and δ are constants.

The given DE is a sixth Painlevé equation. The following result regarding its solution via Riccati equation is proved: If $v(\tau)$ is a solution of the hypergeometric equation

$$\tau(\tau - 1)v'' + [(1 + \alpha_1 + \beta_1)\tau - \gamma_1]v' + \alpha_1\beta_1 v = 0,$$

where

$$\tau = 1/(1-x), \alpha_1 = (2\alpha)^{1/2}, \beta_1 = (-2\beta)^{1/2}, \gamma_1 = \lambda,$$

and if α, β, γ, and δ are chosen according to

$$(2\alpha)^{1/2} - (-2\beta)^{1/2} - 1 \neq 0,\ 2(2\alpha)^{1/2}(-\alpha + 3\beta + \gamma - \delta)$$
$$+ 2(-2\beta)^{1/2}(3\alpha - \beta - \gamma + \delta) + 4(-\alpha\beta)^{1/2}(-\alpha + \beta + \gamma - \delta - 1)$$
$$+ 2(\alpha - \beta - \gamma) + \alpha^2 - 6\alpha\beta - 2\alpha\gamma + 2\alpha\delta + \beta^2 + 2\beta\gamma - 2\beta\delta$$
$$+ \gamma^2 + 2\gamma\delta + \delta^2 = 0,$$

where $2(-\alpha\beta)^{1/2} = (2\alpha)^{1/2}(-2\beta)^{1/2}$, then the function v is related to y by

$$y = -[x(x-1)/(2\alpha)^{1/2}]v'\ [1/(1-z)]/v[1/(1-z)], \alpha \neq 0.$$

The solution of the given DE satisfies the Riccati equation

$$y' = \left[\frac{(2\alpha)^{1/2}}{x(x-1)}\right] y^2 + \left[\frac{\lambda x + \mu}{x(x-1)}\right] y + \frac{(-2\beta)^{1/2}}{x-1}.$$

Special cases when either

$$\alpha = 0 \quad \text{or} \quad (2\alpha)^{1/2} - (-2\beta)^{1/2} - 1 = 0$$

are also considered. Lukashevich and Yablonskii (1967)

21. $$\boldsymbol{y'' - (1/2)[1/y + 1/(y-1) + 1/(y-x)]\,y'^2 + [1/x + 1/(x-1) + 1/(y-x)]}$$
$$\boldsymbol{\times\, y' - \left[y(y-1)(y-x)/x^2(x-1)^2\right]\left[\alpha + \beta x/y^2 + \gamma(x-1)/(y-1)^2\right.}$$
$$\boldsymbol{\left. + \delta x(x-1)/(y-x)^2\right] = 0,}$$

where α, β, γ, and δ are constants.

Painlevé found the general solution of the given DE for $\alpha = \beta = \gamma = \delta = 0$. For $\delta = 0$, $\beta = -\alpha\left[1 \pm (\gamma/\alpha)^{1/2}\right]^2$, there is an obvious constant solution $y = 1 \pm (\gamma/\alpha)^{1/2}$. Lukashevich (1965)

22. $$\boldsymbol{y'' + (\tanh xy)[xy' + y]y' + [1/2(\operatorname{sech} xy)]}$$
$$\boldsymbol{\times \left[x^{-3/4}\sin x^{1/2} + (1/2)x^{-5/4}\cos x^{1/2}\right] y = 0, x > 0.}$$

It is shown that all solutions of the given DE are oscillatory. Blasko, Graef, Hacik, and Spikes (1990)

23. $$\boldsymbol{y'' + f_3(x,y)y'^2 + f_2(x,y)y' + f_1(x,y) = 0,}$$

where f_i $(i = 1, 2, 3)$ are smooth functions.

The given DE has a complete first integral of the form

$$F(x,y)y'(x) + H(x,y) = cE_2(x)/[1 + cE_4(x)],$$

where $F(x,y) = g_y/z, H(x,y) = I/z^2$ with

$$\begin{aligned} g(x,y) &= \int \exp\left[\int f_3(x,y)dy\right] dy, z(x,y) = w(x) + u(x)g(x,y), \\ I(x,y) &= t(x) + \int [z[g_y(f_2+v) - g_{xy}] + g_y z_x]\, dy, \end{aligned}$$

where t, u, v, and w are arbitrary functions ; $E_2(x)$ and $E_4(x)$ are then given by

$$u(x) = \frac{E_4'(x)}{E_2(x)}, v(x) = \frac{E_2'(x)}{E_2(x)}.$$

This is possible only if

$$[f_2F - F_x - 2uFH + vF]_x = [f_1F - uH^2 + vH]_y.$$

The study includes previous methods for finding first integrals as special cases. Finch (1989)

2.18 $y'' + f(y,y') = 0, f(y,y')$ cubic in y'

1. $$y'' - \epsilon(y' - y'^3 + yy'^2) + w_0^2 y = 0,$$

where $\epsilon \ll 1$ and w_0 are constants.

A uniformly valid first approximation is

$$\begin{aligned} y &= a_0 \left[\frac{3}{4}w_0^2a_0^2 + \left\{1 - \frac{3}{4}w_0^2a_0^2 - 4\right\}\exp(-\epsilon x)\right]^{-1/2} \cos\left[w_0 x - \frac{1}{6w_0}\right. \\ &\times \left.\ln\left\{\frac{3}{4}w_0^2a_0^2\exp(\epsilon x) + 1 - \frac{3}{4}w_0^2a_0^2\right\} + \text{ constant}\right] + \cdots. \end{aligned}$$

Nayfeh (1985), p.125

2. $$y'' + y'^3 = 0.$$

Put $y' = p,\ y'' = p\dfrac{dp}{dy}$ so that

$$p\,\frac{dp}{dy} + p^3 = 0.$$

Therefore, either $p = 0$ or $-1/p + y = -c_1$, say. The solutions are then found to be $y = c$ and

$$y = \pm[2x + c_2]^{1/2} + c_1;$$

c, c_1, and c_2 are arbitrary constants. Rabenstein (1972), p.47

3. $$y'' + y'^3 = 0, y'(0) + y(0) = 3(2)^{1/2}/2 = 2.121320, y'(1) = 1/2.$$

Exact solution of the given BVP is $y = (2x+2)^{1/2}$. Roberts (1979), p.364

4.

$$\boldsymbol{y'' + (y^2 + y'^2 - 1)y' + y = 0.}$$

$y = \cos x$ is a solution.

$$y^2 + y'^2 = 1$$

is an isolated closed trajectory in the phase plane—it is the limit cycle. This is stable since all phase paths approach the circle as $x \to \infty$. The given DE can be dealt with easily if we put

$$y = r\cos\theta, y' = r\sin\theta, r^2 = x^2 + y^2;$$

then

$$\frac{dr}{dx} = -r(r^2 - 1)\sin^2\theta, \tag{1}$$

$$\frac{d\theta}{dx} = -1 - (r^2 - 1)\sin\theta \ \cos\theta. \tag{2}$$

Equation (1) clearly shows that r decreases with x outside the limit cycle $r = 1$ while it increases inside it, implying that it is stable; the paths approach it from inside as well as outside. Jordan and Smith (1977), p.19

5.

$$\boldsymbol{y'' - y'^3 = 0.}$$

Putting $Y = x, X = y$ (inverting the role of x and y), we have $\dfrac{d^2y}{dx^2} = -1$, with the solution

$$Y = -X^2/2 + C_1X + C_2$$

or

$$x = -y^2/2 + C_1y + C_2;$$

C_1 and C_2 are constants. Sarlet, Mahomed, and Leach (1987)

6.

$$\boldsymbol{y'' - (y' - 2\alpha)(y' + \alpha)^2 = 0,}$$

where α is a constant.

Putting

$$X = y + \alpha x, Y = y - 2\alpha x \tag{1}$$

in the given DE, we obtain the linear equation

$$Y'' - 3\alpha y' = 0, \ ' = \frac{d}{dX}$$

with the solution

$$Y = C_1 + C_2e^{3\alpha X}. \tag{2}$$

Equation (2) with the help of (1) defines the solution implicitly. Sarlet, Mahomed, and Leach (1987)

7.

$$\boldsymbol{y'' + \epsilon y'^3 + w_0^2 y = 0,}$$

where $\epsilon \ll 1$ and w_0 are constants.

By using the method of multiple scales, a first order uniform solution is

$$y = \frac{a_0}{[1 + (3/4)\epsilon a_0^2 w_0^2 x]^{1/2}} \cos[w_0 x + \beta] + \cdots.$$

Nayfeh (1985), p.113

8.
$$y'' + \epsilon[(y'^3/3) - y'] + y = 0,$$

where ϵ is a small parameter. For the given Rayleigh DE, the solution by the method of multiple scales is

$$y(x) = 2\left[1 - (1 - 4/a^2)e^{-Cx}\right]^{-1/2} \cos x$$

to order ϵ, as long as $x = 0(\epsilon^{-1})$. The limit cycle is given by $a = 2$. Jordan and Smith (1977), p.162

9.
$$y'' + a_1 y' + a_2 y'^3 + a_3 y^3 = 0,$$

where a_i $(i = 1, 2, 3)$ are constants.

Using the method of harmonic balance, an approximate solution is found as $y = A \sin wx$, where

$$w^2 = \left[-\frac{a_1 a_3}{a_2}\right]^{1/2}, A^2 = \frac{4}{3}\left[\frac{-a_1}{a_2 a_3}\right]^{1/2}.$$

Ku and Jonnada (1971)

10.
$$y'' + \epsilon(y^2 + y'^2 - 1)y' + y = 0,$$

where $\epsilon > 0$ is a small parameter. The equivalent linear DE obtained by the harmonic balance method after assuming the solution in the form $y = a \cos wx$ is

$$y'' + \epsilon(a^2 - 1)y' + y = 0, \tag{1}$$

which is easily solved. The linear DE (1) with constant coefficients gives the limit cycles (with $a = 1$) exactly. The nearby spiral paths are also easily obtained for $a > 1$ or $a < 1$. Jordan and Smith (1977), p.119

11.
$$y'' + \epsilon(y^2 + y'^2 - 4)y' + y = 0,$$

where $\epsilon > 0$ is small. The given DE has an exact solution $y = 2\cos x$, which is a limit cycle. It is stable. The slowly varying amplitude method gives the approximate solution of the paths close to the cycle as $y = a\cos\theta$, where

$$a(\theta) = \frac{2}{[1 - (1 - 4/a^2(0))\exp(2\epsilon\theta)]^{1/2}},$$

$\theta = \tan^{-1}(y'/y)$, and the period is $T = 2 + O(\epsilon^2)$. Jordan and Smith (1977), p.118

12.
$$y'' + \epsilon(y^2 + y'^2 - 1)y' + y^3 = 0.$$

Referring to the general equation (2.22.22), all conditions in the theorem therein are satisfied; therefore, the given DE has at least one periodic solution. Jordan and Smith (1977), p.343

13.
$$y'' - \epsilon(1 - ay^2 - by'^2)y' + y = 0,$$

where $a, b,$ and ϵ (small) are positive constants.

Put

$$y = r\cos(x+\psi),\ y' = -r\sin(x+\psi), x+\psi = \theta$$

in the DE and find that

$$\frac{dr}{d\theta} = \frac{r'}{1+\psi'} = \frac{\epsilon[r\sin^2\theta(1 - ar^2\cos^2\theta - br^2\sin^2\theta)]}{\epsilon(\cdots) + O(\epsilon^2)}. \tag{1}$$

Averaging of $O(\epsilon)$ terms over θ on the RHS of (1) produces the DE for the approximate solution r_a:

$$\frac{dr_a}{d\theta} = \frac{1}{2}\,\epsilon r_a\left[1 - ar_a^2/4\ -\ \frac{3br_a^2}{4}\right]$$

with the critical point $2/(a+3b)^{1/2}$. It is shown that a 2π-periodic solution $r(\theta)$ exists, and as the original DE is autonomous, corresponds to a time-periodic solution. Verhulst (1990), p.174

14.

$$\boldsymbol{y'' - \epsilon[1 - \mu_1(y^2 + y'^2/w_0^2)]y' + w_0^2 y = 0,}$$

where ϵ, μ_1, and w_0 are positive constants.

The given DE models a modified Van der Pol oscillator and possesses an exact sinusoidal steady-state solution: $A_0\sin(w_0 x + \phi_0)$, where $A_0 = 1/(\mu_1)^{1/2}$ and ϕ_0 is a constant phase that depends on initial conditions. This solution is shown to be phenomenon-related to a limit cycle. Several numerical illustrations are given. Kaplan (1978)

15.

$$\boldsymbol{y'' + Ay + 2By^3 + \epsilon(z_3 + z_2 y^2 + z_1 y^4 + z_4 y'^2)y' = 0,}$$

where A, B, ϵ, and z_i $(i = 1,2,3,4)$ are constants; $\epsilon \ll 1$ and $A > 0, B > 0$.

The given DE is studied by using the Jacobian elliptic functions with a generalized harmonic balance method. The transitory motion, and consequently the limit cycles and their stability, are also studied qualitatively with a generalized approximation of the Krylov-Bogoliubov slowly varying amplitude and phase type, giving the radius, frequency, and energy of limit cycles. Explicit (approximate) solutions are given for some special cases:

(a) $z_1 = z_2 = 0$.

(b) $z_1 = z_3 = 0$.

(c) $z_2 = z_3 = 0$.

(d) $z_3 = 0$.

(e) $z_2 = 0$.

(f) $z_1 = 0$.

The analytical solutions are compared with numerical results. Garcia–Margallo and Bejarno (1992)

16.

$$\boldsymbol{y'' + \epsilon(y' - 3)(y' + 1)y' + y = 0,}$$

where $\epsilon > 0$ is small. Using energy balance method, the approximate amplitude a of the limit cycle $y = a\cos x$ is 2. The limit cycle is stable. Jordan and Smith (1977), p.118

17.
$$y'' + yy'^3 = 0.$$

Put $y' = p,\ y'' = p\dfrac{dp}{dy}$, etc., and integrate twice. The solution is

$$y^3/3 + cy + a = 2x,$$

where c and a are arbitrary constants. Wilson (1971), p.40

18.
$$y'' + 2yy'^3 + y'^3 = 0,$$

$y(0) = 0, y'(0) = 1.$

The solution is

$$x = y^3/3 + y^2/2 + y,$$

obtained by putting $y' = p,\ y'' = p\dfrac{dp}{dy}$ and integrating twice, etc. Wilson (1971), p.40

19.
$$y'' + (y^2y'^2 + y^2 - 1)y' + y = 0.$$

It is proved that the given DE has at least one nonzero periodic solution. Zuo-huan (1990)

20.
$$y'' + (y'^2 + 1)(y^2 - 4)y' + y = 0.$$

It is proved by a constructive procedure that for the given DE, there does not exist a nonzero periodic solution. Zuo-Huan (1990)

21.
$$y'' + \epsilon(y^2 - 1)y'^3 + y = 0,$$

where ϵ is a small parameter. The amplitude $a(x)$ in the polar form of the solution

$$y = a\cos\theta, y' = a\sin\theta$$

is given for small ϵ, by the first order DE

$$16a'(x) = -a^3\epsilon(a^2 - 6). \tag{1}$$

An averaging method is used to obtain (1). Jordan and Smith (1977), p.119

2.19 $y'' + f(x,y,y') = 0, f(x,y,y')$ cubic in y'

1.
$$y'' - y'^3 - \sum_{j=1}^{3} A_j(y,x)y'^{(3-j)} = 0,$$

where A_j are polynomials in y with coefficients that are analytic functions of x.

Conditions on A_j are found such that the solution of the given DE has stationary transcendental and essential singularities. This is achieved by reducing the DE to a system of Briot–Bouquet equations. First, equations where solutions have algebraic moving singularities are singled out. Bogoslovskii and Ostroumov (1983)

2. $$\mathbf{2y'' + (4/x + x)y' + (2x + x^3){y'}^3 = 0.}$$

Put $P = \dfrac{dy}{dx}$; we get a Bernoulli equation with the general solution

$$\frac{dy}{dx} = P(x, A) = \frac{A}{x(x^2 \exp(x^2/2) - A)^{1/2}}. \tag{1}$$

The general solution of the given DE, therefore, is

$$y = \int \frac{A}{x(x^2 \exp(x^2/2) - A)^{1/2}} dx + B;$$

A and B are constants. Bluman and Reid (1988)

3. $$\mathbf{y'' - (1/x)y' - (1/x){y'}^3 = 0.}$$

Put $x^2 = z$, to get the autonomous equation

$$y'' - 2{y'}^3 = 0, \ ' \equiv \frac{d}{dz}.$$

The solution is obtained by putting $y' = p, \ y'' = p \dfrac{dp}{dy}$, and integrating:

$$x^2 + (y + C_2)^2 = C_1,$$

the equation for circles. C_1 and C_2 are arbitrary constants. Reddick (1949), p.191

4. $$\mathbf{y'' + b(x)y'[{y'}^2 + \epsilon] + a(x)y = 0,}$$

where

$$\hat{b} = (1/\pi) \int_0^\pi b(x)dx \neq 0,$$

ϵ is a small parameter, and $a(x)$ and $b(x)$ are π-periodic real-valued continuous functions. The relation between the qualitative behavior of small oscillations of the given perturbed Hill's equation and the qualitative behavior of solutions of the underlying Hill's equation

$$y'' + a(x)y = 0, a(x) = a(x + T)$$

is considered. Smith (1986)

5. $$\mathbf{y'' + (6x^2 + y){y'}^3 = 0,}$$

$y = y_0$ when $x = x_0$.

The solution curve for which $x > x_0$ is found in the integral form

$$y(x) = y_0 + \int_{x_0}^{x} \frac{dx}{\left[k^2 + 4(x^3 - x_0^3) + 2 \int_{x_0}^{x} y(x)dx\right]^{1/2}},$$

where k is a constant.

The integral curve is such that it is above $y = y_0$. Moreover, for arbitrary k and $b > x_0$,

$$y(b) < x_0 + \int_{x_0}^{b} \frac{dx}{2(x^3 - x_0^3)^{1/2}} < \infty.$$

It follows that $y(\infty) < \infty$. Bartashevich (1973)

6. $$y'' + xyy'^3 = 0.$$

Writing $x = x(y)$, i.e., interchanging the dependent and independent variables, we have

$$\frac{d^2x}{dy^2} - xy = 0. \tag{1}$$

The solution to the Airy equation (1) is

$$x(y) = C_1 \text{ Ai}(y) + C_2 \text{ Bi}(y),$$

where C_1 and C_2 are constants and $\text{Ai}(y)$ and $\text{Bi}(y)$ are Airy functions. Zwillinger (1989), p.264

7. $$y'' + (y'^2 + 1)\,[f(x, y)y' + g(x, y)] = 0.$$

If there exists a potential function $\psi(x, y)$ such that $\psi_x = g$ and $\psi_y = f$, then the given DE has the solution

$$y' + \tan(\psi + C) = 0.$$

The proof is by substitution. Kamke (1959), p.555

8. $$y'' - D(x, y)y'^3 - C(x, y)y'^2 - B(x, y)y' - A(x, y) = 0.$$

(a) The integral solutions of the given DE describe a set of ∞^2 curves of the cubic type. Examples of such systems are the geodesics of a surface S, natural families, isogonal trajectories, and axial systems. Certain interesting projective properties of the system of (integral) curves of the given DE are developed.

(b) See Eqn. (2.16.15). Special cases of this equation having invariance properties are tabulated.

Terracini (1955), Ibragimov (1992)

2.20 $y'' + f(y') + g(x, y) = 0$

1. $$8y'' + 9y'^4 = 0.$$

Writing the given DE as

$$y''/y'^4 = -9/8$$

and integrating twice, we find that

$$(y+C_1)^3 = (x+C_2)^2,$$

where C_1 and C_2 are arbitrary constants. Kamke (1959), p.558 ; Murphy (1960), p.387

2. $$\boldsymbol{y'' + y + \epsilon y'^5 = 0, \epsilon \ll 1.}$$

By the method of multiple scales, a first order uniform expansion is found to be

$$y = a_0 \left[1 + \frac{5}{4}\epsilon a_0^4 x\right]^{-1/4} \cos(x+\beta) + \cdots.$$

Nayfeh (1985), p.119

3. $$\boldsymbol{y'' + (\alpha - \gamma V \beta_1)y' - \gamma V^2 \left[\beta_2(y'/V)^2 - \beta_3(y'/V)^3 \right.}$$

$$\boldsymbol{\left. - \beta_4(y'/V)^4 + \beta_5(y'/V)^5 + \beta_6(y'/V)^6 - \beta_7(y'/V)^7\right] + y = 0,}$$

where $\alpha, \gamma, \beta_1, \ldots, \beta_7$ are real positive constants and $V \geq 0$ is another constant.

The given DE appears in wind engineering in which the motions of a bluff body are modeled by an oscillator. It is shown that it can possess up to three limit cycles in addition to the fixed point $(0,0)$. Novak (1969)

4. $$\boldsymbol{y'' - y'^n + y^m = 0,}$$

where n and m are constants.

It is easy to check that if m, n are positive integers, then for the given DE to have a polynomial solution

$$y = a_0 x^s + a_1 x^{s-1} + \cdots + a_s, a_0 \neq 0,$$

m, n, s must satisfy the equation $n(s-1) = ms$. Clearly, $m < n$ and if m and n are given positive integers, then for the given DE to have a polynomial solution, $s = n/(n-m)$ must be positive integer. Utz (1969)

5. $$\boldsymbol{y'' + w^2 y + \epsilon y'^n = 0,}$$

where $\epsilon \ll 1$ and w are constants, and n is a positive odd integer. Nayfeh (1985), p.121

6. $$\boldsymbol{y'' - (y'-1)^{1/2} = 0.}$$

Here the initial values $x = x_0, y = y_0, y_0' = 1$ are singular because at the point $y' = 1$, the function $(y'-1)^{1/2}$ is not holomorphic. It is easy to check that $y = x + C$ is a singular solution while

$$y = x + (1/12)(x+C_1)^3 + C_2$$

is the general solution not containing the singular solution. Here, C, C_1, and C_2 are arbitrary constants. Through any point (x_0, y_0, y_0') of any singular solution $y_0 = z_0 + C$, $y' = 1$, i.e., for each C, there passes a particular solution obtained from the general solution, because

$$x_0 + C = x_0 + (1/12)(x_0 + C_1)^3 + C_2$$

and

$$y' = 1 + (1/4)(x_0 + C_1)^2$$

for $C_1 = -x_0$ and $C_2 = C$.

The given DE may be written as the system $y' = u$,

$$v = (y' - 1)^{1/2} = (u - 1)^{1/2},$$

so that

$$y' = u, u' = v, v' = 1/2, \tag{1}$$

a system of three equations. In (1), there is no irrationality and there are no finite singular initial points since the system is linear. From the system (1), we easily find that

$$\begin{aligned} v &= x/2 + C_1, u = x^2/4 + C_1 x + C_2, \\ y &= x^3/12 + C_1 x^2/2 + C_2 x + C_3. \end{aligned} \tag{2}$$

This solution, however, must satisfy $v = (u - 1)^{1/2}$; thus,

$$x/2 + C_1 = \left[x^2/4 + C_1 x + C_2 - 1\right]^{1/2}, C_2 = C_1^2 + 1,$$

and hence

$$\begin{aligned} v &= x/2 + C_1, u = x^2/4 + C_1 x + C_1^2 + 1, \\ y &= x^3/12 + C_1 x^2/2 + x + C_1^2 x + C_3. \end{aligned}$$

We must also examine the case $u = 1, v = 0, y = x + C$, where $v' = 1/2$ is not taken into account. This case yields the singular solution of the given DE, which is not a solution of (1). Thus, if we consider the system (1) from the beginning, then we have the solution (2) and there is no singular solution.

If we consider the system (1), we get the integral $v^2 - (u - 1) = C$. If we choose the initial conditions corresponding to $C = 0$, then $v = (u - 1)^{1/2}$ and (1) leads to two equations,

$$y' = u, u' = (u - 1)^{1/2}. \tag{3}$$

Integration of this system gives the general solution

$$u = 1 + \frac{(x + C_1)^2}{4} \quad \text{and} \quad y = x + \frac{(C_1 + x)^3}{12} + C_2.$$

The solution $u = 1, y = x + C$ is the singular solution of (3), which is, however, not a solution of (1) even when taken together with the integral $v = (u - 1)^{1/2}$ because $v' \neq 1/2$.

Thus, a singular solution can appear on transition from (1) to the given DE through the integrals of the system (1). The system (3) is equivalent to the given DE because

$$y'' = u' = (u - 1)^{1/2} = (y' - 1)^{1/2}.$$

The singular solutions are on the boundary of the domain D of existence and uniqueness of solutions of the given DE, and their absence or presence characterizes the behavior of integral curves in the neighborhood of the boundary of D. Erugin (1980)

7. $$\boldsymbol{y'' - (y' - 1)^{1/2} - a = 0.}$$

$y' = 1$ is a singular initial value, but $y = x + C$ is not a solution. If we put $a + (y' - 1)^{1/2} = 0$, then $y' = 1 + a^2$, so that $y = (1 + a^2)x + C$. This forms a solution if $a < 0$ in the given DE.

Putting $u = (y' - 1)^{1/2}$, we get $y' = u^2 + 1$. This leads to

$$dy = (u^2 + 1)dx = (u^2 + 1)[2u/(a + u)]du,$$

i.e.,

$$dx = [2u/(a + u)]du.$$

The solution of this system is

$$\begin{aligned} y &= (2/3)(u + a)^3 - 3a(u + a)^2 + 2(3a^2 + 1)(u + a) - 2a(a^2 + 1)\ln(u + a) + C_2, \\ x &= 2u - 2a\ln(u + a) - C_1 \end{aligned} \tag{1}$$

in a parametric form. C_1 and C_2 are arbitrary constants. Also,

$$y = (1 + a^2)x + C \text{ with } a < 0 \tag{2}$$

is a singular solution.

The given DE is also equivalent to the following third order system with rational right-hand sides:

$$y' = v, v' = a + u, u' = (a + u)/(2u). \tag{3}$$

The system (3) has the integral

$$u^2 - (v - 1) = C; u = (v - 1)^{1/2},$$

and

$$u = (y' - 1)^{1/2}$$

for $C = 0$.

Now the nonstationary singularities of the system (1) are considered, it has no other singularities. The singular solution (2) also has no singularities for $a < 0$; for $a > 0$ there are no singular solutions.

Now turning to solutions of the system (3), we must consider the situation in which $u = -a$ for $a < 0$. Then $v = C_1$ and $y = C_1x + C_2$. Since we must have $u = (y' - 1)^{1/2}$, we conclude that

$$-a = (C_1 - 1)^{1/2}, C_1 = 1 + a^2,$$

and

$$y = (1 + a^2)x + C_2$$

is a singular solution. When $u + a \neq 0$, we find that for (3), x_0 is a singularity if $u \to 0$ when $x \to x_0$ since in this situation, the RHS of the third equation of (3) is not holomorphic at this point.

It may be checked that the general solution (1) and the singular solution

$$y = (1 + a^2)x + C, a < 0,$$

have no finite coincident values of y and y'. Erugin (1980)

8. $$y'' - a(y'^2 + 1)^{1/2} - b = 0,$$

where a and b are constants. The given DE describes motion of a suspension bridge. If we put $y'(x) = p(x)$, we have

$$\frac{dp}{dx} = a(1+p^2)^{1/2} + b,$$

so that

$$x = \int \frac{dp}{a(1+p^2)^{1/2} + b}. \tag{1}$$

Also, with $\tan u = y'(x)$, we get

$$\frac{\sin u}{\cos^2 u(a + b\cos u)} \frac{du}{dy} = 1,$$

with the solution

$$a^2 y = C + \frac{a}{\cos u} + b \operatorname{In} \left| \frac{\cos u}{a + b\cos u} \right|, \quad u = \tan^{-1} p. \tag{2}$$

Equations (1)–(2) give a parametric representation of the solution. Kamke (1959), p.556

9. $$y'' - (1 - y'^2)^{1/2} - y = 0.$$

Put $y' = p(y)$ and then $p^2 = z$; we have

$$\frac{1}{2}\frac{dz}{dy} - (1-z)^{1/2} - y = 0.$$

Now set $1 - z = A^2$, we get

$$A\frac{dA}{dy} + A + y = 0 \quad \text{or} \quad \frac{dA}{dy} = -\frac{A+y}{A}. \tag{1}$$

Equation (1) is homogeneous of first order; hence put $A = vy$ and solve. Wilson (1971), p.24

10. $$y'' - 2y'(y'-1)^{1/2} = 0, y(\pi/4) = 1, y'(\pi/4) = 2.$$

The solution is $y = \tan x$, obtained by putting $y' = p, y'' = p\dfrac{dp}{dy}$, and integrating twice, etc. Wilson (1971), p.40

11. $$y'' - (y'-1)^{1/2} - (y'-2)^{1/2} = 0.$$

The boundary of the domain of existence and uniqueness of solution is formed by the curves $y' = 1$ and $y' = 2$, but $y = x + C$ and $y = 2x + C$ are not solutions.

Setting

$$y' = p, \ y'' = p\frac{dp}{dy} = (p-1)^{1/2} + (p-2)^{1/2}.$$

Therefore,

$$dy = \frac{p\, dp}{(p-1)^{1/2} + (p-2)^{1/2}}, \quad dx = \frac{dp}{(p-1)^{1/2} + (p-2)^{1/2}}.$$

The solution in the parametric form is

$$y = \int^p \frac{p\,dp}{(p-1)^{1/2} + (p-2)^{1/2}} + y_0, \quad x = \int^p \frac{dp}{(p-1)^{1/2} + (p-2)^{1/2}} + x_0, \tag{1}$$

and so $y \to \infty$ and $x \to \infty$ when $p \to \infty$. Putting $u = (p-2)^{1/2}$, we may write (1) as

$$\begin{aligned} x &= u^2 - \frac{2}{3}u^3 + \frac{u^4}{4} + \cdots + x_0, \\ y &= 2u^2 - \frac{4}{3}u^3 + u^4 + \cdots + y_0. \end{aligned} \tag{2}$$

Inverting (2) in the power of $v = (x - x_0)^{1/2}$, we write

$$u = v + \frac{v^2}{3} + \frac{11}{72}v^3 + \alpha_4 v^4 + \cdots,$$

and hence

$$\begin{aligned} y - y_0 &= 2v^3 + \frac{v^4}{2} + \alpha v^5 + \cdots \\ &= 2(x - x_0) + \frac{(x-x_0)^2}{2} + \alpha(x - x_0)^{5/2} + \cdots. \end{aligned}$$

Thus, x_0 is a movable algebraic singularity. Setting

$$\xi = (y' - 1)^{1/2} + (y' - 2)^{1/2}$$

and squaring, etc., we find that

$$\xi^4 + 2\xi^2(3 - 2y') + 1 = 0,$$

so that $\xi \neq 0$. Again calling

$$y' = p = \frac{\xi^2}{4} + \frac{3}{2} + \frac{\xi^{-2}}{4},$$

and using the given equation so that

$$\frac{dp}{\xi} = \frac{dy'}{\xi} = \frac{1}{2\xi}(\xi - \xi^{-3})d\xi = dx,$$

we find by integration that

$$\begin{aligned} x &= (1/2)\int (1 - \xi^{-4})d\xi = (1/2)(\xi + \xi^{-3}/3) + C_1, \\ y &= \frac{1}{2}\int \left[\frac{3}{2} + \frac{\xi^{-2}}{2} + \frac{\xi^2}{4}\right](1 - \xi^{-4})\,d\xi \\ &= \frac{1}{2}\left[\frac{3}{2}\,\xi + \frac{\xi^3}{12} + \frac{\xi^{-3}}{2} + \frac{\xi^{-5}}{20}\right] + C_2, \end{aligned}$$

a parametric representation of the solution, which is evidently algebraic. C_1 and C_2 are arbitrary constants. Erugin (1980)

12. $$\mathbf{y'' - (2y' + 1)^{1/2} + (y' + 2)^{1/2} = 0.}$$

We set $y' = p$ so that

$$\frac{dy'}{dx} = p\,\frac{dp}{dy} = (2p+1)^{1/2} - (p+2)^{1/2}.$$

An integration gives the parametric representation

$$\begin{aligned} y &= \int \frac{p\,dp}{(2p+1)^{1/2} - (p+2)^{1/2}} + C_1 = \int \frac{p[(2p+1)^{1/2} + (p+2)^{1/2}]}{p-1}\,dp + C_1, \\ x &= \int \frac{dp}{(2p+1)^{1/2} - (p+2)^{1/2}} + C_2 = \int \frac{[(2p+1)^{1/2} + (p+2)^{1/2}]}{p-1}\,dp + C_2 \end{aligned} \tag{1}$$

of the general solution. C_1 and C_2 are arbitrary constants. It is clear that $y \to \infty$ and $x \to \infty$ as $p \to 1$. Expanding the integrand and in the expression for y about $p = 1$, using the expression for x, and performing the integration, we have

$$y = x + 2(3)^{1/2}(p-1) + (3^{1/2}/4)(p-1)^2 + \cdots + C. \tag{2}$$

Thus, $y \to x + C$ as $p \to 1$. All solutions approximate $y = x + C$ when $y' = p \to 1$. It follows from the given DE that the RHS is holomorphic about $y' = 1$; therefore, the solution with $y(0) = C$, $y'(0) = 1$ is $y = x + C$, and is not singular. C is a constant.

It may easily be checked from differentiating the given DE, etc., that $y''' \to \infty$ as $y' \to -1/2$ or -2; therefore, $2y' + 1 = 0$ and $y' + 2 = 0$ form the boundary of the domain of existence and uniqueness of solution of the given system. Indeed, putting $u = (2p+1)^{1/2}$ in (1) and assuming u to be small, we find by expansions, etc., that

$$\begin{aligned} y &= (1/6)^{1/2}\left[\frac{u^2}{2} + (2/3)^{1/2}\frac{u^3}{3} + \cdots\right], \\ u &= [6(x - x_0)]^{1/2} + \sum_{k=2}^{\infty} \alpha_k \left[\{6^{1/2}(x - x_0)\}^{1/2}\right]^k; \end{aligned}$$

hence the solution $y = y(x)$ is a two-valued algebraic function in the neighborhood of each point $x = x_0$, when $y' \sim -1/2$; similarly for $y' = -2$. Erugin (1980)

13. $$\mathbf{y'' - a(y'^2 + 1)^{3/2} = 0,}$$

where a is a constant.

With $y' = p$, $y'' = p\,\dfrac{dp}{dy}$, etc., the solution is easily found to be

$$(y - C_1)^2 + (x - C_2)^2 = a^{-2},$$

where C_1 and C_2 are arbitrary constants. Kamke (1959), p.557; Murphy (1960), p.386

14. $$\mathbf{y'' + \sin(y') + 1 = 0, y(0) = 0, y(1) = 1.}$$

The given problem is solved by finite difference methods. Numerical results are tabulated. Kubicek and Hlavacek (1983), p.95

15.
$$y'' - 6y - \sin[y'(x)] = h(x), -1 < x < 1,$$

where

$$h(x) = \begin{cases} 8x^3 - 6x^2 - 21x + 2 - \sin(-4x^2 + 2x + 13/16), & -1 \le x < 0, \\ 12x^3 - 25x - \sin(-6x^2 + 13/6), & 0 < x \le 1, \end{cases}$$

$y(-1) = y(1),\ y'(-1) = y'(1)$.

The solution of the given BVP is constructed iteratively, using a general theorem. Saranen and Seikkala (1989)

16.
$$y'' + \mu \sin y' + y = 0.$$

Using the phase plane analysis, it is shown that the given DE has, for all real μ, an infinite number of limit cycles; the conjecture that the limit cycles crossing the axes roughly every π units are unique and alternatively stable and unstable is demonstrated to be correct. D'heodene (1969)

17.
$$y'' + \eta^{-1} \sin(\theta, y') + \sin y - \sin \theta = 0,$$

where $\eta > 0$ and θ are parameters.

It is proved that all solutions with $\eta \le \cos\theta/2$ are stable and all solutions with

$$\eta \ge [(\pi/2)(\sin\theta + (\pi/2)\cos\theta)]^{1/2}$$

are unstable (stable solutions require $\lim_{x\to\infty} y' = 0$). Bohm (1953)

18.
$$y'' + y''^n \sin(y''^n) - y^3 + y = f(x), 0 < x < 1; n > 0,$$

$y(0) = y(1) = 0$ or $y(0) = y'(1) = 0$, where $f(x) \in$ a Hilbert space. Sufficient conditions for the solvability of the given BVPs are given in suitably chosen spaces. Gaponenko (1983)

19.
$$y'' + e^{y'} = 0.$$

Put $y' = p$, $\dfrac{dp}{dx} = -e^p$. $x = e^{-p} - c$, i.e., $-p = \ln(x + c)$. Hence integrate. The solution is

$$y = a - (x + c)\ln(x + c) + (x + c), x + c > 0.$$

Wilson (1971), p.40

20.
$$y'' + Ky' + \theta\ \mathrm{sgn}(y') + y - y_0 - V^2 F(y) = 0,$$

where K, θ are the coefficients of linear and Coulomb friction damping and $F(y) = (1.3|y| - 1.6y^2)y$. For various choices of the parameters K, θ, y_0, and V^2, it is shown that in some flow velocity intervals, several steady solutions of the given mathematical model (locally stable equilibrium positions and steady vibrations) can exist. The domains of attraction are determined. Tondl (1985)

21.
$$y'' + y - F(y') = 0,$$

where

$$F(y) = \begin{cases} -6(y-1), & |y-1| \le 0.4, \\ -[1 + 1.4 \ \exp\{-0.5|y-1| + 0.2\}] \ \mathrm{sgn}(y-1), & |y-1| \ge 0.4. \end{cases}$$

This is a Coulomb friction-type problem (see Fig. 3.27, p.98 of Jordan and Smith for a phase diagram). Jordan and Smith (1977), p.99

22.
$$y'' + (1/2)\mu y'|y'| + n^2 y = 0,$$

where μ and n are constants.

The given DE represents motion of a particle subject to a linear restoring force and a damping term proportional to the square of the velocity. Assuming that $y(0) = a > 0, y'(0) = 0$ and μa small, the position and time of a first stationary point are estimated. Williams (1951)

23.
$$y'' + (k/m)|y'|y' + (a/m)y + (b/m)y^3 = 0.$$

The given DE describes the motion of a particle in a rough horizontal straight groove under the action of a spring attached to it and to a fixed point on the groove. See Eqn. (2.20.28) for the case $b = 0$. Mickens (1981), p.29

24.
$$y'' + \epsilon(|y'| - 1)y' + y = 0.$$

An approximate solution, which is obtained by averaging methods, is

$$y = a\cos\theta, x = a\sin\theta,$$

where

$$\frac{da}{d\theta} = \left[\frac{4\epsilon}{3\pi}\right] a(a - a^*), a^* = (3/8)(\pi + 1/3)$$

or

$$a(\theta) = \frac{a^*}{[1 - (1 - a^*/a_0)] \ \exp[\epsilon(4a^*/(3\pi))(\theta - \theta_0)]}.$$

This is the path in polar coordinates, satisfying the condition $a = a_0$ when $\theta = \theta_0$. Jordan and Smith (1977), p.110

25.
$$y'' + cy'|y'| + k \sin y = 0,$$

where c and k are constants.

The given DE describes the swinging of a pendulum, immersed in a medium that exerts a force proportional to the square of its angular velocity and in a direction opposite to the velocity. With $y' = v$, we have, in the phase plane,

$$\frac{dv}{dy} = \frac{-k\sin y - cv|v|}{v}. \qquad (1)$$

Equation (1) has singularities at $y = n\pi, v = 0, n$ a positive or negative integer, which are stable spiral points if n is even and saddle points if n is odd. Integral curves for (1) can be drawn in the (y, v) plane.

Scaling (1) by writing $z = v/k^{1/2}$, we have

$$z\,\frac{dz}{dy} + cz|z| + \sin y = 0. \tag{2}$$

Writing $z^2 = \xi$, etc., in (2), we can integrate the resulting linear DE to yield

$$\begin{aligned} z^2 &= c_1 e^{-2cy} + \frac{2}{1+4c^2}\cos y - \frac{4c}{1+4c^2}\sin ny, y > 0, \\ z^2 &= c_2 e^{2cy} + \frac{2}{1+4c^2}\cos y + \frac{4c}{1+4c^2}\sin y, y < 0, \end{aligned}$$

where c_1 and c_2 are arbitrary integration constants.

The phase plane is used to determine the range of initial velocities within which motion with a specified number of full revolutions will occur. Stoker (1950), p.59

26. $$\boldsymbol{y'' + \epsilon\mu_2 y'|y'| + \epsilon \operatorname{sgn} y' + y = 0,}$$

where $\epsilon \ll 1$ and μ_2 are parameters.

By the method of averaging, a first order approximate solution is found to be

$$y = [3/(2\mu_2)]^{1/2}\tan[C - (\epsilon/\pi)(8\mu_2/3)^{1/2}x]\cos(x+\beta) + \cdots,$$

where c and β are arbitrary constants. Nayfeh (1985), p.152

27. $$\boldsymbol{y'' + y + \epsilon\mu_2 y'|y'| + 2\epsilon\mu_1 y' = 0,}$$

where $\epsilon \ll 1, \mu_1$, and μ_2 are parameters.

By the method of averaging, a first order approximate solution is found to be

$$y = a_0\left[e^{\epsilon\mu_1 x} + \frac{4\mu_2 a_0}{3\pi\mu_1}\,(e^{\epsilon\mu_1 x} - 1)\right]^{-1}\cos(x+\beta) + \cdots,$$

where a_0 and β are arbitrary constants. Nayfeh (1985), p.153

28. $$\boldsymbol{y'' + ay'|y'| + by' + cy = 0,}$$

where $a > 0, b \geq 0, c > 0$ are constants.

The given DE describes small vibrations with quadratic damping. With $y(x) = f(z)/2a, z = c^{1/2}x$, it takes the form

$$f'' + (1/2)f'|f'| + Bf' + f = 0, B = b/c^{1/2}. \tag{1}$$

One may build up the solutions of the DE (1) from those of the two DEs

$$f'' \pm (1/2)f'^2 + Bf' + f = 0.$$

If f is the solution of DE with the upper sign, then $\bar{f} = -f$ is the solution of the DE with lower sign. With $B = 0$ in (1), the DE

$$f'' + (1/2)f'|f'| + f = 0$$

may be solved with the help of the solution $f = S(z, a)$ of the IVP

$$f'' - (1/2)f'^2 + f = 0, f(0) = a, f'(0) = 0.$$

If this solution exists for $z > z_0$ and has amplitudes $a_1, a_2, \ldots$ at the points $z_1 < z_2 < \cdots$, where, say, $f(z_1) = -a_1 < 0$, $f(z_2) = a_2 > 0$, etc., then we have

$$f = \begin{cases} S(z - z_1, -a_1) & \text{for } z_0 < z \leq z_1, \\ -S(z - z_1, a_1) = -S(z - z_2, -a_2) & \text{for } z_1 \leq z \leq z_2, \\ S(z - z_2, a_2) = S(z - z_3, -a_3) & \text{for } z_2 \leq z \leq z_3, \\ \ldots\ldots\ldots\ldots\ldots & \end{cases}$$

Kamke (1959), p.552

29. $$y'' + y'|y'| + qy' + y - p^2y^3 = 0,$$

where $p > 0$ and $q \geq 0$ are constants.

It is shown that the origin in the phase plane is stable in the strong sense that solutions near the origin tend to it as $x \to \infty$. The only other singular points $(\pm 1/p, 0)$ are saddle points. It is also shown that for $q = 0, p = 1$, the solutions of the given DE for certain initial values, not close to $(0, 0)$, do not exist over the semi-infinite range (x_0, ∞). Cecconi (1950)

30. $$y'' + y'|y'| - qy' + y - p^2y^3 = 0,$$

where p and q are positive constants.

Using the Poincaré-Bendixson plane it is proved that there exists a $p_0 > 0$ such that if $p < p_0$, the given equation has a periodic solution (different from the trivial solution $y = 0$). Cecconi (1950a)

31. $$y'' - a|y'|^\rho - cy - 1 = 0,$$

where $c > 0, \rho$, and a are constants.

$y = -1/c$ is the trivial solution. Phase plane analysis of the given DE is carried out. Using geometric methods it is shown that for the Neumann problem with $y'(0) = y'(X) = 0$, $y = -1/c$ is the only solution; X is a constant. Chicone (1988)

32. $$\epsilon y'' + f(y') = 0, 0 < x < 1, y(0) = \alpha, y(1) = \beta,$$

where ϵ, α, and β are constants.

The set of values of ϵ for which the given problem has a solution is determined. The shooting method is used. Deshpande and Kasture (1981)

33. $$y'' + F(y') + y = 0.$$

Conditions for the existence of a unique periodic solution are discussed. See Eqn. (2.12.32). Jordan and Smith (1977), p.344

34. $$\epsilon y'' + F(y') + y = 0.$$

Assuming that $f = -F$ is Lipschitzian, zero at zero, increasing on some interval containing zero and decreasing outside and that $\lim_{y\to\infty} f(y) = a$ and $\lim_{y\to-\infty} f(y) = b$, phase portrait properties are studied. Using nonstandard methods, with ϵ small, the following result is proved. If $b \ll a$, all trajectories except the stationary one spiral to infinity; if $b \gg a$, there exists a unique limit cycle, the shape of which is given; finally, if $b \simeq a$, the size of the cycle tends to infinity and eventually disappears. Urlacher (1984)

35. $$y'' + \alpha y' + f(y') + \beta y = 0,$$

where $\beta > 0$ and α are parameters, where

$$f(y') = \sum_{n=1}^{N} a_{2n} y'^{2n}$$

is a nonlinear function of y'.

Two theorems are proved:

(a) When $\alpha = 0$, the unique fixed point at $y = 0, y' = 0$ is a center and the phase plane is densely filled with closed orbits.

(b) When $\alpha > 0$ (resp., < 0) the unique fixed point at $y = 0$, $y' = 0$ is a sink (resp., source) and all orbits spiral around $(0,0)$ and toward it as $x \to \infty$ (resp., $x \to -\infty$).

Holmes (1977)

36. $$y'' + g(y') + cy = 0,$$

where $c > 0$ and $g(z) \le 0$ are real functions, $g(0) = 0$.

It is proved that a solution of the given DE either oscillates or approaches $+\infty$ or $-\infty$ when $x \to \infty$. Utz (1956)

37. $$ay'' + R(y') + cy = 0,$$

where a and c are constants.

The given DE describes damped motion with a general resistance function $R(y')$.

If $a > 0, b \ge 0, c > 0$ are constants and $R(v) = [b + f(v)]v, -\infty < v < \infty$, is a continuous even function, $f(0) = 0$, and $f'(v)$ for $v \ge 0$ exists and is continuous, and for $v > 0$ is greater than 0, then the following is true for the solution of the given DE: Every solution exists in an interval $X < x < \infty$, where, $av^2 + cy^2$ decreases monotonically with increasing x; further, $av^2 + cy^2 \to 0$ for $x \to \infty$ and $\to \infty$ for $x \to X$. If $4ac - b^2 \le 0$, then every solution has at most one zero, that is, there is no oscillation.

For $4ac - b^2 > 0$, every solution has an infinite number of zeros, the amplitude of every solution decreases monotonically to zero. Besides, if $\lim_{v\to\infty} f(v) > a + c - b$, then every solution has a smallest zero and a first maximum, the distance between two zeros and two extreme points is greater than $(a/c)^{1/2}$; there exists a critical solution $y = \eta(x)$ whose amplitudes $\alpha_1, \alpha_2, \ldots$ have the following relation to the amplitudes $a_1, a_2, \ldots$ of every (other) solution:

$$a_1 \geq \alpha_1 > a_2 \geq \alpha_2 > \cdots.$$

Kamke (1959), p.559

38. $\boldsymbol{y'' + g(y') + f(y) = 0.}$

Let $f(z)$ and $g(z)$ be continuous. If $y(x)$ is a solution of the given DE valid for large $x, y(x)$ and $y'(x)$ are bounded, and $zg(z) > 0$ for $z \neq 0$, then if $y'(x)$ is monotonic for all large x, it is shown that $y'(x) \to 0$ as $x \to \infty$. Utz (1969)

39. $\boldsymbol{y'' - F(y') - G(y) = 0,}$

either $y(0) = 0, y(X) = 0$ or $y'(0) = 0, y'(X) = 0; X$ is a constant.

With respect to these boundary value problems, two situations are considered where general existence theorems do not apply and special techniques must be used. The first situation occurs where there is a trivial solution to the BVP so that the general existence theorem guarantees the existence of the trivial solution instead of a solution of interest that satisfies a side condition. The second situation occurs when growth in the derivative of the dependent variable is not bounded by a quadratic function, in which case there may be no solution. Geometric methods are used. Chicone (1988)

40. $\boldsymbol{y'' + g(y') + f(y) = 0, 0 \leq x \leq X,}$

$y'(0) = 0,\ y'(X) = \phi(y(X))$.

The functions $f(y), \phi(y)$, and $g(y)$ are in C^1. The boundary data curve $y' = \phi(y)$ on $0 < y \leq c$ lies in the open fourth quadrant of the (y, y') phase plane. A positive eigenfunction $y(x) > 0$ on $0 \leq x \leq X$ is sought satisfying the two endpoint conditions. The general form applies to chemical processes with a suitable interpretation of the eigenfunctions $y(x, A)$, and their dependedence on A, X, and other parameters, in terms of the physical models.

The existence of the solution is proved in the phase plane, with some smoothness and convexity conditions on f, g, and ϕ. Specifically, it is proved that for each positive $X > 0$, there exists a unique amplitude A on $0 < A < c$ such that the solution $y(x, A)$ initiating at $y(0, A) = A, y'(0, A) = 0$ is the required positive eigenfunction on $0 \leq x \leq X$. Markus and Amundsun (1968)

41. $\boldsymbol{y'' + F(y') + g(y) = 0,}$

where $F(y)$ and $g(y)$ satisfy conditions as follows.

Assume that:

(a) $F(y) = -F(-y); F(y) \in C^1(-\infty, \infty); |F(y)| \leq A$ and $|F'(y)| \leq B$ for all y with $yF(y) > 0$ for y sufficiently small.

(b) $F(y)$ is oscillatory: With $\{y_n\}_{n=1}^{\infty}$ the sequence of positive zeros of $F(y)$, one requires

$$\int_{Y_n}^{Y_{n+1}} |F(y)| dy$$

to be a nondecreasing function of n.

(c) $g(y) = -g(-y); g(y) \in C^1(-\infty, \infty)$, and $2g'(y) \geq r > B$ for all y. Then the given DE has an infinite number of limit cycles.

Sato (1977)

42. $$y'' - F(y') + g(y) = 0,$$

where $F(y')$ and $g(y)$ are known functions with continuous first derivatives. The functions $g(y)$ and $f(y') = dF(y')/dy'$ are required to satisfy

$$\begin{aligned} &\nu \leq \frac{dg}{dy} \leq \mu && \text{for } -\infty < y < \infty, \\ &f(y) \neq 0 && \text{for } -b < y < b, \ \textstyle\int_{-\infty}^{\infty} f_+(y)dy < \infty, \end{aligned} \tag{1}$$

where $f_+(y) = \max[0, f(y)]$, and μ, ν are positive constants.

If the conditions (1) hold and the given DE has a periodic solution $y(x)$ with least period X, then it is proved that

$$(1/2)\mu^{-1/2}\nu X \leq (2 + \nu^{-1}\mu)(J_1 + J_2) + (1 + \nu^{-1}\mu)^2[1 + \max(J_1, J_2)], \tag{2}$$

where

$$J_1 = (b/\sqrt{\nu})^{-1} \int_0^{\infty} f_+(y)dy, J_2 = (b/\sqrt{\nu})^{-1} \int_{-\infty}^{0} f_+(y)dy.$$

Further, if $dg/dy \leq \mu$ for all y, then the least period X of any nonconstant periodic solution of the given DE satisfies $X \geq 2\pi/(\mu)^{1/2}$.

If in addition to conditions (1), we require that

$$g(-y) = -g(y), F(-y') = -F(y')$$

for all y, y', then (2) can be replaced by the sharper inequality

$$(1/2)\mu^{-1/2}\nu X \leq (2 + \nu^{-1}\mu)(J_1 + J_2) + 2(1 + \nu^{-1}\mu).$$

Smith (1971)

43. $$y'' + \phi(y') + \psi(y) = f(x), y(0) = y'(X) = 0.$$

The existence of solutions is examined when $\phi(y)$ is an arbitrary continuous function. Gaponenko (1983)

44. $$y'' + f(y') + b(y)y' + r(y) = 0.$$

A graphical method of Lienard is used for a topological discussion of the behavior of the solution curves near singularities. Cahen (1953)

2.21 $y'' + h(y)f(y') + g(x, y) = 0$

1. $$y'' + (y'^2 + 1)(y^2 - 4)y' + y = 0.$$

It is proved that there does not exist a nonzero periodic solution of the given DE. Zuo-huan (1990)

2. $$y'' + ay(y'^2 + 1)^2 = 0,$$

where a is a constant.

The given DE is autonomous. Put $y'(x) = p(y)$, $y''(x) = pp'$, etc., and integrate twice:

$$x = \int \left[\frac{ay^2 + C_1}{1 - ay^2 - C_1}\right]^{1/2} dy + C_2,$$

an elliptic integral. C_1 and C_2 are arbitrary constants. Kamke (1959), p.555; Murphy (1960), p.385; Ames (1968), p.45

3. $$y'' + f(y)y'^{2n-1} + g(y) = 0,$$

where n is an integer.

It is shown that if $yg(y) > 0$, $f(y) \geq 0$ for all real y and

$$\int_0^y g(y)dy \to \infty \text{ as } |y| \to \infty,$$

then any solution of the given DE defined for all large x is bounded. Utz (1971)

4. $$y'' + f(y)y'^{2n} + g(y) = 0,$$

where n is an integer.

It is proved that if $g(y) > 0$ for $y \neq 0, 0 < b \leq f(y)$ for some real $b, |g(y)| > \epsilon > 0$ for y sufficiently large, and if $\lim_{y \to 0} g(y)/y = a \neq 0$, the given DE has periodic solutions. In particular, any solution of the given DE, with initial condition $y(0) = -m^2, y'(0) = 0$ is periodic. Utz (1971)

5. $$\epsilon y'' + 4yy'^{3/2} = 0, y(0) = -1, y(1) = b,$$

with $b \simeq 1$ and $\epsilon = 0.05$.

Numerical solution is obtained to show significant changes in the position x_0 of a transition layer as $b = 1$ is changed by magnitude of order 10^{-3}. The location point x_0 of the transition point varies from $x_0 = 0.23$ for $b = 1.001$ to $x_0 = 0.89$ for $b = 0.9949$. The results are shown graphically. Bohe (1990)

6. $$y'' - ay(y'^2 + 1)^{3/2} = 0,$$

where a is a constant.

With $p(y) = y'(x)$ we get

$$\frac{pp'}{(p^2 + 1)^{3/2}} = ay. \tag{1}$$

Integration of (1) leads to

$$2(p^2 + 1)^{-1/2} + ay^2 = C,$$

that is, to the DE

$$y' = \pm\frac{1}{ay^2 - C}\left[4 - (ay^2 - C)^2\right]^{1/2}.$$

Now the variables separate. Here C is an arbitrary constant. Kamke (1959), p.557; Murphy (1960), p.386

7. $$m_0 y'' + ky[1-(y'/c)^2]^{3/2} = 0,$$

where m_0 and k are constants.

This is the relativistic equation for an oscillator in the alternative form

$$\frac{d}{dx}\left[\frac{m_0 y'}{\{1-(y'/c)^2\}^{1/2}}\right] + ky = 0,$$

where m_0, c, and k are positive constants. It is easy to check that the phase paths are given by the integral

$$\frac{m_0 C^2}{\{1-(y'/c)^2\}^{1/2}} + \frac{1}{2}ky^2 = \text{ constant.}$$

If $y' = a$ when $y = a$, the period T of the oscillation is given by

$$T = \frac{4}{c\epsilon^{1/2}} \int_0^a \frac{[1+\epsilon(a^2-x^2)]dx}{(a^2-x^2)^{1/2}[2+\epsilon(a^2-x^2)]^{1/2}}, \quad \epsilon = \frac{k}{2m_0c^2}.$$

The constant ϵ is small; therefore, by expanding the integrand in powers of ϵ, etc., we have

$$T \simeq 2\pi\left[(m_0/k)\{1+(3/8)\epsilon a^2\}\right]^{1/2}.$$

Jordan and Smith (1977), p.33

8. $$y'' - [2q^2(q+1)/(q-1)^2]yy'^{(1-1/q)} = 0, y(1) = 0, y'(0) = 1,$$

where q is an odd positive integer.

The given BVP arises in the self-similar analysis of some nonlinear impact problems. Multiplying the given DE by y' and integrating, we get

$$y' = \pm\left[\frac{q(q+1)^2}{(q-1)^2}\right]^{q/(q+1)} (y^2+k^2)^{q/(q+1)},$$

where k is the constant of integration. The solution can now be expressed in terms of beta functions:

$$B\left\{\frac{q-1}{2(q+1)}, 1/2\right\} - B_Y\left\{\frac{q-1}{2(q+1)}, 1/2\right\} = \pm 2k^{(q-1)/(q+1)}\left\{\frac{q(q+1)^2}{(q-1)^2}\right\}^{q/(q+1)} (x-1) \quad (1)$$

where $Y = k^2/(k^2+y^2)$ and the condition $y(1) = 0$ has been used. The constant k is determined by the condition $y'(0) = 1$; we have

$$k = \left[\frac{y(0)}{q}\right]^{1/2} \frac{q-1}{q+1},$$

where $Y(0)$ is the root of the equation

$$B\left\{\frac{q-1}{2(q+1)}, 1/2\right\} - B_{Y(0)}\left\{\frac{q-1}{2(q+1)}, 1/2\right\} = 2[Y(0)]^{(q-1)/2(q+1)}\left\{\frac{q(q+1)^2}{(q-1)^2}\right\}^{1/2}.$$

For negative sign in (1), if we use $y(0) = (q-1)/(2q)$, we have

$$k = \left[\frac{Y(0}{1-Y(0)}\right]^{1/2} \frac{q-1}{2q},$$

where

$$B\left\{\frac{q-1}{2(q+1)}, 1/2\right\} - B_{Y(0)}\left\{\frac{q-1}{2(q+1)}, 1/2\right\}$$
$$= 2\left\{\frac{Y(0)}{1-Y(0)}\right\}^{(q-1)/\{2(q+1)\}}\left[\frac{q-1}{2q}\right]^{(q-1)/(q+1)}\left\{\frac{q/(q+1)^2}{(q-1)^2}\right\}^{q/(q+1)}.$$

Numerical results are depicted graphically. Taulbee, Cozzarelli, and Dym (1971)

9. $$\boldsymbol{y'' - [n^2/(n-1)](y')^{(n-1)/n} y = 0,}$$

$y(\infty) = y'(\infty) = 0$; n is an odd positive integer $\neq 1$.

The given DE arises in the self-similar analysis of some nonlinear impact problems. Writing the given DE as

$$(y')^{1/n} y'' = \left[\frac{n^2}{n-1}\right] yy',$$

integrating twice, and using given initial conditions, we get

$$y = (-1)^{-(n+1)/(n-1)}\left[\frac{(n/2)^n(n-1)}{n+1}\right]^{-1/(n-1)}(x+C)^{-(n+1)/(n-1)},$$

where C is an arbitrary constant to be determined from the condition at $x = 0$, say. Taulbee, Cozzarelli, and Dym (1971)

10. $$\boldsymbol{y'' + y \sec y' = 0.}$$

A first integral is

$$y' \sin y' + \cos y' + y^2/2 = c,$$

where c is an arbitrary constant. A phase plane diagram may be drawn. Wilson (1971), p.40

11. $$\boldsymbol{y'' + y^3 e^{-y'} = 0.}$$

Put $y' = p$, $y'' = p(dp/dy)$, etc., and integrate. A first integral is

$$y'e^{y'} - e^{y'} + y^4/4 = c,$$

where c is an arbitrary constant. The solution may be studied in the phase plane. Wilson (1971), p.40

12. $$\boldsymbol{y'' + k^2 f(y)\,\mathrm{sgn}(y') + w_0^2 y = 0, f = f_0 + |f_1|,}$$

where

$$f_0 = \sum_{\hat{k}=0}^{n} a_{2\hat{k}} y^{2\hat{k}},$$
$$f_1 = \sum_{\hat{k}=0}^{n} a_{2\hat{k}+1} y^{2\hat{k}+1},$$

and w_0 and $\hat{k}$ are constants.

The given system represents a complicated motion under friction (dry friction when f is constant). In the four quadrants f assumes the values of four distinct polynomials. In each quadrant the phase trajectories belong to one of the two families

$$y'^2 + w_0^2 y^2 \pm 2k^2 \int f(y)dy = C,$$

where C is an arbitrary constant. Special attention is given to the case $f_0 = 0, f_1(y) = y$. Voronov (1951)

13. $$y'' + |y'|y + y = 0.$$

A phase diagram shows that except for the equilibrium point $(0, 0)$, there is a loss of energy along every phase path no matter where it goes in the phase plane. From any initial state, the corresponding phase path eventually enters the origin and the motion ceases. Jordan and Smith (1977), p.18

14. $$y'' + \epsilon(|y| - 1)|y'|y' + y + \epsilon y^3 = 0,$$

where ϵ is a small parameter. The method of equivalent linearization and use of the solution $y = a\cos wx$ give the approximate amplitude and frequency of the limit cycle as $8/3$ and $(19/3)^{1/2}$, respectively. Jordan and Smith (1977), p.119

15. $$y'' + |y|^m \operatorname{sgn}(y')|y'|^n + \beta \operatorname{sgn}(y)|y|^p = 0,$$

where m, n, p, and β are constants.

The given DE is invariant under $x = a^A x,\ y = a^B y$ provided that

$$p = p_c \equiv (2m + n)/(2 - n).$$

(a) $p < p_c, n = 1$. We introduce the variables $y = \xi x^k, dx = x^\lambda d\theta$ or $\theta = \int_1^x dx/x^\lambda$ and let $\lambda = (p-1)/(2m)$ with $\lambda < 1$, and $km = -1$. The given DE after some algebra assumes the form

$$\frac{d^2\xi}{d\theta^2} + [1/f(\theta)] \left[|\xi|^m - \frac{p+3}{2m}\right] \frac{d\xi}{d\theta} + \frac{1}{f^2(\theta)} [(1/m)(1/m + 1)\xi - (1/m)\xi|\xi|^m] + \beta \operatorname{sgn}(\xi)|\xi|^p = 0, \quad (1)$$

where $f(\theta) = x^{1-\lambda} = 1 + (1-\lambda)\theta$. It is shown numerically that a limit cycle for (1) exists for $\theta \to \infty$. The results are shown graphically for $p = 5, \beta = 1/15$, and $m = 6, 5, 4, 2$. For $m = 2$, since $p = 5 = p_c$ and $\beta > \beta_c = 1/12$, the solution goes to one of the critical points.

(b) $p > p_c$. Introducing the variables $y = \xi x^k, d\theta = dx/x, x = \exp(\theta)$ and assuming that $mk = -1$ and $n = 1$, one arrives at the equation

$$\frac{d^2\xi}{d\theta^2} + [|\xi|^m + (2k-1)]\frac{d\xi}{d\theta} + k(k-1)\xi + k\xi|\xi|^m + \beta\operatorname{sgn}(\xi)|\xi|^p \exp(-\epsilon/m)\theta = 0, \quad (2)$$

where $\epsilon = p - p_c$. Behavior of the solution of (2) for $m = 2, \beta = 1/5$, and $p = 5, 5.2, 5.5$, and 5.7 is found numerically and depicted. The limit cycle is found for $p = 5$.

Bouquet, Feix, and Leach (1991)

16. $$y'' + \gamma|y|^m \operatorname{sgn}(y')|y'|^n + \beta \operatorname{sgn}(y)|y|^p = 0.$$

The given DE can be rescaled according to

$$\hat{x} = x\gamma^{-(p+1)/D}\beta^{(m+n-1)/D}, \hat{y} = y\gamma^{2/D}\beta^{(n-2)/D}, D = 2m + n + np - 2p$$

to the form

$$\frac{d^2\hat{y}}{d\hat{x}^2} + |\hat{y}|^m(\operatorname{sgn}\hat{y}')|\hat{y}'|^n + (\operatorname{sgn}\hat{y})|\hat{y}|^p = 0, \quad ' \equiv \frac{d}{dx}. \tag{1}$$

Therefore, the form (1) is possible only if $p \neq (2m+n)/(2-n)$. Now see Eqn. (2.21.15). Bouquet, Feix, and Leach (1991)

17. $$y'' + \beta y^3 + y'|y'|^{m-1}|y|^{3-2m} = 0, 0 \leq m \leq 3/2,$$

where β and m are constants.

Introducing the variables $\xi = xy, \eta = x^2y', X = \eta\xi^{-2}$,

$$\frac{dy}{y} = -\frac{XdX}{\beta + 2X^2 + |X|^m \operatorname{sgn} y \operatorname{sgn} X},$$

and rescaled space variables ξ, $w = \eta + \xi, \theta = \ln x$, we finally arrive at the system

$$\frac{d\xi}{d\theta} = w, \quad \frac{dw}{d\theta} = 3w - 2\xi - \xi^3 - (w-\xi)|w-\xi|^{m-1}|\xi|^{3-2m}. \tag{1}$$

The system (1) is studied in the phase space for different values of β and $m = 3/2$. Point attractors and limit cycles are found and depicted. Bouquet, Feix, and Leach (1991)

18. $$y'' = K^2y(1+y'^2)^{3/2} \text{ if } y'' > 0,$$

$$y'' = -K^2y(1+y'^2)^{3/2} \text{ if } y'' < 0;$$

$y(x) \to 0$ as $|x| \to \infty$.

The one-kink solution is given parametrically by

$$y(s) = (2/K)\operatorname{sech}[-K(s-s_0)],$$

$$x(s) = s + (2/K)[\tanh\{-K(s-s_0)\} - 1],$$

where s_0 is an arbitrary constant. Shimizu, Sawada, and Wadati (1983)

19. $$y'' + e^{-|y'|}y + 5y\sin 8y = 10x\sin 2\pi x, 0 < x < 1,$$

$y(0) = y(1), y'(0) = y'(1)$.

Periodic solutions using a general theorem involving iterative procedure are constructed. Saranen and Seikkala (1989)

20. $$y'' + f(y)M(y') + g(y) = 0.$$

It is shown that if $f(y), g(y)$ are continuous, $yg(y) > 0$ for $y \neq 0, f(y) \geq 0$ for all real $y, \int_0^y g(u)du \to \infty$ as $|y| \to \infty$, and $zM(z) \geq 0$ for all real z, then any solution of the given DE valid for all large x is bounded. Utz (1971)

21. $$y'' + h(y)g(y') + h(y)f(y) = 0,$$

where $g(y')$ is an even function, $g(y') > 0$ for $y' > 0, h(y)$ is an odd function for $y > 0$, and $f(y)$ is an even integrable function, $f(y) > 0$ for $y = 0$, and

$$\int_0^y h(u)f(u)du \to \infty \text{ as } y \to \infty.$$

Assuming that the solutions of the given DE are unique, it is shown that if $y = y(x)$ is a solution valid for all x, then $y(x)$ is oscillatory (if $y \not\equiv 0$) and periodic. Utz (1967)

22. $$y'' + f(y)\phi(y') + \psi(y')\eta(y) = 0.$$

The given DE is a generalized Lienard equation. By a series of transformations somewhat like those for Lienard equation, the given DE is transformed to the system

$$y' = h(z) - e(z)F(y), z' = -g(y),$$

where the functions h, e, F, and g are defined in terms of f, ϕ, ψ, and η. Conditions on the function are found that imply the existence of a limit cycle. Huang (1984)

23. $$y'' + f(y)h(y')y' + g(y)k(y') = 0,$$

where f and g are continuous on R and h and k are positive continuous on R.

(a) We define the functions

$$\begin{aligned} F(y) &= \int_0^y f(\xi)d\xi, \\ G(y) &= \int_0^y g(\xi)d\xi, \\ K(y) &= \int_0^y \frac{\eta}{k(\eta)}d\eta. \end{aligned}$$

The following results are proved:

(i) Suppose that $f(y) \geq 0$ for all $y \in R, k(y) = 1$ for all $y \in R$ and $yg(y) > 0$ for all $y \neq 0$. Then all solutions of the given DE are bounded if and only if

$$|F(y)| + G(y) \to \infty \text{ as} |y| \to \infty. \tag{1}$$

(ii) Suppose that $f(y) \geq 0$ for all $y \in R, K(y) \to \infty$ as $|y| \to \infty$ and $yg(y) > 0$ for all $y \neq 0$. If (1) is satisfied, then all solutions of the given DE are bounded.

(b) Define the functions

$$\begin{aligned} F(y) &= \int_0^y f(\xi)d\xi, \\ G(y) &= \int_0^y g(\xi)d\xi, \\ K(y) &= \int_0^y \frac{\eta}{k(\eta)}d\eta. \end{aligned}$$

Suppose that:

(i) $f(y) \geq 0$ for all $y \in R$.

(ii) There exist constants $P > 0$ and $Q > 0$ such that $G(y) \geq -P$ for all $y \in R$,

$$G(y) < \lim_{y \to \infty} \sup G(y) \qquad \text{for all } y \geq Q,$$

$$G(y) < \lim_{y \to -\infty} \sup G(y) \qquad \text{for all } y \leq -Q.$$

(iii) $K(y) \to \infty$ as $|y| \to \infty$.

Then all solutions of the given DE are bounded if and only if there exist sequences $\{y_n\}$ tending to infinity and $\{\tilde{y}_n\}$ tending to $-\infty$ such that $F(\tilde{y}_n) \to \infty$ or $G(y_n) \to \infty$ as $n \to \infty$, and $F(\tilde{y}_n) \to -\infty$ or $G(\tilde{y}_n) \to \infty$ as $n \to \infty$. Burton (1970), Sugie (1987)

2.22 $y'' + f(y, y') = 0$

1. $$y'' + (y^2 + y'^2 - 1)y' + y = 0$$

The given DE has a periodic solution $y = \sin x$. Wilson (1971), p.322

2. $$y'' + \sum_{\alpha} A_{1-\alpha,\alpha} y^{1-\alpha} y'^{\alpha} = 0,$$

where $A_{1-\alpha,\alpha}$ are constants.

The given DE has an integrating factor of the form

$$1/(y'^2 + y \sum_{\alpha} A_{1-\alpha,\alpha} y^{1-\alpha} y'^{\alpha}).$$

Sabata (1980)

3. $$y'' - a(y'^2 + 1)^{1/2} = 0,$$

where a is a constant.

Integrating, we have $y' = \sinh(ax + C_1)$, where C_1 is a arbitrary constant. The solution, therefore, is

$$ay = \cosh(ax + C_1) + C_2,$$

where C_2 is another arbitrary constant. Kamke (1959), p.556; Murphy (1960), p.386; Ames (1968), p.45

4. $$y'' - a(y'^2 + by^2)^{1/2} = 0,$$

where a and b are constants.

Put $p(y) = y'(x)$ so that we have

$$p' = a(p^2 + by^2)^{1/2}/p. \tag{1}$$

Equation (1) may be solved easily, since the RHS is homogeneous in p and y. Kamke (1959), p.557; Murphy (1960), p.386

5. $$\boldsymbol{y'' + y^3 y' - y y'(4y' + y^4)^{1/2} = 0.}$$

There is an obvious singular boundary where

$$4y' + y^4 = 0. \tag{1}$$

In the region bounded by this curve in (y', y) plane, the given DE has a solution

$$y = A\tan(A^3 x + B), \tag{2}$$

where A and B are arbitrary constants. The function

$$y = [(3/4)(x - C)]^{-1/3} \tag{3}$$

satisfies the boundary (1) and

$$y'' + y^3 y' = 0;$$

therefore, it is also a solution of the given DE.

It follows that if y_0 and y_0' are real and satisfy $4y' + y^4 = 0$, then the solution (3) satisfies (1), having one nonstationary singularity C, a three-valued algebraic singularity. If y_0 and y_0' are complex and satisfy $4y' + y^4 = 0$, there can be two solutions, one of type (3) and another of type (2), having an infinite number of nonstationary poles. This is so because the RHS of the given DE

$$y'' = -y^3 y' + y y'(4y' + y^4)^{1/2}$$

is many-valued and thus not holomorphic. Erugin (1980)

6. $$\boldsymbol{y'' + \mu \sin y' + y = 0,}$$

where μ is a constant.

(a) It is proved that the given DE has exactly n limit cycles on the strip $|y'| \leq (n+1)\pi, n = 1, 2, 3, \ldots$ of the (y, y') phase plane.

(b) Put $y = r\cos\theta, y' = r\sin\theta$, so that the given DE is replaced by the system

$$\begin{aligned} r'\cos\theta \quad &- \quad r\sin\theta\,\theta' = r\sin\theta \\ r'\sin\theta \quad &+ \quad r\cos\theta\,\theta' = -r\cos\theta - \mu\sin(r\sin\theta). \end{aligned} \tag{1}$$

Solving for r' and θ' and by division, we obtain the single DE

$$\frac{dr}{d\theta} = \frac{\mu\sin\theta\sin(r\sin\theta)}{1 + (\mu/r)\cos\theta\sin(r\sin\theta)}. \tag{2}$$

A periodic solution of (2) that is equivalent to limit cycle of (1) will have period 2π in the angular variable θ. Integrating (2), we have

$$r(\theta, \mu) = A + \int_0^\theta \frac{\mu\ \sin\phi\sin(r\sin\phi)}{1 + \mu\cos\phi\sin(r\sin\phi)/r}\,d\phi. \tag{3}$$

For r to have a period 2π, we require that $r(2\pi, \mu) = A$, so that

$$\int_0^{2\pi} \frac{\sin\phi\sin(r\sin\phi)}{1 + \mu\cos\phi\sin(r\sin\phi)/r}\,d\phi = 0. \tag{4}$$

In the limit μ small, $r(\theta, 0) = A$ and (4) reduces to

$$\int_0^{2\pi} \sin\,\phi \sin\,(A \sin\,\phi)\,d\phi = 2\pi J_1(A) = 0,$$

showing that the permissible amplitudes associated with each limit cycle must be zeros of $J_1(A)$. The solvability of (4) is guaranteed if the appropriate Jacobian does not vanish. That Jacobian is give by

$$\int_0^{2\pi} \sin^2\phi \cos(A \sin\phi) d\phi = 2\pi[J_0(A) - J_2(A)]. \tag{5}$$

The large zeros of (4) are given by $A_n = n\pi + \pi/4 + O(1/n)$. Therefore, by familiar asymptotic estimates, (5) reduces to $4(-1)^n[2\pi/A_n]^{1/2} + O(1/n)$.

Accordingly, the Jacobian decreases in magnitude as n increases, so that one cannot guarantee a uniform range in μ for which all solutions of the type

$$y = A \cos w(\mu)x + \mu y_1 + \cdots$$

exist.

Using Brouwer's fixed-point theorem, it is shown that there exists a $\bar{\mu} > 0$ such that (2) has, for all $|\mu| < \bar{\mu}$, an infinite number of limit cycles. Hence the results for the given DE follow. Zhang (1980), Hochstadt and Stephan (1967), D'heedene (1969)

7. $$\mathbf{y'' + \epsilon \sin(y^2 + y'^2)\, sgn(y') + y = 0,}$$

where ϵ is a small parameter. The amplitude $a(x)$ in the polar form of the solution $y = a\cos\theta, y' = a\sin\theta$ is given for small ϵ by the first order DE

$$\pi a'(x) = -2a\epsilon\ \sin a^2. \tag{1}$$

An averaging method is used to obtain (1). Jordan and Smith (1977), p.119

8. $$\mathbf{y'' + y + y' - (1/2)(y' - |y'|)\delta(y - y_0) = 0,}$$

where y_0 is a constant and δ is the Dirac delta function. y is small but finite.

A first order uniform solution is

$$y = a(x)\cos[x + \beta(x)], \quad y' = -a(x)\sin[x + \beta(x)],$$

where

$$a'(x) = -\frac{a}{2} + \frac{1}{2\pi}\left[1 - \frac{x_0^2}{a^2}\right]^{1/2}, \quad \beta'(x) = \frac{x_0}{2\pi a^2}.$$

Nayfeh (1985), p.128

9. $$\mathbf{y'' - y - a|y'|^\rho - 1 = 0,}$$

$y(0) = 0, y(1) = 0$, where a and ρ are constants.

It is shown that there is a unique solution for the given boundary value problem when $a < (\pi/2^{1/2})^4 \simeq 24$. Chicone (1988)

10.
$$y'' - a|y'|^\rho - e^y - 1 = 0,$$

$y(0) = 0, y(1) = 0$, where a and ρ are constants.

It is shown that for $\rho = 4$, the stated BVP has a solution for $a < \pi^4 2^{-5}$ and has no solution for $a > \pi^4 2^{-2}$. Geometrical methods are used. Chicone (1988)

11.
$$y'' + \epsilon h(y^2 + y'^2 - 1)y' + y = 0, \epsilon \ll 1,$$

where h is continuous, $h(u) < 0$ for $u < 0, h(0) = 0$, and $h(u) > 0$ for $u > 0$.

The given DE has the periodic solution $y = \cos(x + \alpha)$ for any α. It can be shown by using the method of slowly varying amplitude that this solution is a stable limit cycle when $\epsilon > 0$. Jordan and Smith (1977), p.119

12.
$$y'' + yV(y'/y) = 0,$$

where V is a general function of its argument.

The given DE can be rewritten as

$$\frac{d}{dx}(y'/y) + (y'/y)^2 + V(y'/y) = 0. \tag{1}$$

Now put $z = y'/y$ in (1) to obtain the first order DE

$$z' + z^2 + V(z) = 0,$$

which may be integrated via a quadrature. Sabata (1980)

13.
$$y'' - y^{(k-2)/k} G(y' y^{(1-k)/k}) = 0,$$

where k is a constant and G is an arbitrary function of its argument.

Put $y' = p$ so that

$$p\frac{dp}{dy} = y^{(k-2)/k} G(y^{(1-k)/k} p). \tag{1}$$

Introducing $X = y^{(1-k)/k} p$ and writing its differential, we can express (1) as

$$\frac{dy}{y} = \frac{XdX}{[(1-k)/k]X^2 + G(X)}. \tag{2}$$

The quadrature of (2) is formally performed by introduction of a potential function $H(X)$ such that

$$G(X) = \frac{k-1}{n}X^2 + \frac{XH(X)}{H'(X)}, \tag{3}$$

where $H'(X)$ is the derivative of $H(X)$ with respect to X. The solution of (2) is $y = cH(X)$, where c is a constant. Bouquet, Feix, and Leach (1991)

14.
$$y'' + f'(y)y' + h(y' + f) + y = 0,$$

where h is a function of its argument $y' + f$.

A graphical method of construction of solution of the IVP is given. As examples of this special class of DEs, we have:

(a) Van der Pol's equation:

$$y'' - \mu(1 - y^2)y' + y = 0,$$

with

$$f(y) = -\mu Y + (1/3)\mu y^3,$$

μ a constant and $h(y' + f) = 0$.

(b) Rayleigh's equation:

$$y'' - [\alpha - \beta y'^2]y' + y = 0,$$

with $f(y) = 0$ and $h(y') = -\alpha y' + \beta y'^3$ (α and β are constants).

(c) $y'' + \dfrac{dG}{dy}\, y' + G(y) = 0,$

with $G(y) = f(y) + y, h(y' + f) = y' + f$.

The graphical method makes special use of the curves described by the functions $f(y)$ and $-h(y)$. Nishikawa (1964), p.24

15. $$\boldsymbol{y'' + w^2 y + \epsilon f(y, y') = 0,}$$

where w and ϵ are parameters.

Two methods of obtaining approximate solutions are sketched: (1) Assume a solution of the form $y = a \sin(wx + \phi)$, where a and ϕ are slowly varying coefficients of x; and (2) seek a solution in the form of a power series in ϵ, the coefficients of which are periodic functions. Reissig (1955/56)

16. $$\boldsymbol{y'' + y^3 - \epsilon f(y, y') = 0,}$$

where $0 < \epsilon \ll 1$ and f is a polynomial function of its arguments.

An approximate solution is found by assuming the form

$$y(x) = A(x)\ \mathrm{cn}[wx + \phi(x), \mu^2 = 1/2], \tag{1}$$

where cn is the Jacobi elliptic function, appearing in the solution of $y'' + y^3 = 0$, namely

$$y(x) = A\ \mathrm{cn}(wx + \phi, \mu^2 = 1/2),$$

where $A = w$ and A and ϕ are constants fixed by initial conditions. In (1) $A(x)$ and $\phi(x)$ are functions to be determined such that (a) (1) is a solution of the given DE; and (b) the time derivative of (1) must have the same form as the time derivative of the generating solution (2).

Following the foregoing procedure, and using an averaged form of Jacobi elliptic functions, one finally obtains ODEs for A and ϕ.

A simpler approach using trigonometric functions is as follows. Assume the exact solution of the given DE to be

$$y(x) = A(x) \cos[wx + \phi(x)], \tag{3}$$

where $A(x)$ and $\phi(x)$ are as yet unknown functions of x and w is an unspecified constant. Differentiating (3), we have

$$y' = -wA \sin\psi + [A' \cos\psi - A\phi' \sin\psi], \tag{4}$$

$\psi = wx + \phi$. Now require that

$$A' \cos\psi - A\phi' \sin\psi = 0, \tag{5}$$

so that $y' = -wA\sin\psi$. The second derivative of y is given by

$$y'' = -wA' \sin\psi - wA\phi' \cos\psi - w^2 A\cos\psi. \tag{6}$$

Substituting (6) into the given DE gives

$$A' \sin\psi + A\phi' \cos\psi = -wA\cos\psi + \{3A^3/(4w)\}\cos\psi$$
$$+ [A^3/(4w)]\cos 3\psi - (\epsilon/w) f(A\cos\psi, -wA\sin\psi). \tag{7}$$

Solving (5) and (7) for A' and $A\phi'$, we have

$$\begin{aligned} \frac{dA}{dx} &= -wA\cos\psi\sin\psi + \frac{3A^3}{4w}\cos\psi\sin\psi + \frac{A^3}{4w}\cos 3\psi\sin\psi - (\epsilon/w) f\sin\psi, \\ A\frac{d\phi}{dx} &= -wA\cos^2\psi + \frac{3A^3}{4w}\cos^2\psi + \frac{A^3}{4w}\cos 3\psi\cos\psi - (\epsilon/w) f\cos\psi. \end{aligned} \tag{8}$$

The equations (8) are exact. To get an explicit approximate solution, we average the RHS of (8) over period 2π in the variable ψ. We obtain

$$\begin{aligned} \frac{dA}{dx} &= -\frac{\epsilon}{2\pi w}\int_0^{2\pi} f(A\cos\psi, -wA\sin\psi)\sin\psi\, d\psi, \\ A\frac{d\phi}{dx} &= -\frac{\epsilon}{2\pi w}\int_0^{2\pi} f(A\cos\psi, -wA\sin\psi)\cos\psi d\psi + (A/2)[3A^2/(4w) - w]. \end{aligned}$$

Yuste and Bejarano (1986), Mickens and Oyedeji (1985)

17. $$\boldsymbol{y'' + \sin y - \epsilon f(y, y') = 0,}$$

where $f(y, y')$ is a polynomial, and $0 < \epsilon \ll 1$.

The existence of limit cycles is considered. A special case considered is $f = (a_{01} + a_{11}x + a_{21}x^2)y$. Morozov and Fedorov (1983)

18. $$\boldsymbol{y'' - yF(y, y') = 0,}$$

where $F(y, y') > 0$ for $y^2 + y'^2 \neq 0$.

Assuming that the solutions of the given DE are unique, it is shown that if $y = y(x), y(x) \not\equiv 0$, is a solution of the given DE valid for all $x \geq 0$, then $y(x)$ is eventually montone. If $y(x)$ has a zero, then $y(x) \to \infty$ or $y(x) \to -\infty$ eventually. Utz (1967)

19. $$\boldsymbol{y'' + F(y, y') = 0.}$$

Under the assumptions:

(a) $F : R \times R \to R$ is continuous and the given DE or its equivalent system

$$y' = z, \quad z' = -F(y, z) \tag{1}$$

has the existence and uniqueness property for its solutions.

(b) $F(y, 0)y > 0$ for $y \neq 0$.

(c) There exists $\epsilon > 0$ such that $F(y, 0) \geq F(y, z)$ for $y < 0, z > 0$ and $y^2 + z^2 < \epsilon$, $F(y, 0) \leq F(y, z)$ for $y > 0, z < 0$ and $y^2 + z^2 < \epsilon$, holds and either

$$\begin{aligned}(d_1)F(y,z) &= F(y,-z), \text{ for every}(y,z) \in R^2 \quad \text{or} \\ (d_2)F(y,z) &= -F(-y,z), \text{ for every}(y,z) \in R^2,\end{aligned}$$

then there exists a region in the phase plane of the system that is filled by a family of closed trajectories surrounding the stationary point $(0, 0)$.

Villari and Zanolin (1988)

20. $$\boldsymbol{y'' + f(y, y') + g(y) = 0,}$$

where $f(y, y')$ and $g(y)$ are continuous for all y, y' and $g(y) = -g(-y), g(y) > 0\ \forall y > 0$, and $f(y, 0) = 0$ for all y and the function

$$f(y, y') = f_i(y, y'), i = 1, 2, \ldots, 5$$

satisfy one of the following properties:

(a) $f_1(y, y') = f_1(-y, y') = f_1(y, -y')$,

(b) $f_2(y, y') = -f_2(-y, y') = -f_2(y, -y')$,

(c) $f_3(y, y') = -f_3(-y, y') = f_3(y, -y')$,

or is their combination:

(d) $f_4(y, y') = f_1(y, y') + f_3(y, y')$,

(e) $f_5(y, y') = f_2(y, y') + f_3(y, y')$.

It is shown that there exists a relatively large set of nonlinear, nonconservative systems which contain forces not admitting a potential but which have infinitely many periodic solutions, so that the behavior of such systems is similar to that of conservative oscillatory systems. Phase plane analysis is used. Pivovarov (1981)

21. $$\boldsymbol{y'' + f(y, y') + g(y) = 0,}$$

where f and g are differentiable functions of their arguments.

The following results are proved:

(a) If $f(y, y') \geq 0$ for all $(y, y'), yg(y) > 0$ for $y \neq 0$, and $\int_0^y g(y)dy \to \infty$ as $y \to \infty$, and if $y = y(x)$ is a (nontrivial) solution of the given DE valid for large x, then $y(x)$ is bounded and oscillatory as $x \to \infty$ or $y(x)$ montonically approaches zero as $x \to \infty$.

(b) If $f(y, y') > 0$ except at a discrete set of points, $g(y)$ is an odd function and $y = y(x)$ is an oscillatory solution of the given DE, then the amplitudes of the oscillations of this solution are monotonically decreasing.

Utz (1957)

22.
$$y'' + f(y, y')y' + g(y) = 0,$$

where f and g are continuous.

The given DE has at least one periodic solution under the following conditions:

(a) there exists $a > 0$ such that $f(x, y) > 0$ when $(x^2 + y^2)^{1/2} > a$.

(b) $f(0, 0) < 0$ [hence $f(x, y) < 0$ in a neighborhood of the origin].

(c) $g(0) = 0, g(x) > 0$ when $y > 0$, and $g(y) < 0$ when $y < 0$.

(d) $G(y) = \int_0^y g(u)du \to \infty$ as $x \to \infty$.

Jordan and Smith (1977), p.326

23.
$$y'' + f(y, y')y' + g(y) = 0.$$

The given DE is associated with electronic oscillators. Under various conditions on the functions $f(y, y')$ and $g(y)$, which are not necessarily polynomials, the oscillatory nature of the solution, the amplitudes of maximum and minimum values, and the existence of periodic solutions are demonstrated in an elementary manner. It is, however, assumed that a unique solution exists if y and y' at any $x = x_1$ are prescribed and that this solution may be continued over an infinite x-interval. Graffi (1940)

24.
$$y'' + y'f(y, y') + \phi(y) = 0, y(0) = y_0, y(\infty) = \alpha.$$

It is assumed that f and ϕ are continuous and have continuous derivatives. It is necessary that $\phi(\alpha) = 0$, and by translation may be taken to be zero. A complete treatment is given for the case $y\phi(y) < 0$ for $y \neq 0$ and when for all (y, z), $|f(y, z)| \leq a(y)|z| + b(y)$, where a and b are continuous. With these assumptions it is shown that for each $y_0 \neq 0$, there is a unique z_0 (for which $y_0 z_0 < 0$) such that the solution $y(x)$ satisfying $y(0) = y_0, y'(0) = z_0$ has the property $y(x) \to 0$ as $x \to \infty$. The solution is discussed in the phase plane, and the asymptotic behavior is discussed in a variety of cases. Klokov (1959)

25.
$$y'' + f(y, y')y' + \phi(y, y'^2) = 0.$$

Sufficient conditions on $f(x, y)$ and $\phi(x, y)$ are obtained so that the given DE has periodic solutions. Shimizu (1948)

26.
$$y'' + \phi'(y) + \gamma g(y, y') = 0,$$

$\gamma \geq 0$ is constant, and $g_y(x, y) > 0$.

A typical phase plane study of both cases $\gamma = 0$ and $\gamma \neq 0$ is carried out. The given DE describes the motion of a mass point on a spring. A zero-dimensional shock is envisaged. Antman (1988)

27.
$$y'' + \epsilon h(y, y') + V_y(y) = 0,$$

where V is a double-well potential, and h is dissipative, subject to ϵ-dependent initial conditions

$$y(0) = a_0 + \epsilon a_1 + \cdots,$$
$$\frac{dy(0)}{dx} = b_0 + \epsilon b_1 + \cdots.$$

The solution near the separatrix is represented as a large sequence of perturbed solitary pulses.

The amplitude and phase of the nonlinear oscillator after crossing the separatrix are determined and shown to be sensitively dependent on the initial conditions. Bourland and Haberman (1990)

2.23 $y'' + h(x)k(y)f(y') + g(x, y) = 0$

1. $\boldsymbol{y'' + xyy' - y^2 = 0, \quad y'(0) = -b(3)^{1/2}, y(\infty) = 0}$

The given DE results from the PDE $uu_t = u_{xx}$ by a similarity transformation. Introducing the variables $u = x^2 y, v = x^2(y - xy')$, we transform the given DE to

$$\frac{dv}{du} = \frac{v(2-u)}{3u - v}. \tag{1}$$

A local solution of Eqn. (1) in the neighborhood of singular points is found, and hence the solution of the BVP is found numerically. Dresner (1983)

2. $\boldsymbol{y'' - \alpha p(x)[1 + r(x)]y'^m y^n = 0,}$

$n + m < 1$ or $n + m > 1, x \in \Delta_0 = [a, \infty)$, where

$$0 < p(x) \in C^k, k \geq 1, r(x) = O(1) \in C\Delta_0 : \alpha = \pm 1, n = \frac{n_1}{2n_2 + 1}, m = \frac{m_1}{2m_2 + 1},$$

where n_1, n_2, m_1, m_2 are integers.

The asymptotic behavior as $x \to +\infty$ of those solutions, each of which, in some interval $\Delta_1 = [x_0, +\infty)$, belongs to the class $y(x) \in C^2\Delta_1, y(x) \neq 0, y(x) \not\equiv$ constant, is discussed. The question of whether there actually exist solutions with one or other asymptotic representation is also discussed. Kostin and Evtuhov (1976), Evtuhov (1977)

3. $\boldsymbol{y'' - \alpha g(x) y^n (y')^m = 0, x \in [a, b),}$

where

$$y(x) > 0, y'(x) \neq 0, \ \ \alpha = \pm 1, 0 < g(x) \in C^2[a, b), b \leq \infty.$$

A WKB solution of the given problem is defined and found as follows: A function

$$[\phi'(x_0, x)]^{-1/2} w[\phi(x_0, x)],$$

where $w[\phi(x_0, x)]$ is the solution of the equation

$$\frac{d^2 w}{d\phi^2} = w^n \left(\frac{dw}{d\phi}\right)^m$$

and

$$\phi(x_0, x) = \int_{x_0}^{x} [g(t)]^{2/(3+n-m)} dt, x_0 \in [a, b),$$

is called a WKB solution of the given DE. The existence of WKB solutions and their asymptotes is considered. Aripov and Eshmatov (1988)

4.
$$\mathbf{y'' + x^{-4/3} y^{1/3} (y')^{2/5} = 0, x > 0.}$$

It is shown that the given DE has an oscillatory solution. Grace and Lalli (1987a)

5.
$$\mathbf{y'' + x^{-4/3} y^{1/3} [1 + (y')^{2/5}] = 0, x > 0.}$$

It is shown that the given DE has an oscillatory solution. Grace and Lalli (1987a)

6.
$$\mathbf{y'' + (5/3) x^{-12/7} y'^{2/3} y^{3/7} = 0, x \geq x_0 > 0.}$$

It is proved that the given DE is nonoscillatory. Erbe and Liu (1990)

7.
$$\mathbf{y'' + 5x^{-3/2} y'^{4/5} y^{1/3} = 0, x \geq x_0 > 0.}$$

It is proved that the given DE is nonoscillatory. Erbe and Liu (1990)

8.
$$\mathbf{y'' + [n^2 a^2/(1+n)] x y'^{(2n-1)/n} = 0.}$$

The given DE arises in the flow of non-Newtonian fluid through porous media. Putting $x^2 = z$ and integrating the resulting equation, we get, for $n < 1$,

$$y' = \left[C_1 - \frac{na^2}{2} \frac{1-n}{1+n} x^2\right]^{n/(1-n)}, \tag{1}$$

where C_1 is an arbitrary constant. If we impose the conditions $y(0) = y_0$, say, and $y'(x_1) = 0$, we have

$$C_1 = \frac{na^2}{2} \frac{1-n}{1+n} x_1^2$$

and an integration of (1) then leads to

$$y - y_0 = \left[\frac{na^2}{2} \frac{1-n}{1+n}\right]^{n/(1-n)} x_1^{(1+n)/(1-n)} J_n(x), \tag{2}$$

where

$$J_n(x) = \int_0^{x/x_1} (1 - y^2)^{n/(1-n)} dy; 0 < x/x_1 < 1.$$

Pascal and Pascal (1985)

9.
$$\mathbf{y'' - [n^2 a^2/(1+n)] x (-y')^{(2n-1)/n} = 0,}$$

where n and a are constants.

The given DE with $y' = 0$ when $x = x_1$, say, is integrated once to yield

$$-\frac{dy}{dx} = B^{n/(1-n)} x_1^{2n/(1-n)} (1 - x^2/x_1^2)^{n/(1-n)}, n < 1, \tag{1}$$

where B is a constant.

The solution of (1) is expressible in terms of the function

$$J_n(x/x_1) = \int_0^{x/x_1} (1-\xi^2)^{n/(1-n)} d\xi,$$

where

$$J_n(1) = [\pi^{1/2} n/(1+n)][\Gamma(n/(1-n))]/[\Gamma((1+n)/\{2(1-n)\})],$$

where Γ is the gamma function. Pascal (1991)

10. $$y'' - kx^a y^b y'^c = 0,$$

where k, a, b, and c are constants.

Put $y = x^{(c-a-2)/(b+c-1)}\eta(\xi), \xi = \ln x, b+c \neq 1$ to obtain

$$\begin{aligned}\eta'' &+ [(c-2a-b-3)/(b+c-1)]\eta' - [(a+b+1)/(b+c-1)][(c-a-2)/(b+c-1)]\eta \\ &= k\eta^b[\eta' + [(c-a-2)/(b+c-1)]\eta]^c. \qquad (1)\end{aligned}$$

Equation (1) is autonomous. Put $\eta'(\xi) = p(\eta)$ so that

$$\begin{aligned}pp' &+ [(c-2a-b-3)/(b+c-1)]p - [(a+b+1)/(b+c-1)][(c-a-2)/(b+c-1)]\eta \\ &= k\eta^b[p + \{(c-a-2)/(b+c-1)\}\eta]^c. \qquad (2)\end{aligned}$$

Equation (2) may be solved explicitly only for special choices of a, b, and c. Kamke (1959), p.555; Murphy (1960), p.386

11. $$y'' - \sigma x^p |y|^q |y'|^r = 0, y(0) = 0, y(\tau) = B,$$

where σ, p, q, and r are real; τ and B are prescribed.

The domain of existence of the given problem, i.e., the set of such points (τ, B) for which the solution exists, is determined. Klokov and Stepanov (1980)

12. $$\epsilon y'' - g(x)f(y') = 0, c < x < d, y(c) = a, y(d) = b, a \neq b,$$

where a and b are finite real numbers, ϵ is a fixed positive infinitesimal, and c and d are real numbers.

Using continuity arguments, solutions with a free layer at x_0 are found for all $x_0 \in (c,d)$, and the position of x_0 as a function of ϵ and of the boundary conditions is given. A new phenomenon appears as a consequence of the sensitive dependence of x_0 on the boundary data. In particular, it is proved that for any standard $x_0 \in [c,d]$, there exist a and b such that the solution of the given problem has a jump at x_0. Bohe (1990)

13. $$y'' + p(x)g(y')h(y) = 0.$$

The following qualitative results are obtained:

(a) If $p(x) \geq a > 0, g(y') > 0, yh(y) > o(y \neq 0)$, and $\int_0^{\pm\infty} h(y)dy = +\infty$, then any solution $y(x)$ and its derivative $y'(x)$ are oscillatory and their zeros separate one another.

(b) If $p(x) < 0, g(y') > 0, yh(y) > 0(y \neq 0)$, then any solution $y(x)$ and its derivative $y'(x)$ are strictly monotone.

Some results about the properties of the amplitude and the boundedness of oscillatory solutions when $p(x) > 0$ are given. Asymptotic behavior when $p(x) < 0$ is also given. The theorems proved generalize the work of Bellman (1953). Liang (1966)

14.

$$y'' + a(x)f(y)g(y') = 0,$$

where $a : I \to R^+ = (0, \infty), I = [0, \infty), f : R \to R = (-\infty, \infty)$, and $g : R \to R^+$ are continuous functions, $yf(y) > 0$ for $y \neq 0$. Assuming that every solution of the given DE can be defined on $[\tau, \infty)$ for some $\tau \geq 0$, the stability and boundedness of solutions is considered. Necessary and sufficient conditions are established for the solution of the given DE to be stable, bounded, or vanishing as $x \to \infty$. Liang (1987)

15.

$$y'' + c(x)f(y)h(y') = 0,$$

where $h(s)$ is positive and continuous for all real s, $f(s)$ is continuous for all real s and $sf(s) > 0$ if $s \neq 0$; $c(x)$ is continuous for all real $x \geq x_0$.

Conditions are stated, determining the behavior of a solution of the given DE which are expressed wholly and explicitly in terms of the initial conditions and the given functions $h(s), f(s)$, and $c(x)$. The method of proof for some of the results involves the introduction of two Lyapunov functions which do not require that $c(x)$ be monotone. Petty and Johnson (1973)

16.

$$y'' + p(x)f(y)g(y') = e^{-x}\sin x.$$

If:

(a) $\int_0^\infty p(s)ds = \infty$,

(b) $yf(y) > 0$ for $y \neq 0, df/dy \geq 0$ for all x, and

(c) $g(y') \geq k > 0$ for all y',

then it is shown that every solution of the given DE is oscillatory. Rankin (1976)

2.24 $y'' + f(x, y, y') = 0$

1.

$$y'' - y'^2 - \sum_{j=1}^{2} A_j(y, x)y'^{(k-j)} = 0,$$

where A_j are polynomials in y with coefficients that are analytic functions of x.

It is shown that there are no equations in the given class whose solutions have stationary transcendental and essential singularities. Bogoslovskii and Ostroumov (1983)

2.

$$y'' - a(x) \exp\left\{\sum_{i=1}^{m} b_i(x)y^{2i-1} + \sum_{j=1}^{n} c_j(x)(y')^{2j-1}\right\} = 0,$$

where a, b_i, and c_j are positive continuous functions on $[0, \infty)$.

Explicit sufficient conditions for the existence of some or all of the solutions of the given DE defined on the given interval $[0, \infty)$ are given. The solutions are classified as proper or singular. Usami (1987)

3. $$\boldsymbol{y'' = \sum_n f_n(x, y) y'^n.}$$

It is shown that a necessary condition for exact linearization of this equation is that it be of the form

$$y'' = f_0(x, y) + f_1(x, y)y' + f_2(x, y)y'^2 + f_3(x, y)y'^3.$$

This condition, however, is not sufficient. This is achieved by searching those finite-point transformations (if any), $x_1 = T(x, y), y_1 = S(x, y)$, with nonvanishing Jacobian $T_x S_y - T_y S_x$, which reduce the given equation to a linear second order equation. Aguirre and Krause (1988)

4. $$\boldsymbol{y'' - y'/x - x^3/y' = 0.}$$

One solution by inspection is $y = \pm(1/3)x^3$. Put $y = x^3 z$ and then $x = e^t$ to obtain the autonomous DE

$$\left(\frac{dz}{dt} + 3z\right)\frac{d^2z}{dt^2} + 4\left(\frac{dz}{dt}\right)^2 + 15z\frac{dz}{dt} + 9z^2 - 1 = 0. \tag{1}$$

Now reduce (1) to first order DE by writing

$$\frac{dz}{dt} = p, \quad \frac{d^2z}{dt^2} = p\frac{dp}{dz}, \quad \text{etc.}$$

$$(p^2 + 3zp)\frac{dp}{dz} + 4p^2 + 15zp + 9z^2 - 1 = 0. \tag{2}$$

Equation (2) may be treated in the phase plane. Rabenstein (1966), p.373

5. $$\boldsymbol{(\beta/3)y'' + xy'^{5/3} - \alpha y y'^{2/3} = 0,}$$

where β and α are constants.

An alternative form of the given DE is

$$\beta\frac{d}{dx}\left(\frac{dy}{dx}\right)^{1/3} + x\frac{dy}{dx} - \alpha y = 0.$$

The given DE results from the nonlinear heat equation $u_t = \left[(u_z)^{1/3}\right]_z$ by a similarity transformation. It transforms via $u = xy^{1/2}, v = x(y')^{1/3}$ to the first order DE

$$\frac{dv}{du} = \frac{u(2\beta v - 2v^3 + 2\alpha u^2)}{2\beta u^2 + \beta v^3},$$

which is analyzed in the phase plane and the BVP with $y'(0) = -(q/k)^3, y(\infty) = 0$ is solved by joining the appropriate singular points. Here q and k are some constants.

For the special values $\alpha = -1$ and $\beta = 2/3$, the given DE integrates first to yield $(2/3)y'^{1/3} + xy = 0$, if we choose the constant of integration to be zero, and hence to

$$y = \frac{4}{3(3)^{1/2}}(x^4 + b^4)^{-1/2},$$

where b is the second constant of integration, obtained by a certain conservation condition. Dresner (1983)

6. $$y'' + y - f(x)(1 - y^2 - y'^2)^{1/2} = 0.$$

The given DE arises in connection with the problem of determining a curve by means of its curvature and torsion, expressed in terms of arc length. The solution constructed is rather intricate. Nalli (1947)

7. $$y'' - 2a(y + bx + c)(y'^2 + 1)^{3/2} = 0,$$

where a, b, and c are constants.

Put $u(x) = y + bx + c$ so that $u''[(u' - b)^2 + 1]^{-3/2} = 2au$. Multiplication by u' and integration lead to

$$\left[\frac{bu' - (b^2 + 1)}{(u' - b)^2 + 1}\right]^{1/2} = au^2 + C,$$

where C is an arbitrary constant. Solving for u'^{-1}, we have

$$\frac{dx}{du} = \frac{b}{b^2 + 1} \pm \frac{au^2 + C}{b^2 + 1}\left[b^2 + 1 - (au^2 + C)^2\right]^{-1/2};$$

x can, therefore, be represented as an elliptic integral of u. Kamke (1959), p.557; Murphy (1960), p.386

8. $$y'' - \tanh(y'/x) - (1/x)y' = 0.$$

Put $y' = v$ in the given DE so that $v' - \tanh(v/x) - v/x = 0$. Now put $v/x = z$ and get

$$\frac{dz}{dx} = (1/x)\tanh z. \tag{1}$$

On integration, (1) gives $\sinh z = c_1 x$, or

$$y'/x = \sinh^{-1} c_1 x. \tag{2}$$

On integrating (2), we have $y = \int x \sinh^{-1}(c_1 x)dx + c_2$. c_1 and c_2 are arbitrary constants. Rabenstein (1972), p.47

9. $$y'' = A_1 x^{n_1} y^{m_1} y'^{\ell_1} + A_2 x^{n_2} y^{m_2} y'^{\ell_2},$$

where n_i, m_i, ℓ_i are constants.

The given DE is studied for point transformations which are "closed" on the given class, i.e., that transform the given DE into an equation of similar class:

$$\ddot{u} = B_1 t^{\nu_1} u^{\mu_1} \dot{u}^{\lambda_1} + B_2 t^{\nu_2} u^{\mu_2} \dot{u}^{\lambda_2}.$$

Khakimova (1988)

10. $$y'' - a(xy' - y)^{\nu} = 0,$$

where a and ν are constants.

Put $y = xu(x)$ to obtain Bernoulli's equation for u':

$$xu'' + 2u' = ax^{2\nu}u'^{\nu}.$$

It is linear for $\nu = 1$. It is, in general, solved by putting $(u')^{-\nu+1} = V$, say. Kamke (1959), p.555; Murphy (1960), p.386

11. $$y'' = \alpha(y - xy')^n/\{x^2(y - x)\}, y(a) = a.$$

Proceed as for Eqn. (2.41.10). Forsyth (1959), p.206

12. $$y'' + (y' - y/x)^a f(x, y) = 0,$$

where a is constant.

If $f(x, y) = (1/x)\phi(y/x)$, we may put $y(x) = x\eta(\xi), \xi = \ln x$ so that the given DE becomes

$$\eta'' + e^{\xi} f\{e^{\xi}, e^{\xi}\eta\}\eta'^{a} + \eta' = 0. \tag{1}$$

If f depends only on x, then (1) is of Bernoulli type for $u(\xi) = \eta'$. If f is as assumed, $f(x, y) = (1/x)\phi(y/x)$, then (1) has the form

$$\eta'' + \phi(\eta)\eta'^{a} + \eta' = 0. \tag{2}$$

Equation (2) is autonomous. So put $p(\eta) = \eta'(\xi)$ and obtain the first order DE $pp' + \phi(\eta)p^a + p = 0$. Kamke (1959), p.555

13. $$y'' + n\beta y'^2 + [na^2/(1 + n)]xy'^{(2n-1)/n} = 0,$$

where $n < 1$, and β is small.

The given DE arises in non-Newtonian flows through porous media. Using $\lambda = n\beta$ as a small parameter, a perturbation solution

$$y = y_0(x) + \lambda y_1(x) + \lambda^2 y_2(x) + \cdots$$

is developed. The zeroth order equation is the same as found in Eqn. (2.23.8). The linear equation for $y_1(x)$ is then solved subject to the boundary conditions of the physical problem. For $n = 0.5$ the solution $y = y_0(x) + \lambda y_1(x)$ is explicitly written out for $0 < x < x_1$, say:

$$\begin{aligned} y_0(x) &= y^0 + B^*(x/x_1)[1 - (1/3)(x^2/x_1^2)], x < x_1, \\ y_1(x) &= (B^*)^2 \left[(11/30)(x/x_1) - (1/2)(x^2/x_1^2) + (1/6)(x^4/x_1^4) - (1/30)(x^6/x_1^6)\right], x < x_1, \end{aligned}$$

where $B^* = [(1 - n)a^2/(2n)]^{n/(1-n)} x_1^{(1+n)/(1-n)}$. Pascal and Pascal (1985)

14. $$y'' + y'|y'| + q(x)y' + y - p^2y^3 = 0,$$

where $q(x)$ is positive and differentiable and $q'(x) = O(|x|^{1/2})$ as $x \to \infty$; p is constant.

A region Z of the (y, y') plane is determined such that any integral curve whose initial values for $x = 0$ lie in Z remain in Z for $x > 0$ satisfies the stability condition

$$\lim_{x \to \infty} [y^2(x) + y'^2(x)] = 0.$$

For an integral characterized by the initial values outside Z, it is proved, conversely, that

$$\lim_{x\to X}[y^2(x)+y'^2(x)]=\infty$$

for some finite X. The region Z is bounded by two integral curves that pass, respectively, through the singular points $(-1/p,0)$ and $(1/p,0)$ in the (y,y')-plane. Modona (1953)

15. $$y''+y'|y'|+q(x)y'+y-p^2(x)y^3=f(x),$$

where $p(x)>0$ and $q(x)$ and $f(x)$ are continuous functions with period X.

It is proved that the given DE possesses at least one solution with period X. For the proof, the DE is converted into a nonlinear integral equation whose kernel is a Green's function for the operator $[d^2/(dx^2)+1]$, with periodicity as boundary conditions. A fixed-point theorem is then used. Stoppelli (1953)

16. $$\epsilon y''-f(x,y,y')=0,$$

where $f(x,y,y')=\begin{cases} xy'^2+xyy', & -1\le x\le 0,\\ xy'^2-xyy', & 0\le x\le 1\end{cases}$

and $y(-1,\epsilon)=A, y(1,\epsilon)=B, Ae^{-1}>B>0$.

It is shown that for ϵ sufficiently small, the given BVP has a solution that satisfies

$$A\exp[-(1+x)] - (Ae^{-1}-B)\exp[-Ae^{-1}(2\epsilon)^{-1}x^2]-\epsilon^{1/2}\gamma\le y(x,\epsilon)\le A\exp[-(1+x)]$$

$$B-\epsilon^{1/2}\gamma \le y(x,\epsilon)\le B+(Ae^{-1}-B)\exp[-B(2\epsilon)^{-1}x^2]+\epsilon^{1/2}\gamma,$$

where γ is a positive constant independent of ϵ. Howes (1978)

17. $$y''-f(y',ax+by)=0,$$

where a and $b\neq 0$ are constants.

With $u(x)=ax+by(x)$, the given DE takes the autonomous form

$$u''=bf\left(\frac{u'-a}{b},\ u\right).$$

Hence, put $u'=p, u''=p(dp/du)$, etc., and solve. Kamke (1959), p.558; Murphy (1960), p.386

18. $$y''-yf(x,y'/y)=0.$$

With $y'=u(x)y$, we get a first order DE $u'=f(x,u)-u^2$. Kamke (1959), p.558; Murphy (1960), p.387

19. $$y''-x^{a-2}f(y/x^a,xy'/x^a)=0,$$

where a is a constant.

The given DE is equidimensional with respect to x. Writing $y(x)=x^a\eta(\xi),\xi=\ln x$, we get

$$\eta'' + (2a - 1)\eta' + a(a - 1)\eta = f(\eta, \eta' + a\eta).$$

Kamke (1959), p.558; Murphy (1960), p.387

20.
$$\boldsymbol{y'' + p(x)f(y') + b(x)y = c(x),}$$

where all functions are continuous.

Assuming that:

(a) the given DE has a solution $y_0(x)$ that exists for all $x \geq 0$ and where derivative $y_0'(x)$ is bounded for $x \geq 0$,

(b) $p(x)$ is "integrally" positive,

(c) $f(y)$ is strictly monotone increasing for all y,

(d) $0 < b_1 \leq b(x) \leq b_2$, and either $b'(x) \leq 0$ or $b'(x) \geq 0$,

then it is shown that for every pair of solutions of the given DE, the difference of their derivatives tends to zero as $x \to \infty$. Yoshizawa (1985)

21.
$$\boldsymbol{\epsilon y'' = f(x, y')y' + g(x, y), -1 < x < 1,}$$

$y(-1, \epsilon) = A, y(1, \epsilon) = B$, where ϵ is a small parameter, f and g are smooth functions, $f(0, y) = 0$ for all y, i.e., $x = 0$ is a turning point, and A and B are prescribed.

Existence, for sufficiently small ϵ, of a solution of the given problem that exhibits spike layer behavior and a solution that exhibits nonmonotone transition layer behavior at $x = 0$ is established. De Santi (1987)

22.
$$\boldsymbol{y'' + h(x, y')y' + f^2(y, y')y = 0.}$$

The given DE has importance in many applications to mechanics and radio. It includes Van der Pol's equation as a special case. A nearly periodic solution is found in the form

$$y = \alpha \sin\left(\int_0^x w dx + \epsilon\right),$$

where w and α are functions of x obtainable from

$$w^2 = \pi^{-1} \int_0^{2\pi} f^2(\alpha \sin u, \alpha w \cos u) \sin^2 u du,$$

$$\frac{d\alpha}{dx}\left(1 + \frac{\alpha}{2w}\frac{dw}{d\alpha}\right) = -\frac{\alpha}{2\pi}\int_0^{2\pi} h(\alpha \sin u, \alpha w \cos u) \cos^2 u du,$$

which are found by quadrature; ϵ is assumed small. Kac (1950)

23.
$$\boldsymbol{y'' + \lambda y + q(x)f(y, y') = 0, x > 0, y'(0) = 0, \lim_{x\to\infty} y(x) = 0,}$$

where q and f satisfy:

(a) $q \in C(R_+, R)$ and $\lim_{x\to\infty} q(x) = L, L > 0$.

(b) $f \in C^1(R^2, R)$ with $f(0, 0) = 0$ and grad $f(0, 0) = (0, 0)$.

Conditions are given for the bifurcation and asymptotic bifurcation in L^p of solutions. Bifurcation occurs at the lowest point of the spectrum of the linearized problem. Under stronger hypothesis, there is a global branch of solutions. Stuart (1985)

24. $$y'' - \{w'(x)/w(x)\}y' - f(y, y', w(x), q(x)) = 0.$$

It is proved that if $u_1(x), u_2(x)$ are independent solutions with Wronskian w of the linear equation $u'' = \{w'/w\}u' - q(x)u$, where $w(x), q(x)$ are arbitrarily prescribed functions, then the given DE has the general solution $y = F(u_1, u_2)$ if and only if f has the form $f = -qZ(y) + A(y)y'^2 + C(y)w^2$, where Z, A, C satisfy $Z' - AZ = 1, ZC' + (3 - AZ)C = 0$. The function $F(u_1, u_2)$ is any solution of the system

$$\begin{aligned} F_{uu} &= A(F)F_u^2 + v^2C(F), \\ F_{vv} &= A(F)F_v^2 + u^2C(F), F_{uv} = A(F)F_uF_v - uvC(F), \\ F_u &= u^{-1}(Z(F) - vF_v). \end{aligned}$$

Gergen and Dressel (1965), Herbst (1956)

25. $$y'' + \epsilon h(y, y', X) + V_y(y, X) = 0,$$

where $X = \epsilon x$; $V(y, X)$ is a nonlinear potential that admits periodic solutions when $\epsilon = 0$ in the given DE. The term $h(y, y', X)$ is assumed to be odd in dy/dx in order to represent an arbitrary small nonlinear damping. This study "corrects" and extends some earlier work on the given system by several investigators. Bourland and Haberman (1988)

26. $$y'' + \phi(y, y') + g(y) = e(x).$$

(a) If $e(x) = 0$ and $G(y) = \int_0^y g(u)du$ exists for every finite y and $\to \infty$ with $|y|$ and if $\phi(y, y') \equiv f(y')$, where $uf(u) \geq 0$ for all u, in either case $f(u)$ being integrable over every finite interval, then all solutions are bounded as $x \to \infty$.

(b) If $e(x) \not\equiv 0$ and $g(y) = y$, and if $\phi(y, y')$ satisfies the earlier hypothesis, the same conclusion holds provided that there exists one solution that is bounded as $x \to \infty$. All proofs depend on considerations of the integrated equation or of positive-definite gauge functions. Utz (1956a)

27. $$y'' - F(x, y, y') = 0.$$

It is shown that if $F = A + 3By' + 3Cy'^2 + Dy'^3$ for some functions A, B, C, D of x and y, which satisfy the equations

$$2C_{xy} - B_{yy} - D_{xx} + AD_y + 2A_yD - 3B_xD - 3BD_x - 3B_yC + 6CC_x = 0,$$

$$B_{xy} - C_{xx} - A_xD - AD_x + 3A_yC + 3AC_y + 3BC_x - 6BB_y = 0,$$

then the given DE is linearizable by a point transformation. Grissom, Thompson, and Wilkens (1989)

28. $$y'' - f(x, y, y') = 0.$$

The given DE is linearizable by the point change of variables

$$t = t(x, y), u = u(x, y); \frac{\partial(t, u)}{\partial(x, y)} \neq 0$$

if it has the form

$$y'' + F_3(x,y)y'^3 + F_2(x,y)y'^2 + F_1(x,y)y' + F(x,y) = 0,$$

where the coefficients F_3, F_2, F_1, and F satisfy the compatibility conditions of the auxiliary system

$$\begin{aligned}
\frac{\partial z}{\partial x} &= z^2 - Fw - F_1 z + \frac{\partial F}{\partial y} + FF_2,\\
\frac{\partial z}{\partial y} &= -zw + FF_3 - (1/3)\frac{\partial F_2}{\partial x} + (2/3)\frac{\partial F_1}{\partial y},\\
\frac{\partial w}{\partial x} &= zw - FF_3 - (1/3)\frac{\partial F_1}{\partial y} + (2/3)\frac{\partial F_2}{\partial x},\\
\frac{\partial w}{\partial y} &= -w^2 + F_2 w + F_3 z + \frac{\partial F_3}{\partial x} - F_1 F_3.
\end{aligned}$$

Ibragimov (1992)

29. $\boldsymbol{y'' - f(y, y', x) = 0.}$

Similarity analysis of point transformations is extended to nonpoint transformations (with the inclusion of derivatives), giving an analytic expression for solutions. Meinhardt (1981)

30. $\boldsymbol{y'' - f(y, y', x) = 0,}$

where f is an arbitrary function of its arguments.

It can be shown that a necessary and sufficient condition for the given DE to be invariant under the similarity transformation

$$x = a^A \bar{x}, y = a^B \bar{y}, \text{ with } B/A = k \tag{1}$$

is that we may be able to write

$$f(y, y', x) = x^{k-2} F(\xi, \eta), \tag{2}$$

where ξ and η are elementary invariants of the similarity transformation defined by (1) and given by

$$\xi = x^{-k} y, \eta = x^{1-k} y'. \tag{3}$$

Further, the requirement that f be invariant under translation in x leads to

$$F(\xi, \eta) = \xi^{(k-2)/k} G(X), \tag{4}$$

where G is an arbitrary function of the characteristic $X = \eta\xi^{(1-k)/k}$. Combining (2)–(4), we arrive at the form

$$y'' = y^{(k-2)/K} G(y' y^{(1-k)/k}), \tag{5}$$

which is the most general form of the given second order DE which admits the two symmetries (1) and translation in x. Now see Eqn. (2.22.13). Bouquet, Feix, and Leach (1991)

31. $$y'' - f(y', y, x) = 0,$$

where f is unspecified but may contain integral powers of y' and y.

A nonperturbation approximate method of solution is suggested as follows. Put $\phi = y'/y, y = \exp[\int \phi dx]$, to convert the given DE to a Riccati integrodifferential equation

$$\frac{d\phi}{dx} + \phi^2 - f\left[\phi \exp\left\{\int \phi dx\right\}, \exp\left\{\int \phi dx\right\}, x\right] = 0. \tag{1}$$

The first two terms in (1) are Riccati terms; the f term modifies the form. To approximate this term one may have some idea of y'/y; the term $\exp[\int \phi dx]$ can be estimated and an iteration started by solving the first order inhomogeneous Riccati equation. The entire process is nonperturbative.

Alternatively, one may use the form invariance of the Riccati equation under reciprocation and write $\phi = g/(h + \psi)$, where ψ is the new dependent variable and g and h are arbitrary functions, and obtain the new DE as

$$\begin{aligned}\frac{dg}{dx}(h+\psi) - g\left(\frac{dh}{dx} + \frac{d\psi}{dx}\right) &= (h+\psi)^2 f\left[g(h+\psi)^{-1}\exp\left\{\int g(h+\psi)^{-1}dx\right\};\right. \\ &\left.\exp\left\{\int g(h+\psi)^{-1}dx'; x\right\}\exp\left\{-\int g(h+y_1)^{-1}dx\right\}\right].\end{aligned}$$

The functions g and h are chosen to simplify the equation for a given function f. An iterative method may be used as before by each time evaluating the integral terms approximately and solving the Riccati equation. Burt (1987)

32. $$y'' - F(x, y, y') = 0.$$

Making use of the transformation $y_1 = \lambda^\beta y, x_1 = \lambda x$, $0 < \lambda < \infty$, under which the given DE is invariant, it is possible to discover the form $F = x^{\beta-2}G(u, v)$, where $u = yx^{-\beta}, v = x^{1-\beta}y'$ so that the DE becomes one of first order:

$$\frac{dv}{du} = \frac{G(u, v) - (\beta - 1)v}{v - \beta u}. \tag{1}$$

Equation (1) may be solved in the (u, v) phase plane; if it can be integrated, we obtain v as a function of u, so a first order DE is obtained. Dresner (1983)

33. $$y'' + f(y', y, x) = 0,$$

where the function $f(y', y, x)$ is continuous and single valued.

This general equation, particularly when it possesses oscillatory solutions, may be solved by a graphical method called the delta method. The equation is rewritten as

$$y'' + w_0^2 y + f(y', y, x) - w_0^2 y = 0$$

and introducing $\tau = w_0 x$ and $v = dy/d\tau$, we have

$$\frac{dv}{dy} = -\frac{y + \delta(v, y, \tau)}{v}, \delta(v, y, \tau) = \frac{1}{w_0^2} f(w_0 v, y, \tau/w_0) - y.$$

$\delta(v, y, \tau)$, in general, depends upon all the variables v, y, and τ, but for small changes in these variables it may be assumed to remain constant. This is the basic assumption of the method; hence the name *delta*. Then, a geometrical construction of solution in (y, v) plane is carried out. Nishikawa (1964), p. 32

34. $$\boldsymbol{y'' - f(x, y, y') = 0},$$

where $f(x, y, y')$ is defined in the domain $a \leq y \leq b, c \leq y' \leq d$, $x \in (-\infty, \infty)$, is periodic with respect to x with period X, continuous with respect to all the variables x, y, and y', and satisfies the inequalities

$$|f(x, y, y')| < M, |f(x, y_1, y_1') - f(x, y_2, y_2')| < K_1|y_1 - y_2| + K_2|y_1' - y_2'|,$$

where M, K_1, and K_2 are positive constants depending on a, b, c, d and on the parameters. Furthermore, let

$$b - a \geq MX^2/2, c \leq -5MX/6 \leq 5MX/6 \leq d,$$

$(X/2)[(X/2)K_1 + (5/3)K_2] < 1$. Then the sequence of functions

$$y_{m+1}(x, y_0) = y_0 + L^2[f(x, y_m(x, y_0), y_m'(x, y_0)], \tag{1}$$

$m = 0, 1, \ldots$ which are periodic with respect to x with period X, converges as $m \to \infty$ for

$$x \in (-\infty, \infty), a + MX^2/4 \leq y_0 \leq b - MX^2/4. \tag{2}$$

Moreover, the limit function $y_\infty(x, y_0)$ is defined in the domain (2), is periodic, and is the unique solution of the equation

$$y(x, y_0) = y_0 + L^2 f(x, y(x, y_0), y'(x, y_0)).$$

Here and above,

$$\text{Lf}\ (x) = \int_0^x |f(x) - \overline{f(x)}|dx,$$

where

$$\overline{f(x)} = (1/X)\int_0^x f(x)dx,$$

and therefore

$$\begin{aligned} L^2 f(x) &= L(Lf) \\ &= \int_0^x \left[\int_0^x |f(x) - \overline{f(x)}|dx - \overline{\int_0^x |f(x) - \overline{f(x)}|dx}\right] dx. \end{aligned}$$

A second theorem states that if the given DE with stipulated conditions has a periodic solution $y = \Phi(x)$ with period X, satisfying $\Phi(0) = y_0$ on the interval a $+Mx^2/4 \leq y_0 \leq b - MX^2/4$, then $\Phi(x) = y_\infty(x, y_0)$. Samoilenko (1967)

35. $$\boldsymbol{y'' - f(x, y, y') = 0, a < x < b, y(a) = y(b), y'(a) = y'(b)}.$$

This periodic BVP is solved via an integral equation formulation and Picard iteration. The method is constructive. Contraction mapping is used. Some numerical results are presented. Saranen and Seikkala (1989)

36. $$y'' - f(x, y, y') = 0, y(0) = y(1), y'(0) = y'(1), \quad \text{or}$$

$$u_1 y(0) - u_2 y'(0) = \alpha, v_1 y(1) + v_2 y'(1) = \beta;$$

u_i, v_i $(i = 1, 2)$ are constants.

Here y and f are vectors; $f = A(x, y)y' + f_0(x, y)$, where $y = (y' \cdots y^{(n)})^T$ as well as y' or f_0 are real n-dimensional column vectors and A is a real $n \times n$ matrix.

Given a region Ω in (x, y) space, it is asked whether there exists a solution $y(x)$ of the problem satisfying $[x, y(x)] \in \Omega$. The answer leads to a rather general set of conditions which are sufficient in order that Ω has the desired property. One of these conditions is geometric in nature and depends upon the boundary data only. The second condition can be expressed in terms of inequalities and depends on the value of f in $\partial\Omega$. These inequalities turn out to be common background of a variety of conditions which can be found in literature on boundary value problems, and which in the case of a scalar equation reduce to the well-known properties of upper and lower solutions. Knobloch and Schmitt (1977)

37. $$y'' - f(x, y, y') = 0, y'(a) = A, y(b) + hy'(b) = B,$$

where $a, A, b, h,$ and B are constants.

Let the function $f(x, u, v)$ be continuous in the region $a \le x \le b, -\infty < u < \infty, -\infty < v < \infty$, and have continuous f_u, f_v which satisfy $|f_u(x, u, v)| \le M$ and $f_v(x, u, v) \le K, (K \ge 0)$. Under these assumptions, it is shown that the given problem has exactly one solution if the length $< a, b >$ is smaller than that of the interval in which the solution $w(x)$ of $w' = w^2 + Kw + M, w(0) = 0$, satisfies the inequality $w(x) \le 1/h$, or is smaller than the length of the interval of existence of the solution $w(x)$, according as $h \ne 0$ or $h = 0$. Opial (1960)

38. $$y'' - f(y, y', x) = 0, y(0) = y(1) = 0.$$

A sufficient condition is given for the existence of a unique solution to the given BVP under the assumption that the partial derivative $f_{y'}$ is bounded and the partial derivative $\partial f/\partial y$ is bounded from below. Tippet (1974)

39. $$\epsilon y'' - f(x, y, y') = 0, \; 0 < x < 1, \; y(0, \epsilon) = A, y(1, \epsilon) = B.$$

This singular perturbation problem is studied under the assumption that $f_{y'y'} \ne 0$ in the domain of interest. Solutions are shown to exhibit essentially two types of asymptotic behaviors: (a) boundary layer behavior, and/or (b) smooth transition from one stable reduced root to another. An algorithm for the exact determination of conditions guaranteeing such behavior as well as several illustrative examples are discussed. Differential inequalities using classical Nagumo theory are used. Howes (1978)

40. $$\epsilon y'' - f(x, y, y', \epsilon) = 0, -1 < x < 1, y(-1, \epsilon) = A, y(1, \epsilon) = B.$$

The existence and asymptotic behavior as $\epsilon \to 0^+$ of solutions of the given BVP are studied in the case that $f_{y'y'} = O(1)$, as $|y'| \to \infty$ and $f_{y'y'} = 0$ at $x = 0$. For small values of the perturbation parameter $\epsilon > 0$, the solutions are closely approximated by certain roots of the reduced equation $f(x, y, y', 0) = 0$ throughout most of the x-interval with the possible exception of a shock layer region centered $x = 0$ and/or a boundary layer region at $x = -1$ or $x = 1$ (or both endpoints). Inside such regions a solution changes rapidly either to transfer from one reduced root to another or to satisfy the given boundary data. Howes (1978a)

41. $$\epsilon y'' - f(x, y, y', \epsilon) = 0, -1 < x < 1,$$

$y(-1) = B_1, y(1) = B_2$, for which f possesses a turning point at $x = 0$, i.e., $(\partial f/\partial y') = 0$ at $x = 0$. The parameter ϵ is small and positive.

Sufficient conditions are given for the existence of the solution of the given problem and for studying the behavior of these solutions, especially in the neighborhood of the turning points, as $\epsilon \to 0^+$. The principal assumptions are that the appropriate reduced problems $0 = f(x, u, u', 0), -1 < x < 0,\ 0 < x < 1, -1 < x < 1, u(-1) = B_1$, or $u(1) = B_2$ have smooth solutions and that the function f is of class C^1 and sufficiently well behaved. Howes (1975)

42. $$y'' + y - \epsilon f(y, y', \theta + \delta) = 0, y(0) = A, y'(0) = 0,$$

where $\theta = wx, w$ is a frequency, and δ is a phase shift; A and ϵ are complex numbers. The function f is analytic in its arguments.

Existence results for periodic solutions of the given problem are established. Several specific functions f are chosen. The proof is through changing the given problem to an integral equation and the use of implicit function theorem. Stoker (1950), p.224.

43. $$y'' + c_1 y + c_3 y^3 + \epsilon f(y, y', x) = 0,$$

where c_1, c_3, and $\epsilon \ll 1$ are constants.

An approximate oscillatory solution in terms of Jacobi elliptic functions, namely

$$y(x) = A(x) \text{ cn } (\psi(x), \mu^2(x)), \psi(x) = \int_0^x w(x)dx - \phi(x),$$

where the amplitude A, parameter μ^2, frequency w, and phase ϕ are made time dependent is found. The approximate method is applicable when

$$c_1 > 0, c_3 > 0 \quad \text{or} \quad c_1 < 0, c_3 < 0 \quad \text{or} \quad c_1 < 0 \quad \text{and} \quad c_3 > 0.$$

Transformation properties of elliptic function with respect to their parameters are utilized for improving the previous methods. Yuste and Bejarano (1989)

44. $$y'' + w(y) + \epsilon f(y, y', x) = 0,$$

where $w(y)$ is a strongly nonlinear term.

It is assumed that the unperturbed equation $y'' + w(y) = 0$ has a known solution. A solution of the perturbed equation is found in terms of the known solution of the reduced equation. Autonomous case with $f = f(y)$ is treated more explicitly. The method of strained coordinates is used. Papoulis (1958)

45. $$y'' - R(x, y, y') = 0,$$

where $R(x, y, y')$ is a rational and irreducible function of y and y', with coefficients analytic in x.

It may be proved that a necessary condition for the given DE to have poles as its only movable singularities is that it has the form

$$y'' = A(x, y)y'^2 + B(x, y)y' + C(x, y),$$

where A, B, C are rational functions of y with coefficients analytic in x; more specifically:

(a) All the poles of $A(x, y)$ regarded as functions of y are simple.

(b) The poles of $B(x, y)$ and $C(x, y)$ regarded as functions of y are included among those of $A(x, y)$ and are simple.

(c) Let $D(x, y)$ be the least common denominator of the partial fraction of $A(x, y)$.

Then the degrees of the polynomials $B(x, y)/D(x, y)$ and $C(x, y)/D(x, y)$ in y are at most 4 and 6, respectively. Bureau (1964)

46. $$y'' - R(x, y, y') = 0,$$

where R is a rational function of y' and is algebraic in x and y.

The given DE is changed to the system

$$\frac{dx}{dz} = f_1(x, y, z), \qquad \frac{dy}{dz} = f_2(x, y, z). \tag{1}$$

Wide conditions on f_1, f_2 are given under which (1) has no movable essential singularities. For such a system, if $z = z_0$ is a movable singularity, then for $z \to z_0$, one will have $y \to y_0, x \to x_0(x_0, y_0$ possibly infinite); the method of construction of such a solution is given. This method is stronger than Painlevé's in that it enables one to study equations of the form (1), having multivalued movable singularities, and to study effectively the character of solutions in the neighborhood of a movable singularity. Erugin (1952)

47. $$y'' - f(y', y, x) = 0,$$

where $f(y', y, x)$ is an irrational function. In this case the situation is quite different from the one when $f(y', y, x)$ is a rational function of y' and y, since in the latter case, the system is equivalent to $y' = u, u' = f(u, y, x)$ with rational second members. The irrational $f(y', y, x)$ case leads to a system of $n > 2$ with rational right-hand sides. The study is then fundamentally different from that with $n = 2$. This is why Painlevé began his investigation with the case in which $f(y', y, x)$ is rational. See Eqn. (2.20.6) for an illustration of this situation. Erugin (1980)

48. $$y'' + yF(y'^2, y^2, x) = 0,$$

where $F(u, v, w) \in C\{[0, \infty) \times [0, \infty) \times (-\infty, \infty)\}$.

Suppose that the function $F(u, v, w)$ satisfies for some functions $a(x)$ and $f(y)$ for all u, x, and positive

$$y, yF(u^2, y^2, x) \geq a(x)f(y)(y \geq 0),$$

where $a(x)$ is summable in any finite interval. If there exists a positive concave function $\phi(x)$ such that $\int_0^\infty a(x)\phi(x)dx = +\infty$, then all solutions of the given DE are oscillatory. Chen (1973)

49. $$y'' - x^p y^q y'^r \phi(x, y, y') = 0, \; y(0) = 0, \; \lim_{x \to 0} x^{-\alpha} y(x) = k,$$

where k, p, q, r are real, $\alpha, k > 0, \phi \in C(I \times R^2)$, and $\phi(0, 0, 0) > 0$.

Existence and uniqueness of the solution of IVP are considered. Ad'yutov and Klokov (1989)

2.25 $xy'' + g(x, y, y') = 0$

1. $$\boldsymbol{xy'' + y' - y^4 = 0, y(0) = 1.}$$

The given system has a formal power series solution

$$y = \sum_{n=0}^{\infty} A_n x^n, \text{ where } A_0 = 1,$$

and

$$B_m = \sum_{k=0}^{m} A_k A_{m-k'} A_{n+1}(n+1)^{-2} \sum_{m=0}^{n} B_m B_{n-m}.$$

For n between 40 and 60, the ratio of coefficients can be approximated by

$$A_{n-1}/A_n = 0.79493729 + 0.2648880n^{-1} + 0.118004n^{-2}$$

correct to eight decimal places, implying the radius of convergence to be very close to 0.7949373.

Shift the origin to an arbitrary point h by $x = h(1-u)$; the given DE becomes

$$\frac{d}{du}\left[(1-u)\frac{dy}{du}\right] = hy^4, \tag{1}$$

with the solution

$$y_1 = \left[\frac{10}{9hu^2}\right]^{1/3}\left[1 - \frac{2u}{21} - \frac{11u^2}{882} + \cdots\right], \tag{2}$$

so the equation (1) has a movable singularity. If h is determined so that $y = 1$ when $u = 1$, the point $x = h$ will be a singularity for the given problem. It may be the one that limits the convergence, so $\rho = h$. The leading factor in (2) suggests that we may remove the singularity by writing $y = w^{-2/3}$, which promises to give a rapidly convergent series. So the equation for w is

$$\frac{5x}{3}\left(\frac{dw}{dx}\right)^2 - w\frac{d}{dx}\left(x\frac{dw}{dx}\right) = 3/2, w(0) = 1.$$

When $x = \rho, u = 0, y = \infty$, and $w = 0$, ρ is computed as a root of $w(x) = 0$. The expansion for w is

$$w = 1 - (3/2)x + (3/8)x^2 - (5/48)x^3 + \cdots = \sum_{n=0}^{\infty} a_n x^n,$$

where

$$a_k = \frac{1}{3k^2}\sum_{j=1}^{k-1} j(5k - 8j)a_j a_{k-j}, k > 1, a_1 = -1.5. \tag{3}$$

The series (3) is slowly convergent. D'Alambert's test gives $\rho = 0.795$. First eight coefficients alternate in sign, then they have all negative sign; with 100 terms, $\rho = 0.7949366807$. Jordan (1986)

2. $$\boldsymbol{xy'' + y' + \sin y = 0.}$$

If we put $z = e^{iy}$, z satisfies a particular case of the third Painlevé equation. Kruskal and Clarkson (1992)

3. $$\boldsymbol{xy'' + y' - \sin y = 0.}$$

One interesting result about the given DE is that there exists a $\delta \geq 1$ such that through the point (x_0, π) there pass exactly three bounded solutions if $x_0 > \delta$, and exactly one bounded solution if $0 < x_0 \leq \delta$. Sansone (1962)

4. $$\boldsymbol{xy'' + y' - (1/2)(e^y - e^{-y}) = 0.}$$

The given DE arises from the sinh-Gordon equation by a similarity transformation. Writing $e^y = z$, we obtain $z'' = z'^2/z - z'/x + (z^2 - 1)/(2x)$, which is a special case of P_{III}. Drazin and Johnson (1989), p.189

5. $$\boldsymbol{xy'' + y' - kxe^y = 0, y = 1} \text{ at } \boldsymbol{x = 0,}$$

where k is a constant.

Differentiating the given DE, we have

$$(xy')'' = ke^y(1 + xy'). \tag{1}$$

Setting $v = 1 + xy'$ in (1) and eliminating the exponential with the help of the given DE, we have

$$xv'' = vv'. \tag{2}$$

Using $v'' = v'\dfrac{dv'}{dv}$ and noting that $\dfrac{d}{dv}(xv') = 1 + x\dfrac{dv'}{dv}$, we find from (2) that

$$\frac{d}{dv}(xv') = v + 1, \tag{3}$$

after division by v'. Integration of (3) gives

$$xv' = v^2/2 + v + C. \tag{4}$$

The condition $v(0) = 1$ gives the constant $C = -3/2$. Separating the variables in (4) and integrating gives v as a rational function of x^2; then returning to y' and integrating, we obtain

$$y = \ln\{A(a^2 - x^2)^{-2}\},$$

where A and a are arbitrary constants. The additional condition $y = 1$ when $|x| = 1$ leads to the quadratic in a^2: $e(1 - a^2)^2 = A$ and $A = 8a^2/k$. Thus, selecting the root greater than unity, we have

$$y(x) = \ln\{(8a^2/k)(a^2 - x^2)^{-2}\},$$

with

$$a^2 = 1 + 4/(ke) + \{(1 + 4/(ke))^2 - 1\}^{1/2}.$$

Hart (1980)

6. $$\boldsymbol{xy'' + 2y' + by + cy^2 = 0,}$$

where b and c are constants.

The following results are proved:

(a) Proper solutions of the given DE (i.e., solutions that remain bounded and continuous together with their first derivatives for $0 < x < x_0$) as $x \to 0$ through positive values are such that $xy \to 0$ as $x \to 0$.

(b) There exist improper solutions of the given DE tending to $-\infty$ for finite $x \to 0$. There are no proper solutions y tending to $-\infty$ as $x \to 0$ through positive values.

(c) All solutions of the given DE remaining positive as $x \to 0$ through positive values are proper solutions as $x \to 0$, with y, y' remaining bounded for all finite $x \neq 0$.

Putting $y = z/x, x = e^{-t}$, the given DE goes to

$$\ddot{z} + \dot{z} + be^{-t}z + cz^2 = 0. \tag{1}$$

The leading behavior of (1) is $z = 1/(ct)$ as $t \to \infty$; so we set $z = 1/(ct) + \psi$ to obtain

$$\ddot{\psi} + \dot{\psi} + 2\psi/t = -2/(ct^3) - be^{-t}/(ct) - b\psi e^{-t} - c\psi^2 \equiv f(\psi, t). \tag{2}$$

Equation (2) is solved by variation of parameters to obtain an integral equation for ψ:

$$\psi = A\psi_1 + B\psi_2 + \psi_1 \int_{t_0}^{t} \psi_2 f(\psi, t')/(\dot{\psi}_1\psi_2 - \psi_1\dot{\psi}_2)dt' - \psi_2 \int_{t_0}^{t} \psi_1 f(\psi, t')/(\dot{\psi}_1\psi_2 - \psi_1\dot{\psi}_2)dt',$$

where ψ_1 and ψ_2 are linearly independent solutions of the homogeneous part of (2). A first correction to the leading behavior is found to be $z = 1/(ct) - 2/(ct^2 \ln t) + \cdots$ as $t \to \infty$, or

$$y = -1/(cx \ln x) - 2 \ln(-\ln x)/(cx \ln^2 x) + \cdots \text{ as } x \to 0.$$

Billigheimer (1968)

7. $$\boldsymbol{xy'' + 2y' + y^3 = 0.}$$

The given DE arises from the PDE $u_{\xi\eta} = u^3$ through a similarity ansatz, and can be shown to possess the Painlevé property. Euler, Steeb, and Cyrus (1989)

8. $$\boldsymbol{xy'' + 2y' + \sum_{m=0}^{n} b_m x^m y^m = 0,}$$

where b_m are constants and n is a positive integer.

A first integral of the given DE is

$$I = (xy' + y)^2 + 2\sum_{m=0}^{n} \left[\frac{b_m}{m+1}\right] (xy)^{m+1}.$$

Ranganathan (1992a)

9. $$\boldsymbol{xy'' + 2y' - (7/4)x(y')^2/y + \lambda x - y = 0, y(0) = 1,}$$

where λ is a constant.

For $x < 1$, a solution is found in the form $y = 1 + \sum_n C_n x^n$ with $C_1 = 1/2, C_2 = (15-16\lambda)/96, C_3 = (31-48\lambda)/768, C_4 = (896\lambda^2-1800\lambda+879)/92160$, etc. For $1 < x \ll x_0$, where x_0 is a large number, the solution is found in the form

$$y = \lambda x\{1 + 1/(4x) + \cdots\}. \tag{1}$$

The solution (1) is not found adequate; it is found again by a Padé approximation: $y_{\text{Pade}} = P_{n+1}(x)/P_n(x)$ for different n. Numerical results are found and physically interpreted with reference to electron density, etc. Tu (1991)

10. $$\boldsymbol{xy'' + (1+a)y' - k(y^2 + 2y) = 0,}$$

where a and k are constants.

The given DE arises in boundary layer theory. For $a = -1/2$, it admits the integrating factor $2y'$, so that we may obtain

$$(8k/3)^{1/2}x^{1/2} + B = \int \frac{dy}{(y^3 + 3y^2 + A)^{1/2}}, \tag{1}$$

where A and B are constants. The quadrature in (1) may be evaluated. Burde (1990)

11. $$\boldsymbol{xy'' + ay' + by + cy^2 = 0,}$$

where a, b, c are constants.

It is shown that there exists a one-parameter set of solutions bounded at $x = 0$, having in the neighborhood of the origin the series expansion

$$y = y_0 - \{y_0(b + cy_0)/a\}x + \{y_0(b + 2cy_0)(b + cy_0)/2a(a+1)\}x^2 + \cdots,$$

convergent at least for $|x| < x_1 = |a|/(b + c|y_0|)$, where $y = y_0$ at $x = 0$.

To demonstrate the theorem, one writes the series solution in the form $y = \sum_0^\infty d_n x^n$, where

$$d_{n+1} = -\frac{bd_n + c(d_0 d_n + \cdots + d_n d_0)}{(n+1)(n+a)},$$

which is dominated by the geometric series $y = \sum A(Mx)^n$, where $A = |d_0|, M = (b + c|d_0|)/|a|$. Other qualitative results are as follows:

(a) For $a > 1/2$ there are no oscillatory solutions as $x \to 0$ through positive values of the given DE.

(b) There exist three kinds of proper solutions, i.e., solutions remaining bounded and continuous together with their first derivatives for $0 < x < x_0$:

  (i) Trivial solutions $y \equiv 0, y = -b/c$.

  (ii) Ultimately monotonic solutions tending to $y = C$, where C is a constant, as x tends to zero.

  (iii) Ultimately monotonic solutions tending to $+\infty$ as x tends to 0 through positive values.

(c) There are also improper solutions tending to $-\infty$ for finite positive x.

Billigheimer (1968).

12. $$\boldsymbol{xy'' - yy' = 0.}$$

The given DE is Cauchy-Euler w.r.t. x. So put $t = \ln x$ to obtain

$$\ddot{y} - (y+1)\dot{y} = 0, \quad \cdot \equiv \frac{d}{dt}. \tag{1}$$

A first integral for (1) is

$$y' = C + (y+1)^2/2, \tag{2}$$

where C is an arbitrary constant. Hence the solutions are

$$y = 2C_1 \tan(C_1 \ln x + C_2) - 1; y = C_0;$$

C_i $(i = 0, 1, 2)$ are arbitrary constants. Bender and Orszag (1978), p.153

13. $\boldsymbol{xy'' + (y-1)y' = 0.}$

With $y(x) = \eta(\xi), \xi = \ln|x|$, we get

$$\eta'' - 2\eta' + \eta\eta' = 0. \tag{1}$$

Equation (1) integrates to the first order DE

$$2\eta' = 4\eta - \eta^2 + C, \tag{2}$$

where C is an arbitrary constant. Equation (2) is easily integrated in terms of elementary functions, since the variables separate; the form depends on the sign of C, etc. Kamke (1959), p.563; Murphy (1960), p.388

14. $\boldsymbol{2xy'' - y'^2 + 1 = 0.}$

Put $y' = p, y'' = (dp/dx)$, to get $2x(dp/dx) = p^2 - 1$,

$$\frac{dx}{2x} = \frac{dp}{p^2-1} = (1/2)\left[\frac{1}{p-1} - \frac{1}{p+1}\right] dp. \tag{1}$$

Integration of (1) and then (with $p = dy/dx$) another integration lead to

$$y = -x - (2/c_1)\ln|c_1 x - 1| + c_2.$$

The solutions $y = \pm x + c$ follow from the special first integrals $y' = \pm 1$. c, c_1, and c_2 are arbitrary constants. Rabenstein (1972), p.47

15. $\boldsymbol{xy'' - 2(y'^2 - y') = 0.}$

Put $v = y'$ so that

$$x\frac{dv}{dx} = 2(v^2 - v). \tag{1}$$

The solution of (1) is

$$v = \frac{dy}{dx} = \frac{1}{1 - c_1 x^2}. \tag{2}$$

Equation (2) with $c_1 = a^2$ integrates to $y = \{1/(2a)\}\ln\{|1+ax|/|1-ax|\}$, while with $c_1 = -b^2$, it integrates to $y = (1/b)\tan^{-1} bx + c_2$. Finally, with $v = 0$ and $v = 1$ from (1), we get the solution $y = c$ and $y = x + c$. Rabenstein (1972), p.45

16. $$\boldsymbol{xy'' + yy'^2 = 0, x > 0.}$$

The given DE has a nonoscillatory solution $y(x) = x^{1/2}$. Grace and Lalli (1987a)

17. $$\boldsymbol{y'' - y'/x + y'^2 = 0.}$$

Putting $x^2 = z$, we have $y'' + y'^2 = 0$, $' \equiv d/dz$. An integration gives $y' = e^{-y+C_1}$. A second integration gives

$$y = \ln(x^2 + C_2) + C_1;$$

C_1 and C_2 are arbitrary constants. Reddick (1949), p.191

18. $$\boldsymbol{xy'' - x^2y'^2 + 2y' + y^2 = 0.}$$

A finite transformation $x = A^m X, y = A^n Y$ shows that with $n = -m, xy$ is an invariant. Using this fact, we find that the solutions are

$$x = \exp\left\{\int (C_1 e^{\eta} + 2\eta + 1)^{-1} d\eta + C_2\right\}, \eta = xy, \text{ and } y = 0.$$

C_1 and C_2 are arbitrary constants. Kamke (1959), p.563; Murphy (1960), p.388

19. $$\boldsymbol{xy'' + a(xy' - y)^2 - b = 0,}$$

where a and b are constants.

Putting $y = xu(x)$ leads to a Riccati equation

$$x^2u'' + ax^4u'^2 + 2xu' = b$$

for u'; further, if we write $ax^2u' = v'/v$, we have a linear DE with constant coefficients:

$$v'' = abv.$$

Kamke (1959), p.564; Murphy (1960), p.389

20. $$\boldsymbol{2xy'' + y'^3 + y' = 0.}$$

The given DE is invariant under transformation $x = A^m X$ and $y = A^m Y$. The solution is easily found to be $(y + C_1)^2 = 2C_2 x - C_2^2; C_1$ and C_2 are arbitrary constants. Kamke (1959), p.564; Murphy (1960), p.389

21. $$\boldsymbol{xy'' - y' - 2(x^2 + y'^2)^{1/2} = 0.}$$

We have, with $y' = v, y'' = dv/dx$,

$$x\frac{dv}{dx} - v - 2(x^2 + v^2)^{1/2} = 0$$

or

$$\frac{dv}{dx} = \{v + 2(x^2 + v^2)^{1/2}\}/x. \tag{1}$$

Equation (1) is homogeneous in v and x and hence easily integrated by putting $v = xu$, etc. A second integration then gives

$$y = (c_1/8)x^4 - (1/2c_1)\ \ln|x| + c_2.$$

c_1 and c_2 are arbitrary constants. Rabenstein (1972), p.47

22. $$\boldsymbol{xy'' + y' + x^{10}y^2 - [x^2y\ \cos\ y + y'(\sin\ y')/x + x\ \sin\ x] = 0.}$$

It is shown that every solution of the given DE satisfies the condition

$$\int_{x_0}^{\infty} y^4(x)dx < \infty.$$

Spikes (1977)

23. $$\boldsymbol{xy'' + \exp(-y') - 1 = 0.}$$

Writing in the given DE $y' = v$, we have

$$e^v v'/(e^v - 1) = 1/x. \tag{1}$$

On integration, (1) gives $e^v - 1 = c_1 x$, or

$$\begin{aligned} \frac{dy}{dx} = v &= \ln(1 + c_1 x), \quad y = \int \ln(1 + c_1 x)dx + c_2 \\ &= (1/c_1)\{\ln(1 + c_1 x) - 1\}(1 + c_1 x) + c_2. \end{aligned}$$

c_1 and c_2 are arbitrary constants. Rabenstein (1972), p.47

24. $$\boldsymbol{xy'' - f(y', y, x) = 0, \quad y(0) = \alpha, y'(0) = \beta,}$$

where f is a regular function of its arguments, and α and β are constants.

Conditions are found on the function f such that the given system has a finite integral. It is assumed that $f(\beta, \alpha, 0) = 0$. The conditions arise from reduction of the given system to Briot-Bouquet form. Forsyth (1959), p.193

25. $$\boldsymbol{\mu xy'' - y(1 + y') = 0, y(0) = 1, y(\infty) = 0;}$$

$\mu > 0$ is a constant.

It is proved that the given singular BVP has a unique solution. Hallam and Loper (1975)

26. $$\boldsymbol{\mu xy'' - y(1 + y') = 0, y(0) = 1, y(\phi^{-1}) = 0,}$$

where $0 < \mu, 0 < \phi < 1$ are constants.

It is proved that the given problem has one and only one real solution $y(x, \mu, \phi)$ satisfying the given BC. Numerical solution is also discussed for $\mu = 1, 1/2, 1/5, 1/10, 1/20, 1/50$, and $1/100$ for $\phi \ll 1$. Hallam and Loper (1975)

27. $$\boldsymbol{\mu xy'' - (1 - x)y' - yy' = 0, y(0) = 0, y'(0) = 0,}$$

where μ is a constant.

Neglecting the nonlinear term in the first instance as small, it is possible to integrate once:

$$y' = Cx^{1/\mu}e^{-x/\mu}, \tag{1}$$

where C cannot be found by using the IC. $y'(0)$, is, however, satisfied automatically. Assuming that $\mu = A^{-1}$, where A is a positive integer, it is possible to find a closed-form solution of (1):

$$y = \overline{C}\left[e^{-Ax}\sum_{q=0}^{A}\{(Ax)^q/q!\} - 1\right], \tag{2}$$

where $\overline{C} = -C(A-1)!/A^A$ and the IC $y(0)$ has been satisfied. The solution (2) is used to solve the given IVP numerically. Hallam and Loper (1975)

28. $$\boldsymbol{a^2xy'' - (xy' - y)^2 + c^2 = 0,}$$

where a and c are constants.

Put

$$xy' - y = -(a^2/\eta)\frac{d\eta}{dx}, \tag{1}$$

so that the given DE changes to

$$\frac{d^2\eta}{dx^2} - (c^2/a^4)\eta = 0,$$

with the solution $\eta = Ae^{cx/a^2} + Be^{-cx/a^2}$. Now (1) becomes

$$xy' - y = -c[e^{2cx/a^2} - B]/\{Ae^{2cx/a^2} + B\},$$

which integrates to give the solution of the given DE as

$$y/x = C - c\int\left[\frac{Ae^{2cx/a^2} - B}{Ae^{2cx/a^2} + B}\right](1/x^2)dx. \tag{2}$$

It has two arbitrary constants, C and $B/A = k$, say. The integral can be expanded as

$$y = c(1-k)/(1+k) - \{4k/(1+k)^2\}(c^2/a^2)x\log x + a \text{ regular function of } x.$$

This regular function vanishes with x. The solution is regular if $w(0) = c$, so that $k = 0$ and the logarithmic term vanishes. For all other assigned initial values of y, the term $x\log x$ remains. Forsyth (1959), p.195

2.26 $x^2y'' + g(x, y, y') = 0$

1. $$\boldsymbol{x^2y'' - 2y(1+y)^2 = 0, x > 0, y(0) = -1.}$$

The given problem has a one-parameter family of continuous, uniformly bounded negative solutions on $[0, \infty]$, namely

$$y(x) = -1/(1+cx), c \geq 0.$$

Gross (1963)

2. $$\boldsymbol{x^2y'' - a(y^n - y) = 0,}$$

where a is constant.

With $y(x) = x^{(1/2)(1-k)}\eta(\xi), \xi = x^k, k = (1-4a)^{1/2}$, the given DE becomes

$$\eta'' = \{a/(1-4a)\}\xi^{(n-1)/(2k)-(n+3)/2}\eta^n.$$

Now see Eqn. (2.3.36). Kamke (1959), p.564

3. $$\boldsymbol{x^2y'' + a(e^y - 1) = 0,}$$

where a is constant.

Putting $a = 2/(A-1)$ and $z = y(x) + \log\{(2A-2)/bs^2\}, x = s^{1-A}$ we get

$$sz''(s) + Az'(s) + bse^z = 0.$$

Now see (A) in Eqn. (2.8.81). Kamke (1959), p.564; Murphy (1960), p.389

4. $$\boldsymbol{9x^2y'' + ay^3 + 2y = 0,}$$

where a is a constant.

With $y = x^{1/3}\eta(\xi), \xi = x^{1/3}$, we get

$$\eta'' + a\eta^3 = 0. \tag{1}$$

Multiplying (1) by $2\eta'$ and integrating, we have

$$\eta'^2 + (1/2)a\eta^4 = C, \tag{2}$$

where C is an arbitrary constant. Equation (2) can be solved in terms of elliptic functions. Kamke (1959), p.565; Murphy (1960), p.390

5. $$\boldsymbol{9x^2y'' - (x-2)y - 2y^3 = 0.}$$

Writing $y = \eta w(\eta), \eta = x^{1/3}$ the given DE becomes $d^2w/d\eta^2 = 2w^3 + \eta w$, which is a special case of P_{II}. Drazin and Johnson (1989), p.188

6. $$\boldsymbol{16x^2y'' + 4xy' + 2y + (1/2)y^3 = 0.}$$

The given DE has a Painlevé property since it can easily be shown that it has the Laurent series expansion about an arbitrary movable singularity $x = x_0$:

$$y = \pm 8ix_0 \sum_{j=0}^{\infty} A_j(x - x_0)^{j-1},$$

where

$$A_0 = 1, A_1 = \frac{5}{8x_0}, \ A_2 = \frac{-5}{64x_0^2}, A_3 = \frac{5}{128x_0^3},$$

A_4 is arbitrary.

Actually, the given DE may be integrated exactly: Put $y = x^{1/4}f(x^{1/4})$ so that

$$f''(\eta) + (1/2)f^3(\eta) = 0, \eta = x^{1/4}, \tag{1}$$

where $' \equiv d/d\eta$. Equation (1) has a first integral

$$f'^2 + (1/4)f^4 = I_1. \tag{2}$$

By the scaling $f(y) = 2(I_1/4)^{1/4}g((I_1/4)^{1/4}y)$, (2) can be put in the standard form for integration in terms of an elliptic function:

$$g(\bar{y}) = \pm i2^{1/2}\mathrm{ds}(2^{1/2}(\bar{y} - \bar{y}_0)),$$

where $\bar{y}_0$ is some initial phase and the elliptic function ds has the parameter $m = 1/2$. Fournier, Levine, and Tabor (1988)

7. $\boldsymbol{x^2y'' + xy' - (\alpha/2)\tanh y \cosh^{-2}y - (\gamma/4)x \sinh 2y - (\delta/8)x^2 \sinh 4y = 0,}$

where α, γ, and δ are constants.

The given DE is obtained from Eqn. (2.16.34) by putting $y = \tanh^2 u$, and then replacing u by y. Call $y = Y(x) = Y(x_0, y_0, y_1; x)$ a solution of the given DE satisfying IC

$$y(x_0) = y_0, y'(x_0) = y_1,$$

where x_0 is a positive constant and y_0, y_1 are real constants. Now we assume that

(a) $Y(x)$ is prolonged over R^+ and is a real-valued function.

(b) $|Y(x)| < \infty$ for $x \geq x_0$.

Then it is proved that if $\delta < 0$, then $y(x)$ and $y'(x)$ admit asymptotic representations

$$\begin{aligned} y(x) &= R_1\{1 + O(x^{-1})\}x^{-1/2}\cos[\phi(R_1^2, \theta_1; x)], \\ y'(x) &= -R_1(-\delta/2)^{1/2}\{1 + O(x^{-1})\}x^{-1/2}\sin\{\psi_1(R_1^2, \theta_1; x)\} \text{ as } x \to +\infty. \end{aligned}$$

Here R_1 and θ_1 are real constants depending on (x_0, y_0, y_1) and satisfying $R_1 > 0, 0 \leq \theta_1 < 2\pi$, and ϕ_1 and ψ_1 are real functions defined by

$$\phi_1(r, \theta; x) = (-\delta/2)^{1/2}x - \{(\gamma/4)(-2/\delta)^{1/2} - (-\delta/2)^{1/2}r\}\log x + \theta + O(x^{-1}),$$

$$\psi_1(r, \theta; x) = \phi_1(r, \theta; x) + O(x^{-1}), \text{ as } x \to +\infty.$$

Shimomura (1987a)

8. $$\boldsymbol{x^2y'' + 2xy' + x^\lambda y^n = 0,}$$

where λ and n are parameters.

The given DE is equivalent to the predator-prey equation

$$\begin{aligned} \frac{d\xi}{dt} &= -\xi(1 + \xi + \eta), \\ \frac{d\eta}{dt} &= \eta(\lambda + 1 + n\xi + \eta), \end{aligned}$$

after the change of variables $\xi = xy'/y, \eta = x^{\lambda-1}y^n/y'$, $t = \log|x|$. See Sachdev (1991). Jordan and Smith (1977), p.65

9. $$\boldsymbol{x^2y'' + 2xy' + x^ay^n = 0, a > 0, n \geq 0.}$$

It is proved that the given DE has one and only one solution in an interval $0 < x < x_1$, with the property that $y > 0$ in this interval and $\lim_{x \to +0} y = c, 0 < c < \infty$. The method used is one of successive approximation. It is also proved that the condition $2a - n + 1 > 0$ is necessary and sufficient for $y(x)$ to have a zero at a finite distance. Sansone (1940)

10. $$\boldsymbol{x^2y'' + 2xy' + x^2y - \sum_{k=1}^{N} d_k(x)y^k = 0,}$$

where $d_k(x)(1 \leq k \leq N)$ are continuous functions on $I_a = [1, \infty)$.

It is shown that a constant c can always be chosen such that with $|y(1)| + |y'(1)| \leq c$, the given DE has a solution that satisfies

$$\begin{aligned} y(x) &= x^{-1}\{(\delta_1 + O(1))e^{+ix} + (\delta_2 + O(1))e^{-ix}\}, \\ y'(x) &= ix^{-1}\{(\delta_1 + O(1))e^{ix} - (\delta_2 + O(1))e^{-ix}\}, \end{aligned}$$

where $\delta_k (k = 1, 2)$ are constants, as $x \to \infty$. Pinto (1991)

11. $$\boldsymbol{x^2y'' + 2xy' - \sin 2y = 0.}$$

Put $z = \cos y$ and $r = (\lambda x)^2$, with arbitrary λ, to obtain

$$(1 - z^2)\left(2r^2\frac{d^2z}{dr^2} + 3r\frac{dz}{dr}\right) + 2r^2z\left(\frac{dz}{dr}\right)^2 + z(1 - z^2)^2 = 0.$$

Now see Eqn. (2.40.17). Ferreira and Neto (1992)

12. $$\boldsymbol{x^2y'' + 2xy' - \sin 2y - \beta^2x^2 \sin y = 0,}$$

where β is constant.

To get the behavior for large x, we first note that y tends to an odd or an even integer multiple of π. Writing $y^* = y - N\pi, N$ an integer, and $\sin y \approx (-1)^N y^*$, the dominant terms of the given DE give

$$(x^2y^{*\prime})' - (-1)^N\beta^2x^2y^* = 0$$

with the solution

$$\begin{aligned} y^* &= \lambda e^{-\beta x}/x, \ N \text{ even} \\ y^* &= \eta(1/x)\sin(\beta x + \delta), \ N \text{ odd} \end{aligned} \tag{1}$$

with λ, η, δ constants.

Numerical solutions starting from $y = 0$ and $y = \pi$ at $x = 0$ and having the asymptotic behavior (1) are shown graphically. There are no solutions approaching odd multiples of $\pi/2$ at infinite distance. There are solotonic solutions with finite energy having monotonic behavior, approaching even multiples of π as $x \to \infty$. The solution has $y \to y \pm 2$ symmetry. The values of $y'(0)$ and λ for the solotonic behavior are given for $N = 1, 2, 3$. A suitable formula for λ is derived by integrating the given DE from 0 to ∞ and using conditions at $x = 0$ and (1). The solutions that approach odd multiples of π present an oscillatory behavior at infinity and are finite at the origin. Ferreira and Neto (1992a)

13. $$x^2y'' + 2xy' + x\lambda f(y) = 0, \lambda > 0,$$

where $f(0) = 0, 0 \leq f'(y) \leq L$ for $0 < y < C$, and $f(y)$ is continuous and nondecreasing in $0 \leq y < \infty$; L and C are constants.

The given problem is reduced to a nonlinear Volterra integral equation using a Picard iteration process; the existence and uniqueness of a solution such that $\lim_{x\to 0+} y(x) = c, 0 < c < \infty$, and $y(x) > 0, y'(x) < 0$ in $0 < x \leq x^*$ are proved. A lower bound for zeros of the solution as well as the conditions of their existence are found. Cimino (1953)

14. $$x^2y'' + xy' - F_0(x, e^{-y}, xe^y) - F_1(x, e^{-y}, xe^y) - F_2(x, e^{-y}, xe^y)x^2y'^2 = 0,$$

where $F_j(x, \xi, \eta)(j = 0, 1, 2)$ are holomorphic and bounded in the polydisk $|x| < r_0, |\xi| < r_1, |n| < r_1 (r_0, r_1 > 0)$ and satisfy $F_j(0, 0, 0) = 0$.

The given DE contains transformed forms of P_{III}, P_V and P_{VI} as special cases. The series expansions of the solutions are obtained and their convergence determined. In some special cases, the representations of solutions of Painlevé equations are valid in the domain where the solutions have infinitely many movable poles and zeros. Shimomura (1987)

15. $$x^2y'' + 3xy' - 1/(x^4y^3) = 0.$$

Using the argument of scale invariance with respect to x and y both, we may write $y(x) = u(x)/x$, so

$$x^2u'' + xu' + u = 1/u^3. \tag{1}$$

Equation (1) is equidimensional w.r.t. x. We, therefore, put $x = e^t$ to get

$$u'' + u = 1/u^3, ' \equiv \frac{d}{dt}. \tag{2}$$

Equation (2) is autonomous, so we write

$$u' = p \text{ and } p\frac{dp}{du} = 1/u^3 - u. \tag{3}$$

The variables are separable. Integrating (3), we have

$$\frac{du}{dt} = p(u) = \pm[A - u^2 - 1/u^2]^{1/2}, \tag{4}$$

where A is an arbitrary constant. Integrating the separable equation (4), we get

$$u(t) = \pm[\cosh\ B + \sinh\ B \sin(2t + C)]^{1/2},$$

that is,

$$y(x) = \pm(1/x)[\cosh\ B + \sinh\ B \sin(2\ \ln\ x + C)]^{1/2},$$

where we have used the relations $y(x) = u(x)/x$ and $x = e^t$. Zwillinger (1989), p.295

16. $$x^2y'' + (2n + 1 + k)xy' + (n^2 + kn + 2)y + x^ny^2 = 0,$$

where n and k are constants.

The given DE arises in boundary layer theory. It admits an integrating factor $2x^{-n}y'$, provided that $k = \pm 5/3^{1/2}$ and $n = \mp 2/3^{1/2}$. Then, we obtain, on integration,

$$x^{2-n}y'^2 + (2/3)y^3 + A = 0. \quad (1)$$

If we set $A = 0$, we can integrate (1) again and obtain an explicit solution:

$$2y^{-1/2} + (-2/3)^{1/2}2x^{n/2}/n + B = 0, n = \mp 2/3^{1/2},$$

where B is a constant. Burde (1990)

17. $\boldsymbol{x^2y'' + (1/\beta)xy' + \{\alpha(\alpha-1)/\beta^2\}y + \{1/(\beta^2(N-1))\}y^N = 0,}$

where $\alpha = 2/(1-N)$ and $\beta = 2(1-\alpha)$ and N is a constant.

Writing

$$y = x^{-\alpha/\beta}f(x^{1/\beta}) = x^{1/(N+1)}f(x^{(N-1)/(2(N+1))}),$$

we get

$$f'' + \{1/(N-1)\}f^N = 0, \quad (1)$$

where the prime denotes derivative with respect to $\xi \equiv x^{1/\beta}$. Equation (1) is easily solved. A first integral is

$$f'^2 + \{2/(N^2-1)\}f^{N+1} = C, \quad (2)$$

where C is an arbitrary constant. The solution of (2) is expressible as a quadrature, which may be evaluated for special N values. Levine and Tabor (1988)

18. $\boldsymbol{x^2y'' - xy'(1-y) + y(1-y) = 0.}$

See Eqn. (2.13.6). Webb and Mckenzie (1991)

19. $\boldsymbol{x^2y'' + x(y+1)y' + y = 0.}$

The given DE is Cauchy-Euler type w.r.t. x. Hence put $x = e^t$ to get

$$\ddot{y} + y\dot{y} + y = 0, \quad \cdot \equiv \frac{d}{dt}. \quad (1)$$

Equation (1) is autonomous. Putting $\dot{y} = p(y)$, we have

$$p\frac{dp}{dy} + (p+1)y = 0.$$

The variables separate. A first integral, therefore, is $e^p/(p+1) = Ce^{-y^2/2}$; C is an arbitrary constant. Bender and Orszag (1978), p.34

20. $\boldsymbol{x^2y'' + x(2y+1)y' = 0.}$

The given DE is Cauchy-Euler type w.r.t. x. Put $x = e^t$ to get

$$y'' + 2yy' = 0, \quad ' \equiv \frac{d}{dt}. \quad (1)$$

Equation (1) is exact with the integral

$$y' + y^2 = C.$$

Now the variables separate. Here C is an arbitrary constant. Bender and Orszag (1978), p.34.

21. $$\boldsymbol{x^2 y'' + y(y - xy') = 0.}$$

The given DE results from similarity transformation of the nonlinear Landau damping equation. Put $q = dy/dz, z = \ln x$ to obtain an Abel equation of the second kind,

$$q\frac{dq}{dy} = (y+1)q - y^2, \tag{1}$$

or put $\phi = 1/q$ in (1) to get an Abel equation of the first kind,

$$\frac{d\phi}{dy} = -(y+1)\phi^2 + y^2\phi^3. \tag{2}$$

Equation (1) or (2) may be treated in the phase plane. Webb and Mckenzie (1991)

22. $$\boldsymbol{x^2 y'' - x(y-1)y' - y(y^2 - 1) = 0.}$$

Put $x = e^z$, so that

$$\frac{d^2y}{dz^2} - y\frac{dy}{dz} - y(y^2 - 1) = 0. \tag{1}$$

A particular first integral of (1) is $dy/dz = y^2 - 1$. A general first integral of (1) is found as follows. Put $y^2 - 1 = \phi$ and assume that $dy/dz = \phi(1 + w)$; then (1) reduces to

$$\phi\frac{dw}{d\phi} = -w(2w+3)/\{2(1+w)\}, \tag{2}$$

from which we obtain, on integration,

$$w^2(2w+3) = A^3/\phi^3, \tag{3}$$

where A is a constant, or

$$\left(\frac{dy}{dz} - y^2 + 1\right)^2 \left(2\frac{dy}{dz} + y^2 - 1\right) = A^3. \tag{4}$$

Putting $A = 0$, we have two possibilities; the first one leads to the singular first integral already identified. The second one yields

$$2\frac{dy}{dz} + y^2 - 1 = 0. \tag{5}$$

Equation (5) integrates to give

$$y = \frac{1 - (\mu/2)e^{-z}}{1 + (\mu/2)e^{-z}}, \tag{6}$$

where μ is a constant of integration. McVittie (1933)

23. $$x^2y'' - (1 + 2y/x^2)xy' + 4y = 0.$$

$y = x^2$ is a particular solution. For general solution, write $y = x^2u(t), x = e^t$, and hence solve the resulting autonomous equation. Bender and Orszag (1978), p.34.

24. $$x^2y'' + xy'^2 - yy' = 0.$$

The given DE is invariant under the scaling group $(x, y) \to (ax, ay)$. Symmetry argument shows that the invariants of the given DE are

$$u = \frac{y}{x}, w = \frac{dy}{dx}, \frac{dw}{du} = \frac{x^2 y_{xx}}{xy_x - y}.$$

Introducing u and w as the new variables, we have the first order equation

$$(w - u)\frac{dw}{du} + w^2 = uw. \tag{1}$$

Equation (1) has two families of solutions: $w = u$ or $dw/du = -w$, the latter integrating to $w = ce^{-u}$, where c is a constant. Reverting to original variables, we have two first order homogeneous equations,

$$\frac{dy}{dx} = \frac{y}{x}, \quad \text{or} \quad \frac{dy}{dx} = ce^{-y/x}. \tag{2}$$

The first of (2) has the solution $y = kx$, while the second has the implicit solution

$$\int \left[\frac{1}{ce^{-u} - u}\right] du = \log x + K, \tag{3}$$

where $u = y/x$. Here K is another constant. While (3) gives the general solution, $y = kx$ is a one-parameter family of singular solutions.

Alternatively, we may put $u = y/x, \hat{w} = \log x$, so that

$$y'(x) = (1 + u\hat{w}_u)/\hat{w}_u, xy''(x) = (\hat{w}_u^2 - \hat{w}_{uu})/\hat{w}^3.$$

The given DE takes the form

$$\frac{dz}{du} = (u + 1)z^2 + z, \tag{4}$$

where $z = d\hat{w}/du$. We may solve (4) as a Bernoulli equation or multiply it by $(z + uz^2)^{-1}$ and integrate to find that $u + \log(u + z^{-1}) = c_1$ or $z = (c_1e^{-u} - u)^{-1}$. Recalling that $z = \hat{w}_u = (c_1e^{-u} - u)^{-1}$ and integrating, we again obtain (3). Olver (1986), p.147

25. $$x^2y'' + ayy'^2 + bx = 0,$$

where a and b are constants.

With $y = x\eta(\xi), \xi = \ln x$, we obtain the autonomous form

$$\eta'' + a\eta(\eta' + \eta)^2 + \eta' + b = 0. \tag{1}$$

Hence put $\eta' = p, \eta'' = p(dp/d\eta)$, etc., to reduce (1) to first order. Kamke (1959), p.565; Murphy (1960), p.390

26. $$x^2y'' - [x^2/(y+1)]y'^2 + 2xy' - n^2x^2y(y+1) = 0,$$

where n is a constant.

An even-term power series solution is found and its convergence determined by using the method of majorants. Matsumoto (1953)

27. $$x^2y'' + y'^2 - 2xy' = 0.$$

Put $y' = z$, so that

$$x^2\frac{dz}{dx} + z^2 - 2xz = 0. \tag{1}$$

Put $z = 1/w$ in (1) so that $x^2(dw/dx) + 2xw = 1$. Therefore, on integration, $x^2w = x + c_1$, or $dx/dy = (x+c_1)/x^2$. Thus, $y = x^2/2 - c_1x + c_1^2\ln(x+c_1) + c_2$. c_1 and c_2 are arbitrary constants. Rabenstein (1972), p.47

28. $$x^2y'' - 2xy' + 2y(1-y) + 2x^2y'^2/(1-y) = 0.$$

The solutions of the given DE such that $y(0) = 0$ and $y'(0) \neq \infty$ are $y = (\beta x + \gamma x^2)/\ (1 + \beta x + \gamma x^2)$, where β and γ are arbitrary. The given system may first be made autonomous by writing $y = vx$, etc. Forsyth (1959), p.190

29. $$x^2y'' - 3xy' + y'^2 = 0.$$

Here $y = x^2$ is a special solution. To obtain the general solution, write $y = x^2 + y_1$, say, and integrate twice. The general solution is $y = x^2 + C_1\ln(x^2 - C_1) + C_2$. Reddick (1949), p.191

30. $$x^2y'' - a(xy' - y)^2 - bx^2 = 0,$$

where a and b are constants.

Direct quadrature (after substituting $y = vx$, etc.) shows that the solution is $y/x = \beta - (1/a)\int\{1/(x\eta)\}(d\eta/dx)dx$, where

$$\eta = AJ_0((ab)^{1/2}x) + BY_0((ab)^{1/2}x), \tag{1}$$

β being an arbitrary constant, and J_0, Y_0 denoting Bessel functions of order zero. To get a regular solution about $z = 0$, set $B = 0$ in (1). Forsyth (1959), p.190

31. $$4x^2y'' - x^4y'^2 + 4y = 0.$$

Put $y(x) = x^{-2}\eta(\xi), \xi = \ln x$ to obtain the autonomous form $4\eta'' - \eta'^2 + 4\eta\eta' - 20\eta' + 4\eta(7-\eta) = 0$; hence put $\eta' = p, \eta'' = p(dp/d\eta)$ to get a first order DE. Kamke (1959), p. 565

32. $$x^2y'' + 2xy' + x^2y - \lambda_1(x)y^n - \lambda_2(x)(y')^m = 0, n, m \geq 0, mn \neq 1,$$

where $x^{-(1+n)}\lambda_1(x)$ and $x^{-(1+m)}\lambda_2(x)$ are integrable functions on $[1,\infty)$. The asymptotic behavior of the given DE in terms of the solutions of the homogeneous part equal to zero is discussed. The cases $m, n \leq 1$, and $1 \leq m, n, mn \neq 1$ are discussed separately [see Eqn. (2.27.24)]. Medina and Pinto (1988)

33. $$\boldsymbol{x^2y'' - (ax^2y'^2 + by^2)^{1/2} = 0,}$$

where a and b are constants.

With $y(x) = x\eta(\xi), \xi = \ln x$, we obtain the autonomous form

$$\eta'' + \eta' = \{a(\eta' + \eta)^2 + b\eta^2\}^{1/2}. \tag{1}$$

Now put $\eta'(\xi) = p(\eta), \eta'' = p(dp/d\eta)$ in (1) to get the first order DE

$$p(p' + 1) = [a(p + \eta)^2 + b\eta^2]^{1/2}.$$

Kamke (1959), p.565; Murphy (1960), p.390

34. $$\boldsymbol{x^2y'' - H(xy' - y) = 0,}$$

where H is a given function of its argument. Put $xy' - y = w$ to obtain

$$x\frac{dw}{dx} = H(w). \tag{1}$$

Equation (1) is separable, so that

$$\int \frac{dw}{H(w)} = \log x + c, \tag{2}$$

where c is a constant. Solving (2), we write the explicit form $w = h(x)$. Thus,

$$xy' - y = h(x). \tag{3}$$

An integration of (3) gives $y = x\{\int x^{-2}h(x)dx + k\}$ as the general solution of the given DE. Olver (1986), p.153

35. $$\boldsymbol{x^2y'' + (a+1)xy' - x^a f(x^a y, xy' + ay) = 0,}$$

where $a \neq 0$ is constant.

With $y = x^{-a}\eta(\xi), \xi = x^a/a$, the given DE assumes the autonomous form

$$\eta'' = f(\eta, \eta'); \tag{1}$$

hence put $\eta' = p, \eta'' = p(dp/d\eta)$, etc., and reduce (1) to first order. Kamke (1959), p.564; Murphy (1960), p.389

36. $$\boldsymbol{x^2y'' - f(xy', y, x) = 0,}$$

where f is a regular function of its arguments.

The given system is reduced to Briot–Bouquet form to find conditions such that a solution exists with $y(0) = \alpha$, a constant, in the form

$$y = \alpha + \beta x + x^2(\gamma + v),$$

where v is one component of the vector satisfying the Briot–Bouquet system; β and γ are definite constants. Forsyth (1959), p.187

2.27 $(f(x)y')' + g(x,y) = 0$

1. $$x^{4/3}y'' + (4/3)x^{1/3}y' + [1/(4x)]\cos xy' + (x^2 \sin x)y^{1/3} = 0, x > 0.$$

All solutions of the given DE are shown to be oscillatory. Grace and Lalli (1987)

2. $$x^{4/3}y'' + (4/3)x^{1/3}y' - [1/(4x)](\cos x)y' + (x^2 \sin x)y^{1/3} = 0, x > 0.$$

All solutions of the given equation are shown to be oscillatory. Grace and Lalli (1987)

3. $$x^{3/2}y'' + (3/2)x^{1/2}y' + axy^2 = 0,$$

where a is a constant.

Writing

$$y(x) = x^{-1/2}V(\zeta), \zeta = x^{1/2},$$

we have

$$\ddot{V} + 4aV^2(\zeta) = 0, \quad \cdot \equiv \frac{d}{d\zeta}. \tag{1}$$

Equation (1) has the solution

$$\zeta = 2V_m \int_0^{V/V_m} (1 - z^3)^{-1/2} dz,$$

where $V_m^3 = 3/(32a)$. Hence the solution of the given equation may be found. Rosenau (1984)

4. $$x^{3/2}y'' + (3/2)x^{1/2}y' - \sum_{m=1}^{n} a_m x^m y^{2m} = 0,$$

where a_m are constants and n is a positive integer.

A first integral of the given DE is

$$I = [xy' + (1/2)y]^2 - 2\sum_m [a_m/(2m+1)](x^{1/2}y)^{2m+1}.$$

For the special case with one term, $m = 1$ on the RHS of the given DE, the solution may be found to be

$$y = [Ax^{1/4} + Bx^{3/4}]^{-2}, B^2 = 2a_1/3,$$

where A is an arbitrary constant. Ranganathan (1992a)

5. $$x^m y'' + (\nu + m)x^{m-1}y' + y^s = 0,$$

where $\nu = 0, 1, 2$ and $s > 1$.

We note that the given DE is invariant under a one-parameter group of transformations

$$\Gamma_\lambda : \ y \to \lambda^{-b}y, x \to \lambda x, b = (2-m)/(s-1) \quad (\lambda > 0),$$

where $b \neq 0, \infty$. Now we define $v = x^b y$, $\eta = \ln x$ so that the given DE becomes

$$\frac{d^2v}{d\eta^2} - b_0\frac{dv}{d\eta} + b_1 v + v^s = 0, \tag{1}$$

where $b_0 = -\nu + \{(1-m)s + (3-m)\}/(s-1), b_1 = b(b_0 - b)$. Introducing $z = dv/d\eta$ in (1), we get

$$\frac{dz}{dv} = \{b_0 z - (b_1 v + v^s)\}/z. \tag{2}$$

Equation (2) may be studied in the phase plane. Rosenau (1982)

6. $\qquad \boldsymbol{x^k y'' + kx^{k-1}y' + \lambda x^{-k}|y|^m \operatorname{sgn} y = 0,}$

where k, λ, and m are constants.

The given DE has a first integral

$$x^{2k}y'^2 + \{2/(m+1)\}\lambda|y|^{m+1} = c,$$

where c is constant. Trench and Bahar (1987)

7. $\qquad \boldsymbol{x^k y'' + kx^{k-1}y' + \lambda x^{\{(m+1)k-(m+3)\}/2}|y|^m \operatorname{sgn} y = 0,}$

where k, λ, and m are constants.

A first integral of the given DE is

$$x^{k+1}y'^2 + (k-1)x^k yy' + \{2/(m+1)\}\lambda x^{(m+1)(k-1)/2}|y|^{m+1} = c,$$

where c is constant. Trench and Bahar (1987)

8. $\qquad \boldsymbol{x^k y'' + kx^{k-1}y' + [qx^{k-2} + \beta x^{-k-4\nu}]y + \lambda x^{-k-\nu(m+3)}|y|^m \operatorname{sgn} y = 0,}$

where k, λ, ν, and m are constants.

The given DE has a first integral

$$x^{2(k+\nu)}y'^2 + [\beta x^{-2\nu} + \nu^2 x^{2(k+\nu-1)}]y^2 - 2\nu x^{2k+2\nu-1}yy' + \{2/(m+1)\}\lambda x^{-(m+1)\nu}|y|^{m+1} = c,$$

where c is a constant, provided that $k = 1 - 2\nu$ and $q = \nu^2$. Trench and Bahar (1987)

9. $\qquad \boldsymbol{x^k y'' + kx^{k-1}y' + (\beta + q)x^{k-2}y + \lambda x^{\{(m+1)k-(m+3)\}/2}|y|^m \operatorname{sgn} y = 0,}$

where k, β, q, λ, and m are constants.

The given DE has a first integral

$$x^{k+1}y'^2 + (\beta + q)x^{k-1}y^2 + (k-1)x^k yy' + \{2/(m+1)\}\lambda x^{(m+1)(k-1)/2}|y|^{m+1} = c,$$

where c is a constant. Trench and Bahar (1987)

10. $\qquad \boldsymbol{x^k y'' + kx^{k-1}y' + [qx^{k-2} + \beta x^{3k+4\nu-4}]y + \lambda x^{(m+2)k+(m+3)(\nu-1)}|y|^m \operatorname{sgn} y = 0,}$

where $k, q, \beta, \nu, \lambda$, and m are constants.

The given DE has a first integral

$$x^{2-2\nu}y'^2 + \{\beta x^{2k+2\nu-2} + (k+\nu-1)^2 x^{-2\nu}\}y^2 + 2(k+\nu-1)x^{-2\nu+1}yy'$$

$$+ \{2/(m+1)\}\lambda x^{(m+1)(k+\nu-1)}|y|^{m+1} = c,$$

where c is a constant, provided that $k = 1 - 2\nu$ and $q = \nu^2$. Trench and Bahar (1987)

11. $$r(x)y'' + r'(x)y' - p(x)y^\lambda - f(x) = 0,$$

where $r, p, f \in C([a, +\infty), R)$, $R = (-\infty, \infty)$ with $r > 0$, and $\lambda > 0$ is the quotient of odd integers.

Assume that $p \geq 0, r > 0, r \in C^1[a, \infty)$. In addition, assume that the IVP for the given DE is locally uniquely solvable. Then all solutions of the given equation that exist on $[x_1, \infty)$ for some $x_1 \geq a$ are nonoscillatory provided that either (a) $f(x) \geq 0, x \geq x_1$ or (b) $f(x) \leq 0, x \geq x_1$, where we assume that $f(x) \not\equiv 0$ on any subinterval of $[x_1, \infty)$.

Some other qualitative features of the solution of the given DE with $f(x) \neq 0$ or $f(x) = 0$ are also discussed. Erbe, Sree Hari Rao, and Seshagiri Rao (1984)

12. $$\theta(x)y'' + \theta'(x)y' + \sum_{i=0}^{m} A_{2i+1}(x)y^{2i+1} = 0,$$

where $\theta(x)$ and $A_{2i+1}(x)$ are positive continuous differentiable functions for $x \geq x_0 > 0$. It is shown that if $A_{2i+1}(x)\theta(x) > 0$ for $i = 0, 1, \ldots, m$ and for $x \geq x_0$, then $y(x)$ is bounded as $x \to +\infty$. Butlewski (1945)

13. $$r(x)y'' + r'(x)y' + q(x)y^\gamma = 0,$$

where $\gamma \geq 1$ is the ratio of odd, positive integers; γ and q are positive on a ray $[a, \infty)$ and have two continuous derivatives.

Under the assumptions on γ and q, a local existence and uniqueness theorem holds; moreover, all solutions of the given DE are extendable to $[a, \infty)$. Hence for each choice of initial values $y(a)$ and $y'(a)$, we have a unique solution of the given DE on $[a, \infty)$.

Now, if we assume that

$$k = \int_a^\infty |\eta(r\eta')'|dt < \infty, \tag{1}$$

$$\int_a^\infty 1/(r\eta^2)dt = \infty, \tag{2}$$

where $\eta(t) = [r(t)q(t)]^{-1/(\gamma+3)}$, then it is proved that the given DE has an oscillatory solution. Moreover, for $\gamma > 1$, every solution of the given DE satisfying the inequality $|y(a)| > \eta(a)[2(\gamma+1)k^2]^{1/(\gamma-1)}$ is oscillatory. Hinton (1969)

14. $$p(x)y'' + p'(x)y' + q(x)y^r = 0, x \geq 0,$$

where $p(x)$ is absolutely continuous and positive and $q(x)$ is continuous and nonpositive for $x \geq 0$.

Since $p(x) > 0$, $\int_0^\infty \{1/p(x)\}dx$ is infinite or exists as a finite number. Therefore, it is possible to reduce the given DE to simpler forms.

(a) Suppose that $\int_0^\infty \{1/p(t)\}dt = \infty$. Introducing the Liouville transformation

$$s = \int_0^x \{1/p(\tau)\}d\tau, z(s) = y(x),$$

we reduce the given DE to

$$\frac{d^2}{ds^2}z(s) + p(x)q(x)z^r(s) = 0,\ s \geq 0. \tag{1}$$

(b) Suppose that $\int_0^\infty \{1/p(x)\}dx < \infty$; we introduce the transformation

$$s = \left[\int_x^\infty \{1/p(\tau)\}d\tau\right]^{-1}, z(s) = sy(x)$$

to obtain the given DE in the form

$$\frac{d^2}{ds^2}z(s) + \frac{p(x)q(x)}{s^{r+3}}z^r(s) = 0,\ s > 0.$$

In either case, the transformation maps $x \in [0,\infty)$ onto $s \in [0,\infty)$. To circumvent the difficulty of defining z^r when z is negative and r not an integer, one may write instead of (1) the more realistic equation

$$\frac{d^2y}{dx^2} + a(x)|y|^r \operatorname{sgn} y = 0, x \geq 0.$$

Wong (1975)

15. $$\boldsymbol{m(x)y'' + m'(x)y' + a(x)b(y) = 0,}$$

where $m(x), a(x)$, and $b(y)$ satisfy certain conditions.

The following theorem is proved: Let $M_0 \geq m(x) \geq m_0 > 0$ for some fixed real constants M_0 and m_0. Let $m(x), a(x) \in C^1[0,\infty)$, and let $a(x) \geq a_0 > 0$ for some real constant a_0. Furthermore, suppose that $b(y) \in C(-\infty,\infty), yb(y) > 0$ for $y \neq 0, b(y) = -b(-y)$, and

$$\lim_{|y|\to\infty} B(y) = \int_0^y b(u)du = +\infty;$$

then if $a'(x) \geq 0, m'(x) \geq 0$, the solutions are bounded, oscillatory, and the amplitudes of any solution form a nonincreasing sequence. If $a'(x) \leq 0, m'(x) \leq 0$, the solutions are bounded, oscillatory, and the amplitudes of the solutions form a nondecreasing sequence. In addition, if $a(x)$ is bounded from above, then the absolute value of the greatest lower bound of the amplitudes is positive. Kroopnick (1978)

16. $$\boldsymbol{r(x)y'' + r'(x)y' + p(x)f(y) = 0.}$$

It is shown that if $p(x) > 0$ for large $x, yf(y) > 0$ for $y \neq 0$,

$$\liminf_{y\to\infty} f(y) > 0, \limsup_{y\to-\infty} f(y) < 0 \quad \text{and} \quad \int^\infty r^{-1}(s)ds = \int^\infty p(s)ds = \infty,$$

then all solutions of the given DE are oscillatory. It is assumed that f is continuous for all y and all solutions of the equation can be continued throughout the interval $[0,\infty)$. Tomastick (1967)

17.
$$p(x)y'' + p'(x)y' + q(x)g(y) = f(x), p(x), q(x) \in C[0, \infty),$$
$$p(x) > 0 \text{ for } 0 \leq x < \infty, g(x) \in C(-\infty, \infty).$$

Sufficient conditions for the approach to zero of all nonoscillatory solutions are derived. Londen (1973)

18.
$$p(x)y'' + p'(x)y' + a(x) \text{ sgn } (y)|y|^{\gamma} = e(x), \gamma > 0,$$

where $p(x)$ and $a(x)$ are positive, continuous, real valued, and locally of bounded variation in the interval $[0, \infty]$; $e(x)$ would be required to be locally Lebesgue integrable on $[0, \infty)$. The boundedness of solution is discussed. Dekleine (1971)

19.
$$(p(x)y')' + \sum_{k=1}^{N} a_k(x) f_k(y) = e(x),$$

where $p(x)$ and $a_k(x), k = 1, 2, \ldots, n$ are positive, continuous, real valued, and locally of bounded variation on the interval $[0, \infty)$. The boundedness of all solutions is studied. Dekleine (1971)

20.
$$a^2(x)y'' + 2a(x)a'(x)y' + (N-1)a^2(x)y'/x + \lambda f(y) = 0,$$
$$0 < x < 1, y'(0) = 0, y(1) = 0,$$

where

$$f(0) = 0, f'(0) > 0, \text{sgn } f''(y) = -\text{sgn } y \ \forall y \in R, y \neq 0, \limsup_{\|y\| \to \infty} f(y)/y \leq 0.$$

If $\lambda_n, n \geq 0$ be the $(n+1)$th eigenvalue of the BVP

$$(a^2(x)y')' + (N-1)a^2(x)y'/x + \lambda f'(0)y = 0, x \in (0,1), y'(0) = 0, y(1) = 0,$$

then the main result about the given nonlinear problem is as follows, provided that $x^2a'' + (N-1)xa' \leq (N-1)a$ for $0 < x < 1$.

If $\lambda \in (\lambda_{n-1}, \lambda_n), n \geq 1$, then there are exactly $(2n+1)$ solutions. For $0 < \lambda < \lambda_0$, zero is the only solution. For $\lambda > \lambda_0$, the only stable solutions are the ones with no zeros in $0 < x < 1$. If $C_k, k \geq 0$ denotes the branch of solution of the given problem emanating from the zero solution at $\lambda = \lambda_k$, then C_k is an unbounded continuum and the solutions in it are characterized by having exactly k simple zeros in $0 < x < 1$ for all $\lambda > \lambda_k$. Nascimento (1989)

21.
$$a(x)y'' + a'(x)y'(x) + p(x)f(y)g(y') = r(x).$$

The following assumptions are made: $a(x), p(x)$, and $r(x)$ are continuous in $[0, \infty)$; $f(y)$ and $g(y)$ are continuous in $(-\infty, \infty)$; $a(x) > 0$ and $p(x) > 0$ in $[0, \infty)$ and:

(a) $\int_0^\infty \dfrac{ds}{a(s)} = \infty$.

(b) $\int_0^\infty p(s)ds = \infty$.

(c) $yf(y) > 0$ for $y \neq 0$, $df/dy \geq 0$ for all y.

(d) $g(y') \geq k > 0$ for all y'.

(e) $[\int_b^x (\int_b^s (1/a(w))dw)\, r(s)ds] < M$ for all $x \geq b$ and all $b > 0$.

(f) There exists an increasing sequence $\{b_n\}$ with $\lim_{n\to\infty} b_n = \infty$ such that for each n,

$$\lim_{x\to\infty} \int_{b_n}^x \left(\int_s^t \frac{dw}{a(w)} \right) r(s)ds = L_n.$$

The numbers L_n take on both positive, negative, and possibly zero values for arbitrarily large n.

Under the conditions above, it is shown that every solution $y(x)$ of the given DE is oscillatory. Rankin (1976)

22. $$a(x)y'' + a'(x)y'(x) + q(x)f(y)g(y') = r(x),$$

where $a, q, r : [x_0, \infty) \to R,\ f, g:\ R \to R,\ a(x) > 0, q(x) > 0, g(y') > 0$.

Asymptotic behavior of the given DE is found: By examining the quotient $r(x)/q(x)$ as $x \to \infty$, boundedness and other behavioral results are obtained without requiring the forcing term $r(x)$ to be small in some sense. Indeed, the results allow $r(x)$ to become unbounded as $x \to \infty$. The solutions are either nonoscillatory, oscillatory, or of Z-type, that is, they have arbitrarily large zeros but are ultimately nonnegative or nonpositive. Graef and Spikes (1975)

23. $$p(x)y'' + p'(x)y' - f(x, y, y') = 0, x \geq 0,$$

where the following assumptions on p and f are made:

(a) $P : [0, \infty) \to (0, \infty)$ is continuous and $P(x) = \int_0^x \{1/p(s)\}ds \to \infty$ as $x \to \infty$.

(b) $f : [0, \infty) \times R \times R \to (0, \infty)$ is continuous and $f(x, u, v)$ is nondecreasing in u and v.

(c) Every Cauchy problem for the given DE has a unique solution.

Two types of solutions are defined. Let $y(x)$ be a solution of the given DE with the maximal interval of existence $[0, X_y]$. From the given DE, we have $(p(x)y'(x))' > 0$ so that $p(x)y'(x)$ is increasing on $[0, X_y]$. It turns out that either $X_y < \infty$ and $\lim_{x\to X_y} p(x)y'(x) = \infty$, or else $X_y = \infty$ and $\lim_{x\to\infty} p(x)y'(x)$ exists in $R \cup \{\infty\}$. In the former case $y(x)$ is called a singular solution, and in the latter $y(x)$ is called a proper solution. The class of proper solutions is further divided into four subclasses, depending on the $\lim_{x\to\infty} p(x)y'(x)$. Explicit sufficient conditions are given for the existence of some or all of these classes of proper solutions on $[0, \infty)$ as well as for singular solutions. Usami (1987)

24. $$py'' + p'y' + qy = g(x, y, y'),$$

where $p = p(x)$ is a positive and continuously differentiable function on $I_a = [a, \infty)$ such that $p(a) = 1, q = q(x)$ is a continuous function on I_a, and $g = g(x, u, v)$ is a continuous function on $I_a \times R^2$.

If $\{z_1, z_2\}$ is a fundamental system of solutions of the linear homogeneous equation $(pz')' + qz = 0$, then it is proved that certain solutions of the given equation are defined on all I_a and satisfy, for $x \to \infty$,

$$\begin{aligned} y &= (\delta_1 + O(1))z_1 + (\delta_2 + O(1))z_2, \\ y' &= (\delta_1 + O(1))z_1' + (\delta_2 + O(1))z_2, \end{aligned}$$

where $\delta_i (i = 1, 2)$ are constants. Medina and Pinto (1988)

25. $$\boldsymbol{a(x)y'' + a'(x)y' + q(x)f(y) = r(x, y, y'),}$$

where $yf(y) > 0$ for $y \neq 0$.

Conditions are given such that every solution $y(x)$ of the given DE satisfies $\int_{x_0}^{\infty} y(s)f\{y(s)\}ds < \infty$. Spikes (1977)

2.28 $f(x)y'' + g(x, y, y') = 0$

1. $$\boldsymbol{(a - x)y'' - K(1 + y'^2)^{1/2} = 0,}$$

where K and a are constants.

Putting $y' = p$, we have

$$K(1 + p^2)^{1/2} = (a - x)\frac{dp}{dx},$$

which, being separable, can be solved:

$$\frac{dy}{dx} = -\frac{1}{2}\left[C_1(a - x)^K - \frac{1}{C_1}(a - x)^{-K}\right]. \tag{1}$$

If $K \neq 1$, a solution of (1) is

$$y = \frac{1}{2}\left[\frac{C_1}{1 + K}(a - x)^{K+1} + \frac{1}{C_1(K - 1)}(a - x)^{1-K}\right] + C_2.$$

The case $K = 1$ involves a logarithmic term. Goldstein and Braun (1973), p.76.

2. $$\boldsymbol{(x^2 + 1)y'' + y'^2 + 1 = 0.}$$

Put $y' = p(x), y''(x) = dp/dx$, etc., to obtain

$$\frac{dp}{dx} = -(p^2 + 1)/(x^2 + 1).$$

The variables separate. A first integral is

$$\tan^{-1} p + \tan^{-1} x = C_1, \tag{1}$$

where C_1 is an arbitrary constant. Equation (1) can be written as $(p + x)/(1 - px) = C_2$, where C_2 is another arbitrary constant. $C_2 = 0$ leads to $y = -(1/2)x^2 + C$ with C a constant. $C_2 \neq 0$ yields

$$y = C_3 + C_4 x + (C_4^2 + 1)\log\,|x - C_4|;$$

where C_3 and C_4 are arbitrary constants. Kamke (1959), p.565; Murphy (1960), p.390.

3. $$(1 + x^2)y'' - y'^2 - 1 = 0.$$

Put

$$y' = p, y'' = \frac{dp}{dx}, \frac{dp}{1+p^2} = \frac{dx}{1+x^2}. \tag{1}$$

Two integrations of (1) give $y = Ax - (1 + A^2)\log(1 + x/A) + B$, where A and B are constants. Lawden (1959), p.319

4. $$(x^2 - 1)y'' + 4xy' + 2y - y^3 = 0.$$

The given DE arises from a nonlinear Klein-Gordon equation by a similarity transformation. The local nature of the solution near a movable singularity is found in the form $y \sim A(x - x_0)^{-1}$ as $x \to x_0$. Drazin and Johnson (1989), p.189

5. $$x^3(y'' + yy' - y^3) + 12xy + 24 = 0.$$

This is a special case of Eqn. (2.13.9). The solution is

$$y = (x^3u' + 2)/\{x(x^2u - 1)\},$$

where u is any solution of $u'' = 6u^2$; see Eqn. (2.1.9). The proof is through substitution. Kamke (1959), p.565; Murphy (1960), p.390

6. $$2x^3y'' + (2x^3y + 9x^2)y' - 2x^3y^3 + 3x^2y^2 + axy + b = 0,$$

where a and b are constants.

(a) With $y(x) = \xi'(x)\eta(\xi), \xi = -2/(x)^{1/2}$, we get

$$\xi^3(\eta'' + \eta\eta' - \eta^3) + (2a - 12)\xi\eta - 4b = 0.$$

(b) If $a = 12, b = -6$, the solutions are

$$y = (u' - 1)/(u - x), \tag{1}$$

where $u(x)$ is an arbitrary solution of

$$x^3u' = u^3 + C,$$

where C is an arbitrary constant.

That (1) actually yields solutions is easy to verify by substitution. That (1) also gives all the solutions follows from the fact that among the solutions (1), there exist solutions with arbitrary initial values of y and y' at any point $x \neq 0$. Kamke (1959), p.566; Murphy (1960), p.390

7. $$x^3y'' + x^2y'^2 - y = 1/x.$$

The given DE has a Laurent series solution $y = 1/(6x^2) + 1/(72x^3) + 1/(1440x^4) + \cdots$. Orlov (1980)

8. $$x^3y'' - a(xy' - y)^2 = 0,$$

where a is a constant.

With $y(x) = x\eta(\xi), \xi = \ln x$, one gets a simple DE; the solution is

$$y = (x/a)\ln\{x/(C_1 x + C_2)\},$$

where C_1 and C_2 are arbitrary constants. Kamke (1959), p.566; Murphy (1960), p.390

9. $$\boldsymbol{x^4 y'' - (2/d)y^2 = 0, d \neq 0, x \in [1, \infty).}$$

The given DE has an exact solution $y = dx^2$. More generally, if $|y'(1)| + |y'(1) - y(1)| < |d|/e$, then any solution of the given DE is defined on $[1, \infty)$ and has as $x \to \infty$ the asymptotic form

$$y(x) = \{\delta_1 + O(1)\}x + \delta_2 + O(1),$$

where δ_1 and δ_2 are constants. Pinto (1991)

10. $$\boldsymbol{x^4 y'' + a^2 y^\nu = 0,}$$

where a and ν are constants.

For $\nu = 1$, the DE is linear; for $\nu \neq 1$, one solution is

$$y = [2(\nu - 3)x^2/\{a^2(\nu - 1)^2\}]^{1/(\nu-1)}.$$

See also the solution detail of Eqn. (2.8.40). Kamke (1959) p.567; Murphy (1960), p.391

11. $$\boldsymbol{x^4 y'' - (2xy + x^3)y' + 4y^2 = 0.}$$

With $y(x) = x^2\eta(\xi), \xi = \ln x$ and further $\eta'(\xi) = u(\eta)$, we get $u' = 2(\eta - 1)$ and $u = 0$, implying that $\eta'(\xi) = (\eta - 1)^2 + C$ and $\eta = C$. Therefore, on integration,

$$\eta - 1 = \begin{cases} C_1 \tan(C_1\xi + C_2) \text{ for } C = C_1^2, \\ -C_1 \tanh(C_1\xi + C_2), -C_1 \coth(C_1\xi + C_2), \text{ for } C = -C_1^2. \end{cases}$$

In addition, we have $(\xi + C)(1 - \eta) = 1$ and $\eta = C$. The solution of the original DE is obtained from $y(x) = x^2\eta(\xi), \xi = \log x$. C, C_1, and C_2 are arbitrary constants. Kamke (1959), p.567; Murphy (1960), p.390

12. $$\boldsymbol{x^4 y'' - x^2 y'^2 - x^3 y' + 4y^2 = 0.}$$

The given DE is invariant under $y = A^2 Y$ and $x = AX$. The suggested change of variable is $y = x^2\eta$. Thus,

$$x^2\eta'' + 3x\eta' - 4x\eta\eta' - x^2\eta'^2 = 0. \qquad (1)$$

On integration of (1), we have

$$\eta' = (C_1 e^\eta - 4\eta - 2)/x.$$

Hence, the solutions may be found to be $y = 0$ and

$$x = \exp\left\{\int (C_1 e^\eta - 4\eta - 2)^{-1} d\eta + C_2\right\}, y = x^2\eta,$$

where C_1 and C_2 are arbitrary constants. Kamke (1959), p.567; Murphy (1960), p.390

13. $$x^4y'' + (xy' - y)^3 = 0.$$

With $y(x) = x\eta(\xi), \xi = 1/x$ or with $u(x) = xy' - y$, one obtains a simple DE; the solution is

$$y = C_1x + x\sin^{-1}(C_2/x). \tag{1}$$

Another solution is $y = C_0x$. In (1), we may replace $\cos^{-1}$ by $\sin^{-1}$. $C_i\,(i = 0, 1, 2)$ are arbitrary constants. Kamke (1959), p.567; Murphy (1960), p.390

14. $$2x^2(1-x^2)y'' + 3x(1-x^2)y' + 2x^2yy'^2 + y(1-y^2)^2 = 0.$$

See Eqn. (2.26.11). Ferreira and Neto (1992)

15. $$x^{s+2}y'' - f(xy', y, x) = 0,$$

where s is a positive integer greater than zero and f is a regular function of its arguments.

It may be proved that the point $x = 0$ is an essential singularity of the complete solution, and the equation does not unconditionally possess an integral, which is a regular function of x and vanishes with x. Forsyth (1959), p.220

16. $$4(1 + 2^{2/5}x^{1/5})y'' + 2^{2/5}x^{-22/15}y^{1/3}(1 + y'^{2/5}) = 0, x > 0.$$

The given DE has a nonoscillatory solution $y(x) = x^{1/2}$. Grace and Lalli (1987a)

17. $$(1 + a\cos x)y'' - 2a\sin xy' + \alpha\sin y = 4a\sin x,$$

$$y(0) - y(2\pi) = 0, y'(0) = y'(2\pi) = 0,$$

where $|a| < 1$ and α are arbitrary.

Existence of an odd solution to the periodic boundary value problem is established. Dang (1988)

18. $$f(x)y'' + g(x)y' + ay^n = 0,$$

where a is a constant.

It is shown that the given DE has a solution

$$y = h(x)u(x),$$

where u is a solution of the first order DE

$$fh^{1-n}u'^2 + Au^2 + 2au^{n+1}/(n+1) = C,$$

provided that f and g have the form

$$f(x) = (Bh^2 - A)h^{n+1}/h'^2, g(x) = Ah/h' - (Bh^2 - A)h^{n+1}h''/h'^3$$

for some function h; A, B, and C are constants. Mitrinovitch (1955)

19. $$y'' + p(x)y' - 3h(x)yy' - f(x) - g(x)y - [h'(x) + p(x)h(x)]y^2 + h^2(x)y^3 = 0,$$

where $p(x), h(x)$, and $g(x)$ are arbitrary functions of x.

We show that the change of variable

$$u = e^{-\int y(x)h(x)dx} \tag{1}$$

changes the given second order nonlinear DE into a third order linear DE.

Differentiating (1) three times in succession, we have

$$\begin{aligned} u' &= -yh(x)u, \\ u'' &= (y^2h^2 - y'h - yh')u, \end{aligned} \tag{2}$$

$$u''' = -h(y'' - 3hyy' + 2h'y'/h + h''y/h - 3y^2h' + h^2y^3)u. \tag{3}$$

Substituting y'' from the given DE into (3), we have

$$u''' = -h[(2h'/h - p)y' + (h''/h + g)y + f + (ph - 2h')y^2]u. \tag{4}$$

Now using (2) to eliminate y' in (4), we have

$$u''' + (p - 2h'/h)u'' + hfu = -hyu(h''/h + g - 2h'^2/h^2 + ph'/h).$$

Thus, the transformation (1) changes the given DE to linear third order DE

$$u''' + p_1(x)u'' + q_1(x)u' + r_1(x)u = 0,$$

where

$$p_1(x) = p - 2h'/h, q_1(x) = h''/h + g - 2h'^2/h^2 + ph'/h, r_1(x) = hf.$$

Goldstein and Braun (1973), p.99

20. $\boldsymbol{2f^2y'' - f(3f' - 2fy)y' - (ff'' - 2f'^2 - 2f^3)y - ff'y^2 - 2f^2y^3 = 0, f = f(x).}$

(a) A solution is $y = u(z)v(x)$, where $v = dz/dx, 12v^2 = f(x)$, and $u''(z) + uu'(z) + 12u = u^3$; for the latter, see Eqn. (2.10.3).

(b) Put $y = u'(x)/(u-1), v = 2u'' - (u'f'/f) - (u^2-1)f$; then $(v'/v) - (f'/f) = 2u'/(u-1)$. It follows that $v = C(u-1)^2f, (2u'u''/f) - u'^2f'/f^2 = [C(u-1)^2 + u^2 - 1]u'$,

$$u'^2 = f[C_1(u-1)^3 + (u-1)^2 + C_2]. \tag{1}$$

Equation (1) separates. C, C_1, and C_2 are arbitrary constants. Murphy (1960), p.391

21. $$\boldsymbol{f^2(x)y'' - (3f^3 + 3ff' - f^2y)y' + af^5 - (ff'' - 3f'^2 - 3f^2f' - bf^4)y - (ff' + f^3)y^2 - f^2y^3 = 0, f = f(x),}$$

and a is a constant.

A solution is

$$y(x) = z'u(z),$$

where

$$z = \exp\left(\int f(x)dx\right)$$

and

$$z^3(u'' + uu' - u^3) + (b-2)zu + a = 0.$$

Murphy (1960), p.391

22. $$f^2y'' + ff'y' - \phi(y, fy') = 0, f = f(x).$$

With $y(x) = \eta(\xi), \xi = \int \frac{dx}{f(x)}$, we get the autonomous form

$$\eta'' = \phi(\eta, \eta'). \tag{1}$$

The order of (1) may be reduced by writing $\eta' = p, \eta'' = p(dp/d\eta)$, etc. Kamke (1959), p.569; Murphy (1960), p.391

2.29 $yy'' + G(x, y, y') = 0$

1. $$yy'' + 1 = 0, y(0) = y(1) = 0.$$

A leading behavior argument shows that $y(x) \sim \pm x(-2 \ln x)^{1/2}$, $x \to 0+$, and $y(x) \sim \pm(1-x)\{-2 \ln(1-x)\}^{1/2}$ as $x \to 1-$. The asymptotics are compared with an exact numerical solution of the problem. Bender and Orszag (1978), p.200

2. $$yy'' - 1 = 0, y(0) = y(1) = 0.$$

Using integration by parts, one can show that the given problem has no solution. Bender and Orszag (1978), p.200

3. $$yy'' - 1 = 0, 0 \le x \le 1; y'(0) = -2^{1/2}\gamma, y'(1) = -2\alpha y(1),$$

$\gamma > 0$ is a constant.

The given BVP arises in the theory of chemical reactions. Multiplying it by y' and integrating, we obtain

$$\int_{y(1)}^{y(x)} \frac{du}{(\ln u + A)^{1/2}} = 2^{1/2}(1-x), 0 \le x \le 1. \tag{1}$$

Using the given BC, we obtain

$$A = \gamma^2 - \ln y(0), \tag{2}$$

$$A = 2\alpha^2 y^2(1) - \ln y(1). \tag{3}$$

Equation (3) connects A and $y(1)$. To find $y(1)$, we put $x = 0$ in (1) and consider (2); we get

$$I(\gamma) - I(v) - \alpha \exp(v^2)/v = 0, v = (2\alpha)^{1/2} y(1), \tag{4}$$

where $I(z) = \int_0^z \exp(x^2)dx$; it is clear that $I(z) = \exp(z^2)/(2z)[1+o(1)], z \to \infty$. Equations (1) and (4) provide the solution of the problem. Equation (4) is a transcendental equation whose solutions determine the number of solutions of the given problem. Numerical results are shown graphically. Berman and Vostokov (1982)

4. $$yy'' - a = 0,$$

where a is a constant.

Writing the given DE as $y'y'' = ay'/y$ and integrating twice, we have

$$x = \int (2a \log y + C_1)^{-1/2} dy + C_2,$$

where C_1 and C_2 are arbitrary constants. Kamke (1959), p.569; Murphy (1960), p.392

5. $$\boldsymbol{yy'' + x = 0, y(0) = 0, y'(1) = 0.}$$

The solution is found in the form

$$y(x) = \sum_{n=1}^{\infty} \frac{y^n(0)}{n!} x^n, \tag{1}$$

with

$$y''(x) = \sum_{n=2}^{\infty} \frac{y^n(0)}{(n-2)!} x^{n-2}. \tag{2}$$

The condition $y'(1) = 0$, with the help of (1)-(2), becomes

$$S = \sum_{n=2}^{\infty} \frac{b_n}{S^{2n-3}}, \tag{3}$$

where $b_n > 0$ may be found recursively. Here, $y''(0) = 1/S$ or $y'(0) = S$. Equation (1) may be shown to possess a unique positive root. Approximate values are given by

$$\sigma^3 - \sigma^2 - (1/4)\sigma - 1/9 = 0,$$

where $\sigma = S^2$. This gives $S \sim 9/8$ and $y(1) \sim -0.61$. The numerical solution gives $y(1) = -0.6276$. Callegari and Nachman (1978)

6. $$\boldsymbol{yy'' + x = 0, y'(K) = 0, y(1) = 0,}$$

where $K(0 < K < 1)$ is a constant.

Since

$$y'(x) = -\int_K^x \frac{\xi d\xi}{y} < 0, y''(x) = -x/y < 0,$$

$y'''(x) = -(1 + y'y'')/y < 0$ and by recursive argument $y^{(n+3)}(x) < 0$.

The solution is given by

$$y(x) = \sum_{n=0}^{\infty} \frac{y^n(K)}{n!} (x-K)^n,$$

$$0 = \sum_{n=0}^{\infty} \frac{y^n(K)}{n!} (1-K)^n. \tag{1}$$

Since $g^n(K) < 0, n \geq 1$, Eqn. (1) may be written in the form

$$\alpha = \sum_{n=1}^{\infty} \frac{b_n}{2n-1}, \tag{2}$$

where $y(K) = \alpha$ and $b_n > 0$ may be found recursively. It is proved from (2) that it has a unique positive root. An approximation to $\alpha^2 = \beta$ may be found by solving the equation

$$\beta^2 - (1/6)(1+2K)(1-K)^2\beta - (1/24)K^2(1-K)^4 = 0, 0 < K < 1. \tag{3}$$

Equation (3) gives good results for K near 1 and is in error by about 13% for $K = 0$.

It is also shown that solution to the given problem with $1 < K < 6$ is analytic on $(1, K)$. Callegari and Nachman (1978)

7. $\mathbf{yy'' + (x - \lambda) = 0, 0 < x < 1 + \lambda, y'(0) = 0, y(1 + \lambda) = 0,}$

where λ is a parameter.

The given problem arises in boundary layer theory. It is proved that there is a range of positive values of λ such that the positive continuous solution $y(x)$ of the given BVP is analytic on the closed interval $[0, 1+\lambda]$. The numerical results regarding the nonuniqueness of the solution for $\lambda \le \lambda_c = 0.3541\ldots$ are confirmed analytically. Hussaini, Lakin, and Nachman (1987)

8. $\mathbf{yy'' + 2x = 0, y'(0) = 0, y(1) = 0.}$

It is shown that this BVP has a power series expansion

$$y(x) = \sum_{n=0}^{\infty} \frac{y^{(n)}(0)}{n!} x^n, \tag{1}$$

where the derivatives are given by

$$y^{(n+3)} = -\frac{1}{y}\left[(n+1)y^{(1)}y^{(n+2)} + \frac{1}{2}\sum_{p=2}^{n+1} y(p)y^{(n+3-p)}\left\{\binom{n+1}{n+3-p} + \binom{n+1}{p}\right\}\right]$$

with

$$y^{(r)}(x) = \frac{d^r y}{dx^r} \quad \text{and} \quad \binom{m}{r} = \frac{m!}{r!(m-r)!}.$$

The series (1) is shown to be convergent for $|x| < 1$. It is also proved that the power series for $y(x)$ converges at $x = 1$ to the value $y(1)$. Numerical results for the bounds and exact solutions are provided. Callegari and Friedman (1968)

9. $\mathbf{yy'' + 2x = 0, 0 < x < 1, y'(0) = K \le 0, y(1) = 0.}$

The given BVP arises in boundary layer theory for an incompressible fluid, or a compressible fluid with viscosity inversely proportional to temperature. It is shown that the positive solution of the given problem can be expressed in power series about $x = 0$, which converges on the closed interval $0 \le x \le 1$:

$$y(x) = \sum_{n=0}^{\infty} \frac{y^n(0)}{n!} x^n,$$

where $y(0)$ must satisfy the transcendental equation

$$y(0) + \sum_{n=1}^{\infty} \frac{y^n[0; g(0)]}{n!} = 0. \tag{1}$$

The root $g(0)$ of (1) is shown to be unique and real positive, and is estimated analytically. Callegari and Friedman (1976)

10. $$yy'' - x = 0.$$

The given DE is invariant under $x = \lambda^2 X, y = \lambda^3 Y$, which allows its reduction to a first order DE, which may be studied generally. It has a unique solution $3y^2 - 4x^3 = 0$ for the singular initial condition $y(0) = 0$. A series solution for a general IC is constructed. Levi and Massera (1947)

11. $$yy'' + (1 - x) = 0, y(0+) = 0, y(1-) = 0, \; y(x) > 0 \text{ for } 0 < x < 1.$$

This problem is encountered in boundary layer theory for an incompressible viscous fluid. It is a special case of Eqn. (2.6.20). Lomtatidze (1987)

12. $$yy'' - ax = 0,$$

where $a \neq 0$ is a constant.

For solutions with the initial conditions $y(0) = C_0$, $y'(0) = C_1$, one obtains the following cases:

(a) $C_0 = C_1 = 0, y = \pm[(4a/3)x^3]^{1/2}$.

(b) $C_0 = 0, C_1 \neq 0, y = C_1 x + C_2 x^2 + \cdots$, with the coefficients

$$C_2 = a/2C_1, C_3 = -C_2^2/3C_1, C_4 = 2C_2^3/9C_1^2, C_5 = -17C_2^4/90C_1^3,$$

etc. The series can be shown to converge in a neighborhood of $x = 0$.

(c) $C_0 \neq 0$. The solution in this case can also be found about $x = 0$ in the form of a convergent power series.

Kamke (1959), p.570

13. $$yy'' - ax^2 = 0,$$

where $a \neq 0$ is a constant.

If $y(x)$ is the solution of this DE with IC $y(0) = C_0$, $y'(0) = C_1$, then the following cases arise:

(a) $C_0 = C_1 = 0$: $y = \pm(a/2)^{1/2}x^2$.

(b) $C_0 = 0, C_1 \neq 0$: $y = C_1 x(1 + b_1 x^2 - b_2 x^4 + b_3 x^6 - \cdots)$; the series converges for $|x| < (3C_1^2/|a|)^{1/2}$, the first few coefficients being

$$b_1 = a/6C_1^2, b_2 = (3/10)b_1^2, b_3 = (13/70)b_1^3, b_4 = (25/168)b_1^4.$$

(c) For $a > 0, C_1 > 0, x > 0$, if $y > 0$, then y increases monotonically, and as $x \to \infty$, $y'/(2x) \to (a/2)^{1/2}$, $y'' \to (2a)^{1/2}$.

(d) $C_0 \neq 0$: There exists a solution in the form of a convergent power series about $x = 0$.

Kamke (1959), p.570.

14. $$yy'' + h(x) = 0, k < x < 1, y'(k) = C, k \geq 0, y(1) = 0,$$

where $h(x)$ is continuous, positive, and $\lim_{x\to\infty} h(x)/x > 0$ and k and C are constants.

Existence, uniqueness, and analyticity of the solutions are proved. Numerical results are also proved. The problem arises in boundary layer theory. Vajravelu, Soewono, and Mohapatra (1991)

15. $$yy'' + 2xh(x) = 0, 0 < x < 1, y'(0) = 0, y(1) = 0.$$

It is shown that the given BVP, with $h(x)$ a continuous, nonnegative function on $0 \leq x \leq 1$, has at most one solution in the class of functions that are defined and continuous on $0 \leq x \leq 1$ and positive for $0 \leq x < 1$. Callegari and Friedman (1968)

16. $$yy'' + 2xh(x) = 0, y(0) = a_0, y'(0) = 0,$$

where $h(x)$ is an arbitrary nonnegative continuous function for $0 \leq x \leq 1$.

It is proved that there exists a constant $a_0 > 0$ such that the given IVP has a continuous solution that is positive for $0 \leq x < 1$ and vanishes for $x = 1$.

The proof is carried out by considering the dependence of solutions of the IVP

$$yy'' + 2xh(x) = 0, y(0) = k, y'(0) = 0$$

on the positive parameter k. It is sufficient to show that a value of k exists such that $y(x)$ is continuous for $0 \leq x < 1$, positive for $0 \leq x < 1$, and $y(1) = 0$. Callegari and Friedman (1968)

17. $$yy'' + y^2 - ax - b = 0.$$

Only a graphical solution is indicated by Kamke; however, a power series solution may be constructed to solve an IVP. See Eqns. (2.29.12)–(2.29.13). Kamke (1959), p.570

18. $$(a - y)y'' + c\{y(a - y) - \lambda\} = 0,$$

where $a > 0, c > 0, \lambda > 0$ are constants.

The given DE describes motion of a bar restrained by springs and attracted by a parallel current-carrying conductor. Multiplying it by y' and integrating, we get a first integral

$$(y'^2/2) + (c/2)y^2 + c\lambda \ln (a - y) = C,$$

where C is an arbitrary constant. The given DE has no equilibrium points if $a^2 < 4\lambda$; it has an equilibrium point with index 0 at $y = a/2, y' = 0$, if $a^2 = 4\lambda$. If $a^2 > 4\lambda$, it has equilibrium points at:

(a) $y = (a/2) + (1/2)[a^2 - 4\lambda]^{1/2}, y' = 0$, index -1.
(b) $y = (a/2) - (1/2)[a^2 - 4\lambda]^{1/2}, y' = 0$, index 1.

Jordan and Smith (1977), p.91

19. $$(a - y)y'' + (k/m)\{y(a - y) - \lambda\} = 0,$$

where a, k, m, and λ are constants.

The given DE describes the motion of a current-carrying conductor restrained by springs and subjected to a force from the magnetic field due to another infinitely long fixed parallel wire. Putting $y' = v, y'' = v(dv/dy)$, etc., we have

$$\frac{dv}{dy} = \frac{k}{m}\frac{y^2 - ay + \lambda}{v(a-y)} \quad . \tag{1}$$

Equation (1) has singularities $(y_1, 0)$ and $(y_2, 0)$, where

$$y_{1,2} = a/2 \pm \{a^2/4 - \lambda\}^{1/2}.$$

Three cases arise:

(a) $\lambda < a^2/4$.

(b) $\lambda = a^2/4$.

(c) $\lambda > a^2/4$.

The nature of singular points varies. The integral curves may be drawn and the results physically interpreted. Stoker (1950), p.51

20. $$\boldsymbol{yy'' + f(x)y^2 - \phi(x) = 0.}$$

Certain cases in which the given DE can be transformed into equations that are integrable by quadratures are considered. Leko (1955)

21. $$\boldsymbol{yy'' + f(x)y^2 - \phi(x) = 0,}$$

where f and ϕ are arbitrary functions of x.

It is proved that the general solution of the given DE may be obtained by eliminating t from $\int \phi(x)dx = \int \Phi(t)dt$,

$$y = [\Phi(t)/\phi(x)]^{1/2} z(t, c_1, c_2),$$

where $z = z(t, c_1, c_2)$ represents the general solution of the equation

$$zz'' + F(t)z^2 = \Phi(t), \tag{1}$$

and the coefficients in the given DE and (1) are related by

$$[1/\phi^2(x)]\{f(x) + \Delta_2[\phi(x)]^{-1/2}\} = [1/\Phi^2(t)]\{F(t) + \Delta_2[\Phi(t)]^{-1/2}\},$$

where

$$\Delta_2(u) = \frac{1}{u}\frac{d^2u}{dx^2}.$$

Bandic (1963)

22. $$\boldsymbol{yy'' - xy - 2y^3 = 0.}$$

For this second Painlevé equation, precise asymptotic formulas as $x \to +\infty$ are derived rigorously. Suleimanov (1987)

2.30 $yy'' + ky'^2 + g(x, y, y') = 0,\ k > 0,\ g$ linear in y'

1. $$3yy'' + y'^2 = 0, y(0) = 0, y(1) = 1.$$

The exact solution is $y = x^{3/4}$. The quasilinearization method is illustrated with the help of this example. Roberts and Shipman (1972), p.93

2. $$2yy'' + y'^2 = 0.$$

Writing the given DE as $2y''/y' + y'/y = 0$ and integrating twice lead to $y^3 = (C_1x + C_2)^2$; C_1 and C_2 are arbitrary constants. Reddick (1949), p.191

3. $$2yy'' + y'^2 + 1 = 0.$$

The given DE is a special case of Eqn. (2.32.9). The solution is

$$C_1 \tan^{-1}\{y/(C_1 - y)\}^{1/2} - \{y(C_1 - y)^{1/2}\} = x + C_2,$$

or in the parametric form

$$x = C_1(t - \sin\ t) + C_2, y = C_1(1 - \cos\ t),$$

the equation for cycloids; C_1 and C_2 are arbitrary constants. Kamke (1959), p.576; Murphy (1960), p.396

4. $$2(y - a)y'' + y'^2 + 1 = 0,$$

where a is a constant.

The given DE is autonomous. The solution is

$$2x = C_1 \pm 2\{(y - a + C_2)(a - y)\}^{1/2} \pm C_2 \sin^{-1}\{(2y - 2a + C_2)/C_2\},$$

obtained by putting $y' = p, y'' = p(dp/dy)$, etc., and integrating twice. Here C_1 and C_2 are arbitrary constants. Kamke (1959), p.579

5. $$2yy'' + y'^2 + (2/x)yy' - 3y^2/x^2 = 0.$$

Put $t = xy'/y$ in the given DE to obtain

$$2xt' + 3(t^2 - 1) = 0, t' = \frac{dt}{dx}. \tag{1}$$

For $0 < t < 1$, we have from (1), after integration,

$$x^3 = A(1 + t)/(1 - t). \tag{2}$$

The definition $t = xy'/y$ and (1) yield

$$\frac{1}{y}\frac{dy}{dt} = 2t/[3(1 - t^2)],$$

which, on integration, gives

$$y^3 = B/(1-t^2). \tag{3}$$

Elimination of t from (2) and (3) gives the explicit solution

$$y^3 = (C/x^3)(x^3 + A)^2;$$

A, B, and C are arbitrary constants. Murphy (1992)

6. $$\boldsymbol{yy'' + y'^2 = 0, y(0) = 1, y'(1) = 3/4.}$$

Exact solution of the given BVP is $y = \{3x+1\}^{1/2}$. Numerical solution is obtained by shooting methods and compared with the exact analytic one. Roberts (1979), p.358

7. $$\boldsymbol{yy'' + y'^2 - a^2 = 0,}$$

where a is a constant

(a) With $y' = p$, we have

$$ypp'(y) + p^2 = a^2.$$

A first integral is

$$C^2y^2(a^2 - p^2) = 1;$$

the variables now separate. C is an arbitrary constant.

(b) $y^2 = C_1 + C_2x + a^2x^2$ is another solution with C_1 and C_2 as arbitrary constants. This integral follows from writing the given DE as $(yy')' = a^2$ and integrating twice.

Murphy (1960), p.392

8. $$\boldsymbol{yy'' + y'^2 + y^2 = 0.}$$

With $y'(x) = p(y)$, we have $ypp'(y) + p^2 + y^2 = 0$. So a first integral is $y^2(2p^2 + y^2) = C$; C is an arbitrary constant. Now the variables separate. The solution may be found to be $y^2 = C_1 \sin(2^{1/2}x + C_2)$, where C_1 and C_2 are arbitrary constants. Murphy (1960), p.392

9. $$\boldsymbol{yy'' + y'^2 + 8y(y - 3/4) = 0.}$$

The given DE has a solution

$$y = \begin{cases} \cos^2 x, & |x| < \pi/2 \\ 0, & |x| \geq \pi/2. \end{cases}$$

Satsuma (1987)

10. $$\boldsymbol{yy'' + y'^2 + 2a^2y^2 = 0,}$$

where a is a constant.

With $y'(x) = p(y)$, we have $ypp'(y) + p^2 + 2a^2y^2 = 0$. So a first integral is

$$y^2(p^2 + a^2y^2) = C,$$

where C is an arbitrary constant.

The variables now separate. The solution may be found to be

$$y^2 = C_1 \cos(2ax) + C_2 \sin(2ax),$$

where C_1 and C_2 are arbitrary constants. Murphy (1960), p.392

11. $$\boldsymbol{yy'' + y'^2 + (y^2/2)(y-1) = 0.}$$

The given DE has a pulse-like solution,

$$y = (5/4)\ \mathrm{sech}^2(x/4).$$

Satsuma (1987)

12. $$\boldsymbol{yy'' + y'^2 - y^6 - xy^2 - \alpha = 0,}$$

where α is constant.

The given DE has movable branch points and has solutions in the form

$$y(x) = \sum_{n=0}^{\infty} a_n (x - x_0)^{n-1/2}. \tag{1}$$

However, the given DE is "essentially" Painlevé, since if we make the transformation $z = y^2$ as suggested by the form (1) with half-integer powers on its RHS, we get a second Painlevé equation. Kruskal and Clarkson (1992)

13. $$\boldsymbol{yy'' + y'^2 - y' = 0.}$$

The given DE is autonomous. Writing $y' = p, y'' = p(dp/dy)$, etc., and integrating twice, we have

$$x = y + C_1 \ln|y - C_1| + C_2,$$

where C_1 and C_2 are constants. $y = C$ is another (special) solution. Kamke (1959), p.571; Murphy (1960), p.393

14. $$\boldsymbol{yy'' + y'^2 - y' - 3y^2(y^2 - 1) = 0.}$$

The given DE has a solution with a sharp front,

$$y = \begin{cases} \tanh\ x, & x > 0, \\ 0, & x \le 0. \end{cases}$$

Tanh x itself is also a solution. Satsuma (1987)

15. $$\boldsymbol{yy'' + (y' - 2)(y' + 1) = 0.}$$

The given DE appears in the self-similar analysis of transonic flow with aligned field magnetohydrodynamic nozzle flow.

A solution is easily found to be

$$(y-2x)^2(y+x) = 2\alpha^3,$$

where α is an arbitrary constant. Other special solutions were obtained as $y = 2x + c_1$, and $y = -x + c_2$, where c_1 and c_2 are arbitrary constants. Watanabe, Sichel, and Ong (1973)

16. $$\boldsymbol{yy'' + y'^2 + (c/2)y' + (1/2)y(y-1)(\alpha - y) = 0,}$$

where c and α are constants.

The given DE describes traveling waves for a generalized Fisher's equation. This has a Laurent series expansion about

$$x = x_0, y = 20z^{-2} + (5/12)(1+\alpha) + (1/60)[(5/16)(1+\alpha)^2 - \alpha]z^2 + \cdots,$$

where $z = x - x_0$, provided that $c = 0$. Two other solutions for $c = 0$ are posssible. If we write $y = f/g$, where $f = A + Be^{kx} + Ce^{2kx}$, $g = (1+e^{kx})^2$, then substitution, etc., lead to

(a) $A = 1, B = -2, C = 1$, and $k = 1/5^{1/2}$, provided that $\alpha = 3/5$. Then

$$y = \tanh^2[1/\{2(5)^{1/2}\}]x.$$

(b) $A = 0, B = 5, C = 0$, and $k = i/2$, provided that $\alpha = 0$. Then

$$y = (5/4)\sec^2(x/4).$$

Satsuma (1987)

17. $$\boldsymbol{3yy'' + 3y'^2 + xy' + y = 0.}$$

The given DE describes self-similar solutions $u = t^{-1/3}y(z/t^{1/3})$ of nonlinear heat equation $u_t = (uu_z)_z$. It is immediately integrable once:

$$3yy' + xy = c.$$

For the special case, $c = 0$, the solution is

$$y = (x_0^2 - x^2)/6,$$

where x_0 is constant of integration. Dresner (1983)

18. $$\boldsymbol{yy'' + y'^2 + (1/2)xy' = 0.}$$

The given DE results from the similarity transformation of the nonlinear diffusion equation $u_t = (u_z)_z$. The transformation $u = y/x^2, v = y'/x$ reduces the given DE to one of first order:

$$\frac{dv}{du} = \frac{v(2v+2u+1)}{2u(2u-v)}. \tag{1}$$

A local analysis of (1) in the (u, v) plane gives qualitative features of the solution for a problem in boundary layer flow. The direction field of Eqn. (1) is drawn and a detailed discussion is provided. Dresner (1983)

19. $$yy'' + y'^2 + (1/2)xy' - y = 0.$$

The given DE arises from a similarity reduction of the PDE $u_t = (u_z)_z$, called the Boltzmann equation. The transformation $u = y/x^2, v = y'/x$ changes the given DE to one of first order:

$$\frac{dv}{du} = \frac{2u - 2v^2 - v - 2uv}{2u(v-2u)}. \tag{1}$$

Equation (1) is solved in the phase plane with suitable BCs arising from a consideration of isothermal percolation of a perfect gas through a microporous medium as described by Darcy's law. Dresner (1983)

20. $$3yy'' + 3y'^2 + 2xy' - y = 0.$$

This is a reduced form of the nonlinear diffusion equation $u_t = (uu_z)_z$ obtained by writing $u = t^{1/3}y(z/t^{2/3})$. The given DE is invariant to the associated group $y^1 = \mu^2 y$, $x^1 = \mu x$. An invariant and a differential invariant of this group are $u = y/x^2$ and $v = y'/x$. A short calculation shows that it can be reduced to a first order DE:

$$\frac{dv}{du} = \frac{u - 2v - 3v^2 - 3uv}{3u(v-2u)}. \tag{1}$$

Equation (1) is solved in the phase plane subject to BCs imposed by the physical problem. Dresner (1983)

21. $$yy'' + y'^2 + xy' - [(m-1)/m]y = 0, m > 0.$$

The given DE arises in the self-similar description of flows in mixing layers [see Eqn. (3.5.23) for its analysis]. Diyesperov (1986)

22. $$yy'' + y'^2 - xy' + y = 0.$$

With $y = x^2 u(z), z = \ln x$, and $u'(z) = p$, we have

$$upp'(u) + p^2 + (7u-1)p + u(6u-1) = 0. \tag{1}$$

Equation (1) is of Abel type. Murphy (1960), p.393

23. $$yy'' + y'^2 + axy' = 0,$$

where a is a constant.

With $y = x^2 u(z), z = \ln x, u'(z) = p(u)$, we get the Abel equation

$$upp' + p^2 + (a + 7u)p + 2u(a + 3u) = 0.$$

Murphy (1960), p.393

24. $$\boldsymbol{yy'' + y'^2 - y^2y'/x = 0.}$$

The given DE is Cauchy–Euler type w.r.t. x. Put $x = e^t$ to get the autonomous DE

$$yy'' - (y + y^2)y' + y'^2 = 0, \ ' \equiv \frac{d}{dt}.$$

Now put $y'(t) = p(y)$, etc., to get

$$\frac{dp}{dy} + p/y = 1 + y, p \neq 0, y \neq 0. \tag{1}$$

Integrating (1), we have

$$\frac{dy}{dt} = p = y/2 + y^2/3 + C/y,$$

where C is an arbitrary constant. Now the variables separate. Bender and Orszag (1978), p.34

25. $$\boldsymbol{yy'' + y'^2 - yy'/(1 + x) = 0.}$$

The given DE is Cauchy–Euler type w.r.t. $(1 + x)$. Put $1 + x = e^t$ to get

$$y\ddot{y} - 2y\dot{y} + \dot{y}^2 = 0, \ \cdot \equiv \frac{d}{dt}. \tag{1}$$

Equation (1) is autonomous. Put $dy/dt = p(y)$, etc., we get

$$\frac{dp}{dy} = (2y - p)/y, \text{ if } y \neq 0. \tag{2}$$

Equation (2) is homogeneous and may be integrated by putting

$$p = yv(y)$$

to yield

$$\frac{dy}{dt} = C/y + y,$$

where C is an arbitrary constant. Now the integration is elementary. Bender and Orszag (1978), p.34

26. $\boldsymbol{(yy'' + y'^2) + (N - 1)x^{-1}yy' + \{1/(N + 2)\}xy' + \{N/(N + 2)\}y = 0, N = 1, 2.}$

The given DE arises from similarity reduction of radially symmetric nonlinear diffusion equation. An integration yields

$$x^{N-1}yy' + \{1/(N + 2)\}x^N y + \alpha = 0. \tag{1}$$

It is convenient to introduce the variable $g = y + x^2/\{2(N + 2)\}$ to rewrite (1) as

$$\alpha\frac{dx}{dg} = x^{N-1}\{x^2/(2(N + 2)) - g\}. \tag{2}$$

(a) $N = 1$. Equation (2) is Riccati, so that

$$\alpha \frac{dx}{dg} = x^2/6 - g. \tag{3}$$

Writing $x = -6\alpha q^{-1}(dq/dg)$, we obtain the linear equation $6\alpha^2(d^2q/dg^2) = gq$ with the general solution

$$q = a \text{ Ai}\{(6\alpha^2)^{-1/3}g\} + b \text{ Bi}\{(6\alpha^2)^{-1/3}g\},$$

where a and b are arbitrary constants and Ai and Bi are Airy functions. We therefore get

$$x = -(36\alpha)^{1/3} \frac{\sin\ r \text{ Ai}'\{(6\alpha^2)^{-1/3}(y + x^2/6)\} + \cos\ r \text{ Bi}'\{(6\alpha^2)^{-1/3}(y + x^2/6)\}}{\sin\ r \text{ Ai}\{(6\alpha^2)^{-1/3}(y + x^2/6)\} + \cos\ r \text{ Bi}\{(6\alpha^2)^{-1/3}(y + x^2/6)\}},$$

where $\sin\ r = a/(a^2 + b^2)^{1/2}$.

(b) $N = 2$. Here Eqn. (2) is of Bernoulli type, namely

$$\alpha \frac{dx}{dg} = x^3/8 - xg. \tag{4}$$

Put $x = \xi^{-1/2\alpha}$ in (4) to obtain $\alpha(d\xi/dg) = 2\xi g - 1/4$, with the solution

$$\xi = e^{g^2/\alpha}\{(1/8)(\pi/\alpha)^{1/2} \operatorname{erfc}(g/\alpha^{1/2}) + \beta\},$$

where β is an arbitrary constant. Hence the solution in this case is

$$x = \exp\{-(y + x^2/8)^2/(2\alpha)\}[(1/8)(\pi/\alpha)^{1/2} \operatorname{erfc}\{(y + x^2/8)/\alpha^{1/2}\} + \beta]^{-1/2}.$$

King (1991)

27. $$\mathbf{2yy'' + 3y'^2 + 4y^2 = 0.}$$

Writing $y' = p, y'' = p(dp/dy)$, etc., and integrating, we get

$$y'^2 = -(4/5)y^2 + cy^{-3}, \tag{1}$$

where c is a constant. The only explicit solution is $y \equiv 0$. The solutions of the given DE are seen to have a single extreme value of y (an absolute maximum or minimum), the solutions are bounded, but $|y'| \to \infty$ as $|x| \to \infty$. Utz (1978)

28. $$\mathbf{2yy'' + 3y'^2 - 4y^2 = 0.}$$

Writing $y' = p, y'' = p(dp/dy)$, etc., and integrating, we get

$$y'^2 = (4/5)y^2 + cy^{-3},$$

where c is a constant. For $c = 0$, we easily integrate to get the subfamily of solutions

$$y = c_1 \exp\{a(x + c_2)\}, a^2 = 4/5,$$

where c_1 and c_2 are arbitrary constants. For $c \neq 0$, one may discuss the solutions in the (y', y) plane. Utz (1978)

29. $$yy'' + (3/2)y'^2 - Ay^{-4} + B = 0, y(0) = y_0, y'(0) = 0,$$

where A and B are constants.

The given DE describes the oscillation of a gas sphere that arises from an underwater explosion in an infinite medium.

Writing $v = y^{5/2}$, we have

$$\frac{d^2v}{dx^2} - (5/2)Av^{-7/5} + (5/2)Bv^{1/5} = 0. \tag{1}$$

Equation (1) has a first integral

$$(1/2)\left(\frac{dv}{dx}\right)^2 + (25/4)Av^{-2/5} + (25/12)Bv^{6/5} = \text{constant}.$$

Using the BC $v = v_1, dv/dx = 0$ at $x = 0$, we get

$$\left(\frac{dv}{dx}\right)^2 = (25/2)A(v_1^{-2/5} - v^{-2/5}) + (25/6)B(v_1^{6/5} - v^{6/5})$$

or

$$y^4\left(\frac{dy}{dx}\right)^2 = 2A(y/y_1 - 1) + (2/3)By_1^4(y/y_1 - y^4/y_1^4). \tag{2}$$

Equation (2) is again integrated after several complicated transformations, and the solution is expressed in terms of elliptic integrals of the first and second kinds and some simple algebraic function. Childs (1973)

30. $$y'' + (3/2y)y'^2 - (1/x)y' + y/2x^2 = 0.$$

Put $t = -xy'/y$ to reduce the given DE to one of first order and hence integrate to obtain the exact solution

$$y^5 = C_1x^2/\{1 + \tan^2((1/2)\ln C_2x)\},$$

where C_1 and C_2 are arbitrary constants. Murphy (1992)

31. $$2yy'' + 3y'^2 - (2/x)yy' - y^2/x^2 = 0.$$

Following exactly the steps of Eqn. (2.30.5), we get the explicit solution

$$y^5 = (C/x)(x^3 + D)^2,$$

where C and D are arbitrary constants. Murphy (1992)

2.31 $yy'' + ky'^2 + g(x, y, y') = 0,\ k < 0,\ g$ linear in y'

1. $$\boldsymbol{yy'' - (1/4)y'^2 + my + nx^3 + px^2 + qx + r = 0,}$$

where $m, n, p, q,$ and r are constants.

A solution of the given DE is

$$y = a_1x^{5/2} + a_2x^2 + a_3x^{3/2} + a_4x,$$

where

$$\begin{aligned} a_1 &= \pm(-16n/35)^{1/2},\ a_2 = -4m/13, \\ a_3 &= -8(a_2^2 + ma_2 + p)/21a_1,\ a_4 = -16ma_3/65a_1, \end{aligned}$$

provided that

$$q = -3a_3^2/16 - 9ma_4/13, \quad r = a_4^2/4.$$

Marchenko (1988)

2. $$\boldsymbol{2yy'' - y'^2 - 1 = 0.}$$

Putting $y' = p, y'' = p(dp/dy)$, we have $2yp(dp/dy) = p^2 + 1$. Therefore,

$$\frac{2pdp}{p^2+1} = \frac{dy}{y}, \text{ i.e., } p^2 + 1 = C_1y$$

or

$$\frac{dy}{dx} = \pm(C_1y - 1)^{1/2}.$$

Hence,

$$\begin{aligned} x &= \pm\int \frac{dy}{(C_1y-1)^{1/2}} + C_2 \\ &= \pm 2C_1^{-1}(C_1y - 1)^{1/2} + C_2. \end{aligned}$$

Here C_1 and C_2 are arbitrary constants. Rabenstein (1966), p.373

3. $$\boldsymbol{2yy'' - y'^2 + a = 0,}$$

where a is a constant.

A first integral is $y'^2 - a = Cy, C$ is an arbitrary constant.

Now the variables separate:

$$\int \frac{dy}{(a + Cy)^{1/2}} = x + C_1;$$

C_1 is another arbitrary constant. Kamke (1959), p.576; Murphy (1960), p.396

4. $$\mathbf{2yy'' - y'^2 - 4y^2 = 0}$$

Make the substitution $y = z^2$ and integrate; the resulting linear DE is $z'' - z = 0$. The solution is $y = C_1 \sinh^2(x + C_2); C_1$ and C_2 are arbitrary constants. Reddick (1949), p.191

5. $$\mathbf{2yy'' - y'^2 + f(x)y^2 + a = 0,}$$

where $a > 0$ is a constant.

The given DE is a special case of Eqn. (2.32.9). If u, v are two solutions of the linear DE

$$4y'' + f(x)y = 0$$

that satisfy the condition $(uv' - vu')^2 = a$, then it may be shown that $y = uv$ is a solution of the given DE. Kamke (1959), p.576

6. $$\mathbf{2yy'' - y'^2 - 8y^3 - 4xy^2 = 0.}$$

With $y = \pm u^2$, the DE goes into $u'' \mp 2u^3 - xu = 0$. Now refer to Eqns. (2.4.38) and (2.4.40). Kamke (1959), p.577; Murphy (1960), p.396

7. $$\mathbf{yy'' - (1/2)y'^2 - 4y^3 - 2xy^2 - k = 0,}$$

where k is an integral constant.

The given DE results from the similarity reduction of the Korteweg–deVries equation. It can be shown that if $k = 0$, the given DE is of Painlevé type, having no movable critical point. Putting $y = F^2$, we have $F'' = 2F^3 + xF$, which is a special case of the second Painlevé equation. Tajiri and Kawamoto (1982)

8. $$\mathbf{2yy'' - y'^2 + ay^3 + by^2 = 0,}$$

where a and b are constants.

Putting $y = u^2$, we get

$$4u'' + au^3 + bu = 0;$$

now with $p(u) = u'(x)$, we have the DE

$$2(p^2)' + au^3 + bu = 0$$

with variables separable. Kamke (1959), p.577; Murphy (1960), p.396

9. $$\mathbf{2yy'' - y'^2 + ay^3 + 2xy^2 + 1 = 0,}$$

where $a \neq 0$ is a constant.

Defining u through $y' = 2uy - 1, y(x) \neq 0$, we get

$$ay = -4u' - 4u^2 - 2x \tag{1}$$

and

$$u'' - 2u^3 - xu - a/4 + 1/2 = 0. \tag{2}$$

Therefore, if u is a solution of (2), then one gets a solution of the original DE from (1). For (2), see Eqn. (2.4.38). Kamke (1959), p.577; Murphy (1960), p.396

10. $$\boldsymbol{2yy'' - y'^2 + ay^3 + bxy^2 = 0,}$$

where a and b are constants.

With $y = u^2$, we get $4u'' + au^3 + bxu = 0$; now see Eqn. (2.4.40). Kamke (1959), p.577; Murphy (1960), p.396

11. $$\boldsymbol{2yy'' - y'^2 - 3y^4 = 0.}$$

The given DE is a special case of Eqn. (2.32.9). A first integral is

$$y'^2 = y^4 + Cy;$$

C is an arbitrary constant. Hence the solution can be obtained in terms of elliptic functions. Kamke (1959), p.577; Murphy (1960), p.396

12. $$\boldsymbol{2yy'' - y'^2 - 3y^4 - 8xy^3 - 4(x^2 + a)y^2 + b = 0.}$$

This is the fourth Painlevé equation; see Sachdev (1991) for an elementary discussion of this equation. Kamke (1959), p.578; Murphy (1960), p.396

13. $$\boldsymbol{2yy'' - y'^2 + 3fyy' - 8y^3 + 2(f' + f^2)y^2 = 0, f = f(x).}$$

The given DE permits a first integral

$$(y' + fy)^2 = 4y\left\{y^2 + C\exp\left(-2\int f(x)dx\right)\right\};$$

C is an arbitrary constant. Kamke (1959), p.578; Murphy (1960), p.397

14. $$\boldsymbol{2yy'' - y'^2 + 4y^2y' + y^4 + f(x)y^2 + 1 = 0.}$$

With $y = u'/u$ we get

$$2u'u''' - u''^2 + fu'^2 + u^2 = 0; \tag{1}$$

differentiation of (1) leads to the linear DE

$$u^{(4)} + fu'' + (1/2)f'u' + u = 0.$$

Kamke (1959), p.578; Murphy (1960), p.397

15. $$\boldsymbol{3yy'' - 2y'^2 - ax^2 - bx - c = 0,}$$

where a, b, and c are constants.

Differentiating the given DE thrice, we get

$$3yy^{(5)} + 5y'y^{(4)} = 0;$$

therefore,

$$y^{(4)} = C|y|^{-5/3};$$

C is an arbitrary constant. Now by suitable elimination of the higher derivatives, we get

$$(2Ry' - 3R'y)^2 = 9(b^2 - 4ac)y^2 - 2R^3 + CR|y|^{4/3}, \tag{1}$$

with $R = ax^2 + bx + c$. Introducing u via $u^3y^2 = R^3$, and using (1), we find that

$$\int u^{-1}\{9(b^2 - 4ac) + Cu - 2u^3\}^{-1/2}du \pm \int \frac{dx}{3R} = C_1, \tag{2}$$

where C_1 is an arbitrary constant. We assume that the denominators occurring in (2) do not vanish. Kamke (1959), p.579

16. $$yy'' - (2/3)y'^2 - 12Ny^{2/3} = 0,$$

where N is a constant.

The given DE appears in the discussion of a nonlinear Schrödinger equation for an inhomogeneous medium. It has a special solution,

$$y = (-1)^{3/4}2^3N^{3/4}x^{3/2}.$$

Herrera (1984)

17. $$4yy'' - 3y'^2 + 4y = 0.$$

With $y = \pm u^2$, we have

$$2uu'' - u'^2 \pm 1 = 0.$$

Now see Eqn. (2.31.3). Kamke (1959), p.580; Murphy (1960), p.397

18. $$yy'' - (3/4)y'^2 + (y/x)^2 = 0.$$

Putting $u = y/x^\ell$ and $\xi = \ln x$, the given DE becomes

$$u\frac{d^2x}{d\xi^2} + (\ell/2 - 1)u\frac{du}{d\xi} - (3/4)\left(\frac{du}{d\xi}\right)^2 + (\ell/2 - 1)^2u^2 = 0. \tag{1}$$

We choose $\ell = 2$ to simplify (1) to

$$u\frac{d^2u}{d\xi^2} - \frac{3}{4}\left(\frac{du}{d\xi}\right)^2 = 0. \tag{2}$$

Equation (2), in the form $\frac{d^2u}{d\xi^2}/\frac{du}{d\xi} = (3/4)\frac{du}{d\xi}/u$, is solved easily:

$$4u^{1/4} = C_1\xi + C_2,$$

where C_1 and C_2 are arbitrary constants. Using the definition of u and ξ, we write the explicit solution of the given DE as

$$y = x^2[(C_1 \ln x + C_2)/4]^4.$$

Goldstein and Braun (1973), pp.83-84

19. $$\mathbf{4yy'' - 3y'^2 - 12y^3 = 0.}$$

With $y = \pm u^2$, we get

$$2uu'' - u'^2 \mp 3u^4 = 0.$$

Putting $u' = p, u'' = p(dp/du)$, we have

$$2up\frac{dp}{du} - p^2 \mp 3u^4 = 0. \tag{1}$$

Equation (1) is linear in p^2. Solving for p, taking the square root, and integrating lead to the final solution. Kamke (1959), p.580; Murphy (1960), p.397

20. $$\mathbf{4yy'' - 3y'^2 + ay^3 + by^2 + cy = 0,}$$

where a, b, and c are constants.

With $y = \pm u^2$, we get the simpler form

$$2uu'' - u'^2 \pm (1/4)(au^4 \pm bu^2 + c) = 0.$$

Now proceed as in Eqn. (2.32.9). Kamke (1959), p.580; Murphy (1960), p.397

21. $$\mathbf{4y(y'' + y) - 3(y' - m)(y' + 3m) = 0,}$$

where m is a constant.

The solution of the given DE has a finite Fourier series form: $y = a_0 + a_1 \cos x + a_2 \cos 2x + b_1 \sin x + b_2 \sin 2x$. In fact, it is found to be $y = -m\{2\sin(x+k) + \sin(x+k)\cos(x+k)\} + \nu\{1 + \cos(x+k)\}^2$, where k and ν are arbitrary constants. Roth (1941)

22. $$\mathbf{4yy'' - 3y'^2 + \{6y^2 - 2(f'/f)y\}y' + y^4 - 2(f'/f)y^3 + gy^2 + fy = 0,}$$
$$\mathbf{f = f(x), g = g(x).}$$

The solution is

$$4y = -f(2u' + u^2 - (f'/f) + g/4)^{-1},$$

where $u = v'/v$ and v is a solution of the linear third order DE

$$v''' = (3f'/2f)v'' + \{f''/f - (f'^2/f^2) - g/4\}v' + (1/8)\{(f'/f)g - f - g'\}v.$$

Kamke (1959), p.580

23. $$\mathbf{(y - 2)y'' - y'^2 = 0, 0 \le x \le 1, y(0) = A, y(1) = B.}$$

Separation of variables, integration, and use of the BC leads to the closed-form solution $y = 2 - (2 - A)\{(2 - B)/(2 - A)\}^x$. Baxley (1990)

24. $$\mathbf{yy'' - y'^2 + 1 = 0.}$$

The given DE is autonomous. Putting $y' = p,\ y'' = p(dp/dy)$, etc., and integrating twice, we obtain the solutions

$$C_1 y = \sinh(C_1 x + C_2) \quad \text{and} \quad C_1 y = \sin(C_1 x + C_2);$$

C_1 and C_2 are arbitrary constants. Kamke (1959), p.571

25. $$yy'' - y'^2 - 1 = 0.$$

Putting $y' = p, y'' = p(dp/dy)$, etc., and integrating twice, we obtain $C_1 y = \cosh(C_1 x + C_2)$; C_1 and C_2 are arbitrary constants. Kamke (1959), p.571; Murphy (1960), p.392

26. $$yy'' - y'^2 + k = 0, y(0) = 0, y(1) = 1,$$

where k is a constant.

For $k = 1$, the solution is $y = x$. Sellars (1955)

27. $$yy'' - y'^2 + qy = 0,$$

where q is a real number.

Put $y' = p, y'' = p(dp/dy)$ to obtain the Bernoulli equation

$$yp\frac{dp}{dy} - p^2 + qy = 0,$$

which has the solution

$$y = \{q/(2c_1^2)\}[\cosh\{2^{1/2}c_1(x + c_2)\} - 1], \tag{1}$$

where c_1 and c_2 are constants. The solution (1) is not periodic. To exhibit the periodic solutions, we write the given DE as

$$y(y'' + q) = y'^2 \tag{2}$$

and seek an integrating function $m(x)$ such that

$$y'' + q = my', \tag{3}$$

$$y' = my. \tag{4}$$

Assuming that $m(x)$ exists with a second derivative, we use (4) to eliminate y' from (3). This gives

$$y = -q/m'. \tag{5}$$

Now we seek a tractable equation involving $m(x)$ so as to secure solutions of the given DE by putting $m(x)$ into (5). Combining (4)-(5), we have $m'' + mm' = 0$ with the solution $m(x) = -c_1 \tan\{c_1(x + c_2)/2\}$. Thus, using (5), we have the periodic solutions

$$y = \{2q/c_1^2\} \cos^2\{c_1(x + c_2)/2\}, \tag{6}$$

where c_1, c_2 are arbitrary constants.

The curves (1) in (y, y') are hyperbolas, while the curves (6) are ellipses. Utz (1978)

28. $$yy'' - y'^2 - \beta y - \delta = 0;$$

β and δ are constants.

The given DE is a special case of Painlevé's third transcendent. Putting $y' = u$, we have

$$yu\frac{du}{dy} = u^2 + \beta y + \delta.$$

Integrating twice (putting $u^2 = v$, say), we get the solution in the form of a quadrature:

$$\int \frac{dy}{(C_1 y^2 - 2\beta y - \delta)^{1/2}} = \pm x + C_2,$$

where C_1 and C_2 are arbitrary constants. The integral may be expressed explicitly in terms of elementary functions. Lukashevich (1965)

29. $$\boldsymbol{yy'' - y'^2 + y^2 x = 0.}$$

The given DE is homogeneous in y, y', y''. Therefore, putting $u = y'/y$, we get

$$u' + u^2 - u^2 + x = u' + x = 0. \tag{1}$$

On integration, (1) gives

$$u + x^2/2 = C_1. \tag{2}$$

Hence, on integrating again, (2) (with $u = y'/y$) yields

$$\begin{aligned} \ln y &= \int (C_1 - x^2/2)dx + C_2 \\ &= C_1 x - x^3/6 + C_2 \end{aligned}$$

or

$$y = C_3 \exp(C_1 x - x^3/6),$$

$C_3 = e^{C_2}$; C_1 and C_2 are arbitrary constants. Goldstein and Braun (1973), p.80

30. $$\boldsymbol{yy'' - y'^2 + [x^2/(1-x)]y^2 = 0.}$$

The given DE is equidimensional w.r.t. y. Putting $y = e^{u(x)}$, we have

$$y^2[(1-x)u'' + x^2] = 0.$$

If $y \neq 0$, we get

$$(1-x)u'' + x^2 = 0. \tag{1}$$

Equation (1) is separable. Two integrations give

$$u(x) = x^3/6 + x^2/2 + (x-1)\log(x-1) + Ax + B,$$

where A and B are arbitrary constants. Hence

$$y(x) = e^{u(x)} = (x-1)^{(x-1)} \exp(x^3/6 + x^2/2 + Ax + B).$$

Zwillinger (1989), p.192

31. $$\boldsymbol{yy'' - y'^2 + \{1/(2x^2)\}y^2 = 0.}$$

See Eqn. (2.31.65). Forsyth (1959), p.306

32. $$yy'' - y'^2 + y^3 - 1 = 0.$$

Choosing

$$y(x) = \exp(f(x)), \tag{1}$$

we find that

$$\frac{d^2 f}{dx^2} = \exp(-2f) - \exp(f).$$

Integrating and then putting $g = \exp(f(x))$, we have

$$\left(\frac{dg}{dx}\right)^2 = -1 + kg^2 - 2g^3 = -2(g - m_1)(g - m_2)(g - m_3), m_1 > m_2 > m_3,$$

where R is an arbitrary constant. Using Jacobi's sn function, we can express the solution for f as

$$f(x) = \ln\{m_1 - (m_1 - m_3)\ \mathrm{sn}^2[\{(m_1 - m_3)/2\}^{1/2}x, \{(m_1 - m_2)/(m_1 - m_3)\}^{1/2}]\}.$$

Hence $y(x)$ may be obtained from (1). Kawamoto (1985)

33. $$yy'' - y'^2 - y^3 = 0.$$

Putting $y' = u, y'' = u(du/dy)$, we have

$$\frac{du}{dy} - \frac{1}{y}u = y^2 u^{-1}, u \neq 0, y \neq 0. \tag{1}$$

The Bernoulli's equation (1) can be solved:

$$u = \pm y(2y + c_1)^{1/2}, u \neq 0, y \neq 0, 2y + c_1 > 0,$$

where c_1 is a constant. With $dy/dx = u$, a second integration gives

$$(2/c_1^{1/2}) \log[\{(2y + c_1)^{1/2} - c_1^{1/2}\}/y^{1/2}] = \pm x + c_2, \text{ if } c_1 > 0$$

or

$$\{2/(-c_1)^{1/2}\} \tan^{-1}[\{(2y + c_1)^{1/2}/(-c_1)^{1/2}\}] = \pm x + c_2, \text{ if } c_1 < 0.$$

For $c_1 = 0$, we have $dy/dx = u = \pm 2^{1/2} y^{3/2}$. This integrates to give $(2^{1/2}x - c_2)^2 y = 4, y \neq 0$. We note that $y = 0$ is a particular solution not obtainable from the cases above. c_2 is another arbitrary constant. Tenebaum and Pollard (1963), p.503

34. $$yy'' - y'^2 - (1/2)w_0^2 y^3 = 0,$$

where w_0^2 is a constant.

Put $y' = p, y'' = p(dp/dy)$, etc.

$$yp\,\frac{dp}{dy} - p^2 - (1/2)w_0^2 y^3 = 0. \tag{1}$$

Equation (1) is linear in p^2 and can be solved:

$$y'^2 = w_0^2 y^3 + I_1 y^2, \tag{2}$$

where I_1 is a constant. Writing $y = \{g^2(x)-1\}I_1/w_0^2$, Eqn. (2) reduces to $g' = (1/2)I_1^{1/2}\{1-g^2\}$ with the solution

$$g = \tanh\{(1/2)I_1^{1/2}(x - x_0)\},$$

where x_0 is arbitrary. Thus,

$$y = -(I_1/w_0^2)\ \mathrm{sech}^2\{(1/2)I_1^{1/2}(x - x_0)\}.$$

Parthasarathy and Lakshmanan (1990)

35. $$\boldsymbol{yy'' - y'^2 + \beta y^4 = 0,}$$

where β is a constant.

A first integral is easily found to be $y'^2 = -\beta y^4 + I_1 y^2$, where I_1 is a constant of integration. Putting $y = (I_1/\beta)^{1/2}[2g^2(x) - 1]^{-1}$, we have $g'^2 = (1/4)I_1[g^2(x) - 1]$ with the solution $g(x) = \cosh[(1/2)I_1^{1/2}(x - x_0)]$, thus yielding $f(x) = (I_1/\beta)^{1/2}\ \mathrm{sech}(I_1^{1/2}x)$. Parthasarathy and Lakshmanan (1991)

36. $$\boldsymbol{yy'' - y'^2 - \alpha y^3 - \gamma y^4 = 0,}$$

where α and γ are constants.

The given DE (after some transformations) is a special case of the third Painlevé transcendent. Putting $y' = u$, we get

$$yu\frac{du}{dy} = u^2 + y^3\alpha + \gamma y^4.$$

Putting $u^2 = U$ and integrating twice, we get

$$\int \frac{dy}{y(\gamma y^2 + 2\alpha y + C_1)^{1/2}} = \pm x + C_2, \tag{1}$$

where C_1 and C_2 are arbitrary constants. The integral in (1) may be expressed in terms of elementary functions. Lukashevich (1965)

37. $$\boldsymbol{yy'' - y'^2 - a_0 - a_1 y - (a_2 + a_3 y)y^3 = 0,}$$

where a_i $(i = 0, 1, 2, 3)$ are constants.

Introducing $y' = p, y'' = p(dp/dy)$, etc., write the given DE as

$$py\frac{dp}{dy} - p^2 = a_0 + a_1 y + (a_2 + a_3 y)y^3.$$

Solve this linear DE for $p^2(y)$ to obtain

$$y'^2 = Cy^2 - a_0 - 2a_1 y + (2a_2 + a_3 y)y^3;$$

C is an arbitrary constant. The solution now is expressible in terms of elliptic functions. Murphy (1960), p.392

38. $$\boldsymbol{yy'' - y'^2 - a_0 - a_1 y - a_2 y^2 - a_3 y^3 - a_4 y^4 = 0,}$$

a_i $(i = 0, 1, 2, 3, 4)$ are constants.

With $y'(x) = p(y)$, we have

$$ypp'(y) = p^2 + a_0 + a_1 y + a_2 y^2 + a_3 y^3 + a_4 y^4. \qquad (1)$$

Solving the linear DE (1) for $p^2(y)$, we have

$$y'^2 = Cy^2 - a_0 - 2a_1 y + 2a_2 y^2 \ln y + 2a_3 y^3 + a_4 y^4;$$

now the variables separate. Murphy (1960), p.392

39. $$\boldsymbol{yy'' - y'^2 - a_0 - a_1 y - a_3 y^3 - a_4 y^4 = 0};$$

where a_i $(i = 0, 1, 3, 4)$ are constants.

Writing $y(x) = \beta \exp(\alpha f(x))$, we get

$$\frac{d^2 f}{dx^2} = \frac{a_0}{\alpha\beta^2}\exp(-2\alpha f) + \frac{a_1}{\alpha\beta}\exp(-\alpha f) + \frac{\beta a_3}{\alpha}\exp(\alpha f) + (\beta^2 a_4/\alpha)\exp(2\alpha f). \qquad (1)$$

Integrating (1) once, we have

$$\begin{aligned}\left(\frac{df}{dx}\right)^2 &= \frac{-a_0}{\alpha^2\beta^2}\exp(-2\alpha f) - \frac{2a_1}{\alpha^2\beta}\exp(-\alpha f) + \frac{2\beta a_3}{\alpha^2}\exp(\alpha f)\\ &\quad + (\beta^2 a_4/\alpha^2)\exp(2\alpha f) + K, \end{aligned} \qquad (2)$$

where K is an integration constant. Again by the transformation $g(x) = \gamma \exp(\alpha f(x))$, (2) can be rewritten as

$$\left(\frac{dg}{dx}\right)^2 = A_0 + A_1 g + A_2 g^2 + A_3 g^3 + A_4 g^4 \equiv -\mathcal{F}(g) \geq 0, \qquad (3)$$

where $A_0 = -a_0\gamma^2/\beta^2, A_1 = -2a_1\gamma/\beta, A_2 = \alpha^2 k, A_3 = 2\beta a_3/\gamma, A_4 = \beta^2 a_4/\gamma^2$ with arbitrary constants α, β, and γ. Several special cases of (3) that are related to the one-soliton solution of several soliton equations are treated.

(a) $A_0 = A_1 = A_4 = 0$, so that (3) becomes

$$\left(\frac{dg}{dx}\right)^2 = g^2(A_2 + A_3 g), \qquad (4)$$

where $A_2 > 0, A_3 < 0$. Integration of (4) leads to

$$g_1(x) = -(A_2/A_3)\ \mathrm{sech}^2[A_2^{1/2}(x + x_0)/2]$$

for $0 \leq g(x) \leq -A_2/A_3$,

$$g_2(x) = (A_2/A_3)\ \mathrm{cosech}^2[A_2^{1/2}(x + x_0)/2] \text{ for } g(x) \leq 0,$$

where x_0 is an integration constant.

(b) $A_0 = A_1 = A_3 = 0$, so that

$$\left(\frac{dg}{dx}\right)^2 = g^2(A_2 + A_4 g^2). \tag{5}$$

Equation (5) integrates to give

$$g(x) = \pm(|A_2/A_4|)^{1/2}\ \mathrm{sech}[A_2^{1/2}(x + x_0)],$$

where x_0 is a constant of integration, provided that $A_2 > 0$ and $A_4 < 0$.

(c) $A_0 = A_1 = 0$, so that

$$\left(\frac{dg}{dx}\right)^2 = g^2(A_2 + A_3 g + A_4 g^2), A_2 < 0 \quad \text{and} \quad A_3, A_4 > 0. \tag{6}$$

The solitary wave solutions of (6) are

$$g(x) = \pm\frac{2A_2}{(A_3^2 - 4A_2A_4)^{1/2}}\left\{\frac{2}{\exp(X) - 2A_3/(A_3^2 - 4A_2A_4)^{1/2} + \exp(-X)}\right\},$$

where $X \equiv (|A_2|)^{1/2}(x + x_0)$.

(d) $A_0 = A_4 = 0$ or $A_1 = A_3 = 0$. For example, if $A_0 = A_4 = 0$ so that in the original DE $a_0 = a_4 = 0$, we have

$$yy'' - y'^2 - a_1 y - a_3 y^3 = 0. \tag{7}$$

Introducing $y(x) = [|a_1/a_3|]^{1/2}\exp[i\alpha f(x)]$, we have

$$\frac{d^2 f}{dx^2} = -\left[2\left(|a_1 a_3|\right)^{1/2}/\alpha\right]\sin\alpha f,$$

with the solution

$$f(x) = \pm(4/\alpha)\tan^{-1}[\exp\{2(|a_1 a_3|)^{1/2}\}^{1/2}(x + x_0)].$$

Kawamoto (1985)

40. $$\boldsymbol{yy'' - y'^2 - a_0 - a_1 y - a_3 y^3 - a_4 y^4 - a_5 y^5 = 0,}$$

where a_i are constants.

Put $y(x) = \beta\ \exp(\alpha f(x))$. Integrating once, and then writing $g(x) = \gamma\ \exp\{\alpha f(x)\}$, we obtain

$$\left(\frac{dg}{dx}\right)^2 = A_0 + A_1 g + A_2 g^2 + A_3 g^3 + A_4 g^4 + A_5 g^5, \tag{1}$$

where

$$A_0 = -a_0\gamma^2/\beta^2, A_1 = -2a_1\gamma/\beta, A_2 = \alpha^2 k, A_3 = 2\beta a_3/\gamma, A_4 = \beta^2 a_4/\gamma^2, A_5 = 2\beta^3 a_5/(3\gamma^3),$$

where α, β, and γ are arbitrary constants.

With fifth degree polynomials on the RHS of (1), the integrals of the latter are called hyperelliptic. The following special cases, with $A_5 \neq 0$, may be determined explicitly.

(a) $A_0 = 0$ such that (1) can be written as

$$\left(\frac{dg}{dx}\right)^2 = A_5 g(g+\lambda_1)^2(g+\lambda_2)^2, \tag{2}$$

where λ_1, λ_2 can be found in terms of A_1, A_2, A_3, A_4, and A_5. Taking the square root of (2), we can find the solution in terms of elementary functions.

(b) $A_0 = A_1 = 0, A_5 < 0$, so that (1) can be written as

$$\left(\frac{dg}{dx}\right)^2 = A_5 g^2(g+\lambda_1)^2(g+\lambda_2).$$

Noting that $(g+\lambda_2) \leq 0$, where $A_5, \lambda_1, \lambda_2 < 0$, and $\lambda_1 < \lambda_2$, we have

$$\begin{aligned}\frac{dx}{dg} &= \pm\{1/|A_5|^{1/2}\}[1/\{g(g+\lambda_1)(-g-\lambda_2)^{1/2}\}] \\ &= \pm[1/\lambda_1|A_5|^{1/2}][1/\{g(-g-\lambda_2)^{1/2}\} - 1/\{(g+\lambda_1)(-g-\lambda_2)^{1/2}\}]. \end{aligned} \tag{3}$$

Integrating (3), we have

$$\begin{aligned}\int^g \frac{dg}{g(-g-\lambda_2)^{1/2}} &= \pm\lambda_1|A_5|^{1/2}(x+x_0) + \int \frac{dg}{(g+\lambda_1)(-g-\lambda_2)^{1/2}} \\ &= \pm\lambda_1|A_5|^{1/2}(x+x_0) + \frac{2}{(\lambda_2-\lambda_1)^{1/2}} \arctan \frac{F(x)}{(\lambda_2-\lambda_1)^{1/2}}, \end{aligned} \tag{4}$$

with the integration constant x_0, where

$$F^2(x) = -g(x) - \lambda_2. \tag{5}$$

Since the LHS of (4) is an elementary integral for $g(x)$, we can obtain

$$\begin{aligned} g_1(x) &= -\lambda_2 \operatorname{sech}^2\left[\left(|\lambda_2|^{1/2}/2\right)\{(\pm\lambda_1|A_5|^{1/2}(x+x_0) + G(F_1(x)))\}\right], \\ & \qquad 0 \leq g(x) \leq -\lambda_2, \\ g_2(x) &= \lambda_2 \operatorname{cosech}^2\left[\left(|\lambda_2|^{1/2}/2\right)\{(\pm\lambda_1|A_5|^{1/2}(x+x_0) + G(F_2(x)))\}\right], \\ & \qquad g(x) \leq 0 \end{aligned}$$

where

$$G(F_j(x)) = \frac{2}{(\lambda_2-\lambda_1)^{1/2}} \arctan \left\{\frac{F_j(x)}{(\lambda_2-\lambda_1)^{1/2}}\right\} \quad (j=1,2),$$

and by Eqn. (5),

$$\begin{aligned} F_1^2(x) &= -g_1(x) - \lambda_2 \\ &= -\lambda_2 \tanh^2\left[\left(|\lambda_2|^{1/2}/2\right)(\pm\lambda_1|A_5|^{1/2}(x+x_0) + G(F_1(x)))\right], \\ F_2^2(x) &= -g_2(x) - \lambda_2 \\ &= -\lambda_2 \coth^2\left[\left(|\lambda_2|^{1/2}/2\right)(\pm\lambda_1|A_5|^{1/2}(x+x_0) + G(F_2(x)))\right]. \end{aligned}$$

(c) $A_0 = A_1 = A_2 = 0$, so that (1) becomes

$$\left(\frac{dg}{dx}\right)^2 = A_5 g^3 (g + \lambda_1)^2. \tag{6}$$

The solution of (6) is elementary. The solutions are shown graphically and related to soliton solutions of some model PDE. Kawamoto (1985)

41. $$\boldsymbol{yy'' - y'^2 - a_0 - a_1 y - a_3 y^3 - a_4 y^4 - a_5 y^5 - a_6 y^6 = 0,}$$

where a_i $(i = 0, \ldots, 6)$ are constants.

Conditions on a_i are obtained such that the solutions of the given DE correspond to a stationary solitary wave in sech or a slightly generalized form of this function. Kawamoto (1985)

42. $$\boldsymbol{yy'' - y'^2 + Kx^2 y^4 = 0, y(0) = 1, y'(0) = 0,}$$

where K is a constant.

The given DE may be written as $(\ln y)'' + Kx^2y^2 = 0$. The IVP describes a soliton solution of the nonlinear scalar equation of the unitary quantum theory. It is solved numerically with $K = 16\pi$, and shown graphically. Sapogin and Boichenko (1988)

43. $$\boldsymbol{yy'' - y'^2 + e^{2x}(ay^4 + b) + e^x y(cy^2 + d) = 0.}$$

Put $y(x) = \eta(\xi), \xi = e^x$, to obtain

$$\xi(\eta\eta'' - \eta'^2) + \eta\eta' + \xi(a\eta^4 + b) + c\eta^3 + d\eta = 0.$$

See Ince (1956), Eqn. (XIII). Kamke (1959), p.571; Ince (1956), p.335

44. $$\boldsymbol{yy'' - y'^2 - y^2 \log y = 0.}$$

Putting $u(x) = \ln y$, one may easily obtain the solution as $\log y = C_1 e^x + C_2 e^{-x}$; C_1 and C_2 are arbitrary constants. Kamke (1959), p.571; Murphy (1960), p.393

45. $$\boldsymbol{yy'' - y'^2 - y^2 \ln y + x^2 y^2 = 0.}$$

With $\ln y = u$, we get

$$y^2(u'' - u + x^2) = 0.$$

The solutions are

$$\ln y = C_1 e^x + C_2 e^{-x} + 2 + x^2; y = 0;$$

C_1 and C_2 are arbitrary constants. Murphy (1960), p.393

46. $$\boldsymbol{yy'' - y'^2 + y' = 0.}$$

With $y' = p(y)$, we have $p(yp' - p + 1) = 0$. So a first integral is

$$p = 1 + Cy,$$

where C is an arbitrary constant.

The solution finally is

$$y + C_2 = C_1 e^{x/C_2};$$

C_1 and C_2 are arbitrary constants. Murphy (1960), p.393

47. $$\boldsymbol{yy'' - y'^2 + 2y' = 0.}$$

With $y' = p(y)$, we have $p(yp' + 2 - p) = 0$; so a first integral is $p = Cy + 2$. The solutions are

$$C_1 y + C_2 \exp(C_1 x) + 2 = 0; y = C_0.$$

Here C, C_1, C_2, and C_0 are arbitrary constants. Murphy (1960), p.393

48. $$\boldsymbol{yy'' - y'^2 - y' = 0.}$$

Put $y' = p, y'' = p(dp/dy)$; we have $yp(dp/dy) - p^2 - p = 0$. Therefore, either $p = 0$, i.e., $y =$ constant, or

$$\begin{aligned} y\frac{dp}{dy} &= p + 1, \\ \frac{dy}{y} &= \frac{dp}{p+1}, \quad p + 1 = cy; \\ \frac{dy}{dx} &= cy - 1. \end{aligned} \tag{1}$$

On integration of (1), we have $x = (1/c)\ln(cy - 1) + d$, where c and d are arbitrary constants. Rabenstein (1966), p.373

49. $$\boldsymbol{yy'' - y'^2 - y' - y^2(fy + g') = 0; f = f(x), g(x) = -f'/f.}$$

A first integral is

$$(y' - gy + 1)^2 = 2y^2\left(fy + \int f dx + C\right);$$

C is an arbitrary constant. Murphy (1960), p.393

50. $$\boldsymbol{yy'' - y'^2 - y' + fy^3 + y^2\frac{d}{dx}(f'/f) = 0, f = f(x).}$$

The given DE admits the first integral,

$$[y' + (f'/f)y + 1]^2 + 2y^2\left(yf + \int f dx + C\right) = 0;$$

C is a constant. Kamke (1959), p.571

51. $$\boldsymbol{yy'' - y'^2 + f(x)y' - y^3 - f'(x)y = 0.}$$

The given DE admits a first integral

$$[y' - f]^2 = 2y^2\left(y - \int f dx + C\right);$$

C is an arbitrary constant. Kamke (1959), p.571; Murphy (1960), p.393

52. $$\boldsymbol{yy'' - y'^2 + f'y' - y^4 + fy^3 - f''y = 0, f = f(x).}$$

A first integral of the given DE is

$$(y' - f')^2 - y^2(y - f)^2 = Cy^2;$$

C is an arbitrary constant. Kamke (1959), p.572; Murphy (1960), p.393

53. $$\boldsymbol{yy'' - y'^2 - yy' = 0.}$$

(a) With $y'(x) = p(y)$, we have

$$ypp'(y) = p^2 + yp. \tag{1}$$

One solution of (1) is $p = y\ \ln(Cy)$; hence the solution is

$$y = C_1 \exp(C_2 e^x).$$

(b) $y = C_2$.

C, C_1, and C_2 are arbitrary constants. Murphy (1960), p.392

54. $$\boldsymbol{yy'' - y'^2 + ayy' + by^2 = 0;}$$

a and b are constants.

With $u(x) = y'/y$, we get a first order linear DE

$$u' + au + b = 0.$$

Kamke (1959), p.572; Murphy (1960), p.393

55. $$\boldsymbol{yy'' - y'^2 + ayy' + by^3 - 2ay^2 = 0,}$$

where a and b are constants.

Put $y'(x) = p(y)$ to obtain the Abel DE

$$ypp' - p^2 + apy + by^3 - 2ay^2 = 0.$$

Kamke (1959), p.572

56. $$\boldsymbol{yy'' - y'^2 - (ay - 1)y' - 2b^2y^3 + 2a^2y^2 + ay = 0;}$$

a and b are constants.

The solutions are

$$y = -1/(2a) + e^{2ax}(u^2 + C),$$

where

$$u' = be^{-ax}[(u^2 + C)e^{2ax} - 1/(2a)] \text{ for } a \neq 0;$$

and

$$y = x + u^2 + C,$$

where

$$u' = b(u^2 + C + x) \text{ for } a = 0, b \neq 0.$$

Here $u = u(x)$ and C is an arbitrary constant. For explicit solution, the Riccati equation in u must be solved. Kamke (1959), p.572

57. $$\boldsymbol{yy'' - y'^2 + a(y-1)y' - y(y+1)(b^2y^2 - a^2) = 0;}$$

a and b are constants.

The solution is given by

$$y = -1 + Ce^{-ax}[(1+u^2)/(1-u^2)],$$

where

$$(2/b)u' = Ce^{-ax}(u^2+1) + u^2 - 1; \tag{1}$$

C is an arbitrary constant. Equation (1) is of Riccati form. Kamke (1959), p.572

58. $$\boldsymbol{yy'' - y'^2 + ayy' + by^2(\log y) = 0, y(x) > 0,}$$

where a and b are constants.

Put $\log y = u$ to obtain the linear DE with constant coefficients: $u'' + au' + bu = 0$. Utz (1978)

59. $$\boldsymbol{yy'' - y'^2 + i\alpha yy' + (1/2)w_0^2 y - (1/2)w_0^2 y^3 + i\gamma y^2 \cosh wx = 0,}$$

where α, w_0, and γ are constants and $i = (-1)^{1/2}$.

The given DE is another version of a driven pendulum with viscous damping. To analyze its singularity structure, we assume a Laurent series

$$y = \sum_{j=0}^{\infty} a_i \tau^{j-2}, \quad \tau = (x - x_0) \to 0 \tag{1}$$

about an arbitrary movable singularity, which in order to give a general solution, must have one arbitrary constant in addition to x_0. Substitution of (1) in the given DE leads to the recursion relation

$$\begin{aligned} &\sum_r \{a_{j-r}a_r(j-r-2)(j-2r-1) + i\alpha a_{j-r-1}a_r(j-r-3) \\ &\quad - (1/2)w_0^2 \sum_p a_{j-r}a_{r-p}a_p + i\gamma_p G_{j-r-2}a_{r-p}a_p\} \\ &\quad = -(1/2)w_0^2 a_{j-4}, \ 0 \le p \le r \le j, \end{aligned} \tag{2}$$

where

$$G(x) = \cosh wx, G_n = \frac{1}{n!} \left. \frac{\partial^n G(x)}{\partial x^n} \right|_{x=x_0}.$$

From (2) one obtains

$$\begin{aligned} j &= 0, \quad a_0 = 4/w_0^2, \\ j &= 1, \quad a_1 = -4i\alpha/w_0^2, \\ j &= 2, \quad 0.a_2 + (2\alpha^2 + i\gamma \cosh wx_0)a_0^2 = 0. \end{aligned} \tag{3}$$

It follows from (3) that for a_2 to be arbitrary, $\alpha = 0 = \gamma$. Thus, the given DE is of Painlevé type only if $\alpha = \gamma = 0$. It is easy to check that in this case, it is integrable in terms of Jacobian elliptic functions.

For $\alpha \neq 0$ and $\gamma \neq 0$, we must seek a more general form of the solution to have a_2 arbitrary. We assume the psi series form

$$y = \sum_{j=0}^{\infty} \sum_{k=0}^{\infty} a_{jk} \tau^{j-2} (\tau^2 \ln \tau)^k. \tag{4}$$

Substituting (4) in the given DE leads to the recurrence relation for a_{jk}:

$$\begin{aligned} &\sum_{r,s} \{a_{j-r,k-s}\, a_{rs}(j - r + 2k - 2s - 2)(j - 2r + 2k - 4s - 1) \\ &\quad + a_{j-r-2,k-s+1} a_{rs}[(2j - 2r + 4k - 4s - 5)(k - 2s + 1) - s] \\ &\quad + a_{j-r-4,k-s+2} a_{rs}(k - s + 2)(k - 2s + 1) + i\alpha a_{j-r-1,k-s} a_{rs}(j - r + 2k - 2s - 3) \\ &\quad + i\alpha a_{j-r-3,k-s+1} a_{rs}(k - s + 1) - (1/2) w_0^2 \sum_{p,q} a_{j-r,k-s} a_{r-p,s-q} a_{pq} \\ &\quad + i\gamma \sum_p G_{j-r-2} a_{r-p,k-s} a_{ps} = (-1/2) w_0^2 a_{j-4,k}, 0 \leq p \leq r \leq j, \\ &\quad 0 \leq q \leq s \leq k. \end{aligned} \tag{5}$$

The coefficients a_{00} and a_{10} are found to be $4/w_0^2$ and $-4i\alpha/w_0^2$, respectively. For a_{20}, we have

$$0.a_{20} - a_{01}a_{00} + (2\alpha^2 + i\gamma \cosh wx_0)a_{00}^2 = 0,$$

which means that $a_{01} = 4(2\alpha^2 + i\gamma \cosh wx_0)/w_0^2$.

Thus, in general, the given DE does not have the Painlevé property and has a movable logarithmic singularity, the exceptional case being $\alpha = 0, \gamma = 0$.

It is easy to check that the only closed-form recurrence relations (5) relate to a_{0k}, $k = 0, 1, 2, \ldots$. These satisfy

$$\Sigma_s[\{8(k-s)(k-s-1) - 8s(k-s) + 8s - 4(k-s) + 4\} a_{0,k-s} a_{os} - w_0^2 \Sigma_p a_{0,k-s} a_{0,s-q} a_{0q}] = 0. \tag{6}$$

It is not difficult to check that if we introduce the generating function

$$\theta(z) = \sum_{k=0} a_{ok} z^k, \tag{7}$$

where z is a function of τ, then the following relation for $\theta(z)$ is obtained from (6):

$$8z^2\theta\theta'' - 8z^2\theta'^2 + 4z\theta\theta' + 4\theta^2 - w_0^2\theta^3 = 0, \tag{8}$$

where a prime denotes differentiation with respect to z. In fact, (8) is obtained more easily by the direct substitution

$$y = (1/\tau^2)\theta(z), z = \tau^2 \ln \tau$$

in the given DE. It is easy to check that (8) has the Painlevé property with the local expansion

$$\theta(z) = \sum_{j=0}^{\infty} A_j (z - z_0)^{j-2},$$

in which A_2 and z_0 are the arbitrary constants. Actually, (8) can be explicitly integrated. We write $\theta(z) = \xi^2 f(\xi), \xi = z^{1/2}$ so that (8) becomes

$$ff'' - f'^2 - (1/2)w_0^2 f^3 = 0, \tag{9}$$

where a prime denotes differentiation with respect to ξ. Equation (9) has a first integral

$$f'^2 = w_0^2 f^3 + I_1 f^2, \tag{10}$$

where I_1 is a constant of integration and can be obtained from using the initial conditions as obtained from the recurrence relations

$$I_1 = -3w_0^2 a_{01} = -12(2\alpha^2 + i\gamma \cosh wx_0). \tag{11}$$

On writing $f(\xi) = \{g^2(\xi) - 1\}I_1/w_0^2$, Eqn. (10) becomes

$$g'(\xi) = (1/2)I_1^{1/2}\{1 - g^2(\xi)\}. \tag{12}$$

Equation (12), on integration, yields $g(\xi) = \tanh\{(1/2)I_1^{1/2}(\xi - \xi_0)\}$. Choosing $\xi_0 = 0$ for convenience, we get

$$f(\xi) = -(I_1/w_0^2)\ \mathrm{sech}^2\{(1/2)I_1^{1/2}\xi\}. \tag{13}$$

It is evident from (13) that f has second order poles at

$$\xi_m = (i\pi/I_1^{1/2})(2m+1), m \in Z$$

in the complex ξ-plane, where m denotes a lattice site integer. The corresponding singularity in the complex z plane is situated at

$$z_m = (1/12)\pi^2(2m+1)^2\{(2\alpha^2 - i\gamma \cosh wx_0)/(4\alpha^4 + \gamma^2 \cosh^2 wx_0)\}. \tag{14}$$

The singularity structure in the complex z plane is studied for the parameters $\alpha = 0.3, w_0^2 = 1.0, w = 0.5$, and $\gamma = 0.5$. The analysis in the complex x-plane is also inferred, and depicted. Parthasarathy and Lakshmanan (1990)

60. $$yy'' - y'^2 + (1/x)yy' - \lambda/2 = 0,$$

where λ is a constant.

A solution of the given DE is $y = Ax^a + Bx^b$, where $a = -\beta, b = 2+\beta, \lambda = 8(1+\beta^2)AB$, and A and B are arbitrary constants. Lange and Weinitschke (1991)

61. $$y'' - y'^2/y + (1/x)y' - (1/x)(\alpha y^2 + \beta) - \gamma y^3 - \delta/y = 0,$$

where α, β, γ, and δ are constants.

For the given DE which is P_{III}, an IVP $y(x_0) = y_0, y'(x_0) = y_1$ is solved in the neighborhood of the fixed singular point $x = 0$. It is further indicated that $\{y(x)x^{-w}\}^{\pm 1}, -1 \le w \le 1$ and $\{(\tilde{y}(x))^{-1}x^w\}^{-1}, -1 \le w \le 1$ are bounded on suitably chosen curves tending to zero. Convergent series solutions are constructed in the neighborhood of $x = 0$. Shimomura (1982, 1982a)

62. $$yy'' - y'^2 - yy'/x - 2y^2/x^2 + k = 0,$$

where k is a constant.

The given DE is (inhomogeneous) Cauchy-Euler type with respect to x, which by writing $x = e^t$, etc., suggests writing the solution as $y = (1/x)Y(x)$.

The solution is then found to be $y = (a/x)\cosh(bx^2+\delta)$, where a, b, and δ are constants, such that $4a^2b^2 + k = 0$. Olver (1986), p.201

63. $$(8x-3)(yy'' - y'^2) + 12yy' - [8/(2x-1)]y^2 - y/(2x-1) = 0.$$

The given ODE results from Liouville equation through several transformations. Putting $y = h^2$, we get an equation similar to the given DE:

$$(2x-1)(8x-3)(hh'' - h'^2) + 12(2x-1)hh' - 4h^2 - 1/2 = 0.$$

The given DE has special solutions $y = -1/8, y = -x+3/8$, and $y = -x/2+1/4$. Matsuno (1987)

64. $$yy'' - y'^2 + (\tan x + \cot x)yy' + \{\cos^2 x - \nu^2 \cot^2 x\}y^2 \log y = 0.$$

The solution is $y = \exp[Z_\nu(\sin x)]$; Z_ν is a Bessel function of order ν. Kamke (1959), p.572

65. $$yy'' - y'^2 + f(x)yy' + g(x)y^2 = 0.$$

With $u(x) = y'/y$, we obtain the linear DE $u' + fu + g = 0$. Kamke (1959), p.572; Murphy (1960), p.394

66. $$yy'' - y'^2 + f(x)yy' - \alpha y^3 \exp\left(-2\int f(x)dx\right) = 0,$$

where α is a constant.

Putting $y = e^{z(x)}$, we get $z'' + f(x)z' = \alpha e^z \exp(-2\int f(x)dx)$, which has the general solution $\int (c_1 + 2\alpha e^z)^{-1/2}dz = \theta(x, c_2)$, where $\theta(x, c_2) = c_2 + \alpha^{1/2}\int \exp(-\int f(x)dx)dx$; now see Eqn. (2.32.23). Hence the solution of the original DE via $y = e^{z(x)}$. Bandic (1965)

67. $$yy'' - y'^2 - y^2y' = 0.$$

With $y'(x) = p(y)$ we have $p(yp' - p - y^2) = 0$. So a first integral is $p = y(y + C)$; C is an arbitrary constant. The solutions are

$$y = C_2 e^{C_1 x}(y + C_1), y = C_0;$$

C_2, C_1, and C_0 are arbitrary constants. Murphy (1960), p.393

68. $$yy'' - y'^2 - \alpha y^2 y' - y^3 = 0,$$

where α is constant.

The given DE governs the similarity form of solutions of nonlinear diffusion equations. Put $y = e^p$ so that we get

$$e^p\left(1 + \alpha\frac{dp}{dx}\right) = \frac{d^2p}{dx^2}. \tag{1}$$

Equation (1) integrates to yield

$$\alpha^2 e^p = \alpha\frac{dp}{dx} - \ln\left[\left(1+\alpha\frac{dp}{dx}\right)\beta^{-1}\right],$$

which together with

$$\int_{q_0}^{dp/dx} \frac{dq}{[\alpha q - \ln\{(1+\alpha q)/\beta\}](1+\alpha q)} - x/\alpha^2,$$

where q_0 is an arbitrary constant, gives the solution in implicit form. β is another arbitrary constant. King (1990)

69. $$yy'' - y'^2 + \lambda y^2(xy' + 2y) = 0,$$

where λ is a constant.

The given DE governs the similarity solution of a nonlinear diffusion equation. Putting $y = dh/dx$ and integrating (with constant of integration set equal to zero), we have $(h')^{-1}h'' + \lambda(h + xh') = 0$, so that

$$\frac{dh}{dx} = \lambda e^{-\lambda xh}, \tag{1}$$

where α is an arbitrary constant. King (1990)

70. $$yy'' - y'^2 + (fy^2 + g)y' + f'y^3 - g'y = 0, f = f(x), g = g(x).$$

The solutions of the given DE satisfy Riccati DE $y' + fy^2 + Cy - g = 0, C$ an arbitrary constant, which may be linearized [see Sachdev (1991), p.28]. Kamke (1959), p.572

71. $$yy'' - y'^2 - (a_0y^2 + a_1y + a_2)y' - b_0y^4 - b_1y^3 - b_2y^2 - b_3y - b_4 = 0,$$

where a_i and b_i are analytic functions in a given domain D.

The given DE includes Painlevé's fourth equation as a special case. With a view to studying the multivalued moving singularities, the given DE is transformed according to $y = (A + u)/t, t = x - x_0, x_0 \in D, A \neq 0$, and a solution in the form $u = \sum_{n=1}^{\infty} u_n t^n$ is sought. The given system is first changed to a Briot and Bouquet type, and hence analyzed. Kazakov (1987)

72. $$yy'' - y'^2 - pg(x)y^{p+1}y' - g'(x)y^{p+2} = 0,$$

where $p \neq 0$ is a real constant and $g \neq 0$ is an arbitrary function.

Multiplying the given DE by y^{-2} and integrating once, we have

$$y' - Cy = g(x)y^{p+1}. \tag{1}$$

Equation (1) is of Bernoulli type and is easily solved through a quadrature. It is shown that if $g(x)$ is not of the form

$$g(x) = k_1 e^{k_2 x}(k_3 + k_4 x)^{k_5} \text{ or } g(x) = k_6 e^{k_7 x^2},$$

where k_i $(i = 1, \ldots, 7)$ are real constants, then the given DE admits no nontrivial symmetries but is, nevertheless, integrable by quadratures. Gonzalez–Lopez (1988)

73. $$yy'' - y'^2 - P_2(y, x)y' - P_4(y, x) = 0,$$

where $P_2 = a_0(x)y^2 + a_1(x)y + a_2(x)$,

$$P_4 = b_0(x)y^4 + b_1(x)y^3 + b_2(x)y^2 + b_3(x)y + b_4(x),$$

and $a_i(x)$ and $b_i(x)$ are holomorphic functions of x in some domain D.

Conditions are found on the coefficients in P_2 and P_4 such that the DE reduces to the irreducible form of $\mathrm{P_{III}}$:

$$ww'' - w'^2 + (1/x)ww' - (1/x)w(\alpha w^2 + \beta) - \gamma w^4 - \delta = 0.$$

Kazakov (1986)

74. $$4yy'' - 5y'^2 + ay^3 = 0,$$

where a is a constant.

This is a special case of Eqn. (2.39.38). For $a = -4\alpha^2$, the given DE has the solution

$$y = (\alpha x + C)^{-2};$$

C is an arbitrary constant. Kamke (1959), p.581

75. $$yy'' - (5/4)y'^2 - 4(2 - c^2)y^3 - w^2y^2 = 0,$$

where c and w are constants.

Put $\Omega = y^{-1/4}$ to get a special case of Pinney's equation

$$\Omega'' + (w^2/4)\Omega = (c^2 - 2)\Omega^{-3}. \tag{1}$$

If the initial conditions for (1) are $\Omega(0) = w_0, \Omega'(0) = v_0$, we get the solution

$$\Omega = [\{w_0 \cos \mu x + (v_0/\mu) \sin \mu x\}^2 + (1/w_0)^2 \sin^2 \mu x]^{1/2},$$

where $\mu = w/2, w_0^2 = (1/8)(8 - w^2 - 4c^2)v_0^{-2}$. Hence the solution of the given DE may be found. Levi, Nucci, Rogers, and Winternitz (1988)

76. $$3yy'' - 5y'^2 = 0.$$

The solution of the given DE is easily found by writing it as

$$\begin{aligned} y''/y' &= (5/3)y'/y, \\ y^2 &= (C_1x + C_2)^{-3}, \end{aligned}$$

where C_1 and C_2 are arbitrary constants. Kamke (1959), p.580; Murphy (1960), p.397

77. $$\mathbf{2yy'' - 3y'^2 = 0.}$$

The given DE is easily integrated after division by yy'. A second integration gives

$$y = C_1(x + C_2)^{-2};$$

C_1 and C_2 are arbitrary constants. $y = C$ is another solution. Kamke (1959), p.578; Murphy (1960), p.397

78. $$\mathbf{2yy'' - 3y'^2 + 4y^2 = 0.}$$

Writing $y' = z, z' = (3z^2 - 4y^2)/(2y)$ or $(dz/dy) = (3z^2 - 4y^2)/(2yz)$, that is, $2(dz/dy) - 3z/y = -4y/z$, we have the solution

$$z^2 = y^2(4 + cy), \tag{1}$$

where c is an arbitrary constant. For $c = 0$, we find the solution to be $y = c_1 \exp\{\pm 2(x + c_2)\}$. In general, the phase portraits of (1) are similar to the family of folia of Descartes with $y = \pm 2x$ acting as separatrices. Utz (1978)

79. $$\mathbf{2yy'' - 3y'^2 - 4y^2 = 0.}$$

Choosing y to be the independent variable, we write $p(y) = y'(x)$ and obtain the homogeneous DE in y and p,

$$2ypp' - 3p^2 - 4y^2 = 0,$$

with the solution

$$y \cos^2(x + C_1) = C_2;$$

C_1 and C_2 are arbitrary constants. Kamke (1959), p.578; Murphy (1960), p.397

80. $$\mathbf{2yy'' - 3y'^2 + f(x)y^2 = 0.}$$

With $u(x) = |y|^{-1/2}$, we get the linear DE

$$4u'' = f(x)u.$$

Kamke (1959), p.578; Murphy (1960), p.397

81. $$\mathbf{yy'' - (3/2)y'^2 + ky^4 = 0,}$$

$k > 0$ is a constant.

The given DE has a special solution $y = aq/\{(x + p)^2 + q^2\}, a^2 = 2/k$. Writing $y' = p, y'' = p(dp/dy)$, etc., and integrating, we get

$$y'^2 = y^3(c - 2ky). \tag{1}$$

If $k < 0$, one has the special solutions (with $c = 0$) obtained after integrating (1): $y = 1/\{m \pm (-2k)^{1/2}x\}, k < 0$. The other solutions for $k < 0$ form two families (corresponding to $c < 0$ and $c > 0$) of hyperbola-like curves filling the remainder of the phase plane. Utz (1978)

82. $$\mathbf{yy'' - (3/2)y'^2 + 2a^2y^2 - 2b^2y^4 = 0,}$$

where a and b are constants.

A general solution of the given DE may be found in the form

$$y = (a/2b)[\tanh\{a(x - x_1)\} - \tanh\{a(x - x_2)\}], \tag{1}$$

where x_1 and x_2 are arbitrary constants. One may seek solutions of the form

$$u = c_{1,0}(x)\sigma(\theta(x)) + c_{0,1}(x)\tau(\theta(x)) + c_{0,0}(x)$$

to identify (1). Conte and Musette (1992)

83. $$yy'' - (3/2)y'^2 + \epsilon y' - \epsilon yy' - 2y^2 = 0,$$

where ϵ is a parameter.

The given DE arises from the Van der Pol equation $v'' - \epsilon(1 - v^2)v' + v = 0$ by the transformation $v = y^{-1/2}$. The leading order behavior of the given DE shows that $y \sim 2\epsilon\tau/3, \tau = x - x_0$.

The resonance equation is obtained by substituting $(2/3)\epsilon\tau(1 + \gamma\tau^n)$ for y in the given DE and seeking the coefficient of ν^τ. The latter is $(n + 1)\{2n/3 - 1\}$. The resonances, therefore, are $n = -1, n = 3/2$. The solution may be found in the form

$$y = (2/3)\epsilon\tau(1 - \epsilon\tau + C\tau^{3/2} + \cdots),$$

where C is arbitrary.

The given DE, and hence the Van der Pol equation, do not possess the Painlevé property and hence are not integrable. Ramani, Dorizzi, and Grammaticos (1983)

84. $$yy'' - 2y'^2 = 0, 0 \le x \le 1,$$

where $y(0) = y_0, y(1) = y_1$.

It is shown that there exists a solution $y(x)$ such that $\min\{y(0), y(1)\} \le y(x) \le \max\{y(0), y(1)\}, 0 \le x \le 1$. Knobloch and Schmitt (1977)

85. $$yy'' - 2y'^2 - y^2 = 0.$$

With $y' = yu(x)$, we have $u' = 1 + u^2$, so that $u = \tan(x + C)$, with C an arbitrary constant. Hence the solution is

$$y = C_1 \sec(C + x);$$

C_1 is another arbitrary constant. Murphy (1960), p.394

86. $$yy'' - 2(y'^2 - y^3) = 0.$$

With $y'(x) = p(y)$, we have $ypp'(y) = 2(p^2 - y^3)$. So a first integral is

$$p^2 = y^3(Cy + 4);$$

C is an arbitrary constant. The solution is found to be $(C_1 + C_2x + x^2)y = 1; C_1$ and C_2 are arbitrary constants. Murphy (1960), p.394

87. $$y(y'' - 4y) - 2(y' - m)(y' - 2m) = 0,$$

where m is a constant.

Using the method of Eqn. (2.32.13), the solution of the given DE is

$$y = m\tan 2(x+k) + \nu\sec\{2(x+k)\},$$

where k and ν are arbitrary constants. Roth (1941)

88. $$\boldsymbol{y(y'' + 2y' + 3y - 4m) - 2(y' - m)(y' - 2m) = 0,}$$

where m is a constant.

Using the method of Eqn. (2.32.13), the solution of the given DE is found to be

$$y = (2m/3)\{1 + 2\tanh 2(x+k)\} + \nu e^{x+k}\text{sech}\{2(x+k)\},$$

where k and ν are arbitrary constants. Roth (1941)

89. $$\boldsymbol{y(y'' + 2y' - y - 4m) - 2(y' - m)(y' - 2m) = 0,}$$

where m is a constant.

Using the method of Eqn. (2.32.13), the solution of the given DE is found to be

$$y = -2m\{1 + 1/(x+k)\} + \nu e^{x+k}/(x+k),$$

where k and ν are arbitrary constants. Roth (1941)

90. $$\boldsymbol{yy'' - 2y'^2 + 2xyy' - 2y^2 - 2^{3/2}y' = 0.}$$

The given DE arises from the self-similar form of the Burgers equation and a "reciprocal" transformation. It has an exact solution

$$y = [(2\pi)^{1/2}/(e^R - 1)]\exp(x^2) + (\pi/2)^{1/2}\exp(x^2)\text{ erfc } x,$$

where R is an arbitrary constant and erfc $x = 1-$ erf x is the complementary error function. Sachdev, Nair, and Tikekar (1986)

91. $$\boldsymbol{(y+1)y'' - 3y'^2 = 0.}$$

Put $y+1 = Y$, then $Y' = u, Y'' = u(du/dY)$. Integrating the resulting equation twice, we have

$$(y+1)^{-2} = c_1 x + c_2,$$

where c_1 and c_2 are arbitrary constants. Tenebaum and Pollard (1963), p.504

92. $$\boldsymbol{yy'' - 3y'^2 + (1/2)(3y^2 - 1)y^2 = 0.}$$

With $y = u^{-1/2}$, we get the linear DE $u'' + u - 3 = 0$. The solution is

$$y = (3 + C_1\cos x + C_2\sin x)^{-1/2};$$

C_1 and C_2 are arbitrary constants. Murphy (1960), p.397

93. $$\boldsymbol{2yy'' - 6y'^2 + ay^5 + y^2 = 0,}$$

where a is a constant.

With $p(y) = y'$, we get the Bernoulli equation

$$p' - 3p/y + (ay^4 + y)/2p = 0,$$

with the solution

$$4p^2 = 4ay^5 + y^2 + Cy^6;$$

C is an arbitrary constant. Now, the variables are separable. Kamke (1959), p.579; Murphy (1960), p.397

94. $$yy'' - 3y'^2 + 3yy' - y^2 = 0.$$

With $y' = yu(x)$, we get the Riccati equation $u' = 2u^2 - 3u + 1$ with the solutions

$$(1 - 2Ce^x)u = 1 - Ce^x; \qquad (1)$$

C is an arbitrary constant. Equation (1) integrates to give

$$(2e^x - C_1)y^2 = C_2 e^{2x};$$

$C_1 = C^{-1}$ and C_2 are arbitrary constants. Kamke (1959), p.573; Murphy (1960), p.394

2.32 $yy'' + ky'^2 + g(x, y, y') = 0$, k a general constant, g linear in y'

1. $$yy'' - ay'^2 = 0,$$

where a is a constant.

Division by yy' makes the given DE exact. A first integral is

$$y' = C|y|^a,$$

C a constant, hence

$$y = \begin{cases} |C_1 x + C_0|^{1/(1-a)} & \text{for } a \neq 1, \\ C_1 e^{Cx} & \text{for } a = 1; \end{cases}$$

C_1, C_0, and C are arbitrary constants. Kamke (1959), p.573; Murphy (1960), p.394

2. $$ayy'' + (1 - a)y'^2 = 0,$$

where a is a constant.

Writing the given DE as

$$ay''/y' + (1 - a)y'/y = 0,$$

the general solution is found to be $y(x) = \alpha(x - x_0)^a$; α and x_0 are arbitrary constants. Kruskal and Clarkson (1992)

3. $$(ay + b)y'' + cy'^2 = 0,$$

where a and b are constants.

After dividing by $(ay+b)y'$ and integrating, we obtain

$$ay+b=\begin{cases}(C_1x+C_0)^{a/(a+c)} & \text{for } a+c\neq 0,\\ C_0\exp(C_1x) & \text{for } a+c=0;\end{cases}$$

C_1 and C_0 are arbitrary constants. Kamke (1959), p.582

4. $$\boldsymbol{yy''+a(y'^2+1)=0,}$$

where a is a constant.

For the given autonomous equation, we put $y'=p, y''=p(dp/dy)$ and integrate twice to obtain

$$x=\int (C_1y^{-2a}-1)^{-1/2}dy+C_2;$$

C_1 and C_2 are arbitrary constants. For $a=1,-1,1/2,-1/2$ we get semicircles, catenaries, cycloids, and parabolas. Kamke (1959), p.573; Murphy (1960), p.395

5. $$\boldsymbol{yy''-ay'^2-b=0.}$$

(a) $a=0$; see Eqn. (2.29.4).

(b) $b=0$; see Eqn. (2.32.1).

(c) $a\neq 0, b\neq 0$; put $y'(x)=p$ so that

$$ypp'(y)=ap^2+b.$$

So a first integral of the given DE is

$$(b+ap^2)=Cy^{2a},$$

C is a constant. Now the variables separate. The nature of explicit solution depends on the value of a. Murphy (1960), p.394

6. $$\boldsymbol{yy''-ky'^2+qy=0,}$$

where k and q are constants.

Put $y'=z, z'=(kz^2-qy)/y, y\neq 0$, from which we have the Bernoulli's equation

$$\frac{dz}{dy}=\frac{kz^2-qy}{zy} \quad \text{or} \quad \frac{dz}{dy}-\frac{kz}{y}=-qz^{-1}. \tag{1}$$

Setting $w=z^2$, and solving the equation resulting from (1), we have

$$z^2=w=2qy/(2k-1)+cy^{2k}, k\neq 1/2, \tag{2}$$

$$z^2=w=-2qy\log|y|+cy, k=1/2. \tag{3}$$

For $q\neq 0$, regardless of the sign of q and c, (3) in the (y,z) plane has a cycle and corresponds to a periodic solution there. For example, if $q>0$, then the cycle is in the half-plane $y\geq 0$ between $y=0$ and $y=\exp(e/2q)$. Not all solutions are periodic, as for example, in the half-plane $y\geq 0$ and $q<0$.

For (2), if $k > 1/2$ and $q > 0$, there is, for any $c > 0$, a cycle in $y \geq 0$ between $y = 0$ and $y = \{2q/(c(1-2k))\}^{1/(2k-1)}$. If $q > 0$, there will not be a cycle unless y^{2k} is meaningful for $y < 0$. Similarly, for $0 < k < 1/2$, if $q > 0$, one may take $c > 0$ to secure a cycle in the half-plane $y \geq 0$. If $k < 0$, there are no periodic solutions. Utz (1978)

7. $$\boldsymbol{yy'' + ay'^2 + by^3 = 0,}$$

where a and b are constants.

With $p(y) = y'(x)$, we get $ypp' + ap^2 + by^3 = 0$; the further substitution $p(y) = y^{3/2}u(t)$, $t = \ln y$ leads to the Bernoulli's equation

$$uu' + (a+3/2)u^2 + b = 0; \tag{1}$$

the solution of Eqn. (1) is

$$\begin{aligned} t &= -\int \frac{udu}{(a+3/2)u^2+b} + C \\ &= -\{1/(2a+3)\}\ln\{(a+3/2)u^2+b\} + C; \end{aligned}$$

C is a constant. Therefore,

$$y'^2 = y^3[Cy^{-2a-3} - \{2b/(2a+3)\}].$$

Now, the variables separate. Kamke (1959), p.573; Murphy (1960), p.394

8. $$\boldsymbol{yy'' - ay'^2 - a_0 - a_1y - a_2y^2 - a_3y^3 - a_4y^4 = 0.}$$

With $y'(x) = p(y)$, we have

$$ypp'(y) = ap^2 + a_0 + a_1y + a_2y^2 + a_3y^3 + a_4y^4.$$

Now put $p^2 = P$, and solve the first order linear DE in P. Murphy (1960), p.394

9. $$\boldsymbol{ayy'' + by'^2 + C_4y^4 + \cdots + C_1y + C_0 = 0,}$$

where a, b, and C_i $(i = 0, 1, \ldots, 4)$ are constants.

If we set $y'(x) = p(y)$ and $p^2 = u$, we get a linear DE:

$$(1/2)ayu' + bu + C_4y^4 + \cdots + C_0 = 0. \tag{1}$$

For $a = 1, b = -1$, we find from integrating (1) that

$$y'^2 + C_4y^4 + 2C_3y^3 + 2C_2y^2\log|y| - 2C_1y - C_0 = Cy^2, \tag{2}$$

while for $a = 2, b = -1$, we get

$$y'^2 + (1/3)C_4y^4 + (1/2)C_3y^3 + C_2y^2 + C_1y\log|y| - C_0 = Cy. \tag{3}$$

C in (2) and (3) is an arbitrary constant. If the logarithmic terms are absent, (1), (2), and (3) can be solved in terms of elliptic functions. If $a = -2b$ and $C_1 = 0$, one may proceed as follows. Differentiating the original DE with respect to x and dividing by y, we get

$$ay''' + 4C_4y^2y' + 3C_3yy' + 2C_2y' = 0,$$

so that

$$ay'' + (4/3)C_4y^3 + (3/2)C_3y^2 + 2C_2y + C = 0. \tag{4}$$

Combining (4) with the given DE, we have

$$by'^2 = (1/3)C_4y^4 + (1/2)C_3y^3 + C_2y^2 + Cy - C_0. \tag{5}$$

We can solve (5) in terms of elliptic functions. Kamke (1959), p.581

10. $$\boldsymbol{yy'' - ky'^2 + f(y) = 0,}$$

where k is a positive constant.

Writing $y' = z, z' = \{kz^2 - f(y)\}/y$ and integrating, we have

$$z^2 = y^{2k}\left[c - \int_0^y 2f(u)u^{-2k-1}du\right].$$

Considering the half-plane $y \geq 0$ to avoid restricting k, it is clear that there will be a periodic solution whenever c may be selected to cause

$$\phi(y) = c - \int_0^y f(u)u^{-2k-1}du$$

to satisfy

$$\phi(y) > 0 \text{ for } 0 < y < \alpha \quad \text{and} \quad \phi(\alpha) = 0. \tag{1}$$

Thus, it is proved that if $f(y)$ is continuous for all real y, then for $k > 0$, the given DE has an infinite number of periodic solutions provided that there is a c and $\alpha(c)$ for which (1) holds. Utz (1978)

11. $$\boldsymbol{kyy'' + [k(2-k)/(k-1)]y'^2 + cyy' + (1-k)y^3 + a(k-1)y^2 = 0,}$$

where k, c, and a are constants.

The given DE arises from a degenerate logistic equation. It may be written as the system

$$y' = yw, w' = [(k-1)/k](y-a) - (c/k)w - [1/(k-1)]w^2. \tag{1}$$

A straightforward analysis of (1) shows that $y = a, w = 0$ is always a saddle. Moreover, if $|c| > (ak)^{1/2}$, two additional critical points arise:

$$W_\pm = (y = 0, w = w_\pm),$$

where $W_\pm$ are the solutions of $cw/k + w^2/(k-1) + (k-1)a/k = 0$. If $c < 0, W_+$ is a saddle and W_- is an unstable node. If $c > 0, W_+$ is a stable node while W_- is a saddle. A phase portrait is drawn. Kersner and Mottoni (1990)

12. $$\boldsymbol{yy'' + ay'^2 + byy' + cy^2 + dy^{1-a} = 0,}$$

where a, b, c, and d are constants.

Putting

$$y = \begin{cases} e^u & \text{for } a = -1, \\ u^{1/(a+1)} & \text{for } a \neq -1, \end{cases}$$

we get DEs for u with constant coefficients:

$$u'' + bu' + c + d = 0$$

and

$$u'' + bu' + (a+1)cu = -(a+1)d.$$

Kamke (1959), p.574; Murphy (1960), p.395

13. $$(a_0 y + a_1)y'' + b_0 y'^2 + \{c_0 y + c_1)y' + d_0 y^2 + d_1 y + d_2 = 0,$$

where a_i, b_i, c_i, and d_i are constants.

Changing the variable to $Y = a_0 y + a_1$, the given DE may be rewritten in the form

$$Y(Y'' + aY' + bY + c) + l(Y' - m)(Y' - n) = 0, \tag{1}$$

where m and n are either real or complex-conjugate numbers. Reverting to lower case, we look for a function μ—"an integrating function"—such that when (1) is written as the system

$$y'' + ay' + by + c = \mu\ell(y' - m), y' - n = -\mu y, \tag{2}$$

eliminating y'' from (2), we can write (2) as the equivalent system

$$\{(\ell+1)\mu - a\}y' + (\mu' - b)y = \ell m\mu + c, y' + \mu y = n. \tag{3}$$

On solving for y and y' from (3), we have

$$y' : y : 1 = (\mu' - b)n - (\ell m\mu + c)\mu : [\ell m - (\ell+1)n]\mu + an + c : \mu' - (\ell+1)\mu^2 + a\mu - b$$

or

$$\frac{(\mu' - b)n - (\ell m\mu + c)\mu}{\mu' - (\ell+1)\mu^2 + a\mu - b} = \frac{d}{dx}\left[\frac{[\ell m - (\ell+1)n]\mu + an + c}{\mu' - (\ell+1)\mu^2 + a\mu - b}\right]. \tag{4}$$

If any solution of (4) can be found, the solution y is then obtained from (3) without further integration. For example, if $b = 0$ and $(\ell+1)c + a\ell m = 0$, the solution of (4) is $\mu = \{a/(2\ell+2)\}(1 + \tanh ax)$, and hence y from (3) is

$$y = \frac{\{\ell m - (\ell+1)n\}\{a/(2\ell+2)\}(1 + \tanh ax) + an + c}{\{a^2/(4\ell+4)\}\{2\,\mathrm{sech}^2 ax - (1 + \tanh ax)^2 + 2(1 + \tanh ax)\}}. \tag{5}$$

For the special case

$$\ell m - (\ell+1)n = 0, an + c = 0, \tag{6}$$

(5) gives $y = 0$; but reverting to (3), we find that for $y \not\equiv 0$,

$$\mu' - (\ell+1)\mu^2 + a\mu - b = 0. \tag{7}$$

In this case Eqns. (3) have become identical. Assuming first that $\rho = l + 1 \neq 0$, we have the following cases:

(a) $4b\rho - a^2 < 0$, and solving (7), we get $\mu = a/(2\rho) - (D/\rho)\tanh D(x+k)$, where k is an arbitrary constant, and $D^2 = (a^2 - 4b\rho)/4$. Integration of second of (2) gives

$$\begin{aligned}\exp[\{a/(2\rho)\}(x+k)][\mathrm{sech} D(x+k)]^{1/\rho} y &= n \\ \int \exp\{a/(2\rho)\}(x+k)[\mathrm{sech} D(x+k)]^{1/\rho} dx &+ \nu,\end{aligned}$$

where ν is a constant of integration.

(b) $4b\rho - a^2 = 0$. Equation (7) becomes $\mu' = \rho\{\mu - a/(2\rho)\}^2$ with the solution $\mu = a/(2\rho) - 1/\{\rho(x+k)\}$. The solution y is then found from $\exp\{a/(2\rho)\}(x+k)[x+k]^{-1/\rho}y = n\int \exp\{a/(2\rho)\}[x+k]^{-1/\rho}dx + \nu$.

(c) $4b\rho - a^2 > 0$. Equation (7) gives $\mu = a/(2\rho) + (D/\rho)\tan D(x+k)$, where $(4b\rho - a^2)/4 = D^2$. The solution y in this case is given by

$$\exp\{a/(2\rho)\}(x+k)[\sec D(x+k)]^{1/\rho}y = n \int \exp\{a/(2\rho)\}(x+k)[\sec D(x+k)]^{1/\rho}dx + \nu,$$

where ν is a constant.

Roth (1941)

14. $\quad (\boldsymbol{A_0} + \boldsymbol{B_0 y})\boldsymbol{y''} - \boldsymbol{2B_0}(\boldsymbol{y'^2}) + (\boldsymbol{C_0} + \boldsymbol{D_0 y})\boldsymbol{y'} + \boldsymbol{E_0} + \boldsymbol{F_0 y} + \boldsymbol{G_0 y^2} + \boldsymbol{H_0 y^3} = \boldsymbol{0},$

where $y(0) = \alpha_0, y'(0) = \beta_1, \alpha_0\beta_1 \neq 0$ and each of the coefficient functions $A_0(x), \ldots, H_0(x)$ is expandable in a Taylor's series about

$$x = 0 \text{ and } A_0(0) = B_0(0) = C_0(0) = D_0(0) = G_0(0) = H_0(0) = 0.$$

In particular, the following special case is considered:

$$\begin{aligned} A_0 &= x^2 \sum_{k=0}^{\infty} a_k x^k, \quad B_0 = x^2 \sum_{k=0}^{\infty} b_k x^k, \\ C_0 &= x \sum_{k=0}^{\infty} e_k x^k, \quad D_0 = x \sum_{k=0}^{\infty} d_k x^k, \quad E_0 = \sum_{k=0}^{\infty} e_k x^k, c_0 \neq 0, \\ F_0 &= \sum_{k=0}^{\infty} f_k x^k, f_0 \neq 0, \quad G_0 = x \sum_{k=0}^{\infty} g_k x^k, \quad H_0 = x \sum_{k=0}^{\infty} h_k x^k. \end{aligned}$$

Assuming that the given DE has a power series expansion of the form

$$y = \alpha_0 + \sum_{k=1}^{\infty} \beta_k x^k, \tag{1}$$

and substituting into the given DE, we find that α_0 and β_1 are uniquely determined and the coefficients in (1) have the property that

$$\Delta_p = \begin{vmatrix} \alpha_0 & \beta_1 & \cdots & \beta_p \\ \beta_1 & \beta_2 & \cdots & \beta_{p+1} \\ \vdots & & \vdots & \vdots \\ \beta_p & \beta_{p+1} & \cdots & \beta_{2p} \end{vmatrix} \neq 0, p = 0, 1, 2, \ldots.$$

$$\Gamma_{2p+1} = \begin{vmatrix} \beta_1 & \beta_2 & \cdots & \beta_{p+1} \\ \beta_2 & \beta_3 & \cdots & \beta_{p+2} \\ \vdots & & \vdots & \vdots \\ \beta_{p+1} & \beta_{p+2} & \cdots & \beta_{2p+1} \end{vmatrix} \neq 0, p = 0, 1, 2, \ldots.$$

Then, it is shown that the solution has a continued fraction expansion of the form

$$y = \cfrac{\alpha_0}{1 + \cfrac{\alpha_1 x}{1 + \cfrac{\alpha_2 x}{1 + \cdots}}} \tag{1}$$

Resonance relations for the coefficients α_i are developed using a Padé approximation to (1). Numerical procedure is elaborated. Fair and Luke (1966)

15. $$\boldsymbol{yy'' - \{(\alpha+1)/\alpha\}y'^2 + 2xyy' - 2(1-\alpha j)y^2 - 2^{3/2}y' = 0,}$$

where $\alpha > 0$ is a parameter and $j = 0, 1, 2$.

The solution of the given DE about $x = 0$ may be found to be

$$y = \sum_{n=0}^{\infty} a_n x^n, \tag{1}$$

where

$$\begin{aligned} a_2 &= \{1/a_0\}[\{(\alpha+1)/(2\alpha)\}a_1^2 + (1-\alpha j)a_0^2 + 2^{1/2}a_1], \\ a_{k+2} &= \{2/((k+1)(k+2)a_0)\}[\{(\alpha+1)/(2\alpha)\}(k+1)a_1 a_{k+1} + (1-\alpha j - k)a_0 a_k \\ &+ 2^{1/2}(k+1)a_{k+1} + \sum_{i=1}^{k}\{-(1/2)(k+2-i)(k+1-i)a_i a_{k+2-i} \\ &+ [(\alpha+1)/(2\alpha)](i+1)(k+1-i)a_{i+1}a_{k+1-i} + (1-\alpha j)a_i a_{k-i} - (k-i)a_i a_{k-i}\}]. \end{aligned}$$

Convergence of the series (1) was verified by solving a connection problem over the entire real line for the function $f = (y/\delta^{1/2})^\alpha$, by imposing appropriate conditions at $x = \pm\infty$. See Eqn. (2.13.37). Sachdev and Nair (1987)

16. $$\boldsymbol{yy'' - \{(\beta+1)/\beta\}y'^2 + (1+n)xyy' - (1-n)y^2 - 2y' = 0,}$$

where β and n are parameters.

The given DE arises from similarity reduction and an inverse transformation of the generalized Burgers equation

$$u_t + u^\beta u_x = (\delta/2)(1+t)^n u_{xx}.$$

It has a Taylor series solution about $x = 0$ in the form

$$y(x) = \sum_{r=0}^{\infty} a_r x^r, \tag{1}$$

where

$$a_2 = \{1/(2a_0)\}[\{\{(\beta+1)/\beta\}a_1 + 2\}a_1 + (1-n)a_0^2]$$

and

$$\begin{aligned} a_{r+2} &= [1/\{(r+1)(r+2)a_0\}][\{\{(\beta+1)/\beta\}a_1 + 2\}(1+r)a_{r+1} \\ &+ (1-n)a_0 a_r + \sum_{i=1}^{r}(r+1-i)a_{r+1-i}\{\{(\beta+1)/\beta\}(i+1)a_{i+1} \\ &- (1+n)a_{i-1}\} - \sum_{i=1}^{r} a_i\{(r+2-i)(r+1-i)a_{r+2-i} - (1-n)a_{r-i}\}]. \end{aligned}$$

The convergence of series (1) for different initial conditions and parametric values is proved with the help of a connection problem for Eqn. (2.13.39) over the entire real line.

The solution y_1 of the present equation and y_2 of (2.13.39) are related by $y_2 = y_1^{-\beta}$. The given DE is a special case of the Euler–Painlevé equation (2.32.22). Sachdev, Nair, and Tikekar (1988)

17.
$$(2x-1)(8x-3)\left(yy'' - \frac{p+1}{p-1}y'^2\right) + \left(\frac{8(3p+1)}{p-1}x - \frac{12p}{p-1}\right)yy' - \frac{4(p+1)}{p-1}y^2 - \frac{(p-1)}{2} = 0,$$

where $p \neq 0, 1$ is a constant.

The given DE arises from reduction of a nonlinear Klein–Gordon equation. It has exact solutions:

(a) $y = \pm i(p-1)/[8(p+1)]^{1/2}$.

(b) $y = \pm(ax+b)^{1/2}$, where either

$$a = -(p-1)^2/(p-3), b = (3/8)[(p-1)^2/(p-3)]$$

or

$$a = \frac{-(p-1)^2}{2(p-2)}, \quad b = (1/4)\frac{(p-1)^2}{(p-2)}, p \neq 2, 3.$$

(c) $y = \pm(-3x + 5/4)$ for $p = 3$.

(d) It may be checked that the solutions of the form

$$y = \pm(ax+b)^{(p+1)/4}, p \neq 0, 1$$

do not exist, where $(p+1)/4$ is nonintegral. More general solutions in series form are also discussed.

Matsuno (1987)

18.
$$ayy'' + by'^2 - yy'/(x^2+c^2)^{1/2} = 0,$$

where a, b, and c are constants.

With $y' = yu(x)$, we obtain the Bernoulli equation

$$u' - \frac{u}{a(x^2+c^2)^{1/2}} + (1 + b/a)u^2 = 0;$$

alternatively, dividing the original DE by yy', we have

$$\frac{d}{dx}\left[a \log|y'| + b \log|y| - \log[x + (x^2+c^2)]^{1/2}\right] = 0,$$

so that, on integrating twice,

$$y^{1+(b/a)} = C_1 + C_2[x + (x^2+c^2)^{1/2}]^{1/a}[(x^2+c^2)^{1/2} - ax];$$

C_1 and C_2 are arbitrary constants. Kamke (1959), p.581

19.
$$yy'' + ay'^2 + f(x)yy' + g(x)y^2 = 0,$$

where a is a constant.

Writing $y = u^{1/(a+1)}$, we get a linear DE:

$$u'' + fu' + (a+1)gu = 0.$$

Kamke (1959), p.574; Murphy (1960), p.394

20. $$yy'' + (m-1)y'^2 + f(x)yy' + \Phi(x)y^2 = 0,$$

where m is a constant.

The given DE is related to

$$z\ddot{z} + (m-1)\dot{z}^2 + \Psi(t)z\dot{z} + F(t)z^2 = 0, \tag{1}$$

where $\Psi(t)$ and $F(t)$ are so fixed that (1) is solvable through a quadrature. Changing the variables in (1) according to

$$z = e^{\tau}, \quad \frac{d\tau}{dt} = \xi, \tag{2}$$

we have

$$\frac{d\xi}{dt} + m\xi^2 + \psi(t)\xi + F(t) = 0. \tag{3}$$

Equation (3) can be solved explicitly if we set $F(t) = a$, $\Psi(t) = b$, where a and b are constants. Then (3) can be integrated to give

$$C_1 - t = \int \frac{d\xi}{m\xi^2 + b\xi + a}, \tag{4}$$

where C_1 is an arbitrary constant. The right-hand side of (4) is explicitly integrable, the form depending on $\Delta = 4ma - b^2$; for example, if $\Delta = 0$, we find that

$$\xi = 1/(2m)[2/(t - C_1) - b].$$

Reverting to variables z and t and integrating once again, we have

$$z = C_2(t - C_1)^{1/m}e^{-bt/(2m)}.$$

Now, following the procedure for Eqn. (2.39.4) with $\Theta(x) = 0$, we have the following relations:

$$\int f(x)dx = bt \tag{5}$$

$$y = [b/f(x)]^{1/(2m)}C_2(t - C_1)^{1/m}e^{-bt/(2m)}, \tag{6}$$

$$[1/f^2(x)][\Phi(x) - u'/(2m) + u^2/(4m) - uf(x)/(2m)] = a/b^2, \tag{7}$$

where $u = f'(x)/f(x)$. Now choosing $f(x)$, we find $\Phi(x)$ from (7). For such a choice of $f(x)$ and $\Phi(x)$, the solution of the given DE is given by (5)–(6). Braude (1967)

21. $$yy'' + (1/n - 1)y'^2 + k(x)yy' + m(x)y^2 = 0,$$

where n is a positive integer or infinity and $k(x)$ and $m(x)$ are meromorphic functions.

Necessary and sufficient conditions are derived for all solutions of the given DE to be meromorphic. This is accomplished by writing $y = u^n$, where n is a positive integer, and $y = \exp\left(\int_{x_1}^{x} u\,dx\right)$ when $n \to \infty$, and regarding the given DE to be linear equation with meromorphic coefficients. Necessary and sufficient conditions for y to be entire are derived by studying how solutions of the linear DE have to behave at poles in the finite plane. Jablonskii (1964)

22. $$\boldsymbol{yy'' + ay'^2 + f(x)yy' + g(x)y^2 + by' + c = 0,}$$

where $f(x)$ and $g(x)$ are sufficiently smooth functions and a, b, c are real constants.

The special case $b = c = 0$ was considered by Euler and Painlevé [see Kamke (1959)] and is indeed easily linearized [see Eqn. (2.32.19)]. For $b \neq 0, c \neq 0$, the given DE in general is not linearizable. However, it appears in a natural way from a class of generalized Burgers equations through similarity transformation and a reciprocal relation [see Eqns. (2.33.7), (2.32.15), (2.32.16)]. It was called an Euler–Painlevé transcendent by Sachdev and his co-workers. It includes a large number of equations in the earlier compendium of Kamke (and the present one) as special cases. A simple extension that makes a, b, and c functions of x extends this class much further. Sachdev (1991)

23. $$\boldsymbol{yy'' + ay'^2 + f(x)yy' - \alpha \exp\left(-2\int f(x)dx\right) q(y) = 0,}$$

where a and α are constants.

Introducing the change of variable $x = x(t)$, we get the system

$$y\ddot{y} + a\dot{y}^2 = \alpha q(y), \tag{1}$$

where

$$f(x)\dot{x}^2 - \ddot{x} = 0, \ \ \phi(x)\dot{x}^2 = \alpha, \tag{2}$$

where $\phi(x) = \alpha \exp(-2\int f(x)dx)$, and α is a constant. Putting $\dot{y} = z^{1/2}$, we can transform (1) to the first order linear DE $yz' + 2az = 2\alpha q(y)$, where $z' = dz/dy$ with the solution

$$y^{2a}z = c_1 + 2\alpha \int y^{2a-1}q(y)dy. \tag{3}$$

The general integral of the given DE can now be written as

$$\int y^a(c_1 + 2\alpha \int y^{2a-1}q(y)dy)^{-1/2}dy = \theta(x, c_2),$$

where

$$\theta(x, c_2) = c_2 + \int \exp\left(-\int f(x)dx\right) dx$$

and we have made use of (2) and (3). Bandic (1965)

24. $$\boldsymbol{yy'' + ay'^2 + f(x)yy' - \phi(x)q(y) = 0,}$$

where a is a constant.

Introducing the variable $x = x(t)$, the given DE assumes a similar form

$$y\ddot{y} + a\dot{y}^2 + F(t)y\dot{y} = \Phi(t)q(y), \tag{1}$$

where

$$f(x)\dot{x}^2 - \ddot{x} = \dot{x}F(t); \phi(x)\dot{x}^2 = \Phi(t). \tag{2}$$

On elimination of $\dot{x}, \ddot{x}$ from (1), we get

$$\phi(x)\exp\left[2\int f(x)dx\right] = \Phi(t)\exp 2\left[\int F(t)dt\right],$$

the expression on the left represents a "semi-invariant" of the given DE under the transformation $x = x(t)$. If $y = y(t, c_1, c_2)$ is the general integral of Eqn. (1), then the general integral of the given DE can be obtained in the parametric form

$$y = y(t, c_1, c_2);\ \int [\phi(x)]^{1/2}dx = \int [\phi(t)]^{1/2}dt,$$

where t plays the role of the parameter. Bandic (1965)

25. $$\boldsymbol{yy'' + (\mu - 2)y'^2 + y^2y' = 0,}$$

where μ is a parameter.

The given DE is Painlevé for positive integer values of μ. A first integral for $\mu = 1$ is

$$y^{-1}y' + y = C_1, \tag{1}$$

and the general solution, therefore, is

$$y = \frac{C_1\exp\{C_1(x - C_2)\}}{C_1\exp\{C_1(x - C_2)\} + 1},$$

where $C_1 \neq 0, C_2$ are arbitrary constants. Notice that (1) is a Riccati equation.

For $\mu = 2$, we write $y = 2w$ and hence get the first integral

$$y' + (1/2)y^2 = C_3. \tag{2}$$

Solving the Riccati equation (2), we have

$$y = \{2C_3\}^{1/2}\ \tanh[(C_3/2)^{1/2}(x - C_4)] \text{ for } C_3 > 0$$

and

$$y = \{-2C_3\}^{1/2}\ \tan\{-[(-C_3/2)(x - C_4)]^{1/2}]\} \text{ for } C_3 < 0.$$

For general μ, the first integral of the given DE is

$$y^{\mu-2}\ y' + (1/\mu)y^{\mu} = C_5. \tag{3}$$

For $C_5 = 0$, we have the solution

$$y = \mu/(x - C_6).$$

For $C_5 \neq 0$, the general solution from (3) is

$$x - C_6 = \mu\int \frac{dy}{\mu C_5 y^{2-\mu} - y^2}. \tag{4}$$

The integral (4) is expressible in terms of known functions if $\mu = 1, 2, 3, \ldots$ or $\mu = 1/(1 - m)$ $(m = 2, 3, \ldots)$. For example, for $\mu = 3, C_5 > 0$, an implicit solution is

$$\begin{aligned} x - C_6 &= 3\left[\frac{\log(y^2 + (3C_5)^{1/3}y + (3C_5)^{2/3})}{\{6(3C_5)^{1/3}\}} - \arctan\left(\frac{\{2y + (3C_5)^{1/3}\}/\{3^{5/6}C_5^{1/3}\}}{3^{5/6}C_5^{1/3}}\right)\right.\\ &\left. - \frac{\log\{y - (3C_1)^{1/3}\}}{3(3C_5)^{1/3}}\right]. \end{aligned}$$

In the above, C_i $(i = 1, \ldots, 6)$ are arbitrary constants. Glockle, Baumann, and Nonnenmacher (1992)

26. $$yy'' + ay'^2 + by^2y' + cy^4 = 0,$$

where a, b, and c are constants.

Put $y'(x) = p(y)$ to get

$$ypp' + ap^2 + by^2p + cy^4 = 0.$$

Now write $p(y) = y^2u(t);\ t = \ln y$ to obtain the simple form

$$uu' + (a+2)u^2 + bu + c = 0,$$

with the solution

$$t = -\int \frac{udu}{(a+2)u^2 + bu + c} + C_1;$$

C_1 is an arbitrary constant. The integral here may be evaluated explicitly.

We thus have $y'(x) = y^2u(\ln y)$ with the solution

$$x = \int \frac{dy}{y^2u(\ln y)} + C_2;$$

C_2 is an arbitrary constant. Kamke (1959), p.574

27. $$yy'' + (\mu - 2)y'^2 + xy^2y' + \theta y^3 = 0,$$

where μ and θ are constants.

The given DE arises from similarity reduction of a nonlinear diffusion equation. For $\mu = 1, \theta = 1$, the first integral is

$$y^{-1}y' + xy = C_1; \tag{1}$$

the general solution of the Riccati equation (1) is

$$y = C_1^2/\{C_1x - 1 + C_2C_1^2\exp(-C_1x)\},$$

where C_1 and C_2 are arbitrary constants. For $\mu = 2$, the transformation

$$y = 2(2-\theta)x^{-\theta}W(z), z = x^{2-\theta}$$

changes the given DE to standard form V of Ince (1956), p.334,

$$W'' = -2WW' + q(z)W' + q'(z)W, \tag{2}$$

provided that $\theta = 1$ or $\theta = 1/2$. The function $q(z)$ in (2) is $2/z$ for $\theta = 1$ and $1/(3z)$ for $\theta = 1/2$. On integration of (2), we have

$$C_3 = W' + W^2 - q(z)W,$$

that is,

$$4C_3 = 2xy' + x^2y^2 - 2y, \tag{3}$$

for $\mu = 2, \theta = 1$, and similarly,

$$C_4 = y' + (1/2)xy^2 \tag{4}$$

for $\mu = 2, \theta = 1/2$. Equations (3) and (4), being Riccati, can be solved easily. For arbitrary $\mu \neq 0$, a first integral is

$$C_5 = y^{\mu-2}y' + (1/\mu)xy^{\mu}, \tag{5}$$

provided that $\mu\theta = 1$. For $C_5 = 0$, (5) being Riccati, can easily be integrated:

$$y = 2\mu/(x^2 + C_6).$$

For $\theta = 2/\mu, \mu \neq 0, 1$, we have the first integral

$$C_7 = \mu x y^{\mu-2}\, y' - \{\mu/(\mu-1)\}y^{\mu-1} + x^2y^{\mu},$$

which for $C_7 = 0$ can be integrated:

$$y = \mu(2\mu-1)/(\mu-1)\ \{(C_8x^{1/(1-\mu)} + x^2)\}^{-1}.$$

For $\theta = 1$, a singular integral of the given DE is

$$y = \mu/\{C_9(x - C_9)\};$$

C_i $(i = 1, \ldots, 9)$ are arbitrary constants. There are other special solutions:

(a) $y = a/x^{1/(1-\mu)}$ if $\theta = 1/(1-\mu); a$ is an arbitrary constant.
(b) $y = [2(2\mu-1)/(2-\theta)][1/x^2]$.
(c) $y = 2\mu/(x^2 + c), \theta = 1/\mu, c \neq 0$.
(d) $y = [\mu(2\mu-1)/(\mu-1)][1/(x^2 + cx^{1/(1-\mu)})], \theta = 2/\mu; c$ is an arbitrary but nonzero constant.
(e) $y = -\mu\ c^2/(1+cx); c \neq 0$ is an arbitrary constant and $\theta = 1$.
(f) $y = (3c^2/4)\{1/(x^4 + cx^3)\}$ for $\mu = 3/4$ and $\theta = 3; c \neq 0$ is an arbitrary constant.

Glockle, Baumann, and Nonnenmacher (1992)

28. $\boldsymbol{yy'' - \{(a-1)/a\}y'^2 - fy^2y' + \{a/(a+2)^2\}f^2y^4 - \{a/(a+2)\}f'y^3 = 0,\ f = f(x)}$,

where a is a constant.

Putting $u(x) = y\exp[-\{a/(a+2)\}\int yf dx]$, we get

$$auu'' - (a-1)u'^2 = 0.$$

Now write $u' = uv(x)$ and integrate to obtain

$$u = C|x + C^*|^a,$$

where C^* is a constant. Also, writing $w(x) = 1/y$ in the definition of u, we get a linear DE

$$w' + (u'/u)w + [a/(a+2)]f = 0.$$

The solution may now be written as

$$y = -[(a+2)|x + C_1|^a] / \left\{a\int |x + C_1|^a f(x)dx + C_2\right\};$$

C_2 is an arbitrary constant. Kamke (1959), p.574

29. $\boldsymbol{a(a+2)^2yy''-(a-1)(a+2)^2y'^2-a(a+2)^2f(x)y^2y'-a^2(a+2)f'y^3+a^2f^2y^4=0,}$

where $f = f(x)$, and a is a constant.

With $u(x) = ye^{\phi}, \phi(x) = -\{a/(a+2)\} \int yf(x)dx$, we get

$$u''/u' = \{(a-1)/a\}u'/u,$$

which is easily solved. The solution is

$$y\{aX(x) + C_2\} + (a+2)(x+C_1)^a = 0,$$

where $X(x) = \int (x+C_1)^a f(x)dx$; C_1 and C_2 are arbitrary constants. Murphy (1960), p.398

30. $$\boldsymbol{ayy'' - (a-1)y'^2 + (a+2)fy^2y' + f^2y^4 + af'y^3 = 0,}$$

$f = f(x)$, where a is a constant.

With $y = v(x)/\{\int fvdx\}$, we get $vv'' = \{(a-1)/a\}v'^2$, leading, after two integrations, to

$$v = (C_1x + C_0)^a;$$

C_1 and C_0 are arbitrary constants. Kamke (1959), p.582

31. $$\boldsymbol{yy'' - cy'^2 - u_2(y)y' - u_4(y) = 0,}$$

where c is a constant and u_2 and u_4 are a quadratic and a quartic in y, respectively.

The given DE may be reduced, subject to some conditions, to a "subuniform" equation with no movable algebraic singularities. Forsyth (1959), p.304

2.33 $yy'' + g(x, y, y') = 0$

1. $$\boldsymbol{yy'' + y' + xy - x^2 = 0.}$$

The asymptotic series solution of the given DE for $x \to +\infty$ may be found following Levinson (1970); see Sachdev (1991). Bender and Orszag (1978), p.200

2. $$\boldsymbol{y(y'' + 2y' + 2y - 2m) - m(y' - m) = 0,}$$

where m is a constant.

Using the method of Eqn. (2.32.13), the solution of the given DE is found to be

$$e^x \sec(x+k)y = m \int e^x \sec(x+k)dx + \nu,$$

where k and ν are constants, and the integral may be evaluated by expanding $\sec(x+k) = [1-\sin^2(x+k)]^{-1/2}$ as an infinite series and integrating term by term; the series is uniformly convergent except in the neighborhood of points such that $|\sin(x+k)| = 1$. Roth (1941)

3. $$\boldsymbol{(1+y)y'' + Pr(1-e^{-x})y' - (Re^{-x} - A)y + y'^2 = 0,}$$
$$\boldsymbol{x > 2\ln|m|, y(2\ln|m|) = \beta, y(\infty) = 0,}$$

where Pr, R, A, m, and β are constants.

The given problem arises in flow and heat transfer over a stretching sheet. The following results are proved:

(a) If $0 < A < Pr^2/4$, then for any m and β such that $2\ln|m|$ is large and $|\beta|$ is small, the given BVP has infinitely many solutions.

(b) If $A \le 0$ and $R/m^2 - A \ge 0$, then for any $\beta > -1$ the given BVP has a monotonic solution.

(c) Let $A \le Pr^2/4$, $\lambda \le -1$ and $Pr \cdot \lambda + R > 0$, where λ is the smallest eigenvalue of the matrix

$$\begin{pmatrix} 0 & 1 \\ -A & -Pr \end{pmatrix}.$$

If $|m| \ge 1$ and $\beta > 0$ is small enough, then the given BVP has a solution $y(x)$ satisfying $0 \le y(x) \le pe^{\lambda x}$ for $x \ge 2\ln|m|$, where p satisfies $2p\lambda^2 \le Pr\lambda + R$. Soewono, Vajravelu, and Mohapatra (1992)

4. $$(y+1)y'' - y'^2 = 0, y'(0) + y(0) = 1, y'(1) - y(1) = 1.$$

The exact solution of the problem is $y = -1 + e^x$. Roberts (1979), p.364

5. $$yy'' - y'^2 - ayy' - by^2 \pm y^3 = 0,$$

where a is a constant.

With $y'(x) = p(y)$, we get the Abel equation

$$ypp'(y) = p^2 + ayp + by^2 \mp y^3.$$

Murphy (1960), p.393

6. $$y(y'' + ay' + by - an) + \{n/(m-n)\}(y' - m)(y' - n) = 0,$$

where a, b, n, and m are parameters.

See Eqn. (2.32.13). Roth (1941)

7. $$yy'' - 2(1+\alpha)y'^2 + 2xyy' - 2y^2 - 2^{3/2}y' - 2\lambda = 0,$$

where α and λ are constants.

The given DE arises from a generalized Burgers equation by a similarity and a "reciprocal" transformation, and has a Taylor series representation of the solution about the origin, which converges for certain sets of parameters; see Eqn. (2.13.36) for details. Sachdev, Nair, and Tikekar (1986)

8. $$(y+1)y'' - (1+2x)y'^2 = 0.$$

The given DE has an exact solution

$$y = e^{1/x} - 1 = \frac{1}{x} + \frac{1}{2x^2} + \frac{1}{6x^3} + \cdots.$$

Orlov (1980)

9. $$yy'' + x^2y'^2 + xyy' = 0.$$

The asymptotic behavior for $x \to +\infty$ may be found following Levinson (1970); see Sachdev (1991). It may be noted that the given DE is equidimensional in y and therefore admits exponential asymptotic behavior. $y = C$ is also a solution. Bender and Orszag (1978), p.200

10. $$yy'' - x^3y'^2 = 0.$$

The asymptotic solution as $x \to +\infty$ may be found following Levinson (1970); see Sachdev (1991). It may be noted that the given DE is equidimensional in y and therefore has exponential asymptotic behavior. $y = C$ is a solution. Bender and Orszag (1978), p.200

11. $$yy'' + y'^3 - y'^2 = 0.$$

Put $y' = u$ so that $y'' = u(du/dy)$. Integrating twice, we have $y = c_1 \log(c_2 y) + x$, where c_1 and c_2 are arbitrary constants. $y = c$ is another solution. Tenebaum and Pollard (1963), p.504

12. $$yy'' - iy'^2(y' - 1) = 0,$$

where $i = (-1)^{1/2}$.

The given DE may be written as the system

$$\frac{dz}{dx} = iz^2(z-1)/y, \quad \frac{dy}{dx} = z.$$

A first integral is easily found to be $z = 1/[1 - (i+1)cy^i]$, where c is a constant. A second integral is checked to be $y - cy^{i+1} = x - x_0$, where x_0 is a constant. Thus, we have the general solution of the given DE. Kondratenya (1970)

13. $$yy'' - y'^2(1 - y' \sin y - yy' \cos y) = 0.$$

With $y'(x) = p(y)$, we have

$$p\{yp'(y) - p(1 - p\sin y - yp\cos y)\} = 0.$$

A first integral is

$$1/p = (C/y) + \sin y;$$

C is an arbitrary constant.

The solutions are then found to be

$$x + C_1 = C_2 \ln y - \cos y \quad \text{and} \quad y = C_3;$$

C_i $(i = 1, 2, 3)$ are arbitrary constants. Murphy (1960), p.395

14. $$yy'' - (y'^2 + 1) - 2ay(y'^2 + 1)^{3/2} = 0,$$

where a is a constant.

For $a = 0$, the solution is easily found to be

$$C_1 y = \cosh(C_1 x + C_2),$$

C_1 and C_2 are arbitrary constants. For $a = \pm 1$, we have the solution in the form of an elliptic integral:

$$x = \int \left[\frac{y^2 + C_1}{\{y^2 - (y^2 + C_1)^2\}^{1/2}}\right] dy + C_2;$$

C_1 and C_2 are arbitrary constants. Kamke (1959), p.575

15. $$\boldsymbol{yy'' - \phi(y') = 0.}$$

A geometrical method for solving two-point boundary value problems for the given DE is given. Matos Peixoto (1946)

2.34 $xyy'' + g(x, y, y') = 0$

1. $$\boldsymbol{xyy'' + xy'^2 + yy' = 0.}$$

The given DE is exact. A first integral is

$$xyy' = C_0,$$

so the solution is

$$y^2 = C_1 \ln x + C_2.$$

The other solution is

$$y = C.$$

Here C_i $(i = 0, 1, 2)$ and C are arbitrary constants. Murphy (1960), p.398

2. $$\boldsymbol{xyy'' + xy'^2 + 2yy' = 0}$$

The given DE is exact. A first integral is

$$2xyy' + y^2 = C_0; \tag{1}$$

C_0 is an arbitrary constant.

The solution for y^2 is easily found from (1) as

$$x(C_1 + y^2) = C_2,$$

$y = C$ is another solution; C, C_1, and C_2 are arbitrary constants. Murphy (1960), p.398

3. $$\boldsymbol{xyy'' + xy'^2 - yy' = 0.}$$

Writing the given DE as $x(y^2)'' = (y^2)'$ and integrating twice, we obtain

$$y^2 = C_1 x^2 + C_2;$$

C_1 and C_2 are arbitrary constants. Kamke (1959), p.582; Murphy (1960), p.398

4. $$\boldsymbol{xyy'' + xy'^2 - 3yy' = 0.}$$

In this equation, which is equidimensional with respect to x, write $y(x) = u(z), z = \ln x$, and $u'(z) = p(u)$ to get $p\{up'(u) + p - 4u\} = 0$. A first integral of the given DE, therefore, is

$$\frac{du}{dz} = p = 2u + (C_0/u);$$

C_0 is an arbitrary constant.

The solutions are easily found to be

$$y^2 = C_1 + C_2x^4, y = C;$$

C_1, C_2, and C are arbitrary constants. Murphy (1960), p.398

5. $$\boldsymbol{xyy'' + 2xy'^2 + ayy' = 0,}$$

where a is a constant.

With $u(x) = y^3$, we get a linear DE

$$xu'' + au' = 0.$$

The solution, therefore, is

$$y^3 = C_1 + C_2x^{1-a};$$

C_1 and C_2 are arbitrary constants. Kamke (1959), p.583; Murphy (1960), p.399

6. $$\boldsymbol{xyy'' + xy'^2 + ayy' + f(x) = 0,}$$

where a is a constant.

With $y^2 = u(x)$, we obtain a linear DE,

$$xu'' + au' + 2f(x) = 0$$

with the solution

$$u = C_1 + C_2x^{1-a} - 2\int x^{-a}\left(\int x^{a-1}f(x)dx\right)dx;$$

C_1 and C_2 are arbitrary constants. Kamke (1959), p.582; Murphy (1960), p.398

7. $$\boldsymbol{xyy'' - (1/2)xy'^2 + (1/2)yy' = 0.}$$

The solution is

$$y = C_1(|x|^{1/2} + C_2)^2;$$

C_1 and C_2 are arbitrary constants. Kamke (1959), p.584; Murphy (1960), p.400

8. $$\boldsymbol{xyy'' - xy'^2 + yy' = 0.}$$

It is easily checked that the elements $\ldots x^{-3}, x^{-2}, x^{-1}, 1, x, x^2, x^3, \ldots$ are solutions of the given DE. Chalkley (1960)

9. $$\boldsymbol{xyy'' - xy'^2 + yy' - y^3 - \beta y + x = 0,}$$

where β is a constant.

The given DE is a special case of third Painlevé equation, wherein $\gamma = 0$, so that the y^4 term is absent. Suitable scaling has been carried out to eliminate other parameters α and δ occurring in P_{III}.

The given DE is equivalent to

$$\begin{aligned} \frac{dy}{dx} &= a_0 + \{(a_0 - \beta)/(a_0 x)\}y - \{1/(a_0 x)\}y^2 v, \\ \frac{dv}{dx} &= -a_0 - \{(a_0 - \beta)/(a_0 x)\}v + (1/(a_0 x))yv^2, \end{aligned} \tag{1}$$

where $a_0^2 - 1 = 0$. Eliminating y from (1), we get

$$xvv'' - xv'^2 + vv' - v^3 - pv + x = 0, \tag{2}$$

where $p = \beta - 2a_0$. It follows from (2) and the given equation that if $y_0 = y_0(x, \beta)$ is a solution of the latter for some $\beta = \beta_0$, then the function

$$y_1(x) = \{(a_0 - \beta_0)y_0 + x - a_0 x y_0'\}/y_0^2, a_0^2 - 1 = 0 \tag{3}$$

is a solution of the same with $\beta_1 = \beta_0 - 2a_0$.

It can be shown that to find the general solution of the given equation, it suffices to construct it in the strip [Re β_0, Re β_1], where β_0 is any given number and $\beta_1 = \beta_0 - 2a_0$. As an example, we note that $y_0(x) = x^{1/3}$ is a solution for $\beta_0 = 0$. It follows from (3) that

$$\begin{aligned} y_1(x) &= \{\mp 2 + 3x^{2/3}\}/(3x^{1/3}), \\ y_2(x) &= \{\mp 24x + 20x^{1/3} + 9x^{5/3}\}/\{(2 \mp 3x^{2/3})^2\} \end{aligned}$$

are solutions of the given DE for $\beta_1 = \pm 2$ and $\beta_2 = \pm 4$, respectively. Gromak (1973)

10. $$\boldsymbol{xyy'' - xy'^2 + yy' - x(a_0 + a_1 y^4) - y(a_2 + a_3 y^2) = 0.}$$

With $y(x) = u(z), z = \ln x$, we get

$$uu''(z) = u'^2 + e^{2z}(a_0 + a_1 u^4) + e^z u(a_2 + a_3 u^2).$$

Now see Eqn. (2.31.43). Murphy (1960), p.398

11. $$\boldsymbol{xyy'' - xy'^2 + yy' + axy^4 + by^3 + cy + dx = 0,}$$

where a, b, c, and d are constants.

The given DE is in general not solvable in terms of classical functions. Kamke (1959), p.583; Ince (1956), p.335

12. $$\boldsymbol{xyy'' - xy'^2 + ayy' + bxy^3 = 0,}$$

where a and b are constants.

Writing the given DE as

$$x\frac{d^2}{dx^2}\log y + a\frac{d}{dx}\log y + bxy = 0,$$

we get Eqn. (2.8.81) if we put $u = \log y$. If we set $b = \pm\beta^2$ in the given DE and change to $y(x) = \overline{y}(\overline{x}), \overline{x} = \beta x$, we get

$$xyy'' - xy'^2 + ayy' \pm xy^3 = 0, \tag{1}$$

wherein we have dropped the bars. One solution of (1) is $y = \pm(2a-2)x^{-2}$. In general, writing $y = x^{-2}\eta(\xi), \xi = \log x$ in (1), we get

$$\eta\eta'' - \eta'^2 + (a-1)\eta\eta' \pm \eta^3 + (2-2a)\eta^2 = 0; \tag{2}$$

with $\eta'(\xi) = p(\eta)$, Eqn. (2) goes to

$$\eta pp' - p^2 + (a-1)\eta p \pm \eta^3 + (2-2a)\eta^2 = 0.$$

For further work, follow the analysis in Eqn. (2.8.81). Kamke (1959), p.583; Murphy (1960), p.399

13. $\boldsymbol{xyy'' - 2xy'^2 + yy' = 0.}$

Multiply throughout by x; the given DE becomes equidimensional with respect to x. Put $x = e^z$ to remove explicit dependence of the independent variable; the DE becomes autonomous. The solution then is easily found to be $y^{-1} = c_1 \log x + c_2$, where c_1 and c_2 are arbitrary constants. Tenebaum and Pollard (1963), p.505

14. $\boldsymbol{xyy'' - 2xy'^2 + (y+1)y' = 0.}$

The given DE is equidimensional with respect to x. With $y(x) = \eta(\xi), \xi = \ln|x|$, it becomes autonomous. The solutions are $y = C, y = (1/2)\ln|x|, 2C_1y = \tan(C_2 + C_1\ln|x|)$; C_1 and C_2 are arbitrary constants. Kamke (1959), p.583; Murphy (1960), p.399

15. $\boldsymbol{xyy'' - 2xy'^2 - yy' = 0.}$

With $y' = yu(x)$, we get $xu' - u - xu^2 = 0$, which has the solution

$$u(C - x^2) = 2x;$$

C is an arbitrary constant.

The solution of the given DE is now easily found to be

$$y(C_1 - x^2) = C_2;$$

C_1 and C_2 are arbitrary constants. Murphy (1960), p.399

16. $\boldsymbol{xyy'' - 2xy'^2 + ayy' = 0,}$

where a is a constant.

With $u(x) = 1/y$, we get a linear DE

$$xu'' + au' = 0.$$

The solution, therefore, is

$$\frac{1}{y} = \begin{cases} C_1 + C_2x^{1-a} & \text{for } a \neq 1, \\ C_1 + C_2 \ln x & \text{for } a = 1; \end{cases}$$

C_1 and C_2 are arbitrary constants. Kamke (1959), p.583; Murphy (1960), p.399

17. $$\boldsymbol{xyy'' - 4xy'^2 + 4yy' = 0.}$$

The solution may be found to be

$$y^{-3} = C_1 + C_2x^{-3};$$

C_1 and C_2 are arbitrary constants. Kamke (1959), p.584; Murphy (1960), p.399

18. $$\boldsymbol{xyy'' + ay'(xy' - y) = 0,}$$

where a is a constant.

(a) If $a = -1$, see Eqn. (2.34.8).
(b) If $a \neq -1$, put $y' = yu(x)$ to obtain $xu' - au + (1+a)xu^2 = 0$ with the solution

$$u(x + Cx^{-a}) = 1 \text{ or } y'/y = x^a/[C + x^{a+1}];$$

C is an arbitrary constant. The solution of the given DE is

$$y^{1+a} = C_1 + C_2x^{1+a};$$

C_1 and C_2 are arbitrary constants.

Murphy (1960), p.399

19. $$\boldsymbol{xyy'' + [\{ax/(b^2 - x^2)^{1/2}\} - x]y'^2 - yy' = 0.}$$

Writing $y' = yu(x)$ leads to a Bernoulli equation for u. The solutions may be found to be

$$y = C \text{ and } y = C_1 \exp[(1/a)(b^2 - x^2)^{1/2} + (C_2/a^2)\log\{C_2 - a(b^2 - x^2)^{1/2}\}];$$

C, C_1, and C_2 are arbitrary constants. Kamke (1959), p.584

2.35 $x^2yy'' + g(x,y,y') = 0$

1. $$\boldsymbol{x^2yy'' - 2xyy' + x^2y'^2 + y^2 = 0.}$$

The given DE is equidimensional with respect to x. A dominant balance argument suggests that on putting $y = ux^\theta, \theta = 1, \theta = 1/2$.

(a) With $\theta = 1$, we put $y = ux$, where u is neither zero nor infinite when $x = 0$. The equation for u is

$$xu'' + 2u' + xu'^2/u = 0. \tag{1}$$

The requisite solution of (1) is $u = B$, a constant, so $y = Bx$.

(b) $\theta = 1/2$. Put $x = z^2, y = uz$, so that

$$z\frac{d^2u}{dz^2} - \frac{du}{dz} = -\frac{z}{u}\left(\frac{du}{dz}\right)^2.$$

If $u(0) = \alpha, u'(0) = \beta$, it is easily checked that $\beta = 0$. Writing $u = \alpha + v$, we have

$$z\frac{d^2v}{dz^2} - \frac{dv}{dz} + \frac{z}{\alpha + v}\left(\frac{dv}{dz}\right)^2 = 0. \tag{2}$$

A regular solution of (2) with $v'(0) = 0$ can be found in the form

$$v = a_2z^2 + a_3z^3 + a_4z^4 + \cdots, a_3 = 0, a_4 = -4a_2^2/(8\alpha), \tag{3}$$

etc., where a_2 is arbitrary. It is easy to check that odd powers of z in v have zero coefficients. The solution corresponding to the solution of original equation from (3) is

$$y = x^{1/2}(\alpha + a_2x + a_4x^2 + \cdots),$$

where α and a_2 are arbitrary.

A complete solution of the original DE may be found by simple quadratures as $y = (Ax + Bx^2)^{1/2}$, where A and B are arbitrary constants. Forsyth (1959), p.226

2. $$\boldsymbol{x^2yy'' + (xy' - y)^2 - 3y^2 = 0.}$$

The given DE is equidimensional with respect to x and y. With $y^2 = u(x)$, we get $x^2u'' - 2xu' - 4u = 0$, a linear DE of Cauchy–Euler type. The solution is

$$xy^2 = C_1 + C_2x^5;$$

C_1 and C_2 are arbitrary constants. Murphy (1960), p.400

3. $$\boldsymbol{x^2yy'' - x^2y'^2 + (1/2)xyy' + (1/4)y^2 + (1/4)y^4 = 0.}$$

Put $y = \xi f(\xi)$, where $\xi = x^{1/2}$ to obtain

$$ff'' - f'^2 + \beta f^4 = 0, \ ' \equiv \frac{d}{d\xi}. \tag{1}$$

Integrating (1), we have

$$f'^2 = -\beta f^4 + I_1f^2, \tag{2}$$

where I_1 is a constant of integration. Put

$$f(\xi) = (I_1/\beta)^{1/2}[2g^2(\xi) - 1]^{-1}$$

in (2) to get

$$g'^2(\xi) = (1/4)I_1[g^2(\xi) - 1]. \tag{3}$$

Equation (3) is readily integrated to yield

$$g(\xi) = \cosh[(1/2)I_1^{1/2}(\xi - \xi_0)], \tag{4}$$

where ξ_0 is a constant. Hence, $f(\xi) = (I_1/\beta)^{1/2}\ \mathrm{sech}\{I_1^{1/2}\xi\}$, and so is the solution. Parthasarathy and Lakshmanan (1991)

4. $$\boldsymbol{x^2yy'' - x^2y'^2 + (1/2)xyy' + (1/2)y^2 - (1/8)w_0^2y^3 = 0,}$$

where w_0^2 is a constant.

See Eqn. (2.31.59). Parthasarathy and Lakshmanan (1990)

5. $$\boldsymbol{x^2yy'' - x^2y'^2 + axyy' + by^2(\log y) = 0, y(x) > 0,}$$

where a and b are constants.

Put $\log y = u$ to obtain the linear Euler equation $x^2u'' + axu' + bu = 0$, which is easily solved. Utz (1978)

6. $$\boldsymbol{x^2yy'' - x^2y'^2 + y^2 + \{4xy(xy' - y)^3\}^{1/2} = 0,}$$

where $y(0) = 0, y'(0) = \beta$.

An exact solution is $y = \beta x e^{-x/\{c(x-c)\}}$, where β and c are arbitrary constants. Forsyth (1959), p.242

7. $$\boldsymbol{x^2yy'' - (4/3)x^2y'^2 + (3/4)[(1+\lambda)^2 - 4\mu k - 1]y^2 + (\mu^2/(2\alpha))x^3y^{7/3}y' = 0,}$$

where λ, μ, α, and k are constants.

The given DE arises in the similarity form of solutions of nonlinear diffusion equation. The following set of transformations change it to a first order equation:

$$y = x^{-3/2}u(x),\ \eta = \ln x,\ p = \frac{du}{d\eta} = x\frac{du}{dx},$$

so that

$$p\frac{dp}{du} - \frac{4}{3}\frac{p^2}{u} + \frac{3}{4}[(1+\lambda)^2 - 4\mu k]u + (\mu^2/(2\alpha))\left(p - \frac{3u}{2}\right)u^{4/3} = 0.$$

Dividing the given DE by $x^3y^{7/3}$, we have

$$\left[\frac{y^{-4/3}y'}{x} - \frac{3y^{-1/3}}{x^2}\right]' + \frac{3y^{-1/3}}{4x^3}[(1+\lambda)^2 - 4\mu k - 9] + \{\mu^2/(2\alpha)\}y' = 0,$$

which can be integrated provided that $(1+\lambda)^2 - 4\mu k = 9$.

The first integral in this case is

$$y^{-4/3}y'/x - 3y^{-1/3}/x^2 + [\mu^2/(2\alpha)]y = C_1, \tag{1}$$

where C_1 is the integration constant. If $C_1 = 0$, we may multiply (1) by y^{-1}, put $y^{-4/3} = z$ and integrate. We then have

$$y(x) = [Cx^{-4} + (\mu^2x^2)/(9\alpha)]^{-3/4},$$

where C is an arbitrary constant. Hill (1989)

8. $$\boldsymbol{x^2(1-y)y'' + 2x^2y'^2 - 2x(1-y)y' + 2y(1-y)^2 = 0.}$$

With $(1-y)u(x) = 1$, we have a linear inhomogeneous DE

$$x^2u'' - 2xu' + 2u = 2.$$

Hence the solution may be found to be

$$y(1 + C_1x + C_2x^2) = x(C_1 + C_2x);$$

C_1 and C_2 are arbitrary constants. Murphy (1960), p.400

9. $$\boldsymbol{x^2yy'' - 2x^2y'^2 - axyy' - ay^2 = 0,}$$

where a is a constant.

The given DE is equidimensional in x and y. With $y' = yu(x)$, we obtain the equation $x^2u' = a + xu(a + xu)$ having the solution

$$xu(Cx^{a-1} - 1) = 1 - aCx^{a-1};$$

C is an arbitrary constant. Putting $u = y'/y$ and integrating, we have

$$y(C_1x + C_2x^a) = 1;$$

C_1 and C_2 are arbitrary constants. Murphy (1960), p.400

10. $$\boldsymbol{2x^2yy'' - x^2y'^2 + y^2 = 0.}$$

The given DE is equidimensional in x and y. With $y = xu(z), z = \ln x$, and $p = du/dz$, we get $p[2up' - p] = 0$, which integrates to give

$$p^2 = Cu;$$

C is an arbitrary constant.

The solution on integration is

$$y = x(C_1 + C_2 \ln x)^2;$$

C_1 and C_2 are arbitrary constants.

$$y = C_3x$$

is another solution, with C_3 an arbitrary constant. Murphy (1960), p.401

11. $$\boldsymbol{x^2yy'' - (1/2)x^2(y'^2 + 1) + (1/2)y^2 = 0.}$$

The given DE is a special case of Eqn. (2.31.5) if we divide it by x^2. The solution is

$$y = x[C_1 + (4C_1C_2 - 1)^{1/2} \ln|x| + C_2 \ln^2|x|];$$

C_1 and C_2 are arbitrary constants. Kamke (1959), p.585

12. $\boldsymbol{x^2yy''+x^2y'^2/m+\lambda(4/m+3)xyy'+2\lambda(2\lambda/m+\lambda-1)y^2+[(\lambda+1)^2/2]x^3y' = 0,}$

where m and λ are constants.

It is easy to check that the given DE (which arises from similarity solutions of nonlinear diffusion equations) remains invariant under the one-parameter group $x_1 = e^{\epsilon}x, y_1 = e^{2\epsilon}y$. Therefore, introducing $g = y/x^2$ and $h = y'/x$, we have

$$y' = xh = x^2g' + 2xg.$$

Therefore,

$$\log x = \int \frac{dg}{h(g) - 2g} + \text{ constant.} \tag{1}$$

Further, on using $y'' = xh' + h = (h - 2g)(dh/dg) + h$, it is not difficult to show that the given DE becomes

$$\frac{dh}{dg} = \frac{h^2/m + (4\lambda/m + 3\lambda + 1)gh + 2\lambda(2\lambda/m + \lambda - 1)g^2 + (\lambda + 1)^2h/2}{g(2g - h)}. \tag{2}$$

By postulating simple forms for $h(g)$, it is possible to deduce exact solutions:

(a) $h(g) = a$, a constant, then a valid solution is obtained if $m = -3/2, \lambda = -3$, and $a = 3$. In this case, (1) and $y = x^2g$ yield $y = (3/2)x^2$ + constant.

(b) $h(g) = a + bg$, where a and b are constants, then we find from (2) that

$$a = -m(\lambda + 1)^2/2, b = -2(3 + 2m)a,$$

where $\lambda = -m/(m + 2)$ or $\lambda = -m/(m + 1)$.

(c) $h(g) = ag^n$. Substitution in (2) shows that $a = 12, n = 2$. This case is valid only if $m = -1/x, \lambda = -3/5$. Equation (1) then gives $g = (6 - Cx^2)^{-1}$, where C is a constant. Equation (2) may be analyzed in the phase plane.

Hill and Hill (1990)

13. $$\boldsymbol{ax^2yy'' + bx^2y'^2 + cxyy' + dy^2 = 0,}$$

where a, b, c, and d are constants.

(a) If $a + b = 0$, we get through $y' = yu(x)$, the linear DE

$$ax^2u' + cxu + d = 0.$$

(b) If $a + b \neq 0$, we set $y = u^{\alpha}, u = u(x), \alpha = a/(a + b)$, to obtain the Cauchy–Euler (linear) DE $a\alpha x^2u'' + c\alpha xu' + du = 0$.

Kamke (1959), p.585; Murphy (1960), pp.400, 401

14. $$\boldsymbol{x^2yy'' - [x^2(3y - 1)/2(y - 1)]y'^2 + xyy' + (1/2)(ay^2 + b)(y - 1)^2}$$
$$\boldsymbol{+ (c/2)xy^2 + (d/2)x^2y^2(y + 1)/(y - 1) = 0,}$$

where a, b, c, and d are constants.

The given DE is not integrable in terms of classical transcendental functions [see Ince (1956), p.341]. Kamke (1959), p.590; Murphy (1960), p.403

15. $$\boldsymbol{x^3yy'' + x^3y'^2 + 6x^2yy' + 3xy^2 - a = 0,}$$

where a is a constant.

A first integral is

$$x^3yy' + (3x^2y^2/2) = ax + C_1; \qquad (1)$$

C_1 is an arbitrary constant. Equation (1) is linear in y^2; hence the solution is

$$x^3y^2 = ax^2 + C_1x + C_2;$$

C_2 is another arbitrary constant. Murphy (1960), p.401

16. $$\boldsymbol{x^3y^2y'' - [16/\{3(8-1/x)^{4/3}\} + 10/\{9x(8-1/x)^{4/3}\}]y^2 + 16 = 0,}$$
$$\boldsymbol{0 < x \leq 1, (2/3)y(1) - 7y'(1) = 0.}$$

The exact solution of the given problem is

$$y(x) = (8 - 1/x)^{2/3}.$$

Baxley (1988)

17. $$\boldsymbol{x^3y^2y'' + (xy' - y)^3(y + x) = 0.}$$

The given DE belongs to the type of Eqn. (2.24.12). With $y(x) = x\eta(\xi)$, $\xi = \log x$, we get an autonomous DE,

$$\eta^2\eta'' + (\eta + 1)\eta'^3 + \eta^2\eta' = 0.$$

Further, writing $p(\eta) = \eta'(\xi)$, we get the Riccati DE $\eta^2p' + (\eta + 1)p^2 + \eta^2 = 0$, which may be linearized. Kamke (1959), p.590; Murphy (1960), p.403

2.36 $f(x)yy'' + g(x, y, y') = 0$

1. $$\boldsymbol{x(x+1)^2yy'' - x(x+1)^2y'^2 + 2(x+1)^2yy' - a(x+2)y^2 = 0,}$$

where a is a constant.

With $y' = u(x)y$, we get a linear DE for u:

$$u' + 2u/x = a(x+2)/\{x(x+1)^2\}.$$

The solution may be found to be

$$y = C_1|x+1|^a \exp(C_2/x);$$

C_1 and C_2 are arbitrary constants. Kamke (1959), p.585; Murphy (1960), p.401

2. $$\boldsymbol{8(x^3 - 1)yy'' - 4(x^3 - 1)y'^2 + 12x^2yy' - 3xy^2 = 0.}$$

With $y = \eta^2(\xi)$, $\xi = x^3$, we get the hypergeometric DE

$$\xi(\xi - 1)\eta'' + [(7/6)\xi - (2/3)]\eta' - (1/48)\eta = 0.$$

Kamke (1959), p.585; Murphy (1960), p.401

3. $$\boldsymbol{(a^2 + x^2)^{1/2}(yy'' + by'^2) - yy' = 0,}$$

where a and b are constants.

With $y' = yu(x)$, we have a first order DE

$$(a^2 + x^2)^{1/2}[u' + (1 + b)u^2] = u. \tag{1}$$

Equation (1) is of Bernoulli type, and hence can be integrated. Murphy (1960), p.401

4. $$(a^2 - x^2)^{1/2}(xyy'' - xy'^2 - yy') - bxy'^2 = 0,$$

where a and b are constants.

With $y' = yu(x)$, we get the Bernoulli equation

$$(a^2 - x^2)^{1/2}[xu' - u] = bxu^2.$$

Hence the solution may be found easily. Murphy (1960), p.401

5. $$f(x)yy'' + g(x)y'^2 + h(x)yy' + k(x)y^2 = 0.$$

With $u(x) = y'/y$ we get the Riccati DE

$$fu' + (f + g)u^2 + hu + k = 0,$$

which is linear if $f = -g$. Kamke (1959), p.586; Murphy (1960), p.401

6. $$f_1(x)yy'' - f_1(x)y'^2 + f_2(x)yy' + f_3(x)y^2(\log\ y) = 0, y(x) > 0.$$

Putting $\log y = u$, we get $f_1(x)u'' + f_2(x)u' + f_3(x)u = 0$, which is a linear equation. Utz (1978)

2.37 $f(y)y'' + g(x, y, y') = 0$, $f(y)$ quadratic

1. $$y^2y'' + k = 0,$$

where k is a constant.

Put $y' = u$ so that $y'' = u(du/dy)$ and integrate twice; we have

$$x = \int [y/\{2k + c_1y\}]^{1/2}dy + c_2,$$

where c_1 and c_2 are arbitrary constants. The integral may be evaluated by putting $y = z^2$, etc. Tenebaum and Pollard (1963), p.504

2. $$y^2y'' + x^2/32 = 0,$$

$y(0+) = 0, y(1-) = 0, y(x) > 0$ for $0 < x < 1$.

This problem arises in the investigation of the equilibrium of a membrane with no catenary force on a contour. It is a special case of Eqn. (2.6.20). Lomtatidze (1987)

3. $$y^2y'' - (\lambda^2/8)y^2 + x^2/32 = 0, 0 < x \leq 1,$$

$y(0) = 0, 2y'(1) - (1 + \nu)y(1) = 0$, where $\lambda > 0, \nu > 0$ are constants.

The given BVP arises in nonlinear mechanics. Changing the variable to u so that $y = x^p u$, where $(1+\nu)/2 < p \le 1$, we obtain

$$x^2u'' + 2pxu' + p(p-1)u = -1/(32x^{3p-4}u^2) + \lambda^2/(8x^{p-2}).$$

Further, putting $x = 1/t, 1 \le t < \infty$, we get

$$u'' = -\left\{\frac{2(1-p)}{t}\right\}u' + \left\{\frac{p(1-p)}{t^2}\right\}u - \frac{1}{32t^{6-3p}u^2} + \frac{\lambda^2}{8t^{4-p}}, 1 \le t < \infty. \tag{1}$$

The boundary condition at $x = 1 = t$ becomes

$$\{p - (1+\nu)/2\}u(1) + u'(1) = 0. \tag{2}$$

Making use of the form (1)–(2), the following uniqueness and existence results are proved:

(a) Suppose that $\nu < 1$ and $(1+\nu)/2 < p < 1$. Then the given BVP has at most one positive solution $y(x)$ satisfying $y(x)/x^p \to 0$ as $x \to 0^+$.

(b) Suppose that $\nu < 1$. Then the given BVP has a positive solution $y(x)$ for which $y(x)/x < 1/(2\lambda), 0 < x \le 1$.

Further, there exist constants $C > 0, \beta > 1$ such that $y(x) \ge Cx(\beta - x), 0 < x \le 1, y(x)/x$ is decreasing for $0 < x \le 1$ and $\lim_{x \to 0^+} y(x)/x \le 1/(2\lambda)$. Part (a) leaves open the possibility of multiple solutions satisfying $y(x) \to 0$ as $x \to 0^+$ more slowly than $x^{(1+\nu)/2}$.

A computational approach to finding $\lim_{x \to 0^+} y(x)/x$ is suggested. Numerical results are provided. Baxley (1988)

4. $$y^2y'' - [16/\{3(8-x)^{4/3}\} + 10x/\{9(8-x)^{4/3}\}]y^2 + 16x^2 = 0,$$

$$0 < x \le 1, 7y'(1) - (19/3)y(1) = 0.$$

The exact solution of the given problem is $y(x) = x(8-x)^{2/3}$. Baxley (1988)

5. $$y^2y'' + (1/x)y^2y' - (\alpha y^2 + \beta) = 0, y'(0) = 0, y(1) = 1,$$

where α and β are constants.

This singular BVP describes the equilibrium of neighboring drops at different potentials. On using L'Hospital's rule, it is reduced to $2y'' = \alpha + \beta/y^2$, which is valid near $x = 0$. Two solutions are found numerically for $\alpha = 0, \beta = 0.77295$, giving $y(0) = 0.62773$ and 0.47935. Kubicek and Hlavacek (1983), p.110

6. $$y^2y'' + yy'^2 - a = 0,$$

where a is a constant.

Put $y' = p(y), y'' = p(dp/dy)$, so that the given DE becomes

$$y^2p\frac{dp}{dy} + yp^2 = a. \tag{1}$$

Now put $p^2 = P$ in (1) to get the linear first order DE

$$\frac{dp}{dy} + (2/y)P = 2a/y^2$$

with the solution

$$p^2 = P = (2ay + A)/y^2.$$

Hence the solution is

$$\pm \int \{y/(2ay + A)^{1/2}\}dy = x + B,$$

where A and B are arbitrary constants. Tatarkiewicz (1963)

7. $$\boldsymbol{y^2 y'' + yy'^2 + ax = 0,}$$

where a is a constant.

With $y = \{u'(x)\}^{-1}$, we get

$$-u'u''' + 3u''^2 + axu'^5 = 0.$$

Now interchanging the role of u and x, we get a linear DE with constant coefficients: $x''' + ax(u) = 0$. Kamke (1959), p.586; Murphy (1960), p.401

8. $$\boldsymbol{y^2 y'' + yy'^2 + y^3/x^2 = 0.}$$

Put $t = -xy'/y$ to reduce the given DE to one of first order and hence integrate to obtain the exact solution

$$y^4 = C_1 x/\{1 + \tan^2((1/2)7^{1/2} \ln C_2 x)\},$$

where C_1 and C_2 are arbitrary constants. Murphy (1992)

9. $$\boldsymbol{y^2 y'' + 2yy'^2 - 3y^3 - 8e^{-x}\{1 + 2e^{-4x} - 3e^{-8x}\} = 0.}$$

The given DE has an exact monotonic solution $y(x) = e^x - e^{-3x}$. Graef and Spikes (1987)

10. $$\boldsymbol{y^2 y'' - (cy - x)y'^3 = 0, y(c) = 1, y'(c) = \infty.}$$

Change the role of dependent and independent variables. Rescale the given DE according to $X = x/c, Y = y$ to obtain

$$Y^2 \frac{d^2Y}{dX^2} - (Y - X)\left(\frac{dY}{dX}\right)^3 = 0. \tag{1}$$

Interchanging the dependent and independent variables, (1) goes to the linear DE

$$\frac{d^2X}{dY^2} - X/Y^2 = -1/Y. \tag{2}$$

The solution to (2) subject to the given BC is

$$X = (1/5^{1/2})(Y^\sigma - Y^\tau) + Y,$$

where $\tau = (1/2)(1 + 5^{1/2}), \sigma = (1/2)(1 - 5^{1/2})$. Forsyth (1959), p.197

11. $$(1 + y^2)y'' + w^2 y = 0,$$

where w is a constant.

Rewriting this as $y'' + w^2 y(1 - y^2 + \cdots) = 0$ and introducing the variable $y = \epsilon^{1/2} v$, we have $v'' + w^2 v - \epsilon w^2 v^3 + \cdots = 0$ with a small nonlinear term. A straightforward perturbation gives

$$y = \epsilon^{1/2} a \cos(wx + \beta) + \epsilon^{3/2}[(3/8)wa^3 x \sin(wx + \beta) - (1/32)a^3 \cos(3wx + 3\beta)] + \cdots$$

with a secular term. Renormalization gives a first uniformly valid approximation as

$$y = \epsilon^{1/2} a \cos[\{1 - (3/8)\epsilon a^2\} w^2 x + \beta] + \cdots.$$

Nayfeh (1985), p.93

12. $$(1 + y^2)y'' + y' + C \tan^{-1} y = 0, y(a) = 0, y'(a) = v_0,$$

where C is a constant greater than 1/4.

A constant, $V = (\pi/2)C^{1/2} \exp\{(4C - 1)^{-1/2}(\pi/2 - \tan^{-1}(4C - 1)^{-1/2})\}$, is found such that if $v_0 < V$, the solution of IVP is oscillatory, satisfying $y(x) \to 0, y'(x) \to 0$. If, on the other hand, $v_0 \geq V$, the IVP has a solution that is nonvanishing for $x > a$ and satisfies $|y(x)| \to \infty$ as $x \to \infty$. Benson (1984)

13. $$(1 + y^2)y'' + (a + y'^2)y = 0,$$

where a is a constant.

The given DE describes the motion of a bead sliding on a smooth parabolic wire rotating with constant angular velocity about a vertical axis. Put $y' = p$, etc., and integrate. A first integral is

$$(1 + y^2)(a + p^2) = C;$$

C is an arbitrary constant. Now the variables separate.

Phase plane analysis shows that for $a > 0$, the origin is a center. For $a = 0$, every y is an equilibrium point. For $a < 0$, the origin is a saddle point. Note that $y' = \pm(-a)^{1/2}$ are phase paths. Jordan and Smith (1977), p.31

14. $$(1 + y^2)y'' + 2yy'^2 - y'/x + y/x - (1/x^8)[14 + 8x^4 + x^5] = 0, x \geq x_0 > 0.$$

This forced DE has a nonoscillatory solution $y(x) = 1/x^2$. Grace and Lalli (1990)

15. $$(1+y^2)y''+2yy'^2+\sin x(1+\sin^2 x)y'/x^2+(1+\sin^2 x)(1-\cos x/x^2)y = 0, x > 0.$$

It is shown that all bounded solutions of the given DE are oscillatory. Grace and Lalli (1990)

16. $$(1 + y^2)y'' + (1 - 2y)y'^2 = 0.$$

Writing the given DE as $y''/y + (1 - 2y)y'/(1 + y^2) = 0$ and integrating twice, we find the general solution as

$$y = \tan\{\ln \alpha(x - x_0)\},$$

where α and x_0 are arbitrary constants. The point $x = x_0$ is a nonisolated movable essential singularity. $y(x)$ has no limit point (even when we allow infinity) as $x \to x_0$ unless x is taken along special paths; an infinite number of distinct branches originate from the point x_0, which is a limit point of poles, a branch point, and an essential singularity. Kruskal and Clarkson (1992)

17. $$\mathbf{(1+y^2)y'' - (a+3y)y'^2 = 0,}$$

where a is a constant.

With $y'(x) = p(y)$, we have

$$p[(1+y^2)p' - (a+3y)p] = 0.$$

An integration and substitution of $y = \tan t$, etc., lead to the parametric representation of the solution

$$x = C_1 e^{-at}(\sin t - a\cos t) + C_2; y = \tan t;$$

C_1 and C_2 are arbitrary constants. $y = C_3$, a constant, is another solution. Murphy (1960), p.402

18. $$\mathbf{(y^2+1)y'' - 3yy'^2 = 0.}$$

The given DE is a special case of Eqn. (2.39.38). The solution is

$$\{1 - (C_1x + C_2)^2\}y^2 = (C_1x + C_2)^2;$$

C_1 and C_2 are arbitrary constants. Kamke (1959), p.587; Murphy (1960), p.402

19. $$\mathbf{(1+y^2)y'' + y'(1+y'^2) = 0.}$$

With $y'(x) = p(y)$, we have

$$p[(1+y^2)p' + 1 + p^2] = 0.$$

A first integral of the given DE, therefore, is

$$p(1+Cy) = C - y; \tag{1}$$

C is an arbitrary constant. The solution from (1) is

$$x = C_1 + C_2y - (1+C_2^2)\ln(C_2+y);$$

C_1 and C_2 are arbitrary constants. $y = C_3$, a constant, is another solution. Murphy (1960), p.402

20. $$\mathbf{(1+4p^2y^2)y'' + \Lambda y + 4p^2yy'^2 = 0,}$$

where p and Λ are constants.

The solution for small but finite y is found by first putting $y = \epsilon^\lambda u, \lambda > 0$ such that nonlinearity is small $O(\epsilon)$. Thus, $\lambda = 1/2$, so that the DE becomes

$$u'' + w_0^2 u = -4p^2\epsilon(u^2u'' + uu'^2),$$

where $w_0^2 = \Lambda$. By using the method of multiple scales, a uniformly valid first approximation is

$$y = a\cos\{(1 - p^2a^2)w_0x + \beta_0\} + \cdots,$$

where β_0 is arbitrary. Nayfeh (1985), p.102

21.
$$(\ell^2/12 + r^2y^2)y'' + r^2yy'^2 + gry\cos y = 0,$$

where r, ℓ, and g are constants.

A first order uniform expansion for small but finite y is found by first expanding the equation and retaining only the cubic terms; we obtain

$$y'' + w_0^2y + \alpha_1(y^2y'' + yy'^2) - \alpha_2w_0^2y^3 + \cdots = 0,$$

where $w_0^2 = 12gr/\ell^2, \alpha_1 = 12r^2/\ell^2, \alpha_2 = 1/2$. Now proceed as in Eqn. (2.39.30). Nayfeh (1985), p.105

22.
$$\frac{d^2}{dx^2}(y + \beta y^3) + p\frac{d}{dx}(y + \beta y^3) + q\frac{dy}{dx} + y - r = 0,$$

where β, p, q, and r are constants.

It is easily shown that if $pq = 1$ and $\beta \neq 0$, one solution of the given DE is

$$x = A_2 - p[(3\beta r^2 + 1)\ln|y - r| + \{(3/2)\beta\}(y^2 + 2ry)],$$

where A_2 is an arbitrary constant. Koehler and Pignani (1962)

23.
$$(1 + 2Ky^2)y'' + 2Kyy'^2 + 2(\epsilon + \epsilon_Ly^2)y' + ry^3 + \Omega^2(1 - 2\mu\cos\theta x)y = 0,$$

where $\epsilon, \Omega, \mu, \theta, r, \epsilon_L$, and K are constants.

The following boundedness results are proved: Let y be any solution of the IVP for the given DE in which all constants are nonnegative, and if $K \neq 0$, the inequality $K \leq \epsilon_L/(2\epsilon)$ is satisfied. Then there exists positive constants ζ_0 and η_0 such that $|y| \leq \zeta_0, |y'| \leq \eta_0$ for all nonnegative x. Genin and Maybee (1970)

24.
$$2y(1 - y)y'' - (1 - 2y)y'^2 = 0.$$

With $y'(x) = p(y)$, we have $2y(1 - y)p' = (1 - 2y)p, p' = dp/dy$. A first integral, therefore, is

$$p^2 = Cy(1 - y);$$

C is an arbitrary constant. Now the variables separate.

$$y = C_1, \text{ a constant,}$$

is another solution. Murphy (1960), p.402

25.
$$2y(1 - y)y'' - (1 - 3y)y'^2 = 0.$$

With $y'(x) = p(y)$, we have $p[2y(1 - y)p' - (1 - 3y)p] = 0$. A first integral, therefore, is

$$p^2 = 4C_0^2y(1 - y^2);$$

C_0 is an arbitrary constant.

Hence the solution, by separating the variables, is

$$y = \tanh^2(C_1 + C_2x);$$

C_1 and C_2 are arbitrary constants. $y = C$ is another solution. Murphy (1960), p.402

26. $$\mathbf{2y(y-1)y'' - (2y-1)y'^2 + f(x)y(y-1)y' = 0.}$$

A first integral is

$$y'^2 = Cy(y-1)\exp\left(-\int f dx\right);$$

the variables now separate; C is an arbitrary constant. Kamke (1959), p.588; Murphy (1960), p.402

27. $$\mathbf{2y(y-1)y'' - (3y-1)y'^2 + f(y) = 0.}$$

This is a special case of Eqn. (2.39.38). If $f = 0$, the solutions are

$$y = -\tan^2(C_1x + C_2) \quad \text{and} \quad y = \tanh^2(C_1x + C_2);$$

C_1 and C_2 are arbitrary constants. Kamke (1959), p.588

28. $$\mathbf{2y(y-1)y'' - (3y-1)y'^2 + 4yy'(fy+g) + 4y^2(y-1)(g^2 - f^2 - g' - f') = 0,}$$
$$\mathbf{f = f(x), g = g(x).}$$

With $u = C\exp\left(\int (g-f)dx\right)$, the given DE has a first integral

$$\{y' - 2(f+g)y\}^2 = y(y-1)^2u^2.$$

Kamke (1959), p.588; Murphy (1960), p.402

29. $$\mathbf{2y(y-1)y'' - (3y-1)y'^2 - 4(fy+g)yy' - (y-1)^3(\phi^2y^2 - \psi^2)}$$
$$\mathbf{- 4y^2(y-1)(f^2 - g^2 - f' - g') = 0.}$$

Here f, g, ϕ, and ψ are given functions of x, and $\phi' = 2f\phi, \psi' = -2g\psi$. The given DE is equivalent to the system

$$\begin{aligned} y' &= -2(y-1)u + y(y-1)\phi - 2y(f+g), \\ yu' &= -(y-1)u^2 - 2yug + [(y-1)/4]\psi^2. \end{aligned}$$

Elimination of y gives a DE for $u(x)$ which is equivalent to the Riccati DE:

$$u' + u^2 + (2g - \phi)u = (1/4)\psi^2 + v,$$

where $v' = 2(f-g)v$. Kamke (1959), p.588; Murphy (1960), p.403

30. $$3y(y-1)y'' - 2(2y-1)y'^2 + f(y) = 0.$$

The given DE is a special case of Eqn. (2.39.38). If $f = 0$, a first integral is found to be

$$y'^3 = Cy^2(y-1)^2;$$

C is an arbitrary constant. Now the variables separate. Kamke (1959), p.588; Murphy (1960), p.403

31. $$ay(y-1)y'' - (a-1)(2y-1)y'^2 + f(x)y(y-1)y' = 0,$$

where a is a constant.

A first integral is

$$y' = Cy^{1-(1/a)}(y-1)^{1-(1/a)} \exp\left\{(-1/a)\int f dx\right\};$$

C is an arbitrary constant. Now the variables separate. Kamke (1959), p.589

32. $$y(y-1)y'' - u_1(y)y'^2 - u_3(y)y' - u_5(y) = 0,$$

where u_1, u_3, and u_5 are algebraic polynomials of degree 1, 3, and 5, respectively.

The given DE can be reduced to a "subuniform" one with no movable algebraic singularities, provided that some conditions are satisfied by the coefficients. Forsyth (1959), p.304

33. $$aby(y-1)y'' - [(2ab-a-b)y + (1-a)b]y'^2 + f(x)y(y-1)y' = 0,$$

where a and b are constants.

A first integral is

$$y' = Cy^{1-(1/a)}(y-1)^{1-(1/b)} \exp\left\{(-1/ab)\int f dx\right\};$$

C is an arbitrary constant. Now the variables separate. Kamke (1959), p.589

34. $$(y^2 - 2\alpha y + 2)y'' + (y - 2\alpha)y'^2 = 0,$$

where α is a parameter.

Interchanging the variables y and x, we get

$$(y^2 - 2\alpha y + 2)\frac{d^2x}{dy^2} + (2\alpha - y)\frac{dx}{dy} = 0$$

or

$$(y-\lambda_1)(y-\lambda_2)\frac{d^2x}{dy^2} + (2\alpha - y)\frac{dx}{dy} = 0, \tag{1}$$

where λ_1 and λ_2 are roots of

$$\lambda^2 - 2\alpha\lambda + 2 = 0, \tag{2}$$

which we assume to be real so that $\alpha^2 > 2$. Now put $\tau = (y-\lambda_1)/(\lambda_2-\lambda_1)$ in (1) to obtain

$$\tau(1-\tau)\frac{d^2x}{d\tau^2} - \{(2\alpha-\lambda_1)/(\lambda_2-\lambda_1) - \tau\}\frac{dx}{d\tau} = 0. \tag{3}$$

Since $2\alpha = \lambda_1 + \lambda_2 = \lambda_1 + 2/\lambda_1$ [see (2)], we write (3) as

$$\tau(1-\tau)\frac{d^2x}{d\tau^2} + \{2/(\lambda_1^2 - 2) + \tau\}\frac{dx}{d\tau} = 0, \tag{4}$$

which is a hypergeometric equation having the solution

$$\begin{aligned} x &= A_1 + B_1\{(y-\lambda_1)/(2\lambda_1^{-1} - \lambda_1)\}^{(4-\lambda_1^2)/(2-\lambda_1^2)} \\ &\quad \times 2F_1\{(4-\lambda_1^2)/(2-\lambda_1^2), \lambda_1^2/(2-\lambda_1^2), (6-2\lambda_1^2)/(2-\lambda_1^2); \\ &\quad (y-\lambda_1)/(2\lambda_1^{-1} - \lambda_1)\}, \end{aligned} \tag{5}$$

where A_1 and B_1 are arbitrary constants. Using the formula for the incomplete beta function, namely

$$\beta(p, q, x) = (x^p/p)2F_1\begin{pmatrix} p & 1-q \\ p+1 & x \end{pmatrix},$$

we write (5) as

$$x = A_1 + B_2\beta\{(4-\lambda_1^2)/(2-\lambda_1^2), (2-2\lambda_1^2)/(2-\lambda_1^2), (y-\lambda_1)/(2\lambda_1^{-1} - \lambda_1)\}, \tag{6}$$

where B_2 is an arbitrary constant.

If we have the boundary conditions $x = 0, y = \lambda_1$ and $x = 1, y = \lambda_2 = 2/\lambda_1$, then $A_1 = 0$, and

$$B_2^{-1} = \beta\{(4-\lambda_1^2)/(2-\lambda_1^2), (2-2\lambda_1^2)/(2-\lambda_1^2), 1\}$$

and the solution becomes

$$x = \beta\left\{\frac{4-\lambda_1^2}{2-\lambda_1^2}, \frac{2-2\lambda_1^2}{2-\lambda_1^2}, \frac{y-\lambda_1}{2\lambda_1^{-1} - \lambda_1}\right\} \Big/ \beta\left\{\frac{4-\lambda_1^2}{2-\lambda_1^2}, \frac{2-2\lambda_1^2}{2-\lambda_1^2}, 1\right\}.$$

Sachdev and Philip (1986)

35. $$\frac{d^2}{dx^2}\left[\sum_{k=0}^{3} b_k y^k\right] + y - r = 0,$$

where b_k $(k = 0, \ldots, 3)$ and r are constants.

The general solution of the given DE is

$$x - \int (b_1 + 2b_2 y + 3b_3 y^2) h(y)\, dy = A,$$

where

$$h(y) = [2r(b_1 y + b_2 y^2 + b_3 y^3) - (b_1 y^2 + 4b_2 y^3/3 + 3b_3 y^4/2) + C]^{-1/2}$$

and A and C are arbitrary constants. To prove this result, we put $dy/dx = v$; then the given DE becomes

$$v(b_1 + 2b_2 y + 3b_3 y^2)\frac{dv}{dy} + 2(b_2 + 3b_3 y)v^2 + y = r. \tag{1}$$

Writing $v^2 = t$ in (1), we have

$$\frac{dt}{dy} + \frac{4(b_2 + 3b_3 y)}{b_1 + 2b_2 y + 3b_3 y^2}\, t = \frac{2(r - y)}{b_1 + 2b_2 y + 3b_3 y^2}. \tag{2}$$

Equation (2) becomes exact by multiplying it with $(b_1 + 2b_2 y + 3b_3 y^2)^2$. Thus, on doing this and integrating, we have

$$\frac{dy}{dx} = \frac{[2r(b_1 y + b_2 y^2 + b_3 y^3) - (b_1 y^2 + 4b_2 y^3/3 + 3b_3 y^4/2) + c_1]^{1/2}}{b_1 + 2b_2 y + 3b_3 y^2}, \tag{3}$$

where C_1 is a constant of integration.

Separating the variables in (3) and integrating, we get the desired result. Koehler and Pignani (1962)

36. $\boldsymbol{(a_0 y^2 + a_1 y + a_2) y'' + b_1 y'^2 + (c_0 y^2 + c_1 y + c_2) y' + d_0 y^3 + d_1 y^2 + d_2 y + d_3 = 0,}$

where a_i, b_1, c_i, and d_i are constants.

It is shown that the given DE has a first integral of the type $y' = \mu(x) y + \nu(x)$ if (for $b_1 \neq 0$)μ is a constant given by $a_0\mu^2 + c_0\mu + d_0 = 0$. In addition, the coefficients satisfy the condition $a_0 = 0, c_0 = 0, a_1 = 0$ and $b_1\mu^2 + c_1\mu + d_1 = 0, 2b_1\mu + c_1 = 0, a_2\mu^2 + c_2\mu + d_2 = 0$. Roth (1941)

37. $\boldsymbol{(a_0 y^2 + a_1 y + a_2) y'' + b_0 y y'^2 + (c_0 y^2 + c_1 y + c_2) y' + d_0 y^3 + d_1 y^2 + d_2 y + d_3 = 0,}$

where a_i, b_0, c_i, and d_i are constants.

It is proved that the given DE has a first integral of the form $y' = \mu(x) y + \nu(x)$, where either μ is a constant or else b_0 is zero; and in all cases the general solution of the equation is obtained by quadratures. Roth (1941)

2.38 $f(y)y'' + g(x, y, y') = 0$, $f(y)$ cubic

1. $\boldsymbol{y^3 y'' + 1 = 0.}$

The asymptotic behavior of the given DE as $x \to \infty$ is

$$y(x) \sim bx + c - [1/(2b^3 x)] + [c/(2b^4 x^2)] + \cdots, \tag{1}$$

where $b = 0$ and c are arbitrary constants. This is just the binomial series whose sum has the form

$$y(x) = \pm[\{(C_1 x + C_2)^2 - 1\}/C_1]^{1/2}, \tag{2}$$

where $C_1 \neq 0$ and C_2 are arbitrary constants. Thus, (2) is an exact solution. Another exact solution is

$$y(x) = \pm[2(x + a)]^{1/2}, \tag{3}$$

where a is an arbitrary constant. The solution (3) may be regarded as the singular limit of (2) in which $C_1 \to 0$ and $C_2^2 \to 1$ in such a way that the ratio $(C_2^2 - 1)/2C_1 = a$ remains fixed. Bender and Orszag (1978), p.154

2. $\boldsymbol{y^3 y'' - 2k^{4/3} x^{4n+6} = 0,}$

where k is an arbitrary constant and n is an integer.

One exact solution is

$$y = k^{1/3}x^{(n+2)}, n = 0, -3.$$

Herrera (1984)

3. $$\boldsymbol{y^3y'' + ky - h^2 = 0,}$$

where h and k are constants.

Put $y' = u$ so that $y'' = u(du/dy)$. Two integrations yield

$$\begin{aligned} x &= c_2 \pm (1/c_1)(c_1y^2 + 2ky - h^2)^{1/2} \mp (k/c_1^{3/2}) \log[y + k/c_1 \\ &\quad + \{(c_1y^2 + 2ky - h^2)^{1/2}/c_1\}^{1/2}], \end{aligned}$$

where c_1 and c_2 are arbitrary constants. $y = h^2/k$ is another solution. Tenebaum and Pollard (1963), p.504

4. $$\boldsymbol{y^3y'' + \{A/(4 - x^2)^{1/2}\}y + B = 0,}$$

where A and B are constants.

The problem for different A and B is solved numerically with IC $y(x_0) = 1, y'(x_0) = 0$. Gromov and Nakaryakov (1992)

5. $$\boldsymbol{y^3y'' - A_1y^4 - A^2 = 0,}$$

where A_1 and A are constants.

The given DE arises in the study of nonlinear elastic layer being nonuniformly extended and then twisted. Writing the given DE as $y'' = A_1 + A^2/y^3$, multiplying by y', and integrating, we have $(y')^2 - A_1y^2 + A^2y^{-2} = A_2$, where A_2 is a constant. Solving for y', we have

$$y' = \pm[A_1y^2 + A_2 - A^2y^{-2}]^{1/2}. \tag{1}$$

Introducing $\xi = y^2$ in (1), we get

$$\pm\frac{1}{2}\frac{d\xi}{\{A_1\xi^2 + A_2\xi - A^2\}^{1/2}} = dx.$$

Writing $\Delta = -4A_1A^2 - A_2^2$, different types of solutions are possible for $\Delta > 0$, $= 0$, or < 0, and for different values of A_1, A_2, and A. For example, for $A_1 < 0, \Delta < 0$, we obtain

$$\pm(1/2)\{-1/(-A_1)^{1/2}\} \sin^{-1}\{(2A_1\xi + A_2)/(-\Delta)^{1/2}\} = x + A_3,$$

where A_3 is another constant. Chao, Rajagopal, and Wineman (1987)

6. $$\boldsymbol{2y^3y'' + y^4 - a^2xy^2 - 1 = 0,}$$

where a is a constant.

Evidently, no integral curve has a point in common with the x-axis. Since along with $y(x), -y(x)$ is also a solution, we may confine ourselves to the half-plane $y > 0$. Every integral curve can be continued on both sides to any arbitrary distance. From the DE, it follows easily that no integral curve can decrease monotonically for $x \to \infty$. There is exactly one integral curve that is concave downward for all sufficiently large x (an exceptional integral). For this there is an asymptotic expansion

$$y \sim x^{1/2}\{a_0 + (a_2/x^2) + (a_4/x^4) + \cdots\}.$$

Every other integral curve has an infinite number of points of inflection and has the property that $y(x) - ax^{1/2}$ has an infinite number of maxima and minima which decrease as $x \to \infty$ and whose abscissas tend to infinity; for suitable constants, we have for this integral

$$y = ax^{1/2} + C_1 \sin[x + (C_1^2/12x^2)\ln x - C_1] + O(1/x^{1/2});$$

C_1 is an arbitrary constant. Kamke (1959), p.590

7. $$\boldsymbol{y'' + Q(x)y - \lambda/y^3 = 0,}$$

where λ is a constant.

It is shown that if $y_1(x)$ and $y_2(x)$ are linearly independent solutions of the homogeneous (linear) DE $y''(x) + Q(x)y = 0$, then the solution of the given DE may be written as

$$y = (Ay_1^2 + By_2^2 + 2Cy_1y_2)^{1/2},$$

where A, B, C are constants such that $AB - C^2 = \lambda/W^2$, where W is the constant Wronskian: $W = y_1y_2' - y_2y_1'$. Eliezer and Gray (1976)

8. $$\boldsymbol{2y^3y'' + y^2y'^2 - ax^2 - bx - c = 0,}$$

where a, b, c are constants.

If $y(x) \neq 0$, introduce $\xi = \int \dfrac{dx}{y(x)}, \eta(\xi) = y(x)$ in the given DE to obtain

$$2\eta\eta'' - \eta'^2 = ax^2 + bx + c. \tag{1}$$

Differentiating (1) twice with respect to x, we have

$$2\eta''' = 2ax + b,\ \eta^{(iv)} = a\eta. \tag{2}$$

We have a linear DE for $\eta(\xi)$. If $\eta(\xi)$ is a solution of this DE, then one gets from the first of (2), x as a function of ξ and hence ξ as a function of x; finally, $y(x) = \eta\{\xi(x)\}$. With the functions so obtained, we may find the solution of the given DE. Kamke (1959), p.591; Murphy (1960), p.404

9. $$\boldsymbol{(2/3)\beta^2\ (\cot\alpha)\ (y^3y'' + 3y^2y'^2) + (w - 2Ky^2)y' = 0,}$$

where α, β, w, and k are constants.

The given DE arises in the traveling wave form of nonlinear waves in thin films. A first integral is immediately found as

$$-wy + (2k/3)y^3 - 4A\beta^2y^3y' = C_1, \tag{1}$$

where $A = (1/6)\cot\alpha > 0$ and C_1 is an arbitrary constant. From (1), we immediately find that

$$[k/(6A\beta^2)](x + K) = y + \frac{3w}{2k}\int \frac{y + C_1/w}{y^3 - [3w/(2k)]y - 3C_1/(2k)}\,dy, \tag{2}$$

where K is another arbitrary constant. The form of the explicit solution depends on the roots of the cubic in the denominator under the integral sign:

(a) $y^3 - [3w/(2k)]y - 3c_1/(2k) = (y-a_1)(y-a_2)^2$.
The solution may be found in the form

$$[k/(6A\beta^2)](x+k) = y + [3w/(2k)]\{\epsilon_1 \ln|y-a_1| + \epsilon_2 \ln|y-a_2| - \epsilon_3/(y-a_2)\},$$

where ϵ_i $(i=1,2,3)$ are the constants arising in the partial fraction decomposition of the integrand.

(b) If there is only one real root a_1, say, so that

$$y^3 - [3w/(2k)]y - 3c_1/(2k) = (y-a_1)\{y^2 + a_1 y + [a_1^2 - 3w/(2k)]\},$$

then the solution may be found in the form

$$\begin{aligned}[k/(6A\beta^2)](x+k) &= y + [3w/(2k)]\{\epsilon_1 \ln|y-a_1| \\ &+ (1/2)\epsilon_2 \ln|y^2 + a_1 y + a_1^2 - 3w/(2k)|\} \\ &+ \frac{\epsilon_3 - (1/2)\epsilon_2 a_1}{[3a_1^2/4 - 3w/(2k)]^{1/2}} \arctan\left[\frac{y + a_1/2}{[3a_1^2/4 - 3w/(2k)]^{1/2}}\right],\end{aligned}$$

where ϵ_i appear in the partial fractions in the integrand.

(c) All the roots of the cubic equation are real and distinct,

$$y^3 - [3w/(2k)]y - 3c_1/(2k) = (y-a_1)(y-a_2)(y-a_3).$$

The explicit solution may be found as

$$[k/(6A\beta^2)](x+K) = y + [3w/(2k)]\{\epsilon_1 \ln|y-a_1| + \epsilon_2 \ln|y-a_2| + \epsilon_3 \ln|y-a_3|\},$$

where ϵ_i are the coefficients appearing in the partial fractions in the integrand.

Melkonian and Maslowe (1990)

10. $$(y^3 + y)y'' - (3y^2 - 1)y'^2 = 0.$$

The given autonomous DE has solutions $y = C$ and $(C_1 x - C_0)(y^2+1) = 1; C, C_1$, and C_0 are arbitrary constants. Kamke (1959), p.590; Murphy (1960), p.404

11. $$ny(1-y^2)y'' + y'^2 + (1-y^2)(ny^2 - 1) = 0,$$

where n is a positive integer.

The given DE describes the support function $y(x)$ of a geodesic in the two-dimensional Riemann manifold O_n^2 with a certain metric. It is proved that any constant solution $y(x)$ such that $y^2 + y'^2 < 1$ is periodic and its period is given by the improper integral

$$T = 2\int_{a_0}^{a_1} \frac{dy}{\{1 - y^2 - C(1/y^2 - 1)\}^{\alpha/2}},$$

where $C = (a_0^2)^\alpha(1-a_0^2)^{1-\alpha} = (a_1^2)^\alpha(1-a_1^2)^{1-\alpha}(0 < a_0 < \alpha^{1/2} < a_1 < 1, \alpha = 1/n)$, is the integral constant of the given DE and $0 < C < A = \alpha^a(1-\alpha)^{1-\alpha}$.

Regarding T, it is also stated, with reference to previous work, that $T > \pi$ and $\lim_{C\to 0} T = \pi$ and $\lim_{C\to A} T = 2^{1/2}\pi$. It is shown that $T < 2^{1/2}\pi$ for any $n \geq 3$ and $1 < n < 3$. Ostuki (1974,1974a)

12. $$(4y^3 - g_2 y - g_3)y'' - (6y^2 - g_2/2)y'^2 = 0,$$

where g_2 and g_3 are constants.

For the given autonomous DE, the solution is found in terms of the elliptic function

$$y = \mathcal{P}(C_1 x + C_0, g_2, g_3);$$

C_1 and C_0 are arbitrary constants. Kamke (1959), p.592

13. $$(4y^3 - g_2 y - g_3)[y'' + f(x)y'] - [6y^2 - (g_2/2)]y'^2 = 0,$$

where g_2 and g_3 are constants.

A first integral is

$$y' = C[4y^3 - g_2 y - g_3]^{1/2} \exp\left(-\int f dx\right);$$

C is an arbitrary constant. Now the variables separate. Kamke (1959), p.592

14. $$2y(1-y)(a-y)y'' + \{y(1-y) + y(a-y) - (1-y)(a-y)\}y'^2 - a_0 y^2(1-y)^2(a-y)^2$$
$$- a_1 y^2(1-y)^2 - a_2 y^2(a-y)^2 - a_3(1-y)^2(a-y)^2 = 0,$$

where a, a_i $(i = 0, 1, 2, 3)$ are constants.

A first integral is

$$y'^2 = [(a_0 y + C)y(1-y)(a-y) + a_1 y(1-y) + a_2 y(a-y) - a_3(1-y)(a-y)];$$

C is an arbitrary constant. The solution may be found in terms of elliptic functions. Murphy (1960), p.404

15. $$2(y-a)(y-b)(y-c)y'' - [(y-a)(y-b) + (y-a)(y-c) + (y-b)(y-c)]y'^2$$
$$+ [(y-a)(y-b)(y-c)]^2 + A + B/(y-a)^2 + C/(y-b)^2 + D/(y-c)^2 = 0,$$

where $a, b, c, A, B, C,$ and D are constants.

The given DE is a special case of Eqn. (2.39.38). Its solution can be expressed in terms of elliptic functions [see Ince (1956), p.343]. Kamke (1959), p.591; Murphy (1960), p.404

2.39 $f(y)y'' + g(x, y, y') = 0$

1. $$y^2(a-y)^2 y'' + \alpha m_1(a-y)^2 - \alpha m_2 y^2 = 0,$$

where $\alpha, m_1, m_2,$ and a are constants.

The given DE describes free flight of a space satellite on the line joining, and between, a planet (mass m_1) and its moon (mass m_2), which are at a fixed distance apart, y is the distance of the satellite from the planet, and α is the gravitational constant.

The equilibrium point at $y = am_1^{1/2}/(m_1^{1/2} + m_2^{1/2})$ may be shown to be unstable according to the linear approximation. Jordan and Smith (1977), p.62

2. $\quad (k^2y^2 - 1)(y^2 - 1)y'' + [(k^2 + 1 - 2k^2y^2)y + a\{(k^2y^2 - 1)(y^2 - 1)\}^{1/2}]y'^2 = 0,$

where $a, k \neq 0$ are constants.

The given DE is a special case of Eqn. (2.16.68). The solution is $y = \text{sn}[(1/a)\log(C_1x + C_0), k]$; C_1 and C_0 are arbitrary constants. Kamke (1959), p.592; Murphy (1960), p.405

3. $\quad (1 + \beta y^2)y^3y'' + \beta y^4y'^2 - Fy^4 - B^2\beta^2(-2\beta y^6 + y^4) - B^2 = 0,$

where β, F, and B are constants.

The given DE arises in the study of nonlinear elastic layer being nonuniformly extended and then twisted.

Writing the given DE as

$$y'' - B^2/y^3 - Fy = B^2\beta^2(-2\beta y^3 + y) - \beta yy'^2 - \beta y^2y'',$$

multiplying it by y', and integrating, we have

$$y'^2 + B^2/y^2 - Fy^2 = B^2\beta^2(-\beta y^4 + y^2) - \beta y^2y'^2 + D, \tag{1}$$

where D is a constant.

With $y^2 = \xi$, Eqn. (1) transforms to

$$\{(1 + \beta\xi)/4\}\xi'^2 = B^2\beta^2\xi^2(1 - \beta\xi) + F\xi^2 + D\xi - B^2. \tag{2}$$

Solving for ξ' and integrating, we have

$$\pm\frac{1}{2}\int\left\{\frac{B^2\beta^2\xi^2(1 - \beta\xi) + F\xi^2 + D\xi - B^2)}{(1 + \beta\xi)}\right\}^{-1/2} d\xi = x + E,$$

where E is a constant. We have an implicit solution, expressed as a quadrature. Chao, Rajagopal, and Wineman (1987)

4. $\quad y^{n-2}[yy'' + (m - 1)y'^2 + f(x)yy'] + \phi(x)y^n = \theta(x),$

where m and n are natural numbers.

To solve the given DE, we introduce the variables

$$x = \alpha(t),\ y = \beta(t)z(t),$$

where $\dfrac{d\alpha}{dt} = \beta^{2m}$. Now the given DE becomes

$$z^{n-2}[z\ddot{z} + (m - 1)\dot{z}^2 + \psi(t)z\dot{z}] + F(t)z^n = \phi(t),\ \cdot = \frac{d}{dt}, \tag{1}$$

where the following relations hold:

$$[f(x)]^{-2}[\phi(x) + u^2/(4m) - u'/(2m) - uf(x)/(2m)]$$

$$= [\{\psi(t)\}^{-2}][F(t) + v^2/(4m) - \dot{v}/(2m) - v\psi(t)/(2m)],$$

$$\theta(x)[f(x)]^{(n-4m)/(2m)} = \phi(t)[\psi(t)]^{(n-4m)/(2m)}, \tag{2}$$

$$u = f'(x)/f(x),\ v = \dot{\psi}(t)/\psi(t),$$

and

$$\int f(x)dx = \int \psi(t)dt.$$

Relations (2) follow from the elimination of $\beta(t)$. Relations (2) lead to the solution of the given DE as

$$y = [\psi(t)/f(x)]^{1/(2m)} z(t, C_1, C_2). \tag{3}$$

Here z is the general integral (1) and C_1 and C_2 are arbitrary constants.

Therefore, to obtain the solution of the given DE, we choose functions $F(t), \phi(t)$, and $\psi(t)$ so that (1) can be solved by quadrature. Then substituting these functions into (2), we obtain two relations among $f(x), \phi(x)$, and $\theta(x)$, for which (3) is a solution of the given DE. One of these three functions may be chosen arbitrarily. It is convenient to specify the function $f(x)$ and then find $\phi(x)$ and $\theta(x)$, since this choice enables us to restrict ourselves to elementary functions. Details are similar to those for Eqn. (2.9.36). Braude (1967)

5.
$$y''y^{1/2} - a = 0.$$

For the given autonomous DE, we put $p(y) = y'(x)$ and get $pp' = ay^{-1/2}$, and so

$$y'(x) = p = 2\{ay^{1/2} + C\}^{1/2};$$

now the variables separate. C is an arbitrary constant. Kamke (1959), p.593; Murphy (1960), p.405

6.
$$y''y^{1/2} - 2(a + bx) = 0,$$

where a and b are constants.

A first integral of the given DE is

$$y'^3 - 12y'(a + bx)y^{1/2} + 8by^{3/2} + 8(a + bx)^3/b = C,$$

where C is an arbitrary constant. Murphy (1960), p.405

7.
$$y^{1/2}y'' + 2xy' + 4\lambda(1 - y) = 0, \lambda < 0, y(0) = 0, y(+\infty) = 1,$$

$$0 < y(x) < 1 \text{ for } 0 < x < \infty.$$

This problem results from the Falkner–Skan problem of boundary layer theory by a certain change of variables. Iwano (1977)

8.
$$y^{\lambda}y'' - x^{-2-\alpha} = 0, x \geq 1,$$

where $\alpha > 0$ and λ are constants.

The given DE has exactly one positive decaying solution

$$y(x) = \left[\frac{(\lambda+1)^2}{\alpha(\alpha+\lambda+1)}\right]^{1/(\lambda+1)} x^{-\alpha/(\lambda+1)}, x \geq 1,$$

provided that $[\lambda/(\lambda+1)]\alpha(\alpha+1) \leq 1/4$. Usami (1991)

9. $$y^\lambda y'' - \{(1+\sin x)/x^{2+\epsilon}\}' = 0,$$

where $x \geq 1, \epsilon > \lambda > 0$ are constants.

It is proved that there exists a positive decaying solution of the given DE. Usami (1991)

10. $$y^\lambda y'' - a(x) = 0, x \geq x_0 \geq 0,$$

where $\lambda > 0$. $a(x) : [x_0, \infty] \to R$ is continuous. (See also the following.)

If it is assumed that

$$C_1 x^{-2-\alpha} \leq a(x) \leq C_2 x^{-2-\beta}, x \geq 1$$

or

$$C_1 e^{-\alpha x} \leq a(x) \leq C_2 e^{-\beta x}, x \geq 1,$$

for some positive constants C_1, C_2, α, and β, and $0 < \beta \leq \alpha < (1+1/\lambda)\beta$, then the given DE has at least one positive decaying solution. Usami (1991)

11. $$y^\lambda y'' - a(x) = 0, x \geq x_0 \geq 0,$$

where $\lambda > 0$. $a(x) : [x_0, \infty] \to R$ is continuous. (See also the following.)

Positive decaying solutions ($y \in C^2[x_0, \infty)$) which satisfy the given DE and the asymptotic condition

$$\lim_{x\to\infty} y(x) = \lim_{x\to\infty} y'(x) = 0$$

are considered.

The function $a(x)$ is assumed to be such that:

(a) It has noncompact support.

(b) $A(x) \equiv \int_x^\infty a(s)ds$ converges for all $x \geq x_0$, and $A(x) \geq 0$ on $[x_0, \infty)$.

Conditions are found so that the given DE has positive decaying solutions. Usami (1991)

12. $$y^\lambda y'' - h(x) = 0, \lambda > 0,$$

with either

$$y(a+) = 0, y(b-) = 0$$

or

$$y(a+) = 0, y(b-) = y(x_0)$$

or

$$y(a+) = 0, y'(b-) = 0,$$

where $-\infty < a < x_0 < b < +\infty$. This is a special case of Eqn. (2.6.20). The function $h(x)$ may have singularities at one or both ends. Lomtatidze (1987)

13. $$y^\alpha y'' - \phi(x) = 0,$$

where $\phi \in C(R_+), R_+ = [0, \infty), \phi(x) > 0$, and $\alpha \in (0,1), y(0) = a > 0$ and there exists $x_0 > 0$ such that $y(x_0) = y'(x_0) = 0$.

It is proved that if

$$\int_0^\infty s\phi(s)ds = +\infty, \tag{1}$$

then the given problem has one solution. It is shown that the condition (1) is sufficient but not necessary to ensure the solution of the given problem.

Choose $x_0 > 0$. Integrating the given DE twice and using IC $y(x_0) = y'(x_0) = 0$, we get an integral equation

$$y(x) = \int_x^{x_0} (s - x)\phi(s)/y^\alpha(s)ds \tag{2}$$

over $x \in [0, x_0)$. It follows from (2) that $y'(x) < 0$ and $y''(x) > 0$ for $x \in [0, x_0)$; hence $y(x) \le y(0)(1 - x/x_0), x \in [0, x_0)$, and using this inequality in (2), we obtain

$$y^{1+\alpha}(0) \ge \int_0^{x_0} (1 - s/x_0)^{-\alpha} s\phi(s)ds. \tag{3}$$

It is then clear that if

$$\lim_{x_0 \to \infty} (1 - s/x_0)^{-\alpha} s\phi(s)ds = +\infty, \tag{4}$$

then for each $a > 0$, there is x_0 such that the given problem has a solution. The condition (4) is also not necessary for the given problem to have a solution for arbitrary $a > 0$. Ad'yutov, Klokov, and Mikhailov (1991)

14. $$y^\alpha y'' - \phi(x) = 0, y(\infty) = h > 0,$$

where $\phi \in (R_+), R_+ = [0, \infty), \phi(x) > 0$, and $\alpha \in (0, 1)$.

It is proved that the given problem has a unique solution for arbitrary $h > 0$ provided that $\int_0^\infty s\phi(s)ds < \infty$. An iterative method is used. Ad'yutov, Klokov, and Mikhailov (1991)

15. $$y^m y'' + m y^{m-1} y'^2 + y/m = 0,$$

where m is a constant.

The given DE arises in the similarity solutions for nonlinear diffusions. Multiplying by $2y^m y'$ and integrating, we have a first integral

$$y'^2 y^{2m} = C - 2y^{m+2}/[m(m+2)], \tag{1}$$

from which we deduce that

$$\int^{y(x)} \frac{\eta d\eta}{[C - 2\eta^{m+2}/(m(m+2))]^{1/2}} = \pm(x - x_0), \tag{2}$$

C and x_0 are arbitrary constants in (1) and (2). Hill (1989)

16. $$[(-y)^{1/n}]'' - a^2(\delta y + rxy') = 0,$$

where n, a, δ, and r are constants; here $2r = 1 + \delta(1 - 1/n)$.

The given DE arises in the description of nonlinear effects in a non-Newtonian fluid flow. For $\delta = r = n/(1+n)$, we get

$$\{(-y)^{1/n}\}'' - \{na^2/(1+n)\}(xy)' = 0. \tag{1}$$

On integration, (1) gives $[(-y)^{1/n}]' = c + \{na^2/(1+n)\}xy$, where c is a constant. For $c = 0$, the solution is

$$y = -[c_1 - \{n(1-n)/2(n+1)\}a^2x^2]^{n/(1-n)},$$

where c_1 is another constant. Pascal (1991b)

17. $$my^{m-1}y'' + m(m-1)y^{m-2}y'^2 + cy' + (1/m)y^n = 0,$$

where m and c are constants.

The given DE arises in the traveling wave form of reaction-diffusion equations. Putting $X = y, Y = -(y^m)'$, the given DE is written as

$$\frac{dX}{dx} = -\frac{1}{m}X^{1-m}Y, \frac{dY}{dx} = -(c/m)X^{1-m}Y + (1/m)X^n. \tag{1}$$

The trajectories of the system are solutions of

$$\frac{dY}{dX} = c - X^{m+n-1}Y^{-1}. \tag{2}$$

Admissible trajectories, i.e., solutions of (2), are sought for $X \geq 0$ such that the solution $(X(x), Y(x))$ of (1) is defined for every $x \in R$ and Y vanishes wherever X does. The given DE is invariant under $x \to -x, c \to -c, y \to y$. Phase plane analysis is carried out for $m + n = 2, m + n > 2$, and $m + n < 2$. For $c = 2$, a family of trajectories of (2) for $m + n = 2$ is obtained as $Y - X = k \exp\{X/(Y - X)\}$, where k is a constant. The results are summarized as follows. The given DE allows ("admissible" as noted above) solutions only if $m + n = 2$, in which case there exists a critical value $c^* = 2$, such that:

(a) There exist no solutions for $0 < c < c^*$.

(b) There exists one solution for $c = c^*$.

(c) There exist infinitely many solutions for $c > c^*$.

These solutions are finite and their profiles are decreasing functions $y : R \to [0, \infty)$. Pablo and Vazquez (1991)

18. $$amy^{m-1}y'' + am(m-1)y^{m-2}y'^2 - cy' - by^n = 0, a, b > 0,$$

where m and n are arbitrary parameters.

The given DE describes traveling thermal waves in an absorbing medium. It is proved that a unique solution of the given DE [in the sense of distribution in $(0, \infty)$] satisfying the conditions $y(x) > 0$ for $x > 0$ and $y(0) = 0, (y^m)'(0) = 0$ exists for some choices of the parameters. Detailed analysis is carried out for different values of m and n, as $x \to 0$ and $x \to \infty$. Herrero and Vazquez (1988)

19. $$my^{m-1}y'' + m(m-1)y^{m-2}y'^2 + cy' + (1/m)(1-y)y^n = 0,$$

where m and c are constants; here $0 \leq y \leq 1, y(-\infty) = 1, y(+\infty) = 0$.

The sought solutions may be finite, i.e., $y(x) = 0$ for $x \geq x_0$ or $y(x) = 1$ for $x \leq x_1$, or positive. $[y(x) > 0$ for all $x \in R]$.

Introducing the variables $X = y, Y = [-\{m/(m-1)\}y^{m-1}]'$, $d\tau = (1/m)X^{1-m}dx$, we write the given DE as the nonsingular system

$$\frac{dX}{d\tau} = -XY, \ \frac{dY}{d\tau} = X^p(1-X) - Y(c-Y), p = m+n-2, \tag{1}$$

with critical points $(X,Y) = (0,0),(1,0),(0,c)$. The admissible trajectories are sought from

$$\frac{dY}{dX} = (c-Y)/X - X^{p-1}(1-X)/Y$$

for $0 \leq X \leq 1$, joining the points $(1,0)$ and $(0,B)$ for some $0 \leq B < \infty$. Phase portraits are clearly drawn. The results may be summarized as follows. The given DE allows admissible solutions as defined above with $m > 1, \lambda = 1/m$ only if $p = m+n-2 \geq 0$. Moreover, for $p \geq 0$, there exists a critical speed $c = c^*(p) > 0$, such that:

(a) There exist no solutions for $0 < c < c^*$.

(b) There exists a unique solution for each $c \geq c^*$.

The solution with $c = c^*$ is finite [i.e., $y(x) = 0$ for $x \geq x_0$ or $y(x) = 1$ for $x \leq x_1$] if and only if $n < 1$. Finally, $c^*(p)$ decreases as p increases. In particular, $c^*(0) = 2$, $c^*(1) = 1/2^{1/2}, c^*(p) \to 0$ as $p \to \infty$. Pablo and Vazquez (1991)

20. $$y^n y'' + ny^{n-1}y'^2 + \{1/(2(n+1))\}(2y + xy') = 0,$$

where n is a parameter.

Writing $y = dh/dx$ and integrating, we have

$$\left(\frac{dh}{dx}\right)^n \frac{d^2h}{dx^2} + \frac{1}{(2(n+1))}\left(h + x\frac{dh}{dx}\right) = 0, \tag{1}$$

where we have put the constant of integration equal to zero.

Integration of (1) gives

$$\frac{1}{n+1}\left(\frac{dh}{dx}\right)^{n+1} + \frac{1}{2(n+1)}xh - \alpha = 0,$$

where α is an arbitrary constant. Choosing $\alpha = 0$, we obtain an explicit solution:

$$\begin{aligned} h &= -2[\{n/(2(n+2))\}\{\beta - x^{(n+2)/(n+1)}\}]^{(n+1)/n}, n \neq 0, -2, \\ h &= -2\beta e^{-x^2/4}, n = 0, \\ h &= [-4\ln(x/\beta)]^{1/2}, n = -2 \end{aligned}$$

King (1990)

21. $$y^m y'' + my^{m-1}y'^2 + (x/a)y' - (\alpha + 2/(ma))y = 0,$$

where m, α, and a are constants.

It is easy to find the first integral of the given DE as

$$y^m y' + xy/a = C_1, \tag{1}$$

provided that $a = -(1/\alpha)(1 + 2/m)$, where C_1 is an arbitrary constant. If $C_1 = 0$ in (1), we can integrate the latter to yield

$$y(x) = \{C_2 - mx^2/(2a)\}^{1/m},$$

where C_2 is another arbitrary constant. For $C_1 \neq 0$, (1) can be integrated for special values of m.

(a) $m = -1/2$; we put $\psi = y^{-1/2}$ in (1) to obtain

$$\frac{d\psi}{dx} = x/2a - (1/2)C_1\psi^2, \tag{2}$$

which, with successive substitutions,

$$\psi = (2/(C_1 u))\frac{du}{dx}, x = k\eta$$

yields the Airy equation

$$\frac{d^2u}{d\eta^2} + \eta u = 0, \tag{3}$$

provided that $K^3 = -4a/C_1$. Equation (3) has the general solution

$$u(\eta) = C_3\eta^{1/2}J_{1/3}((2/3)\eta^{3/2}) + C_4\eta^{1/2}J_{-1/3}((2/3)\eta^{3/2}),$$

where C_3 and C_4 denote arbitrary constants. Thus, altogether, we have

$$\frac{1}{(y(x))^{1/2}} = \frac{2}{C_1 x}\left[\frac{1}{2} + \eta^{3/2}\frac{J'_{1/3}((2/3)\eta^{3/2}) + C_2 J_{-1/3}((2/3)\eta^{3/2})}{J_{1/3}((2/3)\eta^{3/2}) + C_2 J_{-1/3}((2/3)\eta^{3/2})}\right],$$

$x = k\eta, k^3 = 2\alpha a, C_2 = C_4/C_3$ is the additional arbitrary constant.

(b) $m = -1, a = 1/\alpha$, so that (1) can be solved as a Bernoulli equation:

$$y(x) = \{C_2 e^{C_1 x} + (C_1 x - 1)/aC_1^2\}^{-1},$$

where C_2 is another arbitrary constant.

(c) $m = -2$. In this case the given DE integrates to (1) only if $\alpha = 0$. Introducing the new variable $\psi = xy$, we get the separable equation

$$x\frac{d\psi}{dx} = -\frac{\psi}{a}(\psi^2 - aC_1\psi - a),$$

which can easily be integrated.

Also, if we multiply the given DE by x, we may easily deduce another first integral

$$xy^m y' + x^2 y/a = C_5 + y^{m+1}/(m+1), \tag{4}$$

provided that $a = -(2/\alpha)(1 + 1/m); C_5$ is an arbitrary constant. If $C_5 = 0$ in (4), we can integrate the latter to give

$$y(x) = \{C_6 x^{m/(m+1)} - [m(m+1)/a(m+2)]x^2\}^{1/m}.$$

If $C_5 \neq 0$, we can integrate (4) for special values of m. For example, if $m = -1/2$, then the substitution $\psi = y^{1/2}$ in (4) gives the Riccati equation

$$x\frac{d\psi}{dx} = (1/2)C_5 + \psi - x^2\psi^2/2a,$$

which, with the successive substitutions

$$\psi = \frac{2a}{xu}\frac{du}{dx}, \quad x = k\eta \text{ with } k^2 = 4a/C_5$$

yields the linear equation

$$\frac{d^2u}{d\eta^2} - \frac{2}{\eta}\frac{du}{d\eta} - u = 0,$$

with the general solution

$$u(\eta) = \eta^{3/2}[C_7 I_{3/2}(\eta) + C_8 I_{-3/2}(\eta)].$$

Altogether, we have the solution

$$y(x) = \{C_5^2/4\eta^2\}\left[\frac{3}{2\eta} + \frac{I'_{3/2}(\eta) + C_9 I'_{-3/2}(\eta)}{I_{3/2}(\eta) + C_9 I_{-3/2}(\eta)}\right]^2,$$

where $C_9 = C_8/C_7$ and prime denotes differentiation with respect to $\eta = (C_5/4a)^{1/2}x$. $C_5 \cdots C_9$ are arbitrary constants. Hill, Avagliano, and Edwards (1992)

22. $\quad \boldsymbol{my^{m-1}y'' + m(m-1)y^{m-2}y'^2 + [1/(m+1)](y + xy') = 0,}$

where $\int_{-\infty}^{\infty} y dx = 1$ and $y(\infty) = 0, m$ is a constant.

The given problem arises from calculation of the current distribution in a superconductor suddenly charged with current. A first integral is

$$(y^m)' + xy/(m+1) = 0, \tag{1}$$

if the constant of integration is chosen to be zero. An integration of (1) gives the solution as

$$y = [(m-1)/\{2m(m+1)\}(x_0^2 - x^2)]^{1/(m-1)}$$

for $0 < x < x_0$ and $y = 0$ for $x > x_0$. The constant x_0 is chosen by the source condition $\int_{-\infty}^{\infty} y dx = 1$. Dresner (1983), p.113

23. $\quad \boldsymbol{y^{n-m}y'' + (n-m)y^{n-m-1}y'^2 - ny^{n-1}y' + b(xy' + y) = 0,}$

$y \to 0$ as $|x| \to \infty$. Here n, m, and b are constants; $n > 0, m - n > 0$.

The given DE is exact and has the first integral

$$y^n - y'y^{n-m} - xy/(2 + n - m) = K, \tag{1}$$

where K is a constant, provided that $b = 1/(2+n-m)$. To satisfy the end conditions, we must choose $K = 0$ so that (1) may be rewritten as

$$y' = y^m - xy^{1+m-n}/(2+n-m). \tag{2}$$

For $m = 2n-1 (m > 1)$, we write (2) as

$$y' = y^{2n-1} - xy^n/(3-n). \tag{3}$$

Introducing the transformation

$$y = \{\alpha/(w + \beta x^2(w))\}^{1/(n-1)},$$

with $\alpha = [1/\{2(1-n)(3-n)\}]^{1/3}$, $\beta = (\alpha/2)(n-1)/(3-n)$, where $1 < n < 3$, into (3), we get

$$\frac{dx}{dw} = -w/(n-1)\alpha^2 - [1/\{2\alpha(3-n)\}]x^2. \tag{4}$$

Equation (4) is Riccati and is linearized by

$$x = 2\alpha(3-n)(1/h)\,\frac{dh}{dw}$$

to the Airy equation

$$\frac{d^2h}{dw^2} - hw = 0,$$

with the general solution

$$h(w) = C_1 \operatorname{Ai}(w) + C_2 \operatorname{Bi}(w),$$

where C_1 and C_2 are arbitrary constants and Ai and Bi are Airy functions. Thus,

$$x(w) = 2\alpha(3-n)\{C \operatorname{Ai}'(w) + \operatorname{Bi}'(w)\}/\{C \operatorname{Ai}(w) + \operatorname{Bi}(w)\},$$

where $C = C_1/C_2$. To satisfy the condition $y \to 0$ as $|x| \to \infty$, we choose, for given C, roots w_i $(i = 1, 2, \ldots)$, each satisfying

$$C = -\operatorname{Bi}(w_i)/\operatorname{Ai}(w_i). \tag{5}$$

The solutions of (5) are computed. It is then concluded that for $C > 0$, the solution is possible only if $w < 0$. For $C < 0$, the domain of validity in (x, w) plane also extends to positive values of w: $0 < w < w_1$, where w_1 is the largest and the only positive root of (5).

The problem arises from the similarity reduction of a nonlinear diffusion-convection equation. Pistiner, Shapiro, and Rubin (1991)

24. $$\boldsymbol{\alpha y^{\alpha-1}y'' + \{(\alpha-1)\alpha\}y^{\alpha-2}y'^2 - (1/2)e^{-2x}y' = 0,}$$

$x \in [0, +\infty), y(0) = 0, \lim_{x\to 0+} [y\alpha(x)]' = 0$, where $\alpha > 1$ is a parameter.

Using the integral equation method, it is shown that the given problem has a unique solution $y(x)$: $[0, +\infty) \to [0, +\infty), y$ is continuous, $y(0) = 0$ and $y(x) > 0$ for $x > 0$. Okrasinski (1989)

25. $$\boldsymbol{\frac{d^2}{dx^2}\left[\sum_{k=0}^{n} b_k y^k\right] + p\frac{d}{dx}\left[\sum_{k=0}^{n} b_k y^k\right] + q\frac{dy}{dx} + y - r = 0,}$$

where b_k, p, q, and r are constants.

It is shown that if $pq = 1$ and $b_n \neq 0$, a general solution is given by $x - p\int h(y)dy = A$, where $h(y) = \left[\sum_{k=1}^{n} kb_k y^{k-1}\right](r-y)^{-1}$ and A is an arbitrary constant of integration.

We indicate a few steps of the proof. Putting $\dfrac{dy}{dx} = \nu$, we have

$$\frac{d}{dx}\left[\sum_{k=0}^{n} b_k y^k\right] = \left[\sum_{k=1}^{n} kb_k y^{k-1}\right]\nu$$

and

$$\frac{d^2}{dx^2}\left[\sum_{k=0}^{n} b_k y^k\right] = \nu^2\left[\sum_{k=1}^{n} k(k-1)b_k y^{k-2}\right] + \nu\left[\sum_{k=1}^{n} kb_k y^{k-1}\right]\frac{d\nu}{dy}.$$

The given DE becomes

$$\frac{d}{dy}\left[\nu\sum_{k=1}^{n} kb_k y^{k-1}\right] + p\left[\sum_{k=1}^{n} kb_k y^{k-1}\right] + q = (r-y)\nu^{-1}. \tag{1}$$

Introducing z in the form $z = \left[p\sum_{k=1}^{n} b_k y^k + qy + B_1\right] + \nu\left[\sum_{k=1}^{n} kb_k y^{k-1}\right]$, where B_1 is a constant, we change (1) to

$$\frac{dz}{dy}\left[z - \left(p\sum_{k=1}^{n} b_k y^k + qy + B_1\right)\right] = (r-y)\sum_{k=1}^{n} kb_k y^{k-1}. \tag{2}$$

It is possible to determine a polynomial solution of Eqn. (2) in the form

$$z = rq + p\sum_{k=1}^{n} b_k y^k + B_1, \tag{3}$$

from which the solution of the given DE can be obtained as follows:

$$\nu = q(r-y)/\sum_{k=1}^{n} kb_k y^{k-1},$$

or

$$\frac{dy}{dx} = q(r-y)/\sum_{k=1}^{n} kb_k y^{k-1},$$

so that $x - p\int\left[\left(\sum_{k=1}^{n} kb_k y^{k-1}\right)/(r-y)\right]dy = A$, where A is a constant; here we use the relation $pq = 1$. Koehler and Pignani (1962)

26. $$\frac{d^2}{dx^2}\left[\sum_{k=0}^{3} b_k y^k\right] + p\frac{d}{dx}\left[\sum_{k=0}^{3} b_k y^k\right] + q\frac{dy}{dx} + y - r = 0,$$

where b_k, p, q, and r are constants.

Putting $z = \left[p\sum_{k=1}^{3} b_k y^k + qy + B_1\right] + \nu\sum_{k=1}^{3} kb_k y^{k-1}$, where B_1 is a constant and $\nu = \dfrac{dy}{dx}$, we change the given DE to

$$\frac{dz}{dy}\left[z - \left(p\sum_{k=1}^{3} b_k y^k + qy + B_1\right)\right] = (r-y)\sum_{k=1}^{n} kb_k y^{k-1}. \tag{1}$$

Assuming that $z = \sum_{k=0}^{n} c_k y^k, n > 3$ and substituting in (1), we have

$$\left[\sum_{k=1}^{n} kc_k y^{k-1}\right]\left[\sum_{k=0}^{3} c_k y^k - \left(p\sum_{k=1}^{n} b_k y^k + qy + B_1\right)\right] = (r-y)\sum_{k=1}^{3} kb_k y^{k-1}. \tag{2}$$

Equating equal powers of y, etc., in (2), we have

$$\begin{aligned}
c_k &= 0, k = 4, 5, \ldots, n, &(3)\\
3c_3^2 &- 3pb_3c_3 = 0, &(4)\\
5c_2c_3 &- 3pb_2c_3 - 2pb_3c_2 = 0, &(5)\\
4c_1c_3 &+ 2c_2^2 - 3(pb_1 + q)c_3 - 2pb_2c_2 - pb_3c_1 = -3b_3, &(6)\\
3c_0c_3 &+ 3c_1c_2 - 3B_1c_3 - 2(pb_1 + q)c_2 - pb_2c_1 = 3rb_3 - 2b_2, &(7)\\
2c_0c_2 &+ c_1^2 - 2B_1c_2 - (pb_1 + q)c_1 = 2rb_2 - b_1, &(8)\\
c_0c_1 &- B_1c_1 = rb_1.
\end{aligned}$$

From (4), we have $c_3 = pb_3 \neq 0$ or $c_3 = 0 \neq pb_3$ or $c_3 = 0 = pb_3$. Thus three cases arise:

(a) $c_3 = pb_3 \neq 0$. Solving the system (4)-(8), we get $c_0 = rq + B_1$, $c_1 = pb_1, c_2 = pb_2, c_3 = pb_3$. Thus assuming that $pq = 1, p \neq 0$, we get

$$z = rq + p\sum_{k=1}^{3} b_k y^k + B_1. \tag{9}$$

(b) $c_3 = 0, pb_3 \neq 0$ and $c_3 \neq pb_3$. In this case, the system (4)-(8) yields $c_0 = prb_1/3 + B_1, c_1 = 3/p, c_2 = 0$, provided that $b_2 = -3rb_3 \neq 0$ and $2(b_1 + rb_2)p^2 + 3pq - 9 = 0$. There is also a solution of degree 1 for z, namely

$$z = B_1 + prb_1/3 + 3y/p, \tag{10}$$

provided that $p \neq 0, b_2 = -3rb_3 \neq 0$ and $2(b_1 + rb_2)p^2 + 3pq - 9 = 0$.

(c) In this case, $c_3 = pb_3 = 0$. Three subcases arise:

(i) $c_3 = b_3 = 0, p \neq 0$.

(ii) $c_3 = p = 0, b_3 \neq 0$.

(iii) $c_3 = b_3 = p = 0$.

From the preceding discussion one may conclude that no polynomial solution of degree greater than 3 for the DE (1) exists. For the case $n = 3$, one may solve explicitly for cases (a) and (b) the first order DE arising from (9) and (10) by the use of (i), respectively. For example, from (9) and (1), we have

$$\frac{dy}{dx} = \frac{q(r-y)}{b_1 + 2b_2y + 3b_3y^2}.$$

Koehler and Pignani (1962)

27. $$\boldsymbol{Ky^{q+1}y'' - x^q(y')^{q+2} = 0,}$$

where K is a constant, and q is a parameter positive, zero, or negative.

Putting $y = ax^\theta$, we have for $\theta \neq 0, \theta^{q+1} - K(\theta - 1) = 0$. Only these roots will be treated as admissible that have their real parts positive.

For $q > 0$, we put $y = ux^\theta$; the equation takes the form

$$xu'' = pu' + xu'^2 P/K,$$

where P is a quotient function, and the character of the equation depends on the value of p. See Forsyth (1959) for a general discussion. Special cases are:

(a) $u = a$, a constant. The solution is $y = ax^\theta$.

(b) $q = 0$. The original DE becomes $ky'' = y'^2/y$ with the solution

$$y = (ax + b)^{K/(K-1)},$$

a and b being arbitrary: $b = 0$ if $y(0) = 0$.

(c) $q = -1$. The original DE becomes $kxy'' - y' = 0$ with the solution

$$y = ax^{1+1/K} + b,$$

a and b being arbitrary constants. $b = 0$ if $y(0) = 0$.

Forsyth (1959), p.223

28. $$(1 + a^2n^2y^{2n-2})y'' + a^2n^2(n-1)y^{2n-3}y'^2 + k(y + a^2ny^{2n-1}) = 0,$$

where $a \neq 0$ is a fixed number, k is a positive constant, and n is a fixed nonnegative integer.

It is proposed that the solutions of the given DE have to be periodic. Amendola and Manes (1987)

29. $$(2 + \sin y)y'' - (\cos y)y'^2 - 8x^{1/4}y'$$
$$+ (2 + \cos x - 2x \sin x)/(2x^{1/2})(y + |y|^\lambda \operatorname{sgn} y) = 0,$$

where $\lambda > 1$, and $x > \pi/2$.

The given DE is shown to have oscillatory solutions. Grace and Lalli (1990)

30. $$(\ell^2 + r^2 - 2r\ell \cos y)y'' + r\ell \sin y(y')^2 + g\ell \sin y = 0,$$

where g, r, and ℓ are constants.

To get a first order uniform expansion for small but finite y, we expand $\sin y$ and $\cos y$ and retain cubic terms only. We have

$$y'' + w_0^2 y + \alpha_1(y^2y'' + yy'^2) - \alpha_2 w_0^2 y^3 + \cdots = 0,$$

where $w_0^2 = g\ell/(\ell - r)^2, \alpha_1 = r\ell/(\ell - r)^2, \alpha_2 = 1/6$. To get a small parameter in the nonlinear terms, we write $y = \epsilon^{1/2}u$. We thus have $u'' + w_0^2 u + \epsilon[\alpha_1(u^2u'' + uu'^2) - \alpha_2 w_0^2 u^3] + \cdots = 0$. A first uniformly valid approximation for u and hence for y is found to be $y = a\cos\{[1 - (1/8)(2\alpha_1 + 3\alpha_2)a^2]w_0 x + \beta\} + \cdots$. Nayfeh (1985), p.105

31. $$(a \sin^2 y + b)y'' + ay'^2 \sin y \cos y + A(a \sin^2 y + c)y = 0,$$

where $a, b, c,$ and A are constants. Choosing y as the independent variable, one gets a linear DE for

$$u(y) = y'^2 : (a \sin^2 y + b)u' + au \sin 2y + 2Ay(a \sin^2 y + c) = 0.$$

The solution is

$$(a \sin^2 y + b)u = C + (A/4)[a \cos 2y + 2ay \sin 2y - (2a + 4c)y^2];$$

C and A are arbitrary constants. Kamke (1959), p.593; Murphy (1960), p.405

32. $$e^y y'' + e^y y'^2 + (N - 1)e^y y'/x + 1 = 0.$$

The solution is easily found to be

$$e^y = -\{1/(2N)\}x^2 + \alpha + \beta x^{2-N}, \text{ if } N \neq 2$$

$$e^y = -\{1/4\}x^2 + \alpha + \beta \ln x, \text{ if } N = 2;$$

α and β are arbitrary constants. King (1990)

33. $$y(1 - \log y)y'' + (1 + \log y)y'^2 = 0.$$

With $u(x) = \log y$, a simpler DE is obtained. The solutions are

$$y = C \quad \text{and} \quad \log y = (x + C_1)/(x + C_2);$$

$C, C_1,$ and C_2 are arbitrary constants. Kamke (1959), p.593; Murphy (1960), p.406

34. $$(|y|^{m-1}y)'' + m\beta xy' - m\alpha y = 0, m > 1,$$

where m, β, α are parameters.

The given DE arises from a similarity reduction of the porous media equation with sign changes. If we set $\tau = \log|x|$, $u(\tau) = xy'(x)/y(x), v(\tau) = x^2|y(x)|^{1-m}$, the given DE reduces to a quadratic autonomous system in the open upper half (u, v)-plane H^+,

$$\dot{u} = u - mu^2 + \alpha v - \beta uv, \ \dot{v} = v\{2 + (1 - m)u\}, \tag{1}$$

where the dot denotes a derivative with respect to τ. The system (1) is analyzed in the (u, v) phase plane. Hulshof (1991)

35. $$k(y)y'' + k'(y)y'^2 + cy' + g(y) = 0 \text{ on } -\infty < x < \infty, y(-\infty) = 0, y(+\infty) = 1,$$

where $g(y) \equiv 0$ on $[0, \theta]$ and $g(y) > 0$ on $(\theta, 1)$, where $0 < \theta < 1$.

The given DE describes traveling waves in combustion models. Here y is the normalized temperature $0 \leq y \leq 1 \cdot k(y)$ is a C^1 function of y, which is strictly positive, $g(y)$ is a renormalized reaction term such that $g(y) = 0$ on $[0, \theta)$ and $g(y) > 0$ on $(\theta, 1)$, where θ is some ignition temperature $0 < \theta < 1$; moreover, $g(1) = 0; c$ is the unknown mass flux of the wave. Existence of the solution of this nonlinear eigenvalue problem is proved for a bounded domain first and then by allowing the domain to tend to infinity. Berestycki, Nicolaenko, and Scheurer (1985)

36. $$f(y)y'' + f'(y)y'^2 + 2xy' = 0, y(0) = 1, y(\infty) = 0,$$

where $x \geq 0, \theta < y \leq 1, \theta < 0$, and $f(y)$ is positive and continuously differentiable.

The given DE arises in many nonlinear diffusion problems. Introducing the transformation $S(y) = \left\{\int_0^1 f(y)dy\right\}^{-1} \int_0^y f(z)dz$, we reduce the given problem to $S'' + 2xF(S)S' = 0, S(0) = 1, S(\infty) = 0$, where $F(S) = [f\{y(S)\}]^{-1}$. Note that $0 \leq S \leq 1$ for $0 \leq y <\leq 1$. Now, proceed to Eqn. (2.13.43). Lee (1971/72)

37. $$f'(y)f(y)y'' + \{f(y)f''(y) - 2f'^2(y)\}y'^2 = 0.$$

The given DE has solutions $y = c$, $f(y) = 1/(c_1x + c_0)$, where c, c_1, and c_0 are constants. Utz (1978)

38. $$f(y)y'' + \alpha f'(y)y'^2 + g(y) = 0.$$

The given DE is of type of Eqn. (2.16.68) if we divide by $f(y)$. Multiplying the given DE by $\pm|f|^{(2\alpha-1)}y'$, (upper or lower sign being chosen according as $f > 0$ or $f < 0$) and integrate:

$$|f|^{2\alpha}y'^2 \pm 2\int |f|^{2\alpha-1}g dy = C;$$

C is an arbitrary constant. Now the variables separate. Kamke (1959), p.594; Murphy (1960), p.409

39. $$k(y)y'' + k'(y)y'^2 - f(x)y' = 0, x > 0, y(0) = 0,$$

where $f : \overline{R}_+ \to R_+$ is a decreasing C^1 function and that $k : \overline{R}_+ \to R_+$ is a positive C^1 function with $k(0) = 0$.

Assuming that $y(x) > 0$ and $\lim_{x\to 0+} y'(x)k(y(x))$ exists, the problem is reduced to an integral equation whose solution is obtained by a monotone iterative scheme. The uniqueness of the solutions is also proved. Okrasinski (1989a)

40. $$2g(y)y'' - g'(y)y'^2 + g(y)f(x)y' = 0.$$

$y = k$ is a solution. A one-parameter family of integrals is

$$y'^2 = cg(y)\exp\left\{-\int f(x)dx\right\},$$

where c is a constant. Utz (1978)

41. $$f(y)y'' - f'(y)y'^2 + g(x)f(y)y' + h(x)f^2(y) = 0.$$

With $y(x) = u(z)$, where z is a solution of

$$z''(x) + g(x)z' + h(x) = 0,$$

we have

$$z'^2 f(u)u''(z) = z'^2 f'(u)u'^2 + h(x)f(u)(u' - f). \tag{1}$$

A solution of Eqn. (1) is $u'(z) = f(u)$. Murphy (1960), p.406

42. $$f(y)y'' - f'(y)y'^2 - f^2(y)\phi\{x, y'/f(y)\} = 0.$$

With $u(x) = y'/f(y)$, we get the first order DE $u' = \phi(x, u)$. Kamke (1959), p.594; Murphy (1960), p.408

2.40 $h(x)f(y)y'' + g(x, y, y') = 0$

1. $$\boldsymbol{xy^2y'' - a = 0.}$$

Putting $y = xu(x)$, we obtain

$$x^4u'^2 = -2a/u + C,$$

a DE with variables separable; C is an arbitrary constant. Kamke (1959), p.589

2. $$\boldsymbol{xy^2y'' - xyy'^2 - (a - y^2)y' = 0,}$$

where a is a constant.

The given DE is equidimensional with respect to x. With $u = \ln x, dy/du, = p = p(y)$, we have $p\{y^2p' - yp - a\} = 0$. A first integral is

$$2py = Cy^2 - a;$$

C is an arbitrary constant.

Now the variables separate. The solution is

$$C_1y^2 = a + C_2x^{C_1};$$

C_1 and C_2 are arbitrary constants. Murphy (1960), p.403

3. $$\boldsymbol{x^2y^2y'' + 3xy^2y' - (\lambda^2/2)x^{2/3}y^2 + x^2/8 = 0,}$$

$y(0) = 1, \lim_{x\to 0} x^{1/3}y'(x) = 2D/3$, where D is a constant.

The given DE appears in the discussion of displacements and stresses in shallow membrane caps. The change of variables

$$\xi = x^{-2/3}y, \eta = x^{1/3}y'$$

produces the system

$$\begin{aligned}\frac{d\xi}{dx} &= \frac{1}{x}(\eta - (2/3)\xi),\\ \frac{d\eta}{dx} &= \frac{1}{x}\left(-\frac{8\eta}{3} - \frac{1}{8\xi^2} + \frac{\lambda^2}{2}\right),\end{aligned}$$

which in the phase plane becomes

$$\frac{d\eta}{d\xi} = \frac{-8\eta/3 - 1/(8\xi^2) + (\lambda^2/2)}{\eta - (2/3)\xi}. \tag{1}$$

The ICs reduce to

$$\lim_{x\to 0} \xi(x) = \infty \text{ since } \lim_{x\to 0} x^{2/3}\xi(x) = \lim_{x\to 0} y(x) = 1 \text{ and } \lim_{x\to\infty} \eta(\xi) = 2D/3. \tag{2}$$

The solution is discussed in the phase plane, with reference to appropriate singularities. It is concluded that if $\lambda^2 \leq (64/3)^{1/3}$, the given problem has a unique solution. It satisfies (1) and (2). Dickey (1989)

4.
$$\boldsymbol{x^2y^2y'' - (x^2 + y^2)(xy' - y) = 0.}$$

With $y = xu(z), z = \ln x$, and $p = du/dz$, we get

$$p(u^2p' - 1) = 0. \tag{1}$$

Equation (1) has the solution

$$p = C - 1/u;$$

C is an arbitrary constant.

The solution of the given DE is therefore

$$\ln x = C_1(y/x) + C_1^2 \ln(y - C_1x) - C_1^2 \ln x + C_2;$$

C_1 and C_2 are arbitrary constants. $y = C_3x, C_3$ a constant, is another solution. Murphy (1960), p.403

5. $\boldsymbol{x^{3/4}y^2y'' + 2x^{3/4}yy'^2 - \{3/(4x^{1/4})\}y^2y' - y^3/\{x(\ln x)^2\} = (\sin x)/\{x(\ln x)^3\}.}$

It is shown that if the solution of the given DE is not eventually monotonic, $y(x) \to 0$ as $x \to \infty$. Graef and Spikes (1987)

6.
$$\boldsymbol{x(1 + y^2)y'' + 2xyy'^2 + (1 + y^2)y' - (x^{1/3} + x^{1/9})yy'}$$
$$\boldsymbol{+ (1/3)x^{-1/3}y^{1/3} = (1/9)x^{-2/3}, x > 0.}$$

The given DE has an exact nonoscillatory solution $y(x) = x^{1/3}$. Grace and Lalli (1987)

7.
$$\boldsymbol{\{(1 + y^2)/x\}y'' + \{(1 + y^2)/x\}'y' + (2/x^3)(y + y^3) = 0, x > 0.}$$

It is shown that the given DE has oscillatory solutions. Grace, Lalli, and Yeh (1984)

8.
$$\boldsymbol{(1/x)(1 + y^2)y'' + (2/x)yy'^2 - (1/x^2)(1 + y^2)y'}$$
$$\boldsymbol{+ (1/x^2)y' + x^2(\cos x)\sinh y = 0, x \geq x_0 > 0.}$$

The given DE is shown to have oscillatory solutions. Grace and Lalli (1990)

9.
$$\boldsymbol{x^{4/3}(1 + y^2)y'' + 2x^{4/3}yy'^2 + (4/3)x^{1/3}(1 + y^2)y'}$$
$$\boldsymbol{+ [1/(4x)](\sin x)y' + (x^2 \sin x)y^{1/3} = 0, x > 0.}$$

All solutions of the given DE are shown to be oscillatory. Grace and Lalli (1987)

10.
$$\boldsymbol{x^{4/3}(1 + y^2)y'' + 2x^{4/3}yy'^2 + (4/3)x^{1/3}(1 + y^2)y'}$$
$$\boldsymbol{- [1/(4x)](\sin x)y' + (x^2 \sin x)y^{1/3} = 0, x > 0.}$$

All solutions of the given DE are shown to be oscillatory. Grace and Lalli (1987)

11.
$$x^{4/3}(1+y^2)y'' + 2x^{4/3}yy'^2 + (4/3)x^{1/3}(1+y^2)y' + [(\sin x)/(4x)]y' + x^{-1/3}y^{1/3}\exp(\sin y) = 0, x > 0.$$

The given DE is shown to have oscillatory solutions. Grace and Lalli (1987)

12.
$$x^{4/3}(1+y^2)y'' + 2x^{4/3}yy'^2 + (4/3)x^{1/3}(1+y^2)y' - [(\sin x)/(4x)]y' + x^{-1/3}y^{1/3}\exp(\sin y) = 0, x > 0.$$

The given DE is shown to have oscillatory solutions. Grace and Lalli (1987)

13.
$$\left[\frac{x}{1+\sin^2(\log x)}\right](1+y^2)y'' + \left[\frac{x}{1+\sin^2(\log x)}(1+y^2)\right]' y' + \frac{y}{x} = 0, x > 0.$$

The given DE has an exact oscillatory solution $y = \sin(\log x)$. It is shown that this DE is generally oscillatory. Grace, Lalli, and Yeh (1984)

14.
$$\left[\frac{1}{1+\sin^2(\log x)}\right](1+y^2)y'' + \left[\frac{1+y^2}{1+\sin^2(\log x)}\right]' y' + \frac{y'}{x} + \frac{y}{x^2} = 0, x > 0.$$

The given DE has an exact solution $y = \sin(\log x)$. It is shown that all bounded solutions of this DE are oscillatory. Grace, Lalli, and Yeh (1984)

15.
$$x(1-y^2)y'' - 2xyy'^2 + (1-y^2)y' + x^{-1/3}y^{1/3} - x^{-1/3}\sin^{1/3}(\ln x) + (3/x)\sin(\ln x)[\cos^2(\ln x)] = 0, x > 0.$$

The given DE has an exact oscillatory solution $y(x) = \sin(\ln x)$. Grace and Lalli (1987)

16.
$$4x^2(1-y^2)y'' + 4x^2yy'^2 + 4x(1-y^2)y' + y(1-y^2)^2 = 0.$$

If we put $x = e^t$, $\dfrac{dy}{dt} = u(y)$, we get a special case of (2) of Eqn. (2.40.17) with $p = q = s = 4$. A first integral is easily obtained and then the solution is written as

$$\int \frac{d(y^2)}{\{y^2(y^2-1)(y^2-C-1)\}^{1/2}} = \int \frac{dx}{x},$$

and is expressible in terms of elliptic integrals of the first kind. Elementary solutions are obtained when the arbitrary constant $C = 0, -1$. Painlevé analysis is also done briefly. Ferreira and Neto (1992)

17.
$$px^2(1-y^2)y'' + qx^2yy'^2 + sx(1-y^2)y' + y(1-y^2)^2 = 0,$$

where p, q, and s are constants.

The given DE is called a sigma model equation. An analysis of the Painlevé type is carried out. For $q = p = 0$, the given DE is solved exactly:

$$y = \{1 - (x/x_0)^{2/s}\}^{-1/2}, \tag{1}$$

where x_0 is an arbitrary constant.

Introducing the variable t through $x = e^t$, and $\dfrac{dy}{dx} = u(y)$, we have

$$p\frac{du}{dy}u(y) + (s-p)u(y) + \{q/(1-y^2)\}yu^2 + y(1-y^2) = 0. \tag{2}$$

For $s = p$, Eqn. (2) becomes linear in u^2 and hence solved. A first integral of the given DE is written in the form

$$\left(x\frac{dy}{dx}\right)^2 = \{1/(2p-q)\}(1-y^2)^2 + C(1-y^2)^{q/p}, 2p-q \neq 0, \tag{3}$$

where C is an integration constant. Equation (3) may be solved in terms of an elliptic function. For $C = 0$, it has the algebraic solution

$$y(x) = \{(\lambda x)^n - 1\}/\{(\lambda x)^n + 1\},$$

where $n = 2/(2p-q)^{1/2}$ and λ is an arbitrary constant. Another simple solution of (3) is found when $p = q$. Choosing now $C = -1/p$, we obtain the solution of (3) as

$$y = 1/\cos\{(1/p)^{1/2}\ln \lambda_1 x\},$$

where λ_1 is an arbitrary constant. For $(2p-q) = 0$, other form for the first integral must be found. Ferreira and Neto (1992)

18. $$\boldsymbol{px^2(1-y^2)y'' + qx^2yy'^2 + sx(1-y^2)y' + y(1-y^2)^2}$$
$$\boldsymbol{+\, \lambda(4/x)(1-y^2)^2[4x^2y'' + 2xy' + y(1-y^2)] = 0,}$$

where p, q, s, and λ are constants.

Painlevé analysis of the given Sigma-Skyrme equation shows that it has, in general, movable transcendental singularities. Characteristic features of physical interest are observed. It is found that the terms multiplying λ are stronger singularities of the equation than the first four terms. The case $\lambda = 0, 1$ are studied separately [see Eqn. (2.40.17) for the case $\lambda = 0$]. Ferreira and Neto (1992)

19. $$\boldsymbol{(x^2-a^2)(y^2-a^2)y'' - (x^2-a^2)yy'^2 + x(y^2-a^2)y' = 0.}$$

With $u(x) = |y^2 - a^2|^{-1/2}y'$, we get $(x^2-a^2)u^2 = C$, i.e., $(x^2-a^2)y'^2 = C(y^2-a^2)$ a DE with variables separable. Kamke (1959), p.590; Murphy (1960), p.403

20. $\boldsymbol{x^\alpha y^{2n}y'' + 2nx^\alpha y^{2n-1}y'^2 + \alpha x^{\alpha-1}y^{2n}y' - x^\beta y^\gamma - (\alpha-2n-2)x^{\alpha-2n-3} + x^{\beta-\gamma} = 0,}$

where $\gamma > 0$ is an even positive integer, n is a nonnegative integer, and either:

(a) $\alpha > 2n+2$, or

(b) $\beta - \gamma > \max\{-2n-3, -1\}$.

The given DE has an exact solution $y(x) = -1/x$. Graef and Spikes (1987)

21. $\mathbf{[2\cosh(\sin x)/(1+y^2)]y'' + [2\cosh(\sin x)/(1+y^2)]'y' + (\sin x\, e^{-\nu \sin x} y^\nu) = 0,}$

where ν is a constant.

The given DE has a nonoscillatory solution $y = e^{\sin x}$. Grace, Lalli, and Yeh (1984)

22. $$\mathbf{np^2(p+1)c^2x^2y^py'' + np^3(p+1)c^2x^2y^{p-1}y'^2 + np(p+1)(3p+2)c^2xy^py' - y' + n(p+1)^2c^2y^{p+1} = 0,}$$

where p, c, and n are constants.

The given DE results from a similarity reduction of a nonlinear diffusion equation in n spatial variables. Branson and Steeb (1983)

23. $$\mathbf{x^{N-1}(y^ny'' + ny^{n-1}y'^2) + (N-1)x^{N-2}y^ny' + \lambda(Nx^{N-1}y + x^Ny') = 0,}$$

where $\lambda = 1/(nN+2)$, and N is a positive integer, n is a parameter.

The given DE arises from a similarity reduction of a nonlinear diffusion equation.

A first integral of the given DE is

$$x^{N-1}y^ny' + \lambda x^N y + \alpha = 0, \tag{1}$$

where α is a constant. For $\alpha = 0$, we have either $y = 0$ or

$$y = \{(\lambda n/2)(\beta - x^2)\}^{1/n}, n \neq 0,$$

$$y = \beta e^{-\lambda x^2/2}, n = 0.$$

For $\alpha \neq 0$ several special cases arise from (1).

(a) $n = -1$. Putting $y = 1/g$, we have from (1)

$$\frac{dg}{dx} - \alpha x^{1-N} g = \lambda x. \tag{2}$$

When $\alpha \neq 0, N \neq 2$, the solution of (2) is

$$g = \lambda \exp[(\alpha x^{2-N})/(2-N)] \int \eta \exp\{(-\alpha\eta^{2-N})/(2-N)\} d\eta,$$

where x_0 is an arbitrary constant. For $N = 1$, we have $g = \beta e^{\alpha x} - (\lambda/\alpha^2)(1+\alpha x)$. For $N = 3, g = \beta e^{-\alpha/x} + (\lambda x/2)(\alpha + x) + (\lambda\alpha^2/2)e^{-\alpha/x}E_i(-\alpha/x)$, where $Ei(x) = \int_z^\infty (e^{-t}/t)dt$ is the exponential integral. For $N = 2$,

$$g = \beta x^\alpha + \{\lambda/(2-\alpha)\}x^2 \text{ if } \alpha \neq 2,$$

$$g = \beta x^2 + \lambda x^2 \ln x \text{ if } \alpha = 2.$$

(b) $n = -2/N$. In this case (1) is invariant under $x \to \beta x, y \to \beta^{-N}y$; the substitution $y = x^{-N}g, \xi = \ln x$ changes (1) to

$$\frac{dg}{d\xi} = Ng - \alpha g^{2/N} - \lambda g^{1+2/N},$$

so that

$$\int_{g_0}^{g} \frac{dg}{Ng - \alpha g^{2/N} - \lambda g^{1+2/N}} = \xi,$$

where g_0 is an arbitrary constant. The constant solution $g = g_0$ is obtained from $Ng_0 - \alpha g_0^{2/N} - \lambda g_0^{1+2/N} = 0$.

(c) $n = -1/2$. Here we put $y = g^2$ in (1) to obtain the Riccati equation

$$2x^{N-1}\frac{dg}{dx} + \lambda x^N g^2 = -\alpha, \tag{3}$$

which on writing $g = (2/\lambda x q)(dq/dx)$ becomes linear:

$$x\frac{d^2q}{dx^2} - \frac{dq}{dx} + \frac{\lambda\alpha}{4}x^{3-N}q = 0. \tag{4}$$

$N = 4$ corresponds to the case discussed in (2), when Eqn. (4) is Cauchy–Euler. Otherwise, we write

$$\begin{aligned} q &= xp, \\ z &= [(\lambda\alpha)^{1/2}/(4-N)]x^{(4-N)/2}, \end{aligned}$$

to obtain

$$z^2\frac{d^2p}{dz^2} + z\frac{dp}{dz} + \{z^2 - 4/(4-N)^2\}p = 0, \tag{5}$$

which is a Bessel equation with the general solution

$$p = AJ_\nu(z) + BY_\nu(z),$$

where A and B are arbitrary constants and $\nu = 2/(4-N)$. The general solution to (3) is

$$\begin{aligned} g &= (\alpha/\lambda)^{1/2}x^{-N/2}[\beta J_{\nu-1}\{((\lambda\alpha)^{1/2}/(4-N))x^{(4-N)/2}\} \\ &\quad + (1-\beta)Y_{\nu-1}\{((\lambda\alpha)^{1/2}/(4-N))x^{(4-N)/2}\}[\beta J_\nu\{((\alpha\lambda)^{1/2}/(4-N))x^{(4-N)/2}\} \\ &\quad + (1-\beta)Y_\nu\{((\lambda\alpha)^{1/2}/(4-N))x^{(4-N)/2}\}]^{-1}. \end{aligned}$$

King (1990)

24. $$\boldsymbol{x(2-\sin y)y'' - x(\cos y)y'^2 + (2-\sin y)y' + (1/3)y' + x^{-1/3}y^{1/3}}$$
$$\boldsymbol{-(1/x^2)[x + 5/3 - \sin(1/x) - (1/x)\cos(1/x)] = 0, x > 0.}$$

The given DE has an exact nonoscillatory solution $y(x) = 1/x$. Grace and Lalli (1987)

25. $$\boldsymbol{(1/x)(2-\sin y)y'' - (1/x)(\cos y)y'^2 - (1/x^2)(2-\sin y)y^2}$$
$$\boldsymbol{+ (\sin x)y' + x^2\cos(xf(y)) = 0,}$$

$x \geq x_0 > 0$, where f can be any of the following functions:

(a) $f(y) = my; y \in R$ for $m > 0$.
(b) $f(y) = ny + |y|^\alpha \operatorname{sgn} y, y \in R$ for $n > 0$ and $\alpha > 0$.
(c) $f(y) = y\ln^2(\mu + |y|); y \in R$ for $\mu > 1$.
(d) $f(y) = ye^{\lambda|y|}; y \in R$ for $\lambda \geq 0$.
(e) $f(y) = \sinh y; y \in R$.

Then the given DE is shown to have oscillatory solutions. Grace and Lalli (1990)

26.
$$(1/x^{1/2})(2 - \sin y)y'' - (1/x^{1/2})(\cos y)y'^2 - (1/[2x^{3/2}])(2 - \sin y)y'$$
$$+ [(\sin x)/(2x^{3/2})]y' + [(2 + \cos x - x \sin x)/(2x^{3/2})]y$$
$$- (1/x^{1/2})[1 - 1/x - (1/2)\cos x + (\sin x)/x - (x/2)\sin x] = 0, x \geq \pi/2.$$

This forced equation has a nonoscillatory solution $y(x) = x$. Grace and Lalli (1990)

27.
$$x^{1/2}(2 - \sin y)y'' - x^{1/2}(\cos y)y'^2 + (1/[2x^{1/2}])(2 - \sin y)y'$$
$$- x^{-1/2}y' + \{(\sin x)/[x(2 - \sin x)]\}y = 0.$$

It is shown that the given DE has oscillatory solutions. Grace and Lalli (1987)

28.
$$x^{4/3}(2 - \sin y)y'' - x^{4/3}(\cos y)y'^2 + (4/3)x^{1/3}(2 - \sin y)y'$$
$$- [1/(4x)](\cos x)y' + (x^2 \sin x)y^{1/3} = 0, x > 0.$$

All solutions of this equation are shown to be oscillatory. Grace and Lalli (1987)

29.
$$x^{4/3}(2 - \sin y)y'' - x^{4/3}(\cos y)y'^2 + (4/3)x^{1/3}(2 - \sin y)y'$$
$$+ [1/(4x)](\cos x)y' + (x^2 \sin x)y^{1/3} = 0, x > 0.$$

All solutions of the given DE are shown to be oscillatory. Grace and Lalli (1987)

30.
$$x^{4/3}(2 - \sin y)y'' - x^{4/3}(\cos y)y'^2 + (4/3)x^{1/3}(2 - \sin y)y'$$
$$+ [(\cos x)/x]y' + x^{-1/3}y^{1/3}\ln(e + y^2) = 0, x > 0.$$

The given DE is shown to have oscillatory solutions. Grace and Lalli (1987)

31.
$$x^{4/3}(2 - \sin y)y'' - (4/3)(\cos y)y'^2 + (4/3)x^{1/3}(2 - \sin y)y'$$
$$- [(\cos x)/x]y' + x^{-1/3}y^{1/3}\ln(e + y^2) = 0, x > 0.$$

The given DE is shown to have oscillatory solutions. Grace and Lalli (1987)

32.
$$[\{x/(2 - \sin(\sin(\ln x)))\}(2 - \sin y)y']' + y/x = 0, x > 0.$$

The given DE has an oscillatory solution $y(x) = \sin(\ln x)$. Grace and Lalli (1990)

33.
$$\{(1 + \sin^2 y)/x\}y'' + \{(1 + \sin^2 y)/x\}'y' + y^3/x^3 = 0, x > 0.$$

It is shown that the given DE has oscillatory solutions. Grace, Lalli, and Yeh (1984)

34.
$$x^{1/2}(1 + \sin^2 y)y'' + [x^{1/2}(1 + \sin^2 y)]'y' + xy'^3 + y^3/x = 0, x > 0.$$

It is shown that the given DE has oscillatory solutions. Grace, Lalli, and Yeh (1984)

35. $$(e^y/x)y'' + (e^y/x)'y' + [1/(x^2 \ln x)]y = 0, x \geq e^e.$$

The given DE has a nonoscillatory unbounded solution $y(x) = \ln x$. All its bounded solutions are shown to be oscillatory. Grace, Lalli and Yeh (1984)

36. $$[x^{1/2} \ln(e + y^2)]y'' + [x^{1/2} \ln(e + y^2)]'y' - x^{-1/2}y' + (1/x)[(1/x) + \sin x](y + y^3) = 0, x > 0.$$

It is shown that all solutions of the given DE are oscillatory. Grace and Lalli (1987)

37. $$[x(1 + |y|^\alpha)y']' - y'/3 + x^{-2/3}|y|^\alpha \operatorname{sgn} y = 4/(3x^2) + (1 + \alpha)/x^{2+\alpha} + x^{-(2/3)-\alpha}$$

for $1/2 \leq \alpha < 1$ and $x > 0$.

The given DE has an exact (nonoscillatory) solution $y(x) = 1/x$. Grace and Lalli (1987)

38. $$(e^{|y|}/x)y'' + (e^{|y|}/x)'y' - y' + [2\{x(x - 1)\}]\sinh y = 0, x \geq e.$$

The given DE has an exact solution $y = \ln x$ which is nonoscillatory. Grace and Lalli (1990)

39. $$\{e^{|y|}/x\}y'' + \{e^{|y|}/x\}'y' + [2/\{x(x - 1)\}]\sinh y = 0, x \geq \pi/2.$$

It is shown that this undamped DE has oscillatory solutions. Grace and Lalli (1990)

40. $$[(1/x)e^{|y|}y']' - (1/x^3)y' + [(1 + x^2)/(x^4 \ln x)]y = 0.$$

It is shown that all bounded solutions of the given DE are oscillatory. Grace and Lalli (1990)

41. $$x^2 g'(y)y'' + x^2 g''(y)y'^2 + kxg'(y)y' + rg(y) - f(x) = 0, y(x) > 0,$$

where k and r are constants and g is an arbitrary C^2 function.

Put $u = g(y), u' = g'(y)y', u'' = g''(y)y'^2 + g'(y)y''$, etc., in the given DE to obtain the linear Euler equation

$$x^2u'' + kxu' + ru = f(x)$$

with an inhomogeneous term. Utz (1978)

42. $$a(x)\psi(y)y'' + (a(x)\psi(y))'y' + q(x)f(y) = 0.$$

Conditions on the functions $a(x), \psi(y), q(x)$, and $f(y)$ are given such that this DE has oscillatory solutions. Grace, Lalli, and Yeh (1984)

43. $$a(x)f(y)y'' + a(x)f'(y)y'^2 + a'(x)f(y)y' + b(x)g(y) = 0,$$

where f, a, g, b are suitable functions.

The given DE is a generalized Emden–Fowler equation. Necessary and sufficient conditions are found such that all solutions of the given DE are oscillatory, bounded, or decaying to zero as $x \to \infty$, and for the trivial solutions of the equivalent system of the given DE to be globally asymptotically stable. Some properties of amplitudes for oscillatory solutions are also discussed. Liang (1987a)

44. $$a(x)\psi(y)y'' + a(x)\psi'(y)y'^2 + a'(x)\psi(y)y' + p(x)y' + q(x)f(y) = 0,$$

where $p, q : [x_0, \infty) \to R = (-\infty, \infty), \psi, f : R \to R$ are continuous, $a(x) > 0$, and $f(y) > 0$ for $y \neq 0$.

Some oscillation theorems for the given DE are proved. Grace and Lalli (1987)

45. $$a(x)\psi(y)y'' + a(x)\psi'(y)y'^2 + a'(x)\psi(y)y' - q(x)f(y) - r(x) = 0,$$

where $q(x) \geq 0$.

Asymptotic properties of solutions of the given DE are discussed, and illustrated with several examples. Graef and Spikes (1987)

2.41 $f(x,y)y'' + g(x,y,y') = 0$

1. $$(y+x)y'' + y'^2 - y' = 0.$$

The given DE is exact and has the first integral

$$(y+x)y' - 2y = C, \tag{1}$$

C an arbitrary constant. Equation (1) may be integrated by writing it in the linear form

$$\frac{dx}{dy} - x/(2y+C) = y/(2y+C).$$

Kamke (1959), p.575; Murphy (1960), p.395

2. $$2(y-x)y'' - y'^2 + 2y' = 0.$$

The solution of the given DE is

$$y - h = a(x-h)^2, \tag{1}$$

where h and a are arbitrary constants. Equation (1) gives vertical parabolas with vertices on the line $y = x$. Reddick (1949), p.29

3. $$(y-x)y'' - 2y'(y'+1) = 0.$$

This is a special case of Eqn. (2.41.6) and has solutions $y = C$; $y = C - x$; $y = C_1 + C_2/(x - C_1)$; C, C_1, and C_2 are arbitrary constants. Kamke (1959), p.575; Murphy (1960), p.395

4. $$(y-x)y'' + (y'+1)(y'^2+1) = 0.$$

This is a special case of Eqn. (2.41.6). One solution is

$$y + x = C_1,$$

C_1 is an arbitrary constant. A first integral is

$$(y-x)^2(y'^2+1) - C_2^2(y'+1)^2 = 0; \tag{1}$$

C_2 is an arbitrary constant. Equation (1) has the solution

$$(x-C_3)^2 + (y-C_3)^2 = C_2^2.$$

Kamke (1959), p.575; Murphy (1960), p.395

5. $$(y-x)y'' - 2y'\{y' - \alpha y'^{1/2}\} - y' = 0, y(1) = 1, y(0) = 0,$$

where α is a parameter.

Put $p = y - x$ to obtain

$$pp'\frac{dp'}{dp} = 2(1+p')^2 - 2\alpha(1+p')^{3/2} + (1+p'). \tag{1}$$

Now write $N^2 = p' + 1$ and integrate to obtain

$$p = C(a-N)^{(a^2-1)/(a(a-b))}(N-b)^{(b^2-1)/(b(b-a))}N^{-1/(ab)}, \tag{2}$$

where a and b are the roots of $\lambda^2 - \alpha\lambda + 1/2 = 0$ and C is an arbitrary constant. It is easily seen that the BCs on p are $p(1) = 0, p(0) = 0$. Since $N^2 - 1 = dp/dx$, we deduce that

$$x = -C\int (a-N)^{-1/(2a^2-1)}(N - a^{-1}/2)^{2a^2/(2a^2-1)}N^{-3}dN. \tag{3}$$

If we write $U = 2\alpha - N^{-1}$, then we conclude from (2) and BCs $p(1) = 0, p(0) = 0$ that $U = 1/a$ at $x = 0$ and $U = 2a$ at $x = 1$. The solution after some manipulation can be written as

$$\begin{aligned} y &= \frac{\int_0^{U-1/a} z^{2a^2/(2a^2-1)}(2a - a^{-1} - z)^{-1/(2a^2-1)}dz}{\int_0^{2a-1/a} z^{2a^2/(2a^2-1)}(2a - a^{-1} - z)^{-1/(2a^2-1)}dz} \\ &= \frac{\beta\{(4a^2-1)/(2a^2-1), 2(a^2-1)/(2a^2-1), (U-a^{-1})/(2-a^{-1})\}}{\beta\{(4a^2-1)/(2a^2-1), 2(a^2-1)/(2a^2-1), 1\}} \end{aligned}$$

in terms of incomplete beta functions, which are tabulated. Freeman (1972)

6. $$(y-x)y'' + f(y') = 0.$$

One solution of the given DE is $y = ax + C_1$, where a is a root of $f(a) = 0$. The remaining solutions are obtained from

$$(y-x)\phi(y') = C_2,$$

where $\phi(u) = \exp\int [\{(u-1)/f(u)\}du]$. These are obtained by first writing $y - x = z$, so that $zz'' + f(z'+1) = 0$ and then putting $z' = p, z'' = p(dp/dz)$; subsequently, put $p+1 = u$ and integrate. Kamke (1959), p.576; Murphy (1960), p.396

7. $$\boldsymbol{x(y+x)y'' + xy'^2 - (y-x)y' - y = 0.}$$

Writing $u(x) = y + x, v = u'/u$, one gets an equidimensional DE whose solution leads to

$$(y+x)^2 = C_1x^2 + C_2;$$

C_1 and C_2 are arbitrary constants. Alternatively, put $u(x) = (y+x)^2$ to obtain the linear DE $xu'' - u' = 0$; hence the solution. Kamke (1959), p.584; Murphy (1960), p.400

8. $$\boldsymbol{x(x+2y)y'' + 2xy'^2 + 4(x+y)y' + x^2 + 2y = 0.}$$

A first integral is

$$x(x+2y)y' + (x^3/3) + 2xy + y^2 = C;$$

C is an arbitrary constant. The solution of the given DE, therefore, is

$$12xy(x+y) = C_1 + C_2x - x^4;$$

C_1 and C_2 are arbitrary constants. Murphy (1960), p.400

9. $$\boldsymbol{x^2(y+x)y'' - (xy'-y)^2 = 0.}$$

Writing $y + x = xu(x)$ and $v = u'/u$, we get $xv' + 2v = 0$, which, on integration, leads to the solution

$$y = -x + xC_1\exp(C_2/x);$$

C_1 and C_2 are arbitrary constants. Kamke (1959), p.584; Murphy (1960), p.400

10. $$\boldsymbol{x^2(y-x)y'' - \alpha(y-xy')^2 = 0, y(a) = a,}$$

where a and α are constants.

Put $x = 1/\zeta, y = (1+v)x$, so that $v(d^2v/d\zeta^2) = \alpha(dv/d\zeta)^2$ with the solution $v = (A\zeta + B)^{1/(1-\alpha)}$, leading to

$$y = x + x^{-\alpha/(1-\alpha)}(A+Bx)^{1/(1-\alpha)}.$$

Forsyth (1959), p.205

11. $$\boldsymbol{x^2(y-x)y'' - a(xy'-y)^2 = 0,}$$

where a is a constant.

With $y - x = xu(x)$, we get $xuu'' - axu'^2 + 2uu' = 0$. Now, putting $v(x) = u'/u$, we obtain the Bernoulli equation

$$xv' + (1-a)xv^2 + 2v = 0.$$

Further, with $w(x) = 1/v(x)$, we get the linear DE

$$w' - (2/x)w = 1 - a$$

with the solutions

$$w = (a-1)x + Cx^2; \tag{1}$$

C is an arbitrary constant. For $a \neq 1$, we get from (1) the solution

$$u(x) = (y - x)/x = \pm|C_0 + (C_1/x)|^{1/(1-a)};$$

C_0 and C_1 are arbitrary constants. For $a = 1$, see Eqn. (2.41.12). Kamke (1959), p.585; Murphy (1960), p.400

12. $$x^2(x - y)y'' + (xy' - y)^2 = 0.$$

With $y = xu(z)$, $z = \ln x$, we have

$$p[(1 - u)p' + 1 - u + p] = 0, \tag{1}$$

where $p = du/dz$. The solution of (1) is

$$p = (u - 1)[C - \ln(u - 1)]; \tag{2}$$

C is an arbitrary constant. The solution of (2) and hence of original DE is

$$C_1(x - y) = x\exp(C_2/x);$$

C_1 and C_2 are arbitrary constants. Murphy (1960), p.400

13. $$(a_1y + a_0)y'' + a_3y'^2 + [(3a_1 + 2a_3)y^2 + (3a_0 + a_1)y + a_5]y'$$
$$+ (a_1 + a_3)y^4 + (a_0 + a_4)y^3 + (a_5 + a_6)y^2 + a_7y + a_8 = 0,$$

where a_i $(i = 0, 1, 2, \ldots, 8)$ are analytic functions of x, in some domain D.

Conditions are found for this equation to have poles, with integral values as their residues, as its only moving singularities. Kolesnikova and Lukashevich (1972)

14. $$(x + y^2)y'' + 2yy'^2 + 2y' - a = 0,$$

where a is a constant

The given DE is exact. A first integral is

$$(x + y^2)y' + y = ax + C_1; \tag{1}$$

C_1 is an arbitrary constant. Equation (1) is also exact. The solution of the given DE is

$$2y(3x + y^2) = 3ax^2 + C_1x + C_2;$$

C_1 and C_2 are arbitrary constants. Murphy (1960), p.402

15. $$(y^2 + x)y'' + 2(y^2 - x)y'^3 + 4yy'^2 + y' = 0.$$

Writing $x = x(y)$, the given DE becomes

$$(y^2 + x)x'' - 2(y^2 - x) - 4yx' - x'^2 = 0. \tag{1}$$

Now put $v(y) = y^2 + x$ in (1) so that $vv'' = v'^2$. The solutions are now easily obtained as

$$y^2 + x = C_1\exp C_2y \quad \text{and} \quad y = C;$$

C_1, C_2, and C are arbitrary constants. Kamke (1959), p.587; Murphy (1960), p.402

16. $$(x^2+y^2)y'' - xy = 0.$$

The asymptotic series solution for $x \to \infty$ may be found following Levinson (1970); see Sachdev (1991). Bender and Orszag (1978), p.200

17. $$(y^2+x^2)y'' - (y'^2+1)(xy'-y) = 0.$$

The first two integrals of the given DE are

$$\tan^{-1} y' - \tan^{-1}(y/x) = C_1 \tag{1}$$

and

$$y' - (y/x) = \{1 + (y'y/x)\}C_2;$$

C_1 and C_2 are arbitrary constants. Introducing polar coordinates $x = r\cos\phi, y = r\sin\phi$ in (1), we get the solution as

$$r = C_3 \exp(C_4\phi);$$

C_3 and C_4 are arbitrary constants. Kamke (1959), p.587; Murphy (1960), p.402

18. $$(y^2+x^2)y'' - 2(y'^2+1)(xy'-y) = 0.$$

The solutions are:

(a) y = Cx.

(b) The upper and lower halves of the circles through the origin $x^2+y^2+C_1x+C_2y = 0$; C, C_1, and C_2 are arbitrary constants.

Kamke (1959), p.587; Murphy (1960), p.402

19. $$\{x/2+4(1-y^2)\}(2xy''+y') + \{y/(1-y^2)\}x^2(y')^2 + xy' + 2y(1-y^2)\{1/4 + (1-y^2)/x\} = 0.$$

The given DE is invariant under the change $y \to -y$. The straight lines $y = +1$, $y = -1$, $x = 0$ and the parabolas $x + 8(1-y^2) = 0$ are singular lines of the given DE and the latter cannot be solved numerically when these points are approached. The boundaries $y = \pm 1$ of the physical interval $-1 \le y \le 1$ can only be crossed for $x = 0$. The behavior of solutions near $x = 0$ is given by

$$y'' + y'/2x + y/4x^2 = 0,$$

so that

$$y \simeq Ax^{1/4}\cos\{(3^{1/2}/4)\log x + \beta\}, x \to 0$$

oscillating with increasing frequency and decreasing amplitude as $x \to 0$. As x increases, almost all solutions oscillate about $y = 0$ in the form

$$y = Bx^{-1/4}\cos\{(7^{1/2}/4)\log x + r\}, x \to \infty. \tag{1}$$

Here A, β, B and r are constants.

There are very special solutions, with $y \to \pm 1$ as $x \to \infty$ in the form

$$y \to 1 + P/x^2,$$

where P is a constant; they also have symmetric form for $-y$. They start from the origin $x = 0$ with the values $y = \pm 1$, behaving like

$$y = -1 + a_1 x + a_2 x^2 + \cdots$$

with sharply defined values of $y'(0) = a_1$ (similarly for $y = 1 + b_1 x + \cdots$, etc.). If the value of a_1 is changed even in a small way, the solution tends to $y = 0$ with generic oscillatory behavior (1) as $x \to \infty$.

A Laurent series solution of the given DE about a moving singularity is also explored. An occurrence of the logarithmic term is investigated. Ferreira and Neto (1992)

20. $$(\boldsymbol{y^2 + 1 + \sin x)y'' + (2yy' + \cos x)y' + (2 + \cos x + 3x \sin x)y^3/(3x^{4/3})}$$
$$\boldsymbol{- (2 + \cos x)/(3x^{13/3}) - (\sin x)/x^{10/3} - 4/x^5 - 2(1 + \sin x)/x^3}$$
$$\boldsymbol{+ (\cos x)/x^2 = 0, x \geq 1.}$$

The given DE has a nonoscillatory solution $y = 1/x$. Blasko, Graef, Hacik, and Spikes (1990)

21. $\boldsymbol{(y^2 + \sin x + 1)y'' + (2yy' + \cos x)y' + (6 + 3 \cos x) + 4x \sin x(y^5 + y)/(4x^{7/4}) = 0.}$

It is proved that all solutions of the given DE are oscillatory. Blasko, Graef, Hacik, and Spikes (1990)

22. $$\boldsymbol{\{x^2/4 + 2(1 - y^2)\}y'' + (x^2/4)yy'^2/(1 - y^2) + xy'/2}$$
$$\boldsymbol{+ 2y(1 - y^2)\{1/4 + (1 - y^2)/x^2\} = 0, y(0) = -1, y(\infty) = +1.}$$

The given DE describes a "hedgehog" description. A series solution near $x = 0$ is found by first introducing $z \equiv x^2$:

$$y = a_0 + a_1 z + a_2 z^2 + a_3 z^3 + \cdots,$$

where $a_0 = -1$,

$$\begin{aligned} a_2 &= -a_1^2(48a_1 + 7)/\{10(16a_1 + 1)\}, a_3 = -\{320a_2^2 a_1 + 8a_2^2 \\ &+ 32a_2 a_1^3 + 18a_2 a_1^2 - 64a_1^5 - 5a_1^4\}/\{28a_1(16a_1 + 1)\}, \end{aligned}$$

which involves an arbitrary constant $a_1 = y''(0)/2$.

The behavior near infinity is obtained by first introducing the variable $u = 1/z$ in the given DE so that

$$\begin{aligned} &4u^2\left[1 + 8u(1 - y^2)\right]\frac{d^2y}{du^2} + 4u^2\{y/(1 - y^2)\}\left(\frac{dy}{du}\right)^2 \\ &+ 2u\left[1 + 24u(1 - y^2)\right]\frac{dy}{du} + 2y(1 - y^2)\{1 + 4u(1 - y^2)\} = 0. \end{aligned} \quad (1)$$

The solution of (1) about $u = 0$ where $y = 1$ is $y = 1 + Pu^2 + Ru^4 + Su^5 + Tu^6 + Vu^7 + Wu^8 + \cdots$, where $R = (5/14)P^2, S = (16/3)P^2, T = (-20R^2 + 24RP^2 + 5P^4)/(88P), V = (-58SR + 15SP^2 + 1696RP^2 + 240P^4)/(130P)$. The numerical results give $a_1 = 0.50377$ and $P = -(1/2)(8.6386)^2$. The numerical results are checked by the following integral formulas obtained from the given DE by suitable manipulation:

$$\int_0^\infty y\{(3/4)x^2y'^2 + 8y'^2(1-y^2) + 2(1-y^2)^2(1/4 + (1-y^2)/x^2)\}dx = 0$$

and

$$\int_0^\infty y'\{x(1-y^2)y' + 2x^2yy'^2 + 8y(1-y^2)^3/x^2 - 4y'\frac{d}{dx}(1-y^2)^2\}dx = 0.$$

Neto, Galain, and Ferreira (1991)

23.
$$\frac{1+y^2}{1+(\sin^2 x)/x^2}y'' + \left[\frac{1+y^2}{1+(\sin^2 x)/x^2}\right]' y' + xy^5$$
$$- \frac{\sin x}{x}\left[\frac{2}{x^2} + \frac{\sin^4 x}{x^3} - 1\right] + \frac{2\cos x}{x^2} = 0, \ x \geq x_0 > 0.$$

The given DE has an exact solution $y(x) = (\sin x)/x$. Oscillatory properties of the given DE are discussed. Yan (1989)

24.
$$2x(x-1)y(y-1)(y-x)y'' - x(x-1)(3y^2 - 2xy - 2y + x)y'^2$$
$$+ 2y(y-1)(2xy - y - x^2)y' - y^2(y-1)^2 - f(x)[y(y-1)(y-x)]^{3/2} = 0.$$

If $\phi(u,x)$ is defined through $u = \int_0^\phi \frac{ds}{s(s-1)(s-x)}$ as an elliptic function with x-dependent periods $2w_1(x)$ and $2w_2(x)$, then the solutions of the given DE are

$$y = \phi(u + C_1w_1 + C_2w_2, x), \tag{1}$$

where $u(x)$ is a solution of the linear DE

$$4x(x-1)u'' + 4(2x-1)u' + u = f(x);$$

C_1 and C_2 are arbitrary constants. If $f \equiv 0$, one may set $u = 0$ in (1). Kamke (1959), p.592; Murphy (1960), p.404

25.
$$2x^2(x-1)^2y(y-1)(y-x)y'' - x^2(x-1)^2(3y^2 - 2xy - 2y + x)y'^2$$
$$+ 2x(x-1)y(y-1)(2xy - y - x^2)y' + ay^2(y-1)^2(y-x)^2$$
$$+ bx(y-1)^2(y-x)^2 + c(x-1)y^2(y-x)^2 + dx(x-1)y^2(y-1)^2 = 0,$$

where a, b, c, and d are constants.

The given DE cannot be solved in terms of classical transcendental functions [see Ince (1956), p.344]. For $a = b = c = 0, d = -1$, see Eqn. (2.41.24). Kamke (1959), p.592; Murphy (1960), p.405

26. $$\mathbf{2x^2(x-1)^2y(y-1)(y-x)y'' - x^2(x-1)^2(3y^2-2xy-2y+x)y'^2 + 2x(x-1)y(y-1)(2xy-y-x^2)y' - 2\alpha y^2(y-1)^2(y-x)^2 - 2\beta x(y-1)^2(y-x)^2 - 2\gamma(x-1)y^2(y-x)^2 - 2\delta x(x-1)y^2(y-1)^2 = 0,}$$

where α, β, γ, and δ are constants.

This is Painlevé's sixth equation. It may be written in the equivalent form

$$\begin{aligned} x(x-1)y' &= ax + (bx-p)y + qy^2 + y(y-1)(y-x)v, \\ x(x-1)v' &= (1/2)(2\alpha - q^2) - (bx-p)v - 2qvy - (1/2)(3y^2 - 2xy - 2y + x)v^2, \end{aligned} \tag{1}$$

where $a^2 + 2\beta = 0, b = r_1 - a, r_1^2 = 2\gamma, p = 1 + a + r_2, r_2^2 = 1 - 2\delta, q = 1 + a - r_1 + r_2$.

It is clear from (1) that for $2\alpha = q^2$, we have a one-parameter class of solutions of the given DE, provided by the general solution of the Riccati equation

$$x(x-1)y' = ax + (bx-p)y + qy^2. \tag{2}$$

This corresponds to $v = 0$ as the special solution of the second equation of system (1).

The given DE may have fixed singular points at $x = 0, x = 1$, or $x = \infty$. To investigate the solution in the neighborhood of the singularities, we put, assuming that $p \neq 0, v = u - (2\alpha - q^2)/2 \equiv u + h$, we obtain the system of the form

$$\begin{aligned} x\frac{dy}{dx} &= -py + F(x, y, u), \\ x\frac{du}{dx} &= -2qhy + pu + \phi(x, y, u), \end{aligned} \tag{3}$$

where F and ϕ are polynomials in their variables, satisfying the property

$$\begin{aligned} F(0,0,0) &\equiv \phi(0,0,0) \equiv \frac{\partial F(0,0,0)}{\partial y} = \frac{\partial \phi(0,0,0)}{\partial y} \\ &= \frac{\partial F(0,0,0)}{\partial u} = \frac{\partial \phi(0,0,0)}{\partial u} = 0. \end{aligned} \tag{4}$$

For $p = 0$, we put $y = x(\xi + k), v = (\eta/x) + h$ and for appropriate choices of the constants k and h, we have the system

$$\begin{aligned} x\frac{d\xi}{dx} &= -\xi + F_1(x, \xi, \eta), \\ x\frac{d\eta}{dx} &= \eta + \phi_1(x, \xi, \eta), \end{aligned} \tag{5}$$

where F_1 and ϕ_1 have the property (4) at $x = \xi = \eta = 0$. Using the systems (3) and (5), it is proved that if $\alpha \neq 0$, all the poles, zeros, and unities (where $y = 1$) of an arbitrary meromorphic solution of the given DE are simple. For cases when $z = 0$ and $z = 1$ points where solutions of the given DE are single valued, the latter may be represented by the meromorphic function

$$y = u(x)/v(x), \tag{6}$$

where $u(x)$ and $v(x)$ are entire functions. We have the following:

(a) If $x = x_0$ is a pole of (6), then the principal part of the polar expansion about x_0 is

$$\mathcal{P}(y) = x_0(x_0 - 1)/[(2\alpha)^{1/2}(x - x_0)]$$

if $\alpha \neq 0$, and

$$\mathcal{P}(y) = a_{-2}/(x - x_0)^2,$$

a_{-2} being arbitrary, if $\alpha = 0$.

(b) About the point of zero, i.e., when $y(x_0) = 0$,

$$y = \{(-2\beta)^{1/2}/(x_0 - 1)\}(x - x_0) + a_2(x - x_0)^2 + \cdots, \; \beta \neq 0.$$

(c) About the point of unity, $y(x_0) = 1$,

$$y = 1 + [(2\gamma)^{1/2}/x_0](x - x_0) + \cdots.$$

(d) Finally, when $y(x_0) = x_0, y = x_0 + b_1(x - x_0) + \cdots$, where $b_1^2 - 2b_1 + 2\delta = 0$.

It is also proved that the necessary and sufficient conditions that the given DE should have a rational solution

$$y(x) = \lambda x + \mu + P_{n-1}(x)/Q_n(x) = P(x)/Q_n(x), \tag{7}$$

where P_{n-1} and Q_n are polynomials of degrees $n-1$ and n, respectively, is that the system

$$\begin{aligned} x^2(x-1)^2(Q_nQ_n'' - Q_n'^2) &= -x(x-1)P'Q_n - x(x-1)^2Q_nQ_n' - 2\alpha P^2 \\ &\quad + [(x+1)\alpha - x(x-1)g']PQ_n - [x(x-1)^2g' + g'']Q_n^2, \end{aligned}$$

$$\begin{aligned} x(x-1)^2(PP'' - P'^2) &= (x^2-1)P'Q_n - x(x-1)Q_nQ_n' - g''P^2 + [(x^2-1)g' \\ &\quad - (\beta+\gamma+\delta)x + (\beta-\gamma-\delta)]PQ_n + [2\beta x - x(x-1)g']Q_n^2, \end{aligned}$$

for $g' = \epsilon, \epsilon$ being a constant, should have a solution in the form of polynomials appearing in (7).

It may be shown that the entire function u or v in the meromorphic solutions (6) is governed by the system

$$x^2(x-1)^2(vv'' - v'^2) + x(x-1)^2yy' + x(x-1)u'v + 2\alpha u^2 - \alpha(x+1)uv = 0,$$

$$x(x-1)^2(uu'' - u'^2) - (x^2-1)u'v + x(x-1)vv' + [(\beta+\gamma+\delta)x + (\beta-\gamma-\delta)]uv - 2\beta xv^2 = 0.$$

Lukashevich (1972)

27. $$(\boldsymbol{x^2 + y^2})^2\boldsymbol{y'' + a^2y = 0},$$

where a is a constant

A first integral of the given DE is

$$(xy' - y)^2 = C + a^2x^2/(x^2 + y^2),$$

where C is an arbitrary constant. Murphy (1960), p.405

28. $$(\boldsymbol{y^2 + ax^2 + 2bx + c)^2 y'' + dy = 0},$$

where a, b, c, and d are constants.

Dividing by the coefficient of y'' and multiplication by $ax(xy' - y) + b(2xy' - y) + cy'$ renders the given DE exact; an integration then gives

$$(ax^2 + 2bx + c)y'^2 - 2(ax + b)yy' + ay^2 + dy^2/(y^2 + ax^2 + 2bx + c) = C, \qquad (1)$$

where C is an arbitrary constant. Equation (1) can be reduced by the transformation $y(x) = u(x)[ax^2 + 2bx + c]^{1/2}$. The special case of the given DE

$$(y^2 + x^2)^2 y'' + ay = 0$$

can be made exact by multiplication by $2x(y^2 + x^2)^{-2}(xy' - y)$. An integration then gives the homogeneous DE

$$(xy' - y)^2 - ax^2/(x^2 + y^2) = C,$$

which is solved by putting $y = vx$, etc. C is an arbitrary constant. Kamke (1959), p.593; Murphy (1960), p.405

29. $$\boldsymbol{(y^2 + x^2)^{1/2} y'' - a(y'^2 + 1)^{3/2} = 0},$$

where a is a constant.

The equation is invariant under $y = AY, x = AX$. The solution in parametric form is

$$x = C_1[(\cos t)/\{\sin s - a\}], y = C_1[(\sin t)/\{\sin s - a\}],$$

with $t = C_2 - \int [\{\sin s\}/\{\sin s - a\}]ds; C_1$ and C_2 are arbitrary constants. Kamke (1959), p.593

30. $$\boldsymbol{\{y^2 + (1 + a \cos x)^2\}^{3/2} y'' + y = 0},$$

where a is constant.

The general solution of the given DE, singular at a point $x = x_0$, has the form

$$y(x) = y_0 + \sum_{k=4}^{\infty} y_k (x - x_0)^{k/5},$$

with $y_0 = i(1+a\cos x_0), y_4 = (625/(128y_0))^{1/5}, y_5 = \mp i \sin x_0$, where x_0 and y_6 are arbitrary corresponding to the resonances -1 and 6. The given DE is said to have a weak Painlevé property, since in the neighborhood of a singularity at x_0, the solution can be expressed in the form

$$y(x) = (x - x_0)^{p/q} \sum_{n=0}^{\infty} y_n (x - x_0)^{n/q},$$

where x_0, y_n are constants, and p and q are integers determined by the leading order analysis.

Turning now to fixed singularities which arise when $1 + a\cos x_0 = 0$, we first expand

$$(1 + a\cos x)^2 = (a^2 - 1)(x - x_0)^2 + o((x - x_0)^2),$$

so that

$$y(x) = y_0(x - x_0)^{2/3} + \{y_2 + a_2 \ln(x - x_0)\}(x - x_0)^{4/3} + o((x - x_0)^{4/3})$$

with $y_0^6 = 81/4, y_2$ arbitrary, and $a_2 = (1/5)(a^2 - 1)$. The constants appear at the "resonances" - 1 and 2/3. Therefore, unless $a^2 = 1$, the logarithmic terms arise in the expansion. It is suggested that this behavior near a fixed singularity is not compatible with integrability and, in fact, numerical solution of the given DE exhibits chaos. Kruskal and Clarkson (1992)

31. $\quad \boldsymbol{(x^2/4 + 2\sin^2 y)y'' + \sin(2y)y'^2 + xy'/2 - \sin(2y)\{1/4 + \sin^2 y/x^2\} = 0,}$

$$\boldsymbol{y(0) = \pi, y(\infty) = 0.}$$

Local solutions near $x = 0$ and $x = \infty$, respectively, are

$$y = \pi + Ax + Bx^3 + Cx^5 + \cdots, \tag{1}$$

where $B = -(2/15)A^3(1 + 2A^2)/(1 + 8A^2), C = (A/7)\{A^4/15 - B(48B - 6A^3 + A)/(1 + 8A^2)\}$, where A is a free parameter to be chosen so that $y(\infty) = 0$, and

$$\begin{aligned} y &= (b/x^2)\left\{1 - \frac{1}{21}\left(\frac{b}{x^2}\right)^2 - (4/3b)\left(\frac{b}{x^2}\right)^3 \right. \\ &\quad \left. + \frac{2}{35 \times 11}\left(\frac{b}{x^2}\right)^4 + \frac{136}{21 \times 13 \times b}\left(\frac{b}{x^2}\right)^5 + \cdots\right\}. \end{aligned} \tag{2}$$

The value of b is obtained as soon as A is obtained. The numerical solution gives $A = -1.003764$ and $b = 8.6386$.

To check the numerical results, the following integral relations are obtained. Write the given DE as

$$\frac{d}{dx}\{(x^2/4 + 2\sin^2 y)y'\} = \sin(2y)\{y'^2 + 1/4 + \sin^2 y/x^2\} \tag{3}$$

and multiply by $1/x^2$ to obtain

$$\begin{aligned} \frac{d}{dx}[y'\{1 + (8/x^2)\sin^2 y\}] &= (4/x^2)\{\sin(2y)[y'^2 + 1/4 + \sin^2 y/x^2] \\ &\quad - (1/x)y'[x^2/2 + 4\sin^2 y]\}. \end{aligned} \tag{4}$$

Integrating (4) from 0 to ∞ and using the series expansions for y developed above, one has

$$-A(1+8A^2) = 4\int_0^\infty (1/x^2)\{\sin(2y)(y'^2+1/4+\sin^2 y/x^2) - (2/x)y'(x^2/4+2\sin^2 y)\}dx. \tag{5}$$

Multiplying (3) by x and integrating by parts leads to the following formula for b:

$$b = 2\pi - 2\int_0^\infty \{(x/4)(xy' + \sin(2y)) + x\sin(2y)(y'^2 + \sin^2 y/x^2)\}dx. \tag{6}$$

Equations (5) and (6) verify the numerical results for A and b. Neto, Galain, and Ferreira (1991)

32. $(x^2/4 + 2\sin^2 y)y'' + (x/2 + 4y' \sin y \cos y)y' - \sin(2y)[y'^2 + \sin^2 y/x^2 + 1/4]$

$$- (1/4)\beta^2 x^2 \sin y = 0,$$

where β is a constant.

The given DE is the Skyrme model. The solution with $y(0) = \pi$, for small x, behaves like

$$y(x) = \pi + Ax + Bx^3 + Cx^5 + \cdots,$$

with

$$\begin{aligned} B &= -\{(2/15)A^3(1 + 2A^2) + \beta^2 A/10\}/(1 + 8A^2), \\ C &= (A/7)\{A^4/15 - B(48B - 6A^3 + A)/(1 + 8A^2)\} + \beta^2(A^6 - 6B)/\{168(1 + 8A^2)\}, \end{aligned}$$

and so on. The value of A for a solotonic solution is -1.131034.

A new form of the given DE with $y^* = \cos y$ and $z = x^2$ gives the polynomial equation

$$\begin{aligned} &[z/2 + 4(1 - y^{*2})]\left\{\frac{dy^*}{dz} + 2z\frac{d^2y^*}{dz^2}\right\} + \frac{y^*}{1 - y^{*2}}z^2\left(\frac{dy^*}{dz}\right)^2 + z\frac{dy^*}{dz} \\ &\quad + 2y^*(1 - y^{*2})\{1/4 + (1 - y^{*2})/z\} + (1/4)\beta^2 z(1 - y^{*2}) = 0. \end{aligned} \tag{1}$$

An expansion of the form

$$y = -1 + a_1 z + a_2 z^2$$

is obtained, where $a_1 = 0.63962$ for a soliton solution. The problem is solved numerically with this value of a_1. A pole-like singularity is located at $z_0 = -4.5902$. Equation (1) is found to be non-Painlevé with a logarithmic branch point superimposed on a pole. Ferreira and Neto (1992a)

33. $$\{(1/4)x^2 + 2\sin^2 y\}y'' + (x/2)y' + \sin 2y(y')^2$$

$$- (1/4)\sin 2y - \{\sin^2 y/x^2\}\sin 2y = 0.$$

The given DE has been introduced as a reasonable model for a soliton description of nucleons—the Skyrme model.

It follows easily that if y satisfies the given DE, so does $\pm y + k\pi$ for any integer k positive or negative. The following general results regarding the given DE are proved.

(a) All solutions of the given DE are bounded.

(b) As $x \to \infty$, every solution converges to either $k\pi$ or $(k+1/2)\pi$, for some integer k positive or negative. As $x \to 0$, every solution converges to either $\ell\pi$ or $(\ell + 1/2)\pi$, for some integer ℓ.

(c) If a solution converges to $k\pi$ as $x \to \infty$, then it is ultimately monotonic and $y - k\pi \sim$ const. x^{-2}. If a solution converges to $(k + 1/2)\pi$ as $x \to \infty$, then it oscillates about $(k + 1/2)\pi$ and $y - (k + 1/2)\pi = O(x^{-1/2})$.
If a solution converges to $\ell\pi$ as $x \downarrow 0$, then it is ultimately monotonic and $y - \ell\pi \sim$ const. r. If a solution converges to $(\ell+1/2)\pi$ as $x \downarrow 0$, then it oscillates about $(\ell + 1/2)\pi$ and $y - (\ell + 1/2)\pi = O(x^{1/2})$.

(d) No solution can have a local maximum for which y lies between $k\pi$ and $(k+1/2)\pi$. If $y(r_0) = \ell\pi$ for some $r_0 \geq 0, \ell \leq k$, and if subsequently y has a local maximum between $(k+1/2)\pi$ and $(k+1)\pi$, then it must converge, as $x \to \infty$, to $(k+1/2)\pi$. [Note that a stationary value at $y = k\pi$ or $y = (k+1/2)\pi$ implies a constant solution.]

No solution can have a local minimum for which y lies between $(\ell+1/2)\pi$ and $(\ell+1)\pi$. If $y(r_1) = k\pi$ for some $x_1 \leq \infty, k \geq \ell$, and if earlier y has a local minimum between $\ell\pi$ and $(\ell+1/2)\pi$, then it must converge as $x \downarrow 0$ to $(\ell+1/2)\pi$.

In particular, any solution going from $\ell\pi$ as $x \downarrow 0$ to $k\pi$ as $x \to \infty$ must be monotonic; and the only solution connecting $k\pi$ to $k\pi$ is constant.

(e) Given any integers k, ℓ, there is at least one solution that connects:

(i) $\ell\pi$ to $k\pi$.

(ii) $(\ell+1/2)\pi$ to $k\pi$.

(iii) $(\ell+1/2)\pi$ to $k\pi$.

(iv) $(\ell+1/2)\pi$ to $(k+1/2)\pi$.

(f) In case (i) of (e), the solution is unique if $|k-\ell| = 0, 1$. In cases (ii), (iii), and (iv), it is not unique.

For the BC $y(0) = 0, y(\infty) = n\pi$, n some positive integer, the existence and uniqueness of solutions are discussed. McLeod and Troy (1991)

34. $$\boldsymbol{X^3(x,y)y'' = 1; \quad X^2(x,y) = a_0 + 2a_1x + a_2x^2 + 2a_3y + 2a_4xy + a_5y^2,}$$

where a_i $(i = 0, \ldots 5)$ are constants. With

$$a_5X^2 = u^2 + X_1^2, u = a_3 + a_4x + a_5y, X_1^2 = b_0 + 2b_1x + b_2x^2,$$

where b_i are functions of a_i, we have

$$u''(x) = a_5^{5/2}(u^2 + X_1^2)^{-3/2} or X_1^3u''(x) = a_5^{5/2}[(u/X_1)^2 + 1]^{-3/2}.$$

A first integral is

$$X_1^4U'^2 = (b_0b_2 - b_1^2)U^2 + 2\int f(U)dU,$$

where $U = X_1^{-1}(x)u(x)$ and $f(U) = a_5^{5/2}(U^2+1)^{-3/2}$. Murphy (1960), p.405

35. $$\boldsymbol{g(x,y)y'' + g_x(x,y)y' + q(x)f(y) - r(x) = 0,}$$

where $q(x)$ is allowed to oscillate.

Sufficient conditions for any solution $y(x)$ of the given DE either to oscillate or to satisfy $\liminf |y(x)| = 0$ are given. Also, sufficient conditions are obtained for all solutions of the given DE to be oscillatory. Blasko, Graef, Hacik, and Spikes (1990)

2.42 $f(y, y')y'' + g(x, y, y') = 0$

1. $$y'y'' - a^2x = 0,$$

where a is a constant.

A first integral of the given DE is

$$y'^2 = C + a^2x^2,$$

C is an arbitrary constant, and hence the solution is

$$2y = a[x(x^2 + C_1^2)^{1/2} + C_1^2 \ln\{x + (x^2 + C_1^2)^{1/2}\}] + C_2;$$

C_1 and C_2 are arbitrary constants. Murphy (1960), p.406

2. $$y'y'' - x^2yy' - xy^2 = 0.$$

The given DE is exact. A first integral is

$$y'^2 - x^2y^2 = C;$$

C is an arbitrary constant. Kamke (1959), p.594; Murphy (1960), p.406

3. $$(1 + \epsilon y')y'' + y = 0.$$

A two-term perturbation solution with period 2π is

$$y = a\cos x + \epsilon a^2[(1/6)\sin 2x - (1/3)\sin x].$$

Jordan and Smith (1977), p.146

4. $$6yy'y'' + 2y'^3 - 1/x^{1/2} = 0.$$

The given DE is the Euler equation of integral functional $J(y) = \int_{x_0}^{x_1}(yy'^3 + y/x^{1/2})dx$. It is conformally invariant under $\overline{x} = (1+\epsilon)x, \overline{y} = (1+\epsilon)^{5/6}y$ with a factor $-1/2$. Logan (1985)

5. $$3yy'y'' - y'^3 - 1 = 0.$$

Put $y' = p, y'' = p(dp/dy)$, so that

$$3yp^2\frac{dp}{dy} = p^3 + 1. \qquad (1)$$

On integration (1) gives

$$p^3 + 1 = c_1y. \qquad (2)$$

Integrating (2) with $p = dy/dx$, we have

$$y = (1/c_1)[1 + \{(2c_1/3)(x + c)\}^{3/2}],$$

where c_1 and c_2 are arbitrary constants. Rabenstein (1972), p.47

6. $$(a^2y'^2-1)y''-(1+ke^{ay}y')y'=0,$$

where a is a constant.

The given DE arises in the similarity reduction of a nonlinear PDE and has a first integral

$$a^2y'^2/2-\log y'-x=(k/a)e^{ay}+C,$$

where C is a constant. Arrigo (1991)

7. $$(y'^2+y^2)y''+y^3=0.$$

Put $y'=yu(x)$ to obtain

$$(1+u^2)u'+1+u^2+u^4=0,$$

with the solution $3^{1/2}u=(1-u^2)\tan(C-3^{1/2}x)$, where C is an arbitrary constant. Hence the solution via $y'/y=u(x)$. Kamke (1959), p.595; Murphy (1960), p.406

8. $$(y^2-y'^2)y''-\ln x=0, y(1)=1, y'(1)=0.$$

A first integral involving a quadrature is easily written by integrating by parts. The problem may be solved numerically. Wilson (1971), p.27

9. $$(1-y^2-y'^2)y''-1=0, y(0)=y'(0)=0.$$

Put $y'=p, p^2=P$ and interchange the role of y and P to obtain the Riccati equation $(dy/dP)+(y^2/2)=1-P/2$. Hence solve by linearizing, etc., and using BCs. Wilson (1971), p.27

10. $$4yy'^2y''-y'^4-3=0.$$

With $y'(x)=p(y)$, we have

$$4yp^3p'=p^4+3. \tag{1}$$

A first integral from (1) is $p^4=Cy-3$, where C is an arbitrary constant. Hence the solution of the given DE is

$$256C_1(y-C_1)^3=243(x-C_2)^4;$$

C_1 and C_2 are arbitrary constants. Murphy (1960), p.407

11. $$y^3y'^2y''+(1/2)=0,$$

$y(0)=y'(0)=1$. Put $y'=p(y), y''=p(dp/dy)$; we have $y^3p^3(dp/dy)=-1/2$. A first integral satisfying the IC is $p^4=y^{-2}$; hence the solution by quadrature satisfying the IC is $y=[(3x/2)+1]^{2/3}$. Bender and Orszag (1978), p.34

12. $$y^5 y'^2 y'' + (5/3) y^4 y'^4 + xy'/12 = 0,$$

where either $y(\infty) = 0, y(-\infty) = 1$, or $y(\infty) = 0, y(0) = 1$.

This problem arises in lubrication theory. The given DE is invariant under the transformation $F(x, \lambda) = \lambda^{-4/7} y(\lambda x)$. Introducing the variables $\tau = \ln|x|, \phi(\tau) = x^{-4/7} y, \psi = x^{3/7} y' = (4/7)\phi + (\partial\phi/\partial\tau)$ sign x, one may obtain an equation for $\psi(\phi)$:

$$\frac{d\psi}{d\phi} = -(5/3)(\psi/\phi) - \{1/(7\psi - 4\phi)\}\{(11\psi/3) + 7/(12\phi^5\psi)\},$$

whose field integral curves in the half-plane $\phi \geq 0$ may be drawn. Some asymptotic results may be obtained. For example, as $\phi \to 0$ corresponding to $x \to 0$, the integral curves admit of the asymptotic representation

$$|\psi| = C_\infty \phi^{-1/3} + o(\phi^{-1/3}), C_\infty^{(o)} \simeq 0.013.$$

Numerical solution of the given problem is presented. Kerchman (1990)

13. $$(a^2 b^2 y'^4 - 1) y'' - (a^2 y + k y^{-3}) y'^4 = 0,$$

where a, b, and k are constants.

The given DE arises in the similarity reduction of a nonlinear wave equation. If $k = -1$, the given equation has solutions

$$y(x) = e^{-x}, e^x, \text{ and } \{2x/(ab)\}^{1/2}.$$

Arrigo (1991)

14. $$(5/3) y'^{2/3} y'' + x^{-11/4} y^{5/3} = 0, x \geq x_0 > 0.$$

It is proved that the given DE is nonoscillatory. Erbe and Liu (1990)

15. $$(y')^{-(2n-1)/n} y'' + [n^2/(n+1)]x = 0, y(0) = 1, y(\infty) = 0,$$

where n is an odd positive integer.

This problem arises in the self-similar analysis of some nonlinear impact problems. A first integral is

$$y' = [C_1 + (1/2)\{n(n-1)/(n+1)\}x^2]^{-n/(n-1)},$$

where C_1 is constant.

A second integration and use of the condition $y(0) = 1$ gives

$$y(x) = \int_0^x [C_1 + (1/2)\{n(n-1)/(n+1)\}\eta^2]^{-n/(n-1)} d\eta + 1. \tag{1}$$

Using the condition $y(\infty) = 0$ leads to the determination

$$C_1 = [(n+1)/\{2n(n-1)\}]^{(n-1)/(n+1)} [B\{1/2, (n+1)/\{2(n-1)\}\}]^{2(n-1)/(n+1)},$$

where $B(a, b)$ is the beta function. Taulbee, Cozzarelli, and Dym (1971)

16. $$(-1/n)(-y')^{1/n-1}y'' + (1/x)(-y')^{1/n} - a^2(\lambda y + \beta x y') = 0,$$

where a, n, λ, and β are constants and $\lambda(1-1/n) = -1 + \beta(1+1/n)$.

The given DE arises in the description of pressure disturbances in a non-Newtonian fluid. For $\lambda = 2\beta$, we have $\lambda = 2n/(3-n)$ and $\beta = n/(3-n)$, and the solution of the given DE may be found to be

$$y = [C - \{(1-n)/(1+n)\}(a^2\beta)^n x^{1+n}]^{1/(1-n)},$$

where C is a constant. Pascal (1991a)

17. $$(1 - a^{m+2}y'^m)y'' - f(y) = 0,$$

where a and m are constants.

The given DE arises in the similarity reduction of a nonlinear wave equation. Integrating once, we have

$$y'^2/2 - a^{m+2}y'^{m+2}/(m+2) = f(y)dy + C,$$

where C is a constant. Using a Legendre transformation, we get the solution in the parmetric form

$$x = (1/t)G^{-1} + \int\{G^{-1}/t^2\}dt, y = G^{-1},$$

where $G^{-1} = G^{-1}\{t^2/2 - a^{m+2}t^{m+2}/(m+2) - C\}$ and $G(t) = \int f(t)dt$. Arrigo (1991)

18. $$[y^{(1+2n)/n}(-y')^{1/n}]' + a^2(\alpha y + \beta x y') = 0,$$

where α and β are related by $\{1 + 2/n\}\alpha = -1 - (1+1/n)\beta$; n is another parameter.

For $\alpha = \beta = -n/(3+2n)$, the given DE can be written as

$$\frac{d}{dx}\{y^{(1+2n)/n}(-y')^{1/n}\} = \{na^2/(3+2n)\}\frac{d}{dx}(xy),$$

which on integration yields

$$y^{(1+2n)/n}(-y')^{1/n} = \{na^2/(3+2n)\}xy + C, \tag{1}$$

where C is a constant.

If we choose $x = x_1$ such that $y(x_1) = 0$, then $C = 0$. Integration of (1) yields

$$y(x) = [C_1 - \{(2+n)/(1+n)\}\{na^2/(3+2n)\}^n x^{1+n}]^{1/(2+n)},$$

where C_1 is another constant. Pascal (1991b)

19. $$(1.2 - \tanh^2 y')y'' + 0.5y' + y - 2.5E = 0.$$

The given equation describes the response of a LRC series circuit, and is solved by a numerical double delta function [see Eqn. (2.24.33)]. Nishikawa (1964), p.37

20. $$(|y'|^{p-2}y')' + f(x,y) = 0, y(0) = y(X) = 0, p > 1;$$

X is a positive real number.

Two existence results, depending on certain conditions on $f(x,y)$, are given for the given Dirichlet BVP. Pino and Manasevich (1989)

21. $$\frac{d}{dx}\left\{h\left(\frac{dy}{dx}\right)\right\}+g\left(x,\frac{dy}{dx}\right)+f(y)=0 \text{ on } (0,\infty),$$

where h, g, and f are functions, like $h(u)=|u|^\alpha u, g(x,u)=(1+x)^\theta |u|^\beta f(u)=|u|^\gamma u$ with $\alpha>-1, \beta>-1$ and $\gamma>-1$.

Precise decay estimates as $x\to\infty$ are derived. Nakao (1989)

22. $$f(y')y''+g(y)y'+h(x)=0.$$

We get a first order DE by integration:

$$\int f(y')dy'+\int g(y)dy+\int h(x)dx=C;$$

C is an arbitrary constant. Kamke (1959), p.596; Murphy (1960), p.407

2.43 $f(x,y,y')y''+g(x,y,y')=0$

1. $$(y'-x)y''-y'=0.$$

A first integral is

$$y'^2/2-xy'=A. \tag{1}$$

Equation (1) is a quadratic in y' and is easily integrated:

$$y=x^2/2\pm\{(x/2)(x^2+c_1)^{1/2}+(c_1/2)\ln|x+(x^2+c_1)^{1/2}|\}+c_2.$$

A, c_1, and c_2 are arbitrary constants. Rabenstein (1972), p.47

2. $$(\alpha y'+\beta y+\gamma x)y''-(ay'+by+cx)=0,$$

$y(0)=y'(0)=0$, where α,β,γ,a,b, and c are constants.

Depending on the value of $\rho=\{1-(1-p)^{1/2}\}/\{1+(1-p)^{1/2}\}$, where $p=4(a\gamma-c\alpha)/(a+\gamma)^2$, the solution may be either regular functions of x and $x^{\rho-1}$ or regular functions of x and $x\log x$ or for a particular relation among coefficients, a regular function of x; the last class involves an arbitrary parameter. For example, for $\rho=2$, the solution has the form $y=\{(a-2\gamma)/(6\alpha)\}x^2+\theta x^3\log x+\sum\sum C_{mn}x^m(x\log x)^n$, the double summation extending over the values of m and n, such that $m\geq 2, m+n\geq 4$; the coefficient θ is arbitrary and enters into the expression of the coefficients C_{mn}; also, since $\rho=2$, the coefficients satisfy the relation $2(a+\gamma)^2=9(a\gamma-c\alpha)$. Forsyth (1959), p.216

3. $$2xy'y''-y'^2-1=0.$$

Writing the given DE as

$$2y'y''/(y'^2+1)=1/x,$$

we integrate to get $y'^2 + 1 = c_1 x$. Hence the solution is

$$y = \pm\{2/(3c_1)\}\{c_1 x - 1\}^{3/2} + c_2;$$

c_1 and c_2 are arbitrary constants. Rabenstein (1972), p.47

4. $$(xy' - y)y'' - a = 0.$$

This is a special case of Eqn. (2.43.18). However, the form of solution therein is not valid here. A solution vanishing at $x = 0$ is easily found to be $y = Ax - (-a)^{1/2} x \log x$, where A is an arbitrary constant. Forsyth (1959), p.207

5. $$(xy' - y)y'' + 4y'^2 = 0.$$

With $u(x) = y'/y$, we get an equidimensional DE. Then put $\eta(\xi) = xu(x), \xi = \log|x|$ to obtain a DE with variables separable:

$$(1 - \eta)\eta' = \eta(\eta + 1)^2.$$

Now setting $t(\xi) = 1/(1+\eta)$, we have $(2t-1)t' = t-1$. Thus, the parametric representation of the solution is

$$x = C_1(t-1)e^{2t}, y = C_2 t e^{-2t};$$

C_1 and C_2 are arbitrary constants. Kamke (1959), p.594

6. $$(xy' - y)y'' - (y'^2 + 1)^2 = 0.$$

The given DE describes involutes of all circles with the origin as the center. Kamke (1959), p.595

7. $$(axy' + \beta y + \gamma x)y'' = \alpha xy' + by + cx, y(0) = 0.$$

A dominant balance argument shows that a regular solution exists in the form

$$y = k_2 x^2 + k_3 x^3 + k_4 x^4 + \cdots,$$

where $k_2 = c/(2\gamma)$ and other coefficients are successively found. The substitution $y = xu$ changes the given DE to

$$xu'' + 2u' = \{c + (a+b)u + axu'\}/\{\gamma + (a+\beta)u + axu'\}.$$

Assuming that $u = p + qx + rx^2 + vx$, where $2q = \{c + (a+b)p\}/\{\gamma + (\alpha+\beta)p\}$ and $3r = q^2\{(2a+b)/(c+(a+b)p) - (2\alpha+\beta)/(\gamma+(\alpha+\beta)p)\}$, we get the form

$$x^2 v'' + 4xv' + 2v = \text{regular function of } xv', v, x, \tag{1}$$

the regular function containing no term in xv' alone or in v alone, or in x alone. Equation (1) has only one solution such that $v(0) = v'(0) = 0$. It is shown that there is only one solution of the given DE that vanishes with x, this solution being regular. Moreover, p is arbitrary and this is the initial value of y'. Thus, a unique formal solution is obtained for a given value of $y'(0)$. Forsyth (1959), p.219

8. $$(y^2+2x^2y')y''+2(x+y)y'^2+xy'+y=0.$$

The given DE is exact; a first integral, therefore, is

$$(y^2+x^2y')y'+xy=C;$$

C is an arbitrary constant. Murphy (1960), p.406

9. $$ax^3y'y''+by^2=0,$$

where a and b are constants.

The given DE is equidimensional with respect to both x and y. With $u(x)=y'/y$, we get the first order DE

$$ax^3u(u'+u^2)+b=0.$$

Now, with $xu=\eta(\xi), \xi=\ln|x|$, we get

$$a\eta(\eta'-\eta+\eta^2)+b=0,$$

which is easily integrated. Kamke (1959), p.595; Murphy (1960), p.406

10. $$f_1y'y''+f_2yy''+f_3y'^2+f_4yy'+f_5y^2=0, f=f(x); f_1\neq 0.$$

If f_1 and f_2 are continously differentiable, we transform the given DE through $\eta(\xi)=y(x)\exp\left(\int (f_2/f_1)dx\right)$,

$$\frac{d\xi}{dx}=\exp\left(\int \{(2f_2-f_3)/f_1\}dx\right)$$

to the form

$$\eta'\eta''+g(\xi)\eta\eta'+h(\xi)\eta^2=0. \tag{1}$$

Now if $h(\xi)=(1/2)g'(\xi)$, the solutions of Eqn. (1) are also those of $\eta'^2+g\eta^2=C$; C is an arbitrary constant. With $u(x)=y'/y$, the original DE changes to an Abel equation. Kamke (1959), p.595; Murphy (1960), p.406

11. $$(ay^2+bxyy'+cx^2)y''-1=0.$$

Substitute $x^2=z^3$ and $y=zu$ to obtain

$$z^2u''=(1/2)u-(3/2)zu'+(9/4)/[\{a+2b/3\}u^2+(2/3)bzu'u+cz], \tag{1}$$

where $u'=du/dz, u''=d^2u/dz^2$. An asymptotic solution of (1) may be developed. Any solution of (1), which is not infinite when $z=0$, leads to a solution of the original DE. Forsyth (1959), p.204

12. $$(2y^2y'+x^2)y''+2yy'^3+3xy'+y=0.$$

The given DE is exact. A first integral is

$$y^2y'^2+x^2y'+xy=C;$$

C is an arbitrary constant. Kamke (1959), p.595; Murphy (1960), p.406

13. $$(y'^2 - x)y'' - y' + x^2 = 0.$$

Writing the given DE as $(x-p^2)p' = x^2 - p, p(y) = y'(x)$, a first integral is found to be

$$p^3 - 3px + x^3 = C; \tag{1}$$

C is an arbitrary constant. Solve (1) for p and integrate with respect to x. Murphy (1960), p.406

14. $$a(y'^2 - 2\beta^2 x^2)y'' - (y' - \beta x) = 0, y(0) = y'(0) = 0;$$

a and β are constants.

A series solution, which may possibly involve a logarithmic term, may be obtained following the general approach of Forsyth (1959). Forsyth (1959), p.218

15. $$(cy'^2 + xy' - y)y'' + 1 = 0, y(0) = y'(0) = 0.$$

Writing the given equation as

$$\frac{dx}{dy'} + xy' = y - cy'^2,$$

we have, using

$$\frac{dy}{dy'} = y'\frac{dx}{dy'}, \frac{d^2x}{dy'^2} + x + y'\frac{dx}{dy'} = \frac{dy}{dy'} - 2cy' = y'\frac{dx}{dy'} - 2cy'.$$

Therefore, $\frac{d^2x}{dy'^2} + x = -2cy'$, so that

$$x = -2cy' + A\sin y' + B\cos y'$$

and

$$\begin{aligned} y &= cy'^2 + xy' + \frac{dx}{dy'} = -2c - cy'^2 + B(y'\cos y' - \sin y') \\ &\quad + A(\cos y' + y'\sin y'). \end{aligned}$$

With the IC $y'(0) = 0, y(0) = 0$, we have $B = 0, A = 2c$. Thus,

$$\begin{aligned} x &= -2c(y' - \sin y'), \\ y &= -2c(1 - \cos y' - y'\sin y') - cy'^2. \end{aligned}$$

These as regular functions of y' are

$$\begin{aligned} x &= 2c\sum_{n=1}^{\infty} \{(-1)^n/(2n+1)!\}y'^{2n+1}, \\ y &= 2c\sum_{n=2}^{\infty} \{(-1)^{n-1}(2n-1)/(2n)!\}y'^{2n}. \end{aligned}$$

We thus have a parametric representation of the solution. Forsyth (1959), p.201

16. $$\{y'^2 + a(xy' - y)\}y'' - b = 0,$$

where a and b are constants.

Differentiating the given DE with respect to x and eliminating $xy' - y$ with its help, one obtains a second order DE for $p(x) = y'(x)$:

$$bp'' + (2p + ax)p'^3 = 0.$$

Now, if p is assumed to be the independent variable, so that $x = x(p)$, we may obtain the linear DE

$$x'' - (a/b)x = (2/b)p,$$

which is easily solved. Kamke (1959), p.596; Murphy (1960), p.407

17. $$(a^2x^2y'^2 - 1)y'' + [a^2xy' - f(y)]y'^2 = 0,$$

where a is a constant.

The given DE arises in the similarity reduction of a nonlinear wave equation. A first integral is found to be

$$a^2x^2y'^2/2 - \log y' = \int f(y)dy + C,$$

where C is a constant. Arrigo (1991)

18. $$(y - xy')^n y'' - a = 0, y(0) = 0,$$

and n is a positive integer.

Since $d/dx(y - xy') = -xy''$, we can integrate the given DE to yield $(y - xy')^{n+1} = -(1/2)(n+1)ax^2$, provided that $y(0) = 0$. A second integration is easily performed:

$$y = Ax - \{(1+n)/(1-n)\}\{-(a/2)(n+1)\}^{1/(n+1)}x^{2/(n+1)},$$

where A is arbitrary. Forsyth (1959), p.207

19. $$\{a(y'^2 + 1)^{1/2} - xy'\}y'' - y'^2 - 1 = 0.$$

Here y does not appear explicitly. The solution in a parametric form may be found to be

$$x = (at + C_1)/v, y = (C_1t - a)/v - C_1 \log(t + v) + C_2,$$

with $v = (t^2 + 1)^{1/2}$; C_1 and C_2 are arbitrary constants. Kamke (1959), p.596

20. $$\frac{1}{x}\frac{d}{dx}\left[xy'/(1 + y'^2)^{1/2}\right] + \omega^2x^2 - k_1\frac{y'^2}{1 + y'^2} + k_2 = 0,$$

where ω, k_1, k_2 are constants.

The given DE arises in electrodynamics. It is first transformed in polar coordinates: An approximate solution is attempted, which is shown to agree reasonably well with exact numerical solutions, and the specific values $\omega = 0.1, R = 0.4$ are chosen. Bohme, Johann, and Siekmann (1974)

21. $$-nx^{2+n}(-y')^{n-1}y'' + (2+n)x^{1+n}(-y')^n + a^2x^3(\alpha y + \beta xy') = 0,$$

where n, α, and β are constants, and $(1-n)\alpha = 1 + 2\beta$.

For $\alpha = 4\beta$, we have $\alpha = -2/(2n-1), \beta = -1/\{2(2n-1)\}$ and the given DE has the exact form

$$[x^{2+n}(-y')^n]' = [a^2/\{2(2n-1)\}](x^4y)'. \tag{1}$$

On integration, (1) gives

$$x^{2+n}(-y')^n = [a^2/\{2(2n-1)\}]x^4y + C, \tag{2}$$

where C is a constant. The solution of (2) such that $y(x_1) = 0$ and $y'(x_1) = 0$ for some $x = x_1$ is

$$y = [C_1 - \{(n-1)/2\}\{a^2/2(2n-1)\}^{1/n}x^{2/n}]^{n/(n-1)}.$$

Pascal and Pascal (1991)

22. $$c^2x^2y'' - (-1)^m x^{2m+2}y'^m(2xy' + x^2y'')$$
$$+ [\{(m+2)/m\}c]^2y + c^2\{(3m+4)/m\}xy' - \lambda^2y = 0,$$

where c, m, and λ are constants.

The given DE arises in the similarity reduction of a nonlinear wave equation and has a particular solution $y = k/x$ provided that $c = \pm m\lambda/2$. Arrigo (1991)

23. $$g(y', y, x)y'' + f(y', y, x) = 0,$$

where $g(y', y, x)$ and $f(y', y, x)$ are continuous and single-valued functions. Adding and subtracting the terms (d^2y/dx^2) and w_0^2y, we rewrite the equation as

$$\{1 + g(y', y, x) - 1\}\frac{d^2y}{dx^2} + w_0^2y + f(y', y, x) - w_0^2y = 0.$$

Introducing the variables $\tau = w_0x$ and $v = dy/d\tau$, we arrive at the first order equation

$$\frac{dv}{dy} = -\frac{y+\delta_1}{v+\delta_2}, \tag{1}$$

where

$$\delta_1 = (1/w_0^2)f(w_0v, y, \tau/w_0) - y,$$
$$\delta_2 = \{g(w_0v, y, \tau/w_0) - 1\}v.$$

In the double delta method, δ_1 and δ_2 are assumed to be constant and Eqn. (1) is integrated to give

$$(y+\delta_1)^2 + (v+\delta_2)^2 = \text{constant} = r^2, \quad \text{say}.$$

The graphical method of solution of the general second order equation is based on this approximation. Nishikawa (1964), p.36

2.44 $f(x, y, y', y'') = 0$, f polynomial in y''

1. $$y''^2 - ay - b = 0,$$

where a and b are constants.

Putting $p(y) = y'(x)$, we obtain $(pp')^2 = ay + b$. An integration leads to the first order DE

$$y'^2 = (4/3a)(ay+b)^{3/2} + C,$$

which itself is expressible in the form of a quadrature. C is an arbitrary constant. Kamke (1959), p.596; Murphy (1960), p.407

2. $$(y'')^2 - A(x)y' = 0,$$

where $A(x)$ is an arbitrary function.

The solution of the given DE is

$$y = (1/4)\int \left\{\int A^{1/2}dx + k_1\right\}^2 dx + k_2,$$

where k_1 and k_2 are arbitrary constants. Cosgrove and Scoufis (1993)

3. $$(y'')^2 - \{A(x)y' + (A'(x) + c_1(c_1x+c_2)^{-1}A(x))y + C(x)\}^2\{c_1(xy'-y) + c_2y' + c_3\} = 0,$$

where $A(x)$ and $C(x)$ are arbitrary functions and c_j $(j = 1, 2, 3)$ are arbitrary constants.

The given DE has a first integral

$$y' = (c_1y - c_3)/(c_1x+c_2) + \{1/(4(c_1x+c_2))\}\{(c_1x+c_2)A(x)y + \int (c_1x+c_2)C(x)dx + K\}^2,$$

which is a Riccati equation. Letting $y = -4(c_1x+c_2)^{-1}A^{-2}w'/w$, we get a second order linear equation for w. Cosgrove and Scoufis (1993)

4. $$y''^2 + y'^2 - 1 = 0, y(\pi/2) = y'(\pi/2) = 1.$$

Put $y' = p, y'' = dp/dx, (dp/dx)^2 = 1 - p^2$. Integrating twice (with p later put equal to dy/dx, etc.) and using IC, we get $y = 1 - \cos x$. Reddick (1949), p.193

5. $$y''^2 - a - by'^2 = 0,$$

where a and b are constants.

Put $y' = p$ so that

$$\left[p\frac{dp}{dy}\right]^2 = a + bp^2. \tag{1}$$

After integrating (1) and then integrating $dy/dx = p(y)$, we obtain $by = a^{1/2}\cosh(C_1 + b^{1/2}x) + C_2; C_1$ and C_2 are arbitrary constants. Murphy (1960), p.407

6. $$(y'')^2 - A(x)\{y'^2 + \delta\} = 0,$$

where $A(x)$ is an arbitrary function and δ is a nonzero constant.

The solution of the given DE may be found to be

$$y = k_1 \int Qdx - (1/4)\delta k_1^{-1} \int Q^{-1}dx + k_2,$$

where $Q(x) = \exp\left\{\int A^{1/2}dx\right\}$, and k_1 and k_2 are arbitrary constants. Cosgrove and Scoufis (1993)

7. $$(y'')^2 - A(x)\{y'^2 + xy' - y\} = 0,$$

where $A(x)$ is an arbitrary function of x.

The solution of the given DE may be found to be

$$y = (1/4)\left\{\int Qdx + k_1\right\}\left\{\int Q^{-1}dx + k_2\right\} - x^2/4,$$

where $Q(x) = \exp\left\{\int A^{1/2}dx\right\}$, and k_1 and k_2 are arbitrary constants. Cosgrove and Scoufis (1993)

8. $$(y'')^2 - A(x)\{y'(xy' - y) + \gamma\} = 0,$$

where γ is a constant.

The solution of the given DE may be found to be

$$y = -2A^{-1}\{(u_1')^2 - \gamma(u_2')^2\} + (1/2)x\{(u_1)^2 - \gamma(u_2)^2\}, \tag{1}$$

where $u_1(x)$ and $u_2(x)$ are two linearly independent solutions of

$$u'' = (1/2)A^{-1}A'u' + (1/4)xAu, \tag{2}$$

subject to the Wronskian being normalized by

$$u_1u_2' - u_2u_1' = \pm A^{1/2}.$$

If $\gamma = 0$ in (1), u_2 drops out and then u_1 is any solution of (2). Cosgrove and Scoufis (1993)

9. $$(y'')^2 - A(x)\{c_1(xy' - y)^2 + c_2y'(xy' - y) + c_3(y')^2 + c_4(xy' - y) + c_5y' + c_6\} = 0,$$

where $A(x)$ is arbitrary and not all of $c_1, c_2, \ldots, c_5$ are zero.

Divide the given DE by $A(x)$, differentiate, and remove the common factor y'' (which itself may give a special solution linear in x). The result is the linear third order equation

$$y''' - (A'(x)/2A(x))y'' = A(x)\{c_1x(xy' - y) + c_2(xy' - y/2) + c_3y' + (1/2)(c_4x + c_5)\}. \tag{1}$$

If $c_1 = c_2 = c_3 = 0$, we can solve (1) by quadratures. Cosgrove and Scoufis (1993)

10. $$\boldsymbol{y''^2 + 2y'^3 - 2g_1xy' + 2g_1y - 2g_2y' + 4g_3 = 0,}$$

where g_i $(i = 1, 2, 3)$ are constants.

See Eqn. (4.4.4) for solution of the given DE. Boiti and Pempinelli (1980)

11. $$\boldsymbol{(y'')^2 + 4(y')^3 - A_1y' - A_2 = 0,}$$

where A_1 and A_2 are arbitrary complex constants.

The given DE is solved by putting $y' = -u(x)$, etc.:

$$y = -\int u dx + K_1, u = \mathcal{P}(x - K_2, A_1, -A_2),$$

where K_1 and K_2 are integration constants and $\mathcal{P}$ is the Weierstrass elliptic function. Cosgrove and Scoufis (1993)

12. $$\boldsymbol{(y'')^2 + 4(y')^3 + 2(xy' - y) = 0.}$$

The given DE is solved by

$$y = (1/2)(u')^2 - 2u^3 - xu, y' = -u,$$

where $u(x)$ satisfies $\mathrm{P_I}$:

$$u'' = 6u^2 + x.$$

Cosgrove and Scoufis (1993)

13. $$\boldsymbol{(y'')^2 + 4(y')^3 + 2y'(xy' - y) - A_1 = 0,}$$

where A_1 is an arbitrary constant.

The given DE is solved by

$$\begin{aligned} y &= (1/2)(u')^2 - (1/2)(u^2 + x/2)^2 - (\alpha + \epsilon_1/2)u, \\ y' &= (-\epsilon_1/2)u' - (1/2)(u^2 + x/2), \\ A_1 &= (1/4)(\alpha + \epsilon_1/2)^2, (\epsilon_1 = \pm 1). \end{aligned}$$

The function $u(x)$ satisfies $\mathrm{P_{II}}$:

$$u'' = 2u^3 + xu + \alpha.$$

Cosgrove and Scoufis (1993)

14. $$\boldsymbol{(y'')^2 + 4(y')^3 - 4(xy' - y)^2 - A_1y' - A_2 = 0,}$$

where A_1 and A_2 are arbitrary complex constants.

The given DE is solved by the following correspondence:

$$\begin{aligned} y &= (1/8)u^{-1}(u')^2 - (1/8)u^3 - (1/2)xu^2 - (1/2)(x^2 - \alpha + \epsilon_1)u \\ &\quad + (1/3)(\alpha - \epsilon_1)x + (\beta/4)u^{-1}, \\ y' &= (-\epsilon_1/2)u' - (1/2)u^2 - xu + (1/3)(\alpha - \epsilon_1), \end{aligned}$$

where $\epsilon_1 = \pm 1, A_1 = (4/3)(\alpha - \epsilon_1)^2 - 2\beta$, and $A_2 = -(4/3)(\alpha - \epsilon_1) - (8/27)(\alpha - \epsilon_1)^3$. The function $u(x)$ satisfies P_{IV}:

$$u'' = (1/2)u^{-1}(u')^2 + (3/2)u^3 + 4xu^2 + 2(x^2 - \alpha)u + \beta u^{-1}.$$

Cosgrove and Scoufis (1993)

15. $$(y'')^2 + 4(c_1 x + c_2)^{-6}\{c_1(xy' - y) + c_2 y' + c_3\}^3 = 0,$$

where c_1, c_2, and c_3 are arbitrary constants.

By linear transformation, etc., the given DE can be changed to the canonical form

$$(y'')^2 = -4(y')^3.$$

The solution may be found to be

$$y = -c_1(c_1 x + c_2)^2/\{1 - K_1(c_1 x + c_2)\} + K_2(c_1 x + c_2) + c_3/c_1 (c_1 \neq 0),$$

$$y = c_2^3(x - K_1)^{-1} + K_2 - (c_3/c_2)x (c_1 = 0, c_2 \neq 0),$$

where K_1 and K_2 are arbitrary constants. Cosgrove and Scoufis (1993)

16. $$(y'')^2 - \{A(x)y' + A'(x)y + C(x)\}^2 y' = 0,$$

where $A(x)$ and $C(x)$ are arbitrary functions.

The given DE has a first integral

$$y' = (1/4)\left\{A(x)y + \int C(x)dx + K\right\}^2,$$

which, being Riccati form, may be transformed to a second order linear DE. Cosgrove and Scoufis (1993)

17. $$a^2 y''^2 - (1 + y'^2)^3 = 0,$$

where a is a constant.

With $y' = p(x)$, we have $a^2 p'^2 = (1 + p^2)^3$. Now the variables separate. The solution of the given DE is

$$(x + C_1)^2 + (y + C_2)^2 = a^2;$$

C_1 and C_2 are arbitrary constants. Murphy (1960), p.407

18. $$y''^2 - xy'' + y' = 0.$$

With $y'(x) = p(x)$, we have $p'^2 - xp' + p = 0$. Solve for p' and integrate twice. The solutions are found to be

$$2y = C_1 x^2 - 2C_1^2 x + C_2; 12y = C_0 + x^3; y = C,$$

where C_1, C_2, C_0, and C are arbitrary constants. Murphy (1960), p.407

19. $$y''^2 - (1/2)x^2y'' + xy' - y = 0.$$

It is easy to check that a general solution of the given DE is

$$y = c_1x - (1/2)c_2x^2 + c_2^2,$$

where c_1 and c_2 are arbitrary constants. Martin and Reissner (1958), p.73

20. $$a^2y''^2 - 2axy'' + y' = 0,$$

where a is a constant

Writing the given DE as $(ay'' - x)^2 = (x^2 - y')$ and putting $y' = u(x)$, we get a first order DE:

$$a\frac{du}{dx} - x = \pm(x^2 - u)^{1/2}.$$

Now put $u = vx^2$. The DE in v is separable. Kamke (1959), p.596

21. $$fy''^2 + 2kxy'' + gy' + hx^2 + \ell x^3 = 0, y(0) = 0,$$

where $f, g, h, k,$ and ℓ are constants.

Put $y = x^3u, y' = x^2v$, where u and v are not zero when $x = 0$. We have $vx^2 = y' = x^2(xu' + 3u)$, so that $xu' + 3u = v$, and $y'' = x(xv' + 2v)$, so that on substituting into the original equation, we get $f(xv' + 2v)^2 + gv + h + 2k(xv' + 2v) + \ell x = 0$. This equation is quadratic in v' and hence can easily be integrated. Forsyth (1959), p.236

22. $$(y'' + \alpha y + \beta)^2 - 4y^2\{y'^2 + \alpha y^2 + 2\beta y + \gamma\} = 0,$$

where $\alpha, \beta,$ and γ are constants.

Taking the square root of the given DE, multiplying by $2y'$ and integrating, we easily obtain the first integral,

$$(y')^2 = y^4 + (k - \alpha)y^2 - 2\beta y - \gamma + (1/4)k^2, \tag{1}$$

where k is a constant. Equation (1) is integrable in terms of Jacobi elliptic functions. Cosgrove and Scoufis (1993)

23. $$(y'' + \alpha y + \beta)^2 - (4y^2/x^2)\{y'^2 + \alpha y^2 + 2\beta y + \gamma\} = 0,$$

where $\alpha, \beta,$ and γ are constants.

The solution of the given DE may be obtained in the form

$$y = (1/2)x\{u^{-1}(u' + \alpha/4) + u\}, \tag{1}$$

where u satisfies P_{III}:

$$u'' = (1/u)u'^2 - (1/x)u' + u^3 + (B_1/\alpha)u^2/x - B_2/(4x) - \alpha^2/(16u), \tag{2}$$

with

$$B_1 = 2\beta - \alpha - 2(\beta^2 - \alpha\gamma)^{1/2},$$

$$B_2 = 2\beta + \alpha + 2(\beta^2 - \alpha\gamma)^{1/2},$$

the square root taking either sign. With $\alpha = 0$ in the given DE, the solution is given by

$$y = (x/2)\{u^{-1}u' + 2\beta u\},$$

where u now satisfies a special case of P_{III}:

$$u'' = (1/u)(u')^2 - (1/x)u' + 4\beta^2 u^3 + 2(\gamma - \beta)u^2/x - 1/(2x). \qquad (3)$$

Equation (3) for $\beta = 1/2$ follows easily from (2) in the limit $\alpha \to 0$ such that $B_1/\alpha \to 2(\gamma - 1/2)$. Cosgrove and Scoufis (1993)

24. $$\boldsymbol{y''^2 - 2y'y'' + 2yy' - y^2 = 0.}$$

The given DE may be written as

$$(y'' - y')^2 - (y' - y)^2 = 0.$$

Therefore, two cases arise on factoring:

$$y'' - 2y' + y = 0. \qquad (1)$$

The linear DE (1) has the solution

$$y = (A + Cx)e^x,$$

where A and C are arbitrary constants.

$$y'' - y = 0. \qquad (2)$$

The linear DE (2) has the solution

$$y = Be^x + De^{-x},$$

where B and D are arbitrary constants. Zwillinger (1989), p.205

25. $$\boldsymbol{(y'' - \beta y' - 2\beta^2 y - 2\beta\gamma)^2 + 4(y^2 + \delta)^2\{y' + \beta y + \gamma\} = 0,}$$

where β, δ, and γ are constants.

The given DE may be shown to have the first integral

$$(y' + \beta y + \gamma)(y' - 2\beta y - 2\gamma)^2 = -(y^3 + 3\delta y + k)^2, \qquad (1)$$

where k is a constant. The solution of the first order DE (1) may be found explicitly for special choices of the parameters. Cosgrove and Scoufis (1993)

26. $$\boldsymbol{\{y'' - \alpha x y' + (\alpha - 2\alpha^2 x^2)y - 2\alpha\nu x\}^2 + 4(y^2 + \delta)^2(y' + \alpha x y + \nu) = 0,}$$

where α, ν, and δ are constants.

The solution of the given DE may be found in the form

$$y = (1/2)\{u^{-1}u' + u + \alpha x + (\nu - \delta + \alpha)u^{-1}\},$$

where $u(x)$ satisfies P_{IV}:

$$u'' = (1/2)u^{-1}(u')^2 - (1/2)u^3 - 2\alpha x u^2 - \{(3/2)\alpha^2 x^2 + \nu + \delta\}u - (1/2)(\nu - \delta + \alpha)^2 u^{-1}.$$

When $\delta = \nu$, there is a limiting solution of the given DE, expressible in terms of parabolic cylindrical functions:

$$y = v^{-1}v' + \alpha x/2, v'' = -\{(3/4)\alpha^2 x^2 + \nu + \alpha/2\}v.$$

Cosgrove and Scoufis (1993)

27. $$\boldsymbol{y''^2 + [a_1(x)y' + a_2(x)y + a_3(x)]y'' + b_1(x)y'^2}$$
$$\boldsymbol{+ b_2(x)yy' + b_3(x)y' + c_1(x)y^2 + c_2(x)y + m(x) = 0,}$$

where a_i, b_j, c_k, and m are functions of x holomorphic in some region G.

First solving for y'', conditions in the form of six theorems are derived under which all solutions of the given "irreducible" DE are single valued in G. Solutions with nonstationary critical points are also discussed.

As a special simple example, if $a_1^2 - 4b_1 \equiv 0, a_1a_2 - 2b_2 \equiv 0, a_1a_3 - 2b_3 \equiv 0$, in G, then the DE has the one-parameter family of solutions

$$y = y_1 + (x - x_0)\left[r_0 + \sum_{k=2}^{\infty} \phi_k(x - x_0)^{k/2}\right]^2,$$

with a nonstationary algebraic critical point. Here y_1, x_0, r_0, and ϕ_k are constants. Sulima (1973)

28. $$\boldsymbol{\{y'' + A(x)y' + B(x)y + C(x)\}^2 + D(x)(y')^2 + E(x)yy' + F(x)y^2}$$
$$\boldsymbol{+ G(x)y' + H(x)y + I(x) = 0.}$$

The given DE, which may be written as $S^2 = R$, where R is not a square, is called an inhomogeneous Appell equation if any of the following equivalent conditions hold:

(a) The solutions of the first order equation $R(x, y, y') = 0$ are singular integrals of the given DE.

(b) There exists $\lambda(x)$ such that $(dR/dx) - \lambda R$ is divisible by S.

(c) The general solution of the given DE, $y(x; k_1, k_2)$, is rational in the constants of integration.

It is shown that the given DE may be reduced to a second order linear equation. Cosgrove and Scoufis (1993)

29. $$\boldsymbol{(y'' + y^3y')^2 - y^2y'^2(4y' + y^4) = 0.}$$

The given DE has the general solution

$$y(x) = \alpha\tan(\alpha^3 x + \beta),$$

where α and β are arbitrary constants. Kruskal and Clarkson (1992)

30. $$\sum_{p,q=0}^{2} f_{pq}(x) y^{(p)} y^{(q)} = 0.$$

The solutions can be expressed as

$$y = C_1^2 u_1 + C_1 C_2 u_2 + C_2^2 u_3,$$

where u_1, u_2, u_3 are solutions of a linear third order DE. There are also singular solutions that satisfy a linear second order DE. Kamke (1959), p.597; Murphy (1960), p.408

31. $$xy''^2 - 2y'y'' + ax = 0,$$

where a is a constant

With $y'(x) = p(x)$, we have

$$xp'^2 - 2pp' + ax = 0. \tag{1}$$

A solution of (1) is

$$2Cp = a + C^2x^2;$$

C is an arbitrary constant. The solution of the given DE is

$$2C_1 y = (C_1^2 x^3/3) + ax + C_2,$$

C_1 and C_2 are arbitrary constants. For $a = 1$, we have another solution,

$$2y = C_0 + x^2,$$

where C_0 is an arbitrary constant. Murphy (1960), p.407

32. $$a^2 xy''^2 + yy'^2 y'' - b^2 x = 0, y(0) = 0,$$

where a and b are constants.

There are two forms of the solution,

$$y = \pm a x^{1/2} P(x^{1/2}), y = \pm(1/2)(b/a) x^2 Q(x),$$

where $P(x)$ and $Q(x)$ are regular functions of x, each of which is unity when $x = 0$. This may be checked by a dominant balance argument. A series solution may be obtained. Forsyth (1959), p.234

33. $$a(a - 2x)y''^2 + 2ay'y''(1 + y'^2) + (1 + y'^2)^3 = 0,$$

where a is a constant.

The solution of the given DE is

$$(y - \alpha)^2 + (x - \beta)^2 = 2ax, \tag{1}$$

where α and β are arbitrary constants. Equation (1) can be interpreted as the doubly infinite family of circles. Forsyth (1959), p.256

34. $$x^2(y'')^2 - y(y - 2k/\rho^{n+1}) = 0,$$

where k, ρ, and n are constants.

The solution is approximated by that of $xy'' - y = 0$; the explicit solution of the latter is obtainable in terms of Bessel functions. The error in the approximation is studied. Mignosi (1964)

35. $$x^2(y'')^2 + 4(xy' - y)y'^2 - A_1(xy' - y)^2 - A_2(xy' - y) - A_3y' - A_4 = 0,$$

where A_j $(j = 1, \dots, 4)$ are arbitrary complex constants.

The given DE is solved by the following correspondence, provided that A_1 and A_2 are not both zero:

$$\begin{aligned}
y &= (1/(4u))\{xu'/(u-1) - u\}^2 - (1/4)\{1 - (2\alpha)^{1/2}\}^2(u-1) \\
&\quad - (\beta/2)(u-1)/u + (\nu/4)x(u+1)/(u-1) + (\delta/2)x^2u/(u-1)^2, \\
y' &= -\{x/(4u(u-1))\}\{u' - (2\alpha)^{1/2}u(u-1)/x\}^2 - (\beta/2x)(u-1)/u \\
&\quad - \gamma/4 - (\delta/2)xu/(u-1), \\
A_1 &= -2\delta, A_2 = (\gamma^2/4) + 2\beta\delta - \delta\{1 - (2\alpha)^{1/2}\}^2, \\
A_3 &= \beta\gamma + (\gamma/2)\{1 - (2\alpha)^{1/2}\}^2, \\
A_4 &= (\gamma^2/8)\{(1 - (2\alpha)^{1/2})^2 - 2\beta\} - (\delta/8)\{(1 - (2\alpha)^{1/2})^2 + 2\beta\}^2,
\end{aligned}$$

where $u(x)$ satisfies P_V with suitable constants.

When $A_1 = 0, A_2$ is unrestricted, the given DE is solved by the following correspondence:

$$\begin{aligned}
y &= (1/4)u^{-2}\{xu' - u\}^2 - (1/16)\alpha u^2 - (1/8)(\beta + 2\alpha^{1/2})u + (1/8)\gamma xu^{-1} + (1/16)\delta x^2u^{-2}, \\
y' &= -(1/4)\alpha^{1/2}u' - (1/8)x^{-1}(\alpha u^2 + \beta u), \\
A_2 &= -(1/16)\alpha\delta, A_3 = (1/16)\gamma(\beta + 2\alpha^{1/2}), \\
A_4 &= (1/256)\alpha\gamma^2 - (1/256)\delta(\beta + 2\alpha^{1/2})^2.
\end{aligned}$$

The function $u(x)$ satisfies

$$u'' = u^{-1}u'^2 - x^{-1}u' + (1/4)x^{-2}(\alpha u^3 + \beta u^2) + (\gamma/4)x^{-1} + (\delta/4)u^{-1}. \tag{1}$$

The change of variables $\overline{x} = x^2, u = \overline{x}\,\overline{u}$ transforms (1) to P_{III}:

$$u'' = u^{-1}(u')^2 - (\overline{x})^{-1}u' + \alpha u^3 + (\overline{x})^{-1}(\beta u^2 + \gamma) + \delta u^{-1},$$

where the prime denotes $d/d\overline{x}$, and the bar on u has been dropped. Cosgrove and Scoufis (1993)

36. $$(xy'' - y')^2 - 1 - y''^2 = 0.$$

With $y'(x) = p(x)$, we have

$$(1 - x^2)p'^2 + 2xpp' + 1 - p^2 = 0. \tag{1}$$

Equation (1) integrates to yield

$$(p - Cx)^2 = 1 + C^2;$$

C is an arbitrary constant. Hence the solution of the original DE is

$$2y = x(1 - x^2)^{1/2} + \sin^{-1} x + C_0,$$

where C_0 is an arbitrary constant. Other solutions of the given DE are

$$y = (C_1/2)x^2 \pm (1 + C_1^2)^{1/2}x + C_2, y = x + C_3,$$

where C_1, C_2, and C_3 are arbitrary constants. Murphy (1960), p.407

37. $$3x^2y''^2 - 2(3xy' + y)y'' + 4y'^2 = 0.$$

We obtain either

$$y = C_1^2x^2 + C_1C_2x + C_2^2$$

or the solutions

$$y = Cx^{1\pm(2/3^{1/2})}$$

of the linear Cauchy-Euler DE

$$3x^2y'' - 3xy' - y = 0.$$

The given DE is equidimensional in both x and y. C_1, C_2, and C are arbitrary constants. Kamke (1959), p.597; Murphy (1960), p.408

38. $$2(x^2 + 1)y''^2 - x(4y' + x)y'' + 2(y' + x)y' - 2y = 0.$$

Differentiating the given DE once, we have

$$\{4(x^2 + 1)y'' - 4xy' - x^2\}y''' = 0.$$

Hence the solution is either

$$16y = (X + C)^2 + 2x(X + C)(x^2 + 1)^{1/2} - 3x^2,$$

with $X = \ln\{x + (x^2 + 1)^{1/2}\}$ or

$$y = C_1x^2 + C_2x + 4C_1^2 + C_2^2;$$

C, C_1, and C_2 are arbitrary constants. Kamke (1959), p.597; Murphy (1960), p.407

39. $$x^3y''^2 + 18yy' = 0.$$

$y = C/x^2$ is an exact solution of the given DE, where C is a constant. Orlov (1980)

40. $$(9x^3 - 2x^2)y''^2 - 6x(6x-1)y'y'' - 6yy'' + 36xy'^2 = 0.$$

We have either

$$y = C_1^2 x^3 + C_1 C_2 x + C_2^2$$

or the solutions

$$y = CEx(4x-1)^{1/2} \quad \text{and} \quad (C/E)x(4x-1)^{1/2},$$

where $E = \exp\left(\int [\{2(9x^2-2x)^{1/2}\}/(4x^2-x)]dx\right)$ of the linear DE

$$(9x^3 - 2x^2)y'' - 3x(6x-1)y' - 3y = 0.$$

Here C_i $(i = 1, 2, 3)$ and C are arbitrary constants. The given DE is equidimensional in y.
Kamke (1959), p.597; Murphy (1960), p.408

41. $$x^2(x-1)^2(y'')^2 + 4y'(xy'-y)^2 - 4(xy'-y)y'^2 - A_1 y'^2 - A_2(xy'-y) - A_3 y' - A_4 = 0,$$

where A_j $(j = 1, \ldots, 4)$ are arbitrary complex constants.

The solution is found by Bäcklund correspondence,

$$\begin{aligned} y &= [x^2(x-1)^2/\{4u(u-1)(u-x)\}]\{u' - u(u-1)/(x(x-1))\}^2 \\ &\quad + (1/8)\{1-(2\alpha)^{1/2}\}^2(1-2u) - (\beta/4)(1-2x/u) \\ &\quad - (\gamma/4)\{1-2(x-1)/(u-1)\} + \{1/8-(1/4)\delta\}\{1-2x(u-1)/(u-x)\}, \\ y' &= -\{x(x-1)/(4u(u-1))\}\{u' - (2\alpha)^{1/2}u(u-1)/(x(x-1))\}^2 \\ &\quad - (\beta/2)(u-x)/((x-1)u) - (\gamma/2)(u-x)/(x(u-1)), \end{aligned}$$

where $(2\alpha)^{1/2}$ can take either sign and

$$\begin{aligned} A_1 &= \alpha - \beta + \gamma - \delta - \sqrt{2\alpha} + 1, \\ A_2 &= (\beta+\gamma)\left(\alpha + \delta - \sqrt{2\alpha}\right), \\ A_3 &= (\gamma - \beta)\left(\alpha - \delta - \sqrt{2\alpha} + 1\right) + (1/4)\left(\alpha - \beta - \gamma + \delta - \sqrt{2\alpha}\right)^2, \\ A_4 &= (1/4)(\gamma-\beta)\left(\alpha + \delta - \sqrt{2\alpha}\right)^2 + (1/4)(\beta+\gamma)^2\left(\alpha - \delta - \sqrt{2\alpha} + 1\right). \end{aligned}$$

The function $u(x)$ satisfies $\mathrm{P_{VI}}$. Cosgrove and Scoufis (1993)

42. $$a(x)y''^2 + b(x)y'y'' + c(x)yy'' + d(x)y'^2 + e(x)yy' + f(x)y^2 = 0,$$

where $a(x), \ldots, f(x)$ are meromorphic functions of a complex variable x on a region ω of the complex plane, the left member $Q(x, y, y', y'')$ of the given DE has no nontrivial factorization, and $a(x) \not\equiv 0$.

The given DE can, under certain conditions, be written in the equivalent form

$$(2ay'' + by' + cy)^2 + A_2 y'^2 + A_3 yy' + A_4 y^2 = 0. \tag{1}$$

Setting $D = A_3^2 - 4A_2A_4$, we must have $D \not\equiv 0$ because $Q(y, y', y'')$ has no nontrivial factorization.

Thus, when the given DE can be put in the form (1) such that $A_2 \not\equiv 0$,

$$\begin{aligned} b &\equiv [(2A_2' + 2A_3)/A_2 - D'/D]a, \\ c &\equiv [(A_3' + 2A_4)/A_2 - A_3 D'/(2A_2 D)]a, \end{aligned}$$

and a, A_2, A_3, A_4 are meromorphic functions such that $a \not\equiv 0, A_2 \not\equiv 0, D \not\equiv 0$, then the problem of solving the given DE is reduced to solving a Riccati equation and a first order linear homogeneous DE. The conditions for the solutions of the given DE to be free from movable branch points are also found. Chalkley (1987)

43. $$yy''^2 - ae^{2x} = 0,$$

where a is a constant.

With $y = x^{4/3}u$, we get

$$\{x^2u'' + (8/3)xu' + (4/9)u\}u^{1/2} = \pm a^{1/2}e^x.$$

For the IC $y(0) = y'(0) = 0$, a power series solution of the given DE can be found. Kamke (1959), p.598

44. $$yy''^2 - 2y'^2y'' - y^3 = 0.$$

Attempt $v = y'/y$ to solve the given homogeneous DE in y, y', and y''. Forsyth (1959), p.305

45. $$yy''^2 - 2y'^2y'' + (1+c)yy'^2 - cy^3 = 0,$$

where $c = 0, 1$ or $0 < c < 1$.

Attempt $v = y'/y$ to solve the given DE, which is homogeneous in y, y', and y''. The equation for v is

$$v' = [(v^2 - 1)(v^2 - c)]^{1/2},$$

which is solvable in terms of elliptic functions. Forsyth (1959), p.305

46. $$[yy'' - y'^2]^2 + 4yy'^3 = 0.$$

The given DE is of degree 4; the general solution is easily found in the exponential form

$$y = \alpha \exp\{(x - x_0)^{-1}\},$$

where α and x_0 are arbitrary constants. Kruskal and Clarkson (1992)

47. $$(1 + y'^2 + yy'')^2 - (1 + y'^2)^3 = 0.$$

With $y'(x) = p(y)$, we have $ypp' = (1+p^2)[(1+p^2)^{1/2} - 1]$ with the solution in terms of p^2, $(C-y)^2p^2 = y(2C - y)$; C is an arbitrary constant. Now the variables separate and the solution of the given DE is

$$(x - C_1)^2 + (y - C_2)^2 = C_2^2;$$

C_1 and C_2 are arbitrary constants. Murphy (1960), p.408

48. $$(a^2y^2 - b^2)y''^2 - 2a^2yy'^2y'' + (a^2y'^2 - 1)y'^2 = 0,$$

where a and b are constants.

Introducing $p(y) = y'(x)$, with y as the independent variable, and differentiating the resulting DE with respect to y, we get

$$\{(a^2y^2 - b^2)p' - a^2yp\}p'' = 0;$$

hence the solutions are

$$y = C_1e^{C_2x} \pm (1/a)(b^2 + C_2^{-2})^{1/2}$$

and

$$y = \pm(b/a)\cos[(x + C)/b];$$

C_1, C_2, and C are arbitrary constants. Kamke (1959), p.598; Murphy (1960), p.408

49. $$[\gamma(yy'' + y'^2) + yy']^2 - y(y' - 2\alpha y)^2(2\gamma y' + y) = 0,$$

where γ and α are constants.

The given DE is equivalent to

$$yy' + zz' = \alpha z^2, z'(x) = \beta y(x), \alpha\beta \neq 0,$$

where $\gamma = 2\alpha/\beta^2$. For the analysis of equivalent system and the BVP $y(0) = 1, y'(1) = 0$. Kuznetskii (1973)

50. $$(x^2yy'' - x^2y'^2 + y^2)^2 = 4xy(xy' - y)^3.$$

With $y = xu(x)$ we get the autonomous DE

$$(uu'' - u'^2)^2 = 4uu'^3;$$

further, putting $v = u'/u$, we have $v'^2 = 4v^3$. Hence it follows that $(x + C)^2v = 1$ and so $y/x = u = C_0\exp[1/(C - x)]; C_0$ and C are arbitrary constants. Kamke (1959), p.598; Murphy (1960), p.409

51. $$y''^3 - a(y - b)y^4 = 0.$$

Follow Eqn. (2.44.1). Forsyth (1959), p.305

52. $$y''^3 - c(y - a)\{y - (1/3)(2a + b)\}^3(y - b)^5 = 0.$$

Follow Eqn. (2.44.1). Forsyth (1959), p.305

53. $$fy''^3 + gy' + hx^2 + kxy'' + \ell y = 0,$$

where f, g, h, k, and ℓ are constants, $y(0) = 0, y'(0) = 0$.

It is shown that there are three kinds of solutions; the first two are such that $y''(0) = 0, y'''(0) = \infty$, while for the third, $y''(0) = 0, y'''(0) = -2h/(g+2k)$. Forsyth (1959), p.239

54. $$y''^3 - 12y'(xy'' - 2y') = 0.$$

With $y' = p(x)$, we have

$$p'^3 - 12xpp' + 24p^2 = 0. \tag{1}$$

The solutions of the given DE with the help of (1) may be found to be

$$y = C_1(x - C_1)^3 + C_2; 9y = x^4 + C_0; y = C,$$

where C_1, C_2, C_0, and C are arbitrary constants. Murphy (1960), p.409

55. $$(yy'' - y'^2)^3 + (27/2)yy'^5 = 0.$$

An exact solution is $y = \alpha e^{(x-c)^{-1/2}}$, where α and c are arbitrary constants; $y = c$ is an essential singularity. Forsyth (1959), p.277

56. $$(2yy'' - y'^2)^3 + 32y''(xy'' - y')^3 = 0.$$

The solutions of the given DE are

$$C_1C_2^3y = (C_1^2x + 1)^2 + 2C_2^2 \text{ and } y^2 = 8x;$$

C_1, C_2, and C_3 are arbitrary constants. Kamke (1959), p.598; Murphy (1960), p.409

57. $$fy''^m + gy' + hx^2 + kxy'' + \ell y = 0,$$

where m is an integer greater than 3, and $y(0) = y'(0) = 0$; f, g, h, k, and ℓ are constants. Forsyth (1959), p.241

58. $$\sum_{\nu=1}^{m} a_\nu (y'')^{2\nu-1} + by = 0, a_\nu \geq 0, a_1 > 0,$$

where m is a positive integer.

It may be proved that all solutions of the given DE in the neighborhood of the origin are periodic. Amel'kin and Gaishun (1971)

59. $$\phi(y'') + ay' + by = 0,$$

where

$$\phi(z) = \sum_{\nu=1}^{m} a_\nu z^{2\nu-1}, a_\nu \geq 0, a_1 > 0 \tag{1}$$

and a and b are constants.

For this equation it is shown that:

(a) If $a > 0, b > 0$, then the point $(0, 0)$ is stable in the large.

(b) If $a = 0, b > 0$, all solutions of the given system are periodic in the neighborhood of the origin.

(c) $a = 0, b = 0$, then $y = C_1 x + C_2$, where C_1 and C_2 are arbitrary constants.

If $m = \infty$ in (1), we must also assume that the series converges in some neighborhood of the origin. Amel'kin and Gaishun (1971)

60. $$F(y'', y', y) = 0,$$

where F is a homogeneous polynomial in its arguments.

The substitution $v = y'/y$ may change it to more convenient form, such as the Briot–Bouquet equation. Forsyth (1959), p.305

61. $$F(y'', y', y, x) = 0,$$

where F is a homogeneous polynomial in the arguments y'', y', y and the coefficients are analytic functions of x.

Conditions are stated for the solution of the given DE to possess no movable algebraic singularities. Forsyth (1959), p.305

62. $$F(y'', y', y, x) = 0,$$

where F is a polynomial in y, y', and y''.

To obtain majorants for arbitrary positive solutions of the given DE on the ray $[0, \infty)$, it is assumed to be in the form $F(x, y, y'/y, (y'/y)') = 0$. The majorants are expressed in terms of the coefficients of the DE. Strelitz (1981)

2.45 $f(x, y, y', y'') = 0$, f not polynomial in y''

1. $$(ay''^2 + by'^2)^{1/2} + cyy'' + dy'^2 = 0,$$

where a, b, c, and d are constants.

With $p(y) = y'(x)$, we get

$$(ap'^2 + b)^{1/2} + cyp' + dp = 0.$$

Squaring, etc., and solving for p', we may separate the variables p and y and hence integrate. Kamke (1959), p.598

2. $$\phi(y'') + by = 0, b > 0,$$

where $\phi(z)$ is C^1 for $z \in (-\infty, \infty)$ such that $\phi(z) \geq \delta > 0, b$ is a real number, and $\phi(0) = 0$.

It is shown that all solutions of the given DE are periodic in the neighborhood of the origin. Amel'kin and Gaishun (1971)

3. $$\phi(y'') + by + \mu\Gamma(x) = 0,$$

where $\phi(z)$ is holomorphic in the neighborhood of the origin and $\phi'(z) \geq \delta > 0$; μ is a small parameter and

$$\Gamma(x) = \int_0^x \gamma(\tau)d\tau + k$$

is C^1 with the positive period w, k is a constant, and $\int_0^w \gamma(\tau)d\tau = 0$.

Introducing $y'(x) = z(x), z'(x) = p$, we get the system

$$z' = p, p' = -(bz + \mu\gamma(x))/\phi'(p). \tag{1}$$

Introducing the variable

$$z_1 = z, z_2 = \phi(p),$$

the system (1) becomes

$$\frac{dz_1}{dx} = \phi^{-1}(z_2), \frac{dz_2}{dx} = -bz_1 - \mu\gamma(x), \tag{2}$$

where ϕ^{-1} is the inverse of the function ϕ. The system (2) is shown first to have a singularity of the center type for $\mu = 0$. By further argument, the existence of a periodic solution is established for $\mu \neq 0$. Amel'kin and Gaishun (1971)

4. $$\phi(y'') + ay' = 0.$$

Putting $y' = z$, the order of this DE can be reduced. Amel'kin and Gaishun (1971)

5. $$\phi(y'') + ay' + by = 0,$$

where $\phi(z)$ is C^1 defined for $z \in (-\infty, \infty)$ such that $|\phi'(x)| \geq \delta > 0$. For definiteness it is assumed that $\phi'(z) \geq \delta > 0, z \in (-\infty, \infty)$, a and b are real numbers, and $\phi(0) = 0$. The condition $\phi'(z) > 0$ ensures that the solution of the given DE can be continued indefinitely in both directions.

Introducing $y' = z(x)$ or $y = \int z(x)dx$, we obtain

$$\phi(z') + az + b\int z(x)dx = 0. \tag{1}$$

Differentiating this relation with respect to x and setting $z' = p$, we obtain the system

$$z' = p, p' = -bz - ap/\phi'(p). \tag{2}$$

For $a > 0$ and $b > 0$, the zero solution of the given equation are shown to be stable in the large. This is accomplished by using the Lyapunov function

$$v = (1/2)bz^2 + \int_0^p p\phi'(p)dp$$

for the system (2). For $a < 0$ and $b > 0$, the zero solution of the given DE is shown to be unstable. Amel'kin and Gaishun (1971)

6. $$\boldsymbol{f(y'') + xy'' - y' = 0.}$$

With $y'(x) = p(x)$, we have

$$p = xp' + f(p'). \tag{1}$$

The solution of Clairaut's equation (1) is

$$p = C_1 x + f(C_1);$$

C_1 is an arbitrary constant. Hence the solution of the given DE is

$$2y = C_1 x^2 + 2x f(C_1) + C_2;$$

C_2 is another arbitrary constant. Murphy (1960), p.409

7. $$\boldsymbol{f(x, y'') = 0.}$$

Put $y' = p(x)$ so that we have a first order DE $f(x, (dp/dx)) = 0$ with the dependent variable missing [see Murphy (1960), p.10]. Murphy (1960), p.409

8. $$\boldsymbol{f(y, y'') = 0.}$$

With $y' = p(y), y'' = p(dp/dy)$, we get $f(y, p(dp/dy)) = 0$, a first order DE in p. Murphy (1960), p.409

9. $$\boldsymbol{y' f(y''/y') - y'^2 + yy'' = 0.}$$

With $y'(x) = p(y)$, we have $p[p - yp'(y) - f(p')] = 0$ with one solution

$$p = C_1 y + f(C_1);$$

C_1 is an arbitrary constant. Hence the solution of the given DE is

$$C_1 x = \ln[C_1 y + f(C_1)] + C_2;$$

C_2 is another arbitrary constant. $y = C$, a constant, is another solution. Murphy (1960), p.409

10. $$\boldsymbol{F[x - \{(y'^2 + 1)/y''\}y', y + \{(y'^2 + 1)/y''\}, \{(y'^2 + 1)\}^{3/2}/y''] = 0.}$$

This is Clairaut's DE of second order. If the (arbitrary) numbers a, b, r belong to the region of definition of the function $F(u, v, w)$ and satisfy the relation $F(a, b, r) = 0$, then one solution is the circle

$$(x - a)^2 + (y - b)^2 = r^2.$$

Kamke (1959), p.599

11. $$\boldsymbol{f(y', y'') = 0.}$$

Put $y' = p(x), y'' = dp/dx$, to get the first order DE

$$f\left(p, \frac{dp}{dx}\right) = 0.$$

Murphy (1960), p.409

12. $$\boldsymbol{F\{y+y'^2/y'',x+(y'-y'^3)/(2y'')\}=0.}$$

The solution of the given DE is

$$(y-a)^2=2C(x-b)+C^2,$$

where $F(a,b)=0$. Whether it is possible to obtain all solutions by following this method needs further investigation. Kamke (1959), p.599

13. $$\boldsymbol{F[y'',y'-xy'',y-xy'+(1/2)x^2y'']=0.}$$

The solutions of the given DE are

$$y=(1/2)ax^2+bx+c,$$

where $F(a,b,c)=0; a,b,$ and c are arbitrary constants. Kamke (1959), p.599; Murphy (1960), p.409

14. $$\boldsymbol{F(y'',y',y,x)=0.}$$

Suppose that the given DE is homogeneous of degree k in the variables y,y',y'' so that

$$F(ty'',ty',ty,x)=t^kF(y'',y',y,x)$$

for any number t. Hence if we put $t=1/y$, we get

$$F(y''/y,y'/y,1,x)=[1/(yu)]F(y'',y',y,x).$$

Thus, the given DE can be written as

$$F(y''/y,y'/y,1,x)=0. \tag{1}$$

Now we introduce $u=y'/y$ so that $y''/y=u'+u^2$. Equation (1) now changes to the first order DE

$$F(u'+u^2,u,1,x)=0. \tag{2}$$

Suppose that Eqn. (2) can be solved and its integral is

$$f(u,x)=C_1; \tag{3}$$

on solving for u from (3), we have

$$y'/y=u=g(x,C_1). \tag{4}$$

Integrating (4), we have

$$\ln y=\int g(x,C_1)dx+C_2.$$

Goldstein and Braun (1973), p.79

15. $\boldsymbol{F(y'', y', y, x) = 0}$.

The given DE is said to be isobaric if there exist numbers ℓ and k such that

$$F(t^{\ell-2}y'', t^{\ell-1}y', t^{\ell}y, tx) = t^k F(y'', y', y, x) = 0.$$

In particular, upon replacing t by $1/x$, we get

$$F(y''/x^{\ell-2}, y'/x^{\ell-1}, y/x^{\ell}, 1) = (1/x^k)F(y'', y', y, x) = 0. \tag{1}$$

Now, if we introduce in Eqn. (1) the new variables

$$u = x^{-\ell}y$$

and

$$\xi = \ln x,$$

we get

$$F\left[\frac{d^2u}{d\xi^2} + (2\ell - 1)\frac{du}{d\xi} + \ell(\ell-1)u, \frac{du}{d\xi} + \ell u, u, 1\right] = 0. \tag{2}$$

The independent variable does not appear explicitly in (2), so this second order DE can be changed to first order by putting $du/d\xi = p$, $d^2u/d\xi^2 = p(dp/du)$, and hence solved. Goldstein and Braun (1973), p.82

16. $\boldsymbol{F(y'', y', x) = 0}$.

Put $y' = p$ so that $F((dp/dx), p, x) = 0$. We may solve this first order equation in the form

$$f(x, p) = C_1, \text{ say.} \tag{1}$$

Now, two cases arise:

(a) Equation (1) can be solved for p:

$$\frac{dy}{dx} = p = g(x, C_1), \text{ say.} \tag{2}$$

Then, by integration, we have

$$y = \int g(x, C_1)dx + C_2.$$

(b) Equation (2) can be written as

$$x = h(p, C_1). \tag{3}$$

Then, substituting from (3) in $p = (dy/dp)(dp/dx)$, we have

$$\frac{dy}{dp} = p\frac{dx}{dp} = p\frac{dh}{dp}.$$

Therefore,

$$\begin{aligned} y &= \int p\frac{dh}{dp}dp + C_2 \\ &= ph(p, C_1) - \int h(p, C_1)dp + C_2. \end{aligned} \tag{4}$$

Equations (3) and (4) yield a solution of the given DE parameterically, p being the parameter. C_1 and C_2 are the arbitrary constants. Goldstein and Braun (1973), p.74

2.46 $y'' + f(y) = a\ \sin(\omega x + \delta)$

1. $$y'' + y^2 = t\ \sin\ x, y(0) = y(\pi) = 0,$$

where t is a parameter.

It is proved that given $N_0 \in N$ there exists $t_{N_0} > 0$ such that for all $t \geq t_{N_0}$, the given problem has at least N_0 solutions. Lupo, Solimini, and Srikanth (1988)

2. $$y'' + y + \mu y^2 = \epsilon\ \cos\ wx (\epsilon \neq 0),$$

where μ, w, and ϵ are constants.

Using topological methods, the existence of $2\pi w^{-1}$ periodic solutions for all sufficiently small ϵ and for $w > 2\pi$ is proved. Ezeilo (1982)

3. $$y'' + 9y + \epsilon y^2 = \Gamma\ \cos\ x,$$

where $\epsilon \ll 1$ and Γ are constants. The 2π-periodic perturbation solution to $O(\epsilon^2)$ is

$$\begin{aligned} y &= a_0 \cos 3x + b_0\ \sin\ 3x + (1/8)\Gamma \cos x - (1/18)\epsilon(a_0^2 + b_0^2) \\ &\quad - (\epsilon/40)\Gamma(a_0 \cos 2x + b_0 \sin 2x) + \cdots. \end{aligned}$$

Jordan and Smith (1977), p.149

4. $$y'' + \mu(y + by^2) = e\ \cos\ x,$$

where μ, b, and e are constants.

Solutions with period equal to 2π are given for $e \ll 1$. Using numerical methods, even, 2π-periodic solutions of the given DE for arbitrary values of the parameters μ, b, and e are obtained. Stability of these solutions in the linear approximation is considered. Zlatoustov and Shipovskikh (1984)

5. $$y'' + \Omega^2 y + \epsilon y^2 = \Gamma\ \cos\ x, \epsilon \ll 1,$$

where Ω, ϵ, and Γ are parameters.

The first few harmonics in a 2π-periodic solution are

$$\begin{aligned} y &= -\epsilon\Gamma^2/\{2\Omega^2(\Omega^2 - 1)^2\} + \{\Gamma/(\Omega^2 - 1)\} \cos x \\ &\quad + \{\epsilon\Gamma^2/2(\Omega^2 - 1)^2(\Omega^2 - 4)\} \cos 2x + \cdots. \end{aligned}$$

The expansion fails near integral values of Ω. Jordan and Smith (1977), p.148

6. $$y'' + \Omega^2 y - \epsilon - y^2 = \Gamma\ \cos\ x;$$

$\epsilon > 0, \Omega$, and Γ are constants.

A 2π-periodic perturbation solution is sought in the form $y = y_0(x) + \epsilon y_1(x) + \cdots$. A perturbation solution without secular terms to $O(\epsilon^2)$ is found to be

$$\begin{aligned} y(\epsilon, x) &= \{\Gamma/(\Omega^2 - 1)\} \cos x + [\epsilon\Gamma^2/\{2\Omega^2(\Omega^2 - 1)^2\}] \\ &\quad + [\epsilon\Gamma^2 \cos 2x/\{2(\Omega^2 - 1)^2(\Omega^2 - 4)\}], \end{aligned}$$

provided that $\Omega^2 \neq 1, 4$. For $\Omega \approx 1$, we write $\Omega^2 = 1 + \beta, \Gamma = \epsilon\gamma$. Now, a perturbation solution to $O(\epsilon^2)$ is found to be

$$y(\epsilon, x) \approx (\gamma/\beta)\cos x + \epsilon[(1/2)\gamma^2/\beta^2 - ((1/6)\gamma^2/\beta^2)\cos 2x + a_1 \cos x + b_1 \sin x].$$

Jordan and Smith (1977), p.136

7. $$\mathbf{y'' + (y/4) + 0.1y^3 = \cos x}$$

The 2π-periodic solution of the given DE by a regular perturbation scheme is

$$y(\epsilon, x) = -(4/3)\cos x - \epsilon[(64/27)\cos x + (64/945)\cos 3x] + O(\epsilon^2).$$

With $\epsilon = 0.1$, we have

$$y(\epsilon, x) \simeq -1.57\cos x - 0.0068\cos 3x.$$

Jordan and Smith (1977), p.127

8. $$\mathbf{y'' + y^3 = \sin x.}$$

Since the equation is invariant under $x \to -x$ and $y \to -y$, the periodic solutions are sought in the form

$$y(x) = \sum_{k=1}^{8} b_{2k-1}\sin(2k-1)x,$$

where $b_1 = 1.431189037$,

$$\begin{aligned} b_3 &= -0.126911530, b_5 = 0.009754734, b_7 = -0.000763601, \\ b_9 &= 0.000059845, b_{11} = -0.000004691, b_{13} = 0.000000368, \\ b_{15} &= -0.000000029. \end{aligned}$$

Using the coefficients, the ratios of the coefficients (except the first and last) are found to be close to -7.84×10^{-2}.

The "trigonometric polynomial" approximate periodic solution to the given DE is

$$y(x) = \sum_{k=1}^{8} b_{2k-1}\sin(2k-1)x,$$

where

$$\begin{aligned} b_1 &= 1.431189037, b_3 = -0.126911530, b_5 = 0.009754734, \\ b_7 &= -0.000763601, b_9 = 0.000059845, b_{11} = -0.000004691, \\ b_{13} &= 0.000000368, b_{15} = -0.000000029. \end{aligned}$$

A general theorem on the bounds of the Fourier coefficients is also proved and illustrated. Mickens (1988)

9. $$\mathbf{y'' + y^3 = p + h \sin wx,}$$

where p, h, and w are parameters.

This special case is considered extensively as an application of the more general equation (2.46.38). Cekmarev (1948)

10. $$y'' + ky + y^3 = B \cos x,$$

where k and B are constants.

For the specific choices $(k, B) = (0.1, 12.0)$, two types of steady phenomena occur, depending on the initial conditions, one of which represents a deterministic process and the other a randomly transitional process. The solutions are depicted in (t, x, y) space as well as in the (x, y) plane as Poincaré maps.

By using analog and digital computers, the change of attractors and of average power spectrum of the random process under the variation of the system parameters are studied. Random process is elucidated with the help of many figures. Ueda (1979)

11. $$y'' + 100y^3 = 100p \cos \Omega x, p = 0.10, 0.01; \Omega = 10.$$

Finite element and perturbation solutions are compared graphically. Burton (1984)

12. $$y'' - (1/10)y^3 + (3/5)y = \cos x.$$

The first two-harmonic 2π-periodic solution of the given DE is

$$3.2 \cos x - 0.027 \cos 3x.$$

Jordan and Smith (1977), p.146

13. $$y'' - (1/2)y^3 + (1/4)y = \cos x.$$

The first two-harmonic 2π-periodic solution of the given DE is

$$-\cos x + \cos 3x.$$

Jordan and Smith (1977), p.146

14. $$y'' + (1/2)(3y - y^3) = -\sin x, y(0) = y(2\pi), y'(0) = y'(2\pi).$$

It is proved that the given problem has a solution. Neito (1988)

15. $$y'' + y + \epsilon y^3 = \epsilon p \cos(\Omega x), y(0) = a, y'(0) = 0,$$

where ϵ, p, and Ω are constants.

Writing $X = \Omega x$, the equation is changed to

$$\Omega^2 y'' + y + \epsilon y^3 = \epsilon p \cos X, \ ' \equiv \frac{d}{dX}.$$

Using a modified Linstedt–Poincaré procedure the solution is found in the form

$$\begin{aligned} y(X) &= [1 - \alpha/8 + (\alpha^2/64)((4p/a^3) - 1)] \cos X + a[\alpha/8 - \alpha^2 p/(16a^3)] \cos 3X \\ &\quad + (\alpha^2 a/64) \cos 5X + O(\alpha^3), \\ \Omega^2 &= [1/(1 - 3\alpha)][1 - (4P\alpha)/a^3 + (P/2a^3 - 3/8)\alpha^2 + O(\alpha^3)], \end{aligned}$$

where $\alpha = \epsilon a^2/(-4 + 3\epsilon a^2)$. Burton (1984)

16. $$y'' + cy + dy^3 = e \sin wx,$$

where c, d, e, and w are constants.

It is shown that if $d \neq 0$, the given DE has solutions with periods exceeding a given constant if $|c| \neq 0$ and $|e|$ are sufficiently small. Shimizu (1954)

17. $$y'' + \alpha y + \beta y^3 = K \cos x, y(-\pi) = y(\pi), y'(-\pi) = y'(\pi),$$

where α, β, and K are parameters.

The solution of the given problem is sought in the Fourier series form

$$y(x) = \sum_{j=0}^{\infty} A_{2j+1} \cos\{(2j+1)x\}, \tag{1}$$

which satisfies the given BC. The substitution of (1), etc., in the given DE leads to an infinite system of nonlinear algebraic equations for the Fourier coefficients. This system is then solved using perturbation series in the amplitude K of the forcing term. Solution profiles of high accuracy and phase plane orbits are presented. The existence of limiting values of the forcing amplitude is discussed and points of nonlinear resonance are identified. Forbes (1987)

18. $$y'' + n^2 y - \beta y^3 = \beta F_0 \cos \lambda x,$$

where n, β, F_0, and λ are constants with $|\beta|$ small.

The solution is sought in the form

$$y = A\cos(nx - w) + \beta y_1 + \beta^2 y_2 + \cdots + \beta^N y_N. \tag{1}$$

For $\beta = 0$, (1) yields the usual general solution $y = A\cos(nx - w)$, where A and w are arbitrary constants. For $\beta \neq 0$, variation in each of A and w is permitted. These variations are determined successively to increasing orders in powers of β together with the additive corrections $y_1, y_2, \ldots, y_N$. Thus, the variation of parameters and power series expansions are both used in the same process. The nonresonant solution to first order in β is found to be

$$y = A\cos(nx - w) - \{\beta A^3/(32n^2)\}\cos(3nx - 3w) + \{\beta F_0/(n^2 - \lambda^2)\}\cos \lambda x,$$

where $w = w_0 + \{3\beta A^2/(8n)\}x$ with w_0 and A each constant. The resonant case leads to a coupled system of equations for w and A, which is discussed in great detail in the phase plane. Struble and Yionoulis (1962)

19. $$y'' + \alpha y + \beta y^3 = k \cos 2x, 0 < x < 1,$$

where α, β, k are real constants and $\alpha > 0$. $y(0) = y(1), y'(0) = y'(1)$.

Periodic solutions are constructed using a general theorem. Numerical results for $\alpha = 5$, $\beta = 1$, $k = 7$ are also depicted. Saranen and Seikkala (1989)

20. $$y'' \pm w^2 y + ry^3 = K \sin \Omega x, y(0) = y(X/2) = 0,$$

where w, r, K, Ω, and X are constants.

The solutions are obtained using Rayleigh–Ritz–Galerkin approximations. Error bounds are obtained. Bazley (1983)

21. $$\boldsymbol{y'' + (9 + \epsilon\beta)y - \epsilon y^3 = \Gamma \cos x,}$$

where ϵ is small and β, Γ are not too large.

Write the given DE as

$$y'' + 9y = \Gamma \cos x + \epsilon(y^3 - \beta y)$$

and seek the solution in the form $y(\epsilon, x) = y_0(x) + \epsilon y_1(x) + \cdots$. The zeroth order solution is

$$y_0(x) = a_0 \cos 3x + b_0 \sin 3x + (\Gamma/8) \cos x,$$

where the 2π-periodicity requirement leads to $b_0 = 0$, while a_0 is given by

$$a_0[(3/4)a_0^2 - \beta + 6\Gamma^2/16^2] + (2\Gamma^3/16^3) = 0.$$

Jordan and Smith (1977), p.133

22. $$\boldsymbol{y'' + (1 + \epsilon\beta)y - \epsilon y^3 = \Gamma \cos 3x,}$$

where ϵ, β, and Γ are constants This is a forced Duffing equation. Its periodic solution of period $2\pi/3$ is approximately

$$y = (1/8)[\Gamma - (\epsilon/8)\{(3\Gamma^3/256) - \beta\Gamma\}] \cos 3x.$$

Jordan and Smith (1977), p.148

23. $$\boldsymbol{y'' + w_0^2 y + \epsilon y^4 = 2\epsilon K \cos \Omega x,}$$

where $\epsilon \ll 1, w_0, K$, and Ω are constants.

Different cases arise. The solution is expressed in the form

$$y = a\cos(w_0 x + \beta) + 2\Lambda \cos \Omega x, \Lambda = K/(w_0^2 - \Omega^2)$$

with a and β as follows:

(a) $\Omega = 4w_0 + \epsilon\sigma$;

$$\begin{aligned} \frac{da}{dx} &= -\{\Lambda/(2w_0)\}a^3 \sin(\epsilon\sigma x - 4\beta), \\ a\frac{d\beta}{dx} &= \{\epsilon\Lambda/(2w_0)\}a^3 \cos(\epsilon\sigma x - 4\beta). \end{aligned}$$

(b) $\Omega = 2w_0 + \epsilon\sigma$;

$$\begin{aligned} \frac{da}{dx} &= -\{\epsilon\Lambda a/w_0\}(\Lambda^2 + a^2) \sin(\epsilon\sigma x - 2\beta), \\ a\frac{d\beta}{dx} &= \{2\epsilon\Lambda a/w_0\}(3\Lambda^2 + a^2) \cos(\epsilon\sigma x - 2\beta). \end{aligned}$$

(c) $4\Omega = w_0 + \epsilon\sigma$; then

$$\begin{aligned} \frac{da}{dx} &= -\{\epsilon\Lambda^4/w_0\}\sin(\epsilon\sigma x - \beta), \\ a\frac{d\beta}{dx} &= \{\epsilon\Lambda^4/w_0\}\cos(\epsilon\sigma x - \beta). \end{aligned}$$

(d) $2\Omega = w_0 + \epsilon\sigma$; then

$$\begin{aligned} \frac{da}{dx} &= -\{\epsilon\Lambda^2/w_0\}\{(3/2)a^2 + 4\Lambda^2)\sin(\epsilon\sigma x - \beta), \\ a\frac{d\beta}{dx} &= \{\epsilon\Lambda^2/w_0\}\{(9/2)a^2 + 4\Lambda^2)\cos(\epsilon\sigma x - \beta). \end{aligned}$$

(e) $3\Omega = 2w_0 + \epsilon\sigma$; then

$$\begin{aligned} \frac{da}{dx} &= -\{2\epsilon\Lambda^3 a/w_0\}\sin(\epsilon\sigma x - 2\beta), \\ a\frac{d\beta}{dx} &= \{2\epsilon\Lambda^3/w_0\}\cos(\epsilon\sigma x - 2\beta). \end{aligned}$$

(f) $2\Omega = 3w_0 + \epsilon\sigma$; then

$$\begin{aligned} \frac{da}{dx} &= -\{3\epsilon\Lambda^2 a^2/(2w_0)\}\sin(\epsilon\sigma x - 3\beta), \\ a\frac{d\beta}{dx} &= \{3\epsilon\Lambda^2 a^2/(2w_0)\}\cos(\epsilon\sigma x - 3\beta). \end{aligned}$$

Nayfeh (1985), p.175

24. $$\boldsymbol{y'' + w_0^2(y - \mu^2 y^3 + \nu^4 y^5) = P \sin \Omega x,}$$

where w_0, μ, ν, P, and Ω are constants.

Starting with zeroth order solution $y_0 = C\sin\Omega x$, the first order solution, by substitution in the undifferentiated terms and two integrations, is found to be

$$\begin{aligned} y_1 = {} & (1/\Omega^2)\{-P + w_0^2 C - (3/4)\mu^2 w_0^2 C^3 + (5/8)\nu^4 w_0^2 C^5\}\sin\Omega x \\ & + (1/\Omega^2)\{(1/36)(\mu^2 w_0^2 C^3 - (5/144)\nu^4 w_0^2 C^5)\sin 3\Omega x \\ & + (\nu^4/400)(w_0^2/\Omega^2)C^5 \sin 5\Omega x. \end{aligned}$$

Hagedorn (1988), p.232

25. $$\boldsymbol{y'' + \mu y - \lambda\mu y/(1 + y^2)^{1/2} = K \cos x,}$$

where μ, λ, and K are constants.

The given DE describes the motion of a spring-mass system, when a mass is attached by two springs to two opposing parallel walls. A series solution, with period 2π, is sought:

$$y(x) = \sum_{j=0}^{\infty} A_{2j+1}\cos\{(2j+1)x\}. \tag{1}$$

The coefficients of the series (1) are themselves expressed as perturbation series in the amplitude of the forcing term. Accurate nonlinear solutions and phase plane portraits are presented. Forbes (1989)

26. $$\mathbf{y'' + \sin y = \delta \sin\{(x + x_0)/\epsilon\}, 0 < \delta \ll 1, \epsilon > 0.}$$

It is a perturbed pendulum equation, with the time-dependent Hamiltonian

$$H = \{(1/2)y'^2 + 1 - \cos y\} + \delta[-y \sin\{(x + x_0)/\epsilon\}].$$

The orbits that connect the two saddle points of the unforced ($\delta = 0$) pendulum are called separatrices. If $\epsilon = O(1)$, then one can use Melnikov's method to show that these separatrices can split for weak forcing ($\delta \ll 1$) and that the perturbed system is chaotic. If $\epsilon \ll 1$, the Melnikov's method fails because the perturbation term is not analytic in ϵ at $\epsilon = 0$. It is shown that for $\delta \ll 1, \epsilon \ll 1$, the solution of the perturbed problem exhibits a symmetry to all orders in an asymptotic expansion. From the asymptotic expansion it follows that the separatrices split by an amount that is at most transcendentally small. Chang and Segur (1991)

27. $$\mathbf{y'' + k^2 \sin y = \lambda \sin wx,}$$

where k, λ, and w are constants.

Periodic solutions satisfying $y(0) = y(X)$ or $y'(0) = y'(X)$ are considered. A sequence of period-doubling bifurcations for the forced oscillations of a pendulum for $k^2 = 1$, $w = 1.3$ was tried. As a generating X-solution ($X = 2/w$), the stable equilibrium state $\lambda_0 = 0, y_0 = 0, y_0' = 0$ was chosen. Numerical methods were used to study the transition from regular to stochastic mode. The transition occurs in regions of phase space characterized by a strong local instability. Results are depicted in the phase plane. Gulyayev, Zubritskaya, and Koshkin (1989)

28. $$\mathbf{y'' + \mu \sin y = e \sin x, y(0) = y'(\pi/2) = 0,}$$

where μ and e are constants.

The class of symmetric periodic solutions satisfying the given BC is determined when e is small. The sketches of these solutions are drawn and the latter are extended to the region of large amplitude of the acting force by using numerical methods. The branching and stability of 2π-periodic solutions are investigated. Zlatoustov, Sazonov, and Sarycherv (1979)

29. $$\mathbf{y'' + \Omega^2 \sin y = \cos x,}$$

where Ω is a constant.

Write the DE as

$$y'' + \epsilon\Omega^2 y + \epsilon\Omega^2(\sin y - y) = \cos x,$$

where $\epsilon = 1$. The perturbation method then gives the leading terms of the solution of the given DE as

$$y = \{1/(\Omega^2 - 1)\}[1 + \Omega^2 - 2\Omega^2 J_1\{1/(\Omega^2 - 1)\}] \cos x + \{2/(\Omega^2 - 9)\} J_3\{1/(\Omega^2 - 1)\} \cos 3x,$$

where Ω is not close to an odd integer and J_{2n+1} is the Bessel function of order $2n+1$. Jordan and Smith (1977), p.146

30. $$y'' + g(y) = E \cos x,$$

where $g(y)$ is a piecewise linear restoring force given by

$$g(y) = \begin{cases} k_2^2 y + (k_1^2 - k_2^2), & y \geq 1, \\ k_1^2 y, & |y| \leq 1, \\ k_2^2 y + (k_2^2 - k_1^2), & y \leq -1. \end{cases}$$

The numbers k_1^2 and k_2^2 are arbitrary nonnegative numbers, $k_1^2 \neq k_2^2$. The parameter E is not small. It is shown that the given system has, at least for some value of E, a 2π-periodic solution that is odd-harmonic. As E varies, this family of solutions undergoes various types of branching; the reasons for these branchings are explained. Loud (1968)

31. $$y'' + y + g(y) = -20 \sin 3x,$$

where

$$y(0) = y'(\pi/6) = 0$$

and

$$g(y) = \begin{cases} \sin y, & 0 \leq y \leq \pi/2, \\ 0, & y \geq \pi/2. \end{cases}$$

A constructive numerical procedure is given for the given BVP. Tables compare the results with those by quasilinearization and shooting methods. Chandra and Fleishman (1972)

32. $$y'' + \Omega^2 y + \epsilon f(y) = \Gamma \cos x,$$

where Ω is not close to an odd integer and $f(x)$ is an odd function of x, with expansion

$$f(a \cos x) = -a_1(a) \cos x - a_3(a) \cos 3x - \cdots.$$

A perturbation solution of period 2π to order ϵ is

$$y = \frac{\Gamma + \epsilon a_1}{\Omega^2 - 1} \cos x + \frac{\epsilon a_3}{\Omega^2 - 9} \cos 3x.$$

Jordan and Smith (1977), p.146

33. $$y'' + y + g(y) = a \sin wx,$$

where $g(y)$ is odd and Lipschitz continuous with Liptschitz constant β; further, $g(y) > 0$ for $y > 0$ and the constants w and a satisfy $w > 1$ and $a < 0$.

It is pointed out that the given DE possesses a unique periodic solution that has least period $2\pi/w$ and is exactly out of phase with $a \sin wx$ provided that $\beta\tau < \cos\tau$, where $\tau = \pi/2w$. This solution is called a harmonic solution of the given DE. A special technique is given for the approximate representation of periodic solution; the method consists in approximating $g(y)$ arbitrarily closely by a multistep relay and then to solve the resulting multistep problem. Some numerical results are presented. Chandra and Fleishman (1972)

34. $$\boldsymbol{y'' + N(y) = \sin x, y(0) = y(\pi) = 0,}$$

with N odd, $N(y) \leq 0, N'(y) \leq 0$ for all $y \geq 0$.

It is proved that the given BVP has exactly one solution. Brüll and Hölters (1986)

35. $$\boldsymbol{y'' + N(y) = \sin x, y(0) = y(\pi) = 0,}$$

where the function $N(y)$ is odd and $\lim_{y\to\infty} N(y)/y = +\infty$. It is proved that the given BVP has infinitely many solutions. Brüll and Hölters (1986)

36. $$\boldsymbol{y'' + N(y) = K \sin \Omega x, y(0) = y(\pi/\Omega) = 0,}$$

where $N(y)$ is odd and $\lim_{y\to\infty} N(y)/y < \infty$.

It is proved that for sufficiently large Ω, the given BVP has exactly one solution. If $y(\Omega, x)$ is this unique solution and $z(\Omega, x)$ the unique solution of the (averaged) approximate system

$$z'' + \{N(\beta)/\beta\}z = \sin\Omega x, y(0) = y(\pi/\Omega) = 0,$$

where $\beta^2 = \{\Omega/\pi\} \int_0^{\pi/\Omega} z^2(s)ds$, then it is proved that

$$||z(\Omega) - y(\Omega)||_\infty = O(1/\Omega^8).$$

Brüll and Hölters (1986), Hale (1969)

37. $$\boldsymbol{y'' + g(y) = K \sin x, y(0) = y(\pi) = 0,}$$

where g is an odd function $g(y) = f(y^2)y$.

An averaging procedure is proposed which while destroying the fine structure of the problem will reveal several qualitative properties. The idea is to replace the argument y^2 of f by its mean value over $[0, \pi]$. Thus, the problem reduces to

$$y'' + \{g(\beta)/\beta\}y = K \sin x, \tag{1a}$$

$$y(0) = y(\pi) = 0, \tag{1b}$$

$$\beta^2 = (1/\pi) \int_0^\infty y^2(x)dx. \tag{1c}$$

The problem (1a)–(1c) is still a nonlinear problem, but it can be handled as a linear BVP with a nonlinear constraint. It is easy to check that if $g(\beta)/\beta < 0$, no solution of (1a) with $K = 0$ satisfies (1b). For $g(\beta)/\beta > 0$, the general solution of (1a) is $y(x) = A\sin(\{g(\beta)/\beta\}^{1/2}x) + B\cos(\{g(\beta)/\beta\}^{1/2}x) + K\beta\sin x/\{g(\beta) - \beta\}$, provided that $g(\beta) \neq \beta$. The BCs require that $B = 0, g(\beta)/\beta = m^2 (m = 2, 3, 4, \ldots)$, while (1c) yields $\beta^2 = A^2/2 + K^2\beta^2/\{2(g(\beta) - \beta)^2\}$. Hence the solutions have the form $y(x) = 2^{1/2}\beta\sin x$, where β satisfies

$$|g(\beta) - \beta| = K/2^{1/2}. \tag{2}$$

Equation (2) includes, because of the absolute value, the solution of the case $g(\beta)/\beta < 0$. Brüll and Hölters (1986)

38. $$\boldsymbol{y'' + f(y) = p + h \sin wx,}$$

where p and h are constants.

An approximate solution in the form $y = b + a\sin wx$ is sought leading to the relations for a and b from the variational equation. Several special cases are considered. Cekmarev (1948)

2.47 $y'' + ay' + g(x, y) = a\sin(\omega x + \delta)$

1. $$y'' + 0.04y' + |y^3|y = 0.06 \cos x + 0.01.$$

Fixed points, invariant curves, and domains of attraction of singular points of the given DE are discussed and depicted. Subharmonics of order 1/4 and 1/3 occur in the Fourier representation of solutions. Hayashi, Ueda, and Kawakami (1969)

2. $$y'' + 0.05y' + y^3 = 0.14 \cos x + 0.005.$$

Harmonic and subharmonic solutions of the given DE are found using the method of harmonic balance, and are depicted. Nishikawa (1964), p. 78

3. $$y'' + 0.1y' + y^3 = 0.15 \cos x.$$

Fixed points, invariant curves, and domains of attraction of singular points of the given system are discussed and depicted. Subharmonic oscillations of order 1/3 predominantly occur. Hayashi, Ueda, and Kawakami (1969)

4. $$y'' + 0.2y' + y^3 = 0.3 \cos x.$$

The periodic solutions related to three fixed points of the given DE are found in the form

$$\begin{aligned} y_{01} &= 0.671 \sin x - 0.744 \cos x + 0.026 \sin 3x + 0.022 \cos 3x + \cdots, \\ y_{02} &= 0.067 \sin x - 0.310 \cos x + 0.001 \sin 3x - 0.001 \cos 3x + \cdots, \\ y_{03} &= 0.988 \sin x + 0.684 \cos x + 0.021 \sin 3x - 0.061 \cos 3x + \cdots. \end{aligned}$$

Another bounded solution is also indicated. Fixed points and invariant curves of the given DE are depicted schematically [see Eqn. (2.49.50)]. Hayashi, Ueda, and Kawakami (1969)

5. $$y'' + 0.2y' + y^3 = 3 \cos x.$$

Periodic solutions related with the fixed points of the given DE are found. Fixed points and invariant curves of the given DE are discussed. Domains of attraction containing two of the fixed points are demarcated. Hayashi, Ueda, and Kawakami (1969)

6. $$y'' + 0.2y' + y^3 = 5.5 \cos x.$$

Fixed points and invariant curves of the given system are discussed. Periodic solutions about the fixed points are found in the form of Fourier series. Hayashi, Ueda, and Kawakami (1969)

7. $$y'' + 0.2y' + y^3 = B \cos x + B_0,$$

where B and B_0 are constants.

For different choices of B and B_0, fixed points of the given DE and their successive multiplication under varying system parameters are discussed and depicted. Hayashi, Ueda, and Kawakami (1969)

8. $$y'' + 0.2y' - y + y^3 = 0.3\cos(1.29x).$$

The solution of the given DE is found both numerically and by harmonic balance approximation. The numerical solution is found to be chaotic, and is depicted in the phase plane and as a Poincaré projection. Tongue (1986)

9. $$y'' + 0.2y' - y + y^3 = 0.3\cos(1.4x).$$

Phase plane solutions obtained by using harmonic balancing and numerical methods are depicted and compared. The solution is periodic. Tongue (1986)

10. $$y'' + 0.25y' - y + y^3 = q + 0.4\cos x,$$

where q is a constant.

It is easy to check that without the forcing term, the oscillator has three equilibrium positions if only the inequality $q < 0.3$ is satisfied. The equilibria are two sinks and a saddle. Numerical solutions show that with periodic term present, we have an infinite sequence of period-doubling bifurcations that finally lead to chaos. For $q = 0.004657$, 0.1, and 0.114, period doubling is shown. The strange attractor appears at $q = 0.12$. Awrejcewicz (1991)

11. $$y'' + 0.7y' + y^3 = 0.75\cos x.$$

This special Duffing equation is rewritten as

$$\frac{dv}{dy} = -\frac{y + \delta(v, y, \tau)}{v}, \text{ where } \delta(v, y, \tau) = -y + y^3 + 0.7v - 0.75\cos\tau,$$

$\tau = x$, and $v = dy/d\tau$. The graphical solution using the delta method is found in the (y, v) and (y, x) planes. Nishikawa (1964), p. 34

12. $$y'' \pm 0.1y' + \epsilon y + y^3 = G\cos wx, \epsilon = -0.1, 0,$$

where G and w are constants.

The given DE is solved numerically for different (G, w) set of values to identify regions of chaos. The experiments are carried out to demonstrate that while it is easier to produce chaos in a system with multiple equilibria, due to the small required force level, chaos can also occur in systems without multiple equilibria, albeit at higher forcing levels. Tongue (1986)

13. $$y'' + w_0^2 y + 2\epsilon^2 \mu y' + \epsilon\alpha_2 y^2 = 2\epsilon^2 f\cos\Omega x,$$

where $\epsilon \ll 1$, and $\Omega = w_0 + \epsilon^2\sigma$. $W_0, \mu, \epsilon, \alpha_2, f$, and Ω are constants.

A second order approximate solution is

$$y = a\cos(w_0 x + \beta) + \{\epsilon\alpha_2/(6w_0^2)\}a^2\{\cos(2w_0 x + 2\beta) - 3\} + \cdots,$$

where a and β are given by

$$\frac{da}{dx} = -\epsilon^2\mu a + \{\epsilon^2 f/w_0\}\sin(\epsilon^2\sigma x - \beta),$$
$$a\frac{d\beta}{dx} = -5\epsilon^2\alpha_2^2 a^3/12w_0^3 - \{\epsilon^2 f/w_0\}\cos(\epsilon^2\sigma x - \beta).$$

Nayfeh (1985), p.183

14. $$\epsilon^2 y'' + \delta\epsilon^2 y' - y^2 = -\cos x - 1 - C,$$

where $\delta \geq 0, 0 < \epsilon << 1$ and $|C|$ is small; i.e., $o(\epsilon)$ as $\epsilon \to 0$.

The given DE describes resonant oscillation in shallow-water equations. Initial conditions $y(0)$ and $y'(0)$ are sought that give rise to a periodic solution with $y(x+2\pi) = y(x)$. A solution is sought in the form $y = Y(\tau, x)$, where $\tau = f(x)/\epsilon$. The given ODE goes to the PDE

$$f'^2 Y_{\tau\tau} + \epsilon[f'' Y_\tau + 2f' Y_{\tau X} + f' Y_\tau] + \epsilon^2[Y_{xx} + \delta Y_x] = Y^2 - \cos x - 1 - C.$$

To order ϵ^0: $f'^2 Y_{0\tau\tau} = Y_0^2 - 1 - \cos x$. f is chosen such that the first approximation to Y satisfies an equation whose solution is periodic in τ with period independent of x. Thus, $f' = 2^{1/2}(1+\cos x)^{1/4} = 2^{3/4}|\cos(x/2)|^{1/2}$ so that $Y_0 = (1+\cos x)^{1/2} X_0$, where X_0 satisfies $2X_{0\tau\tau} = X_0^2 - 1$, giving $X_{0\tau}^2 = (1/3)X_0^3 - X_0 +$ constant.

However, the above two-time approach breaks down. A matched asymptotic expansion is obtained. Asymptotic and numerical methods are used to show that solutions which are bounded for all time can exist only if $C > -2^{-2/3}\epsilon^{4/3}(1.4664)$. Byatt–Smith (1988a)

15. $$y'' + w_0^2 y + 2\epsilon^2\mu y' + \epsilon\alpha_2 y^2 = \epsilon^{1/2} f \cos\Omega x,$$

where $2\Omega = w_0 + \epsilon^2\sigma$, and $\epsilon \ll 1$.

A uniform expansion is found in the form

$$y = a\cos(w_0 x + \beta) + \{\epsilon^{1/2} f/(w_0^2 - \Omega^2)\}\cos\Omega x + \{\epsilon\alpha_2 a^2/(6w_0^2)\}\{\cos(2w_0 x + 2\beta) - 3\} + \cdots,$$

where

$$\frac{da}{dx} = -\epsilon^2\mu a - [\alpha_2\epsilon^2 f^2/\{4w_0(w_0^2 - \Omega^2)^2\}]\sin(\epsilon^2\sigma x - \beta),$$
$$a\frac{d\beta}{dx} = -\{5\alpha_2^2\epsilon^2/(12w_0^2)\}a^3 + [\alpha_2\epsilon^2 f^2/\{4w_0(w_0^2 - \Omega^2)^2\}]\cos(\epsilon^2\sigma x - \Omega).$$

Nayfeh (1985), p.184

16. $$y'' + w_0^2 y + 2\epsilon\mu y' + \epsilon\alpha y^2 = K\cos\Omega x, \epsilon \ll 1,$$

where $w_0, \epsilon \ll 1, \mu, \alpha, K$, and Ω are constants. When $2\Omega \simeq w_0$, a first order perturbation solution is

$$y_0 = A(T_1)\exp(iw_0 T_0) + \Lambda\exp(i\Omega T_0) + \text{ c.c.},$$

where $\Lambda = K/\{2(w_0^2 - \Omega^2)\}, T_0 = x, T_1 = \epsilon x$, and

$$A = a_0\exp(-\mu T_1) - [\alpha\Lambda^2/2iw_0(\mu + i\sigma)]\exp(i\sigma T_1),$$

where $2\Omega = w_0 + \epsilon\sigma$ and a_0 is a constant. When $\Omega - w_0 \simeq w_0$, the function $A(T_1)$ is given by $A = B\exp(i\sigma T_1/2)$, where $B = B_r + iB_i$ and $B_r = b_r\exp(\lambda T_1), B_i = b_i\exp(\lambda T_1)$; nontrivial b_r and b_i exist only if

$$\lambda = -\mu \mp \{\alpha^2\Lambda^2/w_0^2 - (1/4)\sigma^2\}^{1/2}.$$

If $\{\alpha^2\Lambda^2/w_0^2 - (1/4)\sigma^2\}^{1/2} > \mu$, then A is unbounded, and the solution is unstable. Nayfeh (1985), p.162

17. $$y'' + cy' + y^3 = q + F\cos wx,$$

where $c = 0.15, w = 1.0, F = 0.16$.

Local bifurcations of periodic orbits in the q parameter line are investigated. The flow of the given DE is strictly contracted, which implies only that the saddle-node or period-doubling local bifurcations exist. Hopf bifurcations are excluded. For $0.03 \le q \le 0.05$ there exist two independent (main resonance and 1/2 subharmonic) solutions. The 1/2 resonance bursts suddenly into chaos, with an increase of q beyond 0.042385. Starting with the 32π period, the system is extremely sensitive to very small change of q, making further visualization very hard. For $q = 0.037$, chaos is found; for $q = 0.038029$, a 32π-periodic orbit is born. Awrejcewicz (1989)

18. $$y'' + ky' + y^3 = B\cos x.$$

This system is symmetric: $y \to -y$ and $x \to x + \pi$ leave the system invariant. Using the method of harmonic balance, the subharmonic response is found to be

$$y = x_1\sin x + y_1\cos x + (x_{1/3})\sin(x/3) + (y_{1/3})\cos(x/3),$$

where

$$\begin{aligned} x_1 &= k(9R_1 + R_{1/3})/(9B), y_1 = -(9A_1R_1 - A_{1/3}R_{1/3})/(9B), \\ x_{1/3} &= r_{1/3}\cos\theta_{1/3}, r_{1/3}\cos(\theta_{1/3} + 2\pi/3), r_{1/3}\cos(\theta_{1/3} + 4\pi/3), \\ y_{1/3} &= r_{1/3}\sin\theta_{1/3}, r_{1/3}\sin(\theta_{1/3} + 2\pi/3), r_{1/3}\sin(\theta_{1/3} + 4\pi/3), \end{aligned}$$

where

$$\begin{aligned} &R_1 = r_1^2 = x_1^2 + y_1^2, R_{1/3} = r_{1/3}^2 = x_{1/3}^2 + y_{1/3}^2, \\ &\cos 3\theta_{1/3} = -4(A_{1/3}x_1 - ky_1)/(9R_1r_{1/3}), \\ &\sin 3\theta_{1/3} = -4(kx_1 + A_{1/3}y_1)/(9R_1r_{1/3}). \end{aligned}$$

Two possibilities arise: Either $R_{1/3} = 0$ and there is no subharmonic response or $A_{1/3}^2 + K^2 - (81/16)R_1R_{1/3} = 0$. Stability of periodic solutions is discussed. Nishikawa (1964) p.66

19. $$y'' + ky' + y^3 = B\cos x + B_0,$$

where B and B_0 are constants.

The given DE is unsymmetric: $y \to -y$ and $x \to x + \pi$ does not leave the system invariant. The periodic solution might be assumed to take the form:

(a) $y_0(x) = x_1\sin x + y_1\cos x + z_0$ for harmonic response, and

(b) $y_0(x) = x_{1/2}\sin(x/2) + y_{1/2}\cos(x/2) + x_1 \sin x + y_1 \cos x + z_0$, or

(c) $y_0(x) = x_{1/3}\sin(x/3) + y_{1/3}\cos(x/3) + x_1 \sin x + y_1 \cos x + z_0$, for subharmonic response, z_0 has been added since the given system is unsymmetrical.

Case (a): In this case the coefficients are found to be $x_1 = kR_1/B, y_1 = -A_1R_1/B$, where $A_1 = 1 - (3/4)(R_1 + 4Z_0), R_1 = r_1^2 = x_1^2 + y_1^2, Z_0 = z_0^2$. R_1 and Z_0 are obtained from $(A_1^2 + k^2)R_1 = B_1^2, (3R_1/2 + Z_0)Z_0 = B_0$.

Case (b): Subharmonic response with 1/2 harmonics. Here it is found by the method of harmonic balance that

$$\begin{aligned} x_1 &= [k(4R_1 + R_{1/2})]/(4B), y_1 = -(4A_1R_1 - A_{1/2}R_{1/2})/(4B), \\ x_{1/2} &= r_{1/2}\cos\theta_{1/2}, r_{1/2}\cos(\theta_{1/2} + \pi), \\ y_{1/2} &= r_{1/2}\cos\theta_{1/2}, r_{1/2}\sin(\theta_{1/2} + \pi), \end{aligned}$$

where

$$\begin{aligned} A_1 &= 1 - (3/4)(R_1 + 2R_{1/2} + 4Z), \\ A_{1/2} &= 1/2 - (3/2)(2R_1 + R_{1/2} + 4Z), \\ R_1 &= r_1^2 = x_1^2 + y_1^2, R_{1/2} = r_{1/2}^2 = x_{1/2}^2 + y_{1/2}^2, Z = z_0^2, \\ \cos 2\theta_{1/2} &= -(kx_1 + A_{1/2}y_1)/(6R_1z_0), \\ \sin 2\theta_{1/2} &= (A_{1/2}x_1 - ky_1)/(6R_1z_0), \end{aligned}$$

in which the unknown quantities $R_1, R_{1/2}$, and Z may be determined by solving the simultaneous equations

$$\begin{aligned} &(4A_1R_1 - A_{1/2}R_{1/2})^2 + k^2(4R_1 + R_{1/2})^2 - 16B^2R_1 = 0, \\ &A_{1/2}^2 + k^2 - 36R_1Z = 0, (3/2)(R_1 + R_{1/2} + 2Z/3)z_0 + (A_{1/2}R_{1/2})/(8z_0) - B_0 = 0. \end{aligned}$$

Case (c): Subharmonic response with 1/3 harmonics. In this case, the method of harmonic balance gives

$$\begin{aligned} x_1 &= [k(9R_1 + R_{1/3})]/(9B), y_1 = -(9A_1R_1 - A_{1/3}R_{1/3})/(9B), \\ x_{1/3} &= r_{1/3}\cos\theta_{1/3}, r_{1/3}\cos(\theta_{1/3} + 2\pi/3), r_{1/3}\cos(\theta_{1/3} + 4\pi/3), \\ y_{1/3} &= r_{1/3}\sin\theta_{1/3}, r_{1/3}\sin(\theta_{1/3} + 2\pi/3), r_{1/3}\sin(\theta_{1/3} + 4\pi/3), \end{aligned}$$

where

$$\begin{aligned} A_1 &= 1 - (3/4)(R_1 + 2R_{1/3} + 4Z), \\ A_{1/3} &= 1/3 - (9/4)(2R_1 + R_{1/3} + 4Z), \\ R_1 &= r_1^2 = x_1^2 + y_1^2, R_{1/3} = r_{1/3}^2 = x_{1/3}^2 + y_{1/3}^2, Z = z_0^2, \\ \cos 3\theta_{1/3} &= -4(A_{1/3}x_1 - ky_1)/(9R_1r_{1/3}), \\ \sin 3\theta_{1/3} &= -4(kx_1 + A_{1/3}y_1)/(9R_1r_{1/3}), \end{aligned}$$

in which the unknowns $R_1, R_{1/3}$, and Z may be found by solving the simultaneous equations

$$\begin{aligned} (9A_1R_1 - A_{1/3}R_{1/3})^2 + k^2(9R_1 + R_{1/3})^2 - 81B^2R_1 &= 0, \\ A_{1/3}^2 + k^2 - (81/16)R_1R_{1/3} &= 0, \\ (3/2)(R_1 + R_{1/3} + 2Z/3)z_0 - B_0 &= 0. \end{aligned}$$

Nishikawa (1964), p.74

20.
$$y'' + ky' + y + y^3 = F \cos \Omega x,$$

where k, F, and Ω are constants.

Higher harmonic oscillations are investigated by harmonic balance method and by an analog computer analysis. Stability of the harmonic solution is also considered. Resonance curves of first and third harmonic oscillation for various parameters are graphically depicted. Jump and hysteresis effects are observed. Szemplinska–Stupnicka (1968)

21. $$y'' + \epsilon y' + cy + dy^3 = h \sin(2\pi x/p).$$

See Eqn. (2.49.42). For certain conditions on the constants ϵ, c, d, h, and p, periodic solutions of certain types are found. Shimizu (1951)

22. $$y'' + cy' - 0.5(1 - y^2)y = P \cos wx,$$

where c, P, and w are constants.

With $c = 0.1$, the bifurcation parameter P is varied to obtain numerical solutions—basic and successive harmonics—in the period-doubling bifurcation leading to chaos. P values are 0.07, 0.083, 0.0864, 0.08718, 0.08732. The strange attractor as a limit of the period-doubling sequence appears for $P = 0.088$. Awrejcewicz (1991)

23. $$y'' + ky' + 0.3y + 0.7y^3 = B \cos 2x + B_0,$$

where k, B, and B_0 are constants.

A large number of 1/2 harmonic solutions (steady and transient) for varied values of k, B, and B_0 are found in the phase plane. Nishikawa (1964), p.44

24. $$y'' + 2\delta y' + y(1 - y^2) = \epsilon \sin wx.$$

The solution of the given DE is approximated by $y = a_0 + a_1 \sin wx$, and $a_1 = a_1(w; \delta, \epsilon)$ is determined both when $a_0 = 0$ and $|a_0| > 0$. Stability of the solution is discussed. A formal expansion of the solution in terms of Lamé functions is also developed. An extended Fourier calculation based on numerical collocation is also reported. Miles (1989)

25. $$y'' + ky' + y + \beta y^3 = F \cos wx,$$

where $k \geq 0, \beta \geq 0$, and F are constants.

Numerical studies of the given Duffing equation show the peculiar behavior of accurate response curves—the existence of higher harmonic resonances of odd order, where the response curves have loops, and those of even order, where the response curves have branches accompanied by loops. As k approaches zero, the loops expand infinitely. As k increases, the loops change into simple maxima which vanish finally along with the branches. Monguchi and Nakamura (1983)

26. $$y'' + \gamma\rho y' + y + \gamma A y^3 = \gamma\epsilon \cos wx,$$

where $w^{-1} = 1 + \gamma T, \nu$ small; ρ, A, and ϵ are constants.

It is known that for certain values of the parameters ρ, A, and T, the amplitude of a stable periodic solution may depend discontinuously on the forcing amplitude ϵ. It is shown that such discontinuities really arise if ϵ is a function of x which varies sufficiently slowly, and similarly if ϵ is fixed but T is varied slowly. Mizohata (1953)

27. $$y'' + \delta y' - y + y^3 = r \cos x,$$

$\delta \geq 0, r \geq 0$ are parameters.

The given DE is called a Duffing equation with negative stiffness. Dependence of the character of the fixed points on the parameter r is investigated in great detail numerically, and is depicted in the phase plane.

Writing $r = \epsilon^{-3}$, the analysis for large r is carried out for the given DE written as $y''+\delta y'-y+y^3 = \epsilon^{-3}\cos x, \delta > 0, \epsilon \ll 1$. Finding inner and outer expansion, and hence their matching, lead to a periodic solution; it is compared with numerical results. Byatt–Smith (1987)

28. $$y'' + Ay' + By + Cy^3 = D\cos(wx + \phi),$$

$y(0) = \alpha_0, y'(0) = \beta_1; A > 0, B, C, D, w,$ and ϕ are constants.

Specifically, it is assumed that $A = 0.2, B = 5.0, C = 10.0, D = \alpha_0 = 1.0, \beta_1 = \phi = 0$. Changing to the variable u, where $y = 1 - 7x^2u$, we have for the chosen coefficient, $7x^2u'' + (28x + 1.4x^2)u' + (14 + 2.8x + 245x^2)u - 1740x^4u^2 + 3430x^6u^3 - 15 + \cos wx = 0, u(0) = 1.0$. Replacing $\cos wx$ by a polynomial approximation that is accurate to five decimals for $0 \leq w \leq 1$, "rational approximation" solutions are constructed and computed. Special values $w = 0, 1$ are discussed in detail [see Eqn. (2.32.14)]. Fair and Luke (1966)

29. $$y'' + hy' + w_0^2 y - \alpha y^3 = P\cos\nu x,$$

where $h > 0, \alpha = 4$, and $w_0^2 = 1; P$ and ν are other constants.

With reference to this softening-type Duffing's oscillator, the phenomenon of bifurcations leading to chaos in the principal resonance zone is studied with the aid of approximate analytic methods and the digital computer simulation. The stability analysis of the first approximate harmonic solution predicts bifurcation of the symmetric into asymmetric solution and then to period-doubled solutions, provided that the higher order instability of the variational Hill's-type equation is examined. The suitability of the harmonic solution for explanation and prediction of the strange behavior of the system is confirmed by the surprisingly good coincidence of the two critical forcing parameter value necessary for the bifurcation to occur with that theoretical value which brings "crisis" into resonance curve structure. The concept "near chaos," which refers to a solution showing some features of chaos but still remaining very close to a periodic solution, is discussed. Szemplinska–Stupnicka (1988)

30. $$y'' + \delta y' - \beta y + \alpha y^3 = f\cos wx,$$

where δ, β, α, and w are positive constants, and $f \geq 0$ is another parameter.

The given DE is a Duffing system with negative linear stiffness and a periodic forcing term. It is a model nonlinear oscillator and is studied in great detail for the global nature of attracting motions arising as a result of bifurcations. For small and large f, the behavior is as expected and is obtained by averaging theorems yielding acceptable results. For a wide range of moderate f, extremely complicated nonperiodic motions arise–called strange attractors or chaotic oscillations. These are intimately connected with homoclinic orbits arising as a result of global bifurcation. The analysis includes:

(a) Study of the unperturbed system with $f = 0$ in the phase plane.

(b) Lyapunov function approach to find the asymptotic stability of sinks $\{\pm(\beta/\alpha)^{1/2}, 0\}$ with eigenvalues

$$\lambda_{1,2} = (1/2)\{-\delta \pm (\delta^2 - 8\beta)^{1/2}\}.$$

(c) Proof of the preservation of the global structure of the perturbed system for small and large f.

(d) Possible stable and unstable manifolds of the saddle points and use of Melnikov theorem on transversal intersection.

(e) Poincaré maps and evaluation of f_e, the value of f for which transversal homoclinic intersection occurs and then persists, and its comparison with analytic value by Melnikov approach.

(f) Bifurcations of fixed points—the sinks—for different f values. After sufficient number of bifurcations and period doubling, the motion becomes irregular. However, attracting points of very high period may be present and the rate of attraction may be low. Computational results merely suggest that for $1.08 \leq f \leq 2.45$, a strange attractor exists. The structure of these attractors is studied.

Holmes (1979)

31. $$y'' + \alpha y' + w_0^2 y + \beta y^3 = F \cos\Omega x.$$

For the given Duffing oscillator, α is a damping parameter, $\beta > 0$ is the coefficient of the nonlinear term, w_0 is the natural frequency of the oscillator when $\beta = 0$, and $F\cos\Omega x$ is the driving force.

Analog solutions are obtained. Parametric dependencies are discussed in detail and qualitative comparisons are made with predictions of harmonic linearization. Frehlich and Novak (1985)

32. $$y'' + Ky' + \Omega^2 y - \epsilon y^3 = \Gamma \cos x,$$

where K, Ω, ϵ, and Γ are constants.

This is the pendulum equation with small damping. Assuming that $\Gamma = \epsilon\gamma$ and $K = \epsilon k(\gamma, k > 0)$ are small, and $\Omega^2 = 1 + \epsilon\beta$, wherein ϵ is small, we may rewrite the equation as

$$y'' + y = \epsilon(\gamma\cos x - ky' - \beta y + y^3). \tag{1}$$

A 2π-periodic perturbation solution of Eqn. (1) is found in the form

$$y = y_0 + \epsilon y_1 + \epsilon^2 y_2 + \cdots,$$

where $y_0 = a_0\cos x + b_0\sin x$ and a_0 and b_0 are given by

$$\begin{aligned} ka_0 - b_0\{\beta - (3/4)(a_0^2 + b_0^2)\} &= 0, \\ kb_0 + a_0\{\beta - (3/4)(a_0^2 + b_0^2)\} &= \gamma. \end{aligned} \tag{2}$$

The zeroth order solution is found in terms of the amplitude of the generating solution, $r_0 = (a_0^2 + b_0^2)^{1/2}$, found from (2) by squaring and adding, etc.: $r_0^2\{k^2 + (\beta - (3/4)r_0^2)^2\} = \gamma^2$. Jordan and Smith (1977), p.128

33. $$y'' + y + \epsilon(ky' + \beta y + y^3) = \epsilon\gamma \cos x, \epsilon << 1,$$

where ϵ, γ, k, and β are parameters.

The 2π-periodic solution of the given DE is

$$\begin{aligned} y &= a_0 \cos x + b_0 \sin x + \epsilon[a_1 \cos x + b_1 \sin x - (1/32)a_0(a_0^2 - 3b_0^2)\cos 3x \\ &\quad - (1/32)b_0(3a_0^2 - b_0^2)\sin 3x] + O(\epsilon^2), \end{aligned} \tag{1}$$

where a_0 and b_0 are given by

$$ka_0 - b_0\{\beta + (3/4)(a_0^2 + b_0^2)\} = 0,$$

$$kb_0 + a_0\{\beta + (3/4)(a_0^2 + b_0^2)\} = \gamma.$$

The constants a_1 and b_1 are determined by the requirement that the next order solution $y_2(x)$ is 2π-periodic.

Squaring and adding the equations in (1), we obtain

$$\gamma_0^2\{k^2 + (\beta + 3\gamma_0^2/4)\} = \gamma^2; \tag{2}$$

when $\gamma_0^2 = a_0^2 + b_0^2$ is solved for from (2), a_0 and b_0 are obtained from (1). Jordan and Smith (1977), pp.128,138

34. $$y'' + w_0^2 y + 2\epsilon^2 \mu y' + \epsilon\alpha_2 y^2 + \epsilon^2 \alpha_3 y^3 = 2\epsilon f \cos \Omega x,$$

where $\epsilon \ll 1$, and $\Omega = 2w_0 + \epsilon^2\sigma$; $w_0, \mu, \alpha_2, \alpha_3, f$, and σ are constants.

A second order approximation is

$$y = a\cos(w_0 x + \beta) + \epsilon[\{\alpha_2 a^2/(6w_0^2)\}\{\cos(2w_0 x + 2\beta) - 3\} + [2f/(w_0^2 - \Omega^2)\cos \Omega x] + \cdots,$$

where a and β are given by

$$\begin{aligned} \frac{da}{dx} &= -\epsilon^2 \mu a + \{\epsilon^2 \alpha_2/(3w_0^3)\} f a \sin(\epsilon^2 \sigma x - 2\beta), \\ a\frac{d\beta}{dx} &= \{(9w_0^2 \alpha_3 - 10\alpha_2^2)/(24w_0^3)\}a^3 - \{(\epsilon^2 \alpha_2)/(3w_0^3)\} f a \cos(\epsilon^2 \sigma x - 2\beta). \end{aligned}$$

Nayfeh (1985), p.183

35. $$y'' + 2\epsilon y' + cy + dy^3 - f\int_{-\infty}^{x} e^{-\mu(x-s)} y(s)ds = E \sin wx,$$

where $\epsilon, c, d, f, \mu, E$, and w are constants.

It is stated that for a given $\epsilon > 0$, the given DE has a periodic solution if f is sufficiently small and u sufficiently large. Proofs are not provided. Shimizu (1953)

36. $$y'' + \delta y' + (1 + \beta \cos \nu x)y + \alpha y^3 = \gamma \cos wx,$$

where δ, β, ν, and γ are small positive parameters and $w \simeq 1$.

Averaging methods are used to establish the existence of transverse homoclinic invariant tori. Further by applying Melnikov technique to the averaged system, it is shown that there exist transverse homoclinic orbits resulting in chaotic dynamics. Numerical results are also given to demonstrate the theoretical results. Yagasaki, Sakata, and Kimura (1987); Yagasaki (1990)

37. $$y'' + (\gamma/m)y' + (2k/m)\{1 - 1/(b^2 + y^2)^{1/2}\}y = (f_0/m) \sin wx,$$

where γ, m, k, b, f_0, and w are constants.

The given DE describes the snap-through oscillations of a one-hinged arch. When only cubic nonlinearities are retained, the given DE assumes the form for the two-well potential Duffing oscillator; see Eqn. (2.47.30). Moon (1987), p.79

38. $$y'' + \delta y' + \alpha \sin y = f \cos wx,$$

where δ, α, f, and w are constants.

The given DE describes damped sinusoidally driven pendulum. It is numerically investigated, and the effects of fractal boundaries of the basins of attraction on the functional form, noise sensitivity, and noise scaling of low-frequency power spectra are considered. Gwinn and Westervelt (1985)

39. $$y'' + \alpha y' + w_0^2 \sin y = \gamma \cos wx,$$

where α, w_0, and γ are constants.

The given DE describes the motion of a driven pendulum with viscous damping.

Introducing $z = e^{iy}, \tilde{x} = -ix$, we get

$$zz'' - z'^2 + i\alpha zz' + (1/2)w_0^2 z - (1/2)w_0^2 z^3 + i\gamma z^2 \cosh \sim wx = 0, \tag{1}$$

where $' \equiv d/d\tilde{x}$. Now see Eqn. (2.31.59) for a singularity analysis of (1). Parthasarathy and Lakshmanan (1990)

40. $$y'' + \alpha y' + \beta e^{-y}(1 - e^{-y}) = \gamma \cos wx,$$

where α, β, γ, and w are constants.

The given DE is damped and driven Morse oscillator. Put $z = e^{-y}$ to obtain

$$zz'' - z'^2 + \alpha zz' - \beta z^3 + \beta z^4 + \gamma z^2 \cos wx = 0. \tag{1}$$

The singularity structure of Eqn. (1) is analyzed. Writing the Laurent series

$$z = \sum_{j=0}^{\infty} a_j \tau^{j-1}, \quad \tau = x - x_0 \to 0, \tag{2}$$

and substituting into (1), we obtain the recurrence relation

$$\begin{aligned} &\sum_r a_{j-r} a_r (j - r - 1)(j - 2r - 1) + \alpha \sum_r a_{j-r-1} a_r (j - r - 2) \\ &- \beta \sum_{r,p} a_{j-r-1} a_{r-p} a_p + \beta \sum_{r,p,m} a_{j-r} a_{r-p} a_{p-m} a_m \\ &+ \gamma \sum_{r,p} G_{j-r-2} a_{r-p} a_p = 0, \; 0 \le m \le p \le r \le j, \end{aligned} \tag{3}$$

where

$$G(x) = \cos wx \quad \text{and} \quad G_n = (1/n!) \left. \frac{\partial G(x)}{\partial x^n} \right|_{x=x_0}.$$

From (3) we obtain

$$\begin{aligned} j = 0: &\quad a_0 = \pm i/\beta^{1/2}, \quad j = 1: \quad a_1 = (1/2)(1 - \alpha a_0), \\ j = 2: &\quad 0 \cdot a_2 + \alpha a_0 (1 - \alpha a_0) + \gamma a_0^2 \cos wx_0 = 0. \end{aligned} \tag{4}$$

Equation (4) gives the compatibility condition for the arbitrariness of a_2. Thus, if $\gamma \neq 0$, in general, the series (2) will not yield a general solution and must be modified. For $\alpha = 0,\ \gamma = 0$, the solution is meromorphic and the given DE is easily integrated. For $\gamma \neq 0$, one must write

$$z = \sum_{j=0}^{\infty}\sum_{k=0}^{\infty} a_{jk}\tau^{j-1}(\tau^2 \ln \tau)^k \tag{5}$$

and substitute into the given DE, etc., to get $a_{00} = \pm i/\beta^{1/2}$ and $a_{10} = (1/2)(1 - \alpha a_{00})$. For a_{20} to be arbitrary, one now obtains the condition $0.\ \ a_{20} + a_{01}a_{00} + \alpha a_{00}(1 - \alpha a_{00}) + \gamma a_{00}^2 \cos wx_0 = 0$, which requires that $a_{01} = [\alpha(\alpha a_{00} - 1) - \gamma a_{00} \cos wx_0]/3$. Equation (5) indicates that in the general case, x_0 is no longer a movable pole but a movable logarithmic branch point. To find the structure of this branch point, we put $z = (1/\tau)\theta(\xi)$, where $\xi = \tau^2 \ln \tau$, in the given DE to obtain

$$4\xi^2\theta\theta'' - 4\xi^2\theta'^2 + 2\xi\theta\theta' + \theta^2 + \beta\theta^4 = 0. \tag{6}$$

It is easy to check that (6) has Painlevé property. In fact, putting $\theta(\zeta) = f(\zeta)$, where $\zeta = \zeta^{1/2}$ into (6), we have

$$ff'' - f'^2 + \beta f^4 = 0, \tag{7}$$

where a prime denotes differentiation with respect to ζ. A first integral of (7) is

$$f'^2 = -\beta f^4 + I_1 f^2, \tag{8}$$

where I_1 is determined to be $6\beta a_{01}a_{00} = 6[\gamma \cos wx_0 - \alpha(\alpha + i\beta^{1/2})]$. By the simple change of variable $f(\zeta) = (I_1/\beta)^{1/2}[2g^2(\zeta) - 1]^{-1}$, Eqn. (8) goes to $g'^2(\zeta) = (1/4)I_1[g^2(\zeta) - 1]$ with the solution $g(\zeta) = \cosh[(1/2)I_1^{1/2}(\zeta - \zeta_0)]$, where ζ_0 is arbitrary. Choosing $\zeta_0 = 0$, we easily find that

$$f(\zeta) = (I_1/\beta)^{1/2}\operatorname{sech}(I_1^{1/2}\zeta). \tag{9}$$

Equation (9) shows that $f(\zeta)$ has simple poles which are situated at the discrete points $\zeta_m = \{i\pi/(2I_1^{1/2})\}(2m + 1), m \in Z$ in the complex ζ plane, where m denotes the lattice set integer. The location and structure of singularities in the complex plane are studied. Parthasarathy and Lakshmanan (1991)

41. $$\boldsymbol{y'' + ky' + |y|y = B \cos 2x + B_0,}$$

with $k = 0.20,\ B = 1.50$, and $B_0 = 0.50$.

A numerical solution with $y = y' = 0$ at $x = 0$ is found, giving both transient and steady states. Nishikawa (1964), p.30

42. $$\boldsymbol{y'' + ky' + f(y) = B \cos 2x + B_0,}$$

where $f(y) = y|y|$ for analog computer analysis, and $\sim c_1 y + c_3 y$ for phase plane analysis. B, B_0, c_1, and c_3 are constants.

Some general conclusions are drawn regarding the relationship existing between the initial conditions and the resulting 1/2-harmonic responses. Nishikawa (1964), p.60

43. $$\boldsymbol{y'' + y + 2\epsilon\mu y' + \epsilon y|y| = 2\epsilon f \cos \Omega x, \epsilon \ll 1,}$$

where $\Omega = 1 + \epsilon\sigma; \mu, f$, and σ are constants.

A first order solution is $y = a\cos(x + \beta)$, where

$$\frac{da}{dx} = -\epsilon\mu a + \epsilon f \sin(\epsilon\sigma x - \beta),$$
$$a\frac{d\beta}{dx} = \{4/(3\pi)\}\epsilon a^2 - \epsilon f \cos(\epsilon\sigma x - \beta).$$

Nayfeh (1985), p.182

44. $$y'' + ay' + g(y) = E \sin x,$$

where

$$g(y) = \begin{cases} (25/4)y - 4, & y \geq 1, \\ (9/4)y, & |y| \leq 1, \\ (25/4)y + 4, & y \leq -1; \end{cases}$$

a and E are parameters.

For $a = 0.01$, stable 2π-periodic solutions for different values of E are computed and graphically depicted. Loud (1968)

45. $$y'' + 2\zeta y' - F(y) = A \cos wx,$$

where ζ, A, w are parameters, and $F(-y) = -F(y)$.

It is shown that the given DE cannot have "inversion-symmetric" attractors of even periods. For the special case of a Duffing equation with $F = -y + 4y^3, \zeta = 0.2, A = 0.115$, this conclusion is confirmed by numerical solutions. Raty, Von Boehm, and Isomaki (1984)

46. $$y'' - \sigma h(x)y' + g(y) = E \cos x.$$

Here σ is real and h is a 2π-periodic continuous function satisfying some generic conditions. The existence of a periodic solution is proved. Ladeira and Taboas (1986)

2.48 $y'' + f(y, y') = a\sin(\omega x + \delta)$

1. $$y'' + y + 2\epsilon y^2 y' = K \cos \Omega x, \epsilon \ll 1,$$

where K and Ω are constants.

A first order solution is $y = a\cos(x + \beta) + 2\Lambda \cos \Omega x + \cdots$, where $\Lambda = K/\{2(1 - \Omega^2)\}$ and a and β are solutions of

(a)
$$\frac{da}{dx} = -2\epsilon\{\Lambda^2 + (1/8)a^2\}a - (2/3)\epsilon\Lambda^3 \cos(\epsilon\sigma x - \beta),$$
$$a\frac{d\beta}{dx} = \{-2/3\}\epsilon\Lambda^3 \sin(\epsilon\sigma x - \beta), \text{ when } 3\Omega = 1 + \epsilon\sigma x.$$

(b)
$$\frac{da}{dx} = -2\epsilon\{\Lambda^2 + (1/8)a^2\}a - (1/2)\epsilon\Lambda a^2 \cos(\epsilon\sigma x - 3\beta),$$
$$a\frac{d\beta}{dx} = \{-1/2\}\epsilon\Lambda a^2 \sin(\epsilon\sigma x - \beta), \text{ when } \Omega = 3 + \epsilon\sigma x.$$

Nayfeh (1985), p.182

2. $$y'' + 2\epsilon y^2 y' + y = 2\epsilon K \cos \Omega x,$$

where $\Omega = 1 + \epsilon\sigma; \epsilon \ll 1, K$, and σ are constants.

Using the method of averaging and multiple scales, a first order approximation is found to be $y = a\cos(x + \beta) + \cdots$, where $da/dx = -(1/4)\epsilon a^3 + \epsilon K \sin(\epsilon\sigma x - \beta)$,

$$a\frac{d\beta}{dx} = -\epsilon K \cos(\epsilon\sigma x - \beta).$$

Nayfeh (1985), p.174

3. $$y'' - \epsilon(1 - y^2)y' + y = U_0 \cos \delta x,$$

where ϵ, U_0, and δ are parameters.

The numerical solutions of the given DE are studied in the important nonlinear case when one is interested in the evolution and the deformation of the limit cycle, when beats appear. Chappaz (1968)

4. $$y'' + k(y^2 - 1)y' + y = kb\mu \cos\mu x,$$

where k and b are positive real numbers; μ is another constant.

It is shown by numerical experimentation that for $0 < b < 2/3$ and k large, the given DE has subharmonic solutions of order $(2n + 1)$, where $n \simeq O\{(2/3 - b)k\}$ as $k \to \infty$. El-Abbasy (1985)

5. $$y'' + k(y^2 - 1)y' + y = \mu k B \cos(\mu x + \alpha),$$

where k is the tuning parameter, and B, μ, and α are the (normalized) forcing amplitude, frequency, and phase, respectively.

Stable response of the given forced (Van der Pol) oscillators are considered. By numerically computing the rotation number of stable oscillations for various values of the forcing amplitude and oscillator tuning, descriptions of regions of phase locking, successive bifurcation of stable subharmonic and almost periodic oscillations, and overlap regions where two distinct stable oscillations can coexist are obtained. The parameters considered are $\alpha = 0, \mu = 1, 0 < B < 0.8$, and $0 < 1/k < 0.2$. Flaherty and Hoppensteadt (1978)

6. $$y'' + \nu(y^2 - 1)y' + y = (\alpha\nu + \beta) \cos x,$$

where ν is a large constant, $0 < \alpha < 2/3$. β is another constant.

Singular perturbation methods are used for the following cases:

(a) When the solution does not completely cross the unstable region $|y| < 1$ but merely dips through it.

(b) When the solution starts slowly through the unstable region, then abruptly slips toward $y = -2$.

(c) The critical case in which the solution stays in the unstable region for a full period of the forcing term.

Grasman (1981)

7. $$\rho^2 y'' + \epsilon\rho(y^2 - 1)y' + y = K \cos x,$$

where ρ, ϵ, and K are constants.

A perturbation series solution, using multiple scales, is found. Numerical results are presented and graphically shown. Information is obtained about the location and nature of the singularities in the complex ϵ-plane that limit the convergence of the series. Three pairs of singularities whose location changes as the forcing-term amplitude is varied are discussed. In particular, when the forcing amplitude exceeds a certain value, a pair of singularities along the imaginary axis becomes dominant. Dadfar and Geer (1990)

8. $$y'' - \epsilon(1 - y^2)y' + y = U_0 e^{i\delta x},$$

where C, δ, and U_0 are parameters and $i = (-1)^{1/2}$.

The numerical solutions of the given DE are studied in the important nonlinear case when one is interested in the evolution and deformation of the limit cycle, and beats appear. Chappaz (1968)

9. $$y'' + k(y^2 - 1)y' + y = b\lambda k \cos(\lambda x + \mu),$$

where b, λ, k, μ are parameters.

It is proved that for $b > 2/3$, there exists $k_0(b, \lambda, \mu)$ such that for $k \geq k_0$, the given DE has a unique periodic solution. This solution has period $2\pi\lambda^{-1}$ and all solutions converge to it. Lloyd (1972)

10. $$y'' + \mu(y^2 - 1)y' + y = B \cos \nu x,$$

where $\mu(> 0), B$, and ν are constants.

Fixed points and invariant curves for the given DE are discussed and depicted for $\mu = 1$ and varied choices of B and ν. Hayashi, Ueda, and Kawakami (1969)

11. $$y'' + \epsilon(y^2 - 1)y' + y = F \cos wx,$$

where ϵ and F are constants. This is Van der Pol's DE with a forcing term. Putting $wx = \tau$, we have

$$w^2\ddot{y} + \epsilon w(y^2 - 1)\dot{y} + y = F\cos\tau, \quad \cdot = \frac{d}{d\tau}.$$

(a) The case of hard excitation (F is not small) and far from resonance so that w is not close to an integer. The 2π-periodic solution by regular perturbation is

$$y(\epsilon, \tau) = [F/(1 - w^2)]\cos\tau + O(\epsilon).$$

(b) Soft excitation (F small) near resonance.

Writing $F = \epsilon\gamma$ and $w = 1 + \epsilon w_1$, a regular perturbation gives

$$y_0(\tau) = a_0\cos\tau + b_0\sin\tau,$$

where

$$\begin{aligned} 2w_1 a_0 - b_0[(r_0^2/4) - 1] &= -\gamma, \\ 2w_1 b_0 + a_0[(r_0^2/4) - 1] &= 0 \end{aligned} \tag{1}$$

where $r_0 = (a_0^2 + b_0^2)^{1/2} > 0$.

Squaring and adding the equations in (1), we obtain

$$r_0^2\{4w_1^2 + (r_0^2/4 - 1)^2\} = \gamma^2,$$

which gives the possible amplitude r_0 of the response. Jordan and Smith (1977), p.142

12. $$\boldsymbol{y'' + \alpha(y^2 - 1)y' + w^2 y = -2\alpha w \sin 3wx, \alpha > 0,}$$

where α and w are constants.

The given DE admits the solution $y = 2\cos wx$. This solution is shown to be unstable for small values of α. Colombo (1953)

13. $$\boldsymbol{y'' + k(1 - y^2)y' + cy = bk \cos x,}$$

$0 < b < 2/3$ and k is large.

Ignoring first the terms without k, an approximate equation is $(1 - y^2)y' = b\cos x$ with the solution $y = Y_0(x)$ given by $Y_0(x) - (1/3)Y_0^3(x) = b\sin x, |Y_0(x)| < 1$ for all x. First a uniqueness theorem for periodic solutions with $|y(x)| < 1$ is proved. Then it is shown that $Y_0(x)$ is the first term in a formal power series solution

$$y = \sum_{n=0}^{\infty} k^{-n} Y_n(x) \tag{1}$$

and that the partial sums $S_n(x) = \sum_{k=0}^{N-1} k^{-n} Y_n(x)$ of (1) are approximations to a genuine periodic solution when $k > k_N(b)$. Cartwright and Reuter (1987)

14. $$\boldsymbol{y'' - y + y^3 - \epsilon y^2 y' + \delta y' = \gamma \cos x,}$$

where ϵ, δ, and γ are small parameters.

Global perturbation techniques are used to study the bifurcation behavior. The case concentrated upon is that in which the unforced system possesses a one-parameter family of periodic orbits limiting on a homomoclinic orbit. Greenspan and Holmes (1984)

15. $$\boldsymbol{y'' - \epsilon(1 - y^2)y' + y = \epsilon K \cos \Omega(x, \tau),}$$

where K is the amplitude and Ω is the angular displacement of excitation; $d\Omega/dx = w + \epsilon\sigma(\tau)$, $\sigma = \sigma_0 + \epsilon\alpha x$, $\tau = \epsilon x$; ϵ, σ_0, and α are constants.

The results obtained, time histories, and phase plane plots reveal rich structure. Depending on the parameters, periodic, stable, and unstable motions are found, which sometimes lead to chaos. Tran and Evan-Iwanowski (1990)

16. $$\boldsymbol{y'' - (\beta - \sigma y^2)y' + \alpha y + \mu y^3 = q + \eta^2 \cos \eta x;}$$

$\beta, \sigma, \alpha, \mu, q$, and η are constants.

Periodic and chaotic solutions for $\beta < 0.1$ and $\beta = 0.1$, respectively, are found numerically. Awrejcewicz (1991)

17. $$\boldsymbol{y'' - (\beta - \delta y^2)y' + dy + \mu y^3 = q + \eta^2 \cos \eta x,}$$

where $\beta, \delta, \alpha, \mu, q$, and η are constants.

The given DE describes the motion of oscillators for which the rotation of an unbalanced disk or wheel becomes the source for the exciting force. An approximate perturbation analysis first gives bufurcation curves for the one-frequency solution. Subsequently, evolution of strange attractors at certain critical values of the parameters is studied. Long transitional chaotic phenomena and sudden qualitative changes in chaotic dynamics with evolution of chaotic attractors are discussed and illustrated. Awrejcewicz and Grabski (1989)

18.
$$y'' - k(1 + 2cy - y^2)y' + y = bk\mu \cos \mu x,$$

$c > 0, k$ large and $0 < 1/100 < b < 2/3 - 1/100$.

Possible stable subharmonic solutions of the given DE are discussed for the special case $c = 0$. The variation of the solution with increasing value of c is also considered. Cartwright (1985)

19.
$$y'' - (\alpha + \beta y - \gamma y^2)y' + w^2 y = Ew_1^2 \sin w_1 x,$$

where $\alpha, \beta, \gamma, w, w_1$, and E are constants.

This DE is analyzed in great detail for certain ranges of the parameters involved. It is assumed in particular that α, β, γ, and $w - w_1$ are small compared to w. The results are presented without proof. They consist of a classification of various types of periodic and quasiperiodic solutions. Numerical diagrams give a full qualitative and quantitative picture of the different cases. Cartwright (1948)

20.
$$y'' + (k + \beta y^2 + \epsilon y^4)y' + y + \mu y^3 = \begin{cases} \cos \eta x \\ \eta^2 \cos \eta x, \end{cases}$$

where K, β, ϵ, η, and μ are constants.

The solution of the given DE with both excited and self-excited vibration components at resonance are investigated. Tondl (1982)

21.
$$y'' - \epsilon(1 - y^{2n+2})y' + y^{2n+1} = \epsilon a \cos wx,$$

where ϵ is small, $n \geq 1$ is an integer, and $a > 0$ and $w > 0$ are independent of ϵ.

The given DE generalizes, for $n \geq 1$, the Van der Pol equation that corresponds to $n = 0$.

A preliminary study of the given DE for a general n and ϵ, and a more detailed study of its special case, when $n = 1$ and ϵ is small, is carried out. Results on the exact number of periodic oscillations, the relations between the stationary values (amplitudes) and least periods of periodic oscillations, and the ways some of the periodic oscillations appear or disappear as the parameters a, w vary are detailed. Obi (1976)

22.
$$y'' + (1 + \sin y)y' + y^{1/3} \sin^2 x = \sin x.$$

The given DE occurs in circuit theory. It is proved that it possesses at least one 2π-periodic solution. Mehri (1989)

23.
$$y'' + \alpha(1 + r \cos y)y' + \sin y = I \sin wx,$$

where α, r, I, and w are constants.

The given DE is invariant under $y \to -y, x \to x + (2n+1)\pi/w$, where n is an integer. Through an extensive numerical simulation it is shown that the symmetry properties of the equation of motion are clearly related with the development of chaotic solutions. Palmero and Romero (1991)

24.
$$y'' + \beta(1 + \epsilon \cos y)y' + \sin y = \rho + A \sin wx,$$

where β, ϵ, ρ, and w are constants.

The given DE describes phase locking in coherent communication theory, charge density waves, and dynamical behavior of a Josephson junction. For $\epsilon = 0, \beta > 2$, the problem is solved by reduction to a circle mapping. The conclusion for $0 \le \epsilon < 1, \beta > 2$ is that no chaotic motion takes place, and every trajectory is asymptotically periodic or quasiperiodic. Previous studies show chaos for β, ρ, and A small. Min, Xian, and Jinyan (1988)

25.
$$y'' - (1/10)y'^2 + (1/2)y = \cos x.$$

The first two-harmonic 2π-periodic solution of the given DE is

$$2 \cos x + 1.2 \sin x - 0.024 \sin 3x.$$

Jordan and Smith (1977), p.146

26.
$$y'' + p^2 y(1 + ay^2 + by'^2) = \Omega^2 \cos wx,$$

where p, a, b, Ω, and w are constants.

The given mechanical system includes the Duffing equation as a special case. An approximate solution is obtained by the method of Krylov and Bogoliuboff. Kane (1966)

27.
$$y'' - \epsilon(1 - y'^2)y' + y^3 = B \cos \nu x,$$

where $\epsilon = 0.2$; B, and ν are constants.

It is shown that (a) if $B \ne 0$, the given DE has a limit cycle at $\nu_0 = 1.0463$, and (b) if B is sufficiently small and nearly equals an integer multiple or submultiple of ν_0, one witnesses a phenomenon of synchronization. Bifurcation diagrams in the (B, ν) parameter plane are drawn to show generation and extinction of a pair of $2\pi m/\nu$ periodic solutions, period doubling, Hopf bifurcation, and degenerate bifurcation due to invariance of the given DE under the transformation $(y, z, x) \to (-y, -z, x + \pi/\nu), z = y'$. Extinction of a strange repeller and distortion of a strange attractor at $\nu = 0.3$ as B increases from 1 to 1.15 are exhibited. The paper is mostly numerical in nature. Kawakami and Sakai (1988)

28.
$$y'' + (c_2/m)(y' - v)^3 - (c_1/m)(y' - v) + (k_1/m)y$$
$$+ (k_2/m)y^3 + (g\mu/m) \operatorname{sgn}(y' - v) = (P_0/m) \cos wx,$$

where $c_1, m, v, c_2, k_1, k_2, g, \mu, P_0$, and w are constants.

The given DE describes a nonlinear oscillator on a conveyor belt. Phase plane diagrams, Poincaré maps, and time histories are obtained numerically for a set of parameters. A period-doubling route to chaos is observed. Narayanan and Jayaraman (1991)

29. $$y'' + \epsilon y'|y'| + y = 2\epsilon K \cos \Omega x, \epsilon \ll 1,$$

where $\Omega = 1 + \epsilon\sigma; K$ and σ are constants.

First order solution by the method of multiple scales and averaging is $y = a\cos(x + \beta) + \cdots$, where

$$\begin{aligned} \frac{da}{dx} &= -4\epsilon a^2/(3\pi) + \epsilon K \sin(\epsilon\sigma x - \beta), \\ a\frac{d\beta}{dx} &= -\epsilon K \cos(\epsilon\sigma x - \beta). \end{aligned}$$

Nayfeh (1985), p.171

2.49 $y'' + g(x, y, y') = p(x), p$ periodic

1. $$y'' + y + y^3 = B \cos wx + (A^3 \cos 3wx)/4,$$

where $B = A - Aw^2 + 3A^3/4$ and B, A, and w are constants.

An exact solution of the given DE is $y = A\cos wx$. A stability analysis of this solution is carried out, and the results are depicted in an amplitude-frequency plane. Turrittin and Culmer (1957)

2. $$y'' + 2y^3 = \lambda \cos x + \mu \cos 2x,$$

where λ and μ are parameters.

Computations for a 2π-periodic solution of the given DE are reported. Schmitt (1972)

3. $$y'' + 2y^3 = p(x),$$

where $p(x)$ is periodic and is piecewise continuous.

It is shown that all solutions of the given DE are bounded; it is not necessary to have a small parameter multiplying $p(x)$. Further, there is a B, depending only on $\max|p(x)|$, such that every solution satisfies

$$\sup\{(y'(x))^2 + y^4\}^{1/4} - \inf\{(y'(x))^2 + y^4(x)\}^{1/4} \leq B.$$

Morris (1976)

4. $$y'' + 2y^3 = p(x),$$

where $p(x)$ is periodic but not necessarily even.

It is proved that if $p(x)$ is C^1, has least period 2π, and $\int_0^{2\pi} p(x)dx = 0$, then for any positive integer m, the given DE has an infinity of periodic solutions with a least period $2m\pi$. Morris (1965)

5. $$y'' + w_0^2 y + \epsilon y^3 = K_1 \cos(\Omega_1 x + \theta_1) + K_2 \cos(\Omega_2 x + \theta_2),$$

where w_0, K_n, and $\theta_n, n = 1, 2$ are constants, and $\epsilon \ll 1$.

It is assumed that Ω_n is away from w_0. The solution is found in the form

$$y = A(\epsilon x)\exp(iw_0x) + \Lambda_1\exp(i\Omega_1 x) + \Lambda_2\exp(i\Omega_2 x) + \text{ c.c.},$$

where

$$\Lambda_n = K_n e^{i\theta_n/2}(w_0 - \Omega_n^2)$$

and where $A(\epsilon x)$ is governed by some complicated equations depending on different relations between Ω_n and w_0. Nayfeh (1985), p.190

6. $$y'' + w^2 y - \nu^2 y^3 = p(x),$$

where $w > 0$ and $\nu > 0$ are constants, $p(x)$ is a continuous periodic function of period X; moreover, $\int_0^X p(x)dx = 0$.

It is proved that if $X > 0$ is sufficiently small, then the given DE must have at least three periodic solutions of period X. Marlin and Ullrich (1968)

7. $$y'' + ay + by^3 = p(x),$$

where $a \geq 0, b > 0$ are constants and $p(x)$ is continuous and periodic. The solutions of the given DE are shown to be bounded. Moser (1973)

8. $$y'' + y^{2n+1} = -p(x), n \geq 1.$$

It is shown that for a continuous $p(x+1) = p(x)$, every solution of the given DE is bounded. Moreover, for every integer $n \geq 1$, there are infinitely many solutions of the given DE. Dieckerhoff and Zehnder (1987)

9. $$y'' + \beta y^{2n+1} + (a_1 + \epsilon a(x))y = p(x), n \geq 1,$$

where $a(x)$ and $p(x)$ are continuous and 1-periodic functions in $x \in R$, a_1 and β are positive constants, and ϵ is a small parameter.

It is proved that all solutions of the given DE are bounded for $x \in R$ and that there are infinitely many quasiperiodic solutions and an infinity of periodic solutions of minimal period m for each positive integer m. Bin (1989)

10. $$y'' + y^{2n+1} + p_{2n}(x)y^{2n} + \cdots = -p_0(x),$$

where $p_i(x)$ are 2π-periodic and sufficiently smooth in x.

It is proved that every solution of the given DE is bounded for x in R. Dieckerhoff and Zehnder (1985)

11. $$y'' - 1/y^\nu + \beta y = h(x),$$

where h is continuous, X-periodic, and $\nu \geq 1; \beta$ is a constant.

It is proved that the given DE possesses an X-periodic solution if $\beta \neq \mu_k/4$ for all $k = 0, 1, \ldots$, where $\mu_k = (2\pi k/X)^2$. Pino, Manasevich, and Montero (1992)

12. $$y'' + \Gamma^2 y + B^2 y/(1+y^2)^{1/2} = F(x),$$

where Γ and B are constants and $F(x)$ is an odd-periodic function with period 2π.

It is proved that:

(a) If $0 < \Gamma \leq 1/2$ and $0 < B^2 < \Gamma^2$, then the given DE has a unique periodic solution with period 2π.

(b) If Γ is positive and not equal to an integer, there exists a positive number B_0 such that if $-B_0 \leq B \leq B_0$, the DE has a unique periodic solution with period 2π.

Manacorda (1948)

13.

$$y'' + \sigma^2 \sin y = p(x),$$

where σ is a real constant, $p(x + X) = p(x)$ and $\int_0^X p(x)dx = 0$.

It is proved that the given DE possesses at least one periodic solution of period X, provided that $X < \{\pi/(2M)\}^{1/2}$, where $M = \sigma^2 + \sup|p(x)|$. Marlin and Ullrich (1968)

14.

$$y'' + k \sin y = e(x) + c, k > 0,$$

where y is X-periodic, $e = e(x)$ is X-periodic, and $\int_0^X e(s)ds = 0$.

The following result is proved: Assume that $k < w^2$. For every $e = e(x) \in C[0, X]$, X-periodic, with $\int_0^X e(x)dx = 0$, and satisfying $|||e||| \leq \pi/4$, there exist two numbers $d = d(e) < 0 < D = D(e)$ such that:

(a) If $c \notin [d, D]$, then the given problem has no solution.

(b) If $c = d$ or $c = D$, then the given problem has exactly one true solution.

(c) If $c \in (d, D)$, then the given problem has exactly two true solutions: $y_1(x) < y_2(x)\ \forall x \in [0, X]$.

Furthermore, given $\xi \in (-\pi/2, \pi/2)$, if

$$||e||_L 2 < w^2(1 - \sin|\xi|)^{1/2}(2X)^{1/2}(1 - k/w^2),$$

then the given problem with $c = k \sin \xi$ has exactly two true solutions:

$$y_1(x) < y_2(x)\ \forall x \in [0, X].$$

In the above, we define

$$|||e||| = \min\{(X/12)(1 - k/w^2)^{-1}\ ||e||_{L^1}, (X^{1/2}/(2(3)^{1/2}w)\ (1 - k/w^2)^{-1}\ ||e||_{L^2}\}.$$

A solution in the above is called true if

$$(1/X) \int_0^X y(s)ds \in [0, 2\pi).$$

Tarantello (1989)

15.

$$y'' + (1/m)f(y) = e(x),$$

where $e(x)$ consists of two harmonics, and m is a constant.

An approximate solution is found. Assuming y to be a linear combination with varying coefficients of harmonics with the same frequency as $e(x)$, a system for the coefficients is obtained in which all periodic terms are replaced by their mean values over a period;

stability of these solutions is also discussed. As an example, the case $f(y) = \alpha y + \beta y^3$ is worked out. Nitkin (1953)

16. $$y'' + g(y) = p(x),$$

where $g(y)$ and $p(x)$ are real-valued functions continuous on the real $R, p(x + 2\pi) = p(x)$, and the solutions of the given DE are uniquely determined by their initial conditions.

It is proved that if for some integer $n \geq 0, |g(x) - n^2 x|$ is bounded on R, and if a suitable Fourier coefficient of $p(x)$ has sufficiently large absolute value, then all solutions of the given DE are bounded on $[0, \infty)$. Seifert (1990)

17. $$y'' + w(y) = \epsilon e(x),$$

where $e(x)$ is periodic in x with a period X.

It is assumed that a periodic solution of the unperturbed equation $y'' + w(y) = 0$ with period X is known. A periodic solution $y(x, \epsilon)$ is sought, with the same period as that of $e(x)$, i.e., X, in terms of the solution of the unperturbed equation. Papoulis (1958)

18. $$y'' + g(y) = p(x),$$

where $p(x)$ is a continuous and 2π-periodic function of x in R and $g(y)$ is C^1 in R.

It is proved that:

(a) Every solution of the given DE is bounded for x in R if the superlinear conditions $g(y)/y \to \infty$ as $|y| \to \infty$ is satisfied.

(b) Every solution of the given DE is bounded for x in R if the sublinear condition $g(y)/y \to 0$ as $|y| \to \infty$ and $g(y)\ \mathrm{sgn}(y) \to \infty$ as $|y| \to \infty$ are satisfied.

Ding (1988)

19. $$y'' + g(y) = p(x),$$

where g and p are continuous and p is periodic with period X.

Assume that the function $g(y)$ is Lipschitzian and satisfies $yg(y) \geq 0$, and $|g(y)| \to \infty$ as $|y| \to \infty$. The function p is continuous and periodic and hence is bounded. It is then proved that the given DE has a periodic solution with period X, provided that X is sufficiently small. Marlin and Ullrich (1968)

20. $$y'' + g(y) = e(x), \lim_{|y| \to \infty} g(y)/y = +\infty,$$

where $g(y)$ and $e(x)$ are continuous in $(-\infty, \infty), g(y)$ is Lipschitz, $e(x)$ is periodic with minimum period 2π and even, $e(-x) = e(x)$.

Existence of even and periodic solution is proved. If $e(x)$ is odd and $g(y)$ is odd in y, the given DE has odd and periodic solutions. Nakajima (1990)

21. $$y'' + f(y) = Pg(x) = Pg(x + X),$$

$y(0) = Y \neq 0, 0 < P < \infty, -\infty < x < \infty$, where $g(x)$ is a cosine-like function of period $Y; P > 0+, X$ and Y are constants. A necessary condition for the existence of steady-state forced vibration is that $f(y)$ be everywhere analytic, odd, monotonically increasing with

y, and $yf(y) > 0, (y \neq 0)$. The definition cosine-like generalizes the properties of a cosine function $s(x)$: $s(x)$ is analytic everywhere in $-\infty < x < \infty; s(0) = 1, s'(0) = 0; s(x) = s(x + X)$, where $X =$ constant is the least period of s:

$$s(X/4) = 0;\; s(x) = -s(X/2 - x); s(x + \epsilon) < s(x)$$

for $0 \leq x \leq x + \epsilon \leq X/4$. Rosenberg (1966)

22. $$y'' + y - g(y, p) = \mu f(x),$$

where p, μ are small parameters, f is an even-continuous $(2\pi/m)$-periodic function, g is an odd function of y, sufficiently smooth, and $m \geq 2$ is an integer.

It is shown that under certain conditions on g and f, the small 2π-periodic solutions of the given DE maintain some symmetry properties of the forcing term $f(x)$, when $\mu \neq 0$. Bifurcation curves are found. The changes in the number of such solutions as (p, μ) varies in a small neighborhood of the origin are described. Fürkotter and Rodrigues (1991)

23. $$y'' + cy' + y - \beta y^2 - h(x) = 0,$$

$c > 0$ is a constant, where

$$h(x) = \sum_{p_1, p_2} H_{p_1 p_2} e^{i(p_1 w_1 + p_2 w_2)x},$$

where p_1, p_2 are certain positive or negative integers; β is a constant.

The existence of a special type of combination oscillation in the form of certain almost periodic solutions is proved. Stoker (1950), p.235

24. $$y'' + w_0^2 y + 2\epsilon^2 \mu y' + \epsilon \alpha_2 y^2 = F_1 \cos \Omega_1 x + \epsilon F_2 \cos \Omega_2 x,$$

where $\Omega_2 + \Omega_1 = w_0 + \epsilon\sigma, \epsilon \ll 1; w_0, \epsilon, \alpha_2, F_1, F_2, \Omega_1$, and Ω_2 are constants.

A second order approximate solution is

$$\begin{aligned} y \;=\; & a\cos(w_0 x + \beta) + 2\Lambda_1 \cos\Omega_1 x + \epsilon\{2\Lambda_2 \cos\Omega_2 x \\ & + \{\alpha_2 a^2/(6w_0^2)\}\cos(2w_0 x + 2\beta) + \{2\alpha_2\Lambda_1^2/(4\Omega_1^2 - w_0^2)\}\cos 2\Omega_1 x \\ & + [2\alpha_2\Lambda_1 a/\{\Omega_1(\Omega_1 + 2w_0)\}]\cos[(w_0 + \Omega_1)x + \beta] \\ & + [2\alpha_2\Lambda_1 a/\{(\Omega_1(\Omega_1 - 2w_0)\}]\cos\{(w_0 - \Omega_1)x + \beta\} - \alpha_2 a^2/(2w_0^2) \\ & - 2\alpha_2\Lambda_1^2/(w_0^2)\} + \cdots, \end{aligned}$$

where a and β are given by

$$\begin{aligned} w_0\frac{da}{dx} &= -\epsilon^2 w_0 \mu a - 2\epsilon^2\alpha_2\Lambda_1\Lambda_2\sin\gamma, \\ w_0 a\frac{d\beta}{dx} &= 2\epsilon^2\alpha_2^2\Lambda_1^2[2/(\Omega_1^2 - 4w_0^2) - 1/w_0^2]a - 5\alpha_2^2\epsilon^2 a^3/(12w_0^2) + 2\epsilon^2\alpha_2\Lambda_1\Lambda_2\cos\gamma, \\ \gamma &= \epsilon^2\sigma x - \beta, \Lambda_n = F_n/\{2(w_0^2 - \Omega_n^2)\}, n = 1, 2. \end{aligned}$$

Nayfeh (1985), p.210

25. $$y'' + w_0^2 + 2\epsilon^2\mu y' + \epsilon\alpha_2 y^2 = \epsilon^{1/2}F_1 \cos\Omega_1 x + \epsilon^{1/2}F_2 \cos\Omega_2 x,$$

where $\Omega_2 + \Omega_1 = w_0 + \epsilon^2\sigma, \epsilon \ll 1$.

A second order solution is found to be

$$\begin{aligned} y &= a\cos(w_0x+\beta)+2\epsilon^{1/2}\Lambda_1\cos\Omega_1x+2\epsilon^{1/2}\Lambda_2\cos\Omega_2x \\ &\quad +\{\epsilon\alpha_2/(6w_0^2)\}[\cos(2w_0x+2\beta)-3]+\cdots, \end{aligned}$$

where a and β are given by

$$\begin{aligned} w_0\frac{da}{dx} &= -\epsilon^2 w_0\mu a-2\epsilon^2\alpha_2\Lambda_1\Lambda_2\sin\gamma, \\ w_0a\frac{d\beta}{dx} &= -\{5\alpha_2^2\epsilon^2/(12w_0^2)\}a^3+2\epsilon^2\alpha_2\Lambda_1\Lambda_2\cos\gamma, \\ \gamma &= \epsilon\sigma x-\beta, \Lambda_n=F_n/\{2(w_0^2-\Omega_n^2)\}, n=1,2. \end{aligned}$$

Nayfeh (1985), p.211

26. $$y''+hy'+y+y^3=B\cos wx+(A^3\cos 3wx)/4,$$

where h, B, A, and w are real constants, h is small and positive; $B=A-Aw^2+3A^3/4$.

The given DE is rewritten as

$$\frac{d^2y}{dz^2}+\frac{h}{w}\frac{dy}{dz}+\frac{y+y^3}{w^2}=\frac{B}{w^2}\cos z+\frac{A^3}{4w^2}\cos 3z, z=wx.$$

A solution with IC $y(0)=A+p, y'(0)=r$ is sought in the form

$$y(z)=\sum_{i,j,k=0}^{\infty} C_{ijk}(z)p^ir^jh^k. \qquad (1)$$

The triple series (1) converges if $|z|\le 6$ provided that p, r, and h are sufficiently small, say $|p|<\delta, |r|<\delta$, and $|h|<\delta$. The coefficients $C_{ijk}(z)$ are obtained by substitution in the given DE and solving the resulting linear ODEs sequentially. Periodic solutions of the given DE near $y=A\cos wx$ for small h in powers of h are also found. A curious result is that periodic solutions of the given DE are shown to exist with an even-harmonic component when the parmeters h, w, and A are suitably chosen. Turrittin and Culmer (1957)

27. $$y''+ky'+y+\alpha y^3=R_1\cos wx+R_2\cos 2wx,$$

where k is a small damping coefficient, α is a small nonlinearity parameter, R_1 and R_2 are amplitudes of the exciting harmonics, and w is the exciting frequency.

Using the method of small parameter and harmonic balance, the solution is found in the form

$$y=A+B\cos(wx-\psi_1)+C\cos(2wx-\psi_2)+D\cos(3wx/2-\psi_3),$$

where A, B, C and ψ_1, ψ_2, ψ_3 are constants, found by substitution in the given DE, etc. The solution thus obtained is compared with analog computer results. Tomas (1971)

28. $$y''+cy'+y+\beta y^3=p(x),$$

where $c>0, \beta>0$ are constants, and $p(x)$ is a continuous, odd-harmonic period function of $x, \max|p(x)|=1$.

It is proved by using a fixed-point argument that the given DE has an X-periodic solution. Li and Wang (1980)

29. $$y'' + \epsilon\mu y' + p^2 y + \epsilon\beta y^3 = \epsilon F_0 \cos wx + \epsilon G_0 \sin wx,$$

where $\epsilon, \mu, p, \beta, F_0, w$, and G_0 are parameters.

Periodic solutions are found using truncated point mappings, suitably combined with perturbation methods; their stability is also discussed. Flashner and Guttalu (1989)

30. $$y'' + cy' + a(x)y^\alpha = p(x),$$

where:

(a) $a(x)$ and $p(x)$ are real, continuous, and periodic with period X for some $X > 0$.

(b) $\int_0^x p(u)du = 0$.

(c) c is a real number.

(d) $\alpha > 1$ and α is a rational number that can be written as $\alpha = m/n$ with $m > n$ and m, n both positive odd integers.

y^α represents the real-valued function on R, the set of real numbers; it is assumed that $yy^\alpha > 0$ whenever $y \neq 0$.

In addition to the foregoing assumptions, if for all x either

$$a(x) \geq 0 \quad \text{or} \quad a(x) \leq 0,$$

and if

$$||a|| \leq \min[\{1/(3\alpha)\}\{1/(\rho\alpha)\}^{\alpha-1}, \{|c|/(3X\alpha)\}\{1/(\rho\alpha)^{\alpha-1}\}],$$

then the given DE has at least one X-periodic solution y satisfying $||y - \phi|| \leq ||\phi|| + \rho$, where ϕ is a X-periodic solution of $y'' + cy' = p(x)$ and $\rho = 2||\phi||/(\alpha - 1)$. Here

$$||y|| = \max_{0 \leq x \leq X} |y(x)|.$$

Chang (1975)

31. $$y'' + \lambda y'(x) + k \sin y = e(x) + c,$$

where $e(x) \in C[0, X]$ is a X-periodic function with $\int_0^X e(s)ds = 0$ and $\lambda, k, c \in R$ and $k > 0$; an X-periodic solution is sought.

A simple integration from 0 to X shows that a necessary condition for the given DE to have a solution is that there exists $\xi \in [-\pi/2, \pi/2]$ such that $c = k \sin \xi$. On the other hand if $|\xi| = \pi/2$, i.e., $c = \pm k$, then the given DE has a solution if and only if $e = 0$. The following results are proved:

(a) For every $e = e(x), X$-periodic, with $\int_0^x e(s)ds = 0$, there exists a positive $\lambda_0 = \lambda_0(k, e)$ depending on k and e, such that $\forall\ \lambda :\ |\lambda| >_0$ there exist two numbers $d = d(e, \lambda) < 0 < D = D(e, \lambda)$ with the following properties:

  (i) If $c \notin [d, D]$, then the given problem has no solution.

  (ii) If $c = d$ or $c = D$, then the given problem has exactly one true solution.

(iii) If $c \in [d, D]$, then the given problem has exactly two true solutions, y_1 and y_2, with $y_1(x) < y_2(x)\ \forall x \in [0, X]$.

(b) Let $e = e(x)$ as above and $\xi \in (-\pi/2, \pi/2)$. There exists a positive number $\lambda_1 = \lambda_1(k, e, \xi)$ depending on k, e, and ξ such that $\forall \lambda : |\lambda| > \lambda_1$; the given problem with $c = k \sin \xi$ has exactly two true solutions $y_1(x) < y_2(x)\ \forall x \in [0, X]$. $y = y(x)$ is called a true solution for the given problem if $(1/X)\int_0^X y(s)ds \in [0, 2\pi)$.

Observe that if $\lambda \neq 0$ and $e = 0$, then the given problem admits exactly two true (constant) solutions $y_1 < y_2$ if $c = k \sin \xi$ and $\xi \in (-\pi/2, \pi/2)$ and exactly one true (constant) solution if $c = \pm k$. Tarantello (1989)

32. $$y'' + \alpha y' + (1 - \epsilon\lambda \cos wx) \sin y = \beta + \epsilon\mu (\cos wx - w \sin wx),$$

where $\alpha, \epsilon, \lambda, w, \beta$, and μ are parameters.

For α sufficiently large, some parametric regions for which the given system is in chaos are obtained by using the Melnikov function method. Sun (1988)

33. $$y'' + cy' + g(y) = f(x) \equiv f(x + X),$$

where $g : R \to R$ is continuous, $f : R \to R$ is continuous and X-periodic, and $c \in R; R$ denotes real numbers and $X > 0$.

With $m \equiv X^{-1}\int_0^X f(x)dx$, the following result is proved. Suppose that:

(a) There is a number $r \geq 0$ such that $(g(y) - m)y \geq 0$ whenever $|y| \geq r$.

(b) $c \neq 0$.

Then there is at least one X-periodic solution of the given DE. A similar result (with possibly different r) holds when $(g(y) - m)y \leq 0$. Ward (1980)

34. $$y'' + \lambda(y^2 - 1)y' + y = a \cos \nu_1 x + b \cos \nu_2 x,$$

where λ, a, b, ν_1, and ν_2 are constants. Numerical solutions of this quasiperiodically forced Van der Pol equation are found using Galerkin's methods. Mitsui (1977)

35. $$y'' - 2\lambda(1 - \beta y^2)y' + w_0^2 y = F \cos wx \cos \Omega x,$$

where $\lambda, \beta, w_0, F, w$, and Ω are constants.

Using numerical integrations, domains of strange chaotic and strange nonchaotic attractors of the given quasiperiodic forced Van der Pol equation are shown in the $\Omega/2\pi$-F plane. It is also stated, using a general theorem, that the given DE has a simple quasiperiodic solution

$$y = a \cos[(\Omega - w)x + \phi_1] + b \cos[(\Omega + w)x + \phi_2],$$

provided that the constant

$$C = \max[(|F|/2)\{1/|w_0^2 + (\Omega - w)^2| + 1/|w_0^2 - (\Omega + w)^2|\},\\ (|F|/2)\{(\Omega - w)/|w_0^2 - (\Omega - w)^2| + (\Omega + w)/|w_0^2 - (\Omega + w)^2|\}]$$

satisfies the inequality

$$C \leq \{1/(52\xi)^{1/2}\}\{(w_0^2 - \lambda^2)/(2 + 2\lambda)\}^{1/4}.$$

An averaging approach is used to find an approximate solution in the form

$$y \sim a(x) \cos wx \cos \Omega x + b(x) \cos wx \sin \Omega x,$$

where $a(x)$ and $b(x)$ are slowly varying amplitudes. Brindley and Kapitaniak (1991)

36. $$\boldsymbol{y'' - 2\lambda(1 - \beta y^2)y' + w_0^2 y = F \cos wx \cos \Omega x,}$$

where $\lambda < 1, w \ll \Omega, w \ll 1$, and w, Ω are incommensurate.

Numerical results show that there are substantial parameter ranges in which nonchaotic strange attractor exists. The striking differences between the behavior of trajectories near a nonchaotic attractor and those near a chaotic attractor are graphically illustrated. The regions of quasiperiodic, nonchaotic, and chaotic responses, deduced by calculation of Lyapunov exponents, are indicated. Brindley, Kapitaniak, and Naschie (1991)

37. $$\boldsymbol{y'' + w_n^2 y + \{(A + Cy^2)y' + A_2 y^2 + A_3 y^3\} = e \cos wx + f \sin wx}$$

with $w = 2w_n$; here A, C, A_2, A_3 are constants.

It is shown that the conditions for the given DE to have a periodic solution of period $2\pi/w_n$ are $4ACw_n^2 + M < 0$,

$$A^2 w_n^2 (w_n^2 - w^2) M < MK^2 A_2^2 + 2K^2 A^2 C^2 w_n^4,$$

where

$$M = \frac{2k^2(C^2 w_n^2 + aA_3^2)}{(w_n^2 - w^2)^2} \quad \text{and} \quad k^2 = e^2 + f^2$$

and the requirement that a certain determinant should not vanish; a perturbation method is used. Rosenblatt (1945)

38. $$\boldsymbol{y'' + f(y)y' + y = e(x),}$$

where $\int_0^L e(x)dx = 0$ and $f(y)$ is positive, except possibly at discrete points, and is piecewise continuous. It is shown that if for some $a > 0$,

$$\int_0^a f(\xi)d\xi = 3M \quad \left(\text{or} \int_{-a}^0 f(\xi)d\xi = 3M\right),$$

where

$$M = \max \int_0^x |e(\xi)|\, d\xi,$$

then the given DE has a unique periodic solution; the solution is of period L and all other solutions approch it as $x \to +\infty$. The second theorem states that the same conclusion can be drawn if it is known that the given DE has a solution which together with its first derivative is bounded for $x > 0$. Reissig (1955)

39. $$\boldsymbol{y'' + \alpha f'(y)y' + \beta y = E'(x),}$$

where α and β are constants; $f'(y)$ is continuous and such that $0 < a \leq f'(y) \leq b$. $E(x)$ is assumed real, periodic, and absolutely continuous with $E'(x)$ belonging to L_2, the class of Lebesgue measurable and square-integrable functions.

The existence of a periodic solution is proved by a constructive iterative scheme. Swartz (1958)

40. $$y'' + f(y)y' + g(y) = p(x),$$

where $f(y)$ is an even function, $g(y)$ is an odd function, and $p(x)$ is periodic with period X and is odd harmonic.

It is proved that this Duffing equation with damping does have an odd-harmonic steady-state response to an odd-harmonic term. Loud (1955)

41. $$y'' + f(y)y' + g(y) = e(x),$$

where $f, g, e : R \to R$ are continuous functions and $e(x)$ is periodic.

The existence of periodic solutions to this periodically forced scalar Lienard equation, with the same period as the period of the forcing term $e(x)$, is studied. Defining

$$F(y) = \int_{y_0}^{y} f(u)du, G(y) = \int_{y_0}^{y} g(u)du, E(x) = \int_{x_0}^{x} e(s)ds,$$

continuous and X-periodic, the analysis is carried out in the Lienard plane: $y' = z - F(y) + E(x), z' = -g(y)$. It is assumed that $F, g, E : R \to R$ are continuous functions and $E(x)$ is periodic with period $X = X_E > 0$. Omari, Villari, and Zanolin (1987)

42. $$y'' + f(y)y' + g(y) = h(x),$$

where $f(y) \geq c > 0; g(0) = 0, (dg/dy) > 0$ and $|g(y)| \to \infty$ as $|y| \to \infty$; $h(x)$ is a periodic function of period p.

It is shown that the given DE has at least one periodic solution of period p. Shimizu (1951)

43. $$y'' + g'(y)y' + f(y) = e(x),$$

where:

(a) f and g are differentiable.

(b) e is periodic.

(c) There exist positive constants p and q such that

$$p > q^2, p + 4q^2 \geq qG(x) \geq F(x) \geq p,$$

where $F(x) = f(x)/x, G(x) = g(x)/x$.

Under these conditions it is shown that the given DE has at least one periodic solution. Urabe (1950)

44. $$y'' + \mu f(y)y' + g(y) = \mu p(x), \mu \gg 1.$$

The solutions of the given DE are investigated in the (y, z) plane defined by the equivalent pair of first order equations $y' = \mu[z - F(y) + P(x)], z' = -g(y)/u$, where $F(y) = \int_0^y f(y)dy$ and $P(x) = \int^x p(\tau)d\tau$. With certain restriction on the functions involved, it is shown that all solutions of the system (1) enter and remain in a bounded, simply connected region of the plane. This region R_1 then contains all periodic solutions (if, in fact, any exist). By restricting the amplitude of $P(x), R_1$ is divided into subregions, one of which is shown to contain an unstable harmonic solution, for $P(x)$ periodic, and another to contain all other periodic solutions. Ponzo (1967)

45.
$$y'' + \{K + h(y)\}y' + f(x, y) = p(x),$$

where h is a continuous function, and f, p are continuous and periodic with respect to x of period w.

It is proved by using Leray–Schauder fixed-point technique that the given DE possesses at least one periodic solution of period w. Mehri (1989)

46.
$$y'' + w_0^2 y - \epsilon y'^2 = K_1 \cos \Omega_1 x + K_2 \cos \Omega_2 x, \epsilon \ll 1,$$

when $\Omega_2 \pm \Omega_1 \simeq w_0$ and Ω_1 is away from zero; K_1 and K_2 are constants.

A first order approximation is

$$y = a \cos(w_0 x + \beta) + 2\Lambda_1 \cos \Omega_1 x + 2\Lambda_2 \cos \Omega_2 x,$$

where $\Lambda_n = K_n / \{2(w_0^2 - \Omega_n^2)\}, n = 1, 2$. When $\Omega_2 \pm \Omega_1 = w_0 + \epsilon\sigma$, σ a constant, we have

$$\frac{da}{dX_1} = +2w_0^{-1}\Omega_1\Omega_2\Lambda_1\Lambda_2 \sin(\sigma X_1 - \beta),$$
$$a\frac{d\beta}{dX_1} = \pm 2w_0^{-1}\Omega_1\Omega_2\Lambda_1\Lambda_2 cos(\sigma X_1 - \beta),$$

where $X_1 = \epsilon x$. Nayfeh (1985), p.187

47.
$$y'' + y + \phi(y') = f(x),$$

where $f(x)$ is periodic.

The existence of at least one solution with the same period as $f(x)$ is proved assuming (besides the usual smoothness requirements) only that

$$\lim_{y\to\infty} \inf \phi(y) > \max |f(x)|, \lim_{y\to-\infty} \sup \phi(y) < -\max |f(x)|.$$

If, in addition, $\phi(y)$ is increasing for $|y| \le R_0$, where R_0 depends on the properties of $\phi(y)$ and $f(x)$, then the periodic solution is unique and stable. The method of proof is topological. Ascari (1952)

48.
$$y'' + F(y') + G(y) = E(x),$$

where $E(x)$ is periodic with period X.

It is shown that under suitable conditions on E, F, G, the given DE has a periodic solution with period X. A Brouwer fixed-point theorem is used. Reissig (1956)

49.
$$y'' + f(y, y')y' + g(y) = e(x),$$

where $e(x)$ is periodic of period L.

The given DE is a generalization of the Duffing equation. In the method of mapping used here, periodic solutions are correlated with fixed points or periodic points in the phase plane. The behavior of invariant curves of mapping reveals the global aspect of the solution in the transient state. This work is a survey report, dealing with the phase plane potraits of second order differential equations. In particular, attention is directed towards the appearance of homoclinic structure of indefinite number of subharmonic and ultrasubharmonic responses. Hayashi (1980)

50. $$y'' + f(y, y')y' + g(y) = e(x),$$

where $e(x)$ is periodic with period L.

Reduce the given DE to the system $y' = z, z' = -f(y,z)z - g(y) + e(x)$. Mapping procedure based on the transformation theory of DEs is used to find periodic solutions which in this theory correspond to a fixed point. In this paper, fundamental theory of the mapping method is explained and then applied to obtain the solution of nonlinear DEs. This method, when combined with the use of a computer, provides an effective means of finding various types of solutions which may occur, depending on the parameters of these equations. Illustrative examples are given [see Eqns. (2.47.1) and (2.48.10)]. Hayashi, Ueda, and Kawakami (1969)

51. $$y'' + f(y, y')y' + g(y) = p(x),$$

where the continuous functions $f(y,z), g(y)$ satisfy the following conditions:

(a) $f(y,z) \geq 0$ for all y and z.

(b) $yg(y) > 0$ for all $y \neq 0$.

(c) $\lim_{|y| \to \infty} |g(y)| = +\infty$.

(d) The function $p(x)$ is continuous and periodic with period w: $p(x+w) = p(x)$ for all x.

Then, some theorems regarding the existence of periodic solutions are proved. Opial (1961)

52. $$yy'' + ky' + y^3 = B \cos nx, n = 1, 3,$$

where k is a parameter; B is a constant.

Using Van der Pol's method, transient solutions are obtained and depicted graphically. Hayashi (1951)

2.50 $\vec{y}\,' = \vec{f}(x, \vec{y}), \vec{f}$ polynomial in y_1, y_2

1. $$y' = 1 + y + z^2, \quad z' = 2 + x + y,$$

where $y(1) = 1, z(1) = 2$.

Writing $y = 1 + \sum_{j=1}^{\infty} a_j x^j, z = 2 + \sum_{j=1}^{\infty} b_j x^j$, substituting in the given system and equating coefficients of equal powers of x on both sides, the coefficients a_j and b_j may be found. The solution is

$$\begin{aligned} y &= 1 + 6(x-1) + 11(x-1)^2 + (41/3)(x-1)^3 + \cdots, \\ z &= 2 + 4(x-1) + (7/2)(x-1)^2 + (11/3)(x-1)^3 + \cdots. \end{aligned}$$

The radius of convergence may be investigated. Martin and Reissner (1958), p.162

2. $$y' = -\beta - 2yz - C_1(y - a_0 z), \quad z' = -\delta_1 + y^2 - z^2 - C_1(a_0 y + z).$$

It is possible to solve this system exactly by introducing the complex variable $w = z + iy$ so that we have $w' = -K_1 - K_2 w - w^2$ with the solution

$$w = -(1/2)K_2 - (\zeta)^{1/2}\tan(\zeta^{1/2}x + i\rho),$$

where

$$\begin{aligned} K_1 &= \delta_1 + i\beta, K_2 = C_1(1 - ia_0), \zeta \equiv K_1 - (1/4)K_2^2 = \xi + i\eta, \text{ and} \\ &\xi(\Omega, c) = \delta_1 - (1/4)C_1^2(1 - a_0^2), \eta(\Omega, c) = \beta + (1/2)a_0 C_1^2. \end{aligned}$$

Landman (1987)

3. $$y' = -a_{31} + (a_{33} - a_{11})y - a_{21}z + yz,$$

$$z' = -a_{32} - y + (a_{33} - a_{22})z + z^2,$$

where $a_{31}, a_{33}, a_{11}, a_{21}, a_{32}, a_{22}$ are analytic functions of x.

Eliminating y, we get the second order DE

$$\begin{aligned} \frac{d^2z}{dx^2} &= 3z\frac{dz}{dx} + (2a_{33} - a_{22} - a_{11})\frac{dz}{dx} - z^3 \\ &\quad - (2a_{33} - a_{22} - a_{11})z^2 + \left(\frac{da_{33}}{dx} - \frac{da_{22}}{dx} + a_{32}\right. \\ &\quad \left. + a_{11}a_{33} - a_{11}a_{22} - a_{33}^2 + a_{22}a_{33} + a_{21}\right)z + a_{31} \\ &\quad - \frac{da_{32}}{dx} + a_{32}a_{33} - a_{11}a_{32}. \end{aligned} \tag{1}$$

With $z = -u$, (1) reduces to one of the 50 equations of Ince (1956, p.449):

$$\begin{aligned} \frac{d^2u}{dx^2} &= -3u\frac{du}{dx} - u^3 + (2a_{33} - a_{22} - a_{11})\left(\frac{du}{dx} + u^2\right) \\ &\quad + \left(\frac{da_{33}}{dx} - \frac{da_{22}}{dx} + a_{32} + a_{11}a_{33} - a_{11}a_{22} - a_{33}^2\right. \\ &\quad + \left. a_{22}a_{33} + a_{21}\right)u - a_{31} + a_{11}a_{32} - a_{32}a_{33} + \frac{da_{32}}{dx}. \end{aligned} \tag{2}$$

Further, if we put $u = v + (1/3)(2a_{33} - a_{22} - a_{11})$ and $v = w^{-1}(dw/dx)$ in (2), we may obtain a third order linear DE in w:

$$\begin{aligned} \frac{d^3w}{dx^3} &= \left[\frac{da_{11}}{dx} - \frac{da_{33}}{dx} + a_{32} + a_{21} + (1/3)(a_{11}^2\right. \\ &\quad \left. + a_{22}^2 + a_{33}^2 - a_{11}a_{22} - a_{11}a_{33} - a_{22}a_{33})\right]\frac{dw}{dx} \\ &\quad + \left[-(1/3)\left(2\frac{d^2a_{33}}{dx^2} - \frac{d^2a_{22}}{dx^2} - \frac{d^2a_{11}}{dx^2}\right)\right. \\ &\quad + (2/27)(2a_{33} - a_{22} - a_{11})^3 + (1/3)(2a_{33} \end{aligned}$$

$$- a_{22} - a_{11}\Big)\left(\frac{da_{33}}{dx} - \frac{da_{22}}{dx} + a_{32} + a_{11}a_{33} - a_{11}a_{22}\right.$$
$$\left. - a_{33}^2 + a_{22}a_{33} + a_{21}\right) - a_{31} + a_{11}a_{32} - a_{32}a_{33} + \left.\frac{da_{32}}{dx}\right] w. \quad (3)$$

Korzyuk (1986)

4.
$$\dot{x} = -y - bx^2 - (2c+\beta)xy - dy^2 + \mu h_1(t),$$
$$\dot{y} = x + ax^2 + (2b+\alpha)xy + cy^2 + \mu h_2(t),$$

where a, b, c, d, α, and β are constants; $h_1(t)$ and $h_2(t)$ are continuous and periodic with smallest positive period 2π; and the coefficients satisfy the condition that $(0,0)$ is a center for the given system with $\mu = 0$, and not even one of the following conditions hold:

(a) $a + c = b + d = \alpha + 4b = \beta + 4c = 0$.

(b) $a = d = \alpha + 3b = \beta + 3c = 0$.

(c) $(b+d)/(a+c) = k, 6(b+d) + \alpha = 6(a+c) + \beta = a(k^2-1) + (5d+3b)k = d(k^2-1) - (5a+3c)k = 0$.

(d) $(b+d)/(a+c) = k, 10(b+d) + 3\alpha = 10(a+c) + 3\beta = (b+4d)k^2 - 6ak - 3d = 3ak^2 + 6dk - 4a - c = 0$.

(e) $a = c = \beta = d = 6b + \alpha = 0$.

(f) $a = c = \beta = \alpha - 10d = b + 4d = 0$.

Then the given system has at least one periodic solution of period 2. Amelkin (1968)

5.
$$y' = -a_{31} + (a_{33} - a_{11})y - a_{21}z + (y + a_{23}z)y,$$
$$z' = -a_{32} - a_{12}y + (a_{33} - a_{22})z + (y + a_{23}z)z,$$

where $a_{ij} \neq 0$ are analytic functions of x in a domain D of complex plane. Eliminating y and putting $z = 1/w + a_{12}$, the given system can be reduced to a second order DE, which is one of 50 equations in Ince (1956), p.437.

It may easily be found that in the neighborhood of a movable singularity (which is only a first order pole in the present case),

$$\begin{aligned} y &= \{1/(x-x_0)\}\left\{y_0 + \textstyle\sum_{j=1}^{\infty} \phi_j^{(1)}(x-x_0)^j\right\}, \\ z &= \{1/(x-x_0)\}\left\{z_0 + \textstyle\sum_{j=1}^{\infty} \phi_j^{(2)}(x-x_0)^j\right\}, \end{aligned} \quad (1)$$

where $|y_0| + |z_0| \neq 0$, and

$$y_0 + a_{230}y_0z_0 + y_0^2 = 0, z_0 + y_0z_0 + a_{230}z_0^2 = 0,$$

which implies that since $|y_0| + |z_0| \neq 0$,

$$1 + y_0 + a_{230}z_0 = 0. \quad (2)$$

Here $a_{230} = a_{23}(x_0)$.

Thus, if y_0 is determined by (2), z_0 remains an arbitrary constant in the expansion (1), apart from x_0. Korzyuk (1986)

6. $$\boldsymbol{y' = 1 + x^2 + yz^2, \quad z' = 2 + xy^2 + z, y(0) = 0, z(0) = 0.}$$

A series solution, by writing $y = \sum_{j=1}^{\infty} a_j x^j, z = \sum_{j=1}^{\infty} b_j x^j$, substituting in the given system, equating coefficients of equal powers of x on both sides, may be found to be

$$\begin{aligned} y &= x + (1/3)x^3 + x^4 + (4/5)x^5 + \cdots, \\ z &= 2x + x^2 + (1/3)x^3 + (1/3)x^4 + (1/15)x^5 + \cdots. \end{aligned}$$

The radius of convergence may be investigated. Martin and Reissner (1958), p.162

7. $$\boldsymbol{y' = -z(r + y^2 + z^2), \quad z' = -z/x + y(y^2 + z^2 - 1/r),}$$

where r is a parameter and $y'(0) = 0, z(0) = 0$, and $\lim_{x\to\infty}\{y(x), z(x)\} = (0, 0)$.

The following results are proved:

(a) There exists $r_0 \geq 1$ with the property that for each $r \geq r_0$, there is an $\alpha_0 > 0$ such that if $\{y(0), z(0)\} = (\alpha_0, 0)$, then $z(0) = 0$ and $\lim_{x\to\infty}\{y(x), z(x)\} = (0, 0)$. Furthermore, $y > 0$ and $z > 0$ on $(0, \infty)$.

(b) Let $N \geq 1$ be a positive integer. There is a value $r_N > r_0$ with the property that if $r \geq r_N$ and $1 \leq i \leq N-1$, then there exists $\alpha_i > 0$ such that the solution of the given system is with $\{y(0), z(0)\} = (\alpha_i, 0)$ satisfies $\lim_{x\to\infty}(u, v) = (0, 0)$, and there exists a discrete set $0 < \delta_1 < \delta_2 < \ldots < \delta_i < \infty$ such that $y(\delta_j) = 0$ for $1 \leq j \leq i$, and $y(x) \neq 0$ for $x \notin \{\delta_1, \ldots, \delta_i\}$.

Troy (1989)

8. $$\boldsymbol{v' = -v/x - (\gamma - u^2 - v^2)u, \quad u' = -(1/\gamma + u^2 + v^2)v,}$$

where x is a positive radial coordinate and γ is an arbitrary parameter in the semiopen interval $(0, 1]$.

Physical solutions demand that real continuous solutions exist such that energy density w satisfies $(\partial w/\partial x) < 0$, and $\epsilon = \int_0^\infty w(\rho)\rho d\rho$ is bounded where $w^2 = u^2 + v^2$. This system arises in the self-trapping of light beams. The system, which has approximate modified Bessel function behavior, is first changed into an integral equation system; an iterative process gives an approximate solution of the nonlinear system. Some qualitative analytic and numerical solutions are found; the latter are, however, found to be unstable. Writing the equation for $w = u^2 + v^2$, namely

$$(1/2)w' + v^2/x + (\gamma + 1/\gamma)uv = 0,$$

we easily deduce that:

(a) $u(x), w(x)$ have extrema for $v = 0$.

(b) The function $xv(x)$ has extrema for $u = 0, w = \gamma$.

Hilton (1988)

9.
$$\dot{x} = y[\Omega - (x^2 + y^2)] + \epsilon(\delta x - x(x^2 + y^2) + \gamma x \cos t),$$
$$\dot{y} = -x[\Omega - (x^2 + y^2)] + \epsilon[\delta y - y(x^2 + y^2)],$$
where Ω is a fixed parameter, δ and γ vary and $0 < \epsilon << 1$ is a small scaling parameter.

The given system is transformed according to $x = (2I)^{1/2} \sin\theta, y = (2I)^{1/2} \cos\theta$, yielding

$$\begin{aligned} \dot{I} &= \epsilon[2\delta I - 4I^2 + 2\gamma I \sin^2\theta \cos t], \\ \dot{\theta} &= \Omega - 2I + \epsilon[\gamma \sin\theta \cos\theta \cos t]. \end{aligned}$$

A perturbation solution $I = \Omega/2 - 1/4 + (\epsilon)^{1/2} h$, $\theta = t/2 + \phi$ is found when the equations for h and θ are averaged. The averaged system is analyzed. Greenspan and Holmes (1984)

10. $$y' = -z^3/2 + a_2(x)z + a_3(x), \quad z' = -b_0(x)y^3/2 + b_2(x)y + b_3(x),$$

where $a_2(x), a_3(x), b_0(x), b_2(x)$, and $b_3(x)$ are holomorphic functions in some domain D.

For the given system to possess only algebraic nonstationary singularities, the functions b_0, b_2, and a_2 must have the form

$$\begin{aligned} b_0(x) &= 1/(C_1x + C_2)^2, b_2(x) = C_3/(C_1x + C_2)^{3/2}, \\ a_2(x) &= C_4/(C_1x + C_2)^{1/2}, \end{aligned}$$

where C_i $(i = 1, 2, 3, 4)$ are arbitrary constants, for which $|C_1| + |C_2| \neq 0$; the functions $a_3(x)$ and $b_3(x)$ are arbitrary. Bogoslovskii and Yablonskii (1967)

11. $$y' = -y^2/3 + a_2(x), \quad z' = -2b_0(x)y^5/3 + \sum_{j=2}^{5} b_j(x)y^{5-j},$$

where the functions $a_2, b_j, j = 0, 2, 3, 4, 5$ are holomorphic in some domain D.

It is shown that for the given system to have only algebraic movable singularities,

$$b_0(x) = 1/(C_1x + C_2)^3, b_3 = C_3/(C_1x + C_2)^2,$$

where C_λ $(\lambda = 1, 2, 3)$ are arbitrary constants for which $|C_1| + |C_2| \neq 0$; the remaining coefficients in the DE are arbitrary. Bogoslovskii and Yablonskii (1967)

12. $$y' = \sum_{j=0}^{p} a_j(x)z^{p-j}, \quad z' = \sum_{j=0}^{k} b_j(x)y^{k-j},$$

where $(p - 1)(k - 1) \geq 1$ and the coefficient functions $a_j(x), b_j(x)$ are holomorphic in some domain D.

It is proved that if $k+1 \neq M(p+1)$, where M is a positive integer, then all nonstationary singular points of the given DE in the domain D are algebraic. The system is reduced to Briot–Bouquet form and hence analyzed. Bogoslovskii and Yablonskii (1967)

13. $$y' = \sum_{k=q}^{\tilde{q}} a_k(x)y^k z^{n-k}, \quad z' = \sum_{k=t}^{\tilde{t}} b_k(x)y^k z^{m-k},$$

where $q, \tilde{q}, t, \tilde{t}, n, m$ are nonnegative integers, with $q \leq \tilde{q} \leq n, t \leq \tilde{t} \leq m, n > m > 1$, and $a_k(x), b_k(x)$ are functions holomorphic in some domain D. The class of systems of the given form which admits movable algebraic singularities are considered; certain systems with nonalgebraic singularities are also obtained. Specifically, formal solutions with algebraic singularities at x_0 in the forms

$$
\begin{aligned}
y &= \sum_{k=0}^{\infty} \alpha_k (x - x_0)^{(s+k)/p}, \\
z &= \sum_{k=0}^{\infty} \beta_k (x - x_0)^{(r+k)/p}, \alpha_0\beta_0 \neq 0 \text{ if } sr \neq 0; \\
y &= y_0 + \sum_{k=1}^{\infty} \alpha_k (x - x_0)^{(\ell+k)/p}, \\
z &= \sum_{k=0}^{\infty} \beta_k (x - x_0)^{(r+k)/p}, y_0\alpha_1\beta_0 \neq 0
\end{aligned}
$$

if $s = 0, r < 0$; and

$$
\begin{aligned}
y &= \sum_{k=0}^{\infty} \alpha_k (x - x_0)^{(s+k)/p}, \\
z &= z_0 + \sum_{k=1}^{\infty} \beta_k (x - x_0)^{(\ell+k)/p}, \\
& \alpha_0\beta_1 z \neq 0, \text{ if } s < 0, r = 0,
\end{aligned}
$$

were constructed.

In each of these cases, $p > 0, \ell > 0$; each solution has its own value for p, but this value is the same for both components of the solution. Through four lemmas, each with different conditions, the following theorem is proved: Each lemma defines a family of solutions $(y, z)(yz \neq 0)$ of the given system having an algebraic singularity at x_0 and having the property that y/z tends to zero or to infinity as $x \to x_0$; all solutions having an algebraic singularity at x_0, for which y/z has this behavior as $x \to x_0$, are contained in these families.

The convergence of the series solutions thus constructed is proved; in each case, the system is first transformed to the Briot–Bouquet class for which standard results are used. Yablonskii (1967)

14. $$y' = \sum_{i+j=0}^{n} a_{ij}(x) y^i z^j, \quad z' = \sum_{i+j=0}^{n} b_{ij}(x) y^i z^j.$$

It is proved that a necessary condition that the given system has no movable critical point is that it is expressible in the form given in Eqn. (2.51.18). Lukashevich (1967)

15. $$y' = \sum_{i=0}^{n} \sum_{k=0}^{n_i} P_{i,k}(x) z^{n_i-k} y^{n-i}, \quad z' = \sum_{j=0}^{m} \sum_{\ell=0}^{m_j} Q_{j,\ell}(x) y^{m_j-1} z^{m-j},$$

where $P_{i,k}(x)$ and $Q_{j,\ell}(x)$ are functions of x that are holomorphic in some domain D_1.

It may be remarked that any normal system of two DEs whose right-hand sides are polynomials in y and z and holomorphic with respect to x can be written in this form. Solutions of the given system are sought such that $y(x)$ and $z(x)$ tend to ∞ as $x \to x_0$, so that $x = x_0$ is an algebraic singularity:

$$y(x) = (x - x_0)^{r/\mu}[\alpha_0 + u(x)], z(x) = (x - x_0)^{s/\mu}[\beta_0 + v(x)],$$

where $r < 0, s < 0$, and $\mu > 0$, are integers, $\alpha_0\beta_0 \neq 0$, while $u(x)$ and $v(x)$ are analytic functions of x satisfying $u(x) \to 0, v(x) \to 0, (x-x_0)u'(x) \to 0, (x-x_0)v'(x) \to 0$ as $x \to x_0 \in D_1$. In order to simplify the arguments, only such points x_0 in D_1 are considered for which the conditions $P_{i,0}(x_0) \neq 0,\ i = 0, 1, \ldots, n; Q_{j,0}(x_0) \neq 0\ (j = 0, 1, \ldots, m)$ are satisfied. This will lead to the condition that $r/\mu = s/\mu$ for any point $x_0 \in D$, where D is a subdomain of D_1.

The given system is transformed into a Briot–Bouquet system, which gives conditions for the solution to have only algebraic singularities. Alternatively, the given system is shown to have a solution in the form

$$y = (x-x_0)^{r/\mu}\left[\alpha_0 + \sum_{\sigma_1+\sigma_2=1} c^{(\sigma_1,\sigma_2)}(x-x_0)^{(\sigma_1+\sigma_2\lambda_2)/\mu}(\ln(x-x_0)^{1/\mu})^{\sigma_2}\right],$$

$$z = (x-x_0)^{s/\mu}\left[\beta_0 + \sum_{\sigma_1+\sigma_2=1} d^{(\sigma_1,\sigma_2)}(x-x_0)^{(\sigma_1+\sigma_2\lambda_2)/\mu}(\ln(x-x_0)^{1/\mu})^{\sigma_2}\right]$$

satisfying $y(x) \to \infty, z(x) \to \infty$ as $x \to x_0$ at least along paths with bounded argument along which $x \to x_0$. Kondratenya and Yablonskii (1968)

2.51 $\vec{y}\,' = \vec{f}(x, \vec{y}), \vec{f}$ not polynomial in y_1, y_2

1. $\dot{x} = \gamma x(1 - x/k) - yx^n/(a + x^n), \quad \dot{y} = y\{\mu x^n/(a + x^n) - D\}, n = 1, 2,$

where γ, k, a, μ, and D are constants.

It is proved that there are three types of global structures of the given predator-prey system when the given system has:

(a) A unique stable limit cycle in the domain $\Omega\{(x, y)|x > 0, y > 0\}$, and all the trajectories of the given system in the domain Ω approach this limit cycle.

(b) A stable singular point in the domain Ω, and all the trajectories of the given system in the domain Ω terminate at the singular point.

(c) The given system does not have any singular point in the domain Ω, and all the trajectories in the domain Ω terminate at the point $(k, 0)$.

The types of global structure of the given system under the various conditions on the parameters γ, a, μ, k, and D are detailed for $n = 1, 2$. Sunhong (1989)

2. $$\boldsymbol{y}' = \sum_{k=1}^{M} \frac{\boldsymbol{A}_k(\boldsymbol{x})}{\boldsymbol{z} - \boldsymbol{\alpha}_k(\boldsymbol{x})}, \quad \boldsymbol{z}' = \sum_{k=1}^{m} \frac{\boldsymbol{B}_k(\boldsymbol{x})}{\boldsymbol{z} - \boldsymbol{\beta}_k(\boldsymbol{x})},$$

where $\alpha_i(x) \not\equiv \alpha_j(x), \beta_i(x) \not\equiv \beta_j(x), i \neq j, 1 \leq M \leq 4, 1 \leq< m \leq 4, A_k(x) \not\equiv 0, B_k(x) \not\equiv 0$.

It is proved that among systems of the given form with $(M-1)(m-1) \neq 0$, one and only one system, which can be reduced to the form

$$\frac{dy}{dx} = 1/(z - \alpha_1) + 1/(z - \alpha_2),$$
$$\frac{dz}{dx} = 1/(y - \beta_1) - 1/(y - \beta_2),$$

does not have any moving critical singular point. For the case $(M - 1)(m - 1) = 0$, say, $M = 1$, the given system can be reduced to the form

$$\frac{dy}{dx} = 1/(z - \alpha), \frac{dz}{dx} = \sum_{k=1}^{m}\{(1 - n_k)/n_k\}\{1/(y - \beta_k)\}, \tag{1}$$

where some of the n_k may also equal infinity. The system (1) has the first integral

$$z - \alpha = C\Pi_{k=1}^{m}(y - \beta_k)^{-(1-1/n_k)},$$

where C is an arbitrary constant. By lowering the order of the system (1), we obtain the equation

$$\left(\frac{dy}{dx}\right)^N = P(y), \tag{2}$$

where $P(y)$ is a polynomial. Conditions on m and n_i are found so that (2) belongs to the Briot–Bouquet class. For some cases, the given system can be integrated in terms of the elementary and elliptic functions. These cases are:

(a) $m = 1, n_1 = -1$, any positive integer n and ∞.

(b) $m = 2, n_1$ and n_2 may take, respectively, $n, -n$ (any integer); ∞, ∞; 2,2; 2, ∞; 2,3; 2,6; 3,6; 2,4; 4,4; 3,3.

(c) $m = 3, n_1, n_2$, and n_3 may take, respectively, the values: 2,2, ∞; 2,3,6; 2,4,4; 3,3,3; 2,2,2.

(d) $m = 4, n_1 = n_2 = n_3 = n_4 = 2$.

Yablonskii (1967)

3.
$$y' = \frac{P_1(z, x)}{Q_1(z, x)}, \quad z' = \frac{P_2(z, x)}{Q_2(z, x)},$$

where $P_1(z, x), P_2(z, x), Q_1(z, x)$, and $Q_2(z, x)$ are polynomials in the variable u with coefficients analytic in x, holomorphic in a certain region D; P_1/Q_1 and P_2/Q_2 are irreducible rational functions. Conditions on the system are found so that both $y(x)$ and $z(x)$ do not contain moving critical singular points. Yablonskii (1967a)

4.
$$\dot{x} = \frac{X_1(t, x, y)}{X_2(t, x, y)}, \quad \dot{y} = \frac{Y_1(t, x, y)}{Y_2(t, x, y)},$$

in the real region $G\{t, x, y\}$ with boundary $g\{t^*, x^*, y^*\}$. Singular points of solutions whose graphs are spirals in the neighborhood of infinity are investigated. Cylindrical coordinates $x = r\cos\phi, y = r\sin\phi$ are introduced; and special forms of functions X_i, Y_i in the variables t, r, and ϕ are assumed. Artykov, Rabinkov, and Rozet (1984)

5. $$y' = \frac{P_1(y,z,x)}{Q_1(y,z,x)}, \quad z' = \frac{P_2(y,z,x)}{Q_2(y,z,x)},$$

where P_1, P_2, Q_1, and Q_2 are polynonmials in y and z, with coefficients that are analytic functions of x in a finite region D. It is assumed that the given system has a first integral

$$R(x,y) = M(y,z)/N(y,z) = \text{ constant}, \tag{1}$$

where M and N are polynomials in x and y with no common factors. By virtue of the identity

$$\frac{\partial R(y,z)}{\partial y}\frac{P_1(y,z,x)}{Q_1(y,z,x)} + \frac{\partial R(y,z)}{\partial z}\frac{P_2(y,z,x)}{Q_2(y,z,x)} = 0,$$

the given system assumes the form

$$\begin{aligned}\frac{dy}{dx} &= \frac{P(y,z,x)}{Q(y,z,x)}\frac{R_1(y,z)}{S_1(y,z)},\\ \frac{dz}{dx} &= \frac{P(y,z,x)}{Q(y,z,x)}\frac{R_2(y,z)}{S_2(y,z)},\end{aligned} \tag{2}$$

where $P(y,z,x)$ and $Q(y,z,x)$ have no common factors $f(y,z)$ that are polynomials in y and z, and neither $R_1(y,z)$ nor $S_1(y,z)$ has a factor of the form $M(y,z) - CN(y,z)$ or of the form $N(y,z) - CM(y,z)$, where C is arbitrary.

Systems of the form (2) with an integral (1) are sought such that their solutions have only algebraic nonstationary singularities. Pisarenok (1981)

6. $$y' = (1/x)[yz(y+z)]^{1/2}, z' = [1/(1-x)][yz(y+z)]^{1/2}.$$

The given system is the limit of a system encountered in Riemann's problem concerning the construction of a simple class of functions with a given monodromy group. Qualitative study of this system is carried out in the neighborhood of the boundary of the domain D in which the existence and uniqueness of the solution are assured. This system has right-hand sides that are not single valued.

Introducing $u = [yz(y+z)]^{1/2}$, the given system is transformed to

$$\begin{aligned}\frac{dy}{dx} &= u/x, \frac{dz}{dx} = u/(1-x),\\ \frac{du}{dx} &= -(1/2)\{(2yz+z^2)/x + (2yz+y^2)/(1-x)\}\end{aligned} \tag{1}$$

in the neighborhood of the initial point (y_0, z_0) in the complex plane. For the given system, we can take one of the branches of the function $[y_0z_0(y_0+z_0)]^{1/2}$, for example, $[y_0z_0(y_0+z_0)]^{1/2} > 0$ if $y_0z_0(y_0+z_0) > 0$ as the initial value of u. The following results are proved:

(a) The system (1), whose initial values conform to $u_0 = [y_0z_0(y_0+z_0)]^{1/2}$, satisfies the relation $u = [yz(y+z)]^{1/2}$. This is because $u^2 - yz(y+z) = C$ is an integral and $C = 0$ if $u = u_0, y = y_0, z = z_0$.

(b) Every solution of the given system with arbitrary initial values y_0, z_0, u_0, and $x_0 \neq 0, 1$, is regular in the neighborhood of x_0.

(c) Some asymptotic solutions are found:

(i) $y \to y_0 = 0, z \to z_0 \neq 0$ when $x \to x_0 \neq 0, 1, x_0 > 0$; then

$$\begin{aligned} y &= (1/2)\{(z_0^2/x_0)(x-x_0)^2/2!\} + \cdots \\ z &= z_0 + (1/2)\{z_0^2/x_0(1-x_0)\}\{(x-x_0)^2/2!\} + \cdots \\ u &= (1/2)\{z_0^2(x-x_0)/x_0\} + \cdots. \end{aligned}$$

(ii) $y \to y_0 = 0, z \to z_0 = 0$ when $x \to x_0 \neq 0, 1$; then

$$\begin{aligned} y &= y_0 + y_0^2/\{2x_0(1-x_0)\}\{(x-x_0)^2/2!\} + \cdots \\ z &= (1/2)\{y_0^2/(1-x_0)^2\}\{(x-x_0)^2/2!\} + \cdots \\ u &= (1/2)\{y_0^2(x-x_0)/(1-x_0)\} + \cdots. \end{aligned}$$

(iii) $y \to a, z \to a$, when $x \to x_0 = 0, 1, u_0 = 0$; then

$$\begin{aligned} y &= a - [a^2/\{2x_0^2(1-x_0)\}][(x-x_0)^2/2!] + \cdots \\ z &= -a - [a^2/\{2x_0(1-x_0)^2\}][(x-x_0)^2/2!] + \cdots \\ u &= -a^2(x-x_0)/\{2x_0(1-x_0)\} + \cdots. \end{aligned}$$

(iv) It is shown that if $x_1 \neq 0, 1, \infty$ is a singularity of the given system; then $y \to \infty$ and $z \to \infty$ as $x \to x_1$.

Several other asymptotic results are derived when x tends to either a finite or an infinite limit. Erugin (1974)

7. $\quad \boldsymbol{y' = (2/x)[P(y,z)]^{1/2}, \quad z' = [2/(1-x)][P(y,z)]^{1/2}},$

where $P(y, z)$ is a multinomial in y and z.

Putting $u = [P(y,z)]^{1/2}, (\partial P/\partial y) = P_1(y,z), (\partial P/\partial z) = P_2(y,z)$ we get the third order system

$$\frac{dy}{dx} = \frac{2u}{x}, \frac{dz}{dx} = \frac{2u}{1-x}, \frac{du}{dx} = \frac{P_1(y,z)}{x} + \frac{P_2(y,z)}{1-x}. \tag{1}$$

If $y = y_0, z = z_0$, is such that $P(y_0, z_0) = 0$, the given DE has the solution $y \equiv y_0, z \equiv z_0$. If, further, $P_1(y_0, z_0) = P_2(y_0, z_0) = 0$, then this constant solution is the only solution with these initial conditions. If either $P_1(y_0, z_0) \neq 0$ or $P_2(y_0, z_0) \neq 0$, then besides the constant solution $y = y_0$ and $z = z_0$, there is one more solution with these initial values, regular at x_0. This solution is contained in the general solution in the neighborhood of (x_0, z_0, y_0), and the solution $y \equiv y_0, z \equiv z_0$ is singular because it is not contained in the general solution in the neighborhood of (x_0, y_0, z_0). In fact, if $P_2(y_0, z_0) \neq 0$, then $P_2(\overline{y}, \overline{z}) \neq 0$ at points $\overline{y}, \overline{z}$ in the neighborhood of (y_0, z_0). Erugin (1980)

8. $\quad \boldsymbol{y' = (2/x)[P(y,z)]^{1/2}, \quad z' = [2/(1-x)][P(x,y)]^{1/2}},$

where $P(y, z)$ is a polynomial. In particular, the case arising from a problem of Riemann has either

$$P(y,z) = yz(y+z) + a_{20}y^2 + 2a_{11}yz + a_{02}z^2 + a_1y + a_2z + a_0$$

or

$$P(y,z) = a_{20}y^2 + 2a_{11}yz + a_{02}z^2 + a_1y + a_2z + a_0,$$

where a_{kl} and a_i are constants. First assume that $P(y, z)$ is a general polynomial. Here y, z, x and the coefficients a_{kl} are complex, and $[P(y, z)]^{1/2}$ is one of the branches of this two-valued function, which can be continued, together with $P(y,z)$, by means of analytic continuation.

Introducing $u = [P(y,z)]^{1/2}$, where (y,z) are solutions of the given system, we get a system equivalent to the original one:

$$\begin{aligned} \frac{dy}{dx} &= \frac{2u}{x}, \quad \frac{dz}{dx} = \frac{2u}{1-x}, \\ \frac{dy}{dx} &= \frac{P_1(y,z)}{x} + \frac{P_2(y,z)}{1-x}, \end{aligned} \tag{1}$$

where

$$P_1(y,z) = \frac{\partial P}{\partial y}, \quad P_2(y,z) = \frac{\partial P}{\partial z}.$$

Particular solutions of (1) are sought that satisfy the relation $u = [P(y,z)]^{1/2}$.

The following results are proved:

(a) A solution of (1) whose initial values satisfy the relation $u_0 = [P(y_0,z_0)]^{1/2}$ satisfies $u = [P(y,z)]^{1/2}$.

(b) The solutions of the given system cannot have movable, singular points x_0 that are essential singularities, i.e., are such that as $x \to x_0 \neq 0, 1$, either y or z has no limiting value.

(c) If $x_0 \neq 0, 1, \infty$ is a singular point of the given system, then $y \to \infty, z \to \infty$ as $x \to x_0$.

(d) Asymptotic solutions are found such that:

(i) $y \to \infty, z \to \infty$ as $x \to x_1 < 1$.

(ii) $y \to y_0$ (finite), $z \to z_0$ (finite) as $x \to 0, x \to 1$, or $x \to \infty$.

(iii) $y \to \infty, z \to z_0 \neq 0$ as $x \to 1-0$.

(iv) $z \to \infty, y \to y_0$ as $x \to 1-0$.

(v) $y \to y_0 - 0, z \to z_0 + 0$ as $x \to \infty$.

(vi) $y \to \infty, z \to z_0 + 0$ as $x \to \infty$.

(vii) $y \to \infty, z \to -\infty$ as $x \to \infty$.

Erugin (1974)

9. $\boldsymbol{y_1' = -2y_1 + y_2^2 + e^{-x}\cos y_2, \quad y_2' = -y_2 + y_1^2\cos x + e^{-x}y_1,}$

The given sytem is shown to have a solution tending to zero provided that a certain norm condition is satisfied. Daniel and Moore (1970), p.33

10. $\boldsymbol{x' = -\ell + a_{11}(t)e^x + a_{12}(t)e^y, \quad y' = m + a_{21}(t)e^x + a_{22}(t)e^y,}$

where ℓ, m are constants and $a_{ij}(t)$ are continuous functions that are bounded for $t \geq t_0 > 0$. Under an elaborate set of conditions, it is shown by the method of successive approximations that there exists a one-parameter family of solutions for which

$$\lim_{t\to\infty} \frac{dx}{dt} = -\ell \text{ and } \lim_{t\to\infty} \frac{dy}{dt} = m.$$

Peyovitch (1947)

11. $$\dot{x} = \{1 + h_1(t)\} - \{1 + h_3(t)\}e^{\mu y}, \quad \dot{y} = -\{1 + h_2(t)\} + \{1 + h_4(t)\}e^{x},$$

where $h_i : R \to R, i = 1, \ldots, 4$ are continuous and T-periodic functions.

Conditions are given such that the given system has at least $2n$ different T-periodic solutions. A number of zeros of solutions are also determined. Hausrath and Manasevich (1991)

12. $$y' = Da(1 - y)\exp\{z/(1 + \epsilon z)\},$$

$$z' = Da \cdot B(1 - y)\exp\{z/(1 + \epsilon z)\} - \beta(z - z_c),$$

$(1 - \mu)y(1) = y(0), (1 - \lambda)z(1) = z(0)$, where Da, ϵ, B, β, and z_c are constants.

The given system describes temperature and conversion profiles in a nonisothermal, nonadiabatic tubular reactor with recycle.

For the special case $\beta = 0$ and $\mu = \lambda$, one obtains an intermediate integral $z = By$, and the system reduces to

$$\frac{dz}{dx} = Da \cdot B(1 - z/B)\exp[z/(1 + \epsilon z)].$$

More generally, the problem is solved numerically using a "parameter mapping" method. Results are shown graphically for some parametric values. Kubicek and Hlavacek (1983), p.246

13. $$\dot{y}_1 = y_2, \quad \dot{y}_2 = 2 + (2|t + 0.5| + 0.025 - |y_1 - 0.2|)^{1/2},$$

$y_1(0) = 0, y_1(1) = 0$.

This BVP is solved numerically by changing it to several single problems and using an IV method. Polovko (1986)

14. $$\dot{x} = a(t)|y|^{r_1} \operatorname{sgn} y, \quad \dot{y} = -b(t)|x|^{r_2} \operatorname{sgn} x,$$

where $a(t) > 0, b(t) > 0$ are continuous, $r_1, r_2 > 0$, and $a(t)/b(t)$ is of locally bounded variation.

Oscillation criteria are established employing energy function techniques; these depend crucially on whether $r_1 r_2 > 1, r_1 r_2 = 1$, or $r_1 r_2 < 1$. The results obtained include many known nonoscillation theorems for the classical Emden-Fowler equations as special cases. Erbe and Liu (1990)

15. $$y_1' = a_1(x)|y_2|^{\lambda_1} \operatorname{sgn}(y_2), \quad y_2' = a_2(x)|y_1|^{\lambda_2} \operatorname{sgn}(y_1)$$

where $a_i(x) : [0, +\infty) \to [0, \infty)$ are locally summable and $\lambda_i > 0$ $(i = 1, 2)$.

The asymptotic solution for $x \to \infty$ is investigated, as for the Emden-Fowler equation. Both regular and singular solutions are considered. The singular solutions are defined as follows:

(a) A nontrivial solution y_1, y_2 of the given DE on $[x_0; +\infty)$ is so called if there exists $x_1 \in (x_0; +\infty)$ for which $|y_1(x)| + |y_2(x)| \equiv 0$ for $x \geq x_1$.

(b) A nontrivial solution y_1, y_2 of the given DE on $[x_0; x_1)$ is so called if $x_1 < +\infty$ and

$$\lim_{x \to x_1} \{|y_1(x)| + |y_2(x)|\} = +\infty.$$

Previous results by the same author imply that if $\lambda_1\lambda_2 > 1(\lambda_1\lambda_2 < 1)$, then the given system has no singular solutions of the first (second) kind, and if $\lambda_1\lambda_2 = 1$, then all solutions of the given system are regular.

The following result is proved: Assume that $\lambda_1\lambda_2 > 1(\lambda_1\lambda_2 < 1)$ and that $a_1(x)$ is positive, continuous, and such that

$$a_2(x)\left(\int_0^x a_1(\tau)d\tau\right)^{1+\mu} \geq \delta a_1(x)$$

for sufficiently large x, where δ is positive and $\mu = \lambda_2$ $(\mu = 1/\lambda_1)$. Then a solution $y_1(x,\gamma), y_2(x,\gamma)$ of the given system with $\gamma \neq \gamma_0(\gamma = \gamma_0)$ is a singular solution of the second (first) kind. Here the initial conditions $y_1(x_0) = y_{10}, y_2(x_0) = \gamma$ are taken into account. Mirzov (1987)

16. $$y' = z, \quad z' = f(y,x) + \phi(y,x)z$$

in the region defined by the inequalities $|y| < H, |z| < H, x \geq 0$ $(H = \text{constant} > 0)$; here f and ϕ are holomorphic functions of y, for $|y| < H$, with coefficients that are bounded, continuous functions of $x(x \geq 0)$, and $|f(y,x)| \leq K, |\phi(y,x)| \leq K(K = \text{constant} > 0)$ for $x \geq 0$ and all complex y with $|y| \leq H$, and $f(0,x) = 0$ for $x \geq 0$.

Under some further conditions, it is shown that the trivial solution of the given system is stable, uniformly with respect to the initial condition at $x = x_0 \geq 0$, and every solution $y(x), z(x)$ of the given system with sufficiently small initial values $y(x_0), z(x_0)$ satisfies $\lim_{x\to\infty} y(x) = c$, the constant c depending on the choice of the solution, $\lim_{x\to\infty} z(x) = 0$, and $\int_0^\infty |z(x)|dx < \infty$. Krechetov (1975)

17. $$\dot{x} = -y - P(x,y) + \mu h_1(t), \quad \dot{y} = x + Q(x,y) + \mu h_2(t),$$

where $P(x,y)$, and $Q(x,y)$ are holomorphic functions in some neighborhood of the origin, containing terms not less than degree 2; μ is a small parameter, and the functions $h_1(t)$ and $h_2(t)$ are continuous and periodic with smallest positive period w. We assume that the reduced system with $\mu = 0$ has the origin as center. It is then proved that if $w < 2\pi$, then there exists $\mu_0 > 0$ such that the given system has at least one periodic solution of period w for every $|\mu| \leq \mu_0$. Amelkin (1968)

18. $$y' = f_0(x,y) + f_1(x,y)z, \quad z' = \phi_0(x,y) + \phi_1(x,y)z + \phi_2(x,y)z^2.$$

Eliminating the variable z from this system (by differentiating the first equation, etc.), we obtain

$$\frac{d^2y}{dx^2} = R\left(x, y, \frac{dy}{dx}\right), \tag{1}$$

where $R(x,y,(dy/dx))$ is a rational function of y and (dy/dx) with coefficients that are holomorphic with respect to x. Necessary conditions for the general solution of (1) to have no movable critical points are discussed in Ince (1956). Assuming that these conditions are satisfied, the variable $z = z(x)$ also will not have a movable critical point. Ince also shows that the solution of (1) can be expressed in terms of either elementary functions, known classical functions, or transcendental functions, which are solutions of the six irreducible Painlevé equations.

In fact, it can be shown that each of six Painlevé equations can be written in the form that is a particular case of the given system. For example, the fifth Painlevé equation is equivalent (within a nonsingular linear fractional transformation) to

$$\begin{aligned} x\frac{dy}{dx} &= -a - (a+c)y - yz - y^2 z, \\ x\frac{dz}{dx} &= \gamma x + \delta x^2 + (a+c)z + 2\delta x^2 y + (1/2)z^2 + yz^2, \end{aligned}$$

where a, c, ν, and δ are constants, while the sixth Painlevé equation is equivalent to the system

$$\begin{aligned} x(x-1)\frac{dy}{dx} &= \lambda x + (ax+b)y + cy^2 + ky(y-1)(y-x)z, \\ x(x-1)\frac{dz}{dx} &= p - [2cy + (ax+b)]z - (1/2)k(3y^2 - 2xy - 2y + x)z^2, \end{aligned} \qquad (2)$$

where λ, a, b, c, p, and $k \neq 0$ are constants. Lukashevich (1967)

19. $$y' = \sum_{j=0}^{k} \phi_j(z,x) y^{k-j}, \quad z' = \sum_{j=0}^{k-1} \psi_j(z,x) y^{k-1-j},$$

where $k > 1$.

Suppose that (z_0, x_0) is a point where the functions $\phi_j(z,x), \psi_j(z,x)$ are holomorphic. Then the changes of variables

$$y = [y_0 + u(t)]/t, z = z_0 + v(t), t = (x - x_0)^{1/(k-1)}$$

changes the given system to the Briot–Bouquet form

$$\begin{aligned} tu' &= \alpha_0 + \alpha_1 u + \alpha_2 v + F_1(u,v,t), \\ tv' &= \beta_0 + \beta_1 u + \beta_2 v + F_2(u,v,t), \end{aligned} \qquad (1)$$

with coefficients

$$\begin{aligned} \alpha_0 &= y_0 + (k-1)y_0^k \phi_0(z_0,x_0), \quad \beta_0 = (k-1)y_0^{k-1}\psi_0(z_0,x_0), \\ \alpha_1 &= 1 + k(k-1)y_0^{k-1}\phi_0(z_0,x_0), \quad \beta_1 = (k-1)^2 y_0^{k-2}\psi_0(z_0,x_0), \\ \alpha_2 &= (k-1)y_0^k \frac{\partial}{\partial z}\phi_0(z_0,x_0), \quad \beta_2 = (k-1)y_0^{k-1}\frac{\partial}{\partial z}\psi_0(z_0,x_0); \end{aligned}$$

the functions $F_j(u,v,t)$ and $\dfrac{\partial F_j}{\partial u}, \dfrac{\partial F_j}{\partial v}$ are zero at $u = v = t = 0$.

The system (1) has the property that $u \to 0, v \to 0$, as $t \to 0$ under the assumption that $\alpha_0 = \beta_0 = 0$, which holds if

$$\phi_0(z_0,x_0) \neq 0, \psi_0(z_0,x_0) = 0 \qquad (2)$$

and then

$$y_0^{k-1} = -1/[(k-1)\phi_0(z_0,x_0)].$$

Thus, it follows that the given system under the hypothesis (2) has solution with the property $y \to \infty, z \to z_0$ as $x \to x_0$, and the structure of the solution depends on the characteristic roots of the system (1); $\lambda_1 = -(k-1), \lambda_2 = \beta_2(y_0, z_0, x_0)$ and it changes with β_2.

Assuming that in the given system, $\phi_0(z,x) = A =$ constant, $\psi_0(z,x) = B =$ constant, $AB \neq 0$, so that (1) has the form

$$\begin{aligned} tu' &= -(k-1)u + F_1, \\ tv' &= -B/A - [(k-1)B/(y_0 A)]u + F_2. \end{aligned} \tag{3}$$

For $B \neq 0$, it is shown that if ϕ_j and ψ_j in the given system have the polynomial form

$$\begin{aligned} \phi_j(z,x) &= \sum_{\nu=0}^{k_j} a_{j\nu}(x) z^{k_j-\nu}, \; j = 1, \ldots, k, \\ \psi_j(z,x) &= \sum_{\nu=0}^{m_j} b_{j\nu}(x) z^{m_j-\nu}, \; j = 1, \ldots, k-1, \end{aligned}$$

which have coefficients holomorphic in the region D, where D is a complex plane from which the singular points of $a_{j\nu}, b_{j\nu}$ have been deleted together with point $x = \infty$; then at each point $x_0 \in D$ the given system has a solution with the property

$$x \to x_0 + o(y^{-k+1}), z \to z_0 + O(\ln y), \tag{4}$$

which remains bounded as the argument $y \to \infty$ along a certain path L in the complex plane.

If A and B in (3) are polynomials in z, then there is no solution of the given DE with the property (2). Bogoslovskii (1974)

20. $$y' = -\lambda z + Y(y,z,x), \quad z' = \lambda y + Z(y,z,x),$$

where Y and Z are real holomorphic functions of y and z and are continuous w-periodic functions of x; it is also assumed that the expansions of Y and Z in powers of y and z contain no terms of degree lower than the second. It is further assumed that the real positive number λ is such that $\lambda w/\pi$ is irrational.

Changing the variable to r and θ via $y = r\cos\theta, z = r\sin\theta$, the given DE becomes

$$\frac{dr}{dx} = r^2 R(x,\theta,r), \quad \frac{d\theta}{dx} = \lambda + r\phi(x,\theta,r), \tag{1}$$

where R and ϕ are holomorphic functions of r whose coefficients can be expressed as polynomials in sines and cosines of integral multiples of θ with coefficients periodic in x.

Setting $r = z_1 + u^{(2)} z_1^2 + \cdots + v_1 z_1^m$, we transform (1) to

$$\begin{aligned} \frac{dz_1}{dx} &= g z_1^m + \textstyle\sum_{j=m+1}^{\infty} a_j(x,\theta) z_1^j, \\ \frac{d\theta}{dx} &= \lambda + \textstyle\sum_{j=1}^{\infty} b_j(x,\theta) z_1^j, \end{aligned} \tag{2}$$

where the right-hand sides have the same properties as the RHSs of (1). A family of solutions of (2) is constructed such that $z_1 \to 0, \theta \to +\infty$ as $x \to +\infty$. Hence, if the trivial solution of the system is asymptotically stable, we obtain a general solution of the given system in the neighborhood of $y = 0, z = 0$ for all $x \geq x_0$, where x_0 is arbitrary. When the $(0,0)$ solution of the given system is asymptotically stable for $x \to -\infty$, we obtain a general solution of this system in the neighborhood of $(0,0)$ for all $x \leq x_0$. Grudo (1971)

21. $$\dot{x} = X(t, x, y), \quad \dot{y} = Y(t, x, y),$$

where X and Y are analytic with respect to x and y and periodic with respect to t with common period T. A theory corresponding to Poincaré theory for the autonomous system with $X = X(x, y), Y = Y(x, y)$ is constructed. Analogous to a singular (point) solution of the latter, a singular solution of the given system is defined for which x and y are periodic with period T. It is shown that in the ordinary cases, these period solutions can be classified in three types, which present definite analogies with nodes, foci, and saddle points occurring in the Poincaré theory. Analytic aspects of the analog are detailed, the geometrical aspects being treated briefly. Amerio (1951)

22. $$y' = F_1(y, z, x), \quad z' = F_2(y, z, x),$$

where $F_i(y, z, x)$ $(i = 1, 2)$ are meromorphic functions of y and z in some domain $\Omega_y \times \Omega_z$ with coefficients analytic in x. The functions F_i may have the specific form

$$F_1 = \frac{\phi_1(y, z, x)}{[z - a(x)]^n}, \quad F_2 = \frac{\phi_2(y, z, x)}{[y - b(x)]^m}, \tag{1}$$

where n and m are nonnegative integers not both zero; $\phi_i(y, z, x)$ $(i = 1, 2)$ are holomorphic functions in some domain $\Omega = \Omega_y \times \Omega_z \times D$ of the (y, z, x) space and $a(x)$ and $b(x)$ are functions holomorphic in D. Let $x_0 \in D$. Introduce the notation $z_0 = a(x_0), y_0 = b(x_0)$. We shall assume that $\phi_i(y, z, x)$ $(i = 1, 2)$ are finite and nonzero at (y_0, z_0, x_0). Writing $y - b(x) = \overline{y}, z - a(x) = \overline{z}$, the system (1) reduces to

$$\begin{aligned} \overline{z}^n \frac{d\overline{y}}{dx} &= \sum_{i=0}^{\infty} \sum_{j=0}^{\infty} \sum_{k=0}^{\infty} p_{ijk} \overline{y}^i \overline{z}^j (x - x_0)^k, \\ \overline{y}^m \frac{d\overline{z}}{dx} &= \sum_{i=0}^{\infty} \sum_{j=0}^{\infty} \sum_{k=0}^{\infty} q_{ijk} \overline{y}^i \overline{z}^j (x - x_0)^k, \end{aligned}$$

where $p_{000} q_{000} \neq 0$. Now proceed to Eqn. (2.53.1). Here the bar does not indicate complex conjugate. Markina and Yablonskii (1972)

23. $$y' = \frac{P(y, z, x)}{R(y, z, x)}, \quad z' = \frac{Q(y, z, x)}{S(y, z, x)},$$

where P, Q, R, and S are entire functions of y and z and holomorphic functions of x in some region D, while the pairs P and R, Q and S, and R and S have no common factors with respect to y or z for arbitrary $x \in D$. Further, it is assumed that

$$R(y, z, x) = H_1(z, x) H_2(y, z, x), S(y, z, x) = G_1(y, x) G_2(y, z, x),$$

where H_1 and H_2 are entire functions of z and of y and z, respectively, and the equation $H_2(y, z, x) = 0$ has no roots with respect to z independent of y; similarly, G_1 and G_2 are entire functions of y and of y and z, respectively, and the equation $G_2(y, z, x) = 0$ has no roots in y, independent of z.

Under these conditions, various situations arise: the solution in the neighborhood of a point (y_0, z_0, x_0) may be analytic about x_0 or holomorphic about x_0 or may have algebraic singularities about x_0. The existence of movable essential singularities is also discussed. Kondratenya (1970)

24. $$y' = \frac{P(y,z,x)}{R(y,z,x)}, \quad z' = \frac{Q(y,z,x)}{S(y,z,x)},$$

where P, Q, R, and S are entire functions of y and z and holomorphic functions of x in some region D.

It is shown that if for $x \to x_0 \in D$, a solution $y = f(x), z = \phi(x)$ of the given system is such that $f(x) \to y_0$ (a finite number) and $\phi(x)$ is indeterminate, and the given system satisfies the condition $Q(y, z, x_0) \not\equiv 0, R(y_0, z, x_0) \not\equiv 0$, and $S(y, z, x_0) \not\equiv 0$, then $\phi(x)$ takes all values except $\infty, C_1, C_2, \ldots$ in any neighborhood of x_0, where $C_1, C_2, \ldots$ are zeros of $Q(y_0, z, x_0)$ and $R(y_0, z, x_0)$.

Thus, $z_0 \in D$ is an essential singularity of the function $z = \phi(x)$ and behavior of solution $y = f(x)$ and $z = \phi(x)$ is found about the singularity. Kondratenya (1971)

25. $$y' = f(x, y, z), \quad z' = g(x, y, z),$$

where f and g are given functions of x, y, and z, and without loss of generality, $y(0) = 0, z(0) = 0$.

Assuming that f and g are analytic in $|x| < a, |y| < b$, and $|z| < c$, and in this parallelopiped $|f| < M$ and $|g| < N$, we can write the solution in the form

$$y = \sum_{j=1}^{\infty} a_j x^j, \quad z = \sum_{j=1}^{\infty} b_j x^j. \tag{1}$$

Writing

$$\begin{aligned} f(x, y, z) &= A_{000} + A_{100}x + A_{010}y + A_{001}z + A_{200}x^2 + A_{020}y^2 \\ &\quad + A_{002}z^2 + A_{110}xy + A_{101}xz + A_{011}yz + \cdots \\ &\equiv \sum_{k=0}^{\infty} \sum_{m=0}^{\infty} \sum_{n=0}^{\infty} A_{kmn} x^k y^m z^n, \end{aligned}$$

and similarly,

$$g(x, y, z) = \sum_{k=0}^{\infty} \sum_{m=0}^{\infty} \sum_{n=0}^{\infty} B_{kmn} x^k y^m z^n,$$

we have the following formula for A_{kmn} and similarly for B_{kmn}:

$$A_{kmn} = \frac{1}{k!m!n!} \left[\frac{\partial^{k+m+n} f}{\partial x^k \partial y^m \partial z^n} \right]_{x=0, y=0, z=0}.$$

Substituting for y, z, and f and g their series form in the given DE, we have

$$\sum_{j=1}^{\infty} ja_jx^{j-1} = \sum_{k=0}^{\infty}\sum_{m=0}^{\infty}\sum_{n=0}^{\infty} A_{kmn}x^k\left(\sum_{j=1}^{\infty} a_jx^j\right)^m\left(\sum_{j=1}^{\infty} b_jx^j\right)^n,$$

$$\sum_{j=1}^{\infty} jb_jx^{j-1} = \sum_{k=0}^{\infty}\sum_{m=0}^{\infty}\sum_{n=0}^{\infty} B_{kmn}x^k\left(\sum_{j=1}^{\infty} a_jx^j\right)^m\left(\sum_{j=1}^{\infty} b_jx^j\right)^n.$$

Now we must carry out indicated multiplication on the RHSs and equate different powers of x to obtain a_j recursively in terms of A_{kmn} and B_{kmn}.

Under the conditions mentioned in the beginning on f and g, the series solution (1) may be shown to possess intervals of convergence the magnitudes (lengths) of which are bounded below by a definite positive expression depending on a, b, c, M, and N. Martin and Reissner (1958), p.161

2.52 $h_i(x,y_1,y_2)y_i' = f_i(x,y_1,y_2)\,(i=1,2)$, f_i polynomial in y_i

1. $$(z-x)y' + yz' = 0, \quad y' + (z-x)z' = 0.$$

The given system describes long waves in a channel, after similarity reduction. This homogeneous system in y' and z' is compatible if $(z-x)^2 - y = 0$. The solution is then easily found as $z = 2(x - h_0^{1/2})/3, y = \{(x+2h_0)^{1/2}\}^2/9$, where h_0 is an arbitrary constant. Dresner (1983). p 82

2. $$3y' + 4xz' = 2z, \quad 6zz' + 4xy' = 3y,$$

$y(0) = -1, y^2(A) = z^3(A), 16A^2 = 9z(A)$.

The given system decribes waves in a superelastic wire. The transformation $Y = y/x^3, Z = z/x^3$, reduces it to an autonomous form and hence to the (Y,Z) phase plane:

$$\frac{dY}{dZ} = \frac{12Y + 4Z^2 - 18YZ}{3Y + 8Z - 12Z^2}. \tag{1}$$

The local analysis of (1) leads to the solution, described by the separatrix starting from a saddle point and moving to infinity. The waves are headed by a shock. Dresner (1983), p.82

3. $$(z-x)y' + yz' = -y, \quad y' - y(x-z)z' = -\sigma z,$$

$y(\infty) = 0, y(0) = 1, z(0) = z_0; \sigma$ is a constant.

The given system arises in gas dynamics when reaction forces are present. For the case $\sigma = 0$ (no reaction), it has solutions

$$\begin{aligned} z - z_0 &= 1/\{2(2^{1/2})\}\ln\{(z-x+2^{1/2})/(z_0+2^{1/2}) \\ &\quad - (z_0 - 2^{1/2})/(z-x-2^{1/2})\}, \\ y &= [\{(z-x)^2 - 2\}/\{z_0 - 2\}]^{1/2}, 1 < z_0 < 2^{1/2}, 2^{1/2} < z_0, \\ z &= x + 2^{1/2}, y = \exp(-2^{1/2}x), z_0 = 2^{1/2}. \end{aligned}$$

The cases $\sigma = 0$ and $\sigma \neq 0$ are both found numerically and depicted. Gordeyev, Kudryashov, and Murzenko (1985)

4. $$x' + a(xy)' = -k_1 x, \quad y' + a(xy)' = -k_2 y, \quad ' = \frac{d}{dt},$$

where from chemical-biological context, a is a large parameter, and $x(0) > 0, y(0) > 0; k_1$ and k_2 are constants.

A Taylor series solution about $t = 0$ or series in a^{-1} can be found easily but are seen to be slowly convergent. We may rewrite the system as

$$\begin{aligned} x' &= -k_1 x + (k_1 + k_2)xy/(\epsilon + x + y), \\ y' &= -k_2 y + (k_1 + k_2)xy/(\epsilon + x + y), \epsilon = 1/a. \end{aligned} \tag{1}$$

For $\epsilon = 0$, the system (1) is easily integrated by a separation of variables:

$$\{x^2 - c^2/\nu\}^{(\nu+1)/(2\nu)}/x = Ce^{-k_2 t}, y = c^2/x,$$

where c and C are constants and $\nu = k_1/k_2$. To get the asymptotic results for $a = 1$ and large t, we first change the independent variable to $\tau = k_2 t$ and write the given system as $x' + (xy)' = -\nu x, y' + (xy)' = -y$. For different values of ν, we get the following asymptotic forms:

(a) ν is irrational:

$$\begin{aligned} x &\sim \lambda e^{-\nu t} - (1+\nu)\lambda\mu e^{-(\nu+1)t} \\ &\quad + (1/2)(1+\nu)(2+\nu)\lambda\mu^2 e^{-(\nu+2)t} + (2 + 1/\nu)\lambda^2\mu e^{-(2\nu+1)t} + \cdots, \end{aligned}$$

$$\begin{aligned} y &\sim \mu e^{-t} - (1 + 1/\nu)\lambda\mu e^{-(\nu+1)t} + (2+\nu)\mu^2 e^{-(\nu+2)t} \\ &\quad + (1/2)(1+1/\nu)(2+1/\nu)\lambda^2\mu e^{-(2\nu+1)t} + \cdots, \end{aligned}$$

where μ and λ are arbitrary constants.

(b) ν is rational:

$$x \sim \sum_{m=0}^{q-1} \sum_{n=0}^{\infty} a_{mn} e^{-(m/q+n)t}, \quad y \sim \sum_{m=0}^{q-1} \sum_{n=0}^{\infty} b_{mn} e^{-(m/q+n)t}, \tag{2}$$

where a_{mn}, b_{mn} may be found recursively from

$$\{m/q + n\}\{a_{mn} + c_{mn})\} = (p/q)a_{mn}, \{m/q + n\}\{b_{mn} + c_{mn})\} = b_{mn},$$

obtained by substituting (2) in (1).

(c) $\nu = 1$:

$$\begin{aligned} x &\sim \lambda e^{-\lambda} - 2\lambda\mu e^{-2t} + 3\lambda\mu(\lambda+\mu)e^{-3t} + \cdots, \\ y &\sim \mu e^{-t} - 2\lambda\mu e^{-2t} + 3\lambda\mu(\lambda+\mu)e^{-3t} + \cdots, \end{aligned}$$

where λ and μ are arbitrary constants. Asymptotic nature of the series is rigorously proved.

Bihari and Fenyes (1968)

5. $xy' = (\alpha\epsilon - 1)y + xv + \epsilon xy^2, \quad xyv' = \beta y - x + (\alpha\epsilon - 2)yv + xv^2, \quad \epsilon^2 = 1/\gamma = 1.$

The given system in y and v is equivalent to the third equation of Painlevé. Gromak (1975)

6. $xy' = -y^2v - yv - (a + c)y - a, \quad xv' = yv^2 + v^2/2 + 2\delta x^2 y + (a + c)v - \gamma x + \delta x^2,$

where a, c, γ, and δ are constants.

The given system in y and v is equivalent to the fifth Painlevé equation via the following transformations. Eliminating v, we get the second order equation

$$y'' = [(1 + 2y)/\{2y(1 + y)\}]y'^2 - (1/x)y' + \theta(y, x),$$

where

$$\begin{aligned}\theta(y, x) &= [(a + c)/x^2](a + ay + cy) - (1/x)(y^2 + y)(\delta x - \gamma + 2\delta xy)\\ &\quad - [(1 + 2y)/\{2x^2(y^2 + y)\}](a + ay + cy)^2.\end{aligned}$$

Now putting $y = -w/(w - 1)$, we get Painlevé's fifth equation:

$$\begin{aligned}w'' &= [(3w - 1)/\{2w(w - 1)\}]w'^2 - w'/x + (\alpha/x^2)w(w - 1)^2\\ &\quad + (\beta/x^2)\{(w - 1)^2/w\} + (\gamma/x)w + \{\delta w(w + 1)/(w - 1)\},\end{aligned}$$

with the relations $2\alpha = c^2$ and $2\beta = -a^2$. Gromak (1976)

7.
$$xy' = -a - (a + c)y - yz - y^2 z,$$
$$xz' = x + \delta x^2 + (a + c)z + 2\delta x^2 y + (1/2)z^2 + yz^2,$$

where a, c, γ, and δ are constants.

The given system is equivalent to the fifth Painlevé equation, within a bilinear transformation. It is a special case of Eqn. (2.51.18) and therefore has poles as its only movable singularities. Lukashevich (1967)

8.
$$x(x - 1)y' = \lambda x + (ax + b)y + cy^2 + ky(y - 1)(y - x)z,$$
$$x(x - 1)z' = p - [2cy + (ax + b)]z - (1/2)k(3y^2 - 2xy - 2y + x)z^2,$$

where λ, a, b, c, p, and $k \neq 0$ are constants.

The given system is equivalent to the sixth Painlevé equation, within a bilinear transformation. It is a special case of Eqn. (2.51.18) and therefore has poles as its only movable singularities. Lukashevich (1967)

9.
$$x(x - 1)y' = ax + (bx - p)y + qy^2 + y(y - 1)(y - x)z,$$
$$x(x - 1)z' = (1/2)(2\alpha - q^2) - (bx - p)z - 2qyz - (1/2)(3y^2 - 2xy - 2y + x)z^2,$$

where a, b, α, p, and q are constants.

The given system is equivalent to the sixth equation of Painlevé; the constants a, b, α, p, and q are suitably related to the parameters in the second order form of P_{VI}. Lukashevich (1972)

2.53 $h_i(x, y_1, y_2)y_i' = f_i(x, y_1, y_2)$ $(i = 1, 2)$, f_i not polynomial in y_i

1.
$$zy' = \sum_{i=o}^{\infty} \sum_{j=o}^{\infty} \sum_{k=o}^{\infty} p_{ijk} y^i z^j (x - x_o)^k,$$

$$yz' = \sum_{i=o}^{\infty} \sum_{j=o}^{\infty} \sum_{k=o}^{\infty} q_{ijk} y^i z^j (x - x_o)^k,$$

where the ratio p_{000}/q_{000} is not rational but

$$\mathrm{Re}\{p_{000}/(p_{000} + q_{000})\} > 0, \mathrm{Re}\{q_{000}/(p_{000} + q_{000})\} > 0. \tag{1}$$

Introducing the variable u and v according to

$$\begin{aligned} y &= (x - x_0)^{p_{000}/(p_{000}+q_{000})}(\alpha + u), \\ z &= (x - x_0)^{q_{000}/(p_{000}+q_{000})}(\beta + v), \end{aligned} \tag{2}$$

where α and β are constants related by

$$\alpha\beta = p_{000} + q_{000} \tag{3}$$

and calling

$$u_1 = (x - x_0)^{p_{000}/(p_{000}+q_{000})}, \quad v_1 = (x - x_0)^{q_{000}/(p_{000}+q_{000})}, \tag{4}$$

we have in view of (1), $(u_1, v_1) \to (0, 0)$ as $x \to x_{0^-}$, and the given system becomes

$$\begin{aligned} (x - x_0)\frac{du}{dx} &= -[p_{000}/(p_{000} + q_{000})]u - [p_{000}/\beta^2]v \\ &\quad + [p_{100}\alpha/\beta]u_1 + p_{010}v_1 + F_1(u, v, u_1, v_1, x), \\ (x - x_0)\frac{dv}{dx} &= -[q_{000}/\alpha^2]u - [q_{000}/(p_{000} + q_{000})]v \\ &\quad + q_{100}u_1 + [q_{010}\beta/\alpha]v_1 + F_2(u, v, u_1, v_1, x), \end{aligned} \tag{5}$$

where F_1 and F_2 are holomorphic functions of all their arguments in a neighborhood of the point $(0, 0, 0, 0, x_0)$ and they and their partial derivatives vanish with u, v, u_1, v_1 at this point. It is easy to check that u_1 and v_1 satisfy the system

$$\begin{aligned} (x - x_0)\frac{du_1}{dx} &= [p_{000}/(p_{000} + q_{000})]u_1, \\ (x - x_0)\frac{dv_1}{dx} &= [q_{000}/(p_{000} + q_{000})]v_1. \end{aligned} \tag{6}$$

The system (5)–(6) with $(u, v, u_1, v_1) = (0, 0, 0, 0)$ at $x = x_0$ is shown to have a two-parameter family of solutions

$$
\begin{aligned}
u &= \sum u_{klm}(x-x_0)^{k+lu_3+mu_4}, \\
v &= \sum v_{klm}(x-x_0)^{k+lu_3+mu_4}, \\
u_1 &= A(x-x_0)^{\mu_3}, v_1 = B(x-x_0)^{\mu_4}, \qquad (7)
\end{aligned}
$$

where A and B are arbitrary constants and the series converge. Use (3) to put $A = B = 1$. In view of the relation (3), one of the two parameters α and β is arbitrary. Tracing back, it is shown that there exists a one-parameter family of the original system for which $(y,z) \to (0,0)$ as $x \to x_0$, according to (2). Markina and Yablonskii (1972)

2.
$$
z^n y' = \sum_{i=0}^{\infty}\sum_{j=0}^{\infty}\sum_{k=0}^{\infty} p_{ijk} y^i z^j (x-x_0)^k,
$$

$$
y^m z' = \sum_{i=0}^{\infty}\sum_{j=0}^{\infty}\sum_{k=0}^{\infty} q_{ijk} y^i z^j (x-x_0)^k,
$$

where $p_{000}q_{000} \neq 0$.

Conditions are found under which the point x_0 is an algebraic singular point for the solution $y(x), z(x)$ defined by the IC $y \to 0, z \to 0$ as $x \to x_0$. A formal solution is found in the form

$$
y = \sum_{k=0}^{\infty} \alpha_k (x-x_0)^{(k+r)/p_1}, z = \sum_{k=0}^{\infty} \beta_k (x-x_0)^{(k+s)/p_1}, \qquad (1)
$$

where r, s, and p_1 are positive integers and $\alpha_0, \beta_0 \neq 0$. The substitution of (1) in the given DE gives $r+ns-p_1 = 0, mr+s-p_1 = 0$. It is easy to see that if $m = 0, n \neq 0$ $(n = 0, m \neq 0)$ or $m = 1, n \neq 1$ $(n = 1, m \neq 1)$, then a solution of the form (1) does not exist. It is shown that for $n = m = 1$, the necessary and sufficient condition for the convergence of the series in (1) and hence the existence of a formal one-parameter family of solutions is that $r/s = p_{000}/q_{000}$. This is accomplished by transforming the given DE to a Briot–Bouquet system and using the standard theory for the latter.

The case $(n-1)(m-1) \geq 1$ is also considered. Markina and Yablonskii (1972)

3.
$$
y^m z^n y' = \sum_{i+j+k=0}^{\infty} a_{ijk} y^i z^j (x-x_0)^k,
$$

$$
y^q z^\ell z' = \sum_{i+j+k=0}^{\infty} b_{ijk} y^i z^j (x-x_0)^k,
$$

where m, n, q, and ℓ are nonnegative integers, $m+n+q+\ell \neq 0$, and the series converge when $|y|, |z|$, and $|x-x_0|$ are small.

It is shown that if $m+1 > q$ and $\ell+1 > n$ in the given DE, then all nonstationary singularities of solutions with the properties

$$
y \to 0, z \to 0 \text{ for } x \to x_0, \qquad (1)
$$

or

$$
y \to 0, z \to \beta_0 \neq 0 \text{ for } x \to x_0, \qquad (2)
$$

or

$$
y \to \alpha_0 \neq 0, z \to 0 \text{ for } x \to x_0, \qquad (3)
$$

where $|\alpha_0|, |\beta_0|$ are sufficiently small are algebraic if

$$a_{000}b_{000} \neq 0, (a_{000} + a_{010}\beta_0)(b_{000} + b_{010}\beta_0) \neq 0, (a_{000} + a_{010}\alpha_0)(b_{000} + b_{010}\alpha_0) \neq 0.$$

For example, in the case $a_{000}b_{000} \neq 0$, there are solutions with the property (1) of the form

$$y = \sum_{k=0}^{\infty} y_k(x - x_0)^{(k+r)/p}, z = \sum_{k=0}^{\infty} z_k(x - x_0)^{(k+s)/p}, \tag{4}$$

where $y_0 z_0 \neq 0, r > 0, s > 0$ and $p > 0$ are integers if

$$\begin{aligned} \ell + 1 &= n, m + 1 = q, \quad \text{or} \\ \ell + 1 &> n, m + 1 > q, \quad \text{or} \\ \ell + 1 &< n, m + 1 < q. \end{aligned} \tag{5}$$

In case (5) there is a solution of the form (4) if and only if a_{000}/b_{000} is positive and rational: $a_{000}/b_{000} = r/s$. Here y_0, z_0 satisfy the relation

$$r y_0^q z_0^n = p a_{000} \text{ or } s y_0^q z_0^n = p b_{000},$$

so that either y_0 or z_0 is an arbitrary parameter, while y_k, z_k $(k \geq 1)$ are determined uniquely.

Other cases are treated similarly in a brief manner. Markina (1973)

4. $$y' + (2/3)xz' = (4/3)z, \quad z' + (2/3)z^{1/2}xy' = z^{1/2}y,$$

where $y(0) = -1, y(\infty) = 0, z(\infty) = 0$.

The given BVP describes waves in a non-Hookian wire. It is easy to check that the change $Y = x^3 y, Z = x^4 z$ makes the given system autonomous so that we get, in the (Y, Z) plane, the first order DE

$$\frac{dY}{dZ} = \frac{3Y - 2YZ^{1/2} + (4/3)Z}{4Z - (8/3)Z^{3/2} + YZ^{1/2}}. \tag{1}$$

The analysis of (1) in (Y, Z) plane leads to the solution curves of the BVP, joining the singular points. This happens to be a separatrix. Dresner (1983), p.75

5. $$\left[\sum A y_1^{\alpha_1} y_2^{\alpha_2} x^{\beta}\right] \frac{dy_1}{dx} = \sum A^1 y_1^{\alpha_1^1} y_2^{\alpha_2^1} x^{\beta^1},$$

$$\left[\sum A y_1^{\alpha_1} y_2^{\alpha_2} x^{\beta}\right] \frac{dy_2}{dx} = \sum A^2 y_1^{\alpha_1^2} y_2^{\alpha_2^2} x^{\beta^2},$$

where summations are over indicated indices.

The paper details when this system has integrals of the form

$$\begin{aligned} y_1 &= \nu_1 x^{\mu_1} e^{-k(\log 1/x)^m}, \\ y_2 &= \nu_2 x^{\mu_2} e^{-k(\log 1/x)^m}, \end{aligned}$$

where $0 < m < 1, \mu_1 > 0, \mu_2 > 0$, and where ν_1, ν_2 approach definite nonzero limits when x approaches zero along a path that lies within a certain positive angle α. Rosenblatt (1939)

6. $$xy' = f(x, y, z), \quad xz' = g(x, y, z),$$

where it is assumed that:

(a) Both $f(x, y, z)$ and $g(x, y, z)$ are functions holomorphic and bounded in (x, y, z) in $|x| < a, |y| < b, |z| < b$, and vanish at $x = y = z = 0$.

(b) The functions $f(0, y, z)$ and $g(0, y, z)$ have uniformly convergent expansions of the form $f(0, y, z) = y^{m+1}z^n \left(\alpha + \sum \alpha_{k\ell}y^k z^\ell\right), g(0, y, z) = y^m z^{n+1}(\beta + \sum \beta_{k\ell}y^k z^\ell)$, and m and n are nonnegative integers not simultaneously zero.

(c) α and β are nonzero complex constants such that $\mathrm{Re}\{\alpha/(m\alpha + n\beta)\} > 0$, $\mathrm{Re}\{\beta/(m\alpha + n\beta)\} > 0$. A series solution is constructed that depends on two arbitrary constants.

Iwano (1966/67)

7. $$x^{s-1}y' = f(x, y, z), \quad xz' = g(x, y, z),$$

where f and g are holomorphic at $(0, 0, 0)$ and s is a positive integer.

The case with $x = 0$, an irregular singular point, is studied. More precisely, assuming that $f(0, 0, 0)$, $\mathrm{Re}\ f_y(0, 0, 0) > 0$ and $g(0, 0, z) = z^{m+1}(\alpha + \alpha' z^m)$, where $\alpha\alpha' \neq 0$ and m is a positive integer, existence of the solution in a formal series in terms of $V(x)$ and $W(x)$ is proved, where $V(x)$ is any solution of $xv' = v^{m+1}(\alpha + \alpha' v^m)$ and $W(x)$ is any solution of $xw' = w^2(\alpha + \alpha' w)$. Hsieh and Przybylski (1983)

8. $$x^2y' = (\mu + \alpha x)y + xf(x, y, z), \quad x^2z' = (-\nu + \beta x)z + xg(x, y, z),$$

where:

(a) μ and ν are positive constants.

(b) $f(x, y, z)$ and $g(x, y, z)$ are holomorphic functions of (x, y, z) in a domain defined by $|x| < r, |y| < \rho, |z| < \rho$ and satisfy $f(x, 0, 0) = g(x, 0, 0) = 0$, $f_y(x, 0, 0) = f_z(x, 0, 0) = g_y(x, 0, 0) = g_z(x, 0, 0) = 0$. The point $x = 0$ is an irregular singular point.

(c) $\mathrm{Re}\ (\alpha\nu + \beta\mu) > 0, \alpha$ and β are complex constants.

Then, for every positive constant $\epsilon (< \pi)$, the given system possesses a general solution of the form $y = \phi(x, U(x), V(x)), z = \psi(x, U(x), V(x))$, provided that the triple $(x, U(x), V(x))$ belongs to the domain of convergence of $\phi(x, u, v)$ and $\psi(x, u, v)$. Here:

(a) $(U(x), V(x)) = (C_1 x^\alpha e^{-\mu/x}, C_2 x^\beta e^{\nu/x})$ is a general solution of the equation

$$\begin{aligned} x^2\frac{du}{dx} &= (\mu + \alpha x)u, \\ x^2\frac{dv}{dx} &= (-\nu + \beta x)v. \end{aligned}$$

(b) $\phi(x, u, v)$ and $\psi(x, u, v)$ are holomorphic functions in a domain defined by $D(r', \rho') = \{(x, u, v); 0 < |x| < r', |\arg x - \pi/2| < \pi - \epsilon, |u| < \rho', |v| < \rho'\}$, where r' and ρ' are sufficiently small positive constants. Moreover, these functions are expanded into uniformly convergent series

$$\phi(x, u, v) = u + \sum_{j+k\geq 2} p_{jk}(x)u^j v^k, \psi(x, u, v) = v + \sum_{j+k\geq 2} q_{kj}(x)u^j v^k,$$

in the domain $D(r', \rho')$, where the coefficients $p_{jk}(x)$'s and $q_{jk}(x)$'s are holomorphic in the sector $0 < |x| < r', |\arg x - \pi/2| < \pi - \epsilon$ and admit asymptotic expansions in powers of x as x tends to zero.

Shimomura (1983)

9. $$y' = 1 + yP(x, y) + y^k u, \quad yu' = pu + h(x, y) + yH_1(x, y, u),$$

where p is a constant, $k \geq 0$ an integer, $P(x, y)$ a polynomial in y of degree $\leq k-2$, and H is a holomorphic function of y in a neighborhood of $y = 0$ and that $P(x, y), h(x, y), H_1(x, y, u)$ are analytic functions of x in a given domain D.

In order that the system above has only single-valued solution in the neighborhood of a movable singularity, it is necessary:

(a) That p be an integer positive, negative, or zero.

(b) If $p = 0$, that $h(x, 0) \equiv 0$.

Bureau (1964)

10. $$z^n y' = \sum_{\nu=0}^{m} p_\nu(z, x) y^{m-\nu}, \quad z' = \sum_{\nu=0}^{M} q_\nu(z, x) y^{M-\nu},$$

where $n > 1$, and $p_\nu(z, x)$ and $q_\nu(z, x)$ are holomorphic functions in some domain B. Let $(z = 0, x = x_0) \in B$. In the neighbourhood of this point the functions p and q can be represented by

$$p_\nu(z, x) = \sum_{i=0}^{\infty} \sum_{j=0}^{\infty} p_{\nu ij} z^i (x - x_0)^j,$$
$$q_\nu(z, x) = \sum_{i=0}^{\infty} \sum_{j=0}^{\infty} q_{\nu ij} z^i (x - x_0)^j,$$

where we assume that $p_{000} q_{000} \neq 0$.

The formal solution for which $y \to \infty, z \to 0$, as $x \to x_0$ is sought in the form

$$y = \sum_{k=0}^{\infty} \alpha_k (x - x_0)^{(k-r)/p_1}, \quad z = \sum_{k=0}^{\infty} \beta_k (x - x_0)^{(k+s)/p_1}, \tag{1}$$

where r, s, and p_1 are positive integers, and $\alpha_0 \beta_0 \neq 0$. Substitution of (1) in the given system, etc., leads to

$$s = (M - m + 1)\ell/\lambda, r = (n - 1)\ell/\lambda, p_1 = (Mn - m + 1)\ell/\lambda,$$

where $m < M + 1, \lambda = (M - m + 1, n - 1)$, and ℓ is an arbitrary positive integer. The convergence of (1) is discussed by reducing the given system to the Briot–Bouquet form. Markina and Yablonskii (1972)

3

THIRD ORDER EQUATIONS

3.1 $y''' + f(y) = 0$ and $y''' + f(x, y) = 0$

1. $$y''' + y + y^2 = 0$$

The solution may be found in a finite form:

$$y(x) = -30p(p^2 + p') - (15^{1/2} + 3)/6,$$

$$p(x) = 1/[12(15)^{1/2}P(x - x_0; 0, g_3)], g_3 = -1/[12^2 \times 15].$$

Santos (1989)

2. $$y''' - y^{-2} = 0, y(0) = 1, y'(0) = 0, y'' \to 0 \text{ as } x \to -\infty.$$

The unique value of $y''(0)$ that solves the given problem is found numerically to be $y''(0) = 1.2836$; in this case $y''(\infty) = 2.1591$. Tuck and Schwartz (1990)

3. $$y''' - y^{-2} + 1 = 0, y \to 1 \text{ as } x \to -\infty.$$

A formal linear solution about $y = 1$ satisfying the given boundary condition is

$$y = 1 + a \exp\{2^{-2/3}x\} \cos\{3^{1/2}2^{-2/3}x\}, \tag{1}$$

where a is a constant. A numerical algorithm for solving the given DE subject to (1) generates a one-parameter family of solutions satisfying the given BC. The case $a = 0.004$ is depicted. Tuck and Schwartz (1990)

4. $$y''' - (1 + \alpha)/(y^2 + \alpha) + 1 = 0, y = 1 \text{ as } x \to -\infty,$$

where α is a parameter.

The "linearized" solution about $y = 1$ as $x \to -\infty$ is found to be

$$y = 1 + ae^{qx} \cos\{q(3^{1/2})x\}, \tag{1}$$

where $q = 2^{-2/3}(1 + \alpha)^{-1/3}$. The origin of x for the given autonomous equation is arbitrary. The parameter a in (1) is free but must be taken to be small. For every given α, the entire family of solutions that asymptotes to $y = 1$ as $x \to -\infty$ is generated by allowing a to vary within a certain range. Thus, there is really a two-parameter family of solutions. As x increases, the function $y(x)$ oscillates within a growing envelope until the curve meets the axis $y = 0$. Tuck and Schwartz (1990)

5. $$y''' - (1 + \delta + \delta^2)y^{-2} + (\delta + \delta^2)y^{-3} + 1 = 0,$$

$y \to 1$ as $x \to -\infty$ and $y \to \delta$ as $x \to \infty$; δ is a parameter.

Even though the given DE is of third order, the given data provide a complete specification of a boundary value problem, since the absence of x in the equation implies that the origin of x can be specified arbitrarily, eliminating the need for a third BC. An approximate solution as $x \to -\infty$ about $y = 1$ is found to be

$$y = 1 + ae^{qx}\cos\{q(3^{1/2})x\}, \tag{1}$$

where $q = (2 - \delta - \delta^2)^{1/3}/2$. An initial value problem with (1) and a choice of a by trial and error helps solve the given problem. Numerical results suggest that solution is unique if $\delta \le 1$. The results are depicted for various δ values. A small δ limit is considered in some detail. Writing

$$y = \delta Y, \quad x = x_0 + \delta X \tag{2}$$

in the given DE leads to the formal limiting equation

$$Y''' = Y^{-2} - Y^{-3}. \tag{3}$$

Equation (3) is solved subject to $(y, y') = (\delta, 0)$, i.e., $(Y, Y') = (1, 0)$. So the problem posed for (2) is $Y \to 1$ as $X \to +\infty$. The behavior as $X \to -\infty$, namely,

$$Y \to -X(3\log|X|)^{1/3} + O(X\log^{-5/3}|X|),$$

helps to solve the problem uniquely by the use of numerical methods. Tuck and Schwartz (1990)

6. $$y''' - f(y) = 0,$$

where $f(y)$ may assume one of the following forms:

(a) $-1 + y^{-2}$.
(b) $-1 + (1 + \delta + \delta^2)y^{-2} - (\delta + \delta^2)y^{-3}$.
(c) $y^{-2} - y^{-3}$.
(d) y^{-2}.
(e) $-1 + (1 + \alpha)/(y^2 + \alpha)$.
(f) $-1 + (1 + \alpha)/(y^2 + \alpha y)$.

The given forms occur in the description of drainage and coating flows; solutions satisfying $y \to 1$ as $x \to -\infty, y \to \delta$ as $x \to +\infty$ are discussed numerically. Tuck and Schwartz (1990)

7. $$y''' - q(x)|y|^{\alpha}\,\mathrm{sgn}\, y = 0, \alpha \ne 1,$$

where $q(x)$ is a positive continuous function that is locally of bounded variation on $[0, \infty)$.

The following results are proved:

(a) Let $\alpha > 1$ and set $\sigma = 3 + h(\alpha - 1), h = 1 - 1/3^{1/2}$. Assume that

$$\frac{d}{dx}\{x^\sigma q(x)\} \leq 0 \quad \text{and} \quad \int_1^\infty x^{\sigma-1} q(x)\{\ln x\}^{3(\alpha-1)/2} < +\infty.$$

Then the given DE is nonoscillatory.

(b) Let $0 < \alpha < 1$ and assume that $\lim x \to \infty\ x^{2+\alpha} q(x) = +\infty$. Then the given DE is nonoscillatory.

Erbe and Rao (1987)

8.
$$y''' - p(x)f(y) = 0,$$

where $p(x)$ is continuous and positive in the interval $I : [a,b]$ and for all real $y, yf(y) > 0$ for $y \neq 0$.

It is shown that under the given conditions no solution $y \not\equiv 0$ of the given DE in (a,b) can be tangent to $y = 0$ more than once. It is also proved that no solution of the given DE is periodic. Utz (1967a)

9.
$$y''' + f(x,y) = 0, y(0) = y(a) = 0, 0 < a \leq \alpha,$$

where $f(x,y), f_y(x,y) \geq 0, f_{yy}(x,y) > 0$ are continuous functions of $x \in [0,\alpha]$ and $|y| \leq \beta$.

It is proved that there is exactly one solution of the given IVP. Gregus (1987), p.233

10.
$$y''' - f(x,y) = 0, y(0) = y'(0) = y''(0) = 0, 0 \leq x \leq a, 0 \leq a \leq \alpha,$$

where $f(x,y), f_y(x,y) \geq 0, f_{yy}(x,y) > 0$ are continuous functions of $x \in [0,\alpha]$ and $|y| \leq \beta$.

It is proved that there is exactly one nontrivial solution of the given IVP. Gregus (1987), p.233

3.2 $y''' + f(x,y)y' + g(x,y) = 0$

1.
$$y''' + y' + y^2/2 - 1 = 0,$$

$y' > 0, \forall\ x \in (-\infty, \infty), y(-\infty) = -2^{1/2}, y(\infty) = 2^{1/2}$.

It is proved that the given problem has no monotonic solution. Several other qualitative properties, including oscillation at $x = \pm\infty$, are discussed. Troy (1990a)

2.
$$y''' + y' + y^2/2 - c^2 = 0, -\infty < x < \infty, c > 0.$$

The following results are proved:

(a) For small $c > 0$, there exists at least one odd periodic solution.

(b) For each $c > 0$, there exists no solution of the given DE that satisfies $y' > 0$ $\forall x \in (-\infty, \infty)$ with the $\lim_{x\to\pm\infty} y(x) = c(2^{1/2})$.

(c) There exists $\bar{c} > 0$ such that for each $\bar{c} \in (0, c)$ there is an odd-periodic solution of the given DE.

The given DE describes steady-state solutions of the Kuramoto–Sivashinsky equation. Jones, Troy, and MacGillivray (1992)

3. $$y''' + y' + (1/2)y^2 - c^2 = 0, -\infty < x < \infty,$$

where c is a parameter.

It is shown that for large c, the given DE has a unique (up to translation) bounded solution. This solution $y(x)$ is an odd function of x, tends to $\lim_{x\to\pm\infty} y(x) = \mp c(2)^{1/2}$, and vanishes only at $x = 0$. The integral $\int_0^x y(\tau)d\tau$ thus has a conical form with a single maximum at $x = 0$ and slopes $\pm 2^{1/2}c$ as $x \to \mp\infty$. Topological properties of the set of bounded solutions are discussed. Numerical computations for intermediary values of c^2 suggest that below $c^2 \simeq 1.6$ for every speed there is a continuum of odd quasiperiodic solution or a Cantor set of chaotic solutions wrapped by an infinite sequence of conic solutions. Michelson (1986)

4. $$y''' + y' + y^2 + cy = 0,$$

where c is a constant.

Solution of the given DE involving two arbitrary constants is sought in the finite form

$$y = (A_{-1}p^{-1} + A_0 + A_1 p)p' + B_0 + B_1 p + B_2 p^2 + B_3 p^3, \tag{1}$$

where

$$p'^2(x) = p^2 + b_3 p^3 + p^4, \tag{2}$$

where A_i, B_i, and b_3 are constants.

Using the fact that if we put $y(x) = v(x) - c/2$ in the given DE, we get the equation

$$v''' + v' + v^2 - c^2/4 = 0, \tag{3}$$

so that v is an odd function of x, a special solution of the given DE is sought in the form

$$y = (A_{-1}p^{-1} + A_0)p' + B_0, \tag{4}$$

where

$$p'^2 = b_2 p^2 + p^3. \tag{5}$$

Substituting Eqns. (4) and (5) in the given DE shows that (4) is a solution provided that

$$\begin{aligned} A_{-1} &= -15/19, A_0 = -15/2, \\ B_0 &= -15(d/19^3)^{1/2}, b_2 = d/19, \\ d &= -1 \text{ or } 11, k = (b_2/4)^{1/2}, \\ c &= -(30)d^{1/2}/19^{3/2} \text{ and } p(x) = -b_2 \operatorname{sech}^2 k(x - x_0). \end{aligned} \tag{6}$$

Equation (6) follows from an integration of (5).

The solution is written out explicitly as

$$y(x) = A\{1 + \tanh(kx)[1 - B \operatorname{sech}^2(kx)]\},$$

where $A^2 = 225d/19^3$ and $B = d/2$. Santos (1989)

5. $$y''' + y' + 2y^2 - \mu y = 0,$$

where μ is a constant.

Different kinds of solutions are sought:

(a) $y(+\infty) = y(-\infty) = 0, y'(+\infty) = y''(+\infty) = y'''(+\infty) = y'(-\infty) = y''(-\infty) = y'''(-\infty) = 0$; this gives a solitary wave.

(b) $y(+\infty) \neq y(-\infty)$; this is a shock solution.

(c) $y(x) = y(x+\tau)$, where τ is the period of a periodic solution.

The analytic form of these solutions is constructed for small μ. Chang (1986)

6. $$y''' + y' + \{(1/x^2 + 6/x^4)/(1 + \sin x + 1/x)^r\}y^r = 0.$$

The given DE has the exact nonoscillatory, nonmonotonic solution $y(x) = 1 + \sin x + 1/x$. Heidel (1968)

7. $$y''' + y' - \exp(y) = 0,$$

where $x^2[y + \ln(-x)] \to -2$ as Re $x \to -\infty$.

The following result for complex x is proved. There exists $R > 0$ such that the given problem has a unique solution in the region $|x| > R$, Re $x \leq 0$. This solution satisfies the asymptotic behavior not only as Re $x \to -\infty$, but also as $|x| \to \infty$ along any ray path in the half-plane Re $x \leq 0$.

One particular formal solution for large $|x|$ with no free constants is

$$y(x) = -\ln(-x) - 2x^{-2} + (50/3)x^{-4} - (6104/15)x^{-6} + O(x^{-8}).$$

This expansion diverges for all x, but truncation of the series at any finite order yields an approximate solution of the given DE. Kruskal and Segur (1991)

8. $$\epsilon y''' + y' + y^2 - 1 = 0, y(+\infty) = 1, y(0) = 0, y''(0) = 0.$$

The problem has no solution for $\epsilon > 0$; the solution of the given DE satisfying first two conditions, namely $y(+\infty) = 1, y(0) = 0$, is such that $y''(x) < 0$ for $0 \leq x < \infty$. Amick and McLeod (1990)

9. $$\lambda y''' + y' + y^2 - 1 = 0, \lim_{x\to\pm\infty} y = \pm 1,$$

where λ is a constant.

For general $\lambda > 0$, results are not generally known. However, if $\lambda = 11(15)^2/19^3$, then the solution is known exactly:

$$y(x) = \tanh(19x/30)\{1 - (11/2)\,\mathrm{sech}^2(19x/30)\}.$$

For $\lambda = 0, y = \tanh x$ is a solution. For $\lambda \geq 2/9$, it may be shown that there is no solution of the BVP with $y' \geq 0$ on R. It is also known that for large positive λ, there exists a unique, nonconstant, bounded solution of the given DE and this solution satisfies the other boundary condtions possible, namely $\lim_{x\to\pm\infty} y(x) = \mp 1$. [See also Michelson (1986)]. Toland (1988)

10. $$\lambda y''' + y' - f(y) = 0, \lim_{x\to\pm\infty} y(x) = \pm\alpha,$$

where the function f is such that $f(\pm\alpha) = 0$ and $f > 0$ on $(-\alpha, \alpha)$; λ is a constant.

Suppose that $f : R \to R$ is Lipschitz continuous and has $f(0) > 0, f(\pm\alpha) = 0$ and $f(y) \neq 0, y \neq \pm\alpha$. Then it is shown that for all $\lambda \in [-4/(27B^2), 0]$, where

$$B = \sup\{f(y)/(\alpha - y), f(y)/(y + \alpha), \quad y \in (-\alpha, \alpha)\},$$

there exists a unique monotone solution of the given BVP. Also, if f is C^1 and B is either $-f'(\alpha)$ or $f'(-\alpha)$, then this is the maximal λ-interval for the existence of monotone solutions. Moreover, there exists $\epsilon > 0$ such that for all $\lambda \in [-4/(27B^2) - \epsilon, 0]$, there exists a solution of the BVP if f is even and C^1.

Also, if f is even and C^1 on R and nonincreasing on $(0, +\infty)$, then for each $\lambda < 0$, there exists a unique solution of the given DE with $\{y(x) : x \in R\}$ bounded. This solution satisfies $\lim_{x\to\pm\infty} y(x) = \pm\alpha$ and is odd about its unique simple zero on R. If $\lambda = 0$, then clearly, there exists a solution of BVP. If $\lambda > 0$, the general results are not fully known. Toland (1988)

11. $$sy''' + dy' + (1/2)y^2 - Vy - C = 0,$$

where $s > 0, d, V$, and C are constants; $y \to y_{1,2}$ as $x \to \pm\infty$.

With boundary conditions at $x \to \pm\infty$, we find that $C = (1/2)y_2^2 - Vy_2$ and $V = (1/2)(y_1 + y_2)$. Let

$$y = (V - y_2)z + V, p = [(V - y_2)/(2d)]x.$$

Then the given DE goes to

$$\gamma z_{ppp} + z_p + z^2 = 1, \tag{1}$$

where $\gamma = s(V - y_2)^2/(4d^3)$. If $z(p)$ is a solution of (1), then so is $-z(-p)$. Therefore, $z \to 1$ as $x \to \infty$ and $z \to -1$ as $x \to -\infty$. Two integral identities are easily derived from (1):

$$\int_{-\infty}^{\infty} (1 - z)^2 dp = 2, \quad -\gamma \int_{-\infty}^{\infty} z_{pp}^2 dp + \int_{-\infty}^{\infty} z_p^2 dp = 4/3. \tag{2}$$

When $\gamma > 0$ (i.e., $d > 0$), the second of (2) suggests that solutions may not exist as γ increases. Indeed, assuming that $z^2 < 1$ and $z_p \geq 0$, it is easily seen from (1) that $|z_p| \leq \gamma^{-1/2}$, and then it follows from (1) that solutions do not exist for $\gamma > 9/4$. However, this is an upper bound and the numerical solutions indicate that solutions do not exist for $\gamma < \gamma_c$, where $\gamma_c \approx 0.1$. Note that the argument above does not apply if $\gamma < 0$. Numerical solutions of the given DE starting from asymptotic behavior at $+\infty$ are, however, found to be inconclusive with regard to their solving the correct "connection" problem. Hooper and Grimshaw (1985)

12. $$\epsilon y''' + y' - \cos y = 0,$$

$0 \leq x < \infty, \epsilon > 0$ is a parameter, $y(+\infty) = \pi/2, y(0) = 0$.

It is proved that the given problem possesses a unique monotonic solution. For this solution, $y''(x) < 0$ if $0 \leq x < \infty$ and in particular $y''(0) \neq 0$. It has an asymptotic estimate

$$y''(0) \sim -N\epsilon^{-5/4} \exp\{-(1/2)\pi\epsilon^{-1/2}\} \text{ as } \epsilon \downarrow 0$$

for some positive constant N. Amick and McLeod (1990)

13. $$\epsilon^2 y''' + y' - \cos y = 0, -\infty < x < \infty, 0 < \epsilon \ll 1,$$

$y \to \pm\pi/2$ as $x \to \pm\infty$; ϵ is a parameter.

The given problem describes a geometric model of crystal growth. A needle crystal solution satisfies BC: $y(x, \epsilon) \to \pm\pi/2$ as $x \to \pm\infty$. For $\epsilon = 0$ the system admits a "needle crystal" i.e., monotonic, solution $y(x; 0) = -\pi/2 + 2\tan^{-1} e^x$; for small ϵ, an asymptotic expansion in powers of ϵ^2 is found to be valid to all orders, but it provides no information about whether the given DE admits a true needle crystal for $\epsilon > 0$. This is an example of a problem in which one must go "beyond all orders" of an asymptotic expansion to determine whether a solution even exists. Kruskal and Segur demonstrate and prove the validity of a method to obtain information that lies beyond all orders of an asymptotic expansion. It is proved that the given problem has no needle crystal for any small nonzero ϵ. Kruskal and Segur (1991)

14. $$\epsilon y'''(x) + y'(x) - \cos y = 0, -\infty < x < \infty, y(\infty) = \pi/2,$$

where ϵ is a positive constant.

It is shown that for each prescribed $\epsilon > 0$, a "primary" solution y exists such that $y(0) = 0, y'(x) > 0, y''(x) < 0, 0 \leq x < \infty$. It is also shown that all nontrivial solutions of the given DE take either of the forms $y(x + x_0)$ or $\pi - y(x + x_0)$, where x_0 is any real constant. Quantitative data on $y'(0)$ and $-y''(0)$ as functions of ϵ are given for the primary solution. Hammersley and Mazzarino (1989)

15. $$y''' + (1/\delta^2)y' - (1/\delta^2)\cos y = 0,\ \delta \approx 1,$$

$y(0) = y''(0) = 0, y' > 0\ \forall x \in (-\infty, \infty), y(\infty) = \pi/2, y(-\infty) = -\pi/2.$

It is proved that there is no solution of the given problem if $|\delta - 1| > 0$ is sufficiently small. Troy (1990a)

16. $$\epsilon^2 y''' + y' - (\cos y)/(1 + \alpha \cos 4y) = 0, -\infty < x < \infty; 0 < \alpha < 1$$

when $y + \pi/2 \to 0$ as $x \to -\infty$ and $y(0) = 0$.

The needle-crystal problem is posed thus: For fixed ϵ $(0 < \epsilon \ll 1)$, find $\alpha(\epsilon)$ so that the solution of the given problem increases monotonically and is antisymmetric in x. It is proved that there exist solutions of the given problem for a large, discrete set of values of α. Successive values of α at which needle crystals occur differ approximately by ϵ, so the number of these values is $O(\epsilon^{-1})$ as $\epsilon \to 0$. Asymptotic expansions in ϵ^2 are written and the solution determined. Kruskal and Segur (1991)

17. $$y''' + (2/x^2)y' + q(x)y^r = 0,$$

where $q(x) > 0$ and r is a ratio of odd integers.

It is shown that all nonoscillatory solutions of the given DE are monotone. Heidel (1968)

18. $$y''' - (K/x^2)y' + q(x)y^r = 0, x > 0,$$

where

$$q(x) = -(K - 2)/\{x^3(\log x)^{2-r}\} + 6/\{x^3(\log x)^{3-r}\} + 6/\{x^3(\log x)^{4-r}\}.$$

The given DE has an exact solution $y = 1/\{\log x\}$. Heidel (1968)

19. $$y''' + p(x)y' + q(x)y^r = 0,$$

where $p(x)$ and $q(x)$ are continuous and real valued on a half-axis $[a, \infty)$ and r is the quotient of odd positive integers.

The behavior of the nonoscillatory solutions and the existence of oscillatory solutions of the given DE are studied. The two cases $p(x), q(x) \le 0$ and $p(x), q(x) \ge 0$ are discussed. Heidel (1968)

20. $$y''' + p(x)y' + q(x)y^r = 0,$$

where $p(x), q(x) \in C\ [a, \infty), a > 0; q(x)$ is not identically zero for large x and the exponent r is a quotient of odd positive integers. This last condition ensures that solutions with real initial conditions are real and also that the negative of a solution of the given DE is also a solution.

The following results are proved, after defining a continuable solution: A solution of the given DE is continuable if it exists on $[a_1, \infty)$ for some $a_1 \ge a$. A nontrivial solution of the given DE is called oscillatory if it is continuable and has zeros for arbitrarily large x. A nontrivial solution is called nonoscillatory if it is continuable and not oscillatory.

(a) If $r \le 1$ and (x_0, b) is any compact interval with $a \le x_0$, then any solution of the given DE existing at x_0 extends to $[x_0, b]$.

(b) If $p(x) \ge 0, q(x) \ge 0$ and there exists a continuous derivative $p'(x) \le 0$, then every nonextendable solution of the given DE has infinitely many zeros in a finite interval.

Several other results regarding $y(x)$ and $y'(x)$ at finite and infinite values of x are proved. Gregus (1987), p.240

21. $$y''' + q(x)y' + p(x)y^\beta - f(x) = 0,$$

where p, q, and f are real-valued continuous functions on $[0, \infty)$ such that $p(x) \le 0, q(x) \le 0, f(x) \ge 0$, and $\beta > 0$ is ratio of odd integers.

Some qualitative and asymptotic results regarding solutions of the given DE are proved. In particular, conditions are given such that $\lim_{x\to\infty}\ y(x) = 0$. Parhi (1981)

22. $$y''' + p(x)y' + q(x)f(y) = 0,$$

where $p(x)$ and $q(x)$ are continuous and nonnegative and $f(y)/y \ge \alpha > 0$ for some α.

If $\alpha q(x) - p'(x)$ is positive and if $\int^\infty x\{\alpha q(x) - p'(x)\}dx = \infty$, then any continuable solution of the given DE that has a zero is oscillatory. Waltman (1966)

23. $$y''' + yy' + By - 2B^2x - c = 0,$$

where B and c are constants.

The given DE arises in the similarity reduction of the Boussinesq equation. It may be transformed according to

$$y = B^{2/3}Y(X) + Bx + c/(2B), \quad X = -[B^{1/3}x + (c/2)B^{-5/3}]$$

to

$$\frac{d^3Y}{dX^3} + Y\frac{dY}{dX} - \left(2Y + X\frac{dY}{dX}\right) = 0.$$

Now, proceed to Eqn. (3.2.25). Clarkson and Kruskal (1989)

24. $$\mathbf{y''' + [y - (1/3)x]y' - (2/3)y = 0.}$$

The given DE results from a similarity reduction of the Korteweg–deVries equation. Writing $y = (dw/dx) - (1/6)w^2$, we get

$$w^{\text{iv}} - ww'''/3 - w(w')^2/3 - w^2w''/6 + w^3w'/18 - xw''/3$$
$$+ xww'/9 - 2w'/3 + w^2/9 = 0. \tag{1}$$

Equation (1) can be written as

$$(D_x - w/3)(w''' - w^2w'/6 - xw'/3 - w/3) = 0.$$

Therefore, every solution of the "modified" third order equation

$$w''' - w^2w'/6 - xw'/3 - w/3 = 0 \tag{2}$$

gives rise to a solution of the given DE. Equation (2) integrates to yield the second Painlevé equation

$$w'' = w^3/18 + xw/3 + K \tag{3}$$

for some constant K. Olver (1986), p.198

25. $$\mathbf{y''' + (y - x)y' - 2y = 0.}$$

The given DE arises in the similarity reduction of the Boussinesq equation. There is a one–to–one correspondence between the solution of this equation and the second Painlevé equation $v'' = 2v^3 + xv + \alpha$, namely, $y(x) = -6(v'(x) + v^2(x))$ or $v(x) = [y'(x) + 6\alpha]/[2y(x) - 6x]$. Clarkson and Kruskal (1989)

26. $$\mathbf{y''' + 2yy' = 0, y(\pm 0) = 0,}$$

$y'(+0) = y'(-0)$, $y''(+0) = cy''(-0)$, $\lim_{x\to\infty} y'(x) = a$; $\lim_{x\to-\infty} y'(x) = b (c > 0, a \geq 0, b \geq 0, a \neq b)$.

The given problem with $a = 1, b = 1 + \lambda(-1 \leq \lambda \leq 1), c = 1$ arises in the treatment of displacement of two gaseous streams. The following results are proved:

(a) If $y'(x) > 0 (-\infty < x < \infty)$, then both limits $\lim_{x\to+\infty} y'(x), \lim_{x\to-\infty} y'(x)$, exist and are bounded.

(b) Let $y(x)$ be a solution of the given DE and $\lim_{x\to+\infty} y'(x) = a > 0, \lim_{x\to-\infty} y'(x) = b > 0, y(0) = c_0, y'(0) = c_1, y''(0) = c_2$. Then, for arbitrary $\epsilon > 0$ it is possible to find a $\delta > 0$ such that as soon as

$$|y_0 - c_0| < \delta, |y_0' - c_1| < \delta, |y_0'' - c_2| < \delta,$$

we have

$$\max_{-\infty < x < +\infty} |y'(x) - y'(x, y_0, y_0', y_0'')| < \epsilon.$$

(c) Let $y'(0) = \alpha$ and $y''(0) = \beta$. For arbitrary $a \geq 0$ and $\alpha \geq 0$, there exists a $\beta(\alpha)$ such that $y(x)$ satisfies the given DE and $\lim_{x\to+\infty} y'(x) = a$. In addition, if $a - \alpha \geq 0$, then $\beta(\alpha)/\alpha$ is a continuous, monotonically decreasing function of $\alpha; a - \alpha < 0$; then, having defined $\overline{\beta}(\alpha)$ as the exact upper bound of those $y''(0)$ for which $\lim_{x\to+\infty} y'(x) = a$, we obtain that $\overline{\beta}(\alpha)\alpha^{-1}$ is a nondecreasing function of α.

Scerbina (1961)

27. $$\boldsymbol{y''' + 2yy' - 2\alpha xy' - 4\alpha y = 0,}$$

where α is a constant.

This DE results from the similarity reduction of the Korteweg–deVries equation. Using the transformation

$$F = (-y + 3\alpha x)/(6\alpha^{2/3}), \quad \xi = -\alpha^{1/3}x,$$

we get

$$FF'' - (1/2)F'^2 - 4F^3 - 2\xi F^2 - k = 0,$$

where k is an integration constant. Now go to Eqn. (2.31.7). Tajiri and Kawamoto (1982)

28. $$\boldsymbol{y''' + (6y - x)y' - 2y = 0.}$$

Put $y = (dV/dx) - V^2, V = V(x)$ in the given DE and integrate twice, ignoring the constants of integration to arrive at P_{II}:

$$V'' - xV - 2V^3 = 0.$$

Drazin and Johnson (1989), p.18

29. $$\boldsymbol{y''' + [(3\alpha - 4)/12]y + (\nu\phi - x/3 - 6y)y' - [(3\alpha + 2)(3\alpha + 4)/3^3 \cdot 2^4]x - C = 0,}$$

where α, ν, ϕ, and C are constants.

The given DE is a reduced form of the Kadomtsev–Petviashvili equation. There are other variants of this equation for which the numerical coefficients of terms are different. David, Levi, and Winternitz (1989)

30. $$\boldsymbol{y''' - 12yy' = 0.}$$

The given DE has a first integral

$$y'' = 6y^2 + k,$$

where k is a constant. The solutions for $y(x)$ are $0, x^{-2}$, or $\mathcal{P}(x; g_2, 0)$, where $\mathcal{P}$ is the Weierstrass function and $g_2 = -2k$. Bureau (1964a)

31. $$\boldsymbol{y''' - 12yy' - 6y + 6x = 0.}$$

Putting $y = u'$ and integrating with respect to x, we get

$$u''' - 6u'^2 = 6u - 3x^2 + 6K,$$

where K is an arbitrary constant; now replace $u + K$ by u to obtain

$$u''' - 6u'^2 = 6u - 3x^2.$$

This is another canonical form [see Eqn. (3.3.5)]. Bureau (1964a)

32. $$\boldsymbol{y''' - 12yy' + 6Ky + 6K^2x - K_1 = 0,}$$

where K and K_1 are constants.

By introducing X in place of $x - K_1/(6K^2)$, one may assume that $K_1 = 0$. Further, the transformations $x = \alpha_1 X, y = \beta_1 Y$, where $\alpha_1^2\beta_1 = 1$ and $K\alpha_1^3 = -1$, reduce the given DE to the canonical form

$$y''' - 12yy' - 6y + 6x = 0.$$

Now go to Eqn. (3.2.31). Bureau (1964a)

33. $$\boldsymbol{y''' + [\{4\alpha/(v-1)\}y - 4]y' = 0,}$$

where α and v are constants.

The given DE arises from a traveling wave solution of the Benjamin–Bona–Mahony equation $u_t + u_x + uu_x - u_{xxt} = 0$. It can be integrated to yield

$$y'' + \{2\alpha/(v-1)\}y^2 - 4y = 0, \tag{1}$$

if we set constant of integration equal to zero. Equation (1) has a solitary wave solution

$$y = \{3(v-1)/\alpha\} \operatorname{sech}^2 x. \tag{2}$$

Hereman, Banerjee, Korpel, and Assanto (1986)

34. $$\boldsymbol{y''' - [(8\alpha/v)y - 8]y' = 0,}$$

where α and v are constants.

The given DE arises from a traveling wave solution of the Korteweg–deVries equation $u_t + \alpha uu_x + u_{xxx} = 0$. It has a first integral

$$y'' - (4\alpha/v)y^2 + 8y = 0 \tag{1}$$

if we put constant of integration equal to zero. Equation (1) has an exact solitary wave solution

$$y = \{3v/(2\alpha)\} \tanh^2 x. \tag{2}$$

The solution (2) may be obtained by a series method; see Eqn. (4.2.19). Hereman, Banerjee, Korpel, and Assanto (1986)

35. $$\boldsymbol{\mu y''' + (y - \lambda)y' - \epsilon f(y) = 0,}$$

where μ, λ, and ϵ are constants and $f(y) = -(y - 0.5y^2 + 0.05556y^3)$.

The given DE is written in the form of a system $u' = v, v' = w, w' = \mu^{-1}[\lambda v - uv + \epsilon f(u)]$ and solved numerically with IC $u(0) = u_0, v(0) = v_0, w(0) = w_0$ at fixed λ. Other ICs are also used. Phase portraits and time solutions are computed and depicted for different sets of ϵ values. Some unstable solutions are also discussed. Engelbrecht (1991)

36. $$\boldsymbol{y''' + (\lambda - 1 + y^2)y' = 0,}$$

where λ is a constant.

The given DE describes a nondissipative, nonlinear oscillator that is not unlike simple oscillators for which every motion is periodic with finite amplitude. A first integral is easily found to be

$$y'' + (\lambda - 1)y + (1/3)y^3 = b, \tag{1}$$

where b is constant of integration. The positions of equilibrium are given by

$$(\lambda - 1)y + (1/3)y^3 = b. \tag{2}$$

Analysis of this cubic shows that for $\lambda > 1$, it has only one real root whatever b is, while for $\lambda < 1$, it has one real root only when $|b|$ is greater than $b_c = (2/3)(1-\lambda)^{3/2}$. For $\lambda < 1$ and $|b| < b_c$ there are three positions of equilibrium, one lying in the interval $|y| < (1-\lambda)^{1/2}$ and one on either side of the interval.

Writing $y = y_0 + z$, z small, we have

$$\frac{d^2z}{dx^2} + (\lambda - 1 + y_0^2)z = 0. \tag{3}$$

We may check from (3) that an equilibrium point in the region $|y| < (1-\lambda)^{1/2}$ is unstable while the equilibrium points outside this interval are stable. We can then infer that in the phase plane $(y, (dy/dx))$, the case $\lambda < \lambda_c$ will have two centers with a saddle point between them, while for the case $\lambda > \lambda_c$, we have simply one center. Here $\lambda = \lambda_c$ is that value of λ for which (2) has coincident roots, and we have higher–order singularities at the position of equilibrium.

Further integration of (1) gives

$$y'^2 + (\lambda - 1)y^2 + y^4/6 - 2by = 2E.$$

The solution curves with one or three equilibrium positions are drawn. All solution curves are bounded and closed. All solutions except one are periodic and of finite amplitude. The exception corresponds to the figure-of-eight curve that enters the saddle point, and represents an asymptotic approach to unstable equilibrium. Moore and Spiegel (1966)

37. $\boldsymbol{y''' + [6y^2 - c]y' = 0, y, y', y'' \to 0}$ **as** $\boldsymbol{|x| \to \infty}$.

Integrating three times (employing a simple integrating factor after the first integration) and using the given conditions as $|x| \to \infty$, the solution is easily found to be

$$y = \pm c^{1/2}\,\mathrm{sech}(c^{1/2}x + A)$$

for all $c \geq 0$, where A is an arbitrary constant. Drazin and Johnson (1989), p.33

38. $\boldsymbol{y''' + 6y^2y' - Uy' = 0,}$

where U is a constant.

The given DE is traveling wave form of the modified Korteweg–deVries equation. It can be written as the system

$$u_x = v, v_x = w, w_x = Uv - 6u^2v, \tag{1}$$

which has integrals

$$(1/2)(v^2 + u^4 - Uu^2 - 2Bu) = A, w + 2u^3 - Uu = B.$$

It is not difficult to see that on surfaces of constant B, the system (1) is Hamiltonian. McIntosh (1990)

39. $$\boldsymbol{y''' + [6(c+y)^2 - \alpha]y' = 0,}$$

where c and α are constants.

Two integrations with a suitable choice of the constants of integration lead to

$$(1/2)y'^2 + (1/2)(c+y)^4 - (1/2)\alpha y^2 = 2c^3 y + (1/2)c^4.$$

Put $g = 1/y$ to obtain

$$g'^2 = (\alpha - 6c^2)g^2 - 4cg - 1. \tag{1}$$

A special solution of (1) is $g = -\{1/(4c)\}\{4c^2x^2 + 1\}$ if $\alpha = 6c^2$. Hence a special solution of the given DE for this case is

$$y = -4c\{4c^2x^2 + 1\}^{-1}.$$

Drazin and Johnson (1989), p.33

40. $$\boldsymbol{y''' - [(9/2)y^2 - 9c_1xy + 3c_1^2x^2 - 18c_1A]y' + 6c_1y^2 - 9c_1^2xy - 36c_1^2A = 0,}$$

where c_1 and A are constants.

It may be checked that any solution of the special Riccati equations

$$y' = \pm(3^{1/2}/2)y^2 \mp 3^{1/2}c_1xy \mp 6(3)^{1/2}c_1A \tag{1}$$

is a solution of the given DE. Equation (1) may easily be linearized. Kawamoto (1984)

41. $$\boldsymbol{y''' - [(9/2)y^2 + 9c_1xy + 3c_1^2x^2 - 3k_1]y' - 6c_1y^2 - 9c_1^2xy + 6c_1k_1 = 0,}$$

where c_1 and k_1 are constants.

It may be checked that any solution of the special Riccati equations

$$y' = \pm(3^{1/2}/2)y^2 \pm 3^{1/2}c_1xy \mp 3^{1/2}k_1 \tag{1}$$

is a solution of the given DE. Equations (1) may easily be linearized. Kawamoto (1984)

42. $$\boldsymbol{y''' - 6y^2y' = 0.}$$

The given DE has a first integral

$$y'' = 2y^3 + K;$$

K a constant, hence y can be expressed in terms of elliptic functions [see Eqn. (2.1.64)]. Bureau (1964a)

43. $$\boldsymbol{y''' - [6y^2 + K_1x + K_2]y' - K_1y = 0, K_1 \neq 0,}$$

where K_1, K_2 are constants.

It may be shown that the solutions of the given DE have two sets of simple poles as their only movable singularities. Bureau (1964a)

44.

$$\boldsymbol{y''' - [6y^2 + (-2K^2x^2 + K_1x + K_2)]y' - 2Ky^2}$$
$$\boldsymbol{- (-4K^2x + K_1)y - K(-2K^2x^2 + K_1x + K_2) = 0,}$$

where K, K_1, and K_2 are constants.

Use of linear transformation in x shows that we may assume that $K_1 = 0$ without loss of generality. Further, by the transformation $x == \alpha X, y = \beta Y$, where $\alpha\beta = -1, K\alpha^2 = 1$, and $K_2\alpha^2 = -4h$ with h an arbitrary constant, one obtains

$$y''' - 6y^2y' + (2x^2 + 4h)y' + 2y^2 + 4xy - 2x^2 - 4h = 0. \tag{1}$$

Substituting $y = u + x$ in (1), we have

$$u''' = 6u^2u' + 12xuu' + 4(x^2 - h)u' + 4u^2 + 4xu. \tag{2}$$

Equation (2) is also obtained by differentiating with respect to x the DE

$$2uu'' = u'^2 + 3u^4 + 8xu^3 + 4(x^2 - h)u^2 + 2K \tag{3}$$

so that (3) is an integral of (2). Bureau (1964a)

45.

$$\boldsymbol{y''' - cy^2y' = 0,}$$

where c is a constant.

A transformation $y \to \alpha Y, c\alpha^2 = 6$, takes the given DE to Eqn. (3.2.42). Bureau (1964a)

46.

$$\boldsymbol{y''' + \alpha y^2y' - vy' = 0,}$$

where α and v are constants.

The given DE describes traveling waves for the modified Korteweg–deVries equation and has the solitary wave solution

$$y = (6v/\alpha)^{1/2} \operatorname{sech}[v^{1/2}x + \delta],$$

where δ is an arbitrary phase shift. A first integral of the given DE is

$$y'' + (\alpha/3)y^3 - vy + c = 0,$$

where c is a constant. Choosing $c = 0$, we may directly integrate it, but an alternative interesting approach is given in Eqn. (2.1.56). Hereman, Banerjee, Korpel, and Assanto (1986)

47.

$$\boldsymbol{y''' - \{\alpha/(v - \beta)\}y^2y' - y' = 0,}$$

where α, v, and β are constants.

The given DE arises from the traveling wave form of the modified Korteweg–deVries equation

$$u_t - \alpha u^2u_x + u_{xxx} + \beta u_x = 0.$$

Integrating the given DE and setting the constant of integration equal to zero, we have

$$y'' - [\alpha/\{3(v - \beta)\}]y^3 - y = 0. \tag{1}$$

Multiplying (1) by y', integrating and putting the constant of integration equal to zero, we have

$$y'^2 - [\alpha/\{6(v - \beta)\}]y^4 - y^2 = 0. \tag{2}$$

A solitary wave solution of (2) is $y = \pm\{6(v - \beta)/\alpha\}^{1/2}$ cosech x. Hereman, Banerjee, Korpel, and Assanto (1986)

48. $$\epsilon^2 y''' + (3y^2 - 1)y' - y(y - 1) = 0, 0 < x < 1,$$

$y(0, \epsilon) = A, y(1, \epsilon) = 0, y'(1, \epsilon) = C; \epsilon \ll 1, A$, and C are constants.

The key to solution of the given problem is first to derive an asymptotically equivalent problem

$$\epsilon^2 y'' = y^3 - y, 0 < x < 1, y(0, \epsilon) = A, y'(1, \epsilon) = C.$$

Boundary layers, etc., are fitted where needed. Howes (1983)

49. $$y''' - \{1/(r\mu^2)\}(\alpha y^2 - v)y' - \{1/(r\mu^3)\}\{\epsilon_0 + \epsilon_1 y + \epsilon_2 y^2 + \epsilon_3 y^3) = 0,$$

where $r, \mu \neq 0, \alpha, \epsilon_i$ $(i = 0, 1, 2, 3)$ are constants.

The given DE has an exact solution $y = a_0 + a_1 \tanh x$, where

$$\begin{aligned} a_0 &= \mp(\epsilon_3/2\alpha)(6r/\alpha)^{1/2}, \quad a_1 = \pm\mu(6r/\alpha)^{1/2}, \\ \mu &= \pm\left[-\epsilon_1/(6r\epsilon_3) - 3\epsilon_3^2/(4\alpha^2) \mp (\epsilon_2/\alpha)\{\alpha/(6r)\}^{1/2}\right]^{1/2}. \end{aligned}$$

The consistency condition for the parameters $\epsilon_0, \epsilon_1, \epsilon_2$, and ϵ_3 is

$$\begin{aligned} &\pm(\epsilon_2/\alpha)\{\alpha/(6r)\}^{1/2}\left[-9\epsilon_3^2 r/\alpha + 8\epsilon_1\alpha/(3\epsilon_3) + 12\epsilon_3^2 r/\alpha^2\right. \\ &\quad \left.\pm\, \epsilon_2(6r/\alpha)^{1/2} \pm (2r\epsilon_2/\alpha)\{2/(6r)\}^{1/2}\right] \pm \{\epsilon_1\alpha/(6r\epsilon_3)\}\left[\epsilon_2(6r/\alpha)^{1/2}\right. \\ &\quad \left.+\, (2r\epsilon_2/\alpha)\{\alpha/(6r)^{1/2}\}\right] \pm \{3\epsilon_3^2/(4\alpha^2)\}\left[\epsilon_2\{6r/\alpha\}^{1/2}\right. \\ &\quad \left.+\, (2r\epsilon_2/\alpha)\{\alpha/(6r)\}^{1/2}\right] - 3\epsilon_3\epsilon_1/2 + 4\epsilon_1^2\alpha/(9\epsilon_3^2 r) \\ &\quad +\, 4\epsilon_1\epsilon_3/\alpha - 27\epsilon_3^4 r/(4\alpha^3) + 9r\epsilon_3^4/\alpha^4 - \epsilon_1\epsilon_3/(2\alpha) \\ &\quad -\, 3\epsilon_3^4 r/(4\alpha^4) = \mp\{\epsilon_0\{\alpha/(6r)\}^{1/2} \mp \epsilon_2\epsilon_3^2/(4\alpha^2)\}(6r/\alpha)^{1/2}. \end{aligned}$$

The given DE arises from a modified Korteweg–deVries equation with a particular background interaction. Lan and Wong (1989)

50. $$y''' + [3b(\beta/\gamma)^2 y^2 - (\beta^3/6)2x]y' + (\beta^3/6)y = 0,$$

where b, β, γ are constants.

This DE results from the similarity reduction of modified Kadomtsev–Petviashvili equation. Tajiri and Kawamoto (1983)

51. $$y''' - [12\delta y^2 + 6y + c]y' = 0,$$

where δ and c are constants.

Two integrations of the given DE lead to $y'^2 = 2F(y)$, where

$$F(y) = \delta y^4 + y^3 + (c/2)y^2 + Ay + B,$$

with A and B arbitrary constants. The solution is therefore expressed as

$$x = \int [1/\{2F(y)\}^{1/2}]dy + \text{ const.}$$

It may be shown by geometrical arguments or otherwise that periodic solutions may exist for all δ; that if $\delta > 0$, then either a solitary wave or a kink (a topological soliton) may exist; and that if $\delta < 0$, then a solitary wave may exist but not a kink. Drazin and Johnson (1989), p.37

52. $$\boldsymbol{y''' + [(\alpha/\gamma)(1+\beta y)y - (v/\gamma)]y' = 0,}$$

where α, γ, and β are constants.

The given DE arises from the traveling wave form of a combined Korteweg–deVries and modified Korteweg–deVries equation. Integrating and choosing the constant of integration to be zero, we have

$$y'' + \{\alpha\beta/(3\gamma)\}y^3 + \{\alpha/(2\gamma)\}y^2 - (v/\gamma)y = 0. \tag{1}$$

Scaling (1) according to $y = \{2v/\alpha\}\tilde{y}$, we get

$$\tilde{y}'' + (\sigma_2 v/\gamma)\tilde{y}^3 + (v/\gamma)\tilde{y}^2 - (v/\gamma)\tilde{y} = 0,$$

where $\sigma_2 = \{4\beta/(3\alpha)\}v$. Now go to Eqn. (2.1.62). Coffey (1990)

53. $$\boldsymbol{y''' + [(\alpha/\delta)y^3 + (\beta/\delta)y^2 + (\gamma/\delta)y]y' - (c/\delta)y' = 0,}$$

where $\alpha, \beta, \gamma, \delta$, and c are constants.

If we use vanishing boundary conditions y and $y' \to 0$ as $|x| \to \infty$, we obtain, after two integrations of the given DE,

$$(1/2)y'^2 = -(\alpha/20\delta)y^5 - (\beta/12\delta)y^4 - (\gamma/6\delta)y^3 + (c/2\delta)y^2 \equiv -F(y) \geq 0. \tag{1}$$

Although (1) generally has a solitary wave solution, as a special case of a hyperelliptic integral we may write it as

$$y'^2 = -(\alpha/10\delta)y^2(y + A_\pm)^2(y + B), \tag{2}$$

where

$$\begin{aligned} A_\pm &\equiv (1/3)[5\beta/(3\alpha) \pm \{\{5\beta/3\alpha\}^2 - 10\gamma/\alpha\}^{1/2}], \\ &= [1/(4\alpha\beta)][(2\alpha\gamma + 5\beta^2/3) \pm \{(2\alpha\gamma - 5\beta^2/3)^2 + 72\alpha^2\beta c\}^{1/2}], \end{aligned} \tag{3}$$

$$B \equiv 5\beta/(3\alpha) - 2A_\pm. \tag{4}$$

If $-\alpha/(10\delta) > 0$ and $(y + B) > 0$, Eqn. (2) can be written as

$$\frac{dx}{dY} = k\left[\frac{1}{Y^2 - B} - \frac{1}{Y^2 + (A_+ - B)}\right], \tag{5}$$

where

$$\begin{aligned} Y^2 &= y(x) + B, \\ K &= \pm(2/A_+)(-10\delta/\alpha)^{1/2}. \end{aligned}$$

Assuming that $y = 0$ at $|x| = \infty$ and $B > 0$, Eqn. (5) can be integrated to give

$$y(x) = \begin{cases} -B^{1/2}\tanh^{-1}\{(B^{1/2}/K)(x - x_0) + (1/2)G(Y(x))\} \equiv Y_1(x), \\ \quad \text{for } Y > B^{1/2} \text{ or } Y < -B^{1/2}, \\ -B^{1/2}\tanh\{(B^{1/2}/K)(x - x_0) + (1/2)G(Y(x))\} \equiv Y_2(x), \\ \quad \text{for } -B^{1/2} < Y < B^{1/2}, \end{cases}$$

where x_0 is the integration constant and

$$G(Y(x)) = 2\{B/(A_\pm - B)\}^{1/2} \arctan\{Y(x)/(A_\pm - B)^{1/2}\}$$

for $(A_+ - B) > 0$.

In view of the relation $Y^2(x) = y(x) + B$, we may write the solution $y_1(x)$ and $y_2(x)$ corresponding to $Y_1(x)$ and $Y_2(x)$ as

$$\begin{aligned} y_1(x) &= B \operatorname{cosech}^2\{(B^{1/2}/K)(x - x_0) + (1/2)G(Y_1(x))\}, \\ y_2(x) &= -B \operatorname{sech}^2\{(B^{1/2}/K)(x - x_0) + (1/2)G(Y_2(x))\}. \end{aligned}$$

Kawamoto (1984)

54. $$\boldsymbol{y''' + [(\alpha/\gamma)(1 + \beta y^2)y^2 - (v/\gamma)]y' = 0,}$$

where α, γ, v, and β are constants.

The given DE arises from traveling wave solutions of a Korteweg–deVries like equation with fifth degree nonlinearity.

A first integral of the give DE is

$$y'' + \{\alpha\beta/(5\gamma)\}y^5 + \{\alpha/3\gamma\}y^3 - (v/\gamma)y - K_1/\gamma = 0,$$

where K_1 is a constant, which we may choose to be zero. Putting $y = \eta\tilde{y}$, where $v = \alpha\eta^2/3$, we have

$$\gamma\tilde{y}'' + \sigma_1 v\tilde{y}^5 + v\tilde{y}^3 - v\tilde{y} = 0,$$

where $\sigma_1 = \{9\beta/(5\alpha)\}v$. Now proceed to Eqn. (2.1.77). Coffey (1990)

55. $$\boldsymbol{y''' + [(n+1)(n+2)y^n - c]y' = 0, n = 1, 2, \ldots,}$$

c being a constant, where $y, y', y'' \to 0$ as $|x| \to \infty$.

Integrating once, then using simple integrating factors and the conditions as $|x| \to \infty$, the solution is easily found to be

$$y^n = (c/2)\operatorname{sech}^2(nc^{1/2}x + A),$$

where A is an arbitrary constant. Drazin and Johnson (1989), p.33

56. $$y''' - y'f(y) = 0.$$

Put $y' = p(y), y''' = p^2(d^2p/dy^2) + p(dp/dy)^2$ into the given DE to obtain

$$p\frac{d^2y}{dy^2} + \left(\frac{dp}{dy}\right)^2 = f(y). \tag{1}$$

Integrating (1) twice with respect to y, we have

$$p^2/2 = \int F(y)dy + C_1 y + C_2, \tag{2}$$

where $F(y) = \int f(y)dy$ and C_1 and C_2 are arbitrary constants. Using (2) in $dy/dx = p$, the solution can be found by quadrature. Tatarkiewicz (1963)

3.3 $y''' + f(x, y, y') = 0$

1. $$y''' + 6y'^2 = 0.$$

Put $y' = z$ so that

$$z'' + 6z^2 = 0. \tag{1}$$

The solutions of (1) are $0, -1/x^2$, or $-\mathcal{P}(x)$, the Weierstrass $\mathcal{P}$ function; hence $z(x)$ and $y(x)$ have poles as their only movable singularities [see Eqn. (2.1.9)]. Bureau (1964a)

2. $$y''' + 6y'^2 - K = 0.$$

If we set $y' = z$, we have the standard second order equation

$$z'' + 6z^2 = K.$$

The given DE has simple poles as its only movable singularities [see Eqn. (2.1.9)]. Bureau (1964a)

3. $$y''' + 6y'^2 + x = 0.$$

Put $y' = z$, so that

$$z'' + 6z^2 + x = 0.$$

See Eqn. (2.4.11). Bureau (1964a)

4. $$y''' + 6y'^2 - K_1 x - K_2 = 0,$$

where K_1 and K_2 are constants.

By a simple change of variables the given DE may be simplified to the canonical equation (3.3.2) or (3.3.3). Bureau (1964a)

5. $$y''' + 6y'^2 - 6y - 3x^2 - K_1 x - K_2 = 0,$$

where K_1 and K_2 are constants.

The given DE has simple poles as the only movable singularities in its solution. Bureau (1964a)

6. $$y''' + 6y'^2 - K_2 y - (K_2/2)x^2 - K_3 x - K_4 = 0,$$

where $K_2, K_3,$ and K_4 are constants.

By a simple change of dependent and independent variables the given DE may be reduced to the canonical form (3.3.5). It has simple poles as its only movable singularities. Bureau (1964a)

7. $$y''' + 6y'^2 - 12xy - 12x^4 - K_3 x - K_4 = 0,$$

where K_3 and K_4 are constants.

The given DE has simple poles as the only movable singularities in its solution. Bureau (1964a)

8. $$y''' + 6y'^2 - K_1 xy - (K_1^2/12)x^4 - K_3 x - K_4 = 0,$$

where $K_1, K_3,$ and K_4 are constants.

By a simple change of variable, it may be reduced to the canonical form (3.3.7). The given DE has simple poles as the only movable singularities in its solution. Bureau (1964a)

9. $$y''' + 6y'^2 + 4xy' - 2y = 0.$$

The given DE results from the similarity reduction of the "potential" Korteweg–deVries equation. Multiplying the given DE by y'' and integrating, we have

$$y''^2 + 4xy'^2 + 4y'^3 - 4yy' = \mu^2, \tag{1}$$

where μ is an arbitrary constant. Eliminating y from (1) and the given DE, we get

$$2y'y''' - y''^2 + 4xy'^2 + 8y'^3 + \mu^2 = 0. \tag{2}$$

Equation (2), apart from some obvious changes, is one of 50 equations of Painlevé in y'. Actually, its solution is

$$y'(x; \mu^2) = -2^{-1/3}(V_z(z; \mu) + V^2(z; \mu) + z/2),$$

where $V(z; \mu)$ satisfies the second Painlevé equation

$$V_{zz}(z; \mu) = 2V^3(z; \mu) + zV(z; \mu) + \mu - 1/2 \text{ and } z = 2^{1/3}x.$$

Boiti and Pempinelli (1979)

10. $$y''' + 6y'^2 - (6/x^2)(y' + y^2) - K_3/x - K_4 x^2 = 0,$$

where K_3 and K_4 are constants.

The given DE is equivalent to the system

$$y' = -y^2 + u,$$

$$4uy^2 + 2yu' - u'' - 4u^2 + (6/x^2)u + K_3/x + K_4x^2 = 0.$$

Hence it may be shown that the given DE has poles as the only movable singularities in its solution. Bureau (1964a)

11. $\boldsymbol{y''' + 6y'^2 - (6/x^2)(y' + y^2) - 18x^3y - 6x^8 - K_4x^2 - K_3/x = 0,}$

where K_3 and K_4 are constants.

The given DE has poles as the only movable singularities in its solution. It is equivalent to the system

$$y' = -y^2 + u,$$

$$4y^2u + 2y(u' + 9x^3) - u'' - 4u^2 + 6u/x^2 + 6x^8 + K_4x^2 + K_3/x = 0. \quad (1)$$

The assertion about the singularities may be proved with the help of (1). Bureau (1964a)

12. $\boldsymbol{y''' + 6y'^2 - (6/x^2)(y' + y^2) - K_2x^3y - (K_2^2/54)x^8 - K_4x^2 - K_3/x = 0,}$

where K_2, K_3, and K_4 are constants.

The given DE has poles as the only movable singularities in its solution. See Eqn. (3.3.11) for its canonical form, brought about by a simple change of variables. Bureau (1964a)

13. $\boldsymbol{y''' + 6y'^2 - (6/x^2)(y' + y^2) - 2y/x - 1/x^2 - K_4x^2 - K_3/x = 0,}$

where K_4 and K_3 are constants.

The given DE has poles as the only movable singularities in its solution. Bureau (1964a)

14. $\boldsymbol{y''' + 6y'^2 - (6/x^2)(y' + y^2) - K_1y/x^2 - K_1^2/(4x^2) - K_4x^2 - K_3/x = 0,}$

where K_1, K_3, and K_4 are constants.

The given DE has poles as the only movable singularities in its solution. By a simple change of variables, it can be brought to the canonical equation (3.3.13). Bureau (1964a)

15. $\boldsymbol{y''' - y'^2 + 1 = 0, y(0) = y'(0) = 0, y'(\infty) = 1.}$

The exact solution may be found to be

$$y = x + 2(3)^{1/2} - 3(2)^{1/2}\tanh\{x/2^{1/2} + c\},$$

where $\tanh c = (2/3)^{1/2}$. Brodie and Banks (1986)

16. $\boldsymbol{y''' - y'^2 - y' = 0.}$

With $y' = u(x)$, we have the second order autonomous DE

$$u'' = u(1 + u);$$

see Eqn. (2.1.64). Murphy (1960), p.427

17. $$y''' - 12y'^2 - 72y^2y' - 54y^4 = 0.$$

It is shown that the given DE has movable "critical points." Exton (1971)

18. $$y''' - (3/11)[9 + 7(3)^{1/2}]y'^2 - (6/11)[20 + 7(3)^{1/2}]y^2y' - (3/11)[9 + 7(3)^{1/2}]y^4 = 0.$$

It may be proved that the given DE does not have all its critical points fixed. Exton (1971)

19. $$y''' - by'^2 = 0,$$

where b is a constant.

By making the change $y \to \alpha y$, etc., one may choose $b = -6$, without loss of generality; now see Eqn. (3.3.1). Bureau (1964a)

20. $$y''' + 6v\{(\lambda v - 3)/v\}^{1/2}y'^2 + \{12 - 4v\lambda/(\lambda v - 3)\}y' = 0,$$

where v and λ are constants.

The given DE arises from the traveling wave form of Hirota's shallow-water equation $\lambda u_t - u_{xxt} + 3u_x(1 - u_t) = 0$.

The given DE is of second order in $y' = p$; therefore,

$$p'' + \{12 - 4v\lambda/(\lambda v - 3)\}p' + 6v\{(\lambda v - 3)/v\}^{1/2}p^2 = 0. \tag{1}$$

The solitary wave solution for (1) is

$$y = \{(\lambda v - 3)/v\}^{1/2}\{1 \pm \tanh x\}$$

provided that $\lambda v = 4$. Hereman, Banerjee, Korpel, and Assanto (1986)

21. $$y''' + y'^3 = 0.$$

Put $z_1 = y, z_2 = y', z_3 = y''$; we obtain

$$\frac{dz_1}{z_2} = \frac{dz_2}{z_3} = \frac{dz_3}{-z_2^3} .$$

The system admits two integrals,

$$z_3^2 + z_2^4/2 = c_1^2 \tag{1}$$

and

$$z_1 + (1/2^{1/2})\sin^{-1}(z_3/|c_1|) = c_2, \tag{2}$$

where c_1 and c_2 are arbitrary constants. The solution for fixed $c_1 \neq 0$ are curves which are intersections of cylinders defined by (1) and (2). Roberts and Belford (1971)

22. $$y''' - \epsilon y'^3 - (v + F)y' = 0,$$

where ϵ, v, and F are constants.

The given DE arises from the traveling wave form of a modified Korteweg–deVries equation, called a Calogero–Degasperis–Fokas modified Korteweg–deVries equation. Changing to $\tilde{y} = (1/\eta)y$, where $\eta^2 = 1/\epsilon$, and seeking a solution in the form

$$\tilde{y} = \sum_{n=1}^{\infty} a_n g^n(x), g(x) = e^{-Kx}, K^2 = v + F,$$

we obtain a_1 as arbitrary, $a_2 = 0$, and

$$a_{2n+1} = \frac{a_1^{2n+1}}{(2n+1)2^{3n}}, \quad a_{2n} = 0, n \geq 0.$$

Thus,

$$\tilde{y}(x) = 2^{3/2} \sum_{n=0}^{\infty} (dg)^{2n+1}/(2n+1), \tag{1}$$

where $d \equiv a_1/2^{3/2}$. The series in (1) can be summed up as

$$\tilde{y}(x) = 2^{1/2} \ln\{(1+dg)/(1-dg)\}, |dg| < 1.$$

Setting the phase shift $\Delta \equiv -(1/2) \ln d$, we have

$$\tilde{y}(x) = 2^{1/2} \ln\{\coth(Kx/2 + \Delta)\}. \tag{2}$$

The result (2) is valid for the interval (x_0, ∞), where $x_0 \equiv (1/K) \ln d$ $(a_1 > 0)$. Furthermore, $\tilde{y}$ as given by (2) and its first three derivatives vanish at infinity. Coffey (1990)

23. $$y''' - a^2(y'^5 + 2y'^3 + y') = 0,$$

where a is a constant.

The given DE is a special case of Eqn. (N.1.15). The parametric form of its solution is

$$ax = \int \Omega(u)du + C_2, \quad ay = \int u\Omega(u)du + C_3,$$

where $1/\Omega = \{C_1 + (1/3)(u^2+1)^3\}^{1/2}$ and C_i $(i = 1, 2, 3)$ are arbitrary constants. Kamke (1959), p.600

24. $$\epsilon y''' - f(x, y, y', \epsilon) = 0,$$

where $\epsilon > 0$ is a parameter; $y'(0) = 0, y(1) = 0, y'(1) = 0$ or $y''(0) = 0, y(1) = 0, y'(1) = 0$.

Existence, uniqueness, and asymptotic estimates of the solutions of the given singularly perturbed boundary value problem are shown. Weili (1990)

25. $$y''' - F(x, y, y') = 0$$

Invariants of the given DE are sought under the transformation $\overline{x} = \overline{x}(u), \overline{y} = \lambda(x)y + \mu(x)$. Equivalent simpler systems are found in some special cases. Petrescu (1949)

3.4 $y''' + ay'' + f(y, y') = 0$

1. $$y''' + y'' + y'/16 + y^2 - cy = 0,$$

where c is a constant.

The solution may be found in the following finite form:

$$\begin{aligned} y(x) &= -30p^3 - (15/32a)(8a + 135)p^2 - (15/128a)(16a + 45)p + c/2 \\ &\quad - (4a + 33)/384 - p'[30p + (15/32a)(8a + 45)], \end{aligned} \tag{1}$$

where

$$p'^2(x) = p^4 + (45/16a)p^3 + (1/8)p^2 + (a/720)p, \tag{2}$$

where $a^2 = 15(256c^2 - 1)$. Equation (2), on integration, yields

$$p(x) = a/\{2880\mathcal{P}(x - x_0; g_2, g_3) - 30\}, \tag{3}$$

where $\mathcal{P}(x - x_0; g_2, g_3)$ is the Weierstrass function with parameters $g_2 = 1/3072, g_3 = (75 - 8a^2)/66355200$. Equations (1)–(3) constitute the solution. Santos (1989)

2. $$y''' + y'' + (73/256)y' + y^2 \pm (45/2048)y = 0.$$

The solution may be found in a finite form,

$$y(x) = -30p^3 + (15/2)p^2 - (45/64)p \pm 45/4096 - p'[30p - 45/(256p)],$$

where $p(x) = 1/\{16[(15)^{1/2}\cosh(x/16 - x_0) + 4]\}$. Santos (1989)

3. $$y''' + y'' + (47/144)y' + y^2 + (5/144)y = 0.$$

The solution may be found in a finite form,

$$y(x) = -30p^3 + 6p^2 - (p/2) \pm 5/288 - p'[30p + 1/2 + 5/(24p)],$$

where $p(x) = 5/\{12[\pm 12\cosh(x/12 - x_0) + 13]\}$. Santos (1989)

4. $$y''' + y'' + (T - R + Ry^2)y' + Ty = 0,$$

where T and R are constants.

The given model equation describes unstable oscillations in fluids that rotate, have magnetic fields or are compressible, and have thermal dissipation. This system has the same basic importance as the Lorenz system.

Writing $\tau = R^{1/2}x, s = \delta^{-1/2}y$, where $\delta = 1 - T/R$, we get the DE

$$s''' - \delta(1 - s^2)s' = -R^{-1/2}[s'' + (1 - \delta)s], \quad ' \equiv \frac{d}{d\tau}. \tag{1}$$

Equation (1) is equivalent to the system

$$s'' = \theta - (1 - \delta)s, \tag{2}$$

$$\theta' - (1 - \delta s^2)s' = -R^{-1/2}\theta. \tag{3}$$

In the limit $R \to \infty$, when the motion becomes adiabatic, Eqns. (1) and (3) become

$$s''' - \delta(1 - s^2)s' = 0, \tag{4}$$

$$\theta' - (1 - \delta s^2)s' = 0. \tag{5}$$

Equation (4) can be integrated twice to yield

$$(2\delta)^{-1}s'^2 + (1/12)s^4 - (1/2)s^2 + Bs = E, \tag{6}$$

where B [see Eqn. (7)] and E are constants. Equation (5) integrates to give

$$s - (1/3)\delta s^3 - \theta = \delta B, \text{ say.} \tag{7}$$

Solution curves (6) in the (s, s') plane are drawn for various values of B and E. Apart from a possible curve that enters a saddle point, every pair of values (E, B) corresponds to a periodic solution. The dependence of the solutions on B and E is discussed. Actually, if $B = 0$ and we put

$$s_0^2 = 3[1 + \{1 + (4/3)E\}^{1/2}], s_2^2 = 3[\{1 + (4/3)E\}^{1/2} - 1],$$

we may write (6) as $(2\delta)^{-1}s'^2 = (1/12)(s_0^2 - s^2)(s_2^2 + s^2)$, with the solution having $s = 0$, when $\tau = 0$, as

$$s = s_0 \ \mathrm{sn}\{s_2\delta^{1/2}\tau/6^{1/2}, is_0/s_2\},$$

or in real form

$$s = \frac{s_0 s_2}{(s_0^2 + s_2^2)^{1/2}} \ \mathrm{sd}\{(\delta/6)^{1/2}(s_0^2 + s_2^2)^{1/2}\tau\},$$

with the modulus k defined by $k^2 = (1/2)[1 + (1 + 4E/3)^{-1/2}]$ and the period $P_{\text{sym}} = [4(6)^{1/2}\delta^{-1/2}/(s_0^2 + s^2)^{1/2}]K(k)$. The solution is shown graphically.

The (given) full system is solved numerically subject to the periodicity conditions. A technique for finding the periodic solution is explained and is employed for the given system. It is also shown that periodic solutions for $R \to \infty$ are unstable and lead to aperiodicity. A large number of parametric values are chosen to bring out the role of instability in causing aperiodic solutions. Baker, Moore, and Spiegel (1971)

5. $$\boldsymbol{y''' + \beta y'' + y' - \mu y(1 - y) = 0,}$$

where β and μ are constants.

For $\beta = 0.1$, the evolution of the strange attractor through values $\mu = 0.50, 0.56, 0.57, 0.58$ is computed and shown graphically. Coullet, Tresser, and Arnéodo (1979)

6. $$\boldsymbol{y''' + \beta y'' + y' - \mu y(1 - y^2) = 0, \beta = 0.1, \mu = 0.44.}$$

It is shown that after two distinct Rossler-like attractors symmetric with respect to $(0, 0, 0)$, one constructs only one attractor, which, like a Lorenz attractor, exhibits the symmetry of the differential system from which it is obtained. The strange attractor is shown graphically. Coulet, Tresser, and Arneodo (1979)

7. $$\boldsymbol{y''' + d_2 y'' + d_1 y' + c_1 y + c_2 y^2 = 0,}$$

where d_2, d_1, c_1, c_2 are constants.

A finite solution is found in the form

$$y(x) = \sum A_n p^n(x) p'(x) + \sum B_m p^m(x), [p'(x)]^2 = \sum b_j p^j(x);$$

n, m, j finite. The details are rather complicated. Santos (1989)

8. $$\boldsymbol{y''' + (\delta/\gamma)y'' + (\beta/\gamma)y' + (1/2)(\alpha/\gamma)y^2 - (\lambda/\gamma)y + C_0/\gamma = 0,}$$

where $\delta, \gamma, \beta, \alpha, \lambda$, and C_0 are constants.

The given DE arises from a traveling wave form of a modified Korteweg–deVries Burgers equation with a fourth order derivative. A solution is found in the form

$$y = a_0 + a_1 \tanh \mu x + a_2 \tanh^2 \mu x + a_3 \tanh^3 \mu x. \tag{1}$$

On insertion of (1) in the given DE and equating coefficients of powers of tanh to zero, we get

$$\begin{aligned}
a_3 &= 120\gamma\mu^3/\alpha, \\
a_2 &= -15\delta\mu^2/\alpha, \\
a_1 &= \{(120)(114)\gamma\mu^3 - 360\beta\mu + 22 \times 5\delta^2\mu/\gamma\}/(-114\alpha), \\
a_0 &= (2/\alpha)\left[2\delta\mu^2 - \frac{30 \times 114\mu^2\beta\gamma - 960 \times 114\delta\gamma^2\mu^4}{120 \times 114\gamma^2\mu^2 - 360\beta\gamma + 22 \times 5\delta^2}\right] \\
&\pm \left\{2\delta\mu^2 - \frac{30 \times 114\gamma\beta\mu^2 - 960 \times 114\delta\gamma^2\mu^4}{120 \times 114\gamma^2\mu^2 - 360\beta\gamma + 22 \times 5\delta^2}\right\}^2 \\
&- 2\left\{720\gamma^2\mu^6 - 30\delta^2\mu^4 + (1/114)(2\gamma\mu^3 - \beta\mu)\right. \\
&\times \left.(120 \times 114\gamma\mu^3 - 360\beta\mu + 22 \times 5\delta^2\mu/\gamma)\right\}^{1/2}(2/\alpha) \\
\mu^2 &= \left\{(3780/114)\beta^2\gamma^2 - (1113 \times 75/114)\delta^2\beta\gamma + (54 \times 84375\delta^4/114)\right\} \\
&\times \left\{1800\beta\gamma^3 - 112 \times 5\gamma^2\delta^2\right\}^{-1} \\
\lambda &= \alpha a_0 - 2\delta\mu^2 + \frac{30 \times 114\delta\mu^2\beta\gamma - 960 \times 114\delta\gamma^2\mu^4}{120 \times 114\gamma^2\mu^2 - 360\beta\gamma + 22 \times 5\delta^2}
\end{aligned}$$

and one constraint equation for the parameters β, γ, and δ:

$$\begin{aligned}
&A_1\delta^{12} + B_1\beta\gamma\delta^{10} + C_1\gamma^2\delta^{10} + D_1\gamma^4\delta^8 + E_1\gamma^2\beta^2\delta^8 F_1\beta\gamma^3\delta^8 + G_1\gamma^3\beta^3\delta^6 \\
&+ H_1\beta\gamma^3\delta^6 + I_1\beta^2\gamma^4\delta^4 + J_1\beta^4\gamma^4\delta^4 + K_1\beta^2\gamma^6\delta^4 + L_1\beta^3\gamma^5\delta^4 \\
&+ M_1\beta^4\gamma^6\delta^2 + N_1\beta^3\gamma^7\delta^2 + P_1\beta^5\gamma^7 + Q_1\beta^4\gamma^8 = 0,
\end{aligned}$$

where $A_1, B_1, \ldots, Q_1$ are rather long expressions in terms of $A, B, \ldots, Q, S$, which are some definite numbers. Huibin and Kelin (1990)

9. $$\boldsymbol{y''' - ay'' - by' - g(y) = 0,}$$

where $a \leq 0, b \geq 0, g \in C^1(-\infty, \infty), g(0) = 0, g'(y) > 0$ for $y \neq 0$ and $|g(y)| \to \infty$ as $|y| \to +\infty$.

The following result is proved: There exists a continuous function $\phi : R^2 \to R^1$ such that if $y(x)$ is a solution of the given DE with $[y(0), y'(0), y''(0)]$ contained in

$$G_1 = \left\{[\phi(\lambda, \mu), \lambda, \mu]/(\lambda, \mu) \in R^2\right\},$$

then $\overline{Y}(x) = [y(x), y'(x), y''(x)]$ is defined on $[0, \infty)$ and tends to $[0,0,0]$ as $x \to +\infty$. If $[y(0), y'(0), y''(0)] \notin G_1$, then $|\overline{Y}(x)| \to +\infty$ as $x \to +\infty$ if $Y(x)$ is defined on $[0, \infty)$. The behavior as $x \to -\infty$ is also considered, and its unboundedness on $(-\infty, \infty)$ is proven. Anderson (1970)

10. $$y''' + ay'' + by' + h(y) - p(x) = 0,$$

where $a > 0, b > 0$ are constants and $h(y) \in C(R^1)$ and $p(x) \in C[0, \infty)$.

Boundedness and decay of solutions of the given DE are studied. Jan (1989)

11. $$y''' + ay'' + by' + h(y) - p(x) = 0,$$

where a and b are constants, p is continuous and bounded for all x.

Referring to the space of variables $y, y' = z, z' = p$ as E_3, a surface in E_3 will be said to have the property p^* if it is bounded and if every directed line drawn from the origin of coordinate axes in E_3 meets the surface in one point only. The following result is proved. Suppose that $a > 0, b > 0$, and that further:

(a) $h(0) = 0, h(y)/y \geq c > 0 (y \neq 0)$.

(b) $h'(y)$ exists and is continuous and $|h'(y)| \leq C$ for all y, where $ab - C^2/c > 0$.

(c) $|p(x)| \leq A_0 < \infty$ for all x.

Then there is a surface $\sum = \sum(a, b, c, C, A_0)$ in the space E_3 having the property p^* and such that all trajectories of $y' = z, z' = p, p' = -ap - h(y) + p(x)$ cross it only inward. Ezeilo (1962)

12. $$y''' + ay'' + by' + h(y) - Q(x) = 0,$$

where a and b are constants > 0, and $h(y)$ and $Q(x)$ are continuous functions.

Among the theorems proved are the following. If:

(a) $|h(y)| \leq H_0, |Q(x)| \leq Q_0$ for all y, x,

(b) $h(y) \operatorname{sgn} y \geq 0$ for $y \geq h$, and

(c) $|\int_0^x Q(s)ds| \leq Q$, for all x,

then all solutions of the given DE are bounded. Under conditions (a) and (c) and $yh(y) > 0$ for $y \neq 0$, each solution $y(x)$ of the given DE is either oscillatory or satisfies $\lim_{x \to \infty} y(x) = 0$. Voracek (1965)

13. $$\delta y''' - \epsilon y'' + yy' - cy' = 0,$$

δ, ϵ, and c are positive constants.

The given DE describes traveling wave solutions of a Korteweg–deVries Burgers equation. For fixed positive values of ϵ and δ it is proved that the given DE has a unique bounded solution $y = y(x) = y(x; \epsilon, \delta)$ such that $y_L = \lim_{x \to -\infty} y(x)$ and $y_R = \lim_{x \to \infty} y(x)$. The function y also satisfies the additional conditions

$$\lim_{|x| \to \infty} y^{(j)}(x) = 0, j = 1, 2, 3, \ldots,$$

where $y^{(j)}$ denotes the jth derivative of y w.r.t. x. The two asymptotic states y_L and y_R are restricted by the requirements that $c > y_R$ and $y_R + y_L = 2c$.

For $\epsilon^2 \geq 4\delta r$ where $r = c - y_R, y$ resembles the familiar montone traveling wave solution. If $\epsilon^2 < 4\delta r$, then y has an oscillatory character as $x \to -\infty$. For fixed positive δ, y converges to the solitary wave solution $y_R + 3r \operatorname{sech}^2[(r/4\delta)^{1/2}x]$ of Korteweg–deVries equation, as $\epsilon \downarrow 0$.

For fixed $\epsilon > 0$, the limiting form of y as $\delta \downarrow 0$ is $y_R + r\{1 - \tanh(r/\epsilon)x\}$. If ϵ and δ are both allowed to tend to zero, but in such a way that the quotient δ/ϵ^2 remains bounded, then y tends to the step function

$$X(x) = \begin{cases} y_L & \text{for } x < 0, \\ y_R & \text{for } x > 0. \end{cases}$$

Bona and Schonbek (1985)

14. $$\boldsymbol{y''' + \epsilon y'' - 6yy' - Uy' = 0,}$$

where ϵ and U are constants.

The given DE is a traveling wave form of the Korteweg–deVries Burgers equation. A first integral of the given DE is

$$y'' + \epsilon y' - 3y^2 - Uy - B = 0, \tag{1}$$

where B is a constant. Writing $y = z^2 r + b, z = e^{\mu x}$, where $\mu = -\epsilon/5, b = -U/6 - \mu^2$ in (1), setting $B = -b(3b + U)$, and integrating, we find that

$$r_z^2 = 2\mu^{-2} r^3 - \alpha^3. \tag{2}$$

With $\alpha = 0$, we get a rational solution of (2) that leads to bounded solutions of the given DE:

$$y - b = 2\mu^2 e^{2\mu x}(1 + e^{\mu x})^{-2},$$

provided that $\mu\epsilon = 2^{1/2}$. McIntosh (1990)

15. $$\boldsymbol{y''' + \{r/(\beta\mu)\}y'' + \{1/(\beta\mu^2)\}\{\alpha y - v\}y' - \epsilon_0/(\beta\mu^3) = 0,}$$

where $\beta, \mu \neq 0, r, \alpha, v$, and ϵ are constants.

The given DE arises from a Korteweg–deVries Burgers equation with a background force. An exact solution is $y = a_0 + a_1 \tanh x + a_2 \tanh^2 x$, where

$$a_1 = 12r\mu/5\alpha, a_2 = -12\beta\mu^2/\alpha, \mu = \pm[\{r^2\alpha/(5\beta)\} - \{6\alpha^2 r^2/(25\beta)\}\{36\beta - 40/\alpha\}]^{1/2},$$

where the parameters β and ϵ_0 satisfy

$$\begin{aligned}\beta &= (2880\alpha - 2400)(2130\alpha^2 - 1385\alpha - 325)^{-1},\\ \epsilon_0 &= \{72\alpha^2 r^3/(125\beta) - 12\alpha r^3/(25\beta)\}\{r^2\alpha/(5\beta)\\ &\quad - 6\alpha^2 r^2/(25\beta)\}\{36\beta - 40/\alpha\}^{-1} + \{72\beta r/5 - 12\beta r/\alpha\}\{r^2\alpha/(5\beta)\\ &\quad - 6\alpha^2 r^2/(25\beta)\}^2\{36\beta - 40/\alpha\}^{-2},\end{aligned}$$

and a_0 is arbitrary. Lan and Wong (1989)

16. $$\boldsymbol{y''' + \epsilon y'' + 6y^2 y' - U y' = 0,}$$

where ϵ and U are constants.

The given DE has a first integral,

$$y'' + \epsilon y' + 2y^3 - Uy - B = 0, \tag{1}$$

where B is a constant. Painlevé analysis of (1) shows that the only case where equation (1) can be reduced to an equation whose solutions have no movable critical points is when $B = 0$ and $U = -2\epsilon^2/9$. In this case the substitution $y = zr, z = e^{\mu x}$, $\mu = -\epsilon/3$ transforms (1) into

$$z^3(\mu^2 r_{zz} + 2r^3) = 0. \tag{2}$$

An integration of (2) (for $z \neq 0$) yields

$$\mu^2 r_z^2 = -(r^4 - \alpha^4), \tag{3}$$

where α is a constant. The solution of (3) is $r(z) = \alpha \operatorname{cn}(2^{1/2}\alpha\mu^{-1}z, 1/2^{1/2})$ where $\operatorname{cn}(s,k)$ is one of the Jacobi elliptic functions, with modulus k. For α real, we obtain unbounded real-valued solutions of the given DE:

$$y(x) = \alpha e^{\mu x} \operatorname{cn}(2^{1/2}\alpha\mu^{-1}e^{\mu x}, 1/2^{1/2}).$$

In the (u, v) phase plane, these solutions correspond to a family of curves spiraling into a nodal point at the origin as $x \to \infty$ (for $\epsilon > 0$). McIntosh (1990)

17. $$\boldsymbol{y''' + \epsilon y'' - 6y^2 y' - U y' = 0,}$$

where ϵ and U are constants.

A first integral of the given DE is

$$y'' + \epsilon y' - 2y^3 - Uy + B = 0, \tag{1}$$

where B is a constant. Choosing $B = 0$, and putting $y = zr, z = e^{\mu x}, \mu = -\epsilon/3, U = -2\mu^2$, Eqn. (1) transforms, after an integration, to

$$\mu^2 r_z^2 = r^4 - \alpha^4. \tag{2}$$

For $\alpha = 0$, Eqn. (2) integrates to yield $r(z) = \mp\mu(z + c)^{-1}$ for arbitrary constant c, which may be normalized to $c = 1$. Thus the given DE has bounded solutions

$$y(x) = \mp\mu e^{\mu x}(1 + e^{\mu x})^{-1}.$$

In the phase plane of (1), these solutions correspond to saddle node connections where the node lies at the origin while the saddles lie on the y-axis at $\pm\epsilon/3$. McIntosh (1990)

18. $$\boldsymbol{y''' + a y'' + f(y) y' + g(y) - p(x) = 0,}$$

where a is a constant and $f(y), g(y)$, and $p(x)$ are continuous.

Suppose that $a > 0$ and:

(a) $g(y)$ sgn $y \to +\infty$ as $|y| \to \infty$.

(b) There are constants $\delta_1 > 0, \delta_2 > 0$ such that a $\delta_1 - \delta_2 > 0$ and such that $f(y) \geq \delta_1$ and $g'(y) \leq \delta_2$ for $|y| \geq \xi_0 > 0$.

(c) $p(x)$ satisfies either $\int_0^x p(\tau)d\tau \leq A_0 < \infty$ for all x considered or $|p(x)| \leq A_1 < \infty$ for all x considered.

Then every solution of the given DE satisfies $|y(x)| \leq D_1, |y'(x)| \leq D_1, |y''(x)| \leq D_1$, where D_1 is a constant whose magnitude depends only on $\delta_0, \delta_1, \delta_2, A, f$, and g. Ezeilo (1968)

19. $$\boldsymbol{y''' + ky'' + y'^2 - 1 = 0,}$$

where k is a constant.

The given problem arises in boundary layer theory. Put $y' = z^n w(z) + 1, z = e^x$ to obtain

$$z^2 w'' + (2n + 1 + k)zw' + (n^2 + kn + 2)w + z^n w^2 = 0. \tag{1}$$

For $k = \pm 5/3^{1/2}, n = \mp 2/3^{1/2}$, Eqn. (1) admits an integrating factor $2z^{-n}w'$ and hence reduces to

$$z^{2-n}w'^2 + (2/3)w^3 + A = 0. \tag{2}$$

Setting the arbitrary constant $A = 0$, we can integrate (2) and obtain

$$y' = z^{\mp 2/3^{1/2}} w(z) + 1 = 1 - 2[(B \exp(\pm x/3^{1/2}) + 1)]^{-2}, \tag{3}$$

where B is another constant; hence, the solution by quadrature. Burde (1990)

20. $$\boldsymbol{y''' + y'' - \epsilon(1 - y^2)y' + \epsilon y'^3 + y = 0,}$$

where ϵ is a small parameter.

Putting $y = A\cos wx$, one gets $w^2 = 1$ and $A^2 = 2$ for $\epsilon = 1$. Writing the given DE as the equivalent system

$$\begin{aligned} &\dot{x}_1 = x_2, \quad \dot{x}_2 = x_3, \\ &\dot{x}_3 = -x_1 - \epsilon x_2^3 + \epsilon(1 - x_1^2)x_2 - x_3, \quad \cdot \equiv \frac{d}{dt}, \end{aligned}$$

it is shown that the given DE is of Van der Pol type and exhibits limit cycle in the three-dimensional phase space. Digital computer results confirm this conclusion. Results are shown graphically as projections.

For the limit cycle, letting $x_1 = A\cos t, x_2 = -A\sin t, x_3 = -A\cos t$, and writing $r^2 = x_1^2 + x_2^2 + x_3^2 = 3 + \cos 2t$, we easily check that

$$(1/2)\frac{d}{dt}r^2 = -\sin 2t; \tag{1}$$

here we assume that $A^2 = 2$. Equation (1) is verified for different ICs. Ku and Jonnada (1971a)

21. $$\boldsymbol{y''' + a_1 y'' + a_2 y' + a_3 y'^3 + a_4 y = 0,}$$

where a_i $(i = 1, \ldots, 4)$ are constants.

The method of harmonic balance gives $y = A\sin wx$, where

$$A^2 = (4/3)(a_4 - a_1a_2)/(a_3a_4); w^2 = a_4/a_1.$$

For $a_1 < 0, a_4 < 0, a_2 < 0, a_3 > 0, A^2$ and w^2 are both positive. Ku and Jonnada (1971a)

22. $$y''' + a_1y'' + a_2y' + a_3y^2y' + a_4y'^3 + a_5y = 0,$$

where a_i $(i = 1, 2, \ldots, 5)$ are constants.

Using the method of harmonic balance, an approximate solution is found as $y = A\sin wx$, where

$$w^2 = a_3/a_4, A^2 = \{w^2 - a_2\}/a_3 = (a_5 - a_1a_2)/(a_1a_3) = (a_1|a_2| + a_5)/a_1a_3$$

if $a_2 < 0$, where $a_1a_3 = a_4a_5$. So $w > 0$ if $a_1 > 0, a_3 > 0, a_4 > 0$, and $a_5 > 0$ while $a_2 < 0$ gives $A > 0$. Ku and Jonnada (1971a)

3.5 $y''' + ayy'' + f(x, y, y') = 0$

1. $$y''' + (1/4)yy'' + (1/2)y'^2 = 0,$$

$y(0) = \alpha, y'(0) = \beta, y' \to 0$ as $x \to \infty$, and $\int_0^\infty (y-\alpha)y'^2 dx = 1$. Here α and β are constants.

The given system describes a wall jet problem. New branches of solutions are found numerically. Needham and Merkin (1987)

2. $$y''' + (1/2)yy'' = 0, y(0) = 0, y'(0) = 0, y'(\infty) = 1.$$

(a) The given problem is due to Blasius and arises in boundary layer theory. The given DE is invariant under the associated group $y^1 = \mu^{-1}y, x^1 = \mu x$. Therefore,

$$\frac{dy^1}{dx^1} = \mu^{-2}\frac{dy}{dx} \text{ and } \frac{d^2y^1}{dx^2} = \mu^{-3}\frac{d^2y}{dx^2} \text{ so } \frac{d^2y(0)}{dx^2} \Big/ \left[\frac{dy(\infty)}{dx}\right]^{3/2}$$

is an invariant. The first numerical integration with any guess for $y''(0)$ determines the value of this invariant. This value is 0.3320. Thus $y''(0)$ becomes known; hence the numerical solution.

(b) $y''' + (1/2)yy'' = 0$. Invariance properties of the given Blasius equation are discussed. Two simple exact solutions are $y = C$ and $y = 6/(x + \alpha)$, where C and α are constants.

Dresner (1983), Bluman (1990)

3. $$y''' + (2/3)yy'' - (1/3)y'^2 = 0, y(0) = 0, y''(0) = 0,$$

$y(x) \simeq (1/2)\lambda x^2 + o(x^2)$ as $x \to \infty$.

The given BVP is changed to an IVP by the transformation

$$y = |y'(0)|^{1/2}\overline{y}(\overline{x}),\ \overline{x} = |y'(0)|^{1/2}x; \qquad (1)$$

$$\bar{y}''' + (2/3)\bar{y}\,\bar{y}'' - (1/3)y'^2 = 0,$$
$$\bar{y}(0) = 0, y''(0) = 0, \bar{y}'(0) = \pm 1, \tag{2}$$

with an unknown asymptotic property

$$\bar{y} = (1/2)a(\bar{x} + b)^2, \text{ as } \bar{x} \to \infty.$$

It is shown that there are two solutions of the IVP, corresponding to the $\pm$ sign in (2), given either $a = 0.489, b = 1.132$ for $y'(0) = 1$ or $a = 6.082, b = -3.048$ for $\bar{y}'(0) = -1$. Numerical evidence supports the nonuniqueness of the original problem. Smith (1984)

4. $$\boldsymbol{y''' + yy'' = 0, \ y(0) = y'(0) = 0, y'(\infty) = 1.}$$

(a) The equation is rewritten as $y''' + y''y^\delta = 0$ and a solution is sought in the form

$$y = y_0(x) + \delta y_1(x) + \delta^2 y_2(x) + \cdots.$$

The value of $y''(0)$ obtained by this method, supplemented by Padé approximation, has 8.7% error when compared with the exact value. The zeroth order solution

$$y_0(x) = x - 1 + e^{-x}$$

is a good qualitative approximation to exact solution.

(b) The given problem with a factor of 2 multiplying y''' in the given DE is solved numerically and the value of $y''(0) = 0.33205734$ is found to be very accurate. Some approximate analytic results are also obtained by integrating the given DE in the form

$$y'' = y''(0)\exp\left[(-1/2)\int_0^x y dx\right],$$

and obtaining from $y'(\infty) = 1$ the condition

$$1 = y''(0)\int_0^\infty \left(\exp(-1/2)\int_0^x y d\bar{x}\right) dx.$$

The difficulties with formal series solution of Blasius are analyzed.

(c) This is Blasius problem, which in view of invariance $x \to z/c, y \to cf$, may be reduced to an IVP in f and z:

$$f''' + ff'' = 0, \tag{1}$$

$f(0) = f'(0) = 0, f''(0) = 1,\ ' \equiv d/dz$, where $f''(0)$ results from the assumption that $y''(0) = \alpha$. The invariance leads to $f'(\infty) = \beta$, where $\alpha = \beta^{-3/2}$.

Thus, one is led to solve the IVP for (1). Writing the series form

$$y(z) = \sum_{n=0}^{\infty}(-1)^n a_n z^{3n+2} \tag{2}$$

with $a_0 = 1/2$, we obtain on substitution in the given DE, etc., the recurrence relation

$$3n(3n+1)(3n+2)a_n = \sum_{m=0}^{n-1} a_m a_{n-m-1}(3m+1)(3m+2), n = 1, 2, \ldots. \tag{3}$$

It was found previously that the series (2) has only a limited radius of convergence; one needs $f'(\infty)$ in the boundary layer theory where the problem arises. Analytic continuation poses serious difficulties. So an alternative form is sought. Since $f''(z) \neq 0$ for $0 \le z < \infty$ as is seen from the given DE, the boundary conditions give $f''(z) > 0$; i.e., $y'(z)$ is a monotonically increasing function of z that is analytic in a neighborhood of the positive real axis. Using the inverse function theorem and writing $f'(z) = p(z)$, we deduce that $z(p)$ exists and is analytic for real p with $0 \le p < \beta$, while $z(p) \to \infty$ as $p \to \beta-$. Inverting the representation (2), we may write, at least in some neighborhood of the origin $p = 0$,

$$z'(p) = \sum_{n=0}^{\infty} b_n p^{3n}, \tag{4}$$

where the initial conditions require that $b_0 = 1$. The given DE takes on the form

$$z''(p) = [z'(p)]^2 \int_0^p pz\prime(p)dp. \tag{5}$$

The substitution (4) in (5) leads to the recurrence relation

$$3nb_n = \sum_{m=0}^{n-1} \{b_m/(3m+2)\} \sum_{k=0}^{n-m-1} b_k b_{n-m-k-1} \text{ for } n = 1, 2, \ldots. \tag{6}$$

Using $b_0 = 1$, we get all the coefficients:

$$b_1 = 1/6, b_2 = 1/30, b_3 = 1/144, b_4 = 2099/1425600, \text{ etc.}$$

Evidently, for $b_n > 0$ for all n, it follows that one of the singularities of smallest modulus of $z'(p)$ which restricts the convergence of the series (4) must lie on the positive real axis. But we know that $z'(p)$ is analytic on the positive real axis for $0 \le p < \beta$ while it has a singularity at $p = \beta$. This singularity must, therefore, lie on the circle of convergence of the series (4) and therefore yields the radius of convergence of the series:

$$\beta^{-1} = \overline{\lim}_{n\to\infty} b_n^{1/3n}.$$

Thus, the representation (4) is valid over the entire range of interest and the quantity is determined by the asymptotic property of the coefficients. Lower and upper bounds for α are also found. Bender et al. (1989); Parlange, Braddock, and Sander (1981), Richardson (1973)

5. $$\boldsymbol{y''' + yy'' = 0, y(0) = -C, y'(0) = k, y' \to 1 \text{ as } x \to \infty,}$$

where $k \ge 0; C$ may be positive or negative constant.

Using $y'' = g$ as the dependent variable and $u = y'$ as independent variable, the problem reduces to

$$g(u)g''(u) + u = 0, g'(k) = C, k \ge 0, g(1) = 0.$$

Now, see Eqn. (2.29.14). Vajravelu, Soewono, and Mohapatra (1991)

6. $$\boldsymbol{y''' + yy'' = 0, y(0) = 0, y'(0) = K, y'(\infty) = 1,}$$

where either $0 < K < 1$ or $1 < K < 6$.

The given BVPs arise in the description of a boundary layer behind a rarefaction or shock wave traveling down and perpendicular to a flat plate.

Putting $z = y'(x), g(z) = y''(x)$, we get the BVP

$$gg'' + z = 0, g'(K) = 0, g(1) = 0.$$

Callegari and Nachman (1978)

7. $$\boldsymbol{y''' + yy'' = 0, y(0) = 0, y'(0) = 1, y'(\infty) = 0.}$$

This BVP arises in boundary layer theory. By putting $u = y', g(u) = y''$, we get the BVP

$$gg'' + u = 0, \ ' = \frac{d}{du},$$

$g(0) = 0, g'(1) = 0$. Callegari and Nachman (1978)

8. $$\boldsymbol{y''' + yy'' = 0, \ y'(-\infty) = 0, y'(+\infty) = 1.}$$

It is shown that the given problem has exactly one solution such that $y(0) = 0$. Hastings and Siegel (1972)

9. $$\boldsymbol{y''' + yy'' = 0.}$$

The given DE is transformed by $\xi = yy'/y'', \eta = y'^2/(yy''), t = \log|y'|$ to the system

$$\begin{aligned} \frac{d\xi}{dt} &= \xi(1 + \xi + \eta), \\ \frac{d\eta}{dt} &= \eta(2 + \xi - \eta). \end{aligned}$$

See Sachdev (1991). Jordan and Smith (1977), p.66

10. $$\boldsymbol{y''' + yy'' - 1 = 0, y(0) = y'(0) = 0.}$$

Numerical results were first used to conclude that there was a unique solution with the behavior $y \simeq -x(2\log x)^{1/2}$ as $x \to \infty$, and it was found that $y''(0) = -1.544$ and that y'' tended monotonically to zero as x tended to infinity. These results were later justified analytically with several qualitative features. For example, it was shown that all solutions for which y'' at any stage becomes positive must be such that $y \simeq x(2\ln x)^{1/2}$ as $x \to \infty$. The cases with $y''(0) = r, r \geq 0$ or $r < 0$, were treated separately. Brown (1966)

11. $$\boldsymbol{y''' + yy'' + y'^2 - 1 = 0, y(0) = y'(0) = 0, y' \to -1 \text{ as } x \to \infty.}$$

The given problem was solved numerically, giving the principal result of physical interest $y''(0) = -1.0866$. Burggraf, Stewartson, and Belcher (1971)

12. $$\boldsymbol{y''' + yy'' + y'^2 - 1 = 0, y(0) = y'(0) = 0, y'(\infty) = -1.}$$

The given problem is of some interest in boundary layer theory. It is proved that the given problem has a unique solution; that is, there exists one and only one function $y(x)$ satisfying the given DE on $(0, \infty)$ and the prescribed BC. This solution also satisfies $-\infty < y(x) < 0, x \in (0, \infty); -1 < y'(x) < 0, x \in (0, \infty); y''(x) < 0, x \in [0, \infty)$. Furthermore, $-2/3^{1/2} < y''(0) < -1$.

Choosing $y(x) = -f(x)$, the given DE becomes a special case of the Falkner–Skan equation. Gabutti (1984)

13. $$\boldsymbol{y''' + yy'' - (1 - y'^2) = 0, y(0) = 2(2)^{1/2}, y'(0) = 0,}$$

$y'(\infty) = 1.$

Exact solution of this problem arising in boundary layer theory can easily be found to be $y(x) = 2/\{x + 2^{1/2}\} + x + 2^{1/2}$. Brügner (1985)

14. $$\boldsymbol{y''' + yy'' + y'^2 - 1 = 0, y(0) = y'(0) = 0, y''(0) = k,}$$

where k is an arbitrary real number.

For the IVP, the following results are proved. Let $\hat{k} = z''(0) < 0$, where $z(x)$ is the unique solution of $z''' + zz'' + z'^2 - 1 = 0, z(0) = z'(0) = 0, z'(\infty) = -1$. Thus:

(a) When $k < \hat{k}$, the unique solution of the given IVP is defined only in a finite interval $[0, x_E), x_E > 0$; moreover, y, y', y'' are unbounded for $x \to x_E$.

(b) When $k \in (\hat{k}, 0)$, the unique solution of the given IVP has bounded derivatives $y'(x)$ on $[0, \infty)$. The slope $y'(x)$ first decreases and has a relative minimum, after which it increases and has a relative maximum for some value of x when $y'(x) > 1$. After this y' decreases monotonically and $\lim_{x\to\infty} y'(x) = 1$.

(c) When $k \geq 0$, the unique solution of the given problem has bounded derivative on $[0, \infty)$. The slope y' increases up to a maximum point $y'(x^*) > 1$; then it decreases monotonically with $\lim_{x\to\infty} y'(x) = 1$.

Furthermore, for cases (b) and (c), we have

$$y(x) = x + k - (1 + k^2/2)x^{-1} + o(x^{-1}).$$

Gabutti (1984)

15. $$\boldsymbol{y''' + yy'' - (1 - y'^2) = 0.}$$

The given DE arises in boundary layer theory for a specific parametric value $\lambda = -1$ therein. Put $y = 2u'/u$; then

$$y' + (1/2)y^2 = 2u''/u. \tag{1}$$

The given DE can be written as

$$2(u''/u)'' = 1. \tag{2}$$

Therefore,

$$u'' - (x^2/4 + ax + b)u = 0, \tag{3}$$

where a and b are arbitrary constants of integration. By replacing x by $x - 2a$, which does not alter the given DE, we may suppose that $a = 0$ in (3). Then the latter equation is just Weber's equation with the two linearly independent solutions

$$u_1 = D_\nu(x), u_2 = D_\nu(-x), \tag{4}$$

where $\nu = -b - 1/2. D_\nu$ are parabolic cylindrical functions. Since every solution of (3) is a linear combination of u_1 and u_2, we obtain the general solution of the given DE as

$$y = 2\{D'_\nu(x - x_0) - \tau D'_\nu(x_0 - x)\}/\{D_\nu(x - x_0) + \tau D_\nu(x_0 - x)\},$$

where ν, x_0, and τ are arbitrary complex constants ($\tau = \infty$ also being admitted). If y is real for real values of x, then ν, x_0, and τ must be taken real; the general solution is a meromorphic function. Coppel (1961)

16. $$\boldsymbol{y''' + yy'' + y'^2 - 3y^2y' = 0.}$$

The given DE has a first integral

$$y'' = -yy' + y^3 + K, \tag{1}$$

where K is an arbitrary constant.

It may be shown that Eqn. (1) has poles as its only movable singularities only when $K = 0$ [see Eqn. (2.10.4)]. Bureau (1964a)

17. $$\boldsymbol{y''' + yy'' - y'^2 = 0, y(0) = 0, y'(0) = 1, y' \to 0 \text{ as } x \to \infty.}$$

The given system arises from a reduction of Navier-Stokes equations. An exact solution of the given problem is

$$y = 1 - e^{-x}. \tag{1}$$

The given system may be written as

$$\left\{y'' \exp\left(\int_0^x y(t)dt\right)\right\}' = y'^2 \exp\left(\int_0^x y(t)dt\right).$$

The following results are proved:

(a) Solution (1) is the only solution of the problem for which $y' \geq 0$.

(b) For any second solution of the given problem, y', y'', y''' all vanish precisely once at x_1, x_2, x_3, say, with $x_1 < x_2 < x_3$.

Also, ultimately, $y > 0, y' < 0, y'' > 0, y''' < 0$. Therefore, writing $z = -\log y, u = y'/y^2$, we get

$$u'' + [u'^2/u - 7u' - u'/u] + 6u + 1 = 0, \; ' \equiv \frac{d}{dz}. \tag{2}$$

Equation (2), being autonomous, can be reduced to single first order DE in the $(dy/dx, u)$ plane. This was done and the nonexistence of a second solution of the given problem was proved. With $du/dz = \overline{y}, u = \overline{x}$, (2) has the form

$$\frac{d\overline{y}}{d\overline{x}} = -\frac{6\overline{x}^2 - 7\overline{x}\,\overline{y} + \overline{y}^2 + \overline{x} - \overline{y}}{\overline{x}\,\overline{y}}. \tag{3}$$

Equation (3) is discussed in Hilbert's sixteenth problem with reference to the maximum number of limit cycles that such an equation may possess. McLeod and Rajagopal (1987)

18. $$\boldsymbol{y''' + yy'' - y'^2 = 0.}$$

An exact solution of the given DE is $y = a + be^{-ax}$, where a and b are arbitrary constants. McLeod and Rajagopal (1987)

19. $$\boldsymbol{y''' + yy'' + \lambda y'^2 = 0,}$$

where $y(0) = 0, y'(0) = 1; y' \to 0$ as $x \to \infty$. Here λ is a parameter.

For $\lambda = -1$, we have the closed-form solution $y = 1 - e^{-x}$. For $\lambda = 1$, we have the closed-form solution

$$y = 2^{1/2}[(1 - e^{-2^{1/2}x})/(1 + e^{-2^{1/2}x})]. \quad (1)$$

For $\lambda = 2$, it is proved that the BVP has no solution. In this case, multiplication of the given equation by y' and integration give

$$yy'' - (1/2)y'^2 + y^2y' = \text{ constant.}$$

The constant of integration cannot be chosen such that the boundary conditions on $x = 0$ and $x = \infty$ are satisfied simultaneously.

More generally, we put $\phi = C - y$ and $p = dy/dx$; the given DE changes to

$$\frac{d}{d\phi}\left(p\frac{dp}{d\phi}\right) + (\phi - C)\frac{dp}{d\phi} + \lambda p = 0. \quad (2)$$

A series solution of (2) is found as

$$p = A_1\phi + A_2\phi^2 + A_3\phi^3 + A_4\phi^4 + A_5\phi^5 + \cdots,$$

where

$$\begin{aligned} A_1 &= C, A_2 = -(1+\lambda)/4, A_3 = (1-\lambda^2)/(72C), \\ A_4 &= (1-\lambda^2)(1+2\lambda)/(576C^2), \\ A_5 &= (1-\lambda^2)(11+81\lambda+88\lambda^2)/(86400C^3), \text{ etc.} \end{aligned}$$

The condition $p = 1$ when $\phi = C$ gives the value of C through

$$\begin{aligned} C^2\{1 - (1+\lambda)/4 + (1-\lambda^2)/72 + (1-\lambda^2)(1+2\lambda)/576 \\ + (1-\lambda^2)(11+81\lambda+88\lambda^2)/86400 + \cdots\} = 1. \end{aligned} \quad (3)$$

It is shown that C found from (3) gives a good estimate of the physical quantities such as skin friction in the boundary layer theory, where the DE appears. Merkin (1984)

20. $\mathbf{y''' + yy'' + \mu y'^2 = 0, y(0) = y'(0) = y'(\infty) = 0,}$

where μ is a real parameter.

It is proved that:

(a) The given BVP has only the trivial solution $y(x) \equiv 0$ if $\mu > 1, \mu \neq 2,$ or $\mu \leq 0$.

(b) If $0 < \mu < 1$, then for every $K \geq 0$, there is a solution $y(x)$ of the given BVP such that $\lim_{x\to\infty} y(x)/\{x^{1/(1+\mu)}\} = K$, and conversely, every solution of the given BVP has this form.

(c) If $\mu = 1$, then the solutions of the given BVP are precisely those for which $y''(0) \geq 0$ and then $\lim_{x\to\infty} y^2(x)/x = 2y''(0)$.

(d) If $\mu = 2$, then for every $K \geq 0$, there is a solution of the given BVP such that $\lim_{x\to\infty} y(x) = K$, and conversely, every solution of the given BVP has this form.

Heidel and Jones (1975)

21. $$y''' + yy'' + \mu y'^2 = 0, y(0) = a, y'(0) = b, y'(\infty) = 0,$$

where μ, a, and b are constants.

The problem arises in several contexts in fluid mechanics. It is proved that if $a \geq 0, b > 0$, and $\mu < 0$, then the given problem has exactly one solution y such that $y'(x) > 0$ for $x \geq 0$. If $a \geq 0, b = 0$, and $\mu < 0$, then the given problem has the unique solution $y(x) \equiv a$. Moreover, it is proved that if $a \geq 0, b > 0$, and $\mu < 0$, and if $y(x)$ is a unique solution of the given problem such that $y'(x) > 0$ for $x > 0$, then $y(x)/x^\alpha = \infty$ for $\alpha < 1$.

For $\mu > 0$, the following results are proved. If $0 < \mu \leq 1$ and $y(x)$ is a solution of the given DE such that $y(0) \geq 0, y'(0) \geq 0$, and $y''(0) \geq -y(0)y'(0)$, then $y'(x) \geq 0(y'(x) > 0$ (if $y \neq$ constant) for $x > 0$ and $y'(\infty) = 0$. Also, suppose that $\mu > 1$ and $b > 0$; if a is sufficiently large and positive, then the given BVP has a solution (in fact, a whole family of solutions). Heidel (1973)

22. $$y''' + yy'' + \lambda y'^2 = 0,$$

$y(0) = 0, y''(0) = -1; y' \to 0$ as $x \to \infty$. λ is a parameter.

There is a special solution when $\lambda = -1$, namely $y = 1 - e^{-x}$. It can be shown that for $\lambda = 1$, there is no solution to the BVP. The series solution is found as for Eqn. (3.5.19). To satisfy the BC, we write an integral form of the given DE and hence find that

$$y''(0) = (\lambda - 1) \int_0^\infty y'^2 dx = (\lambda - 1) \int_0^C p d\phi.$$

Thus, we have, by using the BC $y''(0) = -1$, the equation

$$(1 - \lambda) \int_0^C p d\phi = 1.$$

Substituting the expansion for p in terms of ϕ [see Eqn.(3.5.19)], we have the following equation for the constant C:

$$\begin{aligned}(1 - \lambda)C^3[1/2 - (1 + \lambda)/12 + (1 - \lambda^2)/288 + (1 - \lambda^2)(1 + 2\lambda)/2880 \\ + (1 - \lambda^2)(11 + 81\lambda + 88\lambda^2)/518400 + \cdots] = 1.\end{aligned}$$

With this choice of C, the solution again gives good results for skin friction, etc. Merkin (1984)

23. $$y''' + yy'' - [(m - 1)/m]y'^2 = 0, m > 0,$$

with BC $y = bx^m + \cdots,\ x \to +\infty, b > 0, m > 0, y(0) = 0$ and $y'(x) \to 0$ as $x \to -\infty$.

The given problem arises in mixing-layer theory. Other boundary conditions arise in other physical contexts. The order of the given DE can be reduced if we set

$$f = \frac{dy}{dx}, \text{ so that } y'' = f\frac{df}{dy}, \ y''' = f\left\{\left(\frac{df}{dy}\right)^2 + f\frac{d^2 f}{dy^2}\right\}, \text{ etc.}$$

We thus have

$$ff'' + f'^2 + yf' - [(m - 1)/m]f = 0, \tag{1}$$

where

$$\begin{aligned} f &= -c(y-c) - [1/(4m)](y-c)^2 + O\{(y-c)^3\}, y \to c < 0, \\ f &= mb^{1/m}y^{(m-1)/m} + [m(m-1)(m-2)/(m+1)]b^{2/m}y^{-2/m} + O(y^{-1-3/m}) \\ &\quad + D_1 y^{k_1} \exp[-b^{-1/m}/(m+1)y^{(m+1)/m}] + \cdots \text{ as } y \to \infty; \\ k_1 &= -(2m^2+4m-4)/[m(m+1)], \end{aligned}$$

$D_1 =$ constant. Equation (1) is reduced further if we set

$$f = y^2 F(y), y\frac{dF}{dy} = \psi.$$

The DE in the (F, ψ) plane is found to be

$$F\psi\frac{d\psi}{dF} = -\left(\psi^2 + 7F\psi + 6F^2 + \psi + \frac{m+1}{m}F\right). \tag{2}$$

If $m = 1, 2$, then $\psi = -2F$ and $\psi = -(3/2)F$ are, respectively, exact solutions of (2). The latter has three singular points $A(0,0), B(0,-1), C\{-(m+1)/(6m), 0\}$ in the finite part of the (F, ψ) plane and three at infinitely remote parts. For $F \to 0$, the following asymptotic relations hold:

$$\begin{aligned} \psi &= \mu(F) + D_0 F^{k_0} \exp[-m/(m+1)F^{-1}] + \cdots, \\ \mu(F) &= -[(m+1)/m]F - [(m-1)(m-2)/m^2]F^2 + O(F^3), \\ k_0 &= (3m^2+4m-5)/(m+1)^2, \end{aligned}$$

$D_0 =$ constant. For $m = 1, 2, \mu(F)$ may conveniently be taken to be the exact solution $-2F$ and $-(3/2)F$, respectively.

A thorough singular point analysis of (2) is carried out. A comprehensive qualitative theory for solution of the BVP is also provided. Various parametric values of m in the ranges $0 \le m \le 1$ and $m > 1$ are considered. This analysis may be found helpful for this class of third order DEs. Diyesperov (1986)

24. $$y''' + yy'' - \lambda y'^2 = 0, \lambda > 0.$$

The given DE does not have an explicit exact solution except when $\lambda = 1$ [see Eqn. (3.5.18)]. However, it can be transformed into the phase plane, following the steps listed for the special case, $\lambda = 1$. The main result is that for $\lambda = 8/7$, the rest point changes from unstable to stable and merges with the limit cycle. For $\lambda > 8/7$, the limit cycle disappears and we have a spiraling motion to the stable rest point. McLeod and Rajagopal (1987)

25. $$y''' + yy'' + \lambda(1 - y'^2) = 0,$$

$y = \alpha, y' = \beta$ at $x = 0, y' \to 1$ as $x \to \infty; 0 < \lambda \le 1/2$.

It is proved that the given BVP of boundary layer has a solution for any nonnegative value of the constants α, β. Moreover, the second derivative y'' is positive, zero, or negative throughout the interval $0 \le x < \infty$ according as β is less than, equal to, or greater than 1. With this restriction on y'', the solution is shown to be unique.

Asymptotic behavior and many other qualitative properties of the given DE are also proved for different values of λ. The discussion for $\lambda > 1/2$ is less complete. Coppel (1960)

26. $$y''' + yy'' + \lambda(1 - y'^2) = 0,$$

$y(0) = \alpha, y'(0) = \beta, y'(\infty) = 1$, as well as the side conditions $0 \leq \beta < y' < 1$ for $x > 0, y'' > 0$ for $x > 0, 0 < y' < 1$ for $x > 0$.

This Falkner–Skan system appears in boundary layer theory. The continuity and monotonicity of functions of the parameters determining the range of existence of solutions are considered. Physically important properties of the solutions are investigated. Hartman (1972)

27. $$y''' + yy'' + \lambda(1 - y'^2) = 0,$$

$y(0) = y'(0) = 0, y'(\infty) = 1$.

This Falkner–Skan equation with $\lambda < 0$ is discussed. Two types of solutions, one for $y'(\infty) = 1$ and another for $y'(\infty) = -1$, are considered. Several types of solutions are found numerically. It is shown that solution branches with $y'(\infty) = 1$ and those with $y'(\infty) = -1$ tend toward a common limit curve as N, the number of zeros of $y' - 1$, tends to infinity. Periodic solutions are also found for $\lambda < -1$. Oskam and Veldman (1982)

28. $$y''' + yy'' + \beta(1 - y'^2) = 0, \beta = -1,$$

$y(0) = 0, y'(0) = 0, y'(+\infty) = 1$.

It is easy to check that the given DE may be integrated twice to obtain the Riccati equation
$$y' + y^2/2 = x^2/2 - \lambda x, y(0) = 0,$$
which is parametrized by $\lambda = -f''(0)$.

The existence of a critical value $\lambda_0 = 1.0863757$ is proved such that:

(a) $y'(x) \to 1$ algebraically as $x \to +\infty$ if $0 \leq \lambda < \lambda_0$.

(b) $y'(x) \to -1$ as $x \to +\infty$ if $\lambda = \lambda_0$.

(c) y and y' blow up to $-\infty$ at a finite value of x if $\lambda > \lambda_0$.

Many other qualitative results are proved. The series form indicating the foregoing asymptotics are given. For example,

(a)
$$\begin{aligned} y &\sim x - \lambda - \{(2 + \lambda^2)/2\}\{1/(x - \lambda)\} - \cdots, \\ &\quad y'(x) \sim 1 + \{(2 + \lambda^2)/2\}\{1/(x - \lambda)^2\} + \cdots \\ &\quad \text{for } 0 < \lambda < \lambda_0 \text{ as } x \to \infty. \end{aligned}$$

(b)
$$\begin{aligned} y(x) &= -(x - \lambda_0) - \{1 - (\lambda_0^2/2)\}/\{x - \lambda_0\}^2 - \cdots, \\ &\quad y'(x) \sim -1 + \{1 - (\lambda_0/2)^2\}\{x - \lambda_0\}^2 + \cdots \\ &\quad \text{for } \lambda = \lambda_0 \text{ as } x \to +\infty. \end{aligned}$$

Numerical results are depicted for various sets of β and λ.

Brauner and Lainé (1982).

29.
$$y''' + yy'' + \beta(1 - y'^2) = 0,$$

$y(0) = y''(0) = 0, y'(\infty) = 1$, where β is negative; $-1 < y' < 1$ on $[0, \infty)$ and $1 - y'(x) = o(e^{-x})$ as $x \to \infty$.

It is proved that for $|\beta|$ sufficiently small, there is a unique solution of the given problem such that $y'(0) < 0$. Hastings and Siegel (1972)

30.
$$y''' + yy'' + \beta(1 - y'^2) = 0,$$

$-1 < y' < 1$, on $[0, \infty)$ and $1 - y'(x) = o(e^{-x})$ as $x \to \infty$, where β is a small parameter; β is positive.

It is proved that for β sufficiently small, there is a unique solution of the given problem such that $f''(0) < 0$. Hastings and Siegel (1972)

31.
$$y''' + yy'' + \beta(1 - y'^2) = 0,$$

$y(0) = y''(0) = 0, y'(0) = \alpha, -1 < \alpha < 0$.

The following results are proved:

(a) For any $\beta < -1$, there is an $\overline{\alpha}$ in $(-1, 0)$ such that the solution $y_{\overline{\alpha}}$ is periodic with some period P and $y'_{\overline{\alpha}}$ has exactly one local maximum in $(0, P)$. If $-1 \leq \beta \leq 0$, then there is no periodic solution of the given problem except $y \equiv 0$.

(b) If $\beta < -1$, then there is a sequence $\{\alpha_j\} \subset (-1, \overline{Z}), j \geq 0$, tending to $\overline{\alpha}$ from below such that $y'_{\alpha_j}(x)$ tends to -1 as x tends to infinity and y'_{α_j} has exactly j local maxima in $0 < \eta < \infty$.

Hastings and Troy (1987)

32.
$$y''' + yy'' + \beta(1 - y'^2) = 0,$$

where β is a constant.

This Falkner–Skan equation is shown to have a periodic solution for $\beta > 1$. It is also shown that the given DE with $y(0) = y'(0) = 0, y'(\infty) = 1$ has:

(a) A solution with at least N relative minima for any $\beta > 1$ and any integer $N \geq 0$.

(b) For $2 < \beta < 2.05$, at least two periodic solutions. Many qualitative properties are discussed.

Hastings and Troy (1988)

33.
$$y''' + yy'' + \beta(1 - y'^2) = 0,$$

where $|\beta| \leq 1$, a constant, $y(0) = 0, y'(0) = -\lambda, \lambda > 0, y'(\infty) = 1$.

The given DE is called Falkner–Skan, and the BCs arise in boundary layer theory. Extensive numerical work for $|\beta| \leq 1$ over a range of positive and negative reveals rich solution behavior. For $-1 \leq \beta \leq 0$, dual solutions (no solution) are found below (above) a critical value $\lambda_m(\beta)$. When $\lambda > 0$, one observes triple solutions for $0 < \beta \leq 0.14$, unique solutions for $0.14 \lesssim 3 \lesssim 0.5$, and dual solutions for $0.5 < \beta \leq 1$.

For $\beta = -1$, an exact solution becomes possible. Integration of the given DE with $y(0) = 0$ yields

$$y'' + yy' = x + y''(0). \tag{1}$$

A second quadrature satisfying $y'(0) = -\lambda$ furnishes the Riccati equation

$$y' + y^2/2 = x^2/2 + y''(0)x - \lambda. \tag{2}$$

From the BCs at ∞, we have the asymptotic behavior $y(x) \sim x + C$ as $x \to \infty$; substitution of this in (1) yields $C = y''(0)$. An evaluation of (2) in the asymptotic limit $x \to \infty$ then provides the link between λ and $y''(0)$:

$$\lambda = -\{1 + (1/2)(y''(0))^2\}. \tag{3}$$

Put $\lambda = -(1 + \alpha)$ and use (3) to write (2) as

$$y' + y^2/2 = x^2/2 + (2\alpha)^{1/2}x + 1 + \alpha. \tag{4}$$

The general solution of (4) is found by writing

$$y(x) - x - (2\alpha)^{1/2} = 2v'(x)/v(x) \tag{5}$$

so that

$$v'' + (x + (2\alpha)^{1/2})v' = 0. \tag{6}$$

The error function solution of (6), when substituted in (5), yields

$$y(x) = x + (2\alpha)^{1/2} + \frac{2(2^{1/2})\exp\{-(1/2)(x + (2\alpha)^{1/2})^2\}}{\pi^{1/2}\text{erf}\{(x + (2\alpha)^{1/2})/(2^{1/2})\} + B},$$

which on using the BC $y(0) = 0$ gives

$$B = -[\pi^{1/2}\text{erf}(\alpha^{1/2}) + \{2/\alpha^{1/2}\}\exp(-\alpha)].$$

Asymptotic analysis is employed to elucidate the solution behavior for $\beta = 0$ as $\lambda \to 0$ and for $\beta = 1$ as $\lambda \to 1$. Results are depicted and interpreted. Riley and Weidman (1989)

34. $$\mathbf{y''' + yy'' + \lambda(1 - y'^2) + \mu(1 - y') = 0,}$$

$\lambda > 0, \mu > 0$ are constants, $y(0) = y'(0) = 0; y'(\infty) = 1$.

The given problem arises in magnetohydrodynamics boundary layer theory. Existence and uniqueness of the solutions of the given system are proved using a geometrical argument of Coppel and Iglisch for the Falker–Skan equation. Srivastava and Usha (1983/84)

35. $$\mathbf{5y''' + 6yy'' + 3y'^2 = 0,}$$

$y(0) = y'(\infty) = 0, y''(0) = -1$.

The given problem arises in thermal capillary flows in viscous layers. The numerical solution gives $y'(0) = 1.824$ and $y(\infty) = 1.543$. Batyshchev (1991)

36.
$$\boldsymbol{y''' + (4/3)yy'' - (2/3)y'^2 - (64/9)y^2y' - (8/9)y^4 - (10/3)y'y''/y + (20/9)y'^3/y^2 = 0.}$$

By putting
$$z = -(y'' - 2yy')/(y' - y^2) + (4/3)(y'/y - y)$$
and using the given DE, we may obtain
$$y = -(z'' - 2zz')/(z' - z^2) + (1/2)(z'/z - z). \tag{1}$$
It may be shown that z satisfies
$$z''' - 2zz'' - 2z'^2 + (1/2)(z' - z^2)^2 = 0,$$
which, according to Chazy (1911), is of Painlevé type, and hence via (1) is the given DE. Martynov (1985a)

37.
$$\boldsymbol{y''' + 2yy'' + 2\lambda(1 - y'^2) = 0, y'(+\infty) = 1, \lambda \in (-0.5, 0).}$$

It is proved that the given problem has one of the following asymptotic forms as $x \to \infty$:

(a)
$$\begin{aligned} y(x) &= x[1 + C_1x^{-1}\{1 + O(x^{-1})\} + (1/4)C_3x^{-2\lambda-3} \\ &\quad \times \{1 + O(x^{-1})\}\exp(-x^2 - 2C_1x)], \\ y'(x) &= 1 - (1/2)C_3x^{-2\lambda-1}\{1 + O(x^{-1})\}\exp(-x^2 - 2C_1x), \\ y''(x) &= C_3x^{-2\lambda}\{1 + O(x^{-1})\}\exp(-x^2 - 2C_1x). \end{aligned}$$

(b)
$$\begin{aligned} y(x) &= x[1 + C_1x^{-1} + \{C_3/(1 + 2\lambda)\}x^{2\lambda}\{1 + O(x^{2\lambda})\}], \\ y'(x) &= 1 + C_2x^{2\lambda}\{1 + O(x^{2\lambda})\}, \\ y''(x) &= 2\lambda C_2x^{2\lambda-1}\{1 + O(x^{2\lambda})\}, \end{aligned}$$

where $C_3 > 0, C_2 < 0$, and C_1 are constants.

Iwano (1977a)

38.
$$\boldsymbol{y''' + 2yy'' + y'^2 + y^2y' = 0.}$$

We seek a solution of the given DE in the form
$$y(x) = \sum_{j=-k}^{\infty} \alpha_j(x - x_0)^j. \tag{1}$$
Residues of a pole are obtained from $\alpha_{-1}^2 - 5\alpha_{-1} + 6 = 0$. Choosing $\alpha_{-1} = 2$, we find that α_1 in (1) is arbitrary and the rest of the coefficients can be expressed uniquely in terms

of α_{-1}, α_1, and x_0. To prove that the series (1) is convergent when $\alpha_{-1} = 2$, we write the given DE as the system

$$y' = v, v' = u, u' = -2yu - v^2 - y^2 v. \tag{2}$$

Put

$$\begin{aligned} y &= (x - x_0)^{-1}[2 + \phi_1(x)], \\ v &= (x - x_0)^{-2}[-2 + \phi_2(x)], \\ u &= (x - x_0)^{-3}[4 + \phi_3(x)], x - x_0 = \tau; \end{aligned} \tag{3}$$

then $\phi_j(\tau), j = 1, 3$, and we obtain the Briot–Bouquet system

$$\begin{aligned} &\tau\frac{d\phi_1}{d\tau} = \phi_1 + \phi_2, \quad \tau\frac{d\phi_2}{d\tau} = 2\phi_2 + \phi_3, \\ &\tau\frac{d\phi_3}{d\tau} = -(\phi_3 - 2\phi_1^2 + \phi_2^2 + 4\phi_1\phi_2 + 2\phi_1\phi_3 + \phi_1^2\phi_2), \end{aligned} \tag{4}$$

whose characteristic equation has roots $\lambda_1 = -1, \lambda_2 = 1, \lambda_3 = 2$. Hence the system (4) has a one-parameter family of holomorphic solutions such that $\phi_j \to 0, j = 1, 3$, for $\tau \to 0$, which generates for the given DE a polar solution with residue 2.

To find an integral of the given DE, we put

$$y = \nu\lambda'/\lambda, \tag{5}$$

where ν is a constant, to obtain

$$\begin{aligned} &\lambda^3\lambda^{\text{iv}} + 2(\nu - 2)\lambda^2\lambda'\lambda''' + (\nu - 3)\lambda^2\lambda''^2 \\ &\quad + (\nu^2 - 8\nu + 12)\lambda\lambda'^2\lambda'' + \lambda'' - (\nu^2 - 5\nu + 6)\lambda'^4 = 0. \end{aligned} \tag{6}$$

For $\nu = 2$, (6) simplifies to

$$\lambda\lambda^{\text{iv}} - \lambda''^2 = 0, \tag{7}$$

while for $\nu = 3$, we get

$$\lambda^2\lambda^{\text{iv}} + 2\lambda\lambda'\lambda''' - 3\lambda'^2\lambda'' = 0. \tag{8}$$

Equation (7) integrates to yield

$$\lambda\lambda'' - \lambda'^2 = C_1 x + C_2, \tag{9}$$

while (8) gives

$$\lambda(\lambda\lambda'' - \lambda'^2) = C_1 x + C_2, \tag{10}$$

where C_1 and C_2 are constants. To solve (9), we use the definition $\lambda y = 2\lambda'$ to obtain

$$yy' + y'' - (x - x_0)^{-1}y' = 0. \tag{11}$$

Equation (11) is easily verified to be an integral of the given DE. Put $y = (x - x_0)^{-1}z$ and $t = \ln(x - x_0)$ in (1), to obtain

$$\frac{d^2z}{dt^2} = (4 - z)\frac{dz}{dt} + z^2 - 3z, \tag{12}$$

which transforms via $z' = w$ into the Abel equation

$$w\frac{dw}{dz} = (4 - z)w + z^2 - 3z. \tag{13}$$

Equation (13) has the solution $w = z$. Moreover, (7) also has an elementary solution $\lambda = a\exp(bx)$, where a and b are arbitrary constants.

It may be noted that the transformation $\lambda = \mu^{3/2}$ changes Eqn. (7) to (8). Hence the latter may be solved. Sidorevich and Lukashevich (1990)

39. $$\boldsymbol{y''' + 2(yy'' + y'^2) - c_0(x)(y' + y^2) - d_0(x) = 0,}$$

where $c_0(x)$ and $d_0(x)$ are analytic functions of x.

The given DE may be cast in either of the forms

$$y' + y^2 = u, \quad u'' = c_0(x)u + d_0(x) \tag{1}$$

or

$$y = t'/t, t'' = ut \quad u'' = c_0(x)u + d_0(x). \tag{2}$$

Here making use of properties of linear and Riccati equations in (1) or (2), we conclude that the given DE has poles as the only movable singularities in its solution. Bureau (1964a)

40. $\boldsymbol{y''' + 2(yy'' + y'^2) - c_1(x)(yy' + y^3) - c_0(x)y' - d_2(x)y^2 - d_1(x)y - d_0(x) = 0,}$

where c_1, c_0, d_2, d_1, and d_0 are analytic functions of x.

Conditions are found on the functions c_0, c_1, d_0, d_1, and d_2 such that the given DE has solutions with only one set of simple movable poles. Bureau (1964a)

41. $$\boldsymbol{y''' + 2yy'' + 4y'^2 + 2y^2y' - 2c_0(x)y' - c_0'(x)y = 0,}$$

where $c_0(x)$ is an analytic function.

The given DE may be shown to have two sets of simple poles as the only movable singularities in its solution: To integrate it, we multiply it by $2y$ and rewrite it as

$$\frac{d}{dx}(2yy'' - y'^2) + \frac{d}{dx}(4y^2y' + y^4) - 2\frac{d}{dx}(c_0y^2) = 0,$$

and so

$$y'' = y'^2/(2y) - 2yy' - y^3/2 + c_0(x)y + K/(2y), \tag{1}$$

where K is an arbitrary constant. It may be reduced by the transformation

$$x = \alpha X, \quad y = \beta Y, \quad \alpha\beta = 1, \quad k\alpha^4 = -1,$$

to the canonical form

$$y'' = y'^2/(2y) - 2yy' - y^3/2 + \alpha^2 c_0 y - 1/(2y). \tag{2}$$

Equation (2) is reducible to a linear equation of order 4 by the transformation $y = w'/w$, and hence differentiation:

$$w^{\text{iv}} - 2\alpha^2 c_0 w'' + w = 0.$$

Bureau (1964a)

42. $$\mathbf{y''' + 2yy'' + (1 - y'^2) = 0, y(0) = y'(0) = y'(\infty) = 0.}$$

Differentiating the given DE, we have

$$y^{\text{iv}} + 2yy''' = 0. \tag{1}$$

Equation (1) is invariant under the transformation $y(x) \to ky(kx)$. Let us denote by $y = f(x)$ that solution of (1) which has the initial values $f(0) = f'(0) = 0, f''(0) = 1, f'''(0) = -\beta$ and which satisfies the third order DE

$$f''' + 2ff'' + (\beta^2 - f'^2) = 0. \tag{A}$$

Then from (1),

$$f''' = -\beta^2 e^{-2\int_0^x f(t)dt} = -\beta^2 e^{-\int_0^x (x-t)^2 f''(t)dt}. \tag{2}$$

Defining $g(x) = f''(x)$, we write (2) as

$$g(x) = 1 - \beta^2 \int_0^x \exp\left\{-\int_0^\sigma (\sigma - t)^2 g(t)dt\right\} d\sigma \equiv 1 - \beta^2 \int_0^x e^{-G(\sigma)} d\sigma. \tag{3}$$

The constant β can be found by using the condition $g(\infty) = 0$:

$$\beta^2 = \frac{1}{\int_0^\infty e^{-G(\sigma)} d\sigma}.$$

Thus,

$$g(x) = \frac{\int_x^\infty e^{-G(\sigma)} d\sigma}{\int_0^\infty e^{-G(\sigma)} d\sigma}. \tag{4}$$

Introducing the operator T as

$$T\{g\} = \frac{\int_x^\infty e^{-G(\sigma)} d\sigma}{\int_0^\infty e^{-G(\sigma)} d\sigma},$$

we have

$$g = T\{g\}. \tag{5}$$

The integral equation (5) can be solved numerically: $g_{n+1} = T\{g_n\}$ with $g_1 = 1.T\{g\}$ may be shown to have the properties $0 < T\{g\} \le 1$, and $T\{g\} \ge T\{g^*\}$ if $g \le g^*$. The solution of the integral equation can be used to determine $f(x) = \int_0^x (x - t)g(t)dt$, which solves the third order equation (A). With this result we have $y(\zeta) = kf(x), \zeta = x/k$, where the constant k is selected so that $y'(x) \to 1$ as $x \to \infty$, hence $k = 1/\beta^{1/2}$. Numerical solution by the foregoing iterative scheme is found and depicted. Siekmann (1962)

43. $$\mathbf{y''' + 2yy'' - 6y'^2 - 16y^2y' - 8y^4 - k_1x - k_2 = 0,}$$

where k_1 and k_2 are constants.

Replace y by $3y/4$ first. The resulting equation may then be shown to be equivalent to the system

$$y' = -(3/4)y^2 + z, \tag{1}$$

$$z'' = 6z^2 + \hat{k}_1 x + \hat{k}_2, \tag{2}$$

where $\hat{k}_1$ and $\hat{k}_2$ are some other constants. The system has fixed critical points; Eqn. (2) is P_I, while (1) is of Riccati type. Exton (1971)

44.
$$y''' + 2yy'' + 2\lambda(1 - y'^2) = 0, \lambda < 0,$$

$y(0) = 0, y'(0) = 0, y'(+\infty) = 1, 0 < y'(x) < 1$ for $0 < x < +\infty$.

First, it is shown that the given problem is equivalent to

$$y'' + (1/y^{1/2})\{2xy' + 4\lambda(1 - y)\} = 0,$$

$\lambda < 0, y(0) = 0, y(+\infty) = 1, 0 < y(x) < 1$ for $0 < x < +\infty$, where a prime now denotes differentiation with respect to a new variable x. Now see Eqn.(2.13.35). Long-time behavior is also found. Iwano (1977)

45.
$$y''' + 2yy'' + 2\lambda(k^2 - y'^2) = 0,$$

for $x \geq 0, y(0) = y'(0) = 0, y'(x) \to k$ for $x \to \infty$.

Introducing $w = y(x)$ and $p = y'(x)$ as the new variables and assuming that $y''(x) > 0$, a uniqueness theorem is proved. The proof with the hypothesis $y''(x) > 0$ requires separate discussion for the ranges $0 \leq \lambda \leq 1/2$ and $1/2 < \lambda < 1$. Furuya (1953)

46.
$$y''' + 3yy'' + (1/2)y'^2 + 2y^2y' = 0.$$

The given DE arises in magnetohydrodynamics. It has some invariance properties. Heidel (1973)

47.
$$y''' + 3yy'' + 2y'^2 + (10/3)y^2y' + (4/9)y^4 = 0.$$

The given DE has special solutions $y = 3/x, 3/(2x)$. It is shown to possess the "weak Painlevé" property. Duarte, Euler, Moreira, and Steeb (1990)

48.
$$y''' + 3yy'' + 3y'^2 + 3y^2y' - c_0(x)y' - c_0'(x)y - d_0(x) = 0,$$

where $c_0(x)$ and $d_0(x)$ are analytic functions of x.

It may be shown that the given DE has two sets of simple movable poles as the only movable singularities in its solution. Indeed, setting $d_0 = q'(x)$, we integrate the given DE to obtain

$$y'' = -3yy' - y^3 + c_0y + q + K,$$

where K is an arbitrary constant. Now, setting $y = v'/v$, we obtain a linear DE

$$v''' - c_0(x)v' - [q(x) + K]v = 0$$

of third order. Bureau (1964a)

49.
$$y''' + 3yy'' + (7/2)y'^2 - 4y^2y' - y^4/2 - c_0(x)(y^2 + y') - d_0(x) = 0,$$

where $c_0(x)$ and $d_0(x)$ are arbitrary analytic functions of x.

If we choose $c_0(x) \equiv 0$, put $z = -(3/2)y$, and then $z = -(3/2)w'/w$, we get

$$ww^{\text{iv}} - w'w''' + w''^2/2 - d_0(x)w^2 = 0,$$

which on differentiation yields the linear DE

$$w^{\mathrm{v}} - 2d_0(x)w' - d_0'(x)w = 0.$$

Thus, the given DE is stable in the sense that its critical points are all fixed. Exton (1971)

50. $$\boldsymbol{y''' - yy'' = 0.}$$

The given DE occurs in boundary layer theory. It has two closed-form solutions $y_1(x) = ax + b, y_2(x) = -3/(x + c)$, where a, b, c are constants. It is suggested that by using point transformations, new solutions may be generated. Meinhardt (1981)

51. $$\boldsymbol{y''' - yy'' + (1/2)y'^2 + 2s^2 = 0, -\infty < x < \infty,}$$

where s is a constant.

It is shown that if there exist constants A, B, and C such that

$$y(x) = Ax^2 + Bx + C + o(1), x \to \infty, A > 0,$$

then

$$\begin{aligned} y(x) &= Ax^2 + Bx + C, -\infty < x < \infty, B^2 - 4AC = -4s^2, \\ y(x) &= A\{x + B/(2A)\}^2 + s^2/A. \end{aligned}$$

Writing

$$F(x) = \int_0^x y(x)dx \sim Ax^3/3, x \to \infty, \tag{1}$$

we have from the given DE, by differentiating and multiplying by $\exp(-F)$, etc.,

$$[\exp(-F)y''']' = 0,$$

that is,

$$y'''(x) = D\ \exp(F(x)), \tag{2}$$

where $D \neq 0$ is an arbitrary constant.

It follows from (1) that y''' and hence y blows up exponentially as $x \to \infty$, since $A > 0$. It contradicts the hypothesis. Therefore, $D = 0, y''' = 0$, and $y = y(x)$ is a parabola. Hence the result easily follows. Dijkstra (1980)

52. $$\boldsymbol{y''' - yy'' + y'^2 = 0.}$$

Two invariants of the given DE are

$$\psi = y'/y^2, \phi = y''/y^3. \tag{1}$$

Compute

$$\frac{d\phi}{d\psi} = \frac{y'''/y^3 - 3(y'y''/y^4)}{y''/y^2 - 2(y'^2/y^3)} = \frac{yy''' - 3y''y'}{y(yy'' - 2y'^2)}. \tag{2}$$

Eliminating all the derivatives of y from the given DE and (1)-(2), we get the first order Abel equation for $\phi(\psi)$:

$$(\phi' - 1)(\phi - 2\psi^2) + 3\phi\psi - \psi^2 = 0. \tag{3}$$

From the solution $\phi = f(\psi)$ of (3), we get $y''/y^3 = f(y'/y^2)$, an equation that is easily solvable. Symmetries of the given DE are derived and used. Alternatively, we first set $s = x, t = y$, $s' = 1/y', s'' = -y''/y'^3$ to obtain the transformed form of the given DE as

$$s''' - ts's'' - 3s''^2/s' - s'^2 = 0, s' = \frac{ds}{dt}. \quad (4)$$

Now put $s' = ve^{2u}, t = e^{-u}$ to obtain the second order equation

$$u'' = u'^3(v^2 - 6v) + u'^2(v - 7) - 3u'/v, u' = \frac{du}{dv}. \quad (5)$$

Equation (5) is Abel's equation of first order in u'. After solving it, one may go back to retrieve $s'(t)$ and hence $s(t)$. Stephani (1989), p.88

53. $$\mathbf{y''' - yy'' - 2y'^2 - 2y^2y' = 0.}$$

Multiplying the given DE by 2y and integrating, we have

$$y'' - y'^2/(2y) - yy' - y^3/2 - K/y = 0, \quad (1)$$

where K is an arbitrary constant. To integrate (1), set $12w = y' - y^2$ to obtain

$$12w' = 72w^2/y + K/y$$

or

$$y = (72w^2 + K)/(12w'). \quad (2)$$

Eliminating y between (1) and (2), we have

$$w'' = -6w^2 - K/12. \quad (3)$$

Equation (3) is solvable in terms of elliptic functions; hence y can be found from (2). Bureau (1964a)

54. $$\mathbf{y''' - yy'' - 5y'^2 + y^2y' = 0.}$$

Setting $P = y'' - yy' - y^3$ and using the given DE, we have

$$P' = 4y'(y' - y^2). \quad (1)$$

Multiplying (1) by P and noting that

$$\frac{d}{dx}(y' - y^2)^2(2y' + y^2) = 6y'(y' - y^2)P,$$

we have, by integration,

$$P^2 = (4/3)(y' - y^2)^2(2y' + y^2) + K, \quad (2)$$

where K is an arbitrary constant. Therefore, y satisfies the equation

$$(y'' - yy' - y^3)^2 = (4/3)(y' - y^2)^2(2y' + y^2) + K. \quad (3)$$

It may be shown that (3) has poles as its only movable singularities. We consider the special case $K = 0$. For convenience we put $b \equiv (2/3)^{1/2}$, so that the square rooting of (3) in this case leads to

$$y'' - yy' - y^3 = 2b(y' - y^2)(y' + y^2/2)^{1/2}. \tag{4}$$

Now, set $y' + y^2/2 = z^2y^2$ and note that

$$y'' - yy' - y^3 = 2zy^2(z' + zyQ),$$

where $Q = z^2 - 1/b^2$. Therefore, (4) is equivalent to the system

$$y' = y^2(z^2 - 1/2), \quad z' = -Qy(z - b). \tag{5}$$

Now, we obtain the DE for z by eliminating y from (5). Taking the logarithmic derivation of second of (5) and using both the equations, we find that

$$\begin{aligned} z''/z'^2 &= (2z^2 - 2zb - 1)/\{(z^2 - 1/b^2)(z - b)\} \\ &= 2(z + 1/b - b)/\{(z + 1/b)(z - b)\} \\ &= (6/5)\{1/(z - b)\} + (4/5)\{(z + 1/b)\}. \end{aligned}$$

An integration gives

$$z'^5 = K_1^5(z - b)^6(z + 1/b)^4. \tag{6}$$

Putting $t^5 = (z + 1/b)/(z - b)$, etc., we may integrate (6) to find that

$$t = -(K_1/5)(b + 1/b)x + K_2 \quad \text{and} \quad z = (1/b + bt^5)/(t^5 - 1).$$

Here K_1 and K_2 are arbitrary constants. Hence y may be found by solving

$$y' + y^2/2 = z^2y^2.$$

Bureau (1964a)

55. $$\boldsymbol{y'''' - yy'' - 5y'^2 + y^2y' - 3c(x)y' - c'(x)y - c''(x) = 0,}$$

where $c(x)$ is given by

$$c'^2 = (1/3)c^3 + K_2c + K_3$$

and is, therefore, an elliptic function of x. K_2 and K_3 are arbitrary constants.

It may be shown that the given DE has two sets of simple poles as the only movable singularities in its solution. It may be shown to have a first integral

$$\begin{aligned} (y'' - yy' - y^3 + cy + c')^2 &= (8/3)(y' - y^2)^2(y' + y^2/2 + 3c/2) \\ &+ 4(y' - y^2)(2cy^2 + c'y + c'') + 4c^2y^2 + 4cc'y + 2c'^2 + K, \end{aligned}$$

where K is a constant. Bureau (1964a)

56. $$\boldsymbol{y''' - yy'' + \lambda(1 + y'^2) = 0,}$$

$0 < \lambda < 1/2, y(0) = 0, y'(0) = 0$; y is algebraically large only as $x \to \infty$.

It is proved that there exists a solution of the given boundary value problem, which satisfies $y \geq 0, y' \geq 0, y'' > 0, y''' < 0$, and $y^{\text{iv}} \geq 0$ for all x. It is also proved that if a solution of the BVP satisfies the restriction $y' \geq 0$, then as $x \to \infty$, we have $y \sim x^{1/(1-\lambda)}, y' \sim$

$x^{\lambda/(1-\lambda)}, y'' \sim x^{-(1-2\lambda)/(1-\lambda)}$, where the statement $f \sim x^{1/(1-\lambda)}$ means that there exist positive constants A and B such that for x sufficiently large, $Ax^{1/(1-\lambda)} < y < Bx^{1/(1-\lambda)}$.

The solution is, moreover, unique if we insist that $y' \geq 0$ for all x. The solutions are compared with those of the Falkner-Skan equation. McLeod (1972)

57.
$$\mathbf{y''' - yy'' + (6/n^2)y'^2 - (6/n^2)y^2y' + (2/n^2)y^4 - 3y'y''/y - (2/n^2 - 2)y'^3/y^2 = 0.}$$

Putting $y = -u'/u$ and $u' = v^n$, we find that

$$v''' - 3v'v''/v = 0. \tag{1}$$

Therefore, it may be checked that (1) has solutions

$$v = \alpha(x - x_0)^{-1} + \sum_{i=1}^{\infty} \alpha_{4i+3}(x - x_0)^{4i+3} \tag{2}$$

or $v = C_1x + C_2$, where $\alpha, x_0, \alpha_3, C_1$, and C_2 are arbitrary constants. It follows that the function u is single valued only if $n \neq 0, \pm(4\ell + 1)$ with $\ell = 0, 1, \ldots$ in the given DE; hence the given DE has Painlevé property if $n \neq 0, \pm(4\ell + 1); \ell = 0, 1, \ldots$. Martynov (1985a)

58.
$$\mathbf{y''' - 2yy'' + 3y'^2 = 0.}$$

It may be verified that the given DE has poles as its only movable singularities. Bureau (1964a)

59.
$$\mathbf{y''' - 2yy'' + 3y'^2 - \{4/(36 - k^2)\}(6y'^2 - y^2)^2 = 0,}$$

where k is an integer greater than 6.

The solution of the given DE is parametrically given by

$$x = z_1(t)/z(t), y = 6z(t)\dot{z}(t)/\{z(t)\dot{z}_1(t) - z_1(t)\dot{z}(t)\},$$

where $z(t)$ and $z_1(t)$ are two linearly independent solutions of the hypergeometric equation

$$t(1 - t)\ddot{z} + (1/2 - (7/6)t)\dot{z} + \{(1/4)k^{-2} - 1/144\}z = 0,$$

where a dot denotes differentiation with respect to t. The given DE includes an equation due to Chazy as a special case; for this equation, $k \to \infty$. Cosgrove and Scoufis (1993)

60.
$$\mathbf{y''' - 2yy'' - 2y'^2 = 0.}$$

The given DE has an integral

$$y' = y^2 + Kx + K_1, \tag{1}$$

where K and K_1 are constants; Eqn. (1) is Riccati and hence may be solved through exact linearization. Bureau (1964a)

61.
$$\mathbf{y''' - 2yy'' - 4y'^2 + 2y^2y' = 0.}$$

Multiplying the given DE by $2y$ and integrating, we obtain

$$y'' - y'^2/(2y) - 2yy' + y^3/2 - K/y = 0, \tag{1}$$

where K is an arbitrary constant. To integrate (1), set $y' = y^2 + h + 2vy$, where h is given by $h^2 + 2K = 0$; therefore, $v' + v^2 = h/2$. Now putting $v = u'/u, y = -z'/z$, we get the system

$$u'' - (h/2)u = 0, \quad z'' - 2(u'/u)z' + hz = 0, \tag{2}$$

equivalent to (1); (2) can be solved sequentially. Bureau (1964a)

62. $$\mathbf{y''' - 3yy'' - 3y'^2 + 3y^2y' = 0.}$$

The given DE has a first integral

$$y'' - 3yy' + y^3 - K = 0, \tag{1}$$

where K is an arbitrary constant. For Eqn. (1), see Eqn. (2.13.16). Bureau (1964a)

63. $$\mathbf{y''' - 7yy'' + 11y'^2 = 0,}$$

where a is a constant.

By using the Painlevé's method of small parameters or otherwise, one may show that the given DE has logarithmic singularities in its solutions. Bureau (1964a)

64. $$\mathbf{y''' - (a/2)(2yy'' - 3y'^2) = 0,}$$

where a is a constant.

By making the change of variable $y \to \alpha y$, etc., one may choose $a = 2$ without loss of generality; now see Eqn. (3.5.58). Bureau (1964a)

65. $$\mathbf{y''' - (a/7)(7yy'' - 11y'^2) = 0,}$$

where a is a constant.

Letting $y \to \alpha y$, etc., one may choose $a = 7$ without loss of genarality; now see Eqn. (3.5.63). Bureau (1964a)

66. $$\mathbf{y''' - a(yy'' - 2y'^2) = 0,}$$

where a is a constant.

It is easily checked that the given DE has no movable poles in its general integral. Bureau (1964a)

67. $$\mathbf{y''' + \beta^{-1}yy'' - y'^2 + 1 = 0,}$$

$y(0) = y'(0) = 0, y' \to 1$ as $x \to \infty$; the coefficient β is assumed to be large.

A perturbation series $y(x) = y_0 + \beta^{-1}y_1 + \cdots$ leads to the following (first) two systems for y_0 and y_1:

$$\begin{aligned}
&y_0''' + 1 - y_0'^2 = 0,\\
&y_0(0) = y_0'(0) = 0, y_0'(\infty) = 1;\\
&y_1''' + y_0y_0'' - 2y_0'y_1' = 0,\\
&y_1(0) = y_1'(0) = 0, y_1'(\infty) = 0.
\end{aligned}$$

These systems are recursively solved to obtain

$$y_0(x) = x + 2(3)^{1/2} - 3(2)^{1/2}\tanh(x/2^{1/2} + c),$$

where $\tanh c = (2/3)^{1/2}, c = 1.146216$,

$$\begin{aligned} y_1(\theta) &= -2^{1/2}(\theta - c + 6^{1/2})/(5S^2) \\ &+ [3(2)^{1/2}S^2/2 - 12(2)^{1/2}/5]\int_c^\theta yT(y)dy - \{6(2)^{1/2}/5\}\{(6^{1/2} - c)\log S\} \\ &+ 161(2)^{1/2}\theta/40 + 7(2)^{1/2}T/(20S^2) - 167(2)^{1/2}T/140 \\ &- 13(2)^{1/2}\theta T^2/40 + K_2T^2/2 + [K_1 - \{8(2)^{1/2}/5\}\log S][3\theta/4 \\ &+ T/(8S^2) - 15\theta S^2/16 - 15T/16] + K_3, \end{aligned}$$

where $\theta = x/2^{1/2} + c$, $S = \operatorname{sech}\theta, T = \tanh\theta$, and the constants K_i $(i = 1, 2, 3)$ are

$$K_1 = [2(2)^{1/2}/5][4(\log 2 + 6^{1/2} - c) - 7] = 0.557587,$$

$$\begin{aligned} K_2 &= [K_1 - (4(2)^{1/2}/5)\log 3][15c/8 + 9(6)^{1/2}/8] \\ &+ 11(3)^{1/2}/2 + 13(2)^{1/2}c/20 = 1.748655, \end{aligned}$$

$$\begin{aligned} K_3 &= 2^{1/2}c(\log 3 - 457) - K_2/3 + 3(6)^{1/2}K_1/16 - 197(3)^{1/2}/60 \\ &- [9(3)^{1/2}\log 3]/10 = 2.112126. \end{aligned}$$

The skin friction as given by $\beta^{-1/2}y''(0)$ is obtained to good accuracy using this perturbation series solution. The results are compared with an exact numerical solution for $0.5 \le \beta \le 100$. Brodie and Banks (1986)

68.
$$\boldsymbol{y''' + \{(m+1)/(2m)\}yy'' - y'^2 + 1 = 0,}$$

$y(0) = 0, y'(0) = 0, y \sim x + C_0$ as $x \to \infty$, where m is a constant.

The given problem arises in mixed convection on a vertical surface. It is solved numerically for $m = 3/5$, yielding $y''(0) = 1.25913$ and $C_0 = -0.61782$. Merkin, Pop, and Mahmood (1991)

69.
$$\boldsymbol{y''' - a(yy'' + y'^2) = 0,}$$

where a is a constant.

Letting $y \to \alpha y$, etc., one may choose $a = 2$ without loss of generality; now see Eqn. (3.5.60). Bureau (1964a)

70.
$$\boldsymbol{y''' + cyy'' + my'^2 = 0,}$$

$y(0) = 0, y'(0) = -1, y'(\infty) = 0$, where c and m are constants and $c \neq 0$. A formal solution in the form

$$F(x) = r/c + r\sum_{i=1}^{\infty} b_i a^i e^{-irx} \tag{1}$$

when substituted in the given DE leads to the recursive formula

$$b_i = \{1/i^2(i-1)\}\sum_{\ell=1}^{i-1}[c\ell^2 + m\ell(i-\ell)]b_\ell b_{i-\ell}, i = 2, 3, \ldots .$$

If $|a| < 1$ and $|b_i| < 1, i = 1, 2, \ldots$, then it is shown that the series (1) converges absolutely for any $r > 0$ for $x = -\epsilon$; $\epsilon = -(\ln|a|/r + \delta) > 0$, where $\delta > 0$ is a sufficiently small number depending on a and r, and then this series also converges absolutely and uniformly on the half-axis $x > -\epsilon$.

To solve the given BVP it is sufficient that the equations

$$\begin{aligned} r/c + r\sum_{i=1}^{\infty} b_i a^i &= 0, \\ r^3 \sum_{i=1}^{\infty} i^2 b_i a^i &= -1 \end{aligned}$$

be satisfied for $r > 0$ and $|a| < 1$.

The convergence proof requires that $|c + m| \leq 4$, and $|c| \leq 2$.

A special case $c = 2/3, m = -1/3$ was computed for which all b_i were found to be positive. The series form of the solution gives a convenient way of estimating the error. Kravenchko and Yablonskii (1965)

71. $$\boldsymbol{y''' + Q[Ayy'' - y'^2] - \beta = 0, x \in [0, 1],}$$

where $A \in [0, \infty), Q > 0$, β real and $y(0) = y(1) = y''(1) = y''(0) + 1 = 0$.

The given system describes a similarity solution of equations for rectangular cavities and disks.

Changing the independent variable to $z = 1 - x$, we get the problem

$$y''' - Q[Ayy'' - y'^2] + \beta = 0, 0 \leq z \leq 1, \quad ' = \frac{d}{dz}, \tag{1}$$

$y(0) = y''(0) = y(1) = y''(1) + 1 = 0$.

The following specific results are proved:

(a) $\beta = 0$. Given $A \geq 0$, there exists at least one $Q \geq 0$ such that there exists at least one solution of the BVP (1).

(b) For given $Q > 0$ and for given $A \in [1, 2]$, there exist at least one number β and a convex function solving the given problem.

(c) Suppose that $A = 2$. Given $Q \geq 0$, there exists a unique number β such that the BVP (1) has a unique solution.

(d) Any nonconcave solution of the BVP (1) corresponds to $\beta > 0$.

(e) Suppose that $1 \leq A \leq 2$ and that $y(x)$ are convex downward, and solve (1). As $Q \to \infty$, then $y(x) \to 0$ and $y'(x) \to 0$ uniformly in $[0, 1]$, and $y''(x) \to 0$ uniformly in $[0, x_0]$ for any $x_0 \in (0, 1)$.

(f) For any $A \in [1, 2]$, there exists a number $Q_0 = Q_0(A)$ such that if the solutions of (1) are convex and $Q > Q_0$, then $\beta < 0$. In particular, $Q_0(2) \leq 6(e^6 - 1)$.

Shooting arguments are used. Lu (1990)

72. $$\boldsymbol{5y''' + b^3(6yy'' + 3y'^2) - 5\alpha b^2 = 0,}$$

$y(0) = y(1) = y'(1) = 0, y''(0) = -1$, where α is a known parameter and b is a constant.

The given problem arises in capillary flows in thin layers. It was solved numerically. For each α, two solutions were found; for one of these $b > 0$ and for the second $b < 0$. An asymptotic formula for relating b and α was found to be $b \approx (1.5/\alpha)^{1/2}$ as $\alpha \to \infty$. An asymptotic formula for $\alpha \to 0$ was constructed, assuming that $b > 0$. Batyshchev (1991)

73. $$\mathbf{5y''' + b^3(8yy'' - y'^2) - 5\alpha b^2 = 0,}$$

$y(0) = y(1) = y'(1) = 0, y''(0) = -1$, where α is a specified number and b is to be determined.

The given problem arises in thermal capillary flows in viscous layers and was solved numerically. The asymptotic formulas $b \approx \pm\{1.5\alpha^{-1}\}^{1/2}$ as $\alpha \to \infty$ and $b \approx 0.5574\alpha^{-1/3}$ as $\alpha \to 0$ (for $b > 0$) were found. Batyshchev (1991)

74. $$\mathbf{y''' + R(y'^2 - yy'') - K = 0,}$$

$y(0) = 0, y''(0) = 0, y'(1) = 0, y(1) = 1$, where $R < 0$ and K are constants.

The given problem arises in laminar flow in a uniformly porous channel with injection.

Several a priori estimates are obtained. It is then proved that the given problem has a unique solution for all $R < 0$. Numerical results are depicted. Shih (1987)

75. $$\mathbf{y''' - R(yy'' - y'^2) - Rk_1 = 0,}$$

$y(0) = y''(0) = y'(1) = 0,\ y(1) = 1; R$ and k_1 are constants.

The given system arises in laminar flow in channels with porous media. For $R \to \infty, y = x$ is a reasonable approximation in most channels, except very near the walls, where the boundary layer would slow up the fluid.

An approximate solution is obtained first by integrating the given DE as

$$y'' = R\left(yy' - 2\int_0^x y'^2\,dx + k_1 x\right), \tag{1}$$

where we use $y''(0) = 0, y(0) = 0$. Replacing the term yy' with xy' and $\int_0^x y'^2 dx$ with x, since $y = x$ is a good first approximation, we get from (1) the linear DE

$$y'' - R[xy' + (k_1 - 2)x] = 0, y(0) = y'(1) = 0, y(1) = 1. \tag{2}$$

The solution of (2) is

$$y = [\{x - I(x)\}/\{1 - I(1)\}] \text{ with } k_1 = [\{1 - 2I(1)\}/\{1 - I(1)\}], \tag{3}$$

where

$$I(x) = \int_0^x \exp[R(x^2 - 1)/2]dx.$$

For large R, it may be shown that

$$I(x) \le I(1) = R^{-1} + R^{-2} + \cdots. \tag{4}$$

Examining the approximations made in the DE, it is seen that terms of order R^{-1} have been neglected in comparison with terms of order unity. Thus, (3) is a good approximation provided that R is sufficiently large. Sellars (1955)

76. $$\mathbf{y''' - R(yy'' - y'^2) - k = 0, k < 0,}$$

$y(0) = y''(0) = y'(1) = 0, y(1) = 1$, where R is a given positive constant and k is real.

The given BVP describes incompressible laminar fluid flow in a two-dimensional channel with porous walls.

Choosing $k = k_1$, writing $\eta = Bx$ and $y = Au(\eta)$, the new form of DE is

$$u''' = uu'' - u'^2 - 1, ' \equiv \frac{d}{d\eta}, \tag{1}$$

provided that $AR = B, k_1 = -AB^3, A$ and B are nonzero constants. The problem for (1) is solved by using the IC

$$u(0) = u''(0) = 0, u'(0) = \nu, \tag{2}$$

where ν is some real number such that for some

$$\eta = \eta_* > 0, y'(\eta_*) = 0.$$

Conversely, if for some $\nu \in R$, the problem (1)–(2) has a solution with the property $y'(\eta_*) = 0$, then there exist R and k such that the given problem is solvable. To prove this, we note that η_* and $u(\eta_*)$ depend on ν; i.e., $\eta_* = \eta_*(\nu), u(\eta_*) = u_* = u_*(\nu)$. Then using the scaling w.r.t. A and B, etc., and the boundary conditions, we find that

$$A = 1/u_*(\nu), B = \eta_*(\nu), \tag{3}$$

$$R = \eta_*(\nu)u_*(\nu), k_1 = -[\eta_*(\nu)]^3/u_*(\nu). \tag{4}$$

With these values of R and k_1, the solution of the given problem can be found numerically using A and B from (3).

We also note from (1) and (2) that

$$u'' = -\exp[F(\eta)] \int_0^\eta [1 + u'^2(s)] \exp[-F(s)] ds, \tag{5}$$

where $F(\eta) = \int_0^\eta u(s)ds$; therefore, $u'' < 0$ for any $\eta > 0$. Moreover, for any $\nu \in (0, +\infty)$, the problem (1)–(2) has just one extremum. It also follows from (5) that for $\nu \leq 0$, a solution has no extremum.

For large ν, we make the substitution

$$u(\eta) = \nu^{1/2} V(\nu^{1/2}\eta), \nu^{1/2}\eta = z, V \in C^2[0, +\infty]$$

and find that $V(z)$ is a solution of the IVP

$$V''' = VV'' - V'^2 - 1/\nu^2, \tag{6}$$

$$V(0) = V''(0) = 0, V'(0) = 1. \tag{7}$$

Denoting the coordinates of the extremum of the solution of the problem (6)–(7) by $z_* = z_*(\nu), V(z_*) = V_*(\nu)$, we get from (4), etc., that

$$R = z_*(\nu)V_*(\nu), k = -[z_*(\nu)]^3/[\nu^2 V_*(\nu)].$$

The problems (1)–(2) for $0 \leq \nu \leq 1$ and (6)–(7) for $\nu \geq 1$ were solved numerically and depicted. Bespalova (1984)

77. $$\boldsymbol{y''' - Re(yy'' - y'^2) - k_1 = 0, k_1 > 0,}$$

where $Re > 0$ is a given constant.

Writing $\eta = Bx, y(x) = Au(\eta)$, we get

$$u''' = uu'' - u'^2 + 1, \quad ' \equiv \frac{d}{d\eta}, \tag{1}$$

provided that $ARe = B, k_1 = +AB^3 (A, B \neq 0$ are constants). Considering the IVP for (1) with

$$u(0) = u''(0) = 0, u'(0) = \nu, \nu \in R, 0 \leq \nu \leq 1, \tag{2}$$

and denoting the coordinates of the extremum of the solution by $\eta = \eta_*, u = u_*$ where $u'(\eta_*) = 0$, we may from the scaling relations obtain

$$Re = \eta_*(\nu)u_*(\nu), k_1 = [\eta_*(\nu)]^3/u_*(\nu).$$

Thus, if the BVP has a solution, then IVP (1)–(2) for some $\nu \in R$ necessarily has a point where $u'(\eta_*) = 0$. Conversely, if for some $\nu \in R$, the problem (1)–(2) has a solution with the foregoing property, then there exist Re and k_1 such that the original problem has a solution. To prove this, we note that $\eta_* = \eta_*(\nu), u(\eta_*) = u_*(\nu)$; then from the scaling relations and boundary conditions, we have

$$A = 1/u_*(\nu), \quad B = \eta_*(\nu),$$
$$Re = \eta_* u_*(\nu), \quad k_1 = [\eta_*(\nu)]^3/u_*(\nu).$$

Relations $y(0) = y''(0) = y'(1) = 0, y(1) = 0$ are parametric equations of a curve in the (Re, k_1) plane, for which the given problem is solvable. If $\nu > 1$, we transform according to

$$u(\eta) = \nu^{1/2}V(\nu^{1/2}\eta), \nu^{1/2}\eta = z, V \in C^2[0, \infty),$$

so that

$$V''' = VV'' - V'^2 + 1/\nu^2, \tag{3}$$

$$V(0) = V''(0) = 0, V'(0) = 1 \tag{4}$$

and $Re = z_{1*}(\nu)V_{1*}(\nu), k_1 = [z_{1*}(\nu)]^3/[\nu^2 V_{1*}(\nu)]$, where $z_{1*}(\nu), V_{1*}(\nu)$ are the coordinates of the extremum of the solution of the problem (3)–(4).

We find that if $\nu = \pm 1$, there are special solutions of the system (1)–(2), namely, $u = \pm\eta$, for which there are no extrema. If we write $F(\eta) = \int_0^\eta u(s)ds$, we reduce (1)–(2) to

$$u''(\eta) = \exp[F(\eta)] \int_0^\eta \exp[-F(s)][1 - u'^2(s)]ds, \tag{5}$$

$$u'(\eta) = \nu + \int_0^\eta u''(s)ds. \tag{6}$$

It follows that the IVP (1)–(2) for $\nu \in (-\infty, -1)$ has no extremum, while if $\nu \in (+1, +\infty)$, it has one extremum. Numerical results show that for $\nu \in (-1, 0)$, the solution has two extrema, and for $\nu \in [0, 1]$, one extremum.

We can also deduce from (5) and (6) that if $u''(\eta_1) < 0, u'(\eta_1) < -1$, for some $\eta = \eta_1$, then $u''(\eta) < 0, u'(\eta) < -1$ for any $\eta > \eta_1$, and the solution decreases rapidly, tending to $-\infty$. The existence of such values of η_1 has been confirmed numerically.

Computer results are given. Bespalova (1984)

78. $$\boldsymbol{y''' + \lambda(yy'' - y'^2) + 1 - y = 0,}$$

where λ is a positive parameter and either $y(0) = 0, y'(0) = 0, y(\infty) = 1$ or $y(0) = 0, y''(0) = 0, y(\infty) = 1$.

(a) The given BVP arises from a similarity solution of the quasigeostrophic potential vorticity equation for a one-layer ocean solution. Using geometric techniques from qualitative theory of differential equations, the existence of at least one solution of each of the BVP is proved.

(b) The given problem arises in physical oceanography. A thorough numerical study is presented. The analysis gives weakly nonlinear solutions for small λ in powers of λ for (1) and (2); the limit of large positive λ is also presented by using matched asymptotic expansions. Comparison with exact numerical solution is depicted.

Dunbar (1993), Ierley and Ruehr (1986)

79. $$y''' + (2/m)yy'' - y'^2/m + 2k/m = 0,$$

$0 < x < \Delta, y(0) = y'(0) = 0, y(\Delta) = 1/2, y'(\Delta) + my''(\Delta) = 1$; Δ and m are constants.

The given system describes flow of a viscous gas in the neighborhood of the critical line of a blunted axisymmetric body when a hypersonic gas flows around it.

An iterative process, requiring integral equation formulation, is proposed to solve the problem numerically for all $m > 0$, and $k > 0$, which is fairly large. Convergence of the process is proved under certain assumptions. Titov (1980)

80. $$y''' + byy'' - ky'^2 + kh^2 = 0,$$

where b, k, and h are constants.

The given DE appears in the boundary layer theory and has an exact solution

$$y = -(2/k)x^{-1} + hx$$

for $b = -k$. Burde (1990)

81. $$y''' - (p - Dy)y'' + qy' + Ky'^2 - Myy' = 0,$$

where $\lim_{x\to\infty} y'(x) = 0$. p, q, D, K, and M are parameters.

A formal series solution is found in the form

$$y = r_1 + \{6r^2/(Dr + M)\} \sum_{\ell=1}^{\infty} b_\ell a^\ell e^{-\ell rx}, \tag{1}$$

where $r \neq -M/D$ and a are parameters. Substituting (1) in the given DE, we have

$$\sum_{\ell=1}^{\infty} [-r^2\ell^3 - pr\ell^2 - q\ell + (Dr\ell^2 + M\ell)r_1]b_\ell a^\ell e^{-\ell rx}$$

$$+ \{6r^2/(Dr + M)\} \sum_{\ell=2}^{\infty} \sum_{k=1}^{\ell-1} [Drk^2 + Krk(\ell - k) + Mk]b_k b_{\ell-k} a^\ell e^{-\ell rx} = 0. \tag{2}$$

Hence, with $\ell = 1$,

$$r_1 = \frac{r^2 + pr + q}{Dr + M} \tag{3}$$

if b_1 is fixed and r and a are arbitrary parameters. Using (3) in (2), we get

$$b_\ell = \frac{6r\sum_{k=1}^{\ell-1}\left[Dk^2r + Kk(\ell-k)r + Mk\right]b_k b_{\ell-k}}{\ell(\ell-1)[D\ell r^2 + M(\ell+1)r + Mp - Dq]}, \ell = 2, 3, \ldots.$$

We thus have a formal two-parameter family of solutions of the given DE in the form (1). It is known that if the series in (1) converges absolutely when $r > 0$ for some x_0, this series converges absolutely and uniformly in the half-plane Re $x \geq$ Re x_0 and represents an analytic $2\pi i/r$ periodic function $y = y(x)$ such that $\lim_{\mathrm{Re}\, x\to+\infty} y'(x) = 0$. The range of values of parameters occurring in the given DE is determined for which the solution exists. This family can then be used to determine the range of admissible initial values $y(0) = a, y'(0) = b$ [or $y''(0) = b$]. Kravchenko and Yablonskii (1972)

82. $$y''' - (p - Dy)y'' + qy' + Ky'^2 - Myy' = 0,$$

where p, q, D, K, and M are constants:

First we note some special simple solutions:

(a) If $p = q = M = 0, D = -2$, and $K = 3$, the given DE has special solutions $y = A/(x - x_0)^2 - 6/(x - x_0)$, where A and x_0 are constants.

(b) If $p = q = M$ and $D = K$, the given DE integrates twice to give

$$y' = -Ky^2/2 + Cx + r^2/2,$$

which for $C = 0$ and $K = 1$, has the two-parameter family of solutions

$$y = r(1 + ae^{-rx})/(1 - ae^{-rx}),$$

$rx_n = \ln|a| + i(\arg a + 2\pi n)$ are the poles of the solution, where a is arbitrary.

(c) If $p = q = M = D = 0$, and $K = 1$, the given DE becomes, with $y' = u$,

$$u'^2 + (2/3)u^3 = C_1. \tag{1}$$

If $C_1 = 0$, the given DE has the two-parameter family of solutions

$$y = 6/(x + C_2) + C_3.$$

If $C_1 \neq 0$, we can reduce (1) to

$$w'^2 = 4w^3 - b, \tag{2}$$

where $y' = -6w, b = -C_1/36$: solving this DE and using the fact that $y' = -6w$, we obtain the solution of the given DE,

$$y = -6\int P(x + C_2; 0, b)dx + C_3,$$

where the elliptic function $P(z)$ is a doubly periodic function, having a second order pole with residue zero in each period parallelogram.

The given system is equivalent to

$$\begin{aligned} y_1' &= y_2, y_2' = y_3, \\ y_3' &= py_3 - qy_2 - Dy_1y_3 + My_1y_2 - Ky_2^2. \end{aligned} \tag{3}$$

Substituting $x = x_0 + z, y_j = [1/(x - x_0)^j][A_j + \phi_j](j = 1, 2, 3)$, where x_0 is an arbitrary complex number and $A_1 = 6/(2D+K), A_2 = -6/(2D+K), A_3 = 12/(2D+K), 2D+K \neq 0$, we get the following Briot–Bouquet system in ϕ_j $(j = 1, 2, 3)$:

$$\begin{aligned} z\frac{d\phi_1}{dz} &= \phi_1 + \phi_2, \\ z\frac{d\phi_2}{dz} &= 2\phi_2 + \phi_3, \end{aligned} \tag{4}$$

$$z\frac{d\phi_3}{dz} = -2A_1D\phi_1 + 2A_1K\phi_2 + (3 - A_1D)\phi_3 + f(z, \phi),$$

where

$$f(z, \phi) = -K\phi_2^2 - D\phi_1\phi_3 + (2A_1p - MA_1^2 - MA_1\phi_1 + MA_1\phi_2 + p\phi_3 + M\phi_1\phi_2)z + (A_1q - q\phi_2)z^2.$$

The system (4) is studied in detail with regard to the nature of solutions about movable singularities, and hence conclusions with respect to the original system are drawn. In particular, if D and K are real constants, and the (D, K) plane is divided into sectors by the lines:

(1) $2D + K = 0$,
(2) $4(2 + 6^{1/2})D + (7 + 6^{1/2})K = 0$,
(3) $8D + 7K = 0$, and
(4) $4(6^{1/2} - 2)D - [7 - 2(6)^{1/2}]K = 0$,

then the given DE, for fixed x_0 and with D and K in the sector:

(a) Between the lines (1) and (2) $(\alpha_1 < 0, \alpha_2 = 0)$ has a unique solution, holomorphic in the neighborhood of $x = x_0$.

(b) Between the lines (2) and (3) $(\alpha_1 \leq 0, \alpha_2 \neq 0)$ has two one-parameter families of solutions

$$y_j = \left\{1/(x - x_0)^j\right\} \left\{A_j + \sum_{\ell+p>0} a_{\ell p}^{(j)}(x - x_0)^{\ell+p\lambda_r}\right\}$$

$(j = 1, 2, 3, r = 1, 2)$ of the equivalent system (3) in y_j.

(c) Between the lines (3) and (4) $(\alpha_1 > 0, \alpha_2 \neq 0)$ has a two-parameter family of solutions

$$y_j = \left\{1/(x - x_0)^j\right\} \left\{A_j + \sum_{\ell+p+s>0} a_{\ell ps}^{(j)}(x - x_0)^{\ell+p\lambda_2+s\lambda_3}\right\} \tag{A}$$

$(j = 1, 2, 3)$ for the equivalent system (1).

(d) Between the lines (1) and (4) ($\alpha_1 > 0, \alpha_2 = 0$) has a two-parameter family of solutions (A), but logarithms can appear at a countable set of straight lines

$$D + [\{n^2 - 7n + 6\}/\{2(n^2 - 4n + 6)\}]K = 0, n = 1, 2, 3, \ldots$$

in the representaion of solutions. α_1 and α_2 are real and imaginary parts in the eigenvalues λ_3 of the linearized system for $y_j : \lambda_1 = -1, \lambda_{2,3} = \alpha_1 \mp i\alpha_2 = (1/2)[7 - A_1D \pm \{(7 - A_1D)^2 - 24\}^{1/2}]$.

Shemyakina (1975)

83.

$$y''' - ayy'' - ay'^2 - 3a^2y^2y' = 0,$$

where a is a constant.

The transformation $y \to \alpha Y, a\alpha = -1$, takes the given DE to Eqn. (3.5.16). Bureau (1964a)

84.

$$y''' - ayy'' - 2ay'^2 + (a^2/2)y^2y' = 0,$$

where a is a constant.

Changing the variable according to $y \to \alpha y$, etc., one may choose $a = 2$ without loss of genarality; now see Eqn. (3.5.61). Bureau (1964a)

85.

$$y''' - ayy'' - 2ay'^2 - 2a^2y^2y' = 0,$$

where a is a constant.

Transforming the given DE according to $y \to \alpha y, a\alpha = 1$, etc., one may take $a = 1$ without loss of generality. Now, see Eqn. (3.5.53). Bureau (1964a)

86.

$$y''' - ayy'' - 5ay'^2 + a^2y^2y' = 0,$$

where a is a constant.

Transforming according to $y \to \alpha Y, a\alpha = 1$, etc., takes the given DE to Eqn. (3.5.54) without loss of generality. Bureau (1964a)

87.

$$y''' + \{6/v^{1/2}\}yy'' + (12/v)y^2y' + (6/v^{1/2})y'^2 - 4y' = 0,$$

where v is a constant.

The given DE arises from a traveling wave form of Sharma–Tasso–Olver equation $u_t + 3u_x^2 + 3u^2u_x + 3uu_{xx} + u_{xxx} = 0$. It has a first integral

$$y'' + \{6/v^{1/2}\}yy' + (4/v)y^3 - 4y = 0, \tag{1}$$

if we set constant of integration equal to zero. Equation (1) has a solitary wave solution

$$y = \{v^{1/2}/2\}\{(\tanh x)^{\mp 1} \mp 1\}. \tag{2}$$

Hereman, Banerjee, Korpel, and Assanto (1986)

88.

$$y''' - ayy'' - by'^2 - cy^2y' - dy^4 = 0,$$

where a, b, c, and d are constants.

Conditions on the constants a, b, c, and d are found such that the DE has poles as its only movable singularities. Bureau (1964a)

89.
$$y''' - Ayy'' - By'^2 - Dy^2y' - Ey^4 = 0,$$

where A, D, B, and E are constants.

It is proved that for the given DE to be invariant under

$$T : (x,y) \to [(ax+b)/(cx+d); \{\Delta/(cx+d)^2\}w - 6c/(cx+d)],$$

where a, b, c, and d are arbitrary constants such that $\Delta = ad - bc \neq 0$, it is necessary that $A = 2, D + 12E = 0$, and $B + 3(D+1) = 0$.

The proof is by direct substitution. The given DE is transformed to

$$w''' - \Delta(2ww'' - 3w'^2) - \Delta E(6w' - \Delta w^2)^2 = 0.$$

Seeking solutions of the given DE in the form

$$y(x) = \sum_{j=-k}^{\infty} \alpha_j (x - x_0)^j$$

by substituting the latter in the former, we find that either $k = 0$ or $k = 1$; hence the series may be found. Sidorevich and Lukashevich (1990)

3.6 $y''' + f(x,y,y')y'' + g(x,y,y') = 0$

1.
$$3y''' + (n+5)xy'' - (4n+2)y' = 0,$$

$y'(\infty) = y(0) = 0, y''(0) = \pm 1$, where n is an integer.

For $n = 0$, numerical solutions of the given problem give $y'(0) = -0.7743, +0.7743$ for $y'' = +1, -1$, respectively. See Eqns. (3.5.72) and (3.5.73) for the nonlinear counterparts. Batyshchev (1991)

2.
$$\left(\frac{d^2}{dx^2} + \frac{2}{x}\frac{d}{dx}\right)\left(\frac{d}{dx} + \frac{2}{x}\right)y + \left(\frac{d}{dx} + \frac{2}{x}\right)y = -y^2,$$

$y(0) = 0, \lim_{r\to\infty} y(r) = 0$.

The given BVP is treated quite extensively using elementary methods. Write

$$y(x) = \sum_{i=0}^{\infty} a_i x^i, \tag{1}$$

since $y(0) = 0; a_1$ is the only parameter. Numerical solution of the problem is depicted graphically. In the (u, u') plane where $u = yx, u' = (yx)'$, it can be shown that the spiraling solutions converge slowly to zero. Numerical experimentations suggest that there is an interval $a_1 \in [a_{1\,\min}, a_{1\,\max}], a_{1\,\min} = -0.3402, a_{1\,\max} = 0.3306$ for which the solution tends to 0 as $x \to \infty$, while for a_1 outside this interval, it blows in a finite interval.

The series (1) is derived and bounds on a_n are obtained. Thus, $y(x_0)$ for small $x_0 (x_0 \simeq 2)$ can be obtained; a rigorous bound on the truncation error is found. For large $x \geq$

$x_1 (x_1 = 80)$, a three-parameter family of asymptotic expansions of the form $y_{\text{asymp}} = \sum_{n=1}^{\infty} x^{-n} \sum_{k=-n}^{n} a_{nk}(\rho) e^{ikx}, \rho = \log x$ exists. Here the coefficients a_{nk} for $|k| \leq 1$ satisfy a first order system of ODEs, while for $|k| > 1$ they satisfy algebraic equations. The three real parameters are the real number $a_{10}(\rho_1)$ and the complex number $a_{11}(\rho_1) = \bar{a}_{1,-1}(\rho_1)$, where $\rho_1 = \log x_1$. The asymptotic expansion is a divergent one, but an asymptotic formula is obtained and effective bounds for the truncation error are also determined.

This is a large paper with extensive figures, computer programs, etc. Michelsen (1990)

3. $$\boldsymbol{y''' - (1/x)y'' - 12yy' + (6/x)y^2 + 4Kxy - K_1/x + (3/2)K^2x^3 = 0,}$$

where K and K_1 are constants.

The transformation $x = \alpha_1 X, y = \beta_1 Y$, where $\alpha_1^2 \beta_1 = 1, K\alpha_1^4 = -2$, reduces the given DE to the canonical form

$$y''' - 12yy' - (1/x)(y'' - 6y^2 - K) - 8xy + 6x^3 = 0,$$

where K is another constant. Now go to Eqn. (3.6.4). Bureau (1964a)

4. $$\boldsymbol{y''' - (1/x)y'' - 12yy' + (1/x)(6y^2 + K) - 12xy + 6x^3 = 0,}$$

where K is a constant.

The given DE is equivalent to the system

$$\begin{aligned} y'' &= 6y^2 + K + z, \\ xz' &= z + 12x^2y - 6x^4 \end{aligned} \tag{1}$$

in y and z. The system (1) and hence the original DE may be shown to possess poles as the only moving singularities in its solution. Bureau (1964a)

5. $$\boldsymbol{y''' + \{r'(x)/r(x)\}y'' + \{q(x)/r(x)\}(y')^{\gamma} + \{p(x)/r(x)\}y^{\beta} - f(x)/r(x) = 0,}$$

where r, p, q, and f are real-valued functions on $[0, \infty)$ such that $r(x) > 0, p(x) \leq 0, q(x) \leq 0, f(x) \geq 0; \beta > 0$ is ratio of odd integers and $\nu > 0$.

Real solutions are considered on the half-line $[X, \infty)$, where $X \geq 0$ depends on the particular solution, and are nontrivial in any neighborhood of infinity. Sufficient conditions are given such that all solutions of the given DE are nonoscillatory. Asymptotic behavior of the nonoscillatory solutions is also studied. For example, it is shown that if $p(x) \equiv 0$, all solutions of the given DE are nonoscillatory. Several such theorems are given. Parhi (1981)

6. $$\boldsymbol{y''' + R(y'^2 - yy'') - M^2y' - K = 0,}$$

$y(0) = 0, y''(0) = 0, y(1) = 1, y'(1) = 0$, where R, M, and K are constants.

The given DE arises in a laminar flow in a uniformly porous channel with an applied transverse magnetic field. A series solution of the given system for small R and arbitrary M is

$$\begin{aligned} y = {} & (A/M)(\sinh Mx - Mx \cosh M) \\ & + \{RA^2 \cosh M/(4M^2)\}\{5AM \cosh M(\sinh Mx - x \sinh M) \\ & + M(6 - x^2) \sinh Mx + 7x \cosh Mx - x(5M \sinh M + 7 \cosh M)\}, \end{aligned}$$

where

$$K = AM^2 \cosh M + (RA^2/4)[4 + (11 + 5M^2 + 5AM^2 \cosh M) \cosh^2 M + 5M \sinh M \cosh M] \quad \text{and} \quad A = M/\{\sinh M - M \cosh M\}.$$

Shrestha (1968)

7. $$y''' + (x/2 + y)y'' + y'(1 - y') = 0,$$

$y(0) = y''(0) = y'(\infty) = 0$.

The given problem describes the asymptotic form of solutions for a Darcian free convective flow in some parametric domain. It is easy to see that an infinite number of solutions of this problem can exist, since, in general, as $x \to \infty, y' \sim Ax^{-2} + Bxe^{-x^2}$ for some constants A and B.

Considering, however, the physical nature of the problem, solution with $A = 0$ only are sought, so that they decay exponentially as $x \to \infty$. To find the specific solution, $y'(0) = k$, say, is specified, and the given DE is then solved using the Runge–Kutta–Merson method, starting at $x = 0$ and integrating to a sufficiently large value of x so that the constant A may be determined. The value of k is adjusted until the required solution for which $A = 0$ is obtained. This value is found to be $k = 0.355129$ and gives $y(\infty) = 0.559830$. Ingham and Pop (1988)

8. $$y''' + (Qy - 1)[y''/x - y'/x^2] - Qx(f'/x)^2 - \beta x = 0,$$

$y(0) = y(1) = 0$ and $(y'/x)'|_{x=0} = (y'/x)'|_{x=1} - 1 = 0$, where Q and β are constants.

The given problem arises from flows in a cylindrical floating zone. Following earlier numerical solutions for $0 \le Q \le 32.7$ and $Q \ge 1749$, existence of at least one solution of the given problem for $[0, Q_0)$, with Q_0 sufficiently small, is proved by using the Schauder fixed-point theorem. The nonautonomous nature of the given DE makes the proof more complex and delicate. All the solutions are such that $y(x) < 0$ on $(0, 1)$ with y' vanishing just once on $(0, 1)$. Lu and Kazarinoff (1989)

9. $$y''' + 3y^2y'' + 6yy'^2 = 0.$$

A second integral of the given DE is

$$y^3 + y' + c_1x + c_2 = 0,$$

where c_1 and c_2 are arbitrary constants. Wilson (1971), p.33

10. $$y''' + 3(y^2y'' + 3yy'^2) + 3y^4y' + Vy' = 0,$$

where V is a constant.

Ignoring the constant solution, we first note that if y is a solution, so is $-y$.

Integrating once, we find that

$$2yy'' - y'^2 + 6y^3y' + y^6 + Vy^2 + B = 0, \tag{1}$$

where B is an arbitrary (integration) constant. Now we set

$$y(x) = f(x)/[2F(x)]^{1/2}, \tag{2}$$

with

$$F'(x) = f^2(x)$$

and obtain

$$2f''f - f'^2 + Vf^2 + 2BF = 0. \tag{3}$$

On differentiation, (3) yields the linear ODE $f''' + Vf' + Bf = 0$. The general solution of this equation is

$$f(x) = \sum_{j=1}^{3} A_j \exp(p_j x), \tag{4}$$

where the three parameters p_j are the roots of the cubic $p^3 + Vp + B = 0$, so that the following relations hold:

$$\begin{aligned} p_1 + p_2 + p_3 &= 0, \\ p_1p_2 + p_2p_3 + p_3p_1 &= V, \\ p_1p_2p_3 &= -B. \end{aligned} \tag{5}$$

It also follows from (2) and (4) that

$$F(x) = A_0^2 + \sum_{j=1}^{3} \sum_{k=1}^{3} \left[\frac{A_jA_k}{p_j + p_k} \exp[(p_j + p_k)x]\right],$$

or

$$F(x) = A_0^2 - 2\sum_{j=1}^{3} \left[\frac{A_{j+1}A_{j+2}}{P_j} \exp[(-p_jx)]\right] + \frac{1}{2}\sum_{j=1}^{3}\left[\frac{A_j^2}{P_j} \exp[(2p_jx)]\right]. \tag{6}$$

To get this second form we use first of (5) and cyclic convention $A_{j+3} \equiv A_j$. In the above, we assume that none of the quantities p_j vanishes. Insertion of the expression (6) for $F(x)$ and the corresponding expression (4) of $f(x)$ in (3) yields, on use of $p^3 + Vp + B = 0, BA_0^2 = 0$. So two cases arise: Either $B = 0$ and A_0 is an arbitrary, nonvanishing constant, or $A_0 = 0$ and B is arbitrary.

(a) $B = 0, A_0 \neq 0$. In this case one of the p_j's, say p_3, vanishes and we have $p_1 = -p_2 = p$, say, and $p = (-V)^{1/2}$. The solution of the original equation becomes

$$y(x) = [B_1 \exp(px) + B_2 \exp(-px) + B_3]/\{2F_1(x)\}^{1/2}, \tag{7}$$

with

$$\begin{aligned} F_1(x) &= 1 + (B_3^2 + 2B_1B_2)x + (2p)^{-1}\{B_1^2 \exp(2px) \\ &\quad - B_2^2 \exp(-2px) + 4B_3[B_1 \exp(px) - B_2 \exp(-px)]\}. \end{aligned}$$

The solution (7) depends on the three arbitrary constants $B_j, B_j = A_j/A_0, j = 1, 2, 3$, and the parameter $p = (-V)^{1/2}$.

(b) $A_0 = 0$; in this case , it is possible to write the solution of the original equation directly as

$$y(x) = \sum_{j=1}^{3} A_j \exp(p_jx) \left[2\sum_{j=1}^{3}\sum_{k=1}^{3} \frac{A_jA_k}{p_j + p_k} \exp[(p_j + p_k)x]\right]^{-1/2}.$$

Here, as before, p_j are the three roots of the cubic $p^3 + Vp + B = 0$. This solution for arbitrary V depends also on the three arbitrary constants, namely, B and two ratios of any two of the three A_j's to the third. Calogero (1987)

11. $$\boldsymbol{y''' + 3(y^2y'' + 3yy'^2) + 3y^4y' + xy'/3c + y/6c = 0.}$$

A first integral, after multiplication by y, can be found to be

$$2yy'' - y'^2 + 6y^3y' + y^6 + xy^2/3c = B/3c,$$

where B is constant of integration. Setting

$$y(x) = g(x)/[2G(x)]^{1/2}, \tag{1}$$

with

$$G'(x) = g^2(x),$$

we obtain

$$xg^2 + 3c(2g''g - g'^2) = BG. \tag{2}$$

Equation (2) on differentiation and use of (1) yields the linear equation

$$6cg''' + 2xg' + (1 - B)g = 0. \tag{3}$$

Equation (3) may be solved by the Fourier or Laplace transform method. In the special case $B = 1$, we immediately write

$$g'(x) = h[-(3c/2)^{-1/3}x],$$

with $h(z)$ an Airy function satisfying the second order linear ODE

$$h''(z) = (z/2)h(z). \tag{4}$$

For the special case $B = 0$, it is convenient to solve (2) directly, by setting $g(x) = \{h[-(6c)^{-1/3}x]\}^2$ and getting again for $h(z)$ the Airy equation (4). Calogero (1987)

12. $$\boldsymbol{y''' - y^ny'' + (n + 1)y^{n-1}y'^2 = 0,}$$

where n is a positive integer.

The given DE, due originally to Chazy, has an asymptotic expansion about a movable logarithmic branch point:

$$y \sim [\{\log(x - x_0)\}/(x - x_0)]^{1/n} \sum_{j=0}^{\infty} \sum_{k=0}^{j} a_{jk}\{\log\log(x - x_0)\}^k/[\log(x - x_0)]^j$$

as $x \to x_0$ in some sector $\beta < \arg(x - x_0) < \gamma$. Here x_0 and a_{10} are arbitrary constants, while the remaining coefficients a_{jk} are determined by a recursion relation. To get the third constants, one must use the α-test, requiring a complicated analysis. Cosgrove and Scoufis (1993)

13. $$y''' + [af(y) + b]y'' + af''(y)y'^2 + a[bf(y) - \alpha]y' + \beta f(y) = 0,$$

where a, b, α, and β are constants.

Subject to conditions $f(0) = 0, f'(y) > 0, yf''(y) \geq 0, y^{-1}f(y) \to \infty$ as $y \to \infty, ab^2 - 4\beta \neq 0$ and $a(af'(0) + b)(bf'(0) - \alpha) - \beta f'(0) < 0$, a rigorous proof of the existence of a periodic solution is given. The method is mainly topological. Colombo (1950)

14. $$y''' - P_1(x,y)y'' - P_2(x,y)y'^2 - P_3(x,y)y' - P_4(x,y) = 0,$$

where $P_n(x, y)$ is a polynomial in y of degree n with analytic coefficients in x.

All equations of this form are determined, whose general integral has no movable critical points. Bureau (1964)

15. $$y''' + 6y'y'' - g_1 x - g_2 = 0,$$

where g_1 and g_2 are constants. Boiti and Pempinelli (1980)

16. $$y''' + y''(1/x + 3y'/y) - y'/x^2 - y'/y^3 + (3/c)y'/y = 0.$$

The given DE appears as descriptor of solitary waves in two-dimensional magma dynamics. A straightforward integration yields

$$y^3(y'' + y'/x) + 3\int_x^\infty (1/r)y^2 y'^2 dr = (1/c)(1 - y^3) - (1 - y)$$

if $y \to 1$, and y' and $y'' \to 0$ as $x \to \infty$. Multiplying by y', dividing by y^2, and integrating, we get, after interchanging two integrations,

$$y'^2 + \int_x^\infty y'^2\{1 - 3y^2(r)/y^2(x)\}(1/r)dr = [(y-1)^2/y^2]\{c - (2y+1)\}/c.$$

We have used the conditions at $x = \infty$. Another integral is obtained by dividing the given differential equation by y^2 and integrating:

$$yy'' + y'^2 + yy'/x + \int_x^\infty (y'^2/r)dr + (y-1)(3/c - 1/y) = 0.$$

It is shown that if $c \geq 3$, and the solution of the original system exists, then $y \geq 1$. Barcilon and Lovera (1989)

17. $$y''' + y''(2/x + 3y'/y) - 2y'/x^2 - y'/y^3 + (3/c)y'/y = 0,$$

where c is a constant.

The given DE describes solitary waves in three-dimensional magma dynamics. This DE can be integrated twice. Follow Eqn. (3.6.16). Barcilon and Lovera (1989)

18. $$y''' - y'y''/y - gyy' = 0,$$

where g is a constant.

By a suitable change of variables, we may transform the given DE to one with $g = 6$:

$$y''' - y'y''/y - 6yy' = 0. \tag{1}$$

Equation (1) has a first integral

$$y'' = y'^2/y + 2y^2 + C/y, \tag{2}$$

where C is an arbitrary constant. Equation (2), and hence, the given DE, may be shown to possess only fixed critical points. For $C = 0$ in (2), see Eqn. (2.32.7). Exton (1973)

19. $$\mathbf{y''' - y'y''/y - gy^2y' = 0,}$$

where g is a constant.

By a simple change of variables, we may transform the given DE such that $g = 4$;

$$y''' - y'y''/y - 4y^2y' = 0. \tag{1}$$

Equation (1) has a first integral

$$y'' = 2y^3 + Cy, \tag{2}$$

where C is an arbitrary constant. Equation (2) may be shown to have fixed critical points in its solution; the same is, therefore, true of the given DE. See Eqns. (2.4.40) for the solution of (2). Exton (1973)

20. $$\mathbf{y''' - (y'/y + 3)y'' + y'^2/y + 2y' = 0.}$$

The given DE is equivalent to the system described in Eqn. (3.17.1). Whittaker (1991)

21. $$\mathbf{y''' - (y + y'/y)y'' - 4y^2y' + 2y^4 = 0.}$$

Putting $y = -u'/u$, we check after one integration that

$$u''' - Cu^2u' = 0, \tag{1}$$

C an integration constant. Equation (1), after integrating once, or otherwise, can be shown to have a meromorphic solution; hence the same is true of the given DE since

$$y = -u'/u.$$

Martynov (1985a)

22. $$\mathbf{y''' - (y + y'/y)y'' - 2y'^2 = 0.}$$

The given DE may be shown to have moving critical points and is therefore not of Painlevé type. Martynov (1985a)

23. $$\mathbf{y''' - (2y + y'/y)y'' - y^2y' + y^4 = 0.}$$

If we put $z = y - y'/y$, we obtain the system

$$y = -(z'' - 2zz')/(z' - z^2), \tag{1}$$

$$z''' - zz'' - 2z'^2 - 2z^2z' = 0. \tag{2}$$

Equation (2), according to Chazy (1911), has the Painlevé property. Therefore, so has y as a solution of the given DE, according to (1). Martynov (1985a)

24. $\boldsymbol{y''' - [(4-p)y + y'/y]y'' - (2p-4)y^2y' = 0, p = -2, 1, 3, 4, 6.}$

Integrating, we have

$$y'' = (4-p)yy' + (p-2)y^3 + Cy.$$

This belongs to the class considered by Chazy (1911) and has the Painlevé property only for $p = 1$ and 4. Martynov (1985a)

25. $\boldsymbol{y''' - (ey + y'/y)y'' - 2e^2y^2y' = 0,}$

where e is a constant.

By a suitable change of variables, we may transform the given DE such that $e = -1$;

$$y''' - y'y''/y + yy'' - 2y^2y' = 0. \tag{1}$$

Equation (1) has a first integral

$$y'' + yy' = y^3 + Cy, \tag{2}$$

where C is an arbitrary constant. Equation (2) may be shown to have fixed critical points in its solution; the same is, therefore, true of the given DE. See Eqn. (2.10.3) for the solution of (2). Exton (1973)

26. $\boldsymbol{y''' - (ey + y'/y)y'' - 4e^2y^2y' + 2e^3y^4 = 0,}$

where e is a constant.

By suitable scaling we may transform the given DE to one for which $e = -1$:

$$yy''' - y'y'' + y^2y'' - 4y^3y' - 2y^5 = 0. \tag{1}$$

With $y' = uy^2, u' = yt$, Eqn. (1) is equivalent to the system

$$t\frac{dt}{du} + 6ut + 4u^3 + t + 2u^2 - 4u - 2 = 0, \tag{2}$$

$$\frac{u''}{u'^2} - \frac{1}{t}\frac{dt}{du} + u/t. \tag{3}$$

Putting $t = 1/v$ in (2), we have an Abel equation:

$$\frac{dv}{du} = (6u+1)v^2 + (4u^3 + 2u^2 - 4u - 2)v^3. \tag{4}$$

Equation (4) does not have solutions with movable poles only as their singularities. Tracing back, we conclude that the same is true of the given DE. Exton (1973)

27. $\boldsymbol{y''' - (ey + y'/y)y'' - 2ey'^2 = 0,}$

where e is a constant.

The given DE has a first integral

$$y'' = y'^2/y + eyy' + c/y, \quad (1)$$

where c is an arbitrary constant. Equation (1) may be shown to have movable critical points. The same is, therefore, true of the original DE. For $c = 0$ in (1), see Eqn. (2.31.70). Exton (1973)

28. $\mathbf{y''' - [(4-p)y + 2y'/y]y'' + (4-p)y'^2 - (p-2)y^2y' - 2y'y''/y = 0, p = 3, 4, 6.}$

On integration we get

$$y'' = (4-p)yy' + (p-2)y^3 + Cy^2, \quad (1)$$

where C is an arbitrary constant.

It may be checked that (1) has the Painlevé property only when $p = 44$. For $p = 3, 6$, it may be shown that there are relations between the coefficients of the Laurent expansion of the function y and so for $p \neq 4$, (1) does not have the Painlevé property. Martynov (1985a)

29. $$\mathbf{y''' - 3y'y''/(2y) + 3y'^3/(4y^2) - 6yy' - Ky' = 0,}$$

where K is a constant.

A first integral is

$$y'' = 3y'^2/(4y) + 3y^2 + Ky + C, \quad (1)$$

where C is an arbitrary constant. Equation (1), and hence the given DE, may be shown to have fixed critical points in its solutions. For $K = C = 0$ in (1), see Eqn. (2.32.7). Exton (1975)

30. $$\mathbf{y''' - 3y'y''/(2y) + 3y'^3/(4y^2) - gyy' = 0,}$$

where g is a constant.

By a simple scaling, this DE may be changed to one with $g = 6$:

$$y''' - 3y'y''/(2y) + 3y'^3/(4y^2) - 6yy' = 0. \quad (1)$$

Equation (1) has a first integral

$$y'' = 3y'^2/(4y) + 3y^2 + C, \quad (2)$$

where C is an arbitrary constant. Equation (2) may be shown to possess only fixed critical points; the same is, therefore, true of the given DE. For $C = 0$ in (2), see Eqn. (2.32.7). Exton (1975)

31. $$\mathbf{y''' - (3/2)(y + y'/y)y'' - (3/2)y'^2 + (3/4)y^2y' + (3/4)y'^3/y^2 = 0.}$$

Integrating once, we get

$$y'' = (3/2)yy' - (1/4)y^3 + (3/4)y'^2/y + C, \quad (1)$$

where C is an arbitrary constant. Equation (1) belongs to the class considered by Ince (1956), and has the Painlevé property; hence so does the original DE. Martynov (1985a)

32. $$y''' - 2y'y''/y - 2y^4 = 0.$$

Putting $z = y - y''/(2y^2)$, we find that

$$y^2 = y' - z' \text{ and } y' = (y-z)^2 + C_1;$$

C_1 a constant; hence $z' = z^2 - 2yz + C_1$ and

$$z'' - z'^2/(2z) + z^3/2 + C_1 z + C_1^2/(2z) = 0. \tag{1}$$

Analysis of (1) shows that z, and hence y, are single valued. Martynov (1985a)

33. $$y''' - 2y'y''/y - hy^4 = 0,$$

where h is an integer.

The given DE may be shown to possess moving critical points in its solution. Exton (1973)

34. $$y''' - 2y'y''/y - gy^2y' = 0,$$

where g is a constant.

By a suitable change of variables, we may transform the given DE such that $g = 1$;

$$y''' - 2y'y''/y - y^2y' = 0. \tag{1}$$

Equation (1) has a first integral

$$y'' = y^3 + Cy^2, \tag{2}$$

where C is an arbitrary constant. Equation (2) may be shown to have fixed critical points in its solution; the same is, therefore, true of the given DE. See Eqn. (2.1.64) for the solution of (2). Exton (1973)

35. $$y''' - 2y'y''/y + y'^3/y^2 - gyy' = 0,$$

where g is a constant.

By a suitable change of variables, we change the given DE to one with $g = 4$:

$$y''' = 2y'y''/y - y'^3/y^2 + 4yy'. \tag{1}$$

Equation (1) has a first integral

$$y'' = y'^2/y + 2y^2 + C, \tag{2}$$

where C is an arbitrary constant. Equation (2) may be shown to possess fixed critical points in its solution. For $C = 0$ in (2), see Eqn. (2.32.7). Exton (1973)

36. $$y''' - 2y'y''/y + y'^3/y^2 - gy^2y' = 0,$$

where g is a constant.

By a suitable change of variables, the given DE may be changed such that $g = 3$:

$$y''' - 2y'y''/y + y'^3/y^2 - 3y^2y' = 0. \tag{1}$$

Equation (1) has a first integral

$$y'' = y'^2/y + y^3 + C, \tag{2}$$

where C is an arbitrary constant. Equation (2) may be shown to have fixed critical points in its solution; the same is, therefore, true of the given DE. Exton (1973)

37. $$\mathbf{y''' - 2(y + y'/y)y'' + 2y'^2 = 0.}$$

Solution of the given DE may be shown to have movable critical points; it is therefore not of Painlevé type. Martynov (1985a)

38. $$\mathbf{y''' - (y + 2y'/y)y'' - y'^2 + y'^3/y^2 = 0.}$$

The given DE has a first integral

$$y'' = y'^2/y + yy' + 2C_1,$$

where C_1 is an arbitrary constant. Now putting $2z = y - y'/y$, we obtain the system

$$y = -C_1/z', \quad z' = z^2 + C_1x + C_2, \tag{1}$$

where C_1 and C_2 are arbitrary constants. Since the second of (1) is a Riccati equation, z has the Painlevé property. Therefore, it follows from the preceding equation that y has the same property. Martynov (1985a)

39. $$\mathbf{y''' - (ey + 2y'/y)y'' + ey'^2 = 0,}$$

where e is a constant.

A series expansion for the solution about an arbitrary moving pole shows that it involves only one other arbitrary constant. It proves that the general integral of the given DE is not free of algebraic critical points. Exton (1973)

40. $$\mathbf{y''' - (ey + 2y'/y)y'' + ey'^2 - e^2y^2y' = 0,}$$

where e is a constant.

A general series development about an arbitrary pole may be constructed to show that the given DE has algebraic critical points in its solution. Exton (1973)

41. $$\mathbf{y''' - (ey + 2y'/y)y'' + (3e/2)y'^2 - (e^2/2)y^2y' + (e^3/8)y^4 = 0,}$$

where e is a constant.

It may be shown that the given DE has movable critical points in its solution. Exton (1973)

42. $$\mathbf{y''' - [(3 - p)y + 2y'/y]y'' - (2p - 3)y^2y' + y'^3/y^2 = 0,}$$

where $p = -3, 1, 2, 3, 6$.

Integrating the given DE once, we obtain

$$y'' - (1/2)y'^2/y - (3-p)yy' + (p-3/2)y^3 - Cy = 0, \tag{1}$$

where C is a constant of integration. Equation (1) is included in Ince (1956) and has the Painlevé property if and only if $p = 1$ and $p = 3$. Martynov (1985a)

43. $$\mathbf{y''' - (ey + 2y'/y)y'' + y'^3/y^2 + (e^2/4)y^2y' = 0,}$$

where e is a constant.

By a suitable change of variables, the given DE may be transformed such that $e = -2$:

$$y''' - 2y'y''/y + y'^3/y^2 + 2yy'' + y^2y' = 0. \tag{1}$$

Equation (1) has a first integral

$$y'' - y'^2/(2y) + 2yy' + (1/2)y^3 - Cy = 0, \tag{2}$$

where C is an arbitrary constant. Equation (2) may be shown to possess only fixed critical points in its solution. The same is, therefore, true of the given DE. Exton (1973)

44. $$\mathbf{y''' - (ey + 2y'/y)y'' + y'^3/y^2 - ey'^2 = 0,}$$

where e is a constant.

By a simple change of variables we may find a transformed equation in which $e = 1$:

$$y''' - 2y'y''/y + y'^3/y^2 - (y'^2 + yy'') = 0. \tag{1}$$

Equation (1) has a first integral

$$y'' - y'^2/y - yy' - C = 0, \tag{2}$$

where C is an arbitrary constant. Equation (2) may be shown to possess no movable critical points in its solution; the same is, therefore, true of the original equation. For $C = 0$ in (2), see Eqn. (2.31.70) Exton (1973)

45. $$\mathbf{y''' - (1/3)[2y + 7y'/y]y'' - y'^2 - (1/3)y^2y' + (4/3)y'^3/y^2 = 0.}$$

If we put $z = -(1/3)(y'/y - y)$, then some manipulation shows that

$$y = -z''/z' + 2z$$

and z satisfies

$$z''' = zz'' + 2z'^2 + 2z^2z'. \tag{1}$$

According to Chazy (1911), Eqn. (1) has the Painlevé property. Hence the given DE is of Painlevé type. Martynov (1985a)

46. $$\mathbf{y''' - 3y'y''/y + (3/2)y'^3/y^2 - gy^2y' = 0,}$$

where g is a constant.

By a suitable change of variables, the given DE may be transformed such that $g = 3/2$:

$$y''' - 3y'y''/y + (3/2)y'^3/y^2 - (3/2)y^2y' = 0. \tag{1}$$

Equation (1) has the first integral

$$y'' = y'^2/(2y) + (3/2)y^3 + Cy^2, \tag{2}$$

where C is an arbitrary constant. Equation (2) may be shown to possess only fixed critical points in its solution. The same is, therefore, true of the given DE. Exton (1973)

47. $$\boldsymbol{y''' - [(4-p)y + 3y'/y]y'' - (2p-8)y'^2 = 0.}$$

The given DE has a first integral

$$y'' = (4-p)yy' + Cy^3, \tag{1}$$

where C is an arbitrary constant. Equation (1) may be shown to possess the Painlevé property only if $p = 4$; the same is true, therefore, of the original DE. Martynov (1985a)

48. $$\boldsymbol{y''' - [(3-p)y + 3y'/y]y'' - (p-3)y'^2 + (p-3/2)y^2y' + (3/2)y'^3/y^2 = 0,}$$

where p is a constant, equal to 2, 3, or 6.

On integration of the given DE, we get

$$y'' - (3-p)yy' - (p-3/2)y^3 - (1/2)y'^2/y - Cy^2 = 0, \tag{1}$$

where C is an arbitrary constant. Equation (1) is included in the class considered by Ince (1956) and has Painlevé property only if $p = 3$; the same, therefore, holds for the original DE. Martynov (1985a)

49. $$\boldsymbol{y''' - 5y'y''/(2y) + 5y'^3/(4y^2) - 2yy' - Ky' = 0,}$$

where K is a constant.

A first integral of the given DE is

$$y'' = 5y'^2/(4y) + y^2 + Ky + C,$$

where C is an arbitrary constant. Replacing y by $1/z$, we get

$$z'' = 3z'^2/(4z) - Cz^2 - Kz - 1. \tag{1}$$

Equation (1), and hence the given DE, may be shown to possess fixed critical points in its solutions. Exton (1975)

50. $$\boldsymbol{y''' - 5y'y''/(2y) + 5y'^3/(4y^2) - gyy' = 0,}$$

where g is a constant.

By a simple scaling, the given DE may be changed to one with $g = 2$, namely

$$y''' - 5y'y''/(2y) + 5y'^3/(4y^2) - 2yy' = 0,$$

with the first integral

$$y'' = 5y'^2/(4y) + y^2 + C, \tag{1}$$

where C is an arbitrary constant. Putting $z = 1/y$ in (1), we have

$$z'' = 3z'^2/(4z) - Cz^2 - 1. \tag{2}$$

Equation (2), and hence the given DE, may be shown to possess only fixed critical points in its solutions. For $C = 0$ in (1), see Eqn. (2.32.7). Exton (1975)

51. $$\boldsymbol{y''' - 5y'y''/(2y) + 3y'^3/(2y^2) - 3yy' - Ky'/y = 0,}$$

where K is a constant.

A first integral of the given DE is

$$y'' = 3y'^2/(4y) + 3y^2 + Cy - K, \tag{1}$$

where C is an arbitrary constant. Equation (1), and hence the given DE, may be shown to have fixed critical points in its solutions. For $K = C = 0$ in (1), see Eqn. (2.32.7). Exton (1975)

52. $$\boldsymbol{y''' - 5y'y''/(2y) + 3y'^3/(2y^2) - gyy' = 0,}$$

where g is a constant.

By a simple scaling, the given DE may be changed to one with

$$g = 3 : y''' - 5y'y''/(2y) + 3y'^3/(2y^2) - 3yy' = 0,$$

with the first integral

$$y'' = 3y'^2/(4y) + 3y^2 + Cy, \tag{1}$$

where C is an arbitrary constant. Equation (1), and hence the given DE, may be shown to possess only fixed critical points. For $C = 0$ in (1), see Eqn. (2.32.7). Exton (1975)

53. $$\boldsymbol{y''' - 7y'y''/(2y) + 5y'^3/(2y^2) - yy' - Ky'/y = 0,}$$

where K is a constant.

If we put $y = 1/z$, we get the same form as Eqn. (3.6.51); hence the given DE has fixed critical points in its solutions. Exton (1975)

54. $$\boldsymbol{y''' - 7y'y''/(2y) + 5y'^3/(2y^2) - gyy' = 0,}$$

where g is a constant.

By a simple scaling, this DE may be changed to one with $g = 1$, namely

$$y''' - 7y'y''/(2y) + 5y'^3/(2y^2) - yy' = 0,$$

with the first integral

$$y'' = 5y'^2/(4y) + y^2 + Cy, \tag{1}$$

where C is an arbitrary constant. Putting $z = 1/y$ in (1), we have

$$z'' = 3z'^2/(4z) - Cz - 1. \tag{2}$$

Equation (2), and hence the given DE, may be shown to possess only fixed critical points in its solutions. For $C = 0$ in (1), see Eqn. (2.32.7). Exton (1975)

55. $$\boldsymbol{y''' - 4y'y''/y + 3y'^3/y^2 - gy^2y' = 0,}$$

where g is a constant.

We may, by a suitable transformation, change the given DE such that $g = 1$:

$$y''' - 4y'y''/y + 3y'^3/y^2 - y^2y' = 0. \tag{1}$$

Equation (1) has the first integral

$$y'' = y'^2/y + y^3 + Cy^2, \tag{2}$$

where C is an arbitrary constant. Equation (2) may be shown to possess only fixed critical points. The same is, therefore, true of the given DE. For $C = 1$ in (2), see Eqn. (2.31.57). Exton (1973)

56. $$\boldsymbol{y''' - [(1 + 1/n)y + (3 - 1/n)y'/y]y'' + (3/n^2)y'^2 - (3/n^2 - 1/n)y^2y'}$$
$$\boldsymbol{+ (1/n^2)y^4 - (1/n - 1)(2 + 1/n)y'^3/y^2 = 0,}$$

where n is an integer, $n \neq 0, -3\ell - 1$, with $\ell = 0, 1, 2, \ldots$. Putting $y = -u'/u$ and $u' = v^n$, we get

$$v''' - 2v'v''/v = 0. \tag{1}$$

Equation (1) has solutions

$$v = \alpha(x - x_0)^{-2} + \sum_{i=1}^{\infty} \alpha_{6i+4}(x - x_0)^{6i+4}$$

and

$$v = C_1x + C_2,$$

where $\alpha, x_0, \alpha_4, C_1$, and C_2 are arbitrary constants. Clearly, u' is a meromorphic function whose power series, for n as given, does not contain $(x - x_0)^{-1}$; hence u is meromorphic and y has the Painlevé property. Martynov (1985a)

57. $$\boldsymbol{y''' - [(1 + 2/n)y + (3 - 2/n)y'/y]y'' + (2/n)y^2y'}$$
$$\boldsymbol{- (2/n - 2)y'^3/y^2 = 0, n \neq 0, -2\ell - 1, \ell = 0, 1, \ldots.}$$

Integrating once, we get

$$y'' = (1 - 1/n)y'^2/y + (1 + 2/n)yy' - (1/n)y^3 + Cy,$$

where C is a constant of integration.

If we put $y = -u'/u$ and $u' = v^n$, we may verify that the resulting DE has solution

$$v = \sin(C_1x + C_2), C_1 = (-C/n)^{1/2},$$

where C_2 is another arbitrary constant. It may also be verified that if $n \neq 0, -2\ell - 1, \ell = 0, 1, \ldots$, then u has movable poles, and hence y has the Painlevé property. Martynov (1985a)

58.
$$y''' - (ey + by'/y)y'' - cy'^3/y^2 + 2ey'^2 = 0,$$

where b, c, and e are constants.

The given DE may be shown to possess moving critical points in its solution. Exton (1973)

59.
$$y''' + (\tan y)y'y'' - Ay'/\cos y = 2C \cos^2 y,$$

where A and C are positive constants.

Writing $y' = [z(x)]^{1/2}$ or $z = y'^2$, so that $2y'y'' = z^{1/2}\dot{z}$, $y'' = \dot{z}/2, y''' = \ddot{z}(z)^{1/2}/2; \ddot{z} + (\tan y)\dot{z} - 4C\cos^2 y/z^{1/2} = 2A/\cos y$, $\dot{} = d/dy$. Now putting $z(y) = \eta(\xi), \xi = \sin y$, we get the equation

$$\eta^{**} - 4C(\eta)^{-1/2} = 2A/\cos^3 y, \quad {}^* \equiv \frac{d}{d\xi}. \tag{1}$$

Equation (1) is approximated by a linear one, using some physical argument, and is then solved explicitly. Panayotounakos and Theocaris (1986)

60.
$$y''' - a(y)y'y'' - b(y)y' = 0.$$

Put $y'' = q(y(x))$ in the given DE to obtain

$$\frac{dq}{dy} - a(y)q - b(y) = 0. \tag{1}$$

The solution may be obtained by first solving the linear first order DE (1) and hence solving $y'' = q(y(x))$. $y = C$, a constant, is another solution. Tatarkiewicz (1963)

61.
$$y''' + \{f_1(y)y' + f_2(y)\}y'' + \{f_3(y)y'^2 + f_4(y)y' + f_5(y)\}y' + f_6(y) = 0.$$

It is proved that the sufficient conditions for the given nonlinear DE to be equivalent to

$$Y''' + PY'' + QY' + RY = 0, \quad ' \equiv \frac{d}{dX}$$

where P, Q, R are constants, under the transformations

$$Y = \int_0^{\infty} f_2(\tau)\exp\left\{\int_0^{\tau} [f_4(\theta)/f_2(\theta)]d\theta\right\} d\tau + K, \frac{dX}{dx} = f_2(y) \neq 0,$$

are

$$\begin{aligned} f_1(y) &= \{3f_4 - f_2'(y)\}/f_2, f_3(y) = \{f_4^2 - 2f_4f_2'(y) + f_2f_4'(y)\}/f_2^2, \\ f_5(y) &= Af_2^2, f_6(y) = \left[B\int_0^y f_2(\tau)\exp\left\{\int_0^{\tau}\{f_4(\theta)/f_2(\theta)\}d\theta\right\} d\tau + C\right] f_2^2 \\ &\quad \times \exp\left\{\int_0^y -(f_4(\tau)/f_2(\tau))d\tau\right\}, \end{aligned}$$

where A, B, and C are arbitrary constants. The theorem is proved by direct substitution, etc.

Some subclasses of the given DE are identified that admit easier linearization: In each such case, one of the functions f_i $(i = 1, 2, \ldots, 6)$ is equal to zero. Then there is freedom to choose one or two of other f_i arbitrarily. Dasarathy (1973)

62. $$\boldsymbol{y''' + (\delta y'^2 + 1)y'' + y' + \epsilon y = 0;}$$

$\epsilon > 0$ and δ are parameters.

A two-timing perturbation solution is found in the form

$$y' = -\mu a(\mu x)\sin\psi(x) + \mu b(\mu x)\exp[-\psi(x)], \mu = \epsilon - 1,$$

where c, b, and ψ are governed by a system of ODEs with suitable IC. Numerical and analytic results for several values of μ and δ are presented. Tarbell (1980)

63. $$\boldsymbol{y''' - (1 + \epsilon y'^2)y'' - \gamma y' + k_1 = 0,}$$

$y(0) = 0, y \to 1$ as $x \to \infty$, where γ, ϵ, and k_1 are constants. The given problem arises in the flow of a non-Newtonian fluid past an infinite porous plate.

A first integral of the given DE is

$$y'' - y' - \gamma y - (\epsilon/3)y'^3 + k_1 x + k_2 = 0, \tag{1}$$

where k_2 is constant. First it is shown that if the given problem has a solution, then $k_1 = 0$ in (1) and $k_2 = \gamma$. In this situation, the problem is rewritten as

$$\begin{aligned} y' &= v, \\ v' &= v + \gamma(y - 1) + (\epsilon/3)v^3, \\ y(0) &= 0, y \to 1, v \to 0 \text{ as } x \to \infty. \end{aligned} \tag{2}$$

It is then shown that problem (2) has a unique solution for which $v(0) \geq 0$. A perturbation solution of the given problem is sought in the form

$$y(x, \epsilon) = y_0(x) + \epsilon y_1(x) + \cdots + \epsilon^n y_n(x) + \cdots.$$

The resulting DE and BV are

$$-y_0''' + y_0'' + \gamma y_0' = 0,$$

$$-(y_{n+1})''' + (y_{n+1})'' + \gamma(y_{n+1})' = -\sum_{j=0}^{n} \left\{ \sum_{i=0}^{j} (y_i' y_{j-i}') \right\} y_{n-j}'',$$

$$n = 0, 1, 2 \ldots, y_0(0) = 0, y_0 \to 1, y_0' \to 0 \text{ as } x \to \infty;$$

$$y_{n+1}(0) = 0, y_{n+1} \to 0, (y_{n+1})' \to 0 \text{ as } x \to \infty.$$

The solutions for various order terms are found to be $y_0(x) = 1 - e^{mx}, m \equiv (1/2)\{1 - (1 + 4\gamma)^{1/2}\} < 0$ since $\gamma > 0$. Similarly,

$$y_1(x) = \alpha_{11}(e^{3mx} - e^{mx}),$$

$$y_2(x) = \alpha_{22}e^{5mx} + \alpha_{21}e^{3mx} - (\alpha_{21} + \alpha_{22})e^{mx},$$

where

$$\alpha_{11} = -m^4/\{(3m)^3 - (3m)^2 - 3m\gamma\},$$

$$\alpha_{21} = [-3m^4/\{(3m)^3 - (3m)^2 - 3m\gamma\}]\alpha_{11},$$

$$\alpha_{22} = [15m^4/\{(5m)^3 - (5m)^2 - 5m\gamma\}]\alpha_{11}.$$

It is easy to verify that the solution at the nth order has the form

$$\begin{aligned} y_n(x) &= \alpha_{nn}e^{(2n+1)mx} + \alpha_{n(n-1)}e^{\{2(n-1)+1\}mx} \\ &+ \alpha_{n1}e^{3mx} - (\alpha_{nn} + \alpha_{n(n-1)} + \cdots + \alpha_{n1})e^{mx}. \end{aligned}$$

The perturbation series in ϵ is convergent for small values of ϵ and γ. Numerical and perturbation solutions are given in tabular form. The former are found by a shooting method. Rajagopal, Szeri, and Troy (1986)

64. $$y''' + g(y')y'' + by' + f(y) - p(x) = 0,$$

where $b > 0$ is a constant, g, f and p are continuous in their respective arguments, and $|f(y)| \leq M < \infty$ for all y.

It is shown that if:

(a) $g(y) \geq A_1 > 0$ for all y,

(b) $f(y)$ sgn $y \geq m > 0$ for all $|y| \geq 1$,

(c) $|p(x)| \leq A_2, |P(x)| \equiv |\int_0^x p(\tau)d\tau| \leq A_2$ for all $x \geq 0$,

then there exists a positive constant D, depending only on b, A_1, A_2, M, and g, such that every solution $y(x)$ of the given DE satisfies $|y(x)| \leq D, |y'(x)| \leq D, |y''(x)| \leq D$ for all sufficiently large x. Ezeilo (1961)

65. $$y''' + f(y')y'' + g(y') + cy = p(x),$$

where $c > 0$ is a constant.

Conditions on f, g, and h are given such that the following properties hold:

(P_1) Existence of a unique solution $y(x)$ which, together with its two derivatives $y'(x)$ and $y''(x)$, is bounded in R.

(P_2) Existence of a solution which is globally exponentially stable, together with its two derivatives.

(P_3) Existence of a solution which is periodic (respectively, almost periodic), together with its two derivatives holds if $p(x)$ is periodic (almost periodic).

Afuwape (1986)

66. $$y''' + f(y')y'' + g(y)y' + cy = p(x),$$

where $c > 0$ is a constant.

Conditions on f, g, and h are given such that the following properties hold:

(P$_1$) Existence of a unique solution $y(x)$ which, together with its two derivatives $y'(x)$ and $y''(x)$, is bounded in R.

(P$_2$) Existence of a solution which is globally exponentially stable, together with its two derivatives.

(P$_3$) Existence of a solution which is periodic (almost periodic), together with its two derivatives, holds if $p(x)$ is periodic (respectively, almost periodic).

Afuwape (1986)

67.

$$y''' + f(y')y'' + g(y') + h(y) = p(x).$$

Under certain conditions on f, g, h, and p, the given DE has the following properties:

(P$_1$) Existence of a unique solution $y(x)$ which, together with its two derivatives $y'(x)$ and $y''(x)$, is bounded in R.

(P$_2$) Existence of a solution which is globally exponentially stable, together with its two derivatives.

(P$_3$) Existence of a solution which is periodic (respectively, almost periodic), together with its two derivatives.

Afuwape (1986)

68.

$$y''' + f(y')y'' + g(y)y' + h(y) = Q(x).$$

Special cases of the given DE, namely

(a) $g(y) \equiv b$,
(b) $f(y') \equiv a$, and
(c) $f(y') \equiv a, g(y) \equiv b$,

and asymptotic properties of their solutions for $x \to \infty$ are studied.

The following theorem is one of the more important results referring to case (a). Let $f, h, Q \in C^o(R)$ and let b be a positive constant; suppose that $\liminf_{|y|\to\infty} yh(y) > 0$ and that there are positive constants $\alpha, \beta, \gamma, \delta$ such that $f(z) \geq \alpha, |h(y)| \leq \beta, |Q(x)| \leq \gamma, |\int_0^x Q(s)ds| \leq \delta$; then there exists a positive constant D such that every solution $y(x)$ of the given DE can be extended for $x \to \infty$ and the inequality

$$\lim_{x\to\infty} \sup(|y(x)| + |y'(x)| + |y''(x)|) \leq D$$

holds. Voracek (1966)

69.

$$y''' - B(x,y,y')y'' - C(x,y,y') = 0,$$

where B and C are rational in y, y' with coefficients analytic in x, and where B has one pole regarded as a function of y, and no poles as a function of y'.

Conditions are found such that the given DE has fixed critical points. Exton (1973)

3.7 $y''' + f(x, y, y', y'') = 0, f$ not linear in y''

1.
$$\boldsymbol{y''' - [(a_0 + a_1 y + a_2 y' + a_3 y'')y'' + b_0 + b_1 y + by' + b_{11}y^2 + b_{12}yy' + b_{22}y'^2] = 0, x \in (0, \infty),}$$

where $x_0, a, b, a_0, \ldots, a_3, b_0, \ldots, b_{22}$ are real $y(0) = y_0, y'(0) = a$, and $y'(\infty) = b$.

Existence of a solution of the given BVP is considered. Bespalova and Klokov (1988)

2.
$$\boldsymbol{y''' - (3/2)y''^2/y' = 0.}$$

Using the symmetries of the given DE, we get two first integrals,

$$y''^2 = 4y'^3/\phi_2, (y - \phi_1)^2 = \phi_2 y', \tag{1}$$

where ϕ_1 and ϕ_2 are constants. From the second of (1), we get on integration the solution $y = \phi_1 - \phi_2/(x - \phi_3)$, where ϕ_3 is another constant. Stephani (1989), p.91

3.
$$\boldsymbol{y''' - (3/2)y''^2/y' + (y'^3/2)\{-1/y + (1 - 1/k^2)/(y - 1) + 1 - (1/n^2)\}[1/\{y(y - 1)\}] = 0.}$$

This is the Schwartz equation. If, in the special system discussed in Eqn. (3.15.11), we put $u = y/x$, then u satisfies the present Schwartzian equation. This equation has a nonstationary singular line if $1/k + 1/n < 1$. In the notation of Eqns. (3.15.11) and (3.15.6), if $\lambda = 0, \alpha = 1/2$, the system equivalent to Schwartz equation, namely, Eqn. (3.15.11), has the solution

$$\begin{aligned} x_1 &= 1/\{2(ct + d)^2\} + \{1/(ct + d)^2\} \sum_{k=1}^{\infty} \alpha_k \exp\{k(at + b)/(ct + d)\}, \\ y_1 &= \{1/(ct + d)^2\} \sum_{k=1}^{\infty} \beta_k \exp\{k(at + b)/(ct + d)\}, \end{aligned}$$

hence

$$y = y_1/x_1 = \sum_{k=1}^{\infty} \hat{y}_k \exp\{k\{(at + b)/(ct + d)\}\}, ad - bc = -1,$$

where the coefficients y_k are determined by the recurrence relations

$$\begin{aligned} \hat{y}_k &= 2\beta_k - \sum_{m=1}^{k-1} \alpha_m \hat{y}_{m-k}, k = 2, 3, \ldots, \\ \hat{y}_1 &= 2\beta_1. \end{aligned}$$

Martynov and Yablonskii (1979)

4. $\boldsymbol{y''' - (1/2)(y'' - 2yy')^2/(y' - y^2) - (3 - p/2)yy'' - (p - 2)y^2 y' - y'y''/y = 0,}$

where $p = 4, 6$.

On integration of the given DE, we have

$$y'' = (4 - p)yy' + (p - 2)y^3 + 2Cy(y' - y^2)^{1/2}, \tag{1}$$

where C is constant of integration. Equation (1) may be replaced by the system

$$\begin{aligned} y' &= y^2 + [4/(2-p)^2](v-C)^2, \\ v' &= [(2-p)/2]vy. \end{aligned} \tag{2}$$

Eliminating y from (2), we get an equation for v:

$$v'' = [1 - 2/(p-2)]v'^2/v + [2/(2-p)]v(v-C)^2. \tag{3}$$

Equation (3) is studied by Ince (1956) and has the Painlevé property only for $p = 4, 6$. It follows from the second of (2) that y has the same property for these values of p. Martynov (1985a)

5. $\boldsymbol{y''' - (1/2)(y'' - 2yy')^2/(y' - y^2) - (3-p)yy'' - (2p-4)y'^2 - 2y'y''/y = 0.}$

The given DE has a first integral

$$y'' - (4-2p)yy' - (2p-2)y^3 - Cy^2(y' - y^2)^{1/2} = 0, \tag{1}$$

where C is an arbitrary constant. Bureau (1972) has shown that (1) is not of Painlevé type; the same is true of the original DE. Martynov (1985a)

6.
$$\begin{aligned} &\boldsymbol{y''' - (1/2)[(y'' - 2yy')^2/(y' - y^2)] - (1 + 1/n)yy''} \\ &\boldsymbol{- [(3/2)(1 - 1/n^2)y'^2] - [3/(2n^2) - 1/n]y^2y' + [1/(2n^2)]y^4} \\ &\boldsymbol{- (2 - 1/n)y'y''/y - (1/2)(1/n - 1)(3 + 1/n)y'^3/y^2 = 0,} \end{aligned}$$

where n is an integer, $n \neq 0, -2\ell - 1$ with $\ell = 0, 1, \ldots$.

Putting $y = -u'/u$ and $u' = v^n$, we get

$$v''' = v''^2/(2v') + v'v''/v. \tag{1}$$

Equation (1) may be shown to have poles as its only movable singularities; hence it follows that u' and hence y has the same property. Martynov (1985a)

7. $\boldsymbol{y''' - (2/3)(y'' - 2yy')^2/(y' - y^2) - (4/3)yy'' + 2y'^2 - 8y^2y' + (8/3)y^4 = 0.}$

Writing

$$y = -3z - (z'' - 2zz')/(z' - z^2), \tag{1}$$

we have

$$[-3z - (z'' - 2zz')/(z' - z^2)]' - [3z + (z'' - 2zz')/(z' - z^2)]^2 = z' - z^2. \tag{2}$$

Therefore,

$$z''' + 4zz'' + 2z'^2 - 8z^2z' - 8z^4 = 0.$$

Putting $z = -u/2$, we get

$$u''' - 2uu'' - 2u'^2 + (u' - u^2)^2 = 0. \tag{3}$$

From Chazy (1911), it follows that (3) has no movable critical points. Therefore, all solutions of the given DE are meromorphic. Martynov (1985a)

8. $$y''' - (2/3)(y'' - 2yy')^2/(y' - y^2) - (5/3)yy'' - (2/3)y'^2 - y'y''/y = 0.$$

Solution of the given DE may be shown to possess movable critical points; it is therefore not of Painlevé type. Martynov (1985a)

9. $$y''' - (4/5)(y'' - 2yy')^2/(y' - y^2) - (6/5)yy'' - 2y'^2 - (6/5)y'y''/y + (4/5)y'^3/y^2 = 0.$$

Solution of the given DE may be shown to possess movable critical points. Hence it is not of Painlevé type. Martynov (1985a)

10. $$y''' - (y'' - 2yy')^2/(y' - y^2) - 2yy'' - (4 - p)y'^2 - (p - 2)y^2y' = 0, p \neq 2.$$

The given DE has a first integral

$$y'' = (4 - p)yy' + (p - 2)y^3 + C(y' - y^2), \tag{1}$$

where C is an arbitrary constant. Equation (1) is discussed by Ince (1956) and is shown to have single-valued solution if and only if $p = 1$. Martynov (1985a)

11. $$y''' - (y'' - 2yy')^2/(y' - y^2) - (3 - 2/p)yy'' - (4/p)y'^2 = 0.$$

For $p = 2$, we easily find the integral

$$y' - y^2 = C_1 \exp(Ct), \tag{1}$$

where C and C_1 are arbitrary constants. Equation (1) is Riccati; therefore, the given DE has meromorphic solutions.

For $p = 1$, we put

$$z = (y'' - 2yy')/(y' - y^2) - y; \tag{2}$$

then

$$z' = -yz + y' - y^2. \tag{3}$$

We check that $z'' = 0$; therefore, $z = 2C_1'x + C_2'$, where C_1' and C_2' are arbitrary constants. If we put $y = -v'/v$, then $v''' = zv''$; hence $v'' = \exp(C_1'x^2 + C_2'x + C_3)$, where C_3 is another constant. The function y is clearly meromorphic.

With $p = -2$, the given DE becomes

$$y''' = (y'' - 2yy')^2/(y' - y^2) + 4yy'' - 2y'^2. \tag{4}$$

It is easily checked that if y is a solution of (4), then so is

$$z = (y'' - 2yy')/[4(y' - y^2)].$$

Here, if we put

$$u = -(y'' - 6yy' + 4y^3)^2/\{16(y' - y^2)^3\},$$

then

$$y = (1/2)[u''/u' - (2u-1)u'/\{2u(u-1)\}],$$

so that u may be shown to satisfy

$$u''' = (3/2)u''^2/u' - (u'^2/2)\{-3/(4u) + 1/(u-1) + 1\}[1/\{u(u-1)\}]. \quad (5)$$

Equation (5) can be shown to have a movable critical point; the same is therefore the case with the original DE. Martynov (1985a)

12. $$\boldsymbol{y''' - (y'' - 2yy')^2/(y' - y^2) - (1 + 1/n)yy'' - 3y'^2}$$
$$\boldsymbol{+ (1/n)y^2y' - (1 - 1/n)y'y''/y - (1/n - 1)y'^3/y^2 = 0.}$$

Putting $y = -u'/u$ and $u' = v^n$, we obtain

$$v''' = v''^2/v',$$

with the solutions $v = (1/C_1)\exp(C_1x) + C_2$ and $v = C_3x + C_4$, where C_i $(i = 1, 2, 3, 4)$ are arbitrary constants. It follows that the function u is single valued if and only if n is a positive integer. Therefore, the given DE has the Painlevé property when n is a positive integer. Martynov (1985a)

13. $$\boldsymbol{y''' - (y'' - 2yy')^2/(y' - y^2) - 4y'^2 - 2y'y''/y + 2y'^3/y^2 = 0.}$$

The given DE has the solution

$$y = -[\exp(C_1x + C_2) + x + C_3]^{-1},$$

where C_i $(i = 1, 2, 3)$ are arbitrary constants. Martynov (1985a)

14. $$\boldsymbol{y''' - 3(y'' - 2yy')^2/[2(y' - y^2)] - 6y'^2 = 0.}$$

Putting $z = \{y'' - 6yy' + 4y^3\}/\{y' - y^2\}$, we get

$$z''' - 6z'z''/z + 6z'^3/z^2 = 0. \quad (1)$$

Equation (1) has the solution

$$z = (C_1x^2 + C_2x + C_3)^{-1}, \quad (2)$$

where C_i $(i = 1, 2, 3)$ are arbitrary constants. The given DE may now be shown to possess the solution

$$y = z'/(2z) - z/4.$$

Martynov (1985a)

15. $$\boldsymbol{y''' - (1 - 1/\nu)(y'' - 2yy')^2/(y' - y^2) - yy'' - (3 - 3/\nu)y'^2}$$
$$\boldsymbol{- (1 + 2/\nu)y'y''/y + (1 + 1/\nu)y'^3/y^2 = 0.}$$

If we put $y = -u'/u$, we get

$$u^{iv} = (1 - 1/\nu)u'''^2/u'' + (1 + 2/\nu)u''u'''/u' - (1 + 1/\nu)u''^3/u'^2.$$

If $u'' \neq 0$, we find that either

$$u' = \exp(C_1 x + C_2)$$

or

$$u' = \exp[(C_1 x + C_2)^{\nu+1} + C_3].$$

where C_i $(i = 1, 2, 3)$ are arbitrary constants. It follows that u is an entire function, so $y = -u'/u$ has movable poles. Hence, the given DE has the Painlevé property. Martynov (1985a)

16. $$y''' - (1 - 1/\nu)y''^2/y' - (1 + 2/\nu)y'y''/y + (1 + 1/\nu)y'^3/y^2 = 0.$$

The given DE has the general solution

$$y = \exp[(C_1 x + C_2)^{\nu+1} + C_3]$$

and the singular solution

$$y = \exp(C_4 x + C_5),$$

where C_i $(i = 1, 2, 3, 4, 5)$ are arbitrary constants. Martynov (1985a)

17. $$y''' - (1 - 1/\nu)(y'' - 2yy')^2/(y' - y^2) - (1 + 2/\nu)yy''$$
$$- (4 - 1/\nu)y'^2 + (3/\nu)y^2 y' - [1/(\nu(\nu + 1))](y' - y^2)^2 = 0.$$

Writing

$$-y = -\nu z - [(z'' - 2zz')/(z' - z^2)], \tag{1}$$

we find the equation for z:

$$z'' - 3zz' + z^3 = 0. \tag{2}$$

The solution for (2) is

$$z = -(2C_1 x + C_2)/(C_1 x^2 + C_2 x + C_3), \tag{3}$$

where C_1, C_2, and C_3 are arbitrary constants. With $y = -u'/u$, we get from (1)

$$u''' = -\nu z u'',$$

so that using (3), we have

$$u'' = C(C_1 x^2 + C_2 x + C_3)^{\nu}. \tag{4}$$

It follows from (4) and $y = -u'/u$ that when ν is a positive integer, y is a rational function; when $\nu < 0$, the function u has critical points. Therefore, ν must be a positive integer for the given DE to have solutions with no movable critical points. Martynov (1985a)

18. $$y''' - (1 - 1/\nu)(y'' - 2yy')^2/(y' - y^2) - [2 + ((2 - p)/\nu)]yy''$$
$$- (4 - p)y'^2 - (p - 2)(1 + 2/\nu)y^2 y' = 0,$$

where $p = -2, 1, 3, 4, 6$ and $\nu = 1, 2$.

It may be verified that the given DE has a first integral

$$\{y'' + (p - 4)yy' + (2 - p)y^3\}^{\nu} = C(y' - y^2)^{\nu-1}, \tag{1}$$

where C is an arbitrary constant [$p = 2$ is also acceptable in the given DE and (1)].

For $\nu = 2$ and $p \neq -2$, (1) can be written as the system

$$\begin{aligned} y' &= y^2 - (p/2)(u-y)^2, \\ u' &= [(p+2)/2]yu - (p/2)u^2 + C_1, \end{aligned} \tag{2}$$

C_1 a constant, from which the equation for u may be found to be

$$\begin{aligned} u'' &- \left(1 - \frac{p-2}{p+2}\right)\frac{u'^2}{u} - \frac{p-6}{2}\left(-\frac{p}{p+2}u + \frac{2C}{(p+2)u}\right)u' \\ &- \frac{p(p-2)}{2(p+2)}u^3 + \frac{4p}{p+2}Cu - \frac{C^2(2-p)}{(p+2)u} = 0. \end{aligned} \tag{3}$$

Equation (3) belongs to the class considered by Ince (1956). Its solutions have no movable critical points only when either $p = 2$ or $p = 6$. If we can solve (3) for u, we may get from (2)

$$y = (2/(p+2))[u'/u + (p/2)u - C/u]. \tag{4}$$

If $p = -2$ and $\nu = 2$, the second of (2) has no term with y. We may then show that the pole of u is not a critical point of y. It follows that if $p = -2$ and $\nu = 2$, then y is meromorphic. For $\nu = 1$, we may conclude from (1) that y is single valued for $p = -2, 1, 2$, and 4. Martynov (1985a)

19. $$\boldsymbol{y''' - (1 - 1/\nu)(y'' - 2yy')^2/(y' - y^2) - [1 + 1/\nu]yy'' - 4y'^2 + (2/\nu)y^2y' = 0, \nu = 1, 2.}$$

The given DE may be written as the system

$$\begin{aligned} &y' - y^2 + \nu(z' - z^2) + (\nu+1)yz = 0, \\ &y - \nu z + \nu(z'' - 2zz')/(z' - z^2) = 0. \end{aligned} \tag{1}$$

Eliminating y from (1), we get

$$z''' - (1-\nu)(z'' - 2zz')^2/(z' - z^2) - (\nu+1)zz'' - 4z'^2 + 2\nu z^2 z' = 0. \tag{2}$$

If $\nu = 1$, Eqn. (2) becomes

$$z''' - 2zz'' - 4z'^2 + 2z^2z' = 0. \tag{3}$$

Equation (3) has a first integral

$$z'' - z'^2/(2z) - 2zz' + z^3/2 - C/z = 0, \tag{4}$$

where C is an arbitrary constant. We may note that (1) was derived by using the ansatz

$$y = \nu z - \nu(z'' - 2zz')/(z' - z^2). \tag{5}$$

We conclude that if z is a solution of (3), then y given by (5) is also a solution of the same. The given DE has no solution with movable critical points. Martynov (1985a)

20. $$\boldsymbol{y''' - (1 - 1/\nu)(y'' - 2yy')^2/(y' - y^2) - yy'' - [4 + 1/\nu]y'^2 + (1/\nu)y^2y' = 0, \nu = 1, 2, 4, -4.}$$

Putting

$$(y'' - 2yy')/(y' - y^2) - y = -2(\nu + 1)z \tag{1}$$

in the given DE, we obtain

$$y = 2(\nu + 1)z - \nu(z'' - 2zz')/2(z' - z^2). \tag{2}$$

Eliminating y from (1) and (2), we get

$$\begin{aligned} z''' &= (1 - \nu/2)(z'' - 2zz')^2/(z' - z^2) + (2\nu + 2)zz'' \\ &+ (10 + 4/\nu)z'^2 - (4/\nu)(\nu + 1)^2 z^2 z'. \end{aligned} \tag{3}$$

For $\nu = 1$, the given DE [and hence (3)], has no movable critical singularities, since it has the form discussed by Chazy (1911). For $\nu = 2$, Eqn. (3) is reduced by substitution $z = u/3$ to

$$u''' = 2uu'' + 4u'^2 - 2u^2u'. \tag{4}$$

Equation (4) has a first integral

$$u'' = u'^2/2u + 2uu' - u^3/2 + C/u, \tag{5}$$

where C is an arbitrary constant. Equation (5) has no solution with critical moving singularities. Hence, in this case, the given DE also has no solution with movable critical point [see (1) above]. Martynov (1985a)

21. $$\boldsymbol{y''' - (1 - 1/\nu)(y'' - 2yy')^2/(y' - y^2) - [2 + 4/\nu]yy'' - (2\nu + 10)y'^2 + 2[(\nu + 2)^2/\nu]y^2y' = 0, \nu = 1, 2.}$$

See the discussion of (3) in Eqn. (3.7.20) wherein the values $\nu = 1, 2$ interchange their roles in the evaluation of coefficients. Martynov (1985a)

22. $\boldsymbol{y''' - (1 - 1/\nu)(y'' - 2yy')^2/(y' - y^2) - 2(1 - \nu)y'^2 - 2(\nu + 2)y^2y' = 0, \nu = 1, 2.}$

Writing

$$-\nu z = (y'' - 2yy')/(y' - y^2) + \nu y, \tag{1}$$

we easily verify that

$$z' - z^2 = y' - y^2. \tag{2}$$

Eliminating y from (1)-(2), we find that z satisfies the given DE.

For $\nu = 1$, the given DE becomes

$$y''' - 6y^2y' = 0. \tag{3}$$

Integrating (3), we get

$$y'' = 2y^3 + C, \tag{4}$$

where C is an arbitrary constant. Equation (4) may be integrated after multiplication by y' etc. For $\nu = 2$, we put

$$u = y' - y^2$$

in the given DE and obtain

$$u''' = 3u'u''/(2u) - 3u'^3/(4u^2) - 8uu'; \tag{5}$$

therefore, on integration

$$u'' = 3u'^2/(4u) - 4u^2 + C_1, \tag{6}$$

where C_1 is an arbitrary constant. Equation (6) has been discussed by Ince (1956) and has meromorphic solutions. When u is determined by (4), it may be checked following the argument for Eqn. (3.7.20) that the pole of u is not a critical point of y. Martynov (1985a)

23.
$$\begin{aligned} &\boldsymbol{y''' - (1 - 1/\nu)(y'' - 2yy')^2/(y' - y^2)} \\ &\boldsymbol{- yy'' - 2y'^2 - 2y^2y' = 0, \nu = 1, 2.} \end{aligned}$$

Putting $y' - y^2 = -(12/\nu)u^\nu, \nu = 1, 2$, we find that

$$u''' = (12/\nu)u^\nu u'. \tag{1}$$

Now an integration gives

$$u'' = (12/[\nu(\nu + 1)])u^{\nu+1} + C, \nu = 1, 2, \tag{2}$$

where C is an arbitrary constant. The solution of (2) can be expressed in terms of elliptic functions and is therefore meromorphic. The function y may be found from $y = -u''/u'$. Martynov (1985a)

24.
$$\boldsymbol{y''' - (1 - 1/\nu)(y'' - 2yy')^2/(y' - y^2) - 2yy'' - 2y'^2 = 0,}$$

where ν is a constant.

The given system is equivalent to

$$y' = y^2 + u, \quad u'' = (1 - 1/\nu)u'^2/u, \nu \geq 1.$$

Hence $u = (C_1x + C_2)^\nu, \nu \geq 1$. Hence the given DE has no solution with moving critical points for all positive integral ν. Martynov (1985a)

25.
$$\boldsymbol{y''' - (1 - 1/\nu)(y'' - 2yy')^2/(y' - y^2) - (4 + 8/\nu)yy'' + (2 + 16/\nu)y'^2 = 0,}$$

where ν is a parameter.

It may be shown that for all positive ν, the given DE has a moving critical line. The special case $\nu = 1$ was investigated by Chazy (1911). Martynov (1985a)

26.
$$\begin{aligned} &\boldsymbol{y''' - (1 - 1/\nu)(y'' - 2yy')^2/(y' - y^2) - (4 - 4/\nu)y'^2} \\ &\boldsymbol{- (2 + 4/\nu)y'y''/y + (2 + 4/\nu)y'^3/y^2 = 0,} \end{aligned}$$

where ν is a positive integer.

This DE may be written as

$$\{(y''-2yy')/(y'-y^2)-2y'/y\}'=-(1/\nu)\{(y''-2yy')/(y'-y^2)-2y'/y\}^2. \quad (1)$$

On integrating (1) we find that

$$(y''-2yy')/(y'-y^2)-2y'/y=\nu C_1/(C_1x+C_2), \quad (2)$$

where C_1 and C_2 are arbitrary constants. Equation (2) has the solution

$$y=-[(\hat{C}_1x+\hat{C}_2)^{\nu+1}+x+\hat{C}_3]^{-1}.$$

Therefore, y has movable poles. Here $\hat{C}_i$ $(i=1,2,3)$ are arbitrary constants. Martynov (1985a)

27.

$$\boldsymbol{y'''-(1-1/\nu)(y''-2yy')^2/(y'-y^2)-2yy''}$$
$$\boldsymbol{-2y'^2-8(1+2/\nu)(y'-y^2)^2/(1-k^2)=0,}$$

where ν and k are constants.

The given DE is equivalent to the system

$$y' = y^2+u, \quad (1)$$
$$u'' = (1-1/\nu)u'^2/u+du^2, \quad (2)$$

provided that $d=8(1+2/\nu)/(1-k^2)$.

It is easily seen that the solutions of (2) are meromorphic for $\nu=1,2,4$. Let $x=x_0$ be a movable pole of (2); then $x=x_0$ can be a critical singular point for y. Putting $y=-v'/v$, we obtain a linear DE for v, namely

$$v''+uv=0,$$

for which $x=x_0$ is a regular point of u. The characteristic roots for v are $\rho_1=(1+k)/2$ and $\rho_2=(1-k)/2$. Since $\rho_1-\rho_2=k$ with k an integer, it follows that v has two linearly independent solutions of the form

$$v_1 = \sum_{i=1}^{\infty}\delta_i(x-x_0)^{i+\rho_1}+\sigma v_2\ln(x-x_0),$$
$$v_2 = \sum_{i=1}^{\infty}\beta_i(x-x_0)^{i+\rho_2}.$$

It is easily seen that $\sigma=0$ only when

$$k\neq(2+4/\nu)\mu, \quad (3)$$

μ being a positive integer. Thus, under (3), the function v is single valued; hence y is single valued under the condition (3). The given DE therefore has no moving critical singularities in this case. (The case $\nu=\infty$ is, however, not excluded.) Martynov (1985a)

28.

$$\boldsymbol{y'''-(1-1/\nu)(y''^2/y')-a_1y'y''/y-b_1y'^3/y^2=0,}$$

where ν, a_1, and b_1 are constants.

If we put $y = u^n$, we get

$$u''' - (1 - 1/\nu)u''^2/u' - \mu u'u''/u = 0, \tag{1}$$

where $\mu = a_1 n - (n-1)(1+2/\nu)$ and $(n-1)(n-2) - (1-1/\nu)(n-1)^2 - a_1 n(n-1) - b_1 n^2 = 0$. Equation (1) implies that

$$u' = (C_1 u^{\mu+1} + C_2)^{\nu/(\nu+1)}, \tag{2}$$

where C_1 and C_2 are arbitrary constants. Furthermore, if $a_1 = 1 + 2/\nu, b_1 = -1 - 1/\nu$, the given DE has the general solution

$$y = \exp[(\hat{C}_1 x + \hat{C}_2)^{\nu+1} + \hat{C}_3]$$

and the singular solution $y = \exp(\hat{C}_1 x + \hat{C}_2)$, where $\hat{C}_i$ $(i = 1, 2, 3)$ are constants. Martynov (1985a)

29.
$$\begin{aligned} &\boldsymbol{y''' - (1 - 1/\nu)(y'' - 2yy')^2/(y' - y^2)} \\ &\boldsymbol{- (2 + 2/\nu)yy'' - 3y'^2 + (1 + 5/\nu)y^2 y' - (1/\nu)y^4 = 0,} \end{aligned}$$

where ν is a parameter.

Putting

$$-\nu z = (y'' - 2yy')/(y' - y^2) - y, \tag{1}$$

we easily check that

$$z' - z^2 = 0. \tag{2}$$

Now, let $y = -u'/u$, so that (1) gives

$$u''' = -\nu z u''. \tag{3}$$

The solution for (3), where z is given by (2), is

$$u = \frac{C_1}{(\nu+1)(\nu+2)}(x - x_0)^{\nu+2} + C_2 x + C_3, \nu \neq -2, -1, 0,$$

where C_1, C_2, and C_3 are constants.

We conclude that the given DE has no critical points for $\nu \neq -2, -1, 0$. Martynov (1985a)

30. $\boldsymbol{y''' - (1 - 1/\nu)(y'' - 2yy')^2/(y' - y^2) - ayy'' - by'^2 - cy^2 y' - dy^4 = 0,}$

where ν is an integer, $\nu \neq 0, -1, \infty$ and a, b, c, and d are constants.

This is a very simple example, in the sense of Painlevé, of a third order DE $y''' = F(x, y, y', y'')$ with right-hand side rational in y'', y', y.

It is shown that if $a = 4 + 8/\nu, b = d - 2 - 16/\nu, d = \{32(1 + 2/\nu)^3\}/\{4(1 + 2/\nu)^2 - k^2\}$, and $c = -2d$, and further, if $1/k + 1/(\nu + 2) < 1/2, k \neq 6\ell$, and ℓ is an integer, then the given DE has a moving singular line. If, however, $1/k + 1/(\nu + 2) > 1/2$, then the general solution of the given DE is a rational function. Martynov (1985a)

31.
$$\begin{aligned} &\boldsymbol{y''' - (1 - 1/\nu)(y'' - 2yy')^2/(y' - y^2) - ayy'' - by'^2} \\ &\boldsymbol{- cy^2 y' - dy^4 - a_1 y'y''/y - b_1 y'^3/y^2 = 0, a_1^2 + b_1^2 \neq 0,} \end{aligned}$$

where ν is an integer distinct from 0 and -1. Conditions on the constants a, b, c, d, a_1, and b_1 are found such that the DE has no moving critical points. In particular, for the following sets of parametric values, the DE has the Painlevé property:

(a)

$$\begin{aligned} a_1 = {} & 2, \quad b_1 = (3/2)(1/n^2 - 1), a = 1, b = (3/2)(\ell - 3/n^2), \\ c & = 9/(2n^2), d = -3/(2n^2), n \neq \pm(3\ell + 1), \\ \ell & = 0, 1, 2, \ldots . \; \nu \neq 0, -1. \end{aligned}$$

(b)

$$\begin{aligned} \nu = {} & 3, \quad a_1 = (1/3)(5 - 2/n), b_1 = (2/3)(1/n - 1)(2 + 1/n), \\ a & = 1 + 2/(3n), b = 2 - 8/(3n^2), c = 8/(3n^2) - 2/(3n), \\ d & = -2/(3n^2), n \neq 0, -2\ell - 1, \ell = 0, 1, \ldots . \end{aligned}$$

(c)

$$\begin{aligned} \nu = {} & 5, \quad a_1 = (1/5)(7 - 1/n), b_1 = (1/5)(1/n - 2)(3 + 1/n), \\ a & = 1 + 1/(5n), b = (3/5)(4 - 1/n^2), c = 1/n, \\ d & = -1/n^2, n \neq 0. \end{aligned}$$

(d)

$$\begin{aligned} \nu = {} & \infty, \quad a_1 = a = 1, b_1 = 1/(4\ell^2) - 1, b = 3(1 - 1/4\ell^2), \\ & c = 3/(4\ell^2), d = -1/(4\ell^2), \ell = 1, 2, \ldots . \end{aligned}$$

Martynov (1985a)

32. $$y''' - \{A/(y' + ay^2) + B/(y' + by^2)\}y''^2 = 0,$$

where $A, B, a,$ and b are constants.

Equations of the given form are considered such that they have invariant critical points. Verholomov (1950)

33. $$y''' - R(y', y, x)y''^2 = 0.$$

Equations of the given form are found such that they have invariant critical points. Verholomov (1950)

34. $$y''' - A(y', y, x)y''^2 - B(y', y, x)y'' - C(y', y, x) = 0,$$

where A, B, C are rational functions in y', y and analytic in x.

A special simplified version of the given DE is constructed such that it does not change its form under the substituition $x, y; x_0 + \lambda x, \lambda^{-1}y$. This DE is

$$\begin{aligned} (y' - y^2)y''' - (1 - 1/k)y''^2 - a_1yy'y'' - a_2y^3y'' \\ - a_3y'^3 - a_4y^2y'^2 - a_5y^4y' - a_6y^6 = 0, \end{aligned} \tag{1}$$

where a_i $(i = 1, \ldots, 6)$ and k are constants. It is further shown that an equation of the form (1) which is invariant with respect to the transformation of the variables

$$y = f'(x)Y(\tau) + \phi(x), \tau = f(x), f'' = 2f'\phi \tag{2}$$

is necessarily of the form

$$\begin{aligned} y''' \ - \ & (1 - 1/k)(y'' - 2yy')^2/(y' - y^2) - 4(1 + 2/k)yy'' \\ & + 2(1 + 8/k)y'^2 - 4(2 + 4/k - a)(y' - y^2)^2 = 0. \end{aligned} \tag{3}$$

Conditions are established under which (3) has rational solutions; also, the conditions for it to have a moving singular curve are derived. Martynov (1979)

35. $\boldsymbol{y''' - H(x, y, y')y''^2 - G(x, y, y')y'' - F(x, y, y') = 0.}$

The integral solutions of the given DE describe a system of ∞^3 curves. Certain interesting projective properties of this system of curves are developed. Terracini (1955)

36. $\boldsymbol{y''' + F(r)y'' + F(r)y' + y = 0,}$

where $F(r) = 1 - \epsilon f(r), f(r) = 1 - r^2, r^2 = y^2 + y'^2 + y''^2$ and $\epsilon \geq 0$ is a scalar parameter. The given system has applications in control theory. It is first replaced by the equivalent first order system $y' = z, z' = p, p' = -F(r)(p + z) - y$.

For small nonlinearity $\epsilon \ll 1$, the existence of a limit cycle is established by a fixed-point technique, the approach to limit cycle is approximated by averaging methods, and the periodic solution is harmonically represented by perturbation.

Computer solutions are provided to reinforce analysis. One approximate solution is

$$y(x) = a(x)\cos[x + \phi(x)],$$

where

$$\begin{aligned} a^2(x) &= (2/3)[1 + e^{-\epsilon x}(2a_0^{-2}/3 - 1)]^{-1}, \\ \phi(x) &= \phi(0) - (\epsilon/8)\int_0^x a^2(s)ds, \end{aligned}$$

where $a_0 = a(0)$. If $a_0^2 = 2/3$, then $a^2(x) = 2/3$ and $\lim_{x\to\infty} a^2(x) = 2/3$. Mulholland (1971)

37. $\boldsymbol{y''' + (1 - \mu)(1 - r^2)y'' + y' + y = 0,}$

where $r^2 = y^2 + y'^2 + y''^2$, and μ is a parameter.

Periodic solutions using perturbation methods are found. The approximate value of the frequency is $w = 1 - \mu(1/2 - 7A^2/8) + o(\mu^2)$, where μ is a small parameter. El-Owaidy and Zagrout (1981)

38. $\boldsymbol{y''' + F(r)y'' + F(r)y' + y = 0,}$

where $F(r) = 1 - \epsilon f(r), f(r) = 1 - r^2, r^2 = y^2 + (y')^2 + (y'')^2$ with $\epsilon \geq a$ scalar parameter.

The given DE is rewritten as the system

$$\dot{x} = y, \dot{y} = z, \dot{z} = -F(r)(y+z) - x. \tag{1}$$

Singular-point analysis in three dimensions is carried out. A degenerate solution is $y = -x, z = x$. It is stated with reference to previous work that the given system has a limit cycle. Other qualitative features are obtained using perturbation analysis, an averaging technique, and computer solutions. Results are depicted graphically. Mulholland (1971).

39. $$y''' - \epsilon(1 - y'^2)y'' + \epsilon y''^3 + y' + ay = 0,$$

where ϵ and a are constants.

The limit cycle is obtained by first writing the DE as an equivalent system

$$\begin{aligned} \dot{x}_1 &= x_2, \quad \dot{x}_2 = x_3, \\ \dot{x}_3 &= -ax_1 - x_2 - \epsilon x_3^3 + \epsilon(1 - x_2^2)x_3. \end{aligned}$$

For $a = 1, \epsilon = 1$, the limit cycle is written as $x_1 = A \sin t, x_2 = -A \cos t, x_3 = A \sin t$, and then for

$$r^2 = x_1^2 + x_2^2 + x_3^2, \quad (1/2)\frac{d}{dt}r^2 = \sin 2t. \tag{1}$$

The existence of a limit cycle in three-dimensional space and the formula (1) are verified numerically. The results are depicted. Ku and Jonnada (1971a)

40. $$y''' + y(y'')^{2-\alpha} = 0,$$

$y(0) = 0, y'(0) = 0, y'(\infty) = 1$, where α is a parameter. The given problem arises in the steady-state two-dimensional boundary layer flow of a power law fluid past a flat plate.

The DE and first two conditions are invariant under the transformation

$$y = A^{\alpha_1}\tilde{y}, \quad x = A^{\alpha_2}\tilde{x},$$

provided that

$$\alpha_1 = (2\alpha - 1)/3, \quad \alpha_2 = (\alpha - 2)/3. \tag{1}$$

The condition at ∞ becomes $A^{\alpha_1 - \alpha_2}\tilde{y}'(\infty) = 1$, or using (1), we get

$$A = [\tilde{y}'(\infty)]^{-3/(\alpha+1)}. \tag{2}$$

The condition $y''(0)$ is chosen to be A. The transformed condition $\tilde{y}''(0)$ becomes 1. Solving the IVP with $\tilde{y}(0) = 0, \tilde{y}'(0) = 0, \tilde{y}''(0) = 1$, one finds $\tilde{y}'(\infty)$. A is then found from (2). Since A, α_1, α_2 are known, one easily reverts to the original variables. The calculations are performed for a given value of α. Numerical results are given. Kubicek and Hlavacek (1983), p.235

41. $$y''' + f(y'') + y' + y = 0,$$

where $f(\eta)$ is continuous and satisfies a Lipschitz condition for all $\eta, f(0) = 0, \eta f(n) > \eta^2$ for $\eta \neq 0$. The given DE is written as the system

$$z' = w - f(z), \quad w' = r - z, \quad r' = -z, \quad ' \equiv \frac{d}{dx}. \tag{1}$$

Many properties of the system (1) are stated without proof. As an example: Let p be the domain $z = 0, w > 0, r > 0, w^2 + r^2 < R$ (a certain R). If there exists $z_0 \geq 0$ and $\epsilon > 0$ such that $f'(z) > 1 + \epsilon$ for $z \geq z_0$, then a necessary and sufficient condition for the stability of the origin in the large is the nonexistence of a periodic solution cutting P in just one general point. An example is also given of an $f(z)$ for which there is a periodic solution. Pliss (1956)

42. $$y''' + p(x)y'' + q(x)f(y') + r(x)h(y, y'') = 0.$$

Sufficient conditions on p, q, r, and h are found that the solutions and the first two derivatives of the given DE are bounded in $[a, \infty)$. Conditions are also found that $y'(x)$ has an infinite number of zeros. Moravsky (1975)

43. $$y''' - f(x, y, y', y'') = 0,$$

where $f(x, y_1, y_2, y_3) : (a, b) \times R^3 \to R$ is continuous and f satisfies the Lipschitz condition

$$|f(x, y_1, y_2, y_3) - f(x, z_1, z_2, z_3)| \leq \sum_{i=1}^{3} k_i |y_i - z_i|$$

for each $(x, y_1, y_2, y_3), (x, z_1, z_2, z_3) \in (a, b) \times R^3$. Interval-length bounds on subintervals of (a, b) in terms of Lipschitz coefficients $k_i, i = 1, 2, 3$, are found on which certain two-point and three-point boundary value problems for the given DE have unique solutions. A contraction mapping principle is used. Henderson (1987)

44. $$y''' - F(x, y, y', y'') = 0.$$

Invariants of the given DE are sought under the group of transformations $\overline{x} = \overline{x}(x), \overline{y} = \lambda(x)y + \mu(x)$. Equivalent simpler systems are found in some special cases. Petrescu (1949)

3.8 $f(x)y''' + g(x,y,y',y'') = 0$

1. $$xy''' + y'' - (1 + x^8 + 4x^3)y'^3 - (x^{-3} + 1)y^3 - x^2 + x^{-3} = 0, x \geq 2.$$

It is shown that all solutions of the given DE are ultimately nonoscillatory. One such exact solution is $y(x) = x^{-1}$. Parhi (1981)

2. $$xy''' + y'' + (1/2)yy'' - (1/4)y'^2 = 0,$$

$y' \to 0$ as $x \to \infty$ and $y \to 0$ as $x \to 0$.

The given BVP describes natural convection from a vertical cylinder at a very large Prandtl number. When x is small, the solution is $y \sim Ax(\log x + B - 1) + O(x^2(\log x)^2)$, where $A = -30.90, B = 4.61$. When x is large, $y = 2(D + 1) + \sum_{n=1}^{\infty} (A_n/x^{nD})$, where the constant D is determined numerically. The asymptotic expression near $x = 0$ is compatible with the value $D = 0.539$. A few coefficients A_n are tabulated. Crane (1976)

3. $$\boldsymbol{x^2y''' + xy'' + (2xy - 1)y' + y^2 - f(x) = 0.}$$

The given DE is exact and has the first integral

$$x^2y'' - xy' + xy^2 = \int f(x)dx.$$

Kamke (1959), p.601

4. $$\boldsymbol{x^2y''' + x(y - 1)y'' + xy'^2 + (1 - y)y' = 0.}$$

If we divide the given DE by x^2, we get an exact form; hence the first integral is

$$xy'' + (y - 1)y' = Cx,$$

where C is an arbitrary constant. Kamke (1959), p.601

5. $$\boldsymbol{r(x)y''' + r'(x)y'' + q(x)(y')^{\beta} + p(x)y^{\alpha} - f(x) = 0,}$$

where p, q, r, f are realvalued continuous functions on $[0, \infty), r(x) > 0, f(x) \geq 0$, and both $\alpha > 0$ and $\beta > 0$ are ratios of odd integers.

Sufficient conditions for nonoscillation of solutions to the given DE are found when:

(a) $p(x) \geq 0, q(x) \leq 0$,

(b) $p(x) \leq 0$

with any $q(x)$. Some results concern the asymptotic behavior and existence of a positive increasing solution of the given DE. Parhi and Parhi (1986)

6. $$\boldsymbol{f(x)y''' - yy'' + p(y') = 0,}$$

where $f(x)$ is a real differentiable function, $\lim_{x\to\infty} f(x) = c, 0 < c < \infty, f'(x)$ is bounded as $x \to \infty$; $p(x)$ is a real differentiable function and $0 < a < \infty, p(a) = 0, p'(a) > 0$. If $y(x)$ is a solution of the given DE for which $\lim_{x\to+\infty} y'(x) = a$, then $y = ax + b$, where b is a constant. Utz (1956b)

3.9 $f(x, y)y''' + g(x, y, y', y'') = 0$

1. $$\boldsymbol{(a + y)y''' + 3y'y'' = 0.}$$

The given DE is exact. A first integral is

$$(a + y)y'' + y'^2 = C; \tag{1}$$

C is an arbitrary constant. Equation (1) is autonomous. The solution of (1) is

$$y(2a + y) = x(C_1x + C_2) + C_3;$$

C_1, C_2, and C_3 are arbitrary constants. Murphy (1960), p.427

2. $$yy''' - y'y'' - 6y^2y' - By' = 0,$$

where B is a constant.

Integrating the given DE with respect to x, we have

$$y'' = y'^2/y + 2y^2 + C/y + B, \tag{1}$$

where C is an arbitrary constant. Equation (1), and hence the original DE, may be shown to possess only fixed critical points in its solution. Exton (1973)

3. $$yy''' - y'y'' - 6y^2y' - (Ax + B)y' = 0,$$

where $A \neq 0$ and $B \neq 0$ are constants.

The given DE may be shown to possess moving critical points. Exton (1973)

4. $$yy''' - y'y'' + y^3y' = 0.$$

Division by y^2 makes the the given DE exact and hence gives

$$y''/y + (1/2)y^2 = C. \tag{1}$$

Multiplying (1) by yy' makes it exact again and hence we have

$$y' = \pm[C_0 + Cy^2 - (1/4)y^4]^{1/2}. \tag{2}$$

Equation (2) is solved by quadrature. C, C_0 are arbitrary constants. Kamke (1959), p.602

5. $$yy''' - y'y'' + (3/\alpha)y''y^2 + (2/\alpha^2)y^3y' = 0,$$

where α is a parameter.

If we put $y = \alpha T'/T$, the given DE assumes the form

$$T'^2T^{\text{iv}} - T'T''T''' = 0. \tag{1}$$

Now put $T' = v$ in (1) so that

$$v^2v''' - vv'v'' = 0. \tag{2}$$

For treatment of Eqn. (2), see Eqn. (3.9.44). Kozulin and Lukashevich (1988)

6. $$yy''' - y'y'' - y^3f(y') = 0.$$

Put

$$\frac{dy}{dx} = q(z), \quad \frac{dz}{dx} = y \tag{1}$$

in the given DE to obtain the second order autonomous equation

$$q''(z) - f(q) = 0. \tag{2}$$

When Eqn. (1) is solvable, $y(x)$ may be obtained with the help of (1). Tatarkiewicz (1963)

7. $$yy''' - y'y'' + (1+\nu)^{-1}\alpha y'' = 0,$$

where $y(0) = y'(1) = 0, y(1) = \pm(1-\alpha); \nu$ and α are constants. The given DE appears in the description of self-similar flows in channels with permeable walls.

With the change of variable, $\phi(x) = y(x)/y(1)$ the given problem becomes

$$\phi\phi''' - \phi'\phi'' + \gamma\phi'' = 0, \gamma = (1+\nu)^{-1}\alpha/y(1), \tag{1}$$

$$\phi(0) = \phi'(1) = 0, \phi(1) = 1. \tag{2}$$

An approximate solution is found in the form

$$\phi(x) = (3/2)x - x^3/2 + a_7(2x - 3x^3 + x^7) + a_9(3x - 4x^3 + x^9) + \cdots. \tag{3}$$

Additional conditions are required to get the coefficients a_7, a_9, etc. One of these conditions,

$$\phi(1)\phi'''(1) + \gamma\phi''(1) = 0, \tag{4}$$

is obtained from (1) and (2) by putting $x = 1$. A second condition is obtained by integrating (1) over $[0, 1]$. Thus,

$$\left|\phi\phi'' - \phi'^2 + \gamma\phi'\right|_0^1 = 0. \tag{5}$$

Substituting (3) in (4) and (5), we obtain the following for the coefficients a_7 and a_9:

$$\begin{aligned} a_7 &= C + Da_9, E^2a_9^2 - 2Ga_9 - H = 0, \\ C &= (1/8)(1+\gamma)/(8+\gamma), D = -2(10+\gamma)/(8+\gamma), \\ E &= -2(16+\gamma)/(8+\gamma), G = (2320 + 97\gamma - 41\gamma^2 - 2\gamma^3)/(8+\gamma)^2, \\ H &= (287 + 1218\gamma + 371\gamma^2 + 28\gamma^3)/4(8+\gamma)^2. \end{aligned} \tag{6}$$

Since $4G^2 \gg E^2H$ for $0 \leq \gamma \leq 1$, a good approximation to the smallest root of the second equation of (6) is

$$a_9 = -[H/(2G)][1 - E^2H/(4G^2)]. \tag{7}$$

The values of a_7 and a_9 obtained from (6)-(7) when substituted in (3) yield the solution of the problem.

The given problem is also solved numerically. Barashkov and Spriridonov (1988)

8. $$yy''' - y'y'' + y^2y'' - 2y^3y' = 0.$$

The given DE can be shown to possess fixed critical points in its solution. Exton (1973)

9. $$yy''' - y'y'' + (3/\alpha)y^2y'' + (2/\alpha^2)y^3y' = 0,$$

where α is an arbitrary parameter.

It can be shown that the solutions of the given DE can be expressed in terms of quadratures involving transcendental functions. Kozulin and Lukashevich (1988)

10. $$\boldsymbol{yy''' - 2y'y'' = 0.}$$

The given DE has solutions

$$y = \alpha(x - x_0)^{-2} + \sum_{i=0}^{\infty} \alpha_{6i+4}(x - x_0)^{6i+4}$$

and

$$y = (C_1 x + C_2),$$

where C_1, C_2, x_0, α, and α_4 are arbitrary constants. Martynov (1985a)

11. $$\boldsymbol{yy''' - 2y'y'' - 2y^3y' - Ky' = 0,}$$

where K is a constant.

An integration w.r.t. x gives

$$y'' = 2y^3 + Cy^2 - K/2, \tag{1}$$

where C is an arbitrary constant [see Eqn. (2.1.64)]. Equation (1), and hence the given DE, have fixed critical points in their solutions. Exton (1973)

12. $$\boldsymbol{yy''' - 2y'y'' + (2/\alpha)y^2y'' - (3/\alpha)yy'^2 - (2/\alpha^2)y^3y' - (1/\alpha^3)y^5 = 0,}$$

where α is a parameter.

If we put $y = \alpha T'/T$, the given DE becomes

$$T'T^{\text{iv}} - 2T''T''' = 0. \tag{1}$$

Now put $T' = v$ in (1) so that

$$vv''' - 2v'v'' = 0. \tag{2}$$

For treatment of Eqn. (2), see Eqn. (3.9.44). We conclude that if the general solution of (2) is $v = \phi(x, c_1, c_2, c_3)$, the general solution of the given DE has the form

$$y = \alpha\phi(x, c_1, c_2, c_3)/\{c_4 + \int \phi(x, c_1, c_2, c_3)dx\}.$$

Kozulin and Lukashevich (1988)

13. $$\boldsymbol{yy''' - 3y'y'' = 0.}$$

This DE has solutions

$$y = \alpha(x - x_0)^{-1} + \sum_{i=0}^{\infty} \alpha_{4i+3}(x - x_0)^{4i+3}$$

and

$$y = (C_1 x + C_2),$$

where $\alpha, x_0, \alpha_3, C_1$, and C_2 are arbitrary constants. Martynov (1985a)

14. $$yy''' - 3y'y'' - K^2yy' = 0,$$

where K is a constant.

The given DE may be solved by Hirota's method. Writing

$$y = f(x) = \sum_{n=0}^{\infty} \epsilon^n f_n, f_0 \equiv 1, \tag{1}$$

where ϵ is an arbitrary parameter. Substituting (1) in the given DE, we have

$$\sum_{n=0}^{\infty} \sum_{m=0}^{n} [-K^2 f_{n-m} f_m' + f_{n-m}''' f_m - 3f_{n-m}' f_m'']\epsilon^n = 0. \tag{2}$$

By equating separately the coefficients of the powers of ϵ to zero and noting the simplifications arising for $m = 0$ and $m = n$, we obtain the recurrence relation

$$f_n''' - K^2 f_n' = -\sum_{m=1}^{n-1} (-K^2 f_{n-m} f_m' + f_{n-m}''' f_m' - 3f_{n-m}' f_m''). \tag{3}$$

Equation (3) has a solution

$$f_j = (1/2^{j-1})e^{jpx}, \tag{4}$$

where $p = \pm K$, as may be verified directly. Putting (4) into (1), we have

$$f(x) = 1 + 2\sum_{\ell=1}^{\infty} \{(e^{px}/2)\epsilon\}^\ell = 1 + \frac{(2\epsilon)^{1/2} e^{px/2}}{(2/\epsilon)^{1/2} e^{-px/2} - (\epsilon/2)^{1/2}\ e^{px/2}}. \tag{5}$$

If we set the phase shift $\ell = (1/2)\ln(\epsilon/2)$, we have

$$y = -\coth(\pm Kx/2 + \Delta).$$

Coffey (1990)

15. $$yy''' - 3y'y'' + (1/\alpha)y^2y'' - (6/\alpha)yy'^2 - (6/\alpha^2)y^3y' - (2/\alpha^3)y^5 = 0,$$

where α is a parameter.

If we put $y = \alpha T'/T$, the given DE assumes the form

$$T'T^{\text{iv}} - 3T''T''' = 0. \tag{1}$$

Now put $T' = v$ in (1) so that

$$vv''' - 3v'v'' = 0. \tag{2}$$

For treatment of Eqn. (2), see Eqn. (3.9.44). We conclude that if the general solution of (2) is $v = \phi(x, c_1, c_2, c_3)$, the general solution of the given DE has the form

$$y = \alpha\phi(x, c_1, c_2, c_3)/\{c_4 + \int \phi(x, c_1, c_2, c_3)dx\}.$$

Kozulin and Lukashevich (1988)

16. $$yy''' - 3y'y'' + (1/\alpha)y^2y'' - (6/\alpha)yy'^2 - (6/\alpha^2)y^3y' - (2/\alpha^3)y^5 = 0,$$

where α is an arbitrary parameter.

It can be shown that the solutions of the given DE can be expressed in terms of quadratures involving transcendental functions. Kozulin and Lukashevich (1988)

17. $$\boldsymbol{yy''' - ay'y'' - hy^2y'' - pyy'^2 - qy^3y' - fy^5 = 0,}$$

where a, h, p, q, and f are constants.

Putting $y = \alpha T'/T$, we obtain

$$T^{\text{iv}}/T''' = aT''/T' + (h\alpha + 4 - a)T'/T, \tag{1}$$

provided that $f\alpha^3 - q\alpha^2 + (2h+p)\alpha + 6 - 2a = 0, q\alpha^2 - (3h+2p)\alpha + 5a - 12 = 0, p\alpha - 3a + 3 = 0$.

It follows from (1) that the solution y is single valued only when $a = 1$ and $h\alpha = 0, 1, 2$, or 3. Kozulin and Lukashevich (1988)

18. $$\boldsymbol{x^3yy''' + 3x^3y'y'' + 9x^2yy'' + 9x^2y'^2 + 18xyy' + 3y^2 = 0.}$$

The given DE is exact and has the first integral

$$x^3yy'' + x^3y'^2 + 6x^2yy' + 3xy^2 = C_1. \tag{1}$$

Equation (1) is again exact. The solution is easily obtained as

$$x^3y^2 = C_1x^2 + C_2x + C_3;$$

C_i $(i = 1, 2, 3)$ are arbitrary constants. Murphy (1960), p.428

19. $$\boldsymbol{y^2y''' + 1/3 = 0.}$$

(a) The given DE has a very implicit "exact" solution—of little practical value. Its asymptotic solution for $x \to +\infty$ is

$$y(x) \sim ax^2 + bx + c + \frac{1}{18a^2x} - \frac{b}{36a^3x^2} + \frac{3b^2 - 2ac}{180a^4x^3} + \cdots. \tag{1}$$

In the limit $a \to 0$, the behavior of (1) is

$$\begin{aligned} y(x) \quad \sim \quad & x(\ln x)^{1/3}[1 + \{A/(\ln x)\} - \{A^2 + 10/27\}/\{(\ln x)^2\} \\ & + \{(5/3)A^3 + (50/27)A\}/\{(\ln x)^3\} + \cdots], \end{aligned} \tag{2}$$

where A is a parameter. See also Sachdev (1991).

(b) $y(0) = a > 0, y'(0) = b, y''(0) = 0.$

It may be shown that the solution of the given IVP reaches $y = 0$ for some finite $x > 0$. Bender and Orszag (1978), pp.155–9

20. $$\boldsymbol{y^2y''' - 1 = 0.}$$

Let x and y be represented in terms of a parameter t such that $dx/dt = y$. Then

$$x = \int y dt \quad \text{and} \quad \ln y = \int z dt,$$

where $z = dy/dx$. It is evident that $d/dx = (1/y)(d/dt)$, so that

$$\frac{d^3y}{dx^3} = \frac{d}{dx}\left(\frac{1}{y}\frac{dz}{dt}\right) = \frac{1}{y^2}\frac{d^2z}{dt^2} - \frac{z}{y^2}\frac{dz}{dt}.$$

Therefore, z satisfies

$$\frac{d^2z}{dt^2} - z\frac{dz}{dt} = 1,$$

which, on integration, yields the Riccati equation

$$\frac{dz}{dt} - (1/2)z^2 = t \tag{1}$$

if we put constant of integration equal to zero. The substitution

$$z = \frac{d}{dt}\{\ln u^{-2}(t)\} \tag{2}$$

carries (1) to the (linear) Airy's equation

$$\frac{d^2u}{dt^2} + (1/2)tu = 0,$$

with the general solution

$$u(t) = t^{1/2}[C_1 J_{1/3}\{(1/3)t(2t)^{1/2}\} + C_2 J_{-1/3}\{(1/3)t(2t)^{1/2}\}]. \tag{3}$$

Using (2), we can write

$$y = e^{\int z dt} = e^{\int \frac{d}{dt}\{\ln u^{-2}(t)\}dt} = u^{-2}(t).$$

Hence the solution in a parametric form can be written as

$$y = u^{-2}(t), x = \int u^{-2}(t)dt,$$

where $u(t)$ is given by (3) [see also Eqn. (3.1.2)]. Ford (1992)

21. $$\boldsymbol{y^2y''' - yy'y'' + y^3y'' - 2y^4y' - Kyy' - (K/2)y^3 = 0,}$$

where K is a constant.

The given DE can be shown to possess fixed critical points in its solution, provided that $K = 0$ [see Eqn. (3.9.8)]. Exton (1973)

22. $$\boldsymbol{y^2y''' - yy'y'' - 4y^4y' - K_1y^3 - K_2yy' = 0,}$$

where K_1 and K_2 are constants.

If $K_1 = 0$, we may integrate the given DE to obtain $y'' = 2y^3 + Cy - K_2$, where C is an arbitrary constant.

When $K_1 \neq 0, K_2 \neq 0$, integration with respect to x yields $y'' = 2y^3 + (K_1x + C)y - K_2$, which is equivalent, by a simple transformation, to

$$y'' = 2y^3 + xy + K_3, \tag{1}$$

where K_3 is another constant. Equation (1) is P_{II}. Exton (1973)

23. $$\boldsymbol{y^2y''' - 2yy'y'' + y'^3 - 4y^3y' - Ky' = 0,}$$

where K is a constant.

Integrating with respect to x, we get

$$y'' = y'^2/y + 2y^2 + C - K/y, \tag{1}$$

where C is an arbitrary constant. Equation (1), and hence the given DE, may be shown to possess only fixed critical points. For $C = 0$, $K = 0$ in (1), see Eqn. (2.32.7). Exton (1973)

24. $$\boldsymbol{y^2y''' - 2yy'y'' + y'^3 - 3y^4y' - Ky' = 0,}$$

where K is a constant.

On integration w.r.t. x, we get

$$y'' = y'^2/y + y^3 - K/y + C, \tag{1}$$

where C is an arbitrary constant. Equation (1) has fixed critical points in its solutions; so has the given DE. See Eqn. (2.32.26) for the special case of (1) with $K = C = 0$. Exton (1973)

25. $$\boldsymbol{y^2y''' - 2yy'y'' + y'^3 + y^2y' + y^3y'' = 0.}$$

This is a special case of Eqn. (3.9.26) with $K_1 = K_2 = 0$. Exton (1973)

26. $$\boldsymbol{y^2y''' - 2yy'y'' + y'^3 + (y^2y' + y^3y'') - K_1y^2 - K_2y' = 0,}$$

where K_1 and K_2 are constants.

Integrating the given DE with respect to x, we have

$$y'' = y'^2/y - yy' + K_1x - K_2/y + C, \tag{1}$$

where C is an arbitrary constant. Equation (1) may be shown to possess only fixed critical points in its solution only if $K_1 = K_2 = 0$. Exton (1973)

27. $$\boldsymbol{y^2y''' - 2yy'y'' + y'^3 + 2y^3y'' + y^4y' - Ky' = 0,}$$

where K is a constant.

Integrating w.r.t. x, we get

$$y'' - y'^2/(2y) + 2yy' + (1/2)y^3 + K/(2y) - Cy = 0, \tag{1}$$

where C is an arbitrary constant. Equation (1) may be shown to have fixed critical points in its solutions and so has the given DE. For $K = C = 0$ in (1), see Eqn. (2.32.26). Exton (1973)

28. $$\boldsymbol{y^2y''' - 3yy'y'' + (3/2)y'^3 + (3/2)y^4y' - Ay^2y' - K_1y' = 0,}$$

where A and K_1 are constants.

Integrating w.r.t. x, we have

$$y'' = y'^2/2y - (3/2)y^3 + Cy^2 - Ay - K_1/(3y), \tag{1}$$

where C is an arbitrary constant. Equation (1), and hence the given DE, may be shown to possess fixed critical points in its solutions. Exton (1973)

29.

$$\begin{aligned} &y^2y''' - 3yy'y'' + (3/2)y'^3 + (3/2)y^4y' - g(x)y^2y' \\ &- K_1y' - K_2y^4 + [g'(x)/2]y^3 - K_2y^2 = 0, \end{aligned}$$

where K_1 and K_2 are constants.

It is easy to check that a series (solution) development about an arbitrary pole is possible only if $g = A$ (constant) and $K_2 = 0$; now see Eqn. (3.9.28). Exton (1973)

30.

$$y^2y''' - 3yy'y'' + 2y'^3 + \{1/\alpha\}y^3y'' = 0,$$

where α is a parameter.

If we put $y = \alpha T'/T$, the given DE assumes the form

$$T'^2T^{\text{iv}} - bT'T''T''' - dT''^3 = 0, \tag{1}$$

where b and d are constants. Now, put $T' = v$ in (1) so that

$$v^2v''' - bvv'v'' - dv'^3 = 0. \tag{2}$$

For treatment of Eqn. (2), see Eqn. (3.9.44). We conclude that if the general solution of (2) is $v = \phi(x, c_1, c_2, c_3)$, the general solution of the given DE has the form

$$y = \alpha\phi(x, c_1, c_2, c_3)/\{c_4 + \int \phi(x, c_1, c_2, c_3)dx\}.$$

Kozulin and Lukashevich (1988)

31.

$$\begin{aligned} &y^2y''' - 3yy'y'' + \{2(N^2-1)/N^2\}y'^3 + (1/\alpha)y^3y'' \\ &- \{6/(\alpha N^2)\}y^2y'^2 - \{6/(\alpha^2N^2)\}y^4y' - \{2/(\alpha^3N^2)\}y^6 = 0, \end{aligned}$$

where $N \in Z, N \neq 0$, and α is a parameter.

If we put $y = \alpha T'/T$, the given DE assumes the form

$$T'^2T^{\text{iv}} - bT'T''T''' - dT''^3 = 0, \tag{1}$$

where b and d are constants. Now, put $T' = v$ in (1) so that

$$v^2v''' - bvv'v'' - dv'^3 = 0. \tag{2}$$

For treatment of Eqn. (2), see Eqn. (3.9.44). We conclude that if the general solution of (2) is $v = \phi(x, c_1, c_2, c_3)$, the general solution of the given DE has the form

$$y = \alpha\phi(x, c_1, c_2, c_3)/\{c_4 + \int \phi(x, c_1, c_2, c_3)dx\}.$$

Kozulin and Lukashevich (1988)

32. $$y^2y''' - 4yy'y'' + 3y'^3 - y^4y' - Ky' = 0,$$

where K is a constant.

On integration w.r.t. x, we have

$$y'' = y'^2/y + y^3 + Cy^2 - K/(3y), \tag{1}$$

where C is an arbitrary constant. Equation (1), and hence the given DE, may be shown to possess no critical moving points in its solutions. Exton (1973)

33. $$y^2y''' - 4yy'y'' + 3y'^3 - y^4y' - Kyy' = 0,$$

where K is a constant.

The given DE may be shown to possess fixed critical points. Exton (1973)

34. $$y^2y''' - 4yy'y'' + 3y'^3 - y^4y' - Kyy' + (K^2/4)y' = 0,$$

where K is a constant.

The given DE may be shown to possess fixed critical points in its solutions only if $K = 0$. Exton (1973)

35. $$y^2y''' - 4yy'y'' + 3y'^3 - y^4y' - K_1(y'^2 - y^4 + y^2) - K_2y' = 0,$$

where K_1 and K_2 are constants.

The given DE may be shown to possess fixed critical points in its solutions only if $K_1 = 0$. Exton (1973)

36. $$y^2y''' - 4yy'y'' + 3y'^3 - g(y)y' = 0.$$

Put

$$y' = p, y'' = p\frac{dp}{dy}, y''' = p^2\frac{d^2p}{dy^2} + p\left(\frac{dp}{dy}\right)^2$$

in the given DE and simplify; we get

$$\frac{d}{dy}\left(p\frac{dp}{dy}\right) - (4p/y)\frac{dp}{dy} + 3p^2/y^2 - g(y)/y^2 = 0. \tag{1}$$

Writing $p^2 = P$ in (1), we have the linear DE

$$y^2\frac{d^2P}{dy^2} - 4y\frac{dP}{dy} + 6P = 2g(y). \tag{2}$$

The LHS of (2) is a Cauchy–Euler type and when equated to zero is exactly solvable:

$$P(y) = C_1y^2 + C_2y^3,$$

where C_1 and C_2 are arbitrary constants. For a given $g(y)$, Eqn. (2) is easily solved and hence, with the help of $dy/dx = p = P^{1/2}$, we can obtain the solution of the original DE in terms of quadratures.

The special cases, $g(y) = 0$ and $g(y) = 2\alpha y^4, \alpha$ a constant, are easily integrated. Tatarkiewicz (1963)

37. $$y^2y''' - (9/2)yy'y'' + (15/4)y'^3 = 0.$$

The given DE can be written as

$$-8y^{7/2}\frac{d^3}{dx^3}y^{-1/2} = 0;$$

hence we have

$$y = C \quad \text{and} \quad y^{-1/2} = C_2x^2 + C_1x + C_0,$$

where C and C_i $(i = 0, 1, 2)$ are arbitrary constants. Kamke (1959), p.602

38. $$y^2y''' - 5yy'y'' + (40/9)y'^3 = 0.$$

The given DE becomes exact if we multiply it by $y^{-11/3}$; therefore, we have the first integral,

$$9y^{-5/3}y'' - 15y^{-8/3}y'^2 = C.$$

More easily, however, if we put $u(x) = y^{-2/3}$, we get

$$u''' = 0;$$

therefore, we have

$$y^2 = (C_2x^2 + C_1x + C_0)^{-3}.$$

C and C_i $(i = 0, 1, 2)$ are arbitrary constants. Kamke (1959), p.602

39. $$y^2y''' - \{3(N-1)/N\}yy'y'' + \{(N-1)(2N-1)/N^2\}y'^3 + \{(N+3)/N\}y^3y''$$
$$+ \{3/N^2\}y^2y'^2 + \{3(N+1)/N^2\}y^4y' + y^6/N^2 = 0,$$

where $N \in Z, N \neq 0$.

The given DE has the general solution

$$y = \frac{[(x + c_1)^2 + c_2]^N}{c_3 + \int[(x + c_1)^2 + c_2]^N dx},$$

where $N \in Z$, and c_i $(i = 1, 2, 3)$ are arbitrary constants. Kozulin and Lukashevich (1988)

40. $$y^2y''' - \{3(N-1)/N\}yy'y'' + \{(N-1)(2N-1)/N^2\}y'^3$$
$$+ \{(N+3)/(\alpha N)\}y^3y'' + \{3/(\alpha N^2)\}y^2y'^2 + \{3(N+1)/(\alpha^2N^2)\}y^4y'$$
$$+ \{1/(\alpha^3N^2)\}y^6 = 0, \alpha = -1, N \in Z, N \neq 0.$$

The given DE is shown to have no movable critical points. Kozulin and Lukashevich (1988)

41. $$y^2y''' - \{(3N-1)/N\}yy'y'' + \{(N-1)(2N+1)/N^2\}y'^3 + \{(N+1)/(\alpha N)\}y^3y''$$
$$+ \{3/(\alpha N^2)\}y^2y'^2 + \{(N-3)/(\alpha^2N^2)\}y^4y' - \{1/(\alpha^3N^2)\}y^6 = 0,$$

where $N \in Z, N \neq 0$ and α is a constant.

If we put $y = \alpha T'/T$, the given DE assumes the form

$$T'^2 T^{iv} - bT'T''T''' - dT''^3 = 0, \tag{1}$$

where b and d are constants. Now, put $T' = v$ in (1) so that

$$v^2 v''' - bvv'v'' - dv'^3 = 0. \tag{2}$$

For treatment of Eqn. (2), see Eqn. (3.9.44). We conclude that if the general solution of (2) is $v = \phi(x, c_1, c_2, c_3)$, the general solution of the given DE has the form

$$y = \alpha\phi(x, c_1, c_2, c_3)/\left\{c_4 + \int \phi(x, c_1, c_2, c_3)dx\right\}.$$

Kozulin and Lukashevich (1988)

42.
$$\begin{aligned} &\mathbf{y^2 y''' - \{(3N-2)/N\}yy'y'' + \{2(N-1)/N\}y'^3} \\ &\mathbf{+ \{(N+2)/(\alpha N)\}y^3 y'' + \{2/(\alpha^2 N)\}y^4 y' = 0,} \end{aligned}$$

where $N \in Z, N \neq 0$ and $\alpha \neq 0$ is a constant.

If we put $y = \alpha T'/T$, the given DE assumes the form

$$T'^2 T^{iv} - bT'T''T''' - dT''^3 = 0, \tag{1}$$

where b and d are constants. Now put $T' = v$ in (1) so that

$$v^2 v''' - bvv'v'' - dv'^3 = 0. \tag{2}$$

For treatment of Eqn. (2), see Eqn. (3.9.44). We conclude that if the general solution of (2) is $v = \phi(x, c_1, c_2, c_3)$, the general solution of the given DE has the form

$$y = \alpha\phi(x, c_1, c_2, c_3)/\{c_4 + \int \phi(x, c_1, c_2, c_3)dx\}.$$

Kozulin and Lukashevich (1988)

43.
$$\begin{aligned} &\mathbf{y^2 y''' - \{(3N+1)/N\}yy'y'' + \{(2N+1)/N\}y'^3} \\ &\mathbf{+ \{(N-1)/(\alpha N)\}y^3 y'' - \{1/(\alpha^2 N)\}y^4 y' = 0,} \end{aligned}$$

where $N \in Z, N \neq 0$, and α is a constant.

If we put $y = \alpha T'/T$, the given DE assumes the form

$$T'^2 T^{iv} - bT'T''T''' - dT''^3 = 0, \tag{1}$$

where b and d are constants. Now put $T' = v$ in (1) so that

$$v^2 v''' - bvv'v'' - dv'^3 = 0. \tag{2}$$

For treatment of Eqn. (2), see Eqn. (3.9.44). We conclude that if the general solution of (2) is $v = \phi(x, c_1, c_2, c_3)$, the general solution of the given DE has the form

$$y = \alpha\phi(x, c_1, c_2, c_3)/\left\{c_4 + \int \phi(x, c_1, c_2, c_3)dx\right\}.$$

Kozulin and Lukashevich (1988)

44. $$y^2 y''' - byy'y'' - dy'^3 = 0,$$

where b and d are constants.

Put $y' = p, p^2 = w$ in the given DE; we get

$$y^2 w'' - byw' - 2dw = 0. \tag{1}$$

We seek solution of (1) in the form $w = y^\rho$, so that

$$\rho^2 - (b+1)\rho - 2d = 0. \tag{2}$$

Therefore, the general solution of (1) is

$$w = c_1 y^{\rho_1} + c_2 y^{\rho_2} \tag{3}$$

if $\rho_1 \neq \rho_2$ and

$$w = (c_1 + c_2 \ln y) y^{\rho_1} \tag{4}$$

if $\rho_1 = \rho_2$. For (4), we have

$$\frac{dy}{dx} = y^{(\rho_1)/2}(c_1 + c_2 \ln y)^{1/2}$$

or

$$\int \frac{dy}{y^{(\rho_1)/2}\{c_1 + c_2 \ln y\}^{1/2}} = x + c_3. \tag{5}$$

For the special case $\rho_1 = \rho_2 = 2$, the solution (5) becomes

$$y = \exp\{(c_2/4)(x + c_3)^2 - c_1/c_2\}, \tag{6}$$

which is an entire function. For $\rho_1 \neq \rho_2$, (3) leads to

$$\int \frac{dy}{y^{(\rho_1)/2}\{y^{\rho_2 - \rho_1} + H\}^{1/2}} = c_2^{1/2}(x + c_3), H \equiv c_1/c_2. \tag{7}$$

In particular, if $\rho_1 = 0$ and $\rho_2 \neq 0$, we obtain

$$\int \frac{dy}{\{y^{\rho_2} + H\}^{1/2}} = (c_2)^{1/2}(x + c_3). \tag{8}$$

It is clear from (8) that y is single valued if $\rho_2 = 1, 2, 3, 4$. Now suppose that $\rho_1 \neq \rho_2, \rho_1\rho_2 \neq 0$, and $\rho_1 \neq 2$. Putting $y = z^{2/(2-\rho_1)}$ in (7), we get

$$\int \frac{dz}{\{z^k + H\}^{1/2}} = \{(2 - \rho_1)/2\} c_2^{1/2}(x + c_3), k \equiv 2(\rho_2 - \rho_1)/(2 - \rho_1).$$

Therefore, z is single valued if $k = 1, 2, 3, 4$. Correspondingly y is single valued only if $\rho_1 = 2(N-1)/N, N \in Z, N \neq 0$. It follows from (2) that $\rho_1 + \rho_2 = b + 1, \rho_1\rho_2 = -2d$.

Therefore, we conclude that the coefficients must satisfy the following conditions if the solution is to be single valued:

(a) $b = 3(N-1)/N, d = -(N-1)(2N-1)/N^2$, if $k = 1$.

(b) $b = (3N-2)/N, d = -2(N-1)/N$, if $k = 2$.

(c) $b = (3N-1)/N, d = -(N-1)(2N+1)/N^2$, if $k = 3$.

(d) $b = 3, d = -2(N^2-1)/N^2$, if $k = 4$.

Finally, if $\rho_1 \neq \rho_2, \rho_1\rho_2 \neq 0, \rho_1 = 2$, we put

$$y = (z - H)^{1/\{\rho_2 - 2\}} \tag{9}$$

in (7) and obtain

$$\int \frac{dz}{(z-H)z^{1/2}} = (\rho_2 - 2)c_2^{1/2}(x + c_3).$$

Therefore, in this case z is single valued. For y in (9) to be single valued, it is necessary that $\rho_2 = (2N+1)/N, N \in Z$. Kozulin and Lukashevich (1988)

45. $$\boldsymbol{y^2y''' - byy'y'' - dy'^3 - hy^3y'' - py^2y'^2 - qy^4y' - fy^6 = 0,}$$

where b, d, h, p, q, and f are constants.

Necessary and sufficient conditions are found such that the given DE has no moving critical points. Special cases thus arising from them are treated separately and solved explicitly when it is possible. Kozulin and Lukashevich (1988)

46. $$\boldsymbol{y^2y''' - a_1yy'y'' - b_1y'^2 - (a_2yy'' + b_2y'^2)(a_3yy'' + b_3y'^2)^{1/2} = 0,}$$

where $a_k, b_k, k = 1, 2, 3$ are constants, $a_3 \neq 0$.

Necessary and sufficient conditions for the uniqueness of the general solution of the given DE are provided. Martynov (1985a)

47. $$\boldsymbol{(x + y^2)y''' + 6yy'y'' + 3y'' + 2y'^3 = 0.}$$

The given DE is exact and has the first integral

$$(x + y^2)y'' + 2yy'^2 + 2y' = C, \tag{1}$$

C is an arbitrary constant. Equation (1) is again exact. The solution is easily obtained as

$$2y(3x + y^2) = C_1 + C_2(1 + 3x^2) + 6C_3x;$$

C_i $(i = 1, 2, 3)$ are arbitrary constants. Murphy (1960), p.428

48. $$\boldsymbol{y^3y''' + 3y^2y'y'' - y' + (3/c)y^2y' = 0,}$$

where $y \to 1$ as $x \to \pm\infty$; c is a constant.

The given system describes solitary waves in magma dynamics. Implicit solution of the problem is

$$|x| = (A + 1/2)^{1/2}\left\{2(A-y)^{1/2} - \frac{1}{(A-1)^{1/2}}\ln\frac{(A-1)^{1/2} - (A-y)^{1/2}}{(A-1)^{1/2} + (A-y)^{1/2}}\right\}, \tag{1}$$

where A which is always greater than 1, stands for the amplitude of the solitary wave at the origin. It may be noted that two integrations and boundary conditions of the given system lead to

$$y'^2 = [(y-1)^2(c - 1 - 2y)/(cy^2)], c = 2A + 1,$$

which results in (1). Barcilon and Lovera (1989)

3.10 $f(x, y, y', y'')y''' + g(x, y, y', y'') = 0$

1. $$y'y''' - x^4 yy'' = 0.$$

The asymptotic behavior for $x \to +\infty$ may be found following Levinson (1970); see Sachdev (1991). The given DE is equidimensional in y and therefore admits exponential asymptotic behavior. $y = C$ is also a solution. Bender and Orszag (1978), p.200

2. $$y'y''' - y''^2 = 0.$$

This DE has solutions

$$y = (1/C_1)\exp(C_1 x + C_2)$$

and

$$y = C_3 x + C_4,$$

where C_i $(i = 1, \ldots, 4)$ are arbitrary constants. Martynov (1985a)

3. $$y'y''' - y''^2 - y'^3 f(y) = 0.$$

Put

$$y' = p(y), y'' = p\frac{dp}{dy}, y''' = p^2\frac{d^2p}{dy^2} + p\left(\frac{dp}{dy}\right)^2$$

so that the given DE becomes linear:

$$\frac{d^2p}{dy^2} = f(y). \tag{1}$$

Equation (1) may be integrated twice and the solution obtained by using $dy/dx = p$, etc. $y = C$, a constant, is another solution. Tatarkiewicz (1963)

4. $$2y'y''' - 3y''^2 = 0.$$

The given DE has the solution

$$y = (Ax + B)/(Cx + D)$$

for arbitrary $A, \ldots, D$. Kamke (1959), p.602

5. $$y'y''' - 2(y'')^2 = 0.$$

Writing the given DE as $y'''/y'' = 2y''/y'$ and integrating twice, etc., a general solution may be found to be $y = A\log(x - x_0) + B$, where A, x_0, and B are constants. Drazin and Johnson (1989), p.188

6. $$y'y''' - 2y''^2 + y'^2 = 0.$$

With $y' = p(y)$, we have

$$p^2(pp'' + 1 - p'^2) = 0;$$

see Eqn. (2.31.25). The solution is easily found to be either $y =$ constant or $x = \ln\{\sec(C_1 + C_2 y) + \tan(C_1 + C_2 y)\} + C_3$, where C_i $(i = 1, 2, 3)$ are arbitrary constants. Murphy (1960), p.428

7. $$\boldsymbol{y'y''' - 3y''^2 - axy'^5 = 0.}$$

Interchanging the dependent and independent variables, we get the linear DE

$$x'''(y) + ax = 0.$$

Murphy (1960), p.428

8. $$\boldsymbol{(2x^3y' - x^4y')y''' - x^3y'^2 + 3xyy' - 2y^2 = 0.}$$

It is shown by the use of a general theorem that the general solution of the given DE is given parametrically by

$$x = c_2 \exp\left(\int w(z, c_1)dz\right), y = c_3 \exp\left(\int (z+1)w(z, c_1)dz\right),$$

where z is the parameter and w is governed by the Abel equation

$$\frac{dw}{dz} = z^3w^3 + 3zw^2, \tag{1}$$

which has the solution

$$w(z^2w + 2) = c_1(z^2w + 1)^2;$$

c_i $(i = 1, 2, 3)$ are constants. Bandic (1965a)

9. $$\boldsymbol{yy'y''' - y'^2y'' + y(y'')^2 = 0.}$$

Put $r(y) = y'(x)y''(x)$ so that

$$r'(y(x))y'(x) = [y''(x)]^2 + y'(x)y'''(x) = y'r/y.$$

Therefore, either $y =$ constant or $r = Cy$, where C is an arbitrary constant. Solving $r \equiv y'y'' = Cy$, we have

$$\frac{dy}{dx} = \{(3C/2)y^2 + D\}^{1/3},$$

where D is another arbitrary constant. Hence the solution is found in terms of a quadrature. Tatarkiewicz (1963)

10. $$\boldsymbol{(y'^2 + 1)y''' - 3y'y''^2 = 0.}$$

With $y' = z$, the given DE goes to Eqn. (2.37.18). The solutions are the circles

$$(y - C_1)^2 + (x - C_2)^2 = C_3^2,$$

where C_i $(i = 1, 2, 3)$ are arbitrary constants. Kamke (1959), p.602

11. $$\boldsymbol{(y'^2 + 1)y''' - (3y' + a)y''^2 = 0,}$$

where a is a constant.

With $y' = p(x)$ the given DE has the form of the Eqn. (2.16.68). Therefore, one has

$$p'(x) = C(p^2 + 1)^{3/2} \exp(a \tan^{-1} p).$$

Now put $t = \tan^{-1} p$ and hence obtain the parametric representation of the solution

$$\begin{aligned} x &= C_2 + C_1 e^{at}(a \cos t + \sin t), \\ y &= C_3 + C_1 e^{at}(a \sin t + \cos t), \end{aligned}$$

which represents logarithmic spirals for $a \neq 0$ and circles for $a = 0$. C_i $(i = 1, 2, 3)$ are arbitrary constants. Kamke (1959), p.603

12. $$\{(\delta_1/2)y'^2 + (\delta_2/2)y'\}y''' - \{(3\delta_1/4)y' + \delta_2\}y''^2 + Dy'^3 = 0,$$

where δ_1, δ_2, and D are constants. The given DE arises in the study of nonuniform extension of a slab of Mooney–Rivlin material.

Writing the DE as

$$(-\delta_1/4)\{y'^2(y'')^{-1}\}\,\{y''^2 y'^{-3}\}' - (\delta_2/4)\{y'^2 y''^{-1}\}\{y''^2(y')^{-4}\}' = D,$$

multiplying by $y''(y')^{-2}$, and integrating, we have

$$(\delta_1/4)(y''^2/y'^2) + (\delta_2/4)(y''^2/y'^3) = Cy' + D, \tag{1}$$

where C is an arbitrary constant. Rewriting (1) as

$$(y''/y')^2 = y'(Ey' + G)/(\delta_1 y' + \delta_2), \tag{2}$$

where $E = 4C$ and $G = 4D$, we have

$$y'' = \pm y'[\{y'(Ey' + G)\}/(\delta_1 y' + \delta_2)]^{1/2}. \tag{3}$$

Putting $y' = u$ in (3) and integrating, we have

$$\int [(\delta_1 u + \delta_2)/\{u^3(Eu + G)\}]^{1/2} du = \pm x + H, \tag{4}$$

where H is another arbitrary constant. The integral on the LHS of (4) can be expressed in terms of elliptic functions. Rajagopal, Troy, and Wineman (1986)

13. $$y'^3 y''' - 1 = 0.$$

With $y'(x) = u(x)$, we have

$$u^3 u'' = 1. \tag{1}$$

Writing Eqn. (1) as $u'u'' = u'/u^3$ and integrating twice, etc., we get

$$2yC_1^{1/2} + C_3 = (C_2 + x)R(x) + C_1^2 \ln(C_2 + x + R),$$

where $R(x) = \{(C_2 + x)^2 + C_1^2\}^{1/2}$ and C_i $(i = 1, 2, 3)$ are arbitrary constants. Murphy (1960), p.428

14. $$(y')^5 y''' - 3(y')^4(y'')^2 - (y'')^3 = 0.$$

Put $y' = v$, to obtain

$$v^5 v'' = 3v^4(v')^2 + (v')^3. \tag{1}$$

Now put $z = v, w = v'(x)$ in (1); we get

$$z^5 \frac{dw}{dz} = 3z^4 w + w^2. \tag{2}$$

Equation (2) is Riccati, which is simplified by writing $t = -1/z$ and $\phi = w/z^3$; we thus have $d\phi/dt = \phi^2$, whose solution is $\phi = 1/(c-t)$, where c is a constant, or $w = z^4/(cz+1)$; that is,

$$\frac{dv}{dx} = v^4/(cv+1). \tag{3}$$

Equation (3) integrates to yield the implicit relation

$$6(x - c_1)v^3 + 3cv + 2 = 6(x - c_1)(y')^3 + 3cy' + 2 = 0. \tag{4}$$

Solving (4) for $y'(x)$, one can obtain the general solution in terms of a quadrature involving three arbitrary constants. Olver (1986), p.157

15. $$(g(y'))'' + yy'' = 0 \text{ on } (-\infty, \infty),$$

where $y(-\infty) = a$ and $y'(\infty) = 1; \lim_{x\to\infty} \{xy'(x) - y(x)\} = 0$, where a is real; $g(s) = 2\lambda s + |s|s, \lambda \geq 0$.

Some qualitative properties of the given problem are proved, and asymptotic results are obtained. Van Duijn and Peletier (1992)

16. $$y''y''' - 2 = 0.$$

The given DE is immediately integrable. The solution is

$$15y = 8(C_1 + x)^{5/2} + C_2 x + C_3,$$

where C_i $(i = 1, 2, 3)$ are arbitrary constants. Murphy (1960), p.428

17. $$y''y''' - a(b^2 y''^2 + 1)^{1/2} = 0,$$

where a and b are constants.

With $y''^2 = u$, the given DE is written as

$$u' = 2a(b^2 u + 1)^{1/2}.$$

Hence the solution is obtained in parametric form:

$$\begin{aligned} x &= C_1 + v/(ab^2), \\ y &= C_2 + C_3 v/(ab^2) + u^3/(6a^2b^2) + u/(2a^2b^4) - \{1/(2a^2b^5)\} \log(bu + v), \end{aligned}$$

where $v = (b^2 u + 1)^{1/2}$. Kamke (1959), p.603

18. $$2xy''y''' - y''^2 + a^2 = 0,$$

where a is a constant.

With $y''(x) = u(x)$, we get

$$2xuu' = u^2 - a^2. \tag{1}$$

Equation (1) integrates to give

$$u^2 = Cx + a^2.$$

The solution of the given DE is now immediate:

$$15C_1 y = 4(C_1^2 a^2 + x)^{5/2} + C_2 x + C_3,$$

where C_i $(i = 1, 2, 3)$ are arbitrary constants. Murphy (1960), p.429

19. $$(x^3 y'' - 2xy)y''' - 4yy'' + 2x^2 y''^2 = 0.$$

Using a general theorem, it is shown that the general solution of the given DE is given parametrically by

$$x = c_2 \exp\left(-\int \{w/\epsilon^2\} dz\right), y = c_3/\epsilon,$$

where ϵ is the parameter and w is governed by the Abel equation

$$\frac{dw}{dz} = (1/\epsilon^2)(w^3 + w^2),$$

whose general solution is $\ln c_1\{(w + 1)/w\} = 1/w + z/\epsilon^2, c_i$ $(i = 1, 2, 3)$ are constants. Bandic (1965a)

20. $$a_0(x)yy''' + a_1(x)y'y''' + a_2(x)y''y''' + a_3(x)y''^2$$
$$+ a_4(x)y'y'' + a_5(x)yy'' + a_6(x)y'^2 + a_7(x)yy' + a_8(x)y^2 = 0,$$

with coefficients a_j $(j = 0, 1, \ldots, 8)$ analytic in x in some domain D. Conditions are found when the solutions have only single valued polar singularities.

Putting $y' = vy$, we get the equation

$$\begin{aligned}(a_2v' + a_2v^2 + a_1v + a_0)v'' + (3a_2v + a_3)v'^2 + \{4a_2v^3 \\ + (3a_1 + 2a_3)v^2 + (a_4v + a_5 + 3a_0v)\}v' \\ + a_2v^5 + (a_1 + a_3)v^4 + (a_0 + a_4)v^3 + (a_5 + a_6)v^2 + a_7v + a_8 = 0.\end{aligned} \tag{1}$$

It is shown that for the given equation to have only single valued polar singularities, it is necessary and sufficient that any solution of (1) has only poles with integer residues as their only movable singularity.

A necessary condition for (1) to have poles as their only movable singularity, with integer residues, is that $a_2 = 0$. Then (1) becomes

$$\begin{aligned}(a_1v + a_0)v'' &= -a_3v'^2 - [(3a_1 + 2a_3)v^2 + (3a_0 + a_4)v + a_5]v' \\ &\quad - (a_1 + a_3)v^4 - (a_0 + a_4)v^3 - (a_5 + a_6)v^2 - a_7v - a_8.\end{aligned} \tag{2}$$

Two distinct cases arise: (I) $a_1 = 0$ and (II) $a_1 \neq 0$. In each of these cases, subcases are considered such that (2) can be made to correspond to Painlevé's equations and hence necessary and sufficient conditions obtained so that (2) has poles as its only singularities, with integral values as residues.

In (I) the subcases are (a) $a_4 = -a_0$, (b) $a_4 = 0$, (c) $a_4 = -3a_0$, and (d) $a_4 = -2a_0$. In (II) the subcases are:

(a) $a_4 = a_8 = 0, a_5 = -a_6, (a_5/a_1)' = -a_7/a_1$.

(b) $a_4 = a_5 = a_8 = 0, a_7/a_1 + [\{(a_5 + a_6)/a_1\}'/\{(a_5 + a_6)/a_1\}]' = 0$.

(c) $a_4 = a_5 = a_6 = 0, a_8 = -a_1, a_7 = -a_1(z + \alpha), \alpha$ a constant.

(d) $a_3 = (1 - m)/m, m > 1$ an integer.

Another case considered briefly is $a_1 \neq 0, a_0 \neq 0$. It is reduced to the case $a_1 \neq 0, a_0 = 0$, by suitable transformation. Kolesnikova and Lukashevich (1972)

21. $$(1 - y''')(1 + y''^2)^{1/2} - y''y''' = 0.$$

With $y''(x) = u(x)$, we have

$$u'(u + R) = R, \tag{1}$$

where $R = (1 + u^2)^{1/2}$. The solution of (1) is

$$2u = x + C - \{1/(x + C)\};$$

C is an arbitrary constant. The solution of the original DE is now immediate:

$$12y = (x + C_1)^3 + C_2(x + C_1) - 6(x + C_1)\ln(x + C_1) + C_3,$$

where C_i $(i = 1, 2, 3)$ are arbitrary constants. Murphy (1960), p.429

22. $$n|y''|^{n-1}y''' + yy'' + Cy'^2 = 0, y(0) = 0, y'(0) = 0,$$

$y'(\infty) = 0$, where n and C are constants. The given DE arises in the similarity prediction of wall jets past axisymmetric bodies for power law fluids.

It is shown that for each n there is a unique value of C for which the given problem has a physically meaningful solution [with a bounded increasing function $y(x)$ and $\max f'(x) = 1$]. The values of $C, y''(0)$, etc., are tabulated, from numerical solution; the function y' is also depicted graphically. Filip, Kolar, and Hajek (1991)

23. $$u(x, y, y', y'')y''' - v(x, y, y', y'') = 0,$$

where u and v are integral rational functions of the indicated arguments, homogeneous with reference to y, y', and y'', u being of mth degree and v of nth degree. The equation is also assumed to be bidimensional in the sense that all the terms of the function u are of dimension p, and the terms of the function v are of dimension q if we suppose that the variables x and y have dimension 1. The given DE is assumed to be of third order of quasihomogeneity (m, n) and bidimensionality (p, q).

Some general results corresponding to various bidimensionality and quasihomogeneity of the given DE, leading to integrals in the form of Abel's differential equations, are proved. The solution is then expressed parametrically. Bandic (1965a)

3.11 $f(x, y, y', y'', y''') = 0, f$ nonlinear in y'''

1. $$(1-x^2)y'''^2 + 2xy''y''' - y''^2 + 1 = 0.$$

With $y''(x) = u(x)$, we get

$$(1-x^2)u'^2 + 2xuu' - u^2 + 1 = 0. \tag{1}$$

The solution of (1) is easily obtained:

$$u = Cx \pm (C^2+1)^{1/2};$$

C is an arbitrary constant. The solution of the original DE is now immediate:

$$6y = C_1x^3 \pm 3x^2(C_1^2+1)^{1/2} + C_2x + C_3,$$

where C_i $(i = 1, 2, 3)$ are arbitrary constants. Murphy (1960), p.429

2. $$(yy''' + 3y'y'')^2 - 16y(y'')^3 = 0.$$

The general solution of the given DE is

$$y(x) = \alpha(x+\beta)\exp\{(\gamma x + \delta)/(x+\beta)\},$$

where α, β, δ, and γ are arbitrary constants; it has a movable critical singularity at $x = -\beta$. Kruskal and Clarkson (1992)

3.12 $f(x, y, y', y'', y''') = p(x), p$ periodic

1. $$y''' + cy' + f(x,y) = e(x),$$

where c is a constant, $f(x, y)$ and $e(x)$ are continuous and w-periodic functions in x, and $\int_0^w e(t)dt = 0$.

Conditions on f are given such that the given DE has at least one solution y satisfying the periodic boundary conditions $y(0) = y(w), y'(0) = y'(w), y''(0) = y''(w)$. Mehri (1980/81)

2. $$y''' + ay'' + by' + h(y) = e(x),$$

where a, b are real numbers, and $h(y), e(x) \in C^0[R^1], h(y)$ is oscillatory for all y and $e(x)$ is w periodic for all x.

Two sets of conditions are given to ensure that the given DE has at least one w-periodic solution. Andres (1987)

3. $$y''' + ay'' + by' + h(y) - p(x) = 0,$$

where a and b are constants, and $p(x)$ is a continuous periodic function of x with a least period w. The function $h(y)$ is assumed continuous for all y considered so that solutions of the given DE exist satisfying any assigned IC.

Two cases arise:

(a) $|h(y)| \to \infty$ as $|y| \to \infty$. Moreover:

(i) $a > 0, b > 0$.

(ii) $h(y)$ sgn $y > 0, |y| \geq 1$.

(iii) $h'(y)$ exists and is continuous for all y, and

$$h'(y) \leq \begin{cases} c, & ab - c \equiv \delta > 0 \quad \text{for} \quad |y| \geq 1 \\ C, & ab < C < \infty \quad \text{for} \quad |y| \leq 1. \end{cases}$$

(iv) $p(x) \leq A_1, |\int_0^x p(\tau)d\tau| \leq A$, for all x.

(b) $|h(y)| \leq M < \infty$ for all y. Moreover:

(i) $a > 0, b > 0$.

(ii) $h(y)$ sgn $y \geq m > 0, |y| \geq 1$.

(iii) $|p(y)| \leq A_1$, and

$$\left| \int_0^x p(\tau)d\tau \right| \leq A_1 \text{ for all } x.$$

It is also shown that for case (a) that if $p(x)$ has a period w in x, then there exists at least one solution of the given DE with a least period w. A similar result is not proved for case (b). Ezeilo (1959)

4. $\boldsymbol{y''' - \lambda y' - 2yy' - \mu_1 (Dy')(x) = \mu_2 \sin x, y(x + 2\pi) = y(x), \int_{-\pi}^{\pi} u(x)dx = 0,}$

where $(D(y(x)) = (2/\pi)^{1/2} \int_0^\infty y(x - s)s^{-1/2}ds$ for all functions y with period 2π and mean zero. λ is regarded as a control parameter. The given integrodifferential system models resonant sloshing. Bifurcation of harmonic solutions is discussed. Reynolds (1989)

5. $$\boldsymbol{y''' + \psi(y')y'' + \{k^2 + \phi(y)\}y' + f(x, y) = e(x),}$$

where $\phi(y), \psi(y')$ are continuous, and $f(x, y), e(x)$ are continuous and periodic in x of period w.

It is proved that the given DE admits at least one w-periodic solution if:

(a) $0 < k \leq \pi/w$.

(b) $\int_0^w e(x)dx = 0$; i.e., $E(x) = \int_0^x e(t)dt$ is w-periodic.

(c) $|\Psi(y)| = \int_0^y \psi(\tau)d\tau \leq M$ for all y, where $\Psi(y) = \int_0^y \psi(y)dy$.

(d) $|\phi(y)|/|y| = |\int_0^y \phi(\tau)d\tau|/|y| \to 0$ as $|y| \to \infty$.

(e) $|f(x, y)|/|y| \to 0$ as $|y| \to \infty$, uniformly in x.

(f) $f(x, y)$ sgn $y \geq 0(|y| \geq h)$.

Mehri (1990)

3.13 $\vec{y}\,' = \vec{f}(\vec{y}); f_1, f_2, f_3$ linear and quadratic in y_1, y_2, y_3

1. $$\dot{x} = -(y+z), \quad \dot{y} = x + 0.2y, \quad \dot{z} = 0.2 + z(x - 5.7).$$

This prototype, similar to the Lorenz model of turbulence, contains just one second order nonlinearity (xz). Numerical solutions with IC $x(0) = 0, y(0) = -6.78, z(0) = 0.02$, and $t(\text{end}) = 339.249, x(\text{end}) = -7.8366, y(\text{end}) = -4.1803, z(\text{end}) = 0.014385$ were found. The flow is found to be nonperiodic and structurally stable, even though all trajectories are unstable. The flow in state space allows for a folded Poincaré map (horseshoe map). The given system, though simpler than Lorenz system, enjoys most of the features of the Lorenz system. The system illustrates a more general principle for the generation of spiral-type chaos. Results are shown graphically. Several applications in astrophysics, economics, and biology are indicated. Rössler (1976)

2. $$\dot{x} = Q(y-x), \quad \dot{y} = -rxz + rx, \quad \dot{z} = -z,$$

where Q and r are constants.

This is an approximate form of the Lorenz system. See Eqn. (3.13.11) for its solution. Fiszdon and Sen (1988)

3. $$\dot{x} = -y - z, \quad \dot{y} = x + ay, \quad \dot{z} = b + zx - cz,$$

where $\alpha = (a, b, c) \in R^3$.

The given system is called Rössler. A complete analytic study of the fixed points of the given system and their local stability condition as functions of the parameter values (a, b, c) is carried out to identify necessary and sufficient conditions for the occurrence of Hopf bifurcations. In particular, two domains of the (a, b) plane are identified for which period-doubling cascades and transition to chaotic attractors occur. Numerical results are also given. Gardini (1985)

4. $$\dot{x} = -y - z, \quad \dot{y} = x + ay, \quad \dot{z} = b - cz + xz,$$

where a, b, and c are positive constants.

Such a system is said to be dissipative in the sense of Levinson if there exists a compact subset K of the phase space such that every point moving along the trajectory must be in a set K after a sufficiently long time. It is shown that the given system is not dissipative in the sense of Levinson. Leonov and Reitmann (1986)

5. $$\dot{x} = -(y+z), \quad \dot{y} = x + ay, \quad \dot{z} = b + z(x - c),$$

where a, b, and c are constants.

Using the given Rössler system as an example, it is shown that some numerical methods may produce discrete dynamical systems that are not chaotic, even when the underlying continuous dynamical system is thought to be chaotic. In the present case it is found that the transition to chaos from false stability mimics the transition to chaos that has been observed previously as parameters were changed in the Rössler system. Parametric values chosen are $a = 0.2, b = 0.4, c = 5.7$. A backward Euler scheme was used. Corless, Essex, and Nerenberg (1991)

6. $$\dot{x} = y, \quad \dot{y} = xz, \quad \dot{z} = x^2 - z^2.$$

A phase space analysis is carried out. Kapitanov (1984)

7. $$\dot{x} = -3(x - y), \quad \dot{y} = -xz + rx - y, \dot{z} = xy - z,$$

where r is a parameter. For the given (special) Lorenz system, r is the only parameter. If $r < 1$, the only critical point is $(0, 0, 0)$ and is stable. Thus, for $r < 1$, the system cannot exhibit random behavior if $x(0), y(0), z(0)$ are small. If $r > 1$, we have three critical points: $(0, 0, 0), \{\pm(r-1)^{1/2}, \pm(r-1)^{1/2}, (r-1)\}$. The additional critical points are stable if $1 < r < 21$. For $r > 21$, all three critical points are unstable. The trajectories are drawn for the following sets of parameters:

(a) $r = 17, x(0) = z(0) = 0, y(0) = 1$. A slow and regular oscillatory approach to the stable critical point at $x = -4, y = -4, z = 16$ is observed.

(b) $r = 17$, IC as in (a). The graph of $y(t)$ versus $z(t)$ for $1 \leq t \leq 50$ is drawn. There is a slow spiral approach to the critical point $(-4, -4, 16)$.

(c) $r = 17$, IC as in (a). The graph of $y(t)$ versus $x(t)$ is drawn.

(d) $r = 26, x(0) = z(0) = 0, y(0) = 1$. The graph of $y(t)$ versus t exhibits random behavior, with intermittent and irregular oscillations.

(e) $r = 26$, IC as in (d). The graph of $z(t)$ versus $y(t)$ $(2.5 \leq t \leq 30)$ shows randomness as haphazard jumping back and forth from the neighborhoods of the unstable critical points $(5, 5, 25)$ and $(-5, -5, 25)$.

(f) $r = 26$, IC as in (d). The graph of $y(t)$ versus $x(t)$ shows randomness.

Bender and Orszag (1978), pp.192–197

8. $$\dot{x} = \sigma(y - x), \quad \dot{y} = rx - y - xz, \quad \dot{z} = xy - bz,$$

with

(a) $\sigma = 16, r = 315, b = 4, x(0) = -3.0, y(0) = -29.4, z(0) = 259, t(\text{end}) = 15$.

(b) $\sigma = 16, r = 300, b = 4.1, x(0) = -70, y(0) = 40, z(0) = 400$, (i) $t(\text{end}) = 21$, (ii) $t(\text{end}) = 109$.

The results of numerical solutions of the given Lorenz system are shown graphically. They depict Lorenzian chaos, horseshoe map chaos of the walking stick type, and the "original" horseshoe map chaos. Rössler (1977)

9. $$\dot{x} = \sigma y, \dot{y} = -rxz + rx, \quad \dot{z} = xy,$$

where σ and r are constants.

This is an approximate form of the Lorenz system. See Eqn. (3.13.11) for its solution. Fiszdon and Sen (1988)

10. $$\dot{x} = Kx - \lambda y - yz, \quad \dot{y} = x, \quad \dot{z} = -z + y^2,$$

where K and λ are constants.

The given system appears in the modeling of double magnetoconvection. It is numerically treated; bifurcation diagrams, periodic solutions, homoclinic explosions, period-doubling cascade, and chaos are depicted. The results are compared with those for the Lorenz system. Rucklidge (1992), Proctor and Weiss (1990)

11. $$\dot{x} = \sigma(y - x), \quad \dot{y} = -xz + rx - y, \quad \dot{z} = xy - z,$$

where σ and r are constants.

Early behavior of the solutions to the given Lorenz system is considered for large r for initial conditions far from its equilibrium points:

$$(0,0,0), [(r-1)^{1/2}, (r-1)^{1/2}, (r-1)]$$

and

$$[-(r-1)^{1/2}, -(r-1)^{1/2}, (r-1)].$$

Transforming according to

$$\xi = x/(r-1)^{1/2}, \eta = y/(r-1)^{1/2}, \zeta = z/(r-1),$$

we have

$$\dot{\xi} = \sigma(\eta - \xi), \dot{\eta} = -r\xi\zeta + r\xi - \eta, \dot{\zeta} = \xi\eta - \zeta, \tag{1}$$

where the approximation $r \gg 1$ is made; the critical points $(0,0,0), (1,1,1), (-1,-1,1)$ of the phase space (ξ, η, ζ) are now independent of r. Exact numerical results for $\sigma = 10$ and $r = 100$ are obtained with $\xi_i = \eta_i = \zeta_i = 10$. The initial trajectory of the solution in the phase space is seen to be quite different from and essentially independent of the long-time behavior. It can be divided into two stages. For short times, we may approximate (1) by the conservative system

$$\dot{\xi} = \sigma\eta, \dot{\eta} = -r\xi\zeta + r\xi, \dot{\zeta} = \xi\eta. \tag{2}$$

System (2) has the first integral

$$2\sigma\zeta - \xi^2 = C_1, \eta^2 + r\zeta^2 - 2r\zeta = C_2,$$

where C_1 and C_2 are arbitrary constants. Using the first of these with C_1 determined from the initial conditions, the system (2) can be reduced to

$$\ddot{\xi} + (r/2)[\xi^2 + (C_1 - 2\sigma)]\xi = 0, \tag{3}$$

which is a nonlinear oscillator and can be written as a Hamiltonian pair

$$\dot{\xi} = +\frac{\partial H}{\partial \eta}, \quad \dot{\eta} = -\frac{\partial H}{\partial \xi},$$

where

$$H(\xi, \eta) = (\sigma/2)\eta^2 + \{r/(4\sigma)\}[\xi^4/2 + (C_1 - 2\sigma)\xi^2]$$

is related to C_2 through $C_2 = (2/\sigma)H - (rC_1/4\sigma^2)(4\sigma - C_1)$. Equation (3) can be studied in the phase plane (ξ, η); it has the period of oscillation

$$T = (8/r^{1/2}) \int_0^{\pi/2} \frac{d\theta}{[4\sigma(\zeta_i - 1) - \xi_i^2 \cos^2\theta]^{1/2}}$$

when $\zeta_i = \zeta(t_i)$, the initial value.

The approximate system (2) gives an accurate frequency representation of the solution for a short time. This solution is examined in some detail.

For intermediate time, we may approximate (with experience from numerical results) the system (1) by

$$\dot{\xi} = \sigma(\eta - \xi), \quad \dot{\eta} = -r\xi\zeta + r\xi, \dot{\zeta} = -\zeta. \tag{4}$$

The last of (4) can be integrated: $\zeta = \zeta_0 e^{-t}$, and by substitution in the first two, we have $\ddot{\xi} + \sigma\dot{\xi} + \sigma r(\zeta_0 e^{-t} - 1)\xi = 0$. Using

$$\xi = \exp(-\sigma t/2)u, \;\; s = 2\exp(-t/2)(r\zeta_0\sigma)^{1/2}, \;\; \lambda = (\sigma^2 + 4r\sigma)^{1/2},$$

we get a Bessel equation of order λ:

$$s^2u'' + su' + (s^2 - \lambda^2)u = 0, \;\; ' \equiv \frac{d}{ds}.$$

The validity of the approximation in this intermediate stage is checked by comparison with numerical solution in $1 < t < 2$, with $t_i = 1, \sigma = 10, r = 100$. Exact numerical and approximate solutions are represented graphically. Fiszdon and Sen (1988)

12. $$\dot{x} = yz, \quad \dot{y} = xz + C, \quad \dot{z} = y + N,$$

where $x(0) = N, y(0) = -N, y(1) = 0$, where C and N are constants.

The given system occurs in the theory of semiconductor devices. An existence and uniqueness theorem is proved for any positive constants C and N. Some a priori estimates are also obtained. It is also shown that the problem has no solution for $C \leq 0$. Gudkov and Klokov (1982)

13. $$\dot{x}_1 = x_1 - x_1x_2 - x_3, \quad \dot{x}_2 = x_1^2 - ax_2, \quad \dot{x}_3 = bx_1 - cx_3,$$

where a, b, and c are constants.

This is a Rössler system. We have the following:

(a) The origin is always an equilibrium point. Linearization about it gives the characteristic equation $(s + a)\{s^2 + (c - 1)s + (b - c)\} = 0$, leading to the local asymptotic stability conditions $a > 0, b > c > 1$.

(b) If $d = a(1 - b/c) > 0$, then there are two more singular points at $\{\pm d^{1/2}, 1 - b/c, \pm b(d)^{1/2}/c\}$. Symmetry of these points suggests that we may consider only one.

Considering the first, we have a matrix of the linearized system:

$$A = \begin{bmatrix} b/c & -(d)^{1/2} & -1 \\ 2(d)^{1/2} & -a & 0 \\ b & 0 & -c \end{bmatrix}.$$

The characteristic equation, therefore, is

$$\det(sI - A) = s^3 + (a + c - b/c)s^2 + a(c + 2 - 3b/c)s + 2a(c - b) = 0,$$

whence, using $d > 0$, the Routh–Hurwitz conditions for local asymptotic stability become $a > 0(< 0), c > b(< b), a > b/c - c, a(c-b)\{(a+c-b/c)(c+2-3b/c) - 2(c-b)\} > 0$. Cook (1986), pp.45, 193

14. $$\dot{x} = R - zy - \nu x, \quad \dot{y} = \sigma(z-y), \quad \dot{z} = xy - z,$$

where R, ν, and σ are constants.

This is a modified dynamo system. The system has three steady solutions $(x, y, z) = (R/\nu, 0, 0)$ in the zero-field state of the dynamo. When $R < \nu$, all solutions approach the zero-field state. At $R = \nu$, $(R/\nu, 0, 0)$ loses stability, and the critical points $\{1, \pm(R-\nu)^{1/2}, \pm(R-\nu)^{1/2}\}$ branch from it. This pair of "cell" solutions describe the steady dynamo or steady convection. Although the cell solutions are linearly stable for $R < R_c = \sigma\nu(\sigma+\nu+3)/(\sigma-1-\nu)$, instabilities of this system occur well below the point of overstability. As the driving is increased above R_c, the cell solutions become oscillatorily unstable.

Numerical results for a series of well-defined transitions from steady motion to highly nonperiodic behavior are presented. Bifurcation at the critical points $(1, \pm(R-\nu)^{1/2}, \pm(R-\nu)^{1/2})$ is studied. Robbins (1977)

15. $$\dot{x} = \sigma(y-x), \quad \dot{y} = -xz + Rx - y, \quad \dot{z} = xy - Bz,$$

where $\sigma, R = r$, and $B = b$ are real constants.

(a) For the given Lorenz system, a complete Painlevé analysis is carried out. When the Painlevé analysis in terms of the Laurent series indicates that the Lorenz system has at least one integral, the following cases arise:

(i) $\sigma = 1/2, B = 1, R = 0$. It is possible to obtain the two integrals $x^2 - z = C_0 \exp(-t), y^2 + z^2 = A^2 \exp(-2t)$. The third integral is effected by quadrature and the solution is obtained in terms of Jacobi elliptic functions.

(ii) $\sigma = 1, B = 2, R = 1/9$. A first integral is obtained as $x^2 - 2z = C_1 \exp(-2t)$. The given system can be reduced to

$$\ddot{x} + 2\dot{x} + (8/9)x + (x/2)\{x^2 - C_1 \exp(-2t)\} = 0,$$

which itself can be transformed to P_{II}.

(iii) $\sigma = 1/3, B = 0, R$ arbitrary. It is possible to change the given system to P_{III}.

(iv) $B = 1, R = 0, \sigma$ arbitrary. A first integral is obtained as $y^2 + z^2 = C_2 \exp(-2t)$. Hence the system is reduced to second order. In this case the Painlevé property is not satisfied.

(v) $B = 2\sigma, R$ is arbitrary. There exists a time-dependent integral $x^2 - 2\sigma z = C \exp(-2\sigma t)$, but the Painlevé property does not hold.

Levin and Tabor made a psi-series analysis of the given system in powers of $t - t_0 = \tau$ and $(\tau^2 \ln \tau)$. From this analysis they obtained a rescaled set of equations of motion which constitutes the most dominant behavior in the psi series. The analysis of the rescaled equations lead to the identification of the following integrals:

(vi) $\sigma = 1, B = 1, R$ arbitrary. $(-Rx^2 + y^2 + z^2)e^{2t} = I_1$, a constant.

(vii) $\sigma = 1, B = 4, R$ arbitrary. $\{4(1-R)z + Rx^2 - 2xy + y^2 + x^2 z - x^4/4\}e^{4t} = I_2$, a constant.

(viii) $B = 6\sigma - 2, R = 2\sigma - 1, \sigma$ arbitrary. $[\{(2\sigma-1)^2/\sigma\}x^2 + \sigma y^2 - (4\sigma-2)xy + x^2 z - \{1/(4\sigma)\}x^4]e^{4\sigma t} = I_3$, a constant.

Thus explicit solution may sometimes be provided following the procedure outlined above—even in the "chaotic regime"—in the neighborhood of the singularity.

(b) Introducing the variables

$$\tilde{x} = \frac{\epsilon x}{(2\sigma)^{1/2}}, y = \frac{\epsilon^2}{(2)^{1/2}}(y - x), v = \epsilon^2\left(z - \frac{x^2}{2\sigma}\right), \tilde{t} = \frac{\sigma^{1/2}}{\epsilon}t,$$

a new system is obtained (where the tilde has been dropped):

$$\frac{dx}{dt} = y, \ \frac{dy}{dt} = -(x^2 + v - 1)x - \lambda y, \ \frac{dv}{dt} = -\alpha v + \beta x^2,$$

where $\alpha = \epsilon b \sigma^{-1/2}$, $\beta = \epsilon(2\sigma - b)\sigma^{-1/2}, \lambda = \epsilon(\sigma+1)\sigma^{-1/2}$, $\epsilon = (r-1)^{1/2}$.

Referring to some previous numerical results indicating the generation of a singular attractor after a bifurcation of a loop of a separatrix of the saddle point $(0,0,0,)$, the following theorem is proved rigorously: For each $\alpha =$ constant > 0, there is, in the region of positive parameters β and λ, a bifurcation curve $\{\rho(\alpha,\beta,\lambda) = 0\}$, beginning at $(0,0)$ and going to infinity for $\beta \to \infty$, corresponding to a loop of the separatrix of the saddle point $O(0,0,0)$ of the Lorenz system. The result is depicted graphically.

(c) The solutions are $x = a\sin(t+b), y = a\cos(t+b), z = (1-a^2)t + c$, where a, b, c are arbitrary constants. The system has no singular points. With $a = 1$, we have infinitely many orbits, filling out the cylinder $x^2 + y^2 = 1$.

(d) This Lorenz system, with the usual notation, has σ, r, and b as parameters. Eliminating y and z, one may obtain the equation

$$\begin{aligned}\frac{1}{\sigma} - (x\dddot{x} - \dot{x}\ddot{x}) + \left(1 + \frac{1}{\sigma} + \frac{b}{\sigma}\right)x\ddot{x} - \left(1 + \frac{1}{\sigma}\right)\dot{x}^2 + b\left(1 + \frac{1}{\sigma}\right)x\dot{x} \\ + \frac{1}{\sigma}x^3\dot{x} + x^4 - b(r-1)x^2 = 0. \end{aligned} \tag{1}$$

Equation (1) has three stationary solutions $x = 0, \pm x_0$, where $x_0 = [b(r-1)]^{1/2}$ if $r > 1$, and $x = 0$ only if $r \leq 1$. Writing

$$A = \ddot{x} + (1+\sigma)\dot{x} - \sigma(r-1)x + \frac{1}{2}x^3, \tag{2}$$

we may put (1) in the form

$$x\dot{A} - \dot{x}A + bxA = (b/2 - \sigma)x^4, \tag{3}$$

with the solution

$$A = (b/2 - \sigma)x\int_0^t x^2(t-\tau)\exp(-b\tau)d\tau + \text{ const. } x\exp(-bt). \tag{4}$$

Equations (2) and (4) constitute a first integral of the Lorenz system. If we assume that $x(t)$ remains finite and $b \gtrsim 1$, the physically interesting case, then the contribution of the integral in (4) comes mainly from $0 \leq \tau \lesssim 1/b$. Furthermore, if we consider long-time behavior, we can expand $x(t-\tau)$ in Taylor series in τ, and approximate Eqn. (2) for $b \geq 1$ by

$$\ddot{x} + [(1+\sigma) - (2/b^2)(\sigma - b/2)x^2]\dot{x} - \sigma(r-1)x + (\sigma/b)x^3 = 0, \quad (5)$$

where the terms that decay exponentially with t as well as terms $O(b^{-2})$ have been omitted. The following cases arise for (5):

(i) If $b \geq 2\sigma$, every solution ultimately settles down to one of the stationary states with or without oscillation due to the large energy dissipation. No chaotic behavior can be expected.

(ii) If $b < 2\sigma$ and $r < 1$, Eqn. (5) can be transformed to

$$\xi'' + \epsilon(1-\xi^2)\xi' + \alpha^2\xi + \xi^3 = 0, \quad ' = \frac{d}{d\tau}, \quad (6)$$

where

$$\begin{aligned} \xi &= x/x_c, x_c = [b^2(1+\sigma)/(2\sigma - b)]^{1/2}, \\ \alpha &= [(2\sigma - b)(1-r)/\{b(1+\sigma)\}]^{1/2}, \\ \epsilon &= [(2\sigma - b)(1+\sigma)/(\sigma b)]^{1/2}, \tau = t/t_0, \\ t_0 &= [(2\sigma - b)/\{(\sigma b)(1+\sigma)\}]^{1/2}. \end{aligned}$$

For certain ranges of parameters b and σ, ϵ is sufficiently small so that solution of (6) is

$$\xi = a(\tau)\ \mathrm{cn}\{w\tau + \phi(\tau), k\},$$

where $w = (a^2 + \alpha^2)^{1/2}, k = a/[2(a^2+\alpha^2)]^{1/2}$, and cn is the Jacobian elliptic function. Both a and ϕ are slowly varying functions of time [i.e., $a', \phi' = O(\epsilon)$].

(iii) $b < 2\sigma, r > 1$; we normalize x by $x_0 = [b(r-1)]^{1/2}$ and t by $t_0 = 1/[\sigma(r-1)]^{1/2}$, and obtain

$$\eta'' + \epsilon^*(1 - \beta^2\eta^2)\eta' - \eta + \eta^3 = 0, \quad ' \equiv \frac{d}{d\tau}, \quad (7)$$

where $\eta = x/x_0, \tau = t/t_0, \epsilon^* = (1+\sigma)/[\sigma(r-1)]^{1/2}$,

$$\beta = [(2\sigma - b)(r-1)/\{b(1+\sigma)\}]^{1/2}.$$

Equation (7) can be solved in terms of Jacobian elliptic functions if ϵ^* and $\epsilon^*\beta^2$ are small.

Many parametric ranges are considered; solutions are found and depicted graphically. The case $b \simeq 2\sigma$ and $r \gg 1$ so that $\epsilon^* \ll 1$ and $\beta^2 = O(1)$ is considered in detail. Levine and Tabor (1988); Belykh (1984); Plaat (1971), p.240; Sano (1983); Takeyama (1978,1980)

16. $\dot{x} = \sigma(y - x), \quad \dot{y} = rx - xz - ay, \quad \dot{z} = -bz + (1/2)(x^*y + xy^*),$

where the parameters are defined as $b > 0, \sigma > 0, r = r_1 + ir_2, r_1 > 0, r_2 > 0, a = 1 - ie, e > 0$.

The given system is a complex Lorenz system, x and y being complex while z is real. The real Lorenz system is recovered by putting $r_2 = e = 0$ and considering real $x(t), y(t)$, since $z(t)$ is in any case real. The given system has the Painlevé property for:

(a) $\sigma = 1/2, r_1 = e^2/2, b = 1, r_2 = e/2, e$ arbitrary.

(b) $\sigma = 1, r_1 = e^2/4 + 1/9, b = 2, r_2 = 0, e$ arbitrary.

(c) $\sigma = 1/3, r_1 =$ arbitrary, $b = 0, r_2 = e, e$ arbitrary.

Writing $x = x_1 + ix_4, y = x_2 + ix_5, z = x_3$, we get the real system

$$\begin{aligned}
\dot{x}_1 &= -\sigma x_1 + \sigma x_2, \\
\dot{x}_2 &= -x_2 - ex_5 - r_2 x_4 + (r_1 - x_3)x_1, \\
\dot{x}_3 &= -bx_3 + x_1 x_2 + x_4 x_5, \\
\dot{x}_4 &= -\sigma x_4 + \sigma x_5, \\
\dot{x}_5 &= -x_5 + ex_2 + r_2 x_1 + (r_1 - x_3)x_4.
\end{aligned} \tag{1}$$

The system (1) can be treated for the Painlevé property in the complex time plane. A table of parameters is given for which the given system has a constant of motion of the type

$$F(x, y, z) = \exp(c_0 t)\{a_0 xx^* + a_1(yy^* + z^2) + a_2 z - (1/2)ia_3(x^* y - xy^*)\},$$

where c_0, a_1, a_2, and a_3 are constants depending on σ, b, r_1, r_2, and e, where $e = \sigma r_2/(1 - \sigma) + r_1(1 - \sigma)/r_2, a_2 = (1 - \sigma)e - 2\sigma r_2, r_2 = (1 - \sigma)e/(2\sigma)$.

Three quartic invariants of the form

$$\begin{aligned}
F(x, y, z, t) &= \exp(c_0 t)\{-(xx^*)^2 + a_4 xx^* z + a_5 xx^* + a_6 yy^* \\
&+ (1/2)a_7(xy^* + x^* y) - (i/2)a_8(x^* y - xy^* + a_9 z)\},
\end{aligned}$$

with $c_0, a_4, a_5, a_6, a_7, a_8$, and a_9 constants depending on σ, b, r_1, r_2, and e, are found. A brief discussion of the regularity of solutions, when constants of motion exist, is appended.

In another formulation, the given complex system may be rewritten in the equivalent form

$$\begin{aligned}
\ddot{x} &+ (\sigma + a)\dot{x} + \sigma(a - r)x = -\sigma xz, \\
\dot{z} &+ bz = |x|^2 + \frac{d}{dt}(x^* x)\{1/(2\sigma)\}, \\
y &= \dot{x}/\sigma + x.
\end{aligned} \tag{2}$$

From the second of (2), we deduce that

$$z(t) = C\exp(-bt) + |x|^2/(2\sigma) + \exp(-bt)\{1 - b/(2\sigma)\} \int |x|^2 \exp(bt)dt, \tag{3}$$

where C is an arbitrary constant. For $C = 0$ and $b = 2\sigma$, (3) reduces to

$$z(t) = |x|^2/(2\sigma). \tag{4}$$

Therefore, the first of (2) reduces to

$$\ddot{x} + (\sigma + a)\dot{x} + \sigma(a - r)x = -x|x|^2/2. \tag{5}$$

Introducing the transformation $x(t) = u(\xi)v(t),\ \xi = \xi(t), u'(\xi) = du/d\xi$ in (5), we get

$$u''\dot{\xi}^2 v + u'[\ddot{\xi}v + 2\dot{\xi}\dot{v} + (\sigma + a)\dot{\xi}v] + u[\ddot{v} + (\sigma + a)\dot{v} + \sigma(a - r)v] = -u|u|^2|v|^2 v/2. \tag{6}$$

To simplify (6), we choose

$$\ddot{\xi}v + 2\dot{\xi}\dot{v} + (\sigma + a)\dot{\xi}v = 0, \quad \ddot{v} + (\sigma + a)\dot{v} + \sigma(a - r)v = 0, (\dot{\xi})^2 = |v|^2. \tag{7}$$

From (7) we have

$$\begin{aligned} \xi(t) &= -\{3/(\sigma+1)\}\exp\{-(\sigma+1)t/3\}, \\ v(t) &= \exp[t\{ie/2-(\sigma+1)/3\}], \end{aligned}$$

so that

$$u''(\xi) = -u(\xi)|u(\xi)|^2/2, \tag{8}$$

provided that $r_1 = 1-\{8(\sigma+1)^2-9e^2\}/(36\sigma)$, $r_2 = e(1-\sigma)/(2\sigma)$, $\sigma < 1$, $9e^2+20\sigma-8\sigma^2-8 > 0$ (we require that $r_1 > 0, r_2 > 0$). Let $u(\xi)$ be real. The solution of (8) is

$$u(\xi) = (4G)^{1/4}\,\mathrm{cn}[G^{1/4}(G_1-\xi), 2^{1/2}],$$

where G and G_1 are positive constants; hence

$$\begin{aligned} x(t) = {} & (4G)^{1/4}\exp[t\{ie/2-(\sigma+1)/3\}]\,\mathrm{cn}[G^{1/4}(G_1 \\ & + \{3/(\sigma+1)\}\exp\{-(\sigma+1)t/3\}), 2^{1/2}]. \end{aligned}$$

The functions $y(t)$ and $z(t)$ may then be obtained easily from (2) and (4). This solution, however, is a particular solution since it does not involve a sufficient number of arbitrary constants. Roekaerts (1988)

17. $\quad \dot{x} = s(y-xy+x-qx^2), \quad \dot{y} = (1/s)(fz-y-xy), \quad \dot{z} = w(x-z),$

where s, q, w, and f are constants. The given system represents a Field–Noyes model for the Belousov–Zhabotinskii reactions. The nature of singular points is studied. It is shown that the solutions are oscillatory, of finite amplitude, and that at least one of them is periodic. Hastings and Murray (1975)

3.14 $\vec{y}\,' = \vec{f}(\vec{y})$; f_1, f_2, f_3 all quadratic in y_1, y_2, y_3

1. $\quad \dot{x} = yz, \quad \dot{y} = 1-xz, \quad \dot{z} = \eta\dot{x},$

where η is a constant.

The given system is a reduced form of the normalized system in Eqn. (3.14.13) when $\epsilon = 0$; here $\eta = \sigma/\sigma_1$ is a parameter. This system has two first integrals $z - \eta x = E$, $z\exp[-\eta(x^2+y^2)/2] = F$, where E and F are constants. If the initial point is taken to be in the plane $y = 0$, then E and F are given by $E = z(0)-\eta x(0)$, $F = z(0)\exp(-\eta x^2(0)/2)$.

If $z > 0$, it is easy to get an equation for z and hence for $Z = \log z$:

$$\frac{\ddot{z}}{z} - \left(\frac{\dot{z}}{z}\right)^2 = -z^2 + Ez + \eta$$

and

$$\ddot{Z} = -e^{2Z} + Ee^Z + \eta. \tag{1}$$

Multiplication of (1) by $\dot{z}$ and integration give

$$\dot{Z}^2 = -e^{2Z} + 2Ee^Z + 2\eta Z + H,$$

where H is an arbitrary constant, $H = z(0)^2 - 2Ez(0) - 2\eta \ln z(0)$. The solution can now be found by quadrature:

$$\int \frac{dz}{z(-z^2 + 2Ez + 2\eta \ln z + H)^{1/2}} = \int dt + M, \tag{2}$$

where M is an arbitrary constant. The function

$$f(z) = -z^2 + 2Ez + 2\eta \ln z + H$$

can be shown to have two roots $z(0)$ and $z_{\max}$; therefore, the solution (1) is periodic, the half-period being given by

$$t_0/2 = \int_{z(0)}^{z_{\max}} \frac{dz}{z(-z^2 + 2Ez + 2\eta \ln z + H)^{1/2}}.$$

Lu (1989)

2. $\quad \boldsymbol{\dot{x} = rx + \delta y + z - 2y^2, \quad \dot{y} = ry - \delta x + 2xy, \quad \dot{z} = -2z - 2zx,}$

where r and δ are constants.

The given system describes the problem of three interacting quasisynchronous waves in a plasma with quadratic nonlinearities. The first step in Painlevé analysis shows that there are two possible leading order behaviors near a movable singularity at $\tau = t - t_0$:

(a) $x \sim A_1\tau^{-1}, y \sim A_2\tau^{-1}, z \sim A_3\tau$.
(b) $x \sim B_1\tau^{-1}, y \sim B_2\tau^{-1}, z \sim B_3\tau^{-2}$.

Examining first a singularity of type (a), it is easy to discover the Laurent series expansions for all values of r and δ:

$$\begin{aligned} x &= -1/(2\tau) + (-r/4 + i\delta/4) + a_1\tau + \cdots, \\ y &= i/(2\tau) + (-\delta/4 + ir/4) + b_1\tau + \cdots, \\ z &= A_3\tau - 2A_3(1 - r/4 + i\delta/4)\tau^2 + \cdots, \end{aligned}$$

where t_0, A_3, and a_1 (or b_1) are the three free constants. Thus, the Painlevé property holds near this singularity. To analyze the other singularity, we make the translation $y' = y - \delta/2$ and rewrite the system as

$$\begin{aligned} \dot{x} &= rx - \delta y + z - 2y^2, \\ \dot{y} &= ry + r\delta/2 + 2xy, \\ \dot{z} &= -2z - 2zx, \end{aligned} \tag{1}$$

where, for simplicity, we drop the prime on y. The Laurent series in the present case may be written as

$$\begin{aligned} x &= 1/\tau - (r/2 + 1) + \cdots, \\ y &= -(r\delta/2) + b_2\tau^2 + \cdots, \\ z &= -1/\tau^2 - r/\tau + c_0 + \cdots, \end{aligned} \tag{2}$$

where b_2 is a (second) free constant provided that the following compatibility condition holds:

$$r\delta = 0. \tag{3}$$

Two cases arise:

(a) $\delta \neq 0, r = 0$. Here the asymptotic conditions are the same as in (2), c_0 being the third free constant to be determined by the initial conditions. In this case, the third Painlevé property again holds, and the system is integrable. Indeed, one integral of the system (1) is easily found to be

$$zy = Ce^{-2t}, \tag{4}$$

C being an arbitrary constant. Using (4) and second of (1) in the first of (1), we obtain

$$\ddot{y} = \dot{y}^2/y - 4y^3 + 2Ce^{-2t} - 2\delta y^2. \tag{5}$$

Equation (5) with the substitution

$$Y = e^t y, \quad T = e^{-t} \tag{6}$$

transforms to P_{III}. Hence the given system in the present case is completely integrable.

(b) $\delta = 0, r \neq 0$. In this case, the asymptotic expansion (2) possesses a third free constant, C_0, if and only if $r(r+1) = 0$. Since we have already considered the case $r = 0$, we concentrate on $r = -1$. For this case, we note that there exists one first integral

$$x^2 + (y + \delta/2)^2 + z = De^{-2t}, \tag{7}$$

with D an arbitrary constant, valid for all δ. In the special case $\delta = 0$, a second integral is obtained by combining the second and third equations of the system (1):

$$zy = Ce^{-3t}, \tag{8}$$

where C is an arbitrary constant. Using (7) and (8) and the transformation (6), we are led, after some algebra, to the single equation

$$\left(\frac{dy}{dT}\right)^2 = 4Y^2D - 4Y^4 - 4CY, \tag{9}$$

which can be integrated completely in terms of elliptic functions. For $\delta = 0, r$ free, one integral of system (1), $zy = Ce^{(r-2)t}$, exists; no further integration seems possible.

Returning now to the "partially integrable" case $r = -1, \delta$ arbitrary, for which only one integral (7) has been found: Violation of the compatibility condition at the first higher order forces us to introduce logarithmic terms in the asymptotic expansions near the singularity (b). The first few terms of the expansions are

$$\begin{aligned} x &= 1/\tau - 1/2 + a_1\tau + \cdots, \\ y &= (\delta/2)\tau - \delta\tau^2 \ln\tau + b_2\tau^2 + \cdots, \\ z &= -1/\tau^2 + 1/\tau + c_0 + \cdots, \end{aligned}$$

with t_0, b_2, and C_0 (or a_1) the three free constants and higher powers of $\tau(\ln\,\tau)$ entering the higher order terms. Bountis, Ramani, Grammaticos, and Dorizzi (1984)

3. $$\dot{x} = \gamma x + \delta y + z - 2y^2, \quad \dot{y} = \gamma y - \delta x + 2xy, \quad \dot{z} = -2z - 2zx,$$

where γ and δ are parameters. The system describes a "reduced" three-wave interaction problem.

An integral of motion exists in the following cases.

(a) $\gamma = 0, \delta$ arbitrary with the integral $I = z(y - \delta/2)e^{2t}$.

(b) $\gamma = -1, \delta$ arbitrary with the integral $I = (x^2 + y^2 + z)e^{2t}$.

In the special case $\delta = 0$, a second integral is found for case (b), namely $I = zye^{3t}$. Realizing that all these integrals are linear in z, a form $I = \{a_1(x, y) + a_2(x, y)z\}e^{\alpha t}$ is assumed. Substituting this into the equation $dI/dt = 0$, using the given system, we find after considerable analysis that there exist other integrals:

(c) $I = yze^{(2-\nu)t}$, with ν arbitrary but $\delta = 0$.

(d) $I = (y^2 + x^2 + 2yz/\delta)e^{4t}$, where $\gamma = -2$ and δ is arbitrary.

Giacomini, Repetto, and Zandron (1991)

4. $$\dot{x} = z + \delta y + \gamma x - 2y^2, \quad \dot{y} = -\delta x + \gamma y + 2xy, \quad \dot{z} = -2z(x + 1),$$

where δ and γ are parameters.

The given system describes the structure of wave turbulence in dissipative media. It has two singular points: the origin and $x = -1, y = \delta/(2 - \gamma), z = \gamma[1 + \delta^2/(2 - \gamma)^2]$. Both these points are unstable, being saddle foci. A simple phase picture results if $\delta = 0$ and contains a steady limit cycle corresponding to a stable single-period motion. Phase portraits on the planes $Y = 0$ and $Z = 0$ are also drawn both when $\delta = 0$ and when $\delta \neq 0$. The onset of disordered behavior as a result of the loss of stability of only doubly periodic motion is confirmed by computer experiments. Vyshkind and Rabinovich (1976)

5. $$\dot{x}_1 = -wx_2 + 2x_1x_3, \quad \dot{x}_2 = wx_1 + 2x_2x_3, \quad \dot{x}_3 = x_3^2 - x_1^2 - x_2^2 + D,$$

where w and D are constants.

The behavior of the trajectories for various values of w and D is studied. For $D \leq 0$, the system is linearized by a conformal transformation. For $D > 0$ one of the system's trajectories is a circle and another is the straight line L: $x_1 = 0, x_2 = 0$. The remaining trajectories lie on two-dimensional tori with the same rotational number $D^{1/2}/w$.

Depending on whether this number is rational or irrational, all trajectories are closed or dense on this torus, respectively. A further description of the situation for an irrational rotation number and for the case where the vector field is linearizable is detailed. Golubyatnikov and Pestov (1983)

6. $$\dot{x} = hy - \nu_1 x - yz, \quad \dot{y} = hx - \nu_2 y + xz, \quad \dot{z} = -z + xy,$$

where h, ν_1, and ν_2 are positive constants.

The given system arises in a magnetoactive nonisothermal plasma. It is invariant under the transformation $x \to -x, y \to -y, z \to z$. The phase volume shrinks uniformly:

$$\frac{\partial \dot{x}}{\partial x} + \frac{\partial \dot{y}}{\partial y} + \frac{\partial \dot{z}}{\partial z} = -(1 + \nu_1 + \nu_2),$$

so that the attractor should have a zero Lebesgue measure. If we put $u = 2x^2+y^2+(z-3h)^2$, then it is easy to check that

$$\dot{u} \le -ku + 9h^2, k = \min(2\nu_1, 2\nu_2, 1),$$

so that all trajectories are contained in the ellipsoid $u \le 9h^2k^{-1}$. Numerical results show that as the parameter h is increased, a strange attractor (analogous to the known Lorenz attractor) is produced and corresponds to stochastic self-oscillations of the wave amplitudes. Pikovskii, Rabinovich, and Trakhtengerts (1978)

7. $$\dot{x} = y, \quad \dot{y} = -x + z + \epsilon(x^2y - \delta y), \quad \dot{z} = \epsilon(r - z + y^2),$$

with parameters $(r, \delta) \in R^2$; ϵ is a small positive number.

For $\epsilon = 0$, the given system is autonomous Hamiltonian with energy $H(x, y; z) = y^2/2 + x^2/2 - zx = h$. The family of unperturbed periodic orbits in action angle variables is given by $x = z + (2I)^{1/2}\sin\theta, y = (2I)^{1/2}\cos\theta, z = z$.

This example is used to illustrate how the Hopf and saddle-node bifurcations for diffeomorphisms can interact, and how the periodic orbits detected by Melnikov theory can be connected to those created in sub- and supercritical bifurcation from an equilibrium point for three-dimensional flow. Wiggins and Holmes (1987)

8. $$\dot{x} = -\mu x + \beta y + yz, \quad \dot{y} = -\mu y - \beta x + xz, \quad \dot{z} = -xy + \alpha,$$

where μ, α, and β are constants.

The given system is called a Rikitake two-disk dynamo model and describes the time variation of the earth's magnetic field. The leading order behavior of the solution near a singularity at $\tau = (t - t_0) = 0$ is $x \sim i/\tau, y \sim -i/\tau, z \sim 1/\tau$.

Painlevé analysis shows that two more free constants appear simultaneously at the second order. The compatibility conditions for these free constants to enter with only integer powers of τ yield (a) $\alpha = 0$ and (b) either $\beta = 0$ or $\mu = 0$.

The two cases possessing the Painlevé property can easily be integrated. For the first with $\alpha = 0$ and $\beta = 0$, one can find a first integral

$$x^2 - y^2 = Ce^{-2\mu t}, \tag{1}$$

where C is an arbitrary constant. Introducing the transformations $x + y = Cue^{-\mu t}$, $T = e^{-\mu t}$, and using (1), the given system reduces to a single second order DE,

$$\frac{d^2u}{dT^2} = \frac{1}{u}\left(\frac{du}{dT}\right)^2 - \frac{1}{T}\frac{du}{dT} - \frac{C^2}{4\mu^2}(u^3 - 1/u). \tag{2}$$

Equation (2) is a special case of the third Painlevé equation.

In the second case with $\alpha = 0$ and $\mu = 0$, we multiply the first and second equations of the given system by $\dot{x}$ and $\dot{y}$, respectively, add and subtract, and integrate to obtain

$$\begin{aligned} x^2 + y^2 + 2z^2 &= C, \\ x^2 - y^2 + 4\beta z &= D. \end{aligned} \tag{3}$$

Equations (3) can be used to obtain the solutions of the given system in terms of elliptic functions.

Now we investigate the consequences of "partial satisfaction" of Painlevé conditions (a) and (b). For example, if we assume that $\beta = 0$ but $\alpha\mu \neq 0$, the integral (1) still exists and the motion as $t \to \pm\infty$ (for $\mu \gtrless 0$) converges to any one of the four limit cycles $x_1 = \epsilon_1 y, (z - \epsilon_1\mu)^2 + y^2 = 2\epsilon_2\alpha \ln|y|$ $(\epsilon_i = \pm 1)$.

On the other hand, if $\mu = 0$, one integral also exists for all α and β, namely, $x^2 - y^2 + 4\beta(z - \alpha t) = D$, but no further integration seems possible.

It may be noted that pole-like behavior at the leading order, accompanied by only $\ln \tau$ terms at higher orders, may yield more than partial integrability. For example, if both μ and β vanish, two integrals may be shown to exist:

$$x^2 - y^2 = D, x^2 + y^2 + 2z^2 - 4\alpha \ln|x + y| = E,$$

where D and E are arbitrary constants. Bountis, Ramani, Grammaticos, and Dorizzi (1984); Grammaticos, Moulin–Ollagnier, Ramani, Strelcyn, and Wojciechowski (1990)

9.

$$\begin{aligned} \dot{x} &= 2wx, \\ \dot{y} &= \{c_0\Omega + \rho\}/(1 + c_0^2) - [\{c_0 + \rho\}/(1 + c_0^2)]x - 2yw, \\ \dot{w} &= \{\Omega - c_0\rho\}/(1 + c_0^2) - [\{1 - c_0\rho\}/(1 + c_0^2)]x - w^2 + y^2, \end{aligned}$$

where c_0, Ω, and ρ are constants.

It is easy to check that the given system, which describes spatial flows of the Ginsburg–Landau equation, is volume preserving in (x^2, y, w) space; that is,

$$\frac{\partial\{(x^2)^{\cdot}\}}{\partial x^2} + \frac{\partial \dot{y}}{\partial y} + \frac{\partial \dot{w}}{\partial w} = 0.$$

Further, it is found that y and w are odd, while x is even. Choosing the IC $w(0) = 0 = y(0), x(0) = x_0$, say, a numerical solution is found for different sets of parameters. Trajectories are determined in the $w = 0$ Poincaré section. Periodic solutions are found. Sirovich and Newton (1986)

10. $\dot{x} = bx^2 - bxy + cxz + ax, \dot{y} = bxy - by^2 + [(n-2)/(n-3)]cyz + ay, \dot{z} = -bxz + byz,$

where a, b, c are parameters and $n = 7/3$ or $5/2$. It is shown that the given system has solutions with no moving critical points. The solution is related to $\mathrm{P_{III}}$. Leonovich (1984)

11. $\dot{x} = bx^2 - bxy + cxz + ax, \dot{y} = bxy - by^2 + [(n-3)/(n-4)]cyz + ay, \dot{z} = bxz - byz,$

where a, b, c are parameters and $n = 7/3$ or $5/2$.

It is shown that the given system has solutions with no critical points. These solutions are related to $\mathrm{P_{III}}$. Leonovich (1984)

12.

$$\begin{aligned} \dot{x} &= cx^2 + [(n-5)/(3n-11)]cxy + [4(n-4)/(3n-11)]k_1cxz + ax, \\ \dot{y} &= [3(n-3)/(3n-11)]cxy + [(n-3)/(3n-11)]cy^2 \\ &\quad + [4(n-3)/(3n-11)]k_1cyz + ay, \\ \dot{z} &= -[(n-4)/(3n-11)][c/k_1]x^2 - [(n-4)/(3n-11)][c/k_1]xy \\ &\quad - [(n-5)/(3n-11)]cxz - [(n-3)/(3n-11)]cyz - [a/(2k_1)]x \\ &\quad - [a/(2k_1)]y, \end{aligned}$$

where c, k_1, and a are arbitrary parameters and $n = 7/3$ or $5/2$.

It is shown that the given system has solutions with no critical points. The solutions are related to $\mathrm{P_{III}}$. Leonovich (1984)

13. $\dot{x} = yz - a, \quad \dot{y} = R - y - xz, \quad \sigma^{-1}\dot{z} = x - z + \sigma_1^{-1}\dot{x}.$

The given system describes a homopolar disk dynamo that exhibits nonperiodic reversals. When $\sigma_1 \to \infty$, this system reduces to Lorenz system.

This system is studied numerically in its normalized form,

$$\begin{aligned} \dot{X} &= YZ - \epsilon X, \\ \dot{Y} &= 1 - XZ - \epsilon Y, \\ \sigma^{-1}\dot{Z} &= \sigma_1^{-1}\dot{X} + \epsilon(X - Z), \text{ now } \cdot \equiv \frac{d}{dT}; t = T/R^{1/2}, \end{aligned} \tag{1}$$

where $\epsilon = 1/R^{1/2}$ is arbitrarily small. Depending on the choice of the parameters, the system (1) shows chaotic behavior, as evidenced by numerical results. For $\epsilon = 0$ in (1), see Eqn. (3.14.1). Lu (1989)

14. $\dot{x} = -y^2 - z^2 - ax + af, \quad \dot{y} = xy - bxz - y + g, \quad \dot{z} = bxy + xz - z,$

where a, b, f, and g are constants.

The given system, a "new" Lorenz system, models general circulation. It is shown that it may possess one or two steady-state solutions, one or two stable periodic solutions, or (irregular) aperiodic solutions.

Detailed (numerical) work shows period doubling bifurcations to chaos, crisis, reverse period doubling, periodic windows, hysterisis, and coexistence of periodic attractors. Poincaré section, Lyapunov spectrum, and correlation dimension analysis are used to demonstrate the above-noted effects. Lorenz (1984); Masoller, Sicardi, and Schifino (1992)

15. $\dot{x} = hy - \nu_1 x + yz, \quad \dot{y} = hx - \nu_2 y - xz, \quad \dot{z} = \nu_3 z + xy,$

where h, ν_1, ν_2, and ν_3 are constants.

(a) The given system is called a Rabinovich three-wave interaction system. The leading order behavior about a moving singularity is found to be

$$x \sim -i/\tau, \ y \sim 1/\tau, \ z \sim i/\tau (\tau = t - t_0). \tag{1}$$

Writing a Laurent series solution starting with (1), one finds that the two remaining constants (apart from t_0) enter at the second order. This will happen with integer powers of τ and no ln τ terms, provided that certain compatibility conditions are satisfied. Thus, Painlevé property holds and the given system is integrable in the following three cases:

A. $h \neq 0$ (by scaling $h = 1$) and either (a) $\nu_1 - \nu_2 = 2i, \nu_3 = 0$ or (b) $\nu_1 + \nu_2 = \nu_3, \ \nu_1 - \nu_2 = 2i\nu_1\nu_2$.

B. $h = 0$ and either (a) $\nu_2 - \nu_1 = \nu_3$ or (b) $\nu_j = \nu_k, \nu_\ell = 0$ (j, k, ℓ unequal).

C. $\nu_1 = \nu_2 = \nu_3 = 0$. This case is completely integrable (for all h) in terms of elliptic functions.

Consider, for example, the case A(a). Multiplying the second of the given system by i and adding to the first one, we get

$$\dot{u}/u = i - \nu_1 - iz, \tag{2}$$

where

$$u = x + iy = \rho \exp(i\theta). \tag{3}$$

Differentiating (2) once and using (3) and the third of the given system, we obtain, upon separating real and imaginary parts,

$$\rho = C\exp(t), \ddot{\phi} = -C^2 \exp(2t) \sin\phi \ (\phi \equiv 2\theta), \tag{4}$$

where C is an arbitrary constant. It is interesting to note that by the change of variable $u = \exp(i\theta), T = (C^2/8)\exp(2t)$, the ϕ-equation of (4) can be transformed to the third Painlevé equation. In an entirely similar way, we can, using (3), integrate B(b), with $\nu_1 = \nu_2$, say, and $\nu_3 = 0$.

Turning to partially integrable cases, when the given system cannot be fully integrated, we obtain the following:

(i) $h \neq 0, \nu_1 = \nu_2 = \nu, \nu_3 = -2\nu$, as suggested by the completely integrable case A(b), we may obtain the integral $x^2 + y^2 - 4hz = C\exp(-2\nu t), \nu > 0$, whence all solutions tend (as $t \to \infty$) to a paraboloid of revolution around the $z > 0$ axis.

(ii) $h \neq 0, \nu_1 = \nu_2 = \nu_3 = \nu$, suggested by the corresponding completely integrable case with $\nu = 0$. A first integral here is $x^2 - y^2 - 2z^2 = C\exp(-2\nu t), \nu > 0$, and all motions collapse, as $t \to \infty$, on a double ellipsoidal cone about the x-axis with apex at the origin.

(iii) $h = 0, \nu_1 = \nu_2 = \nu$, suggested by the completely integrable case B(b). In this case, one integral is $x^2 + y^2 = C\exp(-2\nu t), \nu > 0$. The motion is globally damped and all solutions are attracted to the origin as $t \to \infty$.

It may be checked that all the partially integrable cases quoted above have a movable pole at leading order, with $\ln(t - t_0)$ terms [and integer powers of $(t - t_0)$] needed at higher order to capture the constants for the general asymptotic expansions of the solutions.

(b) Noting that the three integrals above are linear in x^2, an ansatz of the following form is assumed:

$$I = [a_1(y, z) + a_2(y, z)x^2]e^{\nu t}.$$

After writing $dI/dt = 0$, using the given DEs, and further analysis, the following new integrals were found.

D. $I = y^2 + (h - z)^2$ with $\nu_2 = \nu_3 = 0$, and h and ν_1 arbitrary.

E. $I = x^2 - (z + h)^2$, with $\nu_1 = \nu_3 = 0$, h and ν_2 arbitrary.

F. $I = (y^2 + z^2)e^{2\nu t}$ with $\nu_2 = \nu_3, h = 0, \nu_1$ and ν_3 arbitrary.

G. $I = (x^2 - z^2)e^{2\nu t}$ with $\nu_1 = \nu_3, h = 0, \nu_2$ and ν_3 arbitrary.

Bountis, Ramani, Grammaticos, and Dorizzi (1984); Giacomini, Repetto, and Zandron (1991)

16.
$$\begin{aligned}
\dot{x} &= (k_1/2)(n-4)x^2 + (n-4)[k_2 + (k_1/2)]xy + (n-4)xz + ax, \\
\dot{y} &= (k_1/2)[(n/2) - 1]xy + [k_2(n-3) + (k_1/2)(n-4)]y^2 + (n-3)yz + ay, \\
\dot{z} &= -k_1[k_1(n-4) + k_2(n-3)]xy - k_2[k_1(n-4) + k_2(n-3)]y^2 \\
&\quad - (k_1/2)(n-4)xz - [(k_1/2)(n-4) + k_2(n-3)]yz - ak_1x - ak_2y,
\end{aligned}$$

where k_1, k_2, a, and c are arbitrary parameters and $n = 7/3$ or $5/2$.

It is shown that this system has solutions with no critical points. The solutions are related to P_{III}. Leonovich (1984)

17.
$$\begin{aligned}\dot{x} &= (k_1/2)(n-4)x^2 + (n-4)[k_2 + (k_1/2)]xy + (n-4)xz + ax,\\ \dot{y} &= (k_1/2)(n/2)-1)xy + [k_2(n-3) + (k_1/2)(n-4)]y^2 + (n-3)yz + ay,\\ \dot{z} &= -k_1[k_1(n-4) + k_2(n-3)]x^2 - k_2[k_1(n-4) + k_2(n-3)]xy\\ &\quad - (k_1/2)(n-4)xz - [(k_1/2)(n-4) + k_2(n-3)]yz - ak_1x - ak_2y,\end{aligned}$$

where k_1, k_2, a, and c are arbitrary parameters and $n = 7/3$ or $5/2$.

It is shown that the given system has solutions with no critical points. The solutions are related to P_{III}. Leonovich (1984)

18.
$$\begin{aligned}\dot{x} &= (k_1/2)cx^2 + [(k_1/2) + k_2]cxy + cxz + ax,\\ \dot{y} &= \frac{n-1}{2(n-3)}k_1cxy + \left[\frac{k_1}{2} + \frac{n-2}{n-3}k_2\right]cy^2 + \frac{n-2}{n-3}cyz + ay,\\ \dot{z} &= -k_1\left[\frac{n-2}{n-3}k_2 - k_1\right]cx^2 - k_2\left[\frac{n-2}{n-3}k_2 - k_1\right]cxy,\\ &\quad (k_1/2)cxz - \left(\frac{1}{2}k_1 + \frac{n-2}{n-3}k_2\right)cyz - ak_1x - ak_2y,\end{aligned}$$

where k_1, k_2, a, and c are arbitrary parameters and $n = 7/3$ or $5/2$. It is shown that the given system has solutions with no critical points. The solutions are related to P_{III}. Leonovich (1984)

19.
$$\begin{aligned}\dot{x} &= (k_1/2)cx^2 + [(k_1/2) + k_2]cxy + cxz + ax,\\ \dot{y} &= \frac{n-1}{2(n-3)}k_1cxy + \left[\frac{k_1}{2} + \frac{n-2}{n-3}k_2\right]cy^2 + \frac{n-2}{n-3}cyz + ay,\\ \dot{z} &= -k_1\left[\frac{n-2}{n-3}k_2 - k_1\right]cxy - k_2\left[\frac{n-2}{n-3}k_2 - k_1\right]cy^2\\ &\quad - (k_1/2)cxz - \left(\frac{1}{2}k_1 + \frac{n-2}{n-3}k_2\right)cyz - ak_1x - ak_2y,\end{aligned}$$

where k_1, k_2, a, and c are arbitrary parameters and $n = 7/3$ or $5/2$.

It is shown that the given system has solutions with no critical points. The solutions are related to P_{III}. Leonovich (1984)

20. $\dot{x}_1 = -\mu_1x_1 + (x_3 + \alpha_0)x_2, \dot{x}_2 = -\mu_2x_2 + (x_3 - \alpha_0)x_1, \dot{x}_3 = q - x_1x_2 - \epsilon x_3,$

where μ_1, μ_2, α_0, q, and ϵ are positive constants [see Eqn. (4.8.3)]. Cook (1986), pp.42,193

21. $\dot{x}_1 = 1 - Bx_1 - x_1x_2^2 - Ex_1x_2 + x_3, \dot{x}_2 = A(x_1x_2^2 - x_2 + G), \dot{x}_3 = F(Ex_1x_2 - x_3),$

where B, E, A, G, and F are constants.

The given system describes the behavior of coupled cells each of which exhibits a complex pattern. Numerical solutions (and Lyapunov exponent) show the presence of a hyperchaotic regime at low coupling strengths followed by a periodic window at intermediate values, followed by synchronization in a chaotic state. The interesting observation of complete synchronization of individual cells even when the coupled dynamics show chaos is of considerable significance. Badola, Ravikumar, and Kulkarni (1991)

22. $\dot{x} = 2wx, \dot{y} = -\beta + \gamma x - 2yw - c_1(y - a_0 w), \dot{w} = -\delta_1 + \delta_2 x + y^2 - w^2 - c_1(a_0 y + w),$

where $\beta, \gamma, c_1, a_0, \delta_1, \delta_2$ are constants.

The given system describes spatial structure of quasisteady solutions of the Ginzburg–Landau equation, whose amplitude is bounded away from zero. There are exact solutions of the given system:

(a) $c_1 = 0. x = \lambda^2 L^2 \operatorname{sech}^2 \lambda t$, $w = -\lambda \tanh \lambda t$, $y = \nu w$, where $\lambda^2 = \delta_1/(\nu^2 - 1), L^2 = -3\nu/\gamma$, and $\beta/\delta_1 = 2\nu/(1 - \nu^2)$, where ν satisfies $\nu^2 + 3\nu\delta_2/\gamma - 2 = 0$. This is a breather solution and always exists if $\delta_2 < 0$; it exists for $\delta_2 > 0$ only when $\delta_1 > 0$.

(b) $x = L^2(1 - \tanh \lambda t)^2$, $y = k - \nu\lambda \tanh \lambda t$, $w = -\lambda(1 + \tanh \lambda t)$, where $\lambda^2 = \delta_1/(8 - 9a_0^2)$, $L^2 = 3\nu\lambda^2/\gamma, \beta/\delta_1 = 18a_0/(9a_0^2 - 8), c_1 = 6\lambda, k = \lambda(3a_0 + \nu)$, where ν satisfies $\nu^2 + 3\nu\delta_2/\gamma - 2 = 0$. This is a shock solution and exists for a large range of the coefficients of the given system.

(c) $c_1 = 0. x = L^2\lambda^2 \tanh^2 \lambda t$, $y = -\nu\lambda \tanh \lambda t, w = 2\lambda/\sinh 2\lambda t$, where $\lambda^2 = \delta_1/2, L^2 = 3\nu/\gamma$, and $\beta/\delta_1 = 3\nu/2$, where ν satisfies $\nu^2 + 3\nu\delta_2/\gamma - 2 = 0$. This is called a hole solution.

A considerable numerical study of the given system is carried out, with relation to the foregoing exact solutions and other solutions arising from different sets of parameters. Attractors of higher period and perhaps chaotic attractors are also discovered. The solutions are interpreted physically. Landman (1987)

23. $\dot{x}_1 = \lambda x_1 + C x_1 x_2 + x_1 x_3, \dot{x}_2 = \mu x_2 + A x_2 x_3 + x_2 x_1, \dot{x}_3 = \nu x_3 + B x_3 x_1 + x_3 x_2,$

where λ, μ, ν, A, B, and C are constants. The given system is a third order Lotka–Volterra system.

One may investigate for what values of the parameters the given system possesses Painlevé property and is therefore expected to be completely integrable. The leading order behavior leads to the first necessary condition, $\lambda = \mu = \nu$. Now scaling t so that $\lambda = 1 (= \mu = \nu)$, introducing the new variables

$$x = x_1 e^{-t}, y = x_2 e^{-t}, z = x_3 e^{-t}, t' = e^t,$$

and dropping the prime on t, we have a new system,

$$\dot{x} = Cxy + xz, \quad \dot{y} = Ayz + yx, \quad \dot{z} = Bxz + zy. \tag{1}$$

Examining the leading order behavior of (1) near a movable singularity $x = x_0$, one distinguishes two possible singular behaviors:

(a) x, y, z all behave like $\tau^{-1} (\tau = t - t_0)$.

(b) Two of x, y, z diverge as τ^{-1} and the third behaves like $\tau^p (p > -1)$ at leading order.

Three cases arise for (b). Writing $x \sim B_1\tau^p, y \sim \tau^{-1}, z \sim \tau^{-1}$ in the first case and $y \sim B_2\tau^q, z \sim B_3\tau^s$, respectively, in the other two cases, one finds immediately on substituition in (1) that

$$p = -C - 1/A, q = -A - 1/B, s = -B - 1/C, \tag{2}$$

where B_k is in each case ($k = 1, 2, 3$) the second free constant of the asymptotic expansions (the first one is t_0). Here $ABC + 1 = 0$.

The next necessary conditions for the Painlevé property requires that only integer powers of τ are permitted in the series solution. For example, for type (a) singularity, we may assume that

$$x = A_1/\tau + \sum_{r\geq 0} a_r\tau^r, y = A_2/\tau + \sum_{r\geq 0} b_r\tau^r, z = A_3/\tau + \sum_{r\geq 0} c_r\tau^r$$

($A_k \neq 0, k = 1, 2, 3$). This amounts to solving, at each order, three linear equations for a_r, b_r, c_r. The condition that the determinant of the coefficient matrix of these equations vanishes is that

$$(r+1)\{r^2 - r - A_1A_2A_3(1 + ABC)\} = 0. \tag{3}$$

Two cases arise:

α. $1 + ABC = 0$. The free constants center at $r = -1$ (corresponding to the arbitrariness of t_0) and at $r = 0, +1$. The resonance $r = 0$ implies that one of the A_k's must be arbitrary. The compatibility condition at the leading order implies that a solution exists if and only if we also have $A = 1/(1 - C)$ and $B = (C - 1)/C$, with C arbitrary.

β. $1 + ABC \neq 0$. In this case both kinds of singularities, (a) and (b), must be considered separately. For the Painlevé property, first we must insist that roots of the quadratic equation in the braces in (3) and p, q, s in (2) must all be integers. This implies that $M \equiv A_1A_2A_3(1 + ABC) = m(m - 1)$ for integer m. Therefore, the values of p, q, s can be deduced from the relation $1/M = 1/(p + 1) + 1/(q + 1) + 1/(s + 1) - 1$. All possible values satisfying the requirements above are as follows:

p	q	s	m
1	1	1	2
1	2	2	3
1	2	3	4
1	2	4	6

The second condition of the Painlevé property that the asymptotic expansions possess three free parameters near at least one singularity type is thus verified for both cases α and β. In all other cases, there will be either logarithmic or algebraic branch-point singularities in the expansions destroying the Painlevé property. This does not imply, however, that for non-Painlevé cases, no completely integrable cases exist. Consider, for example, case α with $1 + ABC = 0$. In this case, two simple integrals always exist: $x - Cy - z/B = D, xy^{-1/A}z^{-C} = E$, where D and E are arbitrary constants. Now, the solution can be obtained in terms of a single quadrature. Taking $A = 1, BC = 1$, with C arbitrary, one finds two integrals:

$$(x - Cy)^2 z^C/xy = D^2$$

and

$$D(x - Cy) - C\int \frac{D^2dz}{\{D^2 + 4Cz^C\}^{1/2}} = E, \tag{4}$$

where D and E are arbitrary constants. For certain values of C, the integration in (4) can be performed and the solutions explicitly obtained. These solutions can be identified by introducing the variable $(t - t_0)^{1/2}$ and then carrying out the Painlevé analysis. These values of C are $C = -4, -1, 1, 4$; and the fractional power $\tau^{1/2}$ is naturally suggested by an examination of r, p, q, and s, at which the constants enter the asymptotic expansions.

We note that the second of (4) simplifies, leading to logarithmic terms for $C = 2, A = 1, B = 1/2$. The result is

$$2^{1/2}(x - 2y) = E' + D\sinh^{-1}\{2(2)^{1/2}z/D\}.$$

Even in this case, the problem is completely integrable since it can now be reduced with the help of the first of (4) to a single quadrature. The singularities, however, are logarithmic since even at the leading order, $x \sim -1/\tau,\ y \sim -1/(2\tau),\ z \sim F/(\tau \ln \tau)$ with F arbitrary.

A more complete discussion of the given system with regard to integrability may be found in Grammaticos et al. (1990). Bountis, Ramani, Grammaticos, and Dorizzi (1984); Grammaticos, Moulin–Ollagnier, Ramani, Strelcyn, and Wojciechowski (1990)

24.
$$\begin{aligned}
\dot{x} &= c_1 x^2 + (\alpha + c_2)xy + c_3 xz + yz,\\
\dot{y} &= -x^2 + (b + c_1)xy + (c + c_2)y^2 - xz + (\lambda + c_3)yz,\\
\dot{z} &= z(c_1 x + c_2 y + c_3),
\end{aligned}$$

where the constant coefficients are apparent.

It is established that the integral curve may enter the origin either along a characteristic or along a cone whose vertex is at the singular point or at a point at which both directions may coincide. Kajumov (1978)

25. $\dot{x} = -\xi x + \epsilon x(k_1 x^2 + k_2 y^2), \dot{y} = -\zeta y + \epsilon y(P_1 x^2 + P_2 y^2), \dot{z} = w + \epsilon(Q_1 x^2 + Q_2 y^2),$

where $\xi, \epsilon, w, k_1, k_2, \zeta, P_1, P_2, Q_1$, and Q_2 are constants.

This system arises from a more general system describing damped nonlinear oscillating via the Krylov–Bogoliubov–Mitropolsky method. It is solved numerically for different parametric values, with ICs $x(0) = 0.5, y(0) = 1, z(0) = 0$. The results are graphically shown and compared with approximate analytic results. Bojadziev (1983)

26.
$$\begin{aligned}
\dot{x} &= \alpha_1 x + \alpha_2 y + \alpha_3 z + y(p_1 y + p_2 z),\\
\dot{y} &= \beta_1 x + \beta_2 y + \beta_3 z + x(q_1 x + q_2 y + q_3 z),\\
\dot{z} &= \gamma_1 x + \gamma_2 y + \gamma_3 z + x(r_1 y + r_2 z),
\end{aligned}$$

where $\alpha_i, p_i, \beta_i, q_i, \gamma_i$, and r_i are real constants.

For special choices of the parameters, the given system has the Lorenz system and several other systems in mechanics and physics as special cases. The first integrals of several special cases are found, and hence the system is reduced to Painlevé second or third equations. Backlund transformation is given to generate solutions from known special solutions. Apart from the Lorenz system, the system

$$\dot{x} = ax + by + z - 2y^2, \quad \dot{y} = ay - bx + 2xy, \quad \dot{z} = -2z - 2zx$$

is discussed [see Eqn. (3.14.4)]. Gromak and Tsegelnik (1991)

27.

$$\begin{aligned} \dot{x} &= a_1x + a_2y + a_3z + x(a_4x + a_5y + a_6z), \\ \dot{y} &= b_1x + b_2y + b_3z + y(b_4x + b_5y + b_6z), \\ \dot{z} &= c_1x + c_2y + c_3z + z(c_4x + c_5y + c_6z), \end{aligned}$$

where a_i, b_i and c_i $(i = 1, \ldots, 6)$ are complex numbers.

Conditions are established such that the given system has solutions which are meromorphic functions of the complex variable t. Konopljannik (1979)

28.

$$\begin{aligned} \dot{x} &= x(a_1x + a_2y + a_3z + a), \\ \dot{y} &= y(b_1x + b_2y + b_3z + a), \\ \dot{z} &= P(x, y, z) + c_1x + c_2y + c_3z, \end{aligned}$$

where

$$P(x, y, z) = a_{11}x^2 + a_{12}xy + a_{13}xz + a_{22}y^2 + a_{23}yz + a_{33}z^2,$$

with a, a_i, b_i, c_i, and a_{ij} as constants.

Conditions are found such that for either $a_3 \neq b_3$ or $a_3 = b_3$, the given system has no movable critical point in its solution. Leonovich (1984)

29.

$$\begin{aligned} \dot{x} &= x(a_1x + a_2y + a_3z + a), \\ \dot{y} &= y(b_1x + b_2y + b_3z + a), \\ \dot{z} &= P(x, y, z) + c_1x + c_2y + c_3z, \end{aligned}$$

where

$$P(x, y, z) = a_{11}x^2 + a_{12}xy + a_{22}y^2 + a_{13}xz + a_{23}yz + a_{33}z^2,$$

and a, a_j, b_j, c_j, and a_{ij} $(i, j = 1, 2, 3)$ are constants.

It is shown that if $b_1 = a_1 - c, b_2 = a_2 + c, b_3 = a_3 = d$, and $a_{33} = -d$, and if the coefficients in the RHS of dz/dt have one of the forms:

(a) $a_{11} = (1/d)(c^2 + a_1c - 2a_1^2), a_{12} = (1/d)(2a_2c - 2a_1c - 4a_1a_2)$,
$a_{22} = -(1/d)(2a_2^2 + 3a_2c)$,
$a_{13} = c - 3a_1, a_{23} = -2c - 3a_2, c_1 = (a/d)(c - 3a_1)$,
$c_2 = -(a/d)(3a_2 + 2c), c_3 = -2a$,

(b) $a_{11} = (1/d)[-(1/4)c^2 + (3/2)a_1c - 2a_1^2]$,
$a_{12} = (1/d)[(5/2)c^2 - 4a_1a_2 + (5/2)a_2c - (5/2)a_1c]$,
$a_{22} = -(1/d)[(5/4)c^2 + (7/2)a_2c + 2a_2^2]$,
$a_{13} = (3/2)c - 3a_1, a_{23} = -(5/2)c - 3a_2$,
$c_1 = (a/d)[(3/2)c - 3a_1], c_2 = -(a/d)[(5/2)c + 3a_2], c_3 = -2a$,

(c) $a_{11} = (1/d)[-(1/4)c^2 + (3/2)a_1c - 2a_1^2]$,
$a_{12} = (1/d)[(97/18)c^2 - 4a_1a_2 - (13/2)a_1c + (5/2)a_2c]$,
$a_{22} = -(1/d)[(27/4)c^2 + (15/2)a_2c + 2a_2^2]$,
$a_{13} = (3/2)c - 3a_1, a_{23} = -(13/2)c - 3a_2$,
$c_1 = (a/d)[(3/2)c - 3a_1], c_2 = -(a/d)[(13/2)c + 3a_2], c_3 = -2a$,

(d) $a_{11} = (1/d)[-(1/9)c^2 + a_1c - 2a_1^2]$,
$a_{12} = (1/d)[(20/9)c^2 - 4a_1a_2 - 2ca_1 + 2ca_2]$,
$a_{22} = -(1/d)[(10/9)c^2 + 3a_2c + 2a_2^2]$,
$a_{13} = c - 3a_1, a_{23} = -2c - 3a_2, c_1 = (a/d)[c - 3a_1]$,
$c_2 = -(a/d)[2c + 3a_2], c_3 = -2a$,

where $a, a_1, a_2, c,$ and d are parameters, then the general solution of the given system in each case is given by

$$x = u'/S, y = uu'/S, z = (u'' - Tu'^2 - au')/du',$$

where $S = cu(u-1)$ and $T = (S' + P)/S, P = a_1 + a_2 u$, and the function u is a solution of the equation

$$u'^2 = C_1(2u-1) + C_2 u^3(u-2) - Ku^2, \tag{1}$$

where $C_1, C_2,$ and K are arbitrary constants. Equation (1) is solvable in terms of elliptic functions. Leonovich (1984)

30.
$$\begin{aligned} \dot{x} &= x(a_1 x + a_2 y + a_3 z + a), \\ \dot{y} &= y(b_1 x + b_2 y + b_3 z + a), \\ \dot{z} &= P(x, y, z) + c_1 x + c_2 y + c_3 z, \end{aligned}$$

where $a_3 = b_3 = -a_{33} = d \neq 0$ and

$$P(x, y, z) = a_{11}x^2 + a_{12}xy + a_{22}y^2 + a_{13}xz + a_{23}yz + a_{33}z^2,$$

and $a, a_i, b_i, c_i,$ and a_{ij} $(i, j = 1, 2, 3)$ are constants. Moreover,

$$\begin{aligned} a_{11} &= (1/d)[(3/2)c^2 + (1/2)a_1 c - 2a_1^2], \\ a_{12} &= (1/d)[(9/4)c^2 - 4a_1 a_2 - (3/2)a_1 c + (3/2)a_2 c], \\ a_{22} &= (1/d)[(-11/4)c^2 + a_1 c - (3/2)a_2 c - 2a_2^2], \\ a_{13} &= (c/2) - 3a_1, a_{23} = (-3/2)c - 3a_2, \\ c_1 &= (a/d)[(1/2)c - 2a_1], c_2 = -(a/d)[2a_2 + c], c_3 = -a, \end{aligned}$$

wherein $a, a_1, a_2, c,$ and d are parameters; $b_1 = a_1 - c, b_2 = a_2 + c$.

The solution of the given system is given by

$$x = u'/S, y = uu'/S, z = (u'' - Tu'^2 - au')/(du'),$$

where $S = cu(u-1)$ and $T = (S' + P)/S, P(u) = a_1 + a_2 u$ and the function u satisfies

$$u'' = [1/u + 1/2(u-1)]u'^2 + Ku^{1/2}(u-1)e^t, \tag{1}$$

where K is a constant. Equation (1) via

$$u = \left(\frac{1+v}{1-v}\right)^2, \tau = e^t,$$

reduces to the following special case of a fifth Painlevé equation:

$$v'' = [1/(2v) + 1/(v+1)]v'^2 - v'/\tau + Kv/\tau, \quad ' \equiv \frac{d}{d\tau}.$$

Hence the given system is of Painlevé type. Leonovich (1984)

31.
$$\begin{aligned} \dot{x} &= x(a_1 x + a_2 y + a_3 z + a), \\ \dot{y} &= y(b_1 x + b_2 y + b_3 z + a), \\ \dot{z} &= P(x, y, z) + c_1 x + c_2 y + c_3 z, \end{aligned}$$

where $P(x, y, z) = a_{11}x^2 + a_{12}xy + a_{22}y^2 + a_{13}xz + a_{23}yz + a_{33}z^2$, and $a, a_i, b_i, c_i,$ and a_{ij} $(i, j = 1, 2, 3)$ are constants:

$$\begin{aligned} a_{11} &= (1/d)[a_1c - 2a_1^2], \\ a_{12} &= (1/d)[c^2 - 4a_1a_2 - 2a_1c + 2a_2c], \\ a_{22} &= (-1/d)[c^2 + 3a_2c + 2a_2^2], a_{13} = c - 3a_1, \\ a_{23} &= -2c - 3a_2, c_1 = (a/d)[(1/2)c - 2a_1], \\ c_2 &= -(a/d)[c + 2a_2], c_3 = -a, \end{aligned}$$

wherein a, a_1, a_2, c, and d are parameters. It is assumed that $a_3 = b_3 = -a_{33} = d \neq 0$ and $b_1 = a_1 - c, b_2 = a_2 + c$.

The general solution of the given system is given by

$$x = u'/S, y = uu'/S, z = (u'' - Tu'^2 - au')/(du'),$$

where $S = cu(u-1), T = (S' + P)/S, P(u) = a_1 + a_2u$, and u satisfies

$$u'' = (1/2)[1/u + 1/(u-1)]u'^2 + Ku(u-1)e^t,$$

which via

$$u = 1/(1-v), \quad \tau = e^t,$$

transforms into the following special case of fifth Painlevé equation:

$$v'' = [1/(2v) + 1/(v-1)]v'^2 - v'/\tau + Kv/\tau, \ ' \equiv \frac{d}{d\tau},$$

where K is a constant. Hence the given system is of Painlevé type. Leonovich (1984)

32.
$$\begin{aligned} \dot{x} &= x(a_1x + a_2y + a_3z + a), \\ \dot{y} &= y(b_1x + b_2y + b_3z + a), \\ \dot{z} &= P(x,y,z) + c_1x + c_2y + c_3z, \end{aligned}$$

where $P(x,y,z) = a_{11}x^2 + a_{12}xy + a_{22}y^2 + a_{13}xz + a_{23}yz + a_{33}z^2$, and a, a_i, b_i, c_i, and a_{ij} $(i,j = 1,2,3)$ are constants, given either by:

(a) $a_{11} = (1/d)[(2/3)a_1c - 2a_1^2]$,
$a_{12} = (1/d)[(5/3)c^2 - 4a_1a_2 - (8/3)a_1c + (5/3)a_2c]$,
$a_{22} = (-1/d)[(5/3)c^2 + (11/3)a_2c + 2a_2^2]$,
$a_{13} = (2/3)c - 3a_1, a_{23} = (-8/3)c - 3a_2$,
$c_1 = (a/d)[(2/3)c - 2a_1], c_2 = -(a/d)[(5/3)c + 2a_2], c_3 = -a$

or

(b) $a_{11} = (1/d)[(3/2)c^2 + (1/2)a_1c - 2a_1^2]$,
$a_{12} = (1/d)[(-3/2)c^2 - 4a_1a_2 - (5/2)a_1c + (3/2)a_2c]$,
$a_{22} = (-1/d)[c^2 + a_1c + (9/2)a_2c + 2a_2^2]$,
$a_{13} = c/2 - 3a_1, a_{23} = (-5/2)c - 3a_2$,
$c_1 = (a/d)[(1/2)c - 2a_1], c_2 = -(a/d)[(3/2)c + 2a_2], c_3 = -a$,

where a, a_1, a_2, c, and d are parameters. $b_1 = a_1 - c, b_2 = a_2 + c$ in each of cases (a) and (b); $a_3 = b_3 = d = -a_{33}$.

The general solution in each case is given by

$$x = u'/S, y = uu'/S, z = (u'' - Tu'^2 - au')/(du'),$$

where $S = cu(u-1), T = (S' + P)/S, P(u) = a_1 + a_2 u$, and u satisfies

$$u'' = (u'^2/u) + Ku^{1/3}(u-1)e^t,$$

K = constant, for case (a) and

$$u'' - (u'^2/u) - Ku^{1/2}(u-1)e^t = 0,$$

K = constant, for case (b). Leonovich (1984)

3.15 $\vec{y}\,' = \vec{f}(y)$; f_1, f_2, f_3 homogeneous quadratic in y_1, y_2, y_3

1. $$\dot{x}_1 = x_2 x_3, \quad \dot{x}_2 = x_1 x_3, \quad \dot{x}_3 = x_1 x_2.$$

The given system is "algebraically completely-integrable." The first integrals are $I_1 = x_1^2 - x_2^2, I_2 = x_1^2 - x_3^2$. Steeb, Euler, and Mulser (1991)

2. $$\dot{x} = yz, \quad \dot{y} = -2xz, \quad \dot{z} = xy.$$

The given system describes the rotational motion of a book tossed in the air. It is easy to verify that it has a first integral $x^2 + y^2 + z^2 = C$. It has six critical points $(\pm 1, 0, 0)$, $(0, \pm 1, 0)$, $(0, 0, \pm 1)$. The former four are centers, while the latter are saddles. The trajectories of the given system on the surface of a phase sphere are drawn. Bender and Orszag (1978), p.202

3. $$\dot{x} = -3yz, \quad \dot{y} = 3xz, \quad \dot{z} = -xy.$$

Multiplying the first of this system by x, the second by $2y$, and the third by $3z$ and adding, we have

$$x\dot{x} + 2y\dot{y} + 3z\dot{z} = 0,$$

with the integral $x^2 + 2y^2 + 3z^2 = C$, where C is an arbitrary constant. A second integral is found by mulptiplying the first DE by x, the second by y, adding, and integrating:

$$x^2 + y^2 = D,$$

where D is a constant. A third integral is found by quadrature:

$$\frac{dx}{dt} = \pm 3^{1/2}[(D - x^2)(C + x^2 - 2D)]^{1/2}.$$

Zwillinger (1989), p.239

4. $$\dot{x} = yz - x^2, \quad \dot{y} = zx - y^2, \quad \dot{z} = xy - z^2.$$

The given system is a special case of Einstein's equations of gravitation. It is cyclic in x, y, z. Introducing the variables

$$\begin{aligned} u &= x + y + z, \\ v &= xy + yz + zx, \\ w &= xyz, \end{aligned} \tag{1}$$

we get the transformed system as

$$\begin{aligned} &\dot{u} + u^2 = 3v, \\ &\dot{v} = 0, \\ &\dot{w} + 3uw = v^2. \end{aligned} \tag{2}$$

The second of (2) shows that $xy + yz + zx = C$. Now, putting

$$\begin{aligned} U &= x + y + z, \\ V &= x + wy + w^2 z, \\ W &= x + w^2 y + wz, \end{aligned} \tag{3}$$

where w is cubic root of unity, we have

$$\begin{aligned} UV &= x^2 + wy^2 + w^2 z^2 - w^2 xy - yz - wzx, \\ VW &= x^2 + y^2 + z^2 - xy - yz - zx, \\ WU &= x^2 + w^2 y^2 + wz^2 - wxy - yz - w^2 zx. \end{aligned}$$

Adding the equations of the given system, multiplying them by 1, w, and w^2, respectively, and adding, we reduce the given system to

$$\begin{aligned} \frac{dU}{dt} &= -VW, \\ \frac{dV}{dt} &= -UV, \\ \frac{dW}{dt} &= -WU. \end{aligned} \tag{4}$$

From the last two of (4), we get $dV/V = dW/W$, giving $V = k^2 W$, where k^2 is a convenient arbitrary constant. The system (4) is equivalent to

$$\begin{aligned} \frac{dU}{dt} &= -k^2 W^2, \\ \frac{dW}{dt} &= -WU, \\ V &= k^2 W. \end{aligned} \tag{5}$$

Eliminating U from the first two of (5), we have

$$-k^2 W^2 = \frac{-W\dfrac{d^2W}{dt^2} + \left(\dfrac{dW}{dt}\right)^2}{w^2}.$$

Putting $dW/dt = p, d^2W/dt^2 = p(dp/dW)$, etc., and integrating twice, we get $W = (c/k)$ sec $c(t+b)$, where b and c are constants.

The last two equations of (5) then give

$$V = ck \sec c(t+b), \quad U = -c\tan c(t+b).$$

Hence, reverting to original variables via (3), we have

$$\begin{aligned} x &= [(k^2+1)/(3k)]c \sec c(t+b) - (c/3)\tan c(t+b), \\ y &= [(w^2k^2+w)/(3k)]c \sec c(t+b) - (c/3)\tan c(t+b), \\ z &= [(wk^2+w^2)/(3k)]c \sec c(t+b) - (c/3)\tan c(t+b). \end{aligned} \tag{6}$$

Here k is another arbitrary constant. So (6) represents the general solution of the given system. If we write $a_1 = (k^2+1)/(3k), a_2 = (w^2k^2+w)/(3k), a_3 = (wk^2+w^2)/(3k)$, then a_1, a_2, a_3 satisfy the relations $a_1 + a_2 + a_3 = 0, a_1a_2 + a_2a_3 + a_3a_1 = -1/3, a_1a_2a_3 = (k^6+1)/(27k^3)$; hence a_1, a_2, a_3 are roots of the cubic

$$a^3 - (1/3)a - (k^6+1)/(27k^3) = 0. \tag{7}$$

The solution (6) of the given system may, therefore, be written as $x, y, z = a_ic$ sec $c(t + b) - (c/3)\tan c(t+b), i = 1, 2, 3$, where a_i are roots of the cubic equation (7). Reddick (1949), p.226

5. $$\begin{aligned} x' &= x^2 - (5/2)(x-y)^2 + (9/4)(x-z)^2 - (5/2)(y-z)^2, \\ y' &= y^2 - (1/4)(x-z)^2, \\ z' &= z^2 - (5/2)(x-y)^2 + (9/4)(x-z)^2 - (5/2)(y-z)^2. \end{aligned}$$

It follows from the second equation of the given system that

$$(y'' - 2yy')/(y' - y^2) = 2(x+z). \tag{1}$$

Therefore, from (1) and the given system,

$$16xz = [(y'' - 2yy')/(y' - y^2)]^2 + 16(y' - y^2). \tag{2}$$

Differentiating (1) and using the given system, together with (1) and (2), we have

$$\begin{aligned} [(y'' - 2yy')/(y' - y^2)]' &= -8(x+z)^2 + 16xz - 20y^2 + 20y(x+z) + 9(x-z)^2 \\ &= -[(y'' - 2yy')/(y' - y^2)]^2 + 10y(y'' - 2yy')/(y' - y^2) - 20y'. \end{aligned}$$

This simplifies to

$$y''' = 12yy'' - 18y'^2. \tag{3}$$

It is known that Eqn. (3) has a nonstationary singular line [see Eqn. (3.5.64)]. In the interior of the region bounded by this line, y, xz, and $x+z$ have no critical points; however, $x - z$, and thus x and z have branch points in this region. Martynov (1981)

6. $\dot{x} = 2x^2 - 2xy + 2xz, \dot{y} = 2yz, \dot{z} = -x^2 + (1 + 1/k^2 - 1/n^2)xy + (1/n^2)y^2 + z^2$,

where k and n are positive integers for which $1/k + 1/n < 1$.

Following Eqn. (3.15.11), we have $P(u) = 2 - 2u, Q(u) \equiv 0, R(u) = -1 + (1 + 1/k^2 - 1/n^2)u + (u/n)^2$. The equation $Q(u) - uP(u) = 0$ has the roots $\lambda_1 = 0, \lambda_2 = 1$. For the root $\lambda_1 = 0$, we obtain $p = -p_1 = 2, g = g_1 = 0, r = -1, r_1 = 1 + 1/k^2 - 1/n^2$ so that conditions (2) of Eqn. (3.15.11) are satisfied. The solution of the given system has the representation (3) in Eqn. (3.15.11); the coefficients therein may be obtained as described. Martynov and Yablonskii (1979)

7. $$\dot{x} = Cxy + xz, \quad \dot{y} = Ayz + yx, \quad \dot{z} = Bxz + zy,$$

where C, A, and B are constants.

See Eqn. (3.14.23). Bountis, Ramani, Grammaticos, and Dorizzi (1984); Grammaticos, Moulin–Ollagnier, Ramani, Strelcyn, and Wojciechowski (1990)

8. $$\dot{x} = a_{12}xy + a_{13}xz, \quad \dot{y} = b_{23}yz + b_{22}y^2, \quad \dot{z} = c_{33}z^2,$$

where a_{ij}, b_{ij}, c_{ij} are constants.

It is proved that the origin is a saddle-node singularity for the given system. Rusin (1974)

9. $$\dot{x} = a_{12}xy + a_{13}xz + a_{22}yz, \quad \dot{y} = b_{22}y^2, \quad \dot{z} = c_{33}z^2,$$

where a_{ij}, b_{ij}, c_{ij} are constants.

It is proved that the origin is a saddle-node singularity for the given system. Rusin (1974)

10.
$$\begin{aligned}
\dot{x}_1 &= x_1(a_{12}x_2 + a_{13}x_3), \\
\dot{x}_2 &= x_2(a_{21}x_1 + a_{22}x_2 + a_{23}x_3), \\
\dot{x}_3 &= x_3(a_{32}x_2 + a_{13}x_3) + c_1x_1^2 + c_2x_2x_3, c_1 \neq 0,
\end{aligned}$$

where a_{ij} $(i = 1, 2, 3, j = 1, 2, 3), c_1$ and c_2 are constants.

It is shown that for this system to have a single valued solution, it is necessary that $a_{13} = 0$. Martynov (1986)

11.
$$\begin{aligned}
\dot{x} &= a_1x^2 + a_2xy + a_3y^2 + 2xz, \\
\dot{y} &= b_1x^2 + b_2xy + b_3y^2 + 2yz, \\
\dot{z} &= c_1x^2 + c_2xy + c_3y^2 + z^2,
\end{aligned}$$

where $a_i, b_i, c_i, i = 1, 2, 3$ are constants.

First, special solution of the form

$$\begin{aligned}
x &= \alpha + \sum_{k=1}^{\infty} \alpha_k \exp(-kt), \\
y &= \lambda\alpha + \sum_{k=1}^{\infty} \beta_k \exp(-kt), \qquad (1) \\
z &= -1/2 + \sum_{k=1}^{\infty} \gamma_k \exp(-kt),
\end{aligned}$$

are sought. It is found by substitution etc. that the necessary conditions for (1) to satisfy the given system are

$$\begin{aligned}
\alpha p &= 1, \quad g = \lambda p, \quad g_1 = \lambda p_1, \quad p^2 + 4r = 0, \\
\alpha_1 &= -r_1\ell, \quad \beta_1 = (2r - \lambda r_1), \quad \gamma_1 = (pr_1 - p_1r)\ell,
\end{aligned} \qquad (2)$$

where ℓ is a parameter and

$$\begin{aligned} p &= P(\lambda) = a_1 + a_2\lambda + a_3\lambda^2, \\ g &= Q(\lambda) = b_1 + b_2\lambda + b_3\lambda^2, r = R(\lambda) = c_1 + c_2\lambda + c_3\lambda^2, \\ p_1 &= \left[\frac{d}{du}P(u)\right]_{u=\lambda}, \ g_1 = \left[\frac{d}{du}Q(u)\right]_{u=\lambda}, \ r_1 = \left[\frac{d}{du}R(u)\right]_{u=\lambda}. \end{aligned}$$

Higher coefficients $\alpha_k, \beta_k, \gamma_k, k > 1$, can then be found recursively. It is then shown that the series (1) converges absolutely for all t for which $\text{Re}\, t > \text{Re}\, t_0$. The method of majorant series is used.

It may be checked that if $x_1(t)$, $y_1(t)$, and $z_1(t)$ form a special solution of the given system, then the functions

$$\begin{aligned} x &= \{1/(ct+d)^2\}x_1\{-(at+b)/(ct+d)\}, \\ y &= \{1/(ct+d)^2\}y_1\{-(at+b)/(ct+d)\}, \\ z &= \{1/(ct+d)^2\}z_1\{-(at+b)/(ct+d)\} - c/(ct+d) \end{aligned}$$

also form a solution of the same, provided that the arbitrary constants a, b, c, d satisfy the condition $ad - bc = -1$. Using this fact and (1), it is then shown that

$$\begin{aligned} x &= \alpha/(ct+d)^2 + \{1/(ct+d)^2\}\sum_{k=1}^{\infty} \alpha_k \exp[k(at+b)/(ct+d)], \\ y &= \lambda\alpha/(ct+d)^2 + \{1/(ct+d)^2\}\sum_{k=1}^{\infty} \beta_k \exp[k(at+b)/(ct+d)], \qquad (3) \\ z &= -c/(ct+d) - 1/\{2(ct+d)^2\} + \{1/(ct+d)^2\}\sum_{k=1}^{\infty} \gamma_k \exp[k(at+b)/(ct+d)] \end{aligned}$$

is also a solution of the given system. Martynov and Yablonskii (1979)

12.
$$\begin{aligned} \dot{x} &= a_1(x-\mu z)^2 + a_2(x-\mu z)(y-\nu z) + a_3(y-\nu z)^2 + 2xz - \mu z^2, \\ \dot{y} &= b_1(x-\mu z)^2 + b_2(x-\mu z)(y-\nu z) + b_3(y-\nu z)^2 + 2yz - \nu z^2, \\ \dot{z} &= c_1(x-\mu z)^2 + c_2(x-\mu z)(y-\nu z) + c_3(y-\nu z)^2 + z^2, \end{aligned}$$

where a_k, b_k, c_k $(k = 1, 2, 3)$ are parameters; μ and ν take the values 0 and 1.

Conditions are found under which the given system has a nonstationary singular line and has no critical singularities in the domain of existence of x, y, and z. If $x_1(t), y_1(t), z_1(t)$ is a solution of the given system, then it may be easily verified that

$$\begin{aligned} x &= \{1/(Ct+D)^2\}x_1[-(At+B)/(Ct+D)] - \mu C/(Ct+D), \\ y &= \{1/(Ct+D)^2\}y_1[-(At+B)/(Ct+D)] - \nu C/(Ct+D), \\ z &= \{1/(Ct+D)^2\}z_1[-(At+B)/(Ct+D)] - C/(Ct+D) \end{aligned}$$

is also a solution, where A, B, C, and D are arbitrary constants with $AD - BC = -1$. Martynov (1981)

13.

$$\dot{x}_k = x_k(a_{k1}x_1 + a_{k2}x_2 + a_{k3}x_3), \quad k = 1, 2,$$

$$\dot{x}_3 = x_3(a_{31}x_1 + a_{32}x_2 + a_{33}x_3) + cx_1x_2,$$

where a_{kj} $(k = 1, 2, 3, j = 1, 2, 3), c_1$, and c_2 are constants.

Conditions (which are rather elaborate) on a_{ij} are found such that the given system has unique solutions. Martynov (1986)

14.

$$\dot{x}_k = P_k(x_1, x_2) + a_{k3}x_kx_3, \quad k = 1, 2, 3,$$

where $a_{13} \neq a_{23}$ and $P_k(x_1, x_2)$ are quadratic forms in x_1 and x_2; a_{kj} are constants.

First it is shown that using a nonsingular linear transformation of the function x_k, it is possible to reduce the given system to Eqn. (3.15.10) or (3.15.13); numbers α and β can be found such that when the condition $P_1(0, 1) = P_2(1, 0) = 0$ is satisfied, the nonsingular linear transformation is

$$x_k = y_k, k = 1, 2, x_3 = \alpha y_1 + \beta y_2 + y_3,$$

so that the given system reduces to Eqn. (3.15.10) or (3.15.13). Martynov (1986)

15.

$$\dot{x}_k = \sum_{m=1}^{3} a_{km}x_kx_m, \quad k = 1, 2, 3,$$

where a_{km} are constants, and where the determinant

$$\Delta = \begin{vmatrix} a_{11} & a_{12} & a_{13} \\ a_{21} & a_{22} & a_{23} \\ a_{31} & a_{32} & a_{33} \end{vmatrix} = 0,$$

$a_{mm} \neq 0, m = 1, 2, 3; a_{23} - a_{13} \neq 0, a_{12} - a_{32} \neq 0, a_{31} - a_{21} \neq 0$. Under these conditions, we can find constants u_m $(m = 1, 2, 3)$ such that

$$\sum_{k=1}^{3} a_{km}\mu_k = 0, \mu_1^2 + \mu_2^2 + \mu_3^2 \neq 0. \tag{1}$$

When this condition is satisfied, it can be shown that the given system has the first integral

$$x_1^{\mu_1}\, x_2^{\mu_2}\, x_3^{\mu_3} = C, \tag{2}$$

where C is an arbitrary constant. By using (2), we can eliminate one of x_k from the given system and obtain a second order system for the remaining functions x_m and x_n, say. It may be shown that if $\Delta \neq 0, a_{kn} \neq a_{mn}$, and $A_{kk} = 0$ or $a_{kk} = 0$, there are no systems of the given form with the P-property. (Here A_{kk} are the cofactors of a_{kk}.) Martynov (1984)

16.

$$\dot{x}_k = \sum_{m=1}^{3} \sum_{m=1}^{3} a_{km}x_kx_m, \quad k = 1, 2, 3,$$

where a_{km} are constants.

The coefficients a_{km} are determined such that the given system has the Painlevé property. In particular, it is shown that if the matrix $\{a_{km}\}$ is defined by

$$\begin{bmatrix} a & (1 - r_{12})b & (1 - r_{13})c \\ (1 - r_{21})a & b & (1 - r_{23})c \\ (1 - r_{31})a & (1 - r_{32})b & c \end{bmatrix},$$

where a, b, and c are nonzero parameters, and the numbers $1 - r_{km}$ are as tabulated, the given system has the Painlevé property. Martynov (1984)

17.
$$\dot{x}_k = \sum_{m=1}^{3} a_{km} x_k x_m, \quad k = 1, 2, 3,$$

where $\Delta = \det |a_{km}| = 0$ or when $\Delta \neq 0$, but at least one of the relations

$$a_{kn} = a_{mn} \tag{1}$$

holds with distinct k, m, and n. In these cases it is possible to show that the system enjoys the Painlevé property. For example:

(a) If

$$\begin{aligned} a_{11} &= a_{22} = -a_{21} = -a_{32} = -1, \\ a_{12} &= a_{13} = a_{23} = a_{33} = a_{31} = 0, \end{aligned}$$

then $\Delta = 0$, and it is possible to find the first integral $x_1 x_2 x_3 = 2C_1$, and the solutions are given by

$$\begin{aligned} x_1 &= 2C_1/(2C_1 t + C_2), \\ x_2 &= (2C_1 t + C_2)/(C_1 t^2 + C_2 t + C_3), \\ x_3 &= C_1 t^2 + C_2 t + C_3, \end{aligned}$$

where C_1, C_2, and C_3 are arbitrary constants.

(b)

$$\begin{aligned} a_{11} &= -a_{22} = 1/m, a_{21} = 1/m - 1, a_{22} = -1 - 1/m, \\ a_{31} &= -a_{32} = -p/m, a_{33} = -c/n, a_{13} = a_{23} = C \neq 0, \end{aligned}$$

where m is an integer, $m \neq 0$, and n is a positive integer.

Here $p = 0, 1, 2, 3$ for $n = 1; p = 0, 1, 2$ for $n = 2; p = 0, 1$ for $n = 3, 5, \infty; p = 0$ for the remaining cases. In these cases at least one of the conditions (1) is satisfied. For such a system, the relation (3) derived in Eqn. (3.15.18) becomes

$$u''' = (1 - 1/n)(u''^2/u) + a(u''u'/u) + b(u'^3/u^2),$$

where

$$\begin{aligned} a &= (1 + 2/n)(1 - 1/m) + p/m, \\ b &= (1/m - 1)[1 + p/m + 1/n - 1/(mn)], \end{aligned}$$

and has the Painlevé property. We may also obtain

$$\begin{aligned} x_1 &= u'/[u(u-1)], x_2 = u'/(u-1), \\ x_3 &= (1/c)/[u''/u' - \{(m-1)/m\}(u'/u)], \end{aligned}$$

where $u = x_2/x_1$. Martynov (1984)

18.
$$\dot{x}_k = \sum_{m=1}^{3} a_{km} x_k x_m, \quad k = 1, 2, 3,$$

where a_{km} are constant coefficients such that $a_{23} = a_{13} \neq 0$, and $S = a_{21} - a_{11} + (a_{22} - a_{12})u \not\equiv 0$, where $u = x_2/x_1$, we have

$$u' = Sux_1. \tag{1}$$

Differentiating (1) and using the given system, we find that

$$a_{13}\, x_3 = u''/u' - Tu', \tag{2}$$

where $T = \left(u\dfrac{dS}{du} + P + S\right) / (uS)$, $P = a_{11} + a_{12}u$. Differentiating (2) and using the given system, we obtain

$$u''' = (1 + a_{33}/a_{13})(u''^2/u') + f_1(u)u'u'' + \phi_1(u)u'^3, \tag{3}$$

where $f_1(u) = (1 - 2a_{33}/a_{13})T + (a_{31} + a_{32}u)/(uS)$,

$$\phi_1(u) = \frac{dT}{du} + \frac{a_{33}}{a_{13}} T^2 - \frac{a_{31} + a_{32}u}{uS} T.$$

Equation (3) has been studied by Carton–Lebrun (1969) and the functions $f_1(u)$ and $\phi_1(u)$ can be found such that (3) has the Painlevé property. The following cases can be studied similarly:

(a) $a_{23} = a_{13} = 0$.
(b) $a_{23} = a_{13} \neq 0, S \equiv 0$.
(c) $a_{23} = a_{13} = 0, S \equiv 0$.
(d) $a_{12} = a_{32}$.
(e) $a_{21} = a_{31}$.

Martynov (1984)

3.16 $\vec{y}' = \vec{f}(\vec{y}); f_1, f_2, f_3$ polynomial in y_1, y_2, y_3

1.
$$\dot{x} = y, \quad \dot{y} = xz, \quad \dot{z} = x^2 + z^3.$$

A phase space analysis is carried out. Kapitanov (1984)

2.
$$\dot{x} = y, \quad \dot{y} = xz, \quad \dot{z} = x^2 - z^3.$$

A phase space analysis is carried out. Kapitanov (1984)

3.
$$\epsilon\dot{x} = -x^3 + yx - z, \quad \dot{y} = 1 - 0.1\, x - y, \quad \dot{z} = (1 - 0.1x - y)x + x + \lambda,$$

where $\epsilon = 0.05$ and λ is a parameter.

The given system models the electric activity of nerve cells treated with barium. Computations show that it exhibits its first period doubling at λ = -0.555916. Chaotic solutions are found near λ = - 0.555846. Grasman (1987), pp.65, 67

4. $$\epsilon\dot{x} = y - (1/3)x^3 + x, \quad \dot{y} = -x - x^2 z, \quad \dot{z} = (x+a)z^2,$$

where a is a constant.

The trajectories are found numerically and projected on the (x, z) and (y, z) planes. In the limit $\epsilon \to 0$, the solution jumps from $x = 1$ to $x = -2$ and from $x = -1$ to $x = 2$. For $a \in (1.0, 2.0)$, repeated period doubling and chaos are found. A representative chaotic solution for $a = 1.7$ and $\epsilon = 0.05$ is shown graphically. The Lyapunov exponents of this chaotic solution are $\lambda_1 = 0.09, \lambda_2 = 0, \lambda_3 = -26.45$. A positive exponent indicates chaotic behavior. Grasman (1987), p.103

5. $$\dot{y} = w, \quad \dot{w} = -\mu w - zy + y - ry^3, \quad \dot{z} = -Az - Byw,$$

where μ, r, and A are positive numbers and $B \in R^1$. Assuming that $B \geq 0,\ \mu > A$, conditions are established for the global asymptotic stability of this Lorenz system; then each of the solutions tends for $t \to \infty$ to an equilibrium position.

Referring to the special case

$$\begin{aligned} \frac{dx_1}{dt} &= -\sigma_1(x_1 - y_1), \\ \frac{dy_1}{dt} &= -x_1 z_1 + r_1 x_1 - y_1, \\ \frac{dz_1}{dt} &= -b_1 z_1 + x_1 y_1, \end{aligned}$$

the theory developed is applied to show that for $\sigma_1 = 10, b_1 = 8/3$, there is global asymptotic stability provided that $r_1 \leq 4.5$. Leonov (1986)

6. $$\dot{x}_1 = r[f_\beta(x_2 - x_1) - f_\alpha(x_1)], \quad \dot{x}_2 = -x_3 - f_\beta(x_2 - x_1), \quad \dot{x}_3 = x_2 - \rho x_3,$$

where $f_\alpha(v) = -\alpha_1 v + \alpha_3 v^3, f_\beta(v) = \beta_1 v + \beta_3 v^3$, and $\alpha_1, \alpha_3, \beta_1, \beta_3, r$, and ρ are parameters.

The given system describes a modified Van der Pol oscillator. Degenerate bifurcations are found that control the global dynamics of the system. A two-dimensional parameter space for the oscillator is studied and the measured loci of critical points, where a qualitative change of behavior is observed, are given in detail. It is shown explicitly by analysis of expermental data that chaos arises through the Silnikov mechanism. Healey et al. (1991)

7. $$\dot{x} = y, \quad \dot{y} = -\mu y - xz - \phi(x), \quad \dot{z} = -Az - Bxy,$$

where $\phi(x) = -x + \gamma x^3$, and μ, γ, A, and B are constants. The given system is equivalent to Lorenz system through a simple change of variables.

The simplest bifurcation parameters corresponding to loop separatrices of a saddle equilibrium state of the given system and their expression in terms of elementary functions are determined. Such estimates in the space of parameters (involved) single out domains of absence of a loop separatrix of the saddle point $(0, 0, 0)$. Comparison systems are used to obtain cases of global asymptotic stability. The Hausdorff dimensions of attractors of the given system are also determined. Leonov (1989)

8. $\dot{x} = xz - Wy, \quad \dot{y} = xW + yz, \quad \dot{z} = P + z - (1/3)z^3 - (x^2 + y^2)(1 + Qx + Ez),$

where W, P, Q, and E are parameters.

The given system is called a Langford model. For $W = 10, E = 0.5, Q = 0.7$, and the bifurcation parameters $P = 1.2, 0.95, 0.7$, phase portraits are obtained from the numerical integration. A complicated limit cycle is identified. Mullin (1991)

9. $$\dot{x} = y, \quad \dot{y} = x - 2x^3 + \alpha y + \beta x^2 y - \nu yz, \quad \dot{z} = -\gamma z + \delta x^2,$$

with $\gamma = \pm 1$, and $\alpha, \beta, \gamma, \delta$ are parameters.

It is shown that the given system has a transitive attractor similar to that of the geometric model of the Lorenz equations. It is proved that such an attractor results if a double homoclinic connection of a fixed point with a resonance condition among the eigenvalues is broken in a careful manner. Robinson (1989)

10. $\dot{x} = \alpha(x - x^3/3 + w), \quad \dot{y} = -\epsilon(\gamma x + w - z - s) - \epsilon^2 \mu(\beta x + z), \quad \dot{z} = -\epsilon^2 \mu(\beta x + z),$

where α, s, ϵ, and μ are constants.

The given system describes bursting in biochemical and bioelectrical systems. With ϵ small, a small-amplitude solution is obtained by a perturbation series. The analysis shows that there probably are no periodic solutions. Tu (1989)

11. $$\begin{aligned} \dot{x}_1 &= x_1 - x_1 x_2 - x_2^3 + x_3(x_1^2 + x_2^2 - 1 - x_1 + x_1 x_2 + x_2^3), \\ \dot{x}_2 &= x_1 - x_3(x_1 - x_2 + 2x_1 x_2), \\ \dot{x}_3 &= (x_3 - 1)(x_3 + 2x_3 x_2^2 + x_3^2). \end{aligned}$$

The given DE has five critical points $(0,0,0), (\pm\, 1, 0, 1), (1/2, \pm\, 3^{1/2}/2, 1)$. It is easily checked that the planes $x_3 = 0$ and $x_3 = 1$ satisfy the third equation of the system and reduce the first two to a two-dimensional system and therefore form invariant sets. In particular, with $x_3 = 1$, we have

$$\dot{x}_1 = x_1^2 + x_2^2 - 1, \quad \dot{x}_2 = x_2(1 - 2x_1). \tag{1}$$

A phase plane analysis of system (1) shows that it has no periodic solution. Verhulst (1990), p.24

12. $$\dot{x} = -x + x^3 y^2 - x^2 y^3 z, \quad \dot{y} = -y + z^3, \quad \dot{z} = -z + x^4 - z^4.$$

The singular point $(0,0,0)$ is a stable critical point: The linearized system about this point is $\dot{x} = -x, \dot{y} = -y, \dot{z} = -z$, having the solution

$$x(t) = x(0)e^{-t}, y(t) = y(0)e^{-t}, z(t) = z(0)e^{-t}.$$

This behavior persists for the nonlinear form also. In fact, it can be shown that when t is large and $|x(0)|, |y(0)|$, and $|z(0)|$ are sufficiently small, the solution has the form

$$\begin{aligned} x(t) &\sim \textstyle\sum_{n=1}^{\infty} a_n e^{-nt}, \quad y(t) \sim \sum_{n=1}^{\infty} b_n e^{-nt}, \\ z(t) &\sim \textstyle\sum_{n=1}^{\infty} c_n e^{-nt}, \quad t \to +\infty. \end{aligned} \tag{1}$$

The coefficients are $a_2 = b_2 = c_2 = a_3 = c_3 = 0, b_3 = -(1/2)c_1^2$, etc.

Drawing the graph of $x(t)/y(t)$ and $y(t)/z(t)$ with $x(0) = y(0) = z(0) = 1$, it is shown that, in accordance with (1), these ratios approach the constants a_1/c_1 and b_1/c_1, respectively, as $t \to +\infty$. Bender and Orszag (1978), p.186

13. $$\dot{x} = -(1/2)x^3 y, \quad \dot{y} = -x^4 + x^6\rho^2, \quad \dot{z} = 1 - x^6,$$

where ρ is a constant.

The given DE describes an unperturbed two-body system. It has a constant of motion

$$H(x, y, \rho) = (1/2)y^2 + (1/2)x^4\rho^2 - x^2.$$

The flow of the system is depicted graphically. Xia (1992)

14. $$\dot{x} = 2axz^m, \quad \dot{y} = -2bz + 2ayz^m, \quad \dot{z} = 2by - 2a(x^2 + y^2)z^{m-1},$$

where m is a natural number, and a and b are positive numbers.

It is easily checked that $x^2 + y^2 + z^2 =$ constant. Therefore, attention is confined to spheres. It is shown that the topological nature of the trajectories is determined basically by the parity of m. For odd (even) $m < 3 (m > 4)$, the dynamical structure of the system is identical to that of the case $m = 3 (m = 4)$ and is only slightly different from that of the case $m = 1 (m = 2)$. Liao (1987)

15. $$\dot{x} = P(x, y, z), \quad \dot{y} = Q(x, y, z), \quad \dot{z} = R(x, y, z),$$

where P, Q, R are polynomials in Euclidean space E^3.

For behavior near each of isolated critical points, it is assumed that the system is insensitive or structurally stable. This means that the characteristic roots have nonzero real parts. Such critical points are classified as (1) node, (2) focus, (3) saddle, (4) saddle focus, depending on whether (1) all roots are real and of the same sign; (2) there is exactly one real root but all real parts of the same sign; (3) all roots are real but not all of the same sign; and (4) there is exactly one real root but real parts of roots not all of the same sign, respectively. To determine the global behavior of the solutions, E^3 is compactified to the projective space P^3.

A number of examples, with figures for solutions, illustrate the theory. The stability of the solutions at infinity is also considered. Minc (1955)

3.17 $\vec{y}\,' = \vec{f}(\vec{y}); f_1, f_2, f_3$ not polynomial in y_1, y_2, y_3

1. $$\dot{x} = -y, \quad \dot{y} = y + z, \quad \dot{z} = 2z - yz/x.$$

It is easy to see that $(d/dt)(xz + y^2) = 2(xz + y^2)$, so that

$$x(t)z(t) + (y(t))^2 = (x_0 z_0 + y_0^2)e^{2t}, \tag{1}$$

where $x_0 = x(0), y_0 = y(0), z_0 = z(0)$. Assuming that $x_0, z_0 > 0$, we observe that

$$\frac{d}{dt}(z/x) = (x\dot{z} - z\dot{x})/x^2 = \{x(2z - yz/x) - z(-y)\}/x^2 = 2(z/x).$$

Therefore,

$$z(t)/x(t) = (z_0/x_0)e^{2t}. \tag{2}$$

Putting $\mu^2 = (x_0/z_0)(x_0z_0 + y_0^2)$, we eliminate e^{2t} from (1) and (2), to obtain $z/x = (1/\mu^2)(xz + y^2)$. This enables us to solve for z with the result

$$z = xy^2/(\mu^2 - x^2), \tag{3}$$

showing that the trajectory lie on the rational surface (3) in 3-space. Using (1), (2) and the given system, it is possible to find the complete solution of the given system as

$$\begin{aligned} x(t) &= \mu\ \cos\{(z_0/x_0)^{1/2}(e^t - 1)/\mu + \text{arc}\ \cos(x_0/\mu)\}, \\ y(t) &= (z_0/x_0)^{1/2}e^t \sin\{(z_0/x_0)^{1/2}(e^t - 1)/\mu + \arccos(x_0/\mu)\}, \\ z(t) &= \{z_0/(\mu x_0)\}e^{2t}\cos\{(z_0/x_0)^{1/2}(e^t - 1)/\mu + \arccos(x_0/\mu)\}. \end{aligned}$$

When t assumes values such that $(z_0/x_0)^{1/2}(e^t-1)/\mu+\arccos(x_0/u) = 2n\pi$ for some integer n, we will have $x(t) = \mu, y(t) = 0, z(t) = \{z_0/(\mu x_0)\}e^{2t}$.

For any prescribed values of x_0 and μ, we shall choose y_0 and x_0 so that

$$\begin{aligned} (z_0/x_0)^{1/2} &= \mu\ \arccos(x_0/\mu), \\ y_0 &= \mu\ (\mu^2 - x_0^2)^{1/2}\arccos(x_0/\mu), \end{aligned}$$

where x_0, z_0 are required to behave like signs and $y_0 > 0$. The trajectories will accordingly assume the simpler form

$$\begin{aligned} x(t) &= \mu\cos\{e^t\ \arccos(x_0/u)\}, \\ y(t) &= \mu\{\arccos(x_0/\mu)\}e^t\sin\{e^t\ \arccos(x_0/\mu)\}, \\ z(t) &= \mu\{\arccos(x_0/\mu)\}^2e^{2t}\cos\{e^t\ \arccos(x_0/\mu)\}. \end{aligned}$$

The presence of chaotic behavior in these trajectories is then exhibited by reducing the system to a certain iterative map. Whittaker (1991)

2. $\quad \dot{x}_1 = -[2\pi/\{1 + r^2 + x_3\}]x_2, \quad \dot{x}_2 = [2\pi/\{1 + r^2 + x_3\}]x_1, \quad \dot{x}_3 = -x_3,$

where $r^2 = x_1^2 + x_2^2$.

The general solution of the given system is easily found to be

$$x_1 = a\cos\theta, x_2 = a\sin\theta, x_3 = be^{-t},$$

where $\theta = \{2\pi/(1+a^2)\}\ \log|b+(1+a^2)e^t|$; a and b are arbitrary constants. Urabe (1967), p.119

3. $\quad \dot{x} = 2xy, \quad \dot{y} = z^2/x^2 - y^2 + \alpha - \gamma x, \quad \dot{z} = (\beta - \gamma x)x,$

where α, β, and γ are constants. The system describes the spatial structure of time-periodic solutions of the Ginsburg–Landau equation.

It is easy to check that the given system is volume preserving. It is also equivalent under $(x, y, z, t) \to (x, -y, -z, -t)$. Spatially periodic and quasiperiodic solutions as well as heteroclinic orbits are shown to exist. Chaotic solutions are also speculated upon. Holmes (1986)

4. $$\dot{x} = x(\alpha_1 - x - \beta_1 y), \quad \dot{y} = y(\alpha_2 - \beta_2 x - y), \quad \dot{z} = z\{1 - z/(x+y)\},$$

$x(0) = x_0 > 0, y(0) = y_0 > 0, z(0) = z_0 > 0.$

The given system describes a deterministic model of two competing donor species and one recipient species that simulates the dynamics of interactions in estuaries or oceans. Equilibria of the system, their stability, and the asymptotic behavior of the solutions are described. Some numerical results are shown graphically for a few sets of parameters. Sarkar and Roy (1990)

5. $\dot{x} = 1 - x - Axy/(a+x), \dot{y} = Axy/(a+x) - y - Byz/(b+y), \dot{z} = Byz/(b+y) - z,$

where A, B, a, and b are constants.

The given system describes a microbial model. Adding the three equations, we get

$$\dot{x} + \dot{y} + \dot{z} = 1 - x - y - z,$$

so that, on integration, we have

$$x + y + z = 1 + Ce^{-t},$$

where C is an arbitrary constant.

Since in the given problem $x, y, z > 0$, the asymptotic behavior of the given system is considered on the plane $x + y + z = 1$; the given system then reduces to

$$\begin{aligned} \dot{y} &= A(1-y-z)y/(a+1-y-z) - y - Byz/(b+y), \\ \dot{z} &= Byz/(b+y) - z. \end{aligned} \tag{1}$$

The system (1) is studied in detail with respect to its singular points and their stability. Previous results on the given system are quoted. Kot, Sayler, and Schultz (1992)

6. $$\begin{aligned} \dot{x}_1 &= x_1(\lambda_1 - k\alpha_1 x_2/z), \\ \dot{x}_2 &= x_2(-\lambda_2 - k\alpha_2 x_1/z) + \gamma k x_1 x_3/z, \\ \dot{x}_3 &= x_3[-\lambda_3 + (k\alpha_3 x_1/z)(1 - x_3/\theta)] + S, \end{aligned}$$

$x_i > 0$ $(i = 1,2,3); 0 < x_3 \leq \theta$, where $z = 1 + k(x_1 + x_2 + nx_3)$, and α_i, λ_i $(i = 1,2,3), k, \gamma, \theta, n$, and S are positive constants. The given system was introduced as a model for the immune response in an animal to invasion by active, self-replicating antigens.

Introducing the variable $s, s = \int_0^t z^{-1}(w)dw$, the given system is reduced to $dx/ds = E(x)$ where $x = (x_1, x_2, x_3)^T$ and

$$E(x) = \begin{bmatrix} x_1(\lambda_1 z - k\alpha_1 x_2) \\ x_2(-\lambda_2 z - k\alpha_2 x_1) + \lambda k x_1 x_3 \\ x_3[-\lambda_3 z + k\alpha_3 x_1(1 - x_3/\theta)] + Sz \end{bmatrix}. \tag{1}$$

Two necessary conditions for the given system to have a periodic solution in the positive orthant are that

$$\alpha_1 > \lambda_1 \quad \text{and} \quad S/\lambda_3 < \theta. \tag{2}$$

It is first shown that if for any given positive α_i, λ_i $(i = 1,2,3), k, n, S, \gamma$, and θ such that (2) holds, and $[\lambda_1 c_1 + k(\alpha_1 - \lambda_1)S/\lambda_3 + kn\lambda_1 c_1 S/\lambda_3]^2 \geq 4k\lambda_1\theta(1+kn\theta)[\alpha_1 c_2 + (\alpha_1 - \lambda_1)c_1], \lambda_1 c_1^2(1+kn\theta) < k\alpha_1 c_2 S/\lambda_3$, where $c_1 = \alpha_2/\gamma, c_2 = \lambda_2/\gamma$ $(c_1, c_2 > 0)$, then the given system has at least two critical points (steady states) in R_3^+. Using Hopf's theorem on bifurcating periodic solutions and a stability criterion of Hsu and Kazarinoff, existence of a family of unstable periodic solutions bifurcating from one steady state of a reduced 2 × 2 form of the 3 × 3 system is obtained. It is further shown that no periodic solutions bifurcate from the other steady state.

The existence of a family of periodic solutions of the full 3 × 3 system is proved and the stability criterion for the same exhibited. Two specific examples are treated:

(a) $\alpha_3 = 8, \alpha_1 = \lambda_3 = \theta = 2, \lambda_1 = n = k = S = 1, c_1 = 1/8$, and $c_2 = 1/32$. This system has an equilibrium state (2,4,1). The linearized matrix of the system about (2,4,1),

$$A(\gamma) = \begin{bmatrix} 2 & -2 & 2 \\ 3\gamma/8 & -5\gamma/8 & 15\gamma/8 \\ 3 & -1 & -17 \end{bmatrix},$$

cannot have two pure imaginary eigenvalues if $\gamma > 0$. Hence there are no periodic solutions bifurcating from (2,4,1).

(b) $\alpha_3 = 5, \theta = 2, \lambda_3 = 5/4, \alpha_1 = k = n = S = 1, \lambda_1 = 10^{-1}, c_1 = 3, c_2 = 10^{-1}$. In this case, the given system has (1/4,1/4,1) as an isolated equilibrium point. The linearized system about this point has the matrix

$$A(\gamma) = \begin{bmatrix} (1/40) & -(9/40) & (1/40) \\ (9/40)\gamma & -(41/40)\gamma & (9/40)\gamma \\ (9/4) & -(1/4) & -(27/8) \end{bmatrix},$$

and det $A(\gamma) < 0$ for all $\gamma > 0$. The characteristic equation for A is $\nu^3 + [(41\gamma+134)/40]\nu^2 + [(1133\gamma - 45)/320]\nu + 9\gamma/64 = 0$. The condition that there be two purely imaginary eigenvalues at $\gamma_c > 0$ are $1133\gamma > 45$ and $[(41\gamma + 134)/40][(1133\gamma - 45)/320] = 9\gamma/64$.

The unique value $\gamma_c \simeq 0.0401882473$ satisfies all the conditions above. Using the general theorem, it is then shown that there exists a family of asymptotically orbitally stable periodic solution of the given system bifurcating from the steady state (1/4,1/4,1) for values of γ in some neighborhood below $\gamma_c \simeq 0.0401882473$, with the remaining constants assuming the given values. By continuity, this statement holds in some unknown neighborhood of $(1, 1, \ldots, 10^{-1})$ in $(\alpha_1, k, \ldots, c_2)$ space. Hsü and Kazarinoff (1977)

7. $$y' = 2u/x, z' = 2u/(1-x), u' = \{P_1(y,z)\}/x + \{P_2(y,z)\}/(1-x),$$

where $P_1(y,z) = \dfrac{\partial P(y,z)}{\partial y}, P_2(y,z) = \dfrac{\partial P(y,z)}{\partial z}$, P being a polynomial in y and z. Erugin (1974)

8. $$\begin{aligned} \dot{x} &= rx\{1 - x/K\} - (m_1/y_1)\{yx/(a_1+x)\} - (m_2/y_2)\{zx/(a_2+x)\}, \\ \dot{y} &= \{m_1 yx/(a_1+x)\} - D_1 y, \\ \dot{z} &= m_2 zx/(a_2+x) - D_2 z, \end{aligned}$$

$x(0) = x_0 > 0, y(0) = y_0 > 0, z(0) = z_0; m_i, y_i, a_i$ $(i = 1,2)$, and r are constants.

The given problem arises in competing predators. The behavior of solutions of this system of DE is studied to answer the biological question: Under what conditions will neither, one, or both species of predators survive? Numerical results are also referred to. Several asymptotic and qualitative results are derived. Hsu, Hubbell, and Waltman (1978)

9.
$$\begin{aligned} \dot{y}_1 &= -y_2 - \alpha y_1 y_3/\{y_1^2 + y_2^2\}^{1/2}, \\ \dot{y}_2 &= -y_1 - \alpha y_2 y_3/\{y_1^2 + y_2^2\}^{1/2}, \\ \dot{y}_3 &= \alpha y_1/\{y_1^2 + y_2^2\}^{1/2}, \end{aligned}$$

where α is a parameter.

One trajectory of the given system is

$$y_1 = (2 + \cos \alpha t) \cos t, y_2 = (2 + \cos \alpha t) \sin t, y_3 = \alpha \sin t.$$

This trajectory lies on the surface of a torus in E^3 obtained by rotating the circle $(y_1 - 2)^2 + y_2^2 = 1$ about the y_3 axis. It is easy to check that if α is irrational, the trajectory continually winds about the torus without ever intersecting itself. Daniel and Moore (1970), p.9

10.
$$\begin{aligned} y' &= (3w - 2y)/(3z - 2x), \\ z' &= (z - 3\mu x p^{-3})/(3z - 2x), \\ w' &= (w - 3\mu y p^{-3})/(3z - 2x), \end{aligned}$$

where $p = (x^2 + y^2)^{1/2}$ and μ is a constant.

The given system has the first integrals

$$xw^2 - yzw - \mu x/p = \text{ constant}, \quad yz^2 - xzw - \mu y/p = \text{ constant}.$$

Prince and Eliezer (1981)

11. $\dot{x} = K - \sigma\phi(x, y), \dot{y} = q_1\sigma\phi(x, y) - \sigma\eta(y, z), \dot{z} = q_2\sigma\eta(y, z) - k_s z,$

where k, σ, q_1, q_2, and k_s are constants and

$$\phi(x, y) = \{x(1 + x)(1 + y)^2\}/\{L_1 + (1 + x)^2(1 + y)^2\}$$

and

$$\eta(y, z) = y(1 + z)^2/\{L_2 + (1 + z)^2\};$$

L_1 and L_2 are constants. Here x, y, z represent concentration in the biochemical context; the physical space is restricted to $x \geq 0, y \geq 0, z \geq 0$.

Numerical study shows chaotic behavior of the system for certain sets of parameters. Schichtel and Beckmann (1991)

12. $\dot{x} = -x + axy + uy + \{b/(1 + z^p)\}(1 - x), w\dot{y} = x - axy - vy, \dot{z} = d(y - z),$

where $p \geq 1, v > u, a < \epsilon_0$, and ϵ_0 is a fixed small positive number; w, b, and d are also constants. The given system appears in the analysis of an isothermal chemical reaction occurring in a volume bounded by a membrane and immersed in a reservoir of reactants and products at fixed concentration.

The existence of periodic solutions is proved, using the Brouwer fixed-point theorem. The proof also shows that essentially every solution must oscillate. Uniqueness and global asymptotic stability of periodic solutions are proved for small values of the parameters. Dai (1979)

13. $\dot{x} = xy \sin z, \dot{y} = x^2 \sin z, \dot{z} = x\{x/y + 2y/x\} \cos z.$

The given system appears in the description of electrostatic structures in plasmas. Two first integrals are easily obtained:

$$I_1 = x^2 - y^2, I_2 = x^2 y \cos z.$$

This suggests the introduction of a new variable,

$$u = x^2 - x_0^2 = y^2 - y_0^2,$$

for which an "energy law"

$$\frac{1}{2}\left(\frac{du}{dt}\right)^2 + V(u) = 0$$

is obtained with the conservative potential

$$V(u) = -2\left[(u + x_0)^2(u + y_0^2) - \Gamma^2\right], \Gamma = x^2 y \cos z.$$

The solution corresponding to IC $x_0 = y_0, z_0 = 0\,(\Gamma = x_0^3)$ is found by a quadrature:

$$t_0 - t = \pm\,\{3^{-1/4}/2x_0\}F[\arccos\{(3^{1/2} - 2 - u/x_0^2)/(3^{1/2} + u/x_0^2)\}; \sin(\pi/12)],$$

where F is an elliptic integral of the first kind. For $x_0 = y_0, z_0 = \pi/2$, we have $x = (x_0^{-1} - t)^{-1}$, etc. Bauer and Schamel (1992)

14. $\dot{x} = 2x + 4xy \sin z, \dot{y} = -2\gamma z - 2xy \sin z, \dot{z} = 2(y - x) + 2(2y - x) \cos z - 2\delta,$

where γ and δ are constants.

The given system arises in the study of two-stream instability

$$\frac{\partial \dot{x}}{\partial x} + \frac{\partial \dot{y}}{\partial y} + \frac{\partial \dot{z}}{\partial z} = -2(\gamma - 1).$$

Therefore, divergence of this flow is constant and is negative if $\gamma > 1$. Therefore, the volume in space varies with time according to $V(t) = V(0)\exp[-2(\gamma - 1)t]$.

The stationary equilibria and their stability are examined. It is shown that the system can exhibit a wealth of characteristic dynamical behavior, including Hopf bifurcation to periodic orbits, periodic-doubling bifurcations, chaotic solutions, strange attractors, tangent bifurcations from chaotic to periodic solutions, transient chaos, and hysterisis. Results are depicted graphically. One-dimensional maps are used. Russel and Ott (1981)

15.
$$\begin{aligned} \dot{x} &= 2x - 2x(x+y) - xy(\cos z + c_2 \sin z), \\ \dot{y} &= 2y - 2y(2x + 3y/4) - 2xy(\cos z - c_2 \sin z) - 2k^2 y, \\ \dot{z} &= c_2(2x - y/2) + (2x+y)\sin z + c_2(2x-y)\cos z + 2c_1 k^2, \end{aligned}$$

where c_1, c_2 are constants, and $k = \pi/\ell, x \geq 0, y \geq 0$. This system arises from a diffusion process in several physical contexts.

It is easy to see that

$$2\dot{x} + \dot{y} = 2(2x+y) - (2x+y)^2 - y^2/2 - 2k^2 y - 4xy(1+\cos z) \leq 2(2x+y) - (2x+y)^2.$$

Therefore, $2x+y \leq \xi$, where $\xi(t)$ is the solution of the DE $\dot{\xi} = 2\xi - \xi^2, \xi(0) = 2x(0) + y(0) \geq 0$. Since it may be checked that $\xi(t)$ is bounded, while $x \geq 0, y \geq 0$, each of the functions $x(t), y(t)$ is also bounded. Also,

$$\Omega \equiv \frac{\partial \dot{x}}{\partial x} + \frac{\partial \dot{y}}{\partial y} + \frac{\partial \dot{z}}{\partial z} = 4 - 2k^2 - 8x - 5y;$$

the quantity Ω does not depend explicitly on the parameters C_1, C_2 or on the variable z. For $k > 2^{1/2}$, the system is dissipative everywhere. For $k < 2^{1/2}$, a domain where $\Omega > 0$ appears close to the origin of phase space. In it there are neither stable singular points nor stable limit cycles.

The main attention is focused on the attractor with $c_1 = 7, c_2 = -6, k = 1$. Its projection on the (x,y) plane is shown graphically. It lies wholly in the domain where the system is dissipative. The trajectory can clearly hit the neighborhood of the saddle $x = 1, y = 0$, and can therefore pass close to it for a long time. The mean time T of one revolution about the central domain is roughly 1.63, while the mean value of Ω is -4.2. Hence the phase volume decreases to roughly 1/900 during one revolution.

Nonperiodic solutions and Poincaré maps are studied in detail. Akhromeyeva and Malinetskii (1987)

16. $\dot{x} = r_0 x + \alpha xy \sin z, \dot{y} = r_1 y + \beta x^2 \sin z, \dot{z} = x\{\beta x/y + 2\alpha y/x\} \cos z,$

where r_0, r_1, α, and β are real parameters.

For $\alpha = 0.44, \beta = 0.17, r_0 = 0.47, r_1 = 0.28$, and IC $x(0) = y(0) = 10^{-3}, z(0) = 0$, a numerical solution is shown that blows up at $t_0 \simeq 17$. It is easy to check that in the neighborhood of the pole $t = t_0$,

$$x \sim (t_0 - t)^{-1}, y \sim (t_0 - t)^{-1}, \text{ and } z \to \pi/2,$$

in agreement with the numerical solution. For the special choice $r_0 = r_1 \equiv r$, we may introduce the independent variable $\tau = r^{-1}[1 - \exp(-rt)]$ and the dependent variables as

$$x = (\alpha\beta)^{-1/2} \exp(rt)\hat{x}, y = \alpha^{-1} \exp(rt)\hat{y},$$

and the system changes to the one given in Eqn. (3.17.13). Bauer and Schamel (1992)

17. $\dot{x}_1 = vx_2 - x_1 x_3, \dot{x}_2 = -vx_1 - x_2 x_3, \dot{x}_3 = \ln(x_1^2 + x_2^2)^{1/2},$

where v is a constant.

Defining $y = (x_1^2 + x_2^2)^{1/2}$, we have $\dot{y} = -x_3 y,\ \dot{x}_3 = \ln y$. Therefore, $x_3\dot{x}_3 + (1/y)(\ln y)\dot{y} = 0$, leading to

$$(\ln y)^2 + x_3^2 = c^2, \tag{1}$$

where c is an arbitrary constant. For any fixed c, Eqn. (1) describes a torus of noncircular cross section on which the trajectories lie. Also, defining $\theta = \arctan(x_2/x_1)$, we get $\theta = -v$, so that $\theta = \alpha - vt$, for some constant α. Thus, with a suitable choice of time origin, the solution can be written as

$$\begin{aligned} x_1 &= e^{c\cos t}\cos(\alpha - vt), \\ x_2 &= e^{c\cos t}\sin(\alpha - vt), \\ x_3 &= c\sin t. \end{aligned}$$

Hence, the solutions are periodic in t, so that the trajectories form closed curves if and only if v is a rational number; otherwise, they are almost periodic. Cook (1986), pp.44, 192; Jacobs (1977)

18. $\dot{x} = \{1/(RC_1)\}(y-x) - (1/C_1)g(x), \dot{y} = \{1/(RC_2)\}(x-y) - z/C_2, \dot{z} = -y/L,$

$g(x) = m_0 x + (1/2)(m - m_0)|x+b| + (1/2)(m_0 - m_1)|x-b|$, where R, L, C_1, and C_2 are constants.

The given DE describes autonomous nonlinear circuits, with $g(x)$ defined above involving the constants m_i $(i = 0, 1)$ and b. For small voltages the nonlinear resistance is negative; the equilibrium position $(x, y, z) = (0, 0, 0)$ is unstable and oscillations occur. Chaotic oscillations were found for $1/C_1 = 9, 1/C_2 = 1, 1/L = 7, m_0 = -0.5, m_1 = -0.8$, and $b = 1$ in a set of consistent units. A chaotic time history is shown which has the same character as the Lorenz system. Moon (1987), p.110

19. $\dot{x} = Pr\{-F(x) + (\cos\alpha)y - (\sin\alpha)(z - Ra)\}, \dot{y} = -xz - y + Rax, \dot{z} = xy - z,$

where $F(x)$ is a nonlinear friction law; Pr, Ra, and α are constants.

The given system generalizes Lorenz system and reduces to the latter when $\alpha = 0$ and $F(x) = Cx; C$ is a constant. Moon (1987), p.115

20. $$\dot{x} = y - z - \phi(y), \dot{y} = x - y, \dot{z} = y - z.$$

Introducing the variables $u = x - z, v = x - 2y + z, y = y$, the given system takes the form

$$\frac{du}{dt} = -\phi(y), \frac{dv}{dt} = -2v - \phi(y), \frac{dy}{dt} = (1/2)(u+v).$$

Define the region $Q = \{y \geq M, yu^\lambda \geq 1, u + 2v \geq 0\}$, where M is sufficiently large, $\lambda > 2/\alpha - 1$. In the domain Q,

$$\begin{aligned} \frac{dy}{dt} &= (1/2)(u+v) \geq (1/4)u > 0, \\ \frac{d}{dt}(yu^\lambda)\Big|_{y=u^{-\lambda}} &= [(1/2)(u+v)u^\lambda - \lambda u^{\lambda-1}y\phi(y)]\Big|_{y=u^{-\lambda}} \\ &= u^{\lambda-1}[(1/4)u^2 - \lambda y\phi(y)]\Big|_{y=u^{-\lambda}} = u^{\lambda-1}[(1/4)y^{-2/\lambda} - \lambda y\phi(y)] > 0, \\ \frac{d}{dt}(u+2v)\Big|_{v=u/2} &= -\phi(y) + 2u - 2\phi(y) \geq 2y^{-1/\lambda} - 3\phi(y) > 0; \end{aligned}$$

the trajectory of the given system that begins in Q remains in Q. Since the ordinate y increases monotonically on Q, in that domain, the given system has neither states of equilibrium nor closed trajectories nor recurrent trajectories. Hence Q consists entirely of unbounded trajectories. Balitinov (1968)

21. $$\dot{x} = -ax - \phi(y) - f(z), \dot{y} = x - by, \dot{z} = y - bz,$$

where $a + 2b > 0, b\phi(y)/y + f(z)/z + ab^2 > 0,$

$$(a+b)\phi(y)/y - f(z)/z + 2b(a+b)^2 > 0.$$

It is shown that if:

(a) $b > 0, a \geq 0,$
(b) $(a+b)f(z)/z + bf'(z) + ab^2(a+b) \geq 0$ for $z \neq 0,$
(c) $(a+b)\phi(y)/y - f'(z) > 0$ for $y \neq 0,$

then the trivial solution $(0,0,0)$ of the given system is stable in the large. Other sets of conditions for the boundedness of the solution are also established. Balitinov (1968)

22. $$\begin{aligned} \dot{x}_1 &= a_{11}x_1 + \phi_2(x_2) + \phi_3(x_3), \\ \dot{x}_2 &= a_{21}x_1 + a_{22}x_2 + a_{23}x_3, \\ \dot{x}_3 &= a_{31}x_1 + a_{32}x_2 + a_{33}x_3, \end{aligned}$$

where $\phi_2(x_2)$ and $\phi_3(x_3)$ satisfy certain conditions that ensure existence and uniqueness of solutions, and where $\phi_2(0) = \phi_3(0) = 0$. Assuming further that $a_{31} = 0, a_{32} \neq 0$, a linear transformation leads this system to one in Eqn. (3.17.21). Balitinov (1968)

23. $$\dot{x} = z, \quad \dot{y} = x + by, \quad \dot{z} = f(x,y) + az.$$

Stability of the $(0,0)$ point is considered. Sufficient conditions (rather elaborate) for stability are given. Gaishun (1969)

24. $$\begin{aligned} \dot{x}_1 &= a_{11}x_1 + a_{12}x_2 + a_{13}x_3, \\ \dot{x}_2 &= a_{21}x_1 + a_{22}x_2 + a_{23}x_3, \\ \dot{x}_3 &= f_1(x_1, x_2) + a_{33}x_3, \end{aligned}$$

where a_{ij} are constants. We assume that the function $f_1(x_1, x_2)$ is such that the conditions of some theorem for existence and uniqueness of the solutions are satisfied for all real x_1, x_2 and that this system has a unique equilibrium point $O(0,0,0)$ and therefore, $f_1(0,0) = 0$; we also assume that $f \in C^1$ with respect to x_1 and x_2. Further, let $a_{23} = 0, a_{13}a_{21} \neq 0$. Then the nonsingular transformation $x = x_1, y = (1/a_{21})x_2, z = a_{11}x_1 + a_{12}x_2 + a_{13}x_3$ can be used to reduce the given system to

$$\frac{dx}{dt} = z, \frac{dy}{dt} = x + by, \frac{dz}{dt} = f(x,y) + az,$$

where $a = a_{11} + a_{33}, b = a_{22}, f(x,y) = a_{13}f_1(x, a_{21}y) + (a_{21}a_{12} - a_{33}a_{11})x + a_{12}a_{21}(a_{22} - a_{33})y$. Now refer to Eqn. (3.17.23). Gaishun (1969)

25. $$\dot{x} = y, \quad \dot{y} = Y(x,y,z), \quad \dot{z} = Z(x,y,z),$$

where

$$\begin{aligned} Y(x,y,z) &= f_1(x) + f_2(x)y + f_3(x)z + P(x,y,z), \\ Z(x,y,z) &= g_1(x) + g_2(x)y + g_3(x)z + g_4(x)y^2 + Q(x,y,z), \end{aligned}$$

with

$$\begin{aligned} P(x,y,z) &= y^2P_1(x,y)+yzP_2(x,y,z)+z^2P_3(x,y,z),\\ Q(x,y,z) &= y^3Q_1(x,y)+yzQ_2(x,y,z)+z^2Q_3(x,y,z),\\ f_i(x) &= x^{\alpha_i}[a_i+F_i(x)], g_j(x)=x^{\beta_i}[b_j+G_j(x)],\\ F_i(0) &= 0, G_j(0)=0; P_i, Q_i, F_i, G_j \in C^1, i=1,2,3; j=1,2,3,4; \end{aligned}$$

α_i and β_j are integers and $\alpha_1, \beta_2 \geq 2; \alpha_2, \beta_2, \alpha_3, \beta_3 \geq 1; \beta_4 \geq 0$.

We assume that the parameters α_i, β_j where

$$\begin{aligned} 3\alpha_1 &> 2\beta_1+3, 3\alpha_2 > \beta_1, 3\beta_2+3 > 2\beta_1,\\ 3\beta_3 &> \beta_1, 3\beta_4+6 > \beta_1, \end{aligned}$$

β_1 is even, $\alpha_3 = 1, b_1a_3 \neq 0$.

Under these conditions, a qualitative investigation of properties of the trajectories of the given system in the neighborhood of the origin is carried out. The method used is to reduce the singularity by a special transformation of the phase space, in which the singularity is replaced by a complete invariant manifold. Kapitanov (1984)

26. $$\dot{x} = y, \quad \dot{y} = -\frac{\partial V}{\partial x}(x,z) - \epsilon u y, \quad \dot{z} = -\epsilon\{z+g(x)\}, 0 < \epsilon \leq 0.1,$$

where $V(x,z)$ is thought of as a potential function, and ϵ and u are parameters. The Moore–Spiegel oscillator [see Eqn. (3.2.36)] and Lorenz system may be shown to be special cases of this system. The generic form of V is chosen to be

$$V(x,z) = x^m/m - \sum_{k=1}^{m-2} \alpha_k(z)x^k/k.$$

However we choose $\alpha_k(z)$ and $g(x)$, we find that

$$\frac{\partial \dot{x}}{\partial x} + \frac{\partial \dot{y}}{\partial y} + \frac{\partial \dot{z}}{\partial z} = -(\epsilon + \epsilon\mu),$$

which is a negative constant. Hence the divergence of the velocity of the phase flow is everywhere negative, and the volumes of clouds of phase space points will shrink continuously to zero; ultimately, the clouds will condense onto a structure of zero volume called attractors. Numerical experiments show that the appearance of a strange attractor coincides with aperiodic behavior of the solutions, as in the special cases of the given system, namely the Lorenz system and Moore–Spiegel oscillator. Special choices of parameters and functions considered in detail are $m = 4$, case (1): $\alpha_1(z) = z, \alpha_2 =$ constant; case (2): $\alpha_1 =$ constant, $\alpha_2 = z$. These two cases include Lorenz and Moore–Spiegel systems, after suitable transformations.

It is suggested that basic form of the system describes a family of strange attractors; application of the result to overstable fluid dynamical systems is discussed. Computational results are shown and discussed in detail. Marzec and Spiegel (1980)

27. $$\dot{x} = X(x,y,z), \dot{y} = \lambda y + Y(x,y,z), \dot{z} = \mu z + Z(x,y,z),$$

where $\lambda < 0 < \mu$ are constants, and X, Y, Z are analytic functions in some neighborhood of the origin, and their power series expansions at the origin contain no terms of degree less than two.

Local phase portraits near the origin are obtained and the directions of rays through the origin in the limit $t \to \infty$ or $t \to -\infty$ delineated. Appropriate theorems and figures for various cases are provided. Couper (1971)

28. $$\dot{x} = P(x, y, z), \quad \dot{y} = Q(x, y, z), \quad \dot{z} = R(x, y, z),$$

where P, Q, and R are holomorphic in (x, y, z) vanishing at $(0, 0, 0)$.

The characteristic roots of the matrix of the coefficients of the linear terms in P, Q, R are assumed to form a triangle whose interior holds 0. The shape of the integral varieties in the neighborhood of the origin is studied. Urabe (1952)

3.18 $\vec{y}\,' = \vec{f}(x, \vec{y})$

1. $$\dot{x} = \{1/(4t)\}(x + \sigma y), \quad \dot{y} = \{1/(4t)\}(2y + xz), \quad \dot{z} = \{1/(4t)\}(2z + xy),$$

where σ is a constant.

A first integral of the given system is

$$x^2 - 2\sigma z = 0. \tag{1}$$

Writing $x = t^{1/4} f(t^{1/4}) \equiv \xi f(\xi)$, we may reduce the given system to the single equation

$$f'' - (1/2) f^3 = 0. \tag{2}$$

The solution of (2) can be expressed in terms of Leminiscatic elliptic functions [see Eqn. (2.1.42)]. Levine and Tabor (1988)

2. $$\dot{x} = (1/2t)(x + \sigma y), \quad \dot{y} = (1/2t)(2y - xz), \quad \dot{z} = (1/2t)(2z + xy).$$

The given system arises as rescaled from Lorenz system. It has a first integral

$$x^2 - 2\sigma z = \alpha t. \tag{1}$$

Setting the value of the constant α of integration equal to $6i\lambda\sigma$, we may, with the help of (1), reduce the given system to second order in x:

$$4t^2 \ddot{x} - 2t\dot{x} - 3i\lambda\sigma t x + (x/2)(x^2 + 4) = 0. \tag{2}$$

Put $x = t^{1/2} f(t^{1/2}) \equiv \hat{y} f(\hat{y})$ in (2), then

$$f''(\hat{y}) - 3i\lambda\sigma f(\hat{y}) + (1/2) f^3(\hat{y}) = 0. \tag{3}$$

Equation (3) integrates to yield

$$f'^2 - 3i\lambda\sigma f^2 + (1/4) f^4 = I, \tag{4}$$

where I is a constant. The solution of (4) is

$$f(\hat{y}) = a/\{\mathrm{sn}(ia\hat{y}; m)\},$$

where $a^2 = 12i\lambda\sigma$ and

$$m = \frac{\{3\lambda\sigma + (2\lambda^2\sigma^2 + 20i\gamma)^{1/2}\}}{\{3\lambda\sigma - (2\lambda^2\sigma^2 + 20i\gamma)^{1/2}\}},$$

provided that $I = 7\sigma^2\lambda^2 - 20i\gamma$. Levine and Tabor (1988)

3. $\epsilon\dot{x} = x + y - qx^2 - xy, \quad \dot{y} = -y + 2hz - xy, \quad \dot{z} = (x - z)/p + A_z \cos w_z t,$

where ϵ, q, h, p, A_z, and w_z are constants.

This is the Oregonator. The effect of several parameters is studied. It is found that the parameter space (A_z, w_z) is divided into two regions: a regular region and a nonregular region. Chaotic modes appear in the study so that prediction using the strange attractor is found to be difficult. Yamaguchi, Yamamoto, and Imaeda (1989)

4.
$$\begin{aligned} \epsilon\dot{x} &= -\alpha x - \beta y - qx^2 - xy + A_x \cos wt, \\ \dot{y} &= -\gamma x - \delta y + 2hz - xy + A_y \cos wt, \\ p\dot{z} &= x - z + A_z \cos wt, \end{aligned}$$

where $\epsilon, \alpha, \beta, q, A_x, w, \gamma, \delta, h, A_y, p$, and A_z are constants.

The given Oregonator system with an external periodic force is studied numerically. It is found that the system performs bifurcation of multiperiod oscillation mode and chaotic motion which are similar to those observed in the flow experiments. Several sets of parameters are studied. Imaeda and Tei (1986)

5.
$$\begin{aligned} \dot{x} &= -a_{11}x - y - z - a_{41} + (x + a_{44})x, \\ \dot{y} &= -a_{12}x - a_{22}y - a_{32}z - a_{42} + (x + a_{44})y, \\ \dot{z} &= -a_{13}x - a_{23}y - a_{33}z - a_{43} + (x + a_{44})z, \end{aligned}$$

where $a_{ij}(t)$ are analytic in a domain D of the complex plane.

Solving for y from the first equation of the system, we have

$$y = -\frac{dx}{dt} - a_{11}x - z - a_{41} + (x + a_{44})x. \tag{1}$$

Differentiating the first of the given system, and using the given system and (1), we have

$$\begin{aligned} z &= \frac{1}{a_{32} + a_{33} - a_{22} - a_{23}} \left[\frac{d^2x}{dt^2} - 3x\frac{dx}{dt} + (a_{11} + a_{22} - 2a_{44} + a_{23})\frac{dx}{dt} + x^3 \right. \\ &\quad - (a_{11} + a_{22} - 2a_{44} + a_{23})x^2 + \left(\frac{da_{11}}{dt} - \frac{da_{44}}{dt} - a_{12} - a_{13} \right. \\ &\quad \left. + (a_{11} - a_{44})(a_{22} - a_{44} + a_{23}) - a_{41} \right) x - a_{42} - a_{43} \\ &\quad \left. + \frac{da_{41}}{dt} + a_{41}(a_{22} - a_{44} + a_{23}) \right]. \end{aligned} \tag{2}$$

Differentiating the first equation of the given system twice and using it and (1)–(2), we obtain a third order DE in x:

$$\begin{aligned} \frac{d^3x}{dt^3} &= 4x\frac{d^2x}{dt^2} + A(t)\frac{d^2x}{dt^2} + 3\left(\frac{dx}{dt}\right)^2 - 6x^2\frac{dx}{dt} - 3A(t)x\frac{dx}{dt} \\ &\quad + B(t)\frac{dx}{dt} + x^4 + A(t)x^3 - B(t)x^2 + R(t)x - Q(t), \end{aligned} \tag{3}$$

where $A(t), B(t), R(t)$, and $Q(t)$ are rather complicated expressions. Now writing $x = -(1/u)(du/dt)$, we get a fourth order linear DE:

$$\frac{d^4u}{dt^4} = A(t)\frac{d^3u}{dt^3} + B(t)\frac{d^2u}{dt^2} + R(t)\frac{du}{dt} + Q(t)u. \tag{4}$$

Equation (4), at an arbitrary point t_0 of a domain D in which there are no singular points of the coefficients and which does not contain $t = \infty$, has solutions with simple, double, and triple zeros.

If (4) has a solution with a simple zero, then the solution of the given system may be written as

$$\begin{aligned} x &= \{1/(t-t_0)\}\left[-1+\sum_{j=1}^{\infty} x_j(t-t_o)^j\right], \\ y &= \{1/(t-t_0)\}\left[y_0+\sum_{j=1}^{\infty} y_j(t-t_0)^j\right], \\ z &= \{1/(t-t_0)\}\left[z_0+\sum_{j=1}^{\infty} z_j(t-t_0)^j\right], \end{aligned}$$

where y_0 and z_0 are arbitrary parameters. If the solution of (4) has a double zero, then the given system has the solution

$$\begin{aligned} x &= \{1/(t-t_0)\}\left[-2+\sum_{j=1}^{\infty} x_j(t-t_0)^j\right], \\ y &= \{1/(t-t_0)^2\}\left[2-z_0+\sum_{j=1}^{\infty} y_j(t-t_0)^j\right], \\ z &= \{1/(t-t_0)^2\}\left[z_0+\sum_{j=1}^{\infty} z_j(t-t_0)^j\right], \end{aligned}$$

where z_0 is a parameter.

If (4) has a triple zero, the solution of the given system is

$$\begin{aligned} x &= \{1/(t-t_0)\}\left[-3+\sum_{j=1}^{\infty} x_j(t-t_0)^j\right], \\ y &= \{1/(t-t_0)^3\}\left[-z_0+\sum_{j=1}^{\infty} y_j(t-t_0)^j\right], \\ z &= \{1/(t-t_0)^3\}\left[z_0+\sum_{j=1}^{\infty} z_j(t-t_0)^j\right], \end{aligned}$$

where z_0 is a parameter. In each series above, x_j, y_j, and z_j $(j = 1, 2, \ldots)$ are constants. Yablonskii and Korzyuk (1989)

6.
$$\begin{aligned} \dot{x} &= -a_{11}x - a_{12}y - a_{31}z - a_{41} + (a_{14}x + a_{24}y + a_{34}z + a_{44})x, \\ \dot{y} &= -a_{12}x - a_{22}y - a_{32}z - a_{42} + (a_{14}x + a_{24}y + a_{34}z + a_{44})y, \\ \dot{z} &= -a_{13}x - a_{23}y - a_{33}z - a_{43} + (a_{14}x + a_{24}y + a_{34}z + a_{44})z, \end{aligned}$$

where $a_{ij}(t)$ are analytic in a domain D of the complex plane.

The given system can be reduced to a third order DE and shown to have no movable critical points. In particular, if $a_{14} \neq 0, a_{24} \equiv a_{34} \equiv 0, a_{21} \neq 0$, and $a_{31} \neq 0$, then $x(t)$ has a first order pole in the neighborhood of a moving singular point $t = t_0$, and y and z can have poles of first, second, or third order [see Eqn. (3.18.5) for detailed forms]. Yablonskii and Korzyuk (1989)

7. $\dot{x} = -4x + x^2 y + x^2 z, \dot{y} = -4y + z - 1 + y^2 + z^2 + 4 \sin t, \dot{z} = -y - 4z + x^2 y + 4 \cos t.$

The given system has a solution $x = 0, y = \sin t, z = \cos t$, which is asymptotically stable. Wilson (1971), p.323

8.
$$\begin{aligned} \dot{x} &= 1 + \epsilon \sin wt - x - Axy/(a + x), \\ \dot{y} &= Axy/(a + x) - y - Byz/(b + y), \\ \dot{z} &= Byz/(b + y) - z, \end{aligned}$$

where A, a, B, b, ϵ, and w are constants.

The given system, a forced third order system, describes microbial chaos. We find that

$$\dot{x} + \dot{y} + \dot{z} = 1 + \epsilon \sin wt - x - y - z,$$

so that there exists a first integral

$$x + y + z = 1 + \epsilon/(1 + w^2)^{1/2} \sin(wt - \phi) + ce^{-t},$$

where $\phi = \tan^{-1} w$, and c is constant of integration. One may eliminate one of x, y, z and reduce the given system to one of second order involving time. Quasiperiodic flow on a torus and chaos are found numerically. Kot, Sayler, and Schultz (1992)

9.
$$\begin{aligned} \dot{x}_1 &= -x_1 + x_2^m x_3^n - e^{-rt} + 1, \\ \dot{x}_2 &= -Ax_2 - x_2^{C/A} x_3^2 + x_3^3, \\ \dot{x}_3 &= -Cx_3 - x_1^2 + x_2^2 - e^{-2At} - 2e^{-t} + e^{-2t} + 1, \end{aligned}$$

for all $t \in [0, 1], x_1(0) = 0, x_2(0) = 1, x_3(0) = 1$, with parameters $m > 1, n \geq 2, C > A > 0, r = mA + nC$.

The exact solution of the given problem is

$$x_1 = 1 - e^{-t}, x_2 = e^{-At}, x_3 = e^{-Ct}.$$

The given IVP was also solved numerically. Bauch (1982)

10. $\dot{x} = y, \quad \dot{y} = -\sin x + \epsilon(z - \delta y), \quad \dot{z} = \epsilon(-rz + y \cos t),$

where ϵ and δ are parameters.

The given system governs a pendulum subject to weak damping and variable torque. The torque is supplied by a servomotor that is also driven by an external periodic perturbation.

For $\epsilon = 0$, the unperturbed system is Hamiltonian with energy $H(x, y) = y^2/2 + (1 - \cos x)$ which is identical on each $z = z_0$ constant slice. The unperturbed orbits are given by

$$x(t) = 2 \sin^{-1}[\text{sn}\{K(k)(t + t_0)/\pi\}], y(t) = (2/k)\text{dn}\{K(k)(t + t_0)/\pi\},$$

$k \in (0,1), z(t) = z_0$ where sn and dn are the Jacobi elliptic functions and K is the complete elliptic integral of the first kind with modulus k, provided that $kK = \pi$.

It is shown that there exists a countable set of bifurcation curves in (r, δ)-space given by

$$r\delta = \{\pi/(8E(k))\} \operatorname{sech}\{\pi m K'(k)/K(k)\}, kK(k) = \pi m, m = 1, 2, \ldots,$$

near which for ϵ sufficiently small (depending on m), the saddle-node bifurcations to pairs of $2\pi m$-periodic orbits occur for the Poincaré map of the given system. Wiggins and Holmes (1987)

11.
$$\begin{aligned} \dot{x} &= a_{11}x + a_{12}y + a_{13}z + p_1(t, x, y, z), \\ \dot{y} &= f(x) + a_{22}y + a_{23}z + p_2(t, x, y, z), \\ \dot{z} &= a_{31}x + a_{32}y + a_{33}z + p_3(t, x, y, z), \end{aligned}$$

where the functions p_i are bounded. It is shown that the system is dissipative if $f(x)$ satisfies inequalities of Hurwitz type. [A system of DE $w' = f(t, w)$ is defined as dissipative if there exists a constant $C > 0$ such that for every solution $w(t)$ of this system $\lim_{t\to\infty} \sup|w(t)| < C$.]

It is then shown that if the functions p_i have period T, there exists a solution of period T. Anitova (1964)

3.19 $y_1' = f_1(x, \vec{y}), y_2'' = f_2(x, \vec{y})$

1.
$$\ddot{x} + (\dot{y}/y)\dot{x} + (x/y)(\ddot{y} - \dot{y}^2/y) = 0, 2\dot{x} + (\dot{y}/y)x = 0.$$

Solving the second DE and hence the first, one easily obtains the solution

$$x = cy^{-1/2}, y = (c_1 t + c_2)^{-2},$$

where c, c_1, and c_2 are constants.

The given system arises in the description of waves in a channel of variable cross section. Nazarov and Puchkova (1991)

2.
$$y'^2 = z, \quad z'' + yz' - 2y'z = 0,$$

$y(0) = y_0, z_0 \cos\phi - z'(0) \sin\phi = 1, z(\infty) = 0, 0 \le \phi \le \pi/2$.

The exact solution of the problem, which arises in convection in porous media, is

$$\begin{aligned} y &= 2m - (2m - y_0)e^{-mx}, \\ z &= m^2(2m - y_0)^2 e^{-2mx}, \end{aligned}$$

where m is a postive root of

$$\begin{aligned} 8(\sin\phi)m^5 + 4(\cos\phi - 2y_0\sin\phi)m^4 + 2y_0(y_0\sin\phi \\ - 2\cos\phi)m^3 + y_0^2 m^2\cos\phi - 1 = 0. \end{aligned} \tag{1}$$

Equation (1) has only one positive root when $y_0 < 0$, and one or three positive roots, depending on the magnitude of ϕ. For $y_0 = 1, \phi = 0, m = 1$ is the only positive root. Ramanaiah and Malavizhi (1991)

3.
$$\dot{y} = z, 2\ddot{z} + (1 + \lambda)y\dot{z} - 2\lambda\dot{y}z = 0,$$

where $y(0) = y_0, z(0)\cos\phi - \dot{z}(0)\sin\phi = 1, z(\phi) = 0; \lambda, y_0$, and $0 \le \phi \le \pi/2$ are constants.

The given system arises in convection in porous media. The exact solution of the problem for $\lambda = 1$ is

$$y = m - (m - y_0)e^{-mx}, z = m(m - y_0)e^{-mx},$$

where m is a positive root of the cubic

$$(\sin\phi)m^3 + (\cos\phi - y_0 \sin\phi)m^2 - y_0(\cos\phi)m - 1 = 0. \tag{1}$$

Since $0 \le \phi \le \pi/2$, Eqn. (1) has only one positive root by the Descartes rule, for $y_0 = \phi = 0, \dot{z}(0) = -1$, and $m = 1$. Ramanaiah and Malavizhi (1991)

4. $$\boldsymbol{y' = m\xi + z, \quad (1 + \xi x)z'' + (\xi + y)z' - zy' = 0,}$$

$y(0) = y_0, z_0 \cos\phi - z'(0)\sin\phi = 1, z(\infty) = 0$, where $m, \xi, 0 \le \phi \le \pi/2$, and y_0 are constants.

The given problem arises in a combined study of free and forced convection in a porous medium. The exact solution is

$$y = y_0 + m\xi x + (m - y_0 - 2\xi)(1 - e^{-mx}), z = m(m - y_0 - 2\xi)e^{-mx},$$

provided that m is the positive root of the cubic

$$(\sin\phi)m^3 + [\cos\phi - (y_0 + 2\xi)\sin\phi]m^2 - (\cos\phi)(y_0 + 2\xi)m - 1 = 0.$$

Ramanaiah and Malavizhi (1991)

5. $$\boldsymbol{\ddot{x} = \mu\dot{x} - \nu x + y, \quad \dot{y} = -y + x^2,}$$

where μ and ν are constants.

The given system arises in thermosolutal convection. The divergence of the flow is

$$\frac{\partial}{\partial x}\dot{x} + \frac{\partial}{\partial y}\dot{y} + \frac{\partial}{\partial z}\dot{z} = -1 + \mu,$$

where $z = \dot{x}$; therefore, there are no stable attractors for $\mu > 1$. There is a supercritical Hopf bifurcation from the trivial solution at $\mu = 0$ for $\nu > 0$. At $\nu = 0$ an eigenvalue passes through zero and for $\mu < 0$ all eigenvalues are real. There is a stationary bifurcation at $\nu = 0$. Bifurcation diagrams are drawn. The parametric ranges for which the system displays chaos are identified. Proctor and Weiss (1990)

6. $$\boldsymbol{y'' + (\alpha y^2 - \sigma)y' = z - y, \quad z' = -\lambda y + \beta y^2,}$$

where $\alpha, \sigma, \lambda, \beta$ are constants.

It is shown that according to a first order calculation, there is a Hopf bifurcation on crossing the line $\sigma = 0$. This implies the existence of solutions that are asymptotically doubly periodic with frequencies 1 and $\lambda/(2)^{1/2}$. Longford et al. (1980)

7. $$\boldsymbol{\epsilon^2 y'' - \epsilon\theta y' = f(y) + z, \; \theta z' = y,}$$

where $f(y) = y(y - a)(y - 1); 0 < a < 1, \epsilon$ is small, and $\theta = O(1)$. Further conditions are that y is defined and bounded for $-\infty < x < \infty$, and y is periodic in x. The solitary wave is included among the periodic solutions as the limiting case when the period tends to infinity. Matched asymptotic solutions are found. Casten, Cohen, and Lagerstrom (1975)

8. $$y'' - cy' + y(1-y)(y-a) - (b/c)z = 0, \quad z' = y,$$

where a, b, and c are real numbers with $0 < a < 1, b \geq 0$, and $c > 0$. The given Nagumo system appears in biology.

Numerical results of the system indicate the following situation: Consider a fixed $b > 0$. There is a number a_b with $0 < a_b < 1/2$ such that if $0 < a < a_b$, then there exist precisely two positive values of c such that the given system has a nonconstant bounded solution, while if $a = a_b$, then there is one such value. When such a solution exists, it tends to the origin (in phase space) at plus and minus infinity. If $a_b < a < 1$, then there are no nonconstant bounded solutions. It is also found numerically that when $c > 1/2^{1/2}$, then no nonconstant solution exists for any $b \geq 0$.

Some other results are proved analytically. For example:

(a) If $1/2 < a < 1$, then no bounded nonconstant solution can have origin in its negative limit set. If any such solution exists, then there must be a bounded solution that remains bounded away from the origin on $(-\infty, \infty)$.

(b) If $1/2 < a < 1$ and $b/c^2 > f'(u)$ for all u, then there is no nonconstant bounded solution; here $f(u) = u(1-u)(u-a)$.

It is easily verified that the given system, when expressed as a system of three first order DEs, has only the origin as a singularity; this singularity is a saddle point, with a one-dimensional unstable manifold and a two-dimensional stable manifold. Hastings (1972)

9. $$y' = -cy - c\lambda z + c(y_{-\infty} + \lambda z_{-\infty}),$$

$$z'' + \alpha c z' = -(1/2)(z - z^3) - y,$$

where $c, \lambda, \alpha, y_{-\infty}$, and $z_{-\infty}$ are constants.

The given system describes traveling waves in a solidification problem. Singular points of the system are found and local analysis about the points is carried out. Appropriate singular points are then connected to solve some boundary value problems. Wilder (1991)

10. $$y'' = -(1/m)z, \quad gz'^3 + cz' - ey' - \epsilon r y'^3 = -z + dy + \epsilon s y^3,$$

where $m, g, c, d, \epsilon, s, e$ are constants.

The given system models an oscillating nonlinear mechanical elastic systems with internal friction and relaxation. It is solved numerically subject to some specific values of the parameters and initial conditions. Bojadziev (1983)

11. $$y'' + vy' = ky^\alpha z^\beta, \quad vz' = y^\alpha z^\beta,$$

where v, α, k, and β are constants; $y(0) = 1, y(X) = 0, z(0) = 0, z(X) = 1$. It is assumed that $z(x) > 0$ for $x > 0$. The parameter β is such that $0 < \beta < 1$, while $X > 0$.

The given problem arises in reaction diffusion in a moving medium. It is proved that the given problem, for each $X > 0$, has at least one nonnegative solution. Bobisud (1993)

4

FOURTH ORDER EQUATIONS

4.1 $y^{\text{iv}} + f(x, y, y') = 0$

1. $\quad y^{\text{iv}} - (1/2)y^2 + y = 0, \quad y' = y'' = y''' = 0$ **at** $|\, x \,| = \infty.$

The given DE describes traveling wave solutions of the fifth order Korteweg–deVries equation. The characteristic equation of linearized form near $x = \infty$ is

$$\mu^4 + 1 = 0.$$

Therefore, it has eigenvalues $\mu = \pm 2^{1/2}/2 \pm i2^{1/2}/2$, suggesting that the solitary wave has an oscillatory structure at $|\, x \,| = \infty$. The given DE is symmetric about $x = 0$. The oscillatory solitary wave is obtained by numerical integration and is shown graphically. Kawahara and Takaoka (1988)

2. $\quad y^{\text{iv}} - \{1/2\}y^2 + \nu y = 0,$

where ν is a constant.

The given DE arises as the traveling wave form of a fifth order Korteweg–deVries type of equation. Numerical solution of the given DE shows that solitary wave solution exists for an arbitrary value of ν. Il'ichev (1990)

3. $\quad y^{\text{iv}} - y^2/2 + \lambda y - C = 0,$

where λ and C are constants.

Assuming the solution in the form

$$y = AP^2(x), \tag{1}$$

where A is arbitrary and P is the Weierstrass P-function satisfying

$$P'^2 = 4P^3 - g_2 P - g_3 \tag{2}$$

with two invariants g_2 and g_3, assumed to be real and satisfying $g_2^3 - 27g_3^2 > 0$, we may, by substituting (1) in the given DE and using (2), find that $g_2 = \lambda/168, g_3 = 0, A = 1680$, and $C = 5\lambda^2/56$. The exact periodic solution then becomes

$$y(x) = 1680P^2(x + \delta; g_2, 0) - 140g_2, \tag{3}$$

where δ is an integration constant of (2), generally complex. Although the solution of the given DE has poles lined in the real axis of a $x+\delta$ complex plane, we may be able to choose δ in such a way as to shift the poles half a period above the real axis. Thus $P(x+\delta)$ in (3) may be replaced by $e_3 + (e_2 - e_3)\,\mathrm{sn}^2[(e_1-e_3)^{1/2}x + \delta']$, with an arbitrary real constant δ', where e_1, e_2, and e_3 are roots of $4e^3 - g_2 e - g_3 = 0$ and $e_1 > e_2 > e_3$. Thus the exact bounded periodic solution of the given DE is

$$y(x) = (15/2)\lambda\,\mathrm{cn}^4[(\lambda/28)^{1/4}x + \delta'] - 5\lambda/2.$$

Kano and Nakayama (1981)

4. $$\boldsymbol{y^{\mathrm{iv}} - f y^2 = 0,}$$

where f is a constant.

The solutions of the given DE are not single-valued except in the special case $f = 0$. It has a special solution $y = A(x-x_0)^{-4}, A = 840/f$. If we change this DE by the transformation $y = \alpha Y$, we may, by suitable choice of α, put it in the form $y^{\mathrm{iv}} = 840y^2$. In this case, the special solution is

$$y = (x - x_0)^{-4}.$$

Bureau (1964a)

5. $$\boldsymbol{y^{\mathrm{iv}} - (1/3)y^3 + (\lambda - \alpha)y - C = 0,}$$

where α, λ, and C are constants.

It is assumed that $y(x) = AP(x)$, where A is an arbitrary constant and $P(x)$ is the Weierstrass P-function, satisfying

$$\left(\frac{dp}{dx}\right)^2 = 4P^3 - g_2P - g_3, \tag{1}$$

with two invariants g_2 and g_3, which are assumed to be real and which satisfy the condition

$$g_2^3 - 27g_3^2 > 0. \tag{2}$$

Substituting $y = AP(x)$ into the given DE, after some calculations we find that $A = \pm 6(10)^{1/2}, g_2 = (\lambda - \alpha)/18$. The restriction on λ is that it should always be greater than α so that it satisfies (2). Thus, the exact periodic solution of the given DE is

$$y(x) = \pm 6(10)^{1/2}P(x+\delta; (\lambda-\alpha)/18, g_3), \tag{3}$$

where g_3 is required to satisfy the condition $[(\lambda-\alpha)/18]^3 > 27g_3^2$ and δ is a constant of integration arising from (1). Krishnan (1984)

6. $$\boldsymbol{y^{\mathrm{iv}} - y^3 + \lambda y - C = 0,}$$

where λ and C are arbitrary constants.

Following the procedure in Eqn. (4.1.3), the exact solution of the given DE may be found to be

$$y(x) = \pm 2(30)^{1/2}P(x+\delta; \lambda/18, g_3),$$

where g_3 is an arbitrary constant, provided that $g_3^2 < (\lambda/18)^3/27$. Here P is the Weierstrass P-function. Kano and Nakayama (1981)

7. $$y^{\text{iv}} - x^{\beta} \mid y \mid^{\alpha} \text{ sgn } y = 0, x > 1,$$

where β is a real number and $\alpha > 1$.

It is shown that the given DE has an oscillatory solution if and only if $\beta + (3\alpha + 5)2 \geq 0$. Moreover:

(a) There exist nonoscillatory solutions for which

$$y(x)y'(x) < 0, y(x)y''(x) > 0, y(x)y'''(x) < 0$$

for sufficiently large x.

(b) There exist nonoscillatory solutions for which

$$y(x)y'(x) > 0, y(x)y''(x) > 0, y(x)y'''(x) < 0$$

for sufficiently large x iff $\alpha + \beta + 3 \leq 0$.

(c) There exist nonoscillatory solutions for which

$$y(x)y'(x) > 0, y(x)y''(x) > 0, y(x)y'''(x) < 0$$

for sufficiently large x iff $3\alpha + \beta + 1 < 0$.

Kura (1983)

8. $$y^{\text{iv}} - p(x) \mid y \mid^{\alpha} \text{ sgn } y = 0,$$

where $\alpha > 1$ is a constant and $p(x)$ is a positive continuous function on $[x_0, \infty), x_0 > 0$.

A nontrivial real-valued solution $y(x)$ of the given DE is called proper if it exists on some half-line $[X_y, \infty) \subset [x_0, \infty)$.

The following results are proved.

(a) Let $p'(x) \leq 0$ for $x \geq x_0$ and $\int_{x_0}^{\infty} x^{1+2\alpha} p(x)dx < \infty$. Then every proper solution of the given DE is nonoscillatory.

(b) Suppose that there exist positive constants ϵ and K such that $p(x)x^{(3\alpha+5+\epsilon)/\epsilon} \geq K$ and $\frac{d}{dx}\{p(x)x^{(3\alpha+5+\epsilon)/2}\} \leq 0$ for $x \geq x_0$. Then every proper solution of the given DE is nonoscillatory.

(c) Let $p(x)$ be a positive continuous function on $[x_0, \infty), x_0 > 0$. Suppose that there exists a positive constant ϵ such that $\frac{d}{dx}\{p(x)x^{(3\alpha+5-\epsilon)/2}\} \geq 0$ for $x \geq x_0$. Then every proper solution $y(x)$ of the given DE such that $y(x_0) \equiv y'(x_0) = 0$ is oscillatory.

Kura (1983)

9. $$y^{\text{iv}} - \lambda f(x,y) = 0, x \in (0,1), y(0) = y'(0) = y(1) = y'(1) = 0.$$

Suitable conditions on the nonlinearity of f are obtained for the existence of the positive solutions of the given problem. The main tool used is that of upper and lower solutions. Dunninger (1987)

10. $$y^{\text{iv}} - e_1 y y' = 0,$$

where e_1 is a constant.

A first integral of the given DE is

$$y''' = (e_1/2)y^2 + K, \tag{1}$$

where K is an arbitrary constant. Equation (1) may be shown to have a multiple-valued solution except when $e_1 = 0$. Bureau (1964a)

11. $$y^{\text{iv}} - \epsilon y'^2 - Ra \cdot y = 0,$$

where Ra and $\epsilon << 1$ are constants.

$$y(0) = y(1) = 0, y''(0) = y''(1) = 1.$$

Several perturbation solutions are presented and compared with the numerical solution. One such expansion is

$$y = \epsilon^{-1/2} y_0 + y_1 + \epsilon^{1/2} y_2 + \cdots,$$

leading to the boundary conditions $y_0(0) = y_0(1) = y_0''(0) = y_0''(1) = 0, y_1(0) = y_1(1) = 0, y_1''(0) = y_1''(1) = 1$.

The first two terms in the expansion are found to be

$$\begin{aligned}
y_0 &= D_0 \sin(Ra^{1/4}x), D_0 = \pm 3^{1/2} \\
y_1 &= \{7/(5Ra^{1/2})\}[\cosh(Ra^{1/4}x) + \{(1 - \cosh Ra^{1/4}) \\
&\quad \times \sinh(Ra^{1/4}x)/\sinh Ra^{1/4}\}] + D_1 \sin(Ra^{1/4}x) \\
D_1 &= [21(\cosh Ra^{1/4} - 1)/(25Ra^{1/2}\sinh Ra^{1/4})] + 3/(40Ra^{1/4}).
\end{aligned}$$

D_1 is found from the solution for y_2 with appropriate BC. Becket (1980)

4.2 $y^{\text{iv}} + ky'' + f(x,y,y') = 0$

1. $$y^{\text{iv}} + y'' + 4yy' - \mu y' = 0,$$

where μ is a constant.

A first integral of the given DE is

$$y''' = \mu y - 2y^2 - y'$$

if we ignore constant of integration. Now see Eqn. (3.2.5). Chang (1986)

2. $$y^{\text{iv}} + y'' + (1/2)y'^2 - c^2 = 0, -\infty < x < \infty,$$

where c is a constant.

The given DE arises from the Kuramoto–Sivashinsky equation. With $y' = v$, it becomes

$$v''' + v' + (1/2)v^2 - c^2 = 0.$$

Now see Eqn. (3.2.3). Michelson (1986)

3. $$\boldsymbol{y^{\text{iv}} + y'' + (1/2)(y')^2 - E = 0,}$$

where E is a constant.

Calling $\tau = x, X = \dfrac{dy}{d\tau}$, etc., the given system is written as

$$\frac{dX}{d\tau} = Y, \frac{dY}{d\tau} = Z, \frac{dZ}{d\tau} = E - (1/2)X^2 - Y. \tag{1}$$

For periodic orbits of (1), the relation

$$E = (1/\ell)\int_0^\ell (X^2/2)d\tau$$

holds. Large-amplitude turbulent flows are found. Lau (1992)

4. $$\boldsymbol{(2/15)y^{\text{iv}} + (1/3)y'' - [(c^2 - 1) - (1/2)y(3c - y)]y = 0,}$$

where c is a constant.

The given DE describes traveling wave solutions of a continuous Hamiltonian system, occurring in nonsymmetric gravity waves on water of finite depth. It has a first integral

$$(2/15)(y'y''' - y''^2/2) + (1/6)y'^2 - (1/2)y^2[(c^2 - 1) - y(c - y/4)] = C,$$

where c and C are constants. A general computational study of the original system was carried out to find nonsymmetric periodic solutions, which appear via spontaneous symmetry breaking bifurcations from the symmetric waves. Zufiria (1987)

5. $$\boldsymbol{y^{\text{iv}} - 52y'' + 1352(\alpha/\beta^2)y^2 + 576y = 0.}$$

The given DE arises from a traveling wave form of the Korteweg–deVries equation with an additional fifth order dispersive term. It has a solitary wave solution

$$y = (-105/169)(\beta^2/\alpha)\ \text{sech}^4 x.$$

Hereman, Banerjee, Korpel, and Assanto (1986)

6. $$\boldsymbol{y^{\text{iv}} - \xi y'' - (1/2)y^2 + y = 0,}$$

where ξ is a constant. The given DE describes steady pulse solution of a fifth order Korteweg–deVries equation.

For a pulse type of solution, one may consider $|\, x \,| >> 1$ and linearize the given DE. The characteristic equation for the linearized equation is

$$\mu^4 - \xi\mu^2 + 1 = 0.$$

Five cases arise:

(a) $\xi > 2, \mu = \pm[\xi \pm (\xi^2-4)^{1/2}]^{1/2}$.
(b) $\xi = 2, \mu = \pm 1$.
(c) $2 > \xi > -2, \mu = \pm \exp(\pm i\theta/2)$, where θ is defined by $\cos\theta = \xi/2$.
(d) $\xi = -2, \mu = \pm i$.
(e) $\xi < -2, \mu = \pm i[\{-\xi + (\xi^2-4)^{1/2}\}/2]^{1/2}$.

For $\xi \leq -2$, the value of μ is pure imaginary; no pulse solutions are expected.

The symmetry of the given DE w.r.t. x leads to symmetry of the pulse solution, and its structure changes from oscillatory to montone as ξ increases.

The shape of the tail part of the pulse when $x \to -\infty$ or $+\infty$ is given as follows for cases (a)–(c):

Case (a): $y(x) \sim A\,(\exp \pm[\{\xi+(\xi^2-4)^{1/2}\}/2]^{1/2}x) + B\,(\exp \pm[\{\xi-(\xi^2-4)^{1/2}\}/2]^{1/2}x)$,
Case (b): $y(x) \sim (A+Bx)\exp(\pm x)$,
Case (c): $y(x) \sim A\exp[\pm\cos(\theta/2)x]\cos[\sin(\theta/2)x+B]$,

where A and B are constants.

The pulse solution obtained by numerical integration of the given DE for $\xi = 13/6$, $5/2$, $52/(51)^{1/2}, 2, 0, -1$ are computed and depicted.

In fact, the given DE admits exact solutions for $\xi = 13/6, 5/2$, and $52/(51)^{1/2}$.

For $\xi = 13/6, y = (35/12)\,\text{sech}^4[(x-x_0)/(2(6)^{1/2})]$

$$= \sum_n \frac{1680}{n[x-x_0-i(6)^{1/2}\pi(2n-1)]^4} - \frac{140}{3}\sum_n \frac{1}{[x-x_0-i(6)^{1/2}\pi(2n-1)]^2}.$$

For general values of ξ, Painlevé analysis is carried out. It is found that, to leading order,

$$y(x) = A(x-x_*)^p,$$

where $A = 1680, p = -4$. The resonances

$$y(x) = (x-x_*)^{-4}[1680 + B(x-x_*)^r]$$

occur when:

Case (a): $r = -1, 12, [11 \pm i(159)^{1/2}]/2, B$ is arbitrary.

Case (b): $r = 2, B = -(280/13)\xi$.

It is concluded that the given DE does not possess the Painlevé property. For the case (a), an expansion

$$y(x) = [1680 + P[(x-x_0)^{r_c}] + Q[(x-x_0)^{\bar{r}_c}]\left[\sum_{n=0}^{\infty}(x-x_0)^{n-4}\right]$$

is found where

$$r_c = (11+i(159)^{1/2})/2, \bar{r}_c = (11-i(159)^{1/2})/2,$$

and P and Q are polynomial functions.

Distribution of singularities in the complex x plane for different values of ξ and the formation of fractal natural boundaries are discussed. Takaoka (1989)

7. $$y^{\text{iv}} - \xi y'' - (1/2)y^2 - y = 0,$$

where ξ is a constant.

This DE arises from the steady form of a Korteweg–deVries equation of fifth order. A detailed analysis may be carried out as for Eqn. (4.2.6). Takaoka (1989)

8. $$\boldsymbol{\mu^2 y^{\text{iv}} + y'' + y^2 - y = 0, y'(0) = y'''(0) = 0,}$$

$y(\pm\infty) = 0, \mu$ is a small parameter.

It is shown that there are no nontrivial solutions of the given problem. Amick and Kirchgasoner (1988)

9. $$\boldsymbol{y^{\text{iv}} + \alpha y'' + y^2 + wy - c = 0,}$$

where α, w, and c are constants.

The given DE results from a fifth order Korteweg–deVries equation, when traveling wave solutions are sought.

Put $y = v(x) + \beta, w_1 = 2\beta + w$, where β is a root of $\beta^2 - w\beta + c = 0$. The given DE then goes to

$$v^{\text{iv}} + \alpha v'' + v^2 + w_1 v = 0;$$

now proceed to Eqn. (4.2.10). Santos (1989)

10. $$\boldsymbol{y^{\text{iv}} + \alpha y'' + y^2 + wy = 0,}$$

where α and w are constants.

The given DE results from a fifth order Korteweg–deVries equation, when traveling wave solutions are sought.

Solutions may be found in the finite form

$$y = B_0 + B_1 p(x) + B_2 p^2(x), \tag{1}$$

$$p'^2(x) = b_0 + b_2 p^2(x) + p^3(x). \tag{2}$$

Substituting (1) in the given DE and using (2), we find that (1) is a solution of the given DE, provided that

$$\begin{aligned} B_0 &= -w, B_1 = 35(\alpha + 4b_2)/13, B_2 = -105/2, \\ b_0 &= -(714\alpha^3 + 4225\alpha w)/1373125, \\ b_2 &= -14/65\alpha, \end{aligned}$$

where $p(x)$ may be found from (2) in terms of elliptic functions:

$$\begin{aligned} p(x) &= 4\mathcal{P}(x - x_0; g_2, g_3) - b_2/3, g_2 = b_2^2/12, \\ g_3 &= \alpha(31\alpha^2 - 7 \times 78^2 g_2)/(80 \times 39^3), \end{aligned}$$

where $\mathcal{P}(x - x_0; g_2, g_3)$ is a Weierstrass function with invariants g_2 and g_3. For a bounded solution this function may be expressed as

$$p(x) = p_1 + (p_2 - p_1)\text{sn}^2[\{(p_3 - p_1)^{1/2}(x - x_0)\}/2], \tag{3}$$

where p_1, p_2, and $p_3 (p_1 < p_2 < p_3)$ are the roots of $b_0 + b_2 p^2 + p^3 = 0$ and $\text{sn}(x)$ is the Jacobi elliptic function with modulus $q \equiv [(p_2 - p_1)/(p_3 - p_1)]^{1/2}$. In the limit b_0, p_2, and p_3 tending to zero, and $b_2 = \alpha/13$, we obtain, instead of (3), the solitary wave solution

$$p(x) = -(\alpha/13)\ \text{sech}^2[(\alpha/52)^{1/2}(x - x_0)]. \tag{4}$$

Santos (1989)

11. $$y^{\text{iv}} + (\sigma/B)y'' + \{1/(2B^2)\}y^2 - \{C/B^5\}y = 0,$$

where $\sigma, B \neq 0$, and C are constants.

The given DE arises from a fifth order Korteweg–deVries equation. It has an exact solution $y = -(105/169)\ \text{sech}^4 x$ provided that $B = (-\sigma/52)^{1/2}$ and $C = -(36/169)(-\sigma/52)^{1/2}$. A second exact solution is $y = -(105/169)\ \text{cosech}^4 x$ for a suitable choice of parameters B and C. Huang, Luo, and Dai (1989)

12. $$y^{\text{iv}} + (\sigma/B^2)y'' + \{1/(2B^4)\}y^2 - \{C/B^5\}y + K/B^5 = 0,$$

where $\sigma, B \neq 0, C$, and K are constants.

The given DE arises from a higher order Korteweg–deVries equation. It has an exact solution $y = A_0 + A_1\ \text{sech}^4 x$ where $A_0 = 36/169 + \delta_1[(36/169)^2 + K']^{1/2}$, $A_1 = -105(\sigma^2/169)$, $B = (-\sigma/52)^{1/2}$, $C = \delta_1[\{(36/169)^2 + K'\}(-\sigma/52)]^{1/2}$, $\delta_1 = \pm 1$, and $K' = 2K/B$. Huang, Luo, and Dai (1989)

13. $$\beta y^{\text{iv}} - \alpha y'' + \lambda y - (3/4)y^2 = 0, y = y' = y'' = y''' = 0$$

as $x \to -\infty$, where β, α, and λ are constants.

The given system arises from a fifth order (modified) Korteweg–deVries equation and describes the traveling wave form for the latter. It is suitably normalized by writing $y = \lambda u, x = \{\beta / \mid \lambda \mid\}^{1/4} z$, so that a simplified form results, involving only one parameter. For special cases of the normalized form, see Eqns. (4.2.14)–(4.2.17). Kawahara (1972)

14. $$y^{\text{iv}} - \epsilon y'' - y + 0.75y^2 = 0, y = y' = y'' = y''' = 0$$

as $x \to -\infty$, where ϵ is a parameter.

The given problem arises from a fifth order Korteweg–deVries equation. The linearized (asymptotic) solution as $x \to -\infty$ is $y = A\ \exp[\{\{\epsilon + (\epsilon^2 + 4)^{1/2}\}/2\}^{1/2} x]$ for all ϵ, where A is a parameter. The linearized solution provides the initial conditions for integration from $x = -\infty$. The numerical solution is obtained and depicted. Kawahara (1972)

15. $$y^{\text{iv}} + \epsilon y'' - y + 0.75y^2 = 0, y = y' = y'' = y''' = 0$$

as $x \to -\infty$, where ϵ is a parameter.

The given problem arises from a fifth order Korteweg–deVries equation. The linearized (asymptotic) solution as $x \to -\infty$ is $y = A\ \exp[\{\{-\epsilon + (\epsilon^2 + 4)^{1/2}\}/2\}^{1/2} x]$ for all ϵ, where A is a parameter, and provides initial conditions for integration from $x = -\infty$. The numerical solution is found and depicted. Monotone solitary waves are found. Kawahara (1972)

16. $$y^{\text{iv}} - \epsilon y'' + y - 0.75y^2 = 0, y = y' = y'' = y''' = 0$$

as $x \to -\infty$ where ϵ is a parameter.

The given problem arises from a fifth order Korteweg–deVries equation. The asymptotic (linear) forms of the solution of the given DE as $x \to -\infty$ are:

(a) $y = K\ \exp\{(\epsilon + 2)^{1/2} x/2\} \cos\{(2 - \epsilon)^{1/2} x/2 + \theta\}, \epsilon < 2.$

(b) $y = (A + Bx)\ \exp(x), \epsilon = 2.$

(c) $y = A\exp[\{(\epsilon+2)^{1/2} + (\epsilon-2)^{1/2}\}x/2] + B\exp[\{(\epsilon+2)^{1/2} - (\epsilon-2)^{1/2}\}x/2\}], \epsilon > 2,$
where $K, A, B,$ and θ are arbitrary constants.

The linear solution is used to provide the behavior at $-\infty$ as well as initial conditions for integration. The integration for cases (b) and (c) is started at $x = 0$ since the solution is even in x and so $y' = y''' = 0$ at $x = 0$. A local analysis shows that $y = y_0, y' = 0, y'' = -y_0(1 - y_0/2)^{1/2}$, $y''' = 0$ at $x = 0; y_0$ is, therefore, the only parameter instead of the four parameters that appear in the solutions as $x \to -\infty$. A numerical solution is obtained and depicted. Oscillatory waves are found. Kawahara (1972)

17. $$\boldsymbol{y^{\text{iv}} + \epsilon y'' + y - 0.75y^2 = 0, y = y' = y'' = y''' = 0,}$$

as $x \to -\infty$ where ϵ is a parameter.

The given problem arises from a fifth order Korteweg–deVries equation. The asymptotic (linear) forms of the solution as $x \to -\infty$ are:

(a) $y = K\exp\{(2 - \epsilon)^{1/2}x/2\}\cos\{(\epsilon + 2)^{1/2}x/2 + \theta\}, \epsilon < 2.$

(b) There is no solution for $\epsilon \geq 2$ because the characteristic equation in this case has four pure imaginary roots.

Here K and θ are parameters. The linear solution provides the behavior at $-\infty$ as well as initial conditions for integration. The numerical solution is obtained and depicted. An oscillatory solitory wave is obtained. Kawahara (1972)

18. $$\boldsymbol{y^{\text{iv}} + \epsilon y'' - y^2/2 + \lambda y - C = 0,}$$

where $\epsilon, \lambda,$ and C are constants.

Following the procedure in Eqn. (4.1.3), we may find an exact solution by putting

$$y(x) = AP'' + BP, \tag{1}$$

where A and B are arbitrary constants, and P satisfies

$$P'^2 = 4P^3 - g_2P - g_3, \tag{2}$$

where g_2 and g_3 are two invariants, which are real and satisfy $g_2^3 - 27g_3^2 > 0$. By substitution of (1) in the given DE and using (2), we finally obtain the exact solution as

$$y(x) = 280[6P^2 + (\epsilon/13)P - g_2/2],$$

where

$$g_2 = -31\epsilon^2/[13^2(84)] + \lambda/28, g_3 = \epsilon(-112g_2 + \lambda)/(13 \times 720).$$

Kano and Nakayama (1981)

19. $$\boldsymbol{y^{\text{iv}} + \beta y'' + (\alpha/2)y^2 - vy = 0,}$$

where $\beta, \alpha,$ and v are constants.

The given DE describes traveling wave solution of the Korteweg–deVries equation with additional fifth order dispersion, namely $u_t + \alpha uu_x + \beta u_{xxx} + u_{xxxxx} = 0$, after one integration has been performed. Other methods show that it has a sech^4 type of solitary wave solutions.

Here we describe a simple algebraic method. The linear form of the given DE (i.e., with $\alpha = 0$) allows real exponential solutions $\exp\{\pm K(v)x\}$ for two different values of K, namely $K_{1,2} = [(1/2)\{-\beta \pm (\beta^2 + 4v)^{1/2}\}]^{1/2}$ with $\beta < 0, -(1/4)\beta^2 < v < 0$ as sufficient conditions. Anticipating that the final solution may be built up as a sum of powers of only one decaying exponential solution $g(x) \stackrel{\text{def}}{=} \exp(-Kx)$, we look for two integers $M_{1,2}$ satisfying $K = K_1/M_1 = K_2/M_2$. For computational convenience, we rescale the coefficients in the given DE by $y = -(v/18\alpha)\tilde{y}$ and substitute the expansion $\tilde{y} = \sum_{n=1}^{\infty} a_n g^n$ into the rescaled nonlinear equation; we get

$$\sum_{n=1}^{\infty}(n^4K^4 + n^2\beta K^2 - v)a_n g^n - \{v/36\}\sum_{n=2}^{\infty}\sum_{\ell=1}^{n-1} a_\ell a_{n-\ell} g^n = 0, \tag{1}$$

where we use Cauchy's rule for the double product in the nonlinear term. It follows from (1) and the definition of $K_{1,2}$ that $a_1 = 0$. For a nontrivial solution built up of the mixing of two decaying exponentials $g_{1,2} \stackrel{\text{def}}{=} \exp(-K_{1,2}x)$, we require two of the coefficients a_n to be arbitrary. An obvious choice is a_2 and a_3, so that from (1), we have $16K^4 + 4\beta K^2 - v = 0, 81K^4 + 9\beta K^2 - v = 0$. Solving for v and K in terms of β, we have $v = -36\beta^2/169, K = \{-(1/13)\beta\}^{1/2}$. Using these expressions in the definition of $K_{1,2}$ and $K = K_1/M_1 = K_2/M_2$, we find that $M_1 = 2, M_2 = 3$, so that $K_1 = 2\{-(1/13)\beta\}^{1/2}, K_2 = 3\{-(1/13)\beta\}^{1/2}$. Now, the recursion relation simplifies to

$$(n^2-4)(n^2-9)a_n + \sum_{\ell=1}^{n-1} a_\ell a_{n-\ell} = 0, n \geq 2. \tag{2}$$

The first few coefficients are $a_4 = -a_2^2/84, a_5 = -a_2a_3/168, a_6 = a_2^3/36288 - a_3^2/864$.

Note from (2) that if a_n is a solution, so is $a_n a^n$, with $a > 0$, a constant. We may, thus, find that

$$a_n = b(-1)^{n+1}n(n-1)(n+1)a^n, \tag{3}$$

where the constants a and b depend on a_2 and a_3. To get these, we substitute (3) in (2) and using some summation formulas, we find that $b = 140$, and the equations for $n = 2, 3$ give $a = -a_3/(4a_2) > 0$, provided that $a_2^3 = -(105/2)a_3^2$. Finally, substituting $a_n = 140(-1)^{n+1}n(n^2-1)a^n$ in $\tilde{y} = \sum_{n=1}^{\infty} a_n g^n$, we find that $\tilde{y} = -840(ag)^2/(1+ag)^4$, where we have used the relation $(1/6)\sum_{n=2}^{\infty}(-1)^n n(n^2-1)x^n = x^2/(1+x)^4, \mid x \mid < 1$. Hereman, Banerjee, Korpel, and Assanto (1986)

20. $$\boldsymbol{y^{\text{iv}} + (1 + C^2)y'' + y - 30y^3 = 0,}$$

where C is a constant.

An exact (kink) solution is found to be

$$y = (1/30)(10S - C^2 - 1) - 2R_2 - 2(R_1 - R_2)\text{cn}^2[x(R_1 - R_2)^{1/2}, m],$$

where $R_1 \geq R_2 \geq R_3$ are real roots of the cubic $R^3 + (1/2)SR^2 - (1/2)bR - (1/4)d = 0, m^2 = (R_1 - R_2)/(R_1 - R_3), b = -(1/360)(60S^2 + C^4 + 2C_0^2 - 9), d = (1/5400)(4 + 2C^2 - 3C^4 - C^6 + 45S - 10SC^2 - 5SC^4 - 100S^3)$, and S is a constant.

The solution is obtained by relating the given DE to a certain Riccati equation. Kudryashov (1991)

21. $$\boldsymbol{y^{\text{iv}} - \beta y'' - 30y^3 - \alpha y^2 + Cy - q = 0,}$$

where β, α, C, and q are constants.

The solution of the given DE (arising as a traveling wave form of Kawachara equation) is

$$y = C + 2R(x), C = (1/90)(3\beta + 30S - \alpha), \quad (1)$$

where $R(x)$ is the solution of the anharmonic oscillator

$$R'^2 = -4R^3 + 2SR^2 + 2bR + d \quad (2)$$

with $b = (1/3240)\alpha^2 - (1/360)(\beta^2 - 10C + 60S^2), d = -(1/2160)\{\alpha C + 90q - 20CS + 40S^3 + 2C\beta + 2S\beta^2\} - (1/291600)\alpha^2(\alpha - 30S + 3\beta) + (1/5400)\beta^3$, where S is an arbitrary constant.

The solution of (1) is a periodic (cnoidal) wave, provided that $\alpha = \beta = 1, q = 0, 26/225 \leq C < 131/576 - (7/64)(7/15)^{1/2}$ or $131/576 + (7/64)(7/15)^{1/2} < C$, and R satisfies (2). This solution degenerates into a solitary wave $y(x) = (1/15)\,\text{sech}^2\{x/30^{1/2}\}$, for $C = 26/225, S = -1/15$. Kudryashov (1991)

22. $\quad y^{\text{iv}} + \{\beta/(\gamma K^2)\}y'' + \{\alpha'/(\gamma\nu' K^4)\}y^{\nu'} + \{\alpha/(\gamma\nu K^4)\}y^{\nu} - \{w/(\gamma K^5)\}y = 0,$

where $\beta, \gamma, K, \alpha', \nu', \alpha, \nu$, and w are constants.

Exact solitary wave solutions in the form $y(x) = A/\{\lambda + \cosh x\}^n$ can be found in the following two cases:

(a) $\nu' = (1/2)(\nu + 1)$.
(b) $\nu' = (1/4)(3\nu - 1)$.

Dai and Dai (1989)

4.3 $y^{\text{iv}} + ayy'' + f(x,y,y') = 0$

1. $\quad y^{\text{iv}} + yy'' + y'^2 + By' - 2B^2 = 0,$

where B is a constant.

The given DE arises in a similarity reduction of the Boussinesq equation. Integrating once, we get

$$y''' + yy' + By = 2B^2x + c,$$

where c is a constant [see Eqn. (3.2.23)]. Clarkson and Kruskal (1989)

2. $\quad y^{\text{iv}} + yy'' + y'^2 + Axy' + 2Ay - 2A^2x^2 = 0.$

The given DE results from a similarity reduction of the Boussinesq equation. It is a special case of Eqn. (4.3.3) and may, therefore, be related to Painlevé's fourth equation. Clarkson and Kruskal (1989)

3. $\quad y^{\text{iv}} + yy'' + y'^2 + (Ax + B)y' + 2Ay = 2(Ax + B)^2,$

where A and B are arbitrary constants.

It is proved that the given DE has poles as its only movable singularities. This DE arises in a similarity reduction of the Boussinesq equation. If $A \neq 0, B \neq 0$, we make the transformation $y \to (4A/3)^{1/2}y, x \to \{3/(4A)\}^{1/4}x - B/A$, to obtain

$$y^{\text{iv}} + yy'' + y'^2 + (3/4)xy' + (3/2)y = (9/8)x^2. \tag{1}$$

The solution of (1) can be written as

$$\begin{aligned} y &= -[3(3)^{1/2}/2]\left[\frac{dw}{dY} + w^2(Y) + 2Yw(Y) + 3Y^2\right] + [9(3)^{1/2}/8]\alpha - 3^{1/2}, \\ Y &= 3^{1/4}x/2, \end{aligned}$$

where w is the solution of the fourth Painlevé equation,

$$\frac{d^2w}{dY^2} - \frac{1}{2w}\left(\frac{dw}{dY}\right)^2 - (3/2)w^3 - 4Yw^2 - 2(Y^2 - \alpha)w - \beta/w = 0.$$

Here α and β are arbitrary constants. Clarkson and Kruskal (1989)

4. $$\boldsymbol{y^{\text{iv}} + yy'' + y'^2 + f(x)y' + g(x)y = h(x),}$$

where $f(x), g(x)$, and $h(x)$ are analytic.

It is shown that the most general form of the given DE having the Painlevé property, that is, having no solution with movable singularities other than poles, is given by

$$y^{\text{iv}} + yy'' + y'^2 + (Ax + B)y' + 2Ay = 2(Ax + B)^2,$$

where A and B are arbitrary constants. Clarkson and Kruskal (1989)

5. $$\boldsymbol{y^{\text{iv}} - (yy')' - A = 0, y = 0 \text{ when } x = \pm 1, y'(0) = 0,}$$

where A is a constant.

Integrating the given DE twice and using the BC at $x = \pm 1$, we get

$$y'' = (1/2)y^2 - (A/2)(1 - x^2). \tag{1}$$

Consider the following cases of (1):

(a) $A = 0$. Integrating (1) and using $y'(0) = 0$ and $y(0) = y_0$, we have

$$y'^2 = (1/3)(y^3 - y_0^3). \tag{2}$$

Writing $w = y/y_0$ and integrating (2), we have

$$\int_0^w \frac{dw}{(1 - w^3)^{1/2}} = (-y_0/3)^{1/2}(1 + x),$$

where w is an elliptic function. Since $w = 1$ at $x = 0$, the value of y_0 is given by

$$y_0 = -3\left(\int_0^1 \frac{dw}{(1 - w^3)^{1/2}}\right)^2 = -(1/3)[\Gamma(1/3)\Gamma(1/2)/\Gamma(5/6)]^2 = -5.89835.$$

(b) Solution for small A is easily found to be

$$\begin{aligned} y &= (1/24)(x^4 - 6x^2 + 5)A + (1/384)(x^{10}/270 - x^8/14 \\ &\quad + 23x^6/45 - 5x^4/3 + 25x^2/6 - 2.94)A^2 + O(A^3). \end{aligned} \tag{3}$$

An integral equation formulation of (1) also leads to iterative solution, which is in close agreement with (3).

(c) Solutions for intermediate values of A, symmetric about $x = 0$, are found numerically. Two solutions for $A = 16, 100, 400, 1000$ are found. One agreeing with the iterative solution was selected. For $A = 1000$, four solutions are obtained.

(d) Solutions for large A were found by matched asymptotic methods. Inner solution requires solving first Painlevé transcendent. For finite but large values of A, there are four solutions that satisfy $y'(0) = 0$. It is conjectured that for large values of A, more and more solutions will be found that satisfy the necessary boundary conditions.

Turcotte, Spence, and Bau (1982)

6. $$\boldsymbol{y^{\text{iv}} - 6yy'' - (1/3)xy'' - (1/2)y' - 6y'^2 = 0.}$$

The given DE arises from a two-dimensional Korteweg–deVries equation via a similarity transformation. Integrate the DE once and suppose that $y, y', \ldots \to 0$ somewhere. Now multiply by y, integrate again with the same assumption and then put $y = v^2$. The equation satisfied by v is $v'' = v^3 + xv/12$, which is a special case of $\mathrm{P_{II}}$. Drazin and Johnson (1989), p.189.

7. $$\boldsymbol{y^{\text{iv}} - 12(yy'' + y'^2) = 0.}$$

The given DE may be integrated to yield

$$\begin{aligned} y''' &= 12yy' + K, \\ y'' &= 6y^2 + Kx + K_1, \end{aligned} \tag{1}$$

sequentially, where K and K_1 are arbitrary constants. Equation (1) is $\mathrm{P_I}$; hence the original DE has poles as the only movable singularities in its solution. Bureau (1964a)

8. $$\boldsymbol{y^{\text{iv}} - 12(yy')' + 6Ky' + 6K^2 = 0,}$$

where K is a constant.

On integrating once, we find that

$$y''' - 12yy' + 6Ky + 6K^2x - K_1 = 0.$$

Now see Eqn. (3.2.32). Bureau (1964a)

9. $$\boldsymbol{y^{\text{iv}} - 12(yy')' + 6xy' + 12y + 6x^2 = 0.}$$

The given DE is obtained from Eqn. (4.5.14) by the transformation $x = \alpha X, y = \beta Y$, where $\beta\alpha^2 = 1, K_1\alpha^4 = 1$, and putting $K_2 = 0$. Now setting $y = z'$, the given DE goes to

$$z^{\text{v}} = 12(z'z'')' - 6xz'' - 12z' - 6x^2. \tag{1}$$

Integrating (1), we have

$$z^{\text{iv}} = 12z'z'' - 6xz' - 6z - 2x^3 + K.$$

Now see Eqn. (4.4.8). Bureau (1964a)

10. $$\boldsymbol{y^{\text{iv}} - \lambda yy'' = 0, y(0) = V_0, y(1) = V_1, y'(0) = y'(1) = 0,}$$

where λ is a parameter; v_0 and v_1 are given constants.

The given BVP arises from a model for the axisymmetric flow of an incompressible fluid contained between infinte porous disks. The existence of the solution is established using the Schauder fixed-point theorem. Uniqueness is also proved. It is also shown that $y''(x) = o(1)$ as $\lambda \to \infty$ on $[\delta, 1/2]$ for any $0 < \delta < 1/2$ in the case $V_0 < 0$; for $V_0 > 0, y''(x) = y''(0) + O(1/\lambda^{1/2})$ as $\lambda \to \infty$, on $[0, \delta]$ for any $\delta < 1/2$. Elcrat (1976)

11. $$\boldsymbol{y^{\text{iv}} - cyy'' + (c^2/15)y^3 = 0,}$$

where c is a constant.

The given DE is not stable in the sense that not all its critical points are fixed. Exton (1971)

12. $$\boldsymbol{y^{\text{iv}} - c(yy'' + y'^2) = 0,}$$

where c is a constant.

By a suitable transformation $y = \alpha Y$, we may choose $c = 12$. Now see Eqn. (4.3.7). Bureau (1964a)

13. $$\boldsymbol{y^{\text{iv}} + \{16\alpha/v^2\}(yy'' + y'^2) + 8y'' = 0,}$$

where α and v are constants.

The given DE arises from the traveling wave form of the second order Benjamin–Ono equation $u_{tt} + 2\alpha(uu_{xx} + u_x^2) + \beta u_{xxxx} = 0$. Integrating the given DE twice and setting the constant of integration equal to zero each time, we have

$$y'' + [8\alpha/v^2]y^2 + 8y = 0. \tag{1}$$

Multiplying (1) by y', integrating, and putting the constant of integration equal to zero, we have

$$y'^2 + [16\alpha/\{3v^2\}]y^3 + 8y^2 = 0. \tag{2}$$

A solitary wave solution of (2) is $y = -\{3v^2/(4\alpha)\}\tanh^2 x$. Hereman, Banerjee, Korpel, and Assanto (1986)

14. $$\boldsymbol{y^{\text{iv}} - \{4\alpha/(v^2 - 1)\}(yy'' + y'^2) - 4y'' = 0,}$$

where α and v are constants.

The given DE arises from the traveling wave form of the Boussinesq and Good equations. On integration and setting the constant of integration equal to zero, we have

$$y''' - \{4\alpha/(v^2 - 1)\}yy' - 4y' = 0. \tag{1}$$

A second integration of (1), with constant of integration again zero, yields

$$y'' - \{2\alpha/(v^2 - 1)\}y^2 - 4y = 0. \qquad (2)$$

Multiplying (2) by y' and integrating, with constant of integration set equal to zero, gives

$$y'^2 - [4\alpha/\{3(v^2 - 1)\}]y^3 - 4y^2 = 0. \qquad (3)$$

Equation (3) has a solitary wave solution $y = \{3(1 - v^2)/\alpha\}\ \text{sech}^2 x$. Hereman, Banerjee, Korpel, and Assanto (1986)

15. $\boldsymbol{y^{\text{iv}} + [(3b^2 - aw)/a^4 + (6/a^2)y]y'' + (6/a^2)y'^2 = 0,}$

where a, b, and w are constants.

The given DE describes traveling wave solutions of the two-dimensional Korteweg–deVries equation. Integrating twice with respect to x, we get

$$a^4 y'' + 3a^2 y^2 - (aw - 3b^2)y = Ax + Ba^2, \qquad (1)$$

where A and B are constants of integration. Imposing the conditions $y, y', y'', y''' \to 0$ as $x \to \pm\infty$, we have $A = B = 0$ in (1). Another integration gives

$$(1/2)y'^2 = (1/a^2)y^2(C/2 - y), \qquad (2)$$

where $C = (aw - 3b^2)/a^2$. The final solution on integration of (2) is

$$y(x) = (C/2)[1 + \tan^2\{(-C)^{1/2}/(2a)\}(x - x_0)] \text{ if } C < 0$$

and $y(x) = (C/2)\ \text{sech}^2[\{C^{1/2}/2a\}(x - x_0)]$ if $C \geq 0$. Here x_0 is an integration constant. If in (1) we put $A = 0$ and integrate, we get

$$(1/2)y'^2 = (1/a^2)\{-y^3 + (C/2)y^2 + By + D\} \equiv (1/a^2)F_1(y), \text{say},$$

where D is a constant of integration. If $F_1(y)$ has three distinct real simple zeros, say, $y_1 > y_2 > y_3$ and $y_2 \leq y \leq y_1$, we get the explicit cnoidal wave solution

$$\begin{aligned} -(2^{1/2}/a)(x - x_1) &= \int_y^{y_1} \frac{dy}{[F(y)]^{1/2}} = \int_y^{y_1} \frac{dy}{\{(y_1 - y)(y - y_2)(y - y_3)\}^{1/2}} \\ &= [2/(y_1 - y_3)^{1/2}]\ \text{sn}^{-1}(\sin\phi, k) = [2/(y_1 - y_3)^{1/2}]F(\phi, k), \end{aligned}$$

where $y(x_1) = y_1$ and $\phi = \text{sn}^{-1}[(y_1 - y)/(y_1 - y_2)]^{1/2}$, $k^2 = (y_1 - y_2)/(y_1 - y_3)$, and $\text{sn}^{-1}(\sin\phi, k) = F(\phi, k)$ is the normal elliptic integral of the first kind with modulus k. If we let $v = \text{sn}^{-1}(\sin\phi, k)$, then we get the cnoidal wave solution

$$\begin{aligned} y(x) &= y_1 - (y_1 - y_2)\ \text{sn}^2(v, k) \\ &= y_2 + (y_1 - y_2)\ \text{cn}^2(v, k) \\ &= y_3 + (y_1 - y_3)\ \text{dn}^2(v, k) \\ &= y_3 + (y_1 - y_3)\ \text{dn}^2[(1/a)\{(1/2)(y_1 - y_3)\}^{1/2}(x - x_1), k], \end{aligned}$$

where $\text{sn}(v, k) = \sin\phi$, $\text{cn}(v, k) = \cos\phi$, and $\text{dn}(v, k) = [1 - k^2\sin^2\phi]^{1/2}$. Chen and Wen (1987)

16. $\boldsymbol{cy^{\text{iv}} + 3yy'' + \alpha c x^2 y'' + 3y'^2 + 7c\alpha x y' + 8\alpha c y = 0,}$

where c and α are constants.

The given DE arises from analysis of fluid equations by group analysis. It has two special solutions:

$$y_1 = -4\alpha c x^2/3, y_2 = \pm 4x^{-2}.$$

Ames and Nucci (1985)

4.4 $y^{\text{iv}} + f(x, y, y', y'') = 0$

1. $$\boldsymbol{y^{\text{iv}} + \{\alpha/(1-v^2)\}(y^2 y'' + 2yy'^2) - y'' = 0,}$$

where α and v are constants.

The given DE arises from the traveling wave form of the modified improved Boussinesq equation $u_{tt} - u_{xx} + 2\alpha u u_x^2 + \alpha u^2 u_{xx} + \beta u_{xxtt} = 0, \alpha > 0, \beta > 0$. Integrating the given DE twice and setting the constant of integration equal to zero each time, we have

$$y'' + [\alpha/\{3(1-v^2)\}]y^3 - y = 0. \tag{1}$$

Multiplying (1) by y' and integrating, with the constant of integration set equal to zero, we have

$$y'^2 + [\alpha/\{6(1-v^2)\}]y^4 - y^2 = 0. \tag{2}$$

Equation (2) has the solitary wave solution $y = \{6(1-v^2)/\alpha\}^{1/2}\,\text{sech}x$. Hereman, Banerjee, Korpel, and Assanto (1986)

2. $$\boldsymbol{y^{\text{iv}} + [6y^2/a^2 - (aw - 3b^2)/a^4]y'' + (12/a^2)yy'^2 = 0,}$$

where a, b, and w are constants.

The given DE describes traveling wave solutions of the two-dimensional modified Korteweg–deVries equation. Integrating twice with respect to x, we find that

$$a^4 y'' + 2a^2 y^3 - (aw - 3b^2)y = Ax + Ba^2/2, \tag{1}$$

where A and B are constants of integration. If we assume that $y, y', y'', y''' \to 0$ as $x \to \pm\infty$, then $A = B = 0$ and the nontrivial solution of (1) is

$$y(x) = C^{1/2}\,\text{sech}[(C^{1/2}/a)(x - x_0)],$$

where x_0 is a constant of integration and $C = (aw - 3b^2)/a^2 > 0$. If $C < 0$, there is no real solution since $y'^2 = y^2(C - y^2)/a^2$. If we set $A = 0$ in (1) and integrate, we get

$$y'^2 = (1/a^2)[-y^4 + Cy^2 + By + D] \equiv (1/a^2)F(y), \tag{2}$$

where D is an integration constant. If B, C, and D are chosen in such a way that $F(y)$ has four distinct real simple zeros,

$$y_1 > y_2 > y_3 > y_4 \text{ with } y_4 = -y_1, y_3 = -y_2 \text{ and } y_2 \le y < y_1,$$

then (2) integrates to give

$$\begin{aligned} y(x) &= [y_1^2 - (y_1^2 - y_2^2)\,\text{sn}^2(v,k)]^{1/2} \\ &= [y_2^2 + (y_1^2 - y_2^2)\,\text{cn}^2(v,k)]^{1/2} \\ &= y_1\,\text{dn}(v,k), \end{aligned}$$

where $y(x_1) = y_1, v = -(y_1/a)(x - x_1)$, and $k^2 = (y_1^2 - y_2^2)/y_1^2$. Chen and Wen (1987)

3. $$\boldsymbol{y^{\text{iv}} + (y' - c^2 + 1)y'' = 0,}$$

where c is an arbitrary parameter and $y', y'', y''' \to 0$ as $|\,x\,| \to \infty$.

Integrating once and using a BC, we have

$$y''' + y'^2/2 + (1 - c^2)y' = 0. \tag{1}$$

Multiplying (1) by y'', integrating, and using a BC, we get $y''^2/2 + y'^3/6 + (1 - c^2)y'^2/2 = 0$. Now put $y' = p$, and integrate twice, and use a BC to obtain

$$y = y_0 + 6(c^2 - 1)^{1/2}\tanh\{(1/2)(c^2 - 1)^{1/2}x + A\},$$

where y_0 and A are arbitrary constants. Drazin and Johnson (1989), p.33.

4. $\quad \boldsymbol{y^{\text{iv}} + 6y'y'' - g_1 = 0,}$

where g_1 is a constant.

The given DE results from the similarity reduction of the "potential" Boussinesq equation, and is once integrable:

$$y''' + 3y'^2 - g_1 x - g_2 = 0, \tag{1}$$

where g_2 is an arbitrary constant. For $g_1 \neq 0$, Eqn. (1) is transformed by $y' = -2(-g_1/2)^{2/5} V(z)$, $x = (-g_1/2)^{-1/5}z - g_2/g_1$ to the first Painlevé equation $V'' - 6V^2 - z = 0$. Equation (1) is easily integrated to yield

$$y''^2 + 2y'^3 - 2g_1 xy' + 2g_1 y - 2g_2 y' + 4g_3 = 0, \tag{2}$$

where g_3 is an arbitrary constant. Choosing $g_1 = 0$, Eqn. (2) can be reduced to the first order equation for the Weierstrass P-function:

$$P'^2(x) = 4P^3(x) - g_2 P(x) - g_3 \text{ with } y'(x) = -2P(x; g_2; g_3).$$

Boiti and Pempinelli (1980)

5. $\quad \boldsymbol{y^{\text{iv}} + 12y'y'' = 0.}$

Integrating w.r.t. x, we have

$$y''' = -6y'^2 + K,$$

where K is an arbitrary constant. Now, put $y' = z$ so that

$$z'' = -6z^2 + K. \tag{1}$$

The solution of (1) can be expressed in terms of elliptic functions; hence the original DE has solutions with poles as the only movable singularities. Bureau (1964a)

6. $\quad \boldsymbol{y^{\text{iv}} + 12y'y'' - 6y' - 6x - K = 0,}$

where K is a constant.

Setting $y' = z$, we have

$$z''' + 12zz' - 6z - 6x - K = 0; \tag{1}$$

now see Eqn. (3.2.31) to which (1) may be transformed if $K = 0$. Bureau (1964a)

7. $\quad \boldsymbol{y^{\text{iv}} + 12y'y'' - K_0 y' - (1/6)K_0^2 x - K_1 = 0,}$

where K_0 and K_1 are constants.

On replacing x by αx and y by βy, where $\alpha\beta = 1, K_0^3 = 6$, the given DE goes to the canonical form described by Eqn. (4.4.6). Bureau (1964a)

8. $$y^{\text{iv}} + 12y'y'' - 6xy' - 6y - 2x^3 - K = 0,$$

where K is a constant.

The solutions of the given DE have poles as their only movable singularities. Bureau (1964a)

9. $$y^{\text{iv}} + 12y'y'' - (K_0x + K_1)y' - K_0y - [1/(18K_0)](K_0x + K_1)^3 + K_2 = 0$$

where K_0, K_1, and K_2 are constants.

First, let $K_0x + K_1 = X$. The transformation implies that one may assume that $K_1 = 0$ without loss of generality. Then $x = \alpha X$, $y = \beta Y$, where $\alpha\beta = 1, K_0\alpha^4 = \alpha$ take the given DE to the simpler Eqn. (4.4.8). Bureau (1964a)

10. $$y^{\text{iv}} - by'y'' = 0,$$

where b is a constant.

By the transformation $y = \alpha Y$, one may choose $b = -12$; see Eqn. (4.4.5). Bureau (1964)

11. $$y^{\text{iv}} - by'y'' - (b^2/4)yy'^2 - (b^2/8)y^2y'' - (b^3/64)y^3y' = 0,$$

where b is a constant.

On integration, we get

$$y''' - (b/2)y'^2 - (b^2/8)y^2y' - (b^3/256)y^4 = \text{ constant.}$$

By a simple change of variable, the constant can be equated to zero; we get

$$y''' - 12y'^2 - 72y^2y' - 54y^4 = 0.$$

Now see Eqn. (3.3.17). Exton (1971)

12. $$y^{\text{iv}} - (3/2)y'^2y'' + (x/2)y'y'' + (x^2/12)y'' - ry'' + (x/4)y' - r/3 = 0,$$

where r is a constant.

This DE results from the modified Boussinesq equation

$$(1/3)q_{tt} - q_tq_{xx} - (3/2)q_x^2q_{xx} + q_{xxxx} = 0$$

by the similarity transformation

$$\eta = xt^{-1/2}, q(x,t) = r \ln t + \theta(\eta)$$

and change to the notation $\theta \to y, \eta \to x$. Integrating the DE once and then putting $y' = 2w(x) - x/3$ and making the simple scale change

$$w = kw_1, z = \ell x, k = -(i\epsilon/2)^{1/2}, \epsilon^2 = 1, \ell = -(4k)^{-1},$$

we get P_{iv} for w_1 with $r = -\alpha\epsilon i$; the parameter β may be identified to be the constant of integration obtained after performing the first integration. Gromak (1987)

13. $$\boldsymbol{y^{\text{iv}} + y''[(-1/2)y'^2 + (1/2)xy' + \gamma + x^2/4] + (3x/4)y' + \gamma = 0,}$$

where γ is a constant.

The given DE arises in the similarity reduction of the modified Boussinesq equation. If we put

$$y'(x) = -3^{3/4}Q(X) - x, X = 3^{1/4}x/2,$$

then $Q(X)$ satisfies the fourth Painlevé equation

$$\frac{d^2Q}{dX^2} = \frac{1}{2Q}\left(\frac{dQ}{dX}\right)^2 + \frac{3}{2}Q^3 + 4XQ^2 + 2(X^2 - \alpha)Q + \beta/Q$$

with $\alpha = \gamma/[3^{1/2}]$ and β an arbitrary constant. Clarkson and Kruskal (1989)

14. $$\boldsymbol{y^{\text{iv}} + [4\alpha^2/\{\gamma(v - \epsilon w^2)\}](yy'' + y')^2 - (4\alpha/\gamma)y'' = 0,}$$

where α, γ, ϵ, and w are constants.

The given DE appears in the traveling wave form of the Kadomtsev–Petviashvili equation. It has a first integral

$$y''' + [4\alpha^2/\{\gamma(v - \epsilon w^2)\}]yy' - (4\alpha/\gamma)y' = 0 \tag{1}$$

if we set constant of integration equal to zero. Equation (1) can be integrated:

$$y'' + [2\alpha^2/\{\gamma(v - \epsilon w^2)\}]y^2 - (4\alpha/\gamma)y = 0, \tag{2}$$

if we again set constant of integration equal to zero. The given DE has the solitary wave solution $y = -\{3(w^2\epsilon - v)/\alpha\}\ \text{sech}^2 x$. This also satisfies (2) if $\alpha = \gamma$. Hereman, Banerjee, Korpel, and Assanto (1986)

15. $$\boldsymbol{y^{\text{iv}} + f(x)y^3/(y^2 + y''^2) = 0,}$$

where $f(x)$ is positive and continuous in $[0, \infty)$ [see Eqn. (4.4.20)]. Cheng (1980)

16. $$\boldsymbol{y^{\text{iv}} + f(x)yy''^2/(y^2 + y''^2) = 0,}$$

where $f(x)$ is positive and continuous in $[0, \infty)$ [see Eqn. (4.4.20)]. Cheng (1980)

17. $$\boldsymbol{y^{\text{iv}} - x^{-\gamma}(1 - x)^{-\rho}(-y'')^{-\alpha}(y^\beta + 1) = 0, 0 < x < 1,}$$

$y(0) = a \geq 0, y''(1) = b \geq 0, y''(0) = 0, y'''(1) = 0$, with $0 \leq \gamma, \rho, \alpha < 1, \beta \geq 0$, and $\beta < \alpha + 1$.

Existence of the solution of the given BVP is proved. O'Regan (1991)

18. $$\boldsymbol{y^{\text{iv}} - y^{-\alpha} \mid y'' \mid^\beta = 0, 0 < x < 1,}$$

$y(0) = y'(1) = y''(0) = y''(1) = 0$ with $\beta > 0$ and $0 < \alpha < 1/3$.

The existence of the solution of the given BVP is proved. O'Regan (1991)

19. $$y^{\text{iv}} - y^{-\alpha}(|\, y''\,|^{\beta} + 1) = 0, 0 < x < 1,$$

$y(0) = y''(0) = y'''(1) = 0, y'(1) = b \geq 0$ with $0 < \alpha, \beta < 1$. In addition, if $b = 0$, assume that $\alpha < 1/3$.

The existence of the solution of the given BVP is proved. O'Regan (1991)

20. $$y^{\text{iv}} + f(x)g(y, y'') = 0,$$

where f and g satisfy the following:

(a) f is positive and continuous on $[0, \infty)$.

(b) $g(\lambda u, \lambda v) = \lambda g(u, v)$ for every λ, u, v.

(c) $\text{sgn}\ g(u, v) = \text{sgn}\ u$.

(d) $g(u, v)$ is continuous for every u, v and satisfies conditions such that solutions of the given DE are determined uniquely by initial conditions.

An equivalent system $x = y, y = -f(t)g(x, y)$ is considered. Several oscillation and asymptotic results are proved. For example:

(a) Suppose that $y(x)$ is a nonoscillatory solution of the given system. Then there are positive constants C_1 and C_2 such that $C_1 \leq |\, y(x)\,| \leq C_2 x^3$ for large x.

(b) A necessary condition for the given system to have an asymptotically constant solution $y(x)$ is that $\int^{\infty} s^3 f(s) ds < \infty$.

The given DE has several properties in common with $y^{\text{iv}} + f(x)y = 0$. Cheng (1980)

21. $$y^{\text{iv}} - f(x, y, y'') = 0, 0 < x < 1,$$

$$y(0) = 0, y'(1) = b \geq 0, y''(0) = c \leq 0, y'''(1) = 0.$$

The existence of the solution of the given problem is proved when f has a singularity at $y = 0$ but not at $y'' = 0$. More precisely, the following conditions are required:

(a) f is continuous on $[0, 1] \times (0, \infty) \times (-\infty, 0]$ with $f \geq 0$ on $(0, 1) \times (0, \infty) \times (-\infty, \infty)$ and $\lim_{y \to 0^+} f(x, y, q) = \infty$ uniformly on compact subsets of $(0, 1) \times (-\infty, \infty)$.

(b) $0 < f(x, y, q) \leq g(y)\phi(|\, q\,|)$ on $(0, 1) \times (0, \infty) \times (-\infty, 0]$, where $g > 0$ is continuous and nonincreasing on $(0, \infty)$ and ϕ is continuous on $[0, \infty)$.

(c) $u/\phi(u)$ is nondecreasing on $[0, \infty)$.

(d) Suppose that there exist constants $A \geq 0, B \geq 0, 0 \leq r < 1$ such that for all $z \in [0, \infty)$, $\int_0^z g(u) du \leq \int_0^{Az^r + B} [u/\phi(u)] du$.

O'Regan (1991)

4.5 $y^{\text{iv}} + ayy''' + f(x, y, y', y'') = 0$

1. $$\boldsymbol{y^{\text{iv}} + 2yy''' + 6y'y'' = 0.}$$

The given DE is equivalent to the system

$$z''' = 0, \tag{1}$$

$$y' + y^2 = z; \tag{2}$$

therefore, $y(z)$ has poles as its only movable singularities. It can be solved by solving (1) and (2) sequentially. Bureau (1964a).

2. $$\boldsymbol{y^{\text{iv}} + 2yy''' + 6y'y'' - c_0(x)(y'' + 2yy') - e_0(x)(y' + y^2) - f_0(x) = 0,}$$

where c_0, e_0, and f_0 are arbitrary analytic functions of x.

The solutions of the given DE have poles as their only movable singularities. This is easily seen by writing the DE in the equivalent form

$$\begin{aligned} y &= w'/w, \\ w'' &= zw, \\ z''' &= c_0(x)z' + e_0(x)z + f_0(x). \end{aligned} \tag{1}$$

It may be noted that the last of Eqns. (1) is linear. Considering the system backward, the statement regarding the poles as being the only movable singularities follows. Bureau (1964a)

3. $$\boldsymbol{y^{\text{iv}} + 2yy''' + 6y'y'' - c_0(x)(y'' + 2yy') - d_0(x)(y'^2 + 2y^2y' + y^4) - e_0(x)(y' + y^2) - f_0(x) = 0,}$$

where $e_0(x), \ldots, f_0(x)$ are analytic functions of x.

The condition that solutions of the given DE have poles as their only movable singularities is that $d_0 = 0$; now see Eqn. (4.5.2). Bureau (1964a)

4. $$\boldsymbol{y^{\text{iv}} + 2yy''' - 10y'y'' - 16y^2y'' - 32yy'^2 - 32y^3y' - K = 0,}$$

where K is a constant.

Replacing y by $3y/4$, we obtain

$$y^{\text{iv}} + (3/2)yy''' - (15/2)y'y'' - 9y^2y'' - 18yy'^2 - (27/2)y^3y' - K_1^* = 0, \tag{1}$$

where K_1^* is another constant. Equation (1) is equivalent to the system

$$\begin{aligned} y' &= -(3/4)y^2 + z, \\ z''' &= 12zz' + K_1^*. \end{aligned} \tag{2}$$

The first of (2) is of Riccati type. The second of Eqn. (2) may be integrated in terms of Painlevé transcendents. Hence the system may be shown to possess only fixed critical points. Exton (1971)

5. $\boldsymbol{y^{\text{iv}} + 3yy''' + 9y'y'' + 6yy'^2 + 3y^2y'' - c_0(x)(y'' + 3yy' + y^3) - f_0(x) = 0,}$

where c_0 and f_0 are analytic functions of x.

Setting

$$u = y'' + 3yy' + y^3, \tag{1}$$

we obtain

$$u'' = c_0(x)u + f_0(x). \tag{2}$$

The linear equation (2) has fixed singularities. Putting $y = v'/v$ in (1), we get

$$v''' = uv. \tag{3}$$

Thus the given DE is equivalent to the system

$$\begin{aligned} v' &= vy, \\ v''' &= uv, \\ u'' &= c_0(x)u + f_0(x). \end{aligned} \tag{4}$$

Starting from the last of Eqns. (4), which is linear, and checking, we conclude that the solutions of the given DE have poles as their only movable singularities. Bureau (1964a)

6. $$\boldsymbol{\begin{aligned} &y^{\text{iv}} + 3yy''' + 9y'y'' + 6yy'^2 + 3y^2y'' - c(x)y'' - 2c'(x)y' \\ &- c''(x)y - f(x)(y'' + 3yy' + y^3) + c(x)f(x)y - f_0(x) = 0, \end{aligned}}$$

where $c(x)$, $f(x)$, and $f_0(x)$ are analytic functions of x.

Setting

$$u = y'' + 3yy' + y^3 - c(x)y \tag{1}$$

in the given DE, we obtain the linear equation

$$u'' = f(x)u + f_0(x). \tag{2}$$

Now put $y = v'/v$ in (1) to find that

$$v''' - uv - cv' = 0. \tag{3}$$

Equations (2), (3), and $v' = vy$ form a system equivalent to the given system. Equation (2) is linear and therefore has fixed singularities. It is now easy to show that the given DE has solutions with poles as their only movable singularities. Bureau (1964a)

7. $\boldsymbol{y^{\text{iv}} + 3yy''' + 10y'y'' + 4y^2y'' + 8yy'^2 + 2y^3y' - f_0(x) = 0,}$

where $f_0(x)$ is an arbitrary function of x.

Replacing y by $-(2/3)y$, we obtain from the given DE the form

$$\begin{aligned} y^{\text{iv}} &- 2yy''' - (20/3)y'y'' + (16/9)y^2y'' + (32/9)yy'^2 \\ &- (16/27)y^3y' - f_0(x) = 0. \end{aligned} \tag{1}$$

Equation (1) is equivalent to the system

$$y''' - 2yy'' + 3y'^2 - (4/27)(6y' - y^2)^2 - z = 0, \;\; z' - f_0(x) = 0, \tag{2}$$

which may be shown to possess only fixed critical points. Exton (1971)

8.
$$\begin{aligned} &\boldsymbol{y^{\text{iv}} + 4yy''' + 10y'y'' + 6y^2y'' + 12yy'^2 + 4y^3y' - c(x)y''} \\ &\boldsymbol{- 2c(x)yy' - [f(x) + c'(x)]y' - c'(x)y^2 - f'(x)y - f_0(x) = 0,} \end{aligned}$$

where $c(x), f(x)$, and $f_0(x)$ are arbitrary analytic functions of x.

Putting $y = w'(x)/w(x)$, we get a fifth order DE that is shown to be equivalent to the linear system

$$w^{\text{iv}} - c(x)w'' - f(x)w' - zw = 0, z'(x) = f_0(x). \tag{1}$$

Hence the critical points of (1), and therefore those of the given DE are fixed. Exton (1971)

9.
$$\begin{aligned} &\boldsymbol{y^{\text{iv}} + 5yy''' + 10y'y'' + 10y^2y'' + 15yy'^2 + 10y^3y' + y^5} \\ &\boldsymbol{- c_0(x)[y'' + 3yy' + y^3] - e_0(x)(y^2 + y') - f_1(x)y - f_0(x) = 0,} \end{aligned}$$

where c_0, e_0, f_0, f_1 are arbitrary analytic functions of x.

The given DE is equivalent to the system

$$y = w'/w, w^{v} = c_0w''' + e_0w'' + f_1w' + f_0w. \tag{1}$$

The second of (1) is linear. The given system has therefore only fixed critical points. Exton (1971)

10.
$$\begin{aligned} &\boldsymbol{y^{\text{iv}} - yy'''/3 - yy'^2/3 - y^2y''/6 + y^3y'/18} \\ &\boldsymbol{- xy''/3 + xyy'/9 - 2y'/3 + y^2/9 = 0.} \end{aligned}$$

See Eqn. (3.2.24). Olver (1986), p.198

11.
$$\boldsymbol{y^{\text{iv}} - yy''' - 11y'y'' + y^2y'' + 2yy'^2 = 0.}$$

The given DE may be integrated to yield

$$y''' - yy'' - 5y'^2 + y^2y' - K = 0, \tag{1}$$

where K is a constant. Equation (1) has branch points as movable singularities of its solution [see Eqn. (3.5.88)]. Bureau (1964a)

12.
$$\boldsymbol{y^{\text{iv}} - 2yy''' - (20/3)y'y'' + (16/9)y^2y'' + (32/9)yy'^2 - (16/27)y^3y' = 0.}$$

Integrating the given DE, we have

$$y''' - 2yy'' + 3y'^2 - (4/27)(6y' - y^2)^2 - k = 0, \tag{1}$$

where k is a constant. Equation (1) has been shown by Chazy (1911) to have all its critical points fixed. Exton (1971)

13. $$\boldsymbol{y^{\text{iv}} - 3yy''' - 9y'y'' + 3y^2y'' + 6yy'^2 = 0.}$$

Integrating the given DE, we get

$$y''' - 3yy'' - 3y'^2 + 3y^2y' - K = 0, \tag{1}$$

where K is an arbitrary constant. It has two sets of simple poles as movable singularities in its solution. Bureau (1964a)

14. $$\boldsymbol{y^{\text{iv}} - 12(yy'' + y'^2) + 6(K_1x + K_2)y' + 12K_1y + 6(K_1x + K_2)^2 = 0,}$$

where K_1 and K_2 are constants.

It can be proved that the given DE has fixed critical points. Bureau (1964a)

15. $$\boldsymbol{y^{\text{iv}} + R(-yy''' + y'y'') = 0, y(0) = y''(0) = 0,}$$

$y(1) = 1, y'(1) = 0$, where R is a constant.

It is proved that for each R there is at least one solution, and for sufficiently large positive R there are at least three solutions. The asymptotic behavior as $R \to -\infty$ is also studied. The given problem arises in laminar flow in a porous channel. Hastings, Lu, and MacGillivray (1992)

16. $$\boldsymbol{y^{\text{iv}} + \lambda(yy''' - y'y'') - y' = 0,}$$

where λ is a positive constant.

The given DE arises in a nonlinear boundary layer problem in physical oceanography. It has a first integral

$$y''' - \lambda(y'^2 - yy'') - y + 1 = 0$$

if derivatives at ∞ are assumed to be zero and $y(\infty) = 1$. Ierley and Ruehr (1986)

17. $$\boldsymbol{y^{\text{iv}} - a(yy''' - 3y'y'') - c(y^2y'' - 2yy'^2) = 0,}$$

where a and b are constants.

The solutions of the given DE are shown to be single-valued only in the special (trivial) case $a = c = 0$. Bureau (1964a)

18. $$\boldsymbol{y^{\text{iv}} - ayy''' - 2ay'y'' + (2a^2/5)y^2y'' + (3a^2/5)yy'^2}$$
$$\boldsymbol{- (2a^3/25)y^3y' + (a^4/625)y^5 = 0.}$$

The solution of the given DE may be shown to possess only fixed critical points. In particular, if we choose $a = -5$ and put $y = w'/w$, we find that $w^{\text{v}} = 0$, which is easily solved. Exton (1971)

19. $$\boldsymbol{y^{\text{iv}} - 3ayy''' - 9ay'y'' + 3a^2y^2y'' + 6a^2yy'^2 = 0,}$$

where a is a constant.

By the transformation $y = aY$, one may choose $a = 1$; now see Eqn. (4.5.13). Bureau (1964a)

20. $$y^{\rm iv} - ayy''' - 11ay'y'' + a^2y^2y'' + 2a^2yy'^2 = 0,$$

where a is a constant.

By the transformation $y = \alpha Y$, we may change the given DE and choose $a = 1$ without loss of generality; see Eqn. (4.5.11). Bureau (1964a)

21. $$y^{\rm iv} - (ayy''' + by'y'') = 0,$$

where a and b are constants.

A simple change of variable $y = \alpha Y$ transforms the given DE to one with $a = -2, b = -6$; now see Eqn. (4.5.1). Bureau (1964a)

22. $$y^{\rm iv} - ayy''' - by'y'' - cy^2y'' - dyy'^2 - ey^3y' - fy^5 - F(x, y) = 0,$$

where

$$\begin{aligned} F(x,y) &= a_0y''' + (c_1y + c_0)y' + d_0y'^2 + (e_2y^2 + e_1y + e_0)y' \\ &\quad + f_4y^4 + f_3y^3 + f_2y^2 + f_1y + f_0, \end{aligned}$$

and a, b, c, d, e, f with or without subscript are analytic functions of x in a certain domain D. Conditions are found that the given DE has fixed critical points. Bureau (1964a)

23. $$y^{\rm iv} + S\{(y - x)y''' - my'y'' - 3y''\} = 0,$$

where S is a parameter and $m = 0$ or $1; y(0) = y''(0) = 0, y(1) = y'(1) = 1$.

The given problem arises in a self-similar solution describing the sqeezing of a fluid between two plates. If S is small, a perturbation solution in powers of S is found as

$$y = y_0(x) + Sy_1(x) + S^2y_2(x) + S^3y_3(x) + \cdots,$$

where y_i $(i = 0, 1, 2, \ldots)$ with appropriate boundary conditions are found to be

$$\begin{aligned} y_0 &= -(1/2)x^3 + (3/2)x, \\ y_1 &= -(1/560)(x^7 + 35x^5 - 73x^3 + 37x), m = 0, \\ y_1 &= (1/280)(x^7 - 28x^5 + 53x^3 - 26x), m = 1, \\ y_2 &= -(1/140)\{(3/880)x^{11} + (7/72)x^9 + (51/280)x^7 \\ &\quad - (41/20)x^5 + (34901/11088)x^3 - (2551/1848)x\}, m = 0, \\ y_2 &= (1/280)\{(1/330)x^{11} + (1/72)x^9 - (193/70)x^7 + (53/5)x^5 \\ &\quad - (18017/1386)x^3 + (47489/9240)x\}, m = 1. \end{aligned}$$

For S large, we write $1/S = \epsilon^2 << 1$ and the given problem becomes a singular perturbation problem. The interior solution is found in the form

$$y = F_0(x) + \epsilon F_1(x) + \cdots,$$

where F_0 satisfies

$$(x - F_0)F_0''' + (3 + mF_0')F_0'' = 0, F_0(0) = F_0''(0) = 0, F_0(1) = 1. \tag{1}$$

The solution of system (1) is $F_0 = x$. Introducing the stretched variable near $x = 1$, namely

$$\zeta = (1 - x)/\epsilon,$$

and writing the solution as

$$y = 1 - \epsilon g_1(\zeta) - \epsilon^2 g_2(\zeta) - \cdots,$$

we obtain

$$(\zeta - g_1)g_1''' + 3g_1'' + mg_1'g_1'' = g_1^{\mathrm{iv}}, g_1(0) = 0, g_1'(0) = 0, g_1(\infty) \to \zeta. \quad (2)$$

Now writing $g_1 = \zeta + \phi$ in (2) and ignoring higher order terms in ϕ, we get

$$\phi^{\mathrm{iv}} - (3+m)\phi'' = 0,$$

so that

$$\phi \sim C_1 + C_2 e^{-(3+m)^{1/2}}\zeta, \quad (3)$$

where C_1 and C_2 are constants. On integrating the first of (2) once, we get

$$(\zeta - g_1)g_1'' + 2g_1' + \{(m+1)/2\}g_1'^2 - (m+5)/2 = g_1''', g_1(0) = g_1'(0) = 0, g_1'(\infty) = 1. \quad (4)$$

The boundary layer system (4) is integrated numerically with initial values $g_1''(0) = 1.529397$ for $m = 0$ and $g_1''(0) = 1.7320508$ for $m = 1$. The numerical integration shows that

$$C_1 = \begin{cases} -0.6767, & m = 0 \\ -0.57734, & m = 1. \end{cases}$$

The next order interior solution is found to be $F_1 = C_1 x$. Thus, uniformly valid composite solution is

$$y = x + \epsilon C_1 x - \epsilon\{g_1(\zeta) - \zeta - C_1\} + O(\epsilon^2).$$

Numerical and analytic solutions are compared. The original system, with a simple transformation, is solved numerically using shooting arguments. Wang (1976)

4.6 $y^{\mathrm{iv}} + f(x, y, y', y'', y''') = 0$

1. $$\boldsymbol{y^{\mathrm{iv}} + a_1 y''' + a_2 y'' + g(y') + a_4 y = 0,}$$

where a_1, a_2, a_4 are constants and g depends on y' only.

The following result is proved. Suppose that (a) a_1, a_2, a_4, are all positive; (b) $g(0) = 0, g(z)/z \geq a_3 > 0 (z \neq 0)$; (c) $g'(z)$ exists and is continuous and $g'(z) \leq A_3$, for all z, where $\Delta_0 \equiv (a_1 a_2 - A_3) - a_1^2 a_4 > 0$. Then $\Delta_1 \equiv (a_1 a_2 - a_3)a_3 - a_1 a_4^2 \geq \Delta_0$, and every solution $y = y(x)$ of the given DE satisfies $y \to 0$, $y' \to 0, y'' \to 0, y''' \to 0$ as $x \to \infty$, provided that $g'(z) - g(z)/z \leq \delta_1, y \neq 0$, where δ_1 is any constant such that $\delta_1 < 2a_4\Delta_1/(a_1 a_3^2)$.

The given theorem generalizes the result for the linear case for which $g(y') = \alpha_3 y'$, where α_3 is a constant, $g' - g/y \equiv 0 (y \neq 0)$; and thus, by taking $A_3 = \alpha_3 = a_3$, the conditions of the theorem in this special case reduce to $a_1 > 0$, $a_2 > 0, \alpha_3 > 0, a_4 > 0, (a_1 a_2 - \alpha_3)\alpha_3 - a_1 a_4^2 > 0$, which is the well-known Routh's criterion for solution of the linear equation

$$y^{\mathrm{iv}} + a_1 y''' + a_2 y'' + \alpha_3 y' + a_4 y = 0$$

to tend to the trivial solution $y = 0$ as $x \to \infty$. Ezeilo (1962)

2. $\boldsymbol{y^{\text{iv}} + \alpha y''' + (-\beta + 3\gamma y^2 + \Omega^2)y'' + 6\gamma y y'^2 + \Omega^2 \alpha y' - \Omega^2 \beta y + \Omega^2 \gamma y^3 = 0,}$

where α, β, γ, and Ω are constants.

The given DE is equivalent to the system $\ddot{x} + \alpha\dot{x} - \beta x + \gamma x^3 = y, \ddot{y} + \Omega^2 y = 0$, which itself is equivalent to the Duffing equation

$$\ddot{x} + \alpha\dot{x} - \beta x + \gamma x^3 = f \cos \Omega t, \gamma > 0.$$

Considering y and x as complex, one may discuss the Painlevé property of the given DE. To find the dominant behavior one writes $y(x) \sim A(x - x_1)^k$, where x_1 is arbitrary. It is easy to check that $k = -1, A^2 = -2/\gamma$; the leading terms are y^{iv}, $3\gamma y^2 y''$, and $6\gamma y y'^2$. To obtain the resonances, i.e., the powers of $(x - x_1)$ at which arbitrary constants may enter, one writes the series

$$y(x) = A(x - x_1)^{-1} + \sum_{j=1}^{\infty} a_{-1+j}(x - x_1)^{-1+j}, 0 <| x - x_1 |< \epsilon. \tag{1}$$

Substitution, etc., and the usual argument for the Painlevé property shows [see Sachdev (1991)] that the resonances occur at $\gamma_1 = -1, \gamma_2 = 3, \gamma_{3,4} = 4$. Substitution of the Laurent series (1) in the given DE shows that

$$a_0 = \alpha/(3\gamma A), a_1 = \alpha^2/(18\gamma A) + \beta/(3\gamma A).$$

The coefficients a_2 and a_3 are arbitrary, being the constants of integration associated with the resonances γ_2 and γ_3. The multiple resonances at $\gamma_{3,4} = 4$ indicate the presence of logarithmic terms. We therefore obtain

$$\begin{aligned} y(x) &= A(x - x_1)^{-1} + a_0 + a_1(x - x_1) + a_2(x - x_1)^2 + a_3(x - x_1)^3 \\ &\quad + b_1(x - x_1)^3 \ \ln(x - x_1), \end{aligned}$$

where x_1, a_2, a_3, and b_1 are arbitrary complex constants of integration. The general form of the solution is

$$y(x) = A(x - x_1)^{-1} \sum_{k=0}^{\infty} \sum_{j=0}^{\infty} a_{jk}(x - x_1)^j \{(x - x_1)^4 \ln(x - x_1)\}^k,$$

where the presence of logarithmic terms indicates the possibility of chaos. Numerical integration alone can confirm this possibility. Steeb and Kunick (1983)

3. $\boldsymbol{y^{\text{iv}} + \nu y''' + \mu y'' + 2y'y'' = 0,}$

where μ and ν are constants.

Put $y' = u$, and integrate once to obtain

$$u'' + \nu u' + \mu u = C - u^2, \tag{1}$$

where C is a constant. If $u(x)$ is a periodic function with period 2π, we multiply (1) by u' and integrate from 0 to 2π to obtain

$$\nu \int_0^{2\pi} u'^2 \, dx = 0.$$

For $\nu = 0$ and $u(0) = A, u'(0) = 0$ and the condition of periodicity $u'(2\pi) = 0, \int_0^{2\pi} u dx = 0$, we may obtain small amplitude solutions in the form

$$\begin{aligned} u &= Au_1 + A^2u_2 + A^3u_3 + \cdots, \\ C &= A^2C_2 + A^3C_3 + \cdots, \\ \mu &= \mu_0 + A\mu_1 + A^2\mu_2 + \cdots. \end{aligned} \tag{2}$$

Substituting (2) in (1), equating powers of A, and solving the resulting system with appropriate IC following from (2), we have

$$\begin{aligned} \mu_0 &= m^2, u_1 = \cos mx, \\ \mu_1 &= 0, u_2 = \{\cos(2mx) - \cos mx\}/(6m^2), \end{aligned}$$

provided that $c_2 = 1/2$. At the third order,

$$\begin{aligned} u_3'' &+ m^2u_3 = -2\mu_1u_2 - \mu_2u_1 + C_3 \\ &= \{\cos(2mx) + 1 - \cos(mx) - \cos(3mx)\}/(6m^2) - \mu_2\cos(mx) + C_3. \end{aligned}$$

The orthogonality of the right-hand side to $\cos mx$, and $u'(2\pi) = 0$ and $\int_0^{2\pi} u dx = 0$, lead to $\mu_2 = -1/(6m^2), C_3 = -1/(6m^2)$. For further analysis of (1), see Eqn. (2.1.1). Goldshtik, Hussain, and Shtern (1991)

4. $\boldsymbol{y^{\text{iv}} - \epsilon(1-y^2)y''' + 6\epsilon yy'y'' + (1+\Omega^2)y'' + 2\epsilon y'^3 - \Omega^2\epsilon(1-y^2)y' + \Omega^2 y = 0,}$

where ϵ and Ω are constants.

The given DE is obtained from

$$\ddot{x} - \epsilon(1-x^2)\dot{x} + x - y = 0, \ddot{y} + \Omega^2 y = 0$$

by eliminating y and then writing y for x. The latter system is equivalent to the perturbed Van der Pol equation,

$$\ddot{x} - \epsilon(1-x^2)\dot{x} + x = f\cos\Omega t. \tag{1}$$

The leading order behavior of the given DE $y \sim A(x-x_1)^k$ shows that $k = -1/2$ and $A^2 = 3/(2\epsilon)$; the leading order terms are $y^{\text{iv}}, \epsilon y^2y''', 6\epsilon yy'y''$, and $2\epsilon y'^3$. Therefore, the given DE treated in the complex plane has a movable algebraic point of order $-1/2$ and is therefore not of Painlevé type. The same is therefore true of (1). Steeb and Kunick (1983)

5. $$\boldsymbol{y^{\text{iv}} + 12y'y'' - (1/x)(y''' + 6y'^2) - 1/x = 0.}$$

The only moving singularities of the solutions of the given DE are poles. Setting

$$z = y''' + 6y'^2 \tag{1}$$

in the given DE, we obtain $z' = (1/x)z + 1/x$ with the solution $z = Kx - 1$. Therefore, the given DE reduces via (1) to

$$y''' + 6y'^2 - Kx + 1 = 0;$$

Now see Eqn. (3.3.4). Bureau (1964a)

6. $$\boldsymbol{y^{\text{iv}} + 12y'y'' - (1/x)(y''' + 6y'^2) - K_0y' - K_0^2x/12 - K_1/x = 0,}$$

where K_0 and K_1 are constants.

The given DE has poles as the only movable singularities in its solution. It can be transformed to Eqn. (4.6.5) by suitable scaling, if $K_0 = 0$. Bureau (1964a)

7. $$y^{\text{iv}} + 12y'y'' - (1/x)(y''' + 6y'^2) - xy' - x^3/24 - K/x = 0,$$

where K is a constant.

The solutions of the given DE have poles as their only moving singularities. Bureau (1964a)

8. $$y^{\text{iv}} + 12y'y'' - a(y''' + 6y'^2) - 6c(y'' + 2yy') - e_0y' - f_2y^2 - f_1y - f_0 = 0,$$

where a, c, e_0, f_2, f_1, and f_0 are analytic functions of x. It is proved that all solutions of the given DE have fixed critical points. Bureau (1964a)

9. $$y^{\text{iv}} + 12y'y'' - (1/x)(y''' + 6y'^2) - (x + K)y' + (K/x)y - x^3/24 - (1/9)Kx^2 - (1/12)K^2x - K_1/x = 0,$$

where K and K_1 are constants.

The solutions of the given DE have poles as their only moving singularities. Bureau (1964a)

10. $$y^{\text{iv}} + 12y'y'' - (1/x)(y''' + 6y'^2) - (K_0x + K_1)y' + (K_1/x)y - (1/24)K_0^2x^3 - (1/9)K_0K_1x^2 - (1/12)K_1^2x - K_2/x = 0,$$

where K_0, K_1, and K_2 are constants.

The solutions of the given DE have poles as their only moving singularities. The change of variables $K_0x + K \to X, x \to \alpha X, y \to \beta Y$ takes the given DE to Eqn. (4.6.7) if $K_1 = 0$ and to Eqn. (4.6.9) otherwise. Bureau (1964a)

11. $$y^{\text{iv}} - (1 - 1/\nu)y'''^2/y'' - (1 + 2/\nu)y''y'''/y' + (1 + 1/\nu)y''^3/y'^2 = 0.$$

If $y'' \neq 0$, then one may show that either

$$y' = \exp(C_1x + C_2)$$

or

$$y' = \exp[(C_1x + C_2)^{\nu+1} + C_3],$$

where C_i $(i = 1, 2, 3)$ are arbitrary constants. Therefore, y is an entire function. Martynov (1985a)

12. $$y^{\text{iv}} - (1 - 1/\nu)y'''^2/y'' - a_1y''y'''/y' - b_1y''^3/y'^2 = 0,$$

where ν is an integer and a_1 and b_1 are constants.

Solutions of the form

$$y' = [(C_1x + C_2)^{\nu+1} + C_3]^n \tag{1}$$

or

$$y' = (C_1x + C_2)^n, \tag{2}$$

subject to some conditions on the constants may be found. For (2), the condition is $(n-1)(n-2)-(1-1/\nu)(n-1)^2-a_1n(n-1)-b_1n^2=0$. Martynov (1985a)

13. $$y^{\text{iv}}-360x(y'''/120)^{1/3}((\mid y''\mid+\mid y'''\mid)/(\mid 30x^4-6\mid+120\mid x^3\mid))^{2/3}=0,$$

$$y(\pm 1)=y'(\pm 1)=0.$$

The given problem has the solution $y=C(2-3x^2+x^6)$, where C is a real number. Bespalova and Klokov (1989)

14. $$y^{\text{iv}}-A(x,y,y',y'')y'''^2-B(x,y,y',y'')y'''-C(x,y,y',y'')=0.$$

Various problems in geometry and physics that give rise to DEs of the given type are discussed. The topics treated include systems of osculating conics to a curvature element, systems of extremals, and various systems of curves related to the trajectories in a force field. Kasner (1942)

15. $$y^{\text{iv}}=f(x,y,y',y'',y'''),y(0)=a_0,y(\tau)=b_0,y'(0)=a_1,y'(\tau)=b_1,$$

where $x\in I=[0,\tau],\tau>0,a_i,b_i\in R\ (i=0,1),f\in C(I\times R^4)$, and $I_0=(0,\tau)$.

Detailed conditions are given such that the given BVP has a solution. Bespalova and Klokov (1989)

16. $$y^{\text{iv}}=f(x,y,y',y'',y'''),$$

where $f:(a,b)\times R^4\to R$ is continuous and f satisfies the Lipschitz condition

$$\mid f(x,y_1,y_2,y_3,y_4)-f(x,z_1,z_2,z_3,z_4)\mid\le\sum_{i=1}^{4}k_i\mid y_i-z_i\mid$$

for each $(x,y_1,y_2,y_3,y_4),(x,z_1,z_2,z_3,z_4)\in(a,b)\times R^4$.

Optimal-length subintervals of (a,b) are characterized in terms of k_i $(i=1,2,3,4)$ on which two-, three-, and four-point boundary value problems for the given DE have unique solutions. Henderson and McGowier (1987)

4.7 $f(x,y,y',y'',y''')y^{\text{iv}}+g(x,y,y',y'',y''')=0$

1. $$xy^{\text{iv}}+2y'''+R(y'y''-yy''')=0,y(x_0)=-\alpha,y(1)=\beta,y'(x_0)=0,y'(1)=0,$$

where R,α, and β are constants.

The given BVP arises in description of flow through an annulus with porous walls and is solved numerically using quasilinearization. Huang (1974)

2. $$xy^{\text{iv}}+2y'''-S(xy'''+2y''+y'y''-yy''')=0,\lim_{x\to 0}(y/x^{1/2})=0\text{ or }y(0)=0,$$

$$\lim_{x\to 0}(x^{1/2}y'')=0,y'(1)=0,y(1)=1,$$

where S is a parameter.

The given problem describes unsteady sqeezing of a viscous fluid from a tube. A matched asymptotic solution is found for large positive and negative S. For large positive S,

$$y = x + (\epsilon/2^{1/2})x - (\epsilon/2^{1/2})e^{-2^{1/2}(1-x)/\epsilon} + O(\epsilon),$$

where $\epsilon^2 = 1/S \ll 1$. Skalak and Wang (1979)

3. $$\mathbf{(1-x^2)y^{\text{iv}} - 4xy''' + 3y'y'' + yy''' = 0.}$$

The given DE arises in the study of steady spreading of oil on water. Integrating the given DE thrice, we get the Riccati equation

$$(1-x^2)y' + 2xy + (1/2)y^2 = C_1x^2 + C_2x + C_3,$$

where C_i $(i = 1,2,3)$ are constants. Now, if we put $y = 2(1-x^2)u'/u$, we get the linear DE

$$2(1-x^2)^2u'' = (C_1x^2 + C_2x + C_3)u, \tag{1}$$

which is solvable in terms of hypergeometric functions. In particular, if we can write

$$C_1x^2 + C_2x + C_3 = (2k^2 - 1/2)(1-x^2),$$

where k is real or pure imaginary constant, we may solve (1) explicitly and substitute in $y = 2(1-x^2)u'/u$ to get

$$y(x) = \begin{cases} (1-x)[(1+2k)(1+x)^{2k} + C(1-2k)]/[(1+x)^{2k} + C], k \neq 0 \\ (1-x)[1 + 2/\{\ln(1+x) + C\}], k = 0, \end{cases}$$

a solution that remains bounded at $x = 1$. C is an arbitrary constant. Wang (1971)

4. $$\mathbf{yy^{\text{iv}} - 1 = 0,}$$

$y(0) = y''(0) = y(1) = y''(1) = 0$.

The solution of the given problem may be found as a psi series involving x and $\log x$. Bender and Orszag (1978), p.200

5. $$\mathbf{yy^{\text{iv}} - y''^2 = 0.}$$

Integrating the given DE twice, we have

$$yy'' - y'^2 = C_1x + C_2, \tag{1}$$

where C_1 and C_2 are constants. Put $y = 2\lambda'/\lambda$ so that (1) becomes

$$yy' + y'' = y'(x - x_0)^{-1}. \tag{2}$$

Writing $y = (x-x_0)^{-1}z, t = \ln(x - x_0)$ in (2), we have

$$\frac{d^2z}{dt^2} = (4-z)\frac{dz}{dt} + z^2 - 3z. \tag{3}$$

Now put $z' = w$ to obtain Abel's equation

$$w\frac{dw}{dz} = (4-z)w + z^2 - 3z, \tag{4}$$

which has the solution $w = z$. The given DE also has the elementary solution $y = a\exp(bx)$, when a and b are constants. Sidorevich and Lukashevich (1990)

6. $$yy^{\text{iv}} - 2y'y'' + y'^2 - (1/k)(y'' + yy' - y'^2 + 1) = 0,$$

with $y(0) = y'(0) = 0, y(\infty) = 1, y'(\infty) = 0$, where k is a constant.

The given problem arises in the stagnation point flow of a non-Newtonian fluid and is solved numerically, using quasilinearization. Garg and Rajagopal (1991)

7. $$yy^{\text{iv}} + 3y''^2 + 4y'y''' = 0.$$

An exact solution of the given DE is $y = \{2(c+g)\}^{1/2}$, where $g = C_1x^3+C_2x^2+C_3x+C_4$ and C_i $(i = 1, 2, 3, 4)$ are constants. Alidema (1978)

8. $$yy^{\text{iv}} + 3y''^2 + 4y'y''' - \cos 2x = 0.$$

An exact solution of the given DE is $y = \{2(c+g)\}^{1/2}$, where $g = c_1x^3 + c_2x^2 + c_3x + c_4 + (\cos 2x)/16$ and c is a constant. Alidema (1978)

9. $$2yy^{\text{iv}} + 6y''^2 + 8y'y''' - a^2y^2 = 0,$$

where a is a constant.

An exact solution is $y = \{2(u+c)\}^{1/2}$, where c is a constant and u satisfies the linear equation $u^{\text{iv}} - a^2(u+c) = 0$. Alidema (1977)

10. $$yy^{\text{iv}} + y''^2 - 2y'y''' - (1/k)y''' + (1/k)(y'^2 - yy'') = 0,$$

$y'(0) = 1, y(0) = 0, y'(\infty) = 0; k$ is a small parameter.

The given system arises in the flow of a viscoelastic fluid over a stretching sheet. A fourth order DE here has three conditions.

A perturbation solution in the form

$$y = y_0 + ky_1 + k^2y_2 + \cdots$$

leads to

$$y_0''' + y_0y_0'' - y_0'^2 = 0,$$
$$y_1''' + y_0y_1'' + y_0''y_1 - 2y_0'y_1' = s,$$

with

$$s = 2y_0'y_0''' - y_0''^2 - y_0y_0^{\text{iv}}.$$

The BC become $y_0'(0) = 1$,

$$y_0(0) = 0, y_0'(\infty) = 0, y_1'(0) = 0, y_1(0) = 0, y_1'(\infty) = 0.$$

The solution for y_0 is $y_0 = 1 - e^{-x}$, while y_1 is written as $y_1 = y_A + \beta_1 y_B$ with $y_A(0) = y_A'(0) = 0, y_A''(0) = 0, y_B(0) = y_B'(0) = 0, y_B''(0) = 1$; the parameter β_1 is found by solving the problem for y_A and y_B numerically, using BC at infinity: $\beta_1 = -y_A'(\infty)/y_B'(\infty)$. Numerical results are presented. Rajagopal, Na, and Gupta (1984)

11. $$y^2y^{\text{iv}} + 2yy'y''' - 3y'^2y'' = 0.$$

See (8) of Eqn. (3.5.37). Sidorevich and Lukashevich (1990)

12. $\boldsymbol{b^3y^3y^{\text{iv}} + (2ab^2 - 3a^2b + a^3)y'^4 + 6ab(a-b)yy'^2y'' + ab^2y^2(3y''^2 + 4y'y''') = 0,}$

where a and b are constants.

General solution of the given DE is

$$y = (C_1x^3 + C_2x^2 + C_3x + C_4)^{b/(a+b)},$$

where C_i $(i = 1,2,3,4)$ are constants. Alidema (1978)

13. $$\boldsymbol{y'^2y^{\text{iv}} - by'y''y''' - dy''^3 = 0,}$$

where b and d are constants.

Put $y' = v$ to obtain $v^2v''' - bvv'v'' - dv'^3 = 0$. Now see Eqn. (3.9.44). Kozulin and Lukashevich (1988)

14. $$\boldsymbol{3y''y^{\text{iv}} - 5y'''^2 = 0}$$

Put $y'' = z$ and obtain

$$3zz'' - 5z'^2 = 0,$$

the same as Eqn. (2.31.76). Hence the solution can be written as

$$(y + C_1x + C_2)^2 = C_3x + C_4,$$

where C_i $(i = 1,\ldots,4)$ are arbitrary constants. Kamke (1959) p.604

15. $$\boldsymbol{3y'y''y^{\text{iv}} - 4y'y'''^2 - 3y''^2y''' = 0.}$$

Multiplying by $(y''')^{-2}$ and integrating, we have

$$(C + y)y''' + 3y'y'' = 0;$$

see Eqn. (3.9.1). The solution of the original DE is

$$C_1(x - C_2)^2 + (y - C_3)^2 = C_4,$$

where C_i $(i = 1,2,3)$ are arbitrary constants. A singular solution is $y = a + bx + cx^2; a, b,$ and c are arbitrary constants. Murphy (1960), p.429

16. $$\boldsymbol{y'(fy')''' - y''(fy')'' + y'^3(fy')' + 2qy'^2 \sin y + (qy'' - q'y') \cos y = 0,}$$
$$\boldsymbol{f = f(x), q = q(x).}$$

Rewrite the given DE with the notation $y \equiv \theta, x \equiv s$:

$$\theta'(f\theta')''' - \theta''(f\theta')'' + \theta'^3(f\theta')' + 2q\theta'^2 \sin\theta + (q\theta'' - q'\theta')\cos\theta = 0 \tag{1}$$

with $f = f(s), q = q(s), \theta = \theta(s)$:

Multiplying (1) by $\cos\theta/\theta'^2$ makes it exact; therefore, an integration gives

$$(f\theta')''\cos\theta + (f\theta')'\theta'\sin\theta - q\ \cos^2\theta = C_1\theta'. \tag{2}$$

Dividing (2) by $\cos^2\theta$ and integrating, we have (after multiplying by $\cos\theta$)

$$(f\theta')' = C_1\sin\theta + C_2\cos\theta + \cos\theta\int_{s_0}^{s} q(\sigma)d\sigma. \tag{3}$$

Introducing rectangular coordinates through

$$x'(s) = \cos\theta, y'(s) = \sin\theta$$

and integrating with respect to s, we have

$$f\theta' = C_1 y + C_2 x + C_3 + \int_{s_0}^{s} q(\sigma)[x(s) - x(\sigma)]d\sigma.$$

Kamke (1959), p.604

4.8 $\vec{y}\,' = \vec{f}(x, \vec{y})$

1. $\quad \dot{q}_j = p_j, j = 1, 2, \quad \dot{p}_1 = -q_1 - 2q_1q_2, \quad \dot{p}_2 = -q_2 - q_1^2 + q_2^2$

The given system has a simple center at $p_j = q_j = 0$ $(j = 1, 2)$, the linearized system having the first integral,

$$p_1^2 + p_2^2 + q_1^2 + q_2^2 = C,$$

where C is a positive constant. The orbits in the vicinity of this center are almost periodic; others exhibit random behavior.

Poincaré plots are constructed for two sets of ICs. These plots show $p_1(t)$ versus $q_1(t)$ at the discrete times when $q_2(t) = 0$ and $p_2(t) > 0$.

(a) $p_1(0) = 1/3, p_2(0) = 0.1293144, q_1(0) = 1/4, q_2(0) = 1/5$. This datum gives rise to almost periodic behavior.

(b) $p_1(0) = 0.1, p_2(0) = 0.467618, q_1(0) = 0.1, q_2(0) = 0.1$. The solution is chaotic.

The system satisfies the Arnold–Moser theorem, i.e., trajectories originating sufficiently close to center remain close to it for all time. Bender and Orszag (1978), p.188

2. $\quad \dot{x}_1 = y_1, \quad \dot{x}_2 = y_2, \quad \dot{y}_1 = -x_1^2 + x_2^2, \quad \dot{y}_2 = 2x_1x_2.$

The "pathological" behavior of the given system is considered. It has the Hamiltonian $H(x, y) = (1/2) < y, y > +V(x)$, where $y = (y_1, y_2)$ and $x = (x_1, x_2)$ and $< y, y >$ is the inner product; the potential $V(x) = (1/3)x_1^3 - x_1x_2^2$. $H(0, 0) = 0$, and the origin is the only critical point where grad $H = (H_x, H_y) = 0$. The linearized equations about the origin have a coefficient matrix with all four eigenvalues equal to zero. Since $(d/dt)H(x(t), y(t)) = 0$, the Hamiltonian H is an integral of the differential equations implying that the three manifold $H(x, y) = h > 0$ consists of orbits of energy h above that of the degenerate critical point. It is easy to check that if $(x(t), y(t))$ is a solution, then so is $(x(-t), -y(-t))$. Periodic orbits and asymptotic sets are identified. Rod (1973)

3. $$\dot{x}_1 = -\mu_1 x_1 + w_1 x_2, \dot{x}_2 = -\mu_2 x_2 + w_2 x_1,$$

$$\dot{w}_1 = q_1 - \epsilon_1 w_1 - x_1 x_2, \dot{w}_2 = q_2 - \epsilon_2 w_2 - x_1 x_2,$$

where μ_i, q_i, ϵ_i $(i = 1, 2)$ are positive constants.

The given system describes two coupled dynamos. Special cases:

(a) $q_1 = q_2 = 1, \epsilon_1 = \epsilon_2 = 0$. It then follows that $\dot{w}_1 = \dot{w}_2$ so that we can write $w_1 = x_3 + \alpha, w_2 = x_3 - \alpha$, where α is a constant of motion. If we choose $\mu_1 = \mu_2 = \mu$, we have a simplified version of the given system:

$$\begin{aligned} \dot{x}_1 &= -\mu x_1 + (x_3 + \alpha) x_2, \\ \dot{x}_2 &= -\mu x_2 + (x_3 - \alpha) x_1, \\ \dot{x}_3 &= 1 - x_1 x_2. \end{aligned} \tag{1}$$

The system (1) has two equilibrium points, which we may write as $(\beta_1, \beta_2, \gamma)$ and $(-\beta_1, -\beta_2, \gamma)$ by defining

$$\beta_1 = \{(\gamma + \alpha)/\mu\}^{1/2}, \beta_2 = \{(\gamma - \alpha)/\mu\}^{1/2}, \gamma = (\alpha^2 + \mu^2)^{1/2}.$$

Since the model is unaltered by reversing the signs of x_1 and x_2, both singular points have the same dynamical properties. Linearizing around the one in the positive octant, we obtain matrix of linearized form as

$$\begin{bmatrix} -\mu & \gamma + \alpha & \beta_2 \\ \gamma - \alpha & -\mu & \beta_1 \\ -\beta_2 & -\beta_1 & 0 \end{bmatrix},$$

whose eigenvalue equation factorizes as $(s + 2\mu)(s^2 + 2\gamma/\mu) = 0$, so the linearized system is "marginally" stable, with a pair of pure imaginary eigenvalues. Numerical results show strong dependence on μ and α. For some parameter values, a limit cycle exists, though rather distorted and of highly nonplanar form. Projections of these limit cycles on (x_1, x_3) plane are shown graphically. For other values of parameters, the system develops the sort of behavior typical of a strange attractor and is also depicted graphically.

(b) $q_1 \neq q_2$ and $\epsilon_1 = \epsilon_2 = \epsilon > 0$. Now writing $w_1 = x_3 + \alpha(t), w_2 = x_3 - \alpha(t)$, we have $\dot{\alpha} = (q_1 - q_2)/2 - \epsilon\alpha$, so that $\alpha \to \alpha_0 = (q_1 - q_2)/(2\epsilon)$ as $t \to \infty$. In analyzing the equilibria, we assume that this limiting value has been attained, so that the state equations become

$$\begin{aligned} \dot{x}_1 &= -\mu_1 x_1 + (x_3 + \alpha_0) x_2, \\ \dot{x}_2 &= -\mu_2 x_2 + (x_3 - \alpha_0) x_1, \\ \dot{x}_3 &= q - x_1 x_2 - \epsilon x_3, \end{aligned}$$

where $q = (q_1 + q_2)/2$. There is then always a singular point at $(0, 0, q/\epsilon)$, and the characteristic equation of the corresponding linearized model is

$$(s + \epsilon)\{(s + \mu_1)(s + \mu_2) - q_1 q_2/\epsilon^2\} = 0,$$

on taking the values of q and α_0 into account. Also, if $|q| > \epsilon\gamma$, where $\gamma = \{\mu_1\mu_2 + \alpha_0^2\}^{1/2}$, there are two more equilibrium points. Taking q to be positive, for definiteness, these are located at $(\beta_1, \beta_2, \gamma)$ and $(-\beta_1, -\beta_2, \gamma)$, where $\beta_1 = \{(\gamma + \alpha_0)(q - \epsilon\gamma)/\mu_1\}^{1/2}$, $\beta_2 = \{(\gamma - \alpha_0)(q - \epsilon\gamma)/\mu_2\}^{1/2}$ with similar expressions in the case for which q is negative. Linearizing about the first of these points, which are evidently related by symmetry, we obtain the "linearized matrix"

$$A = \begin{bmatrix} -\mu_1 & \gamma+\alpha_0 & \beta_2 \\ \gamma-\alpha_0 & -\mu_2 & \beta_1 \\ -\beta_2 & -\beta_1 & -\epsilon \end{bmatrix},$$

for which the characterisitc equation $\det(sI - A) = 0$ becomes

$$s(s+\epsilon)(s+\mu_1+\mu_2) + (q-\epsilon\gamma)[\{(\gamma+\alpha_0)/\mu_1 + (\gamma-\alpha_0)/\mu_2\}s + 4\gamma] = 0.$$

Since the condition $|q| > \epsilon\gamma$ is equivalent to $q_1 q_2 > \epsilon^2 \mu_1 \mu_2$, the equilibria at $(\beta_1, \beta_2, \gamma)$ and $(-\beta_1, -\beta_2, \gamma)$ exist only when the one at $(0, 0, q/\epsilon)$ is unstable. This can happen if q_1 and q_2 have the same sign and ϵ is sufficiently small. Cook (1986), pp.42, 193

4.

$$\dot{x}_1 = x_2 + 4x_2x_4 + 4x_3^3, \dot{x}_2 = -x_1 - 4x_1x_4,$$

$$\dot{x}_3 = 2x_4 + 2(x_1^2 + x_2^2) + x_3^2, \dot{x}_4 = -2x_3 - 2x_3x_4.$$

It is proved that the given system has a one-parameter family of 2π-periodic solutions emanating from the origin. Sweet (1973)

5.

$$\dot{x}_1 = A - (B+1)x_1 + x_1^2 y_1 + D_1(x_2 - x_1),$$

$$\dot{y}_1 = B_1 x_1 - x_1^2 y_1 + D_2(y_2 - y_1),$$

$$\dot{x}_2 = A - (B+1)x_2 + x_2^2 y_2 - D_1(x_1 - x_2),$$

$$\dot{y}_2 = B x_2 - x_2^2 y_2 + D_2(y_1 - y_2),$$

where A, B, B_i, D_i are constants.

The given system describes the chemical reaction in two coupled stirred-cell reactors. Location of bifurcations, limit points, and primary and secondary Hopf bifurcations for $A = 2, D_1/D_2 = 0.1$ are shown in the B-D_1 plane. Numerical solutions show period doubling. The regions with period 3 are shown. Aperiodic or periodic regime can be reached for the same values of parameters if D_1 is increased or decreased, respectively. Schreiber, Kubicek, and Marek (1980)

6.

$$\dot{x}_1 = x_2, \dot{x}_2 = -(1+\epsilon\nu - 2\epsilon w_1)x_1 + \epsilon\nu x_3 + \epsilon\mu(1-x_1^2)x_2, \dot{x}_3 = x_4,$$

$$\dot{x}_4 = -(1+\epsilon\nu+\epsilon\eta - 2\epsilon w_1)x_1 + \epsilon\nu x_1 + \epsilon\mu(1-x_3^2)x_4,$$

where ϵ, μ, ν, η, and w_1 are parameters.

Periodic solutions with radii $r_1 = (x_1^2+x_2^2)^{1/2}$ and $r_2 = (x_3^2+x_4^2)^{1/2}$ in (x_1, x_2) and (x_3, x_4) planes, repectively, are determined, using "point mapping theory". Numerical results with $\epsilon\mu = 0.10, \epsilon\eta = 0.04$, and $\epsilon\nu = 0.1$ are shown graphically. The corresponding values of w_1 are obtained. Guttalu and Flashner (1989)

7.

$$\dot{x} = z, \dot{y} = w, \dot{z} = -\delta_1 x + \beta y + (\delta_2 x - \gamma y)(x^2+y^2) - c_1(z + a_0 w),$$

$$\dot{w} = -\beta x - \delta_1 y + (\gamma x + \delta_2 y)(x^2+y^2) + c_1(a_0 z - w),$$

where $\delta_1, \beta, \delta_2, \gamma, c_1$, and a_0 are constants.

The given four-dimensional Duffing system arises from the Ginzburg–Landau equation. It has the origin (0,0,0,0) for all values of coefficients, and the ring of fixed points

$$x^2 + y^2 = \beta/\gamma, \Delta \equiv \delta_1 - \delta_2\beta/\gamma = 0$$

for all c_1 as its singular points.

The linearization about zero in (x, y, z, w) space gives two eigenvalues satisfying $\lambda^2 + c_1(1 - ia_0)\lambda + \delta_1 + i\beta = 0$ with the other pair their complex conjugate. For $\delta_1 \neq 0, \beta \neq 0$, as other parameters are varied, the origin will be a double spiral point in general. Therefore, the only bifurcations that can occur from the origin will be of Hopf type, at which a pair of eigenvalues become pure imaginary and a branch of periodic solution is possibly shed.

The stability of bifurcating solutions is considered. A complicated structure of periodic solutions of the given system with parameters appropriate to Poiseuille flow are found numerically. Landman (1987)

8.
$$\dot{x}_1 = -[2\pi/\{1 + r^{1/2}\}]x_2, \dot{x}_2 = [2\pi/\{1 + r^{1/2}\}]x_1,$$
$$\dot{x}_3 = -[\pi/\{1 + r^{1/2}\}]x_4, \dot{x}_4 = [\pi/\{1 + r^{1/2}\}]x_3,$$

where $r = (x_1^2 + x_2^2 + x_3^2 + x_4^2)$.

The general solution of the given system is

$$\begin{aligned} x_1 &= a\cos[2\pi t/\{1 + (a^2 + b^2)^{1/2}\} + \theta_1], \\ x_2 &= a\sin[2\pi t/\{1 + (a^2 + b^2)^{1/2}\} + \theta_1], \\ x_3 &= b\cos[\pi t/\{1 + (a^2 + b^2)^{1/2}\} + \theta_2], \\ x_4 &= b\sin[\pi t/\{1 + (a^2 + b^2)^{1/2}\} + \theta_2], \end{aligned}$$

where a, b, θ_1, and θ_2 are arbitrary constants. Urabe (1967), p.102

9.
$$\dot{x} - X(x, y, z, v) = 0, \dot{y} - Y(x, y, z, v) = 0, \dot{z} - A_1 z + Z(x, y, z, v) = 0,$$
$$\dot{v} - A_2 v + V(x, y, z, v) = 0,$$

where x and X are scalars, y and Y, z and Z, and v and V are column vectors of dimensions n, k, and l; X, Y, Z, and V are holomorphic in the neighborhood of $x = 0, y = 0, z = 0, v = 0$, and their expansions in powers of x, y, z, and v do not contain powers of degree lower than the second. Furthermore,

$$\begin{aligned} X(x, y, 0, 0) &= x^m f(x, y), Y(x, y, 0, 0) = x^m \xi(x, y), \\ Z(x, y, 0, 0) &= x^m \eta(x, y) \text{ and } V(x, y, 0, 0) = x^m \mu(x, y), \end{aligned}$$

where $f(x, y), \xi(x, y), \eta(x, y)$, and $\mu(x, y)$ are holomorphic in the neighborhood of $x = 0, y = 0, f(0,0) = g \neq 0, \eta(0,0) = 0, \mu(0,0) = 0$, and $m \geq 2$ is an integer. It is assumed that the constant matrix A_1 has eigenvalues $\lambda_1, \ldots, \lambda_k$ with negative real parts, while the eigenvalues $\mu_1, \mu_2, \ldots, \mu_\ell$ of A_2 do not vanish and have nonnegative real parts.

Using Lyapunov's method an $(n + k + 1)$-parameter family of solutions of the given system is constructed in the neighborhood of $x = 0, y = 0, z = 0, v = 0$ defined for all $t \geq t_0$. Grudo (1973)

4.9 $y_1'' = f_1(x, \vec{y}), y_2'' = f_2(x, \vec{y})$

1.
$$y'' = y + \alpha z^2, \quad z'' = z + \beta y^2;$$

α, β are constants.

It is shown that if $\zeta = \begin{pmatrix} y \\ z \end{pmatrix}$ is the solution of the given system, then $|\zeta(x)| \leq \max(|\zeta_0|, |\zeta_1|)$, where $\zeta_0 = \zeta(0), \zeta_1 = \zeta(1)$. Knobloch and Schmitt (1977)

2. $$y_1'' = -y_1 - 2by_1y_2, \quad y_2'' = -y_2 - by_1^2 - 3cy_2^2,$$

where b and c are constants.

The given dynamical system is shown to have two first integrals, the Hamiltonian

$$y_1'^2 + y_1^2 + y_2'^2 + y_2^2 + 2b[y_1^2 y_2 + (1/3)y_2^3] = C_1$$

and

$$y_1'y_2' + y_1y_2 + b[y_2^2 y_1 + (1/3)y_1^3] = C_2,$$

provided that $b = 3c$. An intituitive argument is used. Here C_1 and C_2 are arbitrary constants. Sarlet and Bahar (1980)

3. $$y'' - y - \alpha z^2 = 0, \quad z'' - z - \beta y^2 - \gamma x = 0,$$

where α, β, and γ are constants.

The following result is proved: Let $\gamma_0 > 0$ be the radius of the largest disk $\{(y, z) : y^2 + z^2 \leq \gamma_0^2\}$, in which the matrix

$$\begin{bmatrix} 2 & 2(\alpha z + \beta y) & 0 \\ 2(\alpha z + \beta y) & 2 & 1 \\ 0 & 1 & (\pi/2)^2 \end{bmatrix} \geq 0.$$

[Note that for $(y, z) = (0, 0)$, the matrix is positive.] Then for every $\gamma, |\gamma| \leq \gamma_0$, the given system has a solution (y_γ, z_γ) such that

$$\begin{aligned} y_\gamma(0) &= z_\gamma(0) = y_\gamma(1) = z_\gamma(1) = 0, \\ y_\gamma^2(x) &+ z_\gamma^2(x) + [x^2/\sin^2(\pi/2x)]\gamma^2 \leq \gamma^2, 0 \leq x \leq 1. \end{aligned}$$

Knobloch and Schmitt (1977)

4. $$y'' = -w_1^2 y - 2\epsilon yz, \quad z'' = -w_2^2 z + \epsilon(z^2 - y^2),$$

where w_1, w_2, and ϵ are constants.

The given system describes coupled nonlinear oscillators. A straightforward perturbation analysis is used to treat higher order internal resonances. The approximate invariant known to exist for the lowest order internal resonances continues to hold for the higher order internal resonances. These results are used to find a possible relation between energy sharing and the onset of ergodicity for nonintegrable systems. Shivamoggi and Varma (1988)

5. $$\ddot{x} - Ax + 2xy = 0, \quad \ddot{y} - By + gy^2 - x^2 = 0,$$

where A, B, and g are constants.

For $g = 6$, the given system has two first integrals,

$$\begin{aligned} H &= (1/2)(\dot{q}_1^2 + \dot{q}_2^2 + Aq_1^2 + Bq_2^2) - q_1^2 q_2 - 2q_2^3, \\ F &= q_1^4 + 4q_1^2 q_2^2 + 4\dot{q}_1(\dot{q}_1 q_2 - \dot{q}_2 q_1) - 4Aq_1^2 q_2 + (4A - B)(\dot{q}_1^2 + Aq_1^2), \end{aligned} \tag{1}$$

where $q_1 = x/i, q_2 = y$, and the dot in (1) stands for derivative with respect to $\tau = t/i$.

The system (1) has a Lax representation in terms of 2×2 matrices, which allows the system to be explicitly integrated in terms of quasiperiodical solutions, which include elliptic functions as special cases. Christiansen, Eilbeck, Enolskii, and Gaididei (1992)

6. $$y_1'' = -w_1^2 y_1 + \alpha_1 y_1 y_2, \quad y_2'' = -w_2^2 y_2 + \alpha_2 y_1^2,$$

where $w_2 \simeq 2w_1$; w_i, α_i $(i = 1, 2)$ are constants.

The amplitude is considered to be small but finite. The approximate solution is found in the form

$$y_1 = a_1 \cos(w_1 x + \beta_1), y_2 = a_2 \cos(w_2 x + \beta_2),$$

where

$$\begin{aligned}
\frac{da_1}{dx} &= \{\alpha_1/(4w_1)\} a_1 a_2 \sin[(w_2 - 2w_1)x + \beta_2 - 2\beta_1], \\
a_1 \frac{d\beta_1}{dx} &= -\{\alpha_1/(4w_1)\} a_1 a_2 \cos[(w_2 - 2w_1)x + \beta_2 - 2\beta_1], \\
\frac{da_2}{dx} &= -\{\alpha_2/(4w_2)\} a_1^2 \sin[(w_2 - 2w_1)x + \beta_2 - 2\beta_1], \\
a_2 \frac{d\beta_2}{dx} &= -\{\alpha_2/(4w_2)\} a_1^2 \cos[(w_2 - 2w_1)x + \beta_2 - 2\beta_1].
\end{aligned}$$

Nayfeh (1985), p.198

7. $$\ddot{x} = -Ax - 2Dxy, \quad \ddot{y} = -By - Dx^2 + Cy^2,$$

where A, B, C, and D are constants. This is a generalized Hennon–Heiles system. By Lie group analysis, or by a direct method, the following invariants are found.

(a) Multiplying the first of the system by x and second by y, adding, and integrating, we have

$$\dot{x}^2/2 + \dot{y}^2/2 + Ax^2/2 + By^2/2 + Dx^2y - Cy^3/3 = E, \text{ say,}$$

where E denotes the energy invariant.

(b) If $B = A, C = -D$, then one may check that another invariant is

$$\dot{x}\dot{y} + Axy + Dx^3/3 + Dxy^2 = G_1, \text{ say.}$$

(c) If $C = -6D$ and $B = 4A$, one may find the invariant

$$\dot{x}(\dot{y}x - \dot{x}y) + Dx^4/4 + Dx^2y^2 + Ax^2y = G_2, \text{ say.}$$

(d) If $C = -16D, B/A = 16$, we may find that

$$\dot{x}^4/4 + (Ax^2/2 + Dx^2y)\dot{x}^2 - Dx^3\dot{x}\dot{y}/3 + A^2x^4/4 - ADx^4y/3 - D^2x^6/18 - D^2x^4y^2/3 = G_3, \text{ say,}$$

is an invariant.

Various methods—Painlevé, Lie group symmetries, and the direct approach—are detailed with reference to the given system. Abraham–Schrauner (1990)

8. $$\ddot{y}_1 + c_1 y_1 - b y_1^2 + a y_2^2 = 0, \quad \ddot{y}_2 + c_2 y_2 + 2m y_1 y_2 = 0,$$

where c_1, c_2, b, a, and m are constants.

For this generalized Henon–Heiles system, the following first integrals are found; here $v_1 = \dot{y}_1$ and $v_2 = \dot{y}_2$:

(a) $b = -6m (m \neq 0)$

$$\begin{aligned} F_1 &= (1/2)mv_1^2 + (1/2)av_2^2 + (1/2)c_1my_1^2 + (1/2)c_2ay_2^2 \\ &\quad + amy_1y_2^2 + 2m^2y_1^3, \\ F_2 &= y_2v_1v_2 - y_1v_2^2 + \{(4c_2 - c_1)/(4m)\}v_2^2 + c_2y_1y_2^2 + my_1^2y_2^2 \\ &\quad + \{c_2/(4m)\}(4c_2 - c_1)y_2^2 + (a/4)y_2^4. \end{aligned}$$

(b) $b = -m, c_2 = c_1 (m \neq 0)$

$$\begin{aligned} F_1 &= (1/2)mv_1^2 + (1/2)av_2^2 + (1/2)c_1my_1^2 + (1/2)c_1ay_2^2 \\ &\quad + amy_1y_2^2 + (1/3)m^2y_1^3, \\ F_2 &= v_1v_2 + c_1y_1y_2 + my_1^2y_2 + (1/3)ay_2^3. \end{aligned}$$

(c) $a = 0, b = -2m/5, c_2 = 4c_1 (m \neq 0)$.

$$\begin{aligned} F_1 &= (1/2)v_1^2 + (1/2)c_1y_1^2 + (2/15)my_1^3, \\ F_2 &= y_1v_1v_2 - y_2v_1^2 + c_1y_1^2y_2 + (2/5)my_2y_1^3. \end{aligned}$$

(d) $a = 0, b = -2m (m \neq 0)$.

$$\begin{aligned} F_1 &= (1/2)v_1^2 + (1/2)c_1y_1^2 + (2/3)my_1^3, \\ F_2 &= my_2^2v_1^2 - 2my_1y_2v_1v_2 + my_1^2v_2^2 + (c_1 - c_2)y_1v_2^2 - (c_1 - c_2)y_2v_1v_2 \\ &\quad + (1/4m)(c_1 - c_2)(c_1 - 4c_2)v_2^2 - c_2(c_1 - c_2)y_1y_2^2 \\ &\quad + (c_2/4m)(c_1 - c_2)(c_1 - 4c_2)y_2^2 \end{aligned}$$

(e) $m = 0, b = 0$.

$$\begin{aligned} F_1 &= (1/2)v_2^2 + (1/2)c_2y_2^2, \\ F_2 &= (1/2)(c_1 - 4c_2)v_1^2 + 2ay_2v_1v_2 - 2ay_1v_2^2 + (1/2)c_1(c_1 - 4c_2)y_1^2 \\ &\quad + a(c_1 - 2c_2)y_1y_2^2 + (1/2)a^2y_2^4. \end{aligned}$$

(f) $b = -16m, c_1 = 16c_2 (m \neq 0)$

$$\begin{aligned} F_1 &= (1/2)mv_1^2 + (1/2)av_2^2 + 8c_2my_1^2 + (1/2)c_2ay_2^2 \\ &\quad + amy_1y_2^2 + (16/3)m^2y_1^3, \\ F_2 &= (1/4)v_2^4 + my_1y_2^2v_2^2 + (1/2)c_2y_2^2v_2^2 - (1/3)my_2^3v_1v_2 \\ &\quad + (1/4)c_2^2y_2^4 - (1/3)mc_2y_1y_2^4 - (1/3)m^2y_1^2y_2^4 - (1/18)amy_2^6. \end{aligned}$$

The integrability of the given system, in general, is also discussed. Sarlet (1991)

9.

$$x^2y'' = zy + px, \quad x^2z'' = -(1/2)y^2,$$

$y(0) = 0, z(0) = 0, y(1) = [\lambda/(\lambda - 1)]y'(1); z(1) = [\mu/(\mu - 1)]z'(1)$, where p and μ are constants.

The given system is reduced to two integral equations, using Green's functions:

$$\begin{aligned} y(x) &= -\int_0^1 K(x,\xi)(1/\xi^2)yz\,d\xi - \int_0^1 K(x,\xi)(p/\xi)d\xi, \\ z(x) &= (1/2)\int_0^1 G(x,\xi)(1/\xi^2)y^2(\xi)d\xi, \end{aligned}$$

where the kernel functions $K(x,\xi)$ and $G(x,\xi)$ are given by

$$\begin{aligned} K(x,\xi) &= \begin{cases} [(\lambda-1)\xi+1]x, & x \le \xi \\ [(\lambda-1)x+1]\xi, & x > \xi, \end{cases} \\ G(x,\xi) &= \begin{cases} [(\mu-1)\xi+1]x, & x \le \xi \\ [(\mu-1)x+1]\xi, & x > \xi. \end{cases} \end{aligned}$$

Using an iterative scheme,

$$\begin{aligned} y_{n+1}(x) &= -\int_0^1 K(x,\xi)(1/\xi^2)y_n(\xi)z_n(\xi)d\xi + y_1(x), \\ z_n(x) &= (1/2)\int_0^1 G(x,\xi)(1/\xi^2)y_n^2(\xi)d\xi, n = 1, 2, \ldots, \\ y_1(x) &= -\int_0^1 K(x,\xi)(p/\xi)d\xi = px\ln x - \lambda px, \end{aligned}$$

the solution is found in the form

$$\begin{aligned} y(x) &= \sum_{i=1}^{\infty}\sum_{j=0}^{i} A_{ij}x^i(\ln x)^j, \\ z(x) &= \sum_{i=1}^{\infty}\sum_{j=0}^{i} B_{ij}x^i(\ln x)^j, \end{aligned}$$

where A_{ij} and B_{ij} are constants. This solution is shown to be uniformly convergent and continuous in [0,1]. Several other properties of the solution are discussed. Numerical results are provided. Kai–yuan, Xiao–Jing, and Xin–Zhi (1990)

10. $$\boldsymbol{y_1'' = -w_1^2 y_1 - \epsilon(\delta_1 y_1^2 + 2\delta_2 y_1 y_2 + \delta_3 y_2^2) + 2F_1 \cos \Omega x,}$$

$$\boldsymbol{y_2'' = -w_2^2 y_2 - \epsilon(\delta_2 y_1^2 + 2\delta_3 y_1 y_2 + \delta_4 y_2^2),}$$

where $w_2 = 2w_1 + \epsilon\sigma_1$ and $\Omega = w_2 + w_1 + \epsilon\sigma_2$ and $w_i, \delta_i, F_i, \sigma_i, \Omega$, and $\epsilon \ll 1$ are constants.

A first order approximate solution is found to be

$$\begin{aligned} y_1 &= a_1\cos(w_1 x + \beta_1) + 2\Lambda_1\cos\Omega x + \cdots, \\ y_2 &= a_2\cos(w_2 x + \beta_2) + \cdots, \end{aligned}$$

where a_n and β_n $(n = 1, 2)$ are given by

$$\begin{aligned} \frac{da_1}{dx} &= -\{\epsilon\delta_2/(2w_1)\}a_1 a_2 \sin\gamma_1 - \{\epsilon\delta_2\Lambda_1/w_1\}a_2\sin\gamma_2, \\ a_1\frac{d\beta_1}{dx} &= \{\epsilon\delta_2/(2w_1)\}a_1 a_2\cos\gamma_1 + (\epsilon\delta_2\Lambda_1/w_1)a_2\cos\gamma_2, \end{aligned}$$

$$\frac{da_2}{dx} = \{\epsilon\delta_2/(4w_2)\}a_1^2 \sin\gamma_1 - (\epsilon\delta_2\Lambda_1/w_2)a_1 \sin\gamma_2,$$

$$a_2\frac{d\beta_2}{dx} = \{\epsilon\delta_2/(4w_2)\}a_1^2 \cos\gamma_1 + (\epsilon\delta_2\Lambda_1/w_2)a_1 \cos\gamma_2,$$

with $\gamma_1 = \beta_2 - 2\beta_1 + \epsilon\sigma_1 x, \gamma_2 = \epsilon\sigma_2 x - \beta_1 - \beta_2, \Lambda_1 = F_1/(w_1^2 - \Omega^2)$. Nayfeh (1985), p.214

11.
$$\boldsymbol{y_1'' = -w_1^2 y_1 + \epsilon\alpha_1 y_1 y_2 + \epsilon K_1 \cos \Omega_1 x,}$$

$$\boldsymbol{y_2'' = -w_2^2 y_2 + \epsilon\alpha_2 y_1^2 + \epsilon K_2 \cos \Omega_2 x,}$$

where:

(a) $w_2 \simeq 2w_1$ and $\beta_1 \simeq w_1$.

(b) $w_2 \simeq 2w_1$ and $\beta_2 \simeq w_2$.

The solution is sought in the form

$$y_1 = a_1 \cos(w_1 x + \beta_1), \quad y_2 = a_2 \cos(w_2 x + \beta_2),$$

where, for case (a),

$$\frac{da_1}{dx} = \{\epsilon\alpha_1/(4w_1)\}a_1 a_2 \sin\gamma_1 + \{\epsilon k_1/(2w_1)\} \sin\gamma_2,$$

$$a_1\frac{d\beta_1}{dx} = -\{\epsilon\alpha_1/(4w_1)\}a_1 a_2 \cos\gamma_1 - \{\epsilon k_1/(2w_1)\} \cos\gamma_2,$$

$$\frac{da_2}{dx} = -\{\epsilon\alpha_2/(4w_2)\}a_1^2 \sin\gamma_1,$$

$$a_2\frac{d\beta_2}{dx} = -\{\epsilon\alpha_2/(4w_2)\}a_1^2 \cos\gamma_1,$$

where $\gamma_1 = \epsilon\sigma_1 x + \beta_2 - 2\beta_1, \gamma_2 = \epsilon\sigma_2 x - \beta_1, w_2 = 2w_1 + \epsilon\sigma_1$ and $\Omega_1 = w_1 + \epsilon\sigma_2$; and, for case (b),

$$\frac{da_1}{dx} = \{\epsilon\alpha_1/(4w_1)\}a_1 a_2 \sin\gamma_1,$$

$$a_1\frac{d\beta_1}{dx} = -\{\epsilon\alpha_1/(4w_1)\}a_1 a_2 \cos\gamma_1,$$

$$\frac{da_2}{dx} = -\{\epsilon\alpha_2/(4w_2)\}a_1^2 \sin\gamma_1 + \{\epsilon K_2/(2w_2)\} \sin\gamma_2,$$

$$a_2\frac{d\beta_2}{dx} = -\{\epsilon\alpha_2/(4w_2)\}a_1^2 \cos\gamma_1 - \{\epsilon K_2/(2w_2)\} \cos\gamma_2,$$

where $\gamma_1 = \epsilon\sigma_1 x + \beta_2 - 2\beta_1, \gamma_2 = \epsilon\sigma_2 x - \beta_2, w_2 = 2w_1 + \epsilon\sigma_1$ and $\Omega_2 = w_2 + \epsilon\sigma_2$. Nayfeh (1985), p.207

12. $\boldsymbol{\epsilon y'' = y + \delta z, \epsilon z'' = 3y + g(z), y(0) = y(1) = 0, z(0) = 0, z(1) = b > 0,}$

where $g(z) = z - z^3$; δ and ϵ (small) are parameters.

It is proved that the given BVP has a solution with a boundary layer at $x = 1$, provided that $b < 2^{1/2}, |\delta| < 1/3$. Kelley (1984)

13. $$\epsilon y'' - (y - z)(y^2 - 1) = 0, y(0) = \alpha_0, y(1) = \alpha_1,$$
$$z'' + ay = 0, z(0) = \beta_0, z(1) = \beta_1.$$

A solution of the given problem with six transition layers and two boundary layers is found, using some general theorems; it is shown graphically. Depending on the parameters, the system shows a large number of transition layers. Fife (1976)

14. $$\ddot{x} = -x - kx^3 - (x - y)^3, \quad \ddot{y} = -y - ky^3 + (x - y)^3,$$

where k is a constant.

The given system possesses a first integral $(\dot{x}^2 + \dot{y}^2)/2 + V(x, y) =$ constant, where $V = (x^2 + y^2)/2 + (k/4)\ (x^4 + y^4) + (1/4)(x - y)^4$.

The stability of the normal mode $y = x$ is analyzed. Month and Rand (1980)

15. $$\epsilon y'' = y + \delta z(z - 2), y(0) = y(1) = 0,$$
$$\epsilon z'' = y + f(x, z), z(0) = 0, z(1) = 2,$$

where $f(x, z) = 2z(z + x - 3/2)(z - 2)$ and $\delta > 0$.

Existence of solution with an interior layer for small $\epsilon > 0$ is established. The location of the layer and behavior of the solution in the layer are also discussed. Kelley (1984)

16. $$y_1'' = -(\alpha_1/2)y_1 - 2\Gamma_1 y_1^3 - 2y_1 y_2^2,$$
$$y_2'' = -(\alpha_2/2)y_2 - 2\Gamma_2 y_2^3 - 2y_2 y_1^2,$$

where α_i, Γ_i $(i = 1, 2)$ are constants.

The given system arises from the Hamiltonian

$$H = (1/2)\sum_{i=1}^{2}\left\{p_i^2 + (1/2)\alpha_i y_i^2 + \Gamma_i y_i^4\right\} + y_1^2 y_2^2,$$

where $p_i = y_i'$.

The given system has Painlevé property for the following sets:

(a) $\Gamma_1 = \Gamma_2 = 1, \alpha_1, \alpha_2$ arbitrary.
(b) $\Gamma_1 = \Gamma_2 = 1/3, \alpha_1 = \alpha_2$.
(c) $\Gamma_1 = 1/3, \Gamma_2 = 8/3, \alpha_2 = 4\alpha_1$.
(d) $\Gamma_1 = 1/6, \Gamma_2 = 8/3, \alpha_2 = 4\alpha_1$.

The second integrals corresponding to each of the cases are as follows:

$$\begin{aligned} I_1 &= (1/2)(\alpha_1 - \alpha_2)(y_2^4 + y_1^2 y_2^2 + (1/2)\alpha_2 y_2^2 + p_2^2) - (p_1 y_2 - p_2 y_1)^2, \\ I_2 &= p_1 p_2 + y_1 y_2[(1/2)\alpha_1 + (2/3)(y_1^2 + y_2^2)], \\ I_3 &= p_1^4 + 4\{(1/4)\alpha_1 + (1/6)y_1^2 + y_2^2\}y_1^2 p_1^2 - (8/3)y_1^3 y_2 p_1 p_2 \\ &\quad + (2/3)y_1^4 p_2^2 + \alpha_1\{(1/4)\alpha_1 + (1/3)y_1^2 + (2/3)y_2^2\}y_1^4 + (1/9)(y_1^2 + 2y_2^2)^2 y_1^4, \\ I_4 &= p_1(p_2 y_1 - p_1 y_2) + (2/3)y_1^2 y_2^3 + (1/3)y_1^4 y_2 + (1/2)\alpha_1 y_1^2 y_2. \end{aligned}$$

The numerical solution of the given system for α_i and π_i different from (1)–(4) above shows transition from totally regular motion at low energies to irregular motion at higher energies. Poincaré sections are shown for

(a) $\alpha_1 = 1/10, \alpha_2 = 1/10, \Gamma_1 = 3/10, \Gamma_2 = 1/10$, and $E = 1.0 \times 10^{-3}$.

(b) $\alpha_1 = 1/10, \alpha_2 = 1/10, \Gamma_1 = 3/10, \Gamma_2 = 1/10$, and $E = 1.0 \times 10^{-2}$.

(c) $\alpha_1 = 1/10, \alpha_2 = 1.10, \Gamma_1 = 3/10, \Gamma_2 = 1/10$, and $E = 1.8 \times 10^{-2}$.

Baumann (1991)

17.

$$y'' = -w_2^2 y + w_1^2 z - \alpha_1 y^3 - \alpha(y-z)^3,$$

$$z'' = -w_2^2 z + w_1^2 y - \alpha_1 z^3 - \alpha(z-y)^3,$$

where α_1, α, w_1, and w_2 are constants.

The solution is found in the form

$$y = A\cos\ wx, \quad z = B\cos(wx+\beta).$$

By substitution of this form, three modes of vibrations $A = B, A = -B$, and $A \neq B$ or $A \neq -B$ are found and interpreted. Anand (1972)

18.

$$y_1'' = -w_1^2 y_1 + \epsilon(\alpha_1 y_1^2 + \alpha_2 y_1 y_2 + \alpha_3 y_2^2),$$

$$y_2'' = -w_2^2 y_2 + \epsilon(\alpha_4 y_1^2 + \alpha_5 y_1 y_2 + \alpha_6 y_2^3),$$

where $w_2 = 2w_1 + \epsilon\sigma$.

A first order approximate solution is

$$y_n = a_n \cos(w_n x + \beta_n) + \cdots, n = 1, 2,$$

where a_n and β_n are given by

$$\begin{aligned}
\frac{da_1}{dx} &= \{\epsilon\alpha_2/(4w_1)\}a_1 a_2 \sin\gamma, \\
a_1\frac{d\beta_1}{dx} &= -\{\epsilon\alpha_2/(4w_1)\}a_1 a_2 \cos\gamma_1, \\
\frac{da_2}{dx} &= -\{\epsilon\alpha_4/(4w_2)\}a_1^2 \sin\gamma, \\
a_2\frac{d\beta_2}{dx} &= -\{\epsilon\alpha_4/(4w_2)\}a_1^2 \cos\gamma,
\end{aligned}$$

where $\gamma = \beta_2 - 2\beta_1 + \epsilon\sigma x$. Nayfeh (1985), p.212

19.

$$\ddot{x} + \sigma_1^2 x - \epsilon\alpha x - \epsilon A(\cos\ t)x - \epsilon\beta x^3 - \epsilon\gamma x y^2 = 0,$$

$$\ddot{y} + \sigma_2^2 x - \epsilon\delta y - \epsilon B(\cos\ wt)y - \epsilon\mu y^3 - \epsilon\nu x^2 y = 0,$$

where $\epsilon > 0$ is a small parameter, and $\alpha, \beta, \gamma, \delta, \mu, \nu, A, B, \sigma_1$, and σ_2 are real constants and w is a rational number.

It is proved that for $A = B = 0$, and some values of $\alpha, \beta, \gamma, \delta, \mu$, and ν, there are four periodic solutions of the given system, each having the same amplitude as for the case $\epsilon = 0$. Cesari and Hale (1957)

20. $$\ddot{x} = a(t)y^\alpha, \quad \ddot{y} = b(t)t^\beta,$$

where $a, b : R \to R_+$ are continuous functions and α and β are natural numbers. It is shown that under certain conditions on a, b, the solutions of the given system are bounded and $\lim_{t\to\infty} x(t) = 0$, $\lim_{t\to\infty} y(t) = 0$. The main analytic tool is Hölder inequality. Foltynska (1989)

21. $$\ddot{x} - kx/(x^2+y^2)^2 = 0,$$
$$\ddot{y} - ky/(x^2+y^2)^2 = 0,$$

where k is a constant.

Suppose that $x^2 + y^2 \neq 0$; then $(d/dt)[\dot{x}^2 + \dot{y}^2] = (d/dt)\{-k/(x^2+y^2)\}$, so $\dot{x}^2 + \dot{y}^2 = k_1 - k/(x^2+y^2), (d/dt)\{x\dot{x} + y\dot{y}\} = k_1$. From these, it follows that $x\dot{x} + y\dot{y} = k_1 t + k_2, (d/dt)\{x^2+y^2\} = 2k_1 t + 2k_2$ so that $x^2+y^2 = k_1 t^2 + 2k_2 t + k_3$. If the $\lim_{t\to\infty}(x^2+y^2) \neq 0$, then $k_1 = k_2 = 0$ and $x^2 + y^2 = k_3$. If k_1 or $k_2 \neq 0$, x^2+y^2 may become zero for some value of t, but eventually goes to infinity. Laplaza (1970)

22. $$y'' + 1 - y - f(y,z) = 0,$$
$$z'' - (1/d)[kz - f(y,z)] = 0,$$

where $f(y,z) = yz^2/(1+z+z^2)$, and d and k are constants.

Stability analysis and periodic solution of the given system are studied numerically, as well as by the use of some standard theorems. Kazarinoff and Yan (1991)

23. $$\ddot{x} = -Ax + 2k/(x-y)^3, \quad \ddot{y} = -By - 2k/(x-y)^3,$$

where A, B, and k are small positive constants.

When $A = B$, the given system uncouples under the transformation

$$z = x - y, w = x + y \tag{1}$$

into

$$\ddot{w} = -Aw, \quad \ddot{z} = -Az + 4k/z^3 \tag{2}$$

and hence yields the integral

$$(1/2)\dot{z}^2 + (1/2)Az^2 + 2k/z^2 = F = \text{constant}. \tag{3}$$

This is in addition to the Hamiltonian

$$V(x,y) = k/(x-y)^2 + (1/2)(Ax^2 + By^2).$$

It is not difficult to integrate (3): $z^2 = \{1/A^{1/2}\}[F/A^{1/2} + (F^2/A - 4k)^{1/2}\sin\{2A^{1/2}(t-t_0)\}]$, where t_0 is the second arbitrary constant.

The local two-sheeeted solution can be obtained by direct singularity analysis for general A and B:

$$x = \alpha + C_1\tau^{1/2} + \cdots, \quad \dot{y} = \alpha + C_2\tau^{1/2} + \cdots, \tau = t - t_*, \tag{4}$$

where α is a free constant and $C_1 = -C_2 = (-k)^{1/4}$.

Further Painlevé analysis shows that the more complete series form of (4) is

$$\begin{aligned} x &= \alpha + (-k)\tau^{1/2} + d\tau + f\tau^{3/2} + \sum_{n=4}^{\infty} a_n \tau^{n/2}, \\ y &= \alpha - (-k)\tau^{1/2} + d\tau - f\tau^{3/2} + \sum_{n=4}^{\infty} a_n \tau^{n/2}, \end{aligned}$$

where $t_*, a, d,$ and f provide the complete set of the free constants to be specified by the initial conditions of the problem.

Location and arrangement of the singularities in the complex t-plane is studied. Bountis, Drossos, and Percival (1991)

24.

$$\ddot{x} = -Ax + 2kx/(x^2 + y^2 - 1)^3,$$

$$\ddot{y} = -By + 2ky/(x^2 + y^2 - 1)^3,$$

where $A, B,$ and k are all positive constants.

For $A = B$, we introduce the polar coordinates $x = r\cos\theta, y = r\sin\theta$. The Hamiltonian of the given system in this case is

$$H = (1/2)(\dot{x}^2 + \dot{y}^2) + (A/2)(x^2 + y^2) + k/(x^2 + y^2 - 1)^2. \tag{1}$$

Introducing the polar coordinates $x = r\cos\theta, y = r\sin\theta$, we get

$$\frac{1}{2}\left(\frac{dr}{dt}\right)^2 + (A/2)r^2 + J^2/(2r^2) + k/(r^2 - 1)^2 = E, \tag{2}$$

where $J = r^2\dot{\theta}$ = constant is the angular momentum integral and E is the total energy. Introducing $u = r^2 - 1$ and after some manipulation, we have

$$\begin{aligned} t - t_0 &= \pm(1/2)\int_{u_0}^{u} \{u/[2E(u+1)u^2 - A(u+1)^2u^2 - J^2u^2 - 2k(u+1)]^{1/2}\}du \\ &= \pm[i/\{2A^{1/2}\}]\int_{u_0}^{u} \{u/[u^4 + \alpha u^3 + \beta u^2 + \gamma u + \delta]^{1/2}\}du, \end{aligned} \tag{3}$$

with $\alpha = 2 - 2E/A, \beta = 1 - 2E/A + J^2/A, \gamma = 2k/A, \delta = 2k/A$.

The integral in (3) can now be expressed in terms of elliptic integrals of the first and third kind:

$$\begin{aligned} F(\phi, \lambda) &= \int_0^{\sin\phi} [1/\{(1 - x^2)(1 - \lambda^2x^2)\}^{1/2}]dx, \\ \pi(\phi, \nu, \lambda) &= \int_0^{\sin\phi} [1/\{(1 - \nu x^2)\ \{(1 - x^2)(1 - \lambda^2x^2)\}^{1/2}\}]dx, \end{aligned}$$

depending on the location of its upper and lower limit, with respect to the roots of the polynomial in the denominator.

For the special case for which the radical in (3) is a perfect square, we may write

$$\begin{aligned} t - t_0 &= \pm\{i/(2A^{1/2})\}\int_{u_0}^{u} [u/\{(u - \rho_1)(u - \rho_2)\}]du \\ &= \pm\{i/(2A^{1/2})\}[1/(\rho_1 - \rho_2)]\ln[(u - \rho_1)^{\rho_1}/(u - \rho_2)^{\rho_2}], \end{aligned}$$

where $\pm(2kA)^{1/2}\rho_2^2 + k\rho_2 + 2k = 0, \rho_1 = \pm(2k/A)^{1/2}/\rho_2, E = A(1 + \rho_1 + \rho_2), J = \pm[2E - A(1 - \rho_1^2 - \rho_2^2 - 4\rho_1\rho_2)]^{1/2}$ Bountis, Drossos, and Percival (1991)

25.

$$\begin{aligned} \ddot{x} &= -x/r^3, \\ \ddot{y} &= -y/r^3, \\ x(0) &= a(1-e), \ \dot{x}(0) = 0, \\ y(0) &= 0, \dot{y}(0) = [(1+e)/\{a(1-e)\}]^{1/2}, \end{aligned}$$

where $r^2 = x^2 + y^2$.

The given system describes an orbital motion with a as the semimajor axis and e the eccentricity of the orbit. Choosing $a = 1.0$ and $0 < e < 0.9$, the given system was solved numerically, over 100 orbits, using nine different methods; their efficiency was compared. Fox (1984)

26.

$$\ddot{x} = -w^2(t)x + (1/x^3)\{\beta - \alpha(x/y)^4\},$$

$$\ddot{y} = -w^2(t)y + (1/y^3)\{\delta - \gamma(y/x)^4\},$$

where β, δ, α, and γ are constants.

The given system is referred to as a Pinney coupled system. It is a special case of Eqn. (4.9.32). Athorne (1991)

27.

$$y_1'' + y_1/\{y_1^2 + y_2^2\}^{3/2} = 0,$$

$$y_2'' + y_2/\{y_1^2 + y_2^2\}^{3/2} = 0.$$

The given system describes nearly planar motion of the planets about the sun. It has two first integrals, $(1/2)\{y_1'^2 + y_2'^2\} - 1/\{y_1^2 + y_2^2\}^{1/2} =$ constant, $y_1y_2' - y_2y_1' =$ constant, giving conservation of energy and angular momentum, respectively. Daniel and Moore (1970), p.12

28.

$$y'' - \lambda \sin z = 0, \quad z'' - \lambda y \cos z = 0,$$

with two sets of BC:

(a) $y'(0) = z(0) = y(1) = z'(1) = 0$.

(b) $y'(0) = z'(0) = y(1) = z(1) = 0$.

These eigenvalue problems arise in the study of the equilibrium states of a thin rotating rod. A constructive iterative method is developed for the positive solution of this fourth order system. Uniqueness is also discussed. Parter (1970)

29.

$$\ddot{y} = ryte^{-y}e^{-z}, \quad \ddot{z} = -qryze^{-y}e^{-z} + z^2,$$

$y(0) = 1$, and $\dot{y} \to 0$ as $t \to \infty, z \to 0$ as $t \to \pm\infty$, where r and q are constants.

It is proved that for any $r > 0$ fixed, there exists a solution $(y, z) \in (C^\infty(R))^2$ to the given problem such that $y > 0, \dot{y} < 0, y \to \infty$ as $t \to -\infty$, and $y \to 0$ as $t \to +\infty$; moreover, $0 < z < qr/e$. Brauner and Schmidt–Lainé (1987)

30.

$$y_1'' + \rho_1 y_1 + y_1(|y_1|^2 + h|y_2|^2) = 0,$$

$$y_2'' + \rho_2 y_2 + y_2(|y_2|^2 + h|y_1|^2) = 0,$$

where ρ_1, ρ_2, and h are constants.

The given system arises from coupled Schrödinger equations. Assuming the form $y_1 = a \tanh \mu x, y_2 = b \operatorname{sech} \mu x$, substituting in the given system, and equating coefficients of the same powers of sech μx and tanh μx to zero, we get

$$\begin{array}{ll} -2a\mu^2 + \rho_1 a + hab^2 = 0, & 2a\mu^2 + a^3 - hab^2 = 0, \\ b\mu^2 + \rho_2 b + hba^2 = 0, & -2b\mu^2 + b^3 - hba^2 = 0. \end{array}$$

The system (1) admits the solution

$$a = \pm(-\rho_1)^{1/2}, b = \pm[(2\rho_2 - \rho_1)/(h-2)]^{1/2}, \mu = [\rho_1(1+h)/2]^{1/2},$$

where $\rho_1 = 2\rho_2/(h-1)$. Hence the explicit solution. Huibin and Kelin (1990a)

31. $\quad \boldsymbol{\ddot{y} = -w^2(t)y + f(x/y)/(y^2x), \quad \ddot{x} = -w^2(t)x + g(y/x)/(x^2y),}$

wherein w, f, and g are arbitrary functions of their indicated arguments.

The given system is called the Ermakov system. We set

$$x = X_2\overline{x}, y = X_2\overline{y}, s = X_1/X_2, X + w^2(t)X = 0,$$

where X_1, X_2 are linearly independent solutions of this linear equation having unit Wronskian so that the given Ermakov system reduces to the autonomous form

$$\overline{y}_{ss} = f(\overline{x}/\overline{y})/(\overline{y}^2\overline{x}), \overline{x}_{ss} = g(\overline{y}/\overline{x})/(\overline{x}^2\overline{y}) \tag{1}$$

in $\overline{y}(s), \overline{x}(s)$. Introducing

$$p = X_s, q = R_s, z = X/R, \tag{2}$$

where $X = \overline{x}^2, R = \overline{y}^2$, we change (1) to

$$\begin{aligned} (p - zq)z\frac{dp}{dz} - \frac{1}{2}p^2 &= 2z^{1/2}g(z^{-1/2}), \\ (p - zq)\frac{dq}{dz} - \frac{1}{2}q^2 &= 2z^{-1/2}f(z^{1/2}). \end{aligned} \tag{3}$$

Now we set $p = z^{1/2}\pi, q = z^{-1/2}X$ in (3) to obtain

$$\begin{aligned} (\pi - X)\frac{d\pi}{dz} - \frac{1}{2}z^{-1}\pi X &= 2z^{-3/2}g(z^{-1/2}), \\ (\pi - X)\frac{dX}{dz} - \frac{1}{2}z^{-1}\pi X &= 2z^{-1/2}f(z^{1/2}). \end{aligned} \tag{4}$$

On subtraction, the system (4) yields a first integral $p - zq = h(z)$, that is, $z_s = (h/\overline{x}^2)z$, where

$$h(z; I) = 2z^{1/2}\left\{I + \int^z u^{-3/2}g(u^{-1/2})du - \int^z u^{-1/2}f(u^{1/2})du\right\}^{1/2}.$$

Here I is an invariant; $h(z, I)$ is determined once f and g in the Ermakov system are specified. Introducing $\psi = 1/\overline{x}$ in the second of (1) as the dependent variable and z as the independent variable, we get a linear equation

$$hz\psi'' + (hz)'\psi' + g(z^{-1/2})\psi/(hz^{1/2}) = 0, \tag{5}$$

with the solution

$$\psi = \psi(z; I, c_1, c_2) = c_1\psi_1 + c_2\psi_2 = 1/\overline{x},$$

where ψ_1 and ψ_2 are linearly independent solutions of (5). Substitution in the first integral $z_s = (h/\overline{x}^2)z$ yields $z_s = h(z; I)\psi^2(z; I, c_1, c_2)z$, whence on integration we have

$$s = s(z; I, c_1, c_2, c_3). \tag{6}$$

Under suitable conditions (6) can be inverted to yield $z = z(s; I, c_1, c_2, c_3)$ and hence $\overline{x} = \overline{x}(s) = 1/\psi(z(s))$, whence the original Ermakov variable x is given by the nonlinear superposition

$$x = X_2/[c_1\psi_1(X_1/X_2; I, c_1, c_2, c_3) + c_2\psi_2(X_1/X_2; I, c_1, c_2, c_3)]$$

of linearly independent solutions X_1, X_2 of the linear equation $\ddot{X} + w^2(t)X = 0$. The relation $y = y(s) = 1/\{z^{1/2}(s)\psi(z(s))\}$ shows that the original Ermakov variable y is given by the nonlinear superposition

$$\begin{aligned} y \; = \; & X_2/[z^{1/2}(X_1/X_2; I, c_1, c_2, c_3)\{c_1\psi_1(X_1/X_2; I, c_1, c_2, c_3) \\ & + c_2\psi_2(X_1/X_2; I, c_1, c_2, c_3)\}]. \end{aligned}$$

Athorne, Rogers, Ramgulam, and Osbaldestin (1990)

32. $\qquad \boldsymbol{\ddot{x} = -w^2(t)x + x^{-3}f(x/y), \quad \ddot{y} = w^2(t)y + y^{-3}g(x/y).}$

The given system is referred to as Ermakov's system. Let x_1 and x_2 be any two linearly independent solutions of

$$\ddot{x} + w^2(t)x = 0 \tag{A}$$

with unit Wronskian: $x_2\dot{x}_1 - x_1\dot{x}_2 = 1$. Then in the barred variables defined by $x = x_2\overline{x}, y = x_2\overline{y}, s = x_1/x_2$, the given system becomes

$$\frac{d^2\overline{x}}{ds^2} = (1/\overline{x}^3)f(\overline{x}/\overline{y}), \; \frac{d^2\overline{y}}{ds^2} = (1/\overline{y}^3)g(\overline{x}/\overline{y}). \tag{1}$$

The autonomizability of the given system depends crucially upon the fact that the RHSs are homogeneous of weight -3. Using the autonomous nature of the system (1), we introduce the new dependent variables $p = 2\overline{x}(d\overline{x}/ds)$ and $q = 2\overline{y}(d\overline{y}/ds)$ and use $z = \overline{x}/\overline{y}$ as the new independent variable. This leads to the second order system

$$\begin{aligned} (p - qz^2)z\tfrac{dp}{dz} - p^2 &= 4f(z), \\ [(p - qz^2)/z]\tfrac{dq}{dz} - q^2 &= 4g(z). \end{aligned} \tag{2}$$

The invariant

$$I = (1/2)\left(x\frac{dy}{dx} - y\frac{dx}{dt}\right)^2 - \int^z u^{-3}f(u)du - \int^{1/z} u^{-3}g(u)du \tag{3}$$

of the given system written in the new variables becomes $p - qz^2 = zh(z; I)$, where $h^2(z; I) = 8I + 8\int^z (u^{-3}f(u) - ug(u))du$, and (2) must be supplemented by the equation relating z to s,

$$\overline{x}^2(s)\frac{dz}{ds} = (1/2)(p - qz^2)z = (1/2)z^2h(z; I).$$

Using (3), the pair of equations (2) becomes a pair of independent Riccati equations

$$\begin{aligned} z^2h\frac{dp}{dz} - p^2 &= 4f(z), \\ h\frac{dq}{dz} - q^2 &= 4g(z). \end{aligned} \tag{4}$$

Finally, the linearizing transformation $p = -z^2h\psi^{-1}(d\psi/dz)$ and $q = -h\phi^{-1}(d\phi/dz)$ changes (4) to

$$\begin{aligned} &z^4h^2\frac{d^2\psi}{dz^2} + z^2h\frac{d(z^2h)}{dz}\frac{d\psi}{dz} + 4f(z)\psi = 0, \\ &h^2\frac{d^2\phi}{dz^2} + h\frac{dh}{dz}\frac{d\phi}{dz} + 4g(z)\phi = 0. \end{aligned} \tag{5}$$

It is important to note that because $h = h(z;I)$, (5) is really a one-parameter (I) family of linear differential equations. Thus, (5) and the linear DE(A) constitute the full linearization of the given Ermakov system. The method of solution is then explained. The singularity structure of the linear system (5) is analyzed. Athorne (1991)

33. $$\epsilon^2 y'' = f(y,z), \quad z'' = g(y,z),$$

$y(0) = \alpha_0, y(1) = \alpha_1, z(0) = \beta_0, z(1) = \beta_1$.

Asymptotic behavior, for small ϵ, of solutions of two-point BVPs for the given autonomous system is considered. Principal attention is paid to the fact that families of solutions with ϵ as a parameter commonly exist which approach discontinuous functions of x as $\epsilon \to 0$. The solutions when ϵ is small but nonzero, being smooth, exhibits an abrupt but continuously differentiable transition at the location of the limit discontinuity. Fife (1976)

34. $$\ddot{x} = \phi(x,y), \quad \ddot{y} = \psi(x,y),$$

$x(0) = a_0, x(\tau) = b_0, \dot{x}(0) = a_1, \dot{x}(\tau) = b_1$, where $t \in [0,\tau] = I, I_0 = (0,\tau), a_i, b_i \in R$ $(i = 0,1), x, y \in R, \phi, \psi \in C(R^2)$.

Let us set $m_0 = \max(|a_0|,|b_0|), m_1 = \max(|a_1|,|b_1|), m_0^* = m_0 + m_1\tau + p\tau^2/2$, with p a nonnegative constant. The following result is proved:

Let there exist constants $m_2, p \geq 0$ such that the following conditions are fulfilled:

(a) $y\phi(x,y) > 0, |y| > m_2, |x| \leq m_0^*, y\psi(x,y) \geq 0, |y| \geq m_2, |x| \leq m_0^*, \phi(x,y) \to \pm\infty$ as $y \to \pm\infty$ uniformly with respect to $x \in [-m_0^*, m_0^*]$ and $0 \leq \phi(x,m_2) \leq p, 0 \geq \phi(x,-m_2) \geq -p$ for $|x| \leq m_0^*$.

(b) Suppose, further, that there exist constants $A_0, A_1, B_0, B_1, r, q > 0$ such that $|\phi(x,y)| \geq A_0|y|^r - A_1, |\psi(x,y)| \leq B_0|y|^q + B_1$ for $|x| \leq m_0^*, \forall y \in R$, and $q \leq 1 + 2r$.

Then the given problem has a solution. Bespalova and Klokov (1990)

4.10 $y_i'' + g_i(x, y_1, y_2, y_1', y_2') = f_i(x, y_1, y_2), (i = 1, 2);$ g_i linear in y_i

1. $$y' = \lambda\mu^{1/2} f(w), \quad u'' + yu' = -w + u - \mu^{1/2} f(w), \quad w' = (u - w)/\mu,$$

where $f(w) \propto w^2$; it may be a more general function. The BCs are

$$\begin{aligned} y(0) &= q_L(\alpha), u(0) = u_c, u'(0) = (u_c - u_a)\alpha, \\ w(0) &= u_a + (u_c - u_a)/(1 + \mu\alpha), \\ y(L) &= c, u(L) = u_c, u'(L) = B(u_c - u_a)\beta_+ + (1 - B)(u_c - u_a)\beta_-, \\ w(L) &= u_a + B(u_c - u_a)/(\mu\beta_+ + 1) + (1 - B)(u_c - u_a)/(1 + \mu\beta_-). \end{aligned}$$

Here $2\beta_\pm = -(c + \mu^{-1}) \pm [(c + \mu^{-1})^2 - 4(c\mu^{-1} - 1)]^{1/2}$. This is a two-point BVP involving a fourth order system for y, w, u, and u' and four unknown constants α, L, c, and $B[q_L(\alpha)$ is known]; it describes traveling combustion waves in a porous medium. Existence of the solution of the given BVP is proved and some approximate power series solutions are found. The results are depicted graphically. The IVP is well posed at $x = 0$; the conditions at $x = L$ determine the unknown constants. Norbury and Stuart (1988)

2. $$y_1'' = -y_1(1 + y_1^2) + 4y_2(1 + y_2^2) + 2,$$
$$y_1'' + 8y_2'' = -y_2(1 + y_2^2) + 2.$$

Perturbation solution about the equilibrium point $y_1 = 2, y_2 = 1$ is

$$\begin{aligned} y_1 &= 2 + \beta_1 B_1 \cos(w_1 x + \beta_1) + \beta_2 B_2 \cos(w_2 x + \beta_2), \\ y_2 &= 1 + B_1 \cos(w_1 x + \beta_1) + B_2 \cos(w_2 x + \beta_2), \end{aligned}$$

where $B_1, B_2, \beta_1, \beta_2$ are determined from the initial conditions. Hagedorn (1988), p.232

3. $$\ddot{x}_1 + 0.2\dot{x}_1 - 0.1\dot{x}_2 = -x_1^3 - (x_1 - x_2)^3 + A\cos t,$$
$$\ddot{x}_2 - 0.1\dot{x}_1 + 0.1\dot{x}_2 = -(x_2 - x_1)^3.$$

The given fourth order forced system is investigated numerically. It is found that there is an invariant torus bifurcation series that precedes the onset of chaos. Poincaré maps are used to show chaos. Cheng (1991)

4. $$y'' + z' = -2y - \lambda z + (y + z)(y^2 + z^2),$$
$$z'' - y' = -2z + \lambda y - (y - z)(y^2 + z^2),$$

where λ is a constant.

The given system has an exact periodic solution

$$\begin{bmatrix} y \\ z \end{bmatrix} = (s/2^{1/2}) \begin{bmatrix} \cos wx \\ \sin wx \end{bmatrix}, \lambda = s^2/2, w = 2 + (1/2)\{(9 - 2s^2)^{1/2} - 3\}.$$

Here s is a parameter. A perturbation approach gives good approximation to these exact solutions. Cicogna (1987)

5. $$y_1'' - 2y_2' = y_1 - \{(1-\mu)(y_1+\mu)\}/r_1^3 - \mu(y_1 - 1 + \mu)/r_2^3,$$
$$y_2'' + 2y_1' = y_2 - (1-\mu)y_2/r_1^3 - \mu y_2/r_2^3,$$
with $r_1 = \{(y_1+\mu)^2 + y_2^2\}^{1/2}, r_2 = \{(y_1 - 1 + \mu)^2 + y_2^2\}^{1/2}$.

The given system describes the motion in the (y_1, y_2) plane of a small third body that is gravitationally attracted by two large bodies which are themselves rotating in circles about their common center of mass (at the origin) and whose mass ratio is $m_2/(m_1 + m_2) = \mu$.

The given system has a first integral (Jacobi integral)

$$J = (1/2)(y_1'^2 + y_2'^2 - y_1^2 - y_2^2) - (1-\mu)/r_1 - \mu/r_2 = C, \text{ say.}$$

This integral can be used to reduce the given system to one in three-dimensional phase space. Topological characterization of the integral manifolds $J = C$ depends on the value of the constant. Five different cases arise and show diverse behavior. Daniel and Moore (1970), p.126

6. $$y'' + \alpha y' = \beta y - \gamma y^3 + z, \quad z'' = -\Omega^2 z,$$

where α, β, γ, and Ω^2 are constants.

The given system is an equivalent form of the Duffing equation,

$$y'' + \alpha y' - \beta y + \gamma y^3 = f \cos \Omega x, \gamma > 0.$$

Now see Eqn. (4.6.2). Steeb and Kunick (1983)

7. $$y'' = -(1/\gamma)\{(1/3)\beta y^3 + (1/3)\alpha z^3 - wy + (1/2)\lambda y^2 + C_0\},$$
$$z'' + (1/\epsilon_2)\epsilon_1 z' = -(1/\epsilon_2)\{(1/2)\epsilon_0 z^2 + \delta yz - wz + C_1\},$$

where $\gamma, \beta, \alpha, w, \lambda, C_0, C_1, \epsilon_1, \epsilon_2$, and ϵ_0 are constants.

The given ODE system arises from a coupled system of two nonlinear PDEs of third order. For $\epsilon_0 = 0$, the solution has the form $y = a + c \tanh^2 \mu x, z = d + f \tanh^2 \mu x$, where on substitution in the given system, etc., we get

$$\begin{aligned}
c &= -(\alpha/\beta)^{1/2} f \equiv \alpha_0 f, \\
\mu^2 &= \{1/(6\epsilon_2)(\delta\alpha_0 + \epsilon_0/2)f \equiv Ff, \\
a &= (8\epsilon_2/\delta)Ff + w/\delta + Ed, \\
d &= \{\gamma\delta\alpha_0^2/\epsilon_2 + \gamma\alpha_0\epsilon_0/(2\epsilon_2) - \lambda\alpha_0^2/2 - \beta\alpha_0^2 w/\delta + 4\beta\alpha_0^2/(3\delta) \\
&\quad \times (\delta\alpha_0 + \epsilon_0/2)f\}\{(\alpha - \beta\alpha_0^2(\delta\alpha_0 + \epsilon_0)/\delta\}^{-1} \equiv A + Bf,
\end{aligned}$$

with the constraint

$$\begin{aligned}
&f^2[\alpha B^2 + \beta\alpha_0\{EB + (8\epsilon_2/\alpha)F\}^2] + f[2\alpha AB + 2(EA + w/\delta)(EB + 8\epsilon_2 F/\delta) \\
&\quad + \lambda\alpha_0(EB + 8\epsilon_2 F/\delta) - 8\alpha_0\gamma F] - w\alpha_0 + \alpha A^2 \\
&\quad + \beta\alpha_0(EA + w/\delta)^2 + \lambda\alpha_0(EA + w/\delta) = 0.
\end{aligned}$$

Huibin and Kelin (1990a)

8. $$\ddot{x} + \delta\dot{x} = -[1 - q + x^2 + 4y^2]x + Qy,$$
$$\ddot{y} + \delta\dot{y} = -4[4 - q + x^2 + 4y^2]y - Qx,$$

where δ, q, and Q are constants.

The given problem, known as panel flutter, analyzes the vibrations of a buckled plate with supersonic flow on one side of the plate. The analytic-analog computer study of this system predates that of Lorenz system by one year. Kobayashi (1962) remarks that in some unstable region of a moderately buckled plate, only an irregular vibration is observed. Kobayashi (1962)

9.
$$\ddot{x} + \gamma\dot{x} = (1/2)x(1 - x^2) - \beta x y^2 + f_2,$$
$$\ddot{y} + \gamma\dot{y} = -\alpha(1 + \epsilon y^2)y - \beta x^2 y + f_0 + f_1 \cos wt,$$
where $\gamma, \beta, f_2, \alpha, \epsilon, f_0$, and f_1 are constants.

This system describes a two-degree-of-freedom buckled beam. There is experimental evidence for chaotic behavior of the given system; no analysis seems to have been done for the same. Moon (1980)

10.
$$y'' + [1/\{m(C_0^2 - V^2)\}]DVy' = [1/\{m(C_0^2 - V^2)\}][Cy^3 - By^2 - 2Kyz^2 + Ay],$$
$$z'' + [1/\{M(V_0^2 - V^2)\}]EVz' = [1/\{M(V_0^2 - V^2)\}][-2KP(y^2 - y_0^2) - M\Omega_0^2 z],$$
where all the coefficients of y, y', z, and z' are constants.

The given system has solutions of the following form, obtained by substitution in the given system, etc.

(a) $y = \alpha + \beta \tanh ux, \quad z = \gamma + \delta \tanh \mu x$, where

$$\begin{aligned}
\beta &= \{M(V^2 - V_0^2)/K\}^{1/2}\mu \equiv \beta_0\mu, \\
\delta &= [\{C\beta_0^2 - 2m(C_0^2 - V^2)\}/(2K)]^{1/2}\mu \equiv \delta_0\mu, \\
\alpha &= \{2EV\delta_0^2 - DV\beta_0^2 + B\beta_0^3\}/\{3\beta_0^3 C + 6K\beta_0\delta_0^2\}, \\
\gamma &= (EV\delta_0 - 4K\alpha\delta_0\beta_0)/(2K\beta_0^2), \\
\mu &= \pm[(A\alpha - 2K\alpha\gamma^2 - B\alpha^2 + C\alpha^3)/(DV\beta_0)]^{1/2},
\end{aligned}$$

with the constraints

$$\begin{aligned}
\delta AE\alpha &= (2K\alpha\gamma^2 + B\alpha^2 - C\alpha^3)E\delta_0 - D[m\Omega_0^2\gamma + 2K\gamma(U_0^2 - \alpha^2)]\beta_0, \\
&- 2\beta_0\mu^2 m(C_0^2 - V^2) = A\beta_0 - 2K(2\gamma\delta_0\alpha + \beta_0\gamma^2) \\
&- 2B\alpha\beta_0 + 3\alpha^2\beta_0 C, \\
&2\delta_0\mu^2 M(V_0^2 - V^2) = M\delta_0\Omega_0^2 + 2K[2\alpha\beta_0\gamma + \delta_0(\alpha^2 - U_0^2)].
\end{aligned}$$

(b) $y = \alpha + \beta \tanh \mu x, \quad z = \delta \operatorname{sech} \mu x$, where

$$\begin{aligned}
\alpha &= \pm EV/(4[KM(V^2 - V_0^2)]^{1/2}), \\
\beta &= \pm\{M(V^2 - V_0^2)/K\}^{1/2}\mu \equiv \beta_0\mu, \\
\delta &= \pm[\{2m(C_0^2 - V^2) - C\beta_0\}/(2K)]^{1/2}\ \mu \equiv \delta_0\mu, \\
\mu &= \pm[(2B\alpha - A - 3C\alpha^2)/(C\beta_0^2)]^{1/2}, \\
B &= (\alpha/\beta_0^2)\{2m(C_0^2 - V^2) + 2\beta_0^2 C + DV\beta_0/\alpha\},
\end{aligned}$$

and D satisfies $A_3D^2 + B_3D + C_3 = 0$, where

$$\begin{aligned}
A_3 &= -2V^2\alpha^2/(C^2\beta_0^2), \\
B_3 &= V(A + 3C\alpha^2)/(C\beta_0) - 2V\alpha^2 D_3/(C\beta_0) - \{\alpha^2 V/(C\beta_0^3)\} \\
&\quad [4m(C_0^2 - V^2) - C\beta_0^2], \\
C_3 &= A\alpha + C\alpha^3 - \alpha^2 D_3[\{4m(C_0^2 - V^2) - C\beta_0^2\}/(C\beta_0^2)] \\
&\quad + [\alpha(A + 3C\alpha^2)/(C\beta_0^2)][2m(C_0^2 - V^2) - CB_0^2], \\
D_3 &= (\alpha/\beta_0^2)[2m(C_0^2 - V^2) + 2\beta_0^2 C],
\end{aligned}$$

and a constraint on C and A: $-M\Omega_0^2 = \mu^2 M(V_0^2 - V^2) + 2K(\alpha^2 + \beta^2 - U_0^2)$.

(c) $B = 0$. The solution has the form $y = \beta \operatorname{sech} \mu x, \quad z = \gamma + \delta \tanh \mu x$, where

$$\begin{aligned} \beta &= [M(V_0^2 - V^2)/K]^{1/2}\mu \equiv \beta_0\mu, \\ \delta &= \{-D\beta_0^2/(2E)\}^{1/2} \equiv \delta_0\mu, \quad \gamma = -EV\delta_0/(2K\beta_0^2), \\ \mu &= [(A - 2K\gamma^2)/\{m(C_0^2 - V^2) + 2K\delta_0^2\}]^{1/2}, \\ C &= [2K\delta_0^2 - 2m(C_0^2 - V^2)]/\beta_0^2 \text{ with } M\Omega_0^2 = 2KU_0^2. \end{aligned}$$

(d) $D = E = 0$. There are algebraic solutions

$$y = 1/(a + bx^2), \quad z = \alpha + \beta/(a + bx^2),$$

where

$$\begin{aligned} \alpha &= \pm\{A/(2K)\}^{1/2}, \quad \beta = \{-B + (B^2 - 8K\alpha^2 C)^{1/2}\}/(4K\alpha), \\ a &= -3\beta/(4\alpha), \quad b = -(4K\alpha\beta + B)/\{6m(C_0^2 - V^2)\}, \\ V &= [\{\beta^2 - C/(2K)\}V_0^2 - m/(MC_0^2)]^{1/2}/\{\beta^2 - C/(2K) - m/M\}^{1/2}, \end{aligned}$$

with $M\Omega_0^2 = 2KU_0^2$.

Huibin and Kelin (1990)

11. $$y_1'' + \ell_{10}y_1' + \ell_{12}(y_1' - y_2') = -C_{12}(y_1 - y_2)^3 - C_{10}y_1 + P\cos\nu x$$

$$\gamma y_2'' + \ell_{12}(y_2' - y_1') = -C_{12}(y_2 - y_1)^3 - C_{23}y_2^3,$$

where $\ell_{ij}, C_{ij}, \gamma, P$, and ν are constants.

Resonance curves of the given system are found for different sets of parameters. Szemplinska–Stupnicka (1980)

12. $$y_1'' + C_1y_1' = -w_1^2y_1 - \delta_{11}y_1^2 - \delta_{12}y_1y_2 - \delta_{13}y_2^2 - \epsilon_1y_1^3 + P_1\cos\Omega_1x + Q_1\cos\Omega_2x,$$

$$y_2'' + C_2y_2' = -w_2^2y_2 - \delta_{21}y_1^2 - \delta_{22}y_1y_2 - \delta_{23}y_2^2 - \epsilon_2y_2^3 + P_2\cos\Omega_1x + Q_2\cos\Omega_2x,$$

This is a nonlinear damped mechanical system with two degrees of freedom with quadratic and cubic spring characteristics, excited by two external harmonic forces with different frequencies. Here $C_i, w_i, \delta_{ij}, \epsilon_i, P_i$, and Q_i $(i = 1, 2)$ are constants. Steady-state (approximate) differential tones of this system in the form

$$\begin{aligned} y_1 &= A_1\cos(\Omega_1x + \alpha_1) + B_1\cos(\Omega_2x + \beta_1) + U_1\cos(\Omega x + \gamma_1), \\ y_2 &= A_2\cos(\Omega_1x + \alpha_2) + B_2\cos(\Omega_2x + \beta_2) + U_2\cos(\Omega x + \gamma_2) \end{aligned} \tag{1}$$

are written out by the balance method of dominant harmonics; the coefficients A_i, B_i, U_i and α_i, β_i, and γ_i $(i = 1, 2)$ are found by substituting (1) into the given DE. Approximate steady-state solutions and the corresponding Galerkin approximations of high order are obtained and error bounds are given. For a certain frequency the existence of three exact periodic solutions is proved by Urabe's method. Van Dooren (1973)

13. $\boldsymbol{y_1'' - \epsilon(-\mu_1 y_1' + \mu_2 y_2') = -w_{n_1}^2 y_1 + \epsilon(\alpha_1 y_1^3 + \alpha_2 y_1^2 y_2 + \alpha_3 y_1 y_2^2 + \alpha_4 y_2^3) + F_1 \cos \Omega x,}$

$\boldsymbol{y_2'' - \epsilon(\mu_3 y_1' - \mu_4 y_2') = -w_{n_2}^2 y_2 + \epsilon(\alpha_5 y_1^3 + \alpha_6 y_1^2 y_2 + \alpha_7 y_1 y_2^2 + \alpha_8 y_2^3) + F_2 \cos \Omega x,}$

where $w_{ni}, \mu_i, \alpha_i, F_i, \Omega$, with i assuming suitable integral values, and ϵ are constants.

The given system describes vibration absorber. Method of multiple scales and numerical methods are used to obtain large-amplitude almost-periodic vibrations. Shaw, Shaw, and Haddow (1989)

14. $$\boldsymbol{y'' - cy' = z^p, \quad z'' - cz' = y^q,}$$

where c is a parameter; $p > 0$ and $q > 0$.

The given system describes finite traveling wave solutions to a semilinear diffusion system. Nonnegative solutions are sought that possess sharp fronts, i.e., $y(x) = z(x) = 0$ for $x \leq x_0$, for some finite x_0. The following results are proved:

(a) There exist finite (wave) solutions of the given system if and only if

$$pq < 1. \tag{1}$$

Moreover, if (1) holds for any real c, then the corresponding wave profiles are unique up to translation in x.

(b) Assuming that $pq < 1$ and any real c, for which the result (1) holds, the following are true:

(i) $y(x) \simeq Ax^\alpha, z(x) \simeq Bx^\beta$ as $x \to 0+$, where

$$\begin{aligned} \alpha &= 2(1+p)/(1-pq), \quad \beta = 2(1+q)/(1-pq), \\ B &= A^q[\beta(\beta-1)]^{-1}, \quad A^{1-pq} = [(\beta(\beta-1))^p \alpha(\alpha-1))]^{-1}. \end{aligned}$$

(ii) If $c < 0$, then $y(x) \simeq Cx^\gamma$ and $z(x) \simeq Dx^\delta$ as $x \to \infty$, where

$$\begin{aligned} \gamma &= (1+p)/(1-pq), \quad \delta = (1+q)/(1-pq), \\ C^{1-pq} &= [(-c)^{1+p}\delta^p\gamma]^{-1}, \quad D = C^q[(-c)\delta]^{-1}. \end{aligned}$$

(iii) If $c > 0$, the asymptotics for $x \to \infty$ depend on the values of p and q as follows:

If $p < 1, q < 1$, then $y(x) \simeq M_1 e^{cx}, z(x) \simeq N_1 e^{cx}$, where

$$M_1^{1-pq} = \frac{1}{[c^2(1-q)]^p c^2(1-p)}, \quad N_1 = \frac{M_1^q}{c^2(1-q)}.$$

If $p < 1, q = 1$, then $y(x) \simeq M_2 e^{cx}, z(x) \simeq N_2 x e^{cx}$, where

$$M_2^{1-p} = \frac{1}{c^{p+1}} \int_0^\infty s^p e^{c(p-1)s} ds, N_2 = M_2/c.$$

If $p < 1, q > 1$, then $y(x) \simeq M_3 e^{cx}, z(x) \simeq N_3 e^{cqx}$, where

$$M_3^{1-pq} = \frac{1}{[c^2(q-1)q]^p c^2(1-pq)}, \quad N_3 = M_3 q/[c^2(q-1)q].$$

The cases $p > 1, q < 1$ and $p > 1, q = 1$ are obtained in an obvious way by suitably changing the coefficients in case (iii).

Esquinas and Herrero (1990)

15. $$y'' - cy' = (yz)^p, \quad z'' - cz' = (yz)^q,$$

where c, p, and q are constants.

The given system describes traveling wave solutions of a nonlinear diffusion system. The solutions over a semi-infinite line vanishing at plus infinity and at the left (finite) end are mentioned to exist if $p + q < 1$. Local analysis at infinity and at the finite end of the wave may be carried out. For example, for $c < 0$, we have

$$y(x) = \begin{cases} A_1 x^{\alpha_1} \text{ with } \alpha_1 = 2(1-q+p)/[1-(p+q)] & \text{if } x \simeq 0 \\ A_2 x^{\alpha_1/2} & \text{if } x \gg 0, \end{cases}$$
$$z(x) = \begin{cases} B_1 x^{\beta_1} \text{ with } \beta_1 = 2(1-p+q)/[1-(p+q)] & \text{if} x \simeq 0 \\ B_2 x^{\beta_1/2} & \text{if } x \gg 0, \end{cases}$$

where A_1, A_2, B_1, B_2 are positive constants, depending on p and q. Esquinas and Herrero (1990)

16. $$\ddot{x} + V_0\dot{x} = xy^m, \quad \ddot{y} + V_0\dot{y} = -xy^m + Ky^n, -\infty < t < \infty.$$

where V_0, m, n, and K are constants; $x \to 1, y \to 0$ as $t \to \infty$ and $x \to x_s, y \to 0$ as $t \to -\infty$ for some $x_s \geq 0$, which may depend on K and V_0.

The given system describes permanent-form traveling waves in a simple isothermal chemical system. The following qualitative results are proved:

(a) $n, m \geq 1$

- (i) Then $x(t) > 0$ and $y(t) > 0$ for all $-\infty < t < \infty$.
- (ii) $x_s < x(t) < 1$ and $0 < x(t) + y(t) < 1$ for all $-\infty < t < \infty$.
- (iii) $x(t)$ is strictly monotone decreasing.

(b) $n = m \geq 1$

- (iv) The solution exists only if $K < 1$.
- (v) $x_s < K$.
- (vi) $y(t) < (1 - K)$.

(c) $n > m \geq 1$

- (vii) $y(t)$ has a unique turning point which is a local maximum.
- (viii) $x_s = 0$.
- (ix)

$$y(t) < \begin{cases} K^{-1/(n-m)}, & K > 1, \\ 1, & k \leq 1. \end{cases}$$

(d) $m > n \geq 1$.

- (x) The solution exists if $K < P^P/(P+1)^{P+1}$, where $P = m - n$. Under this condition, $y^-(K) < y_{\max}(t) < y^+(K)$, where $y^-(K)$ and $y^+(K)$ are two positive roots of the equation $\psi^P(\psi - 1) + K = 0$.

Needham and Merkin (1991)

17.

$$\ddot{x} + \gamma\dot{x} = -\sin x - k(x - y) + I,$$
$$\ddot{y} + \gamma\dot{y} = -\sin y - K(y - x),$$

where $K > 0, \gamma > 0, \gamma^2 > 5K$, and $I \geq 0$.

Existence of a global attractor and existence of a (restricted horizontal) invariant manifold in the global attractor of the given system (a coupled Josephson junction) are proved. Min, Wenxian, and Jinyan (1990)

18.

$$\ddot{x} + \gamma\dot{x} = -\sin x - K(x - y) + I + A \sin wt,$$
$$\ddot{y} + \gamma\dot{y} = -\sin y - K(y - x),$$

where $K > 1, \gamma > 0, \gamma^2 > 5K$, and $I \geq 0$.

Qualitative properties with respect to the existence of a global attractor are discussed. Min, Wenxian, and Jinyan (1990)

19.

$$\ddot{x} + \gamma\dot{x} = -2 \sin(x/2) \cos y - 2K\{x - I/(2K)\},$$
$$\ddot{y} + \gamma\dot{y} = -\sin y \cos(x/2) + I/2,$$

where γ, K, and I are constants, $K > 0, \gamma > 0, \gamma^2 > 5K$, and $I \geq 0$.

The given system is the same as DE (4.10.17) if we write there $\overline{x} = x - y, \overline{y} = (1/2)(x + y)$ and drop the bars. Min, Wenxian, and Jinyan (1990)

20.

$$\ddot{x}_1 + a\dot{x}_1 - 2\mu w\dot{x}_2 = -[\nu^2 + c(x_1^2 + x_2^2)^q]x_1 - Hw^2 \sin wt,$$
$$\ddot{x}_2 + a\dot{x}_2 + 2\mu w\dot{x}_1 = -[\nu^2 + c(x_1^2 + x_2^2)^q]x_2 + Hw^2 \cos wt,$$

where a, ν, c, q, μ, w, and H are constants subject to various restrictions.

Sufficient conditions for the boundedness of the solutions of the given system are derived by reducing the DEs to a system of Volterra integral equations. Sufficient conditions are given both when $H \neq 0$ and when $H = 0$.

The existence of fundamental solutions, i.e., of the form $x = M\sin(wt + r), y = M\cos(wt + r)$, the number of such solutions (three, two, or one, depending on the coefficients in the system) and their stability in the Liapunov sense are discussed. Musinskaja (1966, 1967)

21.

$$y'' - Pe_y y' = -Pe_y B \cdot Da(1 - z) \exp\{y/(1 + \epsilon y)\},$$
$$z'' - Pe_z z' = -Pe_z Da(1 - z) \exp\{y/(1 + \epsilon y)\},$$

$Pe_y y(0) = y'(0), Pe_y z(0) = z'(0), y'(1) = z'(1) = 0$, where Pe_y, Pe_z, Da, B, and ϵ are constants.

(a) $Pe_y = Pe_z = Pe$, say. In this case, it is possible to find a first integral of the given system, $y = Bz$, and hence a simple equation for y:

$$(1/Pe)y'' - y' + Da(B - y)\exp[y/(1 + \epsilon y)] = 0,$$

$Pe_y(0) = y'(0), y'(1) = 0$. Putting $t = Pe(x - 1)$ and writing $r = Da/Pe$, we get the reduced equation

$$\frac{d^2y}{dt^2} - \frac{dy}{dt} + r(B - y)\exp\{y/(1 + \epsilon y)\} = 0$$

with BC, $y'(0) = 0, y(-Pe) = y'(-Pe)$. This problem is easily solved using initial value methods.

(b) More generally, when $Pe_y \neq Pe_z$, the given problem is solved numerically using initial value methods.

Kubicek and Hlavacek (1983), p.250

22.
$$y'' - Hy' = \beta H(y-1) - BDHze^{-\gamma/y},$$
$$z'' - Mz' = MDe^{-\gamma/y}z, 0 < x < 1,$$

with the BC $y' - H(y-1) = 0$ at $x = 0, y' = 0$ at $x = 1$; $z' - M(z-1) = 0$ at $x = 0, z' = 0$ at $x = 1$, where H, M, B, D, β, and γ are constants.

The given system describes axial dispersion of the nonadiabatic tubular chemical reactor. It is shown that the given system can have arbitrary number of low-conversion steady-state solutions, provided that the reactor parameters are in a suitable range. These solutions exhibit oscillations that grow slowly in amplitude as the reactor is transversed from inlet to outlet. Alexander (1990)

23.
$$y'' - cy' = zf_\epsilon(y), y(-\infty) = 0, y(+\infty) = 1,$$
$$\Omega z'' - cz' = zf_\epsilon(y), z(-\infty) = 1, z(+\infty) = 0,$$

where $\Omega, c > 0$ are constants; the nonlinear term f_ϵ satisfies the following:
$f_\epsilon(s) = 0 \ \forall\ s \in [0, \theta)$ for $0 < \theta < 1, f_\epsilon(s) > 0 \ \forall\ s \in (\theta, 1]$, and f_ϵ is Lipschitz on $[\theta, 1]$.

The given system models premixed laminar flames in the case of a single reactant. The existence and asymptotic behavior of the solutions as some parameter $\epsilon \to 0$ are studied. Berestycki, Nicolaenko, and Scheurer (1983)

24.
$$y'' - cy' = -f(y)z, \quad \Lambda z'' - cz' = f(y)z,$$

on the whole real line, with $y(-\infty) = 0, y(+\infty) = 1, z(-\infty) = 1, z(+\infty) = 0$; Λ is a constant.

The given system describes traveling waves in a combustion model. Here $y, 0 \leq y \leq 1$, is renormalized temperature, z is a renormalized reactant concentration $0 \leq z \leq 1$, and $f(y)$ is a renormalized reaction term such that $f(y) \equiv 0$ on $[0, \theta)$ and $f(y) > 0$ on $(\theta, 1)$, where θ is some ignition temperature $0 < \theta < 1; f(1) > 0$. The existence of a solution of this eigenvalue problem is proved first for a bounded domain and then the domain is allowed to extend to infinity. Berestycki, Nicolaenko, and Scheurer (1985)

25.
$$y'' - Cy' = \lambda g(y) - f(y)z^n, \quad z'' - Cz' = f(y)z^n,$$

$y(-\infty) = 0, z(-\infty) = 1, y(+\infty) = 0, z'(+\infty) = 0$, where $C > 0$ is a parameter, $n > 0$, and $\lambda > 0$ is an eigenvalue. Here f is $C^1[0,1]$ and is nondecreasing. There exists $\theta \in (0,1)$, such that $f = 0$ on $[0, \theta]$ and $f > 0$ on $(\theta, 1)$. g is $C^1[0,1], g(0) = 0, g(1) = 1$; there exist α, β such that $0 < \alpha < g' < \beta$.

Existence of a solution is proved first considering the problem in a boundary domain and using the Leray-Schauder degree theory, and then by taking an infinite domain limit. Giovangigli (1990)

26.
$$\ddot{x} + \beta\dot{y} = \lambda x + u(x,y), \quad \ddot{y} - \beta\dot{x} = \lambda y + \nu(x,y),$$

where the functions $u(x,y)$ and $v(x,y)$ are sufficient to form an analytic function $h(x+iy) = u(x,y) + iv(x,y)$ in the neighborhood of zero, which satisfies the conditions $h(0) = h'(0) = 0$.

It is proved that for every β and λ such that $4\lambda \leq \beta^2$, the given system has a family of real-valued periodic solutions. The period is $2\pi/|\rho_m|$, where ρ_m is the greatest real root of the equation $\rho^2 - \beta\rho + \lambda = 0$. Ifantis (1987)

27.
$$y'' + x^{-1}y' = -(Ax + Bz)y, \quad z'' + x^{-1}z' = yz,$$

$y(x) > 0(0 \leq x < 1), y(1) = 0, y'(0) = 0, z'(0) = 0, z(1) = 1.$

The given problem arises in work on gas discharges. It is proved that it has a solution with $y(x) > 0(0 \leq x < 1)y(1) = 0, y'(0) = 0, z'(0) = 0, z(1) = 1$ if and only if $A^{1/2} < r < (A + B)^{1/2}$, where r is the first zero of the Bessel function J_0. Topological methods are used. Bruijn (1981)

28.
$$y'' + (2/x)y' = (z + k^2)y$$
$$z'' + (2/x)z' = (1/2)y^2, 0 < x < \infty,$$

$y'(0) = z'(0) = 0; y(\infty) = z(\infty) = 0.$

The given problem arises from the motion of an electron in a polar crystal.

A short-distance asymptotic analysis leads to the conditions at $x = x_0 \sim 0$, namely,

$$y'(x_0) - (1/3)x_0y(x_0)[z(x_0) + k^2] = 0, z'(x_0) - (x_0/6)y'(x_0) = 0,$$

instead of singular conditions at $x = 0$. The problem is then solved numerically. Balla (1980)

29.
$$x^2y'' - 2xy' = -2y + 2y^2z,$$
$$x^2z'' - 2xz' = -2z + 2z^2y.$$

On making the substitution $F = y/x, G = z/x$, one gets the rescaled equations

$$F'' - 2F^2G = 0, G'' - 2G^2F = 0.$$

It is easy to show that $\theta = FG$ satisfies the Weierstrass equation

$$\theta'' = 6\theta^2 + C,$$

where C is a constant of integration, and hence that both F and G satisfy the Lame equations

$$F'' - 2\theta F = 0, G'' - 2\theta G = 0.$$

Cariello and Tabor (1991)

30.
$$y'' + (a/x)y' = (\delta/\gamma\beta)y^n \exp[z/(1 + z/\gamma)],$$
$$z'' + (a/x)z' = -\delta y^n \exp[z/(1 + z/\gamma)],$$

$y(1) = 1, z(1) = 0, y'(0) = 0, z'(0) = 0$, where $a, \delta, \alpha, \beta, n$, and γ are constants.

The given problem describes heat and mass transfer accompanied by an exothermal chemical reaction occurring in a porous catalyst. It is solved numerically using the method of "false transients." Kubicek and Hlavacek (1983), p.145

31.
$$\epsilon_1 y'' + a(x)y' = f(x, y, z),$$
$$\epsilon_1\epsilon_2 z'' + b(x)z' = g(x, y, z),$$
$y'(0) = z'(0) = 0, \epsilon_1 y'(1) + y(1) = A, \epsilon_1\epsilon_2 z'(1) + z(1) = B$, where ϵ_1 and ϵ_2 are small positive parameters tending simultaneously to zero, a and b are strictly positive, and a, b, f, and g are infintely differentiable in their arguments.

Asymptotic solutions of the given BVP are formally constructed and shown to be asymptotically correct. Chen and O'Malley (1974)

32.
$$y'' - z = 0, \quad \epsilon z'' + yz' = 0;$$
$y(0, \epsilon) = y(1, \epsilon) = 0, z(0, \epsilon) = z_0, z(1, \epsilon) = z_1$; ϵ is a small parameter, z_0 and z_1 are constants.

Asymptotic behavior, as $\epsilon \to 0^+$, of the solution of the given system is discussed in four cases arising from the different signs of z_0 and z_1. Howes and Shao (1989)

33.
$$\ddot{y} = -2z, \quad \ddot{z} + 2y\dot{z} - 2\dot{y}z = 0,$$
where $y(0) = y_0, z(0)\cos\phi - \dot{z}(0)\sin\phi = 1, \dot{y}(\infty) = 0, z(\infty) = 0; 0 \leq \phi \leq \pi/2$.

The given system arises in convection in porous media. The exact solution is
$$\begin{aligned} y &= m - (m - y_0)e^{-2mx}, \\ z &= 2m^2(m - y_0)e^{-2mx}, \end{aligned}$$
where m is the positive root of the quartic
$$4m^4 \sin\phi + 2(\cos\phi - 2y_0 \sin\phi)m^3 - 2y_0(\cos\phi)m^2 - 1 = 0. \tag{1}$$
Since $0 \leq \phi \leq \pi/2$, Descartes' rule shows that (1) has a unique positive root. Ramanaiah and Malavizhi (1991)

34.
$$y'' + \lambda z + \{(\lambda - 2)/3\}xz' = 0,$$
$$z'' - \lambda zy' + \{(1 + \lambda)/3\}yz' = 0,$$
$z(0) = 1, y(0) = 0, z(\infty) = 0, y'(\infty) = 0$; λ is a parameter.

The given problem arises in flows in a saturated porous medium and is solved numerically. Cheng and Chang (1976)

35.
$$\ddot{x}_1 + (bx_1^2 - a)\dot{x}_1 = \epsilon\, Qx_2 - w_1^2 x_1,$$
$$\ddot{x}_2 + (bx_2^2 - a)\dot{x}_2 = \epsilon\, Qx_1 - w_2^2 x_2,$$
where w_1, w_2, a, b, Q, and ϵ are constants.

The given system represents two coupled Van der Pol oscillators. Assuming that the parameters $a, b, \epsilon, |w_1 - 1|, |w_2 - 1|$ are small, and using stroboscopic methods, approximate description of the solutions is obtained. Minorsky (1954)

36.
$$\ddot{x} + \mu(\beta_1 x^2 - \alpha_1)\dot{x} = -\Omega_1^2 x - m\rho_1\Omega_2^2 y,$$
$$\ddot{y} + \mu(\beta_2 y^2 - \alpha_2)\dot{y} = -\Omega_2^2 y - m\rho_2\Omega_1^2 x,$$
where μ is a small constant, $\Omega_j, \alpha_j, \beta_j, \rho_j$, and m are constants.

For certain conditions on the parameters in the given system, the existence and stability properties of integral manifold of the given system are studied. The manifold may be either an equilibrium point, a closed curve, or a torus. Method of averaging is employed to obtain the results. Banfi (1963/64)

37.

$$\ddot{y} - \epsilon[\ddot{x} + \lambda\delta\dot{x}e^{\delta x} - \dot{y}] = (\epsilon\beta - 1)y,$$

$$\ddot{x} = \beta\dot{y} - \epsilon\lambda\beta(e^{\delta x} - 1) - \epsilon^2\frac{R}{R_0}\beta x,$$

where $\epsilon, \lambda, \delta, \beta, R$, and R_0 are constants.

The system describes classical bipolar transistor oscillator circuits. The two-timing method is used to construct the asymptotic behavior of the solution as $\epsilon \to 0$. These solutions are shown to depend on amplitudes that evolve on a slow time scale, according to amplitude equations. The amplitude equations are analyzed and the results are interpreted physcically. Kreigsman (1989)

38.

$$L_1 y_1'' + M y_2'' + f_1(y_1)y_1' = -y_1/c_1,$$

$$M y_1'' + L_2 y_2'' + f_2(y_2)y_2' = -y_2/c_2,$$

where L_1, L_2, M, c_1, and c_2 are constants.

The given system describes two inductively coupled electric circuits with currents y_1 and y_2. It is proved that if this system possesses a periodic solution, then its period is not less than the smaller one of the periods of the corresponding linear system wherein $f_1(y_1) = f_2(y_2) = 0$. Graffi (1953)

39.

$$\ddot{x} - x\dot{y}^2 + Bx^3\dot{y} = -x + x^5, \quad 2\dot{x}\dot{y} + x\ddot{y} - 4Hx^2\dot{x} = 0,$$

where B and H are constants.

The given system has first integrals

$$\begin{aligned} &x^2\dot{y} - Hx^4 = \Omega, \\ &\dot{x}^2 + x^2\{1 - \Omega(2H - B)\} - (1/3)x^6(H^2 - BH + 1) + (\Omega^2/x^2) = K, \end{aligned} \qquad (1)$$

where Ω and K are constants. The critical points of (1) are studied. Doelman and Eckhaus (1991)

40.

$$\ddot{x} - x\dot{y}^2 = -x + x^3, \quad 2\dot{x}\dot{y} + x\ddot{y} = 0.$$

The given system has first integrals

$$\begin{aligned} x^2\dot{y} &= \Omega, \\ \dot{x}^2 + x^2 - (1/2)x^4 + (\Omega^2/x^2) &= K, \end{aligned} \qquad (1)$$

where Ω and K are constants. The critical points of (1) are studied. Doelman and Eckhaus (1991)

41.

$$y'' - yz'^2 + (k/m)y' = -(c/m)y,$$

$$yz'' + 2y'z' + [ky/m]z' = 0,$$

where c, k, and m are constants.

A first integral of the given system is

$$[y'^2/2 + (y^2/2)z'^2 + (w^2/2)y^2 + \mu yy']e^{2\mu x} = \text{constant},$$

where $w^2 = c/m; \mu = k/(2m)$. Djukic (1973)

42. $$\ddot{x} - x\dot{y}^2 = a(t)x, \quad x\ddot{y} + 2\dot{x}\dot{y} = b(t)x.$$

The given system can be changed to the form

$$\dot{u} = a(t) - u^2 + v^2, \quad \dot{v} = b(t) - 2uv \tag{1}$$

by the transformation $u = \dot{x}/x, v = \dot{y}; a$ and b are continuous on $[t_0, \infty)$.

The system (1) is first considered with a and b as constants and reduced to a complex Riccati equation. Conditions are found such that the trajectories of (1) behave like those of the same system with constant coefficients for large t. Rab (1970)

43. $$(zy' - yz')' + (1/2)xy' = 0,$$

$$z'' + (p/2)xz' + (q/2)xy' = 0,$$

$y(\infty) = 0, z(\infty) = 1, y(0) = 1, z(0) = 1; p$ and q are parameters.

The given DE arises from a self-similar reduction of the problem of infilteration of Dopant into semiconductors. Existence of a solution of the given two-point BVP is proved. In the course of existence proof, it is found that y and z have the following properties:

$$y'(x) < 0, 0 \leq x < \infty,$$

and there exists a number $\hat{x} > 0$ such that

$$z'(x)\begin{cases} < 0 & \text{for } 0 \leq x < \hat{x} \\ > 0 & \text{for } \hat{x} < x < \infty, \end{cases}$$

so that the qualitative behavior of y and z is easily sketched. It is also shown that there exists a constant $P > 0$ such that

$$y(x) \sim Px^{-1}e^{-x/4} \text{ as } x \to \infty$$

and that

$$z(x) \sim \begin{cases} 1 + O(xe^{-px^2/4}) & \text{if } p < 1 \\ 1 - (Pq/4)xe^{-x^2/4} & \text{if } p = 1 \\ 1 - \{Pq/(p-1)\}e^{-x^2/4} & \text{if } p > 1 \end{cases}$$

as $x \to \infty$.

The existence proof uses a shooting argument. Peletier and Troy (1991)

44. $$y\ddot{x} - 2\dot{x}\dot{y} = 0 \quad y\ddot{y} - (\dot{y}^2 - \dot{x}^2) = 0.$$

One solution of the given system is $x(t) = -\tanh\ t, y(t) = \text{sech}\ t$. Ueda and Noguchi (1983)

4.11 $y_i''+g_i(x, y_1, y_2, y_1', y_2') = f_i(x, y_1, y_2)$ $(i = 1, 2)$, g_i not linear in y_i

1. $$y'' = z, \quad \epsilon z'' + y'z' = 0,$$

$y(0) = y(1) = 0, z(0) = A, z(1) = B; A < 0 < B.$

The following result is proved. Let $y(x, \epsilon), z(x, \epsilon)$ be a solution of the given system; then:

(a) If

$$A + B \geq 0, \lim_{\epsilon \to 0^+} y(x, \epsilon) = \lim_{\epsilon \to 0^+} z(x, \epsilon) = 0, 0 < x < 1.$$

(b) If

$$A + B < 0, \lim_{\epsilon \to 0^+} z(x, \epsilon) = (1/2)(A + B), 0 < x < 1.$$

and

$$\lim_{\epsilon \to 0^+} y(x, \epsilon) = (1/4)(A + B)(x^2 - x), 0 \leq x \leq 1.$$

Dorr, Parter, and Shampine (1973)

2. $$y_1'' - \alpha_1 y_1' y_2' = -w_1^2 y_1, y_2'' - \alpha_2 y_1'^2 = -w_2^2 y_2,$$

where $w_2 \simeq 2w_1$ and w_i, α_i $(i = 1, 2)$ are constants.

A first order solution is $y_n = a_n \cos(w_n x + \beta_n)$ $(n = 1, 2)$, where a_n and β_n $(n = 1, 2)$ are given by

$$\begin{aligned}
\frac{da_1}{dx} &= (1/4)\alpha_1 w_2 a_1 a_2 \sin\gamma, \\
a_1 \frac{d\beta_1}{dx} &= -(1/4)\alpha_1 w_2 a_1 a_2 \cos\gamma, \\
\frac{da_2}{dx} &= \{\alpha_2 w_1^2/(4w_2)\} a_1^2 \sin\gamma, \\
a_2 \frac{d\beta_2}{dx} &= \{\alpha_2 w_1^2/(4w_2)\} a_1^2 \cos\gamma,
\end{aligned}$$

with $\gamma = \beta_2 - 2\beta_1 + (w_2 - 2w_1)x$. Nayfeh (1985), p.212

3. $$y'' + (1/z)z'y' - Pe\{y' + (y/z)z'\} = -Pe \cdot Da(y/z)\exp[\gamma(z-1)/z],$$

$$z'' - Pez' = -\beta Pe \cdot Da(y/z) \exp[\gamma(z-1)/z],$$

$y(0) = 1/z(0) + (1/Pe)y'(0), z(0) = 1 + (1/Pe)z'(0),\ z'(1) = y'(1) = 0.$

The given system describes axial heat and mass transfer with an exothermic chemical reaction taking into consideration the temperature dependence of the physicochemical variables.

The problem is solved numerically using the method of false transient. Kubicek and Hlavacek (1983), p.135

4. $$y'' + y'/x - z'^2 y = y/x^2 - y\lambda(y),$$

$$z'' + (1/x + 2y'/y)z' = \Omega - w(y),$$

$y(0) = 0, z'(0) = 0$, and $y \to y(\infty)$ as $x \to \infty, w[y(\infty)] = \Omega$, where:

(a) $\lambda = \lambda(y)$ is defined and continuously differentiable on $0 \le y \le a$ for some $a > 0, \lambda(y) > 0$ for $0 \le y < a, \lambda(a) = 0$ and $\lambda'(a) < 0$.

(b) $w = w(y)$ is defined and continuous for $0 \le \rho \le a$ and there exist $\epsilon \ge 0$ and $\mu > 0$ such that

$$|w(a) - w(y)| \le \epsilon(a-y)^{1+\mu}, 0 \le y \le a.$$

Ω is a constant.

It is shown that for ϵ sufficiently small, there exists a number $\Omega = w(\rho(\infty)) = w(a)$ and functions $y = y(x)$ and $z = z(x)$, twice continuously differentiable on $0 \le x < \infty$, satisfying the given system, and

$$0 < y(x) < a \text{ for } 0 < x < \infty, \tag{1}$$

$$y(x) = \begin{cases} O(x) & \text{as } x \to 0 \\ a + O(x^{-2}) & \text{as } x \to \infty, \end{cases} \tag{2}$$

$$z''(x) = \begin{cases} O(x) & \text{as } x \to 0 \\ cx^{-1} + O(x^{-1-2\mu}) & \text{as } x \to \infty, \end{cases} \tag{3}$$

where

$$c = (1/a^2) \int_0^\infty sy^2(s)[w(a) - w(y(s))]ds.$$

From these results, it can further be inferred that

$$\begin{aligned} y'(0) &> 0, y''(0) = 0, \\ y(x) &= 1 + \frac{1+c^2}{\lambda'(a)x^2} + O(x^{-2}), \end{aligned}$$

$y'(x) = O(x^{-2})$ and $y''(x) = O(x^{-2})$ as $x \to \infty$, $z(x) = c\ln x + \text{constant} + O(x^{-2\mu})$ as $x \to \infty$. If it is further assumed that $\lambda = \lambda(y)$ and $w = w(y)$ are n-times differentiable at $y = a$, for some $n \ge 1$, then the given system has asymptotic series solutions satisfying (1)–(3).

Numerical results are also presented for y and z. The given ODE system arises from a reaction-diffusion system. Cohen, Neu, and Rosales (1978)

5.
$$\ddot{x} + \epsilon\lambda^2 h\dot{x} = Q \sin rt - \lambda^2 x - \epsilon\lambda^2(cy^2 x + \alpha x^3),$$

$$\ddot{y} + \epsilon w^2 F(y, \dot{y}) = -w^2 y - \epsilon w^2 \; [bx^2 y + \beta y^3],$$

where $F(y, \dot{y})$ is nonlinear friction, $F(0,0) = 0; h > 0, \lambda, w, b, \alpha, \beta, Q$, and r are constants. It is assumed that for $b = 0$, the second equation of the system has a stable limit cycle. It is also assumed that $w^2 = r^2 + \epsilon\Delta w^2$ and there is no resonant relation between r and λ.

The solution is sought in the form

$$\begin{aligned} x &= q \sin rt + a_1 \cos\theta_1, q = \theta/(\lambda^2 - r^2), \\ \dot{x} &= rq\cos rt - \lambda a_1 \sin\theta_1, \\ y &= a\cos\theta, \dot{y} = -\Omega a \sin\theta, \end{aligned} \tag{1}$$

where Ω is the frequency of oscillation of coordinate y. By substitution of (1) in the given system and suitable transformation a system of ODE for $a_1, \phi = \theta_1 - \lambda t, a, \psi = \theta - \Omega t$

is written and solved approximately. The particular case $\Omega = r$ is studied in detail. Dao (1975)

6.

$$D_1 y_1'' + \psi(x, y_1, y_2, y_1', y_2') = 0,$$

$$D_2 y_2'' - \psi(x, y_1, y_2, y_1', y_2') = 0,$$

$y_1(0) = A, y_1(x_1) = B, y_2'(0) = 0, y_2'(x_1) = 0$, where $A, B \in R$, and $D_1, D_2 > 0$ are constants, $\psi \in C(I \times R^4)$, and $I = (0, x_1)$.

Such problems arise in biochemistry. The following result is proved: Suppose that ψ satisfies the following conditions:

(a) For some constants h_2 and $H_2(-\infty < h_2 < H_2 < \infty)$, the inequalities

$$\begin{aligned} \psi(x, y_1, H_2, y_1', 0) &\geq 0 \\ \psi(x, y_1, h_2, y_1', 0) &\leq 0 \end{aligned}$$

hold for $\forall x \in I$ and $\forall\ y_1, y_1' \in R$.

(b) If $M > 0$, then there is a constant $B(M)$ such that

$$|\psi(x, y_1, y_2, y_1', y_2')| \leq B(M)(1 + y_1'^2 + y_2'^2),$$

for $\forall(x, y_1', y_2') \in I \times R^2, |y_1| + |y_2| \leq M$. Then the given BVP has a solution for all A and B. Some other variants of this result are also proved.

Bespalova and Klokov (1985)

4.12 $h_i(x, y_1, y_2, y_1', y_2')y_i'' + g_i(x, y_1, y_2, y_1', y_2') = f_i(x, y_1, y_2)$ $(i = 1, 2)$

1.

$$y'' - yz'^2 - \epsilon y^3 z' = \lambda y, \quad 2y'z' + yz'' + 3\epsilon y^2 y' = 0.$$

This system results from the nonlinear Schrödinger equation

$$i\psi_t + \psi_{xx} + \epsilon i(\psi^2\psi^*)_x = 0, \epsilon^2 = 1,$$

by the substitution $\psi(x, t) = \phi(x)\exp(i\lambda t)$ with arbitrary real λ, and $\phi(x) = \sigma(x)\exp[i\theta(x)]$ and then changing the notation: $\sigma \to y, \theta \to z$. It follows from the second of the given system that

$$z' = \alpha/(kw) - (3/4)\epsilon k w, w = k^{-1}y^2, k^2 = 4, \tag{1}$$

where α is an arbitrary constant. Now the first of the given DE gives

$$w'' = w'^2/2w - 3w^3/2 - (\alpha\epsilon - 2\lambda)w + \alpha^2/2w. \tag{2}$$

Equation (2) has a first integral

$$w'^2 = -w^4 - 2(\alpha\epsilon - 2\lambda)w^2 + 4Kw - \alpha^2, \tag{3}$$

where K is a constant of integration. Equation (3) has solution in terms of elementary functions or elliptic functions depending on the zeros of the RHS. Gromak (1987)

2. $$yy'' - y'^2 + zz' = -xyz, zz'' - z'^2 = -2yz.$$

The meromorphic solution of the first Painlevé equation is shown to be expressible as $w(x) = y(x)/z(x)$, where $y(x)$ and $z(x)$ satisfy the given system and are entire functions. Lukashevich (1970)

3. $$yy'' - y'^2 + zz' = -\alpha yz, \quad zz'' - z'^2 = -y^2,$$

where α is a constant.

It is shown that the meromorphic solution of the second Painlevé equation can be written as $w(x) = y(x)/z(x)$, where $y(x)$ and $z(x)$ satisfy the given system and are entire functions. Lukashevich (1970)

4. $$yy'' - y'^2 + 2yz' = 2\beta z^2, \quad zz'' - z'^2 = -2xyz - y^2,$$

where β is a constant.

The meromorphic solution of the fourth Painlevé equation is shown to have the form $w(x) = y(x)/z(x)$, where $y(x)$ and $z(x)$ satisfy the given system and are entire functions. Lukashevich (1970)

5. $$yy'' = -xz^\alpha, \quad (1-\sigma)y'z' + yz'' = -\sigma ay, 0 \le x \le 1,$$

with $y'(0) = 0, z(0) = b, y(1) = 0, z(1) = 1$. Here α, σ, a, and b are given constants. Conforto (1941)

6. $$yy'' - y'^2 - a_0yy' - a_1zz' - a_2yz' - a_3y'z = a_4y^2 + a_5yz + a_6z^2,$$
$$zz'' - z'^2 - b_0yy' - b_1zz' - b_2yz' - b_3y'z = b_4y^2 + b_5yz + b_6z^2,$$

where $a_i = a_i(x), b_i = b_i(x)$ $(i = 1, 2, \ldots, 6)$.

It is shown that the coefficients $a_i(x)$ and $b_i(x)$ $(i = 0, 1, \ldots 6)$ can be chosen such that any solution of a Painlevé equation will be of the form $w(z) = y(x)/z(x)$, where $y(x)$ and $z(x)$, satisfying the given system, can only have singularities coinciding with fixed singularities of Painlevé equations. Lukashevich (1970)

7. $$y'' + (1/2)xyz' - yz'^2 - \epsilon y^3z' = -ky^{-3},$$
$$yz'' - (1/2)xy' + 2y'z' + 3\epsilon y^2y' = (1/4)y.$$

This system arises from the nonlinear Schrödinger equation

$$i\psi_t + \psi_{xx} + \epsilon i(\psi^2\psi^*)_x + kt^{-2}\psi^{-1}\psi^{*-2} = 0, \quad \epsilon^2 = 1,$$

by the ansatz

$$\psi(x,t) = t^{-1/4}\sigma(\eta)\exp[i\theta(\eta)], \quad \eta = xt^{-1/2}$$

and then writing (y, z) for σ and θ and x for η. Multiplying the second of the given system by y and integrating, we have

$$y^2z' = y^2x/4 - 3\epsilon y^4/4 + r, \tag{1}$$

where r is a real constant of integration. Substituting for z' from (1) in the first of the given system, we have

$$y'' + (3/16)y^5 - (1/4)\epsilon x y^3 + [(1/16)x^2 + (1/2)\epsilon r]y + (k - r^2)y^{-3} = 0. \qquad (2)$$

Now, putting $y^2 = \lambda w, x = \mu z, \lambda^4 = -1$, and $\mu = -2\epsilon\lambda$ in (2) we obtain the fourth Painlevé equation

$$w'' = w'^2/(2w) + 3w^3/2 + 4zw^2 + 2(z^2 - \alpha)w + \beta/w$$

when $\alpha = 2r\epsilon\lambda^2$ and $\beta = 8(r^2 - k)$. Gromak (1987)

8. $\qquad (xy')' = -xa(y)zy, \quad (xyz')' = -x(b(y) - c(y)z)z,$

$y'(0) = 0, y(1) = 1, z'(0) = 0, z(1) = 0.$

The given problem arises in gas-flow discharge theory. Here $a, b, c \in C^1$ are complicated (positive) functions for $y > 0$. Only solutions with $y(x), z(x) > 0, 0 \leq x < 1$ are of physical interest. Bounds for solutions of this problem are found. Computational results are quoted. Bushard (1978)

9. $\qquad (xy')' = -xa(y)zy, \quad (xyz')' = -x\{b(y) - c(y)z\}z,$

$y'(0) = 0, y(X) = y_w > 0, z'(0) = 0, z(X) = 0, X$ and y_w are constants, and $a(y) = P/\{y(1+Dy^{1/3})\}, b(y) = A \exp\{-B/y^{4/3}\}/\{y^{1/3}(1+Dy^{1/3})\}, c(y) = C/\{y^{4/3}(1+Dy^{1/3})\}^2$, where A, B, C, D, and P are constants.

The given problem arises in a simplified version of a three-species, three-moment treatment of an axisymmetric electric discharge plasma. The principal result is the identification of a region in which solutions must originate. Some numerical results are also provided. Bushard (1976)

10.
$$x(yy'' - y'^2) + yy' = \delta x z^2 + \beta y z,$$
$$x(zz'' - z'^2) + zz' = -\gamma x y^2 - \alpha y z,$$

where α, β, γ, and δ are constants.

The meromorphic solution of the third Painlevé equation is shown to have the form $w(x) = y(x)/z(x)$, where $y(x)$ and $z(x)$ satisfy the given system and are entire functions. Lukashevich (1970)

11.
$$x^2(yy'' - y'^2) + xyy' + xyz' = -(\gamma x + 2\beta)yz + 2\beta z^2,$$
$$x^2(zz'' - z'^2) + xzz' + xyz' = -2\alpha y^2 + 2\alpha y z,$$

where α, β, and γ are constants.

The meromorphic solutions of the fifth Painlevé equation are shown to have the form $w(x) = y(x)/z(x)$, where $y(x)$ and $z(x)$ satisfy the given system and have singularities coinciding with the fixed singularities of the fifth Painlevé equation. $y(x)$ and $z(x)$ are entire functions. Lukashevich (1970)

12. $$x(x-1)^2(yy''-y'^2)+(x-1)^2yy'+x(x-1)yz'=-(\gamma+\delta)yz,$$

$$x^2(x-1)^2(zz''-z'^2)+x(x-1)^2zz'+x(x-1)y'z=-2\alpha y^2+(\alpha+\beta)(x+1)yz-2\beta xz^2,$$

where α,β,γ, and δ are constants.

The meromorphic solutions of the sixth Painlevé equation are shown to have the form $w(x)=y(x)/z(x)$, where $y(x)$ and $z(x)$ satisfy the given system and have singularities coinciding with the fixed singularities of the sixth Painlevé equation. Lukashevich (1970)

13. $$x^2(x-1)^2(yy''-y'^2)+x(x-1)^2yy'+x(x-1)yz'=-2\alpha z^2+\alpha(x+1)yz,$$

$$x(x-1)^2(zz''-z'^2)-(x^2-1)yz'+x(x-1)yy'=-[(\beta+\gamma+\delta)x+(\beta-\gamma-\delta)]yz+2\beta xy^2,$$

where α,β,γ, and δ are constants.

The given system is equivalent to the sixth Painlevé equation, whose arbitrary meromorphic solution is expressed as the ratio $z(x)/y(x)$, where $z(x)$ and $y(x)$ are entire solutions of the given system. Lukashevich (1972)

14. $$(hy')'=M^2sy-1,\quad (kz')'+h(y')^2=-M^2sy^2,$$

$y'(0)=0, y(1)=-K, z'(0)=0, z(1)=0$, where $s=(1+Nz)^\alpha$, $k=(1+Nz)^\beta, h=(1+Nz)^\gamma; \alpha,\beta,\gamma,N$, and M are constants. The given problem arises in magnetohydrodynamic flows and is solved numerically. Kubicek and Hlavacek (1983), p.96

5

FIFTH ORDER EQUATIONS

5.1 Fifth Order Single Equations

1. $$\boldsymbol{y^{\text{v}} - yy' + xy' + 4y = 0,}$$

$y \to 0$ as $x \to -\infty$ and $y \to 5x$ as $x \to \infty$ or $y \to 5x$ as $x \to -\infty$ and $y \to 0$ as $x \to \infty$.

The given DE describes a self-similar form solution of a fifth order nonlinear dispersive evolution equation. Some numerical experiments on the given DE prove the asymptotic character of these solutions with respect to some sets of ICs. Il'ichev (1990)

2. $$\boldsymbol{y^{\text{v}} + 60y'y'' - 16y' = 0.}$$

The given DE appears as a traveling wave form of a fifth order dispersive equation $u_t + \beta u_x u_{xx} + u_{xxxxx} = 0$ and has a first integral

$$y^{\text{iv}} + 30y'^2 - 16y = 0, \tag{1}$$

if we set the constant of integration equal to zero. Equation (1) has a solitary wave solution

$$y = \operatorname{sech}^2 x. \tag{2}$$

The solution (2) may be obtained using a series method [see Eqn. (4.2.19)]. Hereman, Banerjee, Korpel, and Assanto (1986)

3. $$\boldsymbol{y^{\text{v}} - 52y''' + \{2704/\beta^2\}\alpha yy' + 576y' = 0,}$$

where α and β are constants.

The given DE arises from a traveling wave form of a Korteweg–deVries equation with additional fifth order dispersion term. It has a first integral

$$y^{\text{iv}} - 52y'' + \{1352/\beta^2\}\alpha y^2 + 576y = 0 \tag{1}$$

if we put constant of integration equal to zero. Equation (1) has a solitary wave solution

$$y = -\{105/169\}\{\beta^2/\alpha\} \operatorname{sech}^4 x. \tag{2}$$

The solution (2) may be found by the series method [see Eqn. (4.2.19)]. Hereman, Banerjee, Korpel, and Assanto (1986)

4. $$\boldsymbol{y^{\mathrm{v}} + \{\beta/(\gamma K^2)\}y''' + \{1/(\gamma K^5)\}\,\{\alpha K y^{\nu-1} - w\}y' = 0,}$$

where $\beta, \nu \neq 0, K \neq 0, \alpha, \gamma$, and w are constants.

Exact solitary wave solutions can be found in the form

$$y(x) = A/\{\lambda + \cosh x\}^n, \tag{1}$$

where $n = 2/(\nu - 1), \lambda = \pm 1$,

$$\begin{aligned} A &= \{-(\gamma/\alpha)K^4[1/(\nu-1)]^4[2(\nu+1)(\nu+3)(3\nu+1)]\}^{1/(\nu-1)}, \\ K &= \pm[-(\beta/\gamma)(\nu+3)(\nu-1)^2/\{4\nu+12+(\nu+1)^2(\nu+3)\}]^{1/2}, \\ w &= 4\beta(\nu-1)(\nu+1)^2(-\beta/\gamma)^{3/2}[(\nu+3)/\{4\nu+12+(\nu+1)^2(\nu+3)\}]^{5/2}. \end{aligned}$$

A similar result corresponding to $\lambda = 0$ in (1) may also be derived. Dai and Dai (1989)

5. $$\boldsymbol{y^{\mathrm{v}} + (6r^2/\alpha)yy''' + (6/\alpha)(10\alpha - r^2)y'y'' + (36r^2/\alpha)y^2y' - 16y' = 0,}$$

where r and α are constants.

The given DE arises from a traveling wave form of a fifth order generalized Korteweg–deVries equation

$$u_t + \alpha u^2 u_x + \{(10\alpha - \gamma^2)/\gamma\}u_x u_{xx} + \gamma u u_{xxx} + u_{xxxxx} = 0.$$

It has an exact solitary wave solution

$$y = \mathrm{sech}^2 x, \tag{1}$$

provided that $\alpha = 1$. The solution (1) may be found by the series method [see Eqn. (4.2.19)]. Hereman, Banerjee, Korpel, and Assanto (1986)

6. $$\boldsymbol{y^{\mathrm{v}} - dyy''' - ey'y'' - jy^2y' = 0,}$$

where d, e, and j are constants.

The given DE has a first integral

$$y^{\mathrm{iv}} = dyy'' + [(e-d)/2]y'^2 + (j/3)y^3 + K, \tag{1}$$

where K is an arbitrary constant. Equation (1) may be shown to have movable critical points. The same is true of the original DE. Exton (1971)

7. $$\boldsymbol{y^{\mathrm{v}} + 12(y'y''' + y''^2) - Ky'' - K^2/6 = 0,}$$

where K is a constant.

Integrating once, we get the fourth order DE

$$y^{\mathrm{iv}} + 12y'y'' = Ky' + (K^2/6)x + K_0, \tag{1}$$

where K_0 is another arbitrary constant. Equation (1) can again be integrated to yield

$$y''' + 6y'^2 = Ky + (K^2/12)x^2 + K_0 x + K_1, \tag{2}$$

where K_1 is constant of integration. Equation (2) may be shown to have a solution with fixed critical points. Exton (1971)

8. $$y^{\text{v}} + 12(y'y''' + y''^2) - Kxy'' - K^2x^2/6 = 0,$$

where K is a constant.

The given DE, on integration, yields

$$y^{\text{iv}} + 12y'y'' - Kxy' + Ky - K^2x^3/18 + K_1 = 0, \tag{1}$$

where K_1 is another constant. Equation (1) is known to have fixed critical points in its solutions. Exton (1971)

9. $$y^{\text{v}} - by'y''' + (b^2/15)y'^3 = 0,$$

where b is a constant.

The given DE is equivalent to the system

$$y' = z, z^{\text{iv}} = bzz'' - (b^2/15)z^3. \tag{1}$$

The system (1) has movable critical points, and so has the original DE. Exton (1971)

10. $$y^{\text{v}} - b(y'y''' + y''^2) = 0,$$

where b is a constant.

Integrating once, we have

$$y^{\text{iv}} = by'y'' + K, \tag{1}$$

where K is a constant. Equation (1) can again be integrated:

$$y''' = by'^2/2 + Kx + K_1, \tag{2}$$

where K_1 is another arbitrary constant. Equation (2) and hence the original DE may be shown to have fixed critical points. Exton (1971)

11. $$y^{\text{v}} - by'y''' - (2b/3)y''^2 + (2b^2/27)y'^3 = 0,$$

where b is a constant.

The given DE is equivalent to the system

$$y' = z, z^{\text{iv}} = bzz'' + (2b/3)z'^2 - (2b^2/27)z^3. \tag{1}$$

The system (1) may be shown to have movable critical points, so the same is true of the original DE. Exton (1971)

12. $$y^{\text{v}} - by'y''' - (4b/3)y''^2 + (b^2/9)y'^3 = 0,$$

where b is a constant.

The given DE is equivalent to the system

$$\begin{aligned} y' &= z, \\ z^{\text{iv}} &= bzz''' + (4b/3)z'^2 - (b^2/9)z^3. \end{aligned} \tag{1}$$

The system (1) and hence the given DE may be shown to have movable critical points. Exton (1971)

13.
$$y^{\mathrm{v}} + 2yy^{\mathrm{iv}} + 8y'y''' + 6y''^2 - d_0(x)(y''' + 2yy'' + 2y'^2)$$
$$-g_0(x)(y'' + 2yy') - j_0(x)(y^2 + y') - k_0(x) = 0,$$
where d_0, g_0, j_0, and k_0 are arbitrary analytic functions of x.

The given DE is equivalent to the system

$$y = w'/w, w'' = zw, z^{\mathrm{iv}} = d_0 z'' + g_0 z' + j_0 z + k_0, \tag{1}$$

which, in view of the linearity of the last of Eqns. (1), has fixed critical points; the same, therefore, holds for the given DE. Exton (1971)

14.
$$y^{\mathrm{v}} + 3yy^{\mathrm{iv}} + 13y'y''' + 10y''^2 + 4y^2y''' + 24yy'y'' + 8y'^3 + 2y^3y'' + 6y^2y'^2$$
$$-d_0(x)[y''' + 3yy'' + (7/2)y'^2 + 4y^2y' + y^4/2] - k_0(x) = 0,$$
where d_0 and k_0 are arbitrary analytic functions of x.

The given DE is equivalent to the system

$$y''' + 3yy'' + (7/2)y'^2 + 4y^2y' + y^4/2 = z,$$
$$z'' = d_0 z + k_0,$$

which may be shown to have fixed critical points in its solution. The same is, therefore, true for the original DE. Exton (1971)

15.
$$y^{\mathrm{v}} + 4yy^{\mathrm{iv}} + 14y'y''' + 10y''^2 + 12y'^3 + 6y^2y'''$$
$$+ 36yy'y'' + 4y^3y'' + 12y^2y'^2 = 0.$$

The given DE is equivalent to the system

$$y = w'/w, w^{\mathrm{iv}} - zw = 0, z'' = 0.$$

Hence w, and therefore the original DE, has fixed critical points. Exton (1971)

16.
$$\begin{aligned} &y^{\mathrm{v}} + 4yy^{\mathrm{iv}} + 14y'y''' + 10y''^2 + 12y'^3 + 6y^2y''' + 36yy'y'' + 4y^3y'' \\ &+ 12y^2y'^2 - [f(x) + g(x)]y''' - [2f(x) + 4g(x)]yy'' - [2f'(x) + h(x)]y'' \\ &- [2f(x) + 3g(x)]y'^2 - 6g(x)y^2y' - 4f'(x)yy' \\ &- [f''(x) + 2h'(x) - f(x)g(x)]y' - g(x)y^4 - [f''(x) - f(x)g(x)]y^2 \\ &- [h''(x) - g(x)h(x)]y - k(x) = 0, \end{aligned}$$

where $f(x), h(x), g(x)$, and $k(x)$ are arbitrary analytic functions of x.

The given DE is equivalent to the system

$$y = w'/w, w^{\mathrm{iv}} = fw'' + hw' + zw, z'' = gz + k, \tag{1}$$

which, in view of the linearity of the last of Eqns. (1), has fixed critical points in its solution. The same is, therefore, true of the original system. Exton (1971)

17. $$y^{\mathrm{v}} + 5yy^{\mathrm{iv}} + 15y'y''' + 10y''^2 + 10y^2y''' + 50yy'y'' + 15y'^3 + 10y^3y'' + 30y^2y'^2 + 5y^4y' - d(x)y''' - 3d(x)yy'' - [d'(x) + e(x)]y'' - 3d(x)y'^2 - 3d(x)y^2y' - [3d'(x) + 2e(x)]yy' - [e'(x) + f]y' - d'(x)y^3 - e'(x)y^2 - f'(x)y - k(x) = 0,$$

where d, e, f, and k are arbitrary analytic functions of x.

The given DE is equivalent to the system

$$y = w'/w, w^{\mathrm{v}} = dw''' + ew'' + fw' + zw, z' = k,$$

which, in view of the linearity of second of the system, has fixed critical points in its solution. The same is, therefore, true for the given DE. Exton (1971)

18. $$y^{\mathrm{v}} + 6yy^{\mathrm{iv}} + 15y'y''' + 10y''^2 + 15y^2y''' + 60yy'y'' + 15y'^3 + 45y^2y'^2 + 20y^3y'' + 15y^4y' + y^6 - d_0(x)(y''' + 4yy'' + 6y^2y' + 3y'^2 + y^4) - g_0(x)(y'' + 3yy' + y^3) - j_0(x)(y^2 + y') - k_1(x)y - k_0(x) = 0,$$

where d_0, g_0, j_0, k_1, k_0 are arbitrary analytic functions of x.

The given DE is equivalent to the system

$$y = w'/w, w^{\mathrm{vi}} = d_0w^{\mathrm{iv}} + g_0w''' + j_0w'' + k_1w' + k_0w, \tag{1}$$

which in view of the linearity of second of the system (1) has fixed critical points in its solution. The same is, therefore, true of the given DE. Exton (1971)

19. $$y^{\mathrm{v}} - a(yy^{\mathrm{iv}} + 4y'y''' + 3y''^2) = 0,$$

where a is an arbitrary constant.

The given DE may be shown to have fixed critical points. In particular, choosing $a = -2$ (which results from a simple transformation of the given DE), we have

$$y^{\mathrm{v}} + 2yy^{\mathrm{iv}} + 2[4y'y''' + 3y''^2] = 0. \tag{1}$$

Equation (1) is equivalent to the system

$$y = w'/w, w'' = zw, z^{\mathrm{iv}} = 0,$$

showing that the Eqn. (1) has fixed critical points. Exton (1971)

20. $$y^{\mathrm{v}} - ayy^{\mathrm{iv}} - 4ay'y''' - 3ay''^2 + (2a^2/3)y'^3 + 2a^2yy'y'' + (a^2/3)y^2y''' = 0.$$

By a simple change of variable the given DE is equivalent to an equation with $a = -3$:

$$y^{\mathrm{v}} + 3yy^{\mathrm{iv}} + 12y'y''' + 9y''^2 + 6y'^3 + 18yy'y'' + 3y^2y''' = 0. \tag{1}$$

Equation (1) is equivalent to the system

$$y = w'/w, w''' - zw = 0, z''' = 0. \tag{2}$$

Therefore, the system (2) and hence the original DE have fixed critical points. Exton (1971)

21. $$\boldsymbol{y^{\text{v}} - ayy^{\text{iv}} - 3ay'y''' - 2ay''^2 + (2a^2/5)y^2y''' + 2a^2yy'y''}$$
$$\boldsymbol{+ (3a^2/5)y'^3 - (2a^3/25)y^3y'' - (6a^3/25)y^2y'^2 + (a^4/125)y^4y' = 0.}$$

By a simple change of variables this DE may be shown to be equivalent to one obtained by replacing a by -5; we have

$$\begin{aligned} y^{\text{v}} + 5yy^{\text{iv}} + 15y'y''' + 10y''^2 + 10y^2y''' + 50yy'y'' \\ + 15y'^3 + 10y^3y'' + 30y^2y'^2 + 5y^4y' = 0. \end{aligned} \tag{1}$$

Equation (1) is equivalent to the system

$$y = w'/w, w^{\text{v}} = zw, z' = 0. \tag{2}$$

w has fixed critical points; hence so has y. Exton (1971)

22. $$\boldsymbol{y^{\text{v}} - ayy^{\text{iv}} - (5a/2)y'y''' - (5a/3)y''^2 + (5a^2/12)y^2y''' + (5a^2/3)yy'y'' + (5a^2/12)y'^3}$$
$$\boldsymbol{- (5a^3/54)y^3y'' - (5a^3/24)y^2y'^2 + (5a^4/432)y^4y' - (a^5/7776)y^6 = 0.}$$

A simple change of variables shows that this DE is equivalent to one obtained by replacing a by -6:

$$\begin{aligned} y^{\text{v}} + 6yy^{\text{iv}} + 15y'y''' + 10y''^2 + 15y^2y''' + 60yy'y'' \\ + 15y'^3 + 20y^3y'' + 45y^2y'^2 + 15y^4y' + y^6 = 0. \end{aligned} \tag{1}$$

Equation (1) is equivalent to the system

$$y = w'/w, w^{\text{vi}} = 0 \tag{2}$$

and has fixed critical points. Exton (1971)

23. $$\boldsymbol{y^{\text{v}} - ayy^{\text{iv}} - 4[(7-k^2)/(1-k^2)]ay'y''' - [3(9-k^2)/(1-k^2)]ay''^2 + [(12a^2)/(1-k^2)]}$$
$$\boldsymbol{(y^2y''' + 6yy'y'' + 2y'^3) - [12a^3/(1-k^2)](y^3y'' + 3y^2y'^2) = 0,}$$

where a and k are constants.

If we replace a by $(1-k^2)/2$, we arrive at a DE that is equivalent to the system

$$y' = [(1-k^2)/4]y^2 + u, \tag{1}$$

$$u^{\text{iv}} = 12(uu'' + u'^2). \tag{2}$$

The system (1)–(2) and hence the original DE may be shown to have fixed critical points; we may note that Eqn. (1) is of Riccati type, while (2) may be solved. Exton (1971)

24. $$\boldsymbol{y^{\text{v}} + (1/2)F^{\text{iv}}(y)y'^3 + (3/2)F'''(y)y'y'' + F''(y)y'''/2 = 0,}$$

where $F'(y) = dF/dy$, etc., for a given function F of y.

Integrating the given DE twice and putting the first constant of integration equal to zero, we have

$$2y''' + F''(y)y' = A, \tag{1}$$

where A is a constant. Equation (1) can be integrated to give

$$2y'' + F'(y) = Ax + K, \tag{2}$$

where K is another constant. If we choose $F'(y) = -12y^2$ in (2), and $A \neq 0, K = 0$, we get the first Painlevé equation. Putting $A = 0$ and $K = 0$ in (2) and integrating, we have

$$y'^2 + F(y) = E, \tag{3}$$

where E is an arbitrary constant. Choosing $E = 0$ in (3) and $F(y) = \alpha y^{5/2} - \beta y^{3/2} (\alpha > 0, \beta > 0)$ and integrating, we have

$$y = (\beta/\alpha)\mathrm{cn}^4\left[1/\{2(2^{1/2})\}(\alpha\beta)^{1/4}(x - x_0), (1/2)^{1/2}\right],$$

where $\mathrm{cn}(z, k)$ is a Jacobi cn function of modulus k with a constant x_0.

If in the discussion above, we choose $A = E = K = 0$ and $F(y)$ to be one of the following:

(a) $\gamma y^{5/2} - \delta y^2, \quad \gamma > 0, \delta > 0,$

(b) $\alpha y^3 - \beta y^2, \quad \alpha > 0, \beta > 0,$

(c) $\gamma y^4 - \delta y^2, \quad \gamma > 0, \delta > 0,$

(d) $\epsilon y^4 + \mu y^3 - \nu y^2, \quad \epsilon > 0, \mu > 0, \nu > 0.$

we can again express the solution in terms of elliptic functions.

It may be remarked that all the cases listed above describe steady-state solutions of higher order or modified Korteweg–deVries equations. Yamamoto and Takizawa (1981)

25.
$$\boldsymbol{y^4 y^{\mathrm{v}} + (11k^2 - 6k - 6k^3 + k^4)y'^5 + 10(2k - 3k^2 + k^3)yy'^3 y''}$$
$$\boldsymbol{+ 10ky^3 y'' y''' + 15(k^2 - k)y^2 y' y''^2 + 10(k^2 - k)y^2 y'^2 y''' + 5ky^3 y' y^{\mathrm{iv}} = 0,}$$

where k is a constant.

An exact solution is

$$y = \{C_1 x^4 + C_2 x^3 + C_3 x^2 + C_4 x + C_5\}^{1/(k+1)}.$$

Alidema (1977)

26.
$$\boldsymbol{9y''^2 y^{\mathrm{v}} - 45y'' y''' y^{\mathrm{iv}} + 40y'''^3 = 0.}$$

Put $y'' = z$ to obtain Eqn. (3.9.38). Hence the solution can be found to be

$$(y + C_1 x + C_2)^2 = C_3 x^2 + C_4 x + C_5,$$

where C_i $(i = 1, \ldots, 5)$ are arbitrary constants. Kamke (1959), p.603

5.2 Fifth Order Systems

1.
$$\dot{y}_1 = y_2, \quad \dot{y}_2 = y_3,$$
$$\dot{y}_3 = -cy_1y_3 - ny_2^2 + 1 - y_4^2 + sy_2, \quad \dot{y}_4 = y_5,$$
$$\dot{y}_5 = -cy_1y_2 - (n-1)y_2y_4 + s(y_4 - 1),$$

$y_1(0) = y_2(0) = y_4(0) = 0, y_2(t_f) = 0, y_4(t_f) = 1$. Here $c = (3-n)/2$ and $n = -0.1, s = 0.2, t_f = 3.5$.

The given problem arises in boundary layer theory and is solved numerically, using one-parameter embedding. Kubicek and Hlavacek (1983), p.185

2.
$$\dot{y}_1 = y_2, \quad \dot{y}_2 = y_3, \quad \dot{y}_3 = -\{(3-n)/2\}y_1y_3 - ny_2^2 + 1 - y_4^2 + sy_2,$$

$$\dot{y}_4 = y_5, \quad \dot{y}_5 = -\{(3-n)/2\}y_1y_5 - (n-1)y_2y_4 + s(y_4 - 1),$$

with $y_1(0) = 0, y_2(0) = 0, y_4(0) = 0, y_2(t_f) = 0, y_4(t_f) = 1$, where t_f is the final value of t.

The given BVP with specific value of the constants $n = -0.1, s = 0.2, t_0 = 0, t_f = 11.3$ was found to be sensitive to changes to ICs. It was solved numerically using "continuation methods." Conventional shooting methods fail. Roberts and Shipman (1972), p.158

3.
$$\dot{x} = -Pr(x - y), \quad \dot{y} = -xz + rx - y, \quad \dot{z} = xy - bz + \tilde{x}\,\tilde{y},$$

$$\dot{\tilde{x}} = -Pr(\tilde{x} - \tilde{y}), \quad \dot{\tilde{y}} = -\tilde{x}z + r\tilde{x} - \tilde{y},$$

where Pr, r, and b are constants.

The given five-dimensional system describes the onset of self-pulsing in lasers. It is possible to find an exact integral

$$\dot{\tilde{x}}\,x - \dot{x}\,\tilde{x} = \exp[-(1 + Pr)t] \cdot \text{ const.},$$

which shows that $\tilde{x}\,x = \dot{x}\,\tilde{x}$ as $t \to \infty$. In this limit the ratio $\tilde{x}/x$ is constant, which may be chosen to be zero. Therefore, $\tilde{x} = 0$ in this limit. The penultimate equation of the system then implies that $\tilde{y} = 0$ in the same limit. The system then reduces to the Lorenz system, which proves the mathematical analogy of the given system in the limit $t \to \infty$.

Some inferences regarding the instability of the given system from the Lorenz system are drawn. Graham (1976)

4.
$$\dot{x}_1 = -x_2, \dot{x}_2 = x_1 + x_3x_4 \quad \dot{x}_3 = -x_2x_4, \quad \ddot{x}_4 + \mu^2 x_4 = \alpha x_1,$$

where μ and α are constants.

The given system is referred to as a Jaynes–Cummings model. Painlevé analysis shows that the system has the following expansions about a movable singularity:

$$x_1 = \sum_{j=0}^{\infty} x_{1,j}(t - t_1)^{j-3}, \quad x_2 = \sum_{j=0}^{\infty} x_{2,j}(t - t_1)^{j-4},$$
$$x_3 = \sum_{j=0}^{\infty} x_{3,j}(t - t_1)^{j-4}, \quad x_4 = \sum_{j=0}^{\infty} x_{4,j}(t - t_1)^{j-1},$$

with three arbitrary constants, including t_1. Using the analysis of dominant terms and some inspection, the first two integrals of the given system are obtained:

$$x_1^2 + x_2^2 + x_3^2 = I_1,$$
$$\alpha x_3 - \alpha x_1 x_4 + (1/2)\mu^2 x_4^2 + (1/2)(\dot{x}_4)^2 = I_2.$$

By comparison of solution of the dominant terms with expansions of the elliptic functions, a particular solution for x_4 is found to be

$$x_4 = \hat{x}_4 \,\mathrm{dn}(t, k)$$

where $\hat{x}_4^2 = 16\Omega^2$ and

$$\begin{aligned} k &= 2\{1 + (1/c)(1 - (1+c)^{1/2})\}, \\ \Omega^2 &= c(\mu^2 - 1/3)/[4\{(1+c)^{1/2} - 1\}], \\ c &= -(\mu^2 - 1/3)^{-2}[(4/3)\{\alpha^2 - 4(\mu^2 - 1/9)^3\}^{1/2} + (\mu^2 - 1/9)(\mu^2 - 17/9)]. \end{aligned}$$

The functions x_1, x_2, x_3 may be found from the given system, which is now linear in these variables. Steeb, Euler, and Mulser (1991a)

5. $$y'' - hy' = y^2 + z^2 - 1, \quad z'' - hz' = 0, \quad h' = -y,$$

where $y(0) = h(0) = 0, z(0) = \Omega$, and $y \to 0, z \to 1$ as $x \to \infty$: Ω is a parameter.

The given BVP arises in the boundary layer near the equator of a rotating sphere in a rotating fluid. It is solved numerically in different ranges of the parameter Ω:

$$-2.01 \lesssim \Omega \lesssim 1; \Omega \sim 1, \Omega \sim -2.01; -2.01 \lesssim \Omega \lesssim -1; \Omega \sim -1.$$

The results are shown graphically. In certain regions it is found that at least three solutions exist. The question of nonuniqueness is discussed by referring to the numerically obtained results. Ingham (1982)

6. $$y'' - (Re)^{1/2} wy' = Re(y^2 - z^2 + k),$$
$$z'' - (Re)^{1/2} yz' = 2\, Re \cdot yz,$$
$$w' = -2(Re)^{1/2} y, w(0) = y(0) = 0, z(0) = 1,$$

$w(1) = y(1) = 0, z(1) = s$, where Re and s are parameters.

The given system describes steady-state flow of a viscous incompressible fluid between two coaxial rotating disks and is solved numerically by the method of parametric differentiation. Kubicek and Hlavacek (1983), p.272

7. $y'' - (Re)^{1/2} wy' = Re(y^2 - w^2 + k), z'' + (Re)^{1/2} wz' = 2\, Re \cdot yz, w' = -2(Re)^{1/2} y,$

$w(0) = y(0) = w(1) = y(1) = 0, z(0) = 1, z(1) = s$. Here, k, Re, and s are constants.

The system describes flow of a viscous fluid between two coaxial rotating disks. The number of BCs is one higher than the order of the system and determines the unknown constant k. The problem is solved numerically using embedding procedures. Kubicek and Hlavacek (1983), p.189

8. $$y''' = -z, \quad z'' + \{(3 + \lambda)/4\} yz' - \lambda y' z = 0,$$

$y(0) = y'(0) = 0, z(0) = 1, y''(\infty) = z(\infty) = 0$, where λ is a constant.

The given system holds in a certain domain for the similarity solution for free convection in a vertical plate. The value of λ satisfying this system is found numerically to be 0.856. Merkin (1985)

9. $$2y''' + yy' = -Kxz, \quad (2/Pr)z'' + yz' + zy' = 0, 0 \le x < \infty,$$

$y(0) = y'(0) = 0, y'(\infty) = 1, z(0) = 1, z(\infty) = 0$. Here K and Pr are constants.

The given problem arises in combined forced and free convection flow over a horizontal plate when the wall temperature is inversely proportional to the square root of the distance from the leading edge.

Scaling and shooting arguments are used to solve the given BVP; initial value methods are employed. de Hoog, Laminger, and Weiss (1984)

10. $$y''' - y'^2 = -z, \quad z'' - 3\sigma y'z = 0,$$

$y(0) = y'(0) = 0, z(0) = 1, z(\infty) = y'(\infty) = 0$, where σ is a parameter.

For $\sigma = 1$, a numerical solution is obtained. Matched asymptotic expansions are used to treat cases for large and small σ. Kuiken (1981)

11. $$y''' + (3/4)yy'' - (1/2)y'^2 = -z \quad Dz'' + (3/4)yz' = 0,$$

$y(0) = 0, y'(0) = 0, y(\infty) = 0; z(0) = 1, z(\infty) = 0$. D is a constant.

The given system is a reduced similarity form of the basic system describing a free convection boundary layer. Although the system is invariant to the associated group $y^1 = \mu^{-1}y, z^1 = \mu^{-4}z$, and $x^1 = \mu x$, the simplification thus provided would not advance the solving of this fifth order system to any easier level. The zero boundary conditions at infinity cannot be altered by the permitted scaling. Trial and error in solving the given BVP is found to be necessary. Dresner (1983), p.67

12. $$y''' + yy'' + y'^2 = -z^2 + 1, \quad z'' + yz' = 0,$$

$y(0) = y'(0) = z(0) = 0, y'(\infty) = 0, z(\infty) = 1$.

It is shown that the given system has no solution. Rott and Lewellen (1966)

13. $$y''' + yy'' + (1/2)(z^2 - y'^2) = 1/2, \quad z'' + yz' - zy' = 0,$$

$y'(\pm\infty) = 0, z(\pm\infty) = \pm 1$.

The given system describes axially symmetric flows with constant angular velocities at infinity. It is shown that the given BVP has no solution. Elcrat and Siegel (1983)

14. $$y''' + yy'' - (1/2)y'^2 = (1/2)\Omega_\infty^2 - (1/2)z^2, \quad z'' + yz' - zy' = 0,$$

where $y(0) = a, y'(0) = 0, z(0) = \Omega_0, z'(\infty) = 0, z(\infty) = \Omega_\infty$, $\Omega_0 > 0$, a, and Ω_∞ are constants.

An existence theorem for the given problem arising in swirling flows is proved; shooting arguements are employed. McLeod (1969)

15. $$y'' + \sigma z y' - \lambda\sigma z' y = 0, \quad z''' + z z'' + \beta(1 - z'^2) = 0,$$

where $y(0) = 1, y(\infty) = 0, z(0) = z'(0) = 0, z'(\infty) = 1$.

The given system arises in forced convection. Knowing the behavior of y near $x \sim 0$, a perturbation approach is adopted to solve the BVP for z for β small and β large. Korpela (1991)

16. $$y''' + y y'' - \beta y'^2 = -\beta z, z'' - y z' = 0,$$

$y(0) = y'(0) = 0, y'(\infty) = 1, z(0) = z_w, z(\infty) = 1$, where β and z_w are constants.

The given system describes the boundary layer for a compressible fluid via a similarity variable. It is solved numerically, using parametric differentiation, with β as the parameter. Kubicek and Hlavacek (1983), p.271

17. $$y''' + y y'' - \lambda y'^2 = -\lambda z, \quad z'' + y z' = 0,$$

$y(0) = y'(0) = 0, z(0) = a; y'(\infty) = z(\infty) = 1$, where a is a given positive number; λ is a parameter.

Using the Schauder–Tychnov fixed-point theorem, the following result is proved: For any $a \geq 0$, there is $a_0 < 0$ such that for $\lambda_0 \leq \lambda < 0$, the given problem has at least one solution y, z with $0 < y' < 1$ on $x \in (0, \infty)$. Hastings (1971)

18. $$y''' + y y'' - \beta y'^2 = -\beta(z^2 - \sigma^2), z'' + y z' - 2\alpha y' z = 0,$$

with $y(0) = a, y'(0) = 0, z(0) = w, y'(\infty) = 0$, and $z(\infty) = \sigma > 0; \beta, \sigma$, and α are constants.

The following nonexistence and uniqueness results were proved.

(a) If $a \leq 0, w = -\sigma \neq 0, \alpha = \beta$, and $1/4 \leq \alpha \leq 1/2$, then a solution to the given BVP does not exist.

(b) If $a \leq 0, w = \sigma \neq 0, \alpha = \beta$, and $1/4 \leq \alpha \leq 1/2$, then the obvious solution $(y, z) = (a, w)$ to the BVP is unique.

Bushell (1973/74)

19. $$y''' + y y'' - \beta y'^2 = -\beta(z^2 - w^2), z'' + y z' - 2b y' z = 0,$$

$y(0) = y_0, y'(0) = y_0$ and $z(0) = z_0, y'(\infty) = 0$ and $z(\infty) = w$, with the constraints $0 < \beta < 2b$ and $z_0 > 0, w > 0$ on the constants β, b, z_0, and w.

It is proved that the given BVP under the stated conditions has a solution. Hartman (1972a)

20. $$y''' + y y'' - \lambda y'^2 = -\lambda z^n, \quad z'' + \sigma y z' - \gamma z y' = 0,$$

$y(0) = \alpha, y'(0) = \beta, z(0) = a, y'(\infty) = z(\infty) = 0$, where $\alpha \geq 0, \beta \geq 0, \gamma \geq 0, \lambda > 0, \sigma > 0$, $n > 0$, and $a > 0$, are given constants.

It is proved that the given BVP has at least one solution. Hastings (1970)

21. $$y''' + yy'' - \beta y'^2 - (2-\beta)I\alpha_0 y' = -\beta - (2-\beta)I(\beta_0 z + \alpha_0),$$

$$z'' + yz' - \beta y'z - (2-\beta)I\beta_0 y' = -(2-\beta)I(\beta_0 - \alpha_0 z),$$

$y = y' = z = 0$ at $x = 0$; $y' \to 1$ and $z \to 0$ as $x \to \infty$. Here β, I, α_0, and β_0 are constants.

The given BVP is solved by initial value methods numerically. The problem arises in the consideration of Hall effect in a boundary layer flow. Rao, Mittal, and Nataraj (1983)

22. $$y''' + 2yy'' - y'^2 = 1 - z^2, \quad z'' + 2yz' - 2y'z = 0,$$

$y(0) = y'(0) = 0, z(0) = 1/s, y'(\infty) = 0, z(\infty) = 1$, where s is a parameter.

Asymptotic solution for $x \to \infty$ is found to be

$$\begin{aligned} y &\sim A + e^{Px}\left[\frac{BP+CQ}{P^2+Q^2}\sin Qx + \frac{CP-BQ}{P^2+Q^2}\cos Qx\right] + \frac{C_y}{2P}e^{2Px}, \\ z &\sim 1 + \operatorname{sgn}(s)e^{Px}\{C\sin Qx - B\cos Qx\} + C_z e^{2Px}, \end{aligned}$$

where

$$\begin{aligned} P^2 - Q^2 &= -2AP, PQ = -AQ + \operatorname{sgn}(s), \\ C_y &= (B^2+C^2)(2Q^2-P^2)/[2\{(P^2+Q^2)^2+1)\}], \\ C_z &= (B^2+C^2)(Q^2-5P^2)/[2(P^2+Q^2)\{(P^2+Q^2)^2+1)\}]. \end{aligned}$$

The problem is solved numerically. It is shown that an axially symmetric solution of the Navier–Stokes equations, which describes the rotating fluid above a disk which itself is rotating, is nonunique. Zandergen and Dijkstra (1977)

23. $$y''' + 2(yy')' - 2SI(2y'^2 - yy'') = -6Sz^2, \quad z'' + 2(1+SI)(yz)' = 0,$$

$y(0) = 0, y'(0) = z(0) = 1, y'(\infty) = z(\infty) = 0$, where S and I are constants.

Exact solution of the given system with $I = 1$, which arises in the problem of a swirling flow from an annular orifice, is

$$y' = z = \operatorname{sech}^2\{(1+S)^{1/2}x\}.$$

Dzhaugashtin and Shelepov (1991)

24. $$y''' + 3yy'' - 2y'^2 = -z, \quad z'' + 3Pyz' = 0,$$

where P is a constant; $y(0) = 0, y'(0) = 0, y''(0) = a, z(0) = 1, z'(0) = b, y'(\infty) = 0, z(\infty) = 0$, a and b being constants.

The given Pohlhausen system of equations is solved in the five-dimensional phase space x_i, where $x_1 = y, x_2 = y', x_3 = y'', x_4 = z, x_5 = z'$, with BCs $x_1(0) = 0, x_2(0) = 0, x_3(0) = a, x_4(0) = 1, x_5(0) = b$. Numerical solution is depicted graphically. A Taylor series form for y''' and z'' is also developed. Ku (1966)

25. $$y''' + \{(3+\lambda)/4\}yy'' - \{(\lambda+1)/2\}y'^2 = -z,$$

$$(1/Pr)z'' + \{(3+\lambda)/4\}yz' - zy' = 0,$$

where λ and Pr are constants. $y = y' = 0, z = 1$ at $x = 0$ and $y' \to 0, z \to 0$ as $x \to \infty$.

The given BVP arises in the boundary layer flow on a heated vertical plate. The solution is attempted in powers of λ^{-1} and expressed in terms of a special system. Merkin (1985)

26.

$$y''' + \{(3+\lambda)/4\}yy'' - \{(1+\lambda)/2\}y'^2 = -z,$$

$$(1/Pr)z'' + \{(3+\lambda)/4\}yz' - \lambda zy' = 0,$$

$y = y' = 0, z = 1$, on $x = 0; y' \to 0, z \to 0$ as $x \to \infty$. Here λ and $Pr > 0$ are constants.

The given BVP, which arises in free convection on a vertical plate, is shown to have a solution for $-0.6 \leq \lambda \leq 0.2$. Tam (1993)

27.

$$y''' + \{(4+\mu)/5\}yy'' - \{(3+2\mu)/5\}y'^2 = -z,$$

$$(1/Pr)z'' + \{(4+\mu)/5\}yz' - \{(1+4\mu)/5\}zy' = 0,$$

$y = y' = 0, z' = -1$ at $x = 0; y' \to 0, z \to 0$ as $x \to \infty$. Here μ and $Pr > 0$ are constants.

It is assumed that $y(\infty) = c > 0$, a constant to be determined as part of the solution. It is proved that the given problem has a solution for $-1 < \mu \leq 0$. Tam (1993)

28.

$$y''' + (1/2)(m+1)yy'' - my'^2 = -m - \alpha z,$$

$$z'' + \sigma\{(1/2)(m+1)yz' + (1-2m)y'z\} = 0,$$

where α, m, and σ are constants, and, $y(0) = 0, y'(0) = 0, z'(0) = -1, y' \to 1, z \to 0$ as $x \to \infty$.

The given problem arises in a mixed-convection boundary layer. Numerical solutions for different sets of (m, α) are found. The singular case $m \to 1/5$ is shown to exhibit different behaviors when $\alpha = O(1)$ and when α is small. A large m case is also considered. $\sigma = O(1)$ is generally assumed. Merkin and Mahmood (1989)

29.

$$y''' + \{(3-n)/2\}yy'' - sy' + ny'^2 = 1 - z^2,$$

$$z'' + \{(3-n)/2\}yz' + (n-1)zy' = s(z-1),$$

$y(0) = 0, y'(x_f) = 0, y'(0) = 0, z(x_f) = 1, z(0) = 0; n, s$, and x_f are constants.

The given problem arises in boundary layer theory. The problem is very sensitive to ICs. It was solved numerically using multiple shooting methods, with the parameter values $n = -0.1, s = 0.2, x_f = 6$. Kubicek and Hlavacek (1983), p.229

30.

$$y''' + (1/6)(y'')^2uu' = 0, \quad u'' + (1/2)y'(u')^3 = 0,$$

$y(0) = u(0) = 1, y(1) = 16, u'(0) = -1, u(1) = 0.5$.

The given system is solved using quasilinearization and iteration. Kubicek and Hlavacek (1983), p.117

6

SIXTH ORDER EQUATIONS

6.1 Sixth and Specific Higher Order Single Equations

1.
$$\boldsymbol{y^{\text{vi}} - p_2 y^2 = 0,}$$

where p_2 is a constant.

The given DE may be shown to have movable critical points. Exton (1971)

2.
$$\boldsymbol{y^{\text{vi}} - h_1 y y' = 0,}$$

where h_1 is a constant.

The given DE integrates to give

$$y^{\text{v}} = (h_1/2)y^2 + K, \tag{1}$$

where K is a constant. Equation (1) may be shown to have movable critical points, and the same is therefore true of original equation. Exton (1971)

3.
$$\boldsymbol{y^{\text{vi}} - h y y''' - j y' y'' = 0,}$$

where h and j are constants.

Integrating the given DE once, we have

$$y^{\text{v}} = h y y'' + [(j - h)/2]y'^2 + c, \tag{1}$$

where c is an arbitrary constant. Equation (1) may be shown to have movable critical points. The same is, therefore, true of the given DE. Exton (1971)

4.
$$\boldsymbol{y^{\text{vi}} + 2 y y^{\text{v}} + 10 y' y^{\text{iv}} + 20 y'' y''' = 0.}$$

The given DE is equivalent to the system

$$y = w'/w, w'' = zw, z^{\text{v}} = 0,$$

which in view of last two equations of the system has fixed critical points in its solution. The same holds for the given DE. Exton (1971)

5. $$y^{\text{vi}} + 2yy^{\text{v}} + 10y'y^{\text{iv}} + 20y''y''' - d_0(x)(y^{\text{iv}} + 2yy''' + 6y'y'') - h_0(x)(y''' + 2yy'' + 2y'^2) - \ell_0(x)(y'' + 2yy') - n_0(x)(y' + y^2) - p_0(x) = 0,$$

where $d_0, h_0, \ell_0, n_0, p_0$ are arbitrary analytic functions of x.

The given DE is equivalent to the system

$$y = w'/w, w'' = zw, z^{\text{v}} = d_0 z''' + h_0 z'' + l_0 z' + n_0 z + p_0,$$

which in view of its last two members has fixed critical points. The same is, therefore, true of the given DE. Exton (1971)

6. $$y^{\text{vi}} + 3yy^{\text{v}} + 15y'y^{\text{iv}} + 30y''y''' + 3y^2 y^{\text{iv}} + 24yy'y''' + 18yy''^2 + 36y'^2 y'' = 0.$$

The given DE may be shown to be equivalent to the system

$$y = w'/w, w''' = zw, z^{\text{iv}} = 0,$$

and hence, in view of the last two equations of the system, has no movable critical points. Exton (1971)

7. $$y^{\text{vi}} + 4yy^{\text{v}} + 18y'y^{\text{iv}} + 34y''y''' + 72y'^2 y'' + 36yy''^2 + 48yy'y''' + 6y^2 y^{\text{iv}} + 4y^3 y''' + 36y^2 y'y'' + 24yy'^3 = 0.$$

The given DE is equivalent to

$$y = w'/w, w^{\text{iv}} = zw, z''' = 0,$$

which in view of the last two equations of the system has fixed critical points in its solution. The same holds, therefore, for the given DE. Exton (1971)

8. $$y^{\text{vi}} + 5yy^{\text{v}} + 20y'y^{\text{iv}} + 35y''y''' + 10y^2 y^{\text{iv}} + 70yy'y''' + 50yy''^2 + 95y'^2 y'' + 10y^3 y''' + 90y^2 y'y'' + 60yy'^3 + 5y^4 y'' + 20y^3 y'^2 = 0,$$

which may be shown to be equivalent to the system

$$y = w'/w, w^{\text{v}} = zw, z'' = 0. \tag{1}$$

The system (1), in view of its last two equations, has fixed critical points. The same is, therefore, true of the given DE. Exton (1971)

9. $$y^{\text{vi}} + 6yy^{\text{v}} + 21y'y^{\text{iv}} + 35y''y''' + 15y^2 y^{\text{iv}} + 90yy'y''' + 105y'^2 y'' + 60yy''^2 + 90yy'^3 + 150y^2 y'y'' + 20y^3 y''' + 15y^4 y'' + 60y^3 y'^2 + 6y^5 y' = 0.$$

The given DE is equivalent to the system

$$y = w'/w, w^{\text{vi}} = zw, z' = 0. \tag{1}$$

In view of its last two members, the system (1) has fixed critical points. The same is, therefore, true of the given DE. Exton (1971)

10. $$\boldsymbol{y^{\text{vi}} + 7yy^{\text{v}} + 21y'y^{\text{iv}} + 35y''y''' + 21y^2y^{\text{iv}} + 105yy'y''' + 105y'^2y'' + 70yy''^2}$$
$$\boldsymbol{+ 105yy'^3 + 210y^2y'y'' + 35y^3y''' + 35y^4y'' + 105y^3y'^2 + 21y^5y' + y^7 = 0.}$$

The given DE is equivalent to the system

$$y = w'/w, w^{\text{vii}} = 0,$$

which, in view of its second member, has fixed critical points. The same is, therefore, true of the given DE. Exton (1971)

11. $$\boldsymbol{y^{\text{viii}} - \frac{1}{5}y^5 - \frac{1}{3}\beta\, y^3 - \frac{1}{2}\alpha y^2 + \lambda y - C = 0,}$$

where β, α, λ, and C are constants.

A solution is sought in the form $y = AP(x)$, where $P(x)$ is the Weierstrass P-function satisfying

$$\left(\frac{dp}{dx}\right)^2 = 4P^3 - g_2P - g_3,$$

with two invariants g_2 and g_3, which are assumed to be real and which satisfy

$$g_2^3 - 27g_3^2 > 0.$$

By substitution, etc., the exact bounded periodic form of the solution is

$$y(x) = 6(1400)^{1/4}[e_3 + (e_2 - e_3)\text{Sn}^2\{(e_1 - e_3)^{1/2}x + \delta'\}]$$

and the solitary wave form (with $m = (e_2 - e_3)/(e_1 - e_3) = 1$) is

$$y(x) = 6(1400)^{1/4}[e_1 - (e_1 - e_3)\text{sech}^2\{(e_1 - e_3)^{1/2}x + \delta'\}],$$

where e_1, e_2, and e_3 are the roots of

$$4e^3 - g_2e - g_3 = 0.$$

Krishnan (1984)

6.2 Sixth and Specific Higher Order Systems

1. $$\boldsymbol{\dot{x} = yz, \quad \dot{y} = -xz - \gamma, \quad \dot{z} = \beta,}$$
$$\boldsymbol{\dot{\alpha} = 2z\beta - \gamma y, \quad \dot{\beta} = -2z\alpha + \gamma x, \quad \dot{\gamma} = \alpha y - \beta x.}$$

It is claimed that this special sixth order Euler-Poisson system, by the introduction of some new variables, can be solved explicitly in terms of t. Two first integrals are easily found to be $(x^2 + y^2)/2 + z^2 = \alpha + C_1, \alpha^2 + \beta^2 + \gamma^2 = C_2$, where C_1 and C_2 are arbitrary constants. Doksevic (1980)

2. $$\dot{x}_1 = -Ax_1 + y_1x_3, \quad \dot{y}_1 = 1 - x_1x_3, \dot{x}_2 = -Ax_2 + y_2x_1,$$

$$\dot{y}_2 = 1 - x_2x_1, \quad \dot{x}_3 = -Ax_3 + y_3x_2, \quad \dot{y}_3 = 1 - x_3x_2,$$

where A is a constant.

This system describes three disk dynamos. A correlation dimension of a strange attractor and K_2 entropy of the chaotic motion are obtained. Miura and Kai (1986)

3. $$y_1'' + w_1^2y_1 - \alpha_1y_2y_3 = 0, \; y_2'' + w_2^2y_2 - \alpha_2y_3y_1 = 0, \; y_3'' + w_3^2y_3 - \alpha_3y_1y_2 = 0,$$

where $w_3 \simeq w_1 + w_2$ and the amplitude is small but finite.

The solution is found in the form

$$y_n = a_n \cos(w_nx + \beta_n), n = 1, 2, 3,$$

where

$$\begin{aligned}
\frac{da_1}{dx} &= \{\alpha_1/(4w_1)\}a_1a_3 \; \sin[(w_3 - w_2 - w_1)x + \beta_3 - \beta_2 - \beta_1], \\
a_1\frac{d\beta_1}{dx} &= -\{\alpha_1/(4w_1)\}a_2a_3 \; \cos[(w_3 - w_2 - w_1)x + \beta_3 - \beta_2 - \beta_1], \\
\frac{da_2}{dx} &= \{\alpha_2/(4w_2)\}a_1a_3 \; \sin[(w_3 - w_2 - w_1)x + \beta_3 - \beta_2 - \beta_1], \\
a_2\frac{dbt_2}{dx} &= -\{\alpha_2/(4w_2)\}a_1a_3 \; \cos[(w_3 - w_2 - w_1)x + \beta_3 - \beta_2 - \beta_1], \\
\frac{da_3}{dx} &= -\{\alpha_3/(4w_3)\}a_1a_2 \; \sin[(w_3 - w_2 - w_1)x + \beta_3 - \beta_2 - \beta_1], \\
a_3\frac{d\beta_3}{dx} &= -\{\alpha_3/(4w_3)\}a_1a_2 \; \cos[(w_3 - w_2 - w_1)x + \beta_3 - \beta_2 - \beta_1].
\end{aligned}$$

Nayfeh (1985), p.202

4. $$u'' = (1/\sigma_1)(-v + \alpha uv - \beta u + 2\gamma u^2),$$

$$v'' = (1/\sigma_2)(v + \alpha uv - f\mu w), \quad w'' = (1/\sigma_3)(-2\beta u + \mu w),$$

where $\sigma_i, \alpha, \beta, \gamma, f$, and μ are constants.

Invariance properties of this system are discussed. The system with BC $u' = v' = w' = 0$ at $x = 0, 1$ is solved in the form

$$\begin{aligned}
w(x) &= A_0 + \sum_{p=1}^{\infty} A_p \cos mp\pi x, \\
u(x) &= B_0 + \sum_{p=1}^{\infty} B_p \cos mp\pi x, \\
v(x) &= C_0 + \sum_{p=1}^{\infty} C_p \cos mp\pi x,
\end{aligned}$$

where $m \geq 1$ is free to be chosen. The coefficients A_p, B_p, C_p are themselves found in a suitable perturbation series and computed with good accuracy. Forbes (1990)

5. $$\ddot{x}_1 = x_1(x_2^2 + x_3^2), \quad \ddot{x}_2 = x_2(x_1^2 + x_3^2), \quad \ddot{x}_3 = x_3(x_1^2 + x_2^2).$$

The given system is derivable from Eqn. (3.15.1) by differentiation and substitution for $\dot{x}_1, \dot{x}_2, \dot{x}_3$, etc. It is not algebraically completely integrable and does not have the Painlevé property. Steeb, Euler, and Mulser (1991)

6. $$\ddot{x}_1 = -x_1(x_2^2 + x_3^2), \ddot{x}_2 = -x_2(x_1^2 + x_3^2), \ddot{x}_3 = -x_3(x_1^2 + x_2^2).$$

The given system is derived from the Hamiltonian

$$H(\overline{x}, \dot{\overline{x}}) = (1/2)(\dot{x}_1^2 + \dot{x}_2^2 + \dot{x}_3^2) + (1/2)(x_1^2 x_2^2 + x_1^2 x_3^2 + x_2^2 x_3^2).$$

It is not integrable and shows chaos. Steeb, Euler, and Mulser (1991)

7. $$\ddot{x} = -kx/r^3, \ddot{y} = -ky/r^3, \ddot{z} = -kz/r^3,$$

where $r = (x^2 + y^2 + z^2)^{1/2}$ and $k = 1$, and $x(0) = 1, x(1) = 1.35649, y(0) = 0, y(2) = 0.17807, z(1) = 1.26011, z(2) = 1.31388$.

Numerical solution yields $x'(0) = 0.5, y'(0) = 0.1, z'(0) = 0.4$ for the given two-body equations of motion. Kubicek and Hlavacek (1983), p.230

8. $$\ddot{x} = -kx/r^3, \ddot{y} = -ky/r^3, \ddot{z} = -kz/r^3,$$

where $r = (x^2 + y^2 + z^2)^{1/2}$ and $k = 1$.

The given system describes motion of a two-body problem. $x(0) = 1.076, x(2) = 0, y(0) = 0, y(2) = 0.576, z(0) = 0, z(2) = 0.997661$. The given two-point BVP was solved numerically using shooting techniques. Roberts and Shipman (1972), p.133

9. $$\ddot{x} = 2f(x) - f(y) - f(z), \ddot{y} = 2f(y) - f(z) - f(x), \ddot{z} = 2f(z) - f(x) - f(y),$$

with $f(x) = \exp(-x/2^{1/2})\cos(x/2^{1/2} - \phi)$, where $\phi = 1.4611572$, and $x + y + z = L$, where L is the total periodicity length.

The given system has two first integrals:

$$\dot{x} + \dot{y} + \dot{z} = 0$$

and $(1/2)(\dot{x}^2 + \dot{y}^2 + \dot{z}^2) + V(x, y, z) = \text{constant} = E$ with the potential

$$\begin{aligned} V(x, y, z) &= -3[F(x) + F(y) + F(z)], \\ F(x) &= \int^x f(x)dx = -\exp(-x/2^{1/2})\cos[(x/2^{1/2}) - \phi + \pi/4]. \end{aligned}$$

Numerical solutions of the given system show two types of chaotic changes of the interpulse distances, depending on L. As the deviation of the initial value from a fixed point with center-like singularity increases, the periodic motions show frequency downshifts and lead to chaotic behavior. Motions associated with one stable fixed or three stable fixed points involve chaotic behavior. Perturbation solutions are also found, by the introduction of a small parameter. Kawahara and Takaoka (1988)

10. $$\ddot{x} + p(t)\dot{x} - Ke^{-2F}\rho^{-3}\eta^n = 0; \eta = x/\rho; n \neq -3 \text{ an integer},$$

$$\ddot{u} + p(t)\dot{u} = 0, \quad \ddot{\rho} + p(t)\dot{\rho} + (1/4)e^{-2F}\rho^{-3} = 0, F = \int^t pdt,$$

K is a constant.

The given system is shown to have the following invariants:

$$\begin{aligned} I &= e^{2F}\dot{x}^2u - e^F x\dot{x} - [2K/(n+1)]x^{n+1}u^{-(n+1)/2}, n \neq -1, \\ I &= e^{2F}\dot{x}^2u - e^F x\dot{x} + K\log(ux^{-2}), n = -1. \end{aligned}$$

Ranganathan (1992)

11. $$y''' + (1/2)yy'' - (1/2)xz' = 0,$$

$$z''' + (\sigma/2)(yz'' + y'z') = 0,$$

where $y(0) = y''(0) = 0, y'(\infty) = 1; z''(0) = 0, z(\infty) = 0$; σ is a constant.

The given system arises in steady laminar gravity currents. For small $\beta = z'(0)$, an approximate solution is found to be

$$\begin{aligned} y' &= 1 - [(\beta\pi^{1/2})/(1-\sigma)][(1/\sigma^{1/2})\text{erfc}((1/2)\sigma^{1/2}x) - \text{erfc}(x/2)] + O(\beta^2), \sigma \neq 1, \\ &= 1 - (\beta/2)[xe^{-x^2/4} + \pi^{1/2}\text{erfc}(x/2)] + O(\beta^2), \sigma = 1, \\ z' &= \beta e^{-\sigma x^2/4} + O(\beta^2). \end{aligned}$$

Numerical solutions are obtained for the full problem. Brighton (1988)

12. $$y''' + (2/3)yy'' - (1/3)y'^2 - (1/3)xz' + (2/3)z = 0,$$

$$z''' + \sigma[(2/3)yz'' - (1/3)y'z'] = 0,$$

$y(0) = y'(0) = 0, y''(\infty) = 1, z''(0) = 0, z'(\infty) = 0$, where σ is a constant.

The given system arises from the similarity reduction of the system describing two-dimensional steady laminar gravity currents. Numerical solutions are obtained. Brighton (1988)

13. $$y''' + (y+z)y'' - y'^2 = 0, \quad z''' + (y+z)z'' - z'^2 = 0,$$

$y(0) = z(0) = 0, y'(0) = 1,\ z'(0) = \gamma, 0 < \gamma < 1, y'(\infty) = z'(\infty) = 0$.

The given problem arises in flow due to a stretching of a plate with dilation set equal to zero. Existence of solutions is established using geometrical arguments. Srivastava and Usha (1988)

14. $$y''' + \{1/\alpha\}y' + \{6/v\}yy' + \{2\beta/(\alpha v)\}zz' = 0, z''' + \{3y/v - 1\}z' = 0,$$

where α, β, and v are constants.

The given DE arises from the traveling wave form of coupled Korteweg–deVries equations $u_t - \alpha(u_{xxx} + 64uu_x) - 2\beta ww_x = 0, w_t + w_{xxx} + 3uw_x = 0$.

The first of the given system admits one integration. Solitary solutions of the given system is $y = 2v\ \text{sech}^2 x, z = Av\ \text{sech}\ x$, where $A^2 = -2(1+4\alpha)/\beta$. Hereman, Banerjee, Korpel, and Assanto (1986)

15. $$\epsilon y^{\text{iv}} + yy''' + zz' = 0, \quad \epsilon z'' + yz' - zy' = 0,$$

where $y(\pm 1) = y'(\pm 1) = 0, z(-1) = -1, z(1) = 1$; ϵ is a parameter.

The existence of odd solutions $\{y(x,\epsilon), z(x,\epsilon)\}$ is established for all $\epsilon > 0$. This particular solution is shown to possess the following properties:

(a) $y(x, \epsilon) \leq 0$ for $0 \leq x \leq 1$.

(b) $y'(x, \epsilon)$ has precisely one zero in $0 < x < 1$ with $y'(0, \epsilon) < 0$.

(c) $y''(x, \epsilon)$ has precisely one zero in $0 < x < 1$ with $y''(0, \epsilon) = 0$ and $y''(1, \epsilon) < 0$.

(d) $y'''(x, \epsilon)$ has precisely one zero in $0 < x < 1$ with $y'''(0, \epsilon) > 0$ and $y'''(1, \epsilon) < 0$.

(e) $z'(x, \epsilon) > 0$ for $-1 \leq x \leq 1$.

(f) $z''(x, \epsilon) \geq 0$ for $0 \leq x \leq 1$.

The estimate of the magnitude and behavior of the solution as $\epsilon \to 0$ is obtained. McLeod and Parter (1974)

16. $$\boldsymbol{y^{\text{iv}} + yy''' + zz' = 0, z'' + yz' - zy' = 0,}$$

where $y(0) = y'(0) = y(1) = y'(1) = 0, z(0) = \Omega_0, z(1) = \Omega_1$, Ω_0, and Ω_1 are constants.

It is proved that the given system has a solution provided that either:

(a) $\Omega_1 \in [-\Omega_0, 0]$ and $0 \leq \Omega_0 < C$, or

(b) $\Omega_1 \in (0, \Omega_0]$ and $\Omega_0^2 - \Omega_1^2 < C^2$, where $C = (4/3)(2^{1/2})(e^{1/4} + e^{4/9} - 1)^{-1/4} \doteq 1.5$.

Elcrat (1975)

17. $$\boldsymbol{2y^{\text{iv}} + yy''' + y'y'' \pm \xi xz' = 0, 0 \leq x < \infty, (2/Pr)z'' + yz' = 0,}$$

$y(0) = y'(0) = 0, z(0) = 1, y'(\infty) = 1, y''(\infty) = 0, z(\infty) = 0$; ξ and Pr are parameters. The given problem arises in combined forced and free convection flow over a horizontal plate when the wall temperature is constant.

Scaling and shooting methods are used to solve the given BVP; initial value solutions are employed. de Hoog, Laminger, and Weiss (1984)

18. $$\boldsymbol{\epsilon y^{\text{iv}} + yy''' + zz' = 0, \epsilon z'' + yz' - zy' = 0,}$$

$y(0) = y'(0) = y(1) = y'(1) = 0, z(0) = \Omega_0/\Omega_1, z(1) = 1$ where ϵ, Ω_0, and Ω_1 are constants. The given BVP describes the behavior of a fluid occupying the region $0 \leq x \leq 1$ between two rotating disks, rotating about a common axis perpendicular to their planes, when the disks are rotating in the same sense with speeds $0 \leq \Omega_0 \leq \Omega_1$.

A quantitative study of the given system shows that if $\epsilon > 0$ is sufficiently small and $z(0) \geq 0$, then there cannot exist a solution of the given BVP for which $z' \geq 0$. This has implications in the analysis of existence theory for the given problem. McLeod and Parter (1977)

19. $$\epsilon y^{\text{iv}} + yy''' + zz' = 0, \epsilon z'' + yz' - zy' = 0,$$

where $\epsilon > 0$ is a constant. The boundary conditions are

$$\begin{aligned} y(0,\epsilon) &= y(1,\epsilon) = 0, y'(0,\epsilon) = y'(1,\epsilon) = 0, \\ z(0,\epsilon) &= \Omega_0, z(1,\epsilon) = \Omega_1, |\Omega_0| + |\Omega_1| \neq 0. \end{aligned}$$

"Pathological" solutions (y,z) of the given system satisfying $|y(x,\epsilon)| + |y'(x,\epsilon)| + |z(x,\epsilon)| \leq B, |y(x_0,\epsilon)| \geq \delta, 0 < x_0 < 1, 0 < \delta < B, |\Omega_0(\epsilon)| + |\Omega_1(\epsilon)| \neq 0, \lim_{\epsilon\to 0} \{|\Omega_0(\epsilon)| + |\Omega_1(\epsilon)|\} = 0$ are analyzed with regard to the question of their existence. B is a positive constant; δ and x_0 satisfying the implied conditions are assumed to exist. Kreiss and Parter (1983)

20. $$xy^{\text{iv}} + 2y''' + (1/2)R[y'y'' - yy''' + z^2/x^2] = 0,$$

$$xz'' + (1/2)R(y'z - z'y) = 0,$$

$y(0) = z(0) = 0$, and $\lim_{x\to 0} x^{1/2}y''(x) = 0, y(1) = 1, z(0) = 0, y'(1) = 0$, where R is a constant.

The given problem arises in spiral flow in a porous pipe. Numerical solution is obtained. Terril and Thomas (1973)

21. $$\begin{aligned} \dot{y}_1 &= -(y_1/y_5)[y_6 + 2(y_3 + y_5)], \dot{y}_2 = y_1 y_3(y_3 + y_5), \dot{y}_3 = y_4, \\ \dot{y}_4 &= Re(y_1/y_7)[-2\sigma A y_7^2 + y_4 y_5 + y_3(y_3 + y_5)] + y_4 y_5 y_6/(2Ay_7^2) + 2Re \cdot y_2/y_7, \\ \dot{y}_5 &= y_6, \dot{y}_6 = [y_1/\{(4/3Re)y_7\}][\{-\sigma A y_7^2/y_5\}\{y_6 + 2(y_3 + y_5)\} + y_5 y_6(1-\sigma)] \\ &- y_3 y_5 y_6/(2Ay_7^2) - y_4/2 - \{y_5/(2A)\}\{y_6/y_7\}^2, \dot{y}_7 = -y_5 y_6/(2Ay_7), \end{aligned}$$

with $y_1(t_0) = 0.9617, y_2(t_0) = -0.1018, y_3(t_0) = 0.4078, y_5(t_0) = -0.0212, y_7(t_0) = 0.9998$, and $y_3(t_f) = 1.0, y_5(t_f) = -1.0$, where t_0 and t_f are initial and final values of t.

The constants specifically chosen were $Re = 100, A = 0.515, \sigma = 0.400, t_0 = 0.05, t_f = 0.3303$. The given BVP was solved numerically using "continuation techniques". Roberts and Shipman (1972), p.163

22. $$u^{\text{iv}} - \delta u'' + \zeta w'' - (1/2)(u^2 - w^2) \pm u = 0, \quad w^{\text{iv}} - \zeta w'' - \delta w'' - uw \pm w = 0$$

where δ and ζ are real constants.

If we write $\xi = \delta + i\zeta, v = u + iw$, the given system has the following complex solution for $v = v(x)$:

$$v(x) = \alpha \operatorname{sech}^4(\beta(x - x_0)) + \gamma \operatorname{sech}^2(\beta(x - x_0)),$$

where

$$\alpha = -\frac{70(31)^{1/2}}{5(31)^{1/2} \pm 13i}, \quad \beta = \frac{[-(31)^{1/2}]^{1/4}}{[120(31)^{1/2} \pm 312i]^{1/4}}$$

$$\gamma = \frac{70[(31)^{1/2} \pm i)]}{5(31)^{1/2} \pm 13i}, \quad \xi = -\frac{13[(-31)^{1/2}[(31)^{1/2} \pm 3i]}{(31)^{1/2}[30(31)^{1/2} \pm 78i]^{1/2}}.$$

Here $\pm$ are taken in the given order. For $\xi = -13/6$, the solution is $v = (35/12)\operatorname{sech}^2[i(x - x_0)/(2(6)^{1/2})]$. Takaoka (1989, 1989a)

23. $\dot{y}_1 = y_4, \dot{y}_2 = y_5, \dot{y}_3 = y_6, \dot{y}_4 = K(y_8y_6 - y_9y_5),$
$\dot{y}_5 = K(y_9y_4 - y_7y_6), \dot{y}_6 = K(y_7y_5 - y_8y_4), \dot{y}_7 = y_{11}y_6 - y_{12}y_5,$
$\dot{y}_8 = y_{12}y_4 - y_{10}y_6,\ \dot{y}_9 = y_{10}y_5 - y_{11}y_4, \dot{y}_{10} = F_1, \dot{y}_{11} = F_2, \dot{y}_{12} = F_3,$

$0 \leq t \leq 1$, with BCs $y_1(0) = y_2(0) = y_3(0) = y_5(0) = y_6(0) = 0, y_4(0) = 1, y_1(1) = 0.6, y_2(1) = 0.5, y_3(1) = y_7(1) = y_8(1) = y_9(1) = 0, F = (F_1, F_2, F_3) = (0, 1, -6|y_2|\{1 - (y_6)^2]^{1.5})$.

The given system describes the equilibrium of a thin flexible unexpanded bar of unit length and circular cross section.

A first integral is

$$y_4y_7 + y_5y_8 + y_6y_9 = C.$$

The constant C by IC at $t = 1$ is zero. The BVP was solved numerically, using shooting methods. Gaiduk (1984)

N

GENERAL ORDER EQUATIONS

N.1 General Order Single Equations

1. $$y^{(2k)} - f(y;m+1) + \lambda y - C = 0,$$

where $f(y;m+1)$ is a polynomial function of degree $m+1$ and k and m are arbitrary natural numbers; λ and C are arbitrary constants.

Recall that the $(2n)$th order dervative of the Weierstrass $\mathcal{P}$ function with respect to its argument is an $(n+1)$th degree polynomial of the function $\mathcal{P}$ itself,

$$\mathcal{P}^{(2n)} = (n+1)\text{th degree polynomial of the } \mathcal{P} \text{ function,} \tag{1}$$

where $\mathcal{P}$ satisfies the DE

$$(\mathcal{P}')^2 = 4\mathcal{P}^3 - g_2\mathcal{P} - g_3, \tag{2}$$

with two invariants g_2 and g_3, which are assumed to be real and which satisfiy $g_2^3 - 27g_3^2 > 0$. We assume the solution of the given DE in the form

$$y(x) = A\mathcal{P}^{(2n)}(x),$$

where A is an arbitrary constant; then we get

$$\frac{d^2(k+n)}{dx^{2(k+n)}}\,\mathcal{P} = f[A\mathcal{P}^{(2n)};(n+1)(m+1)]/A - \lambda\mathcal{P}^{(2n)} + C/A. \tag{3}$$

The LHS of (3) is a polynomial of degree $(k+n+1)$ of the $\mathcal{P}$ function whose coefficients involve the invariants g_2 and g_3, and the right-hand side a degree $(n+1)(m+1)$ polynomial whose coefficients involve the quantities A, λ, and C in addition to g_2 and g_3. For a solution of the assumed form to be possible, the two polynomials thus obtained must be identical. Thus, finding the solution of the given DE is reduced to the problem of two identical polynomials in (3). This requires that $k+n+1 = (n+1)(m+1)$, i.e., $nm = k-m$. Special cases may be constructed [see Eqns. (4.1.3), (4.1.6), (4.2.18)]. Kano and Nakayama (1981)

2. $$y^{(n)} + q(x)y^{\gamma} = 0, n > 1, \gamma \neq 1 \textbf{ are constants.}$$

Necessary and sufficient conditions for the solutions of the given DE to be oscillatory are deduced. Licko and Svec (1963)

3. $$y^{(n)} + q(x)y^{r}[\log(1+y)]^{\delta} = 0,$$

where $n \geq 2, 0 < r < 1, 0 < r+\delta < 1$ and $q(x) : [0,\infty) \to R$ is continuous. For $n = 2$, and

$$q(x) = \begin{cases} \sin(\log x)/x^2 \text{ for } \sin(\log x) \geq 0, \quad \text{i.e.,} \quad x \in \cup_{i=0}^{\infty} [e^{2i\pi}, e^{(2i+1)\pi}] \\ \sin(\log x)/x^3 \text{ for } \sin(\log x) \geq 0, \quad \text{i.e.,} \quad x \in \cup_{i=1}^{\infty} [e^{(2i-1)\pi}, e^{2i\pi}]. \end{cases}$$

It is shown that the given DE has infinitely many positive solutions $y(x)$ which exist on $[e,\infty)$ and have asymptotic behavior $\lim_{x\to\infty} y(x)/x = 0, \lim_{\to\infty} y(x) = \infty$. Kusano (1989)

4. $$y^{(n)} - p(x)|y|^{\lambda} \operatorname{sgn} y = 0, \ \lambda > 1, \ n \geq 2, \ x > 0,$$

where $p(x) \geq 0$ is a piecewise-continuous function.

It is shown that the given DE has a regular solution $y(x)$ satisfying the inequalities $0 < y^{(n-1)}(x) \leq$ constant. $y(x)x^{1-n}, x > 0$, if and only if $p(x)$ satisfies the condition

$$\int_0^{\infty} p(\tau)\tau^{(n-1)\lambda} \, d\tau < \infty.$$

It is shown that the regular solution has the form

$$u^{(i)}(x) \sim c(n-1)!x^{n-i-1}/(n-i-1)!, i = 0,1,2,\ldots,(n-1), c > 0,$$

$x \to +\infty$ so that $u^{(n-1)}(x)/u(x) \sim$ constant. x^{1-n} as $x \to +\infty$. Izobov and Rabtsevich (1987)

5. $$y^{(n)} - p(x)|y|^{\lambda} \operatorname{sgn}(y) = 0, \ \lambda > 1, \ x \geq 0,$$

where $p(x) \geq 0$ is a piecewise-continuous function.

It is proved that if $\phi(x) \geq 1$ is monotonic, piecewise-continuous, and unbounded for $x \geq 0$, then there is a piecewise-continuous function $p(x) \geq 0$ satisfying the condition

$$\int_0^{\infty} \tau^{(n-1)\lambda} \, p(\tau) \, \phi^{-1}(\tau) \, d\tau = +\infty \tag{1}$$

for which the given DE has an n-parameter family of regular unbounded solutions $y(x)$ satisfying the condition

$$0 < y^{(n-1)}(x)/y(x) < (n-1)! \, x^{1-n} \, \phi^{\mu}(x), \tag{2}$$

where $x \geq x_1 > 0$.

At the same time no equation of the given type with a piecewise-continuous function $p(x) \geq 0$ satisfying condition (1) has a solution of the form (2) with $\mu < 1$ for all $x > x_y > 0$, where x_y is a constant. Rabtsevich (1986)

6. $$y^{(n)} - \phi(x)|y|^{\lambda} \operatorname{sgn} y = 0,$$

where n is a positive integer and $\lambda > 1$ is a real number.

It is also assumed that $\phi(x)$ is a continuous function defined on some half-line $[a,\infty)$ and that all the solutions are extended on this interval. It is shown that no solutions of the given DE grow exponentially. Tanimoto (1989)

7. $$y^{(n)} + \phi(x)g(y) = 0, \ n \geq 2,$$

where $\phi : [0,\infty) \to (0,\infty)$ is continuous and $g : R \to (0,\infty)$ is continuous and nondecreasing.

The following results are proved:

(a) Suppose that $\int_\delta^\infty [1/g(y)]dy < \infty$ for any $\delta \in R$. Then every solution $y(x)$ of the given DE has the property that $\lim_{n\to\infty} y(x) = -\infty$ if and only if $\int_a^\infty x^{n-1}\phi(x)dx = \infty$.

(b) Suppose that $g(y)/y$ is nondecreasing for all sufficiently large $y > 0$, $h(z) = \inf_{x>0} g(xz)/g(x) > 0$ for $z > 0$, and $\int_0^\delta [1/h(z)]dz < \infty$ for some $\delta > 0$. Suppose, moreover, that $\lim_{y\to-\infty} g(y) = 0$ when n is odd. Then, every solution $y(x)$ of the given DE has the property that $\lim_{x\to\infty} y(x) = -\infty$ if and only if $\int_a^\infty \phi(x)g(\lambda x^{n-1})dx = \infty$ for all $\lambda > 0$.

Kusano, Naito, and Swanson (1988)

8.
$$\boldsymbol{y^{(n)} + \sum_{i=1}^{N} q_i(x)f_i(y) = 0,}$$

where:

(a) $n \geq 2$.

(b) Each $q_i : [0, \infty) \to R, 1 \leq i \leq N$, is continuous.

(c) Each $f_i : [0, \infty) \to (0, \infty), 1 \leq i \leq N$, is continuous and nondecreasing.

The functions $q_i(x)$ are oscillatory, i.e., $q_i(x)$ change in sign in any neighborhood of infinity.

Criteria are found for the existence of a positive solution of the given DE with the property that $\lim_{x\to\infty} y(x)/x^k = 0$ and $\lim_{x\to\infty} y(x)/x^{k-1} = \infty$ for some $k, 1 \leq k \leq n-1$. The existence criteria are formulated in terms of the positive part $(q_i)_+(x)$ and the negative part $(q_i)_-(x)$ of the coefficients $q_i : (q_i)_+(x) = \max\{q_i(x), 0\}, (q_i)_-(x) = \max\{-q_i(x), 0\}, 1 \leq i \leq N$. Kusano (1989)

9.
$$\boldsymbol{y^{(n)} + f(x, y) = 0, \ n \geq 2,}$$

where $f : [0, \infty) \times R \to (0, \infty)$ is continuous and nondecreasing in the second variable: $\lim_{y\to-\infty} f(x, y) = 0$ for each $x \in [0, \infty)$.

It is observed that all solutions of the given DE can be continued indefinitely to the right; more precisely, for any $a \geq 0$ and $(\eta_0, \eta_1, \dots, \eta_{n-1}) \in R^n$, the solution of the given DE satisfying IC $y^i(a) = \eta_i, i = 0, 1, \dots, (n-1)$ exists throughout the interval $[a, \infty)$. Kusano, Naito, and Swanson (1988)

10.
$$\boldsymbol{L_n(y) + v^{n\alpha+1}F(y/v) = 0, \ v = v(x),}$$

where F is an arbitrary function and where the linear part

$$L_n(y) = \sum_{k=0}^{n} a_k(x)y^{(k)}(x)$$

is reducible to one with constant coefficients $M_n(z(t)) = \sum_{k=0}^{n} b_k z^{(k)}(t)$ by the transformation

$$y = vz, dt = v^\alpha dx. \tag{1}$$

Through (1), we can change the given DE to the autonomous form $M_n(z) + F(z) = 0$, which has the solution $y = \rho v$, where $\rho =$ constant satisfies the equation $b_0\rho + F(\rho) = 0$ and v satisfies the nonlinear equation

$$L_n(v) - b_0 v^{n\alpha+1} = 0.$$

If either $a_{n-1} \not\equiv 0$ or $\alpha \neq 2/(n-1)$, then v may be expressed in closed form in terms of elementary functions and quadratures. Berkovich (1971)

11. $$\sum_{k=0}^{n} a_k(x)y^{(k)}(x) + u^n vF(y/v(x)) - \phi(x) = 0, \; a_n = 1,$$

where $a_k(x)$ are real, continuous, and sufficiently differentiable, F and ϕ are arbitrary functions, and $u = u(x)$, $v = v(x)$ are chosen such that the linear part $L_n(y) \equiv \sum_{i=0}^{n} a_k(x)y^{(k)}(x) = 0$ of the given equation is reducible, i.e., it is transferrable by the change of variables

$$y = v(x)z, dt = u(x)dx$$

into an equation with constant coefficients:

$$M_n(z(t)) \equiv \sum_{k=0}^{n} b_k z^k(t) = 0.$$

In these circumstances, the given DE changes to

$$M_n(z) + F(z) = \phi(x(t))u^{-n}v^{-1}. \tag{1}$$

If $\phi(x) \equiv 0$ in (1), we obtain an autonomous transformed equation with the solution $y = \rho v, \rho$ is a constant, where v satisfies $L_n(v) - b_0 u^n v = 0$, where u is a known function of x, and satisfies the equation $b_0\rho + F(\rho) = 0$, as may be checked by direct verification. Berkovich (1971)

12. $$L_n(y) + \sum_{s=1}^{\ell} f_s(x)y^{m_s} - \phi(x) = 0, \; 1 \leq m_1 < m_2 \cdots < m_\ell,$$

where the linear part $L_n(x) = \sum_{k=0}^{n} a_k(x)y^{(k)}(x)$ is reducible to one with constant coefficients

$$M_n(z(t)) = \sum_{k=0}^{n} b_k z^{(k)}(t)$$

by the transformation

$$y = v(x)z, dt = u(x)dx \tag{1}$$

and

$$p_s u^n = f_s(x)v^{m_s - 1}, \; p_s = \text{ constant}.$$

With the given condition, (1) changes the given equation to

$$M_n(z(t)) + \sum_{s=1}^{\ell} p_s z^{m_s} = \phi(x(t))u^{-n}v^{-1}. \tag{2}$$

If $\phi(x) \equiv 0$, the system (2) becomes autonomous and the solution is $y = \rho v$, where $\rho =$ constant satisfies the equation

$$b_0\rho + \sum_{s=1}^{\ell} p_s \rho^{m_s} = 0.$$

The result may easily be checked by direct substitution. Berkovich (1971)

13. $$L_n(y) + u^n F(y, y'/u) - \phi(x) = 0,$$

where $u = u(x)$ and $\phi(x)$ are arbitrary and the linear part

$$L_n(y) \equiv \sum_{k=0}^{n} a_k(x) y^{(k)}(x) = 0, a_n(x) \equiv 1,$$

is reducible by a transformation of the independent variable $y = z, dt = u dx$ to one with constant coefficients, namely

$$M_n(z(t)) \equiv \sum_{k=0}^{n} b_k z^{(k)}(t) \equiv 0.$$

Now the given DE changes to

$$M_n(z(t)) + F(z, z_t) = u^{-n} \phi(x(t)).$$

If $\phi(x) \equiv 0$, the given DE becomes autonomous and has the solution $y = \rho$, where ρ satisfies the equation $b_0 \rho + F(\rho, 0) = 0$. The function u in the transformation $dt = u dx$ may take one of the following forms:

$$u = (a_0)^{1/n}, a_0 \not\equiv 0,$$

or

$$u = \exp\left[(-2/\{n(n-1)\}) \int a_{n-1} dx\right], a_{n-1} \not\equiv 0.$$

Berkovich (1971)

14. $$y^{(n+1)} + F(x, y) y^{(n)} = 0, \; 0 \le x \le a,$$

$y(0) = y^{(1)}(0) = \cdots y^{(n-1)}(0) = 0, y^{(n-1)}(a) = \lambda$.

The given class of equations appears frequently in all areas of applied mechanics. The given BVP is changed to a nonlinear integral equation as follows. Put $u(x) = y^{(n-1)}$. Then by integration and use of first n conditions in the given BVP, we can write

$$y(x) = S(u|x), x \in [0, a], \tag{1}$$

where $S(u|x)$ is the linear operator defined as follows:

$$S(u|x) = \begin{cases} u(x) & \text{for } n = 1 \\ [1/(n-2)!] \int_0^x (x-t)^{n-2} u(t) dt & \text{for } n > 1. \end{cases} \tag{2}$$

Replacing $y^{(n+1)}$ and $y^{(n)}(x)$ in the given DE according to $u(x) = y^{(n-1)}(x)$, we get the second order integrodifferential equation

$$u'' + F(x, S(u|x)) u' = 0, x \in [0, a]. \tag{3}$$

We assume that $F(x, S(u|x))$ is a known function of x, and integrate (3) with the conditions $u(0) = 0, u(a) = \lambda$, which follows from the given BC. This yields the integral equation

$$u(x) = A(u|x),$$

where

$$A(u|x) = \lambda \, \frac{\int_0^x \exp[-B(u|t)]dt}{\int_0^a \exp[-B(u|t)]dt} \tag{4}$$

and

$$B(u|x) = \int_0^x F(t, S(u|t))dt.$$

A solution of (4) is sought as the limit of the sequence

$$u_0(x) \equiv 0, u_{k+1}(x) = A(u_k|x), \; k = 0, 1, \ldots. \tag{5}$$

If the sequence (5) converges to a function $u_*(x)$, then

$$y_*(x) = S(u_*|x).$$

The basis of this method is an elaborate theorem requiring certain conditions on $F(x, y)$ and hence on the Fréchet derivative of $A(u|x)$. The convergence and the rate of convergence of the sequence are discussed. The special case when $F(x, y)$ is positive is also discussed. Pykhteev and Myachina (1972)

15. $$\boldsymbol{y^{(n)} - f(y^{(n-2)}) = 0.}$$

With $u(x) = y^{(n-2)}$, the given DE goes to

$$u'' = f(u). \tag{1}$$

Equation (1), being autonomous, is easily integrated:

$$x = \pm \int \left[\hat{C}_1 + 2 \int f(u)du\right]^{-1/2} du + C,$$

where $\hat{C}_1$ and C are arbitrary constants. After integrating $(n-2)$ times, the equation $u(x) = y^{(n-2)}$, one obtains a parametric representation of the solution:

$$x = \int_{C_0}^{u} \frac{du}{\phi(u)}, \; y = \int_{C_1}^{u} \frac{du_1}{\phi(u_1)} \int_{C}^{u_1} \frac{du_2}{\phi(u_2)} \cdots \int_{C_{n-2}}^{u_{n-3}} \frac{u_{n-2}du_{n-2}}{\phi(u_{n-2})},$$

where C_i $(i = 0, 1, \ldots, n-2)$ are constants and

$$\phi(u) = \pm \left[\hat{C}_1 + 2 \int f(u)du\right]^{1/2}.$$

Kamke (1959), p.605

16. $$\boldsymbol{y^{(2n+1)} + \sum_{i=1}^{\infty} p_{s+1} y^{(2n-s)} y^{(s)} = 0, \; y'(\infty) = 0,}$$

where p_k $(k = 1, 2, \ldots, n+1)$ are real constants, $p_1 \neq 0$.

The solution is found in the form

$$y(x) = \{j/p_1\} + j \sum_{i=1}^{\infty} b_i a^i e^{-ijx}, \tag{1}$$

where j and a are parameters. Substituting (1) in the given DE leads to

$$-b_1 a + b_1 a = 0$$

and

$$b_i = \frac{1}{i^{2n}(i-1)} \sum_{\ell=1}^{i-1} \sum_{s=0}^{\infty} p_{s+1} \ell^{2n-s}(i-\ell)^s b_\ell b_{i-\ell}.$$

It is shown that if $|a| < 1$ and $|b_1| \le 1$, the series (1) is absolutely convergent at $x = -\epsilon$ for each $j > 0$, where $\epsilon = -(\ln|a|/j + \delta)$, where $\delta > 0$ is a sufficiently small number chosen to depend on j and a in such a way that ϵ must always be positive. The series (1) is then uniformly and absolutely convergent for $x > -\epsilon$. It is found that the condition $|b_i| \le 1$ leads to the condition $\sum_{s=0}^{n} |p_{s+1}| \le 1$ and ensures the convergence of the series (1). The two-parameter (a and j) family of solutions is obtained. Bainov (1966)

17. $$\boldsymbol{y^{(n)} + p_0 y y^{(n-1)} + \sum_{\ell=1}^{n-4} \sum_{s=1}^{n-3-\ell} p_{\ell s} y^{(n-2-\ell-s)} y^{(\ell)} y^{(s)} = 0,}$$

$n \ge 6$, with $y'(\infty) = 0$ or $p_0 \ne 0$.

A series solution is sought in the form

$$y = \frac{\rho}{p_0} + \rho \sum_{i=1}^{\infty} b_i a^i e^{-ipx}, \qquad (1)$$

where a, b_1, and $\rho > 0$ are parameters. The parameter p is arbitrary, while the parameters a and b_1 satisfy $|a| < 1, |b_1| \le 1$.

Substituting (1) in the given DE, we get

$$\sum_{i=1}^{\infty} i^n b_i a^i e^{-ipx} - \sum_{i=1}^{\infty} i^{n-1} b_i a^i e^{-ipx} - p_0 \sum_{i=2}^{\infty} \sum_{\nu=1}^{i-1} \nu^{n-1} b_\nu b_{i-\nu} a^i e^{-ipx}$$
$$+ \sum_{\ell=1}^{n-4} \sum_{s=1}^{n-3-\ell} p_{\ell s} \sum_{r=1}^{i-2} \sum_{m=r+1}^{i-1} r^{n-2-\ell-s} \times (m-r)^\ell (i-m)^s b_r b_{m-r} b_{i-m} a^i e^{-ipx} = 0.$$

Setting the coefficients of $a^i e^{-ipx}$ to zero, we obtain

$$b_2 = \frac{b_1^2 p_0}{2^{n-1}},$$
$$b_i = \frac{1}{i^{n-1}(i-1)} \left[p_0 \sum_{\nu=1}^{i-1} \nu^{n-1} b_\nu b_{i-\nu} - \sum_{\ell=1}^{n-4} \sum_{s=1}^{n-3-\ell} p_{\ell s} \times \sum_{r=1}^{i-2} \sum_{m=r+1}^{i-1} r^{n-2-\ell-s}(m-r)^\ell (i-m)^s b_r b_{m-r} b_{i-m} \right], i = 3, 4, \ldots.$$

It is shown that the condition

$$\{2(1/n + 1/3)\} + 8/\{(n-1)(n-2)(n-3)\} + \frac{[(n-3)/2]\{(n-3)/2+1\}}{4(n-2)^2} \times \frac{1}{(n-1)(n-2)}$$
$$+ \frac{1}{2(n-1)^2} + (1/96)[(n-3)/2]\{[(n-3)/2]+1\} - 3/2 \ \max(|p_0|, |p_{ls}|) \le 1,$$

ensures the uniform and absolute convergence of the three-parameter series (1), and hence the solution. Bainov (1967)

18. $$\boldsymbol{y^{(n)} - y' \, f(y^{n-1}) = 0.}$$

Put $y^{(n-1)} = p(y)$ in the given DE to obtain the first order equation

$$\frac{dp}{dy} = f(p). \tag{1}$$

If Eqn. (1) can be solved, the original DE may be solved using

$$y^{(n-1)} = p(y).$$

Tatarkiewicz (1963)

19. $$\boldsymbol{y^{(n)} - f(x, y, y', ..., y^{(n-1)}) = 0, y^{\ell}(x_0) = y_0^{\ell}, \ell = 0, 1, 2, ..., n-1.}$$

Numerical methods involving analytic continuation for an IVP are detailed and illustrated with examples. In particular, when a pole is approached, it is suggested that the analytic continuation be halted and the solution resumed as a truncated Laurent series about the pole. The n number of unknown coefficients in the Laurent's series is determined from $y_n^{\ell}(x_q)$ $(\ell = 0, 1, 2, ..., n-1)$; one of the unknowns is necessarily the location of the pole. After the pole is passed and the criterion for using the Laurent series is no longer satisfied, the continuous analytic continuation is again picked up and continues as before. Simon (1965)

20. $$\boldsymbol{y^{(n+1)} - f\{x, y, y', \ldots, y^{(n)}\} = 0, \;\; x \geq x_0.}$$

Asymptotic behavior, as $x \to \infty$, of complex-valued solutions of the given DE is found. Upper bounds for $|y^{(j)}(x)|, 0 \leq j \leq n$ are obtained by securing the upper bounds $y(x)$ of Bihari-type integral inequalities of the form

$$y(x) \leq p(x) + \sum_{i=0}^{n} y_j(x, s)\phi_j\{y(s)\}ds, x \geq x_0.$$

Beesack (1984)

21. $$\boldsymbol{(xy)^{(2m)} + xF(x, y) = 0, \;\; m \geq 1, \;\; x \geq a > 0.}$$

(a) The given DE has a solution satisfying $\lim_{|x|\to\infty} y(x)/|x|^{k-1} = \text{const.} < 0$ for some $k, 1 \leq k \leq 2m-1$ if and only if $\int^{\infty} t^{2m-k} F(t, -\lambda t^{k-1})dt < \infty$ for some $\lambda > 0$.

(b) The given DE has a solution satisfying $\lim_{|x|\to\infty} y(x)/|x|^k = 0, \lim_{|x|\to\infty} y(x)/|x|^{k-1} = -\infty$ for some odd $k, 0 \leq k \leq 2m-2$, if

$$\int^{\infty} x^{2m-k-1} F(x, -\lambda x^{k-1})dx < \infty \text{ for some } \lambda > 0$$

and

$$\int^{\infty} x^{2m-k} F(x, -\mu x^k)dx = \infty \text{ for all } \mu > 0.$$

Kusano, Naito, and Swanson (1988)

22. $$\boldsymbol{A(x)y^{(m)} - P(x,y) = 0,}$$

where $A(x)$ and $P(x, y)$ are polynomials and

$$\begin{aligned} P(x,y) &= B_{\nu_0}(x)y^{\nu_0} + B_{\nu_1}(x)y^{\nu_1} + \cdots + B_{\nu_n}(x)y^{\nu_n}, \\ &\quad \nu_n > \nu_{n-1} > \cdots > \nu_0 \geq 0, \quad \Pi_{i=0}^{n} |B_{\nu_i}(x)| \neq 0. \end{aligned}$$

It is proved that the given DE cannot have more than $n+1$ polynomial solutions of different degrees for $P(x, y) \not\equiv 0$. Samuilov (1971)

23. $$\boldsymbol{yy^{(n+2)} - y'y^{(n+1)} - by^3y^{(n)} - ay^2y^{(n+1)} = 0,}$$

where a and b are constants.

Put

$$\frac{d^n y}{dx^n} = q(z), \quad \frac{dz}{dx} = y \tag{1}$$

in the given DE to obtain the linear equation in q:

$$q''(z) - aq'(z) - bq = 0. \tag{2}$$

Equation (2) and hence (1) may be solved to obtain $y(x)$. Tatarkiewicz (1963)

24. $$\boldsymbol{yy^{(n+2)} - y'y^{(n+1)} - y^3g\{y^{(n)}, y^{(n+1)}/y\} = 0.}$$

Two special cases of the given DE are easily solved.

(a) $g(v, u) = f(v)$. In this case, we write

$$y^{(n)} = q(z), \quad \frac{dz}{dx} = y, \tag{1}$$

so that the given DE transforms to

$$\frac{d^2q}{dz^2} - f(q) = 0. \tag{2}$$

If (2) can be solved, y may be obtained from (1) by integration, etc.

(b) $g = 0$. Write the given DE as

$$y^{(n+2)}/y^{(n+1)} = y'/y,$$

which yields the linear DE

$$y^{(n+1)} = Cy, \tag{3}$$

where C is an arbitrary constant.

Tatarkiewicz (1963)

25. $$\boldsymbol{y'y^{(n+2)} - y''y^{(n+1)} - y'^3 g\{y, y^{(n)}, y^{(n+1)}/y'\} = 0.}$$

Put $y^{(n)}(x) = q(y)$, so that the given DE transforms to the second order equation

$$\frac{d^2q}{dy^2} = g\left(y, q, \frac{dq}{dy}\right),$$

which may be solved, depending on the form of g. Tatarkiewicz (1963)

26. $$\boldsymbol{y^{(n-2)}y^{(n)} - \{y^{(n-1)}\}^2 = 0.}$$

Write the given DE as

$$y^{(n)}/y^{(n-1)} = y^{(n-1)}/y^{(n-2)}$$

and integrate n times. It follows that the solution of the original DE is

$$y = C_2 \exp(C_1 x) + (C_0 + C_1 x + C_2 x^2 + \cdots + C_{n-3}x^{n-3}),$$

where C_i $(i = 1, 2, \ldots, n-3)$ are arbitrary constants. Murphy (1960), p.430

27. $$\boldsymbol{F(y^{(m)}, y^{(m-1)}, \ldots, y', y, x) = 0.}$$

The given general DE is said to be isobaric if there exist numbers k and ℓ such that

$$F(t^{\ell-m}y^{(m)}, t^{\ell-m+1}y^{(m-1)}, \ldots, t^{\ell-1}y', t^{\ell}y, tx) = t^k F(y^{(m)}, y^{(m-1)}, \ldots, y', y, x)$$

for all values of t.

See Eqn. (2.45.15) for the case $m = 2$. Goldstein and Braun (1973), p.82

N.2 Systems of General Order

1. $$\boldsymbol{\dot{x}_1 = \sum_{j=1}^{N} x_j, \quad \dot{x}_k = \sum_{j=1}^{N} a_{kj}x_j, k = 2, \ldots, N-1, \quad \dot{x}_N = 1 + x_1 x_N,}$$

where a_{kj} are constants.

The given system is called a generalized Rössler system, which for $N = 3$ has been shown to exhibit wildly chaotic behavior for some parametric values. This is perhaps the simplest nontrivial system of its kind with $N-1$ of its equations linear, but the Nth equation with a nonlinear term is sufficient, in general, to render it nonintegrable.

Conditions are sought for the given system to possess as many integrals of motion as possible. First, it is easy to check that the given system has only one type of singularity near which the leading order behavior of x_k's is

$$x_k \sim C_k\tau^{-1}, k = 1, 2, \ldots, N-1, \tau = t - t_0, x_N \sim C_N\tau^{-2}. \tag{1}$$

Inserting (1) in the given system, one immediately finds that $C_1 = -2 = -C_N, C_k = -2a_{k,N}, k = 2, \ldots, N-1$.

To determine at which order free constants arise, one substitutes in the given system

$$\begin{aligned} x_k &= C_k\tau^{-1} + \alpha_k\tau^{-1+r}, k = 1, 2, \ldots, N-1, \\ x_N &= C_N\tau^{-2} + \alpha_N\tau^{-2+r} \end{aligned}$$

and derives, as usual, linear equations for the a_k's whose coefficient matrix turns out to have vanishing determinant if and only if

$$(r+1)(r-1)^{N-2}(r-2) = 0. \qquad (2)$$

This implies that $N-2$ free constants appear at $r = 1$ and one at $r = 2$. It is more direct to proceed in an alternative way. Owing to the linearity of $N-1$ linear equations in the given system, we may guess the form of the integrals as

$$\exp(-\lambda_k t) \sum_{j=1}^{N-1} b_{kj}x_j = C_k, k = 1, 2, \ldots, N-1, \qquad (3)$$

C_k in (3) are arbitrary constants and the b_{kj} are to be determined by substituting (3) in the first $N-1$ equations of the given system. This easily leads to the eigenvalue problem

$$Ab_k = \lambda_k b_k, k = 1, 2, \ldots, N-1, \qquad (4)$$

where $b_k = (b_{k,1}, \ldots, b_{k,N-1})^T$ and A is the $(N-1) \times (N-1)$ matrix

$$A = \begin{bmatrix} 1 & a_{2,1} & a_{3,1} & \cdots & a_{N-1,1} \\ 1 & a_{2,2} & a_{3,2} & \cdots & a_{N-1,2} \\ 1 & & & & \\ 1 & & & & \\ 1 & & & & \\ 1 & a_{2,N-1} & a_{3,N-1} & \cdots & a_{N-1,N-1} \end{bmatrix}. \qquad (5)$$

We must also require that the coefficients of x_N terms resulting from the foregoing substitution vanish. This leads to the following equations for $a_{j,N}$:

$$b_{k,1} + \sum_{j=2}^{N-1} b_{k,j}a_{j,N} = 0, k = 1, 2, \ldots, N-1. \qquad (6)$$

Thus, in general, given the matrix A in (5), the system (4) can be solved for its $N-1$ eigenvectors b_k and eigenvalues λ_k. Thus, we may think of (6) as $N-1$ necessary and sufficient conditions for $a_{j,N}$ so that $N-1$ integrals exist. However, since $a_{j,N}$ to be determined are $N-2$ in number, they can satisfy, in general, $N-2$ of equations (6) and hence only $N-2$ integrals of the form (3) are to be expected. Bountis, Ramani, Grammaticos, and Dorizzi (1984)

2.
$$\begin{aligned} \dot{u}_1 &= u_1(u_2 - u_n), \dot{u}_2 = u_2(u_3 - u_1), \\ &\vdots \\ u_n &= u_n(u_1 - u_{n-1}). \end{aligned}$$

The given system arises from a semidiscretisation of the PDE $U_t + UU_x = 0$. It is first observed that the given system is scaleinvariant under $t \to \epsilon^{-1}t, u_j \to \epsilon u_j$. If we impose the cyclic boundary condition $0 \equiv n, n+1 \equiv 1$, we find that

$$I_1(u) = \sum_{j=1}^{n} u_j \quad \text{and} \quad I_2(u) = \sum_{j=1}^{n} u_j$$

are polynomial first integrals of the given system. For $n \geq 4$, additional first integrals $I_3(u) = \sum_{j=1}^{3} u_{j-1}u_{j+1}$ are obtained. For a given $n, n-1$ polynomial first integrals of the given system are found, and thus the latter is algebraically completely integrable.

For $n = 3$, it is easy to check that Laurent series expansion

$$u_j(t) = \sum_{k=0}^{\infty} a_{jk}\tau^{-1+k}, \tau = t - t_0, u_3(t) = \sum_{k=0}^{\infty} a_{3k}\tau^{2+k}, j = 1, 2$$

holds, where $a_{10} = 1, a_{20} = -1$, and a_{30} is arbitrary. A detailed Painlevé analysis is carried out. Steeb, Louw, and Maritz (1987)

3. $$\boldsymbol{y_k' = y_k(y_{k+1} - y_{k-1}), k = 1, 2, \ldots, y_0 = 0.}$$

It is shown that the given system can be reduced to the system of linear equations

$$\psi_k' = 4\psi_{k+1}, k = 1, 2, 3, \ldots.$$

Finite continued fractions and their expansions are used. Yamazaki (1987)

4. $$\boldsymbol{\dot{y}_k = 2y_k(y_{k+1} - y_{k-1}), k = 1, 2, \ldots, y_0 = 0,}$$

with initial conditions $y_k(0) \to 0 (k \to \infty)$.

The given system is integrated in two different ways—one using finite continued fractions. Asymptotic behavior is also obtained. Yamazaki (1990)

5. $$\boldsymbol{y_k' = \frac{1}{2}\sum_{i+j=k} (i+j)y_i y_j - y_k \sum_{j=1}^{\infty} (k+j)y_j, \quad y_k(0) = \hat{y}.}$$

The solution is found in the form

$$y_k(x) = (1/k)e^{-x}\exp[-M_0(0)k(1-e^{-x})]\sum_{\{n_j\}} \frac{k^m(1-e^{-x})^m}{\Pi_j n_j!}\Pi_j \hat{y}_j^{n_j},$$

where the summation goes over all possible sets of $\{n_j\}$, which satisfy the conditions $\sum_j jn_j = k - 1, m = \sum_j n_j$. Binglin (1987)

6. $$\boldsymbol{y_k' = \frac{1}{2}\sum_{i+j=k} [A(i+j)+B]y_i y_j - y_k \sum_{j=1}^{\infty} [A(k+j)+B]y_j, k = 1, 2, \ldots,}$$

where $y_k(0) = \delta_{kl}$.

The solution of the given system is found in the form

$$\begin{aligned} y_k(x) &= (1/k!)[(2A/B)k + k][(2A/B)k + k - 1] \\ &\times [(2A/B)k + 2](B/2)^{k-1}e^{-Ax}(1 - e^{-Ax})^{k-1} \\ &\times [A + (1/2)BM(1 - e^{-Ax})]^{-[(2A/B+1)k+1]}, \end{aligned}$$

where M is a constant. Binglin (1987)

7.

$$\dot{\boldsymbol{y}}_k = (1/2)\sum_{j=1}^{k-1} \boldsymbol{K}_{j,k-j}\boldsymbol{y}_j\boldsymbol{y}_{k-j} - \boldsymbol{y}_k\sum_{j=1}^{\infty} \boldsymbol{K}_{k,j}\boldsymbol{y}_j,$$

$k = 1, 2, \ldots,$ $K_{i,j}$ are constants. $y_1(0) = 1, y_k = 0$ for $k \geq 2$.

It is proved that if $K_{i,j} \leq (i + j)$ for all i and j, the given system has a unique solution that is analytic along the positive t-axis, and all its moments are analytic, too, thus excluding any kind of singular behavior. Heilmann (1992)

8.

$$\boldsymbol{y}'' + (3\boldsymbol{I}/\boldsymbol{x})\boldsymbol{y}' - \boldsymbol{F}(\boldsymbol{x}, \boldsymbol{y}) = \boldsymbol{0},\ \ 0 < \boldsymbol{x} \leq 1,$$

$$\boldsymbol{y}'(0) = \boldsymbol{0},\ \ \boldsymbol{B}_1\boldsymbol{y}(1) \equiv \begin{bmatrix} 1 & 0 & 0 & 0 \\ 0 & 2/3 & 0 & 1 \end{bmatrix} \boldsymbol{y}'(1) = \boldsymbol{0},$$

where

$$F(x, y(x)) = \begin{bmatrix} y_1(x)y_2(x) - \mu^2 y_2(x) - 2\gamma \\ -(1/2)y_1^2(x) + \mu^2 y_1(x) \end{bmatrix}, Y(x) = (y(x), y'(x))^T$$

I is a unit matrix and μ and γ are problem parameters. The function y is a vector-valued function of dimension 2. Using the general theory for this class of problems, existence, uniqueness, and smoothness of continuous solutions are considered. Weinmüller (1984)

9.

$$\boldsymbol{y}' = \frac{\boldsymbol{y}(\boldsymbol{x})\boldsymbol{A}\boldsymbol{y}(\boldsymbol{x})}{\boldsymbol{x}^2},$$

where $y(x) = \sum_{k=1}^{\infty} C_k x^k$, with C_1 and C_2 given, is an analytic square matrix function of x. A is some given square matrix. The given system has a solution if and only if

$$\begin{aligned} C_1AC_1 &= C_1, \\ C_1AC_2 &= C_2AC_1 = C_2. \end{aligned} \tag{1}$$

The solution is

$$y(x) = C_1x + \sum_{k=2}^{\infty} (C_2Ax)^{k-2}C_2x^2,$$

where $C_k = (C_2A)^{k-2}C_2$ for $k = 2, 3, 4, \ldots$. For $k = 2$,

$$2C_2 = C_1AC_2 + C_2AC_1. \tag{2}$$

Premultiplication of (2) by C_1A and postmultiplication of (2) by AC_1 and use of the first of conditions (1) yield

$$C_1AC_2 = C_1AC_2AC_1 = C_2AC_1 = C_2, \tag{3}$$

which is second of (1). For $k > 2$, we again pre- and postmultiply

$$kC_k = \sum C_jAC_{k-j+1}, k = 1, 2, 3, \ldots$$

by C_1A and AC_1 and use the equations derived from lower values of k to obtain $C_k = C_jAC_{k-j+1}$ for $j = 1, 2, \ldots k$. By induction we find that $C_k = (C_2A)^{k-2}C_2$ for $k = 2, 3, 4, \ldots$. Thus, the solution y is uniquely obtained, in terms of C_1 and C_2. Schweitzer (1986)

10.

$$\epsilon\boldsymbol{y}'' = \boldsymbol{H}(\boldsymbol{x}, \boldsymbol{y}), \boldsymbol{y}(\boldsymbol{0}) = \boldsymbol{A}, \boldsymbol{y}(1) = \boldsymbol{B};$$

A and B are constants. ϵ is a small parameter.

Here y is a vector-valued function. Sufficient conditions are given for the existence of a solution that exhibits boundary or interior layer behavior for small positive values of the parameter ϵ. Kelley (1984)

11. $$\dot{x}_i = G_i(x_1, x_2, x_3, \ldots, x_n), i = 1, 2, \ldots, n,$$

where G_i are quadratic forms in $x_1, x_2, \ldots, x_n$, invariant under the transformations

$$x_i(t) = \frac{1}{(ct+d)^2} X_i(\tau) - \frac{\nu_i C}{ct+d}, \tau = -\frac{at+b}{ct+d}, \quad ad - bc = -1,$$

where the numbers ν_i $(i = 1, 2, \ldots, n-1)$ take the values 0 or 1 and $\nu_n = 1$. Conditions similar to (B) in Eqn. (3.15.11) can be derived so that there are solutions of the form (A) therein. Martynov and Yablonskii (1979)

12. $$\dot{x} = \frac{P(x, y_1, \ldots, y_n)}{Q(x, y_1, \ldots, y_n)}, \quad \dot{y}_i = \frac{R_i(x, y_1, \ldots, y_n)}{S_i(x, y_1, \ldots, y_n)} \quad (i = 1, 2, \ldots, n),$$

where P, Q, R_i, S_i $(i = 1, 2, \ldots, n)$ are polynomials in their arguments:

$$\begin{aligned} P(x, y_1, \ldots, y_n) &= \sum_{\nu=0}^{p} P_\nu(x, y_1, \ldots, y_n), \\ Q(x, y_1, \ldots, y_n) &= \sum_{\nu=0}^{p} Q_\nu(x, y_1, \ldots, y_n), \\ R_i(x, y_1, \ldots, y_n) &= \sum_{\nu=0}^{r_i} R_{i\nu}(x, y_1, \ldots, y_n) \ (i = 1, 2, \ldots, n), \\ S_{i\nu}(x, y_1, \ldots, y_n) &= \sum_{\nu=0}^{s_i} S_{i\nu}(x, y_1, \ldots, y_n) \ (i = 1, 2, \ldots, n). \end{aligned}$$

Here $P, Q, R_{i\nu}, S_{i\nu}$ $(i = 1, 2, \ldots, n)$ are homogeneous polynomials in x, $y_1, \ldots, y_n$ of degree $\nu; p, q, r_i$, and s_i are nonnegative integers.

Solutions of the given system are sought so that at least one of the component solutions tends to infinity when $t \to t_0$. The method employed is to change the given system to a Briot–Bouquet system and use the known theory for the latter. Psi series for different sets of parameters with poles and pseudopoles are found. Bud'ko and Yablonskii (1989)

13. $$\dot{q}_j = p_j, \quad \dot{p}_j = \exp(q_{j-1} - q_j) - \exp(q_j - q_{j+1}),$$

where $j = 1, 2, \ldots, m, m > 1$, and $q_0 = q_m, q_{m+1} = q_1$.

These are known as equations of the Toda lattice. The point $p_1 = p_2 = \cdots = p_m = 0, q_1 = q_2 = \cdots = q_m = 0$, is a simple center. Despite the nonlinearity of the system, it can be shown to possess almost periodic solutions for all m. A numerical solution for $m = 3$ is depicted; the repetitive structure with IC $p_1(0) = -1, p_2(0) = 0.7, p_3(0) = 0.3, q_1(0) = -1, q_2(0) = 0$, and $q_3(0) = 1$ suggests almost periodic behavior. Bender and Orszag (1978), p.187

14. $$\dot{x}_r = X_r(x_1, x_2, \ldots, x_n), r = 1, 2, \ldots, n,$$

where X_r are analytic functions, expansible in convergent series of powers of $x_1, \ldots, x_n$, and where not all functions $X_1, \ldots, X_n$ vanish when $x_1 = \ldots = x_n = 0$. It is proved that the given system possesses $(n-1)$ independent integrals, independent of t, expansible in convergent series of powers of $x_1, x_2, \ldots, x_n$; i.e., functions $\phi_r(x_1, x_2, \ldots, x_n)$ $(r = 1, \ldots, n-1)$ exist such that

$$\frac{d\phi_r}{dt} = \frac{\partial \phi_r}{\partial x_1} X_1 + \cdots + \frac{\partial \phi_r}{\partial x_n} X_n = 0$$

identically.

However, if the origin is a singular point of the given system so that $X_r(x_1, \ldots, x_n), r = 1, \ldots, n$ all vanish at the origin, then unless there is a restriction upon the form of $X_1, \ldots, X_n$, the given system possesses no integrals developable about their singular points. Cherry (1925)

BIBLIOGRAPHY

Abdullaev, A. S. (1983) On the theory of the second Painlevé equation. Sov. Math. Dokl., 28, 726.

Abraham–Schrauner, B. (1990) Lie group symmetries and invariants of the Henon–Heiles equations. J. Math. Phys., 31, 1627.

Abraham–Schrauner, B., Bender, C. M. and Zitter, R. N. (1992) Taylor series and δ-perturbation expansions for a nonlinear semi-conductor transport equation. J. Math. Phys., 33, 1335.

Ackerberg, R. C. (1969) On a nonlinear differential equation of electrohydrodynamics. Proc. Roy. Soc. London A, 312, 129.

Adler, J. (1991) Thermal explosion theory with Arrhenius kinetics: homogeneous and inhomogeneous media. Proc. Roy. Soc. London A, 433, 329.

Ad'yutov, M. M. and Klokov, Yu. A. (1989) An initial value problem for second order ordinary differential equations with singularities (Russian). Diff. Eqns., 25, 1268.

Ad'yutov, M. M., Klokov, Yu. A. and Mikhailov, A. P. (1991) Some problems for an ordinary differential equation arising in gas dynamics. Diff. Eqns., 26, 803.

Ad'yutov, M. M. and Zmitrenko, N. V. (1991) A generalisation of Emden's equation. Diff. Eqns., 27, 767.

Afuwape, A. U. (1986) Frequency-domain approach to nonlinear oscillations of some third order differential equations. Nonlinear Anal. Theory Meth. Appl., 10, 1459.

Aguirre, M. and Krause, J. (1988) Finite point transformations and linearisation of $y'' = f(x,y)$. J. Phys. A Math. Gen. 21, 2841.

Airault, H. (1986) Polynomial invariants for the equation $\frac{d^2v}{d\xi^2} = \lambda(\xi)\frac{dv}{d\xi} + F(\xi,\nu)$. Int. J. Nonlinear Mech., 21, 197.

Airault, H. (1990) Reducible cases for $v'' = h(T)\frac{\partial T}{\partial v}$. J. Fac. Sci. Univ. Tokyo Sec. IA Math., 37(2), 235.

Akhromeyeva, T. S. and Malinetskii, G. G (1987) On the strange attractor in a problem of synergetics. USSR Comp. Math. Math. Phys., 27, 132.

Albrecht, F. and Villari, G. (1987) On the uniqueness of periodic solutions of certain Lienard equations. Nonlinear Anal. Theory Meth. Appl., 11, 1267.

Alexander, R. K. (1984) The multiple steady states of the non-adiabatic tubular reactor. J. Math. Anal. Appl., 101, 12.

Alexander, R. (1990) Spatially oscillatory steady states of tubular chemical reactors. SIAM J. Math. Anal., 21, 137.

Alidema, R. I. (1977) Linearisation of nonlinear ordinary differential equations. Mat. Vestnik, 14, 339.

Alidema, R. I. (1978) The equivalence of certain linear and nonlinear differential equations of Nth order. Publ. Inst. Math., 24(38), 5.

Amel'kin, V. V. (1968) Periodic solutions of a system with a small parameter. Diff. Eqns., 4, 887.

Amel'kin, V. V. and Gaishun, I. V. (1971) Some properties of solutions of a second order equation. Diff. Eqns., 7, 1603.

Amel'kin, V. V. and Zhavnerchik, V. E. (1988) Periodic solutions of a Lienard equation. Diff. Eqns., 24, 1085.

Amendola, G. and Manes, A. (1987) A particular class of second order nonlinear differential equations (Italian, English summary). Riv. Mat. Univ. Parma, 13, 51.

Amerio, L. (1951) Sull'estensióne délle nozióni di "còlle," "nòdo" e "fuòco" ai sistèmi di due equazióni differenziali perìodiche in tre variàbili. Atti Accad. Naz. Lincei Rend. Cl. Sci. Fiz. Mat. Nat., 10, 206; ibid., 289.

Ames, W. F. (1968) Nonlinear Ordinary Differential Equations in Transport Processes. Academic Press, New York.

Ames, W. F. and Adams, E. (1979) Nonlinear boundary and eigenvalue problems for the Emden–Fowler equations by group methods. Int. J. Nonlinear Mech., 14, 35.

Ames, W. F. and Nucci, M. C. (1985) Analysis of fluid equations by group methods. J. Engg. Math., 20, 181.

Amick, C. J. and Kirchgasoner, K. (1988) A theory of solitary water waves in the presence of surface tension. Arch. Rat. Mech. Anal., 105, 1.

Amick, C. J. and McLeod, J. B. (1990) A singular perturbation problem in needle crystals. Arch. Rat. Mech. Anal., 109, 139.

Amick, C. J. and Toland, J. F. (1990) A differential equation in the theory of resonant oscillations of water waves. Proc. Roy. Soc. Edin., 114A, 15.

Anand, G. V. (1972) Natural modes of a coupled nonlinear system. Int. J. Nonlinear Mech., 7, 81.

Anderson, L. R. (1970) Integral manifolds of a class of third order autonomous differential equations. J. Diff. Eqns., 7, 274.

Anderson, N. and Arthurs, A. M. (1982) Pointwise bounds for solutions of a nonlinear problem in heat conduction. Z. Angew. Math. Mech., 62, 701.

Andres, J. (1987) On local w-cycles to certain third order nonlinear differential equations. Fasc. Math., 17, 49.

Andres, J. (1992) Note on the asymptotic behaviour of solutions of damped pendulum equations under forcing. Nonlinear Anal. Theory Meth. Appl., 18, 705.

Anitova, E. S. (1964) On the boundedness of the solutions of a system of differential equations of third order. Vestnik Leningrad Univ. Sec. Meh. Astron., 19, 5.

Antman, S. N. (1988) A zero dimensional shock. Q. Appl. Math., 46, 569.

Aripov, M. and Eshmatov, D. (1988) On WKB solutions of a generalised equation of Emden–Fowler type. Dokl. Akad. Nauk UzSSR, 9, 4.

Armellini, G. (1942) Sópra una classe di equazióni differenziali délla meccànica celèste di cui l'integrale generale tende a zèro. Pont. Acad. Sci. Acta, 6, 387.

Aronson, D. G. and Graveleau, J. (1993) A self-similar solution to the focusing problem for the porous medium equation. Euro. J. Appl. Math. 4, 65.

Arrigo, D. J. (1991) Group properties of $u_{xx} - u_y^m u_{yy} = f(u)$. Int. J. Nonlinear Mech., 26, 619.

Artykov, A. R., Rabinkov, G. A and Rozet, I. G. (1984) Moving and nonmoving singular point solution of some nonautonomous systems. Diff. Eqns., 20, 659.

Ascari, A. (1952) Stùdio asintotico di un equazióne relativa alla dinàmica del punto. Ist. Lumbardo Sci. Lett. Rend. Cl. Sci. Mat. Nat., 16, 278.

Ascoli, G. (1951) Ricerche asintotiche sópra una classe di equazióni differenziali nonlineari. Ann. Scuola Norm. Sup. Pisa (3), 5, 1.

Athorne, C. (1991) Rational Ermakov systems of Fuchsian type. J. Phys. A, 24, 945.

Athorne, C., Rogers, C., Ramgulam, V., and Osbaldestin, A. (1990) On linearisation of the Ermakov system. Phys. Lett. A, 143, 207.

Atkinson, F. V. (1955) On second order nonlinear oscillations. Pacific J. Math., 5, 643.

Atkinson, F. V., Brezis, H. and Peletier, L. A. (1990) Nodal solutions of elliptic equations with critical exponents. J. Diff. Eqns., 85, 151.

Avakumovic, V. G. (1947) Sur l'équation différentielle de Thomas–Fermi. Acad. Serbe Sci. Publ. Inst. Math., 1, 101.

Awrejcewicz, J. (1989) Gradual and sudden transition to chaos in a sinusoidally driven nonlinear oscillator. J. Phys. Soc. Japan, 58, 4261.

Awrejcewicz, J. (1991) Three routes to chaos in simple sinusoidally driven oscillators. Z. Angew. Math. Mech., 71, 71.

Awrejcewicz, A. and Grabski, J. (1989) Chaos in a particular nonlinear oscillator. Acta Mech., 79, 303.

Aymerich, G. (1955) Cicli di prima e di secónda spècie di un sistèma meccànico autosostenuto impulsivamente. Rend. Sem. Fac. Sci. Univ. Cagliari, 25, 25.

Badola, P., Ravikumar, V. and Kulkarni, B. D. (1991) Effects of coupling nonlinear systems with complex dynamics. Phys. Lett. A, 155, 365.

Bailey, P. B., Shampine, L. F. and Waltman, P. E. (1968) Nonlinear Two Point Boundary Value Problems. Academic Press, New York, pp. 13, 28, 45, 64.

Bainov, D. D. (1966) Some remarks on certain odd-order differential equations. Diff. Eqns., 2, 448.

Bainov, D. D. (1967) Some remarks on a certain differential equation of arbitrary order. Diff. Eqns., 3, 2003.

Baker, N. H., Moore, P. W. and Spiegel, F. A. (1971) A periodic behaviour of a nonlinear oscillator. Q. J. Mech. Appl. Math., 24, 391.

Balachandran, K., Thandapani, E. and Balasubramanian, G. (1988) Some series solutions of the Duffing equation. Ind. J. Pure Appl. Math., 19, 429.

Balitinov, M. A. (1968) Behaviour of the solution of a nonlinear system in the large. Diff. Eqns., 5, 209.

Ball, F. K. (1962) An exact theory of simple finite shallow water oscillations on a rotating earth. Proc. Conf. on Fluid Mechanics and Hydraulics, Univ. West Australia. Pergamon Press, Elmoford, N.Y., p. 293.

Balla, K. (1980) Solutions of singular boundary value problems for nonlinear systems for systems of ordinary differential equations. Appl. Sci. Res., 20(4), 100.

Bandic, I. (1963) Sur le critère d'integrabilité d'une équation différentielle nonlinéaire du deuxième ordre qui apparat dans l'électronique. Z. Angew. Math. Mech., 43, 429.

Bandic, I. (1964) Sur les invariantes de quelques équations différentielles nonlinéaires du deuxième ordre qui apparaissent dans la physique théorique. C.R. Acad. Sci. Paris, 258, 4417.

Bandic, M. I. (1965) Sur une classe d'équations différentielles nonlinéaires du deuxième ordre qui apparat en physique théorique. C.R. Acad. Sci. Paris, 260, 6269.

Bandic, M. I. (1965a) On the reduction of some nonlinear differential equations of the third order to Abel's differential equation of the first order. Bull. Calcutta Math. Soc., 55, 97.

Banfi, C. (1963/64) Sulle oscillazióni di un sistèma non lineare in due gradi di libertà. Atti Accad. Sci. Torino Cl. Sci. Fis. Mat. Nat., 98, 418.

Barashkov, N. M. and Spriridonov, F. F. (1988) Non-stationary flows in channels with permeable walls. J. Appl. Math. Mech., 52, 458.

Barbalat, I. (1955) L'allure globale des solutions de certaines équations différentielles nonlinéaires de second ordre. Acad. Repub. Pop. Romine Bul. Stünt. Sec. Sci. Mat. Fiz., 7, 653.

Barcilon, V. and Lovera, O. M. (1989) Solitary waves in magma dynamics. J. Fluid Mech., 204, 121.

Bartashevich, D. A. (1973) The qualitative nature of real solutions of Painlevé's first equation. Diff. Eqns., 9, 714.

Bassom, A. P. et al. (1992) Integral equations and exact solutions for the fourth Painlevé equation. Proc. Roy. Soc. London A, 437, 1.

Batalova, Z. S. and Belyakova, G. V. (1985) Bifurcation diagrams of some periodic solutions of a nonlinear Mathieu equation: differential and integral equations (Russian). Gor'kov. Gos. Univ. Gorki, 124, 28–324.

Batyshchev, V. A. (1991) Self-similar solutions describing thermal capillary flow in viscous layers. J. Appl. Math. Mech., 55, 315.

Bauch, H. (1982) The iterative solution of initial value problems for ordinary differential equations. Appl. Sci. Res., 22(2), 58.

Bauer, H. F. (1966) The response of a nonlinear system to pulse excitation. Int. J. Nonlinear Mech., 1, 267.

Bauer, F. and Schamel, H. (1992) Spatio-temporal structures in collisionless electrostate plasmas. Physica D 54, 235.

Baumann, G. (1991) Integrability and chaos for two copropagating pulses in optical fibers. Phys. Lett., 156, 298.

Baxley, J. V. (1988) A singular nonlinear boundary value problem: membrane response of a spherical cap. SIAM J. Appl. Math., 48, 497.

Baxley, J. V. (1990) Existence theorems for a nonlinear second order boundary value problem. J. Diff. Eqns., 85, 125.

Bazley, N. W. (1983) Approximations to periodic solutions of a Duffing equation. Z. Angew. Math. Phys., 34, 301.

Bebernes, J. and Troy, W. (1987) Non-existence for the Kassoy problem. SIAM J. Math. Anal., 18, 1157.

Becket, P. M. (1980) Combined natural and forced convection between parallel vertical walls. SIAM J. Appl. Math., 39, 372.

Beesack, P. R. (1984) Asymptotic behaviour of solutions of some general nonlinear differential equations and integral inequalities. Proc. Roy. Soc. Edin., 85A, 49.

Behzad, M. and Mehri, B. (1971) On the boundedness property of the solution of certain nonlinear differential equations. Studi. Sci. Math. Hung., 6, 163.

Beklemiseva, L. A. (1962) On a nonlinear second order differential equation. Mat. Sb. (N.S.), 56(98), 207.

Belhorec, S. (1967) On some properties of the equation $y''(x) + f(x)y^{\alpha} = 0$, $0 < \alpha < 1$. Mat. Casopis Sloven. Akad. Vied, 17, 10.

Bellman, R. (1953) Stability Theory of Differential Equations. McGraw-Hill, New York.

Belykh, V. N. (1984) Bifurcation of separatrices of a saddle point of the Lorenz system. Diff. Eqns., 20, 1184.

Belyustina, L. N. (1955) On an equation from the theory of electrical mechines. Pamyati Aleksandra Aleksandrovica Andronova (In memory of Aleksandra Aleksandrovica Andronova). Izv. Akad. Nauk SSR Moscow, 173.

Ben M'Barek, A. and Arino, O. (1988) An integrability criterion for non-forced nonlinear differential equations. Rad. Mat., 4(2), 261.

Bender, C. M. and Orszag, S. A. (1978) Advanced Mathematical Methods for Scientists and Engineers. McGraw-Hill, New York.

Bender, C. M., Milton, K. A., Pinsky, S. S., and Simmons, Jr., L. M. (1989) A new perturbative approach to nonlinear problems. J. Math. Phys., 30, 1447.

Benson, D. C. (1973) Comparison and oscillation theory for Lienard's equation with positive damping. SIAM J. Appl. Math., 24, 251.

Benson, D. C. (1974/75) Comparison theorems for a class of nonlinear ordinary differential equations. Proc. Roy. Soc. Edin., 73A, 77.

Benson, D. C. (1981) Principal solutions for Lienard's equation. SIAM J. Math. Anal., 12, 398.

Benson, D. C. (1984) A Prüfer transformation for Lienard's equation. SIAM J. Math. Anal., 15, 656.

Berestycki, H., Nicolaenko, B. and Scheurer, B. (1983) Sur quelques problèmes asymptotiques avec applications a la combustion. C.R. Acad. Sci. Paris T, 296, Series 1, 105.

Berestycki, H., Nicolaenko, B. and Scheurer, B. (1985) Travelling wave solutions to combustion models and their singular limits. SIAM J. Math Anal., 16, 1207.

Berkovich, L. M. (1971) Transformation of ordinary nonlinear differential equations. Diff. Eqns., 7, 272.

Berman, A. S. and Vostokov, V. V. (1982) Exact solutions of a nonlinear boundary value problem of the theory of chemical reactions. J. Appl. Math. Mech., 46, 411.

Bernussou, J., Liu, H. and Mira, C. (1976) On nonlinear periodic differential equations and associated point mappings. Int. J. Nonlinear Mech., 11, 1.

Bespalova, S. A. (1984) On a nonlinear eigenvalue problem. USSR Comp. Math. Math. Phys., 24(3), 88.

Bespalova, S. A. and Klokov, Yu. A. (1985) Boundary value problem for a system of ordinary differential equations, encountered in biochemistry. Diff. Eqns., 21, 491.

Bespalova, S. A. and Klokov, Yu. A. (1988) A generalisation of the Falkner–Skan equation. I. Current topics in boundary value problems: theory and applications (Russian). Latv. Gos. Univ. Riya, iv, 99–108.

Bespalova, S. A. and Klokov, Yu. A. (1989) Boundary value problems for fourth order ordinary differential equations. Diff. Eqns., 25, 381.

Bespalova, S. A. and Klokov, Yu. A. (1990) Boundary value problems for a system of fourth order. Diff. Eqns., 26, 664.

Bihari, I. (1961) On nonlinear equations $u'' + a(t)u + q(t)f(u^2) = 0$. Magyar Tud. Akad. Mat. Kutato Int. Közl., 6, 287.

Bihari, I. (1962) Extension of a theorem of Armellini–Tonelli–Sansone to the nonlinear equation $y'' + a(x)f(y) = 0$. Magyar Tud. Akad. Mat. Kutato Int. Közl., 7, 63.

Bihari, I. and Fenyes, T. (1968) On a first order nonlinear differential equation system. Stud. Sci. Math. Hung., 3, 257.

Billigheimer, C. E. (1968) Solutions of a nonlinear differential equation: II. Proc. Camb. Phil. Soc., 64, 127.

Bin, L. (1989) Boundedness for solutions of nonlinear Hill's equation, with periodic forcing terms via Moser's twist theorem. J. Diff. Eqns., 79, 304.

Binglin, L. (1987) The exact solution of the coagulation equation with kernel $K_{ij} = A(i+j) + B$. J. Phys. A, 20, 2347.

Blanchard, Ph., Stubbe, J. and Vazquez, L. (1988) Stability of bound states for (1+1)-dimensional nonlinear scalar fields. J. Phys. A. Math. Gen., 21, 1137.

Blaquiere, A. (1956) Equation de Hill nonlinéaire et méthode stroboscopique de N. Minorsky. C.R. Acad. Sci. Paris, 243, 1711.

Blasko, R., Graef, J. R., Hacik, M. and Spikes, P. W. (1990) Oscillatory behaviour of solutions of nonlinear differential equations of the second order. J. Math. Anal. Appl., 151, 330.

Bluman, G. W. (1990) Invariant solutions for ordinary differential equations. SIAM J. Appl. Math., 50, 1706.

Bluman, G. W. and Reid, G. J. (1988) New symmetries for ordinary differential equations. IMA J. Appl. Math., 40, 87.

Bobisud, L. E. (1972) The distance to vertical asymptotes for solutions of second order equations. Michigan Math. J., 19, 277.

Bobisud, L.E. (1993) A steady-state reaction diffusion problem in a moving medium. J. Math. Anal. Appl., 175, 239.

Bobisud, L. E., O'Regan, D. and Royalty, W. D. (1987) Existence and non-existence for a singular boundary value problem. Appl. Anal., 28, 245.

Bobisud, L. E., O'Regan, D. and Royalty, W. D. (1987a) Singular boundary value problems. Appl. Anal., 23, 233.

Bogoslovskii, B. P. (1966) On a class of nonlinear differential equations of the second order with algebraic movable singularities. Diff. Eqns., 2, 414.

Bogoslovskii, B. P. (1972) On fixed singularities of one class of nonlinear equations. Diff. Eqns., 8, 1483.

Bogoslovskii, B. P. (1974) Movable singularities for a system of two nonlinear equations. Diff. Eqns., 10, 262.

Bogoslovskii, B. P. and Ostroumov, S. I. (1981) Algebraic moving singular points of a nonlinear equation: differential and integral equations. Gor'kov. Gos. Univ. Gorki, No. 59–12.

Bogoslovskii, B. P. and Ostroumov, S. I. (1983) Equations with stationary transcendental and essential singularities. Diff. Eqns., 20, 1227.

Bogoslovskii, B. P. and Yablonskii, A. I. (1967) Systems of differential equations with non-stationary algebraic singular points. Diff. Eqns., 3, 1110.

Bohe, A. (1990) Free layers in a singularly perturbed boundary value problem. SIAM J. Math. Anal., 21, 1264.

Bohm, C. (1953) Nuòvi critèri di esistenza di soluzióni periòdiche di una nòta equazióne differenziale nonlinéaire. Ann. Mat. Pura Appl., 35, 343.

Bohme, G., Johann, W. and Siekmann, J. (1974) Note on a differential equation of electrodynamics. Acta Mech., 20, 303.

Boiti, M. and Pempinelli, F. (1979) Similarity solutions of the Korteweg–deVries equation. Nuovo Cimento, 51, 70.

Boiti, M. and Pempinelli, F. (1980) Similarity solutions and Bäcklund transformations of the Boussinesq equation. Nuovo Cimento B, 56, 148.

Bojadziev, G. N. (1983) Damped nonlinear oscillations modelled by a three-dimensional differential system. Acta Mech., 48, 193.

Bona, J. L. and Schonbek, M. E. (1985) Travelling wave solutions to Korteweg–deVries Burgers equation. Proc. Roy. Soc. Edin., 101A, 207.

Bountis, T. C. (1983) A note on the "Painlevé" property of anharmonic systems with an external periodic field. Phys. Lett. A, 97, 85.

Bountis, T., Drossos, L. and Percival, I. C. (1991) Non-integrable systems with algebraic singularities in the complex time. J. Phys. A, 24, 3217.

Bountis, T. C., Ramani, A., Grammaticos, B. and Dorizzi, B. (1984) On the complete and partial integrability of non-Hamiltonian systems. Physica A, 128, 268.

Bountis, T., Tsarouhas, G. and Herman, R. (1988) Normal form solutions of dynamical systems in the basin of attraction of their fixed point. Physica D, 33, 34.

Bouquet, S. E., Feix, M. R. and Leach, P. G. L. (1991) Properties of second order differential equations invariant under time translation and self-similar transformations. J. Math. Phys., 32, 1480.

Bourland, F. J. and Haberman, R. (1988) The modulated phase shift for strongly nonlinear slowly varying, and weakly damped oscillations. SIAM J. Appl. Math., 48, 737.

Bourland, F. J. and Haberman, R. (1990) Separatrix crossing: time invariant potentials with dissipation. SIAM J. Appl. Math., 50, 1716.

Bradley, J. S. (1972) Boundary value problems involving the Thomas–Fermi equation. Conf. on Theory of Ordinary and Partial Differential Equations, Univ. Dundee, Dundee, Scotland, 1972. Lecture Notes Mathematics, Vol. 280. Springer-Verlag, New York, pp. 223–226.

Branson, T. R. and Steeb, W. H. (1983) Symmetries of nonlinear diffusion equations. J. Phys. A, 16, 469.

Braude, S. Ya. (1967) Integrability conditions for certain nonlinear second order equations. Diff. Eqns., 3, 535.

Brauner, C. M. and Lainé, N. B. (1982) Further solutions of the Falkner–Skan equation for $\beta = -1$ and $r = 0$. Mathematika, 29, 231.

Brauner, C. M. and Schmidt–Lainé, Cl. (1987) Existence of solution to a certain plane premixed flame problem with two step kinetics. SIAM J. Math. Anal., 18, 1406.

Brighton, P. W. M. (1988) Similarity solutions for two dimensional steady laminar gravity currents. J. Fluid Mech., 192, 75.

Brindley, J. and Kapitaniak, T. (1991) Analytic predictions for strange non-chaotic attractors. Phys. Lett. A, 155, 361.

Brindley, J., Kapitaniak, T. and El Naschie, M.S. (1991) Analytic conditions for strange chaotic and nonchaotic attractors of the quasiperiodically forced Van der Pol equation. Physica D, 51, 28.

Brodie, P. and Banks, W. H. H. (1986) Further properties of the Falkner–Skan equation. Acta Mech., 65, 205.

Brown, S. N. (1966) A differential equation occurring in a boundary layer. Mathematika, 13, 140.

Brügner, G. (1985) Eine explizite "ähnliche Lösung" der grenzschicht Theorie. Z. Angew. Math. Mech., 3, 190.

Bruijn, N. G. (1981) Topological existence proof for a nonlinear two-point boundary value problem. Philips J. Res., 36, 229.

Brüll, L. and Hölters, H. P. (1986) A geometrical approach to bifurcation for nonlinear boundary value problems. Z. Angew. Math. Phys., 37, 820.

Budd, C. J. (1988) Comparison theorems for radial solutions of semilinear elliptic equations. J. Diff. Eqns., 72, 338.

Budd, C. and Qi, Y. (1989) The asymptotic behaviour of the solutions of the Kassoy problem with a modified source term. Proc. Roy. Soc. Edin., 113A, 347.

Bud'ko, T. S. and Yablonskii, A. I. (1989) General method of finding a solution with components having infinite limit values of the autonomous differential systems with rational right sides. Diff. Eqns., 25, 1291.

Burde, G. I. (1990) A class of solutions of boundary layer equations. Fluid Dynamics, 25, 201.

Bureau, F. J. (1964) Differential equations with fixed critical points. Ann. Mat., 64., 231.

Bureau, F. J. (1964a) Differential equations with fixed critical points. Ann. Mat., 66, 1.

Bureau, F. J. (1972) Integration of some nonlinear systems of ordinary differential equations. Ann. Mat. Pura Appl. (4), 94, 345.

Burggraf, O. R., Stewartson, K. and Belcher, R. (1971) Boundary layer induced by a potential flow. Phys. Fluids, 14, 1821.

Burt, P. B. (1987) Non-perturbative solution of nonlinear field equations. Nuovo Cimento, B, 100, 43.

Burt, P. B. and Reid, J. L. (1973) On the radial distribution function in quantum mechanics. J. Chem. Phys., 58, 2194.

Burton, T. A. (1970) On the equation $x'' + f(x)h(x')x' + g(x) = e(f)$. Ann. Mat. Pura Appl., 85, 277.

Burton, T. D. (1983) On the amplitude decay of strongly nonlinear damped oscillator. J. Sound Vib., 87, 535.

Burton, T. D. (1984) A perturbation method for certain nonlinear oscillators. Int. J. Nonlinear Mech., 19, 397.

Burton, T. D. and Hamdan, M. N. (1983) Analysis of nonlinear autonomous conservative oscillators by a time transformation method. Sound Vib., 87, 543.

Bushard, L. B. (1976) Initial value region for a boundary value problem arising in a gas flow discharge theory. SIAM J. Appl. Math., 31, 547.

Bushard, L. B. (1978) Computationally useful bounds for a singular nonlinear second order boundary value problem. SIAM J. Math. Anal., 9, 651.

Bushell, P. J. (1973/74) On the nonexistence of solution to generalised swirling flow problems. Proc. Roy. Soc. Edin., 72A, 271.

Butlewski, Z. (1945) Sur les intégrales bornèes des équations différentielles. Ann. Soc. Polon. Math., 18, 47.

Byatt–Smith, J. G. (1987) 2π periodic solutions of Duffing's equation with negative stiffness. SIAM J. Math. Anal., 47, 60.

Byatt–Smith, J. G. (1988) On the solutions of a second order DE arising in the theory of resonant oscillations in a tank. Stud. Appl. Math., 79, 143.

Byatt–Smith, J. G. (1988a) Resonant oscillations in shallow water with small mean square disturbances. J. Fluid. Mech., 193, 369.

Byatt–Smith, J. G. (1989) The asymptotic solution of a connection problem of a second order ordinary differential equation. Stud. Appl. Math., 80, 109.

Caginalp, G. and Hastings, S. (1986) Properties of some ordinary differential equations related to free boundary problems. Proc. Roy. Soc. Edin., 104A, 217.

Caginalp, G. and McLeod, B. (1986) The interior transition layer for an ordinary differential equation arising from solidification theory. Q. Appl. Math., 44, 155.

Cahen, G. (1953) Systèmes électromécaniques nonlinéaires. Rev. Gen. Elec., 62, 277.

Callegari, A. J. and Friedman, M. B. (1968) An analytical solution of a nonlinear, singular boundary value problem in the theory of viscous fluids. J. Math. Anal. Appl., 21, 510.

Callegari, A. J. and Friedman, M. B. (1976) An analytic solution for the laminar flow over a flat plate with similarity preserving section. Int. J. Nonlinear Mech., 11, 147.

Callegari, A. and Nachman, A. (1978) Some singular nonlinear differential equations arising in boundary layer theory. J. Math. Anal. Appl., 64, 96.

Calogero, F. (1987) The evolution partial differential equation $u_t = u_{xxx} + 3(u_{xx}u^2 + 3u_x^2 u + 3u_x u^4)$. J. Math. Phys., 28, 538.

Cariello, F. and Tabor, M. (1991) Similarity reductions from extended Painlevé expansions for nonintegrable evolution equations. Physica D, 53, 59.

Carton–Lebrun, C. (1969) Simplifiées de Painlevé dont les solutions sont à points critiques isolés fixes. Acad. Roy. Belg. Bull. Cl. Sci. (5) 55, 883.

Cartwright, M. L. (1948) Forced oscillations in nearly sinusoidal systems. J. Inst. Elec. Engrs. Part III, 95, 88.

Cartwright, M. L. (1985) An unsymmetrical Van der Pol equation with stable harmonies. J. Nonlinear Mech., 20, 359.

Cartwright, M. L. and Reuter, G. E. H. (1987) On periodic solutions of Van der Pol's equation, with sinusoidal forcing term and large parameter. J. London Math. Soc., 36, 102.

Casten, R. G., Cohen, H. and Lagerstrom, P. A. (1975) Perturbation analysis of an approximation to the Hodgkin–Huxley theory. Q. Appl. Math., 32, 365.

Castro, A. and Shivaji, R. (1988) Non-negative solutions for a class of non-positive problems. Proc. Roy. Soc. Edin., 108A, 29.

Caughey, T. K. (1969) Whirling of a heavy string under constant axial tension: a nonlinear eigenvalue problem. Int. J. Nonlinear Mech., 4, 61.

Cecconi, J. (1950) Su di una equazióne differenziale non lineare di secónda órdine. Ann. Scuola Norm. Sup. Pisa (3), 4, 245.

Cecconi, J. (1950a) Su di una equazióne differenziale di rilassamento. Atti. Accad. Naz. Lincei Rend. Cl. Sci. Fis. Mat. Nat. (8), 9, 38.

Cekmarev, A. I. (1948) The influence of a constant force on the oscillations in nonlinear systems. Akad. Nauk SSSR Inzenernyi Sbornik, 4, 80.

Cesari, L. and Hale, J. K. (1957) A new sufficient condition for periodic solutions of weakly nonlinear differential systems. Proc. Amer. Math. Soc., 8, 757.

Chalkley, R. (1960) On the second order homogeneous quadratic differential equation. Math. Ann., 141, 87.

Chalkley, R. (1987) New contributions to the related work of Paul Appell, Lazarus Fuchs, George Hamel, and Paul Painlevé on nonlinear differential equations whose solutions are free of movable branch points. J. Diff. Eqns., 68, 72.

Chan, C. Y. and Du, S. W. (1986) The interior transition layer for an ordinary differential equation arising from solidification theory. Q. Appl. Math., 44, 155.

Chan, C. Y. and Hon, Y. C. (1987) A constructive solution for a generalised Thomas–Fermi theory of ionised items. Q. Appl. Math., 45, 591.

Chan, C. Y. and Hon, Y. C. (1988) Computational methods for generalised Thomas–Fermi models of neutral atoms. Q. Appl. Math., 46, 711.

Chandra, J. and Fleishman, B. A. (1972) Approximate determination of periodic solution of a class of nonlinear differential equation. Int. J. Nonlinear Mech., 7, 207.

Chandrasekhar, S. (1957) An introduction to the study of stellar structure. Dover, New York.

Chang, S. H. (1975) Existence of periodic solutions to second order nonlinear equations. J. Math. Anal. Appl., 52, 255.

Chang, H. (1986) Travelling waves on fluid surfaces: normal form analysis of the Kuramoto–Sivashinsky equation. Phys. Fluids, 29, 3142.

Chang, Y. and Segur, H. (1991) An asymptotic symmetry of the rapidly forced pendulum. Physica D, 51, 109.

Chao, R. M., Rajagopal, K. R. and Wineman, A. S. (1987) Non-homogeneous tension-torsion of neo-Hookean and Mooney–Rivlin materials. Arch. Mech., 39, 551.

Chappaz, G. (1968) Aspects analytiques et analogiques des solutions de l'équation de Van der Pol en régime force sinusoïdal. Int. J. Nonlinear Mech., 3, 245.

Chauvette, J. and Stenger, F. (1975) The approximate solution of the nonlinear equation $\Delta u = u - u^3$. J. Math. Anal. Appl., 51, 229.

Chawla, M. M. and Shivakumar, P. N. (1987) On the existence of solutions of a class of singular nonlinear two point boundary value problems. J. Comp. Appl. Math., 19, 379.

Chazy, J. (1911) Sur les èquations différentielles du troisième ordre et d'ordre supérieur fixes. Acta Math., 34, 1–69 and 317–385.

Chen, Y. (1973) An oscillation criterion for the second order nonlinear differential equation $x'' + xF(x'^2, x^2, t) = 0$. Q. J. Math. (Oxford), 24, 165.

Chen, S. (1987) Asymptotic linearity of solutions of nonlinear differential equations. Bull. Austral. Math. Soc., 35, 257.

Chen, J. and O' Malley, R. E., Jr. (1974) On the asymptotic solution of a two-parameter boundary value problem of chemical reactor theory. SIAM J. Appl. Math., 26, 717.

Chen, Y. and Wen, S. (1987) Travelling wave solution to the two dimensional Korteweg–deVries equation. J. Math. Anal. Appl., 127, 226.

Cheng, S. (1980) On a class of fourth order half-linear differential equations. Czech. Math. J., 30, 84.

Cheng, C. (1991) Invariant torus bifurcation series and evolution of chaos exhibited by a forced nonlinear vibration system. Int. J. Nonlinear Mech., 26, 105.

Cheng, P. and Chang, I. (1976) Buoyancy induced flows in a saturated porous medium adjacent to impermeable horizontal surfaces. Int. J. Heat Mass Transfer, 19, 1267.

Cheng, K. and Hsu, S. (1991) On the singular behaviour of the solution of $v''(x)+xv(x) = 0$. J. Math. Anal. Appl., 163, 20.

Cherry, T. M. (1925) Integrals of systems of ordinary differential equations. Proc. Camb. Phil. Soc., 22, 273.

Chicone, C. (1988) Geometric methods for two point nonlinear boundary value problems. J. Diff. Eqns., 72, 360.

Childs, D. R. (1973) Exact solution of the nonlinear differential equation $R\ddot{R} + (3/2)R^2 - AR^{-4} + B = 0$. Int. J. Nonlinear Mech., 8, 371.

Chisholm, J. S. R. and Common, A. K. (1987) A class of second order differential equations and related first order systems. J. Phys. A, 20, 5459.

Christiansen, P. L., Eilbeck, J. C., Enolskii, V. Z. and Gaididei, Ju. B. (1992) On ultrasonic Davydov solitons and the Henon–Heiles system. Phys. Lett. A, 166, 129.

Cicogna, G. (1987) Simple criteria for stable bifurcating periodic solutions of ODEs. SIAM J. Math. Anal., 18, 34.

Cimino, M. (1953) Sulle soluzióni dell equazióne generale del potenziale Newtoniano di una sfèra flùida in equilìbrio. Boll. Un. Mat. Ital., 8, 164.

Cimino, M. (1956) Una condizióne sufficiènte per l'equilìbrio spontàneo di un flùido sotto l'azióne della pròpria gravità. Boll. Un. Mat. Ital., 11, 499.

Clarkson, P. A. and Cosgrove, C. M. (1987) Painlevé analysis of the nonlinear Schrödinger's family of equations. J. Phys. A, 20, 2003.

Clarkson, P. A. and Kruskal, M. D. (1989) New similarity reductions of the Boussinesq equation. J. Math. Phys., 30, 2201.

Clarkson, P. A. and McLeod, J. B. (1988) A connection problem for the second Painlevé transcendent. Arch. Rat. Mech. Anal., 103, 97.

Clèment, Ph. and van Kan, J. (1981) An example of secondary bifurcation in a nonautonomous two-point boundary value problem. Proc. Roy. Soc. Edin., 91A, 101.

Clèment, Ph and Peletier, L. A. (1985) On a nonlinear eigenvalue problem occurring in population genetics. Proc. Roy. Soc. Edin., 100A, 85.

Coffey, M. W. (1990) On series expansions giving closed form solutions of Korteweg–deVries–like equations. SIAM J. Appl. Math., 50, 1580.

Coffman, C. V. (1972) Uniqueness of the ground state solution for $\Delta u - u + u^3 = 0$ and a variational characterization of other solutions. Arch. Rat. Mech. Anal., 46, 81.

Coffman, C. V. and Ullrich, D. F. (1967) On the continuation of solutions of a certain nonlinear differential equation. Monat. Math., 71, 365.

Coffman, C. V. and Wong, J. S. W. (1972) Oscillation and nonoscillation of solutions of generalised Emden–Fowler equations. Trans. Amer. Math. Soc., 167, 399.

Cohen, D. S, Neu, J. C. and Rosales, R. R. (1978) Rotating spiral wave solutions of reaction diffusion equations. SIAM J. Appl. Math., 36, 536.

Colombo, G. (1950) On the ultimate boundedness of the solutions of differential equations. Math. Rev., 12, 611.

Colombo, G. (1953) Sopra un singolare caso che se presenta in un problema di stabilita in meccanica non lineare. Rend. Sem. Mat. Univ. Padova, 22, 123.

Conforto, F. (1941) Sull'integrazióne di un sistèma di equazióni, relativo alla teorìa dello strato lìmite gassoso. Univ. Roma e Ist. Naz. Alta. Mat. Rend. Mat. e Appl., 2(5), 127.

Conley, C. and Smoller, J. (1986) Bifurcation and stability of stationary solutions of the Fitz–Hugh–Nagumo equations. J. Diff. Eqns., 63, 389.

Conte, R. and Musette, M. (1992) Link between solitary waves and projective Riccati equations. J. Phys. A, 25, 5609.

Cook, P. A. (1986) Nonlinear Dynamical Systems. Prentice Hall, Englewood Cliffs, N.J.

Coppel, W. A. (1960) On a differential equation of boundary layer theory. Phil. Trans. Roy. Soc., 253, 101.

Coppel, W. A. (1961) Note on an equation of boundary layer theory. Proc. Camb. Phil. Soc., 57, 696.

Corless, R. M., Essex, C. and Nerenberg, M. A. H. (1991) Numerical methods can suppress chaos. Phys. Lett., 157, 27.

Cosgrove, C. M. and Scoufis, G. (1993) Painlevé classification of a class of differential equations of second order and second degree. Stud. Appl. Math., 88, 25.

Coullet, P., Tresser, C. and Arnéodo, A. (1979) Transition to stochasticity for a class of forced oscillators. Phys. Lett. A, 72, 268.

Couper, G. (1971) Some analytic critical points in 3-space. J. Diff. Eqns., 10, 27.

Crane, J. L. (1976) Natural convection from a vertical cylinder at very large Prandtl number. J. Engg. Math., 10, 115.

Crespo Da Silva, M. R. M. (1974) A transformation approach for finding first integrals of motion of dynamical systems. Int. J. Nonlinear Mech., 9, 241.

Croquette, V. and Poitou, C. (1981) Cascade of period doubling transitions and large stochasticity in the motions of a compass. J. Phys. (Paris) Lett., 42, 537.

Dadfar, M. B. and Geer, J. F. (1990) Resonance and power series solutions of the forced van der Pol oscillator. SIAM J. Appl. Math., 50, 1496.

Dai, L. S. (1979) On the existence, uniqueness and global asymptotic stability of the periodic solution of the modified Michaelis–Menten mechanism. J. Diff. Eqns., 31, 392.

Dai, X. and Dai, J. (1989) Some solitary wave solutions for families of generalised higher order Korteweg–deVries equations. Phys. Lett. A, 142, 367.

Dang, D. H. (1988) On a second order periodic boundary value problem. Diff. Integ. Eqns., 1, 377.

Daniel, J. W. and Moore, R. E. (1970) Computing and Theory in Ordinary Differential Equations. W.H. Freeman, San Francisco.

Dao, N. V. (1975) Interaction between forced and self-excited oscillation in multidimensional systems. Z. Angew. Math. Mech., 55, 683.

Dao, N. V. (1975a) Interaction between parametric and forced oscillations in multidimensional systems. J. Tech. Phys., 16, 213.

Das, P. C. and Venkatesulu, M. (1984) The alternative method for boundary value problems with ordinary differential equations. Riv. Mat. Univ. Parma, 9, 15.

Dasarathy, B. V. (1973) Equivalent nonlinear and linear third order systems. Int. J. Sys. Sci., 4, 243.

David, D., Levi, D. and Winternitz, P. (1989) Soliton in shallow seas of variable depth and in marine straits. Stud. Appl. Math., 80, 1.

Davies, T. V. and James, E. M. (1966) Nonlinear Ordinary Differential Equations. Addison-Wesley, Reading, Mass.

Davis, H. T. (1962) Introduction to Nonlinear Differential and Integral Equations. Dover, New York.

Davis, R. T. and Alfriend, K. T. (1967) Solutions to Van der Pol's equation using a perturbation method. Int. J. Nonlinear Mech., 2, 153.

de Hoog, F. R., Laminger, B. and Weiss, R. (1984) A numerical study of similarity solutions for combined forced and free convection. Acta Mech., 51, 139.

Dekleine, H. A. (1971) Bounded solutions of a second order nonlinear equation. SIAM J. Math. Anal., 2, 511.

Demekhin, Y. A., Tokarev, G. Y. and Shkadov, V. Y. (1991) Hierarchy of bifurcations of space periodic structures in a number of dissipative media. Physica D, 52, 358.

De Santi, A. J. (1987) Non-monotone interior layer theory for some singularly perturbed quasilinear boundary value problems with turning points. SIAM J. Math. Anal., 18, 321.

Deshpande, M. A. and Kasture, D. Y. (1981) Two-point boundary value problems inimical to singular perturbation theory. Rend. Sem. Mat. Univ. Politec. Torino, 39, 99.

De Simone, J. A. and Pennline, J. A. (1978) A new asymptotic analysis of the nth order reaction-diffusion problem: analytic and numerical studies. Math. Biosci., 40, 303.

de Spautz, J. F. and Lerman, R. A. (1967) Equations equivalent to nonlinear differential equations. Proc. Amer. Math. Soc., 18, 441.

D'heedene, R. N. (1969) For all real μ, $\dot{x}+\mu\sin\dot{x}+x=0$ has an infinite number of infinite cycles. J. Diff. Eqns., 5, 564.

Dickey, R. W. (1989) Rotationally symmetric solutions for shallow water membrane caps. Q. Appl. Math., 47, 571.

Dieckerhoff, R. and Zehnder, E. (1985) Boundedness of solutions via the twist-theorem. Lecture Notes in Mathematics. Springer-Verlag, New York, p.1125.

Dieckerhoff, R. and Zehnder, E. (1987) Boundedness of solutions via the twist theorem. Ann. Dell. Scuola Norm. Sup. Pisa, 14, 79.

Dijkstra, D. (1980) On the relation between adjacent inviscid cell type solutions to the rotating disc equations. J. Engg. Math., 14, 133.

Ding, T. (1988) An answer to Littlewood's problem on boundedness for super-linear Duffing equations. J. Diff. Eqns., 73, 269.

Ditto, W. L. and Pickett, T. J. (1988) Non-perturbative solutions of nonlinear differential equations using continued fractions. J. Math. Phys., 29, 1761.

Ditto, W. L. and Pickett, J. J. (1990) Exact solutions of nonlinear differential equations using continued fractions. Nuovo Cimento B, 105, 429.

Dixon, J. M., Kelley, M. and Tuszynski, J. A. (1992) Coherent structures from the three dimensional nonlinear Schrödinger equation. Phys. Lett. A, 170, 77.

Dixon, J. M., Tuszynski, J. A. and Otwinowski, M. (1991) Special analytic solutions of the damped-anharmonic-oscillator equation. Phys. Rev. A, 44, 3484.

Diyesperov, V. N. (1986) Investigation of self-similar solutions describing flow in mixing layers. J. Appl. Math. Mech., 50, 303.

Djukic, D. S. (1973) A procedure for finding first integral of mechanical systems with gauge invariant Lagrangians. Int. J. Nonlinear Mech., 8, 479.

Doelman, A. and Eckhaus, W. (1991) Periodic and quasi-periodic solutions of degenerate and modulation equations. Physica D, 53, 249.

Doksevic, A. I. (1980) A particular solution of an Euler–Poisson system under Kovalevskaja conditions (Russian). Meh. Tverd. Tela, 12, 16–19, 118.

Dorr, F. W., Parter, S. V. and Shampine, L. F. (1973) Application of the maximum principle to singular perturbation problems. SIAM Rev., 15, 43.

Doyle, J. and Englefield, M. J. (1990) Similarity solutions of a generalised Burgers equation. IMA J. Appl. Math., 44, 145.

Drazin, P. G. and Johnson, R. S. (1989) Solitons: An Introduction. Cambridge Univ. Press, Cambridge, pp. 18, 35, 188, 189.

Dresner, L. (1983) Similarity Solutions of Nonlinear Partial Differential Equations. Research Notes in Mathematics, Vol. 88, Pitman Advanced Publishing Program. Pitman, London.

Duarte, L. G. S., Euler, N., Moreira, I. C. and Steeb, W. H. (1990) Invertible point transformations, Painlevé analysis and anharmonic oscillators. J. Phys. A, 23, 1457.

Dunbar, S. R. (1993) Geometric analysis of a nonlinear boundary value problem from physical oceanography. SIAM J. Math. Anal., 24, 444.

Dunninger, D. R. (1987) Existence of positive solutions for fourth order nonlinear problems. Bull. Un. Mat. Ital. B (7), 1, 1129.

Dym, C. L. and Rasmussen, M. L. (1968) On a perturbation problem in structural dynamics. Int. J. Nonlinear Mech., 3, 215.

Dzhaugashtin, K. E. and Shelepov, A. A. (1991) Solution to the problem of a swirling jet from an annular orifice. Fluid Dynamics, 26, 197.

Eberly, D. and Troy, W. C. (1987) Existence of logarithmic type solutions to the Kapila–Kassoy problem in dimensions 3 through 9. J. Diff. Eqns., 70, 309.

Eckhaus, W. (1983) Relaxation oscillation including a standard chase on French ducks, in Asymptotic Analysis, Vol. II, F. Verhulst, ed. Springer Lectures. Springer-Verlag, New York.

El–Abbasy, E. M. (1985) On the periodic solutions of the Van der Pol oscillator with large damping. Proc. Roy. Soc. Edin., 100A, 103.

Elcrat, A. R. (1975) On the swirling flow between rotating coaxial disks. J. Diff. Eqns., 18, 423.

Elcrat, A. R. (1976) On the radial flow of viscous fluid between porous disks. Arch. Rat. Mech. Anal., 61, 91.

Elcrat, A. R. and Siegel, D. (1983) A nonexistence result for axially symmetric flows with constant angular velocities at infinity. Proc. Roy. Soc. Edin., 93A, 229.

Eliason, S. B. (1972) Vertical asymptotes and bounds for certain solutions of a class of second order equations. SIAM J. Math. Anal., 3, 474.

Eliezer, C. J. and Gray, A. (1976) A note on time-dependent harmonic oscillator. SIAM J. Appl. Math., 30, 463.

Elnaggar, A. and Thana, E. (1982) Harmonic and subharmonic solutions of a weakly nonlinear conservative differential equation with a periodically varying coefficient. Proc. Math. Phys. Soc. Egypt, 52, 11.

El–Owaidy, H. and Zagrout, A. (1981) On analytical methods in nonlinear differential equations. Arab. J. Math., 2, 75.

Engelbrecht, J. (1991) Solutions to the perturbed Korteweg–deVries equation. Wave Motion, 14, 85.

Erbe, L. (1977) Comparison theorems for the second order Riccati equations with applications. SIAM J. Math. Anal., 8, 1032.

Erbe, L. H. and Liu, X. (1990) Non-oscillation criteria for Emden–Fowler systems. Bull. Austral. Math. Soc., 42, 455.

Erbe, L. H. and Rao, V. S. H. (1987) Non-oscillation results for third order differential equations. J. Math. Anal. Appl., 125, 471.

Erbe, L. H., Sree Hari Rao, V. and Seshagiri Rao, K. V. V. (1984) Non-oscillation and asymptotic properties of a class of forced second order nonlinear equations. Math. Proc. Camb. Phil. Soc., 95, 155.

Erugin, N. P. (1952) Analytic theory of nonlinear systems of ordinary differential equations. Akad. Nauk SSSR Prik. Mat. Meh., 16, 465.

Erugin, N.P. (1974) Analytic and qualitative theory and the asymptotics of solutions of the system $\frac{dy}{dx} = (1/x)[yz(y+z)]^{1/2}$, $\frac{dz}{dx} = [1/(1-x)][yz(y+z)]^{1/2}$. Diff. Eqns., 10, 279 & 739.

Erugin, N. P. (1980) The equation $w'' = f(w', w, z)$ with an irrational right side. Diff. Eqns., 16, 148.

Ervin, V. J., Ames, W. F. and Adams, E. (1984) Nonlinear waves in the pellet fusion process, in Wave Phenomena: Modern Theory and Applications, C. Rogers and T. B. Moodie, eds. Elsevier-North Holland, Amsterdam. J. Phys. A Math. Gen., 199.

Esquinas, J. and Herrero, M. A. (1990) Travelling wave solutions to a semilinear diffusion system. SIAM J. Math. Anal., 21, 123.

Euler, N., Steeb, W. H. and Cyrus, K. (1989) On exact solutions for damped anharmonic oscillators. J. Phys. A, 22, L195.

Evtuhov, V. M. (1977) On a nonlinear differential equation of second order. Sov. Math. Dokl., 18, 427.

Exton, H. (1971) On nonlinear ordinary differential equations with fixed critical points. Rend. Mat., 6, 385.

Exton, H. (1973) Nonlinear ordinary differential equations with fixed critical points. Rend. Mat., 6, 419.

Exton, H. (1975) Ordinary differential equations with fixed critical points. Funkcial. Ekvac., 18, 41.

Ezeilo, J. O. C. (1959) On the boundedness of solution of a certain differential equation of the third order. Proc. London Math. Soc. (3), 9, 74.

Ezeilo, J. O. C. (1961) A note on a boundedness theorem for some third order differential equations. J. London Math. Soc., 36, 439.

Ezeilo, J. O. C. (1962) A property of the phase space trajectories of a third order nonlinear differential equation. J. London Math. Soc., 37, 33.

Ezeilo, J. O. C. (1962) A stability result for solutions of a certain fourth order differential equation. J. London Math. Soc., 37, 28.

Ezielo, J. O. C. (1968) On the boundedness of the solutions of the equation $\dddot{x} + a\ddot{x} + f(x)\dot{x} + g(x) = p(t)$. Ann. Mat., 80, 281.

Ezeilo, J. O. C. (1982) A Leray–Schauder technique for the investigation of the equation $\ddot{x} + x + \mu x^2 = \epsilon \cos \omega t \quad (\epsilon \neq 0)$. Acta Math. Acad. Sci. Hung., 39, 59.

Fair, W. G. and Luke, Y. L. (1966) Rational approximation to the generalised Duffing equation. Int. J. Nonlinear Mech., 1, 209.

Fairen, V., Lopez, V. and Conde, L. (1988) Power series approximation to solutions of nonlinear systems of differential equations. Am. J. Phys., 56, 57

Ferreira, E. and Neto, J. A. (1992) Movable singularities in the Sigma and Skyrme models. J. Math. Phys., 33, 1185.

Ferreira, E. and Neto, J. A. (1992a) Singularities in the solutions of the massive Skyrme model. J. Math. Phys., 33, 2626.

Fife, P. C. (1976) Boundary and interior transition layer phenomena for pairs of second order differential equations. J. Math. Anal. Appl., 54, 497.

Filcakova, V. P. (1974) On the question of the construction of a certain Painlevé transcendent: projective-iterative methods for the solution of differential and integral equations (Ukranian). Vidannja Int. Mat. Akad. Nauk Ukrain, RSR, Kiev, pp. 162–191, 205–206.

Filip, P., Kolar, V. and Hajek, R. (1991) Similarity prediction of wall jets past axisymmetric bodies for power-law fluids. Acta Mech., 88, 167.

Finch, M. R. (1989) The first integral of a class of nonlinear second order ordinary differential equations. J. Sound Vib., 130, 321.

Finkelstein, R., Lelevier, R. and Ruderman, M. (1951) Nonlinear spinor field. Phys. Rev., 83, 326.

Fiszdon, W. and Sen, M. (1988) Initial behaviour of solutions to the Lorenz equation. Int. J. Nonlinear Mech., 23, 53.

Flaherty, J. E. and Hoppensteadt, F. C. (1978) Frequency entrainment of a forced Van der Pol oscillator. Stud. Appl. Math., 58, 5.

Flashner, H. and Guttalu, R. S. (1989) Analysis of nonlinear non-autonomous systems by truncated point mappings. Int. J. Nonlinear Mech., 24, 327.

Fokas, A. S., Mugan, V. and Ablowitz, M. J. (1988) A method of linearisation for Painlevé equations: Painlevé IV, V. Physica D, 30, 247.

Fokas, A. S., and Zhou, X. (1992) On the solvability of Painlevé II and IV. Comm. Math. Phys., 144, 601.

Foltynska, I. (1989) Functional integrability of the solutions of the system of nonlinear differential equations of second order. Demonstratio Math., 21, 767.

Forbes, L. K. (1987) Periodic solutions of high accuracy to the forced Duffing equation: perturbation series in the forcing amplitude. J. Austral. Math. Soc. Ser. B, 29, 21.

Forbes, L. K. (1989) A series analysis of forced transverse oscillations in a spring-mass system. SIAM J. Appl. Math., 49, 704.

Forbes, L. K. (1990) Stationary patterns of chemical concentration in the Belousov–Zhabotinskii reaction. SIAM J. Appl. Math., 43, 140.

Ford, W. F. (1992) A third order differential equation. SIAM Rev., 34, 121.

Forsyth, A. R. (1959) Theory of Differential Equations. Part II. Ordinary Equations. Not Linear. McGraw-Hill, New York.

Fournier, J. D., Levine, G. and Tabor, M (1988) Singularity clustering in the Duffing oscillator. J. Phys. A, 21, 33.

Fox, K. (1984) Numerical integration of equations of motion of celestial mechanics. Celestial Mech., 33, 127.

Fraenkel, L. E. (1980) Completeness properties in L_2 of the eigenfunctions of two semilinear differential operators. Math. Proc. Camb. Phil. Soc., 88, 451.

Frankel, M. L. (1991) Qualitative approximation of oscillatory flames in premixed gas combustion by a local equation of front dynamics. SIAM J. Appl. Math., 51, 673.

Freeman, N. C. (1972) Simple waves on shear flows: similarity solutions. J. Fluid Mech., 56, 257.

Frehlich, R. C. and Novak, S. (1985) The Duffing oscillator: analog solution and comparison with harmonic linearisation. Int. J. Nonlinear Mech., 20, 123.

Friedman, A., Friedman, J. and McLeod, B. (1988) Concavity of solutions of nonlinear ordinary differential equations. J. Math. Anal. Appl., 131, 486.

Funato, M. (1958/59) On Duffing's equation. Math. Jap., 5, 29.

Fürkotter, M. and Rodrigues, H. M. (1991) Symmetry and bifurcation to 2π periodic solutions of nonlinear second order equations with $2\pi/m$ periodic forcings. SIAM J. Math. Anal., 22, 169.

Furuya, S. (1953) Notes on a boundary value problem. Comment. Math. Univ. St. Paul, 1, 81.

Furuya, S. (1955) Periodic solutions of a nonlinear differential equation. Comment. Math. Univ. St. Paul, 4, 47.

Gabutti, B. (1984) An existence theorem for a boundary value problem related to that of Falkner and Skan. SIAM J. Math. Anal., 15, 943.

Gagliardo, E. (1953) Sul comportoménte dégli integrali delli equazióne differenzialle non lineare $x'' + f(x)x' + g(x) = 0$ con $g(x)$ aresente e $f(x)$ positive per $|x| > M > 0$. Boll. Un. Mat. Ital., 8, 309.

Gaiduk, V. F. (1984) Some modifications of the shooting method for solving nonlinear two point boundary value problems. USSR Comp. Math. Math. Phys., 24(4), 109.

Gaishun, I. V. (1969) On the stability as a whole of motion determined by a certain nonlinear third order system. Diff. Eqns., 5, 1617.

Gang, X. and Changqing, S. (1985) Multiple scale analysis of a nonlinear ordinary differential equation. J. Math. Phys., 26, 1566.

Gaponenko, Y. I. (1983) Solvability of very simple two point boundary value problems. Diff. Eqns., 20, 1137.

Garcia–Margallo, J. and Bejarno, J. D. (1992) The limit cycles of the generalized Rayleigh–Lienard oscillator. J. Sound Vib., 156, 283.

Gardini, L. (1985) Hopf-bifurcations and periodic doubling transitions in Rossler model. Nuovo Cimento B, 89, 139.

Garg, V. K. and Rajagopal, K. R. (1990) Stagnation point flow of a non-Newtonian fluid. Mech. Res. Comm. 17, 415.

Garner, J. B. and Shivaji, R. (1990) Diffusion problems with a mixed nonlinear boundary condition. J. Math. Anal. Appl., 148, 422.

Gatica, J. A., Oliker, V. and Waltman, P. (1989) Singular nonlinear boundary value problems for second order ordinary differential equations. J. Diff. Eqns., 79, 62.

Geicke, J. (1988) Solitary waves in the $\phi^4 + \lambda\phi^3$ model with and without dissipation. J. Phys. A, 211, 391.

Genin, J. and Maybee, J. S. (1970) Boundedness theorem for a nonlinear Mathieu equation. Q. Appl. Math., 28, 450.

Gergen, J. J. and Dressel, F. G. (1965) Second order linear and nonlinear differential equations. Proc. Amer. Math. Soc., 16, 767.

Giacomini, H. J., Repetto, C. E. and Zandorn, O. P. (1991) Integrals of motion for three dimensional non-Hamiltonian dynamical systems. J. Phys. A, 24, 4567.

Gilding, B. H. (1987) The first boundary value problem for $-u'' = \lambda u^p$. J. Math. Anal. Appl., 128, 419.

Giorgini, A. and Toebes, G. (1971) Mode interaction segregation for nonlinear differential equations. Int. J. Nonlinear Mech., 7, 549.

Giovangigli, V. (1990) Non-adiabatic plane laminar flames and their singular limits. SIAM J. Math. Anal., 21, 1305.

Gisin, B. V. (1990) Localized solutions of a certain nonlinear second order differential equation. USSR Comp. Math. Phys., 30, 232.

Glockle, W. G., Baumann, G. and Nonnenmacher, T. F. (1992) Painlevé test and exact similarity solutions of a class of nonlinear diffusion equations. J. Math. Phys., 33, 2456.

Goldshtik, M., Hussain, F. and Shtern, V. (1991) Symmetry breaking in vortex-source and Jeffrey–Hamel flows. J. Fluid Mech., 232, 521.

Goldstein, M. E. and Braun, W. H. (1973) Advanced Methods for the Solutions of Differential Equations. NASA SP-316. U.S. Govt. Printing Office, Washington, D. C.

Golubyatnikov, V. P. and Pestov, L. N. (1983) Trajectories of a dynamical system defined by a one parameter group of conformal transformations of R^3. Siberian Math. J., 24, 52.

Gonzalez–Lopez, A. (1988) Symmetry and integrability by quadratures of ordinary differential equations. Phys. Lett. A, 133, 190.

Gordeyev, Yu. N., Kudryashov, N. A. and Murzenkov, V. V. (1985) Shock waves in an isothermal gas in the presence of reaction forces. J. Appl. Math. Mech., 49, 129.

Grabmüller, H. and Novak, E. (1988) Nonlinear boundary value problems for the annular membrane: new results on existence of positive solutions. Math. Meth. Appl. Sci., 10, 37.

Grace, S. R. and Lalli, B. S. (1987) Oscillation theorems for second order nonlinear differential equations. J. Math. Anal. Appl., 124, 213.

Grace, S. R. and Lalli, B. S. (1987a) An oscillatory criterion for certain second order strongly sublinear equations. J. Math. Anal. Appl., 123, 584.

Grace, S. R. and Lalli, B. S. (1990) Integral averaging techniques for the oscillations of second order nonlinear differential equations. J. Math. Anal. Appl., 149, 277.

Grace, S. R., Lalli, B. S. and Yeh, C. C. (1984) Oscillation theorems for nonlinear second order differential equation with a nonlinear damping. SIAM J. Math. Anal., 15, 1082.

Grace, S. R., Lalli, B.S. and Yey, C. C. (1988). Oscillation theorems for nonlinear second order differential equation with a nonlinear damping. SIAM J. Math. Anal., 19, 1252.

Graef, J. R. and Spikes, P. W. (1987) On the nonoscillations, convergence to zero and integrability of solutions of a second order nonlinear differential equation. Math. Nachr., 130, 139.

Graef, J. R. and Spikes, P. W. (1975) Asymptotic behaviour of solutions of second order nonlinear differential equations. J. Diff. Eqns., 17, 461.

Graffi, D. (1940) Sópra alcune equazióni differenziali nonlineari della fisica-matemàtica. Math. Rev., 9, 589.

Graffi, D. (1953) Sur la période d'oscillation des systèmes nonlinéaires a plusieurs degrés de liberté. Actes du Colloque International de Vibrations Nonlinéarises, Ile de Porquerolles, 1951, discussion 195. Publ. Sci. Tech. Ministère de l'Air, Paris, No. 281, pp. 189–193.

Graffi, D. (1954) Dario su alcune equazioni differenziali non lineare. Atti Accad. Sci. Ist. Bologna Cl. Sci. Fis. Rend. (11), 1, 57.

Graham, R. (1976) Onset of self-focusing in lasers and Lorenz model. Phys. Lett. A, 58, 440.

Grammaticos, B., Moulin–Ollagnier, J., Ramani, A., Strelcyn, J. M. and Wojciechowski, S. (1990) Integrals of quadratic ordinary differential equations in R^3: the Lotka–Volterra system. Physica A, 163, 683.

Grasman, J. (1981) Dips and slidings of the forced Van der Pol relaxation oscillator. Afdeling Toegepaste Wiskunde 214. Math. Cent. Amsterdam, 1+12 pp.

Grasman, J. (1987) Asymptotic Methods for Relaxation Oscillations and Applications. Applied Mathematical Sciences, Vol. 63. Springer-Verlag, New York, p. 103.

Greenspan, B. and Holmes, P. (1984) Repeated resonances and homoclinic bifurcation in a periodically forced family of oscillators. SIAM J. Math. Anal., 15, 69.

Gregus, M. (1987) Third Order Linear Differential Equations. D. Reidel, Dordrecht, The Netherlands.

Grigorenko, V. P., Smirnova, V. N. and Tai, M. L. (1975) A method for solving nonlinear boundary value problems describing the minority carrier distribution in the base of a semiconductor structure. Appl. Sci. Res., 15(4), 107.

Grissom, C., Thompson, G. and Wilkens, G. (1989) Linearisation of second order ordinary differential equations via Cartan's equivalent method. J. Diff. Eqns., 77, 1.

Grizan, G. P. (1988) The domain of existence of a solution of a boundary value problem for the Emden–Fowler equation. Diff. Eqns., 24, 384.

Gromak, V. I. (1973) Solutions of the third Painlevé equation. Diff. Eqns., 9, 1599.

Gromak, V. I. (1975) Theory of Painlevé equations. Diff. Eqns., 11, pp. 285, 519.

Gromak, V. I. (1976) Solutions of Painlevé's fifth equation. D.H. Eqns., 12, 519.

Gromak, V. I. (1987) Theory of the fourth Painlevé equation. Diff. Eqns., 23, 506.

Gromak, V. I. and Tsegelnik, V. V. (1991) Solution of a system of three differential equations with quadratic nonlinearities. Diff. Eqns., 27, 273.

Gromov, E. M. and Nakaryakov, V. M. (1992) Returnal dynamical states and transmission of intense high frequency pulse through parabolic density barrier. Phys. Lett. A, 163, 266.

Gross, O. A. (1963) The boundary value problem on an infinite interval: existence, uniqueness, and asymptotic behaviour of bounded solution to a class of nonlinear second order differential equations. J. Math. Anal. Appl., 7, 100.

Grossinho, M. R. and Sanchez, L. (1986) A note on periodic solutions of some nonautonomous differential equations. Bull. Austral. Math. Soc., 34, 1986.

Grudo, E. I. (1971) The asymptotic representation of solution of a system of two differential equations. Diff. Eqns., 7, 158.

Grudo, E. I. (1973) An autonomous system of differential equations with a characteristic equation having zero roots. Diff. Eqns., 9, 161.

Grundy, R. E. and Peletier, L. A. (1990) The initial interface development for a reaction-diffusion equation with power-law initial data. Q. J. Mech. Appl. Math., 43, 535.

Gudkov, V. V. and Klokov, Yu. A. (1982) A two point boundary value problem for a third order system. Diff. Eqns., 18, 414.

Guidorizzi, H. L. (1993) Oscillating and periodic solutions of the type $\ddot{x}+f_1(x)\dot{x}+f_2(x)\dot{x}^2+g(x)=0$. J. Math. Anal. Appl., 176, 11.

Gulyayev, V. I., Zubritskaya, A. L. and Koshkin, V. L. (1989) A universal sequence of period doubling bifurcations of the forced oscillations of a pendulum. J. Appl. Math. Mech., 53, 561.

Guo, D. J. (1984) The number of nontrivial solutions of nonlinear two point boundary value problems. J. Math. Res. Exposition, 4, 55.

Guttalu, R. S. and Flashner, H. (1989) Periodic solutions of nonlinear autonomous systems by approximate point mappings. J. Sound Vib., 129, 291.

Gwinn, E. G. and Westervelt, R. M. (1985) Intermittent chaos and low frequency noise in the driven damped pendulum. Phys. Rev. Lett., 54, 1613.

Habets, P. and Metzen, G. (1989) Existence of periodic solutions of Duffing equation. J. Diff. Eqns., 78, 1.

Hagedorn, P. (1988) Nonlinear Oscillations. Clarendon Press, Oxford. pp. 232, 233.

Hagedorn, P. and Schäfer, B. (1980) On nonlinear free vibrations of an elastic cable. Int. J. Nonlinear Mech., 15, 333.

Hale, J. (1969) Ordinary Differential Equations. Wiley, New York.

Hale, J. K. and Rodrigues, H. M. (1977) Bifurcation in the Duffing equation with independent parameters: II. Proc. Roy. Soc. Edin., 79A, 317.

Hallam, T. G. and Loper, D. E. (1975) Singular boundary value problems arising in a rotating fluid flow. Arch. Rat. Mech. Anal., 60, 356.

Hamd–Allah, G. M. (1981) Determination of harmonic and subharmonic synchronisation of weakly nonlinear conservative physical systems. Bull. Fac. Sci. Assiut Univ. A, 10(1), 115.

Hammersley, J. M. and Mazzarino, G. (1989) A differential equation connected with the dendritic growth of crystals. IMA J. Appl. Math., 42, 43.

Hart, V. G. (1980) Exact solutions of two nonlinear equations and hyper-circles estimates. J. Austral. Math. Soc. B, 22, 98.

Hartman, P. (1972) On the existence of similar solutions of some boundary layer problems. SIAM J. Math. Anal., 3, 120.

Hartman, P. (1972a) On the swirling flow problems. Indiana Univ. Math. J., 21, 849.

Hastings, S. P. (1970) An existence theorem for some problems from boundary layer theory. Arch. Rat. Mech. Anal., 38, 308.

Hastings, S. P. (1971) An existence theorem for a class of nonlinear boundary value problems including that of Falkner and Skan. J. Diff. Eqns., 9, 580.

Hastings, S. P. (1972) On a third order differential equation from biology. Q. J. Math., Oxford (2), 23, 435.

Hastings, S. P., Lu, C. and MacGillivray, A. D. (1992) A boundary value problem with multiple solutions from the theory of laminar flow. SIAM J. Math. Anal., 23, 201.

Hastings, S. P. and McLeod, J. B. (1985) The number of solutions to an equation from catalysis. Proc. Roy. Soc. Edin., 101A, 15.

Hastings, S. P. and Murray, J. D. (1975) The existence of oscillatory solutions in the Field–Noyes model for the Belousov–Zhabotinskii reaction. SIAM J. Appl. Math., 28, 678.

Hastings, S. P. and Poore, A. B. (1983) A nonlinear problem arising from combustion theory: Linan's problem. SIAM J. Math. Anal., 14, 425.

Hastings, S. P. and Siegel, S. (1972) On some solutions of the Falkner–Skan equation. Mathematika, 19, 76.

Hastings, S. P. and Troy, W. C. (1987) Oscillating solutions of the Falkner–Skan equation for negative β. SIAM J. Math. Anal., 18, 422.

Hastings, S. P. and Troy, W. C. (1988) Oscillating solutions of the Falkner–Skan equation for positive β. J. Diff. Eqns., 71, 123.

Hastings, S. P. and Troy, W. C. (1989) On some conjectures of Turcotte, Spence, Bau, and Holmes. SIAM J. Math. Anal., 20, 634.

Hausrath, A. R. and Manasevich, R. F. (1991) Periodic solutions of a periodically perturbed Lotka–Volterra equation using the Poincaré–Birkhoff theorem. J. Math. Anal. Appl., 157, 1.

Hayashi, C. (1951) Forced oscillations with nonlinear restoring force. Mem. Fac. Engg. Kyoto Univ., 13, 180.

Hayashi, C. (1980) The method of mapping with reference to the doubly asymptotic structure of invariant curves. Int. J. Nonlinear Mech., 15, 341.

Hayashi, C., Ueda, Y. and Kawakami, H. (1969) Transformation theory as applied to the solutions of nonlinear differential equations of second order. Int. J. Nonlinear Mech., 4, 235.

Healey, J. J., Broomhead, D. S., Cliffe, K. A., Jones, R. and Mullin, T. (1991) The origin of chaos in a modified Van der Pol oscillator. Physica D, 48, 322.

Heckenback, A. and Heimes, K. (1976) Real solutions of $y'' = Ay^2 + 2By + C$. Amer. Math. Monthly, 83, 129.

Heidel, J. W. (1968) Qualitative behaviour of solution of a third order nonlinear differential equation. Pacific J. Math., 27, 507.

Heidel, J. W. (1973) A third order differential equation arising in fluid mechanics. Z. Angew. Math. Mech., 53, 167.

Heidel, J. W. and Hinton, D. B. (1972) The existence of oscillatory solutions for a nonlinear differential equation. SIAM J. Math. Anal., 3, 344.

Heidel, J. W. and Jones, G. D. (1975) Asymptotic characterisation of solutions of a boundary value problem arising in fluid mechanics. Z. Angew. Math. Mech. 55, 191.

Heilmann, O. J. (1992) Analytical solutions of Smluschowski's coagulation equation. J. Phys. A, 25, 3763.

Henderson, J. (1987) Best interval lengths for boundary value problems for third order Lipschitz equations. SIAM J. Math. Anal., 18, 293.

Henderson, J. and McGowier, R. W., Jr. (1987) Uniqueness, existence and optimality for fourth order Lipschitz equations. J. Diff. Eqns., 67, 414.

Herbst, R. T. (1956) The equivalence of linear and nonlinear differential equations. Proc. Amer. Math. Soc., 7, 95.

Hereman, W., Banerjee, P. P., Korpel, A. and Assanto, G. (1986) Exact solitary wave solutions of nonlinear evolution and wave equations using a direct algebraic method. J. Phys. A, 19, 607.

Herrera, J. J. E. (1984) Envelope solitons in inhomogeneous media. J. Phys. A, 17, 95.

Herrero, M. A. and Vazquez, J. L. (1988) Thermal waves in absorbing media. J. Diff. Eqns., 74, 218.

Hill, J. M. (1989) Similarity solutions for nonlinear diffusion: a new integration procedure. J. Engg. Math., 23, 141.

Hill, J. M., Avagliano, A. J. and Edwards, M. P. (1992) Some exact results for nonlinear diffusion with absorption. IMA J. Appl. Math., 48, 283.

Hill, D. L. and Hill, J. M. (1990) Similarity solutions for nonlinear diffusion: further exact solutions. J. Engg. Math., 24, 109.

Hille, E. (1969) Lectures on Ordinary Differential Equations. Addison–Wesley, Reading, Mass.

Hilton, P. (1988) Remark on a system of two nonlinear differential equations. SIAM J. Appl. Math., 48, 286.

Hinton, D. (1969) An oscillation criterion for solutions of $(ry')' + qy^r = 0$. Michigan Math. J., 16, 349.

Hochstadt, H. and Stephan, B. H. (1967) On the limit cycles of $\ddot{x} + \mu \sin \dot{x} + x = 0$. Arch. Rat. Mech. Anal., 23, 369.

Hocking, Z. M., Stewartson, K. and Stuart, J. T. (1972) (with an appendix by Brown, S. N.) A nonlinear instability burst in plane parallel flow. J. Fluid Mech., 51, 705.

Holmes, P. J. (1977) Behaviour of an oscillator with even nonlinear damping. Int. J. Nonlinear Mech., 12, 323.

Holmes, P. J. (1979) A nonlinear oscillator with a strange attractor. Phil. Trans. Roy. Soc. London A, 292, 419.

Holmes, P. (1982) On a second order boundary value problem arising in combustion theory. Q. Appl. Math., 40, 53.

Holmes, P. (1986) Spatial structure of time periodic solutions of the Ginzburg–Landau equation. Physica D, 23, 84.

Holmes, P. and Rand, D. (1980) Phase portraits and bifurcations of the nonlinear oscillator $\ddot{x} + (\alpha + rx^2)\dot{x} + \beta x + \delta x^3 = 0$. Int. J. Nonlinear Mech., 15, 449.

Holmes, P. and Spence, D. (1984) On a Painlevé boundary value problem. Q. J. Mech. Appl. Math., 37, 525.

Hooper, A. P. and Grimshaw, R. (1985) Nonlinear instability at the interface between two viscous fluids. Phys. Fluids, 28, 37.

Horvath, A. J. T. (1975) Periodic solutions of a combined Van der Pol–Duffing equation. Int. J. Mech. Sci., 17, 667.

Howes, F. A. (1975) Singularly perturbed nonlinear boundary value problems with turning points. SIAM J. Math. Anal., 6, 644.

Howes, F. A. (1977) Singular perturbation analysis of a class of boundary value problems arising in catalytic reaction theory. Arch. Rat. Mech. Anal., 66, 237.

Howes, F. A. (1978) The asymptotic behaviour of a class of singularly perturbed nonlinear boundary value problems via differential inequalities. SIAM J. Math. Anal., 9, 215.

Howes, F. A. (1978a) Singularly perturbed nonlinear boundary value problems with turning points II. SIAM J. Math. Anal., 9, 250.

Howes, F. A. (1983) Nonlinear dispersive systems: theory and examples. Stud. Appl. Math., 69, 75.

Howes, F. A. and Shao, S. (1989) Asymptotic analysis of model problems for a coupled system. Nonlinear Anal. Theory Meth. Appl., 13, 1013.

Hsiao, G. C. (1973) Singular perturbations for a nonlinear differential equation with a small parameter. SIAM J. Math Anal., 4, 283.

Hsieh, P. F. and Przybyiski, J. J. (1983) On a degenerate system of two nonlinear differential equations at an irregular type singularity. Bull. Inst. Math. Acad. Sinica, 11, 375.

Hsu, S. B., Hubbell, S. P. and Waltman, P. (1978) Competing predators. SIAM J. Appl. Math., 35, 617.

Hsu, S. and Hwang, S. F. (1988) Analysis of large deformation of a heavy cantilever. SIAM J. Math. Anal. 19, 854.

Hsü, I. and Kazarinoff, N. D. (1977) Existence and stability of periodic solution of a third order nonlinear autonomous system simulating immune response in animals. Proc. Roy. Soc. Edin., 77A, 163.

Huang, C. (1974) Applying quasilinearisation to the problem of flow through an annulus with porous walls of different permeability. Appl. Sci. Res., 29, 145.

Huang, Q. C. (1984) On the existence of periodic solutions of nonlinear oscillation equations (Chinese). Ann. Math. Ser. B, 5, 311.

Huang, G., Luo, S. and Dai, X. (1989) Exact and explicit solitary wave solutions to a model equation for water waves. Phys. Lett., 19, 373.

Huibin, L. and Kelin, W. (1990) Exact solutions for two nonlinear equations: I. J. Phys. A, 23, 3923.

Huibin, L. and Kelin, W. (1990a) Exact solutions for some coupled nonlinear equations: II. J. Phys. A, 23, 4097.

Hulshof, J. (1991) Similarity solutions of the porous medium equation with sign changes. J. Math. Anal. Appl., 157, 75.

Hunter, C. and Tajdari, M. (1990) Singular complex periodic solutions of Van der Pol's equation. SIAM J. Appl. Math., 50, 1764.

Hunter, C., Tajdari, M. and Boyer, S. D. (1990) Lagerstrom's model for slow incompressible viscous flow. SIAM J. Appl. Math., 50, 48.

Hussaini, M. Y., Lakin, W. D. and Nachman, A. (1987) On similarity solutions of a boundary layer problem with an upstream moving wall. SIAM J. Appl. Math., 47, 699.

Huseyin, K. and Lin, R. (1992) A perturbation method for the analysis of vibrations and bifurcations associated with non-autonomous systems: I. Int. J. Nonlinear Mech., 27, 203.

Ibragimov, N. Kh. (1992) Group analysis of ordinary differential equations and the invariance principle in mathematical physics. Russian Math. Surv., 47(4), 89.

Ierley, G. R. and Ruehr, O. G. (1986) Analytic and numerical solutions of nonlinear boundary layer problem. Stud. Appl. Math., 75, 1.

Ifantis, K. I. (1987) Analytic solutions for nonlinear differential equations. J. Math. Anal. Appl., 124, 339.

Ifantis, K. I. (1987a) Global analytic solutions of the radial nonlinear wave equation. J. Math. Anal. Appl., 124, 381.

Iffland, G. (1987) Positive solution of a problem of Emden–Fowler type with a free boundary. SIAM J. Math. Anal., 18, 283.

Il'ichev, A. T. (1990) Theory of nonlinear waves described by fifth order evolution equations. Fluid Dynamics, 25, 247.

Imaeda, K. and Tei, T. (1986) Bifurcation and chaos caused by an external periodic force in the Oregonator of BZ reaction. J. Phys. Soc. Jap., 55, 743.

Ince, E. L. (1956) Ordinary Differential Equations. Dover, New York.

Ingham, D. B. (1982) Non-unique solutions of the boundary layer equations for flow near the equator of a rotating sphere in a rotating fluid. Acta Mech., 42, 111.

Ingham, D. B. and Pop, I. (1988) The sudden melting of a thin vertical flat plate in a Darcian free convection flow. Acta Mech., 71, 77.

Ito, A. (1979) Successive subharmonic bifurcations and chaos in a nonlinear Mathieu equation. Prog. Theor. Phys., 61, 815.

Iwano, M. (1966/67) On a singular point of a system of two ordinary nonlinear differential equations. Publ. Res. Inst. Math. Sci., 2, 17.

Iwano, M. (1977) Application of Nagumo–Hukuhara theory on the boundary value problems for nonlinear ordinary differential equations to the Abrikosov problem and Falkner–Skan problem. Ann. Mat. (4), 113, 303.

Iwano, M. (1977a) Asymptotic solutions of the boundary layer equation. J. Fac. Sci. Univ. Tokyo Sec. 1A Math., 24, 433.

Izobov, N. A. and Rabtsevich, V. A. (1987) A best possible I.T. Kiguradze– G.G. Kiguradze condition for the existence of unbounded regular solutions of an Emden–Fowler equation. Diff. Eqns., 23, 1263.

Jablonskii, A. I. (1964) On differential equations with entire solutions. Vesci. Akad. Navuk BSSR Ser. Fiz.-Tehn. Navuk, 3, 5.

Jacobs, J. A. (1977) The earth's core and geomagnetism. Bull. IMA, 13, 86.

Jan, A. (1989) Structure of the phase-space for the third order nonlinear differential equations belonging to the generalised class D^1 in the sense of Levinson. Fasc. Math., 63.

Jasny, M. (1960) On the existence of an oscillatory solution of the nonlinear differential equation of the second order $y'' + f(x)y^{2n-1} = 0, f(x) > 0$. Casopis Pest. Mat., 85, 78.

Johnson, R. S. (1970) A nonlinear equation incorporating damping and dispersion. J. Fluid Mech., 42, 49.

Jones, C. W. (1953) On reducible nonlinear differential equations occurring in mechanics. Proc. Roy. Soc. London A, 217, 327.

Jones, S. E. (1978) Remarks on the perturbation process of certain conservative systems. Int. J. Mech. Sci., 17, 125.

Jones, S. E. and Ames, W. F. (1967) Similarity variables and first integrals of ordinary differential equations. Int. J. Nonlinear Mech., 2, 257.

Jones, J., Troy, W. C., and MacGillivray, A. D. (1992) Steady solutions of the Kuramoto–Sivashinsky equation for small wave speed. J. Diff. Eqns., 96, 28.

Jordan, W. B. (1986) Radius of convergence of a power series. SIAM Rev., 28, 570, 572.

Jordan, D. W. and Smith, P. (1977) Nonlinear Ordinary Differential Equations. Oxford Univ. Press, New York.

Joseph, D. D. and Lundgren, T. S. (1972) Quasilinear Dirichlet problems driven by positive sources. Arch. Rat. Mech. Anal., 49, 241.

Joshi, N. and Kruskal, M. D. (1988) An asymptotic approach to the connection problem for the first and second Painlevé equations. Phys. Lett., 130, 129.

Kac, A. M. (1950) On the approximate solution of nonlinear differential equations of the second order. Math. Rev., 11, 519.

Kaiyuan, Y., Xiao–Jing, Z. and Xin–Zhi, W. (1990) On some properties and calculation of the exact solution to von Kármán's equation of circular plates under a concentrated load. Int. J. Nonlinear Mech., 25, 17.

Kajumov, D. H. (1978) The behaviour of the characteristics and limit manifold in a neighbourhood of the origin of a system of three differential equations. Izv. Akad. Nauk UzSSR Ser. Fiz. Mat. Nauk, 5, 21.

Kaminogo, T. (1979) On a boundary value problem for a nonlinear Bessel equation. Funkcial. Ekvac., 22, 241.

Kamke, E. (1959) Differential Gleichungen: Lösungsmethoden und Lösungen. Akademische Verlaggesellschaft, Leipzig.

Kane, T. R. (1966) Free and forced oscillations of a class of mechanical systems. Int. J. Nonlinear Mech., 1, 157.

Kano, K. and Nakayama, T. (1981) An exact solution of the wave equation $u_t+uu_x-u_5x=0$. J. Phys. Soc. Jap., 50, 361.

Kapaev, A. A. (1988) Asymptotics of solutions of the Painlevé equation of the first kind. Diff. Eqns., 24, 1107.

Kaper, H. G. (1990) Free boundary problems for Emden–Fowler equations. Diff. Integ. Eqns., 3, 353.

Kaper, H. G. and Kwong, M. K. (1988) A non-oscillation theorem for the Emden–Fowler equation: ground states for semilinear elliptic equations with critical exponents. J. Diff. Eqns., 75, 158.

Kapila, A. K. and Matkowsky, B. J. (1980) Reactive diffusive system with Arrhenius kinetics: the Robin problem. SIAM J. Appl. Math., 39, 391.

Kapitanov, A. Ya. (1984) A qualitative investigation of a system in R^3. Diff. Eqns., 20, 177.

Kaplan, B. Z. (1978) On second order nonlinear systems with conservative limit cycles. Int. J. Nonlinear Mech., 13, 43.

Karreman, G. (1949) Some types of relaxation oscillations as models of all-or-none phenomenon. Bull. Math. Biol. Phys., 11, 311.

Kasner, E. (1942) Differential equations of the type $y^{\text{iv}} = Ay'''^2 + By''' + C$. Univ. Nac. Tucuman Rev. A, 3, 7.

Kath, W. L., Knessl, C. and Matkowsky B. J. (1987) A variational approach to nonlinear singularly perturbed boundary value problems. Stud. Appl. Math., 75, 61.

Kawahara, T. (1972) Oscillatory solitary waves in dispersive media. J. Phys. Soc. Jap., 33, 260.

Kawahara, T. and Takaoka, M. (1988) Chaotic motions in an oscillatory soliton lattice. J. Phys. Soc. Jap., 57, 3714.

Kawakami, H. and Sakai, M. (1988) Synchronization and chaos in a Duffing–Raleigh oscillator (Russian). Asymptotic Methods in Mathematical Physics. Naukova Dumka, Kiev, pp. 83–92, 300.

Kawamoto, S. (1984) Linearisation of the classical Boussinesq and related equations. J. Phys. Soc. Jap., 53, 2922.

Kawamoto, S. (1984a) Solitary wave solutions of the Korteweg–deVries equation with higher nonlinearity. J. Phys. Soc. Jap., 53, 3729.

Kawamoto, S. (1985) Construction of stationary solitary wave solutions. J. Phys. Soc. Jap., 54, 1701.

Kazakov, V. A. (1986) On differential equations which are similar to the third Painlevé equation (Russian). Partial Differential Equations. Leningrad Gos. Ped. Inst. Leningrad, pp. 19–22 (Math. Rev. (1988), 88e: 34011).

Kazakov, V. A. (1987) Multivalued moving singular points for a class of differential equations (Russian). Math. Phys. Leningrad Gos Ped. Inst., Leningrad, pp. 153–157 (Math. Rev. (1989), 89e: 34058).

Kazarinoff, N. D. and Yan, J. G. G. (1991) Spatially periodic steady state solutions of a reversible system at strong and subharmonic resonances. Physica D, 48, 147.

Keckic, J. D. (1975) Addition to Kamke's treatise: V. A remark on the generalised Emden's equation. Univ. Beogrod. Publ. Electrotehn. Fak. Ser. Mat. Fiz., 39, 498–541.

Kelley, W. G. (1984) Boundary and interior layer phenomena for singularly perturbed systems. SIAM J. Math. Anal., 15, 635.

Kelley, W. G. (1989) Uniform approximation of singular perturbation problems having singular regular functions. SIAM J. Math. Anal., 20, 479.

Kerchman, V. I. (1990) Problems in the spreading and extrusion of a layer of nonlinear viscous fluid. J. Appl. Math. Mech., 54, 204.

Kersner, R. and Mottoni, P. de (1990) Support properties of non-negative solution of a degenerate logistic equation. Nonlinearity, 3, 453.

Kevorkion, J. and Cole, J.D. (1981) Perturbation Methods in Applied Mathematics. Springer-Verlag, New York.

Khakimova, Z. N. (1988) Point transformations of a second order nonlinear equation. Partial Differential Equations (Russian). Leningrad Gos. Ped. Inst. Leningrad. pp. 85–90 (Math. Rev., 90b: 34059).

Kiguradze, I. T. (1962) On the conditions for oscillation of solutions of the differential equation $u'' + a(t)|u|^n \operatorname{sgn} u = 0$. Casopis Pest. Mat., 87, 492.

King, J. R. (1990) Exact similarity solutions to some nonlinear diffusion equations. J. Phys. A, 23, 3681.

King, J. R. (1991) Exact solutions to a nonlinear diffusion equation. J. Phys. A Math. Gen., 24, 3213.

Kitada, A. and Umehara, H. (1988) Existence of a stationary solution of the Korteweg–deVries–Burgers equation. J. Phys. Soc. Jap., 57, 1855.

Klaasen, G. A. and Troy, W. C. (1984) The existence, uniqueness and instability of spherically symmetric solutions of a system of reaction-diffusion equations. J. Diff. Eqns., 52, 91.

Klamkin, M. S. and Reid, J. L. (1976) Nonlinear differential equations equivalent to solvable nonlinear equations. SIAM J. Math. Anal., 7, 305.

Klebanov, L. B. (1971) Local behaviour of solutions of ordinary differential equations. Diff. Eqns., 7, 1056.

Klokov, J. A. (1959) A limiting boundary value problem for the equation $\ddot{x} + \dot{x}f(x, \dot{x}) + \phi(x) = 0$. Izv. Vyss. Mcebu Zaved. Mat., 6(13), 72.

Klokov, Yu. A. and Stepanov, A. A. (1980) Domain of existence of a solution to the first boundary value problem for a second order equation. Latv. Mat. Ezhegodnik, 24, 92, 256.

Kluwick, A. (1991) Weakly nonlinear kinematic waves in suspension of particles in fluids. Acta Mech., 88, 205.

Knobloch, H. W. and Schmitt, K. (1977) Nonlinear boundary value problems for system of differential equations. Proc. Roy. Soc. Edin., 78A, 139.

Kobayashi, S. (1962) Two-dimensional panel flutter: 1. Simply supported panel. Trans. Jap. Soc. Aerosp. Sci., 5(8), 90.

Koehler, D. I. and Pignani, T. J. (1962) On the solution of a certain nonlinear differential equation of second order. Math. Student, 30, 171.

Kolesnikova, N. S. and Lukashevich, N. A. (1972) On a class of third order differential equation with fixed critical singularities. Diff. Eqns., 8, 1615.

Kolmogoroff, N., Petrovskii, I. and Piskounoff, N. (1937) Etude de l'équation de la diffusion avec croissance de la quantité de matière son application à un problèm biologique. Bull. Moskov Gos. Univ. A1, 1.

Kolosov, A. I. and Lyubarskii, G. Ya. (1968) An equation of Thomas–Fermi type. Diff. Eqns., 5, 1199.

Kondratenya, S. G. (1970) Essential singularities of solutions of certain systems of differential equations. Diff. Eqns., 6, 914.

Kondratenya, S. G. (1971) Behaviour of solutions of systems of two differential equations in the neighbourhood of an essential singularity. Diff. Eqns. 7, 573.

Kondratenya, S. G. and Prolisko, E. G. (1973) The existence and the form of solutions of Lienard equations with a moving algebraic singularity. Diff. Eqns., 9, 198.

Kondratenya, S. G. and Yablonskii, A. I. (1968) Moving singularities of a system of two differential equations. Diff. Eqns., 4, 508.

Konopljannik, I. A. (1979) A system of third order equations without movable critical points. Dokl. Akad. Nauk BSSR, 23, 777.

Korpela, S. A. (1991) On large Prandtl number convection in Falkner–Skan flow. Z. Angew. Math. Mech., 71, 121.

Korzyuk, A. F. (1986) Third order nonlinear systems with single valued movable singularities. Diff. Eqns., 23, 532.

Kostin, A.V. (1971) On asymptotics of extendable solutions of equations of Emden–Fowler type. Sov. Math. Dokl., 12, 1316.

Kostin, M. D. (1971a) Linear differential equations for the radial distribution function of quantum mechanics. J. Chem. Phys., 54, 2739.

Kostin, A. V. and Evtuhov, V. M. (1976) Asymptotic behaviour of the solutions of a nonlinear differential equation. Sov. Math. Dokl., 17, 1700.

Kot, M., Sayler, G. S. and Schultz, T. W. (1992) Complex dynamics in a model microbial system. Bull. Math. Biol., 54, 619.

Kozulin, A. V. and Lukashevich, N. A. (1988) Third order differential equations of a special form. Diff. Eqns., 24, 1387.

Kranje, A. (1951) Sull' integrazióne dell' equazióne di Ritter–Emden. Math. Rev., 13, 38.

Kravenchko, T. K. and Yablonskii, A. I. (1965) Solution of an infinite boundary-value problem for a third order equation. Diff. Eqns., 1, 248.

Kravchenko, T. K. and Yablonskii, A. I. (1972) A boundary value problem on a semi-infinite interval. Diff. Eqns., 8, 1685.

Krechetov, G. S. (1975) Asymptotic properties of solutions of a system of two differential equations. Diff. Eqns., 11, 600.

Kreigsman, G. A. (1989) Bifurcation in classical bipolar transistor oscillator circuits. SIAM J. Appl. Math., 49, 390.

Kreiss, H. and Parter, S. V. (1983) On the swirling flow between rotating co-axial disks: existence and non-uniqueness. Commun. Pure Appl. Math., 36, 55.

Krishnan, E. V. (1982) On the classical Boussinesq equation. J. Phys. Soc. Jap., 51, 3413.

Krishnan, E. V. (1982a) An exact solution of the classical Boussinesq equation. J. Phys. Soc. Jap., 51, 2391.

Krishnan, E. V. (1984) On the exact solutions of certain nonlinear dispersive wave equations. J. Phys. Soc. Jap., 53, 947.

Krishnan, E. V. (1986) On the Ito-type coupled nonlinear wave equation. J. Phys. Soc. Jap., 55, 3753.

Kroopnick, A. (1978) Oscillation properties of $\{m(t)x'\}' + a(t)b(x) = 0$. J. Math. Anal. Appl., 63, 141.

Kruskal, M. D. and Clarkson, P. A. (1992) The Painlevé–Kowalevski and Poly–Painlevé tests for integrability. Stud. Appl. Math., 86, 87.

Kruskal, M. D. and Segur, H. (1991) Asymptotics beyond all orders in a model of crystal growth. Stud. Appl. Math., 85, 129.

Krzywicki, A. and Nadzieja, T. (1989) Radially symmetric solutions of the Poisson–Boltzmann equation. Math. Meth. Appl. Sci., 11, 403.

Ku, Y. H. (1966) Heat transfer problems solved by the method of nonlinear mechanics. Int. J. Nonlinear Mech., 1, 1.

Ku, Y. H. and Jonnada, R. K. (1971) Bifurcation theorems and limit cycles in nonlinear systems: I. J. Franklin Inst., 292, 19.

Ku, Y. H. and Jonnada, R. K. (1971a) Bifurcation theorems and limit cycles in nonlinear systems: II. J. Franklin Inst., 292, 293.

Kubicek, M. and Hlavacek, V. (1983) Numerical Solutions of Nonlinear Boundary Value Problems with Applications. Prentice Hall, Englewood Cliffs., N.J.

Kudryashov, N. V. (1991) On the types of nonlinear nonintegrable equations with exact solutions. Phys. Lett., 155, 269.

Kuiken, H. K. (1981) A backward free convection boundary layer. Q. J. Mech. Appl. Math., 34, 397.

Kura, T. (1983) Existence of oscillatory solutions of fourth order superlinear ordinary differential equation. Hiroshima Math. J., 13, 653.

Kurtz, J. C. (1981) A class of singular nonlinear boundary value problems. J. Math. Anal. Appl., 83, 26.

Kusa, T. (1982) Oscillation theorems for a second order sublinear ordinary differential equation. Proc. Amer. Math. Soc., 84, 535.

Kusano, T. (1989) On unbounded positive solutions of nonlinear differential equation with oscillating coefficients. Czech. Math. J., 39, 133.

Kusano, T., Naito, M. and Swanson, C. A. (1988) On the asymptotic behaviour of nonlinear differential equations. Czech. Math. J., 38, 498.

Kuznetskii (1973) The solution of two ordinary differential equations. Diff. Eqns., 6, 714.

Kwong, M. K. (1991) Uniqueness results for the Emden–Fowler boundary value problem. Nonlinear Anal. Theory, Meth. Appl., 16, 435.

Kwong, M. K. and Wong, J. S. W. (1983) On the oscillation theorem of Belohorec. SIAM J. Math. Anal., 14, 474.

Ladeira, L. A. C. and Taboas, P. (1986) Periodic solutions of the equation $y'' - \sigma h(x)y' + g(y) = E\cos x$. Q. Appl. Math., 45, 429.

Laetsch, T. (1970) The number of solutions of a nonlinear two point boundary value problem. Indiana Univ. Math. J., 20, 1.

Lakin, W. D. and Sanchez, D. A. (1970) Topics in Ordinary Differential Equations. A Potpourri. Prindle, Weber and Schmidt, Boston, pp.84, 92, 93, 145.

Lamarque, C. H. and Stoffel, A. (1992) Parametic resonance with a nonlinear term: comparison of averaging and the normal form method using a simple example. Mech. Res. Commun., 19, 495.

Lan, H. and Wong, K. (1989) Exact solutions for some nonlinear equations. Phys. Lett. A, 137, 369.

Landman, M. J. (1987) Solutions of the Ginzburg–Landau equation of interest in the shear flow transition. Stud. Appl. Math., 76, 187.

Lange, C. G. (1987) Asymptotic analysis of forced nonlinear Strum–Liouville systems. Stud. Appl. Math., 76, 239.

Lange, C. G. and Weinitschke, H. J. (1991) Singular perturbations of limit points with application to tubular chemical reactors. Stud. Appl. Math., 84, 7.

Laplaza, M. L. (1970) The inverse cube law for a planet. Amer. Math. Monthly, 77, 659.

Lau, Y. (1992) Large period turbulent solutions of the Kuramoto–Sivashinsky equation. Phys. Lett. A, 169, 329.

Lawden, D. F. (1959) Mathematics of Engineering Systems. Wiley, New York.

Lazer, A. C. (1990) On the existence of stable periodic solutions of differential equations of Duffing type. Proc. Amer. Math. Soc., 110, 125.

Leach, P. G. L., Feix, M. R. and Bouquet, S. (1980) Analysis and solution of a nonlinear second order differential equation through rescaling and through a dynamical point of view. J. Math. Phys., 29, 2563.

Lebeau, G. and Lochak, P. (1987) On the second Painlevé equation: the connection formula via a Riemann–Hilbert problem and other results. J. Diff. Eqns., 68, 344.

Lee, C. F. (1971/72) On the existence of solution of a nonlinear differential diffusion system. Proc. Roy. Soc. Edin., A71, 1.

Leko, T. (1955) Uber die Integration der Differentialgleichungen $yy'' + f(x)y^2 = \phi(x)$. Hrvatsko Prirod Drustvo (Serbo–Croatian summary). Glas. Mat. Fiz. Astr. Ser. II, 10, 171.

Len, J. L. and Rand, R. H. (1988) Lie transforms applied to a nonlinear parametric excitation problem. Int. J. Nonlinear Mech., 23, 297.

Leonov, G. A. (1986) Estimation of the location of separatrices of a Lorenz system. Diff. Eqns., 22, 297.

Leonov, G. A. (1989) Asymptotic behaviour of solutions of a Lorenz system. Diff. Eqns., 25, 1492.

Leonov, G. A. and Reitmann, V. (1986) Das Rössler-system ist morn dissipative in Sinne von Levinson. Math. Nachr., 129, 31.

Leonovich, N. S. (1984) Classes of third order systems without movable critical points. Diff. Eqns., 20, 1326.

Lepin, L. A. (1990) Spectrum of eigenfunctions of a semilinear heat equation. Ser. Math. Dokl., 41, 340.

Levi, B. and Massera, J. L. (1947) Study in the large of a differential equation of the second order (Spanish) Math. Notae, 7, 91.

Levi, D., Nucci, M.C., Rogers, C. and Winternitz, P. (1988) Group theoretical analysis of a rotating shallow liquid in a rigid container. Preprint. CRM-1580, Université de Montreal.

Levin, P. W. and Koch, B. P. (1981) Chaotic behaviour of a parametrically damped pendulum. Phys. Rev. Lett. A, 86, 71.

Levine, G. and Tabor, M. (1988) Integrating the non-integrable: analytic structure of the Lorenz system. Physica D, 33, 189.

Levinson, N. (1970) Asymptotic behaviour of nonlinear differential equations. Stud. Appl. Math., 49, 285.

Li, Z. and Wang, M. Q. (1980) On periodic solution of Duffing's equation with damping (Chinese). A translation of Texue Tongbao, 25, 1015.

Liang, Z. (1966) Asymptotic behaviour of the solutions of a class of second order nonlinear differential equations. Shuxue Jin Zhan, 9, 251.

Liang, Z. (1987) Necessary and sufficient conditions of the stability for a class of second order nonlinear oscillations. Funkcial. Ekvac., 30, 45.

Liang, Z. C. (1987a) Asymptotic behaviour of solutions of the generalized Emden–Fowler equation. Ann. Diff. Eqns., 3, 311.

Liao, K. R. (1987) The qualitative analysis of a spin system (Chinese). Acta Math. Sinica, 30, 799.

Liberatore, A. and de Mottoni, P. (1983) A case of bifurcation for a one-dimensional boundary value problem with nonlinearity of concave type (Italian, English summary) Rend. Mat. (7), 3, 593.

Licko, I. and Svec, M. (1963) Le caractère oscillatoire des solutions de l'équation $y^{(n)} + f(x)y^{\alpha} = 0$, $n > 1$. Czech. J. Math., 88, 481.

Lloyd, N. G. (1972) On the non-autonomous Van der Pol equation with large parameters. Proc. Camb. Phil. Soc., 72, 213.

Locker, J. (1970) An existence analysis for nonlinear boundary value problems. SIAM J. Appl. Math., 19, 199.

Logan, J. D. (1985) Similarity solutions of the Euler equation in the calculus of variations. J. Phys. A, 18, 2151.

Lomtatidze, A. G. (1987) Positive solutions of boundary value problems for second order ordinary differential equations with singular points. Diff. Eqns., 23, 1146.

Londen, S. (1973) Some nonoscillation theorems for a second order nonlinear differential equation. SIAM J. Math. Anal., 4, 460.

Longford, W. F. et al. (1980) A mechanism for a soft model instability. Phys. Lett. A, 78, 11.

Lorenz, E.N. (1984) Irregularity: a fundamental property of the atmosphere. Tellus 36A, 98.

Loud, W. S. (1955) On periodic solutions of Duffing's equation with damping. J. Math. Phys., 34, 173.

Loud, W. S. (1968) Branching phenomena for periodic solutions of non-autonomous piecewise linear systems. Int. J. Nonlinear Mech., 3, 273.

Lu, Y. Y. (1989) The shunted homopolar dynamo: an analytic approach to a Poincaré map. Stud. Appl. Math., 80, 239.

Lu, C. (1990) Existence, bifurcation, and limit of solutions of the similarity equations for floating rectangular cavities and disks. SIAM J. Math. Anal., 21, 721.

Lu, C. and Kazarinoff, D. (1989) On the existence of solutions of a two point boundary value problem arising in a cylindrical floating zone. SIAM J. Math. Anal., 20, 494.

Ludeke, C. A. and Wagner, W. S. (1968) The generalised Duffing equation with large damping. Int. J. Nonlinear Mech., 3, 393.

Lukashevich, N. A. (1965) Elementary solutions of certain Painlevé equations. Diff. Eqns., 1, 561.

Lukashevich, N. A. (1967) Functions defined by a system of differential equations. Diff. Eqns., 5, 310.

Lukashevich, N. A. (1970) The theory of Painlevé equations. Diff. Eqns., 6, 329.

Lukashevich, N. A. (1972) On the theory of Painlevé's sixth equation. Diff. Eqns., 8, 1081.

Lukashevich, N. A. and Yablonskii, A. I. (1967) On a set of solutions of the sixth Painlevé transcendents. Diff. Eqns., 3, 264.

Luning, C. D. and Perry, W. L. (1974) Positive solutions of superlinear eigenvalue problems via a monotone iterative technique. J. Diff. Eqns., 3, 359.

Luning, C. D. and Perry, W. L. (1984) Iterative solution of a nonlinear boundary value problem for a rotating string. Int. J. Nonlinear Mech., 19, 83.

Lupo, D., Solimini, S. and Srikanth, P. N. (1988) Multiplicity results for an ODE problem with even nonlinearity. Nonlinear Anal. Theory Meth. Appl., 12, 657.

MacGillivray, A. D. (1990) Justification of matching with the transition expansion of Van der Pol's equation. SIAM J. Math. Anal., 21, 221.

Macki, J. W. (1978) A singular nonlinear boundary value problem. Pacific J. Math., 78, 375.

Mahaffey, R. A. (1976) Anharmonic oscillator description of plasma oscillations. Phys. Fluids, 19, 1387.

Makino, T. (1984) On the existence of positive solutions at infinity for ordinary differential equations of Emden type. Funkcial. Ekvac., 27, 319.

Manacorda, T. (1946) Soluzióni periodiche di una equazióne differenziale nonlineare. Math. Rev., 8, 464.

Manacorda, T. (1948) Vibrazióne forzato di un particolare sistèma oscillante nonlineare. Atti Accad. Naz. Lincei Rend. Cl. Sci. Fis. Mat. Nat. (8), 4, 557.

Manaresi, G. (1954) Sopra alcune limitazioni per l'ampiezzu delle oscillazioni nonlineari. Atti Accad. Sci. Ist. Bologna Cl. Sci. Fis. Rend., 2, 184.

Manic, V. and Tomic, M. (1980) Asymptotics of a generalised Thomas–Fermi equation. J. Diff. Eqns., 35, 36.

Marchenko, A. V. (1988) Long waves in shallow liquid under ice cover. J. Appl. Math. Mech., 52, 180.

Markina, A. M. (1973) Forms of nonstationary singularities of a differential system. Diff. Eqns., 9, 1036.

Markina, A. M. and Yablonskii, A. I. (1972) Algebraically movable singularities. Diff. Eqns., 8, 1085.

Markovich, P. A. (1985) A nonlinear eigenvalue problem modelling the avalanche effect in semi-conductor diodes. SIAM J. Math. Anal., 16, 1268.

Markus, L. and Amundsun, N. R. (1968) Nonlinear boundary value problems arising in chemical reactor theory. J. Diff. Eqns., 4, 102.

Marlin, J. A. and Ullrich, D. F. (1968) Periodic solutions of second order nonlinear diffusion equations without damping. SIAM J. Appl. Math., 16, 998.

Marshall, E. A. (1979) On the exact analytical solution of the Poisson–Boltzmann equation. J. Theor. Biol., 81, 613.

Martin, W. T. and Reissner, E. (1958) Elementary Differential Equations. Addison-Wesley, Reading, Mass., pp. 73, 76, 77, 161, 162.

Martinov, N., Ouroushev, D. and Chelebiev, E. (1986) New types of polarisation following from the nonlinear spherical radial Poisson–Boltzmann equation. J. Phys. A, 19, 1327.

Martynov, I. P. (1979) Properties of the solutions of a third order differential equation. Dokl. Akad. Nauk. BSSR, 23, 780.

Martynov, I. P. (1981) Third order systems with non-stationary singular lines. Diff. Eqns., 17, 154.

Martynov, I. P. (1984) Contributions to the theory of third order systems without moving critical points. Diff. Eqns., 20, 1231.

Martynov, I. P. (1985) Third order equations with no moving critical singularities. Diff. Eqns., 21, 623.

Martynov, I. P. (1985a) Analytic properties of solution of a third order differential equation. Diff. Eqns., 21, 512.

Martynov, I. P. (1986) Third order systems with no moving critical singularities. Diff. Eqns., 22, 151.

Martynov, I. P. and Yablonskii, A. I. (1979) Solution of a class of third order system. Diff. Eqns., 15, 1263.

Marzec, C. J. and Spiegel, E. A. (1980) Ordinary differential equations with strange attractors. SIAM J. Appl. Math., 38, 403.

Masoller, C., Sicardi Schifino, A. C. and Romanelli, L. (1992) Regular and chaotic behaviour in the new Lorenz system. Phys. Lett. A, 167, 185.

Massera, J. L. (1956) Qualitative study of the equation $u''^2 = u + u^1$. Bol. Fac. Ingen. Agrimen. Montevideo, 5, 339–347. Fac. Ing. Agrimen. Montevideo. Publ. Didact. Inst. Mat. Estadist 3, 1–10. (Spanish). (Math. Rev. (1957), 18, 211).

Matlak, R. F. (1969) An autonomous system of differential equations in the plane. Bull. Austral. Math. Soc., 1, 391.

Matos Peixoto, M. (1946) On the solutions of the equation $yy'' = \phi(y')$ which pass through two points of the half-plane $y > 0$. Rev. Un. Mat. Argentina, 11, 84.

Matsumoto, T. (1944) A note on Fowler's differential equation. Math. Rev., 11, 359.

Matsumoto, T. (1953) Note on nonlinear differential equation of catalysis. Mem. Coll. Sci. Univ. Kyoto. Ser. A Math., 27, 267.

Matsuno, Y. (1987) Exact solutions for the nonlinear Klein–Gorden and Liouville equations in four dimensional Euclidean space. J. Math. Phys., 28, 2317.

Matsuno, Y. (1991) Similarity solution of a nonlinear diffusion equation describing Alfven wave propagation. J. Phys. Soc. Jap. 60, 3197.

Mawhin, J. and Willem, M. (1984) Multiple solutions of the periodic boundary value problem for some forced pendulum type equations. J. Diff. Eqns., 52, 264.

McIntosh, I. (1990) Single phase averaging and travelling wave solutions of the modified Burgers–Korteweg–deVries equation. Phys. Lett. A, 143, 57.

McLachlan, N. W. (1951) Nonlinear differential equation having a periodic coefficient. Math. Gaz., 35, 32.

McLachlan, N. W. (1954) On a nonlinear differential equation in hydraulics. Proceedings of Symposia in Applied Mathematics. Vol. V. Wave Motion and Vibration Theory. McGraw-Hill, New York, p. 49.

McLeod, J. B. (1969) Von Kármán's swirling flow problems. Arch. Rat. Mech. Anal., 33, 91.

McLeod, J. B. (1972) The existence and uniqueness of a similarity solution arising from separation at a free stream line. Q. J. Math. Oxford (2), 23, 63.

McLeod, J. B. and Parter, S. V. (1974) On the flow between two counter-rotating infinite discs. Arch. Rat. Mech. Anal., 54, 301.

McLeod, J. B. and Parter, S. V. (1977) The monotonicity of solutions in swirling flows. Proc. Roy. Soc. Edin., 76A, 161.

McLeod, J. B. and Rajagopal, K. R. (1987) Uniqueness of flow due to a stretching boundary. Arch. Rat. Mech. Anal., 98, 385.

McLeod, J. B. and Troy, W. C. (1991) The Skyrme model for a nucleus under spherical symmetry. Proc. Roy. Soc. Edin., 118A, 271.

McVittie, G. C. (1933) The mass particle in an expanding universe. Monthly Notices Roy. Astron. Soc., 93, 325.

Medina, R. (1991) On the boundedness and stability of the solution of semilinear differential equations. J. Math. Anal. Appl., 157, 48.

Medina, R. and Pinto, M. (1988) On the asymptotic behaviour of solutions of certain second order nonlinear differential equations. J. Math. Anal. Appl., 135, 399.

Mehri, B. L. (1980/81) Periodic solutions for a certain nonlinear third order differential equation. Bull. Iranian Math. Soc., 8, 9.

Mehri, B. (1989) Periodic solutions of a second order nonlinear differential equation. Bull. Austral. Math. Soc., 40, 357.

Mehri, B. (1990) Periodic solution for certain nonlinear third order differential equation. Indian J. Pure Appl. Math., 21, 203.

Mehta, B. W. and Aris, R. (1971) A note on a form of the Emden–Fowler equation. J. Math. Anal. Appl., 36, 611.

Meinhardt, J. R. (1981) Symmetries and differential equations. J. Phys. A, 14, 1893.

Melkonian, S. and Maslowe, S. A. (1990) Analysis of a nonlinear diffusive amplitude equation for waves on thin films. Stud. Appl. Math., 82, 37.

Melvin, P. J. (1978) The phase shifting limit cycles of the Van der Pol equation. J. Res. Bur. Stand., 83, 593.

Merkin, J. H. (1984) A note on the solution of a differential equation arising in boundary-layer theory. J. Engg. Math., 18, 31.

Merkin, J. H. (1985) A note on the similarity solution for free convection on a vertical plate. J. Engg. Math., 19, 189.

Merkin, J. H. and Mahmood, T. (1989) Mixed convection boundary layer similarity solutions: prescribed wall heat flux. J. Appl. Math. Phys., Z. Angew. Math. Phys., 40, 51.

Merkin, J. H., Pop, I. and Mahmood, T. (1991) Mixed convection on a vertical surface with a prescribed heat flux: the solution for small and large Prandtl numbers. J. Engg. Math., 25, 165.

Micheletti, A. M. (1967) Le soluzione periodiche dell'equazione differenziale nonlineare $x''(t) + 2x^3(t) = f(t)$. Ann. Univ. Ferrara Sez. VII (N.S.), 12, 103.

Michelsen, D. (1986) Steady solutions of the Kuramoto–Sivashinsky equation. Physica D, 19, 89.

Michelsen, D. (1990) Elementry particles as solutions of the Sivashinsky equation. Physica D, 44, 502.

Mickens, R. E. (1981) An Introduction to Nonlinear Oscillations. Cambridge University Press, Cambridge.

Mickens, R. E. (1987) Nonlinear analysis and applications. Proceedings of the Seventh International Conference held at the University of Texas, Arlington, Tex., July 28–August 1, 1986, pp. 339–344. Lecture Notes in Pure and Applied Mathematics. Marcel Dekker, New York, p. 109.

Mickens, R. E. (1988) Bounds on the Fourier coefficients for the periodic solutions of nonlinear oscillator equations. J. Sound Vib., 124, 199.

Mickens, R. E. and Oyedeji, K. (1985) Construction of approximate analytic solutions to a new class of nonlinear oscillator equations. J. Sound Vib., 102, 579.

Mickens, R. E. and Ramadhani, I. (1992) Investigation of an anti-symmetric quadratic nonlinear oscillator. J. Sound Vib., 155, 190.

Mignosi, G. (1964) Su un equazióne differenziale del second ordine nonlineare che si presenta néllo stùdio di cèrti sistémi fisici oscillanti. Atti. Acad. Sci. Lett. Arti. Palermo Part I (4), 23, 273.

Mihailovic, M. V. (1950) Sur l'intégrale de l'équation différentielle de Thomas–Fermi autour du point $x = 0,\ y = 1$. Acad. Serbe Sci. Publ. Inst. Math., 3, 259.

Miles, J. W. (1978) On the second Painlevé transcendent. Proc. Roy. Soc. London A, 361, 277.

Miles, J. (1989) Resonances and symmetry breaking for a Duffing oscillator. SIAM J. Appl. Math., 49, 968.

Min, Q., Wenxian, S. and Jinyan, Z. (1990) Dynamical behaviour in coupled system of J-J type. J. Diff. Eqns., 88, 175.

Min, Q., Xian, S. W. and Jinyan, Z. (1988) Global behaviour in the dynamical equation. J. Diff. Eqns., 71, 315.

Minc, R. M. (1955) Investigation of the trajectories at infinity of three differential equations. In Memory of Aleksandra Aleksandrovica Andronova. Izdat Akad. Nauk SSSR, Moscow, p.499.

Minorsky, N. (1954) Sur les systèmes nonlineaires a deux degrés de liberté. C.R. Acad. Sci. Paris, 238, 646.

Minorsky, N. (1955) Sur l'interaction des oscillations nonlineaires. Rend. Sem. Mat. Fis. Milano, 25, 145.

Minorsky, N. (1962) Nonlinear Oscillations. Van Nostrand, New York.

Mirzov, D. D. (1987) Asymptotic properties of solutions of an Emden–Fowler system. Diff. Eqns., 23, 1042.

Mitrinovitch, D. S. (1955) Sur l'équation différentielle d'Emden generalisèe. C.R. Acad. Sci. Paris, 241, 724.

Mitrinovitch, D. S. (1956) Compléments an traite de Kamke, II. Bull. Soc. Math. Phys. Serbie, 7, 161.

Mitsui, T. (1977) Investigation of numerical solutions of some nonlinear quasi-periodic differential equations. Publ. RIMS Kyoto Univ., 13, 793.

Miura, T. and Kai, T. (1986) A strange attractor of a system of three disc-dynamos and a geomagnetic attractor: their dimensions and K_2 entropy. J. Phys. Soc. Jap., 55, 2562.

Mizohata, S. (1953) Sur les phénoménes de sants dans certains systèmes nonlineaires. Mem. Coll. Sci. Univ. Kyoto Ser. A Math., 27, 203.

Modona, L. (1953) Su di una equazióne differenziale non lineare del secondo ordine. Boll. Un. Mat. Ital., 8, 428.

Moiseev, E. I. and Sadovnichii, V. A. (1989) Uniqueness of the solution of a boundary value problem for a nonlinear equation (Russian). Math. Modelling, 254–261, 308. Nauka, Moscow.

Monguchi, H. and Nakumara, T. (1983) Forced oscillations of systems with nonlinear restoring force. J. Phys. Soc. Jap., 52, 732.

Month, L. A. and Rand, R. H. (1980) An application of the Poincaré map to the stability of nonlinear normal modes. Trans. ASME J. Appl. Mech., 47, 645.

Moon, F. C. (1980) Experimental models for strange attractor vibration in elastic system, in New Approaches to Nonlinear Problems in Dynamics, D. J. Holmes, ed., SIAM, Philadelphia, Pa.

Moon, F. C. (1987) Chaotic Vibrations: An Introduction for Applied Scientists and Engineers. Wiley-Interscience, New York, pp. 79, 110.

Mooney, J. W. (1979) Constructive existence theorems for problems of Thomas–Fermi type. Math. Meth. Appl. Sci., 1, 554.

Moòre, D. W. and Spiegel, E. A. (1966) A thermally excited nonlinear oscillator. Astrophys. J., 143, 871.

Moraux, M., Fijalkow, E. and Fiex, M. R. (1981) Asymptotic solutions of time dependent anharmonic oscillator equation. J. Phys. A, 14, 1611.

Moravsky, L. (1975) Über die nichlineare Differentialgleichung in der Form $y''' + p(x)y'' + q(x)f(y') + r(x)h(y, y'') = 0$. Acta Fac. Rerum. Nat. Uni. Comerian Math., 23, 91.

Morozov, A. D. (1976) A complete qualitative investigation of Duffing's equation. Diff. Eqns., 11, 164.

Morozov, A. D. (1989) Limit cycles and chaos in equations of pendulum type. J. Appl. Math. Mech., 53, 565.

Morozov, A. D. and Fedorov, E. L. (1983) Equation with one degree of freedom differing only slightly from integral nonlinear equations. Diff. Eqns., 20, 1133.

Morris, G. R. (1965) An infinite class of periodic solutions of $\ddot{x}+2x^3=p(x)$. Proc. Camb. Phil. Soc., 61, 157.

Morris, G. R. (1976) A case of boundedness in Littlewood's problem on oscillatory differential equations. Bull. Austral. Math. Soc., 14, 71.

Moser, J. (1973) Stable and random motions in dynamical systems. Annals. of Mathematics. Princeton University Press, Princeton, N.J.

Moslehy, F. A. and Evan–Iwanowski, R. M. (1991) The effects of non-stationary processes on chaotic and regular responses of the Duffing oscillator. Int. J. Nonlinear Mech., 26, 61.

Mulholland, R. J. (1971) Nonlinear oscillation of a third order differential equation. Int. J. Nonlinear Mech., 6, 279.

Mullin, T. (1991) Finite dimensional dynamics in Taylor–Couette flow. IMA J. Appl. Math., 46, 109.

Murphy, G. M. (1960) Ordinary Differential Equations and Their Solutions. D. Van Nostrand, Princeton, N.J.

Murphy, J. G. (1992) Some new closed form solutions describing spherical inflation in compressible finite elasticity. IMA J. Appl. Math., 48, 305.

Musinskaja, A.(1966) The boundedness of solutions of certain nonlinear systems with two degrees of freedom. Nonlinear Vibration Problems, Vol. 7. Wydawnictwa Nauk. Warsaw, pp. 225–232.

Musinskaja, A. (1967) On fundamental periodic solutions of a certain nonlinear system with two degrees of freedom. Nonlinear Vibration Problems, Vol. 8, Wydawnictwa Nauk, Warsaw, pp. 37–48.

Nakao, M. (1989) Decay estimates for some nonlinear second order differential equations. Bull. Austral. Math. Soc., 40, 25.

Nakajima, F. (1990) Even and periodic solutions of the equation $\ddot{u}+g(u)=e(t)$. J. Diff. Eqns., 83, 277.

Nalli, P. (1947) L'equazióne differenziale $y'' + y = f(x)(1 - y^2 - y'^2)^{1/2}$. Boll. Un. Mat. Ital. (3), 2, 195.

Narayanan, S. and Jayaraman, K. (1991) Chaotic vibration on a nonlinear oscillator with Coulomb damping. J. Sound Vib., 146, 17.

Nascimento, A. S. D. (1989) Bifurcation and stability of radially symmetric equilibria of a parabolic equation with variable diffusion. J. Diff. Eqns., 77, 84.

Nayfeh, A. H. (1985) Problems in Perturbations. Wiley, New York.

Nazarov, G. I. and Puchkova, N. G. (1991) The motion of waves in a channel of variable cross section. Fluid Mech. Sov. Res., 20, 37.

Needham, D. J. and Merkin, J. A. (1987) A note on the wall-jet problem. J. Engg. Math., 21, 17.

Needham, D. J. and Merkin, J. A. (1991) The development of travelling waves in a simple chemical system with general orders of autocatalysis and decay. Phil. Trans. R. Soc. London A, 337, 261.

Nehari, Z. (1960) On a class of nonlinear second order differential equations. Trans. Amer. Math. Soc., 95, 101.

Neto, J. A., Galain, R. M. and Ferreira, E. (1991) Properties of the Skyrme soliton configuration. J. Math. Phys., 32, 1949.

Newton, P. K. and Sirovich, L. (1986) Instabilities of the Ginzburg–Landau equation: periodic solutions. Q. Appl. Math., 44, 49.

Nieto, J. J. (1988) Nonlinear second order periodic boundary value problem. J. Math. Anal. Appl., 130, 22.

Nijenhuis, W. (1949) A note on a generalized Van der Pol equation. Math. Rev. 11, 249.

Nishikawa, Y. (1964) A Contribution to the Theory of Nonlinear Oscillations. Nippon Printing and Publishing Company, Osaka, Japan.

Nitkin, A. K. (1953) Nonlinear oscillations of a system with a disturbing force consisting of two harmonics. Rostov. Gos. Univ. Ucen. Zap. Fiz. Mat. Fak., 18, 55.

Norbury, J. and Stuart, A. M. (1988) Travelling combustion waves in a porous medium. Parts I and II. Existence. SIAM J. Appl. Math., 155, 374.

Norbury, J. and Stuart, A. M. (1989) A model for porous medium combustion. Q. J. Mech. Appl. Math., 42, 159

Novak, M. (1969) Aeroelastic galloping of prismatic bodies. Proc. ASCE J. Mech., 5, 115.

Obi, C. (1953) A nonlinear differential equation of the second order with periodic solutions whose associated limit cycles are algebraic curves. J. London Math. Soc., 28, 356.

Obi, C. (1976) An analytic theory of nonlinear oscillations: IV. The periodic oscillations of the equation $\ddot{x} - \epsilon(1 - x^{2n+2})x + x^{2n+1} = \epsilon a \cos \omega t$, $a > 0$, $\omega > 0$ independent of ϵ. SIAM J. Appl. Math., 31, 345.

Obi, C. (1980) Analytical theory of nonlinear oscillations: X. Some classes of equations $\ddot{x} + g(x) = 0$ with no finite Fourier series solution. Atti Accad. Naz. Lincei Rend. Cl. Sci. Fis. Mat. Nat., 67, 53.

Ockendon, H., Ockendon, J. R. and Johnson, A. D. (1986) Resonant sloshing in shallow water. J. Fluid Mech., 167, 465.

Okrasinski, W. (1989) On subsolutions of a nonlinear diffusion problem. Math. Meth. Appl. Sci., 11, 409.

Okrasinski, W. (1989a) On a nonlinear ordinary differential equation. Ann. Polon. Math., 49, 237.

Olver, P. J. (1986) Applications of Lie Groups to Differential Equations. Springer-Verlag, New York, p. 147.

Omari, P., Villari, G. and Zanolin, F. (1987) Periodic solution of the Lienard equation with one-sided growth restrictions. J. Diff. Eqns., 68, 278.

Onose, H. (1983) On Butler's conjecture for oscillation of an ordinary differential equation. Q. J. Math. Oxford Ser. (2), 34, 235.

Opial, Z. (1960) Sur un problème aux limites pour l'équation différentielle du second ordre. Ann. Polon. Math., 7, 223.

Opial, Z. (1961) Sur l'existence des solutions periodiques de l'équation différentielle $x'' + f(x, x')x' + g(x) = p(t)$. Ann. Mat., 11, 149.

O'Regan, D. (1991) Solvability of some fourth (and higher) order singular boundary value problems. J. Math. Anal. Appl., 161, 78.

Orlov, K. (1980) Effective finding of solutions of the form $\sum_{l=1}^{\infty} a_l/x^l$ for differential equations of finite and infinite order. Mat. Vesnik, 4, 481.

Osipov, A. V. and Pliss, V. A. (1989) Relaxation oscillation in a Duffing equation. Diff. Eqns., 25, 300.

Oskam, B. and Veldman, A. E. P. (1982) Branching of Falkner–Skan solutions for $\lambda < 0$. J. Engg. Math., 16, 295.

Otsuki, T. (1974) On a bound for periods of solutions of a certain nonlinear differential

equation: I. J. Math. Soc. Jap., 26, 206.

Otsuki, T. (1974a) On a bound for periods of solutions of a certain nonlinear differential equation: II. Funkcial. Ekvac., 17, 193.

Otwinowski, M., Paul, R. and Laidlaw, W. G. (1988) Exact traveling solutions of a class of nonlinear diffusion equations by reduction to a quadrature. Phys. Lett., 128, 483.

Pablo, A. and Vazquez, J. L. (1991) Travelling waves and finite propagation in a reaction-diffusion equation. J. Diff. Eqns., 93, 19.

Palmero, F. and Romero, F. R. (1991) Interior crises and symmetries in a driven Josephson junction. Phys. Lett. A, 160, 553.

Panayotounakos, D. E. and Theocaris, P. S. (1986) Exact solution for an approximate differential equation of a straight bar under condition of a nonlinear equilibrium. J. Nonlinear Mech., 21, 421.

Papoulis, A. (1958) Strongly nonlinear oscillations. J. Math. Phys., 37, 147.

Parhi, N. (1981) Non-oscillatory behaviour of solutions of non-homogeneous third order differential equations. Appl. Anal., 12, 273.

Parhi, N. and Parhi, S. (1986) On the behaviour of solutions of the differential equation $(r(t)y'')' + q(t)(y')\beta + p(t)y^{\alpha} + f(t)$. Ann. Polon. Math., 47, 137.

Parlange, J. Y., Braddock, R. D. and Sander, G. (1981) Analytical approximations to solution of the Blasius equations. Acta Mech., 38, 119.

Parter, S. V. (1970) Nonlinear eigenvalue problems for some fourth order equations: I. Maximal solutions. II. Fixed point methods. SIAM J. Math Anal., 1, 458.

Parter, S. V. (1972) Remarks on the existence theory of multiple solutions of a singular perturbation problem. SIAM J. Math. Anal., 3, 496.

Parthasarathy, S. and Lakshmanan, M. (1990) On the analytic structure of the driven pendulum. J. Phys. A, 23, L1223–1228.

Parthasarathy, S. and Lakshmanan, M. (1991) Analytic structure of the damped driven Morse oscillator. Phys. Lett. A, 157, 365.

Pascal, H. (1991) On nonlinear effects in unsteady flows through fractured porous media. Int. J. Nonlinear Mech., 26, 251.

Pascal, H. (1991a) On propagation of pressure disturbances in a non-Newtonian fluid flowing through a porous medium. Int. J. Nonlinear Mech., 26, 475.

Pascal, H. (1991b) On nonlinear effects in unsteady flows of non-Newtonian fluids fractured

porous media. Int. J. Nonlinear Mech., 26, 487.

Pascal, H. and Pascal, F. (1985) Flow of non-Newtonian fluid through porous media. Int. J. Engg. Sci., 23, 571.

Pascal, H. and Pascal, F. (1990) On some self-similar flows of non-Newtonian fluids through a porous medium. Stud. Appl. Math., 82, 1.

Pascal, H. and Pascal, J. P. (1991) Nonlinear effects of non-Newtonian fluids in unsteady flow with circular streamlines. Math. Comp. Modelling, 15, 107.

Pasynkova, I. A. (1981) Existence of limit cycles of the first kind in automatic phase frequency control system (Russian) Oscillations and Stability of Mechanical Systems. Prikl. Mekh. 5. Leningrad Univ., Leningrad, pp. 88–95.

Peletier, L. A. and Troy, W. C. (1988) On non-existence of similarity solutions. J. Math. Anal. Appl., 133, 57.

Peletier, L. A. and Troy, W. C. (1991) Self-similar solutions for infiltration of dopant into semiconductors. Arch. Rat. Mech. Anal., 116, 71.

Pennline, J. A. (1981) Improving convergence rate in the method of successive approximations. Proc. Amer. Math. Soc., 37, 127.

Pennline, J. A. (1984) Constructive existence and uniqueness for two point boundary value problems with a linear gradient term. Appl. Math. Comp., 15, 233.

Petrescu, St. (1949) On point invariants of the differential equation $y''' = F(x, y, y', y'')$. Acad. Repub. Pop. Romane Bul. Sci. Ser. A, 1, 433.

Petrov, V. G. (1991) Mechano-mathematical model of cardiac-fiber length-pressure pulsations. Acta Mech., 87, 239.

Petty, C. M. and Johnson, W. E. (1973) Properties of solution of $u'' + c(t)f(u)h(u') = 0$ with explicit initial conditions. SIAM J. Math. Anal., 4, 269.

Peyovitch, T. (1947) L'existence de solutions asymptotiques de certaines équations différentielles. Acad. Serbe Sci. Publ. Inst. Math., 1, 88.

Pikovskii, A. S., Rabinovich, M. I. and Trakhtengerts, V. Yu. (1978) Onset of stochasticity in decay confinement of parametric instability. Sov. Phys. JETP, 47, 715.

Pino, M. and Manasevich, R. (1989) A homotropic deformation along p of a Leray–Schauder degree result and existence for $(|u'|^{p-2}u')' + f(t, u) = 0,\ u(0) = u(T) = 0, \quad p > 1$. J. Diff. Eqns., 80, 1.

Pino, M., Manasevich, R. and Montero, A. (1992) T-periodic solutions for some second order differential equations with singularities. Proc. Roy. Soc. Edin., 120A, 231.

Pinto, M. (1991) Continuation, nonoscillation and asymptotic formulas of solutions of second order differential equations. Nonlinear Anal. Theory Meth. Appl., 16, 981.

Pisarenok, V. P. (1981) Critical points of a differential equation system. Diff. Eqns., 17, 1039.

Pismen, L. M. and Rubinstein, J. (1991) Motion of vortex lines in the Ginzburg–Landau model. Physica D, 47, 353.

Pistiner, A., Shapiro, M. and Rubin, H. (1991) A new solution for the nonlinear diffusion convection equation. SIAM J. Appl. Math., 51, 1616.

Pivovarov, I. (1981) Non-conservative oscillatory systems with periodic solutions. Int. J. Nonlinear Mech., 16, 187.

Plaat, O. (1971) Ordinary Differential Equations. Holden–Day, San Francisco.

Pliss, V. A. (1956) Investigation of a nonlinear differential equation of third order. Dokl. Acad. Nauk SSSR (N.S.), 111, 1178.

Polovko, Y. K. (1986) A numerical method of solving nonlinear two point boundary value problems. USSR Comp. Math. Math. Phys., 26, 180.

Ponzo, P. J. (1967) Forced oscillations of the generalised Lienard equation. SIAM J. Appl. Math., 15, 75.

Ponzo, P. J. and Wax, N. (1965) On certain relaxation oscillations: Confining regions. Appl. Math., 23, 215.

Ponzo, P. J. and Wax, N. (1965a) On certain relaxation oscillations: Asymptotic solutions. SIAM J. Appl. Math. 13, 740.

Powell, J. and Tabor, M. (1992) Non-generic connections corresponding to front solutions. J. Phys. A, 25, 3773.

Prachar, K. and Schmetterer, L. (1956) Über eine spezielle nichtlineare Differentialgleichung. Osterreich Ing. Arch., 10, 247.

Prince, G. E. and Eliezer, C. J. (1981) On the Lie symmetries of the classical Kepler problem. J. Phys. A, 14, 587.

Proctor, M. R. E. and Weiss, N. O. (1990) Normal forms and chaos in thermosolutal convection. Nonlinearity, 3, 619.

Pykhteev, G. N. and Myachina, L. V. (1972) Successive approximation solution of a boundary value problem for a class of nonlinear ordinary differential equations (Russian) Diff. Eqns., 8, 598.

Rab, M. (1970) The Riccati differential equation with complex-valued coefficients. Czech. Math. J., 20, 491.

Rabenstein, A. L. (1966) Introduction to Ordinary Differential Equations. Academic Press, New York.

Rabenstein, A. L. (1972) Introduction to Ordinary Differential Equations, 2nd ed. Academic Press, New York.

Rabtsevich, V. A. (1986) Unbounded regular solutions of the Emden–Fowler equation with a supplementary differential property. Diff. Eqns., 22, 555.

Rajagopal, K. R., Na, T. Y. and Gupta, A. S. (1984) Flow of a viscoelastic fluid over a stretching sheet. Rheol. Acta, 23, 213.

Rajagopal, K. R., Szeri, A. Z. and Troy, W. (1986) An existence theorem for the flow of a non-Newtonian fluid past an infinite porous plate. Int. J. Nonlinear Mech., 21, 279.

Rajagopal, K. R., Troy, W. C. and Wineman, A. S. (1986) A note on non-universal deformations of nonlinear elastic layers. Proc. Roy. Irish Acad. A, 86, 107.

Ramanaiah, G. and Malavizhi, G. (1991) A note on exact solutions of certain nonlinear boundary value problems governing convection in porous media. Int. J. Nonlinear Mech., 26, 345.

Ramani, A., Dorizzi, B. and Grammaticos, B. (1983) Comment on "Painlevé property of anharmonic systems with an external, periodic field." Phys. Lett. A, 97, 87.

Ranganathan, P. V. (1987) General solution of Kostin's equation arising in quantum mechanics. J. Phys. A Math. Gen., 20, 5935.

Ranganathan, P. V. (1988) Solution of some classes of second order nonlinear ordinary differential equations. J. Nonlinear Mech., 23, 421.

Ranganathan, P. V. (1992) Invariants of a certain nonlinear N-dimensional dynamical system. Int. J. Nonlinear Mech., 27, 43.

Ranganathan, P. V. (1992a) Integrals of the Emden–Fowler equations. Int. J. Nonlinear Mech., 27, 583.

Rankin, S. M. (1976) Oscillations of a forced second order nonlinear differential equation. Proc. Amer. Math. Soc., 59, 279.

Rao, B. N., Mittal, M. L. and Nataraj, H. R. (1983) Hall effect in boundary layer flow. Acta Mech., 49, 147.

Raty, R., Von Boehm, J. and Isomaki, H. M. (1984) Absence of inversion symmetric limit cycles of even periods and the chaotic motion of a Duffing oscillator. Phys. Lett. A, 103, 289.

Reddick, H. W. (1949) Differential Equations. Wiley, New York.

Reid, J. L. (1971) An exact solution of the nonlinear differential equation $\ddot{y} + p(t)y = q_m(t)/y^{2m-1}$. Proc. Amer. Math. Soc., 27, 61.

Reid, J. L. and Cullen, J. J. (1982) Two theorems for time-dependent dynamical systems. Prog. Theor. Phys., 68, 989.

Reissig, R. (1955) Über eine nichtlineare differentialgleichung 2. Ordnung. Nachrichtentechnik, 13, 313.

Reissig, R. (1955/56) Neue methoden der nichtlineare mechanik von Krylow and Bogoliubov. Wiss. Humboldt Univ. Berlin Math. Nat. Reihe, 5, 99.

Reissig, R. (1956) Über eine nichtlineare differentialgleichung 2. Ordnung III. Mathematika, 15, 39.

Reynolds, D. W. (1989) Bifurcation of harmonic solutions of an integrodifferential equation modelling resonant sloshing. SIAM J. Appl. Math., 49, 362.

Richard, U. (1951) Su un equazióne nonlineare del secondo ordine. Univ. Politecnico Torino Rend. Sem. Mat., 10, 305.

Richardson, S. (1973) On Blasius's solution governing flow in the boundary layer on a plate. Proc. Camb. Phil. Soc., 74, 179.

Riley, N. and Weidman, F. D. (1989) Multiple solutions of the Falkner–Skan equation for flow past a stretching boundary. SIAM J. Appl. Math., 49, 1350.

Robbins, K. A. (1977) A new approach to subcritical instability and turbulent transitions in a simple dynamo. Math. Proc. Camb. Phil. Soc., 82, 309.

Roberts, C. E., Jr. (1979) Ordinary Differential Equations: A Computational Approach. Prentice Hall, Englewood Cliffs, N.J.

Roberts, S. M. (1984) Solution of $\epsilon Y'' + YY' - Y = 0$ by a nonasymptotic method. J. Opt. Theory Appl., 44, 303.

Roberts, C. E. and Belford, G. G. (1971) Stability and entering of the origin for real, nonlinear, autonomous differential equations of third order. SIAM J. Math. Anal., 2, 133.

Roberts, S. M. and Shipman, J. S. (1972) Two-Point Boundary Value Problems: Shooting Methods. Elsevier, New York.

Robinson, C. (1989) Homoclinic bifurcation to a transitive attractor of Lorenz type. Nonlinearity, 2, 495.

Robinson, F. N. H. (1987) The modified Van der Pol oscillator. IMA J. Appl. Math., 38, 135.

Rod, D. L. (1973) Pathology of invariant sets in the monkey saddle. J. Diff. Eqns., 14, 129.

Roekaerts, D. (1988) Painlevé property and constants of motion of the complex Lorenz model. J. Phys. A, 21, L495.

Rosales, R. (1978) The similarity solutions for the Korteweg–deVries equation and the related Painlevé transcendent. Proc. Roy. Soc. London A, 361, 265.

Rosen, N. and Rosenstock, H. B. (1952) The force between particles in nonlinear field theory. Phys. Rev., 85, 257.

Rosenau, P. (1982) Thermal equilibrium and stability of an ohmically heated plasma. Phys. Fluids, 25, 148.

Rosenau, P. (1984) A note on integration of the Emden–Fowler equation. J. Nonlinear Mech., 19, 303.

Rosenberg, M. R. (1955) On the stability of a nonlinear non-autonomous system. Proc. 2nd U.S. National Congress of Applied Mechanics, Ann Arbor, Mich. ASME, New York, 1954, p.63.

Rosenberg, R. M. (1966) Steady state forced vibrations. Int. J. Nonlinear Mech., 1, 95.

Rosenblatt, A. (1939) Sur les points singuliers des équations différentielles. Acta Acad. Cl. Lima, 2, 59.

Rosenblatt, A. (1945) On the phenomenon of subresonance: the case of the generalised Van der Pol equation with forced vibrations. Bol. Fac. Ing. Montevideo, 3, 116.

Rössler, O. E. (1976) An equation for continuous chaos. Phys. Lett. A, 57, 397.

Rössler, O. E. (1977) Horseshoe map chaos in the Lorenz equation. Phys. Lett., 60, 392.

Roth, L. (1941) On the solution of certain differential equations of second order. Phil. Mag., 32, 155.

Rott, N. and Lewellen, W. S. (1966) Boundary layers and their interactions in rotating flows, in Progress in Aeronautical Science, Vol.7, D. Kuchemann, ed. Pergamon Press, Elmsford, N.Y., p.111.

Rucklidge, A. M. (1992) Chaos in models of double convection. J. Fluid Mech., 237, 209.

Ruf, B. and Solimini, S. (1986) On a class of superlinear Sturm–Liouville problem with arbitrarily many solutions. SIAM J. Math. Anal., 17, 761.

Rusin, M. M. (1974) The nature of the singular point of special three dimensional differential systems (Russian). Collection of Scientific Articles in the Physical and Mathematical Sciences, Vol. 2. (Publi Grodno State Ped. Inst.) Izdat Vyseisajaskola, Minsk, Russia, pp.26–29.

Russel, D. A. and Ott, E. (1981) Chaotic (strange) and periodic behaviour in instability saturation by the oscillating two stream instability. Phys. Fluids, 24, 1976.

Ryder, G. H. (1967) Boundary value problems for a class of nonlinear differential equations. Pacific J. Math., 22, 477.

Rypdal, K., Rasmussen, J. J. and Thomsen, K. (1985) Similarity structure of wave-collapse. Physica D, 16, 339.

Sabata, H. (1980) A certain type of integrating factors for some single systems. Lett. Nuovo Cimento, 28, 395.

Sachdev, P. L. (1991) Nonlinear Ordinary Differential Equations and Their Applications. Marcel Dekker, New York.

Sachdev, P. L. and Nair, K. R. C. (1987) Generalised Burgers equations and Euler–Painlevé transcendents: II. J. Math. Phys., 28, 997.

Sachdev, P. L., Nair, K. R. C. and Tikekar, V. G. (1986) Generalised Burgers equations and Euler–Painlevé transcendents: I. J. Math. Phys. 27, 1506.

Sachdev, P. L., Nair, K. R. C. and Tikekar, V. G. (1988) Generalised Burgers equations and Euler–Painlevé transcendents: III. J. Math. Phys., 29, 2397.

Sachdev, P. L. and Philip, V. (1986) Invariance group properties and exact solutions of equations describing time-dependent free surface flows under gravity. Q. Appl. Math., 43, 463.

Saito, T. (1978) On bounded solutions of $x'' = t^{\beta}x^{1+\alpha}$. Tokyo J. Math., 1, 57.

Saito, T. (1978a) Solutions of $x'' = t^{\alpha\lambda-2}x^{1+\alpha}$ with movable singularity. Tokyo J. Math., 2, 262.

Sajben, M. (1968) An exact solution for axially symmetric equilibrium electron density distributions. Phys. Fluids, 11, 2501.

Samoilenko, A. M. (1967) Periodic solutions of second order nonlinear equations. Diff. Eqns., 3, 989.

Samuilov, A. Z. (1971) Some properties of polynomial solutions of high order differential equations. Diff. Eqns., 7, 1729.

Sanchez, W. E. and Nayfeh, A. H. (1990) Prediction of bifurcations in a parametrically excited Duffing oscillator. Int. J. Nonlinear Mech., 23, 163.

Sano, O. (1983) An analytic solution of the Lorenz model. J. Phys. Soc. Jap., 52, 466.

Sansone, G. (1940) Sulle soluzióni di Emden dell'equazióne di Fowler. Univ. Roma Ist. Naz. Aita. Mat. Rend. Mat. Appl. (5), 1, 163.

Sansone, G. (1949) Sopra l'equazióne di A. Lienard delle oscillazióni di rilassaménto. Ann. Mat. Pura Appl., 28(4), 153.

Sansone, G. (1962) L'equation de M. Dini $xy''+y' = \sin y$. Bull. Soc. Math. Phys. Serbie, 14, 17.

Santos, E. C. (1989) Application of finite expansion in elliptic functions to solve differential equations. J. Phys. Soc. Jap., 58, 4301.

Sapogin, L. G. and Boichenko, V. A. (1988) On the solution of one nonlinear equation. Nuovo Cimento, 102, 433.

Saranen, J. and Seikkala, S. (1988) Solutions of a nonlinear two-point boundary value problem with Neumann-type boundary data. J. Math. Anal. Appl., 135, 691.

Saranen, J. and Seikkala, S. (1989) Some remarks on nonlinear second order differential equations with periodic boundary conditions. J. Math. Anal. Appl., 139, 465.

Sarkar, A. K. and Roy, A. B. (1990) Asymptotic behaviour of solutions of a donor-recipient system. Appl. Math. Modelling, 14, 433.

Sarlet, W. (1991) New aspects of integrability of generalized Henon–Heiles systems. J. Phys. A, 24, 5245.

Sarlet, W. and Bahar, L. Y. (1980) A direct construction of first integrals for certain nonlinear systems. Int. J. Nonlinear Mech., 15, 133.

Sarlet, W., Mahomed, F. M. and Leach, P. G. L. (1987) Symmetries of nonlinear differential equations and linearisation. J. Phys. A, 20, 277.

Sato, Y. (1977) On the limit cycles of $\ddot{x} + F(\dot{x}) + g(y) = 0$. J. Fac. Sci. Univ. Tokyo, 24, 261.

Satsuma, J. (1987) Explicit solutions of nonlinear equations with density-dependent diffusion. J. Phys. Soc. Jap., 56, 1947.

Sawada, K. and Osawa, T. (1978) On exactly solvable nonlinear ordinary differential equa-

tions of the Lienard type. J. Phys. Soc. Jap., 44, 1730.

Scerbina, G. V. (1961) On a boundary value problem for an equation of Blasius. Sov. Math., 2, 1219.

Schichtel, T. and Beckmann, P. E. (1991) The use of first return maps in the computation of basin boundaries in three dimensional phase space. Phys. Lett., 156, 163.

Schmitt, B. V. (1972) Sur l'existence et la localisation de solutions périodiques de systèmes différentials périodiques: application a l'équation de Duffing. Int. J. Nonlinear Mech., 7, 199.

Schreiber, I., Kubicek, M. and Marek, M. (1980) On coupled cells, in New Approaches to Nonlinear Problems in Dynamics, P. J. Holmes, ed. SIAM, Philadelphia, Pa, pp.496–508.

Schweitzer, P. (1986) On a nonlinear matrix differential equation. SIAM Rev., 28, 241.

Segur, H. and Ablowitz, M. J. (1981) Asymptotic solutions of nonlinear evolution equations and a Painlevé transcendent. Physica D, 3, 165.

Seifert, G. (1990) Resonance in undamped second order nonlinear equations with periodic forcing. Q. Appl. Math., 48, 527.

Sellars, J. R. (1955) Laminar flow in channels with porous walls at high suction Reynolds number. J. Appl. Phys., 26, 489.

Shampine, L. F. (1969) Existence of solutions for certain nonlinear boundary value problems. J. Math. Phys., 10, 1177.

Shaw, J., Shaw, J. W. and Haddow, A. G. (1989) On the response of the nonlinear vibration absorber. Int. J. Nonlinear Mech., 24, 281.

Shekhter, B. L. (1986) On a boundary value problem arising in nonlinear field theory. Ann. Mat. Pura Appl., 144, 75.

Shekhter, B. L. (1989) Properties of solutions of a boundary value problem arising in combustion theory. Diff. Eqns., 25, 566.

Shemyakina, T. K. (1975) Analytic properties of solutions of a class of third order differential equation. Diff. Eqns., 11, 864.

Shih, K. (1987) On the existence of solutions of an equation in the theory of laminar flow in a uniformly porous channel with injection. SIAM J. Appl. Math., 47, 526.

Shilova, G. I. (1967) On the number of periodic solutions of the second kind of the differential equation $\ddot{\phi} - f_1(\phi)\dot{\phi} - f_0(\phi) = 0$. Diff. Eqns., 3, 873.

Shimizu, T. (1948) On the existence of limit cycles for some nonlinear differential equations. Math. Japon., 1, 125.

Shimizu, T. (1951) On differential equations for nonlinear oscillations. Math Japon., 2, 86.

Shimizu, T. (1953) On differential equations for oscillations with after effects and nonlinearity. Proc. 2nd Japan National Congress for Applied Mechanics. Science Council of Japan, Tokyo, p.305.

Shimizu, T. (1954) Subharmonics for nonlinear differential equations. Proc. 3rd Japan National Congress for Applied Mechanics, 1953. Science Council of Japan, Tokyo, p.421.

Shimizu, T., Sawada, K. and Wadati, M. (1983) Determination of the one-kink curve of an elastic wire through the inverse method. J. Phys. Soc. Jap., 52, 36.

Shimomura, S. (1982) Painlevé transcendents in the neighbourhood of fixed singular points. Funkcial. Ekvac., 25, 163.

Shimomura, S. (1982a) Series expansions of Painlevé transcendents in the neighbourhood of fixed singular points. Funkcial. Ekvac., 25, 184.

Shimomura, S. (1983) Analytic integration of some nonlinear ordinary differential equations and the fifth Painlevé equation in the neighbourhood of an irregular singular point. Funkcial. Ekvac., 26, 301.

Shimomura, S. (1987) A family of solutions of a nonlinear ordinary differential equation and its application to Painlevé equations III, V, and VI. J. Math. Soc. Jap., 39, 649.

Shimomura, S. (1987a) On solutions of the fifth Painlevé equation in the positive real axis (II). Funkcial. Ekvac., 30, 203.

Shivamoggi, B. K. and Varma, R. K. (1988) Internal resonances in nonlinearly coupled oscillators. Acta Mech., 72, 111.

Shrestha, G. M. (1968) Heat transfer in laminar flow in a uniformly porous channel with an applied transverse magnetic field. Appl. Sci. Res., 19, 352.

Sidorevich, M. P. and Lukashevich, N. A. (1990) A nonlinear third order differential equation. Diff. Eqns., 26, 450.

Siekmann, J. (1962) Note on an integral equation occurring in stagnation point flow. Z. Angew. Math. Phys., 13, 182.

Simon, W. E. (1965) Numerical technique for solution and error estimate for an initial value problem. Math. Comp., 19, 387.

Sirovich, L. and Newton, K. (1986) Periodic solutions of the Ginzburg–Landau equation. Physica D, 21, 115.

Skalak, F. M. and Wang, C. Y. (1979) On the unsteady squeezing of a viscous fluid from a tube. J. Austral. Math. Soc. B, 21, 65.

Smith, R. A. (1953) On the singularities in the complex plane of the solutions of $y'' + f(y)y' + g(y) = p(x)$. Proc. London Math. Soc., 3, 49.

Smith, R. A. (1961) A simple nonlinear oscillation. J. London Math. Soc., 36, 33.

Smith, R. A. (1970) Periodic bound for autonomous Lienard oscillators. Q. Appl. Math., 27, 516.

Smith, R. A. (1971) Periodic bounds for the generalised Raleigh equation. Int. J. Nonlinear Mech., 6, 271.

Smith, D. R. (1975) A nonlinear boundary value problem on an unbounded interval. SIAM J. Math. Anal., 6, 601.

Smith, F. T. (1984) Non-uniqueness in wakes and boundary layers. Proc. Roy. Soc. London, 391, 1.

Smith, H. L. (1986) On the small oscillations of the periodic Raleigh equation. Q. Appl. Math., 44, 223.

Soewono, E., Vajravelu, K. and Mohapatra, R. N. (1992) Existence of solutions of a nonlinear boundary value problem arising in flow and heat transfer over a stretching sheet. Nonlinear Anal. Theory Meth. Appl., 18, 93.

Spikes, P. W. (1977) On the integrability of solutions of perturbed nonlinear differential equations. Proc. Roy. Soc. Edin., 77A, 309.

Srivastava, V. N. and Usha, S. (1983/84) On a boundary value problem for a magneto fluid dynamic boundary layer. J. Math. Phys. Sci., 18(5), 599.

Srivastava, V. N. and Usha, S. (1988) On the existence of similar solutions for a three dimensional flow problem. Rend. Circ. Mat. Palermo (2), 37, 384.

Steeb, W. H., Euler, N. and Mulser, P. (1991) A note on integrability and chaos of reduced self-dual Yang–Mills equations and Yang–Mills equations. Nuovo Cimento B, 106, 1059.

Steeb, W. H., Euler, N. and Mulser, P. (1991a) Semiclassical Jaynes–Cummings model, Painlevé test, and exact solutions. J. Math. Phys., 32, 3405.

Steeb, W. H. and Kunick, A. (1983) Painlevé property of anharmonic systems with an external periodic field. Phys. Lett. A, 95, 269.

Steeb, W. H., Louw, J. A. and Maritz, M. F. (1987) Singular point analysis, resonances and Yoshida's theorem. J. Phys. A, 20, 4027.

Steinmetz, N. (1983) Über eine Klasse von Painleve'schen Differentialgleichungen. Arch. Math., 41, 261.

Stepanov, A. A. (1981) On the existence and uniqueness of the solution of a class of two-point boundary value problems. Lat. Mat. Ezhegodnik, 25, 88.

Stephani, H. (1989) Differential Equations: Their Solutions Using Symmetries. Cambridge Univ. Press, Cambridge.

Stevens, R. R. (1980) Periodic solutions of $x''+g(t,x)=0$. SIAM J. Math. Anal., 11, 400.

Stoker, J. J. (1950) Nonlinear Vibrations in Mechanical and Electrical Systems. Interscience, New York.

Stoppelli, F. (1953) Su un'equazióne differenziale della meccànica dei fili. Rend. Accad. Sci. Fis. Mat. Napoli, 19, 109.

Strelitz, Sh. (1981) On the growth of solutions of first and second order algebraic differential equations. J. Diff. Eqns., 42, 375.

Struble, R. A. (1962) Nonlinear Differential Equations. McGraw-Hill, New York.

Struble, R. A. and Yionoulis, S. M. (1962) General perturbation solution of the harmonically forced Duffing equation. Arch. Rat. Mech. Anal., 9, 422.

Stuart, C. A. (1985) A global branch of solutions to a semilinear equation in an unbounded interval. Proc. Roy. Soc. Edin., 101A, 273.

Su, R. and Chen, T. (1987) The soliton solution of ϕ^6 field theory at finite temperature. J. Phys. A, 20, 5939.

Sugie, J. (1987) On the generalised Lienard equation without the signum condition. J. Math. Anal. Appl., 128, 80.

Suleimanov, B. I. (1987) The relation between asymptotic properties of solutions of the second Painlevé equation in different directions towards infinity. Diff. Eqns., 23, 589.

Sulima, L. P. (1973) A second order differential equation of the second degree. Diff. Eqns., 9, 1543.

Sun, J. H. (1988) Periodic solutions and chaotic behaviour of a class of nonautonomic pendulum systems with large damping. Appl. Math. Mech. (English Ed.), 9, 1195.

Sunhong, D. (1989) On a kind of predator–prey system. SIAM J. Math. Anal., 20, 1426.

Suris, Y. B. (1987) Some properties of the methods for numerical integration of systems of the form $\ddot{x} = f(x)$. USSR Comp. Math. Math. Phys., 27(5), 149.

Swartz, W. J. (1958) An iterative procedure for certain nonlinear circuits. Proc. Amer. Math. Soc., 9, 533.

Sweet, D. (1973) Periodic solutions for dynamical systems possessing a first integral in the resonance case. J. Diff. Eqns., 14, 171.

Szemplinska–Stupnicka, W. (1968) Higher harmonic oscillations in heteronomous nonlinear systems with one degree of freedom. Int. J. Nonlinear Mech., 3, 17.

Szemplinska–Stupnicka, W. (1980) The resonance vibration of homogeneous nonlinear systems. Int. J. Nonlinear Mech., 15, 407.

Szemplinska–Stupnicka, W. (1988) Bifurcations of harmonic solution leading to chaotic motion in the softening type Duffing's oscillator. Int. J. Nonlinear Mech., 23, 297.

Tajiri, M. (1983) Similarity reductions of the one and two dimensional nonlinear Schrödinger equation. J. Phys. Soc. Jap., 52, 1908.

Tajiri, M. (1985) Note on similarity reduction to the second Painlevé equation and soliton solutions of the Zakharov equation. J. Phys. Soc. Jap., 54, 851.

Tajiri, M. and Kawamoto, S. (1982) Reduction of Korteweg–deVries and cylindrical Korteweg–deVries equations to Painlevé equations. J. Phys. Soc. Jap., 51, 1678.

Tajiri, M. and Kawamoto, S. (1983) Similarity reduction of modified Kadomtsev–Petviashvili equation. J. Phys. Soc. Jap., 52, 2315.

Takaoka, M. (1989) Pole distribution and steady pulse solution of the fifth order Korteweg–deVries equation. J. Phys. Soc. Jap., 58, 73.

Takaoka, M. (1989a) A note on "Pole distribution and steady pulse solution of the fifth order Korteweg–deVries equation." J. Phys. Soc. Jap., 58, 3028.

Takeyama, K. (1978) Dynamics of the Lorenz model of convective instabilities. Prog. Theor. Phys., 60, 613.

Takeyama, K. (1980) Dynamics of the Lorenz model of convective instabilities. Prog. Theor. Phys., 63, 91.

Takuno, K. (1983) A two-parameter family of solutions of Painlevé equation V near the point at infinity. Funkcial. Ekvac., 26, 79.

Taliaferro, S. D. (1978) Asymptotic behaviour of solutions of $y'' = p(t)y^{\lambda}$. J. Math. Anal. Appl., 66, 95.

Taliaferro, S. D. (1981) Asymptotic behaviour of solutions of $y'' = \phi(t)f(y)$. SIAM J. Math. Anal., 12, 853.

Tam, K. K. (1993) On the existence and parameter dependence of similarity solutions for free convection on a vertical plate. Stud. Appl. Math., 88, 141.

Tang, J. C. (1967) A nonlinear differential equation involving characteristic values. Z. Angew. Math. Mech., B, 47, 473.

Tanimoto, S. (1989) Asymptotic behaviour of functions and solutions of some nonlinear differential equations. J. Math. Anal. Appl., 138, 511.

Tarantello, G. (1989) On the number of solutions for the forced pendulum equation. J. Diff. Eqns., 80, 79.

Tarbell, J. M. (1980) A two-timing formalism for autonomous oscillations. Int. J. Nonlinear Mech., 15, 235.

Tatarkiewicz, K. (1963) Sur les équations différentielles ordinaires résolubles par des méthodes élémentines. Colloq. Math., 11, 113.

Taulbee, D. B., Cozzarelli, F. A. and Dym, C. L. (1971) Similarity solution to some nonlinear impact problems. Int. J. Nonlinear Mech., 6, 27.

Tenebaum, M. and Pollard, H. (1963) Ordinary Differential Equations. Harper & Row, New York.

Terracini, A. (1948) Geometric characterizations of equation G subordinate to a differential equation of type F. Univ. Nac. Tucuman Rev. A, 6, 255.

Terracini, A. (1955) Aspetti proiettívi nella teòria delle equazióni differenciali. Rend. Sem. Mat. Messina, 1, 115.

Terril, R. M. and Thomas, P. W. (1973) Spiral flow in a porous pipe. Phys. Fluids, 16, 356.

Tineo, A. (1991) Existence of two periodic solutions for the periodic equation $\ddot{x} = g(t, x)$. J. Math. Anal. Appl., 156, 588.

Tippet, J. (1974) An existence-uniqueness theorem for two point boundary value problems. SIAM J. Math Anal., 5, 153.

Titov, O. V. (1980) Theorems of the existence of a solution for the self-similiar boundary layer problem of viscous hypersonic shock layer. Appl. Sci. Res., 19(3), 149.

Toland, J. F. (1988) Existence and uniqueness of heteroclinic orbits for the equation $u'' + u' = f(u)$. Proc. Roy. Soc. Edin., 109A, 23.

Tomas, J. (1971) Ultra subharmonic resonance in a Duffing system. Int. J. Nonlinear Mech., 7, 625.

Tomastick, E. C. (1967) Oscillations of a nonlinear second order differential equation. SIAM J. Appl. Math., 15, 1275.

Tondl, A. (1982) Zum Problem des gegenseitigen Einflusses von selbsterregten und fremderregten Schwingungen. Z. Angew. Math. Mech., 62, 103.

Tondl, A. (1985) Analysis of a selfexcited system with dry friction. Int. J. Nonlinear Mech., 20, 471.

Tong, J. (1982) The asymptotic behaviour of a class of nonlinear differential equations of second order. Proc. Amer. Math. Soc., 84, 235.

Tongue, B. H. (1986) Existence of chaos in a one-degree-of-freedom system. J. Sound Vib., 110, 69.

Tran, M. H. and Evan–Iwanowski, R. M. (1990) Nonstationary responses of a selfexcited driven system. Int. J. Nonlinear Mech., 25, 285.

Trench, W. F. and Bahar, L. Y. (1987) First integrals for equations with nonlinearities of Emden–Fowler type. Int. J. Nonlinear Mech., 22, 216.

Troy, W. C. (1987) The existence of bounded solutions of a semilinear heat equation. SIAM J. Math. Anal., 18, 332.

Troy, W. C. (1989) Multiple solutions of a nonlinear boundary value problem. Proc. Roy. Soc. Edin., 113A, 191.

Troy, W. C. (1990) Bounded solutions of $\Delta u + |u|^{p-1}u - |u|^{q-1}u = 0$ in the super-critical case. SIAM J. Math. Anal., 21, 1326.

Troy, W. C. (1990a) Non-existence of monotonic solutions in a model of dendritic growth. Q. Appl. Math., 48, 209.

Tsukamoto, I. (1989) On solutions of $x'' = -e^{-\alpha\lambda t}x^{1+\alpha}$. Tokyo J. Math., 12, 181.

Tsukamoto, I., Mishina, T. and Ono, M. (1982) On solutions of $x'' = e^{\alpha\lambda t}x^{1+\alpha}$. Keio Sci. Tech. Rep., 35, 1.

Tu, S. T. (1989) A phase-plane analysis of bursting in the three dimensional Bonhoeffer–Van der Pol equations. SIAM J. Appl. Math., 49, 331.

Tu, K. (1991) Analytic solution to the Thomas–Fermi and Thomas–Fermi–Dirac–Weizsacker equations. J. Math. Phys., 32, 2250.

Tuck, E. O. and Schwartz, L. W. (1990) A numerical and asymptotic study of some third order ordinary differential equations relevant to drainage and coating flows. SIAM Rev., 32, 453.

Turcotte, D. L., Spence, D. A. and Bau, H. H. (1982) Multiple solutions for natural convective flows in an internally heated, vertical channel with viscous dissipation and pressure work. Int. J. Heat Mass Transfer, 25, 699.

Turrittin, H. L. and Culmer, W. J. A. (1957) A peculiar periodic solution of a modified Duffing equation. Ann. Mat., 44, 23.

Ueda, Y. (1979) Random transitional phenomena in the system governed by Duffing's equation. J. Stat. Phys., 20, 181.

Ueda, Y. and Noguchi, A. (1983) A model that realizes the solution and the chaos simultaneously. J. Phys. Soc. Jap., 52, 713.

Urabe, K. (1950) On the existence of periodic solutions for certain nonlinear differential equations. Math. Japon., 2, 23.

Urabe, M. (1952) Certain singularity of ordinary differential equations of three variables. J. Sci. Hiroshima Univ. Ser. A, 16, 57.

Urabe, M. (1967) Nonlinear Autonomous Oscillations: Analytic Theory. Academic Press, New York, p.119.

Urlacher, E. (1984) Équations différentielles du type $\epsilon x'' + F(x') + x$ avec ϵ petit. Proc. London Math. Soc., 49, 207.

Usami, H. (1987) Global existence and asymptotic behaviour of solutions of second order nonlinear differential equations. J. Math. Anal. Appl., 122, 152.

Usami, H. (1991) On positive decaying solutions of singular Emden–Fowler type equations. Nonlinear Anal. Theory Meth. Appl., 16, 795.

Utz, W. R. (1956) A note on second order nonlinear differential equations. Proc. Amer. Math. Soc., 7, 1047.

Utz, W. R. (1956a) Boundedness and periodicity of solutions of the generalized Lienard equation. Ann. Mat. Pura Appl., 42, 313.

Utz, W. R. (1956b) A third order differential equation. Monat. Math., 60, 329.

Utz, W. R. (1957) Properties of solutions of certain second order nonlinear differential equations. Proc. Amer. Math. Soc. B, 8, 1024.

Utz, W. R. (1967) The behaviour of solutions of the equations $x'' \pm xx'^2 \pm x^3 = 0$. Amer. Math. Monthly, 74, 420.

Utz, W. R. (1967a) Non-periodicity of solutions of a third order equation. Amer. Math. Monthly, 74, 705.

Utz, W. R. (1969) Properties of solutions of certain nonlinear differential equations of the form $y'' + g(y)y'^2 + f(y) = 0$. Ann. Mat., 81, 61.

Utz, W. R. (1971) Periodic solutions of $y'' + f(y)(y'^m) + g(y) = 0$. Ann. Polon. Math., 24, 327.

Utz, W. R. (1978) Properties of solutions of the differential equation $yy'' - ky'^2 + f(y) = 0$. Pub. Inst. Math. (N.S.), 24(38), 189.

Vajravelu, K., Soewono, E. and Mohapatra, R. N. (1991) On solutions of some singular, nonlinear differential equations arising in boundary layer theory. J. Math. Anal. Appl., 155, 499.

Van Dooren, R. (1973) Differential tones in a damped mechanical system with quadratic and cubic nonlinearities. Int. J. Nonlinear Mech., 8, 575.

Van Duijn, C. J. and Peletier, L. A. (1992) A boundary-layer problem in fresh-salt groundwater flow. Q. J. Mech. Appl. Math., 45, 1.

Vatsya, S. R. (1987) The existence and approximation of the solution of some nonlinear problems. J. Math. Phys., 28, 1283.

Verholomov, D. F. (1950) Equations of the form $y''' = R(y', y, x)y''^2$ with invariant critical points. Ukrain. Mat. Z., 2(2), 84.

Verhulst, F. (1990) Nonlinear Differential Equations and Dynamical Systems. Springer-Verlag, New York.

Villari, G. (1987) On the qualitative behaviour of solutions of the Lienard equation. J. Diff. Eqns., 67, 269.

Villari, G. and Zanolin, F. (1988) Some remarks on non-conservative oscillatory systems with periodic solutions. Int. J. Nonlinear Mech., 23, 1.

Vinokurov, V. A. and Repnikov, N. F. (1981) An iterational method for solving nonlinear boundary value problems. Appl. Sci. Res., 21(4), 81.

Vlieg–Hultsman, M. and Halford, W. D. (1991) The Korteweg–deVries–Burgers equation: a reconstruction of exact solutions. Wave Motion, 14, 267.

Voracek, J. (1965) Einige Bemerkungen über eine nichtlineare Differentialgleichung dritter Ordnung. Abh. Deutsch. Akad. Wiss. Berlin. Kl. Math. Phys. Tech., 1, 372.

Voracek, J. (1966) Certain third order nonlinear differential equations. Sb. Praci. Prirodoved. Fak. Univ. Palackeho Olomouci Fyz., 21, 109.

Vorobev, A. P. (1965) On rational solutions of the second Painlevé equation. Diff. Eqns., 1, 58.

Voronov, A. A. (1951) Free oscillations of an oscillator with variable friction. Math. Rev., 13, 347.

Vrdoljak, B. (1987) On solutions of the general Lagerstrom equation. Z. Angew. Math. Mech., 67, T456.

Vyshkind, S. Ya. and Rabinovich, M. I. (1976) The phase stochastization mechanism and the structure of wave turbulence in dissipative media. Sov. Phys. JETP, 44, 292.

Waltman, P. (1963) Some properties of the solutions of $u'' + a(t)f(u) = 0$. Monat. Math., 67, 50.

Waltman, P. (1964) On the asymptotic behaviour of solutions of a nonlinear equation. Proc. Amer. Math. Soc., 15, 918.

Waltman, P. (1966) Oscillation criteria for third order nonlinear differential equations. Pacific J. Math., 16, 385.

Wang, Y. W. (1971) Effect of spreading of material on the surface of a fluid: an exact solution. Int. J. Nonlinear Mech., 7, 255.

Wang, C. Y. (1976) The squeezing of a fluid between two plates. Trans. ASME J. Appl. Mech., 48, 579.

Wang, D. (1987) The critical points of the periodic solutions of $x'' - x^2(x - \alpha)(x - 1) = 0\ ,\ \ 0 \leq \alpha < 1$. Nonlinear Anal. Theory Meth. Appl., 11, 1029.

Ward, J. R. (1980) Periodic solutions for a class of ordinary differential equations. Proc. Amer. Math. Soc., 78, 350.

Ward, M. J. (1992) Eliminating indeterminacy in singularly perturbed boundary value problems with translation invariant potentials. Stud. Appl. Math., 87, 95.

Wasow, W. (1965) Asymptotic Expansions for Ordinary Differential Equations. Interscience, New York.

Watanabe, H., Sichel, M. and Ong, R. S. B. (1973) Transonic similarity solution for aligned field nozzle flow. J. Engg. Math., 7, 127.

Webb, G. M. and Mckenzie, J. F. (1991) Similarity solutions of the nonlinear damping equation using Lie group analyis. IMA J. Appl. Math., 47, 99.

Weili, Z. (1990) Singular perturbations of boundary value problems for a class of third order nonlinear ordinary differential equations. J. Diff. Eqns., 88, 265.

Weinmüller, E. (1984) On the boundary value problem for systems of ordinary second order differential equations with a singularity of the first kind. SIAM J. Math. Anal., 15, 287.

Whittaker, J. V. (1991) An analytic description of some simple cases of chaotic behaviour. Amer. Math. Monthly, 98, 489.

Wiggins, S. and Holmes, P. (1987) Periodic orbits in slowly varying oscillators. SIAM J. Math. Anal., 18, 592.

Wilder, J. W. (1991) Travelling wave solutions for an interface arising from phase boundaries based on a field model. Q. Appl. Math., 49, 333.

Williams, J. (1951) Small oscillations with damping. Math. Rev., 12, 707.

Williams, L. R. and Leggett, R. W. (1982) Unique and multiple solutions of a family of differential equations modelling chemical reactions. SIAM J. Math. Anal., 13, 122.

Wilson, H. K. (1971) Ordinary Differential Equations: Introductory and Intermediate Course Using Matrix Methods. Addison-Wesley, Reading, Mass.

Wong, P. (1970) Bounds for solutions to a class of nonlinear second order differential equations. J. Diff. Eqns., 7, 139.

Wong, J. S. W. (1975) On the generalised Emden–Fowler equation. SIAM J. Appl. Math., 17, 339.

Wong, J. S. W. (1988) Second order nonlinear forced oscillations. SIAM J. Math. Anal., 19, 667.

Wong, J. S. W. (1989) A sublinear oscillation theorem. J. Math. Anal. Appl., 139, 408.

Wu, Z. (1985) A free boundary problem for degenerate quasilinear parabolic equations. Nonlinear Anal. Theory Meth. Appl., 9, 937.

Xia, Z. (1992) Melnikov method and transversal homoclinic points in the restricted three point problem. J. Diff. Eqns., 96, 170.

Yablonskii, A. I. (1967) Movable singularities of systems of differential equations. Diff. Eqns., 3, 383.

Yablonskii, A. I. (1967a) Systems of differential equations whose critical singular points are fixed. Diff. Eqns., 3, 237.

Yablonskii, A. I. (1972) Asymptotic properties of regular solutions of Painlevé's first and second problems. Diff. Eqns., 8, 870.

Yablonskii, A. I. and Kozyuk, A. F. (1989) Classes of third order systems with quadratic right sides with no moving critical points. Diff. Eqns., 25, 433.

Yagasaki, K. (1990) Second order averaging and chaos in quasi-periodically forced weakly nonlinear oscillators. Physica D, 44, 445.

Yagasaki, K., Sakata, M. and Kimura, K. (1987) Dynamics of a weakly nonlinear system subjected to combined parametric and external excitation. Trans. ASME J. Appl. Mech., 57, 209.

Yamaguchi, T., Yamamoto, Y. and Imaeda, K. (1989) The structure and order of the appearance of the oscillation mode of an Oregonator exerted by an external periodic force. J. Phys. Soc. Jap., 58, 1550.

Yamamoto, Y. and Takizawa, E. I. (1981) On a solution of nonlinear time evolution equation of fifth order. J. Phys. Soc. Jap., 50, 1421.

Yamazaki, S. (1987) On the system of nonlinear differential equations $\dot{y}_k = y_k(y_{k+1} - y_{k-1})$. J. Phys. A, 20, 6237.

Yamazaki, S. (1990) Semi-infinite systems of nonlinear differential equations $\dot{A}_k = 2A_k (A_{k+1} - A_{k-1})$: the method of integration and asymptotic behavior. Nonlinearity, 3, 653.

Yan, Z. Q. (1988) A free boundary problem for ODEs arising from degenerate parabolic equations. J. Math. Res. Exposition, 8, 579.

Yan, J. (1989) On some properties of solution of second order nonlinear differential equations. J. Math. Anal. Appl., 138, 75.

Yanagida, E. (1991) Uniqueness of positive radial solutions of $\Delta u + g(r)u + h(r)u^p = 0$ in R^n. Arch. Rat. Mech. Anal., 115, 257.

Ye, I. Q. and Wang, X. (1978) Nonlinear differential equations arising in the theory of electron beam focusing. Acta Math. Sinica, 1, 13.

Yoshizawa, T. (1985) Attractivity in non-autonomous systems. Int. J. Nonlinear Mech., 20, 519.

Yuste, S. B. and Bejarano, J. D. (1986) Construction of approximate analytic solutions to a new class of nonlinear oscillator equations. J. Sound Vib., 110, 347.

Yuste, S. B. and Bejarano, J. D. (1989) Extension and improvement to the Krylov–Bogoliubov methods using elliptic functions. Int. J. Control, 49, 1127.

Zakharov, V. E. and Kuznetsov, E. A. (1986) Quasidimensional theory of three-dimensional wave collapse. Sov. Phys. JEPT, 64, 773.

Zandergen, P. J. and Dijkstra, D. (1977) Non-unique solutions of Navier–Stokes equations for Kármán swirling flow. J. Engg. Math., 11, 167.

Zhang, Z. F. (1980) Theorem of existence of n limit cycles on $|x| \leq (n+1)$ for the differential equation $\ddot{x} + u \sin x + x = 0$. Sci. Sinica, 23, 1502.

Zhidkov, E. P. and Shirikov, V. P. (1964) Boundary value problem for ordinary second order differential equation. USSR Comp. Math. Math. Phys., 4(4), 18.

Zhilevich, L. I. (1987) Periodic solution of a Lienard equation. Diff. Eqns., 23, 409.

Zlatoustov, V. A., Sazonov, V. V. and Sarycherv, V. A. (1979) Forced periodic oscillations of a simple pendulum. J. Appl. Math. Mech., 43, 270.

Zlatoustov, V. A. and Shipovskikh, T. A. (1984) Forced 2π-periodic solutions of a system with asymmetric restoring force (Russian). Nauk SSSR Inst. Prikl. Mat. Preprint, 124, 29.

Zufiria, J. A. (1987) Weakly nonlinear non-symmetric gravity waves on water of finite depth. J. Fluid Mech., 180, 371.

Zuo–huan, Z. (1990) Periodic solutions of generalised Lienard equations. J. Math. Anal. Appl., 148, 1.

Zwillinger, D. (1989) Handbook of Differential Equations. Academic Press, New York.